AF817675

PROCESS PLANT EQUIPMENT

PROCESS PLANT EQUIPMENT

Operation, Control, and Reliability

Edited by

MICHAEL D. HOLLOWAY
CHIKEZIE NWAOHA
OLIVER A. ONYEWUENYI

WILEY

A JOHN WILEY & SONS, INC., PUBLICATION

Cover photography: courtesy of Chikezie Nwaoha

Copyright © 2012 by John Wiley & Sons. All rights reserved.

Published by John Wiley & Sons, Inc., Hoboken, New Jersey
Published simultaneously in Canada

No part of this publication may be reproduced, stored in a retrieval system, or transmitted in any form or by any means, electronic, mechanical, photocopying, recording, scanning, or otherwise, except as permitted under Section 107 or 108 of the 1976 United States Copyright Act, without either the prior written permission of the Publisher, or authorization through payment of the appropriate per-copy fee to the Copyright Clearance Center, Inc., 222 Rosewood Drive, Danvers, MA 01923, 978-750-8400, fax 978-750-4470, or on the web at www.copyright.com. Requests to the Publisher for permission should be addressed to the Permissions Department, John Wiley & Sons, Inc., 111 River Street, Hoboken, NJ 07030, 201-748-6011, fax 201-748-6008, or online at http://www.wiley.com/go/permission.

Limit of Liability/Disclaimer of Warranty: While the publisher and author have used their best efforts in preparing this book, they make no representations or warranties with respect to the accuracy or completeness of the contents of this book and specifically disclaim any implied warranties of merchantability or fitness for a particular purpose. No warranty may be created or extended by sales representatives or written sales materials. The advice and strategies contained herein may not be suitable for your situation. You should consult with a professional where appropriate. Neither the publisher nor author shall be liable for any loss of profit or any other commercial damages, including but not limited to special, incidental, consequential, or other damages.

For general information on our other products and services or for technical support, please contact our Customer Care Department within the United States at 877-762-2974, outside the United States at 317-572-3993 or fax 317- 572–4002.

Wiley also publishes its books in a variety of electronic formats. Some content that appears in print may not be available in electronic formats. For more information about Wiley products, visit our web site at www.wiley.com.

Library of Congress Cataloging-in-Publication Data:

Process plant equipment : operation, control, and reliability / edited by
Michael D. Holloway, Chikezie Nwaoha, Oliver A. Onyewuenyi.
p. cm.
Includes index.
ISBN 978-1-118-02264-1 (cloth)
1. Pumping machinery–Maintenance and repair. 2. Pipelines–Maintenance and repair. 3. Valves–Maintenance and repair. 4. Compressors–Maintenance and repair. 5. Storage tanks–Maintenance and repair. 6. Mixing machinery–Maintenance and repair. 7. Boilers–Maintenance and repair. 8. Filters and filtration. I. Holloway, Michael H., 1963– II. Nwaoha, Chikezie, 1984– III. Onyewuenyi, Oliver A., 1952–
TJ900.P725 2012
621.8–dc23

2011028229

Printed in the United States of America

10 9 8 7 6 5 4 3 2 1

For the memory of
Denton Ward
student and friend

CONTENTS

SECTION III PROCESS MEASUREMENT, CONTROL, AND MODELING

CONTRIBUTORS

Mathew Chidiebere Aneke, Department of Built Environment, Northumbria University, Newcastle upon Tyne, England

Alberto R. Betancourt-Torcat, Department of Chemical Engineering, University of Waterloo, Waterloo, Ontario, Canada

Ali Ahammad Shoukat Choudhury, Department of Chemical Engineering, Bangladesh University of Engineering and Technology, Dhaka, Bangladesh

Jim Drago, P.E., Sr. Manager Marketing, Intelligence, GPT, Palmyra, New York

Ali Elkamel, Department of Chemical Engineering, University of Waterloo, Waterloo, Ontario, Canada

Marcelo Ferrara, ITW S.r.l., Innovative Technologies Worldwide, Augusta, Italy

Robert Free, Department of Engineering Physics, University of Oklahoma, Norman, Oklahoma

Michael D. Holloway, Certified Laboratories, NCH Corporation, Irving, Texas

Shaohui Jia, PetroChina Pipeline R&D Center, Langfang, Hebei, China

Gregory Livelli, ABB Measurement Products, Warminster, Pennsylvania

Krupavaram Nalli, Tebodin & Co., Al-Athaibab, Muscat, Oman

Celestine C. G. Nwankwo, Federal University of Technology, Owerri, Nigeria

Chikezie Nwaoha, Control Engineering Asia, Ten Alps Communications Asia, Aladinma, Owerri, Imo State, Nigeria

Okenna Obi-Njoku, Owerri, Nigeria

Oliver A. Onyewuenyi, MISOL Technology Solutions, Katy, Texas

Craig Redmond, The Gorman-Rupp Company, Mansfield, Ohio

L. A. Ricardez-Sandoval, Department of Chemical Engineering, University of Waterloo, Waterloo, Ontario, Canada

Jelenka Savkovic-Stevanovic, Department of Chemical Engineering, Faculty of Technology and Metallurgy, Belgrade University, Belgrade, Serbia

John A. Shaw, Process Control Solutions, Cary, North Carolina

N. Sitaram, Thermal Turbomachines Laboratory, Department of Mechanical Engineering, IIT Madras, Chennai, India

Jayesh Ramesh Tekchandaney, Unique Mixers and Furnaces Pvt. Ltd., Thane, Maharashtra, India

Matt Tones, Director, Marketing Intelligence, Garlock Sealing Technologies, Palmyra, New York

Flora Tong, Dow Chemical (China), Shanghai, China

Jacob E. Uche, Port Harcourt Refining Company, Eleme, Nigeria

Jerry Uttrachi, WA Technology, Florence, South Carolina

Sharad Vishwasrao, Vigilant Plant Services, Yokogawa Engineering, Asia Pte. Ltd., Singapore

Maher Y. A. Younan, Department of Mechanical Engineering, School of Sciences and Engineering, American University in Cairo, Cairo, Egypt

Zaki Yamani Zakaria, Department of Chemical Engineering, Faculty of Chemical Engineering, Universiti Teknologi Malaysia, Skudai, Johor, Malaysia

PREFACE

The nature of human beings has been to control the immediate environment for safety and comfort. Building shelter, finding food and water, and staying warm and dry are the reasons for our success. None of this would or could be possible if we did not develop a means to communicate and retain information. Consider the fact that humans do not possess great physical strength or agility compared to other animals or the ability to withstand harsh environments without the use of an extension of our bodies (clothing and shelter). We truly rely on each other's experiences to help us accommodate to what the world throws at us. This manuscript is an extension of just that—a means by which humans can share ideas and experiences in order to live our lives more comfortably. From petroleum, pharmaceuticals, and various chemicals and food products, to energy and power production, processing plants produce products essential to our survival. Without these plants we would not have the ability to get much older than our prehistoric ancestors. Developing the competence to run these plants efficiently, reliably, safely, and profitably is therefore a prime human objective.

The effort that proceeds over the next several hundred pages is nothing short of a miracle! Rarely can you get one or two people to commit to such a monumental project. This particular undertaking has herded over thirty (yes, thirty!) of the world's top academic and engineers to embark on a project that is encyclopedic in nature. Talented folks from Asia, Africa, Europe, the Middle East, South America, and North America have each put forth the effort to take on a topical chapter in order to build this work, all intended to provide readers with the information that will lead them to understand and implement best practices in process plant equipment operations, reliability, and control.

The initial idea for the book sprang from the musings of a very talented and promising young engineer, Chikezie Nwaoha. His vision to provide a comprehensive text that would enable readers to have access not only to the fundamental information concerning process plant equipment but also to practical ideas, best practices, and experiences of highly successful engineers from around the world. Nwaoha, being a smart man, decided to delegate some of the work. He broke the book into three parts. He edited the first section, Process Equipment Operations, Michael D. Holloway edited the second section, Process Plant Reliability, and the third section, Process Measurement, Control, and Modeling, was edited by Oliver A. Onyewuenyi, a world-renowned engineer.

The work incorporates the latest information, best practices, and trends. The sound foundations of engineering principles for a process facility provided in this book have solid roots from which the tree of productivity is constantly bearing fruit. If the reader chooses to put any of these ideas into practice, improvement in equipment operation, reliability, and control should be witnessed and enhanced safety, profitability, and performance will ensue. Only good things can happen.

Like any experienced bushman on the savanna knows, you can only eat an elephant one bite at a time. It is suggested that you take your time and read and digest each chapter carefully. Feel free to write in the margins, highlight passages, and quote as you see fit (but please use sound judgment concerning copyright laws!). Most important, use this work as a tool. Employed with care, information can develop into knowledge. With proper application and sound judgment, wisdom can spring forth. This work is the beginning of a very wise approach.

CHIKEZIE NWAOHA
MICHAEL D. HOLLOWAY
OLIVER A. ONYEWUENYI

SECTION I

PROCESS EQUIPMENT OPERATION

1

INTRODUCTION

Michael D. Holloway
NCH Corporation, Irving, Texas

A *process* is an amalgamation of machines, methods, materials, and people working in concert to produce something. Generally, the end product is something tangible: fuel, food, textiles, building materials—the list is exhaustive. The end product from a process can also be intangible: a bond, software, laws. It is difficult to say where a person begins and a process ends. Human beings are dependent on processes to live, as we are dependent on water to live. The first known process was probably irrigating fields to grow crops. Many argue that this process began over 20,000 years ago, others that it was closer to 50,000 years ago. Every few years a discovery is made that puts the date back even further as well as the place of origin: Africa, Asia, the Middle East? Needless to say, humans have been trying for a very long time to reduce labor and add comfort through the systematic use of materials and machines to implement a process to achieve a desired goal. Consider the following incomplete list of materials and machines. All required a process.

- Machines
  - *Primary machines*: simple machines that rely on their own structure to complete work: lever, pulley, inclined plane, hammer
  - *Secondary machines*: simple machines that rely on an accompanying machine: screw, wheel, axle, saw
  - *Tertiary machines*: complex machines that require a contribution from a compliant machine: gear, valve, pump, furnace, bearing, engines, boiler
- Materials
  - *Primary materials*: material used in the unprocessed state: water, wood, pitch, clay, stone, sand, wax, bone, fiber
  - *Secondary materials*: material developed from a combination or treatment of primary materials: leather, cement, paint, pigments, cloth, metal, glass
  - *Tertiary materials*: materials made from chemical manipulation: alloys, polymers, semiconductors, composites

A process does not become successful without observation and communication. One of the most important devices developed for a process was the *pump*. The first piston pump was invented by Ctesibius of Alexandria, a Greek physicist and inventor born around 300 B.C. One of his better known engineering efforts was improvement of the water clock. A water clock keeps time by means of dripping water maintained at a constant rate. His ideas of refinement of the water clock allowed for accurate timekeeping. The accuracy of his water clock was not improved upon for 1500 years. The second invention he is noted for is the water organ, the precursor of the hydraulic pump. This was a mechanized device in which air was forced by water through organ pipes to produce sounds. At first glance one would be in error not to think of the vast number of applications such a device could have. There are hundreds of different pumps in any given process plant. The concept of conveying gas or liquids without a pump is unheard of

Process Plant Equipment: Operation, Control, and Reliability, First Edition. Edited by Michael D. Holloway, Chikezie Nwaoha, and Oliver A. Onyewuenyi.
© 2012 John Wiley & Sons, Inc. Published 2012 by John Wiley & Sons, Inc.

today. This invention resulted from observation of one of his first inventions—a counterweighted mirror.

Ctesibius was born the son of a barber, and like many good sons he tried to follow in his father's footsteps. Perhaps it was a good thing that he spent more time thinking about how to improve his father's trade than in clipping bangs. He invented a device: a mirror placed at the end of a tubular pole, with a lead counterweight of the exact same weight placed at the other end that allowed the mirror to be adjusted for each customer. He noticed that when he moved the mirror, the weight bounced up and down while making a strange whistling noise. He theorized that this noise was air escaping from the tube. He tinkered with various dimensions and escape holes, which led to other observations and inventions using the power of pressure, gases, and liquids to achieve certain results. Without these musings the piston pump might never have came into being.

Pumping water for consumption, irrigation, and washing changed human society. If a stable water source was found, the water could be transported with minimal labor—all that was needed was a pump. People no longer had to move repeatedly to new areas to find food and water. They could stay put, farm, and live. In doing so, cities were established. With a high concentration of people, the odds of more improved processes increased exponentially. With the increased demand for improved comfort and greater commercial profits came a higher concentration of thinkers. Some people despise the modern city, but it must be admitted that cities are responsible for generating many of the ideas that make the rest of society flourish.

Mechanical means to move gases and fluids are essential in any process plant, but so is chemical manipulation. Perhaps the first known form of manipulating something chemically would be the cooking of food. With cooking, meats, grains, and vegetables become easier to digest and transport, and spoilage is reduced. Adding heat requires a fuel source and a means to control the thermal output. Being able to heat a substance in a controlled fashion on a larger scale introduced materials such as alloys, glass, and a whole host of chemicals. This process required furnaces and valves, among other devices. The second great feat of chemical manipulation is fermentation followed by distillation. Fermention of grains and berries has been carried out for tens of thousands of years. Humans are not the only creatures to enjoy a good "buzz." Many animals will have a party ingesting fermented berries and fruit. The ethanol produced provides a feeling of euphoria. One cannot blame any creature for wanting to feel better, but hopefully, it doesn't get in the way of the success of a species. To be able to separate alcohol from water requires observing condensation, fashioning a controllable heat source, and qualitative analysis. Alcohol is not just for drinking; it is actually a very valuable solvent, and the principles needed to understand how to make and distill alcohol are the very reasons that humans have become so successful. Without knowledge of the principles of fermentation and distillation, our heat, shelter, clothing, transportation, medicines, food, and materials would not exist as we know them.

The most influential industry to date is petroleum refining. Distillation is the main process in petroleum refining. Pharmaceuticals, building materials, solvents, plastics, and various fuels are all a result of the controlled distillation of crude oil. All this came about from the refinement of fermented grain. In fact, it is fair to say that without fermentation, we would not have progressed much further than the Cro-Magnons. Think about that the next time you sip a beer or enjoy a glass of wine or Scotch.

The effort that unfolds over the next several hundred pages is an undertaking that convinced over thirty of the world's top academic and engineers to embark on a project that is encyclopedic in nature. Talented and practicing experts in process plant engineering from Asia, Africa, Europe, the Middle East, and North America have contributed chapters to this book: all intended to help the reader to understand and implement best practices in process plant equipment operations, reliability, and control. The book is a comprehensive text that will provide the reader with access not only to fundamental information concerning process plant equipment but also with access to practical ideas, best practices, and experiences of highly successful engineers from around the world. The book is divided into three sections: Section I, Process Plant Equipment Operations; Section II, Process Plant Reliability; and Section III, Process Measurement, Control, and Modeling. An overview of the main highlights of the various chapters follows.

Section I: Process Equipment Operation

Chapter 2: Valves This chapter provides an introductory description of control valves, their types, and selection criteria, sizing procedures, operating principles, and maintenance and troubleshooting methods. It also describes common problems suffered by control valves and their remedies. Procedures for preventive and predictive maintenance of control valves and nonintrusive methods for detection of valve stiction are also discussed briefly.

Chapter 3: Pumps Water and other liquids are the lifeblood of many industrial processes. If those fluids are the blood, the plumbing system makes up the veins and arteries, and the pump is the heart. This chapter touches briefly on several types of industrial pumps, but deals primarily with the most common type, the centrifugal pump. Most of the principles apply to other types of pumps, but regardless of the type of pump in use, the pump manufacturer's manual and recommendations should always be followed.

The chapter also provides the following: general terms commonly used in the pump industry; brief information on several different types of pumps that may allow a user to identify what type of pump is either in use or needed for a particular application; basic component descriptions common to centrifugal pumps; instructions on how to read a typical pump performance curve; categories of different types of pump applications; how to size and select a pump properly, including net positive suction head calculations and considerations; proper pump maintenance; and basic pump troubleshooting guidelines.

Chapter 4: Pipes Pipelines are one of the main methods of transporting oil and gas worldwide. Historically, pipelines have been the safest means of transporting natural gas and hazardous liquids. The integrity, safety, and efficiency of a pipeline system is important and key to operators. Based on these considerations, this chapter covers mainly pipe types and pipe selection strategy, including pipe strength, toughness, weldability, and material; pipeline network design; pipe problems; pipeline inspection; and pipe maintenance.

Chapte 5: Cooling Towers Cooling towers are the most basic type of evaporative cooling equipment used primarily for process water cooling purposes in many chemical plants. Their principal task is to reject heat to the atmosphere and they are deemed a relatively inexpensive and reliable means of removing heat from water. Basically, hot water from heat exchangers or other units will be sent to a cooling tower and the water exiting the tower (which is cooler) will be sent back to the heat exchanger for cooling purposes.

Chapter 6: Filters and Membranes Filters and membranes are used in vast industrial processes for the separation of mixtures, whether of raw process media materials, reactants, intermediates, or products—comprising gases, liquids, or solutions. This chapter identifies gas and liquid filtration covering solid–liquid separations, solid–gas separations, solid–solid separations, liquid–liquid separations, and liquid–gas separations. It includes membrane technology such as microfiltration, reverse osmosis, ultrafiltration, and nanofiltration. It is a complete reference tool for all involved in filtration as well as for process personnel whose job function is filtration.

Chapter 7: Sealing Devices This chapter covers a variety of gasket types, compression packing, mechanical seals, and expansion joints. Discussed are materials of construction, principles of operation, and applications of sealing products. Wherever there are pumps, valves, pipes, and process equipment, there are sealing devices. Although relatively low in cost, sealing devices can have huge consequences if they don't work as needed or if they fail. All these devices are used in process industries and are critical to plant safety and productivity.

Chapter 8: Steam Traps A steam trap is a device attached to the lower portion of a steam-filled line or vessel which passes condensate but does not allow the escape of steam. It is also a piece of equipment that automatically controls condensate, air, and carbon dioxide removal from a piping system with minimal steam loss. Hot condensate removal is necessary to prevent water hammer, which is capable of damaging or misaligning piping instruments. Air in the steam system must be avoided, as any volume of air consumes part of the volume that the system would otherwise occupy. Apart from that, the temperature of the air–steam mixture normally falls below that of pure steam. It has been proven that air is an insulator and clings to the pipe and equipment surfaces, resulting in slow and uneven heat transfer. This chapter covers the various types and classification of steam traps and their installation, common problems, sizing, selection strategies, application, and maintenance.

Chapter 9: Process Compressors This chapter deals with compressors used in the process industry. Basic theory with practical aspects is provided in sufficient detail for the use of process industry personnel.

Chapter 10: Conveyors This chapter takes into account the types of conveyors been manufactured by modern industries to meet the current challenges encountered in conveying operations. It enumerates their usefulness, what conveyors are, industries that use them, conveyor selection and types, and safety and maintenance.

Chapter 11: Storage Tanks Storage tanks pose a complex management problem for designers and users. Because of the wide variety of liquids that must be stored, some of which are flammable, corrosive, or toxic, material selection for tanks is a critical decision. This chapter provides general guidelines that will aid in the selection of the proper type of storage to be used in a particular application. Various codes, standards, and recommended practices should be used to supplement the material provided. Manufacturers should be consulted for specific design information pertaining to a particular type of storage.

Chapter 12: Mixers Effective mixing of solids, liquids, and gases is critical in determining the quality of food, pharmaceuticals, chemicals, and related products. It is therefore essential that research and development scientists, process and project engineers, and plant operational personnel understand the mixing processes and equipment. Mixing processes may be batch or continuous and may involve materials in combination of phases such as liquid–liquid, liquid–solid, liquid–solid–gas, liquid–gas, and solid–solid (free-flowing powders and viscous pastes). An understanding of mixing mechanisms, power requirements, equipment design, operation and scale-up, and maintenance will lead to maximizing the mixing performance and enhancing business profitability.

Chapter 13: Boilers A boiler is process equipment comprising a combustion unit and boiler unit, which can convert water to steam for use in various applications. Boilers are of different types and generally work with various fittings, retrofits, and accessories. Boiler efficiency is achieved by skillful maintenance practices, including preventive and repair maintenance, in addition to use of only suitably conditioned water as feed water.

Section II: Process Plant Reliability

Chapter 14: Engineering Economics for Chemical Processes This chapter presents basic tools and methods used traditionally in engineering to assess the viability and feasibility of a project. Presented first are the tools available to represent money on a time basis. Next, the mathematical relationships frequently used to model discrete cash flow patterns are presented. The equivalence between the different discrete models is included on this section. The various indexes available to select the most profitable project between a set of alternatives are then presented. In this section, the payback period, the minimum acceptable rate of return, and the internal rate of return are introduced. An illustrative case study showing the application of these concepts is presented at the end of this section. The methods available to perform cost estimation and project evaluation are presented next, including several examples to show the application of cost estimation techniques. Companies execute engineering projects based on the revenues expected. Accordingly, they invest time and money in the process of selecting the project that would return the maximum revenues and satisfy such project constraints as environmental and government regulations. Therefore, the tools, techniques, and methods presented in this chapter would be used by engineers to assist them in the selection of the most suitable engineering project and to accurately estimate the costs associated with the project.

Chapter 15: Process Component Function and Performance Criteria This chapter explores the basic and advanced concepts of material transfer and conveyance equipment for air, steam, gases, liquids, solids, and powders. Also included are the engineering considerations for the component construction for material transfer. Each component section consists of a portion dedicated to selection specifications, reliability and cost savings, various maintenance approaches, and process development and improvement of transfer systems.

Chapter 16: Failure Analysis and Interpretation of Components This chapter highlights the fact that understanding how a component or device fails is essential in developing a scheme as to how to increase reliability and system robustness and ultimately reduce operational costs. There are essentially only four reasons for failure: the material, the methods, the machine, or the man. To identify the source of failure requires an understanding of the signs of the various sources. This chapter provides a fundamental explanation of failure by helping organize information to make the failure assessment a logical process.

Chapter 17: Mechanical Integrity of Process Vessels and Piping This chapter builds a focused and practical coverage of engineering aspects of mechanical integrity as it relates to failure prevention of pressure boundary components in process plants. Principal emphasis is placed on the primary means of achieving plant integrity, which is the prevention of structural failures and failure of pressure vessels and piping, particularly any that could have significant consequences. It provides practical concepts and applicable calculation methodologies for the fitness-for-service assessment and condition monitoring of process piping systems and pressure vessels.

Chapter 18: Design of Pressure Vessels and Piping This chapter covers the basic principles behind the design equations used in pressure vessels, and piping design codes. The design procedures for vessels and pipes are outlined. Numerical examples have been used to demonstrate some of the design procedures. This chapter is not intended to replace design codes but rather to provide an understanding of the concepts behind the codes.

Chapter 19: Process Safety in Chemical Processes In this chapter risk analysis and equipment failure are provided; process hazard analysis and safety rating are studied; safe process design, operation, and control are highlighted; and risk assessment and reliability analysis of a process plant are examined.

Section III: Process Measurement, Control, and Modeling

Chapter 20: Flowmeters and Measurement There are many different methods of measuring fluid flow, which are useful but can be very confusing. The objective of this chapter is to unravel some of the mysteries of flow technology selection and teach how different flowmeters work and when and when not to use them. This chapter covers the basics, including terminology, installation practices, flow profiles, flow disturbances, verfication techniques, flowmeter selection, and troubleshooting.

Chapter 21: Process Control Process control is used to maintain a variable in a process plant at a set point or cause it to respond to a set point change. The most common method used in process control is the PID (proportional, integral, and derivative) control algorithm. This algorithm and how it is used are discussed in this chapter.

Chapter 22: Process Modeling and Simulation This work serves as a guide and deals with the basic requirements for developing a model of a process. It covers the basic steps necessary for developing either a dynamic or steady-state model of a process. The case studies provided are made as simple as possible and make it possible for students and nonexperts to develop a simple model of a process that will help them investigate the behavior of either the entire process plant or a unit operation of interest.

As any experienced bushman on the Savannah knows, you can only eat an elephant one bite at a time. It is suggested that you take your time and read and digest each chapter carefully. Feel free to write in the margins, highlight passages, and quote as you see fit (but please use sound judgment concerning copyright laws!). Most important, use this work as a tool. Information can develop into knowledge with proper application. With proper application and sound judgment, wisdom can come forth. This work is the beginning of a very wise approach.

2

VALVES

Ali Ahammad Shoukat Choudhury
Bangladesh University of Engineering and Technology, Dhaka, Bangladesh
Chikezie Nwaoha
Control Engineering Asia, Ten Alps Communications Asia, Aladinma, Nigeria
Sharad Vishwasrao
VigilantPlant Services, Yokogawa Engineering, Asia, Singapore

Control valves are the most commonly used actuators or final control elements in process industries. They manipulate the flowing fluids to keep the variables being controlled in the desired positions. A control valve is known as the final control element because it is the element that ultimately manipulates the value of the variable in the control process. It is defined as a mechanism that alters the value of the variable being manipulated in response to the output signal from a controller, whether automatic, manual, or by direct human action. It is the element that implements the decision of the controllers. Controllers can be set in either automatic or manual mode control. A cross-sectional diagram of a typical pneumatic control valve is shown in Fig. 2-1. The purpose of the valve is to restrict the flow of process fluid through the pipe that can be seen at the very bottom of the figure. The valve plug is attached rigidly to a stem that is attached to a diaphragm in an air pressure chamber in the actuator section at the top of the valve. When compressed air is applied, the diaphragm moves up and the valve opens. The spring is compressed at the same time. The valve illustrated in Fig. 2-1 is a fail-closed type of valve because when the air pressure is reduced, the spring forces the valve to close.

A control valve has three basic components:

1. *Actuator*. Most actuators are pneumatic. Usually, an actuator works with the help of a diaphragm and instrument air. This is the device that positions the throttling element (i.e., the valve plug inside the valve body).
2. *Valve body subassembly*. This is the part where the valve plug, valve seats, and valve casing are located. The valve body and the valve plug differ in geometry and material construction. The combined body and plug geometry determines the flow properties of the valve. There are through-flow, blending, and stream-splitting types of configurations. Similarly, valve seats also differ in construction. There are conventional and contoured valve seat types with parabolic and quick-opening plugs whose internals can be inspected only during servicing.
3. *Accessories*. These include positioners, *I/P* (current-to-pressure) transducers, and position sensors.

In the process industries, hundreds or even thousands of control loops are in use to produce marketable end products. Many of these valves are housed in an attractive fashion, as shown in Fig. 2-2.

Typically, a control loop consists of three major elements: a sensor and transmitter, a controller, and a control valve. A feedback control loop is shown in Fig. 2-3. The control loop is a closed system consisting of selected instruments that work together as a unit with the single objective of controlling an identified variable. A loop consists of a sensor that can be an orifice, a thermocouple,

Process Plant Equipment: Operation, Control, and Reliability, First Edition. Edited by Michael D. Holloway, Chikezie Nwaoha, and Oliver A. Onyewuenyi.
© 2012 John Wiley & Sons, Inc. Published 2012 by John Wiley & Sons, Inc.

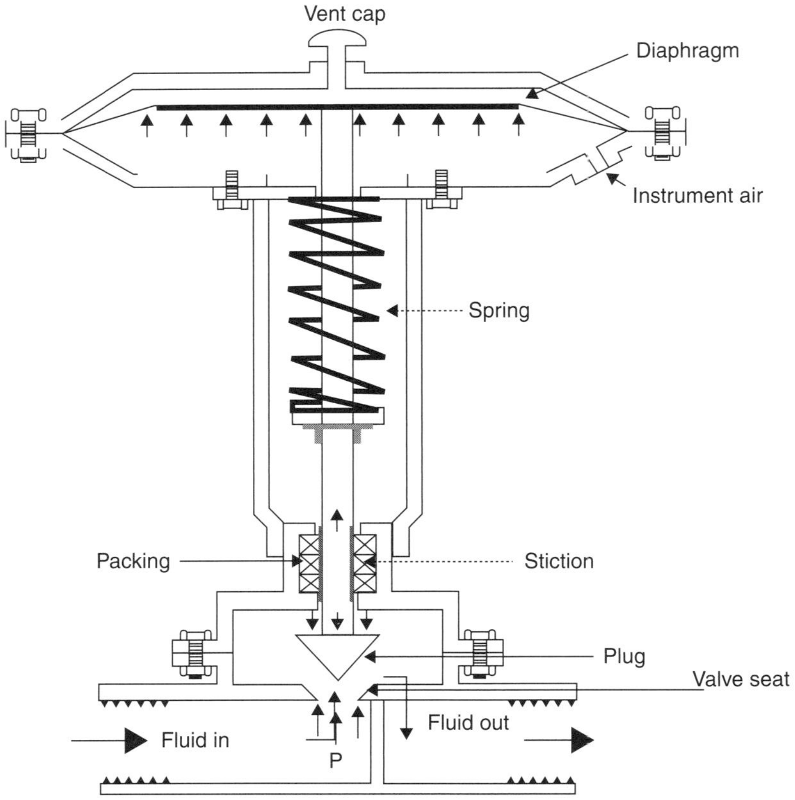

Figure 2-1 Cross-sectional diagram of a pneumatic control valve.

Figure 2-2 Assembly of control valves.

or a venture meter; a transmitter, which can be either a differential pressure electropneumatic or pneumatic transmitter; an indicator, which can be a pressure gauge, a level gauge, or a temperature gauge; and a transducer, which converts the signal reported from the form manipulated to a form understandable to the controller. The controller makes the decision and sends it to an *I/P* converter that converts the electric signal to a pneumatic signal and sends it to the final control instrument, or a positioner that gives proportional positional action to the valve stem so as to position the plug correctly in the valve body and, finally, regulates the flow (Fig 2-3).

2.1 TYPES OF CONTROL VALVES

A variety of types of control valves are used in all sectors of the process industries, depending on the suitability of a valve for a process. Two general types of control valves are based on their motion: linear-motion valves and rotary-motion valves.

2.1.1 Linear-Motion Control Valves

Linear-motion valves have a tortuous flow and low recovery. They can be offered in a variety of special trim designs and can throttle small flow rates. Most linear-motion valves are suitable for high-pressure applications. They are usually flanged or threaded and have separable bonnets. Examples of linear-motion valves are gate valves, diaphragm valves and globe valves.

2.1.1.1 Gate Valves Gate valves are generally used when a straight-line flow of fluid and minimum restriction

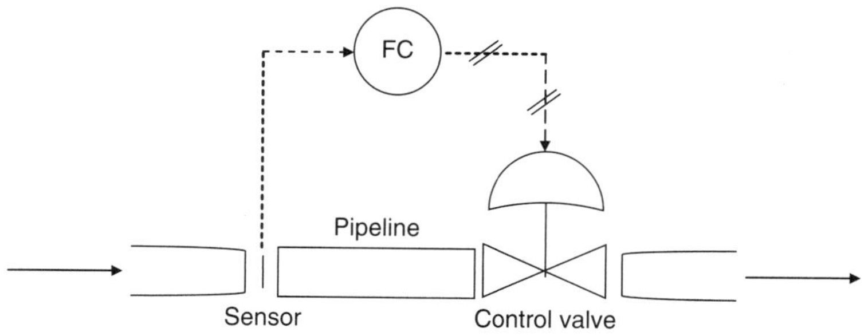

Figure 2-3 Components of a typical control loop arranged in a feedback configuration.

are desired. They are so named because the part that either stops or allows flow through the valve acts somewhat like the opening and closing of a gate. When the valve is wide open, it is fully drawn up into the valve, leaving an opening for flow through the valve of the same size as the pipe in which the valve is installed. Therefore, there is little pressure drop or flow restriction through the valve. Gate valves are not usually suitable for throttling purposes because flow control would be difficult, due to the valve design, and the flow of fluid slapping against a partially open gate can cause serious damage to the valve. Gate valves used in steam systems always have flexible gates [26]. The reason is to prevent binding of the gate within the valve when the valve is in the closed position. When steam lines are heated, they will expand, causing some distortion of valve bodies. If a solid gate fits snugly between the seat of the valve in a cold steam system, when the system is heated and pipes elongate, the seats will compress against the gate, wedging the gate between them and clamping the valve shut. This problem is overcome by the use of a flexible gate. This allows the gate to flex as the valve seat compresses it, thus preventing clamping [27].

2.1.1.2 Diaphragm Valves In a diaphragm control valve, operating air from the pilot acts on the valve diaphragm. The substructure that contains the diaphragm is direct acting in some valves and reverse acting in others. If the substructure is direct acting, the operating air pressure from the control pilot is applied to the top of the valve diaphragm. If the substructure is reverse acting, the operating air pressure from the pilot is applied to the underside of the valve diaphragm [26]. Diaphragm valves are lined to pressures of approximately 50 psi. They are used for fluids containing suspended solids and can be installed in any position. In this valve, the pressure drop is reduced to a negligible quantity. The only maintenance required in this valve is the replacement of the diaphragm, which can be done without removing the valve from the line.

2.1.1.3 Globe Valves These are probably the most common valves in existence. The globe valve derives its name from the globular shape of the valve body. However, positive identification of a globe valve must be made internally because other valve types may also have globular bodies [26]. Globe valve inlet and outlet openings are used extensively throughout the engineering plant and other parts of the ship in a variety of systems. In this type of valve, fluid passes through a restricted opening and changes direction several times. It is used extensively for the regulation of flow.

2.1.2 Rotary-Motion Control Valves

Rotary-motion control valves have a streamlined flow path and high recovery in nature. They have more capacity than that of linear-motion valves. This type of valve has an advantage in handling slurries and abrasives. They are easy to handle because they are flangeless and have an integral bonnet. Rotary-motion valves are designed to have high rangeability. Examples of this type of valve are butterfly valves, ball valves, and plug valves.

2.1.2.1 Butterfly Valves The butterfly valve is used in a variety of systems aboard vessels. These valves can be used effectively in saltwater, lube oil, and freshwater systems [25]. Butterfly valves are light in weight, relatively small, quick acting, provide positive shutoff, and can be used in throttling. This valve has a body, a resilient seat, a butterfly disk, a stem, packing, a notched positioning plate, and a handle. The resilient seat is under compression when it is mounted in the valve body, thus making a seal around the periphery of the disk and both upper and lower points where the stem passes through the seat. Packing is provided to form a positive seal around the stem for added protection in case the seal formed by the seat should become damaged. Butterfly valves are easy to maintain [26]. The resilient seat is held in place by mechanical means, and neither bonding nor cementing is necessary. Because the seat is replaceable, the valve seat does not require lapping, grinding, or machine work.

2.1.2.2 Ball Valves These are stop valves that use a ball to stop or start the flow of fluid [25]. When the valve handle is operated to open the valve, the ball rotates to a point where the hole through the ball is in line with the valve body inlet and outlet. When the valve is shut, which requires only a 90° rotation of the hand wheel for most valves, the ball is rotated so that the hole is perpendicular to the flow openings of the valve body, and flow is stopped. Most ball valves are of the quick-acting type, but many are planetary gear operated [26]. This type of gearing allows the use of a relatively small hand wheel and operating force to operate a fairly large valve but increases the valve operating time. Ball valves are normally found in the following systems: desalination, trim and drain, air, hydraulic, and oil transfer. They are used for general service, high-temperature conditions, and slurries.

2.1.2.3 ***Plug Valves*** These are quarter-turn valves that controls flow by means of a cylindrical or tapered plug with a hole through the center which can be positioned from open to close by a 90° turn. They are used for general services slurries, liquids, vapors, gases, and corrosives [26].

Other types of control valves are used either to control the flow of fluids or to control the pressure of fluids: nonreturn valves and relief valves.

2.1.3 Nonreturn Valves

Also known as *reflux valves* or *check valves*, these valves possess automatic devices that allow water to flow in one direction only (Fig. 2-4). They are made of brass or gun metal. Usually, a valve is pivoted at one end and can rest on a projection on the other end. This valve is provided in the pipeline that draws fluid from the pump [27]. When the pump is operated, the valve is open and the fluid flows through the pipe. But when the pump is suddenly stopped or fails due to a power failure, the valve is closed automatically and the fluid is prevented from returning to the pump [28].

2.1.4 Relief Valves

Relief valves are also known as *pressure relief valves*, *cutoff valves*, or *safety valves* [25]. These are automatic

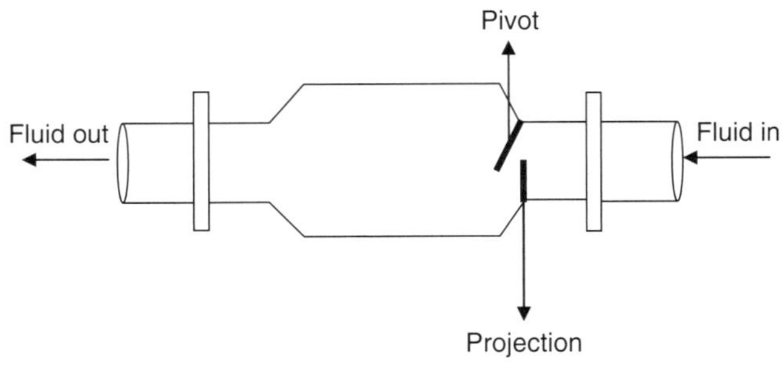

Figure 2-4 Nonreturn valve.

valves used on system lines and equipment to prevent overpressurization. Relief valves normally have a spring, and the power of the spring is adjusted such that a valve always remains in the closed position up to some permissible fluid pressure in the pipeline. When the pressure of the fluid suddenly exceeds the permissible pressure, the valve opens (lifts) automatically and the excess pressure is released instantaneously and then resets (shuts). Thus, the pipeline is protected from bursting. These valves are provided along the pipeline at points where the pressure is likely to increase. Other types of relief valves are high-pressure air safety relief valves (PRVs) and bleed air surge relief valves. Both are designed to open completely at a specified lift pressure and to remain open until a specific reset pressure is reached, at which time they shut [25]. However, the PRV is also the one piece of equipment that we hope never needs to operate. Because the PRV is the last line of defense against the catastrophic failure of a pressurized system, it must be maintained in "like new" condition if it is to provide the confidence necessary to operate a pressurized system.

2.2 CONTROL VALVE ACTUATORS

A control valve, typically outfitted with an actuator, provides the final control element in many process systems. The actuator accepts a signal from an external source and, in response, positions (opens or closes) the valve to the position required or designed. Valve actuators enable remote operation of control valves, which is essential for worker safety in many application environments. Actuators can be moved into position by either hydraulic, air/gas, or electric signals. Typical control valve position commands include more closed, more open, fully closed, and fully open. There are different types of control valve actuators, and they are classified according to the power supply required for activation. Types of valve actuators include pneumatic valve actuators, electric valve actuators, and hydraulic valve actuators.

2.2.1 Pneumatic Valve Actuators

A pneumatic valve actuator is a control valve actuator that can adjust the position of the valve by converting air pressure into rotary or linear motion. Rotary motion actuators are used on butterfly valves, plug valves, and ball valves, and they position from open to closed by a 90° turn [30]. Meanwhile, linear motion actuators are used on globe valves, diaphragm valves, pinch valves, angle valves, and gate valves, and they employ a sliding stem that controls the position of the element (closure). Pneumatic valve actuators can be single- acting, in that air actuates the valve in one direction and a compressed spring actuates the valve in the other direction. Single-acting devices can be either reverse-acting (spring-to-extend) or direct-acting (spring-to-react). The operating force is generated from the pressure of the compressed air. Choosing between reverse-acting and direct-acting is dependent on the safety requirements (in the event of a compressed supply air failure), response/activation time, air supply pressure, and so on. For example, for safety reasons steam valves must close upon failure of the air supply. Pneumatic valve actuators have the advantage of simple construction, requiring little maintenance, and a quick valve response time to changes in the control signal.

2.2.2 Electric Valve Actuators

An electric valve actuaor is a compact valve actuator with a large stem thrust. Electric valve actuators are typically employed in systems where a pneumatic supply is not needed or available. An electric valve actuator is more complex than a pneumatically operating valve actuator. When control valves are spread out over large distances, as is often the case in pipeline applications, an electric valve actuator should be chosen for purely economic reasons (i.e., because electrical energy is cheaper and easier to transport than instrument air and/or hydraulic fluid). Electric valve actuators rely on an electrical power source for their position signal [31]. They employ single- or three-phase ac/dc motors to move a combination of gears to produce the desired level of torque. Subsequently, the rotational motion is converted into a linear motion of the valve stem via a gear wheel and a worm transmission. Electric valve actuators are used primarily on linear motion valves, globe valves, and gate valves. They are also used on quarter-turn valves such as butterfly valves and ball valves. Linear electric valve actuators are installed in systems where tight tolerances are required, whereas rotary electric valve actuators are suitable for use in packaging and electric power. Electric valve actuators have the disadvantage of valve response, which can be as low as 5 s/min.

2.2.3 Hydraulic Valve Actuators

Hydraulic valve actuators usually employ a simple design with a minimum of mechanical parts. Hydraulic valve actuators convert fluid pressure into linear motion, rotary motion, or both. Like electric actuators, they are also used on both quarter-turn and linear valves. In quarter-turn valves, the hydraulic fluid provides the thrust, which is converted mechanically to rotary motion to adjust the valve. For linear valves, the pressure of the hydraulic fluid acts on the piston to provide the thrust in a linear motion, which is a good fit for gate or globe valves [31]. Hydraulic valve actuators are used particularly in situations where a large stem thrust is required, such as the steam supply in turbines or the movement of large valves in chimney flues. In a situation where very large valves are to be actuated, it is often advisable to install the actuators on mechanical gearboxes to provide increased output (torque). There are different types of hydraulic valve actuators that convert linear motion to rotary motion. For example, whereas diaphragm actuators are generally used with linear motion valves, they can also be used for rotary motion valves if they are outfitted with linear-to-rotary motion linkage. Similarly, lever and link actuators transfer the linear motion of a piston. Rack-and-pinion actuators transfer the linear motion of a piston cylinder to rotary motion, and scotch yoke actuators convert linear motion to rotary motion as well. For safety reasons, most hydraulic actuators are provided with fail-safe features: either fail open, fail close, or fail stay put.

For a control system to be effective, the control valve must adjust to its desired position as quickly and efficiently as possible. To achieve this, the right valve actuator must be selected for the application. Therefore, it could be said that the valve actuator specification process is more important than the selection of the control valve itself. To ensure that the right valve actuator is chosen for a given process, critical site information, such as the availability of power supply, hydraulic fluid pressure, and air pressure, must be considered. In addition, the stroke time of the valve, fail-safe position, control signal input, and safety factors must be given due consideration.

2.3 CONTROL VALVE SIZING AND SELECTION

Control valve sizing is a procedure by which the dynamics of a process system are matched to the performance characteristics of a valve. This is to provide a control valve of appropriate size and type that will best meet the needs of managing flow within that process system.

The task of specifying and selecting the appropriate control valve for any given application requires an understanding of the following principles [23]:

- How fluid flow and pressure conditions determine what happens inside a control valve
- How control valves act to modify pressure and flow conditions in a process
- What types of valves are commonly available
- How to determine the size and capacity requirements of a control valve for any given application
- How actuators and positioners drive the control valve
- How the type of valve influences the costs

Selecting the right valve for the job requires that the engineer is able to:

- Ensure that the process requirements are defined properly
- Calculate the flow capacity required over the operating range
- Determine any limiting or adverse conditions, such as cavitation and noise, and know how to deal with them
- Know how to select the valve that will satisfy the constraints of price and maintainability while providing good performance during process control.

Valve sizing involves several steps. They can be described briefly as follows [34]:

1. *Define the system*. To begin, the system should be defined properly, based on information regarding the fluid and its density, the temperature, the pressures, the design flow rate, the minimum flow rate, the operating flow rate, and the pipe diameter.

2. *Define the maximum allowable pressure drop*. The maximum allowable pressure drop across the valve should be determined from the difference between the net positive suction head available and the net positive suction head required. It's important to remember the trade-off: Larger pressure drops increase the pumping cost (operating), and smaller pressure drops increase the valve cost because a larger valve is required (capital cost). The usual rule of thumb is that a valve should be designed to use 10 to 15% of the total pressure drop, or 10 psi, whichever is greater.

3. *Calculate the valve characteristic*.

$$C_v = Q\sqrt{\frac{G}{\Delta P}}$$

where Q = is the design flow rate (gpm), G = the specific gravity relative to water, and ΔP = the allowable pressure drop across a wide-open valve.

4. *Make the preliminary valve selection*. The C_v value should be used as a guide in the valve selection along with the following considerations:

   a. Never use a valve that is less than half the pipe size.

   b. Avoid using the lower 10% and upper 20% of the valve stroke. The valve is much easier to control in the stroke range 10 to 80%.

   Before a valve can be selected, one needs to decide what type of valve will be used. Is it a globe valve or a butterfly valve? Equal percentage or quick opening? Depending on the valve type, the appropriate valve chart supplied by the manufacturer should be used to get the valve size or the diameter.

5. *Check the C_v and stroke percentage at the minimum flow*. If the stroke percentage falls below 10% at the minimum flow rate, a smaller valve may have to be used in some cases. Judgment plays a role in many cases. For example, is the system more likely to operate closer to the maximum flow rates more often than close to the minimum flow rates? Or is it more likely to operate near the minimum flow rate for extended periods of time? It's difficult to find the perfect valve, but you should find one that operates well most of the time. At the minimum flow rate, the C_v value should be recalculated. Then from the valve chart, the stroke percentage should be determined. If the valve stroke is within 10 to 90%, the valve is acceptable. Note that the maximum pressure drop is to be used in the calculation. Although the pressure drop across the valve will be lower at smaller flow rates, using the maximum value gives the worst-case scenario and the conservative estimate. Essentially, at lower pressure drops, C_v would only increase, which would be advantageous in this case.

6. *Check the gain across the applicable flow rates*. Gain is defined as

$$\text{gain} = \frac{\Delta \text{ flow}}{\Delta \text{ stroke or travel}}$$

Valve gains should be calculated for the minimum flow rate, the operating flow rate, and design flow rate conditions. For these three flow conditions, two gains can be calculated, taking the operating flow as basis. The difference of these two gains should not be more than 50% of the higher value. Also, the gain value should never be less than 0.50.

2.3.1 Selecting a Valve Type

When speaking of valves, it is easy to get lost in the terminology. *Valve types* are used to describe the mechanical characteristics and geometry (gate, ball, globe valves). From the flow characteristics, there are three primary types of control valves:

1. *Equal percentage*. Equal increments of valve travel produce an equal percentage in flow change.
2. *Linear*. Valve travel is directly proportional to the valve stoke.
3. *Quick opening*. In this type, a large increase in flow is coupled with a small change in valve stroke.

So how do you decide which control valve to use? Here are some rules of thumb for each:

1. Equal percentage (the most commonly used valve control)
   a. Used in processes where large changes in pressure drop are expected
   b. Used in processes where a small percentage of the total pressure drop is permitted by the valve
   c. Used in temperature and pressure control loops

2. Linear
   a. Used in liquid level or flow loops
   b. Used in systems where the pressure drop across the valve is expected to remain fairly constant (i.e., steady-state systems)
3. Quick opening
   a. Used for frequent on–off service
   b. Used for processes where an "instantly" large flow is needed (i.e., safety systems or cooling water systems)

2.3.2 Sizing and Selection: Letting the Computer Do It All [2]

There are two types of valve sizing software. The first lets you pick the valve type and then gives you the specifics, such as rated C_v, F_L, and F_d values, allowing you to carry out the correct sizing calculations. The second type lets you enter only the flow conditions, and the computer calculates the C_v and selects the right valve, usually the best economical choice. This software is vendor-specific and usually valid for valves from the same manufacturer. When selecting a program, the following things are to be kept in mind:

1. Valve sizing should accord with current ISA standard S75.01 or the corresponding IEC standard.
2. The noise equation should follow ISA standard S75.17 or IEC 534.8.3.
3. The required maximum C_v should not be more than 85% of the rated C_v of the valve selected.
4. The minimum C_v should be greater than the C_v of the valve selected at 5% of valve travel.
5. Pay attention to the cavitation or flashing warnings. They may indicate trouble.

Following is a partial list of vendors offering computer programs for control valve sizing and selection.

H. D. Baumann Inc.
35 Mironal Road
Portsmouth, NH 03801

Eckardt AG
Postfach 50 03 47
D-7000 Stuttgart 50
Germany

Engineered Software, Inc.
P.O. Box 2514
Olympia, WA 98507

Fisher Controls International Inc.
295 South Center Street
Marshalltown, IA 50158

Gulf Publishing Co.
Houston, TX 77252

Instrumentation Software, Inc.
P.O. Box 776
Waretown, NJ 08758

ISA
P.O. Box 12277
Research Triangle Park, NC 27709

Masoneilan Dresser
Dresser Valve and Controls Division
275 Turnpike Street
Canton, MA 02021

Neles-Jamesbury, Inc.
P.O. Box 15004
Worcester, MA 01615

Valtek, Inc.
P.O. Box 2200
Springville, UT 84663

2.4 COMMON PROBLEMS OF CONTROL VALVES

Control valves suffer major problems when in use. The most common problems are described below.

2.4.1 Control Valve Cavitation

Cavitation is a two-stage transformation of the initial formation of vapor bubbles by the flowing fluid and then the reverse: bubbles back to the liquid at high downstream pressure. For the bubbles to form the static pressure of the flowing liquid falls below the fluid vapor pressure. The bubbles eventually collapse at the downstream pressure, which is higher than the vapor pressure of the liquid.

In a pure liquid distribution system involving a high pressure differential and high flow rates, automatic control valves tend to vibrate and make excessive noise. The noise and vibration problems are safety hazards. In valves, cavitation is therefore caused by a sudden and severe fall in pressure below the vapor pressure level and consequent vapor bubble formation as a result of excessive fluid velocity at the seating area. The energy released by the eventual collapse of vapor bubbles eats away the surfaces of the valve plug and seat. This causes loss of flow capacity and erosion damage. The collapse of vapor bubbles can cause local pressure waves of up to 1,000,000 psi. Fluid microjets are also formed, due to the asymetrical bubble collapse. The high-intensity pressure waves combine with the microjet impingement on the valve surface to cause severe damage to the valve.

2.4.1.1 Types of Cavitation It is usually not difficult to determine whether a valve is cavitating. One has merely to listen. However, to determine if the cavitation intensity is high enough to cause damage requires quantifying the intensity and comparing it with available experimental cavitation reference data for the valve of interest. Cavitation intensity can be quantified relative to four levels.

1. *Incipient cavitation* refers to the onset of audible, intermittent cavitation. At this lower limit, cavitation intensity is slight. The operating conditions that foster incipient cavitation are conservative and are seldom used for design purposes.

2. *Critical cavitation*, the next stage, describes the condition when the cavitation noise becomes continuous. The noise intensity is often difficult to detect above the background flow noise. Critical cavitation causes no adverse effects and commonly defines the "no cavitation" condition. This level is referred to as critical because cavitation intensity increases rapidly with any further reduction in σ.

3. *Incipient damage* refers to the conditions under which cavitation begins to destroy the valve. It is usually accompanied by loud noise and heavy vibration. The potential for material loss increases exponentially as σ drops below the value that initiates incipient damage. Consequently, this is the upper limit for safe operation with most valves. Unfortunately, it's the limit that is most difficult to determine, and experimental data are available for only a few valves.

4. *Choking cavitation* is a flow condition in which the mean pressure immediately downstream from the valve is the fluid's vapor pressure. This represents the maximum flow condition through a valve for a given upstream pressure and valve opening. It is a condition that damages both valve and piping. Choking cavitation is an interesting and complex operating condition. Even though the valve outlet is at vapor pressure, the downstream system pressure remains greater. Reducing the downstream pressure increases the length of the vapor cavity but doesn't increase the flow rate. The noise, vibration, and damage occur primarily at the location where cavity collapse occurs.

2.4.1.2 Cavitation Prevention Strategies

1. *Applying two valves in series*. In the case of extreme high-pressure differentials, two valves installed in series effectively mitigate the incidence of cavitation. The second valve acts as a backup when the first valve fails and also ensures pressure reduction to some level [10]. The problems associated with this method are lack of enough space to install two valves and the cost of the second valve.

2. *Applying orifice plates*. Devices that produce back-pressure, such as orifice plates, can be used downstream of a valve to prevent cavitation. The orifice plate has the advantages of low ongoing and installation costs. It has the disadvantage of being only effective within a narrow flow range and can cause reduction of flow capacity within the system. This can possibly cause cavitation, thereby creating a potential for damage to downstream fittings, so absolute care must be taken to follow the manufacturer's specifications [10].

3. *Applying anticavitation valves and trim*. The most effective approach to controlling valve cavitation is to install an anticavitation valve. Where an anticavitation valve is in existence, it should be retrofitted with an anticavitation trim. Equipping valves with anticavitation trim is considered most often in systems where extreme pressure differentials and high-velocity florates are present. The cavitation solution is self-contained either for an existing valve equipped with anticavitation components or for a new valve with trim [10]. This provides a wider range of flow rates and smooth operation with low levels of vibration and noise.

4. *Designing a cavitation-free system*. The best method for preventing cavitation is the inclusion of cavitation prevention measures as an integral part of the design of the distribution system. This involves a complete cavitation study before selection, purchase, and installation of valves in a pipeline. Consulting valve manufacturers is necessary to specify the proper size of a valve equipped with anticavitation trim or another option [10]. This method offers the advantages of lower maintenance cost, fewer equipment failures, less downtime, and optimum efficiency of systems.

2.4.2 Control Valve Leakage

There are two types of valve leakage:

1. *Stem leakage*. A loose or worn stem packing causes external leakage of the process fluids, which may violate V.S. Environmental Protection Agency regulations. On the other hand, tight packing may cause excessive friction, which can make the loop performance unsatisfactory [6].

2. *Valve seat leakage*. Control valves are not shutoff valves. Often, there may be fluid leakage through the valve

seat. Depending on the quantity of fluid that passes through the leakage, valve seat leakages are classified into one of six categories.

Control valves are designed to throttle but they are not shutoff valves, so will not necessarily close 100%. A control valve's ability to shutoff has to do with many factors, such as the type of valve. A double-seated control valve has very poor shutoff capability. The guiding, seat material, actuator thrust, pressure drop, and type of fluid can all play a part in how well a particular control valve shuts off. There are actually six different seat leakage classifications, as defined by ANSI/FCI-70-2-1976 (rev. 1982). An overview of these classifications is provided in Table 2-1.

2.4.3 Control Valve Nonlinearities

There are two types of nonlinearities that may be related to control valves. The first type arises from the nonlinear characteristics of valves, such as their equal percentage, quick opening, and square-root characteristics. Usually, the effect of these types of nonlinearities is minimized during the installation of valves, so that their characteristics are linear. The second type of nonlinearity may appear due to manufacturing limitations or gradual development of faults, Among these faults, deadband, hysteresis, backlash, and stiction are problems commonly found in control valves and other instruments.

Control valves frequently suffer from such problems as stiction, leaks, tight packing, and hysteresis. Bialkowski [1] reported that about 30% of the loops are oscillatory, due to control valve problems. In recent work, Desborough et al. [11,12] reported that control valve problems account for about one-third of the 32% of controllers classified as poor or fair in an industrial survey [1]. If the control valve contains nonlinearities (e.g., stiction, backlash, and deadband), the valve output may be oscillatory, which in turn can cause oscillations in the process output. Among the many types of nonlinearities in control valves, stiction is the most common and a longstanding problem in the process industry. It hinders proper movement of the valve stem and consequently affects control loop performance. Therefore, it is important to learn what stiction is and how it can be detected and quantified. *Deadband, backlash*, and *hysteresis* are often misused and used wrongly in describing valve problems such as stiction. For example, quite commonly a deadband in a valve is referred to as backlash or hysteresis. Therefore, before proceeding to the definition of stiction, these terms are first defined for a better understanding of the stiction mechanism and a more formal definition of stiction.

2.4.3.1 Terms Relating to Valve Nonlinearity In this section we review the American National Standards Institute's (ANSI) formal definition of terms related to stiction. The aim is to differentiate clearly between the key concepts that underlie the ensuing discussion of friction in control valves. These definitions can also be found elsewhere in the literature [13,14]. An ANSI ISA subcommittee report [21] defines the stiction terms as follows:

1. *Backlash*. "In process instrumentation, it is a relative movement between interacting mechanical

TABLE 2-1 Valve Seat Leakage Classifications

Leakage Class Designation	Maximum Leakage Allowable	Test Medium	Test Pressure	Testing Procedures Required for Establishing Rating
I	—	—	—	Not test required
II	0.5% of rated capacity	Air or water at 50–125°F (10–52°C)	45–60 psig or maximum operating differential, whichever is lower	45–60 psig or maximum operating differential, whichever is lower
III	0.1% of rated capacity	As above	As above	As above
IV	0.01% of rated capacity	As above	As above	As above
V	0.0005 mL/min of water per inch of port diameter per psi differential	Water at 50–125°F (10–52°C)	Maximum service pressure drop across valve plug, not to excced ANSI body rating	Maximum service pressure drop across valve plug, not to exceed ANSI body rating
VI	Not to exceed amounts shown above	Air or nitrogen at 50–125°F (10–52°C)	50 psig or max. rated differential pressure across valve plug, whichever is lower	Actuator should be adjusted to operating conditions specified with full normal closing thrust applied to valve plug seat

Source: Adapted from ANSI/FCI-70-2-1976 (rev. 1982).

parts, resulting from looseness, when the motion is reversed."

2. *Hysteresis*. "Hysteresis is that property of the element evidenced by the dependence of the value of the output, for a given excursion of the input, upon the history of prior excursions and the direction of the current traverse.... It is usually determined by subtracting the value of deadband from the maximum measured separation between upscale-going and downscale-going indications of the measured variable (during a full-range traverse, unless otherwise specified) after transients have decayed." Figure 2.5(a) and (c) illustrate the concept. "Some reversal of output may be expected for any small reversal of input. This distinguishes hysteresis from deadband."
3. *Deadband*. "In process instrumentation, it is the range through which an input signal may be varied, upon reversal of direction, without initiating an observable change in output signal.... There are separate and distinct input—output relationships for increasing and decreasing signals" [see Fig. 2-5b]. "Deadband produces phase lag between input and output.... Deadband is usually expressed in percent of span." Deadband and hysteresis may be present together. In that case, the characteristics in the lower left panel of Fig. 2-5 would be observed.
4. *Dead zone*. "It is a predetermined range of input through which the output remains unchanged, irrespective of the direction of change of the input signal.... There is but one input–output relationship" [see Fig. 2-5d]. "Dead zone produces no phase lag between input and output."

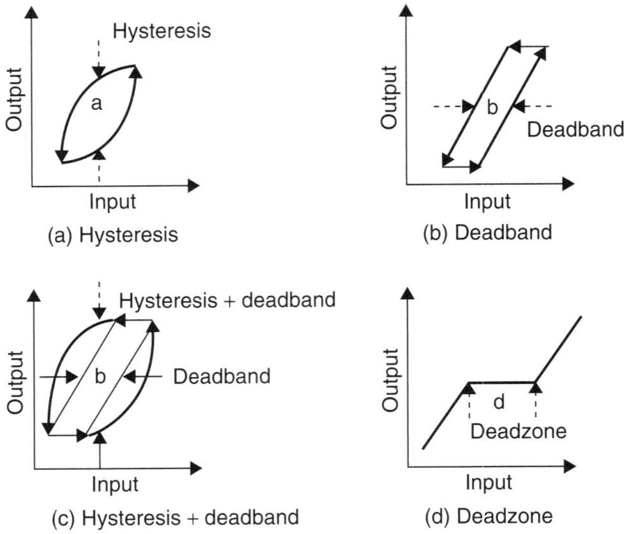

Figure 2-5 Input–output behavior of (a) hysteresis, (b) deadband, (c) hysteresis plus deadband, and (d) deadzonel nonlinearities.

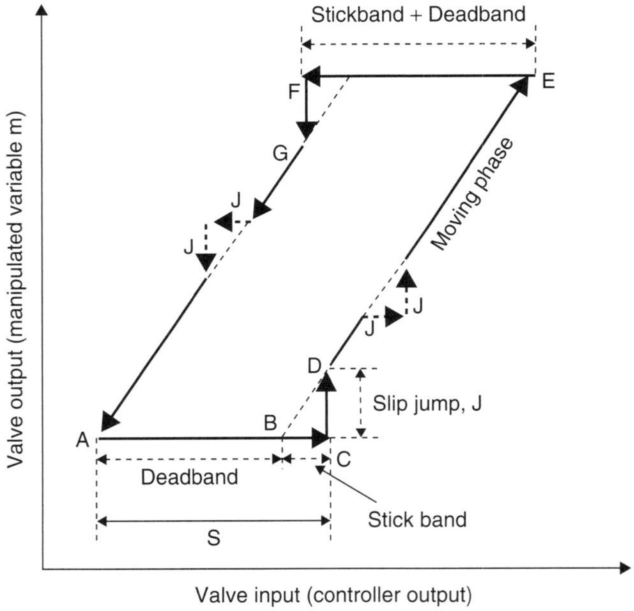

Figure 2-6 Phase plot of a sticky valve.

2.4.3.2 Stiction Different people or organizations have defined stiction in different ways. Some of these definitions have been presented by Choudhury et al., [3–4,7,9]. Based on careful investigation of real process data, a new definition of stiction has been proposed by the authors [8] and is summarized as follows.

The phase plot of the input–output behavior of a valve "suffering from stiction" can be described as shown in Fig. 2-6. It consists of four components: deadband, stickband, slip jump, and the moving phase. When the valve comes to rest or changes direction at point A in Fig. 2-6, the valve sticks, as it cannot overcome the force due to static friction. After the controller output overcomes the deadband (AB) plus the stickband (BC) of the valve, the valve jumps to a new position (point D) and continues to move. Due to very low or zero velocity, the valve may stick again between points D and E in Fig. 2-6 while traveling in the same direction [13]. In such a case, the magnitude of the deadband is zero and only the stickband is present. This can be overcome if the controller output signal is larger than the stickband only. It is usually uncommon in industrial practice.

The deadband and stickband represent the behavior of the valve when it is not moving, although the input to the valve keeps changing. The slip jump phenomenon represents the abrupt release of potential energy stored in the actuator chambers due to high static friction in the form of kinetic energy as the valve starts to move. The magnitude of the slip jump is very crucial in determining the limit cyclic behavior introduced by stiction [24,32]. Once the valve jumps or slips, it continues to move until it sticks again (point E in Fig. 2-6). In this moving phase, dynamic friction is present which may be much lower than the static friction. Therefore, "stiction is a property of an element such that its smooth movement in response to a varying input is preceded by a static part followed by a sudden abrupt jump called *slip-jump*. Slip-jump is expressed as a percentage of the output span. Its origin in a mechanical system is static friction which exceeds the dynamic friction during smooth movement" [5]. This definition has been exploited in the next and subsequent sections to quantify stiction of control valves. In the process industry, stiction is generally measured as a percentage of the valve travel or the span of the control signal [14]. For example, 2% stiction means that when the valve gets stuck, it will start moving only after the cumulative change of its control signal is greater than or equal to 2%. If the range of the control signal is 4 to 20 mA, 2% stiction means that a change in the control signal of less than 0.32 mA in magnitude will not be enough to move the valve.

2.5 DIAGNOSING CONTROL VALVE PROBLEMS

The type of valve diagnostics to be performed determines the cost savings. Plants might be unaware of the diagnostics equipment for testing a variety of valves. It is possible to diagnose problems in control valves and rotary valves (with or without instrumentation) and any valve having an air-operated actuator (spring return or double acting) either by using transducers for different pressure ranges or by preloading with additional supplies. These techniques can be used with valves that lack instrumentation. Using them allows us to examine spring ranges, valve performance, friction, seat load, and actuator leaks.

We should never make the mistake of assuming that a valve is functional just because it's new. When a valve arrives, it should be tested before it is installed so that deficient valves can be returned without setup effort or wasted time installing them. Also, it is important to make certain that valves placed in service are completely healthy. Once they are set up and tested, their properly installed functionality should be documented. But the first step in achieving effective valve diagnostics is to train personnel in control valve function. Technicians who don't understand a valve's function cannot be expected to make the valve operate consistently at peak performance. Then follow the 10 tips for diagnosing control valve problems described below.

1. *Keep good records*. This basic truth is vital to every aspect of a valve wellness program. Record every change and repair as well as details on the valve's history and current performance. During a turnaround, provide copies of these valve records and performance standards to your repair vendors. This ensures that everyone works from the same page without reams of unnecessary paperwork.

When repairs have been completed, test the valve for satisfactory performance and document the results before reinstalling the unit. Update the maintenance history and make sure that the repair shop provides its test data to the plant for archiving.

2. *Make obvious repairs up front*. Before doing anything else, repair obvious problems, such as leaky tubing and gauges. Then run diagnostics to measure three readings: transducer pressure, actuator pressure, and supply pressure. Cylinder actuators (consider these "double-acting" actuators) are a different configuration, comprised of upper and lower cylinder pressures.

3. *Test initial controlling devices*. This is the device that receives the first indication of change or problems in the process flow. Examples include transducers, temperature controllers, pressure controllers, and level controllers. Locating and testing these first may save you time, money, and unnecessary plant shutdowns. If you simply bypass the transducer, you risk overlooking the true problem with the control valve. If the transducer is mounted remotely, extend test leads to reach it.

4. *Check positioner points*. Positioners change the pneumatic pressure to operate an actuator and to overcome friction and pressure imbalance. Positioners put distance between controllers and control valves and increase the speed of a control valve's response to changes. It is imperative that positioners function correctly, especially at high pressures or for complicated chemical processes.

Accessing all positioner points is critical when measuring transducer, actuator, and overall valve performance. To troubleshoot a valve, put the valve in its fail position, disconnect the positioner, and verify that the valve's handwheel is fully backed out. Next, check the positioner and controller for physical damage. Regularly testing properly calibrated diagnostic equipment against functional valves is essential for all of these procedures.

5. *Take accurate travel measurements*. Ensure that travel-measuring centers are properly set and positioned. Diagnostics measure the travel of a control valve with a travel transducer. On globe valves, it is important to use the right scale. It is also important that the correct travel variables are used in determining the length of travel. Ensuring that the correct configuration is vertical will have a major effect on the testing results. There are to be no angles or slacks, which can result in false measurements.

The travel transducer must be placed directly on the valve stem, not on the positioner, linkage, or actuator. Direct placement on the stem effectively determines how far the valve has traveled. The ultimate objective is to find out what the valve is doing.

6. *Use correct inputs*. Ensure that the proper control valve parameters are input into diagnostic machinery. Diagnostics look at valve performance while keeping manufacturer guidelines and specifications in mind. Diagnostics equipment cannot tell if a fluid is flowing through the valve but can determine if a valve has the wrong spring or settings, if the diaphragm settings are correct, if the seals are bad, or if the positioner is producing problems arising from incorrect settings or improper maintenance.

In most cases, valves don't need to be pulled off-line and replaced. The problem may instead be with a transducer and positioner. Diagnostics identify specific problems and evaluate control valve parameters, including transducer, positioner, and actuator performance. Additionally, diagnostics can assess spring rates and bench sets, allowing for intelligent calibration.

Diagnostics can also evaluate the valve body, ascertaining stem friction, packing friction, and the seat load. Although many plant employees are not trained to interpret these data, those who know how to tweak control valve performance to conform to manufacturer guidelines, industry standards, and specific process requirements are as valuable as having an in-house repair team.

The diagnostics equipment makes its evaluation based on the variables you input. For example, effective actuator area, the action of a valve (air to open, or air to close), stem diameter, packing configuration and type, seat diameter, and instrumentation action and characteristics are critical for obtaining a correct evaluation. The final variable needed is the travel distance or rotation.

7. *Do an as-found test*. This diagnostic test determines how well a valve is performing in its current condition. It correlates the transducer's input and output signals and evaluates the positioner's input versus the actuator's applied air pressures. Taking all this into consideration helps determine overall valve performance. Evaluating the data based on manufacturers' standards determines what repairs are needed.

8. *Make control valve repairs*. Performing diagnostics enables you to make necessary replacements and repairs in the field rather than on the bench. This includes work on instrumentation (positioner, transducer, controller, or replacement), actuator (diaphragm, seals), body subassembly (packing, trim components), and adjusting bench set, as well as making necessary implementation repairs.

9. *Perform final diagnostics*. After you make the repairs, always perform final calibration testing on the valve based on the original equipment manufacturer's standards and process conditions. Baseline diagnostic testing evaluates control valves inline, and dynamic testing such as sensitivity and deadband testing further uncovers what's going on inside a valve that is dysfunctional or less than optimal.

10. *Analyze trends*. Take proactive steps toward efficiency by correcting problems early. Put your diagnostic history of valve repair into a database or diagnostic package for future use. This ensures that valves aren't repaired unnecessarily and permits spotting future problems more easily. If the same problems crop up consistently, such records are the only way to spot patterns and diagnose illness.

When vendors and repair shops receive valves with as-found test results, detailed technical notes, and plant performance standards, the expectations for repairs are clearly defined and efforts can focus on a targeted, cost-effective solution to a documented performance program. Providing detailed records minimizes the chance of making mistakes.

Ideally, the objective of a valve wellness program should be efficiency. Correcting problems early and using diagnostics at every opportunity not only saves money and catches a greater percentage of malfunctions, it also saves time by eliminating the need to shut down an entire plant to search blindly for one faulty valve.

The skills of maintenance technicians are maximized when they have a consistent, coherent plan of regular diagnostic maintenance instead of waiting to fix something after it goes bad. As many as 60% of air-operated control valves have serious performance problems, most of which are discoverable only through diagnostics. Many defects can be repaired without removing the valve from the line. Carefully maintained, tuned, and calibrated valves produce more uptime and product. Good documentation and records ensure that valve problems can be isolated faster.

Taken all together, effective valve maintenance can be achieved by strategically applying intelligent diagnostic procedures on a regular basis and keeping good records of trials, successes, and failures. Although contingencies should always be in place, using benchmarks and diagnostics can save countless hours and minimize stress while maximizing efficiency and profits.

2.6 CONTROL VALVE RELIABILITY AND SELECTION

A *Reader's Digest* story once told of a fellow hiking through Japan who came upon a man in a field working on an irrigation sluice diverter plate. The sluices had been blocked, the stem support frame removed, and the handwheel, support bushings, and even the slide rails had been removed and cleaned. The support steel had been repainted. Reassembly was in progress, all done with applications of lubricants to parts kept clean throughout the work.

The hiker complimented the worker on his attention to detail and asked how long the control plate had been in service. The response was a questioned look. The worker explained that he had been doing this annual ritual on the valves in his fields for 10 years, following his father, who had done it for the previous 35 or 36 years, and that he had learned what to do from *his* father, who had done it for an unknown number of years before that.

A little maintenance can make for significant longevity. But not every flow control application provides the conditions available to our Japanese friend. Most valves serve somewhat harsher lives, either from the fluids handled or the demands of service. That's where I hope this chapter will help you. Having experience in the fluid-handling field in a variety of industries and as many service fluids, from benign to aggressive, I offer some considerations that can assist you in selecting new or replacement control units.

1. *Selection isn't simple*. The Japanese example represents a simple system: water in an open channel, limited flow, upstream generally limited by other diversion, low pressure, frequently under observation, and a few other basic criteria. Industrial valve applications are rarely that simple. Although basic applications can be handled easily, sometimes we're called on to select components for systems that few have considered or know about. For example, the Italian engineer who devised the water distribution system for the garden fountains of Tivoli also faced unique challenges. His jump to the next level of technology introduced control by orifices and constant levels.

We can see the game getting more interesting. New applications bring new pressures, temperatures, flow rates, control accuracies, solutions, and mixture conditions that might cause us to paraphrase Thomas Paine: "These are the 'things' that try men's souls." So, let's start making the job a bit easier.

2. *Seek reliability*. The valve has been a reliable flow-management device for several hundred years. The Italians are still uncovering cast plug-type valves that were used on Caesar's barges. Roman aqueducts controlled channel flow. Dams and waterwheels have been part of our own history. Over the ages, we've developed a variety of valves to serve many purposes. The first were basically to stop or start flow.

There are reasons for our choice of valves. When we select a valve, we rely on guidelines we hope will be justified by good operation and long life. A key element in any of these choices is reliability.

3. *Consider the purpose*. A gate valve is a start–stop unit. The fluid must be clean enough to prevent buildup in the cavity where the gate fits. If the gate won't seat, the valve won't stop the flow. This concern led to double-block-and-bleed valve configurations in critical areas. Plug valves, although great for limiting clean fluid flow, have been destroyed by fluids that contain sand or other grit that erodes seat rings. The point is: correctness of application. Vendors spend time and effort asking detailed questions to assist selection decisions, but operations people often recognize that a given piece of data, by omission or commission, can make the difference. Keep the ultimate user in the loop to get better application selections.

4. *Specify suitable materials*. Part of a good selection is materials, both of manufacture and application. Although valves of durable materials may be available, the method of manufacture might also be of concern. An alloy casting process that produces fine-grained valve surfaces should prove beneficial. Corrosion resistance seems to correlate with smaller grains on exposed interior surfaces. Plastic coatings have negated some of this issue, but the surface condition must still be considered.

Using test coupons in a given stream is a good way to confirm metallurgical suitability. Some argue that the method by which the coupons are connected can influence the test results. For a particular chlorinated organic application, engineering references may suggest several suitable materials of construction. The manufacturing process can, however, leave corrosion residue on the coupons. Testing may reveal that what seemed to be a relatively pure stream contained enough corrosion promoter to restrict use of the materials suggested. A final choice might be polymer-lined valves.

5. *Location, location, location*. Valve accessibility is an important variable. Placing one where it cannot help but accumulate even a small amount of solids makes failure probable. Installing a valve in a downward vertical run upstream of an elbow is far superior to placing it in a horizontal run just after an elbow.

Placing a valve far from structural support exposes a line to strains from operation shock. One plant with mostly polymer piping has a corporate guideline stating that valves should be supported rigidly. As a result of this rule, lines between valves need little attention except in longer runs.

6. *Ensure smooth operation*. The manner of valve operation is also relevant. Your choices are more a function of the actual piping layout and arrangement than of the piking and instrumentation diagram (P&ID). This selection is often a field choice during plant construction. Fast-track installation directly from P&IDs means that to make proper choices owner's job-site representative must be cognizant of the needs of operation. Placing an automatic operator close to walls or where movement cannot be observed can lead to production delays and maintenance headaches.

Valves with rising stems can bring hand-pinch problems, and quarter-turn valves can be knuckle busters. Although these issues are rare in well-developed plants, our old friend Murphy tells us that when they arise, it will be in the worst place at the worst time. We have all seen drain valves for dangerous liquids installed so that the discharge points directly into a wall or equipment base where splashback is a serious safety concern.

7. *Pick appropriate seals*. A stem seal or packing is another source of concern because a seal that binds or dissolves in the service medium is an obvious problem. Also, anything that can damage packing or score a shaft is important. This brings us back to the application, and the circle can start again, often with additional possible side branches for more selection questions.

8. *Consider time and frequency*. Frequency of operation can drive a sane choice crazy. Awhile ago I had to consider how much movement there was on a control valve. The first answer was "None. It's in control, so the net movement is zero." But after research and discussions with control valve vendors, it became more reasonable to suggest a 10% movement on a 30-s cycle. This led to a better understanding of how much motive effort was needed to maintain control and an appropriate design for the pneumatic part of the control system.

The frequency of manual operation is also a factor. If once-a-year operation is typical and the location isn't ideal, manual operation might be acceptable, but for weekly or monthly operation you might need to provide another appropriate method. Remote manual operators have been worthwhile investments for some of these situations. They can prevent complicated drop legs and high-pressure drops that absorb energy continuously.

9. *Avoid entrapment*. If you're designing a slurry system, don't place valves near dead-end turns and in long vertical lifts. Maintaining fluid velocity is the goal, but follow-up flushes can prevent some problems.

Clear liquids aren't immune. Consider a vessel feed system where several liquids enter through the same nozzle. If the fluids react or the solutes undergo a phase change, reliability can be aided through use of a block valve at the nozzle or a feed system flush using the bulk fluid in the vessel.

Phase changes can subject valves to significantly different conditions that might not be obvious. Boiler blowdown, for example, often discharges directly to low-pressure lines. The combination of scale buildup, suspended solids, and sonic velocity from flashing discharge can form a grinding medium. A bit from line length downstream from the valve can help retain liquid conditions and reduce velocity through the valve. Such a downstream flash chamber is a lot easier to replace than a blowdown valve.

10. *Consult operators and technicians*. Whether you're upgrading an existing system or designing a new one, get to know the operators. Talk to them about the suitability of the valves currently in use to help determine which types are reliable and which are not. Remember the oft-neglected truism that the most expensive valve is the one that burns up more time each day than it is worth.

Having said all this, it is still difficult to provide a unified checklist for valve selection, because new applications often bring new problems. Vendors are still a great reference source, but operations and plant maintenance personnel are closer to your situation and should be consulted. An unscheduled shutdown can easily overwhelm any savings derived from an inadequate valve. Reliability is the aim, because plants make money only when they run.

2.7 CONTROL VALVE MAINTENANCE

Reactive maintenance is usually defined as running equipment until it fails, with no planned maintenance. With reactive maintenance, no actions or efforts are made to monitor or maintain the equipment, which can decrease equipment life.

Having some form of program in place involving regularly scheduled maintenance is described as *preventive maintenance* or *scheduled maintenance*. This type of maintenance program has specific maintenance actions performed on a time- or machine-runtime-based schedule. Preventive maintenance programs are set up to detect or lessen the degradation of specific components or systems. Overall, equipment performance and life can be lengthened significantly. These programs increase the reliability of the equipment, can be scheduled as part of a routine program, and can increase equipment life while garnering energy savings. A 12 to 18% cost savings can be achieved over a reactive maintenance program. One disadvantage is still the potential for failure that could occur between scheduled maintenance periods.

Predictive maintenance programs incorporate some type of measurement and/or monitoring of equipment in order to observe or predict equipment degradation or failure. These measurements are able to detect the onset of problems or degradation of the equipment or a particular mechanism within the equipment before partial or total failure occurs. Predictive maintenance is performed based on the actual state of the equipment rather than on a preset schedule. Predictive maintenance can increase equipment life and decrease downtime. Through monitoring, preemptive measures can be taken to prevent valve failure, increasing environmental safety as well.

Predictive maintenance is the best way to extend the life of valves. When making repairs on more sophisticated valve types, always make use of available manufacturers' manuals. As soon as you observe a leak, determine its cause and then apply the proper corrective maintenance, and later, follow its predictive maintenance. Maintenance may be as simple as tightening a packing nut or gland. A leaking flange joint may only need to have the bolts tightened or a new gasket inserted. If allowed to collect, dirt and scale will cause leakage.

Whenever you are going to install a valve, be sure you know the function the valve is going to perform: that is, whether it must start flow, stop flow, regulate flow, regulate pressure, or prevent backflow. Always inspect the valve body for the information that is stamped on it by the manufacturer: for example, type of system (oil, gas, or water), operating pressure, operating temperature, and direction of flow.

You should also know the operating characteristics of the valve, the metal from which it is made, and the type of end connection with which it is fitted. Operating characteristics and the materials are factors that affect the length and type of service that a valve will give; end connections indicate whether or not a particular valve is suited to the installation.

When you install valves, ensure that they are readily accessible and allow enough headroom for full operation. Install valves with stems pointing upward if possible. The position of a stem between horizontal and straight up is acceptable, but avoid the stem pointing downward, because if allowed, sediments will collect in the bonnet and score the stem. Also, in a line that is subject to freezing temperatures, liquid that is trapped in the valve bonnet may freeze and rupture the bonnet.

2.7.1 Detecting Control Valve Stiction

There are two types of methods of detecting stiction: invasive and noninvasive. The invasive method requires putting the loop in manual mode and then stroking or traveling the valve over its full travel span when in or out of service. This is called the *valve travel test* in the Instrument Society of America (ISA) standards [19,20]. Using this type of test, stiction can be quantified as the amount of change required in the control signal to move the valve from its stuck position [15]. Since it is neither feasible nor cost-effective to test hundreds of valves at a plant site, the noninvasive method is preferred to the invasive method. Horch's cross-correlation method for detecting stiction is popular among the noninvasive methods reported so far. Horch's method [16–18] detects stiction with the use of the cross-correlation function between the process variable and the controller output. Their method is not applicable for processes containing an integrator (e.g., a level control loop) or for loops carrying a compressible medium (e.g., steam or air). Their method is useful primarily for flow control loops. Even for flow control loops it sometime produces inconclusive results [12]. Also, if a sinusoidal disturbance has entered the control loop, the method falsely detects stiction in the control valve. Moreover, none of the existing methods can quantify stiction. A method based on higher-order statistics has been developed [6] that can detect as well as quantify stiction and is applicable for all types of control loops. The method first examines the presence of nonlinearity in a control loop. If nonlinearity is detected,

TABLE 2-2 Common Control Valve Problems and Solutions

Possible Cause	Remedy
Control Valve Severe Plugging	
Is the flow rate higher than normal?	Adjust the handle stem to achieve the desired/normal output.
Is the pressure gauge functioning?	Replace the faulty gauge.
Is a particle of debris caught in the port under the valve flap?	Check for obstruction. If found, duplicate the cam opening to flush the ports.
Is a pump discharge filter installed?	Confirm that the filter is installed and without fault.
Is the piston/orifice blocked?	Perform a backflush of the orifice/piston until the fluid flows freely and normally.
Excessive Noise	
Is the valve seat faulty or worn out?	Replace the valve seat assembly.
Is the oil viscosity too high?	Check for oil purity, separator efficiency, and status.
Is an improper spring installed behind the valve?	Ensure proper number, proper alignment, and correct thickness of springs.
Is the pressure setting too close to that of another valve in the circuit?	Adjust the pressure gauge.
Improper Flow	
Is the fluid too hot?	Check the status and operation of the fluid cooler.
Is the valve not adjusted properly?	Check for proper installation and adjust properly.
Is the orifice restricted?	Clean the orifice.

the process variable, set point, and controller output data are used to diagnose the possible causes of nonlinearity. A recent book by Jelali and Huang [22] compares eight different nonintrusive methods of detecting valve stiction on a benchmark data set consisting of 89 control loops from various industries. It was found that the bicoherence-based method proposed by Choudhury et al., [6] stood best among these methods.

2.8 CONTROL VALVE TROUBLESHOOTING

Troubleshooting begins with the piping configuration, physical orientation, and application data, such as system pressure, temperature, and the physical properties of the fluid contained. You'll need to address installation issues such as stress from vessel and piping expansion and unsupported discharge lines. In addition, improper valve application can result in mechanical damage from elevated temperature, backpressure, material incompatibility, or incorrect pressure setting.

The first step in troubleshooting a control valve problem is to make sure that it is installed properly. Table 2-2 provides guidance on troubleshooting three common control valve problems [29].

REFERENCES

1. Bialkowski, W. L. Dreams versus reality: a view from both sides of the gap, *Pulp and Paper Canada*, vol. 94, 1992, pp. 19–27.
2. Baumann, H. D. *Control Valve Primer: A User's Guide*, Instrument Society of America, Research Triangle Park, NC, 2009.
3. Choudhury, M. A. A. S., Detection and Diagnosis of Control Loop Nonlinear-ities, Valve Stiction and Data Compression, Ph.D. dessertation, University of Alberta, Edmonton, Alberta, Canada, 2004.
4. Choudhury, M. A. A. S., Jain M., and Shah S. L., Stiction: definition, modelling, detection and quantification, *Journal of Process Control*, vol. 18, 2008, pp. 232–243.
5. Choudhury, M. A. A. S., Kariwala, V. Shah, S. L., Douke, H., Takada, H. and Thornhill, N. F., A simple test to confirm control valve stiction, *16th IFAC World Congress*. Prague, Czech Republic, 2005.
6. Choudhury M. A. A. S., Shah S. L., and Thornhill N. F., *Diagnosis of Process Nonlinearities and Valve Stiction: Data Driven Approaches*, Springer-Verlag, New York, 2008.
7. Choudhury, M. A. A. S., Thornhill, N. F., and Shah, S. L., A data-driven model for valve stiction, *Proceedings of the IFAC Symposium on Advanced Control of Chemical Processes (ADCHEM)*, Hong Kong, 2004.
8. Choudhury, M. A. A. S., Thornhill, N. F., and Shah, S. L., Diagnosis of poor control loop performance using higher order statistics, *Automatica* vol. 40, 2004, pp. 1719–1728.
9. Choudhury, M. A. A. S., Thornhill, N. F., and Shah, S. L. Modelling valve stiction, *Control Engineering Practice*, vol. 13, 2005, pp. 641–658.
10. Cla-val Technical Products Department, Pump. and systems, http://www.pump-zone.com/valves/valves/cavitation-prevention.html, 2008.
11. Desborough, L., Miller, R., and Nordh, P. Regulatory control survey, unpublished manuscript, Honeywell, Morristown, NJ, 2000.
12. Desborough, L., Nordh, P., and Miller, R., Control system reliability: process out of control, *Industrial Computing*, vol. 8, 2001, pp. 52–55.
13. *EnTech Control Valve Dynamic Specification*, Version 3.0, EnTech, Texas, USA 1998.
14. Fisher-Rosemount, *Control Valve Handbook*. Fisher Controls International, Marshalltown, IA, 1999.
15. Gerry, J., and Ruel, M., How to measure and combat valve stiction online, *Proceedings of the ISA International Fall Conference*, Houston, TX, http://www.expertune.com/articles/isa2001/StictionMR.htm, 2001.
16. Horch, A., A simple method for detection of stiction in control valves, *Control Engineering Practice*, vol. 7, 1999, pp. 1221–1231.
17. Horch, A., Condition Monitoring of Control Loops, Ph.D. dissertation, Royal Institute of Technology, Stockholm, Sweden, 2000.
18. Horch, A., Isaksson, A. J., and Forsman, K. Diagnosis and characterization of oscillations in process control loops, *Proceedings of Control Systems*, Victoria, British Columbia, Canada, 2000, pp. 161–165.
19. ISA Committee SP51, *Method of Evaluating the Performance of Positioners with Analog Input Signals and Pneumatic Output*, Technical Report ANSI/ISA-75.13–1996, Instrument Society of America, Research Triangle Park, NC, 1996.
20. ISA Committee SP51, *Control Valve Terminology*, Technical Report ANSI/ISA-75.05.01–2000, Instrument Society of America, Resarch Triangle Park, NC, 2001.
21. ISA Subcommittee SP75.05, *Process Instrumentation Terminology*, Technical Report ANSI/ISA-S51.1–1979, Instrument Society of America, Research Triangle Park, NC, 1979.
22. Jelali, M., and Huang, B. (Eds.), *Detection and Diagnosis of Stiction in Control Loops: State of the Art and Advanced Methods*, Springer-Verlag, New York, 2010.
23. McDonald, D., *Practical Control Valve Sizing, Selection and Maintenance*, IDC Technologies, West Perth, Western Australia, 2008.
24. McMillan, G. K., Improve control valve response, *Chemical Engineering Progress*, vol. 91, 1995. pp. 77–84.
25. Nwaoha, C., Extending control valve life by proper selection and maintenance, *Pipeline and Gas Journal*, vol. 235, no. 11, November 2008, pp. 78–80.
26. Nwaoha, C., Controlling fluid flow through correct control valve operation, *Oil Review Africa*, no. 2, March–April 2009, pp. 62–64.
27. Nwaoha, C., Extending the life of control valve: through proper selection operation, *Steam and Boiler Review*, vol. 3, no. 6, June 2009.

28. Nwaoha, C., Improving valve performance, *Pipeline and Gas Technology*, vol. 8, no. 2, March 2009, pp. 56–59.
29. Nwaoha, C., Maintaining valves the right way, *Valve World*, vol. 14, no 3, April 2009, pp. 75–76.
30. Nwaoha, C., The final element: tips for maintenance, *Control Engineering Asia*, no. 3, March 2009, pp. 28–29.
31. Nwaoha, C., Control valve actuators: key considerations for effective technology selection, *Flow Control Magazine*, vol. xvi, no. 2, February 2010, pp. 34–35.
32. Pipponen, J., Controlling processes with nonideal valves: tuning of loop. and selection of valves, *Proceedings of Control Systems*, Chateau, Halifax, Nova Scotia, Canada, 1996. pp. 179–186.
33. Shahda, J., Consideration of pressure parameters and trim fluid velocity in valve liquid applications, http://www.controleng.com/article/CA492625.html, 2005.
34. The Chemical Engineers' Resource Page, http://www.cheresources.com/valvezz.shtml, accessed March 10, 2011.
35. Tullis, J., P., *Hydraulics of Pipelines: Pumps, Valves, Cavitation and Transients*, Wiley, New York, 1989.

3

PUMPS

CRAIG REDMOND
The Gorman-Rupp Company, Mansfield, Ohio

It has been said that other than electric motors, more pumps are produced than any other mechanical device in the world. Pumps are all around us: bringing irrigation to dry fields, water to put out a fire, pushing petroleum products through a refinery, and bringing drinking water into our homes while pushing sewage back out.

Despite the fact that pumps are so common, it is worth providing a working definition to build on. In the context of this chapter, the term *pump* is used for a mechanical device that moves a liquid from one location to another. Included in the concept of a mechanical device is the fact that it has moving parts and requires some form of energy source to operate.

A few other definitions will prove useful as we discuss pumps.

1. *System*: the entire flow path that a liquid travels from its source to its final destination. This definition includes piping, hose, tubing, valves, filters, and the pump itself. By this definition the same pump can be a part of a countless number of systems in different locations.

2. *Application*: an individual combination of the system and the liquid being pumped, along with all of its physical properties. By this definition the same system can be used for a countless number of applications. For example, the system pumping hot water today could be pumping cold water next week, although these constitute different applications.

3. *Flow rate*: an easily understood term that represents the volume of fluid pumped in a given period. Some typical units of measurement include liters per second, cubic meters per hour, and gallons per minute.

4. *Head*: in its simplest form, a measure of the energy added to the fluid by the pump. In many applications it is easily checked in the field by subtracting the suction gauge reading from the discharge gauge reading (both gauges must be reading in the same units). If the pipe diameters where these readings are taken are of different size, the difference in velocity head also needs to be taken into account. The velocity head is calculated easily by squaring the velocity of the fluid and then dividing by $2g$ ($2 \times$ gravity; the velocity should be in meters per second and $g = 9.81$ m/s^2; conversely, if using English units, the velocity should be in feet per second and $g = 32.2$ ft/s^2). To maintain the same flow rate through two different pipe sizes, the velocity must be higher in the smaller pipe, which means that there is a higher-velocity head in the smaller pipe. A third factor to consider is whether there is an elevation difference between the two gauge readings. If so, this also needs to be taken into account.

5. *Specific gravity*: the ratio of the density of a fluid divided by the density of water. A specific gravity of <1.0 is lighter than water; a specific gravity >1.0 is heavier than water.

There are numerous different types of pumps, but all pumps require an external source of energy. Although this is most often provided by an electric motor, many pumps are driven by alternative means. Pumps can also be driven by hydraulic motors, engines, air, or even by hand. As with every other mechanical device ever invented, the useful output of a pump is always less than the power input. The ratio of the useful output of a pump (often called the *water horsepower*) divided by the input power into the

Process Plant Equipment: Operation, Control, and Reliability, First Edition. Edited by Michael D. Holloway, Chikezie Nwaoha, and Oliver A. Onyewuenyi.
© 2012 John Wiley & Sons, Inc. Published 2012 by John Wiley & Sons, Inc.

pump (often called the *brake horsepower*) is the efficiency rating of the pump. The losses within the pump can be categorized primarily in two ways. Mechanical losses are due to friction, which would include losses from bearings and seals. Hydraulic losses are due to the internal friction losses within the pump body itself.

For all pump types, a variety of factors affect the efficiency. Several of these are under the pump manufacturer's control (the actual pump design itself); others are application related (speed, viscosity, flow rate). Most manufacturers will advertise their pump efficiency based on a set of "standard" conditions, such as at a standard temperature and pressure, pumping clean water.

Although there is no governing body that regulates the construction, installation, or operation of pumps per se, the Hydraulic Institute sets voluntary guidelines. The Hydraulic Institute is a trade association comprised primarily of pump and seal manufacturers that have agreed to work together to define a set of standards. Most of what is included here is in agreement with the Hydraulic Institute's published standards. However, if strict compliance with Hydraulic Institute standards is required, one should go directly to the standards themselves.

Other specific manufacturers that typically target specific process industries may comply with either the American National Standards Institute (ANSI) guidelines or with the American Petroleum Institute (API) guidelines. If there are local or national law or codes that govern the application in question, ensuring that the pump complies with those codes is a necessary step to take.

3.1 TYPES OF PUMPS

Due to the variety of liquids that require pumping, there are numerous types of pumps. A different type of pump is used to move tar than that to move food products than that to move cooling water at a power plant than that to move industrial wastewater at a chemical plant. Pumps in general are categorized as either positive-displacement pumps or dynamic pumps, the most common of which are centrifugal pumps. Although we discuss several types of pumps, we focus on centrifugal pumps. A great many of the principles discussed apply to a broad range of pumps.

3.1.1 Positive-Displacement Pumps

Positive-displacement pumps are named after the fact that they displace nearly the same amount of liquid per cycle regardless of system pressure. Positive-displacement pumps are further broken into two categories: reciprocating and rotary. Depending on the type of positive-displacement pump in question, a cycle can be viewed as either one complete revolution or one complete stroke. To vary the flow rate of a positive-displacement pump, the operator must change the pump speed, and/or the length of the stroke.

Positive-displacement pumps have countless applications but are quite often the best selection for pumping higher-viscosity liquids. Due to the mechanical design of most positive-displacement pumps, there is a much smaller impact due to viscosity on the pump's flow rate and efficiency than on those of a centrifugal pump. It has been said that if you can make the liquid flow into the pump, certain positive-displacement pumps can move it.

When selecting a positive-displacement pump, it is very important to be sure that appropriate consideration is given to proper sizing of the driver. Higher-viscosity liquids often are accompanied by higher-specific-gravity fluids. Both of these factors can greatly increase the power required by the pump. Although some small positive-displacement pumps can rotate at synchronous motor speeds, it is not uncommon to need a gear reduction drive or some other arrangement to match driver speed to pump speed. Consult the pump manufacturer's application information to ensure that there will be enough power to drive the pump for each application.

While all pumps generate pressure, it is especially important with positive-displacement pumps to have an automatic pressure relief valve on the discharge side of the pump if there is any chance at all of the pump ever operating against a closed discharge valve.

3.1.1.1 Reciprocating Pumps Reciprocating pumps use a cyclical vertical motion to impart the energy necessary to cause pumping to occur. These units most often translate rotational motion into linear motion through the use of a cam/eccentric arrangement, but can also be air driven. Consider the motion of a piston within an engine block to help envision how a reciprocating pump operates.

Reciprocating pumps can be compared to engines in other ways, too. Increasing the number of chambers can increase the flow rate, just as more cylinders increase an engine's power output. Increasing the number of chambers, with each cycle out of phase with every other, can help smooth out the flow through the pump, just as a two-cylinder engine runs smoother than a single-cylinder engine. Some pump manufacturers recommend the use of accumulators on either the suction or discharge plumbing (or perhaps both) to help reduce some of the pulsation inherent in reciprocating pumps.

It can be difficult to measure the performance of a reciprocating pump. Gauge readings can fluctuate throughout each cycle and the flow rate can pulsate, making flowmeter readings unreliable. The best way to measure the actual output of a reciprocating pump is to record the change in liquid level over a period of time in either the suction source or the discharge destination. Using this information, one can calculate the volume of liquid displaced over that time span.

Dividing the volume of liquid displaced by the time span will give you the net flow rate of a reciprocating pump.

Although other forms of reciprocating pumps, such as bilge pumps and bellows pumps, exist, just a few of the more common reciprocating pumps are discussed here.

3.1.1.1.1 Diaphragm Pumps Diaphragm pumps (Fig. 3-1) are among the most common types of pumps in the world. Since diaphragm pumps are used with so many different liquids, many different materials of construction are available. The diaphragm itself is the most critical component of the pump, as it must flex with each cycle and be compatible with the liquid being pumped. There must be nonreturn valves on both the suction and discharge side of the pump; otherwise, the motion of the diaphragm would cause a "plunger" motion, with the flow going back and forth within the piping instead of being forced through the piping.

When all components are operating properly, the start of the upward stroke of the diaphragm will simultaneously pull open the suction valve and pulls the discharge valve closed. As the diaphragm pulls upward, liquid flows into the housing until the diaphragm reaches the top of its stroke. Once the diaphragm begins its downward stroke, the suction valve is forced closed and the discharge valve opens. Flow leaves the pump until the diaphragm reaches the bottom of its stroke. This process repeats over and over with each cycle.

3.1.1.1.2 Piston Pumps The operating principle behind a piston pump is identical to that of a diaphragm pump. The primary difference is that rather than a flexible diaphragm flexing with each cycle, a piston pump uses a rigid body to promote flow.

3.1.1.2 *Rotary Pumps* While reciprocating pumps use a linear motion to induce motion in a liquid, rotary pumps rely on rotational motion. Rotary pumps typically generate a much smoother flow than that generated by a reciprocating pump, thus eliminating the need for accumulators. Pressures developed may be lower in some cases, but are still often well above what can be obtained with a similarly sized centrifugal pump.

Figure 3-1 Mechanically driven diaphragm coupled to an electric motor. A diaphragm pump like this has a wide variety of uses, but is typically used on viscous fluids. (Courtesy of The Gorman-Rupp Company.)

Whereas it is very easy to envision the pumping action of a reciprocating pump, rotary pumps pose a bit more of a challenge. Typically, there is a rotating member located in a housing with both a suction and a discharge port. When the rotating member passes initially by the suction port, there is just a small volume of space between it and the housing. As it rotates toward the discharge port, this volume increases, drawing liquid into that volume. This volume reaches a peak, at which point no more flow comes into that volume. Further rotation begins to contract the volume, forcing it out the discharge port. There are usually multiple members on or attached to the rotating member that create separate, discrete volumes.

Some rotary pumps can be operated in either direction with limited impact on a pump's performance. If the pump is to be run in both directions as a regular course of action, it is advisable to ensure that the system has appropriate pressure relief mechanisms in place for both directions of flow.

Although other forms of rotary pumps, such as peristaltic and screw pumps, exist, some of the more common forms are discussed here.

3.1.1.2.1 Gear and Lobe Pumps There are two different types of gear pumps: external tooth and internal tooth. The external tooth design, sometimes referred to as *spur gear design*, involves two identical gears; the internal tooth design has two different gears, one inside the other (Fig. 3-2). Both operate largely on the same principle: that there is a vacuum created as the space between two gear teeth increases and then as two gear teeth come closer together, pressure is generated. In either design, one gear is driven directly by a shaft and the other is forced to rotate due to the mesh of the gear teeth. The driven gear is referred to as the *rotor*, and the driven gear is the *idler*. In an internal tooth gear pump, the rotor is the outer, larger gear. Rotary lobe pumps operate almost identically to external tooth gear pumps.

Gear and lobe pumps rely on very tight tolerances and precise fits to prevent internal leakage, also known as *recirculation*. While all pumps can suffer performance degradation due to abrasive wear, gear and lobe pumps either need to be used on liquids free of abrasives or in conjunction with wear-resistant components to ensure a long useful life for the pump.

3.1.1.2.2 Vane Pumps Vane pumps, often referred to as *rotary vane pumps*, rely on the radial motion of vanes

Figure 3-2 Gears used in an internal gear pump. The outer gear, the rotor, is driven by the pump shaft; the inner gear, the idler, is driven by the rotor. These pumps are used primarily on viscous liquids. (Courtesy of The Gorman-Rupp Company.)

within a rotor that is positioned eccentrically within a housing. As with the other rotary pumps covered here, the pumping action is caused first by an increasing volume on the suction side that draws the fluid in; then, as that area contracts during the rotor's revolution, it forces the fluid out the discharge side.

As one may expect based on the pump's name, the vanes themselves are the most critical part of a vane pump. The material selection for the vanes is important. The presence of abrasives in the pumpage, temperature, the possibility of running dry, and also the operating pressure of the pump are all factors that should be considered when selecting vane material.

3.1.1.2.3 Progressive Cavity Pumps Progressive cavity pumps are used to pump shear-sensitive liquids or fluids containing delicate solids that cannot be damaged during the pumping cycle. This type of pump relies on many tight machined tolerances within the pump to minimize internal leakage. The rotor spins within the housing, similar to a screw being tightened. The fluid is pushed along by the "threads" on the rotor. Although progressive cavity pumps can be relatively expensive up front, there are certain applications where they are the best choice, due to the pumpage. Progressive cavity pumps also create a very smooth flow as the rotor is producing a consistent flow rate throughout each cycle.

3.1.2 Dynamic Pumps

Unlike positive-displacement pumps, which produce a given amount of flow per cycle, dynamic pumps can operate at a range of flows for any given rotational speed. This range of flows is referred to as the pump's *operating range*. Pump manufacturers will typically show the entire range of performance on what is known as the *performance curve*, a plot of head versus flow. Typically as flow increases, head will decrease. Most pump performance curves will show multiple lines representing either various impeller trims or rotational speeds. Additional data typically shown on a pump performance curve include the net positive suction head required (for more on this, see Section 2.4.1 on cavitation), efficiency, and the power required.

One common characteristic of dynamic pumps is that while the discharge pressure measured in kPa or psi varies depending on the specific gravity of the fluid pumped, the discharge head, measured in terms of the fluid being pumped, remains constant. In other words, if the pump is capable of producing a total dynamic head (TDH) of 15 m at 100 L/s when pumping diesel fuel, that same pump is also capable of producing a TDH of 15 m at 100 L/s when pumping water. The relationship between TDH and discharge pressure is that TDH in meters of pumpage is = (discharge pressure in kPa)(0.102)/(specific gravity of the pumpage); conversely, in English units, TDH in feet of pumpage is = (discharge pressure in psi)(2.31)/(specific gravity of the pumpage).

3.1.2.1 Centrifugal Pumps Despite being the most common process pump in most industries, centrifugal pumps are often misunderstood. These devices are relatively simple, but can be tricky in their application. Centrifugal pumps come in many varieties, including submersible, multistage, engine-driven, self-priming, and prime-assisted (Fig. 3-3).

3.1.2.1.1 Basic Components and Ways to Categorize

1. *Volute*: the primary outer portion of the pump. Technically, it is the portion of the pump surrounding the impeller that collects and redirects the flow from the impeller out through the discharge port of the pump. Each volute has at least one *cutwater* (sometimes called a *casing splitter*). Volutes are sometimes also referred to as the pump housing or pump casing.

Volutes can be further categorized into various types. A scroll type of volute is one with a constantly increasing cross-sectional area as one moves from the cutwater toward the pump discharge. A double volute is one with two cutwaters, spaced 180° from each other. The double volute has the added benefit of reducing the radial loads applied to the impeller due to the varying pressures within the volute itself. A circular volute, sometimes referred to as a concentric volute, has a constant cross-sectional area all

Figure 3-3 A centrifugal pump, the most common type of pump used in industry. (Courtesy of The Gorman-Rupp Company.)

360° around the impeller except for the portion that opens to the pump discharge.

Furthermore, volutes can be designed to be self-priming through the use of recirculation and priming ports. Self-priming volutes are often significantly larger than traditional volutes and often appear to be a volute inside a box. Non-self-priming volutes can also be referred to as being straight centrifugal volutes. If automatic priming is desired, another option is to use a priming assist unit. In this situation, a straight centrifugal volute is equipped with an auxiliary vacuum source to draw the fluid into the eye of the impeller.

2. *Diffuser*: portion of the pump immediately around the impeller that allows the fluid to slow down prior to exiting the pump. A diffuser can be viewed similar to a volute with more than two cutwaters.

3. *Impeller*: the rotating member of the pump that induces the actual flow. Flow will typically enter the impeller axially through its eye, which is located at the axis of rotation, then exit the impeller radially. Through imparting kinetic energy to the fluid being pumped, the impeller induces flow through the pump. All impellers have at least one vane that extends at least partially in a direction parallel with the axis of rotation. Unless they are for pumps handling solids, most impellers will have several vanes. Impellers are occasionally referred to as rotors, turbines, or propellers.

Impellers are further categorized in several ways. The most common way is to look at the shrouding on the impeller. If there is no shrouding at all on either side of the vanes, it is categorized as an open impeller. An impeller with a shroud on one side, typically called the *hub*, is called a partially open impeller. An impeller with shrouding on both sides is referred to as an enclosed, or shrouded, impeller (Fig. 3-4). In general, an enclosed impeller is typically associated with higher efficiencies and heads, while an open or semiopen impeller is associated with higher flows and handling solids in the fluid.

Impellers are also categorized based on the number of inlets the impeller may have. Most impellers are single-suction impellers, meaning that the flow enters the impeller from one side only. Double-suction impellers appear like back-to-back single-suction impellers. Typically, the shaft extends through the eye of both sides of the impeller and is supported by a bearing on each side.

Impellers are also categorized based on the type of flow the pump generates. A high-head low-flow pump has a radial flow impeller. A low-head high-flow pump has an axial flow impeller. Heads and flows in between typically have a mixed-flow impeller. These types are further defined by looking at a dimensionless parameter known as the specific speed.

4. *Wear ring*: used only with enclosed impellers, this creates a tight clearance around the impeller eye to reduce internal leakage. Wear rings can be attached to the volute, the impeller, or both. Historically, the wear ring material was made out of something softer than the volute and impeller materials, so it was a sacrificial part. More recent times have led to using harder materials for the wear rings to help reduce the mean time between failures (MTBF). As the wear ring wears and the radial clearance between it and the impeller increases, more recirculation will be allowed, lowering both the pump's head and the efficiency. Wear ring clearances cannot be renewed without replacing the parts.

Figure 3-4 Impellers are classified in many different ways. Two common impeller types are depicted here. On the left is a semiopen impeller and on the right is an enclosed impeller. Different impeller types offer different benefits. (Courtesy of The Gorman-Rupp Company.)

5. *Wear plate*: used with open and semiopen impellers, this creates a tight clearance at the face of the impeller vanes to reduce the internal leakage. The gap between the impeller vanes and the wear plate is known as the *face clearance*. As the face clearance opens up, the head and efficiency will drop. If the pump happens to be a self-priming pump, its priming ability will also begin to decrease as the face clearance increases. Similar to wear rings, wear plates are now being made of harder materials, to reduce the MTBF. On many pumps with wear plates, the clearances can be reset to get additional life out of the parts before they need to be replaced. Some units require disassembly to reset this clearance; others permit it to be done externally.

6. *Seal*: generally used to refer to the component(s) used to prevent the fluid from escaping along the rotating shaft. Seals are one of the most common service items in most applications. Many factors go into proper seal selection, including temperature, pressure, chemical compatibility, and the possibility of the presence of abrasives. A mechanical seal (Fig. 3-5) is generally a spring-loaded arrangement that presses two highly polished seal faces together, one that rotates with the shaft and one that remains stationary. Packing, on the other hand, is braided material that is placed along the shaft in concentric rings. These rings are then squeezed so that they conform to the housing bore but allow a small amount of leakage along the shaft. Packing, then, does not really seal; it reduces leakage. Although some applications may still require packing, in most applications it is considered out-of-date technology.

It is critical to follow carefully the installation directions given in the pump owner's manual, as seals can be very sensitive to assembly problems. In particular, the seal faces themselves need to be kept as clean as possible. If seals are failing repeatedly in a given application, the user should consult with the pump manufacturer to ensure that the proper seal is being used. Sometimes a simple change of material can result in a dramatically increased seal life.

For some highly corrosive applications, magnetic drive pumps can be considered. These units do not have a seal and rely on the magnetic field being strong enough to spin the impeller inside the housing without any direct mechanical contact with that impeller. Unfortunately, due to cost limitations on the magnetic drive concept, the overall performance range of magnetic drive pumps is relatively limited compared to most other types of pumps.

Cartridge, Packing, Single spring

Figure 3-5 Sometimes referred to as the heart of a pump, the mechanical seal is what keeps the fluid being pumped from leaking along the shaft. Depicted from left to right are a cartridge seal, mechanical packing, and a common single-spring seal design. (Courtesy of The Gorman-Rupp Company.)

3.1.2.1.2 Special Variations Just as there are many different variations of automobiles, such as sports cars, passenger vans, and pickup trucks, there are many specialized versions of centrifugal pumps.

1. *Chopper*: as solids pass through the pump, they are cut up into smaller pieces. Used primarily either as part of a process where smaller solids are required downstream or to pump stringy solids that may clog a more traditional pump.

2. *Magnetic drive*: the impeller is driven by a rotating magnetic field, not mechanically. Although relatively expensive, this type of pump eliminates mechanical seals and is generally used on applications that are very tough to seal or those with highly acidic or caustic pumpage, where it is critical to have zero leakage.

3. *Multistage*: when a pump has more than one impeller. Although the vast majority of centrifugal pumps have only one impeller, when discharge pressures required are high, a multistage pump is often required. In such an arrangement, the flow from one impeller is directed internally into the eye of a different impeller. This process repeats until the flow has passed through all the stages. Impellers on most multistage pumps are typically identical and are positioned in alternating orientations, which reduces the axial load on the shaft.

4. *Prime-assist*: a straight centrifugal pump that incorporates an auxiliary device that creates a vacuum (Fig. 3-6). Used primarily due to ease of operation, when long or oversized suction lines need to be evacuated or when a flooded suction is not practical.

5. *Self-priming*: capable of lifting the liquid on the suction side up to the pump automatically (Fig. 3-7). Used primarily due to ease of service or when operating on a flooded suction is not practical.

6. *Slurry*: capable of handling a liquid with a relatively high concentration of solids. Slurry pumps are typically manufactured using hard materials, due to the often abrasive nature of the pumpage.

7. *Submersible*: the entire pump, including the driver, is capable of being fully submerged in the pumpage

Figure 3-6 Although there are multiple ways to create a vacuum, this prime-assist pump uses an air compressor (mounted above the pump shaft) to drive a venturi. (Courtesy The Gorman-Rupp Company.)

Figure 3-7 Self-priming pumps serve many different purposes and can handle more entrained air than can a typical centrifugal pump. The pump depicted is also capable of pumping large solids. (Courtesy of The Gorman-Rupp Company.)

(Fig. 3-8). Submersibles are typically powered by either an electric motor or a hydraulic motor. Used primarily either for portability (no suction plumbing required) or when the net positive suction head (NPSH) dictates.

8. *Vortex*: the face clearance is excessively large such that the impeller itself sits either completely or nearly completely out of the flow path in the pump (Fig. 3-9). Used primarily for pumping either very large solids or stringy solids that may clog a more traditional pump.

3.1.2.1.3 Centrifugal Pumps Contrasted with Positive-Displacement Pumps Following is a list contrasting several different attributes of centrifugal and positive-displacement pumps. This list is not meant to be exhaustive but, rather, to cover several items that may be of interest.

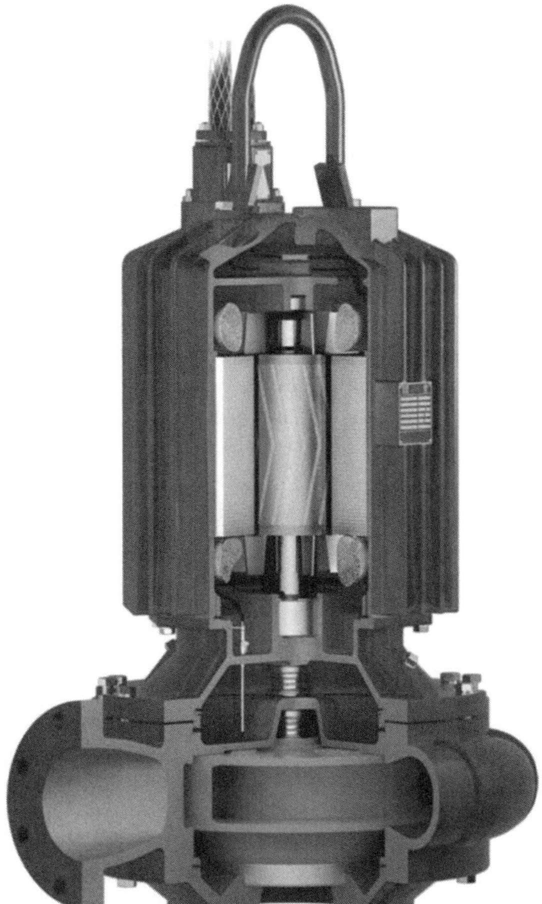

Figure 3-8 Cross section of a submersible pump. Note that the electric motor is incorporated into the pump itself. Since the entire pump is submerged in the pumpage, it is critical that the motor housing be sealed completely. (Courtesy of The Gorman-Rupp Company.)

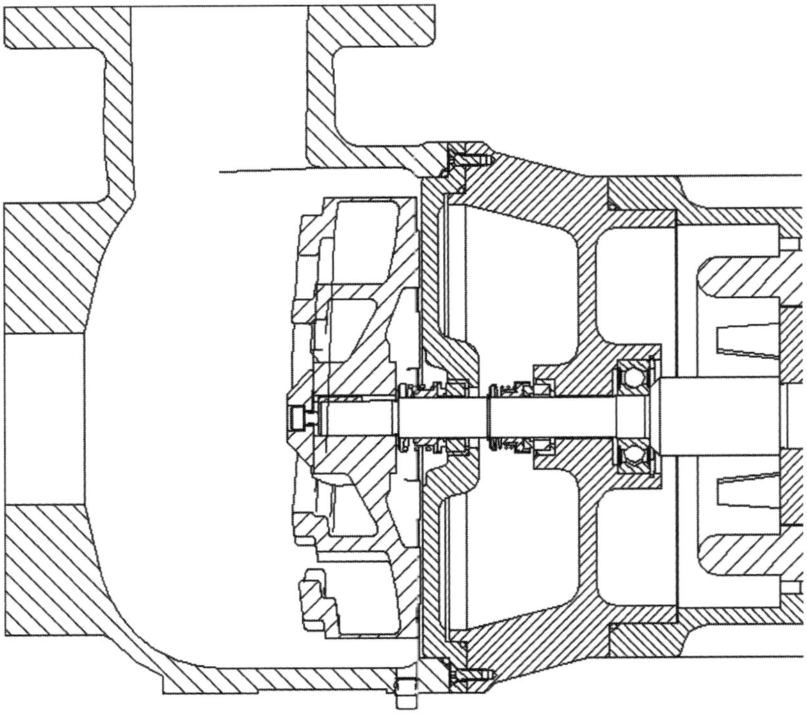

Figure 3-9 In this cross-sectional drawing, one can easily see the flow path that exists inside a vortex pump. While vortex pumps use semiopen impellers, there is not a wear plate against which the impeller spins. (Courtesy of The Gorman-Rupp Company.)

1. A centrifugal pump produces a steady flow; positive-displacement pumps produce flow that is not as steady, especially when reciprocating pumps are considered.

2. Centrifugal pumps generally produce higher flows at lower heads; positive-displacement pumps generally produce lower flows at higher heads. There are exceptions, but this is a general guideline.

3. Centrifugal pumps operate at higher rotational speeds that match motor synchronous speeds (60 Hz: 3450, 1750, 1150 rpm; 50 Hz: 2900, 1450, 960 rpm); most positive-displacement pumps operate much below synchronous motor speeds, with reciprocating pumps typically being below 100 cycles per minute.

4. Centrifugal pumps require an almost constant input torque; positive-displacement pumps require different levels of torque, depending on where it is at in its cycle. This uneven torque requirement can fatigue power transmission components such as couplings quicker than the more constant load of a centrifugal pump.

5. Centrifugal pumps are generally more efficient.

6. Centrifugal pump performance can be affected dramatically by the viscosity of the pumpage; positive-displacement pump performance sees relatively little impact, due to the viscosity of the pumpage.

7. Centrifugal pumps operate such that small changes in system pressure can cause relatively large changes in the flow; positive-displacement pumps generate approximately the same flow, regardless of system pressure.

8. Centrifugal pumps are capable of operating against a closed discharge valve for a very short period of time; positive-displacement pumps should never be run against a closed discharge valve unless other adequate forms of pressure relief are provided in the system.

3.2 PUMP APPLICATIONS

Many documents exist that instruct on how to select and size a pump properly for a given application. Similar to the way in which various pumps can be categorized by certain features of the pump, applications can be categorized to some extent by some of their characteristics. An understanding of what type of application is in question is critical to selecting the correct pump.

3.2.1 Flooded Suction Applications

A flooded suction application is one in which the surface level of the fluid on the suction side of the pump is at a higher elevation than the centerline of the pump suction (Fig. 3-10). Another way to look at this is to ask: If the pump was removed from the system and any valves on the

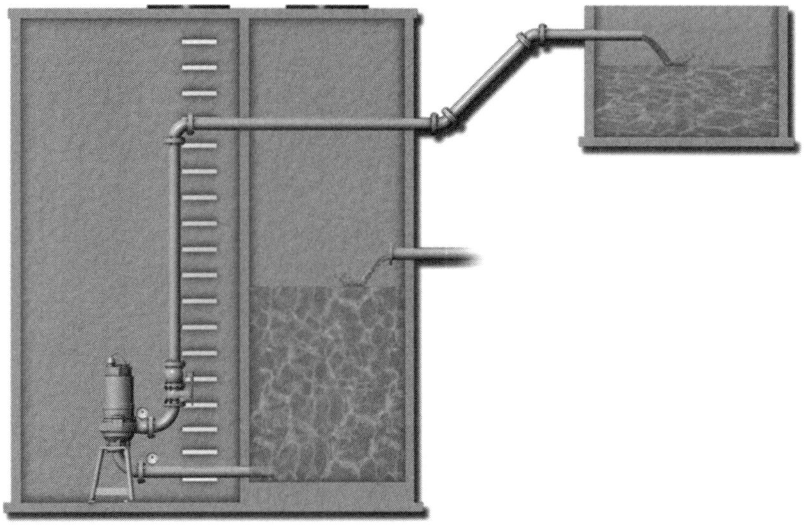

Figure 3-10 Whenever the pump inlet is located lower than the liquid level on the suction side, the application can be classified as flooded suction. Depicted here is a pump located beside a wet well where the pump suction is clearly below the liquid level. (Courtesy of The Gorman-Rupp Company.)

suction plumbing were open, would the fluid escape out and "flood" the area where the pump sits?

Most pumps will operate well on a flooded suction application. It is worth considering, though, that being on a flooded suction does not guarantee that there will be sufficient net positive suction head available to avoid cavitation. Envisioning a long, complicated suction arrangement where the friction losses are significant may help one understand how this can become a concern.

3.2.2 Suction Lift Applications

A suction lift application is one in which the surface level of the fluid on the suction side of the pump is at a lower elevation than the centerline of the pump suction (Fig. 3-11). Another way to look at this is to ask if the liquid needs to be "lifted" up to the pump suction.

Most pumps can also operate on a suction lift, but some pumps are better designed for this than others. A self-priming or priming-assisted pump, for example, can create a vacuum in the suction plumbing that allows the suction piping to fill with liquid. Otherwise, either an auxiliary vacuum source or reliance upon either a foot valve or a check valve to seal correctly is necessary to draw liquid into the pump.

Any suction lift application should have the NPSH verified prior to pump selection. When selecting a self-priming or priming-assist pump, do not assume that the available priming lift is the same as the maximum lift on which the pump should operate. Priming lifts do not take friction, atmospheric conditions, or the vapor pressure of the liquid into consideration, but NPSH calculations do.

It is possible that an application that begins as a flooded suction application can turn into a suction lift application as the pumpage on the suction side is pumped down. These instances are relatively rare, but can occur when the suction of the pump is attached to a tank below the pump's ON level and above the pump's OFF level and the plumbing internal to the tank is arranged such that it is submerged at all times. In such cases it is imperative to be sure that the pump can operate well at the OFF level—this can be verified by performing NPSH calculations for the system at the OFF level.

3.2.3 Staged Pumping

When one pump is not capable of generating enough flow, or perhaps when multiple smaller pumps are preferred over one larger pump, staging of two or more pumps is common (Fig. 3-12). Staging is very common when many different condition points are required. Imagine a situation where there is relatively steady demand for 16 h a day, but for the other 8 h there is a higher level of demand. This is an instance where staging could be considered. One pump could run independently for the 16 h of steady demand and an additional pump (or more) could be run in conjunction with the first pump for the 8 h of higher demand.

Staging is done in two ways: series and parallel. The concepts here are similar to those for an electric circuit. Series systems have pressures that add together like voltage drops in series circuits, while parallel systems have equal head across the circuit, like voltage drop in parallel circuits.

3.2.3.1 ***Series Pumping*** The discharge of one pump is piped directly into the suction of a second pump. This is

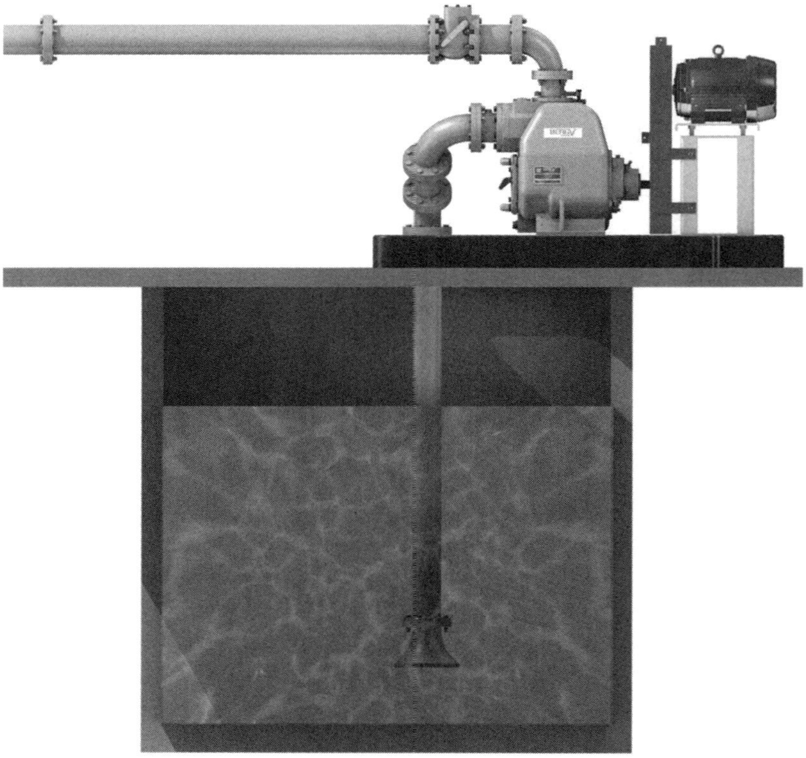

Figure 3-11 Self-priming pump operating on a suction lift where the pump is located above the liquid level. This type of application can make the pump much easier to service. (Courtesy of The Gorman-Rupp Company.)

a very common technique used to meet various condition points that are beyond the capability of one pump by itself. In staging pumps, it is critical to ensure that the maximum operating pressure of the second unit will not be exceeded during operation. To determine the possible performance of two pumps staged together, one must simply add the head of the first pump at a given flow rate to the head of the second pump at the same flow rate. This can be done across a range of flow rates and then plotted to see the possible performance of the two pumps staged. For identical pumps, one can simply double the head on the performance curve. Note that doubling the head will *not* double the flow rate in a given installation, as the pump's operating point is dictated by where the system curve crosses the pump performance curve. In a staged installation, all of the flow goes through both units. If the units are identical, they will both produce an equal amount of head.

3.2.3.2 ***Parallel Pumping*** The suction and discharge of two or more pumps is common. This is another common technique to meet various flow conditions that are outside the normal operating range for one pump by itself. Note that operating in parallel will not increase the potential head that the pumps can generate; rather, it increases the potential flow. To calculate a combined parallel pump performance curve, add the flow rate of one pump at a given head to the flow rate of a second pump at that same head. This can be done across a range of heads and then be plotted to see the potential performance of the pumps in parallel. For identical pumps, one can simply double the flow at each given head. Note that running pumps in parallel will not double the flow in a given installation, as the pump's operating point is dictated by where the system curve crosses the pump performance curve. In a parallel application, both pumps operate at the same head. If the units are identical, including all of the plumbing to and from both units, the flow rates should be exactly equal to each other, and be exactly one-half of the flow where the system curve crosses the parallel performance curve.

3.2.4 Solids-Handling Applications

Many applications require the pumping of liquids with suspended solids within them. These solids can create additional considerations during pump selection. The size of the solids, whether they are abrasive, and whether they are likely to get clogged in the pump are all factors that need to be considered when choosing the proper pump for the application. If a strainer is used on the inlet to the suction pipe, the pump itself needs to be capable of passing

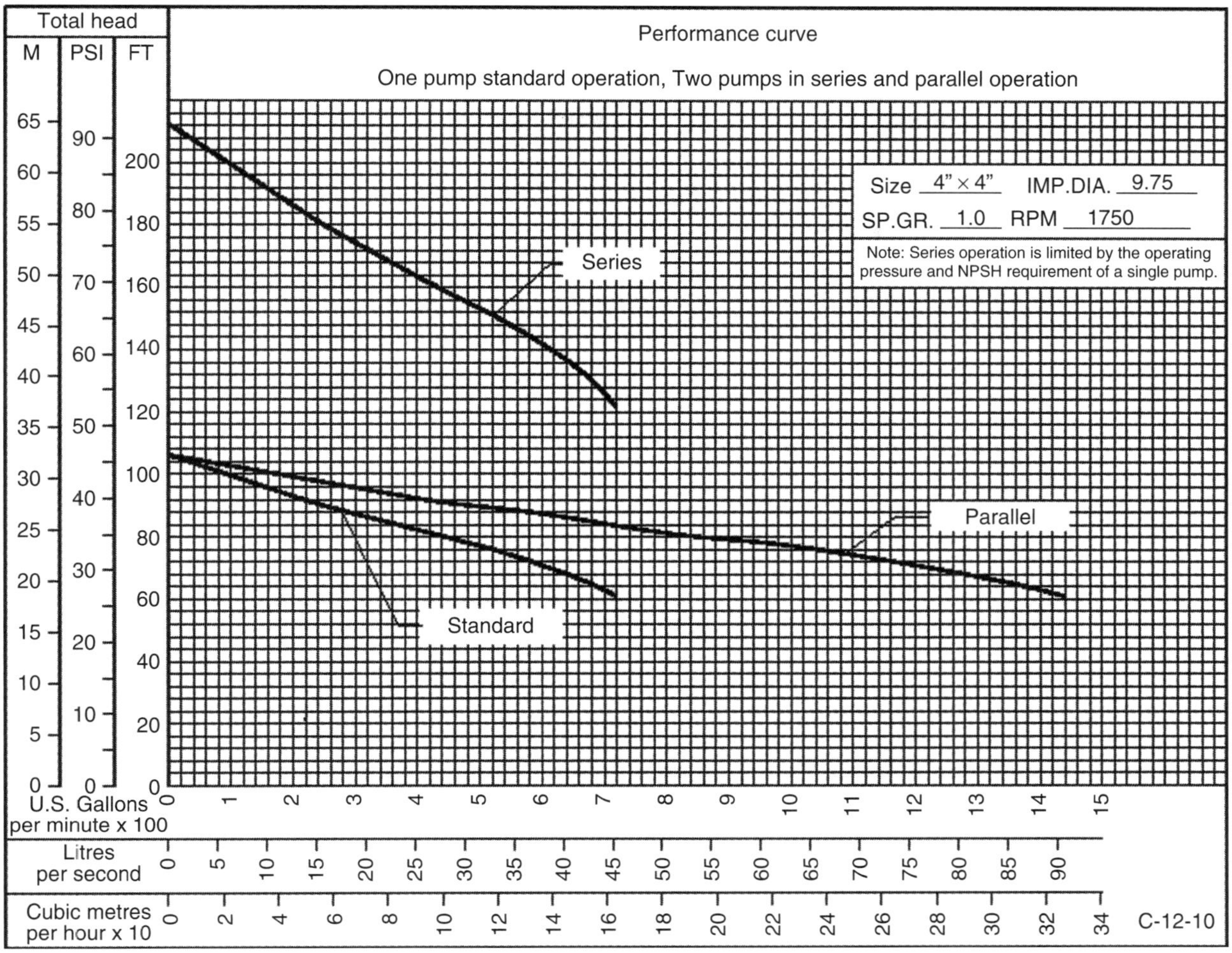

Figure 3-12 Depicted here are three different pump performance curves. The curve marked "Standard" represents one pump running by itself. The curve marked "Series" represents the combined output found by putting two of the standard pumps in series. To find the head generated by each pump, divide the total by 2. The curve marked "Parallel" represents the combined output found by putting two of the standard pumps in parallel. To find the flow generated by each pump, divide the flow by 2. Note that these relationships follow these rules only if both pumps in use are identical, including rpm and impeller size. Consult your pump manufacturer if combining nonidentical pumps for assistance. (Courtesy of The Gorman-Rupp Company.)

any solid that can get past the strainer and into the suction pipe. Such a strainer should also be checked periodically to ensure that it is not clogged with solids. As a strainer clogs, it will increase the friction loss, which can lead to NPSH concerns.

3.3 PUMP SIZING AND SELECTION

Pump sizing and selection must be done in light of the system into which it is going to be installed. Because a pump's operating point in any given application is dictated by where the pump operating curve crosses the system curve, it is recommended that the system head curve be determined first.

3.3.1 System Head Curve

Computing a system head curve (Fig. 3-13) can be a time-consuming process for a new installation. There are many different approaches and theories on how to calculate friction losses through plumbing and fittings. Although we do not go into great detail on this topic, following are some general comments about calculating system curves.

1. System curves have two primary elements:

a. *Static head:* the elevation difference between the surface level of the pumpage on the suction side to where the discharge leaves the piping. This determines the system head curve's value at zero flow, also known as the *y-intercept*.

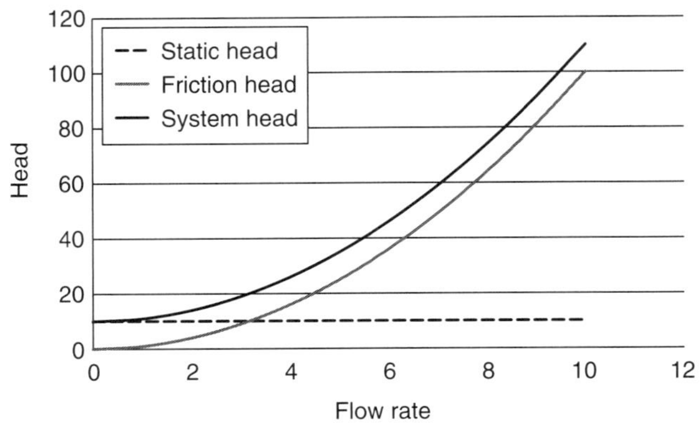

Figure 3-13 Three separate curves are shown, one representing the static portion of the head, which does not change with flow; one representing the friction portion of the head, which increases with flow; and the total head, which is the sum of the static and friction portions. It is the intersection of the system curve and the pump performance curve that dictates what flow and head the pump will generate.

b. *Friction loss:* the amount of additional head that the pump must generate to force the pumpage through the system at varying flow rates. Friction losses always vary based on the square of the flow rate. Therefore, friction losses increase at an increasing rate as the flow rate increases. To double the flow rate in the same system would create four times the friction loss. Friction loss changes if anything in the system changes. For example, throttling a valve that was fully open to a partially closed position will change the system curve by steepening it up.

2. The same flow through a smaller pipe will create much higher friction losses, which leads to a much steeper system curve. As a general comment, then, selecting a larger pipe size will probably lead to greater energy savings over time than the initial added cost of the larger pipe. However, there is a point of diminishing returns from a larger pipe. The best solution is to calculate a system curve for three or four different pipe sizes and consider the additional upfront cost of each larger pipe size and choose accordingly. One other thought to keep in mind is that if there are solids suspended in the pumpage, a minimum velocity of 1 m/s (3 ft/s) is recommended to prevent the solids from settling.

3. More pipe fittings and bends can add significant amounts of additional friction loss, which leads to a much steeper system curve. A single elbow can add as much friction loss as dozens of feet of straight pipe. Obviously, in most systems some fittings and bends are required, but should be kept to a minimum if the desire is to minimize the friction losses.

4. Keep in mind that friction losses represent wasted energy. Since it takes more energy to generate higher heads, reducing friction losses can lead to lower energy costs. Rather than throttling a valve to get the desired flow rate, if the pump is driven by a variable-speed driver such as a variable-frequency drive or an engine, slowing the driver down can also decrease the flow rate without adding unnecessary friction to the system.

5. Any system calculations are subject to error and can change over time. Pipe diameters vary based on any number of factors, including age, wear, schedule of pipe, and pipe material, and can even reduce over time if there is buildup within the pipe. It is not uncommon for system head calculations to be off by 5 to 10% or more initially. For this reason, it is advisable to be sure that there is some variation that is allowed in either the system or the pump to account for this deviation.

6. It is worthwhile to consider calculating the friction loss on the suction and discharge sides of the pump separately. This is because the friction losses on the suction side of the pump become a critical part of the NPSH calculation discussed later.

7. A system head curve for an existing system can be quite simple to determine. From measurements, if one can determine the static head and one condition point (flow at a given TDH) in the system, those two values can be plotted. Subtracting the static head from the TDH at the condition point will give the friction loss at that flow. Double the flow, quadruple the friction loss and add the static head back in. Half the flow, quarter the friction loss, and add the static head back in. You now have four points on the system curve: zero flow (static head), half the condition point, the condition point, and double the condition point. A line can now be sketched in between these points.

3.3.2 Pump Peformance Curves

Once a system curve is developed, the user can determine the necessary head at the flow rate desired. As stated previously, this is the condition point. Although many pump companies now offer computerized pump sizing and selection programs, understanding a pump performance curve is still critical to ensuring the correct pump is selected.

A pump performance curve shows many values that are of interest to operators and those selecting the pumps. They can also be used for troubleshooting problems in the field. Each manufacturer has the freedom to generate their performance curves as they choose, but most contain the same basic information. Here are the main parts of a pump performance curve.

1. *Flow rate*: always on the horizontal (x-axis). Be sure you note the units on the flow rate, as many manufacturers produce pumps for global markets and will place multiple sets of units on this axis.

2. *TDH lines*: these are probably the primary lines of most interest when starting to size a pump. These curves represent the head that the pump can develop at various flow rates. Many pump curves will have multiple TDH lines that represent either different rotational speeds or different impeller diameters. As a rule of thumb, as the pump's rotational speed increases, so does its head. The same is true for the impeller diameter. TDH lines for most centrifugal pumps will have a designated portion of which is labeled the operating range. Operation of the pump beyond these parameters is not recommended for an extended period of time.

3. *Power*: generally slope down and to the right. These lines represent the power required at the impeller shaft to generate the condition points shown on the TDH lines. These data can be used to size the pump driver. At a minimum, the motor should be equal to or greater than the power line to the right of the condition point. More preferred, though, is for the driver to be sized to be nonoverloading across the entire pump curve. To do this, follow the TDH line all the way to its end on the right to select a driver size. Selecting a nonoverloading driver is wise for the following reasons:

a. Any system curve calculation contains uncertainty. If the system curve is not as steep as expected, the condition point will be farther to the right on the pump's operating curve, which will require more power.

b. A nonoverloading driver gives some freedom for future growth if there is ever a desire to increase the flow rate.

c. A nonoverloading driver gives an additional cushion to cover losses, such as transmission losses through V-belts or flexible couplings that are not reflected on the pump performance curve.

d. In general, a driver that is not operating at its maximum capacity will last longer than one that is.

e. With modern high-efficiency electric motors, the overall motor efficiency typically rises as the motor sizes increase. Also, modern high-efficiency motors typically operate efficiently across a very broad range of loads.

4. *Efficiency*: the shape of these lines varies depending on the particular hydraulic design in question. In general, efficiencies start low at low flows, increase as the flow rate increases until you reach the pump's best efficiency point (BEP), and then decrease as the flow rate increases past BEP. Efficiency is a measurement of how well a pump is turning the input power into what is known as *water power:* specific gravity × TDH(in meters) × flow rate (in L/s)/102.1[specific gravity × TDH(in feet) × flow rate (in gal/min)/3960]. Pump efficiency then is water power/input power. In general, the closer the pump operates to BEP, the quieter it will run, with less vibration and with fewer mechanical problems.

5. *Net positive suction head required*: the shape of these lines varies depending on the particular pump hydraulic design, but they generally increase at an increasing rate as the flow rate increases. More on NPSHR later.

3.3.3 Actual Pump Sizing and Selection

Pump sizing and selection involve a balance of many different considerations.

1. Does the system curve cross the pump's TDH curve at the condition point desired? If an application is going to require more than one operating point, all points should be located on the pump performance curve and one needs to ensure that they are all within the recommended operating range of that pump.

2. Is the pump's NPSHR acceptable for the application? Does the NPSH calculation come out favorably for all suction lifts and flow rates required? In almost all cases, NPSH is more critical on higher flow rates and larger suction lifts than lower flow rates and lower suction lifts.

3. Have any additional considerations for the application been considered? Are solids a concern? Are the correct materials selected to increase the life of the pump if abrasives are present? Has the chemical compatibility of the pump materials been verified with the pumpage? If on a suction lift, is there a means to prime the pump automatically, or will that be done manually?

4. If the pump is on a critical process, do users want to stock their own repair parts, or can they rely on the distributor and manufacturer to respond quickly to anything that might cause downtime and the expenses associated with that?

5. Who is going to provide the driver and other accessories, such as discharge check valves or suction strainers? Many pump manufacturers offer numerous options that should be considered.

6. How maintenance friendly is the design? Can the pump be easily repaired by the operator or their maintenance department, or will it require service from the distributor or manufacturer?

7. Although cost is always a factor in any decision, the cost that should be considered is the total cost of ownership. Once the cost of running the driver over the course of the pump's life is considered, the initial purchase price is not a big factor in the total cost over the life of the pump. Add in maintenance expenses and downtime and that initial cost becomes even less of a factor.

3.3.4 Net Positive Suction Head

While some have said that "NPSH" stands for "not pumping so hot," it really is something to take seriously. NPSH calculations are probably the most detailed and complicated part of selecting an appropriate pump. We cannot possibly cover all conceivable liquids, elevations, and so on. However, what is presented here is a basic guideline to follow. The bottom line is that if the NPSHA does not exceed the pump's NPSHR, the pump will experience suction cavitation and a noticeable decline in performance and durability.

Some basic concepts that will help explain NPSH is beginning with the concept that atmospheric pressure at sea level is considered to be 101.3 kPa (14.7 psi). 101.3 kPa (14.7 psi) represents the pressure created by the column of air from seal level up to the outer limits of Earth's atmosphere. A column of water, in contrast, only needs to be 10.3 m (33.9 ft) tall to equal 101.3 kPa (14.7 psi). The height of liquid necessary to equal 101.3 kPa (14.7 psi) will vary from one fluid to the next, based on the fluid's density. Since we already established that specific gravity is a given liquid's density divided by the density of water, specific gravity can be used to determine the height of a column of any liquid needed to equal 101.3 kPa (14.7 psi). Taking 10.3 m (33.9 ft) and dividing by the specific gravity will give you the height of the liquid required to equal 101.3 kPa (14.7 psi). A fluid lighter than water, with a specific gravity below 1.0, will require more than 10.3 m (33.9 ft); a fluid heavier than water, with a specific gravity above 1.0, will require more than 10.3 m (33.9 ft). It is important that any NPSH calculations be done in a consistent set of units, either in meters (feet) of water or in meters (feet) of the fluid being pumped. In this chapter we assume the fluid to be clear water.

3.3.5 Net Positive Suction Head Available

Calculating the NPSHA takes several steps and can lead to some mistakes if one does not take care to ensure they are covering all considerations listed here. To calculate the NPSHA, one must know the elevation of the location compared to sea level, the vapor pressure of the liquid at the temperature it enters the pump, the friction loss in the suction plumbing, the static lift, and how much of a safety factor is desired.

1. *Elevation.* Earlier it was mentioned that 101.3 kPa (14.7 psi), atmospheric pressure at sea level, was equal to the column of air from sea level to the outer atmosphere. For this reason, there is less atmospheric pressure anywhere that the column of air is shorter. Therefore, atmospheric pressure is less on a mountain than in a valley below sea level. Tables giving atmospheric pressure at various elevations are easy to find. NPSHA calculations start with this value.
2. *Vapor pressure.* There are two ways to make a liquid turn into a gas: Heat it up or lower the atmospheric pressure around it. Therefore, it is actually possible to boil water at room temperature. One must simply pull a vacuum over the top of the liquid until it begins to boil. Therefore, the liquid's vapor pressure at its temperature entering the pump must be considered in this calculation. The vapor pressure of the liquid must be subtracted from the subtotal up to this point. Once again, it is important to ensure that units are consistent.
3. *Friction loss.* Earlier it was suggested that the friction losses for the suction and the discharge piping be calculated separately. Since friction losses represent lost pressure, the friction loss for the suction plumbing must also be subtracted from the subtotal.
4. *Static lift.* This is the elevation difference between the surface level of the liquid on the suction side to the centerline of the pump suction. If the centerline of the pump's suction is higher than the liquid, subtract the distance from the subtotal. If the centerline of the pump's suction is lower than the liquid level, add the distance to the subtotal. In situations where the static lift is going to change dramatically, be sure to look at the worst-case scenario—when the liquid level is at its lowest point on the suction side.
5. *Safety factor.* Since there are many things that can vary in an installation (e.g., atmospheric pressure varies depending on weather, the liquid temperature may not always be a constant as it enters the pump), it is recommended that an additional safety factor be included. A good value to use here is 0.6 m (2 ft). This also gets subtracted from the subtotal.

The value left after these computations is the NPSHA for the application. Keep in mind that while many of the variables here do not change with flow rate, overall the NPSHA does change with flow rate since friction loss is one of the factors included. Tables 3-1–3-5 can be used as a guide in establishing NPSH calculations. These calculations are very important to ensure that a proper pump selection is made. As explained in the text, these calculations have multiple steps.

3.4 PUMP MAINTENANCE

The best instructions for how to maintain a pump is simply to follow the manual that was received with the pump. As with any other industrial piece of machinery, preventive maintenance is a wise investment since many pumps are

TABLE 3-1 Net Positive Suction Head Calculations

	Feet	Meters
Atmospheric pressure at sea level	33.96	10.35
Atmospheric pressure correction for elevation of job site (see Table 3-2)		
Atmospheric pressure available at job site		
Total dynamic suction lift (static suction lift + friction and entry losses)		
Vapor pressure of liquid (see Table 3-3)		
Safety factor (covers variation in atmospheric pressure, calculation errors)		
Net positive suction head available		
Net positive suction head required (per pump manufacturer's data)		
Net positive suction head excess (must be > 0 or pump will cavitate)		

TABLE 3-2 Atmospheric Pressure Correction Chart

Sea Level Altitude Above		Reduction	
Feet	Meters	Feet	Meters
0	0	0	0
1000	305	1.2	0.366
2000	610	2.33	0.725
3000	914	3.53	1.076
4000	1220	4.63	1.44
5000	1525	5.71	1.74
6000	1830	6.74	2.054
7000	2134	7.75	2.362
8000	2438	8.74	2.664

TABLE 3-3 Vapor Pressure Correction Chart[a]

Temperature		Reduction	
°F	°C	Feet	Meters
40	4.4	0.28	0.09
60	15.6	0.59	0.18
80	26.7	1.17	0.36
100	37.8	2.19	0.67
120	48.9	3.91	1.19
140	60	6.68	2.03
160	71.1	10.95	3.34
180	82.2	17.35	5.29
200	93.3	26.65	8.12

[a]For water only.

on "mission critical" applications where downtime can be very detrimental and expensive in downtime and lost production. In general, though, the areas described below are among those that should be regularly checked by a qualified technician.

TABLE 3-4 Net Positive Suction Head Calculations[a]

	Feet	Meters
Atmospheric pressure at sea level	33.96	10.35
Atmospheric pressure correction for elevation of job site (see Table 3-2)	3.53	1.08
Atmospheric pressure available at job site	30.43	9.27
Total dynamic suction lift (static suction lift + friction and entry losses)	15.00	4.57
Vapor pressure of liquid (see Table 3-3)	6.68	2.03
Safety factor (covers variation in atmospheric pressure, calculation errors)	2.00	0.61
Net positive suction head available	6.75	2.06
Net positive suction head required (per pump manufacturer's data)	8.00	2.44
Net positive suction head excess (must be > 0 or pump will cavitate)	−1.25	−0.38

[a]Example conditions: 140°F; 3000 ft elevation; 15-ft TDSL; 8-ft NPSHR; cavitates.

TABLE 3-5 Net Positive Suction Head Calculations[a]

	Feet	Meters
Atmospheric pressure at sea level	33.96	10.35
Atmospheric pressure correction for elevation of job site (see Table 3-2)	3.53	1.08
Atmospheric pressure available at job site	30.43	9.27
Total dynamic suction lift (static suction lift + friction and entry losses)	15.00	4.57
Vapor pressure of liquid (see Table 3-3)	1.17	0.36
Safety factor (covers variation in atmospheric pressure, calculation errors)	2.00	0.61
Net positive suction head available	12.26	3.73
Net positive suction head required (per pump manufacturer's data)	8.00	2.44
Net positive suction head excess (must be > 0 or pump will cavitate)	4.26	1.29

[a]Example conditions: 80°F; 3000 ft elevation; 15-ft TDSL; 8-ft NPSHR; should not cavitate.

3.4.1 Bearing Lubrication

Pumps come with a wide variety of methods to keep bearings lubricated. Some units have an oil bath, others have an oil mist system. In either case, follow the pump manufacturer's recommendations as to how often to change the oil and what type of oil to use. Other units have greased bearings that require the periodic addition of more grease. Once again, follow the pump manufacturer's recommendations on how often to add grease and the type of grease to be used.

3.4.2 Seal Maintenance

There is a wide range in mechanical seals, some of the simplest costing only a few dollars, with others well into the

TABLE 3-6 Troubleshooting Problems and Solutions

Concern	Comments	Action
Is the pump operating according to its advertised curve?	Gauge readings, flowmeter readings, and confirmation of rotational input speed are all factors that might help determine where the pump is actually operating. If there is an adjustable discharge valve, one could take multiple readings at various points to see whether or not the pump is operating on its advertised curve.	If the pump is on its advertised curve but performance is low, there is a system problem. Items to review: NPSH; whether there is a problem with the system that is causing more friction loss than expected. If the pump is not on its advertised curve, some of the wear components probably need to be replaced.
Is the pump noisy during operation?	Several factors contribute to a noisy pump, but most common complaints can be traced to one of three things: misapplication, the size of the pump, and noise from other sources.	One of the most common misapplication problems is operating the pump at too low a flow rate. Sometimes this is referred to as being too far to the left on the curve, or outside the operating range, or back toward shutoff. In general, this is caused by either too much static head or too much friction head in the system. Slowing the pump down might help some, but the best solution if the system cannot be revised might be to install a bypass line that allows the pump to operate at a higher flow rate. Another common misapplication is operating with insufficient NPSH. This topic was covered earlier. In general, consider the following steps if NPSH is a problem: 1. Reduce the flow rate by changing the system to reduce the friction. 2. Reduce the flow rate by slowing the pump down or trimming the impeller. 3. Decrease the static suction lift. 4. Cool the fluid before it enters the pump if it is at an elevated temperature. 5. Increase the size of the suction plumbing or remove any unnecessary pipe fittings on the suction side of the pump to reduce the friction. Some noise complaints related to large installed pumps relate to the simple fact that a lot of work is being done inside the pump. Noise is going to be produced by a mechanical device, and as a general rule, this noise is going to get larger as the device gets larger. Pumps are no exception. Frequently, the pump is blamed for what is in reality a noisy system. Every time flow is forced to change directions through an elbow or a fitting, there is noise. Every time flow goes through a restriction such as a valve, there is noise. If the plumbing is not supported rigidly, there can be additional noise that comes from the piping. The pump's driver—whether it be a motor or an engine—can contribute as much or more noise than that of some pumps. Belt drives, chain drives, and even couplings can also add to the noise that surrounds the pump. In rare cases, there might even be vibration from the pump baseplate or a rattle from a guard.

TABLE 3-6 (*Continued*)

Concern	Comments	Action
Are there problems getting the mechanical seals to hold up?	The big key here is to consult with the pump and seal manufacturers, but there are many factors that can lead to premature seal failure. These factors include, but are not limited to the items listed to the right.	1. Chemical compatibility 2. Abrasive content in pumpage 3. Temperature of fluid 4. Lubrication—or lack thereof 5. Improper installation 6. Shaft deflection (may occur if the pump is operated to the left of the operating range)
Are there frequent bearing problems?	Similar to above, consult the pump manufacturer, but here are a few common causes of premature bearing failures are listed to the right.	Improper lubrication—or lack thereof—or contamination Improper installation Improper pump operation 1. Operating on the left side of the operating range 2. Operating while cavitating due to NPSH problems 3. Improper drive alignment or tensioning

thousands of dollars. Most mechanical seals require almost no maintenance, but some have an associated barrier fluid or oil bath. Many seals operate properly without a barrier fluid or an oil bath as long as the seal is not run dry. In any case, the pump manufacturer's recommendations for maintenance should be followed. If there are any signs of a leak, it might be too late for preventive maintenance. It is advised to investigate signs of a seal leak when first noticed, especially if the pumpage contains anything hazardous. Some pump manufacturers recommend that whenever you do maintenance that affects a seal, you install a new seal rather than reusing the existing seal.

Another area of maintenance involves seals equipped with packing, where typically there is a gland that must be retightened periodically to maintain the proper seal. Be careful to follow the pump and/or packing manufacturer's recommendations, as it is possible to overtighten a gland against the packing.

3.4.3 Maintaining Performance

Checking gauge readings on the suction and discharge sides of the pump periodically can be a very effective way to track pump performance over time. For these values to be meaningful, though, they must be taken under identical conditions. The pump speed and all aspects of the system must remain constant from one reading to the next. Charting the readings over weeks and months can help indicate if a pump's performance is degrading.

If degradation is being noticed, it might be time either to reset some clearances, if the pump has an open or semiopen impeller, or to replace some of the wear components. The wear components in most pumps are the impeller, the wear plate, or the wear ring, depending on the impeller style, and in some cases the seal plate. Another short-term fix could be to speed up the pump if it is attached to a variable-speed driver.

It is possible that after resetting clearances or replacing worn components, the pump performance still will not return to its original level. If that is the case, there is probably something that is now different in the system, possibly buildup or wear inside the piping.

3.4.4 Winterizing and Long-Term Storage

Some units are stored when not in use, possibly due to seasonal concerns. If a pump is to be stored for an extended period of time, refer to the pump manual for instructions. If there are no instructions, consider the following steps prior to long-term storage.

1. Drain the pump completely and dry out the internal wetted surfaces with compressed air.
2. Drain and refill any oil cavities with clean oil.
3. Cover the suction and discharge openings.

3.4.5 Cold Temperature Installations

In some climates or processes it may be necessary to provide heat to the pump to ensure that the pumpage does not freeze inside the volute. Notify the manufacturer if the pump will be exposed to extreme cold temperatures, as special modifications such as different elastomers may be required. Allowing pumpage to freeze in the volute has been known to cause the volute to crack. Some alternatives are as follows:

1. If the pump is not going to be in use, winterize it according to the pump manual or the steps in Section 3.3.4.
2. Some pump manufacturers make casing heaters available that can provide additional heat to prevent the pumpage from freezing.
3. Heat tape can be wrapped around the suction and discharge plumbing, and in some cases the volute itself, to provide heat to the volute. This may have limited success, depending on the wall thicknesses of the plumbing and volute.

Other considerations for cold-temperature installations and applications may include using a thinner oil for bearing and seal lubrication and in some cases considering a heating jacket or heat tape for the pump.

3.5 PUMP TROUBLESHOOTING

As with any mechanical device, there may be times when troubleshooting a pump is necessary. Always refer to the pump manufacturer's instruction manual for proper steps to be taken. Table 3-6 is not intended to be an exhaustive list of problems and solutions, but rather as a sampling of the types of problems that operators are most likely to face in actual applications. Use the table only for basic guidance; always refer to the owner's manual for troubleshooting.

4

PIPES

Shaohui Jia
PetroChina Pipeline R&D Center, Langfang, Hebei, China

Pipelines are one of the main methods of transporting oil and gas worldwide. Historically, pipelines have been the safest means of transporting natural gas and hazardous liquids. We describe pipe types, pipe selection, pipe problems, and pipe maintenance.

4.1 TYPES OF PIPES

According to the process of manufacture, there are two primary types of pipes: seamless and welded. Such specifications as American Petroleum Institute (API) 5L and those of the American Society for Testing and Materials are published for steel pipe to provide standards for pipe suitable for use in conveying gas and oil in both the oil and natural gas industries.

4.1.1 Seamless Pipe

Seamless pipe is produced by the seamless process. The seamless process is a process of hot working steel to form a tubular product without a welded seam. If necessary, the hot-worked tubular product may subsequently be cold finished to produce the desired shape, dimensions, and properties.

4.1.2 Welded Pipe

Welded pipe is subdivided into continuous welded pipe, electric welded pipe, laser welded pipe, and submerged-arc welded pipe.

4.1.2.1 Continuous Welded Pipe *Continuous welded pipe* is defined as pipe that has one longitudinal seam produced by the continuous welding process. Continuous welding is a process of forming a seam by heating the skelp in a furnace and pressing the formed edges together mechanically, with successive coils of skelp joined together to provide a continuous flow of steel for the welding mill. This process is a type of butt welding.

4.1.2.2 Electric Welded Pipe *Electric welded pipe* is defined as pipe that has one longitudinal seam produced by the electric welding process. Electric welding is a process of forming a seam by electric-resistance or electric-induction welding, wherein the edges to be welded are pressed together mechanically and the heat for welding is generated by the resistance to flow of the electric current.

4.1.2.3 Laser Welded Pipe *Laser welded pipe* is defined as pipe that has one longitudinal seam produced by the laser welding process. Laser welding is a welding process that uses a laser beam and a keyholing technique to produce melting and coalescence of the edges to be welded. The edges may be preheated. Shielding is obtained entirely from an externally supplied gas or gas mixture.

4.1.2.4 Submerged-Arc Welded Pipe *Submerged-arc welded pipe* is defined as pipe that has one seam produced by an automatic submerged-arc welding process. A submerged-arc weld is a longitudinal or helical seam weld produced by the submerged-arc welding process. Submerged-arc welding is a welding process that produces coalescence of metals by heating them with an arc or arcs between a bare metal consumable electrode or electrodes and the work. The arc and molten metal are shielded by a

Process Plant Equipment: Operation, Control, and Reliability, First Edition. Edited by Michael D. Holloway, Chikezie Nwaoha, and Oliver A. Onyewuenyi.
© 2012 John Wiley & Sons, Inc. Published 2012 by John Wiley & Sons, Inc.

blanket of granular fusible material on the work. Pressure is not used, and part or all of the filler metal is obtained from the electrodes.

4.2 PIPE SELECTION

There are three main metallurgical requirements for steel pipes: strength, toughness, and weldability. They are very important for pipe selection.

4.2.1 Pipe Strength

The *yield strength* or *yield point* of a material is defined in engineering and materials science as the stress at which a material begins to deform plastically. Prior to the yield point the material will deform elastically and will return to its original shape when the applied stress is removed. Once the yield point is passed, some fraction of the deformation will be permanent and nonreversible.

Standard line pipe has grade designations A and B. Stronger-grade standard line pipe is commonly designated X grade in the API 5L specification for line pipe [1]. Table 4-1 demonstrates the pipe strength requirements.

For grades intermediate to X42 and X80, the symbol is X followed by the first two digits of the minimum yield strength specified, measured in kilopounds per square inch (ksi; e.g., X80 for pipe having a minimum yield strength of 80 ksi).

Yield strength testing involves taking a small sample with a fixed cross-sectional area, and pulling it with a controlled, gradually increasing force until the sample changes shape or breaks. Longitudinal and/or transverse strain is recorded using mechanical or optical extensometers. Three testing methods are used to measure the pipe strength.

1. *The flattened strap test*. Strap specimens that have full wall thickness and rectangular cross section are flattened before testing. Although this method is simple, the flattening action changes the measured value of the yield strength so that the true strength is usually underestimated compared with the strength of a ring of pipe material.
2. *The round bar test*. A round bar tensile specimen is used, but the test area represents only a small part of the pipe wall thickness because of the reduced diameter of the specimen in the test length.
3. *The ring tension test*. A full ring of pipe material is tested by applying internal hydraulic pressure. The pipe yield strength measured by the ring tension test is the most accurate, but specialized equipment is required and the test is slow and expensive.

4.2.2 Pipe Toughness

Toughness in steels is the capability of a material to resist brittle fracture (the sudden fracture of materials when a load is applied rapidly, typically with little ductility in the area of the fracture) and tolerate defects, particularly crack-type defects. Two common test methods used to measure toughness are the Charpy impact test and the drop weight tear test (DWTT). The Charpy test was developed as a general test applicable to all types of steel, whereas the DWTT applies specifically to high-pressure pipelines. Although there are differences in the specimen shape, dimensions, and notch dimensions, the principle of the two tests is identical, in that a series of specimens is fractured in impact at a number of different temperatures.

4.2.2.1 The Charpy Impact Test The Charpy impact test (Fig. 4-1) is most commonly used to evaluate the relative toughness or impact toughness of materials and as such is often used in quality control applications, where it is fast and economical. The Charpy test involves striking a

TABLE 4-1 Pipe Strength Requirements

Grade	Symbol	Yield Strength (Minimum) ksi	Yield Strength (Minimum) MPa	Tensile Strength (Minimum) ksi	Tensile Strength (Minimum) MPa	Yield-to-Tensile Ratio (Maximum)	Elongation (Minimum)[a] (%)
A	A	30	207	48	331	0.93	28
B	B	35	241	60	413	0.93	23
X42	X42	42	289	60	413	0.93	23
X46	X46	46	317	63	434	0.93	22
X52	X52	52	358	66	455	0.93	21
X56	X56	56	386	71	489	0.93	19
X60	X60	60	413	75	517	0.93	19
X65	X65	65	448	77	350	0.93	18
X70	X70	70	482	82	565	0.93	17
X80	X80	80	551	90	620	0.93	16

[a] API 5L elongation figures vary with specimen dimensions; those quoted are for a 0.2-in.^{2} specimen.

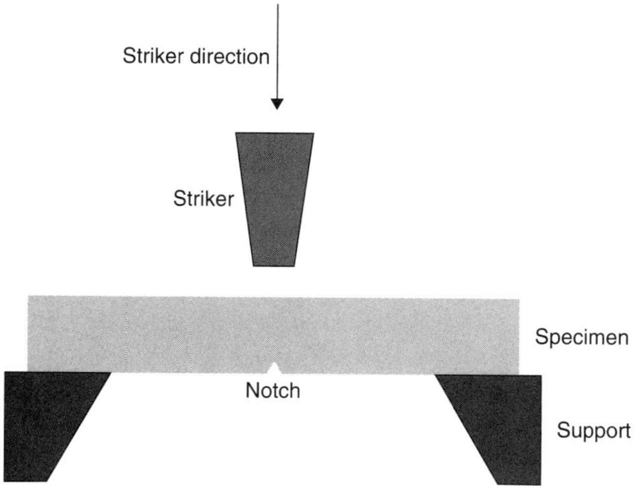

Figure 4-1 Schematic of the Charpy impact test.

suitable test piece with a striker mounted at the end of a pendulum. The test piece is fixed in place at both ends and the striker contacts the test piece immediately behind a machined notch.

The Charpy impact energy of a specimen is affected by several factors:

1. *Yield strength and ductility*. For a given material the impact energy will be seen to decrease if the yield strength is increased: if the material undergoes some process that makes it more brittle and less able to undergo plastic deformation. Such processes may include cold working or precipitation hardening.
2. *Notches*. The notch serves as a stress concentration zone, and some materials are more sensitive than others toward notches. The notch depth and tip radius are therefore very important. Two types of notch are specified [2]:
   a. V-notch of 45°, 2 mm deep with a 0.25-mm radius of curvature at the base of the notch.
   b. U-notch or keyhole notch, 5 mm deep, with a 1-mm radius of curvature at the base of the notch.
3. *Temperature and strain rate*. Most of the impact energy is absorbed by means of plastic deformation during yielding of the specimen. Therefore, factors that affect the yield behavior, and hence ductility, of the material, such as temperature and strain rate, will affect the impact energy. Figure 4-2 illustrates impact strength plotted against temperature.
4. *Fracture mechanism*. Metals tend to fail by one of two mechanisms: microvoid coalescence or cleavage. Cleavage can occur in body-centered-cubic materials, where cleavage takes place along the crystal plane. Microvoid coalescence is the more common fracture mechanism, where voids form as strain increases, and these voids eventually join together and failure occurs. Of the two fracture mechanisms, cleavage involves far less plastic deformation and hence absorbs far less fracture energy.

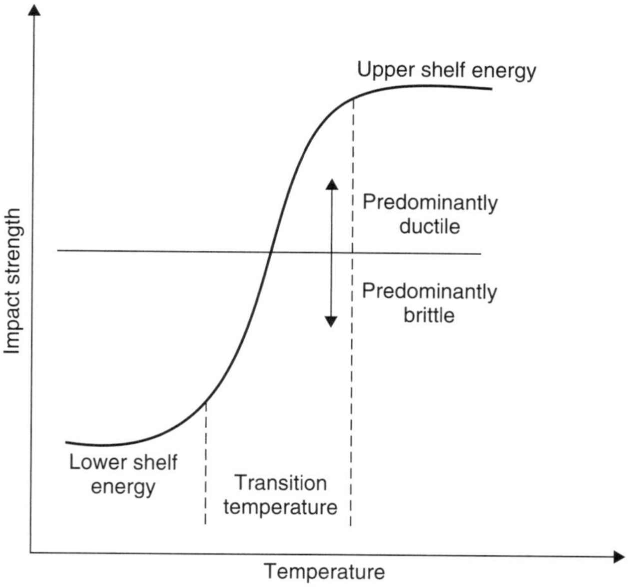

Figure 4-2 Schematic of a Charpy transition curve.

4.2.2.2 The Drop Weight Tear Test The drop weight tear test (DWTT) has been in use for over 40 years as a practical laboratory-scale way of ensuring that steel used in the manufacture of line pipe is not subject to brittle failure when in service. The American Petroleum Institute (API) 5L test standard is used to determine the fracture ductility of metal line pipe. Specimens are cut from sections of pipe soaked at a prescribed temperature and tested within 10 s. ASTM E436, similar to API 5L, is used to establish the temperature range over which ferritic steels undergo a fracture mode transition from ductile to brittle. In both standards a determination regarding ductile-to-brittle behavior is based on visual inspection of the specimens in conjunction with, at times, a calculation to determine the percentage of shear seen in material fracture.

In particular, demands for high operating pressures of line pipes and larger diameters have driven the development of higher-strength steels. Forty years ago the work that led to the drop weight tear test was done on X52 steel (360 MPa yield strength). Improvements in thermomechanical processing have yielded improvements of approximately 10,000 psi per decade, to the point where the state of the art is now X100 steels, and use of X120 steels is being considered.

Depending on the application, the material requirements, as indicated by notch toughness, are divided into three categories [10]:

1. *Category* 1 *requirement:* no notch toughness requirement. Typical application: low-vapor-pressure fluids (water, crude oil, etc.).
2. *Category* 2 *requirement:* notch toughness in the form of energy absorption and fracture appearance. Typical application: buried and aboveground pipelines (-5 to -45°C); high vapor pressure.
3. *Category* 3 *requirement:* proven notch toughness in the form of energy absorption only. Typical application: high-vapor-pressure liquids.

4.2.3 Pipe Weldability

One way to estimate pipe weldability is to use a carbon equivalent (Table 4-2). In welding, equivalent carbon is used to understand how the various alloying elements affect the hardness of the steel being welded. This is then related directly to hydrogen-induced cold cracking, which is the most common weld defect for steel; thus, it is most commonly used to determine weldability.

For pipe grades up to grade X70 inclusive, the carbon equivalent (CE), calculated using product analysis and the following equation, defined in the API 5L standard,

$$\mathrm{CE} = \mathrm{C} + \frac{\mathrm{Mn}}{6} + \frac{\mathrm{Cr} + \mathrm{Mo} + \mathrm{V}}{5} + \frac{\mathrm{Ni} + \mathrm{Cu}}{15} \tag{4.1}$$

must not exceed 0.43%. For this equation, weldability is defined as follows [7]: The *carbon equivalent* is a measure of the tendency of a weld to form martensite on cooling and to suffer hydrogen cracking. When the carbon equivalent is between 0.40 and 0.60, weld preheat may be necessary. When the carbon equivalent is above 0.60, preheat is necessary, postheat may be necessary, and techniques to control the hydrogen content of the weld are required. Preheating the pipe before welding is beneficial in avoiding hydrogen cracking because it dries the pipe and, more important, reduces the cooling rates in the heat affected zone (HAZ).

4.2.4 Piping Material

Piping material is selected by optimizing the basis of design. First, eliminate from consideration those piping materials that [12]:

TABLE 4-2 Pipe Weldability According to Carbon Equivalent

Carbon Equivalent	Weldability
Up to 0.35	Excellent
0.36–0.40	Very good
0.41–0.45	Good
0.46–0.50	Fair
Over 0.50	Poor

- Are not allowed by code or standard
- Are not chemically compatible with the fluid
- Have system-rated pressures or temperatures that do not meet the full range of process operating conditions
- Are not compatible with environmental conditions such as external corrosion potential, heat tracing requirements, ultraviolet degradation, impact potential, and specific joint requirements

The remaining materials are evaluated for advantages and disadvantages, such as capital, fabrication, and installation costs; support system complexity; compatibility in handling thermal cycling; and cathodic protection requirements. The highest-ranked material of construction is then selected. The design proceeds with pipe sizing, pressure-integrity calculations, and stress analysis. If the piping material selected does not meet those requirements, the second-ranked material is used and the pipe sizing, pressure-integrity calculations, and stress analysis are repeated.

The sizing for any piping system consists of two basic components: fluid flow design and pressure integrity design. *Fluid flow design* determines the minimum acceptable diameter of the piping necessary to transfer the fluid efficiently. *Pressure integrity design* determines the minimum pipe wall thickness necessary to handle safely the expected internal and external pressure and loads. After piping materials, design pressure, and sizes have been selected, a stress analysis is performed that relates the piping system selected to the piping layout and piping supports. The analysis ensures that the piping system meets intended service and loading condition requirements while optimizing the layout and support design. The analysis may result in successive reiterations until a balance is struck between stressed and layout efficiency, and stresses and support locations and types. The stress analysis can be a simplified analysis or a computerized analysis, depending on system complexity and the design code.

4.3 PIPELINE NETWORK DESIGN AND OPTIMIZATION

Oil and gas pipelines are in a network rather than single-line configuration. It is more economical to build a gathering or export station for many fields that are separated in a variety of remote areas. Thus, a pipeline is an effective means of oil and gas transportation from field to gathering station or export station. Recently, more researchers have become interested in pipe network design and optimization. The pipeline network design process includes the development of cost estimates for various possible combinations of pipe size, compression equipment, and interstation distances to find the optimal combination that minimizes the transportation cost given the desired flexibility and expandability goals.

We use the following description of pipeline network optimization in western Sichuan Province to explain how to perform pipeline network design and optimization [3]. The natural gas gathering and transmission pipeline network system in western Sichuan is huge and complicated. The optimal design of a pipeline network should be taken as a priority for oil and gas field surface engineering while developing an oil or gas field. The optimal design involves pipeline network disposition not only from wellhead to primary gas gathering station, but also from primary gas gathering station to secondary gas gathering station.

The physical model established for the gas gathering pipeline network in western Sichuan Province is shown in Fig. 4-3. A mathematical model for optimal operation of the gas pipeline network has also been established. In this model the maximum operational benefit is the objective function, and gas throughput, pressure, length of pipe, pipeline strength, coefficient of compression, and pipeline flow are considered as constraint conditions. Meanwhile, a relationship is built from gas production to operation to consumers.

The nonlinear mathematical pipeline network model used in western Sichuan can be described as follows. The objective function is

$$\max F = \sum_{i=1}^{N_n} \int_0^T S_i Q_i dt - \sum_{j=1}^{N_c} C_j N_j \tag{4.2}$$

where Q_i = gas throughput of the i node as a function of time
S_i = cost factor of the i node between gas purchase and gas sale
N_j = power of compressor station j
C_j = cost factor of the power of compressor station j
N_n = total count of nodes
N_c = total count of compressor stations
T = time period

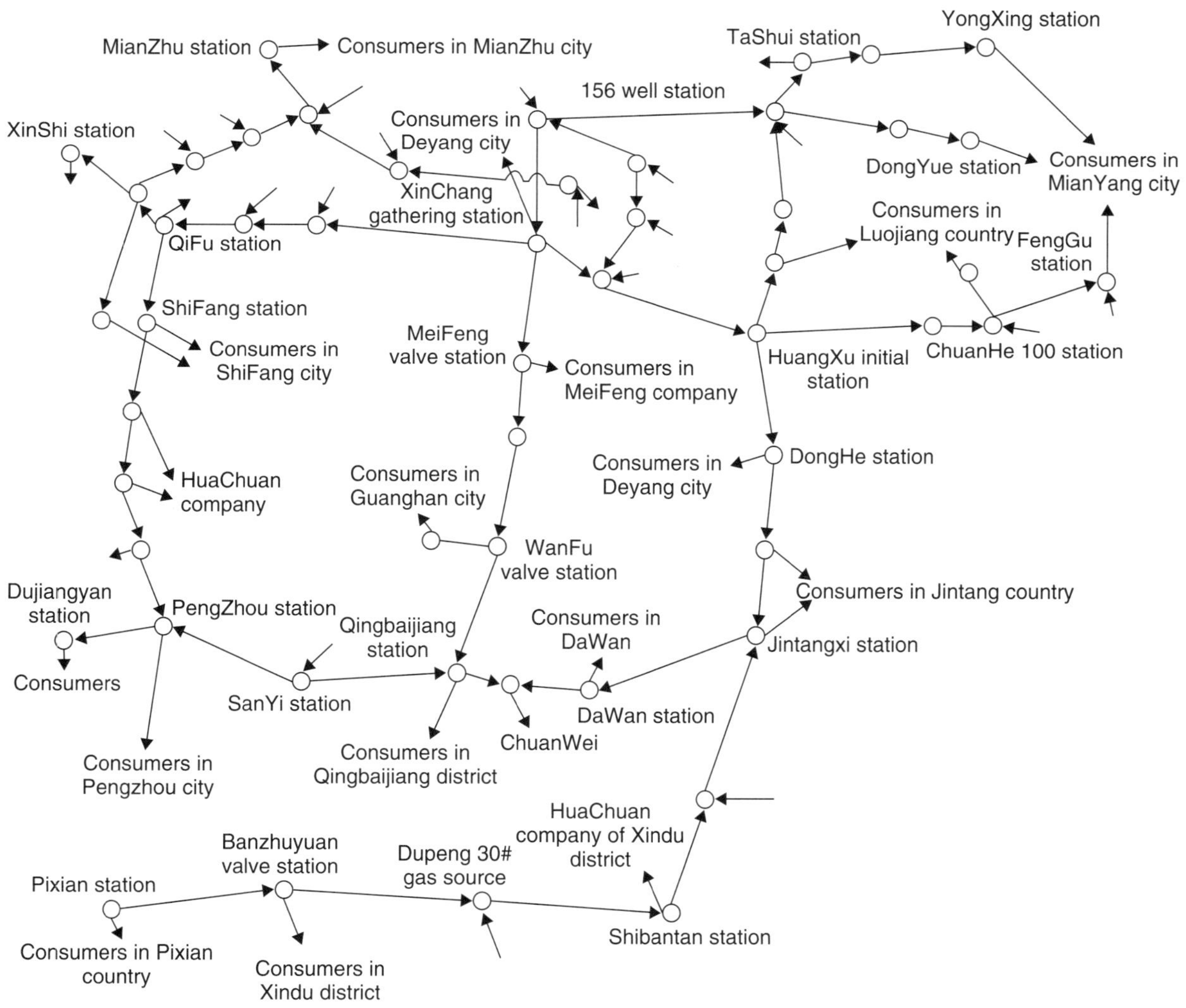

Figure 4-3 Physical model of gas gathering pipeline network in western Sichuan Province.

and the *constraint condition* is

$$Q_{i,\min} \le Q_i \le Q_{i,\max} \qquad (i = 1, 2, \ldots, N_n) \qquad (4.3)$$

where $Q_{i,\min}$ and $Q_{i,\max}$ represent the minimum and maximum gas throughput of the i node, respectively;

$$p_{i,\min} \le p_i \le p_{i,\max} \qquad (i = 1, 2, \ldots, N_n)$$

where p_i = pipeline pressure of the i node $p_{i,\min}$ and $p_{i,\max}$ = minimum and maximum pressure of the i node, respectively;

$$N_{j,\min} \le N_j \le N_{j,\max} \qquad (j = 1, 2, \ldots, N_c) \qquad (4.4)$$

where $N_{j,\min}$ and $N_{j,\max}$ = minimum and maximum power of compressor station j, respectively;

$$p_x \le p_{x,\max} \qquad (x = 1, 2, \ldots, N_p) \qquad (4.5)$$

where p_x = pressure of pipeline x $p_{x,\max}$ = maximum allowable operation pressure;

$$N = Q\frac{k}{k-1}RZ_1T_1[\varepsilon^{(k-1)/k} - 1]\frac{1}{\eta} \qquad (4.6)$$

where N = compressor station power
Q = compressor displacement
T_1 = compressor intake gas temperature
R = gas constant
Z_1 = compressor intake gas compressibility
ε = compressor pressure ratio
η = compressor efficiency
k = compressor absolute heat index;

$$\sum_{k \in C_i} a_{ij}M_{jy} + Q_i = 0 \qquad (4.7)$$

where C_i = components connected with the i node
M_{iy} = absolute value of gas flow from component y to the i node or from the i node to component y (component y is connected to the i node)
Q_i = flow flux of the i node exchanging with others
a_{iy} = a factor whose value is 1 when gas flows from component y to the i node, and -1 when gas flows from the i node to component y;

$$M = \frac{\pi}{4}\sqrt{\frac{[p_Q^2(1 - C_1\Delta h) - p_Z^2]D^5}{\lambda ZRT'L(1 - C_1\Delta h/2)}} \qquad (4.8)$$

where p_Q and p_Z = pipeline pressure at the start and end points, respectively
T = average temperature of gas flow
L = pipeline length
D = pipeline diameter
Δh = difference in elevation along the pipeline
Z = gas compressibility
λ = gas friction factor
N_p = total count of pipelines
M = pipe flow flux

Using simulation analysis, the optimal goals of the planning solution are drawn up:

- Enhancing the gas transmission capacity from the Xinchang and Xiaoquan gas fields to Mianzhu, Shifang, and Pengzhou City.
- Reducing the gas delivery capacity of the Xinqing pipeline while transmitting gas from the Golden pipeline to the MeiFeng Company.
- Increasing the gas transmission capacity from the YuanDong pipeline to Anxian County and the Yongxing Company while reducing the gas delivery capacity of the Yuanyong pipeline.
- Transmitting gas from the Xindu gas field to the Qingbaijiang District of Chengdu City.
- Integrating new pipelines into the gas pipeline network of western Sichuan Province.

In the optimal planning solution, the flow flux and pressure of main pipelines are calculated as shown in Table 4-3.

4.4 PIPELINE FAILURE

According to ASME B31.8S, the causes of pipeline failure are grouped into nine categories of related failure types based on their nature of growth characteristics and are further delineated into time-related defect types (Table 4-4).

The EGIG (The European Gas Pipeline Incident Data Group) is a cooperative of 15 major gas transmission system operators in Europe and owner of an extensive database of pipeline incident data collected since 1970. Through an analysis of incident causes, EGIG provides insight as to which causes effort should be directed. Incidents have been categorized into six groups and are presented in Table 4-5. Figure 4-4 shows the distribution percentage of incidents.

In the United States, the Office of Pipeline Safety (OPS) within the Department of Transportation has gathered pipeline failure data. Their statistics show that the single greatest cause of pipeline accidents is damage from an

TABLE 4-3 Results of Flow Flux and Pressure in Optimal Planning

Pipeline	Diameter (mm)	Length (km)	Starting Pressure (MPa)	End Pressure (MPa)	Flow Flux (10^4 m^3/d)
Xinchang–MeiFeng valve station	273	8.5	2.28	1.8	4
Xinchang–168 Valve Station	325	30	2.28	2.27	26
168 Valve Station–QiFu	325	30	2.27	2.24	55
MaJing–QiFu	325	30	2.25	2.24	19
PengZhou–SanYi	273	21	2.35	2.25	35
DongHe valve station–MeiFeng	273	10	1.88	1.8	79
SanYi–Qinbaijiang	273	15	2.25	1.48	56
HuangXu–DongHe valve station	377	44	2.09	1.88	121
DongHe valve station–Qingjiang valve station	377	44	1.88	1.71	92
Qingjiang valve station–Jintang	377	44	1.71	1.68	82
Jintang–DaWan station	377	15	1.68	1.52	103
Longmenshan–PenZhou station	325	30	2.78	2.35	100
Wenjiang–Pixian	273	15	1.9	1.83	30
Zhongjiang–DongHe valve station	273	35	2.28	1.88	50
YuanDong–TaYong valve station	159	10	1.92	1.91	2.5
YuanDong pipeline	325	37.5	2.13	2.07	4
XInDu–DaWan station	273	15	1.74	1.52	50

TABLE 4-4 Threats to Pipes

Time-dependent	External corrosion	
	Internal corrosion	
	Stress corrosion cracking	
Stable	Manufacturing-related defects	Defective pipe seam
		Defective pipe
	Welding and fabrication related	Defective pipe girth weld
		Defective fabrication weld
		Wrinkle bend or buckle
		Stripped threads/broken pipe/coupling fail
	Equipment	Gasket O-ring failure
		Control/relief equipment malfunction
		Seal/pump packing failure
		Miscellaneous
Time-independent	Third-party damage	Third-party inflicted damage (instantaneous/immediate fail)
		Previously damaged pipe (delayed failure mode)
		Vandalism
	Incorrect operations	Incorrect operation company procedure
	Weather related and outside force	Cold weather
		Lightning
		Heavy rain or floods
		Earth movement

Source: ASME B31.8S.

outside force such as excavation equipment or ships' anchors. Table 4-6 provides a distribution of reported incidents by cause for transmission and gathering pipelines (natural gas and hazardous liquids) from OPS statistics analyzed between 1985 and 1996 [8,13].

The American Gas Association has investigated incidents and causes of pipeline failure between 1984 and 1992 and compared the results, as shown in Table 4-7. Common causes of failure are described next.

4.4.1 Pipe External Corrosion

The external corrosion mechanism of buried pipeline is electrochemical corrosion, and the corrosion rate is dependent on such factors as the nature of the soil. In a moist atmosphere or soil, a thin layer of water film, which induces pipeline corrosion, is formed on the surface of steel pipeline. When the film is neutral, the steel and the water film, which contains oxygen, constitute the

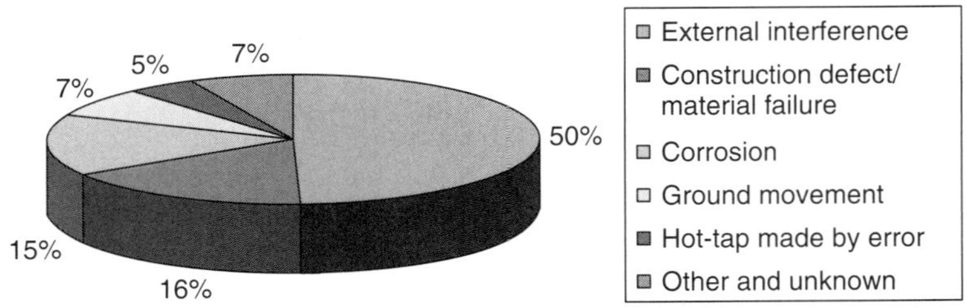

Figure 4-4 Distribution percentage of pipeline incidents.

TABLE 4-5 Causes of Pipeline Failure

Cause	Overall Percentage (%)
External interference	49.6
Construction defect or material failure	16.5
Corrosion	15.4
Ground movement	7.3
Hot tap made by error	4.6
Other and unknown	6.7

TABLE 4-6 Causes of Pipeline Incidents Between 1985 and 1996

Cause	Natural Gas, 1985–1995 (%)	Hazardous Liquids, 1986–1996 (%)
Internal corrosion	9	10
External corrosion	15	32
Third-party damage	36	33
Weather-related	11	3
Defective pipe seam	3	6
Previously damaged pipe	4	8
Unknown	4	0
Defective girth weld	3	4
Defective fabrication weld	2	2
Defective pipe	2	3
Construction damage	1	0
Miscellaneous	8	0
Stress corrosion cracking	1	0

Source: U.S. Office of Pipeline Safety, Department of Transportation.

galvanic cell:

$$Fe - 2e = Fe^{2+} \qquad \text{(cathodic)} \qquad (4.9)$$

$$2H_2O + O_2 + 4e = 4OH^- \qquad \text{(anodic)} \qquad (4.10)$$

When the film is acidic, the steel and the water film, which contains carbon dioxide, constitute the galvanic cell:

$$Fe - 2e = Fe^{2+} \qquad \text{(cathodic)} \qquad (4.11)$$

$$2H^+ + 2e = H_2 \qquad \text{(anodic)} \qquad (4.12)$$

Figure 4-5 shows external corrosion of steel pipes.

TABLE 4-7 Comparison of Pipeline Incidents and Causes Between 1984 and 1992

Cause of Failure	U.S. Gas Pipeline	European Gas Pipeline	Canadian Gas Pipeline
Third-party damage (%)	40.4	52	12.6
Corrosion (%)	20.4	13.91	11.60
Material and construction damage (%)	12.7	19.13	34.3
Operation error (%)	26.4	4.9	41.5
Incident rate (times/kilomiles-year)	0.26	1.85	2.93

Source: American Gas Association.

Figure 4-5 External corrosion of pipes.

4.4.2 Pipe Internal Corrosion

Pipe internal corrosion (Fig. 4-6) is of several types, such as pitting corrosion and groove corrosion. The presence of corrosive agents and conditions inside a pipeline can result in premature failure, loss of service, possible pollution, injury, property damage, or fatalities. Common causes of internal corrosion include:

- Hydrogen sulfide, which is a poisonous corrosive gas and forms sulfuric acid in the presence of water.

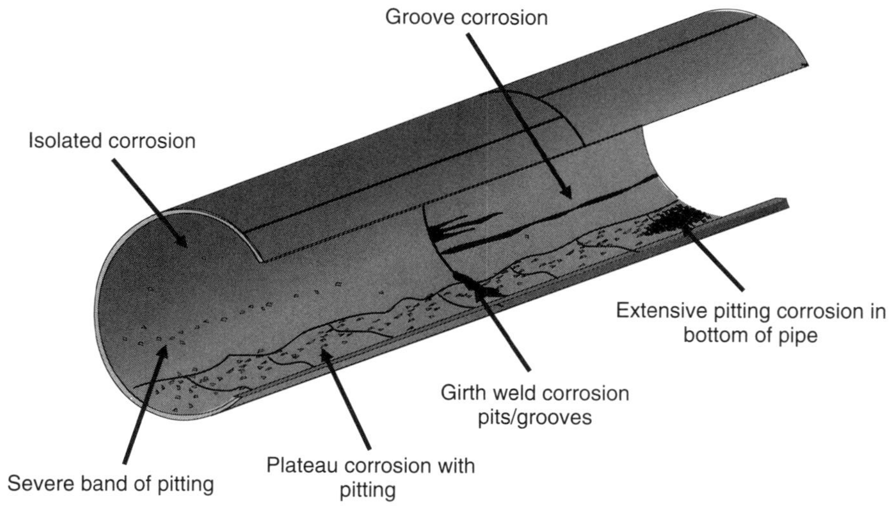

Figure 4-6 Schematic of internal corrosion of pipes.

- Carbon dioxide, which forms carbonic acid in the presence of water.
- Water vapor present in a pipeline will condense to liquid water when cooled sufficiently. Liquid water in a pipeline will contain dissolved corrosive gases as acids and may also contain bacteria responsible for some corrosion reactions. High-velocity flow conditions can remove protective oxide layers on the inside of a pipeline and accelerate the corrosion attack. Low-velocity-flow conditions can allow wetting of the pipe wall to occur as well as near-stagnant conditions, which are ideal for biological corrosion mechanisms.

Take pitting corrosion, for example; it is the localized corrosion of a metal surface confined to a point or small area, which takes the form of cavities. Pitting is one of the most damaging forms of corrosion. Pitting corrosion is caused by the environment (i.e., the chemistry), which may contain aggressive chemical species, such as chloride. Chloride is particularly damaging to the passive film (oxide), so pitting can be initiated at oxide breaks. The environment may also set up a differential aeration cell (e.g., a water droplet on the surface of a steel), and pitting can be initiated at the anodic site (the center of the water droplet).

In a homogeneous environment, pitting is caused by the material, which may contain inclusions (MnS is the major culprit for the initiation of pitting in steels) or defects. In most cases, both the environment and the material contribute to pit initiation. The environment and the material (i.e., metallurgical) factors determine whether or not an existing pit can be repassivated. Sufficient aeration (supply of oxygen to the reaction site) may enhance the formation of oxide at the pitting site and thus repassivate or heal the damaged passive film (oxide)—the pit is repassivated and no pitting occurs. An existing pit can also be repassivated if the material contains a sufficient amount of alloying elements, such as Cr, Mo, Ti, W, and N. These elements, particularly Mo, can significantly enhance the enrichment of Cr in the oxide and thus heal or repassivate the pit.

4.4.3 Stress Corrosion Cracking

Stress corrosion cracking (SCC) is cracking due to a process involving conjoint corrosion and straining of a metal due to residual or applied stresses. SCC may occur through a number of mechanisms. When cracking is clearly a result of hydrogen embrittlement, this term may be used in place of SCC. However, this distinction is rather arbitrary. We are often unsure of the mechanism of SCC, and many failures that are actually due to the effects of hydrogen would conventionally be ascribed to SCC.

Two types of SCC are normally found on pipelines, known as high-pH SCC (9 to 13) and nearly neutral pH SCC (5 to 7). The high-pH SCC caused numerous failures in the United States in the early 1960s and 1970s, whereas nearly neutral pH SCC failures were recorded in Canada during the mid-1980s to early 1990s. The SCC failures have continued throughout the world, including in Australia, Russia, Saudi Arabia, and South America.

4.4.3.1 ***High-pH SCC*** This is classical SCC, originally noted in gas transmission pipelines. It is normally found within 20 km downstream of a compressor station. High-pH SCC normally occurs in a relatively narrow cathodic potential range (−600 to −750 mV Cu/$CuSO_4$) in the presence of a carbonate/bicarbonate environment in a pH window from 9 to 13. Temperatures greater than 40°C are necessary for high-pH SCC susceptibility; growth rates decrease exponentially with temperature.

Intergranular cracking mode generally represents high-pH SCC. A thin oxide layer is formed in the concentrated carbonate–bicarbonate environment which provides protection around the crack surfaces. However, due to changes in loading or cyclic loading, there is crack tip strain, resulting in breakage of oxide film. This results in crack extension due to corrosion. Because of such a stringent environmental requirement for SCC initiation, this is not as prevalent as is nearly neutral pH SCC. This type of SCC has been noted primarily in gas transmission lines (i.e., temperature).

4.4.3.2 Nearly Neutral pH SCC This type of transgranular cracking mode of SCC was noted initially in Canada and has been observed by U.S. operators. The environment primarily responsible is diluted groundwater containing dissolved CO_2. The CO_2 originates (as in high-pH SCC) from the decay of organic matter. Cracking is further exacerbated by the presence of sulfate-reducing bacteria. This occurs primarily due to disbonded coatings, which shield the cathodic current that could reach a pipe surface. There is a free corrosion condition below the coating that results in an environment with a pH of about 5 to 7.

A cyclical load is critical for crack initiation and growth. There are field data which indicate that with a decreasing stress ratio there is an increased propensity for cracking. Hydrogen is considered a key player in this SCC mechanism, where it reduces the cohesive strength at the crack tip. Attempts have been made to relate soil and drainage type to SCC susceptibility; however, limited correlation has been noted.

SCC is an insidious form of corrosion, and it produces a marked loss of mechanical strength with little metal loss. The damage is not obvious to casual inspection and the stress corrosion cracks can trigger mechanical fast fracture and catastrophic failure of components and structures. Several major disasters have involved stress corrosion cracking, including the rupture of high-pressure gas transmission pipes.

Figure 4-7 Pipeline failure due to SCC.

On December 13, 2003 the Williams 26-in. line ruptured near Toledo, Washington due to SCC (Fig. 4-7). The pipeline company that supplies most of Washington's natural gas was ordered on December 19, 2003 to all but shut down its trunkline from Canada to Oregon after federal safety inspectors determined that frailties in the 268-mile pipe would "likely result in serious harm to life, property and the environment."

4.5 PIPELINE INSPECTION AND LEAK DETECTION

4.5.1 Pipeline Inspection

For pipes, some inspection methods detect defects, and some repair methods protect the defect area. Several inspection methods are discussed in this section.

4.5.1.1 Pipeline Inspection Principle Both magnetic flux leakage (MFL) and ultrasonic testing (UT) are well-established in-line inspection technologies. Whereas MFL provides a versatile and reliable method for determining the geometry of metal loss in pipelines, UT allows direct and highly accurate measurements of pipeline wall thickness. However, there are other inspection technologies, such as caliper inspection for pipelines, but they are not discussed in this section.

4.5.1.2 Magnetic Flux Leakage Inspection Magnetic flux leakage (MFL) is a magnetic method of nondestructive testing that is used to detect corrosion and pitting in pipelines. The basic principle is that a powerful magnet is used to magnetize the steel. At areas where there is corrosion or missing metal, the magnetic field "leaks" from the steel. In an MFL tool, a magnetic detector is placed between the poles of a magnet to detect the leakage field. When the tool passes a location where the amount of metal in the pipe wall has been decreased by corrosion or pitting, magnetic flux leakage takes place. Figure 4-8 shows the principle of MFL for pipeline inspection.

MFL technology has evolved to a state that now makes it an integral part of any cost-effective pipeline

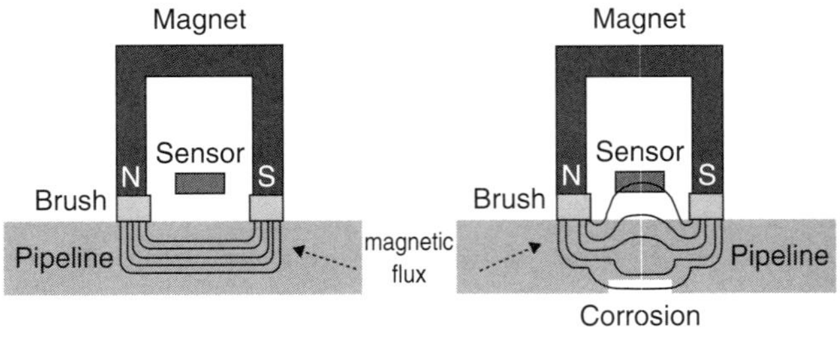

Figure 4-8 Principle of MFL for pipeline inspection.

integrity program. MFL tools, especially high-resolution MFL tools, are designed to detect, locate, and characterize corrosion successfully. However, a pipeline operator should not dismiss the ability of an MFL tool to identify and characterize dents, wrinkles, corrosion growth, mechanical damage, and even some cracks.

4.5.1.3 Ultrasonic Testing Ultrasonic testing (UT) uses high-frequency sound energy to conduct examinations and make measurements. Ultrasonic inspection can be used for flaw detection and evaluation, dimensional measurements, material characterization, and more. A typical UT inspection system consists of several functional units, such as pulser/receiver, transducer, and display devices. A pulser/receiver is an electronic device that can produce high-voltage electrical pulses. Driven by the pulser, the transducer generates high-frequency ultrasonic energy. Sound energy is introduced and propagates through the materials in the form of waves. When there is a discontinuity (such as a crack) in the wave path, part of the energy will be reflected back from the flaw surface. The reflected wave signal is transformed into an electrical signal by the transducer and is displayed on a screen.

Figure 4-9 indicates the principles of ultrasonic testing used for pipe wall thickness measurement:

- If the stand-off increases and the wall thickness decreases, internal metal loss is indicated.
- If the stand-off remains unchanged and the wall thickness decreases, the defect is located on the outer wall surface.

Ultrasonic inspection is a very useful and versatile NDT method. Some of the advantages of ultrasonic inspection that are often cited include:

- It is sensitive to both surface and subsurface discontinuities.
- The depth of penetration for flaw detection or measurement is superior to that of other NDT methods.
- Only single-sided access is needed when the pulse-echo technique is used.
- It is highly accurate in determining reflector position and estimating size and shape.
- Minimal part preparation is required.
- Electronic equipment provides instantaneous results.
- Detailed images can be produced using automated systems.
- It has other uses, such as thickness measurement, in addition to flaw detection.

As with all NDT methods, ultrasonic inspection also has its limitations, which include:

- The surface must be accessible to transmit ultrasound.
- Skill and training are more extensive than with some other methods.
- It normally requires a coupling medium to promote the transfer of sound energy into the test specimen.
- Materials that are rough, irregular in shape, very small, exceptionally thin, or non homogeneous are difficult to inspect.
- Cast iron and other coarse-grained materials are difficult to inspect due to low sound transmission and high signal noise.
- Linear defects oriented parallel to the sound beam may go undetected.
- Reference standards are required for both equipment calibration and the characterization of flaws.

4.5.2 Pipeline Inspection Tools

To achieve the best inspection result, it is very important to perform pipe cleaning before pipeline inline inspection. Unless its condition is already well known, a pipeline should never be inspected before undergoing progressive cleaning. A clean pipeline allows the inspection tool to collect the most accurate data and avoids the possibility of an expensive inspection tool sticking in the pipeline. Regular cleaning is a key component of any long-term integrity program.

GE-PII Company is a service provider for pipeline inspection and integrity services whose products, MagneScan and UltraScan WM [6] we look at next. MagneScan

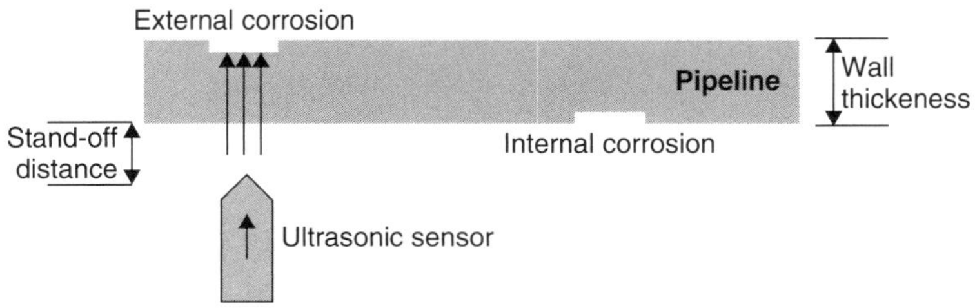

Figure 4-9 Principle of ultrasonic testing for pipeline inspection.

tools offer true high-resolution performance, delivering the detailed data and high confidence levels essential for a productive integrity management program. They can be used in every conceivable type of pipeline inspection: in dry or liquid product, overland or subsea, in diameters from 6 to 56 in. (15 to 42 cm). They record the position of the following pipeline features and anomalies:

- Internal and external pitting and general corrosion on the pipe body
- Metal loss in the vicinity of welds
- Metal loss associated with dents and under casings

MagneScan tools are designed to locate and size areas of metal loss of 10% of wall thickness or greater, and in practice, detect even smaller metal loss. MagneScan tools use the MFL principle to locate metal loss. As the tool travels through a pipe, powerful permanent magnets magnetize the surrounding metal via wire brushes that contact the internal wall. Flux density is driven to the point of saturation. Any change in the thickness of the metal in the pipe wall—a fitting, weld, or patch of corrosion—causes disturbances in the magnetic field. Sensors surrounding the circumference of the tool read these disturbances and record the data onboard.

The characteristic patterns of flux leakage can be interpreted to establish the dimensions of each anomaly. Using a ring of secondary sensors, it is possible to discriminate between internal and external metal loss. Other data are recorded as the tool progresses. Odometer wheels log the distance traveled, and an internal pendulum records the orientation of the tool within the pipeline. A time-based marker system is deployed along the pipeline right-of-way to log the time of passage of the tool.

UltraScan WM can detect and measure precisely mid-wall anomalies such as laminations and inclusions. Many pipeline operators use a baseline UltraScan WM inspection to confirm the quality of new construction before commissioning. To deliver its full potential, ultrasound must be coupled to the pipe wall by a liquid medium. To inspect a dry pipeline, UltraScan WM is run in a liquid batch.

Buried pipeline leak detection can be a frustrating and expensive task. Complete excavation is usually not a practical option, and traditional methods of leak detection such as ultrasonic testing are often costly and inaccurate.

4.5.3 Pipeline Leak Detection

In this section, two new techniques are described in detail.

4.5.3.1 Negative Pressure Wave Leak Detection in Oil Pipelines The negative pressure wave technique is an effective method for paroxysmal oil leakage detection and location. When a leak occurs, the medium at the leak point quickly discharges because of the pressure difference between the inside and outside of the pipeline. Mass loss at the leak point causes the fluid density of that point to decrease, leading to an instantaneous decline of pressure at the leak point. Because of the continuity of fluid, the fluid velocity will not change immediately, but the fluid pressures between the leak point and its adjacent area are different. The difference leads to flow of the high-pressure fluid at two sides of the leak point to the low-pressure region of the leak point. Thus, it causes a reduction in the fluid density and pressure at two sides of the leak point. A negative pressure wave propagates up and down along the pipeline from the leak location. Special transducers installed at both ends of the monitored stretch detect and transform pressure into an electrical signal which is read and analyzed by the computer processing system. Figure 4-10 indicates the principle of the negative pressure wave method.

The leak location can be calculated as

$$X = \frac{L + v\Delta t}{2} \tag{4.13}$$

where X is the distance between the leakage point and the head of the pipeline, L the length of the pipeline, v the propagation velocity of the negative pressure wave, and Δt the time difference between the reception of the negative pressure wave by pressure transducers mounted up- and downstream. In China, the negative pressure wave method has been used for some pipelines, including pipelines in the Shengli oil field. The response time of the leak detection system has been reduced through real-time transmission in the local area network. Moreover, by helping to drill holes effectively it assists pipeline operators in reducing the incidents of stolen oil.

4.5.3.2 Acoustic Leak Detection in Natural Gas Pipelines Acoustic techniques are being employed to detect leakage. When a leak or rupture occurs, the pressure balance of the pipeline is damaged. The gas leaking from the pipeline will cause an acoustic wave due to the friction between the gas and the pipeline. An acoustic emission is a sound wave, or more properly a stress wave, that travels through a material as the result of a sudden release of strain energy. The acoustic energy travels in ultrasonic waves through the material of the structure. Thus, it may be detected by a remote sensor situated either on the containing structure or in the fluid itself. When the pipeline works well, the signals received by the acoustic sensors are treated as background noises. When pipeline leaks, both the acoustic leak signal and the background noises are received by the acoustic sensors. Acoustic monitoring techniques typically utilize acoustic emission sensors to detect leaks based on changes in the background noise pattern, and the acoustic leak detection system will determine whether or not a leak occurs.

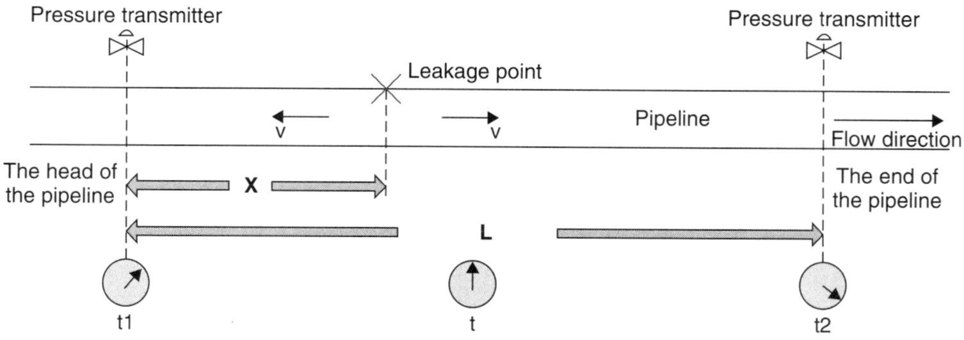

Figure 4-10 Principle of negative pressure wave method.

The advantages of the system include detection of the location of the leaks as well as noninterference with operation of the pipelines. In addition, they are easily ported to various pipe sizes. However, a large number of acoustic sensors is required to monitor an extended range of pipelines. The technology is also unable to detect small leaks that do not produce acoustic emissions at levels substantially higher than the background noise. Attempts to detect small leaks can result in many false alarms [11]. Figure 4-11 is a schematic diagram of acoustic leak detection.

The leak location can be calculated by the formula.

$$S = \frac{L + v\Delta t}{2} \tag{4.14}$$

where S is the distance between the leak point and the upstream acoustic sensor, L the distance between the upstream and downstream acoustic sensors, V the acoustic wave velocity in the pipeline, and Δt the time difference in receiving acoustic signals between the upstream and downstream sensors. The propagation velocity of acoustic waves in the actual pipeline is influenced by medium pressure, temperature, and so on. According to the gas mass conservation and momentum conservation principles, the propagation velocity of an acoustic wave in the rigid pipeline can be expressed as

$$v = \sqrt{\frac{k_v p}{\rho}} = \sqrt{k_v z R T} \tag{4.15}$$

where k_v is the volume insulation index, z the compressibility factor, R the gas constant [kJ/(kg·K)], and T the temperature (K).

Li [9] has carried out experiments based on acoustic leak detection of natural gas pipelines. In the experiment, the researchers analyzed the leak influence of various leaking rates and leaking locations. The results show that the acoustic leak detection method has very good precision and good sensitivity. Following is a brief description of this experiment.

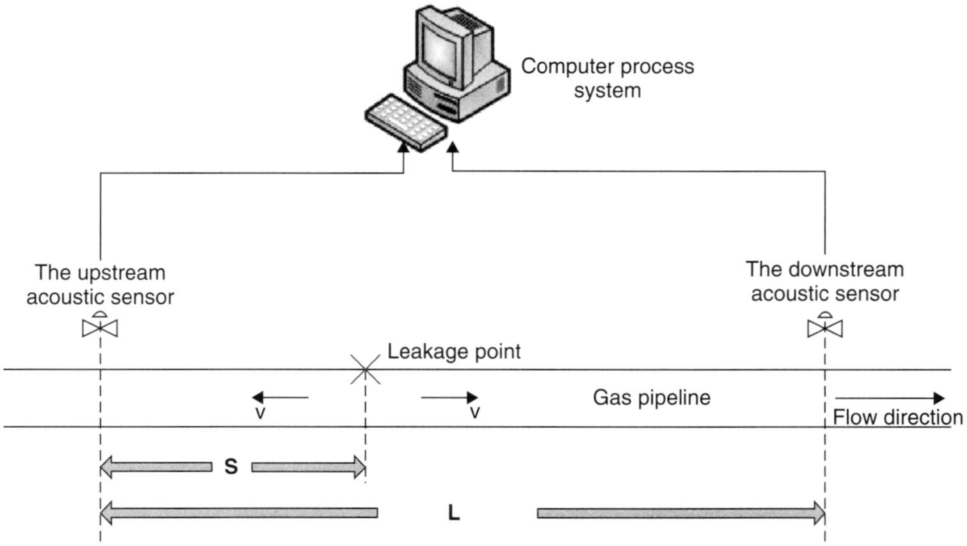

Figure 4-11 Schematic diagram of acoustic leak detection.

Compressed air provided by a compressor was used as a gas source, and water and oil droplets in the gas were removed by a cold and dry machine and filters. After entering a high-pressure buffer tank, the gas entered the test tube sections and finally went into a medium-pressure buffer tank and vents. The full length of the test pipeline was 251.5 m and the test tube section length was 200.8 m. The internal diameter was 10 mm and the maximum pressure was 8 MPa. At both ends of the test pipeline, a pressure transmitter, a differential pressure transmitter, a flow transmitter, temperature transmitters, and other equipment was installed to obtain the parameters used in this experiment. Three simulation leakage points were located 39.8, 88.8, and 49.6 m upstream from the test tube section.

The following conclusions have been drawn from this experiment:

- When the pipeline pressure was higher, the leak hole diameters were larger, and the distance between the leak point and the end of pipeline was greater, it was easier to detect and locate the leakage.
- The locating error of leakage occurring in the middle of the pipeline was greater than it was at the ends of the pipelines, and the locating error of leakage occurring in the end point of the pipeline was smaller than it was at the starting point of the pipeline.
- Under the varying conditions of this experiment, the locating errors were very small and the maximum error was about 1.37%.

Compared with the negative pressure wave method, the acoustic method had greater positioning accuracy and smaller positioning errors in this experiment. According to the results, the acoustic leak detection method had good sensitivity and a high level of locating accuracy in a natural gas pipeline. It was demonstrated that the acoustic leak detection method can achieve leak detecting and locating effectively, which can ensure the safety of natural gas pipelines.

4.6 PIPE MAINTENANCE

4.6.1 Pipeline Coatings

Steel pipelines are used all over the world for cross-country transport of natural gas, crude oil, water, and petrochemical and petroleum products at high pressures over long distances. These pipelines are protected against corrosion by external coating systems and cathodic protection.

Various coating systems have been tried over the past 45 years, and they have evolved with time and with the innovation of new materials. Today, five main coating systems are commonly used for pipelines: three-layer PE (3LPE), three-layer PP (3LPP), fusion-bonded epoxy (FBE or Dual FBE), coal tar enamel (CTE), asphalt enamel, and polyurethane (PUR). The different systems are specified by pipeline owners and consultants based on various factors, including short-term cost, long-term cost, captive usage, regional availability of the coating material, control on handling, transportation and installation of pipelines, and technical reasons [4].

3LPE coating is a multilayer coating composed of three functional components. This anticorrosion system consists of a high-performance fusion-bonded epoxy, followed by a copolymer adhesive and an outer layer of polyethylene, which provides tough, durable protection. 3LPE systems provide excellent pipeline protection for small- and large-diameter pipelines with moderate to high operating temperatures. In Fig. 4-12 the number "1" represents fusion-bonded epoxy, the number "2" represents copolymer adhesive, and the number "3" represents polyethelene.

3LPE coating is dominant worldwide—with a 50% market share—for onshore pipelines, with the exception of North America. The trend is increasing, with a greater number of projects coated with 3LPE in China, India, and the Middle East. The increased acceptance of 3LPE is due to its broad operating temperature range (from −45 to +85°C) and ability to withstand very rough handling and installation practices without damage to the coating.

3LPP coating is similar to the 3LPE coating system. The difference is that instead of using polyethylene as an outer layer, 3LPP uses polypropylene tape. 3LPP provides better impact and abrasion resistance than does 3LPE and is suitable with operation temperatures of more than 110°C. The 3LPP system consists of an epoxy primer and a grafted copolymer PP adhesive to bond the epoxy primer with a PP topcoat. In Fig. 4-13 the number "1" represents fusion-bonded epoxy, the number "2" represents copolymer adhesive, and the number "3" represents polypropylene.

3LPP systems are recognized as excellent systems for offshore projects with elevated operating temperatures

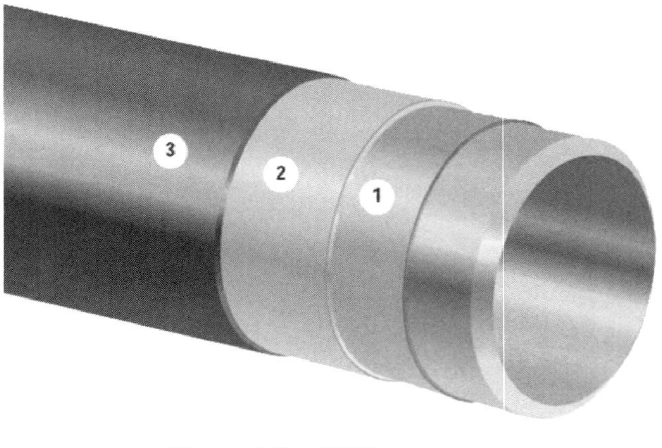

Figure 4-12 3LPE structure.

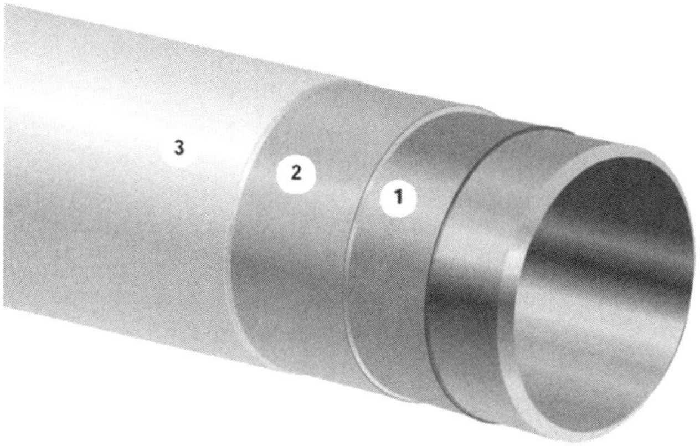

Figure 4-13 3LPP structure.

(0° to +40°C) and extreme mechanical stress on the pipes. Recent projects in the North Sea, Africa, the Gulf of Mexico, and Arabian regions have set new standards for 3LPP coatings, which provide access to deeper gas and oil fields.

FBE or Dual FBE is dominant in North America, the United Kingdom, and a few other countries, but the trend is declining in favor of 3LPE and PP systems. Some pipeline owners have graduated from coal tar coating to dual FBE, as the cost has become quite competitive after increases in coal tar prices. Dual-layer abrasion-resistant FBE systems provide excellent properties for a variety of service applications, which may include directional drilling and antiabrasion for road and river crossings, and elevated temperatures in wet environments and antislip applications. In Fig. 4-14 the number "1" represents anticorrosion coating, and the number "2" represents abrasion-resistant overcoat.

Coal tar and asphalt enamel are both still used in some countries. For many refineries which have their own pipelines, coal tar is the cheapest coating option, being their

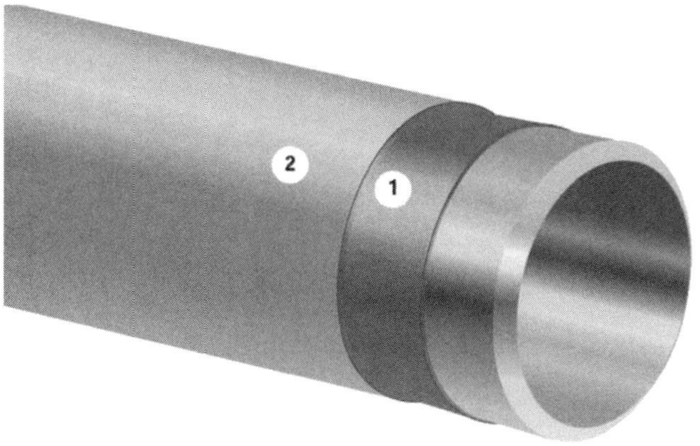

Figure 4-14 Dual FBE structure.

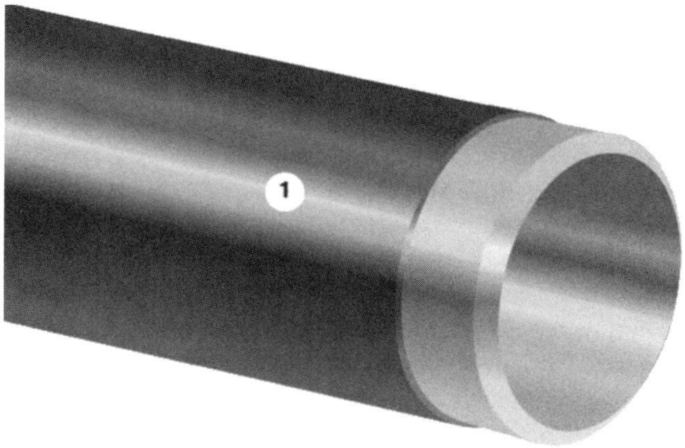

Figure 4-15 Asphalt enamel structure.

own product. Asphalt enamel is a plant-applied durable coating based on modified bitumen (asphalt) that has been used successfully for many years for corrosion protection of steel pipes. Both systems are declining and suffer from health and environmental concerns. In Fig. 4-15, the number "1" represents asphalt enamel.

Polyurethane (PUR) systems are used primarily for pipeline rehabilitation projects or girth weld coating. PUR systems also suffer from concerns as to human health.

4.6.2 Pipeline Repair

Defect repair is the action that most pipeline operators consider as a response to an integrity evaluation. Many types of repair methods deal effectively with such pipeline defects as corrosion and cracking. Next we describe the main pipeline repair methods and the applicability of each technique for defect repair.

4.6.2.1 Dressing Dressing is a procedure that involves careful removal of metal from around a defect to produce a smooth profile that eliminates stress points. It is a highly skilled operation that must be carried out in accord with strict guidelines to avoid the removal of too much metal. Typical defects repaired by dressing are superficial gouges, corrosion, spalling, extremely superficial cracks, and manufacturing faults. However, dressing does not involve the application of any fitting.

4.6.2.2 Weld Repair There are two principal methods of weld repair: weld deposition and welded patch. The concept behind *weld deposition repair* is to deposit weld metal onto the outside diameter of a pipe to replace a pipe wall lost due to corrosion. The most effective technique for depositing a weld metal deposition repair has been found to be a series of perimeter welds that are followed by layers of consecutive parallel fill passes, which should be deposited in the circumferential direction around the pipe rather than the longitudinal weld direction [5]. The first layer must be deposited using a welding heat input that reduces the risk of burning through the corroded area. Higher-welding-heat inputs are then employed for the second and subsequent layers to refine and temper the heat-affected zones of the first and previous passes. Higher heat inputs can be used for the second and subsequent layers because the first layer deposited increases the local wall thickness, which reduces the burn-through risk. For weld metal deposition repairs for extremely corroded pipelines, the second layer may also require a maximum heat input limit until sufficient layers of the weld metal deposition repair are completed. However, this repair has strict limitations relating to the weldability of the pipe material, the remaining wall thickness, and the need for highly qualified welders to perform the operation.

Welded patch repairs are normally circular in shape to reduce stress concentrations and are made from material that is equal to or better than that of the pipeline. These repairs are used to repair nonleaking defects of relatively low severity (minor mechanical damage or low-level corrosion). Patch repairs provide only protection and containment; they do not provide support to the defects. The limitations of welded patch repair is much the same as those of other welded repairs.

4.6.2.3 Shell Repair There are several methods of shell repair: snug-fitting sleeve, stand-off sleeve, hot tapping, epoxy sleeve repairs, and petrosleeve. Epoxy sleeve repairs is are the focus in this section.

Epoxy sleeve repair (ESR) is a mature technology used for pipeline repair. A very stiff material that provides excellent support to a defective area is used to fill in an annular space between the pipe and the steel sleeve. The repair comprises two oversized steel half-shells joined together to encircle the damaged area, leaving an annular gap. This annulus is sealed at each end of the sleeve using a simply applied fast-setting material and then filled at very low pressure with a highly stiff epoxy-based compound. ESR can repair all types of nonleaking defects, including cracks, dents, girth welds, gouges, corrosion, manufacturing defects, and combinations of the above.

4.6.2.4 Replacement For many pipeline operators, replacement is the preferred method of repair. The defective section is removed and replaced by a new, problem free-pipeline. It is one of the most expensive options, however, especially when considering environmental issues and lost production. The major drawback with this alternative is that it is often impossible to achieve. Many pipelines are strategically uninterruptible, and continuity of flow is essential, leaving the operator with few choices.

4.6.2.5 Clock Spring Repair Method Clock spring is the leading permanent pipeline repair solution and it is now recognized as an effective permanent repair alternative for corrosion, mechanical damage, or other defects on high-pressure pipelines. It is not suitable for internal defects, sharp cracklike defects, or, because of its unidirectional structure, girth weld/circumferential defects. A clock spring repair is comprised of eight wraps of composite, the clock spring high-strength filler material, and the clock spring adhesive. The individual wraps of the repair are bonded together, and to the pipe surface, to restore serviceability. This repair can be used as a permanent repair for external blunt metal loss defects with a depth of less than 80% of the nominal wall thickness. The repair system is a high-strength corrosion-resistant E-glass and resin composite sleeve, a high-performance adhesive, and a high-strength filler material. The composite sleeves are shaped and sized uniquely to wrap around pipe diameters of 4 to 56 in. The individual layers are bonded together and to the pipe to fully restore serviceability.

4.7 PIPE TROUBLESHOOTING

In the sections above we discussed pipe types, pipe selection, the problems pipes encounter, and pipe maintenance. However, cost is also an important factor for pipes. For long-distance pipeline systems, the significant cost in terms of capital investment is the cost of the pipe material and installation (Fig. 4-16). Pipeline pressure, grade, installation location, and technique affect the cost and design. Pipe material or grade affect the wall thickness and determine the choice of, and limit on, the welding or installation technique. For a given design pressure and pipe diameter,

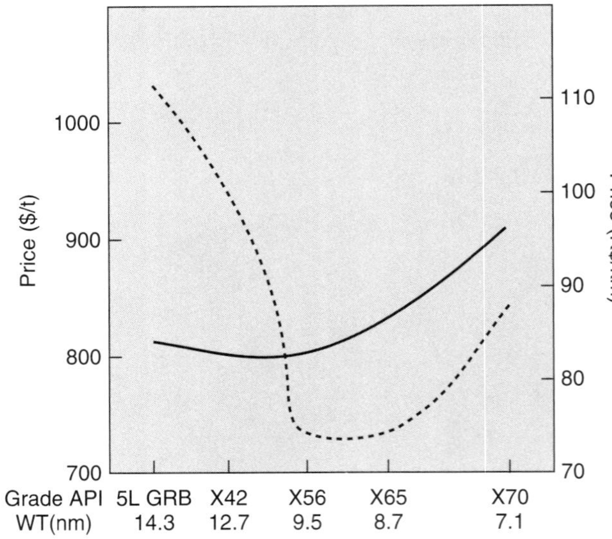

Figure 4-16 Cost of typical pipeline grade.

the wall thickness decreases with a higher-grade material. However, higher grades of steel are usually accompanied by cost premiums and more stringent construction techniques, which translate into higher costs. The location of the pipeline or surrounding environment determines allowable material, labor, and equipment, including construction material requirements.

In the pipeline design, construction, and operational stage, we have to protect the security of the entire pipeline process. To achieve that objective, several advanced technologies are adopted. Currently, a new technology called *pipeline integrity management* is prevails in the pipeline industry. It identifies and analyzes potential threats that can result in pipeline transport incidents and provides a structured and comprehensive means to select and implement risk reduction activities.

REFERENCES

1. American Petroleum Institute, *Specification for Line Pipe*, API 5L, 42nd ed., API, Washington, DC, 2000.
2. British Standards Institution, *Charpy. Impact Test on Metallic Materials, Part 1: Test Method V- and U-Notches*, British Standard EN 10045-1-1990, BSI, London, 1990.
3. Chuan, H., Qinghua, M., and Du, L. et al., Optimal planning studies on natural gas gathering pipeline system of West Sichuan, *Natural Gas Industry*, vol. 26, no. 7, 2006.
4. Dasgupta, C., Developments in steel pipeline coating systems, *Australian Pipeliner*, September, 2007, pp. 17–18.
5. *EWI, In-Service Weld Metal Deposition Repair for Pipelines*, EWI Tech Brief, EWI, Columbus, OH, 2009.
6. *General Electric*, Products, Systems and Services Brochure, GE, Fairfield, CT, 2009.
7. Ginzburg, V. B., and Ballas, R., *Flat Rolling Fundamentals*, CRC Press, Boca Raton, FL, 2000, pp. 141–142.
8. Kiefner, J. F., Kiefner, B. A., and Vieth, P. H., *Analysis of DOT Reportable Incidents for Hazardous Liquid Pipelines: 1986 through 1996*, American Petroleum Institute, Washington, DC, January 7, 1999.
9. Li, Y., Wang, W., Zhao, F., et al., Experimental research on leak detection and location of gas pipeline based on acoustic method, *China International Oil and Gas Pipeline Conference Proceedings*, 2009.
10. Mohitpour, M., Golshan, H., and Murray, A. *Pipeline Design and Construction: A Practical Approach*, 3rd ed., ASME Press, New York, 2006.
11. Sivathanu, Y., *Technology Status Report on Natural Gas Leak Detection in Pipelines*, prepared for the U.S. Department of Energy, En'Urga, Inc., West Lafayette, IN, 2005.
12. U.S. Army Corps. of Engineers, *Engineering and Design Liquid Process Piping*, EM 1110-1-4008, May 5, 1999.
13. Vieth, P. H., Roytman, I., Mesloh, R. E., and Kiefner, J. F., *Analysis of DOT Reportable Incidents for Gas Transmissions and Gathering Pipelines, 1985 through 1994*, PRC International Catalog No. L51745, May 31, 1996.

5

COOLING TOWERS

ZAKI YAMANI ZAKARIA
Universiti Teknologi Malaysia, Skudai, Johor, Malaysia
CHIKEZIE NWAOHA
Control Engineering Asia, Ten Alps Communications, Aladinma, Nigeria

The scope of cooling tower knowledge is too broad to allow complete coverage and comprehension in a single chapter. Hence, the purpose of this section is to provide basic overview knowledge on cooling towers which will facilitate understanding for beginners and users, and can later encourage discussion with manufacturers or cooling tower improvement as an important facility in a processing plant.

Cooling towers are the most basic type of evaporative cooling equipment used (primarily) for process water cooling purposes in many chemical plants. Its prime task is to reject heat to the atmosphere. It is deemed as a relatively inexpensive and reliable means of removing heat from water. Basically, hot water from heat exchangers or other units will be sent to the cooling tower. The incoming hot water will be sprayed down the tower as droplets via the fills and drift eliminator to enable maximum contact with the air while the cooling fans are running. One will thereby experience mist or carryover of moisture around the tower. However, this should be relatively very small because the adequate engineering design will reduce the mist level substantially. The cooled water exiting the cooling tower will be sent back to the heat exchanger as a cooling medium. The makeup water source is used to replenish water lost to evaporation. The temperature drop is dictated by several features, which are explained later in the chapter. A cooling tower is normally located in the utility section and is a large humidification structure consisting of draft fans at the top and a transfer pump at the bottom. Figure 5-1 is a schematic diagram of the cooling tower process.

5.1 COOLING TOWER OPERATION

5.1.1 Cooling Tower Psychrometrics

Psychrometry is the study of cooling by evaporation. Maximum evaporation takes place when water, in the form of tiny droplets, is exposed to maximum airflow for the longest possible time. The process of evaporation through removal of latent heat allows the water to be cooled below the ambient dry-bulb temperature. The dry air enters the cooling tower and begins to gain moisture and enthalpy in an effort to reach equilibrium with the water. The process can be resembled when one cools a hot cup of tea by interspraying it between two cups. The water may be cooled 8°C or more, while the air mass dry-bulb temperature may increase only slightly.

Usually, water is cooled by exposing its surface to available moving air. Cooling of water on the surface of a pond and spraying of water into slow-moving or forced air are other known processes. The cooling of water of varying degrees involves the heat-transfer process in the form of latent heat transfer as well as sensible heat due to the differences in the temperature of the water and air. Cooling by air depends on the moisture

Process Plant Equipment: Operation, Control, and Reliability, First Edition. Edited by Michael D. Holloway, Chikezie Nwaoha, and Oliver A. Onyewuenyi.
© 2012 John Wiley & Sons, Inc. Published 2012 by John Wiley & Sons, Inc.

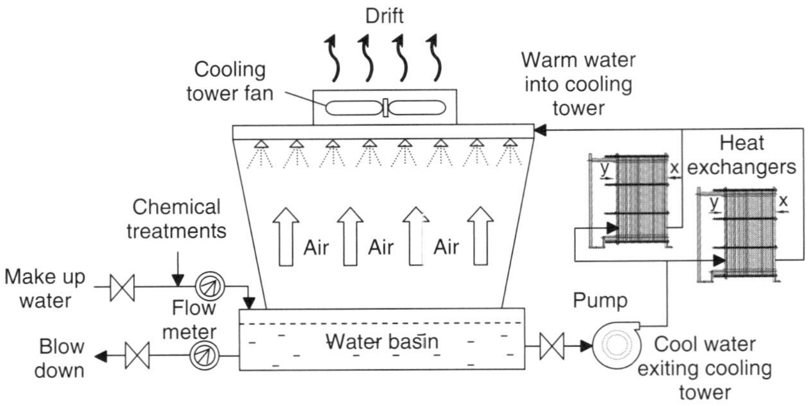

Figure 5-1 Schematic diagram of a cooling tower system.

content of air, indicated by the wet-bulb temperature of the surrounding air. The wet-bulb temperature is the lowest theoretical temperature to which the water can be cooled. A psychrometric chart (Fig. 5-2) may be used to illustrate the relationships between wet- and dry-bulb temperatures. It represents the thermodynamic parameters of moist air at a constant pressure, often equated to an elevation relative to sea level. A psychrometric chart is used to carry out heat or cooling load calculations and find solutions to various cooling-related problems [8].

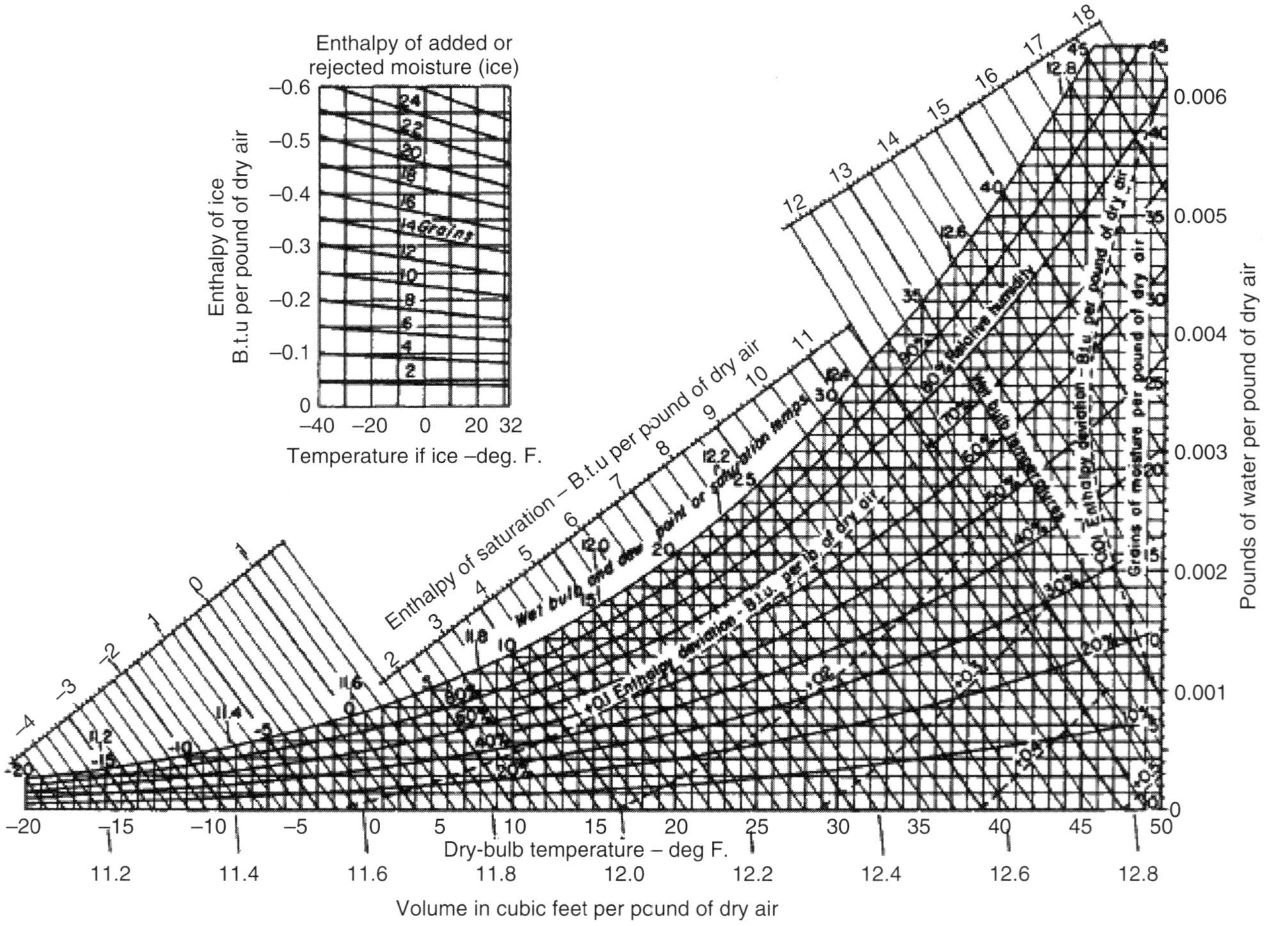

Figure 5-2 Psychrometric chart—low temperatures. Barometric pressure, 29.92 in.Hg. (From [10].)

5.1.2 Principles of Cooling

Before going to the operational part of a cooling tower, it is better to understand its principles. Generally, water can be cooled by exposing its surface to air. The process can be slow or fast, depending on how the water is exposed to the air. A slow process can be seen from the cooling of water on the surface of a lake, whereas a faster process can be observed from the spraying of water to the air.

Heat removal in a cooling tower depends strongly on the temperature and moisture content of air. The moisture content of air can be resembled as the wet-bulb temperature. Wet-bulb temperature is the lowest temperature that can be reached by the evaporation of water. The cooled-water temperature drops close to, but does not equal, the wet-bulb temperature. This is due primarily to the fact that not all water droplets can be in contact with air in the cooling tower. Getting the cool water temperature closer to the wet-bulb temperature will create a good cooling tower. This condition can be achieved from the tower design, which focuses on good air-to-water contact.

The theory behind the cooling water process is based on the enthalpy potential difference as the driving force, developed by Merkel. It is assumed that there is a thin film of air on each particle of water. The enthalpy potential difference between the films on each particle of water and the surrounding provides the driving force for the cooling process. Figure 5-3 illustrates this condition.

The Merkel equation is

$$\frac{KaV}{L} = \int_{T_2}^{T_1} \frac{dT}{h' - h} \tag{5.1}$$

where K = mass-transfer coefficient [lb water/($h \times$ ft^2)]
a = contact area (ft^2/ft^3 tower volume)
V = active cooling volume (ft^3/ft^2 of plan area)
L = water rate [lb/($h \times$ ft^2)]
h' = enthalpy of saturated air at water temperature (Btu/lb)
h = enthalpy of airstream (Btu/lb)
T_1 = entering water temperatures (°F)
T_2 = leaving water temperatures (°F)

The left-hand side of Eq. (5.1) is related to tower specifications, whereas the right-hand side is entirely in terms of air and water properties.

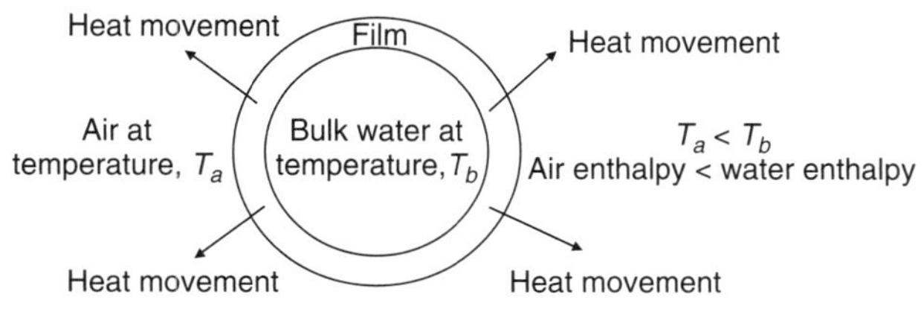

Figure 5-3 Water drop with interfacial film.

Figure 5-4 illustrates a cooling tower process heat balance showing water–air relationships and the driving potential in a counterflow tower. Line AB demonstrates the water operating line, which has fixed inlet and outlet tower water temperatures. Directly beneath B, the air operating line begins at C; this is the enthalpy of the entering wet-bulb temperature, t_{wb}. The t_{wb} of the air entering the cooling tower determines operating temperature levels throughout the plant, process, or system. Line BC represents the initial enthalpy driving force ($h' - h$) and is known as the *cooling tower approach*. In cooling 1°F of water, the enthalpy per pound of air is increased 1 Btu, multiplied by the ratio of pounds of water per pound of air. The liquid/gas ratio L/G is the slope of the straight operating line, which is CD. This line is also the cooling *range* in the tower, which can be known by projecting the line on the temperature scale. Point D represents the air leaving the tower. Projecting the exiting air point toward the water operating line and then onto the temperature axis (DEF) shows the outlet air web-bulb temperature. The area $ABCD$ is the *tower characteristic* and is represented by the integral of Eq. (5.1). When there is a change in L/G, the area $ABCD$ will change; hence, the tower characteristic will change accordingly as well.

For the water operating line, the coordinates can be referred directly to the temperature (x-axis) and enthalpy (y-axis) of any point. However, for the air operating line, only the enthalpy can be referred (y-axis). The corresponding wet-bulb temperature of a point on line CD can be obtained by projecting the point horizontally to the saturation curve, then vertically to the temperature coordinate.

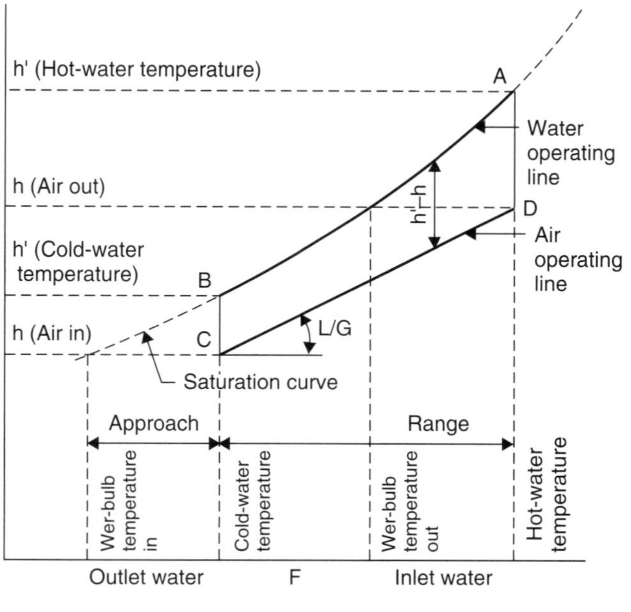

Figure 5-4 Cooling tower process heat balance. (From [10].)

Consider a situation where the entering air wet-bulb temperature increases. This will move point C upward and line CD shifts to the right to maintain constant *KaV/L*. On top of that, it will increase the hot and cold water temperatures as well as cooling range and approach areas. If the cooling range increases, line CD lengthens. At a constant wet-bulb temperature, equilibrium is established by moving the line to the right to maintain a constant KaV/L. A change in L/G changes the slope of CD, and the tower comes to equilibrium with a new KaV/L.

To predict tower performance it is important to know the tower characteristics for fixed ambient and water conditions. The tower characteristic KaV/L can be determined by integration using the *Chebyshev method* for evaluating the integral numerically [10]:

$$\frac{KaV}{L} = \int_{T_2}^{T_1} \frac{dT}{h_w - h_a} \cong \frac{T_1 - T_2}{4}\left(\frac{1}{\Delta h_1} + \frac{1}{\Delta h_2} + \frac{1}{\Delta h_3} + \frac{1}{\Delta h_4}\right) \qquad (5.2)$$

where h_w = enthalpy of air–water vapor mixture at bulk water temperature (Btu/lb dry air)
h_a = enthalpy of air–water vapor mixture at wet-bulb temperature (Btu/lb dry air)
Δh_1 = value of $(h_w - h_a)$ at $T_2 + 0.1(T_1 - T_2)$
Δh_2 = value of $(h_w - h_a)$ at $T_2 + 0.4(T_1 - T_2)$
Δh_3 = value of $(h_w - h_a)$ at $T_1 - 0.4(T_1 - T_2)$
Δh_4 = value of $(h_w - h_a)$ at $T_1 - 0.1(T_1 - T_2)$

For better comprehension on the calculation to obtain the tower characteristic, see Example 5.1.

Example 5.1 The water circulation rate and airflow rate for a given cooling tower is 115,000 lb/min and 97,500 lb of dry air/min, respectively. It was measured that the ambient wet-bulb temperature, T_{wb}, is 80°F, while the inlet and outlet water temperatures are found to be 104°C and 88°F, respectively. Determine the tower characteristic if the cooling tower is altitude at sea level.

Solution: First, organize all information:

$$t_1 = 104^\circ\text{F}$$
$$t_2 = 88^\circ\text{F}$$
$$t_{\mathrm{wb}} = 80^\circ\text{F}$$
$$\text{water flow rate, } L = 115{,}000 \text{ lb/ min}$$
$$\text{airflow rate, } G = 97{,}500 \text{ lb/ min}$$
$$L/G = \frac{\text{water flow rate}}{\text{airflow rate}} = \frac{115{,}000}{97{,}500} = 1.6$$

From the air–water vapor mixture tables, the enthalpy h_1 of ambient air at 80°F wet-bulb temperature is 43.69 Btu/lb.

$$\begin{aligned}\text{enthalpy } h_2 \text{ of leaving air} &= h_1 + (L/G)(t_2 - t_1)\\ &= 43.69 + 1.6(104 - 88)\\ &= 62.5618 \text{ Btu/lb}\end{aligned}$$

Using Eq. (5.2), create and arrange all information as in Table 5-1. Search for the respective enthalpy of water, h_w (column 3), for other temperatures (column 2) as calculated in the table. Calculate the enthalpy of air (column 5). Hence, using Eq. (5.2), the tower characteristic $(KaV/L) = 1.6142$.

If we repeat the calculation above using a lower L/G, the tower characteristic will be smaller under the same design conditions. This means that the airmass in the tower will increase. The decrease in L/G for the same water flow rate means that the decrease of enthalpy in the air side and $1/(h_w - h_a)$ will decrease as well. The exit enthalpy per pound of dry air will decrease and the temperature of the exit air will be reduced.

An alternative method of obtaining the tower characteristic is by using a nomograph prepared by Wood and Betts. It is a fast approximate value of *KaV/L* provided that the values of L/G, t_{wb}, and the cooling range are available [10].

The performance characteristics of various types of towers will vary with height, fill configuration, and flow

TABLE 5-1 Data for Example 5.1

Water Side			Air Side		Enthalpy Difference	
Description	T_w (°F)	h_w (Btu/lb)	Description	h_a (Btu/lb)	$h_w - h_a$	$1/(h_w - h_a)$
t_2	88.0	54.06	h_1	43.69		
$t_{w2} + 0.1 \times$ range	89.6	55.86	$h_{a1} + 0.1 \times L/G \times$ range	45.58	10.283	0.0973
$t_{w2} + 0.4 \times$ range	94.4	61.25	$h_{a1} + 0.4 \times L/G \times$ range	51.24	10.011	0.0999
$t_{w1} - 0.4 \times$ range	97.6	64.84	$h_{a1} - 0.4 \times L/G \times$ range	55.01	9.827	0.1018
$t_{w1} - 0.1 \times$ range	102.4	70.23	$h_{a1} - 0.1 \times L/G \times$ range	60.67	9.555	0.1047
t_1	104	72.03	h_2	62.56		
Sum of $1/(h_w - h_a)$						0.4036
Tower characteristic (KaV/L) = sum of $1/(h_w - h_a)/4 \times$ range						1.6142

arrangement (crossflow or counterflow); however, these factors have been taken into consideration in the preparation of the performance characteristic nomograph. When accurate characteristics of a specific tower are required, the cooling tower manufacturer should be consulted. Performance tests on a cooling tower should be done in accordance with the Cooling Tower Institute (CTI) Acceptance Test Code and the American Society of Mechanical Engineers (ASME) test code.

5.1.3 Heat Exchange

A cooling tower is a specialized heat exchanger in which air and water are brought into direct contact with each other to affect the transfer of heat. Disregarding any negligible amount of sensible heat exchange that may occur, the heat gained by the air must equal the heat lost by the water. Within the airstream, the rate of heat gain is identified by the expression $G(h_2 - h_1)$, where G is the mass flow of dry air through the tower (lb/min), h_1 the enthalpy (total heat content) of entering air (Btu/Ib of dry air), and h_2 the enthalpy of leaving air (Btu/Ib of dry air). For a water stream, the rate of heat loss appears to be $L(t_1 - t_2)$, where L is the mass flow of water entering the tower (lb/min), t_1 the hot water temperature entering the tower (°F), and t_2 the cold water temperature leaving the tower (°F). This derives from the fact that 1 Btu (British thermal unit) is the amount of heat gain or loss necessary to change the temperature of 1 lb of water by 1°F. Since heat lost should be equal to heat gained, the equation can be combined and becomes

$$G(h_2 - h_1) = L(t_1 - t_2)$$

$$L/G = \frac{h_2 - h_1}{t_1 - t_2}$$

The value of L/G is an important indicator for cooling tower performance and is discussed later in the chapter. It is important to note that a change in wet-bulb temperature (due to atmospheric conditions) and a change in cooling range will not change the tower characteristic *(KaV/L)*. Only a change in the L/G ratio will change *KaV/L* [1,12].

Besides *KaV/L* and L/G, the amount of heat transfer and water evaporation in a cooling tower is imperative for cooling tower selection. Heat transfer (heat rejection) and water evaporation rates in a cooling tower can be calculated and are shown in Example 5.2.

Example 5.2 A cooling tower is designed to cool 1100 gal/min of water from 93°F to 84°F. Calculate the heat rejection and evaporation rate.

Solution:

$$\begin{aligned}\text{Heat rejection (Btu/h)} &= 1100\ \text{gal/min} \times 9^\circ\text{F} \\ &\quad \times 8.33\ \text{lb/gal} \times 60\,\text{min/h} \\ &\quad \times 1\ \text{Btu/lb-}^\circ\text{F} \\ &= 1100 \times 9 \times 500 \\ &= 4{,}950{,}000\ \text{Btu/h}\end{aligned}$$

$$\begin{aligned}\text{Evaporation rate (Btu/h)} &= \frac{\text{heat rejection}}{1000\ \text{Btu/lb}} \\ &= \frac{4{,}950{,}000}{1000} = 4950\ \text{Btu/h}\end{aligned}$$

Sometimes, misconception occurs when one assumes that the cooling tower dictates the rate of heat transfer. Actually, a cooling tower gives up the heat it is supplied with. A huge cooling tower may cool water from 90°F to 80°F, whereas a smaller unit can cool the water in the same process from 100°C to 90°C. Under both conditions, the heat transfer and evaporation rates are similar [3].

5.1.4 Components of Cooling Towers

A cooling tower consists of some basic components that enable the evaporative effects of hot water to take place. The following points describe the important building block of a typical cooling tower with its functions. Figure 5-11 illustrates a schematic overview of the cooling tower components.

5.1.4.1 Frame and Casing The frame, crucially the main building block for a cooling tower, supports the exterior casings, motors, fans, fills, louvers, drift eliminators and other components. For a smaller-design cooling tower, the casing will naturally act as the frame. A typical example of a cooling tower is shown in Fig. 5-12.

5.1.4.2 Fill Most cooling towers employ fills made of either plastic or wood to facilitate heat transfer by maximizing water and air contact. There are two types of fill: splash type and film type. With *splash fill*, the hot water falls over successive layers of horizontal splash bars, continuously breaking into smaller droplets which at the same time are wetting the fill surface. Wood fill is employed by relatively older cooling towers (as shown in Fig. 5-5) and can deteriorate with time due to the effect of weather and excessive chemical water treatment. Plastic fill can last longer but is relatively more expensive.

Film fill, on the other hand, typically consists of thin, narrowly spaced plastic surfaces over which the hot water spreads, forming a thin film in contact with the air. These surfaces may be flat, honeycombed, corrugated, or in other patterns.

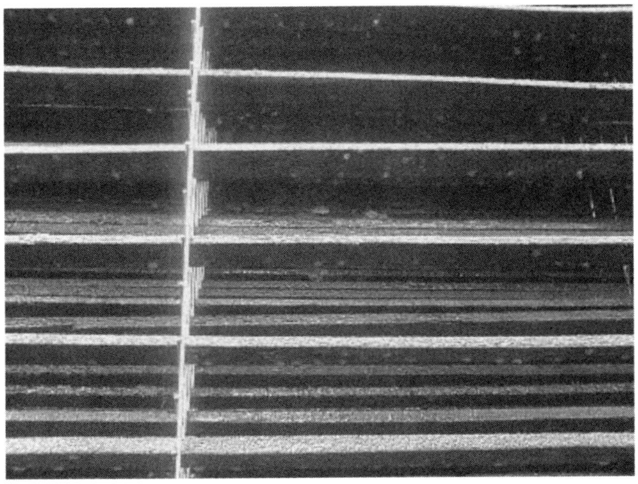

Figure 5-5 Cross-flow wood fill.

Figure 5-6 Wooden drift eliminators.

5.1.4.3 ***Cold Water Basin*** A cold water basin is a pond where the cooled water flows down through the fills and is collected. It is located at or near the bottom of the tower and is normally made of a combination of concrete and cement. There are also basins made of steel, but they are costly and could easily experience corrosion. The basin will have a cold water discharge connection to a few pumps positioned next to the tower to deliver cooled water to the plants. There will be a period where a cooling tower is shut down to allow massive cleaning of the basin. This is due to scale, corrosion, bacteria attack, debris, and various unwanted and unexpected accumulation of sediments at the bottom of the basin.

5.1.4.4 ***Drift Eliminators*** Drift eliminators are an arrangement of flat material or labyrinth passages (made of wood or plastic), such as a baffle device. They are meant to capture water droplets entrapped in the airstream (drift) that would otherwise be lost to the atmosphere. Drift is the circulating water lost from the tower as liquid droplets entrained in the exhaust airstream. A drift eliminator is located next to the fill inside the cooling tower. Without drift eliminators, the percentage of water lost to the atmosphere will increase, resulting in an inefficient cooling tower. Figure 5-6 shows drift eliminators made of wood in a typical cooling tower.

5.1.4.5 ***Air Inlet*** An air inlet is the point of entry for the air entering a cooling tower via the louvers. The inlet may take up an entire side of a tower for a cross-flow design or be located low on the side or the bottom of counter flow designs. Figure 5-7 shows the direction of air inlets via louvers.

5.1.4.6 ***Louvers*** In general, cross-flow towers have inlet louvers made of fiberglass-reinforced polyester (FRP). The purpose of louvers is to equalize airflow into the fill (cooling tower) and retain the water within the tower. In certain cases where the louvers are not aligned or dented properly, airflow into the cooling tower will be restricted and result in a poor evaporative process. Counterflow tower designs do not require louvers. Figure 5-7 shows an example of louvers made of FRP.

Figure 5-7 Cooling tower louvers. (From flickr.com/photos/asbestos_pix/4064198899/, 2010.)

5.1.4.7 ***Nozzles*** A nozzle is a small piece of equipment positioned on top of a tower that provides a water spray effect for the hot water before it enters the fill and wets it. Uniform water distribution at the top of the fill is imperative to attain proper wetting of the entire fill surface. Nozzles, sometimes called target nozzles, can have either round or square spray patterns. The nozzles should have regular checkups to avoid blockage from solid particles that may appear with the hot water. Figure 5-8 demonstrates a hot water deck holding a number of nozzles that will drop into the cooling tower. Figure 5-9 shows examples of target nozzles.

5.1.4.8 Fans The fans used in a cooling tower are axial (propeller) and centrifugal types. In general, propeller fans are used in induced-draft towers; both propeller and centrifugal fans are found in forced-draft towers. A fan having nonautomatic adjustable pitch blades can be used over a wide range of kilowatts with adjustable desired airflow at the lowest power consumption. Automatic variable-pitch blades can change the airflow in response to changing load conditions. The angle of the blade also influences the airflow and the power consumed by the tower. Figure 5-10 shows a cooling tower fan for an induced-draft tower, and Fig. 5-11 allows schematic tower breakups.

Figure 5-10 Cooling tower fans.

Figure 5-8 Hot water deck with a number of nozzles installed.

Figure 5-9 Cooling tower target nozzles.

5.2 TYPES OF COOLING TOWERS

Basically, two major types of cooling towers are used in the process industries: natural-draft cooling towers and mechanical-draft cooling towers. A third type is a combination of natural and mechanical. An additional division is made based on the type of flow: cross-flow, counterflow, or co-current flow. In cross-flow towers, the air moves horizontally across the downward flow of water. In counterflow towers, the air moves vertically upward against the downward fall of the water. In co-current flow, air and water both move in a similar direction.

Cooling towers are also sometimes distinguished by heat dissipation: wet (evaporative cooling), dry, and wet-dry, as well as type of application, either industrial or power plant. Each major type of cooling tower has a distinctive and unique configuration. Figure 5-13 is an overview of the types of cooling towers.

5.2.1 Natural-Draft Cooling Towers

In natural-draft cooling towers, the air that flows through is largely due to the difference in density between the cool inlet air and the warm exit air. The air leaving the stack is lighter than ambient air, and a draft is created by the chimney effect without the need of mechanical fans [11]. Natural-draft designs use very large concrete chimneys to

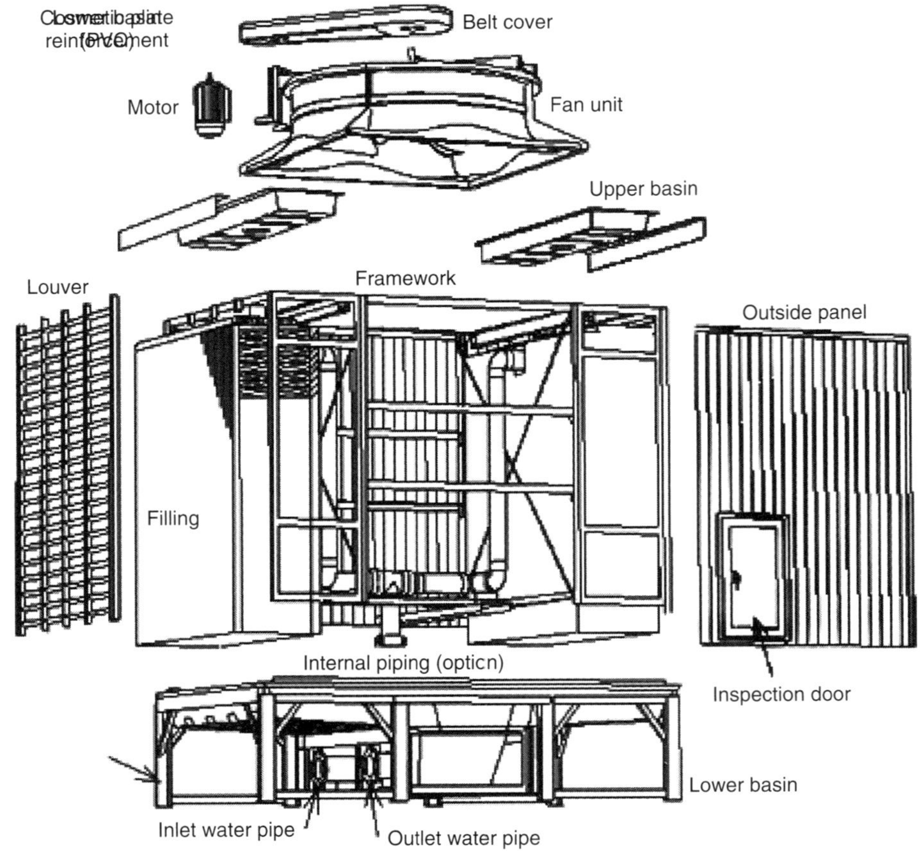

Figure 5-11 Cooling tower schematic breakups. (From [4].)

Figure 5-12 Actual image of cross-flow Induced draft cooling tower. (From http://cae2k.com/db-photos-0/cooling-tower-photos.html).

introduce air through the medium. Due to the tremendous size of these towers (500 ft high and 400 ft in diameter at the base) they are generally used for huge water flow rates: typically above 200,000 gal/min. Usually, these types of towers are used only by utility power stations in the United States.

The water falls downward over fill surfaces, which help to increase the contact time between the water and the air. This helps in maximizing heat transfer between the two. Figure 5-14 shows a schematic diagram of a natural-draft cooling tower. The green flow paths show how the warm water leaves the plant, is pumped, and is distributed onto fills inside the natural-draft cooling tower. The cooled water, including makeup water (in this case from a lake) accounted for in evaporation losses to the atmosphere, is returned to the condenser in the plant. Figure 5-15 shows a natural-draft cooling tower in Callaway, Missouri. Natural-draft cooling towers are classified into four different types: atmospheric cooling, hyperbolic cooling, spray-filled cooling, and wood-filled cooling.

5.2.1.1 ***Atmospheric Spray Towers*** Cooling towers of this type are dependent on atmospheric conditions and the aspirating effects of the spray nozzles. No mechanical devices are used to move the air [10]. This type of cooling tower is used when small sizes are required and when a low level of performance can be tolerated. Figure 5-16 is a schematic diagram of a spray tower.

5.2.1.2 ***Hyperbolic Natural-Draft Towers*** These towers are extremely dependable and predictable in their thermal performance. A chimney or stack is used to induce air

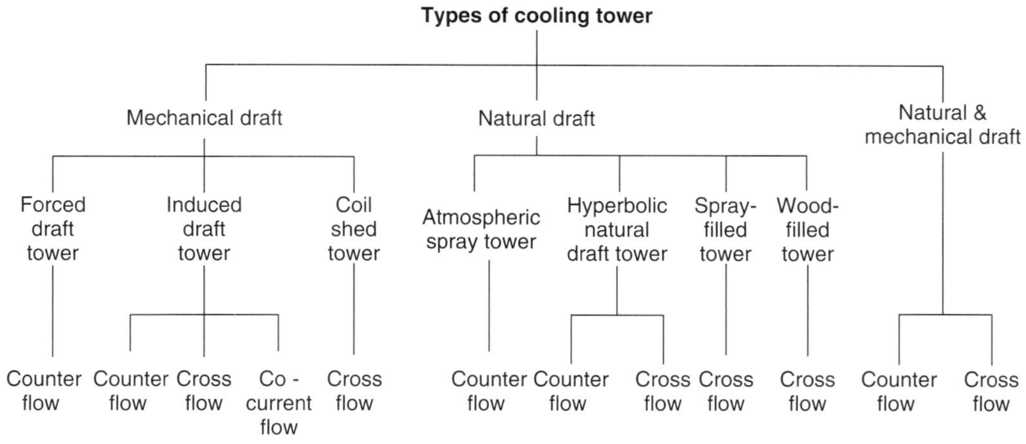

Figure 5-13 Types of cooling towers used in industry.

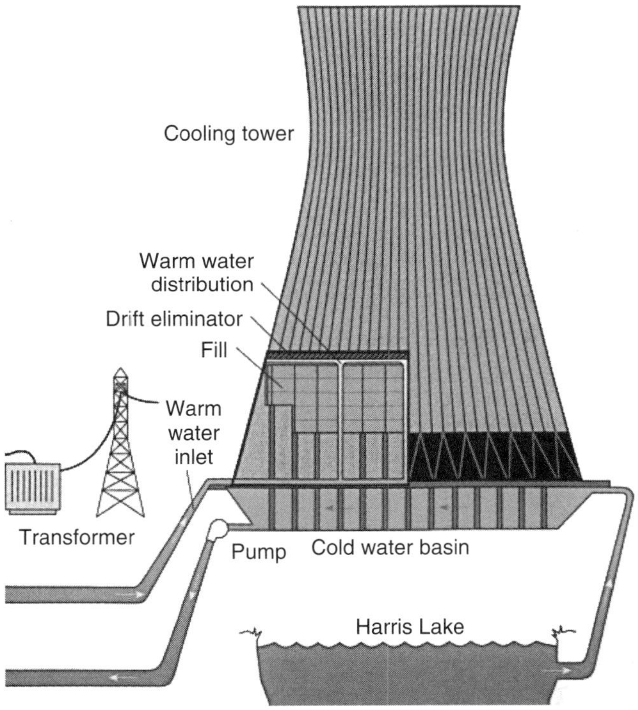

Figure 5-14 Natural-draft cooling tower in the United States. (From [5].)

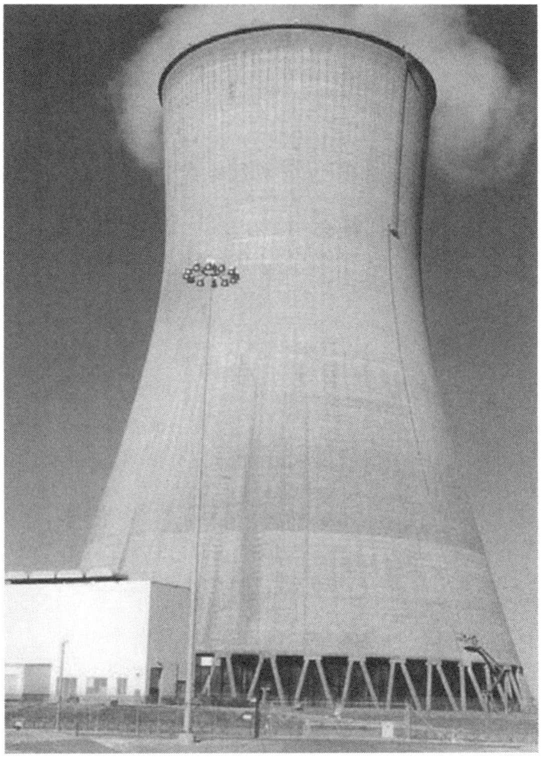

Figure 5-15 Natural-draft cooling tower in Callaway, Missouri (From [6].)

movement through the tower [11]. This type of cooling tower makes use of the stack effect of a chimney above the packing to induce airflow through the packing in counterflow to the water. It has the advantages of low maintenance costs and much better performance than cooling frames but is not suitable for high dry-bulb air temperature because the water inlet temperature is higher than the air dry-bulb temperature. It is used primarily for large cooling duties: for example, in power stations. Figure 5-17 is the schematic of a hyperbolic natural-draft tower.

5.2.1.3 Spray-Filled Towers Spray-filled cooling towers are cooled by evaporative cooling, which requires less area than that required by cooling ponds. A spray-filled cooling tower uses a number of nozzles that distribute water into the available air in the surroundings. Sufficient distance should be provided between the spray nozzle and the basin to keep the spray from being carried over to the sides of the pond basin. This type of cooling tower does not contain packing. There are numerous spray nozzles at the top; air flows crosswise to the vertical fall water. This tower type

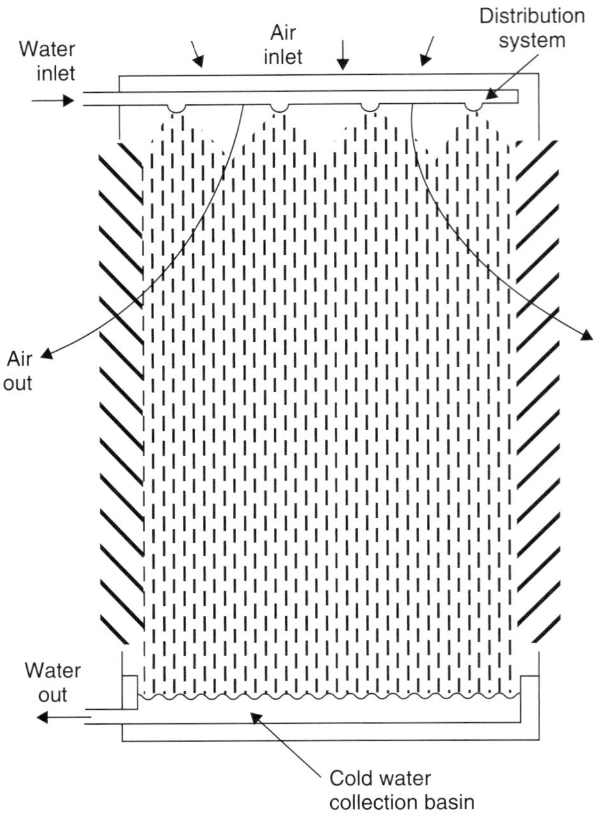

Figure 5-16 Atmospheric spray tower. (From [3].)

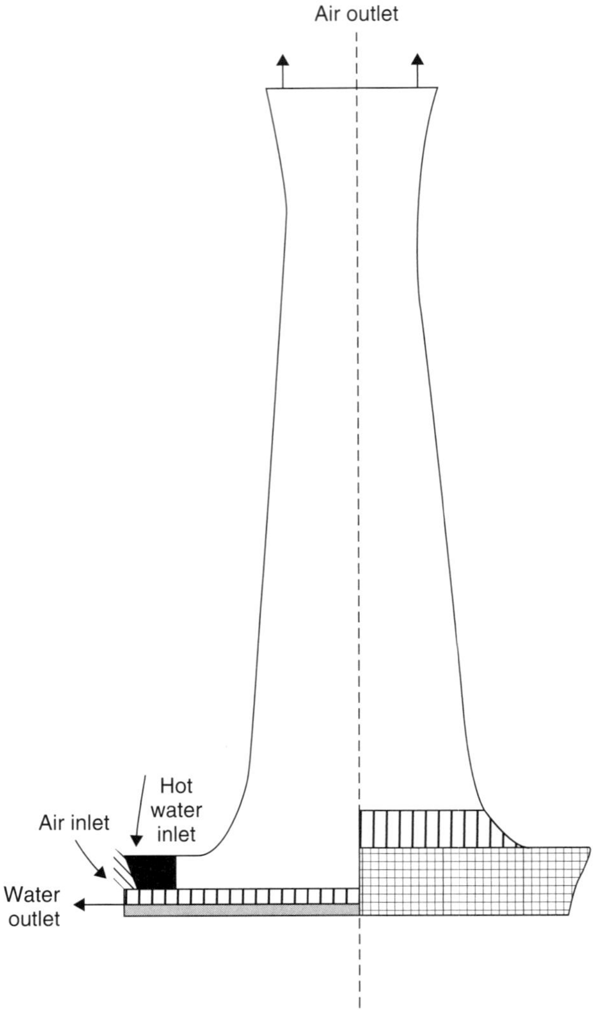

Figure 5-17 Hyperbolic natural-draft tower. (From [3].)

has the advantage of low maintenance costs. It is important to note that spray-filled cooling towers need a high pump head to atomize water through nozzles. However, it could also experience high losses from windage.

*5.2.1.4 **Wood-Filled Cooling Towers*** Wood-filled cooling towers contain baffling to increase the wetted surface for air–water contact. The baffles create a longer contact time as the water drops from deck to deck. Its performance is much better than that of spray-filled frames.

5.2.2 Mechanical-Draft Cooling Towers

Mechanical-draft cooling towers employ fans to move air. The towers have greater stability because it is affected by fewer psychrometric variables. The fans provide a means of regulating airflow. Mechanical-draft cooling towers are subclassified into forced draft and induced draft [11]. Another type, the coiled shed tower, is now receiving less attention. Mechanical-draft towers offer the control of cooling rates through their fan diameter and speed of operation. These towers often contain several areas called *cells*, each with its own fan.

*5.2.2.1 **Forced-Draft Towers*** In forced-draft towers the fan is located on the airstream entering the tower, which is characterized by high air entrance velocities and low exit velocities; therefore, the towers are susceptible to recirculation, thus having less performance stability. The fans can also be subjected to icing under conditions of low ambient temperature and high humidity.

The blow of ambient air into the tower is in the direction across the packing. Its mechanical equipment is near the ground on a firm foundation, thus keeping vibration to a minimum. The fan size is limited; thus, a larger number of smaller fans of higher speed compared with those in an induced-draft arrangement results in more noise, but the tower itself provides some attenuation. Figure 5-18 shows forced-draft cooling towers for counterflow. It is important to note that there is no forced-draft cross-flow cooling tower.

Among the advantages of forced-draft towers are the following:

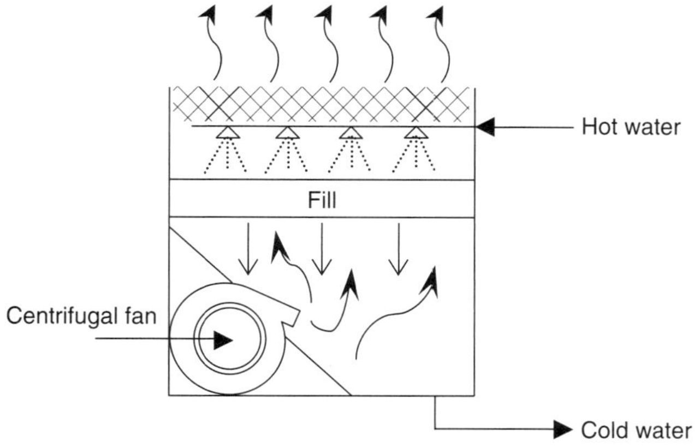

Figure 5-18 Counterflow forced-draft cooling tower. (From [2].)

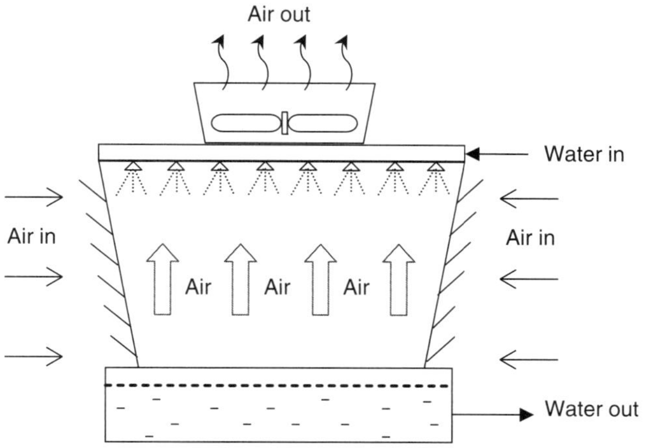

Figure 5-19 Cross-flow induced-draft cooling tower.

- Slightly lower horsepower since the fan is in cold air (horsepower varies directly as the absolute temperature)
- Better accessibility of mechanical components for maintenance
- Easily adaptable for warm air recirculation for cold climates

The disadvantages of forced-draft towers are as follows:

- Poor distribution of air over the section
- Greatly increased possibility of hot air recirculation, due to low discharge velocity from the sections and the absence of a stack
- Low natural-draft capability on fan failure due to the small stack effect
- Total exposure of tubes to sun, rain, and hail

The horizontal section is the most commonly used air-cooled section, and generally the most economical. For a fluid with freezing potential, the tubes should be sloped at least 10 mm/m to the outlet header. Since in most cases no problem will be associated with freezing and it is more costly to design a sloped unit, most coolers are designed with level sections.

Vertical sections are sometimes used when maximum drainage and head are required, such as for condensing services. Angled sections, such as vertical sections, are used for condensing services, allowing positive drainage. Frequently, angle sections are sloped 30° from horizontal.

5.2.2.2 ***Induced-Draft Towers*** The fan is located at the airstream leaving the tower. This causes air exit velocities that are three to four times higher than the air entrance velocities. This improves the heat dispersion and reduces the potential for recirculation. Induced-draft towers require about 1 kW of input for every 18,000 m^3/h of air. This type can be considered one of the most popular options in the processing industry [11]. The fan is situated at the air outlet from the tower, usually at the top, sometimes at the side, or even in ducting. Larger fans can be employed, as it requires low speed and a low noise level. Recirculation of used air is not possible due to the high outlet velocity. An induced-draft tower is more prone to vibration because the fan is mounted on the superstructure and therefore usually has a more compact ground plan than in a forced-draft design, due to the absence of fans on the side. Figures 5-19 and 5-20 illustrate induced-draft cooling towers for cross-flow and counter-flow, respectively.

There are several advantages in using induced-draft towers:

- There is better distribution of air across the section.
- There is less possibility of the hot effluent air recirculating around to the intake of the sections. The hot air is discharged upward at approximately 450 m/min.
- The effect of sun, rain, and hail is smaller, since 60% of the face area of the section is covered.
- There is increased capacity in the event of fan failure, since the natural-draft stack effect is much greater with an induced draft.

The disadvantages of induced-draft towers are:

- The horsepower is higher since the fan is located in the hot air.
- The effluent air temperature should be limited to 95°C to prevent potential damage to fan blades, bearings,

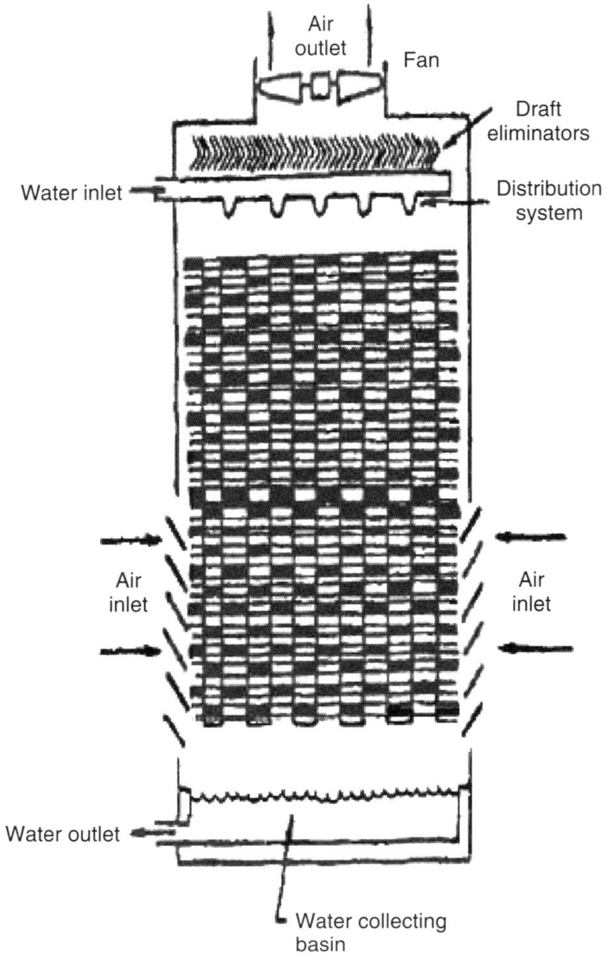

Figure 5-20 Counterflow induced-draft cooling tower. (From [3].)

V-belts, and other mechanical components in the hot airstream.

- The fan drive components are less accessible for maintenance, which may have to be done in the hot air generated by natural convection.
- For inlet process fluids above 175°C, forced-draft design should be used; otherwise, fan failure could subject the fan blades and bearings to excessive temperatures.

5.2.2.3 Coil Shed Towers In coil shed towers, an older cooling tower type, the atmospheric coils or sections are located in the basin of the tower. The sections are cooled by flooding their surfaces with cold water. Reasons for discontinued use were scaling problems, poor temperature control, and construction costs. This type of tower can exist in both mechanical-and natural-draft forms. Figure 5-21 shows a coil shed cooling tower.

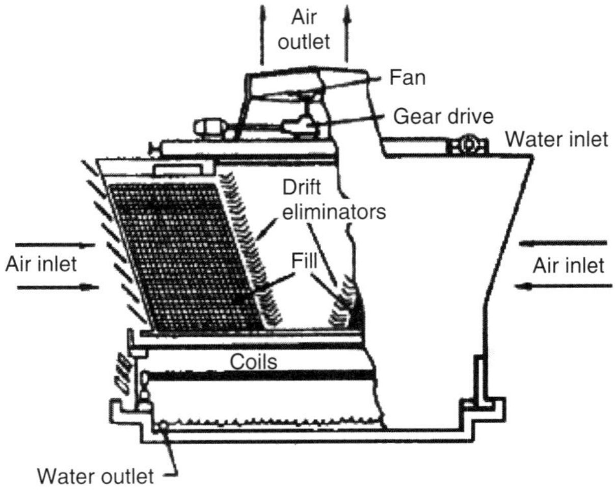

Figure 5-21 Coil shed cooling tower. (From [3].)

5.3 COMMON PROBLEMS OF COOLING TOWERS

Cooling towers are huge pieces of equipment that need to be maintained carefully and regularly. Failure to control or monitor will result in an upset in the process. Some of their common problems are described below.

5.3.1 Scale Deposits

In the evaporation of water from a cooling tower, it leaves deposits on the surface of the fills from the dissolved minerals in that water. This buildup of scales acts as a barrier to heat transfer from the process water to the air. Scales also increase the water pressure drop through the water coolers, condensers, and pipes, and leads to higher energy costs [9]. This is because a cooler will consume up to 2% more energy for each degree rise in the cooler temperature. Scales deposited can provide an ideal environment for the growth of microbial organisms (e.g., harmful *Legionella* bacteria). This represents a health risk to maintenance personnel, and as it becomes airborne, residents living downwind of the cooling tower will inhale the aerosolized cooling tower drift and become infected, as happened in late 2002 in Harnes, France. Figure 5-22 is an example of scaling in a cooling tower.

5.3.2 Delignification of Wood

In some cooling towers, the fills are made of wood. This can be detrimental in the long run, because when the wood is delignified as a result of long-term overfeed plus the effect from chemical water treatment, debris will be released and can foul surfaces and heat exchangers and eventually compromise the cooling tower structure, creating a serious safety problem [9].

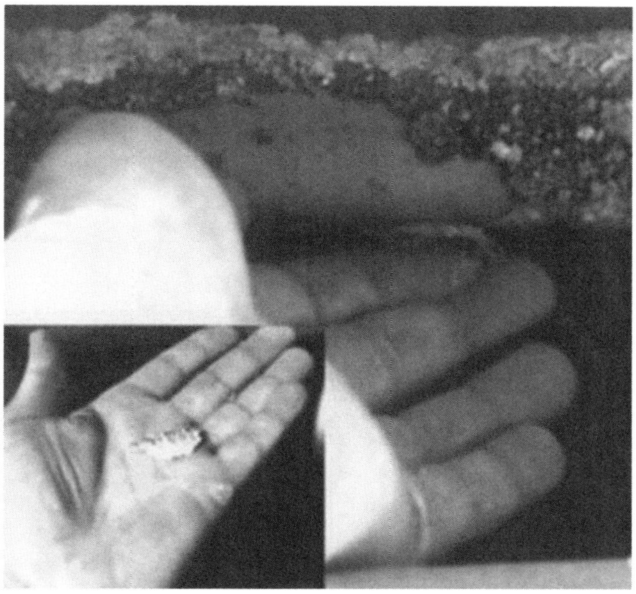

Figure 5-22 Scale deposit in a cooling tower. (From [7].)

5.3.3 Poor Pump Performance

An indirect cooling tower makes use of a cooling tower pump. To achieve optimum heat transfer, proper water circulation is very important. Loose connections, blocked strainers, pump cavitation, and nondesign operating conditions result in reduced water flow, premature equipment failure, and reduced efficiency.

5.3.4 Poor Airflow

Poor airflow through a cooling tower affects the heat transfer between the water and the air. This can be caused by debris at the suction and discharge of the cooling tower or in the fill [9]. Other causes of poor airflow are loose fan and motor mountings, improper fan pitch, and damage to fan blades, among others.

5.3.5 Makeup Water

Makeup water is dependent on the water source, prefiltration techniques, and competence and may contribute to contaminant (sand) buildup if the water is not filtered properly. Installing a sand media filter is an efficient way to clarify to both makeup water and water entering a cooling tower.

5.3.6 Clogging of Distribution Nozzles

Algae and sediment that accumulate in the cooling tower water basin get into the cooling water and clog the distribution nozzles. This causes an uneven distribution over the fill, resulting in uneven air circulation through the fill, and further reduces heat-transfer surface area contact [9]. To prevent such cooling tower problems, a filtration package needs to be installed to check the adverse effects of scale formation, clogging of distribution nozzles, poor pump performance, poor airflow, wood delignification, and so on. Constant monitoring should also be carried out to avoid further deterioration of the cooling tower.

5.4 MEASURING COOLING TOWER PERFORMANCE

Being a very useful piece of utility equipment, it is important to comprehend and assess the performance of a cooling tower. This is to ensure that optimum operation conditions are achieved and to minimize operating cost and unnecessary loss. Following are the parameters used to determine the performance of cooling towers (with reference to Fig. 5-23).

1. *Range* is the temperature difference between the water inlet and exit states. It is not determined by the cooling tower but by the process it serves. The range at a cooling tower is determined by the heat load and water circulation rate through the process:

$$\text{range } (^\circ\text{C}) = \frac{\text{heat load (kcal/h)}}{\text{water circulation rate}}$$

Range can also be measured by the temperature difference between the inlet and outlet water flow of the cooling tower:

$$\text{range } (^\circ\text{C}) = T_{\text{inlet}} - T_{\text{outlet}}$$

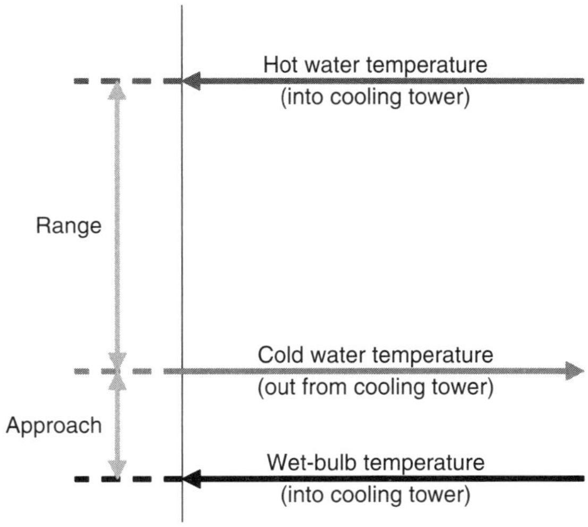

Figure 5-23 Range and approach.

2. *Approach* is the difference in temperature between the cooled water temperature and the entering wet-bulb temperature, t_{wb}. Although both range and approach should be monitored, the latter is a better indicator of cooling tower performance. In general, the closer the approach to the wet-bulb temperature, the more expensive the cooling tower is, due to increased size. Usually, a 2.8°C approach to the design wet-bulb temperature is the coldest water temperature that can be guaranteed by manufacturers:

$$\text{approach } (^\circ\text{C}) = T_{\text{outlet}} - T_{\text{wb}}$$

Figure 5-23 illustrates the relationships among approach, range, wet-bulb temperature, and inlet and outlet temperatures.

3. *Cooling tower effectiveness* (measured as a percentage) is the ratio of the range to the ideal range. It means the difference between cooling water inlet temperature and ambient wet-bulb temperature, which leads to

$$\text{effectiveness} = \frac{\text{range}}{\text{range} + \text{approach}}$$

4. *Cooling capacity* is the heat rejected or heat dissipation (kcal/h), given as a product of mass flow rate of water, specific heat, and temperature difference. Without the cooling capacity, it is meaningless to analyze cooling tower performance.

5. *Evaporation loss* is the water quantity evaporated for cooling duty and, theoretically, for every 1,000,000 kcal heat rejected, evaporation quantity works out to 1.8 m^3. An empirical relation often used is

$$\begin{aligned}\text{evaporation loss } (\text{m}^3/\text{h}) = {} & 0.00085 \times 1.8 \\ & \times \text{circulation rate } (\text{m}^3/\text{h}) \\ & \times (T_1 - T_2)\end{aligned}$$

where $T_1 - T_2$ is the temperature difference between the inlet and outlet water.

6. A cycle of concentration (COC) is the ratio of dissolved solids in circulating water to the dissolved solids in makeup water.

7. A *blowdown loss* is the water discharged from the system to control concentrations of salts or other impurities in the circulating water. The unit involved is % of circulating water rate or gal/min. Blowdown depends on COC and the evaporation losses and is given by the relation

$$\text{blowdown} = \frac{\text{evaporation loss}}{\text{COC} - 1}$$

8. The *liquid/gas* (*L/G*) *ratio* of a cooling tower is the ratio between the water and the airmass flow rates. Against design values, seasonal variations require adjustment and tuning of water and airflow rates to get the best cooling tower effectiveness through such measures as water box loading changes and blade angle adjustments. Thermodynamics also dictates that the heat removed from the water must be equal to the heat absorbed by the surrounding air [2].

5.4.1 Performance Assessment

It is imperative to carry out operational performance assessment for a cooling tower. This is to ensure efficient and effective cooling tower operation. To perform a cooling tower assessment, the typical measurements involved are:

1. Cooling tower design data and curves (the basis)
2. Cooling tower inlet and outlet water temperatures
3. Wet- and dry-bulb air temperatures
4. Exhaust air temperature
5. Water and air flow rates
6. Electrical reading of pumps and fan motors
7. Total dissolved solids (TDS) in the cooling water
8. Cycle of concentration

Since a cooling tower is based on evaporative cooling, the maximum cooling tower efficiency is limited by the wet-bulb temperature of the cooling air. Hence, the cooling tower efficiency can be expressed basically as

$$\text{cooling tower efficiency } (\%) = (t_i - t_o)\frac{100}{t_i - t_{\text{wb}}}$$

where t_i is the inlet temperature of water to the tower (°C, °F), t_o the outlet temperature of water from the tower (°C, °F), and t_{wb} the wet-bulb temperature of air (°C, °F).

To better visualize the performance assessment of a cooling tower, let's look at an example.

Example 5.3 A series of cooling towers are employed to cool down water from a thermal power plant from 42°C to 35.3°C. The wet-bulb air temperature is 27.1°C and the dry-bulb temperature is 38.8°C. The number of cells in operation is 43 of a total of 46. The cooling tower water flow was measured at 68,413 m^3/h and the cooling tower fan flow at 947,521 m^3/h. The design inlet temperature, outlet temperature, and wet-bulb temperature of the air are 41, 31, and 25.2°C, respectively. The density of air is taken to be 1.08 kg/m^3, and the TDS is 2.6. Measure and analyze the performance of the cooling tower.

Solution: Organize all information.

Inlet water temperature = 42°C (rated design = 41°C)

Outlet water temperature = 35.3°C (rated design = 31°C)

Air wet-bulb temperature = 27.1°C (rated design = 25.2°C)

Air dry-bulb temperature = 38.8°C

Total number of cells in cooling tower = 46

Total cells on line with water flow = 43

Cooling water flow = 68,413 m^3/h

Fan flow = 947,521 m^3/h (rated design = 984,300 m^3/h)

Density of air (ρ) = 1.07 kg/m^3

TDS = 2.6

Cooling tower cell = smallest tower subdivision that can function as an independent unit with regard to air and water flow. It is bounded by either exterior walls or partition walls. Each cell may have one or more fans and one or more distribution systems.

Calculation:

Cooling tower water flow/cell = 68,413.34/43 = 1591 m^3/h = 1,591,000 kg/h

Cooling tower fan airflow (average) = 947,521 m^3/h

Cooling tower fan airflow (ρ = 1.07 kg/m^3) = 947,521 × 1.08 = 1,023,322.7 kg/h

L/G ratio of cooling tower (kg/kg) = 1,591,000/1,023,322.7 = 1.55

Cooling tower range = 42°C − 35.3°C = 6.7°C

Cooling tower approach = 35.3°C − 27.1°C = 8.2°C

% Cooling tower effectiveness

$$= \frac{\text{range}}{\text{range} + \text{approach}} \times 100$$

$$= \frac{6.7}{6.7 + 8.2} \times 100 = 44.97\%$$

Rated cooling tower range = 41°C − 31°C = 10°C

Rated cooling tower approach = 31°C − 25.2°C = 5.8°C

Rated % cooling tower effectiveness

$$= \frac{10}{10 + 5.8} \times 100 = 63.29\%$$

Cooling duty handled/cell (kcal) = 1591 × 6.7 = 10,659,753

Evaporation losses (m^3/h) = 0.00085 × 1.8 × 1591(42−35.3) = 16.31 m^3/h

% Evaporation loss = 16.31/1591 × 100 = 1.0251%

Blowdown for site COC of 2.6 = 16.31/2.6−1 = 10.19 m^3/h

Makeup requirement/cell = 16.31 + 10.19 = 26.50 m^3/h

From the solution for Example 5.2 it can be seen that the L/G ratio of the cooling tower is 1.55, which is lower than the rated L/G. But most important, the effectiveness of the cooling tower is only 44.97%, compared to the rated percentage, which is 63.29%. There is a difference of 18.32%. Hence, for several reasons, this cooling tower is not operating at its optimum capacity. The person in charge of the cooling tower should therefore check all the processing parameters related to the tower and correct them. At the same time, a thorough inspection of the tower needs to be carried out to identify problems contributing to the lack of effectiveness [2]. The factors affecting cooling tower effectiveness and maintenance are described next.

5.5 COOLING TOWER MAINTENANCE

Reducing energy costs for a cooling tower is highly dependent on regular maintenance and routine checks. Few routine checks are necessary with towers. The following are the most important cooling tower maintenance activities.

1. *Monitor the process stream*. Monitoring feed and outlet water temperature and feed inlet and outlet water flow rate is crucial. It should be carried out on a daily basis to ensure that the cooling tower is attaining the temperature and flow rate desired. Not only is this imperative for the plant, but it also provides a general indication of cooling tower performance. For a plant employing a *supervisory control and data acquisition* (SCADA) system or a *distributed control system* (DCS), monitoring the aforementioned data would be easier. On top of that, the data are recorded and stored in the server. Hence, the instruments (e.g., resistance temperature detector, flow meter) need to be checked and maintained in good condition.

2. *Monitor and treat the water*. Water quality inside a cooling tower must be maintained to prevent scaling, corrosion, and biological fouling or attack. The chemical properties of water should be checked on a daily basis and the COC should be noted. With the increasing cost of water, efforts to increase COC by cooling water treatment would help to reduce makeup water requirements significantly. In industry, COC improvement is often considered to be a key area of water conservation.

Based on the test results, corrective water treatment action should be taken to maintain the quality of water. Water blowdown should also be carried out to reduce the amount of total suspended solid (TSS) inside the water. Normally, chemical treatments address physical and biological treatments separately.

TABLE 5-2 Typical Problems and Troubleshooting for Cooling Towers

Problem	Possible Causes	Action Recommended
Carryover of water outside the tower	Uneven water distribution on the tower deck, due to some ineffective spray nozzles	Check and clear the nozzle of dirt or blockages. Correct the alignment of the nozzle if disoriented.
	Obstruction within fill pack interrupting smooth water droplets going down	Clear any dirt in the fill or rearrange the fill the way it is supposed to be.
	Defective or displaced drift eliminators	Replace or realign the drift eliminators.
	Excessive circulating water flow	Adjust the water flow rate by regulating valves.
	Excessive airflow	Adjust the fan blade angles.
Excessive absorbed current flow	Drop in voltage	Check the voltage supply; check with the power company.
	Inaccurate angle of fan blades	Measure and adjust the blade angle accordingly.
	Loose or damaged belts on centrifugal fans	Check the belt tightness or replace with new belting.
	Overloading through excessive airflow	Regulate the water flow by means of the valve or adjust the fan blade angle.
	Low ambient air temperature	Cool motor proportionately using inverter to regulate the speed of the fan.
Outlet cooling water temperature increase	Excessive or insufficient cooling water flow	Regulate the cooling water to the flow specified.
	Excessive or insufficient airflow	Improve ventilation in the tower. Adjust the angle of the fan blades.
	Recycling of humid discharge air	Check the air velocity and improve the ventilation.
	Intake of hot air from other sources	Install deflectors.
	Uneven water distribution on the tower deck due to ineffective spray nozzles	Check and clear the nozzle of dirt or blockages. Correct the alignment of the nozzle if disoriented.
	Scaling within joints and fill	Clean or replace the item.
	Blocking of the fill	Clear the blocked section of the fill.
	Loosened or severed belt	Adjust or change the belts.
Cooling water volume decrease	Float valve for makeup water is at the wrong level	Adjust the float to enable correct makeup water.
	Blocking of the sprinkler holes in the upper water basin	Remove dirt and scale.
	Blocking of strainer mesh	Remove the strainer.
	Improper selection of water circulating pump	Replace the pump with one of the proper capacity.
	Lack of equalizing connections	Equalize the basins of towers operating in parallel.
Noise and vibration	Bending of fan shaft	Repair or replace with a new shaft.
	Damage in fan and/or motor	Replace the fan.
	Loose bolts	Tighten loose bolts.
	Belt motor damage	Repair or replace the motor fan.

3. *Check the distribution nozzle*. Nozzles could be blocked and will cause uneven water spray or distribution across a tower, which may result in tower inefficiency. Good distribution is hence very important. Blocked nozzles could result from a fouling effect due to deposits of scales formed by the minerals salts present in water. Blocked nozzles should be freed, and this can be ensured by regular checking.

4. *Monitor the draft fan speed*. The objective of having a fan is to move a specified quantity of air through the system while overcoming the system resistance (pressure loss). The product of airflow and pressure loss is air power developed or work done by the fan. This is termed fan output and input kilowatts, which depend heavily on fan efficiency. Hence, fan efficiency correlates directly with power consumed by the cooling tower. Where draft is

Figure 5-24 Large cooling tower under repair.

provided by the fan mechanism, the motor bearings need to be greased regularly to prevent noise and the belts need to be checked for breakage and looseness. Vibration tests should be performed regularly, as should monitoring of the current (amperage) of the motor fan.

5. *Monitor the louver position*. Loose ends are to be replaced to ensure complete water containment and to prevent water loss. In some designs, louvers are made of wood; others, of metal or FRP. In any case, the interlock bolts should be checked and loose ends tightened. Cross-sectional partitions, louvers, and fence must be checked for breakaway and any loose part must be tightened.

6. *Check for leakage*. Any leakage observed at any points in the inlet or outlet pipeline, or at the pump suction or discharge, needs to be corrected to avoid water, flow rate, and energy loss.

7. *Maintain the water level*. The water contained in the basin needs to be maintained below the water level established. To ensure this, the incoming flow rate and makeup water should be balanced with the drift water, blowdown water, and outgoing water flow rate. Failure to balance this may result in either a decreased water level or an overflow of the basin, which may upset a related process.

8. *Check the transfer pumps*. The flow rate of water needs to be checked from time to time. This can be done by using the existing flowmeter attached to the pipeline. If no flowmeter is available, use of an ultrasonic flowmeter can be a good indicator of the flow rate. The pressure gauge of the pump should also be checked. A higher pressure will indicate that there might be a blockage along the stream or in the suction line of the pump. Immediate attention and corrective maintenance will avoid unnecessary unbalanced flow in the cooling tower.

9. *Shut down the cooling tower periodically*. After a few years of operation, the cooling tower needs to be cleaned and serviced thoroughly. Broken or damaged parts need to be replaced or repaired. In a severe case, brittle treated wood infills and drift eliminators need to be replaced. Under moderate conditions, where the fills and drift eliminators are still in good shape but are dirty due to slime and scale, massive physical cleaning should be carried out using a water jet. Figure 5-24 shows a huge cooling tower under repair and maintenance. Table 5-2 summarizes typical problems faced by cooling towers, possible cause, and action recommended [2].

REFERENCES

1. Baker, D. R., and Shryock, H. A., A comprehensive approach to the analysis of cooling tower performance, *ASME Journal of Heat Transfer*, vol. 83, 1961, pp. 339–350.
2. Bureau of Energy Efficiency, Ministry of Power, India, Cooling towers, in *Energy Efficiency in Electrical Utilities*, 2004.
3. Cheremisinoff, N. P., *Handbook of Chemical Processing Equipment*, Butterworth-Heinemann, Woburn, MA, 2000.
4. http://www.fortepak.com/prototype/kuken.htm, 2010.
5. http://www.nucleartourist.com/systems/ct.htm, 2010.
6. http://www.nucleartourist.com/systems/naturalct.htm, 2010.
7. http://www.zetacorp.com/ctscale.shtml, 2010.
8. McCabe, W. L., Smith, J. C., and Harriott, P., *Unit Operation of Chemical Engineering*, 6th ed., McGraw-Hill, New York, 2001.
9. Nwaoha, C., Cooling tower: improving heat transfer efficiency through filtration and maintenance, *Everything About Water*, no. 4, April 2009, pp. 76–79.
10. Perry, R. H., and Green, D. W., *Perry's Chemical Engineer's Handbook*, 7th ed., McGraw-Hill, New York, 1999.
11. Rakesh, P., In industries, major problem is not proper selection and calculation, *Steam and Boiler Review*, vol. 3, no. 7, July 2009, pp. 6–8.
12. SPX Cooling Technologies, *Cooling Tower Performance: Basic Theory and Practice*, SPX, Overland Park, KS, June 1986.

6

FILTERS AND MEMBRANES

FLORA TONG
Dow Chemical (China), Shanghai, China
CHIKEZIE NWAOHA
Control Engineering Asia, Ten Alps Communications Asia, Alandinma, Nigeria

Filtration of gases and liquids is very critical in all sectors of the fluid-handling industry to maintain fluid cleanliness. Achieving such cleanliness plays an important role in extending the service life of system components. In practice, filtration is the recovery of a material that is either wanted for reuse or unwanted, so that it can readily be withdrawn from the system. It is a separation process in which gas steam or liquid slurry is purified by the use of a filter. That is accomplished when the material passes through a filter medium. The driving force across the medium could be the differential pressure between the medium or gravity or even the concentration; however, there are factors that affect the process, such as the mechanism of operation of the system, the nature of material being handled, the size specification of the filter medium, the operating conditions (cycle, temperature, flow rate, etc.), and the hydraulic properties of the fluid. The choice of a filter medium is often very important for such factors as minimum propensity to bleed, which is the ability to create a solid bridge across the pores in less time; minimum propensity to bind (i.e., the wedging of solids between the interstices of the pores should also be less); minimum resistance to filtrate flow; resistance to attack by the corrosive tendency of the fluid being handled; and resistance to mechanical wear.

Filters are classified as full flow and proportional or partial flow. In a *full-flow filter*, all the fluid that enters the unit passes through the filtering element; in a *proportional-flow filter*, only a portion of the fluid passes through the element. A full-flow filter provides positive filtering action; however, it offers resistance to flow, particularly when the element becomes dirty.

6.1 TYPES OF FILTERS

Filters can be classified by their characteristic applications, for example as air filters, oil filters, flush filters, and sand filters, or by their operational provisions, such as rotary vacuum filters, or by their design principle (e.g., cake gravity, pressure and press filters). Cake filters accumulate appreciable quantities of filtered solids on the surface of a medium; pressure filters operate under superatmospheric pressure at the filtering surface and at that pressure or more downstream. Whereas gravity filters operate at the available hydrostatic pressure of the fluid above, hydraulic filters combine the principles of a filter and a hydraulic press.

Filter media are manufactured from cotton, polymers, asbestos, glass, cellulose. metals, fabrics, refractories, ceramics, sands, and particulate solids, among others. A filter aid is also incorporated in most cases to improve filter performance, especially when filtrate clarity is required. This is mostly in the form of woven metal or light textile material that has a low bulk density, is very porous, and has a high degree of porous cake formation.

One very good thing about a filter is that no matter how big or small it is, the basic principles are the same. When you consider an instrument air filter, for example, you will

Process Plant Equipment: Operation, Control, and Reliability, First Edition. Edited by Michael D. Holloway, Chikezie Nwaoha, and Oliver A. Onyewuenyi.
© 2012 John Wiley & Sons, Inc. Published 2012 by John Wiley & Sons, Inc.

find them small in size, unlike a rotary vacuum filter or a sand filter, but they all operate to get rid of impurities by passing across a medium of special specification. Filters do not require rigorous operations; they are simply prepared, lined up, and commissioned for operation, except when they are observed to be blocked by impurity cake or the filtrate seems to be impure. Then such filters are diverted to standby operation because in most cases process filters are provided with spares.

6.1.1 Gas Filters

Air and other gas filtration represents a less exciting part of the filtration business than liquid filtration. Most gas-phase filtration is used for the removal of a small quantity (i.e., −1% or often much less) of suspended solid particles or liquid droplets from a continuous gas flow. But most important, gas filtration helps to promote environmental protection and personal health and safety and to keep atmospheric and internal air quality as high as possible—not forgetting that it increases process efficiency (i.e., reduces losses). In this section we describe the major types of gas filters.

6.1.1.1 Bag Filters Bag filters are the most commonly used in air filtering. They are manufactured in different filter classes for a great variety of filtering purposes, and they can be divided into coarse mesh and fine filters. Bag filters are made of glass or synthetic fiber. They work by the following principles:

- Dust-laden air enters the bag filter.
- The dust or powder separates inside the bag air filter chamber.
- The air is distributed uniformly, avoiding channeling, while the powder is separated.
- Initially, a coating of material forms on the bag. Subsequently, the coating acts as a filtering medium.
- Dust is accumulates on filter elements as air passes through the filter bags from outside to inside.
- The powder accumulated is dislodged from the bags intermittently by reverse pulse jet air.
- The dislodged powder falls on the bottom cone and is discharged through powder discharge valves.
- The dust-free air is sucked up by an induced-draft fan and exhausted to the atmosphere.

Bag filtration compatibilities are shown in Table 6-1.

6.1.1.2 Cartridge Gas Filters Cartridge filters are developed for remove of dust, rust, and other solid particles from a stream of dry gas or air. Cartridge filters are encased in welded bodies designed in conformance with the standards governing the construction of pressure vessels. Gas flows into the body of the filter through the inlet nozzle. As the gas enters the body of the filter, major impurities and dirt particles will drop to the bottom of the filter body because of the reduced rate of flow. The gas now passes through the filtration materials toward the inside cartridge. Impurities are retained, and the purified gas flows out through the outlet nozzle.

The filter medium consists of radially pleated pulp paper with a pore size of 2 μm and is guaranteed a filtration performance of around 99.9%. The filtration paper is polyester-reinforced and impregnated with phenol resin. All steel parts are galvanized.

Measured across the entire cartridge, the maximum admissible pressure loss is 500 mbar. As soon as pressure-loss readings reach between 200 and 300 mbar, it is recommended that the filter cartridge be replaced. Cleaning a cartridge polluted with dry particles with a jet of pressurized air is only a makeshift process, for even though the surface of the filter may appear clean afterward, the cellulose material is full of deep-seated dust and dirt particles. A filter that has been apparently cleaned in this fashion will have of no more than 50% the cleanliness of that of a new cartridge.

6.1.1.3 Panel Filters Panel filters are used in general ventilation systems as a main filter or prefilter of a multistage filtration. Panel filters are suitable for ventilation units, which have limited space for a filter. Panel filters are manufactured of progressive thermally smoothened synthetic polyester that has a high dust-holding capacity and high constancy for humidity. The filter medium is framed by a rigid polyester, cardboard, or plastic frame, which makes filters completely incinerable.

TABLE 6-1 Bag Filtration Compatibilities

	Compatibility					
Fiber	Weak Acid	Strong Acid	Weak Alkali	Strong Alkali	Solvent	Maximum Temperature (°C)
Polypropylene	Excellent	Excellent	Excellent	Excellent	Fair	95
Polyester	Very good	Good	Good	Poor	Good	150
Nylon	Fair	Poor	Excellent	Excellent	Good	160

6.1.2 Liquid Filters

Liquid filtration is also known as *macrofiltration*. This is characterized by removal of particles between 1 and 100 μm in diameter.

6.1.2.1 Cartridge Liquid Filters Cartridge filters are normally used for fluid polishing, with the particle size to be separated smaller than about 50 μm. In the manufacturing industry, it aids in the removal of contaminants from a more valuable fluid and increases product yield. Surface filtration takes place in pleated filters, while depth filtration is the mechanism operating in string-wound, resin-impregnated, and molded cartridges. Depth filters have traditionally been lower in cost than surface filters. Cartridge filters are typically installed prior to the reverse osmosis units. Cartridge filter types include membrane, nonwoven, carbon, string wound, and metal.

A cartridge filter uses a replaceable filter element, generally cylindrical in shape and long with respect to its diameter, which operates by filtering a fluid from the outside of a cartridge to its inside. It normally comprises a central open-structured core on which the filter medium is placed, and is contained in a cylindrical housing. The medium can be a thin flat sheet, or to maximize the filtration area, a thicker layer of bonded granules or fibers for depth filtration applications. Cartridges are made to a set of generally accepted standard dimensions, so as to be interchangeable as to source.

Cartridge filter types include membrane, nonwoven, carbon, string wound, and metal. Membrane filters are ideal for critical applications where high performance is necessary. Membrane filter materials include vinyl, nylon, nitrile, polypropylene, and poly(tetrafluoroethylene) (Teflon, or PTFE).

Wound filters provide selective particle retention. The winding creates diamond-shaped openings which become progressively smaller toward the center of the cartridge. In this configuration, large particles never reach the center of the filter where the smaller particles are trapped. Cotton, polypropylene, nylon, rayon, Orlon, and glass are among the many materials that may be used as windings. Carbon filters are used to achieve organic purity and are ideal final filters prior to plating or coating applications.

Metal filters maintain high filtration efficiency and structural integrity in most applications through years of continuous use. Metallic media are inherently strong and when supported properly can provide filtration products capable of withstanding very high differential pressure. They can also tolerate high temperatures, with ranges from cryogenic, as low as −450°F (−268°C), to temperatures in excess of 1500°F (815°C). This depends on the metal alloy used, the type of medium selected, and the atmosphere to which the medium will be exposed.

6.1.2.2 Sand (Deep Bed) Filters Sand filters are used for water purification (Fig. 6-1). There are three main types:

1. Rapid (gravity) sand filters
2. Upflow sand filters
3. Slow sand filters

All three methods are used extensively in the water industry throughout the world. The first two require the use of flocculant chemicals to work effectively, while slow sand filters can produce very high quality water free of pathogens, taste, and odor without the need for chemical aids.

A sand bed filter is a type of depth filter. Broadly, there are two types of filter for separating particulate solids from fluids:

- Surface filters, where particulates are captured on a permeable surface
- Depth filters, where particulates are captured within a porous body of material

In addition, there are passive and active devices for causing solid–liquid separation, such as settling tanks, self-cleaning screen filters, hydrocyclones, and centrifuges.

There are several types of depth filter, some employing fibrous material and others employing granular materials. Sand bed filters are an example of a granular loose-media depth filter. They are generally used to separate small amounts (<10 ppm or <10 g/m^3) of fine solids (<100 μm) from aqueous solutions. In addition, they are generally used to purify the fluid rather than to capture solids as a valuable material. Therefore, they find most of their uses in liquid effluent (wastewater) treatment.

Sand bed filters work by providing the particulate solids with many opportunities to be captured on the surface of a

Figure 6-1 Sand filter in operation in a steam production unit. (Courtesy Porty Harcourt Refining Company, Eleme, Nigeria.)

sand grain. As fluid flows through the porous sand along a tortuous route, the particulates come close to sand grains.

In some applications it is necessary to pretreat the effluent flowing into a sand bed to ensure that the particulate solids can be captured. This can be achieved by one of several methods:

- Adjusting the surface charge on the particles and the sand by changing the pH
- Coagulation: adding small, highly charged cations (Al 3+ or Ca 2+ are generally used)
- Flocculation: adding small amounts of charge polymer chains which form a bridge between either the particulate solids (making them bigger) or between the particulate solids and the sand.
- Operating with either upward- or downward-flowing fluids, the latter being much more usual. For downward-flowing devices the fluid can flow under pressure or by gravity alone. Pressure sand bed filters tend to be used in industrial applications and are often referred to as *rapid sand bed filters*. Gravity-fed units are used in water purification, especially in drinking water, and these filters have found wide use in developing countries (slow sand filters).

6.1.2.3 Filter Presses The origins of many, if not most, filter presses (Fig. 6-2) are in areas were clay fields were exploited (i.e., places where semicolloidal, plastic, or poor-draining material had to be dewatered and brought to as low a moisture content as was feasible). With the filter cloths that were available at the time, the only practical solution was to subject the suspension to as much pressure as could be generated. But herein lies a conflict.

Fine and evenly dispersed solids in suspension will inevitably form a tightly packed filter cake, and the more pressure one puts on the cake, the tighter it gets packed until one arrives at a stage where the cake becomes almost impervious and further drainage is barely possible. Obviously, the thicker the layer of solids, the worst it gets, and one soon arrives at a situation where any extra pressure on the cake creates a resistance to draining that is almost equal to the extra pressure. The logical thing to do would be to stop at this point, open the press, remove the cake, and start all over again. This is fine if the press holds a reasonable amount of solids, but if all this effort yields only a thin sheet of cake, it is hardly worth the trouble. In addition, the thinner and thus the lighter the cakes are, the greater is the tendency for them to stick to the cloth.

Figure 6-2 Filter press in use. (From Port Harcourt Refining Company, Eleme, Nigeria.)

Conversely, the thicker the cake, the better the chance that it will drop free by its own weight, leaving a moderately clean filter cloth. The accepted practice therefore is to work with thick (normally 25 to 50 mm) cakes and to keep the pressure up and, if possible, to increase it (although this in itself reduces the drainage capacity) and keep squeezing droplet by droplet until the cake is "dewatered."

Fairly thick cakes are almost inevitable with filter presses and that may mean that a press has to cycle for quite a long period of time before it can suddenly drop a whole load of cakes, which have to be handled as a separate operation. For continuous operations, this clearly creates a bottleneck, and it does little for any meaningful quality control, as the cakes can vary plate to plate as well as within a plate itself.

6.1.2.3.1 Historical Problems with Filter Presses

- The cake has to be scraped manually out of the frames.
- The frames cannot be too thin; otherwise, the feed ports block. This point applies to plate and frame presses. For the rest, almost all filter presses now have recessed or chamber plates.
- If the plates are too large or there are too many of them, the weight is excessive.
- Almost every plate nowadays is made out of a plastic material, usually polypropylene, which also solves many chemical resistance problems, although this does pose limitations on the temperatures of the feed, the wash liquids, and the cleaning liquids.
- If the cake sticks to the cloth, there is no discharge. The much better filter cloths have reduced this problem considerably; in addition, most manufacturers offer a range of cloth scrapers, plate "bumpers," plate shakers, and other mechanical devices to induce cake to drop away.
- If some of the cake does, in fact, stick to the cloth, especially the edges, the next cycle may result in leaks and the plates may distort when closing the plate pack.

- Many manufacturers offer cloth-washing systems with trays to channel the water away, whereas others offer compensators to allow for misalignment.
- If the filter cycle is a bit too short, sloppy cakes with wet centers may result.
- However, advanced electronics allow for a fairly accurate interplay among time, pressure, backpressure, filtrate clarity, and so on, so that this risk is reduced.
- Opening and closing the filter press is time consuming. However, almost all modern presses are equipped with automatic plate moving systems, which select one or several plates, allowing them to discharge the cake.
- If there is only a partial batch remaining in the reactor, the chambers cannot be filled and the result is a partially filtered mess which cannot be washed or dried. If the press is designed with a membrane compression device, this can be overcome.
- Admittedly, this may come at the cost of reduced filtration area and definitely at the cost of reduced chemical resistance, higher capital cost, and more maintenance, but it does resolve the problem.

6.1.2.3.2 Filter Cake Washing Filter press operators frequently claim "extremely well washed" filter cakes, often giving the impression that no other filter would be an equal to this. In some special cases this can be true, but in general, whereas the washing can be good, it is rarely very efficient in terms of time cycle and use of wash fluid.

First, it is almost impossible to visualize a filter press operating a countercurrent washing system. This increases the potential volume of wash fluid immediately by a factor of probably 3 or 4.

It is axiomatic that a filter cake that has been dewatered to its maximum (i.e., out of which no more liquid can be expelled with mechanical means) is also a cake through which one can squeeze a washing liquid. This means that the dewatering cycle has to be stopped well short of its maximum—by default leaving a poorly dewatered cake. This applies to all filters, but obviously, the thicker the overall cake, the greater the residual mother liquor, the greater the risk of back-mixing, and the greater the resistance—all of which translates into a tendency for long washing cycles.

The typical filter press produces not only thick cakes but also "two-sided" cakes with a compacted layer on either side and softer, better-draining matter in the middle. The wash liquor therefore has to be forced (in the opposite direction of the earlier dewatering) through the compacted outer layer into the softer material and then again be forced through the second compacted layer at the other side. The pressure required to force the wash liquid through is therefore at least twice that for a single-sided cake, and it is no surprise that the slightest pinhole, fissure, or shrinkage in the cake will cause bypassing of the wash liquid, quite apart from a tendency for the wash liquid to back-mix instead of doing a displacement wash.

Filters that filter on one side only have it much easier. Not only is the cake thickness normally much less than half that of a filter press but the wash fluid travels in the same direction as the earlier mother liquor, and once it has broken through the final cake layer it emerges as wash filtrate without risk of back-mixing.

Since most filter presses have vertically mounted plates, there is a greater chance of not presenting a homogeneous cake for washing (due to settlement) than would be the case with nonvertical plates. (The much more expensive membrane chamber plates can overcome this problem to an extent by pre-squeezing the cake.)

Taking all these factors together, the normal practice is to "overwash," just to be on the safe side. For difficult or critical products, it is not uncommon to find dewatering cycles of 3 to 4 h being followed by 8 h or more of washing.

6.1.2.4 Pressure Leaf Filters The pressure leaf filter is a MS/SS vertical vessel with filter leaves inside. The leaves are mounted vertically on a common manifold pipe, through which the filtered liquid flows out. On the top, the leaves are held by a vibrating shaft. A mechanical vibrator driven by an electric motor/pneumatic vibrator is provided for vibrating the leaf shaft for cake discharge. A jacket can be offered for hot filtration if desired. Overflow, vent/steam/air charging, a pressure gauge, and a safety valve are provided on the top. The top cover is provided with a devit arm mechanism for raising the lid for cleaning or for removing the leaves. I bolts are provided for quick opening and closing of the top lid. A mechanical jack is provided to lift the top for cleaning and for removing the leaves.

6.1.2.4.1 Applications Pressure leaf filters are used industries:

- *Edible-oil industry*: bleached, winterized, deodorized, hydrogenated, and fractionized oils; dewaxing; catalyst; mineral oil; sulfur
- *Beverage industry*: for glucose, fruit juices, cold drinks, sugar, vinegar
- *Chemical industry*: for organic and inorganic salts, dyes, chemicals, plastizers
- *Pharmaceutical industry*: for pharmaceutical intermediates, syrup, bulk drugs, antibiotics, intravenous solutions
- *Petrochemical industry*: crude oil, liquefied petroleum gas, lubricating oil, sulfur
- Others

- Interesterification
- Herbal extraction plants
- Micronutrient recovery
- Essential oil distillation
- High-vacuum separation by molecular distillation
- Lube oil re-refining

6.1.2.4.2 Working Principle Leaf filters are ideal for solid–liquid separation and work on the principle of precoating and pressure. Precoating is done with the help of suitable filter aid (e.g., bleaching earth, Supercel, Hyflo, Superflo) mixed about 0.5 to 1% with the liquid to be filtered. Unfiltered liquid is pumped into the filter vessel. Initially, the filter aid starts forming a precoat layer on both sides of the filter leaves until cloudy material comes out of the filter. Once the layer is formed, pressure begins to develop, restricting the impurities. Clear liquid flows from both sides into the leaves (filter elements), flows along the tubular channel, and gets discharged from the bottom of the leaf. All the leaves are mounted on a common manifold. The leaves start getting choked on both sides by impurities forming cake, which is in wet form. Once the leaves get chocked completely, the pressure rises to 3 to 4 kg/cm^2 and the output flow almost stops.

The pump is stopped and steam or air pressure is applied from the top (without dropping the filter pressure) to filter the material around the leaves held up in the tank, to squeeze the cake further, and to reduce the liquid retention. The holdup unfiltered liquid in the conical portion is taken back. In pressure leaf filter the cake can be dried by steam or hot air and then discharged from the bottom with the help of mechanical or pneumatic leaves vibrator. The entire operation of cake drying takes 30 to 45 min, and cake discharge takes 5 to 10 min.

6.1.2.4.3 Selection Criteria The most common type of leaf filter used is the vertical leaf filter. They are best selected under the following circumstances:

- When minimum floor space for large filtration areas is required
- When the liquids are volatile and may not be subjected to vacuum
- When there is a risk of environmental hazard from toxic, flammable, or volatile cakes (specially secured discharge mechanisms may be incorporated)
- When high filtrate clarity is required for polishing applications
- When handling saturated brines that require elevated temperatures (the tank may then be steam jacketed)
- When the cake may be discharged either dry or as a thickened slurry

They should be selected with care:

- When the cake is thick and heavy and the pressure is not sufficient to hold it on the leaf
- When coarse mesh screens are used

In the latter case, the filtration step must be preceded by a precoat to retain cakes with fine particles. Precoating with a thin layer of diatomite or perlite is not a simple operation and should be avoided whenever possible.

6.1.2.5 Coalesing Filters The main function of coalescing filters is to trap and, subsequently, remove liquid oil and water from a compressed air or gas stream. The secondary purpose is to remove particulate matter. When the amount of liquid challenging a filter becomes too great, the liquid tends to burst out of the filter medium instead of dropping quietly out of the bottom of the filter. With this bursting or spraying through the medium, some of this liquid can be reentrained into the compressed air and carried downstream.

A low release point is critical to the success of any coalescing filter. If oil and water permeate the medium anywhere in the upper 70% of the cartridge, reentrainment in the airstream is almost certain to occur. According to Fluid Energy, experience has shown that all coalescing filters can have low release points provided that the challenge rate is not excessive. For example, we have tested many coalescing filters that are highly efficient as long as the liquid challenge rate does not exceed 3 ppmw. The most efficient coalescing filter we have found can handle liquid challenge rates up to 50 ppmw and provide a residual carryover of 0.0014 ppmw.

For example, consider three different brands of coalescing filters all rated for a capacity of 1000 scfm at 100 psig with challenge rates of 50, 10, and 3 ppmw, respectively. To put the term ppmw in perspective, Table 6-2 converts ppmw into oz/h, assuming that the liquid contaminant is a typical mixture of water and compressor lubricant. In other words, the best 1000-scfm coalescing filter we have found can remove 3.5 oz/h of liquid and provide a residual carryover of 0.0014 ppmw. Coalescing filters rated at 3 ppmw can remove only 0.2 oz/h and still remain efficient. What this means is that it doesn't matter what the residual carryover rating of a coalescing filter is, provided that it is low. This information is useless without knowing the associated challenge rate. Note that knowing the efficiency of stopping 0.01-μm particles is irrelevant information when evaluating coalescing filter performance.

Coalescing filters must be used according to manufacturers' specifications to keep liquid collected from being reentrained into the airstream. Also, always make sure that the flow direction is according to information on the filter housing. Several companies make these high-efficiency

TABLE 6-2

Filter Capacity at 100 psig and 100°F (scfm)	Challenge Rate (oz/h)		
	50 ppmw	10 ppmw	3 ppmw
1000	3.5	0.7	0.2

air filters, although they are not often applied to everyday circuits.

6.2 MECHANISMS OF FILTRATION

There are four main types of relevant filtration mechanism using nanofiber filter media: depth straining, surface straining, depth filtration, and cake filtration. In practice, the filtration and cake filtration processes often involve a combination of two or more mechanisms, described briefly below.

6.2.1 Depth Straining

Depth straining applies to felts and nonwoven materials that are relatively thick compared to pore diameters, and where the pore diameters are quite variable in their length. The particles penetrate the pores until they reach a necking point, where the diameter becomes smaller than the particle, and at this point the particle is trapped in the pore.

6.2.2 Surface Straining

In surface straining, the particle is larger than the pores and simply cannot pass through. Particles smaller than the pore diameters pass through the medium and are not separated. This type of separation is generally not associated with nonwoven fabrics but, rather, with media that have uniform pore openings. Examples are woven mesh fabrics, screens, and membrane materials where the openings are uniform in diameter.

6.2.3 Depth Filtration

Depth filtration is different from depth straining, It involves mechanisms for removing a particle from a fluid even though the particle may be smaller than the diameter at any point in the pore structure.

6.2.4 Cake Filtration

Cake (or surface) filtration involves the capture of particles on the surface (or near the surface) of a filter medium so that the buildup of particulate matter into a layer of filter cake participates in the filtration process. Surface-modified needlefelts.

6.3 FILTER SELECTION

To use a filter properly to protect a system, it is critical to understand the full scope of the filter's performance characteristics.The following should be considered when choosing a filter: chemical compatibility (hydrophilic or hydrophobic, extractables, filter durability, and adsorption) and accurate pore size for system protection.

6.3.1 Chemical Compatibility

6.3.1.1 Hydrophilic Versus Hydrophobic Filters When determining the chemical requirements of a filter, first determine if you need hydrophobic or hydrophilic filters. Hydrophobic membranes repel water and are inert to aggresive organic solvents, making them ideal for organic solvents. Hydrophilic membranes, which have an affinity for water, are preferable when filtering aqueous samples.

6.3.1.2 Extractables The next point to consider when selecting a filter is the role that extractables can play in compromising your analytical results. Extractables are the substances present in the composition of the filter medium or filter manufacturing process that may be leached into the fluid as it is filtered, thereby affecting the final chromatography results.

6.3.1.3 Filter Durability The third parameter to consider is the filter durability in terms of chemical compatibility. This is the filter's ability to resist select chemicals so that the pore structure is not affected adversely by chemical exposure. Compatibility requires that you consider the effects of temperature, concentration, applied pressure, and length of exposure. This decision is made by consulting the chemical compatibility guide from the filter supplier.

6.3.1.4 Adsorption The final point to consider when selecting a filter is adsorption. As with extractables in the pharmaceutical industry, drug binding can present a challenge when filtering dissolution samples. The purpose of sample filtration is to remove nondissolved solids prior to high-performance liquid chromatographic (HPLC) injection. Nondissolved solids interfere with the resulting chromatography by continuing to dissolve throughout the analysis period and by plugging the HPLC column. Filtration results in more reproducible chromatography and a longer column life. But it also presents drawbacks, including the potential adsorption of activated pharmaceutical ingredients from a drug mixture that can falsely render the product out-of-specification.

6.3.2 Accurate Pore Size

Proper filter selection requires choosing the correct pore size according to the HPLC column packing size. Filtration will extend the life of a column and reduce maintenance due to particulates in the pumping system, thereby giving more analysis per dollar spent, when the pore size of the filter is determined based on the column packing size. The goal is to determine the space size between column packing beads and remove all particles of that size and greater to ensure continuous column use without plugging. Once you select the optimal pore size for your application, you must rely on the filter manufacturer to provide an accurate pore size rating. An improper rating can jeopardize the life span of the column.

6.4 PARTICLE-SIZE MEASUREMENT TECHNIQUES

Why measure particle size? Although many do not realize it, particle size is a major factor in many aspects of processing plant design and performance. It is also critical to the economic return of the plant [12]. Outlined below are some of the ways in which particle-size analysis is so important:

- Recovery of valuable minerals can be compromised by particles outside the optimal size range.
- Excessive grinding increases energy consumption significantly in mills and can reduce throughput.
- Reagent consumption increases with finer particle size.
- Changes in particle size cause unwanted variability in reagent consumption and process operation.
- Filtration and thickening capacity decrease with finer particle size.
- Water recovery and tailings disposal is affected by particle-size distribution.
- Finer products in wet plants can mean higher transport costs and increased cost in downstream processing.

As can be seen, particle size has to be monitored so that it can be controlled. Some of the ways to monitor size is via manual sampling, sample preparation, and analysis using sieves or more modern instruments. Manual methods are labor intensive, however, so many operations reduce size analysis to a minimum. Process control requires frequent measurements (e.g., the reaction time of a grinding circuit is typically 5 to 15 min). Automatic techniques, such as on-stream particle-size measurement, have become common practice in industry because the return on investment is very high [12].

It is well documented that different particle-size measurement techniques will yield different results. This is because each technique measures a different dimension of a three-dimensional particle. For on-stream particle measurement, the focus is on repeatability, precision, and reliability of the equipment.

6.4.1 Image Analysis

There are many different methods of optical image analysis. With the wide size distributions present in mineral processes, they are quite inefficient in achieving the desired volume or weight distribution. This is because the number distribution is so heavily weighted toward fine particles. Significant dilution is required to overcome opaqueness. Errors can be significant if large particles are missed.

6.4.2 Direct Mechanical Measurement

The direct mechanical measurement principle employed in the Outotec 200-psi particle-size instrument [9] has been widely used over the past decade. A representative sample is passed between the ceramic-tipped moving element and the fixed ceramic element, the position of the moving element being measured by an electronic sensor. Random particles are trapped between the elements, where their diameters are measured, at a rate of two measurements per second. The simplicity of this operation means that it has high availability and minimal maintenance. One drawback of this method is that it does not give direct information below the D60 particle size or useful information on bimodal distributions, so alternative techniques have been deployed.

6.4.3 Ultrasonics

Ultrasonic attenuation has been applied in wet mineral processes since the 1970s. Because ultrasonic particle-size measurement is sensitive to air bubbles, the sample has to be deaerated before measurement. Recently, ultrasonics has been combined with gamma-ray transmission and sound velocity measurement for more accurate solids content compensation [9]. The sample volume measured can be relatively large, which is an advantage of this technology. However, the major limitations in mineral processing operations are its sensitivity to air entrainment, flaky particles, solids content, and slurry viscosity changes.

6.4.4 Laser Scattering Technology

Low-angle laser light scattering or laser diffraction has been known as a laboratory technique since the 1960s. The size analysis is based on an intensity distribution measurement of coherent laser light scattered by the particles. The form of the scattering pattern is described by the Mie theory, and the width of the pattern is dependent on the size. When laser light meets a population of particles, volumetric size distribution can be calculated back from the scattered light distribution.

In practice, lase diffraction offers the following advantages:

1. The method is nondistructive and nonintrusive.
2. The measurement exhibits high resolution; up to 20 size fractions can be displayed in addition to specific surface area.
3. The method is absolute, giving volumetric particle size without need for external calibration. This is equal to weight distribution if density is constant.
4. The method is rapid—results are produced in just over 1 min.
5. The technique is applicable over a wide particle-size range with excellent precision and repeatability. The online system described below can achieve 1 to 2% relative precision of the distribution medium over the size range 1 to 500 μm.

6.5 FILTER LOCATION

Filters are the most common device installed in hydraulic systems to prevent foreign particles and contamination from remaining in the system. They may be located in the reservoir, in the return line, in the pressure line, or in any other location in the system where the designer decides they are needed to safeguard the system against impurities. The main objective of this section is to discuss the classification of filters based on location. There are three major areas in a system where filters may be located: suction lines return lines, and pressure lines.

6.5.1 Pressure Line Filters

Pressure line filters are usually located at the point of pump discharge, where system components such as valves have a low tolerance for dirt. This protects system components that are located immediately downstream of the pump, because the filter traps very fine particles from the fluid that leaves the pump. Pressure line filters usually have elements with 1 to 5 μm and have a beta ratio above 50. A pressure line filter and its housing are designed to withstand full system pressure and any cyclic pulses generated by variations in the pump. These requirements makes pressure line filters more expensive than both suction line filters and return line filters. High initial and ongoing cost is the reason for not installing pressure line filters on all systems.

6.5.2 Suction Line Filters

Suction line filters are normally located at the suction line of a pump to protect the pump from particles that can damage and cause problems to downstream components. Such particles may come from debris in the supplying reservoir or rust resulting from corrosion of component internals. When a suction line filter is too fine (has a low micrometer rating, e.g., about 25 μm), it restricts inlet flow, and this restriction increases as the filter clogs and at low fluid temperatures. This therefore creates more drop in pressure (causing cavitation) than can be tolerated at the inlet line. But a low-micrometer-rated filter can be used if the pump is force-fed by another pump. It is advicable that suction line filters have relatively coarse mesh or very large fine mesh, about 150 μm. Finally, suction line filters have the advantage of being an inexpensive filter location option.

6.5.3 Return Line Filters

The return line is another location for filters. Return line filters are located on the return line just before the line enters the reservoir. This is to prevent generated or ingested contaminants from entering the reservoir. They are usually installed in systems that do not have large reservoirs that enable debris to settle out of the fluid. These filters can be offered with ratings from 3 to 25 μm. But a common rating of 10 μm is mostly available. Obviously, if the required cleanliness is 10 μm a filter of 10 μm or less is used. A return line filter has the advantage of achieving high filtering efficiency at an economical cost because it has sufficient pressure available to force the fluid through the filter medium without collapsing the filter housing. This, combined with its very low velocity, makes a return line filter nearly a necessity in most systems.

6.6 MEMBRANE FILTRATION

Membrane filtration utilizes thin sheets of permeable material, made from polymers and materials such as ceramics and metals. Membrane filter media are configured into tubular, hollow fiber, and sheet formats; sheets may be formed into pleated or tubular filters. Many membranes have an asymmetric structure, composed of a thin skin that acts as the surface filter, supported by a thicker layer designed to give mechanical intergrity to the entire structure; the thickness of the membrane may be from greater than 1 μm to several hundred micrometers. Membrane filter media are classified according to the sizes of their pores: membrames with pore sizes between 0.1 and 20 μm are used in microfiltration, between 0.001 and 0.1 μm [molecular weights (MWs) of 500 to 500,000] in ultrafiltration, between 200 and 1000 in nanofiltration and in reverse osmosis.

Membrane filtration has not developed interchangeable products, although a degree of commonality is beginning to emerge. The existence of more choices in membrane filtration is due both to the newness of this application in the process industry and to the fact that the optimum technical choice between different approaches may be a close call, varying from application to application.

6.6.1 Ultrafiltration

Ultrafiltration (UF) is a pressure-driven membrane separation process that separates particulate matter from soluble components in the carrier fluid (such as water). The technology is used to remove particulate and microbial contaminants, but it does not remove ions and small molecules [5]. Pressure drives the process, which typically operates with a feed pressure of 4 to 100 psig. UF plants are automated and have low operational labor requirements. A typical UF system is shown in Fig. 6-3. These systems, can, however require frequent cleaning. UF membranes have a service life of three to five years or longer, which is comparable to that of reverse osmosis membranes.

UF membranes typically have pore sizes in the range 0.01 to 0.10 μm and have a high removal capability for bacteria and most viruses [25], colloids, and silt. The smaller the nominal pore size, the higher the removal capability. Most materials that are used in UF are polymeric and are naturally hydrophobic. Common polymeric materials used in UF include polysulfone, poly(ether sulfone), polypropylene, or poly(vinylidene fluoride). Different membrane materials with the same nominal molecular weight cutoff (MWCO) may have differing solute rejection. Pore-size distribution and uniformity rather than the chemical nature of the membrane material may cause this effect. Because factors other than pore size or MWCO affect the performance of membranes, challenge studies are used to demonstrate membrane performance and benchmark different membranes [8].

UF modules are commercially available in tubular, hollow-fiber, plate and frame, and spiral-wound configurations. Among them, hollow-fiber membrane is most frequently seen in water treatment. The modules contain several small (0.6 to 2 mm diameter) tubes or fibers. The feed solution flows through the open cores of the fibers and the permeate is collected in the cartridge area surrounding the fibers. The filtration can be carried out either "inside out" or "outside in."

Figure 6-3 Large-scale ultrafitration system.

A UF system can also be operated in a pressurized system or an immersed system. In a pressurized system, transmembrane pressure (TMP) is generated in the feed by a pump while the permeate stays at atmospheric pressure. While in an immersed system, membranes are suspended in basins containing the feed and open to the atmosphere. Pressure on the influent side is limited to the pressure provided by the feed column. TMP is generated by a pump that develops suction on the permeate side. In the subsequent sections we use outside-in hollow-fiber UF elements as examples to explain the characteristics, operation and maintenance of a UF system.

6.6.1.1 Configurations of Outside-In Pressurized Hollow-Fiber UF Elements Hollow-fiber UF membranes usually consist of several hundred to several thousand fibers encased in a module, the fibers being bonded at each end with an epoxy or urethane resin (Fig. 6-4). The inner lumens or internal diameters, which are small to avoid fiber collapse when placed under pressure, range from 0.4 to 1.5 mm. The physical strength of these membrane fibers allows them to be backwashed.

Pressurized hollow-fiber membrane systems utilize hollow fibers that are packed into a cylindrical casing, usually with a diameter of 8 in. and a length of 40 or 60 in. These fiber packages, or elements, are then arranged end to end in a pipe or vessel. Most commonly, in UF applications, the vessels are oriented vertically and each vessel houses one element. For large-capacity systems—particularly in Europe—multiple modules can be installed in long vessels in a horizontal configuration similar to that of reverse osmosis systems. Some manufacturers provide their modules with end structures that are already configured to accept the required piping connections, thus eliminating the need for housing. This can save a significant amount of capital cost.

In an outside-in system, the feed water surrounds the membrane, and the filtrate is collected from inside the hollow tube fibers (lumens). The outside-in scheme has the advantage of a larger membrane surface area, which allows for a slightly higher flow than the inside-out model while maintaining the same flux rate and solids concentration.

6.6.1.2 Typical System Design The UF system applies two primary flow designs: dead-end filtration and cross-flow mode. The feed water all becomes permeate after dead-end filtration. But after cross-flow filtration, part of the feed becomes permeate, the other becomes concentrate water with impurities. Dead-end filtration uses less energy

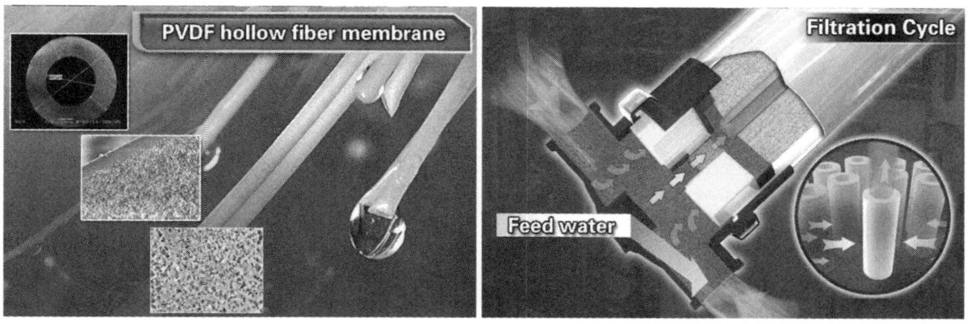

Figure 6-4 Configuration of a typical outside-in pressurized hollow-fiber UF module.

and has a lower operating pressure than cross-flow, therefore reducing operating costs. Alternatively, cross-flow can handle higher solids loading.

Typically, UF is operated at a constant permeate flow. The transmembrane pressure will naturally increase over time and the module can be backflushed or air-scrubbed periodically to remove the fouling layer. Disinfectants and other cleaning agents can be used to fully remove and prevent biological growth as well as other foulants. A typical UF system layout is shown in Fig. 6-5, including the water inlet system, air inlet system, permeate collection system, drainage system, backwash system, and chemical dosing system.

6.6.1.3 Key Terms Used in the Ultrafiltration Process

- *Log removal value* (*LRV*): filtration removal efficiency
- *Recovery*: ratio of filtrate flow to feed flow (%)
- *TMP*: transmembrane Pressure (similar to head loss with conventional granular media filters) (bar or psi)
- *Flux*: permeate flow divided by membrane area ($L/m^2{\cdot}h$ or gal/ft^2-day)
- *Normalized flux*: temperature-corrected, usually to 20°C ($L/m^2{\cdot}h$ or gal/ft^2-day)
- *Permeability*: normalized flux divided by TMP (sometimes called specific flux) ($L/m^2{\cdot}h$ per bar or gal/ft^2-day per psi)
- *BW*: backwash
- *CIP*: clean in place
- *CEB*: chemically enhanced backwash (other terms: mini-clean, enhanced flux maintenance, maintenance clean, chemical wash)
- *Air scouring*: employs the injection of filtered compressed air into the water mains to form ‘slugs’. The air/water combination along with an increase in the flow velocity removes the solids, sludge, particulates and other foreign matter from the internal walls of the pipeline.

6.6.1.4 Operation and Maintenance [11]

6.6.1.4.1 System Startup The following procedures should be followed for the startup of typical pressurized ultrafiltration modules. Manually start the equipment during

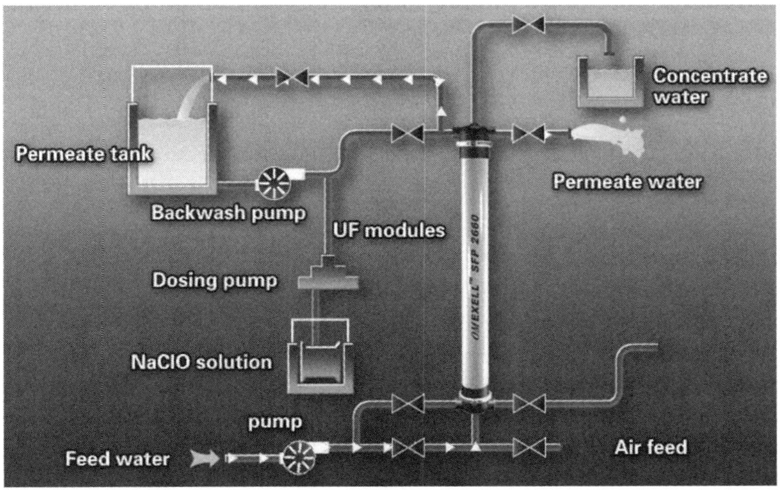

Figure 6-5 Typical UF system flow diagram.

initial operation. Flush the UF parts to remove the preservative used in shipping or the storage solution before starting the equipment. Target a permeate flow of 60% of design during initial operations. After 24 h the permeate flow can be adjusted to design conditions.

Pre-start Checks

1. The UF pretreatment system should operate properly and the UF feed water should meet the design requirements. Ensure that chemical addition points are properly located and that proper mixing of chemicals in the feed streams is possible. Check the addition of pretreatment chemicals.
2. Verify that the drain and waste collection system is functional.
3. Verify that the PLC program is loaded and functioning.
4. Complete an electrical system check. Verify that the instrumentation is working and that the calibration has been completed. Calibrate gauges and meters based on manufacturers' recommendations.
5. Clean and connect interconnecting piping. Flush the system without modules to remove fabrication debris. During the flushing operation, check all pipe connections and valves for leaks. Tighten connections where necessary.
6. Residual air should be removed from the system during startup.

Startup Process Check that all valves are closed and pumps are off before starting the system. Start the equipment by carrying out the following steps:

1. Pumps should be aligned, lubricated, and properly rotated.
2. Start the feed pump.
3. Fill the equipment and start a flush.
4. Start the backwash pump.
5. Set and adjust the backwash pressure.
6. Set and adjust the inlet air pressure.
7. Set the backwash time interval.
8. Set the air scour time interval.
9. Set the backwash sequence.

Module Flushing The ultrafiltration modules should be flushed before operation to remove preservative shipped in the modules. Flushing should be performed until no foam is observed in the wash water. Depending on the treatment application, additional rinsing or disposal of permeate during startup may be required.

6.6.1.4.2 Process Operations The basic operating process for conventional ultrafiltration modules is shown in Fig. 6-6.

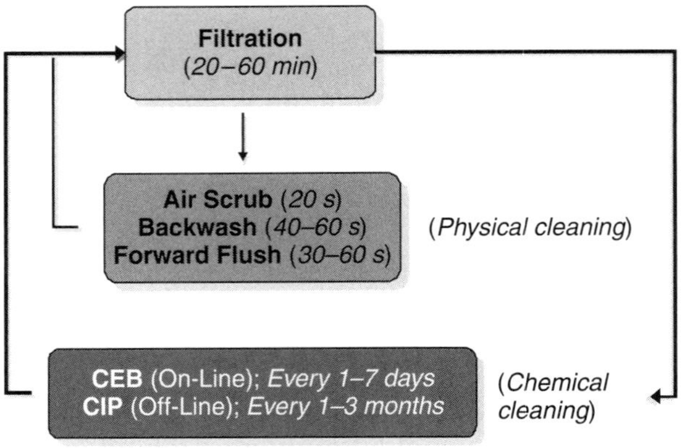

Figure 6-6 Process operation.

Normal Operation Normal operation refers to the routine operating sequence of a system using an ultrafiltration module and includes the operating and backwash steps. At initial startup the modules are flushed using a "forward flush" to remove any residual chemicals or trapped air from the module. The flush occurs on the outside of the fibers and does not filter the feed water to make permeate. After the forward flush is discontinued, the modules can be placed in the operating mode. An operating cycle ranges from 20 to 60 min. While operating, 100% of the feed water is converted to permeate. As contaminants are removed during the operating step, the transmembrane pressure will rise. At the end of the preset operating cycle time, a backwash sequence is triggered.

The backwash mode occurs automatically and may include an air scour but always includes draining, backwash through the top drain, backwash through the bottom drain, and a forward flush. The air scour step, when included, is used to loosen particulates deposited on the outside of the membrane surface. Air is introduced on the outside of the fibers and displaced feed flow or concentrate is allowed to discharge through the top of the module for disposal. After 20 to 30 s of air scour, the module is drained by gravity to remove dislodged particulates. If the air flush is not included, the backwash sequence is started simply by draining the module by gravity to remove the concentrated feed water before starting backwashing.

After draining, the first backwash step is performed. Permeate flow is reversed from the inside of the fiber to the outside and backwash flow is removed from the module housing through the top drain on the module. A top-draining backwash is carried out first to remove contaminants in the area of greatest concentration.

The second backwash step is performed to remove contaminants through the bottom of the module housing. Permeate flow is reversed from the inside of the fiber to the outside, and backwash flow is removed from the module housing through the bottom drain on the module for efficient removal of heavier materials. The two backwash steps can be repeated numerous times depending on the degree of fouling. After backwash is complete, a forward flush is performed to remove any remaining contaminants and remove any air trapped on the outside of the fibers. After a backwash, the modules are returned to the operating mode.

Chemically Enhanced Backwash The frequency of chemically enhanced backwash (CEB) dependent on the feed water quality. On high-quality feed waters, a CEB may not be required. However, a CEB can be performed as frequently as a backwash. The CEB process occurs automatically, but the frequency can be adjusted after gaining operating experience. The CEB is performed using UF permeate, but with either an acid or base combined with an oxidant added, to clean contaminants from the membrane surface more effectively.

A CEB follows the steps of a normal backwash except that a CEB chemical is dosed into the backwash water and a soak step is added after the second backwash step. The soak is performed for 5 to 10 min and allows more time for the oxidant to react with contaminants that have attached to the membrane surface or penetrated the fiber wall. After the soak a forward flush is performed to remove any remaining contaminants on the outside of the fibers. After a CEB and at the start of the operating step, permeate may be wasted to remove residual chemicals held in the fiber or module.

Cleaning in Place Cleaning in place (CIP) includes backwash and chemical recycle to clean the fibers. The CIP is an on-demand operation. The frequency of a CIP depends on the feed water quality but can range from one to three months. Prior to a CIP, the routine backwash steps, including air scour, draining, backwash through the top drain, and backwash through the bottom drain, are performed. The backwash steps are repeated three to eight times to remove contaminants or foulants that do not require chemical removal. After completing the backwash steps, the module is drained by gravity to remove excess water and prevent dilution of the CIP chemicals. CIP chemicals are recycled on the outside of the module for 30 min through the chemical cleaning tank. A small chemical permeate stream will also be collected and recycled to the chemical cleaning tank. Note that the CIP solution can be heated up to 40°C to improve its effectiveness at removing contaminants from the membrane. A soak follows the initial recycle step for 60 min or longer, depending on the degree of fouling that has occurred.

After the soak step, CIP chemicals are again recycled on the outside of the module for 30 min. When the recycle is completed, an air scour is performed and the module is then drained to remove concentrated chemicals. The two steps of backwash and a forward flush are performed to remove any remaining contaminants on the outside of the fibers. After a CIP operation and at the start of the operating step, permeate may be used to remove residual chemicals held in the fiber or module. The CIP steps described above are for a single alkali or acid chemical solution. If an acid and alkali cleaning are required, the CIP steps would be repeated for each chemical solution.

Citric acid or other acid solutions are used to remove scalants, and oxidants may be used to remove biological or inorganic foulants if the membrane material is tolerant of oxidants. The chemicals to be used depend on the type of fouling and the material of membrane construction. Cleaning cycles can be automated and implemented into the system design. A typical cleaning cycle can last from 10 min to 2 to 3 h per bank, including soaking and rinsing times. Cleaning frequency for surface water applications can again vary tremendously, from once per day to about 4 to 6 weeks. Cleaning procedures, durations, and frequencies will vary significantly, depending on feedwater quality, system design and operation, flux rates, and recovery rates. Cleanings may be automated and performed in place, or they may be manual, with the modules actually being removed from operation and cleaned in a special cleaning area. Encased modules are usually cleaned in place but can be removed to a special cleaning rack.

6.6.1.4.3 System Shutdown

1. *Manual shutdown*. A manual shutdown is conducted by opening the concentrate rinse valve and flushing for 15 s. Then the inlet valve is slowly closed.

2. *Equipment shutdown during automatic operation*. The equipment will stop automatically or will not allow automatic operation if the feed pump did not start when operation was initiated or the inlet or permeate pressure was too high to allow operation.

3. *Long-term equipment shutdown*. If the equipment is down for more than 2 days, operation for 30 to 60 min per day can protect the equipment from bacterial fouling. If the equipment is down for more than 7 days, a manual air scrub is carried out before the equipment is turned off, a storage solution (1% $NaHSO_3$) is added, and all valves are closed. During long-term storage, the pH should be checked monthly. Replace the storage solution if the pH is below 3. During shutdown, the UF membrane should always keep moist. The membrane components will be damaged irreversibly after drying.

Integrity Testing Membrane integrity testing ensures that the membrane is still intact. These tests check the membrane structure for wear or damage, which would result in a pore size that is larger than permitted. It is important that these tests be carried out regularly, since finished water quality can be affected adversely by damage to the membrane surface. All tests are nondestructive and are designed to isolate the problem area, should the need arise. Such tests include bubble point, diffusive flow, and pressure hold. The tests are characterized as direct or indirect. Direct tests challenge the physical structure of the membrane utilizing air or pressure-hold tests. Indirect tests monitor filtrate water quality parameters such as turbidity and particle counts to determine membrane integrity.

One of the more popular tests is the *bubble point test*, which uses capillary properties of the membrane to locate any problems. According to capillary theory, the smaller the pore size, the higher the pressure that is required to expel water from the pore. Pressurized air hits the filtrate side of the membrane, while feed water covers the feed side. The feed side is at atmospheric pressure via an open backwash valve. The test is successful when at a specified air pressure no air is released into the feed end for a specified time period. Therefore, the membrane passes the test when the feed water remains in the pores of the membrane at a specified air pressure. This ensures that the pores of the membrane are intact.

Membrane Fouling Four types of fouling are common to UF operations:

1. *Particulate fouling* is caused by suspended solids, colloids, and turbidity that can be reduced by coagulation, sedimentation, clarification, and media filtration. The common cleaning method for particulate fouling is air scour and backwash.

2. *Biological fouling* is caused by the growth of microorganisms that can be reduced by using in-line chemical feed of chlorine or biocide or by elimination of nutrients by using powdered activated carbon (PAC), granular activated carbon (GAC), or coagulation. The common cleaning method for removal of biological fouling is chemically enhanced backwash with oxidizers or biocides (e.g., Cl_2, H_2O_2, SBS).

3. *Inorganic fouling* is caused by the precipitation of inorganics on the membrane that can be reduced by using oxidation/precipitation and filtration as pretreatment to the UF or in some cases using low-hardness water for the alkali chemically enhanced backwash. The common cleaning method for removal of inorganic fouling is chemically enhanced backwash with acid at pH 2 (e.g., HCl, H_2SO_4, citric acid, oxalic acid).

4. *Organic fouling* is caused by organics adsorbing on the membrane (silt, organic acids, humus) that can be reduced by using PAC, GAC, or coagulation. The common cleaning method for removal of organic fouling is CEB with alkali at pH 12 (NaOH).

6.6.1.4.4 Preservation of UF Systems Modules installed during assembly of a skid should not be allowed to dry out. Dry membrane fibers will lose flux irreversibly.

UF systems are designed to run continuously, and membrane systems perform better when operated continuously. However, in reality, a UF system will start up and shut down with some frequency. When the UF system is shut down, the system must be cleaned using air scour and backwashed with permeate water to prevent bacterial growth in the system.

The water used for backwash before shutdown should not contain chemicals. Any feed water and backwash chemical dosing used should be stopped before the last cleaning and shutdown. After cleaning, all valves in the UF system should be closed to seal the system. To avoid leakage in the module housing end caps and clamps, the backpressure in the modules should be controlled when the UF system shuts down, especially in case of nonscheduled shutdowns (e.g., power failure or emergency shutdowns).

When the system is down for more than 48 h, note the following:

- The module should not dry out. Dry membrane fibers will irreversibly lose flux at any time.
- The system should be adequately protected against microbiological growth, flushed 30 to 60 min per day, or operated every 24 h.
- The system should be protected against temperature extremes. The UF system can be shut down for 48 h without adding preservative and taking precautions for microbiological fouling.
- Addition of chemical preservative is necessary for system downtime greater than 48 h.

6.6.2 Microfiltration

Microfiltration (MF) is a membrane-based technology that is very different from nanofiltration, reverse osmosis, and UF but is similar to UF. This process removes suspended solids but virtually no dissolved contaminants. Depending on the configuration and application, UF and MF membranes operate between 30 and 250 psi (2 to 14 bar), have an upper temperature limit of 194°F (90°C), and have a pH range of 1 to 14 [24].

Typically, MF is operated at a constant permeate flow. The transmembrane pressure will increase over time and require a cleaning procedure that includes backflushing and/or air scouring to remove the fouling layer. Disinfectants and other cleaning agents can be used to fully remove and prevent performance loss due to biological growth as well as other foulants [24].

Microfiltration is also a low-pressure cross-flow membrane process for separating colloidal and suspended particles in the range 0.05 to 10 μm. MF is used for fermentation broth clarification and biomass clarification and recovery.

Microfiltration is also a filtration process that operates as a physical sieving separation process. It is best used for the removal of suspended solids, *Giardia*, *Cryptosporidum*, and the reduction of turbidity. It is also used as a pretreatment to desalination technologies such as nanofiltration and reverse osmosis [26].

MF has the largest pore size of the wide variety of membrane filtration systems available. In terms of pore size, MF fills in the gap between ultrafiltration and granular media filtration. In terms of characteristic particle size, this MF range covers the lower portion of the conventional clays and the upper half of the range for humic acids. This is smaller than the size range for bacteria, algae, and cysts, and larger than that of viruses. MF is also typically used for turbidity reduction and to remove suspended solids, *Giardia*, and *Cryptosporidum*. This process requires low transmembrane pressure (1 to 30 psi) to operate and is now used as a pretreatment for such desalination technologies as reverse osmosis, nanofiltration, and electrodialysis, which cannot themselves remove salt [26].

MF membranes can operate in either cross-flow separation or dead-end filtration. In cross-flow separation, only part of the feed stream is treated; the remainder of the water is passed through the membrane untreated. In dead-end separation, all of the feed water is treated. There are also two pump configurations, either pressure-driven or vacuum systems. Pressure-driven membranes are housed in a vessel and the flow is fed from a pump. Vacuum systems consist of membranes submerged in nonpressurized tanks and driven by a vacuum created on the product side. Typical recoveries can range from 85 to 95% [26]. Flux rates range from 20 to 100 gal/ft^2 per day, depending on the application. Backwashes are usually carried out for short durations (3 to 180 s) and occur in relatively frequent intervals (5 min to several hours) [7]. The frequency and duration depend on the specific application. CIP operation can also be performed as a periodic major cleaning technique.

Typical cleaning agents are sodium hypochlorite, citric acid, caustic soda, and detergents. They can be initiated manually or be controlled automatically. CIP operations are used when backwashing and chemically enhanced backwashes are in sufficient.

Factors influencing membrane selection are cost, percent recovery, percent rejection, raw water characteristics, and pretreatment requirements. Factors influencing performance are raw water characteristics, transmembrane pressure, temperature, and regular monitoring and maintenance.

6.6.2.1 Pretreatment A self-backwashing strainer is often necessary to protect the membranes and to moderate particulate loading. Depending on the raw water, a coagulant such as ferric chloride may be added to form pinfloc and help improve rejection.

6.6.2.2 Maintenance It is necessary to monitor filtrate turbidity to give a rough indication of membrane integrity. Membrane integrity can be tested through a pressure decay test. In this test, pressurized air is applied to membranes at a pressure less than would cause the air to flow through the membrane, and the pressure decay is measured. Regular monitoring of membrane performance is necessary to ensure that the membrane system is operating at the most effective loading rate and backwash regime. Membrane life is typically estimated at seven years or more, with manufacturer warranties covering five years [26].

6.6.2.3 Waste Disposal Waste includes pretreatment waste, backwash flow, retentate flow (if applicable), and CIP waste. Waste streams are either discharged to the sewer or treated if discharging to surface waters. Waste streams being discharged to surface waters are typically processed for turbidity removal through settling ponds or other treatment systems. CIP waste is neutralized and usually combined with the remainder of the waste [26].

6.6.2.4 Limitations

- Membrane integrity and testing protocols are still under development.
- Some regulatory agencies are slow to accept MF applications.

6.6.3 Reverse Osmosis

Reverse osmosis (RO) is the finest level of filtration available. The RO membrane acts as a barrier to all dissolved salts and inorganic molecules, as well as organic molecules with a molecular weight greater than approximately 100. Water molecules, on the other hand, pass freely through the membrane, creating a purified product stream. Rejection of dissolved salts is typically 95% to greater than 99%. The uses for RO are numerous and varied and include desalination of seawater or brackish water for drinking purposes, wastewater recovery, food and beverage processing, biomedical separations, and purification of home drinking water and industrial process water. Also, RO is often used in the production of ultrapure water for use in the semiconductor industry, power industry (boiler feed water), and in medical and laboratory applications. Utilizing RO prior to ion exchange dramatically reduces operating costs and regeneration frequency of the an ion-exchange system. Transmembrane pressures for RO typically range from 75 psig (5 bar) for brackish water to greater than 1200 psig (84 bar) for seawater.

From a material standpoint of view, for standard water purification applications, the two most common families of RO membranes are made using polymers of either cellulose acetate or polyamide. A third membrane type, introduced within the last few years, uses charged polysulfone. From a construction point of view, RO membrane is used in four configurations: plate and frame, tubular, hollow fiber, and spiral wound. The most frequently utilized RO membranes are polyamide thin-film composite membranes packed in a spiral-wound configuration.

6.6.3.1 Construction of a Spiral-Wound Configuration Element Spiral-wound designs offer many advantages over the other module designs for most RO applications in water treatment. Typically, a spiral-wound configuration offers significantly lower replacement costs, simpler plumbing systems, easier maintenance, and greater design freedom than other configurations, making it the industry standard for reverse osmosis and nanofiltration membranes in water treatment.

The construction of a spiral-wound membrane element as well as its installation in a pressure vessel is shown schematically in Fig. 6-7. An RO element contains from one to more than 30 membrane leafs, depending on the element diameter and element type. Each leaf is made of two membrane sheets glued together back to back with a permeate spacer in between. There is a side glue line at the feed end and at the concentrate end of the element, and a closing glue line at the outer diameter of the element. The open side of the leaf is connected to and sealed against the perforated central part of the product water tube, which collects the permeate from all leaves. The leaves are rolled up with a sheet of feed spacer between each of them, which provides the channel for the feed and concentrate flow. In operation, the feed water enters the face of the element through the feed spacer channels and exits on the opposite end as concentrate. A part of the feed water, typically, 10 to 20%, permeates the membrane into the leaves and exits the permeate water tube. In membrane systems the elements are placed in series inside a pressure vessel. The concentrate of the first element becomes the feed to the second element, and so on. The permeate tubes are connected with interconnectors (also called couplers), and the combined total permeate exits the pressure vessel at one side (sometimes at both sides) of the vessel.

6.6.3.2 Key Terms In practice, reverse osmosis is applied as a cross-flow filtration process. With a high-pressure pump, feed water is continuously pumped at elevated pressure to the membrane system. Within the membrane system, the feed water will be split into a low-saline and/or purified product, called *permeate*, and a high-saline or concentrated brine, called *concentrate* or *reject*. A flow-regulating valve called a concentrate valve controls the percentage of feed water that is going to the concentrate stream and the permeate that will be obtained from the feed. Key terms used in the reverse osmosis process are defined below.

1. *Recovery*: the percentage of membrane system feed water that emerges from the system as product water (i.e., permeate). Membrane system design is based on expected feed water quality, and recovery is defined through initial adjustment of valves on the concentrate stream. Recovery

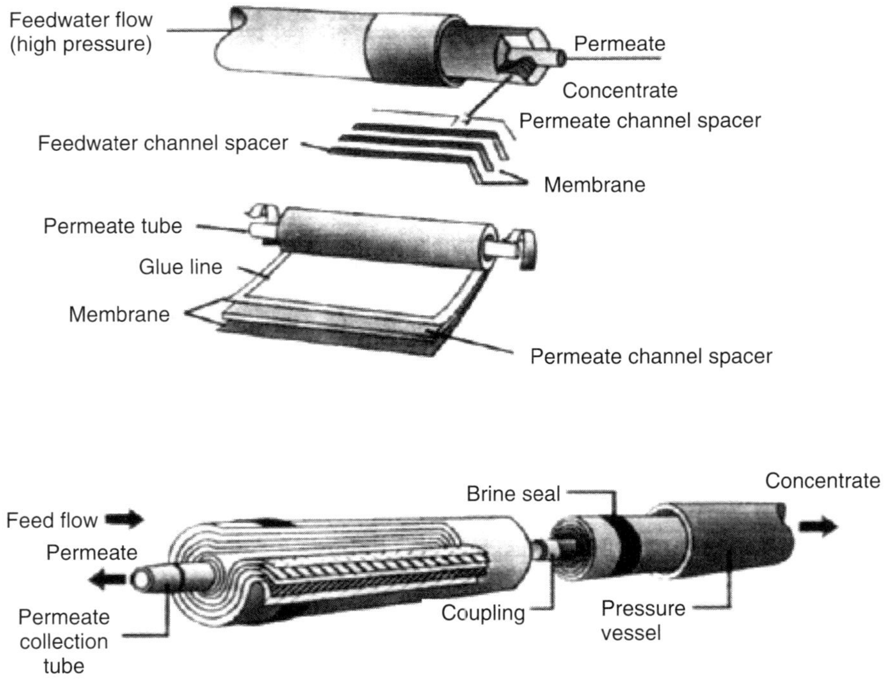

Figure 6-7 Construction of a spiral-wound RO membrane element.

is often fixed at the highest level that maximizes permeate flow while preventing precipitation of supersaturated salts within the membrane system.

2. *Rejection*: the percentage of solute concentration removed from the system feed water by the membrane. In reverse osmosis, a high rejection of total dissolved solids is important, whereas in nanofiltration the solutes of interest are specific (e.g., low rejection for hardness and high rejection for organic matter.

3. *Passage*: the percentage of dissolved constituents (contaminants) in the feed water allowed to pass through the membrane (the opposite of rejection).

4. *Permeate*: the purified product water produced by a membrane system.

5. *Flow*: the rate of feed water introduced to the membrane element or membrane system, usually measured in gallons per minute or cubic meters per hour. Concentrate flow is the rate of flow of nonpermeated feed water that exits the membrane element or membrane system. This concentrate contains most of the dissolved constituents originally carried into the element or into the system from the feed source. It is usually measured in gallons per minute or cubic meters per hour.

6. *Flux*: the rate of permeate transported per unit of membrane area, usually measured in gallons per square foot per day or liters per square meter per hour.

6.6.3.3 Typical System Design Multistage systems are used most frequently in conventional RO systems. Usually, two stages will suffice for recovery up to 75%, and three must be used for higher recovery. These numbers are based on the assumption that standard pressure vessels with six elements are used. For shorter vessels housing only three elements, for example, the number of stages has to be doubled for the same system recovery. Generally speaking, the higher the system recovery, the higher the number of membrane elements that have to be connected in series. To compensate for the permeate that is removed and to maintain uniform feed flow to each stage, the number of pressure vessels per stage decreases in the direction of feed flow. A schematic show of a two-stage system using a staging ratio (pressure vessels in two adjacent stages, upstream vessels/downstream vessels) of 2 : 1 is shown in Fig. 6-8, and a typical RO system is shown in Fig. 6-9.

6.6.3.4 Factors Affecting Reverse Osmosis Performance Permeate flux and salt rejection are the key performance parameters of a reverse osmosis or nanofiltration process. Under specific reference conditions, flux and rejection are intrinsic properties of membrane performance. The flux and rejection of a membrane

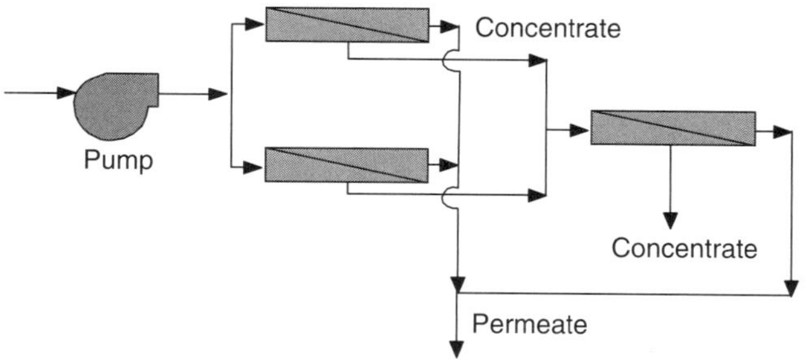

Figure 6-8 Scheme of a two-stage RO system.

Figure 6-9 Typical RO system.

system are influenced primarily by several variable parameters:

1. *Pressure*. With increasing effective feed pressure, the permeate's total dissolved solids will decrease while the flux increases.

2. *Temperature*. If the temperature increases and all other parameters are kept constant, the permeate flux and salt passage will increase.

3. *Recovery*. In the case of increasing recovery, the permeate flux will decrease and stop if the salt concentration reaches a value where the osmotic pressure of the concentrate is as high as the feed pressure applied. The salt rejection will drop with increasing recovery.

4. *Feed water salt concentration*. If the feed water salt concentration increases, the permeate flux and salt rejection will both decrease.

The functions can be understood through with the solution-diffusion model.

6.6.3.5 Operation and Maintenance Successful long-term performance of a membrane system depends on proper operation and maintenance. (For a comprehensive discussion, see DOW Chemical's *Filmtec Reverse Osmosis Membranes Technical Manual*.) This includes the initial plant startup as well as operational startups and shutdowns. Preventing fouling, scaling, plugging, and degradation (e.g. by oxidation) of the membranes is not only a matter of system design, but also a matter of proper commissioning and operation. Record keeping and data normalization are required in order to understand actual plant performance and to enable corrective measures when necessary.

6.6.3.5.1 Initial Startup Before initiating system startup procedures, pretreatment checks, loading of the membrane elements, instrument calibration and other system checks should be completed [30].

Steps in a typical startup sequence are as follows:

1. Before initiating the sequence, rinse the pretreatment section thoroughly to flush out debris and other contaminants, without letting the feed enter the elements. Follow the pre-startup checklist.
2. Check all valves to ensure that settings are correct. The feed pressure control and concentrate control valves should be fully open.
3. Use low-pressure water at a low flow rate to flush the air out of the elements and pressure vessels. Flush at a gauge pressure of 30 to 60 psi (0.2 to 0.4 MPa). All permeate and concentrate flows should be directed to an appropriate waste collection drain during flushing. *Air remaining in the elements and/or in the pressure vessels may lead to excessive force on the element in the flow or radial direction, causing fiberglass shell cracking, if the feed pressure is also ramped up*.
4. During the flushing operation, check all pipe connections and valves for leaks. Tighten connections where necessary.
5. After the system has been flushed for a minimum of 30 min, close the feed pressure control valve.
6. Ensure that the concentrate control valve is open. *Starting against a closed or almost closed concentrate valve could cause the recovery to be exceeded, which may lead to scaling*.
7. Slowly crack open the feed pressure control valve [the feed pressure should be less than 60 psi (0.4 MPa)].
8. Start the high-pressure pump.
9. Slowly open the feed pressure control valve, increasing the feed pressure and feed flow rate to the membrane elements until the design concentrate flow is reached. The feed pressure increase to the elements should be less than 10 psi (0.07 MPa) per second to achieve a soft start. Continue to send all permeate and concentrate flows to an appropriate waste collection drain. *If the feed pressure and/or the feed flow rate are ramped up too quickly, the housing of the elements may be damaged by excessive forces in the flow and/or radial direction (especially if air is in the system), leading to telescoping and/or fiberglass shell cracking*.
10. Slowly close the concentrate control valve until the ratio of permeate flow to concentrate flow approaches, but does not exceed, the design ratio (recovery). Continue to check the system pressure to ensure that it does not exceed the upper design limit.
11. Repeat steps 9 and 10 until the design permeate and concentrate flows are obtained.
12. Calculate the system recovery and compare it to the system's design value.
13. Check the addition of pretreatment chemicals (acid, scale inhibitor, and sodium metabisulfite, if used). Measure the feed water pH.
14. Check the Langelier Saturation Index or Stiff and Davis Stability Index of the concentrate by measuring pH, conductivity, calcium hardness, and alkalinity levels and then making the necessary calculations.
15. Allow the system to run for 1 h.
16. Take the first reading of all operating parameters.
17. Check the permeate conductivity from each pressure vessel to verify that all vessels conform to performance expectations (e.g., vessels with leaking O-rings or other evidence of malfunction are to be identified for corrective action).
18. After 24 to 48 h of operation, review all recorded plant operating data, such as feed pressure, differential pressure, temperature, flows, recovery, and conductivity. At the same time, draw samples of feed water, concentrate, and permeate for analysis of constituents.
19. Compare system performance to design values.
20. Confirm proper operation of mechanical and instrumental safety devices.
21. Switch the permeate flow from drain to the normal service position.
22. Lock the system into automatic operation.
23. Use the initial system performance information obtained in steps 16 to 18 as a reference for evaluating future system performance. Measure system performance regularly during the first week of operation to ensure proper performance during this critical initial stage.

6.6.3.5.2 Operation Startup Once a membrane system has been started up, ideally it should be kept running at constant conditions. In reality, membrane plants have to be shut down and restarted frequently. Start–stop cycles result in pressure and flow changes, causing mechanical stress to the membrane elements. Therefore, the start–stop frequency should be minimized, and the regular operation startup sequence should be as smooth as possible. In principle, the same sequence is recommended as for the initial startup. Most important is a slow feed pressure increase, especially for seawater plants. The normal startup sequence is typically automated through the use of programmable controllers and remotely operated valves. The calibration of instruments, the function of alarms and safety devices, corrosion prevention, and leak-free operation have to be checked and maintained on a regular basis.

6.6.3.5.3 Adjustment of Operation Parameters A membrane system is designed on the basis of a defined set of data, such as the permeate flow, feed water composition, and temperature. In reality, the plant operation has to be flexible to respond to changing needs or changing conditions. The normal way of operating RO membrane plants is to keep the flows and thus the recovery constant at the design values. Changes in the membrane flux (e.g., by temperature or fouling) are compensated by adjusting the feed pressure. However, the maximum feed pressure specified must not be exceeded, nor should too much fouling be tolerated. If the feed water analysis changes such that the scaling potential increases, the system recovery has to be decreased or other measures have to be taken to cope with the new situation.

The most common situation is that the permeate capacity of the plant has to be adjusted to the current needs. Normally, the capacity is designed to meet the peak needs. Operating with overcapacity is generally not recommended. Thus, adjustment means lowering the design permeate output. The easiest way is to shut the plant down when no permeate is needed. A high start–stop frequency, however, can lower membrane performance and lifetime. A permeate buffer tank may be used to allow more constant operation.

Reducing the feed pressure is another way to reduce the permeate flow. Preferably, this is done using a speed-controlled pump in order to save energy. Normally, system recovery is kept constant when the permeate flow is reduced. It has to be ensured by system analysis using the computer program that single-element recoveries do not exceed their limits. During low-flow operation, the system salt rejection is lower than during design flow operation. Also, minimum concentrate flows must be maintained during low-flow operation [7].

6.6.3.5.4 System Shutdown An RO system is designed to be operated continuously. However, in reality membrane systems will start up and shut down with some frequency. When the membrane system is shut down, the system must be flushed preferentially with permeate water or, alternatively, with high-quality feed water to remove the high salt concentration from the pressure vessels until concentrate conductivity matches feed water conductivity. Flushing is done at low pressure [about 40 psi (3 bar)]. A high feed flow rate is sometimes beneficial for a cleaning effect; however, the maximum pressure drop per element and per multielement vessel must not be exceeded. During low-pressure flushing, the vessels of the last stage of a concentrate staged system are normally exposed to the highest feed flow rates, and therefore they show the highest pressure drop.

The water used for flushing must contain no chemicals used for the pretreatment, especially no scale inhibitors. Therefore, any chemical injection (if used) is stopped before flushing. After flushing the system, the feed valves are closed completely. If the concentrate line ends in a drain below the level of the pressure vessels, an air break should be employed in the line at a position higher than that of the highest pressure vessel. Otherwise, the vessels might be emptied by a siphoning effect.

When the high-pressure pump is switched off, and the feed/concentrate side has not been flushed out with permeate water, natural osmosis will cause a temporary permeate reverse flow, sometimes referred to as permeate draw-back or suck-back. Permeate suck-back alone or in combination with a feed-side flush may provide a beneficial cleaning effect. To accommodate permeate suck-back, enough water volume should be available to prevent a vacuum from being drawn or air being sucked back into the membrane element.

If the permeate line is pressurized during operation and the system shuts down, the membrane might become exposed to static permeate backpressure. To avoid membrane damage from backpressure, the static permeate backpressure must not exceed 5 psi (0.3 bar) at any time. Check valves or atmospheric drain valves in the permeate line can be used to safeguard the membrane. These safeguard valves also need to work, especially in case of emergency or nonscheduled shutdowns (e.g., because of a power failure).

When a system must be shut down for longer than 48 h, take care that:

- The elements do not dry out. Dry elements will lose flux irreversibly.
- The system is adequately protected against microbiological growth or that regular flushing is carried out every 24 h.
- When applicable, the system is protected against temperature extremes.

The membrane plant can be stopped for 24 h without preservation and precautions for microbiological fouling. If feed water for flushing every 24 h is not available, preservation with chemicals is necessary for longer stops than 48 h.

6.6.3.5.5 Chemical Cleaning In normal operation, the membrane in reverse osmosis elements can become fouled by mineral scale, biological matter, colloidal particles, and insoluble organic constituents. Deposits build up on the membrane surfaces during operation until they cause a loss of normalized permeate flow, a loss of normalized salt rejection, or both [7].

Occasionally, fouling of the membrane surfaces is caused by:

- Inadequate pretreatment system
- Pretreatment upset conditions
- Improper material selection (pumps, piping, etc.)
- Failure of chemical dosing systems
- Inadequate flushing following shutdown
- Improper operational control
- Slow buildup of precipitates (e.g., barium, silica) over extended periods
- Change in feed water composition
- Biological contamination of feed water

Elements should be cleaned when one or more of the following parameters are applicable:

- The normalized permeate flow drops by 10%.
- The normalized salt passage increases by 5 to 10%.
- The normalized pressure drop (feed pressure minus concentrate pressure) increases by 10 to 15%.

If you wait too long, cleaning may not restore the membrane element performance successfully. In addition, the time between cleanings becomes shorter, as the membrane elements will foul or scale more rapidly. Differential pressure (ΔP) should be measured and recorded across each stage of the array of pressure vessels. If the feed channels within the element become plugged, ΔP will increase. It should be noted that the permeate flux will drop if the feed water temperature decreases. This is normal and does not indicate membrane fouling [4].

Cleaning Equipment The equipment for cleaning is shown in Fig. 6-10. The pH of the cleaning solutions used can be in the range 1 to 13 (depending on the RO product

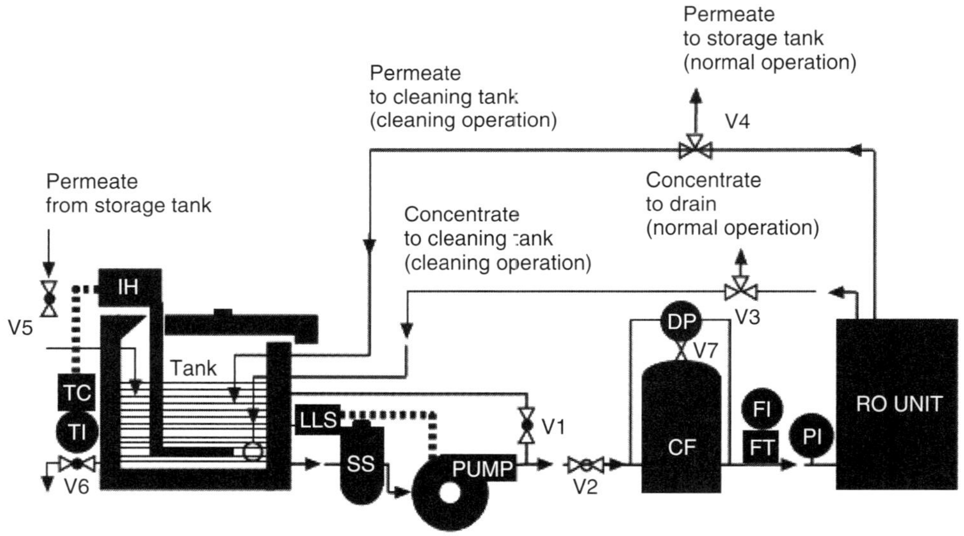

Tank Chemical mixing tank, polypropylene or FRP
IH Immersion heater (may be replaced by cooling coil for some site locations)
TI Temperature indicator
TC Temperature control
LLS Lower level switch to shut off pump
SS Security screen–100 mesh
PUMP Low-pressure pump, 316 SS or non-metallic composite
CF Cartridge filter, 5–10 micron polypropylene with PVC, FRP, or SS housing
DP Differential pressure gauge
FI Flow indicator
FT Flow transmitter (optional)
PI Pressure indicator
V1 Pump recirculation valve, CPVC
V2 Flow control valve, CPVC
V3 Concentrate valve, CPVC 3-way valve
V4 Permeate valve, CPVC 3-way valve
V5 Permeate inlet valve, CPVC
V6 Tank drain valve, PVC, or CPVC
V7 Purge valve, SS, PVC, or CPVC

Figure 6-10 Cleaning system flow diagram.

used), and therefore noncorroding materials of construction should be used in the cleaning system.

Cleaning Procedure There are six steps in the cleaning of elements:

1. Make up the cleaning solution.
2. In low-flow pumping, pump mixed, preheated cleaning solution to the vessel at conditions of low flow rate (about half of that shown in Table 6-3) and low pressure to displace the process water. Use only enough pressure to compensate for the pressure drop from feed to concentrate. The pressure should be low enough that essentially no or little permeate is produced. Low pressure minimizes redeposition of dirt on the membrane. Dump the concentrate, as necessary, to prevent dilution of the cleaning solution.
3. After the process water is displaced, cleaning solution will be present in the concentrate stream. Recycle the concentrate and permeate to the cleaning solution tank and allow the temperature to stabilize. Measure the pH of the solution and adjust the pH if needed.
4. Turn the pump off and allow the elements to soak. Sometimes a soak period of about 1 h is sufficient. For difficult fouling, an extended soak period is beneficial; soak the elements overnight for 10 to 15 h. To maintain a high temperature during an extended soak period, use a slow recirculation rate (about 10% of that shown in Table 6-3).
5. In high-flow pumping, feed the cleaning solution at the rates shown in Table 6-1 for 30 to 60 min. The high flow rate flushes out the foulants removed from the membrane surface by the cleaning. If the elements are heavily fouled, a flow rate 50% higher than that shown in Table 6-1 may aid cleaning. At higher flow rates, excessive pressure drop may be a problem. The maximum recommended pressure drops are 15 psi per element or 50 psi per multielement vessel, whichever value is more limiting. Note that 15 psi per element or 50 psi per multielement vessel should *not* be used as a cleaning criterion. Cleaning is recommended when the pressure drop increases 15%. A pressure drop above 50 psi in a single stage may cause significant membrane damage.
6. RO permeate or deionized water is recommended for flushing out the cleaning solution. Prefiltered raw water or feed water should be avoided, as its components may react with the cleaning solution: precipitation of foulants may occur in the membrane elements. The minimum flush-out temperature is 20°C.

Cleaning Chemicals Table 6-4 lists suitable cleaning chemicals. Acid and alkaline cleaners are the standard cleaning chemicals. Acid cleaners are used to remove inorganic precipitates (including iron), and alkaline cleaners are used to remove organic fouling (including biological matter). Sulfuric acid should not be used for cleaning because of the risk of calcium sulfate precipitation. Specialty cleaning chemicals may be used in cases of severe fouling or unique cleaning requirements. Preferably, RO permeate should be used for the preparation of cleaning solutions; however, prefiltered raw water may be used. The feed water can be highly buffered, so more acid or hydroxide may be needed with feed water to reach the desired pH level, which is about 2 for acid cleaning and about 12 for alkaline cleaning.

6.6.3.5.6 Preservation of RO Systems The procedure for shutting down an RO system was described in Section 6.6.3.5.5. The elements must be preserved any time the plant is shut down for more than a maximum of 48 h to prevent biological growth. Depending on the previous operational history of the plant, it will be necessary in almost all cases to clean the membranes prior to shutdown and preservation. This applies to cases when the membranes are known or assumed to be fouled. After cleaning, the preservation should follow within the next 10 h as follows:

TABLE 6-3 Recommended Feed Flow Rate per Pressure Vessel During High-Flow-Rate Recycle

Cleaning Pressure[a]			Flow per Pressure Vessel	
psig	bar	Element Diameter (in.)	gal/min	m^3/h
20–60	1.5–4.0	2.5	3–5	0.7–1.2
20–60	1.5–4.0	4	8–10	1.8–2.3
20–60	1.5–4.0	6	16–20	3.6–4.5
20–60	1.5–4.0	8[b]	30–40	6.0–9.1
20–60	1.5–4.0	8[c]	35–45	8.0–10.2

[a]Depends on the number of elements per pressure vessel.
[b]Suitable for element with 365-in^2 active area.
[c]Suitable for full-fit element or element with 400- or 400-m^2 active area.

TABLE 6-4 Simple Cleaning Solutions

Foulant	0.1% (W) NaOH and pH 12, 35°C max., or 1.0% (W) Na_4EDTA and pH 12, 35°C max.	0.1% (W) NaOH and pH 12, 35°C max., or 0.025% (W) Na-DSS and pH 12, 35°C max.	0.2% (W) HCl, 25°C, and pH 1–2	1.0% (W) $Na_2S_2O_4$, 25°C, and pH 5	0.5% (W) H_3PO_4, 25°C and pH 1–2	1.0% (W) NH_2SO_3H, 25°C and pH 3–4
Inorganic salts (e.g., $CaCO_2$)	—	—	Preferred	Alternative	Alternative	
Sulfate scales ($CaSO_4$, $BaSO_4$)	OK					
Metal oxides (e.g., iron)	—	—	—	Preferred	Alternative	Alternative
Inorganic colloids (silt)		Preferred				
Silica	Alternative	Preferred				
Biofilms	Alternative	Preferred				
Organic	Alternative	Preferred				

1. Totally immerse the elements in the pressure vessels in a solution of 1 to 1.5% sodium metabisulfite (SMBS), venting the air outside the pressure vessels. Use the overflow technique: Circulate the SMBS solution in such a way that the remaining air in the system is minimized after the recirculation is completed. After the pressure vessel is filled, the SMBS solution should be allowed to overflow through an opening located higher than the upper end of the highest-pressure vessel being filled.
2. Separate the preservation solution from the air outside by closing all valves. Any contact with oxygen will oxidize the SMBS.
3. Check the pH once a week. When the pH becomes 3 or lower, change the preservation solution.
4. Change the preservation solution at least once a month. During the shutdown period, the plant must be kept frost-free and the temperature must not exceed 113°F (45°C). A low temperature is desirable.

6.6.4 Nanofiltration

Nanofiltration (NF) is a special process selected when RO and UF are not the ideal choice for separation. NF can perform separation applications that are not otherwise economically feasible, such as demineralization, color removal, and desalination. In concentrations of organic solutes, suspended solids, and polyvalent ions, the permeate contains monovalent ions and low-molecular-weight organic solutions such as alcohol.

NF is a cross-flow, pressure-driven process that is characterized by a membrane pore size corresponding to a molecular weight cutoff of approximately 200 to 1000 Da, and operating pressures of 150 to 500 psi (10 to 34 bar) [29]. NF is used primarily to separate low-molecular-weight organics and multivalent salts from monovalent salts and water.

Nanofiltration membranes are manufactured using two preparation techniques: polymer phase inversion, resulting in a homogeneous asymmetric membrane, and interfacial polarization of a thin-film composite layer on top of a substrate ultrafiltration membrane or other porous substrate. Cellulose acetate and sulfonated polysulfone are two common materials used for making homogeneous asymmetric NF membranes. Thin-film composite nanofiltration membranes use cross-linked polyamide polymers reacted to carboxylic group or other charged "pendant." Substrate materials commonly used for thin-film composite membranes are polysulfone, poly(ether sulfone), poly(vinyledene fluoride), polyacrylonitrile, and Poly(ether ether ketone) [29]. Recent development of NF membranes with exceptional stability under conditions of very low or high pH, very high temperature, or organic solvent media required membrane manufacturers to seek new materials for membrane manufacturing. The materials used for these innovative membranes are highly cross-linked, to allow long-term stability and practical membrane life in aggressive environments. Nanofiltation membranes have a slightly charged surface [29].

Most NF membranes are packed into spiral-wound elements, although tubular, hollow-fiber, and flat sheet and plate and frame modules are also available.

6.6.4.1 Parameters Affecting the Performance of Nanofiltration Membranes When designing a nanofiltration process, one should consider several operating parameters, the most important being parameters similar to those for most cross-flow filtration processes.

1. *Pressure*. Pressure difference is the driving force responsible for a nanofiltration process. The effective driving pressure is the hydraulic pressure supplied, less the osmotic pressure applied on the membrane by the solutes. Nanofiltration provides good separation at net pressures of 150 psi (10 bar) or higher [29].
2. *Salinity*. The effective pore radius of a charged pore will increase as the ionic strength of the surrounding liquid increases. Therefore, the rejection of monovalent ions will decrease as their concentration in the feed solution increases. The rejection of divalent ions will be affected to a lesser extent [29].
3. *Cross-flow velocity*. Increasing the cross-flow velocity in an NF membrane process increases the average flux due to efficient removal of fouling layer from the membrane surface. However, the mechanical strength of the membrane and the construction of the element and system hardware will determine the maximum cross-flow velocity that can be applied [29]. Running a nanofiltration membrane at too high a cross-flow velocity may cause premature failure of membranes and modules.
4. *Temperature*. Increasing the process temperature increases the NF membrane flux due to viscosity reduction. The rejection of NF membranes does not depend significantly on the process temperature.
5. *pH*. pH affects the performance of nanofiltration membranes in more than one way. The charged sites on an NF membrane surface (i.e., carboxylic group, sulfonic group) are negatively charged at neutral pH or higher but lose their charge at acidic pH. It is well known that most NF and RO membranes have a lower rejection rate at low pH or after an acid rinse. It should be noted, however, that since different membrane manufacturers use different chemistries to produce their thin-film composite layer, the pH dependency of a membrane should be determined for each membrane type [29].

   In addition to the effect of pH on the membrane itself, pH can be responsible for changes in the feed solution, causing changes in membrane performance. Two examples are a change in the solubility of ions at different pH regimes, causing a different rejection rate; and a change in the dissociation state of ions at different pH ranges [29].

6.6.4.2 Applications Industrial applications of NF are quite common in the food and dairy sector, in chemical processing, in the pulp and paper industry, and in texiles, although the chief application continues to be in the treatment of fresh, process, and waste waters. In the treatment of water, NF finds use in the polishing that occurs at the end of conventional processes. It cannot be used for water desalination, but it is an effective means of water softening, as the main hardness chemicals are divalent. NF membranes are also used for the removal of natural organic matter from water, especially tastes, odors, and colors, and in the removal of trace herbicides from large water flows. They can also be used for the removal of residual quantities of disinfectants in drinking water.

Food industry applications are quite numerous. In the dairy sector, NF is used to concentrate whey and permeates from other whey treatments, in the recycling of clean-in-place solutions, and in the concentration of sugar, dextrose syrup, and thin sugar juice while ion-exchange brines are dimineralized. NF is used to degum solutions in the edible-oil processing sector, for continuous cheese production, and in the production of alternative sweeteners.

There are probably as many applications in the entire chemical sector (including petrochemicals and pharmaceuticals) as in the rest of industry put together. NF is a valuable contributor to the totality of the chemicals industry. The production of salt from natural brines uses NF as a purification process, while most chemical processes produce quite vicious wates, from which valuable chemicals can usually be recovered by processes including NF. The high value of many of the products in the pharmaceutical and biotechnical sectors allows the use of NF in their purification processes.

The paper and pulp industry uses a very great quantity of water in its production processes, a quantity that the industry is striving to reduce mainly by "closing the water cycle," a system in which the purification properties of NF have a major role. All of these specifically mentioned applications have been water based, but NF is not restricted to the treatment of aqueous suspensions. Indeed, one of the largest NF plants was installed at a refinery for the dewaxing of oils.

In aqueous systems, NF uses hydrophilic polymeric materials such as poly(ether sulfone), polyamides, and cellulose derivatives. These materials, in contact with organic solvents, quickly lose their stability. Special membranes have therefore been developed to provide the same type of performance as in aqueous systems, and they are now used for solvent exchange, solvent recovery and separation, catalyst recovery, and heavy metal removal.

6.7 FILTER MAINTENANCE

Filters and their elements in any process system should be monitored continuously. This provides users with both routine and critical information about filter performance. One of the most important parameters monitored is the filter pressure, which helps provide an early signal of failure of the filter element and components. For example, when there is a sudden increase in the pressure differential across a pressure line filter from say 1 to 4 bar, this is an indication of an upstream component failure (all other things being equal). Whenever installing a filter, be sure to know the right location for the filter, whether its for the return line, suction line, or pressure line.

In summary, a good understanding of filter location and maintenance goes a long way toward achieving proper fluid filtration.

6.7.1 Preparation for Maintenance

Preparation for maintenance is based on the procedure established for the specific filter: rotary vacuum, sand, and to some extent, clay. These filters are process equipment used directly in-line, unlike the flush oil and air filters used in process utilities.

1. Mostly important, make sure that the filter is isolated. Double blocking and spade blinding are required when any hot work is to be carried out on the filter. Spading is recommended where process flammable fluids are employed.
2. Drain out any hydrocarbon to a closed system.
3. Open the vent and air-purge to the atmosphere. However, flammables are to be flared until no traces are noticed when work or entry is required.
4. Open the manways and authorize entry when safety is guaranteed.

These are only some of the basic maintenance preparation requirements for process filters in diverse use.

REFERENCES

1. American Society for Testing and Materials, *Standard Guide for Record Keeping for Reverse Osmosis Systems*, ASTM D4472-89 (reapproved 2003). ASTM, West Conshokocken, PA, 2003.
2. American Society for Testing and Materials, *Standard Practice for Standardizing Reverse Osmosis Performance Data*, ASTM D4516-00. ASTM, West Conshokocken, PA, 2000.
3. American Society for Testing and Materials, *Standard Guide for Water Analysis for Reverse Osmosis Application*, ASTM D4195-88 (reapproved 2003). ASTM, West Conshokocken, PA, 2003.
4. Amjad, Z., Workman, K. R., and Castete, D. R., Considerations in membrane cleaning, in Amjad, Z. (Ed.), *Reverse Osmosis: Membrane Technology, Water Chemistry, and Industrial Applications*, Van Nostrand Reinhold, New York, 1993, pp. 210–236.
5. Bannerjee, A., Lozier, J., and Carlson, K., An on-line, multisensor, membrane filtration permeate water quality monitoring system, *Proceedings of the AWWA Membrane Technology Conference*, American Water Works Association, Denver, CO, 2001.
6. Byrne, W. (Ed.), *Reverse Osmosis: A Practical Guide for Industrial Users*, 2nd ed., Tall Oaks Publishing, Littleton, CO, 2002, pp. 437–494.
7. Byrne, W., and Bukay, M., How to monitor a reverse osmosis system, *Ultrapure Water*, vol. 1, no. 2, September–October 1984, pp. 32–36.
8. Coffey, B. M., Stewart, M. H., Wattier, K. L., and Wale, R. T. Evaluation of microfiltration for metropolitan's small domestic water systems, *Proceedings of the AWWA Membrane Technology Conference*, American Water Works Association, Denver, CO, 1993.
9. Deitz, K., Fine chemical processing: on-stream particle size analysis, *Filtration + Separation*, vol. 45, no. 4, May 2008, pp. 26–28.
10. Dow Chemical Co., Filmtec products, in *FILMTEC™ Reverse Osmosis Membranes Technical Manual*, Dow Midland, MI,
11. Dow Chemical Co., DOW UF, *Product Manual*, Dow, Midland, MI, April 2011.
12. Hendrix, E., Technical focus—Particle size analysis: technologies used in on-stream particle size analysis, *Outotec*, June 2007, p. 9.
13. http://en.wikipedia.org/wiki/Sand_filter.
14. http://www.chempro.in/equipmentplf.htm.
15. http://www.envirosindia.com/nano-filtration.html.
16. http://www.fluidenergy.com/market/closer_look_coalescer.htm.
17. http://www.globalspec.com/reference/35832/203279/chapter-2-filtration-mechanisms-and-theory.
18. http://www.halton.com/halton/cms.nsf/www/bagfilters.
19. http://www.industrialwaterequipment.co.uk/bag-tech.html.
20. http://www.natural-gas-valves.com/cartirge-filters.php.
21. http://www.niroinc.com/filtration/filtration_technologies.asp.
22. http://www.tpub.com/content/engine/14105/css/14105_146.htm.
23. http://www.yansonsgroup.net/air-pollution-control-equipments.html.
24. Huchler, L., Update your industrial water treatment operations, *Hydrocarbon Processing*, vol. 87, no. 12, December 2008, pp. 83–90.
25. Jacangelo, J. G., and Adham, S. A., Comparison of microfiltration and ultrafiltration for microbial removal, *Proceedings*

of Microfiltration for Water Treatment Symposium, American Water Works Association, Denver, CO, 1994.

26. Jurenka, B., http://www.usbr.gov/pmts/water/publications/primer.html, 2009.
27. Perlmutter, B. A., A Review of Filter Press Basics and Issues Versus Alternative Batch or Continuous Replacement Technologies, BHS-Filtration, Inc., Philadelphia, PA, April 2011.
28. Vacura, K., Cartridge filtration and the world market, *Everything About Water*, April 2009, pp. 38–42.
29. Yacubowicz, H., and Jorge Yacubowicz, J., *Nanofiltration: Properties and Uses*, Koch Membrane Systems, Wilmington, MA, 2007.
30. Youngberg, D. A., Start-up of an RO/DI pure water system, *Ultrapure Water*, March–April 1986, pp. 46–50.

7

SEALING DEVICES

MATT TONES AND JIM DRAGO
Garlock Sealing Technologies and GPT, Palmyra, New York

Fluid sealing devices play a critical role in a wide range of processing industries: chemicals, refining, pulp and paper, and many others. In this role they keep pumps from leaking (Fig. 7-1), valves from releasing emissions, and flanges from spraying fluids, and minimize the possibility of other undesirable and often dangerous conditions [3]. Yet because these devices account for a tiny fraction of the cost of the systems in which they are installed, they often do not receive the attention they warrant.

7.1 GASKETS

7.1.1 Gasket Materials

In this section we outline the features, advantages, and disadvantages of the materials used most in commonly sealing devices, whether they be gaskets, seals, packing, expansion joints, hydraulic components, or other sealing products. It is important to know the properties of the individual constituents in order to have a better understanding of the features and benefits of the sealing device chosen.

7.1.1.1 Basic Elastomers or Rubbers Rubber has three predominant characteristics:

1. *Incompressibility*: can be deformed but not reduced in volume
2. *Extensibility*: can be stretched
3. *Impermeability*: can prevent the passage of gases through the material

The following elastomers are listed alphabetically by common trade or generic name, type, and, in parentheses, the American Society of Testing and Materials (ASTM) designation. They are all man-made except "natural."

Butyl: isobutene–isoprene (IIR)

- Generally resistant to animal and vegetable fats, strong and oxidizing chemicals, and ozone
- Attacked by petroleum solvents, coal tar solvents, and aromatic hydrocarbons (i.e., toluene, benzene, xylene, etc.)
- Excellent weathering properties, low permeability to air, and excellent dielectric properties
- Good physical properties

Ethylene propylene: ethylene propylene–diene terpolymer (EPDM) or ethylene–propylene copolymer (EPM)

- Resistant to animal and vegetable oils, strong and oxidizing chemicals, and ozone
- Attacked by petroleum oils and solvents and aromatic hydrocarbons (i.e., toluene, benzene, xylene, etc.)
- Excellent weathering properties, low-temperature resistance, low permeability to air, and good dielectric strength
- Good bonding properties

Fluoroelastomer: hexafluoropropylene–vinylidene fluoride (FKM or FPM)

Process Plant Equipment: Operation, Control, and Reliability, First Edition. Edited by Michael D. Holloway, Chikezie Nwaoha, and Oliver A. Onyewuenyi.
© 2012 John Wiley & Sons, Inc. Published 2012 by John Wiley & Sons, Inc.

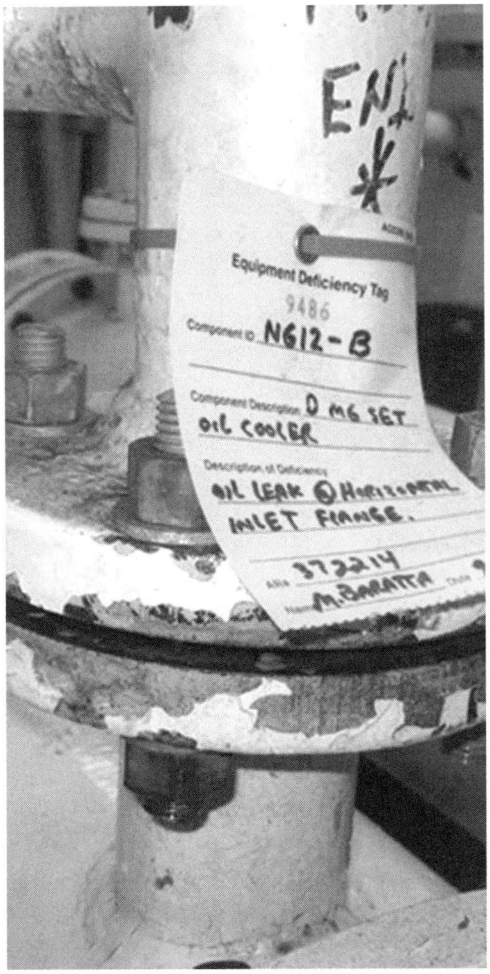

Figure 7-1 Leaking fluid connections pose both operational and environmental problems. (Courtesy of Garlock, an EnPro Industries family of companies.)

- Most common trade names: DuPont's Viton and 3M's Fluorel
- Resistant to aliphatic, aromatic, and halogenated hydrocarbons, acids, and animal and vegetable oils
- Attacked by ketones, low-molecular-weight esters and nitro-containing compounds
- Good weathering properties and resistance to elevated temperatures

Natural: isoprene natural (NR)

- Generally resistant to most moderate wet or dry chemicals, organic acids, and alcohols
- Attacked by strong acids, fats, oils, grease, most hydrocarbons, and ozone
- Excellent physical properties, including abrasion and tear resistance
- Superior tackiness in an uncured state

Neoprene: chloroprene (CR)

- Resistant to moderate chemicals, acids, oils, fats, grease, many solvents, and ozone
- Attacked by strong oxidizing acids, esters, ketones, and chlorinated, aromatic, and nitro hydrocarbons
- Good weathering resistance
- Flame retarding
- Good general physical properties

Nitrile or Buna-N: acrylonitrile butadiene (NBR)

- Generally resistant to fats, oils, greases, and aliphatic hydrocarbons
- Not ozone resistant
- Attacked by ketones, esters, aldehydes, and aromatic, chlorinated, and nitro hydrocarbons
- Good physical properties
- Good abrasion resistance and bonding properties

Polyisoprene: isoprene synthetic (IR)

- Basically, a synthetic version of "natural" SBR or Buna-S—styrene butadiene (SBR)
- Fluid compatibility is similar to that of "natural"
- Good physical properties

Silicone: polysiloxane (VMQ or MQ)

- Generally resistant to moderate or oxidizing chemicals, concentrated sodium hydroxide, and ozone
- Fair resistance to alkalies
- Attacked by many solvents, oils, concentrated acids, and dilute sodium hydroxide
- Very poor wet steam resistance
- Fair to poor physical properties
- Serviceable over a broad temperature range

7.1.1.2 ***Fibers*** The term *fiber* covers a very broad range of natural and man-made materials. Once again, only the most common fibers used in gasketing constructions are reviewed; they are listed alphabetically by generic name.

Acrylic fibers

- Man-made organic fiber
- Relatively low cost
- Not a high-performance fiber

Aramid

- Man-made organic fiber
- Introduced by DuPont in the early 1970s under the trade name Kevlar® E.I. duPont de Nemours and Company
- Chemically known as aromatic polyamides
- Very high tensile strength, high modulus, and low density
- Flame resistant; does not melt
- Starts degrading between 500°F (260°C) and 600°F (315°C), with very little aramid left at 800°F (425°C); withstands short exposures to 1000°F (540°C) in inert atmospheres
- With resins or elastomers, oxidation is retarded further

Carbon

- Rodlike fibers of various diameters and lengths
- Produced by pyrolysis (heat treatment process) to various levels of carbon-content materials (i.e., rayon, acrylonitriles, pitch, etc.)
- Good heat and chemical resistance

Cellulose

- Natural fibers; most common: so-called "vegetable" fibers
- Flax, jute, ramie, and cotton are common forms
- Readily available in huge quantities; low price
- Moderate chemical and general fluid resistance
- Not generally recommended for use over 250°F (121°C)
- Does not melt

Glass and fiberglass

- Man-made material composed mainly of silica (sand)
- Each fiber is a very tiny glass rod
- Excellent heat resistance; incombustible
- Thermal insulator
- Tensile loss above 600°F (315°C)
- Softens at approximately 1500°F (815°C)
- Cannot absorb moisture; won't rot or decay
- Resists acids, oils, many solvents, weather, and corrosive vapors
- Nonconductor of electricity
- Readily available at moderate prices
- Difficult to process because of its brittleness and high wear characteristics

Graphite fibers

- Rodlike fibers of various diameters and lengths
- Produced by special heat treatment of carbon fibers at approximately 5072°F (2800°C)
- Excellent heat and chemical resistance
- Versatile; popularity in the gasketing industry continues to grow

Mineral and rock fibers

- Fibers are amorphous and highly homogeneous because of their metamorphosis from volcanic rock
- Do not burn; chemically inert
- Resistant to high temperatures [above 930°F (500°C)]

Nylon

- Man-made polyamide with good tensile strength
- Good ultimate elongation properties (it can be stretched to a very high degree without breaking)
- Upper temperature limit of about 250°F (120°C)
- Good wet strength
- Resists common solvents, fuels, oils, and greases
- Not recommended against strong alkalies and acids; oxidizing agents, phenol, formic acid

7.1.1.3 Other Materials

Cork

- From the bark of the *Quercus suber* tree, growing mainly in Portugal
- Very light, elastic
- Good insulator to heat, sound, vibration, and electricity
- Free from toxic ingredients; does not impart odor or taste to the products it encloses

- Poor upper temperature limit, about 225°F (110°C)
- High variability in texture
- Limited chemical resistance
- Most often sold as an ingredient with glycerine-saturated vegetable fiber

Graphite

- One of the two crystalline forms of the element carbon (the other is diamond) (human-made carbon, graphite, and diamonds also exist)
- Natural lubricity
- One of the most stable and chemically resistant materials in the world
- Does not melt, but sublimes: changes from the solid to the gaseous state (without going through the liquid state) at temperatures over 5400°F (2980°C); in the presence of oxygen oxidation, above 850°F (450°C)
- Excellent conductor of heat and electricity
- Readily available in various forms; moderately priced to very expensive

Polytetrafluoroethylene (PTFE)

- Most commonly called Teflon® E.I. duPont de Nemours and Company, a DuPont trade name
- Man-made thermoplastic introduced in the late 1940s or early 1950s
- Extremely good chemical inertness and resistance to a wide range of fluids; only a few, very rare fluids will attack it
- Withstands a wide temperature range, from −450°F (−268°C) to +500°F (+260°C)
- A very low static coefficient of friction; very slippery
- Poor heat conductor; good heat insulator
- High coefficient of thermal expansion: swells significantly under heat
- Flows (creeps) under relatively low loads even at room temperature (usually stated as "cold flow")
- Can be blended readily with other materials to improve its features and performance characteristics
- Readily available at moderately high prices

Blocking agents and lubricants As the name suggests, blocking agents are used to fill the internal voids that exist in a braided packing so that fluid cannot wick through the braid. Many blocking agents also function as a lubricant.

PTFE dispersion coatings PTFE dispersed in a water carrier has been a most valuable asset to the packing industry. After yarns or braids are dipped in the coating, the water carrier is evaporated as the yarn or braid is heated in a drying oven. The solid PTFE, particles that have been deposited throughout the packing become a lubricant that possesses all of the features of solid PTFE, such as excellent chemical inertness, a low coefficient of friction, and self-lubricating properties. In addition, the PTFE lubricant will greatly reduce the harshness and abrasive nature of fibers that contact the shaft or sleeve. The main advantage of using PTFE as a blocking agent is that it will remain within the braid even when subjected to temperature and pressure. PTFE dispersion coating may also contain graphite powder, which will increase the thermal conductivity of the coating and allow the packing to withstand higher surface speeds without charring.

Many other types of greases, waxes, and oils can be used as lower-temperature blocking agents and lubricants. These types of coatings are less costly than PTFE dispersion coatings, but they are more easily driven out of the packing by pressure and temperature. They can also be dissolved by strong solvents or broken down by certain chemicals.

Solid lubricants There are a few types of solid lubricants that can be added to a packing material to decrease the friction the packing creates on the surface that is moving against it.

Graphite Powdered graphite or flake graphite are both dry, solid lubricants that are very common in packing materials. They can be combined with other materials in a coating, or they may simply be applied directly to the surface of a braid. Graphite retains its properties as a solid lubricant, even at very high temperatures.

Molybdenum disulfide Molybdenum disulfide powder is another type of solid lubricant. It is often mistaken for graphite because of its similar appearance and behavior.

7.1.1.4 Metals and Alloys Temperature limits for common metals are given in Table 7-1.

Aluminum

- Excellent corrosion resistance to fresh and salt water
- Maximum recommended operating temperature is 800°F (472°C)

Brass

- Copper alloy typically used with nonoxidizing acids and alkaline and neutral salt solutions
- Maximum recommended operating temperature of 500°F (260°C)

TABLE 7-1 Temperature Limits for Common Metals

Material	Minimum °F	Minimum °C	Maximum °F	Maximum °C	Abbrev.
304 stainless steel	−320	−195	1400	760	304
316L stainless steel	−150	−100	1400	760	316L
317L stainless steel	—	—	—	—	317L
321 stainless steel	−320	−195	1400	760	321
347 stainless steel	−320	−195	1700	925	347
Carbon steel	−40	−40	1000	540	CRS
20Cb-3 (Allloy 20)	−300	−185	1400	760	A-20
Hastelloy B 2	−300	−185	2000	1090	HAST. B
Hastelloy C 276	−300	−185	2000	1090	HAST. C
Incoloy 800	−150	−100	1600	870	IN 800
Inconel 600	−150	−100	2000	1090	INC 600

Carbon steel

- Most common metal for double-jacketed gaskets
- Poor resistance to corrosion
- Not recommended for water or diluted acids
- Maximum recommended operating temperature of 1000°F (540°C)

Copper

- Used successfully in acetic acids, nitrates, and many organic chemicals
- Maximum recommended operating temperature of 600°F (316°C)

Hastelloy® Trademark of Haynes International B

- Corrosion-resistant alloy
- Resists hydrochloric acid, phosphoric acid, other halogen acids, and reducing conditions
- Maximum recommended operating temperature of 2000°F (1090°C)

Hastelloy C

- Alloy with exceptional resistance to severe oxidizing conditions in nitric acid, free chlorine, and strong aqueous and acid solutions
- Maximum recommended operating temperature is 2000°F (1090°C)

Inconel® Tradename of International Nickel

- Withstands high temperatures; maximum operating temperature of 2000°F (1090°C)
- Excellent resistance to corrosion by halogen gases and compounds

Monel® Trademark of Special Metal Company

- Excellent resistance to most acids and alkalies, except extremely oxidant acids
- Commonly used in hydrofluoric acid applications
- Maximum recommended operating temperature of 1500°F (820°C)

Nickel

- Resists caustic media and corrosion from neutral and distilled water
- Maximum recommended operating temperature of 1400°F (760°C)

Stainless steel, Type 304

- Widely used for industrial gasketing
- Excellent corrosion resistance
- Maximum recommended operating temperature of 1400°F (760°C)

Stainless steel, Type 316L

- Greater corrosion resistance than 304 SS due to added molybendum
- Maximum recommended operating temperature of 1400°F (760°C)

Stainless steel, Type 321

- Similar to 304 SS, but has titanium added
- Widely used in high-temperature corrosive applications

- Maximum recommended operating temperature of 1400°F (760°C)

Stainless steel, Type 347

- Similar to 304 SS, but has columbium and titanium added
- Good performance in high-temperature corrosive applications, to 1600°F (870°C)

Stainless steel, Type 410

- Martensitic stainless steel
- Heat-treatable, 12% chromium steel
- Good corrosion resistance; high strength
- Maximum recommended operating temperature of 1200°F (650°C)

Titanium

- Good resistance to wet chlorine and chlorine dioxide
- Not suitable for dry chlorine
- Maximum recommended operating temperature of 2000°F (1090°C)

7.1.1.5 Fillers Fillers are materials added to another material either to improve quality or to reduce cost (and sometimes both). Unfortunately, the term *filler* has often taken on the connotation of something used to cheapen products. As will be shown, fillers can be very important ingredients for improving physical properties and chemical resistance, resulting in more efficient and effective sealing constructions. Fillers can take the form of powders or fibers; combinations of fibers and powders are also used as fillers. The following filler materials are listed alphabetically. Temperature limits for filler materials are given in Table 7-2.

Aluminum silicate

- Natural; mined from the earth
- Available in filler or fibrous forms
- Upper temperature range of 1000°F (540°C) to 1200°F (650°C)
- Excellent fluid resistance against water, steam, organic chemicals, fuels, oils, and solvents
- Excellent thermal insulating properties
- Not recommended against strong alkalies, some strong mineral acids, and inorganic fluorides
- Used in rubber products, compressed gasketing, and packing

Barium sulfate

- Commonly called *barytes*
- Natural and mined from the earth or can be manufactured synthetically
- Upper temperature limit of approximately 1500°F (820°C)
- Good resistance against almost all fluids, except for concentrated sulfuric acid
- Used in rubber products, compressed gasketing, and Gylon Style 3510

Calcium carbonate

- Natural and mined from the earth
- Upper temperature limit of approximately 1200°F (850°C)
- Generally good fluid resistance; not recommended for either dilute or strong acids
- Used in rubber products

Carbon black

- Same characteristics as carbon/graphite
- Not recommended for temperatures over 850°F (450°C) in an oxidizing atmosphere or for strong chemical oxidizers
- Used in some rubber products and some compressed gasketing

Ceramic

- Fibrous, aluminosilicate filler material
- Generally used for high-temperature, low-pressure applications, where sealability is not a primary concern
- Does not interlock and densify under compressive stress, unlike other nonasbestos materials
- Used in spiral-wound gaskets; valued for high-temperature capability to 2000°F (1090°C)

Verdicarb-mica-graphite

- First alternative to asbestos in spiral-wound gaskets, developed because of market demands.
- Utility-grade material comprised of mica, graphite, fillers, and an elastomer binder
- Initially rated to 1000°F (540°C), now downrated to 500 to 600°F (260 to 315°C)
- Not successful in the demanding API 607 4th edition fire test

Flexible graphite

- Used in spiral-wound gaskets

TABLE 7-2 Temperature Limits for Filler Materials

	Minimum		Maximum		
Material	°F	°C	°F	°C	Abbrev.
Ceramic	−350	−212	2000	1090	CER
Flexible graphite	−350	−212	950	510	F.G.
PTFE	−400	−240	500	260	PTFE
Verdicarb (mica graphite)	−350	−212	550	288	VC

- Excellent chemical resistance and fire safety
- Temperature capability to 950°F (510°C), depending on grade of flexible graphite and application
- Front runner for a broad range of applications
- Material makeup (such as leachable halogen content) can be controlled; ideal for nuclear applications

PTFE

- Widely used in Gylon and spiral-wound gaskets
- High chemical resistance; low permeability
- Especially successful in hydrocarbon alkylation units in refineries and food-related industries
- Temperature limitation with this material; not fire safe

Silica

- Natural material (sand)
- Many different structures, all with similar properties
- Upper temperature limit of approximately 1500°F (815°C)
- Excellent fluid resistance
- Recommended for most fluids except inorganic fluorides (e.g., hydrofluoric acid) and strong caustics
- Excellent reinforcing filler for black or white rubber products, such as nitriles and SBRs

7.1.2 Gasket and Seal Construction

In this section we outline the features, advantages, and disadvantages of the most commonly used materials found in the most popular metallic and nonmetallic gasketing constructions. It is important to know the properties of the individual constituents in order to have a better understanding of gasket and seal construction. Knowledge of the features, advantages, and disadvantages of individual materials is critical for understanding as well as evaluating various constructions. In this section we show that in many instances, shortcomings in individual materials can be overcome through combination with other materials. The best features of each material can be combined to further improve the final product. Only the most important nonmetallic gasketing constructions are reviewed, with particular emphasis on those sold by Garlock. Products are listed in order of current popularity among industrial maintenance users. Table 7-3 provide data on gasket thicknesses and tolerances.

7.1.2.1 Sheet Gasketing Garlock introduced the first compressed asbestos-replacement gasketing product in 1980. The first material was constructed of aramid fibers, silicate reinforcing agents, and a nitrile binder. The tensile strength was 2000 psi (13.8 MPa) and the compressibility was 7 to 17%. In the manufacture of compressed sheet, various types of fibers are mixed with different types of elastomers and fillers in mixers to form a viscous dough with the introduction of solvent. From the mixers, the dough is fed into special two-roll calenders called *sheeters*. The larger of the two rolls is steam-heated and the smaller is water-cooled. The dough adheres to the hot roll in the form of a thin film. The smaller roll presses against the film being built up on the larger roll; with each revolution of the rolls they are separated a little farther, permitting a continuous buildup of sheet on the hot roll. When a desired thickness is obtained, the sheeter is stopped and

TABLE 7-3 Available Thicknesses and Tolerances (in.)

		Width Limits		
Gasket Thickness	Tolerance	Min.	Max.	Compressed Thickness
0.125	±0.005	$\frac{3}{16}$	1	0.090–0.100
0.175	±0.005	$\frac{1}{4}$	$1\frac{1}{2}$	0.125–0.135
0.250	±0.005	$\frac{5}{16}$	$1\frac{1}{2}$	0.180–0.200
0.285	±0.005	$\frac{5}{16}$	$1\frac{1}{2}$	0.200–0.220

the sheet removed. The constant pressing of the smaller roll against the film being built up on the larger roll causes the film's fibers to orient (position) themselves lengthwise in the direction of roll rotation. Therefore, compressed gasketing always has greater strength in the machine direction (MD). Using the principle that "a chain is as strong as its weakest link," one should only be interested in the minimum tensile strength of the sheet, which is across the machine direction (transverse). The size of the sheet coming off a sheeter is dependent on both the width of the rolls and the circumference of the larger roll.

Compressed gasketing is the single most used gasketing construction in the world. The reasons for its great popularity are:

- Wide range of application and service capabilities
- Relatively low price
- Availability
- Sheet sizes allow the use of one-piece gaskets for large applications
- Ease of cutting into gaskets

7.1.2.2 Polytetrafluoroethylene Garlock manufactures a family of flat polytetrafluroethylene (PTFE) gasketing materials under the trade name Gylon®. Gylon gasketing is manufactured using a proprietary process that imparts unique physical properties, which are not obtainable through conventional manufacturing methods:

- Uniform crystalline structure throughout the sheet to ensure consistent performance
- Uniform filler content throughout the sheet to ensure consistent performance
- Creep and cold flow substantially reduced; uncontrollable in normal PTFE manufacturing
- Equal strength in all directions to ensure consistent performance
- Sheet products without "skive" marks to eliminate high leak rates
- Patented welding process to fabricate gaskets of virtually any size

7.1.2.3 Flexible Graphite Graph-Lock® flexible graphite gasketing is a distinctive flat gasketing material composed of pure exfoliated flake material. Graph-Lock excels in extreme conditions and withstands high temperatures, high pressures, and aggressive chemicals. Flexible graphite is manufactured in such a manner that no organic binders and fillers are introduced (note that very small amounts of naturally occurring inorganic components may be present, depending on the grade of the graphite). The end product is essentially graphite with outstanding physical properties:

- Wide fluid resistance in nonoxidizing media
- Leachable chloride content of homogeneous graphite is less than 100 ppm
- Temperature resistance from cryogenic to over 5400°F (2980°C) in inert or reducing atmospheres, and up to approximately 850°F (455°C) in the atmosphere
- Excellent compressibility and conformability
- Very low creep relaxation

Flexible graphite sheet requires careful handling because, compared with other gasketing materials, it is fairly fragile. The insertion of wire mesh, stainless steel foil, and tanged metal inserts increases strength and ease of handling. This gasketing is very expensive relative to other constructions, however, as is the case in most applications, the price of the gasket is minor compared to the costs incurred by premature failures.

7.1.2.4 Elastomers or Rubbers Rubber sheet is used not only for fabricating gaskets but also for many other applications. The largest market for rubber sheet is in North America. The reasons for its popularity include:

- Wide range of application and service capabilities
- Relatively low price (for most elastomers)
- Ease of cutting into final parts

The most common forms of rubber gasketing and gaskets are:

- Molded to specified shapes (such as the Stress Saver)
- Extruded and spliced
- Lathe cut from sleeves
- Flat die cut

This overview covers flat gasketing and gaskets with or without cloth (fabric) reinforcing. In the manufacture of rubber sheet, various types of basic elastomers are mixed with other ingredients, depending on the final physical properties desired. Other ingredients may include nonblack fillers, curing agents, plastizers, carbon black, and antioxidants—frequently, more than one basic elastomer is used in the final construction. The mixing process and ingredients are at the heart of the manufacture of all rubber goods. After mixing, the uncured material in the form of very roughly formed slabs is ready to make into cured flat sheet of varying widths, thicknesses, surface finishes, and insertion materials. Cloth inserted (C.I.) sheet includes some sort of woven insert, whereas homogeneous sheet has none.

7.1.2.4.1 Curing Processes Garlock accomplishes the curing process by one of three methods:

1. *Flat press cure*. In flat press curing, the calendared sheet is plied to the required thickness and the total slab is inserted into the flat press between side irons, which confine the material and prevent it from flowing over the edges. The two platens are brought together and closed, applying pressure and heat to the uncured material. The length of slab being cured is usually limited by the length of the press. It is possible to cure greater lengths by a method called *step curing*.

2. *Drum cure*. In the drum curing method, the calendered sheet is simply wrapped on a steel drum or mandrel with a fabric liner, which separates the layers of uncured compound so that it does not bond to itself. The material is then inserted in a steam or hot-air vulcanizer and allowed to cure for the necessary period of time. After the material has been cured properly, it is removed from the vulcanizer and unwrapped. The separator or liner (typically, a cotton or Nylon fabric) used in wrapping the material leaves a woven fabric impression on the surface. Some manufacturers use paper, which leaves a smooth surface. This method of curing is considerably more economical than the flat press cure, but generally speaking, does not achieve the same tightness of cure as when pressure is applied to the material on a flat press.

3. *Rotocure*. In rotocure, material is plied up at the press and cured as it passes around the curing drum, between the drum and the endless steel band. Both heat and pressure are applied to the material as it passes through the curing cycle. There are no interruptions in the curing process as is the case with the flat press when longer lengths are needed; nor is there a fabric finish as with drum curing. Rotocure provides a high-gloss smooth finish with a tight cure through a continuous method of curing in long lengths and wide widths. Some rotary presses are equipped to cure up to 72 in. in width; others, 36 or 48 in.

7.1.2.4.2 Physical Properties Any review of constructions available for rubber products should be preceded by an overview of the physical properties. This is crucial for end-user purchasing procedures. All too frequently, users simply state a particular manufacturer's style number followed by the statement "or equal"; or they just give the generic elastomer name, hardness, and color. Both of these purchasing habits can lead to price cutting because suppliers have great latitude in the products they provide. Users can be assured of getting the exact product they want by:

- Specifying a given style along with the term "no equal"
- Listing in detail the physical properties desired
- Specifying the ASTM D-2000 line callout for the material desired

Rubber sheet should be specified on the basis of environmental resistance, consisting of:

- Fluid, temperature, and pressure requirements
- Other important qualities: abrasion, weather, impact, and vibration resistance

It is common practice to blend some elastomers to get the best combination of properties. Blending can also be used to reduce cost, but that can result in reduced environmental resistance. The relevant physical properties are described next.

1. *Hardness.* Hardness determines the ease with which rubber can be deflected. Two hardness scales are used, Shore A and Shore D. The A scale is determined with a blunt indenter, the D scale with a needle point. Soft and medium-hard elastomers are designated by the A scale with readings from 20 through 100. The D scale is graduated so that the needle point registers 100 on a smooth glass surface. A Shore D reading of 50 corresponds approximately to a Shore A reading of 100.

Hardness is usually specified in 10-point increments with a tolerance of >5 points—the higher the number, the harder the elastomer. A typical rubber band is 35 durometer (Shore A) and a rubber heel on a shoe is 80 durometer (Shore A). A bowling ball is approximately 80 to 90 durometer (Shore D). The Shore A instrument is not very accurate over 90 durometer; at this point the Shore D instrument is recommended.

2. *Tensile strength.* Tensile strength is a measure of resistance to rupture, normally expressed in pounds per square inch of cross section. This property indicates the overall character of a rubber. In gasketing, tensile strength is not usually a critical factor; however, in some cases it can be important. It cannot be used as an absolute measure of performance. As an example: rubber rated at 2000 psi (13.8 MPa) will not necessarily last twice as long as rubber rated at 1000 psi (6.9 MPa).

3. *Elongation or stretch.* Elongation is the measure of the ability of rubber to stretch without breaking. More frequently, the term *ultimate elongation* is used, since

this value is expressed as a percentage of original length at the moment of rupture. This value is important when the end-use form must be stretched, then returned to its original shape.

4. *Modulus.* Modulus is a measure of resistance to stretch; high modulus means high tension to stretch, low modulus, the opposite. Natural rubber and neoprene are the only two elastomers that develop low modulus along with high strength without the addition of reinforcing fillers.

5. *Compression set.* Compression set (memory) measures the lack of recovery of rubber after it has been deformed for a period of time at a given temperature, then released. Low compression set is an important property for any gasketing material, particularly where dynamic applications are involved, such as refrigerator door and oven door and hatch cover gaskets, which are intermittently under compression. A gasket that readily takes a permanent set is not preferred.

6. *Resilience.* Resilience is the ability of a material to return to its original shape after it has been deformed. This property is usually applied to rubber materials.

7. *Specific gravity.* Specific gravity is the ratio of the weight of a given substance to the weight of an equal volume of water. Most elastomers have a specific gravity greater than 1.00 (i.e., heavier than water). The range for elastomers normally used in rubber gasketing is from 0.95 to 2.00. Specific gravity plays an important role in rubber sheet because of the economics of selling sheet "by the pound."

7.1.2.5 *Inflatable Seals* The most effective products for sealing between surfaces that move in relation to each other are Cefil'Air® pneumatic seals. Cefil'Air seals can hold, position, or handle objects in a wide range of applications. Cefil'Air seals are homogeneous elastomeric seals with a high modulus of elasticity and considerable tensile strength. The seals are designed to be fitted into grooves and are restricted to low pressures to prevent bursting. They expand and retract with the pressurization and deflation of the seal within the groove. The exact groove and gap dimensions are critical in designing and producing the correct seal for any application.

As a result of its patented design, modern manufacturing techniques, and the most advanced elastomers, Cefil'Air seals can be used in a multitude of sealing, handling, and holding applications. They withstand temperatures from -148°F (-100°C) to $+482^\circ$F ($+250^\circ$C) and pressures from 7 to 150 psig (0.5 to 10.4 bar) in a variety of liquid or gaseous media.

7.1.2.5.1 Profiles

Cefil'Air HP (high-pressure) seals

- Must be captive in slots or grooves within specified dimensions (groove mount)
- Never pressurize or inflate a seal when any one face of the groove is open

Cefil'Air LP (low-pressure) seals

- Secured by their base; work freely outside the confines of a groove (foot mount)
- Maximum pressure cannot be applied until the contact face (grooved/toothed side) is against the surface to be sealed

7.1.2.5.2 Types of Expansion

Axial

- Movement of seal, when pressurized, is perpendicular to the groove and seal plane
- Example: a door seal, where the flat surface of the door, parallel to the bottom of the seal groove, is contacted by the Cefil'Air seal when inflated.

Internal radial

- Movement of seal, when pressurized, is inward with respect to seal placement in the internal groove
- Example: a seal located at the farthest point of the housing or stuffing box of a ship's drive shaft at the exit from the hull. If packing or a mechanical seal located in front of a Cefil'Air seal must be replaced, the seal can be inflated to prevent water from entering the hull.

External radial

- Movement of the seal, when pressurized, is outward with respect to seal placement in an internal groove
- Example: a water- or airtight door, where a Cefil'Air seal is located at the outer edge of the door. The door recesses into the casing when closed; when the seal is inflated, it expands outward to seal against the door casing.

7.1.2.6 *Vegetable Fiber and Blends* Vegetable fiber was one of the first types of gasketing constructions, where processed plant fibers are formed into a homogeneous sheet employing papermaking techniques and glueglycerine saturation. It has limited fluid, temperature, and pressure capabilities; however, it is very economical.

- *Cork–vegetable fiber.* Cork is added to give vegetable fiber more resilience.
- *Cork–vegetable fiber–rubber.* Nitrile rubber added to cork–vegetable fiber increases strength and improves "sealability."

7.1.2.7 *Metal Gaskets*

7.1.2.7.1 Spiral-Wound Gaskets Spiral-wound gaskets are made with an alternating combination of formed metal strip and soft filler materials to form an effective seal when compressed between two flanges. A V-shaped crown centered in the metal strip acts as a spring, giving gaskets greater resiliency under varying conditions. Filler and strip material can be changed to accommodate different chemical compatibility requirements. If the load available to compress a gasket is limited, gasket construction and dimensions can be altered to provide an effective seal. A spiral-wound gasket may include a centering ring, an inner ring, or both. The centering ring centers the gasket within the flange and, to a certain extent, acts as a compression limiter, while the inner ring provides additional radial strength to address concerns of radial buckling. The inner ring also reduces flange erosion and protects the sealing element. Resiliency and strength make spiral-wound gaskets an ideal choice under a variety of conditions and applications. Used widely throughout refineries and chemical processing plants, spiral-wound gaskets are also effective for power generation, aerospace, and a variety of valve and specialty applications.

In response to radial buckling concerns, in 2008 the ASME B16.20 standard was changed to read that all graphite-filled spiral-wound gaskets will be furnished with inner rings unless otherwise specified by the purchaser. To receive a spiral-wound gasket without an inner ring, purchasing must specifically state that no inner ring is required when placing an order. However, even if the gaskets are purchased without the inner ring, the gasket will be stamped in compliance to B16.20. This change affects only graphite-filled spiral wounds, since the inner rings were already required for PTFE-filled spirals.

7.1.2.7.2 The Flexseal® Edge Gasket The Garlock Edge gasket provides improved performance over conventional spiral-wound gaskets; its benefits include:

- Stabl-Lock™ inner wrap construction, a Garlock innovation, prevents the sealing element from flowing (radial buckling) into, and contaminating, the process stream.
- Controlled Density winding, another Garlock innovation, creates a more consistent product, helps prevent bolt torque loss, and requires less compressive force to seal.
- Centering ring with relief ports: allows the gasket to expand outward instead of inward toward the process stream
- Dual flange design option with centering ring: accommodates 150- and 300-lb flanges, reducing inventory costs.

7.1.2.7.3 Graphonic™, Tephonic™, and G.E.T. Gaskets The superior technology of this family of gaskets ensures excellent sealing performance and reliability, even in the most difficult applications. Each of the three styles combines a corrugated metal core with a compressible sealing element of various materials, for resistance to a wide range of harsh conditions, including extreme temperature, corrosive chemicals, and thermal cycling.

- *Graphonic gasket*: corrugated metal core with a flexible graphite sealing element
- *Tephonic gasket*: corrugated metal core with an expanded PTFE (ePTFE) sealing element
- *G.E.T. gasket*: corrugated metal core with a flexible graphite sealing element at the onter diameter (OD) and an ePTFE sealing element at the inner diameter (ID).

7.1.2.7.4 Kammprofile Gasket A high-performance gasket, the Kammprofile gasket, is constructed of a metal core with machined serrations in the sealing area, with graphite laminated to both sides. As a result, the Kammprofile seals fairly easily, is self-centering, resists cold flow, has high-pressure resistance, and withstands extreme temperature and/or pressure cycles. The graphite facing is soft enough to handle mild sealing surface imperfections. The gasket performs well in flanges with high compressive load but less-than-perfect surfaces.

7.1.2.7.5 Double-Jacketed Gaskets Double-jacketed gaskets consist of a metal ring turned up on the ID and OD, with filler material layered in the central area and a flat strip of metal placed on top. The inner and outer edges are then bent over the flat metal, totally encapsulating the filler. Double-jacketed gaskets are the most popular clad gaskets, offering good compressibility and resilience. They are most commonly used in heat exchanger and coker applications. Heat-exchanger gaskets are manufactured in a wide range of sizes and rib configurations. They are used for spiral-wound gaskets not otherwise specified.

7.1.2.8 *Helicoflex® Seals*

7.1.2.8.1 Metal Seals For successful sealing, metal seals should be constructed from a material softer than that of the flange surfaces. The Helicoflex spring-energized seal has an outer jacket that is more ductile (softer) than the mating surfaces. The compression on this seal is usually set with a predetermined groove depth, but seals with limiting rings can be used as well. In the case of a metal O-ring or C-seal, plating and coatings (i.e., silver, PTFE, nickel, etc.) are used to provide this softer lining, which will conform to the irregularities of the flanges. After proper compression, a specific pressure is created by the spring in the Helicoflex spring-energized seals. This pressure maintains the integrity of the seal during normal service conditions. O-rings and C-seals rely on the properties of the base material and internal service pressure to provide this specific pressure.

Helicoflex seals have several advantages over metal O-rings:

- Greater resiliency, due to specific pressure created by the spring
- Greater springback; reacts better to thermal and pressure cycling
- Can be made with an outer jacket of almost any engineering alloy
- Load control in spring design
- Wide choice of nearly any cross-sectional design, geometric shapes, and materials

The most popular seal cross sections account for roughly 80% of applications: plain round cross sections (HN200) and the limiter cross sections (HN208a). Helicoflex seals most often replace spiral-wounds and double-jacketed seals in heat exchangers, reactors, and agitators. Replacements were required because of very high temperatures or significant thermal cycling of the vessel.

Operating Parameters

- *Temperature range*: cryogenic to 1800°F (980°C)
- *Pressure range*: ultrahigh vacuum to 50,000 psi (3450 bar)
- *Media*: strong oxidizers, nitric acid, H_2S and NO_x gases, fluorine services, acetic acids, benzene, paraxylene, methane, phosgene, plant refrigeration systems (NH_4, liquid H_2, liquid N_2)

Applications Helicoflex seals should be specified wherever spiral-wound gaskets or double-jacketed gaskets are failing. For new equipment, Helicoflex seals can be recommended for services above 800°F (430°C) or where strong oxidizers are involved.

7.1.2.8.2 Metal O-Rings The sealing concept for metal O-rings is based on the elastic deformation of the hollow metal tube. When the tubular cross section is compressed, a specific pressure is created at the contact areas of the seal. The pressure on the seal is created by the rigidity of the circular cross section and the wall thickness. The pressure is great enough to force the tubing itself or a softer plating/coating into the voids of the flange faces. Garlock Helicoflex manufactures metallic O-rings mainly from three base materials:

- Stainless steel 321
- Inconel 600
- Inconel® Trademark of International Nickel X-750

and in three styles:

- Nonpressurized or plain
- Self-energized or vented
- Pressure-filled

Self-energizing metal O-rings have holes drilled in the pressure side of the ring to allow the pressure of the process fluid to force the faces of the seal against the flanges. Pressure-filling the metal O-ring allows the seal to keep its resiliency (recovery) even when temperatures are greatly increased or severely cycled. Garlock Helicoflex also has three types of plating or coating that can be applied to the O-rings to improve the sealing levels: nickel, silver, and PTFE.

Metal O-rings are typically used to replace elastomer O-rings when spiral-wounds cannot be used. Metal O-rings have four main advantages over spiral-wounds:

- Greater overall springback
- Less force required to compress
- Will not contaminate process flows
- Can be formed to nearly any contiguous geometric shape

Operating Parameters

- *Temperature range*: cryogenic to 1400°F (760°C)
- *Pressure range*: strong vacuum to 20,000 psi (1380 bar)
- *Media*: any molten plastic, hot liquids, steam, heat-transfer fluids

Applications Metal O-rings are used primarily for:

- Melt-stream plastics: plastic production at refineries, in extruders, and in injection molders
- Replacement of more expensive elastomer O-rings (e.g., Kalrez® Registered trademark of E.I. DuPont DeNemours Co.)
- Body seals in valves and pumps
- Porous metal filters; screen changers and filter packs (tubular and flat)

- Any O-ring groove where temperatures have been elevated over 400°F (205°C); a common occurrence in plastics industries; new polymers require higher melting temperatures and cannot use high-grade elastomer compounds as seals

7.1.2.8.3 Metal C-Seals Sealing with metal C-seals depends on the elastic deformation of Inconel or other alloy material with a C-shaped cross section. When the C cross section is compressed, a specific pressure is created at the contact areas of the seal. The pressure on the seal is created by the rigidity of the C cross section, the material thickness, and the system pressure that energizes the seal. The specific pressure is great enough to force the base material or a softer plating or coating into the voids of the flange faces.

Technetics Helicoflex manufactures metallic C-seals from one base material: Inconel X-750. Other materials are available, such as the 300SS series and the Hastelloys, but the strength and resiliency of Inconel X-750 make it the most feasible choice. Inconel C-seals are put through a heat treatment cycle that gives the seals much better springback (resiliency) characteristics. In some cases, the heat treatment gives better chemical compatibility as well. Available in-house plating and coatings include nickel, silver, and PTFE. Other outside plating and coatings are available through outside vendors.

Metal C-seals are used to replace elastomer O-rings when spiral-wounds and other conventional gaskets cannot be used. Metal C-seals have these advantages over spiral-wound gaskets:

- Greater overall springback
- Less force required to compress
- Will not contaminate process flows

Operating Parameters

- *Temperature range*: cryogenic to 1800°F (980°C)
- *Pressure range*: ultrahigh vacuum to 50,000 psi (3450 bar)
- *Media*: strong oxidizers, corrosive environments, radioactive conditions, any other environment compatible with the base material or plating and coating options

Applications C-seal seals should be recommended wherever spiral-wound gaskets or double-jacketed and/or other conventional gaskets are failing. The plastics and aerospace industries are the two main users of the C-seal.

7.1.2.8.4 Design and Ordering Before designing any Technetics Helicoflex product for a customer, a number of details must be determined: application data such as pressure, temperature, and metal seal. The customer must provide the target leak rates, since no connection system is completely leak free! See the following data on leak rates:

1×10^{-4} atm cm^3/s He = 1 cm^3/3 h	traditional seals
1×10^{-5} atm cm^3/s He = 1 cm^3/24 h	traditional seals
1×10^{-6} atm cm^3/s He = 1 cm^3/2 weeks	traditional seals
1×10^{-7} atm cm^3/s He = 3 cm^3/yr	traditional seals
1×10^{-8} atm cm^3/s He = 1 cm^3/3 yr	Helicoflex O-rings
1×10^{-9} atm cm^3/s He = 1 cm^3/30 yr	Helicoflex C-seals
1×10^{-10} atm cm^3/s He = 1 cm^3/300 yr	Helicoflex O-rings
1×10^{-11} atm cm^3/s He = 1 cm^3/3000 yr	Helicoflex

Note: When dealing with a completely new design, gather all relevant information so that engineers can select the best seal for the application. Note that metal seals are recommended for one-time use only!

7.1.3 Principles of Gasket Operation

7.1.3.1 Gasketed Joints The gasketed joint has three essential parts:

1. The gasket
2. The flange face surfaces
3. A means of clamping the gasket between the sealing surfaces to create and maintain a sufficient seal

When designing gasketed joints or problem solving, you must consider and evaluate these three parts individually before determining correct solutions. The equipment attached to the gasketed joint is also important. Consider standard piping as an example; you must address the following conditions:

- Is sufficient bolt load available?
- Is the piping properly supported so that the sealing surfaces remain parallel? This is usually accomplished with pipe hangers.
- Is axial movement and/or vibration of the piping under control? This is usually done with expansion joints and/or piping "loops."

7.1.3.2 Gaskets The basic function of a gasket is to create and maintain a positive seal between two relatively stationary parts. The gasket must do a number of different jobs to function properly:

- Create an initial seal
- Maintain the seal over a desired length of time
- Be easily replaced

The degree of success depends on how effectively the gasket does the following:

- Remains impervious to the fluid in the system

- Chemically resists the system fluid to prevent serious impairment of its physical properties
- Deforms enough to flow into the imperfections on the gasket seating surfaces to provide intimate contact between the gasket and the seating surfaces
- Withstands system temperatures without serious impairment of its physical properties
- Is resilient enough to maintain an adequate portion of the applied load when joint movements are not fully eliminated by equipment design
- Has sufficient strength to resist crushing under the applied load and to resist blowout under system pressure
- Avoids contaminating the system fluid
- Does not promote corrosion of the gasket sealing surfaces
- Is easily and cleanly removable at the time of replacement

7.1.3.3 Seating Surfaces Gaskets are required because sealing surfaces contain imperfections. If these surfaces were perfectly smooth, flat, parallel, and kept in intimate contact, gaskets would not be needed. But the cost for such a finish would be unacceptable. A properly functioning gasket in a well-designed joint serves the purpose at relatively low cost.

7.1.3.4 Clamping the Gasket Between Seating Surfaces One of the main functions of a flange assembly is to clamp and hold the gasket in place. How the gasket is clamped is also a key factor in proper performance. All gaskets require a certain amount of force or compression to deform to the seating surface and create a seal. Since the force to seal a gasket is directly related to the contact area of the gasket and the available bolt load, it is extremely important that the load be evenly distributed over the entire contact area. The most accurate and economical way to do this is to use a calibrated torque wrench. In addition to using a torque wrench, the force should be applied in numerous steps, in a crossing pattern from one side of the flange to the other to prevent the assembly from tipping. If the flange tips, it can result in higher compression on one side of the gasket and little or no load on the opposing side.

Surface finish is also very important. Nonmetallic gaskets rely heavily on friction between the gasket and flange faces to contain the internal system pressure. Smoother surfaces can be sealed more easily, but the gasket will have a greater tendency to slide or blow out. Rougher surfaces are just the opposite. The rough surfaces will "bite" into the gasket and improve blow out resistance, but if the surface is too rough, the gasket will not conform to the face and affect a seal. Typically, the ideal surface is 125 to 500 μin. RMS with 30 to 55 serrations per inch created with a $\frac{1}{16}$-in.-radius-tipped tool (per ASME B16.5).

7.1.4 Gasket and Metal Seal Applications

7.1.4.1 Flanged Joints in Piping Systems Piping systems are important production arteries in our industrial economy, ranking as the nation's fifth carrier, behind rail, motor transport, air, and water. Piping systems vary widely from a simple gravity feed line for unloading a tankcar to the intricate maze of piping used in chemical or petroleum processing plants.

Selecting the right gaskets and installing them correctly is critical for efficient handling of fluids and transferring heat in piping systems in:

- Power-generating stations
- Chemical processing plants
- Petroleum refining plants
- Papermaking plants
- Food-processing plants
- Primary metal plants
- Commercial and institutional buildings
- Underground transmission lines

7.1.4.1.1 Pipe Joining Methods Pipe joints, often using gaskets or seals, are necessary to:

- Connect pipe to pipe
- Connect pipe to equipment
- Change pipe direction
- Introduce branch lines
- Change pipe size
- Close off a line

Many methods are used for joining pipe or tubing to other lengths, as well as to valves, pumps, heat exchangers, pressure vessels, mixers, agitators, compressors, engines, generators, steam traps, and so on. Design engineers choose one method over another based on:

- Equipment materials of construction
- Operating conditions of pressure and temperature
- Installation and maintenance costs

7.1.4.1.2 Flanged Joints Flanged joints, or a combination of flanged joints and welded joints, are used in most process piping. Flanged joints require gaskets or other sealing devices. The five principal methods of attaching flanges to pipes are:

1. Threaded

2. Slip-on-weld
3. Welding-neck
4. Socket-weld
5. Lap joint

It is important to know how flanges are attached to the piping, as not all flanges are created equal (Fig. 7-2). For example, a "stub end" flange with a lightweight "slip-on" backer flange may have the same contact dimensions as a standard weld neck raised face flange, but the ability to handle higher torque and bending forces in the stub end flange is greatly diminished due to the thinner, lighter-weight design.

Raised face is the most popular facing for metallic or nonmetallic gaskets or seals. The gasket or seal is unconfined. The mating face is flat, but the portion inside the bolt holes is raised $\frac{1}{16}$ in. (1.6 mm) to $\frac{1}{4}$ in. (6.4 mm). The gasket or seal assembly is usually a ring type, which self-centers through its OD contacting the inside edge of the bolts. They may be disassembled easily without springing the flange. Steel flanges that are to mate with cast iron flanges should have flat faces and full-face gaskets to prevent cracking the cast iron during bolt tightening.

Flat face is the second most popular facing for nonmetallic gaskets. Mating faces of both flanges are flat. Depending on the flange construction, the gasket or seal assembly may be of ring type, entirely inside the bolts, or full face (with bolt holes), covering the entire face both inside and outside the bolts. Again, cast iron or other easily damaged flanges should use full-face gaskets.

Tongue-and-groove facing has some popularity, particularly where higher fluid pressures are involved. The gasket or seal assembly is fully confined. The depth of the groove is usually equal to or less than the height of the tongue; the groove is not normally over $\frac{1}{16}$ in. (1.6 mm) wider than the tongue. Gaskets are usually dimensioned the same as the tongue. This joint must be pried apart at disassembly.

A Helicoflex metallic seal, if used, may be located anywhere in the compressed area. Typically, a metal seal is placed in a machined groove or limiter plate, which holds the seal stationary.

7.1.4.2 ***Heat Exchangers*** Heat exchangers transfer thermal energy from one medium to another. Industry typically uses "two-fluid" exchangers. Both fluids can be part of the process stream. If only one process stream fluid is involved, the other fluid is usually steam for heating, or water or air for cooling.

Shell-and-tube heat exchangers are by far the most popular in industry. Figure 7-3 shows a typical shell-and-tube heat exchanger. The exchanger consists of a metal shell with tubes inside, which are typically welded in the end plate, called a *floating head*. Inlet and outlet fittings for the process stream fluid and the heating or cooling of the fluid and metal covers at each end of the floating head keep the fluid in the exchanger. Most frequently, the process stream fluid flows through the tubes and the heating or cooling fluid around the outside of the tubes in the shell. Gaskets or seals are required on the inlet and outlet flanges, which are joined to the piping system and between the shell, floating heads, and covers. The inlet and outlet flanges are of the popular types discussed previously and use ring or full-face gaskets or seal assemblies. The cover gaskets can be complex configurations, depending on the design of the exchanger. These gaskets can be very large, also depending on exchanger design, and could have outside diameters of over 100 in. (2.54 m). The gaskets that seal between the floating heads and the shell are typically ring gaskets.

Note: In most cases, both cover and shell gaskets are sealed with the same bolts. One of the most successful gaskets in heat-exchanger applications is the Graphonic gasket, since it handles the frequent thermal cycling common in this equipment. The oil refining, chemical processing, and power-generation industries are primary users of heat exchangers. Many hundreds of thousands of heat-exchanger gaskets are replaced every year.

7.1.4.3 ***Valves*** A valve is a device that stops, starts, or changes the direction or magnitude of fluid flow, or its pressure and temperature. The history of valves parallels that of piping. In the days of the Roman Empire they used wooden valves. With the advent of metal piping, basic valve designs were developed, such as the gate and

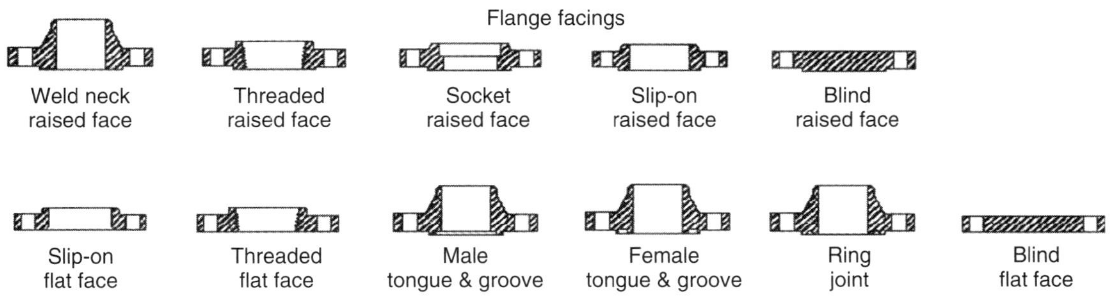

Figure 7-2 Flange facings. (Courtesy of Garlock, an EnPro Industries family of companies.)

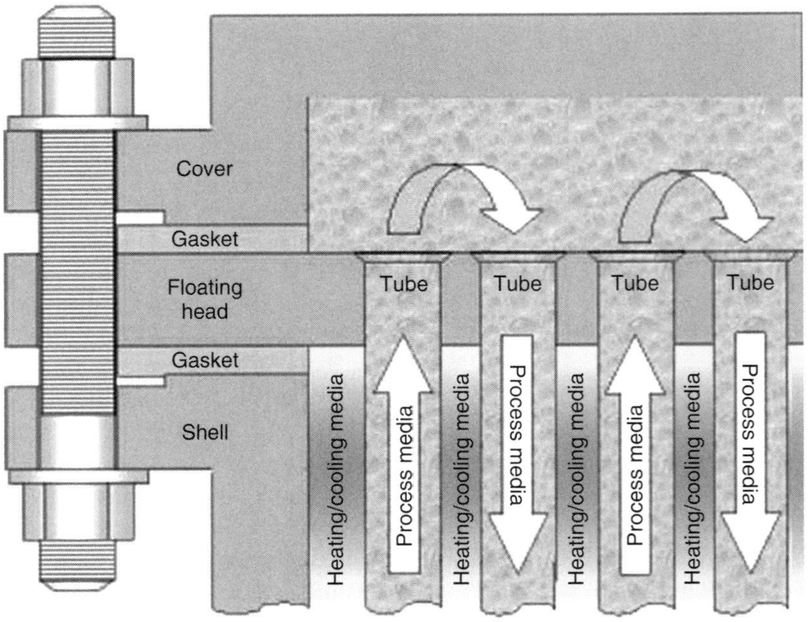

Figure 7-3 Shell-and-tube heat exchangers. (Courtesy of Garlock, an EnPro Industries family of companies.)

globe types shown below. In recent years, valve technology has expanded to keep up with the sophisticated needs of modern energy systems. We now have ball, butterfly, check, diaphragm, pinch clamp, plug, pressure regulator and safety relief valves in addition to gate and globe. There are literally billions of these valves in service, and many of them require two or three gaskets: one each for the inlet and outlet flanges and one for the bonnet (see Fig. 7-4). Valves with threaded fittings would use PTFE pipe thread tape instead of gaskets.

7.1.4.4 ***Valve Actuators*** Almost all valves must be physically operated by some sort of actuator: mechanical, electrical, hydraulic, or pneumatic. The handwheel is the simplest mechanical type; motors and solenoids are electrical; oil-operated cylinders are hydraulic; diaphragm types are pneumatic. We'll focus on the pneumatic type because of the millions of them in service today. They use fabric-inserted rubber, PTFE and Gylon diaphragms. These valve actuators are connected directly to the valve stem—they translate an air signal into valve stem motion. Every industrial plant has this type of valve actuator operating the valves. Some require molded fabric-inserted diaphragms (for longer strokes or movements) but most are cut from flat sheet. Gylon, nylon-inserted nitrile rubber, and nylon-inserted neoprene rubber are some of the major diaphragm constructions in use.

Figure 7-4 Valve. (Courtesy of Garlock, an EnPro Industries family of companies.)

7.1.4.5 ***Pumps*** Pumps are by far the most widely used mechanical devices in industry. Dating from the waterwheel of ancient history, pumps are among the oldest machines used by human beings for transferring energy from one form to another. The three major categories of pumps normally found in industry are centrifugal, rotary, and reciprocating. Gaskets are generally located at the inlet and outlet flanges which join the piping system to the pump.

The horizontally split case pump has a gasket between the casings, and the vertically split pump may have a flat gasket between the main housing and the impeller housing. The chemical processing, power-generation, food-processing, and pulp and paper industries require large quantities of gasketing materials for their pumps. For high-temperature and high-pressure conditions, as well as severe chemical service, Helicoflex metal seals are the best choice.

7.1.4.6 Mixers, Blenders, Reactors, and Agitators Over 100 manufacturers in the United States produce mixing equipment, used in a wide range of industries to blend gases, liquids, and/or solids. Most equipment designs have one common feature: the physical movement of an impeller or a fluid to move the material held in a vessel, tank, vat, or large container.

7.1.4.7 Equipment Requiring Metal Seals The fuel nozzles, cooling systems, and other high-temperature flange connections on turbines require metal seals. Garlock Helicoflex spring-energized seals and C-seals are commonly used in turbines in the power generation industry.

1. *Aerospace equipment.* Electronic instruments, valves, compressors, temperature sensors, and various duct systems can use metal seals. The ideal seal for applications with low seating loads are low-load Helicoflex spring-energized seals and metal C-rings.
2. *Injection molding equipment.* Equipment in the plastics industry includes injection molding equipment, hot runners, runner-less bushings, heated manifolds, filter packs, and other disk filtration systems. Due to temperatures and pressures in these applications, and the need for noncontaminating seals, Helicoflex seals prove superior to conventional gaskets.
3. *Internal combustion engines.* Helicoflex metal seals are often used in cylinder head gaskets in high-performance racing engines. The temperature and pressure cycling in the cylinders of such engines often causes conventional gaskets to leak. The extra resiliency of Garlock Helicoflex metal seals compensates for lift-off and ensures excellent sealing. Exhaust manifolds, which also exhibit high temperatures, use metal seals as well.
4. *Semiconductor process equipment.* Many machines in the semiconductor industry require metal seals, including valves, mass flow controllers, gas delivery systems, vacuum pumps, and special vacuum flanges. Elastomeric seals often cannot be used because of extreme vacuum conditions or high temperatures. Since they are impermeable, metal seals are ideal for vacuum applications.

7.1.4.8 Miscellaneous Equipment The equipment discussed above accounts for the majority of uses for metal and nonmetallic gaskets. Other types of equipment needing gaskets are listed below.

- Blowers
- Boilers
- Centrifuges
- Compressors
- Condensers
- Crystallizers
- Cylinders
- Dehydrators
- Dryers
- Fans
- Filters
- Gauges
- Gear reducers
- Pile drivers
- Steam turbines
- Strainers
- Tanks
- Transformers
- Vaporizers

7.1.5 Gasket Selection

What information is needed for proper gasket or seal selection? When this question was asked years ago of one of our longtime, experienced engineers, his answers came in the form of a "question checklist," comprising four typewritten, single-spaced pages of questions that, ideally, should be answered before any recommendation is given. This checklist is the basis for this section, with other design parameters and criteria added.

7.1.5.1 Conditions Acting on a Gasket Figure 7-5 shows the various forces acting on a gasketed flange joint. This particular application consists of two sealing surfaces with raised faces, a ring gasket, and bolts. The same forces act on any gasketed flange joint. Hydrostatic end force, the fluid pressure trying to separate the flange face, is shown clearly. This force is different from internal pressure, which acts on the inside diameter of the gasket or seal, trying to push or blow it out.

7.1.5.2 Problem-Solving Checklist A gasketed joint is composed of many elements functioning together to achieve a final end result. All of these elements must be considered in gasket selection:

1. *Flange condition* (Since there are two flanges, each should be considered independently.)
   a. Are the flanges new or old?
   b. What is the surface finish?
      - Expressed either in microinch RMS (root mean square) or microinch AARH (arithmetic average roughness height)
   c. What is the lay of the finish?
      - Concentric machined circles, so-called phonographic (spiral) finish, parallel machined lines, or lines curving across the flange?
   d. What is the nature of the finish?

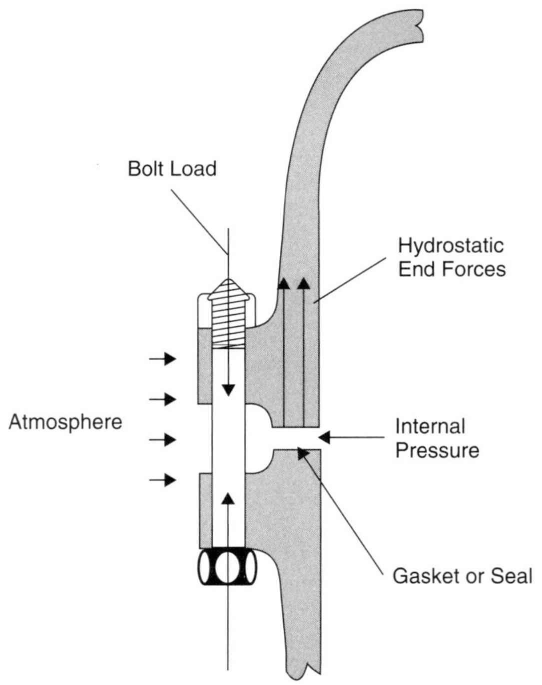

Figure 7-5 Forces acting on a gasketed flanged joint. (Courtesy of Garlock, an EnPro Industries family of companies.)

- Lathe-turned, milled, ground, lapped, or polished?

e. What is the surface contour?
   - Wavy, warped, distorted, or dished so that it is either concave or convex?

f. What is the surface condition?
   - Nicks, dents, voids, cracks, pits, or steam cuts?

g. Is foreign material present?
   - Residue of a former gasket, dirt, scale, oil, grease, tar, gasket cements, or paste?
   - Metal chips or weld spatter?

h. What is the chemical condition; is there evidence of chemical reactions?
   - Rust, corrosion, etching?

Most off-the-shelf standard ASME flanges have a 125- to 500-μin. RMS finish (which means that they are fairly rough) and a serrated finish. This is good, since the rough surface grabs the gasket and keeps it in place. However, as mentioned previously, there is a fine line between a finish that is too smooth and one that is too rough. Again, smoother surfaces are easier to seal (fewer leak paths) but have reduced pressure and crush resistance (lack of friction between the gasket and the flange face), and rougher surfaces are just the opposite. The higher peaks and valleys created by the machining provide more potential leak paths, but they increase the gasket's resistance to pressure and crushing by "biting" into the gasket and holding it in place.

To overcome most of the surface problems in the checklist, if repair of the problem is not feasible, select a thicker gasket and apply more load to make the gasket flow.

Note: In critical services, use of thicker gaskets should be a last resort, since the thicker gaskets tend to have lower blowout resistance and slightly higher creep relaxation. If suitable for the service, a softer gasket, such as Gylon 3545 or Graph-Lock, may be a good choice.

When problems do occur, an examination of the gasket used may provide some explanation if the flanges are not accessible, as the gasket surfaces will provide insight into such things as flange conditions and features as well as the installation process. Compression measurement at various sections of the gasket can confirm if improper bolt loading, flange waviness, or bowing has occurred.

2. *Bolts and nuts*

a. What are the number and size of bolts used?

b. What is the grade?
   - *Example*: ASTM A193 B7, SAE grade 8, ASTM A 307 grade A or B, etc.

c. Are they new or old?

d. What is the thread condition?
   - Clean-cut and free of burrs or galling?
   - Free of rust, dirt, and corrosion?

e. Is lubricant used when tightening?

f. What is the condition of the flanges under the head of the nuts and bolts?

g. Are hardened washers used under the nut, and if so, were they lubricated?

These questions help determine whether or not sufficient load is transmitted from the bolts onto the gasket. For example, according to a flange design reference book, the friction (nut) factor on an alloy steel (A490) bolt is as much as twice as high as on an unlubricated bolt. In other words, to achieve the same amount of compressive load as that created in a lubricated, 1/2-in. bolt at 54 ft-lb, a rusted, unlubricated bolt would have to be tightened to a torque of 108 ft-lb. Therefore, well-lubricated bolts and nuts in good condition can prevent many joint problems by transferring as much load as possible to the gasket or seal.

3. *Joint assembly*

a. What sequence or steps are used to tighten the bolts and nuts?

b. Is a torque wrench used?
   - If yes, how were torque values obtained?
   - If not, how are bolts tightened?

c. Are trained personnel involved in the joint assembly procedure?

d. How accessible is the joint for assembly and service?

These questions help determine the load transmitted to the gasket in the bolt tightening procedure. Most plants do not use torque wrenches for tightening bolts, but this one procedure could save them countless dollars.

4. *Joint environment*
 a. Are there any external mechanical loads on the piping system that could affect the joint?
  - Tension, compression, torsion (twist), shock, impact, or vibration?
 b. Assuming the probability of some external mechanical loads, what provision was made to relieve the loads?
  - Adequate pipe supports and anchors?
  - Expansion joints or expansion loops incorporated in the system?
 c. Are there potentially damaging external atmospheric conditions?
  - Temperature extremes?
  - Corrosive atmosphere (acid rain, submerged pipeline, etc.)?
  - Chemical or solvent spills that might come in contact with the joint?
 d. Are the sealing surfaces of the piping, valves, and so on, aligned properly?
  - An off-center condition prior to bolting, angular misalignment, or twisting?

Any of these conditions will affect the actual operation of the gasket or seal. If the flanges are not aligned properly, the sealing device won't function properly.

7.1.5.3 Human Factors The first two sections covered the mechanical side of gasketing, which is objective. What about the subjective side? What about the needs, desires, and prejudices of the people involved? Consider the following checklist:

1. *Current product*
 a. What is the customer's attitude to the current product?
  - Is the product working?
  - Is the customer happy or unhappy with it?
2. *Price considerations*
 a. How important is price? Does the customer differentiate between cost and price?
  - A higher-priced sealing solution could be more cost-effective, reducing downtime cost, fines for product release into the environment, and the cost of lost product or media
3. *Other factors influencing the customer*
 a. Is quality most important to the customer? Somewhat important? Not at all?
 b. Is availability the first requirement? The second?
 c. Is technical expertise and support a criterion for buying?
 d. Any product prejudices based on previous experiences?
4. *Customer role*
 a. Is the customer a novice or an old hand? With the company? In the job?
 b. Is the contact a key decision maker or only influences the decisions of others?

The human side is frequently the key element for success, not only overall but also in product selection. Garlock is in the replacement business; their products must be replaced periodically, so forming lasting customer relationships, built on earning confidence, is critical for success.

7.2 COMPRESSION PACKINGS

Controlling fluid leakage is essential to the successful operation of many types of mechanical equipment used in fluid-handling. Two common types of fluid-handling equipment are pumps and valves. Pumps provide the driving force to move fluid through a system, while valves are used to control the flow of fluids. To accomplish their purpose, pumps and valves have moving parts that operate in contact with the fluid. In order to move, these internal parts are usually connected to a motor, handwheel, actuator, or other type of force-producing device that is outside the fluid system. In a centrifugal pump, a motor drives a shaft connected to the impeller, which in turn generates pressure to drive the fluid. In valves, a valve stem connects the handwheel or actuator to the internal components of the valve. Where the stem or shaft penetrates the body of the valve or pump, there must be a seal. One type of seal used in these applications is compression packing, which derives its name from the manner in which it performs the sealing function. Made from relatively soft, pliable materials, a compression packing set consists of a number of rings that are inserted into the annular space between the shaft or stem and the body of the pump or valve. The annular space into which the packing is inserted is called a *stuffing box* or *packing gland*. When a load is applied through the gland follower, the packing is compressed, causing the rings to expand radially against the side of the stuffing box and the shaft or stem, creating a seal.

Compression packings find their major use in the process industries, such as petrochemical, paper, and steel mills, and in service industries, such as utilities, marine, water, sewage, food, and nuclear. They seal all types of fluids, including water, steam, acids, caustics, solvents, gases, oil, gasoline, heat-transfer fluids, and other chemicals, over a broad range of temperature and pressure conditions. In addition to their use in many types of pumps and valves, they are also used in linear expansion joints, soot blowers, mixers, screw conveyors, and many other types of equipment.

Compression packings are relatively easy to install and maintain. With proper attention, successful operation can be anticipated. Successful sealing with compression packing requires the following:

- Proper selection of packing materials to meet the specific application requirements, (necessitates careful consideration of equipment type, surface speeds, pressures, temperatures, fluid being sealed, and packing gland dimensions)
- Proper attention to good installation and break-in procedures
- High standards of equipment maintenance

7.2.1 Packing Materials

An understanding of the raw materials used in compression packing can be helpful in many situations: when making packing recommendations, when troubleshooting, and when replacing a competitive product. The performance capabilities and the behavior of various packing materials are highly variable and are dependent on both the ingredients and the construction used to make the packing. Braided packings consist of several different parts. First, the "backbone" of the packing is the yarn. Next, a coating (or blocking agent) can be added to coat the yarns and plug up the areas between the yarns. In addition, many types of liquid, oil, grease, or powder coatings may be added to provide lubrication or corrosion protection.

At this point it is important to clearly define several terms that we use to describe the yarn portion of a packing material.

7.2.1.1 Fibers The yarns used in braided packings may be classified into several groups according to the fibers that make them up. Yarns may be produced using cellulosic (vegetable) fibers, natural mineral fibers, glass fibers, carbon fibers, graphite fibers, or synthetic polymer fibers. Fibers and other materials may also be combined to produce composite yarns that utilize the advantages of several fiber materials while reducing or eliminating their less desirable characteristics.

- *Fiber*: a single strand.
- *Filament*: same as "fiber" above, usually referring to fibers of very small diameter.
- *Continuous filament*: very long fibers. Fishing line is an example of a single continuous filament. Some yarns are made up of twisted bundles of continuous filaments. Continuous filament yarns have a very smooth appearance.
- *Staple fibers*: very short fibers that must be spun together to make a yarn. Wool yarns like those used in a sweater are an everyday example of a spun staple fiber yarn. Yarns made from staple fibers have a characteristic "fuzzy" appearance.
- *Wire*: a single metallic fiber having a diameter of 0.008 in. or greater. Smaller metallic fibers are generally referred to as filaments.
- *Yarn*: one or more bundles of twisted or spun fibers. A single-ply yarn is a single bundle of fibers. A multiple-ply yarn has two or more bundles of fibers that are twisted together.

7.2.1.1.1 Cellulosic (Vegetable) Fibers Cellulosic fibers such as flax, jute, ramie, and cotton are natural fibers used in packings. These are among the oldest fibers used to construct packing materials. Their chief advantage is their low price compared to other fibers, but they have the disadvantages of poor chemical and heat resistance. They have very poor resistance to acids, but fairly good resistance to alkalis.

Flax is a medium-brown colored fiber that is obtained from a plant stalk. Garlock selects high-quality long-fiber flax roving yarns, braids them, and then impregnates them thoroughly with lubricating agents. Flax packings are designed for service in waste and dilute aqueous solutions up to 220°F (104°C) at low to medium pressures. Industries such as wastewater treatment, marine, mining, milling, steel, and pulp and paper specify flax packings for some applications, taking advantage of its tendency to swell slightly as it absorbs water.

7.2.1.1.2 Natural Mineral Fibers Alumina–Silica Alumina–silica fibers receive the highest-temperature rating of any packing materials. These fibers exhibit excellent thermal insulating properties, chemical resistance, and oxidation resistance. Packings made from alumina–silica fibers are not used as a seal. They are normally used as thermal insulators, shielding other materials from exposure to high temperatures.

7.2.1.1.3 Glass Fibers Glass fibers have superior thermal properties, dimensional stability, and tensile strength. They resist most chemicals and can be formulated to resist strong acids. Glass fibers are available in continuous filament yarn, staple fiber yarn, textured yarn, chopped strand, and mats. One hundred percent glass fiber packings have limited use. These packings would be very abrasive to equipment if used in high-speed rotary service. Due to the coarseness of the fibers, glass fiber packings do not function well as a seal. One hundred percent glass fiber packings are typically used as low-pressure door gaskets such as those found on furnaces and wood-burning stoves. These are static services where a tight seal is not required.

7.2.1.1.4 Carbon and Graphite Fibers Carbon and graphite fibers are considered premium-grade materials that are used in some of the most severe packing applications. Their extreme chemical resistance, high-temperature capability, strength, toughness, and thermal conductivity make them suitable for a wide variety of services.

Packings constructed of carbon or graphite materials have excellent resistance to chemical attack. Their chemical resistance covers the pH range 0 to 14. They can also be used with very strong solvents. They are not recommended for use in strong oxidizing media. Carbon and graphite fiber packing also exhibit good abrasion resistance without the wear characteristics that some other abrasion-resistant materials have.

7.2.1.1.5 Synthetic Polymer Fibers Following the elimination of asbestos as a packing material, emphasis was placed on developing and marketing packing configurations that took advantage of the many excellent properties to be found in synthetic polymer fibers. The wide variety of available fibers creates a pool of raw materials that the packing industry can draw upon for sealing materials that will meet the requirements of many applications.

7.2.1.1.6 Acrylic Fibers Acrylic fibers are a relatively low-cost synthetic fiber that can be used in many situations that do not require high-performance materials.

7.2.1.1.7 Aramid Fibers Aromatic polyamid fibers are given the generic name of aramid. Aramid fibers are used in braided pump packings, protective clothing, coated fabrics, belting, hoses, tires, and many other end uses requiring excellent heat resistance, high tensile strength, toughness, and abrasion resistance. The most common reason for using aramid fibers in braided packings is to take advantage of their extreme resistance to abrasion. These fibers are often used either alone or in conjunction with other types of yarn to produce a packing that will withstand the abrasive wear that is encountered in pumps handling liquid–solid mixtures (slurries).

One drawback to aramid packings is that they themselves can be abrasive to the equipment in which they are installed. This is particularly true for equipment operating at high speeds, high pressure, or both. In general, aramid fibers have good chemical resistance, but are not as chemically resistant as fluorocarbon fibers. Strong acids and strong alkalis adversely affect aramid fibers.

7.2.1.1.8 PTFE (Fluorocarbon) Fibers PTFE fibers are commonly referred to under the trade name Teflon. PTFE is known for its exceptionally high resistance to chemicals and low friction. It is also known to have poor mechanical properties and a lack of thermal conductivity that can make it sensitive to heat and high speeds.

It is very important to distinguish between PTFE fiber packings and packings that have some other fiber that is simply PTFE-coated. A PTFE fiber packing should only be used in cases where its extreme chemical resistance is required. Otherwise, some other type of PTFE-coated fiber will often perform with much better long-term results.

7.2.1.2 Flexible Graphite Foil Flexible graphite foil is manufactured by exfoliating, expanding, and then compressing natural graphite flakes into a sheet form of a specific density. Graphite foil is very chemically resistant, the only exception being strong oxidizing chemicals. It is self-lubricating and has excellent high-temperature capability and thermal conductivity. But perhaps its strongest quality is its ability to seal to extremely low levels of leakage. Graphite foil is an ingredient in many of the packings used to seal equipment to the extremely low levels required by the U.S. Environmental Protection Agency (EPA). Some grades of graphite foil also find use in nuclear power applications, where radiation resistance and high chemical purity are required. Since the mid-1990s many methods have been developed to combine graphite foil with other fiberous materials (such as graphite, carbon, stainless steel, and others) to produce graphite foil composite yarns that can be braided in the same manner as other fiberous materials.

7.2.1.3 Blocking Agents and Lubricants As their name suggests, blocking agents are used to fill the internal voids that exist in a braided packing so that fluid cannot wick through the braid. Many blocking agents also function as a lubricant.

PTFE dispersion coatings (PTFE dispersed in a water carrier) have been a most valuable asset to the packing industry. After yarns or braids are dipped in the coating, the water carrier is evaporated as the yarn or braid is heated in a drying oven. The solid PTFE particles that have been deposited throughout the packing become a lubricant that possesses all of the features of solid PTFE, such as excellent chemical inertness, a low coefficient of friction, and self-lubricating properties. In addition, the PTFE lubricant will greatly reduce the harshness and abrasive nature of the fibers that contact the shaft or sleeve. The main advantage of using PTFE as a blocking agent is that it will remain within the braid, even when subjected to temperature and pressure. PTFE dispersion coating may also contain graphite powder, which will increase the thermal conductivity of the coating and allow the packing to withstand higher surface speeds without charring.

Many other types of greases, waxes, and oils can be used as lower-temperature blocking agents and lubricants. These types of coatings are less costly than PTFE dispersion coatings, but they are more easily driven out of the packing by pressure and temperature. They can also be dissolved by strong solvents or broken down by certain chemicals.

7.2.1.4 ***Solid Lubricants*** There are a few types of solid lubricants that can be added to a packing material to decrease the friction the packing creates on the surface that is moving against it.

1. *Graphite.* Powdered graphite or flake graphite are both dry, solid lubricants that are very common in packing materials. They can be combined with other materials in a coating, or they may simply be applied directly to the surface of a braid. Graphite retains its properties as a solid lubricant, even at very high temperatures.
2. *Molybdenum disulfide.* Molybdenum disulfide powder is another type of solid lubricant. It is often mistaken for graphite because of its similar appearance and behavior.

7.2.2 Packing Construction

The raw materials that make up a compression packing can be combined in many different ways. Several types of braiding, twisting, coating, and die-forming processes exist, producing packing materials with widely varying qualities. Brief descriptions of the types of constructions that are commonly used follow.

7.2.2.1 ***Braiding*** The majority of compression packing materials are constructed of yarns that are braided together using a few different methods. These methods vary in the number of yarns that are used, the size of yarns that are used, and the pattern that the yarns follow as they are being braided. There are three different methods of braiding: square braid, Garlock's Lattice Braid® Garlock Sealing Technologies, and round braid.

Square-braided materials are also referred to as square-plaited or simply as plaited. Yarns are processed on equipment where strands pass over and under one another. The resulting packings are usually supplied in square cross section, but rectangular cross sections can also be produced using this method. The resulting packing is usually soft and can carry a large percentage of lubricant. Square-braided packings are often softer and more flexible than a lattice-braided packing made from the same materials. The softness of a square-braided material makes it suitable for use in worn equipment, where the packing is required to conform to irregularities on the stuffing box or shaft surfaces. The flexibility of a square-braided material is an advantage in applications where the packing is bent around a very small radius (as in knife gate valves). Applications that involve large amounts of shaft runout (such as mixers and agitators) also benefit from the flexibility that a square-braided material offers. In these situations, the shaft may compress the packing against the side of the stuffing box and the packing will "bounce back" and maintain a seal.

Garlock's Lattice Braid construction is known generically as *multitrack-braid* construction. Yarns, wires, and other materials, either alone or in combination, can be processed on multitrack braiding equipment where the strands are tightly interwoven to form a solid braid (usually square in cross section). Compared to a square-braided material, multitrack-braided materials use a greater number of smaller-diameter yarns. The strands are more tightly interwoven than in a square-braided material, and the surface of the packing is generally smoother. Each yarn strand is locked together with others to form a solid structure that cannot easily unravel in service. The finished packing is relatively dense, but flexible. The majority of Garlock's braided packing styles are Lattice Braid. Other manufacturers may use their own trade names for this type of braiding construction (such as cross-locked, interlock, multilock, multiple braid, interlace, interbraid, etc.)

One unique feature of Lattice Braid construction is that different yarn materials may be used in different positions in the braid. A common method for improving a braid's resistance to abrasion is to use an abrasion-resistant fiber in the corner positions. The corner positions are the most susceptible to abrasive wear. The rest of the braid may then be constructed of materials that are not as abrasion resistant but may have better sealing characteristics.

Round braiding is a type of construction that can be used in several different ways. A round braider will produce a single tube of braided yarn that is round in cross section. Several variations of round braiding exist:

1. *Braid-over-braid.* Also known as multiple braid or simply as round-braid, a series of round-braided jackets are braided one on top of another. The larger the final packing size, the more layers are required. The finished product is usually supplied in round cross sections, but may also be calendered square or rectangular.
2. *Braid-over-core.* Round-braiding layers of yarns or wires over a core produces braid-over-core products. Many combinations of yarn materials and core materials can be combined in this manner. A core may be made from extruded plastic compounds, rubber, twisted yarns, or braided materials. Braiding fiberous materials over a rubber core produces a very resilient material that can be used in equipment with a high degree of runout. Braiding a wire-reinforced yarn around a core is sometimes done to increase the pressure resistance of a material for use in high-pressure service.
3. *Braided tape.* When a round-braided jacket is constructed without core material, this hollow tube of yarn can

simply be collapsed to form a wide, thin braided tape. This type of construction is used to produce braid that can be used in some flange and tank lid sealing applications.

7.2.2.2 Twisting

Yarns and metallic foils can be twisted together or around a core to create a packing ring.

1. *Twisted yarns.* Single strands (or plies) of yarn are routinely twisted together to form yarns of larger diameter. Usually, these larger-diameter yarns are then used in a braided packing, but in some cases these twisted yarns can be used by themselves as a packing material. In the case of twisted yarns, one size of packing can be used for several stuffing box sizes. Because of its twisted construction, strands from a larger size can be untwisted and removed so that the remaining packing will fit into a smaller stuffing box.
2. *Twisted metal foils.* Metallic foils may be twisted together to form a solid cord, or they can be twisted around a core material (usually, fiberglass braid). After being twisted, the material is then calendered to a square or rectangular cross section. Metallic materials are able to resist high temperatures and the penetration of fluids, and to conform to the irregularities of worn equipment. Twisted foils that do not contain a braided core are very hard and not very elastic. That is, they do not bounce back when they are compressed. Fiberglass braid is used as a core in order to increase the softness and give the packing some elasticity. In some cases, die-formed metallic foil rings are used as a bearing material in applications where shaft runout is a problem.

7.2.2.3 Die-Formed Materials

Many different types of compression packing materials may be die-formed. There are different reasons for die-forming materials, but regardless of the materials, the type of equipment used in the die-forming operation is the same.

1. *Die-formed graphite tape.* This is *the* premium type of sealing material for valve stem applications. Graphite foil is supplied in wide rolls that are slit down to smaller widths, texturized, cut to length, wrapped around the center pin of a die set, and pressed in the die to form a solid ring. After the die-forming process, the rings can be split if desired. Graphite foil rings can be die-formed in various densities and shapes. They are generally used in multiple ring sets in combination with braided end rings.
2. *Die-formed braid.* Braided materials may be die-formed to increase their density, to compress them to a specific shape, or to form them to a specific size. Braided materials vary in the effect that die-forming has on them. Braids that have a relatively high PTFE content and braids that contain wire generally respond well to die-forming. That is, they retain the shape into which they are formed. Some braided materials are unaffected by the die-forming process. Still others are actually damaged by die-forming.
3. *Die-formed twisted foil.* Twisted foil packings can be supplied as individual die-formed rings. This process makes them very hard and dense and gives them a smooth surface finish. Rings made using this process are often used simply as bushings (spacers).
4. *Die-formed metallic tinsel.* Metal tinsel packing is provided in die-formed rings only. Copper, lead, and aluminum tinsels can be pressed into rings. These rings are typically used as spacer bushings, or as bearings in low-speed rotating equipment.

As we have seen, there are a wide variety of raw materials and types of construction that can be used to create a packing material. With such a wide variety of products to choose from, how does one choose the correct packing for a specific application?

7.2.3 Packing Selection

The first step toward making a good packing recommendation is to collect several key pieces of information. The acronym "STAMPS" can be used to help remember what information to look for: **S**, *Size*: What are the stuffing box dimensions? **T**, *Temperature*: What is the temperature of the fluid? **A**, *Application*: What type of equipment is it? **M**, *Media*: What is the fluid being sealed? **P**, *Pressure*: What is the pressure of the fluid? **S**, *Shaft Speed*: What is the surface speed of the shaft?

7.2.3.1 Size

The size of the stuffing box will determine the size of packing that is used. To determine the size of braided stock to use for a given application, the equipment should be measured and then the following equation should be used to determine the packing size:

$$\text{packing size} = \text{OD} - \frac{\text{ID}}{2} \qquad (7.1)$$

where OD represents the bore diameter and ID the stem/shaft diameter. If a die-formed ring set is to be used, the packing size does not need to be calculated. Simply specify the OD and ID dimensions and a die-formed set will be made to fit.

7.2.3.2 Temperature

It is very important to know the temperature that the packing will be exposed to in a given application. It is recommended that temperature limits for all packing styles should be adhered to. One important item to note is that there are two different temperature ratings for the carbon and graphite materials, depending on the media that they are sealing. In a steam or inert gas environment, these materials can be used up to 1200°F (649°C). In an air environment (or one that contains free oxygen) the temperature rating is lowered to 850°F (454°C). This is particularly important in power-generation applications, where it is not uncommon to encounter steam temperatures over 900°F (483°C). At temperatures over 1200°F (649°C), very few options remain. Ceramic fiber packings can be used at these temperatures; however, their sealing capability is severely limited. They are used as either a thermal insulator in a stuffing box or as a furnace door gasket, where absolute sealing is not critical.

*7.2.3.3 **Application*** Many different types of equipment use compression packing as a seal. Each piece of equipment operates with a certain type, frequency, and speed of motion. Equipment also differs in the amount of leakage that is considered acceptable. Pump leakage may be measured in terms of drops per minute or gallons per day, whereas valve leakage may be undetectable or detectable only on a very fine level. These characteristics of the equipment (as well as others) have a significant effect on the type of packing materials that would be recommended.

*7.2.3.4 **Media*** When determining whether a packing material will be suitable for use with the fluid being sealed, there are several questions that should be answered.

*7.2.3.5 **Pressure*** When determining the suitability of a packing for a specific pressure, one simply needs to compare the system pressure with the pressure rating that the manufacturer gives for that style. It is important to note that for packing materials that can be used in pumps and valves, the pressure rating for rotary service (pumps) is usually much lower than the pressure rating for the same packing in valve service.

*7.2.3.6 **Shaft Speed*** In rotary equipment applications, one issue of concern is the frictional heat that is generated between the rotating shaft and the packing. Some packing materials can dissipate this heat quickly, while other materials can withstand high temperatures without melting or charring. The ability of a packing to dissipate and withstand frictional heat is reflected in the shaft speed rating.

In Garlock's product literature the shaft speed ratings are given in units of feet per minute, a measure of the speed of a point on the surface of the shaft. It should not be confused with rpm (rotations per minute), which is a measure of the rotational speed of the shaft. It is common for pump users to know the rotational speed of their pumps. The two most common electric motor speeds are 1750 and 3600 rpm. Unfortunately, customers rarely know the surface speed of the shaft in the packing area. Therefore, it is necessary to convert from rpm to fpm. The following equation is used to make this conversion:

$$\text{surface (ft/min)} = 0.26 \times \text{shaft diameter (in.)} \times \text{rpm speed} \quad (7.2)$$

If the pump has a wear sleeve over the shaft, the outside diameter of the sleeve should be substituted for the shaft diameter in Eq. (7.2). For example, for a pump with a shaft sleeve OD of 2.000 in. operating at 3600 rpm,

$$\begin{aligned}\text{surface speed} &= 0.26 \times 2.000 \text{ in.} \times 3600 \text{ rpm}\\ &= 1872 \text{ ft/min}\end{aligned} \quad (7.3)$$

Note that Eq. (7.3) requires that the shaft diameter be expressed in inches.

7.2.4 Packing Installation

Different packing materials can be chosen for the many different services. In the same way, installation methods also vary depending on the type of equipment. In the following text we describe two different installation methods that will cover nearly all of the applications that you may encounter. One set of instructions is for centrifugal pumps and other rotating equipment, such as mixers, agitators, and other equipment with relatively high shaft speeds. These types of equipment may generate high amounts of frictional heat on the shaft–packing interface, and the installation methods need to account for this.

The second installation method is intended for use in valves and other very low-speed or static services. In these types of equipment the pressures are typically much higher, frictional heat on the packing surface is negligible, and the packing is often required to provide a much tighter seal.

One of the most important things to know about compression packing is the difference between pump packing service and valve packing service. When packing a pump, the main objective will be to Control the leakage while keeping frictional heat generation at a minimum. When packing a valve, the main objective is simply to Cut Off leakage.

The instructions that follow are very general guidelines that can be applied to almost any packing. For many of our packing styles, installation instructions are included in the box. These instructions may offer more detailed information that is specific to that particular style.

*7.2.4.1 **Packing a Pump*** The following steps are required for packing a pump:

1. Remove the old packing as well as the lantern ring if one has been used.

2. Inspect the pump shaft and stuffing box for cleanliness and wear. Replace worn parts and clean the stuffing box as necessary.
3. Determine the packing size and number of rings required by measuring the stuffing box ID, OD, depth, and any other applicable dimensions.

The same equation as that given earlier should be used to determine the packing size:

$$\text{packing size} = \frac{\text{OD} - \text{ID}}{2} \tag{7.1}$$

where OD represents the bore diameter and ID the stem/Shaft diameter. To determine the number of packing rings in the set, divide the stuffing box depth by the packing size and round down to the lowest whole number.

1. Cut rings to the proper length by using a spare sleeve or packing cutter.
2. Lubricate the inside diameter of each ring of packing with a break-in lubricant prior to installation.
3. Install rings individually, seating each ring all the way to the bottom of the box before the next one is installed. Be sure to rotate the joint of each successive ring by 90°.
4. Tighten the gland stud nuts fingertight after the last ring is installed.
5. Start the pump and allow the packing to leak for several minutes. *Note*: For media containing solids, leakage may need to be reduced more quickly.
6. Adjust the packing to arrive at an acceptable leakage rate. (*Make adjustments that are proportional to the leakage. Do not overtighten!*)

If the packing is leaking excessively, it may be tightened to reduce leakage to a steady small stream. Once leakage is reduced to a small stream or less, make only slight adjustments and leave time between adjustments.

7.2.4.2 Packing a Valve The following steps are required for packing a value:

1. Remove all old packing.
2. Inspect the stuffing box for wear and cleanliness. Replace worn parts and clean as required.
3. Determine the packing size by measuring ID, OD, box depth, and any other applicable dimensions.

Equation (7.1) should be used to determine the packing size.

1. Determine the number of rings required to provide adequate sealability. Consult the installation instructions that come with the packing to determine specific requirements for the amount of compression and the number of rings. Take into account the need for bushings when the stuffing box is very deep.
2. Cut rings to the proper length by using a spare stem or packing cutter.
3. Do not lubricate the packing rings. Install each ring individually, being sure to fully seat each ring in the bottom of the stuffing box before the next ring is installed. Stagger the joints 90°.
4. Apply compression to the lower rings of the packing set when possible.
5. After the packing is installed and compressed, check to make sure that there is a minimum of $\frac{1}{8}$ in. (3.2 mm) penetration of the gland follower into the stuffing box, and that there is sufficient gland follower length remaining for future adjustment.
6. When possible, a cycle and adjustment procedure can help to reduce the amount of retightening that may be required later.

7.2.4.3 Cycle and Adjust Procedure for Valves This procedure will help to ensure more even compressive loading throughout the packing set, and reduce the amount of packing relaxation that may occur over the life of the packing.

1. Install and compress the packing set, following normal packing procedures.
2. Raise the valve stem to the full open (up) position.
3. Check the torque on the gland studs. This can be done with a torque wrench, or simply by "feel" with a hand wrench.
4. Actuate the stem through two or three revolutions of the hand wheel in the closing direction.
5. Check the torque on the gland studs. If any loss has occurred, retighten the gland studs to the tightness measured in step 3.
6. Repeat steps 4 and 5 at least three times, or until there is no significant torque loss after actuating the stem.

7.3 MECHANICAL SEALS

Everyone in the industrialized world uses mechanical seals. Various pieces of rotary or rotating equipment, pumps in particular, depend on mechanical seals to control leakage. Familiar rotary equipment devices include automobile water pumps, washing machines, dish washers, compressors, swimming pool pumps, and farm service pumps. Mechanical seals are used anywhere that liquid and gases are transferred by rotating equipment.

Pumps are one of the most widely sold pieces of equipment in industry, second only to electric motors. Pumps are found in all industries, large buildings and households, farming operations, mining, construction, and city services. Although some of the small, inexpensive pumps are disposable (i.e., automotive water pumps), most pumps require packing or mechanical seals to control leakage between the rotating elements and stationary housings. These packings and seals must be serviced to ensure that controlled leakage is maintained.

Maintenance of pumping systems is what we focus on in the following sections. We note that some products are allowed to drip without concern, whereas others must not leak more than a few parts per million and need to be monitored closely with electronic detection devices.

7.3.1 Considerations for Using Mechanical Seals

7.3.1.1 How Pumps Work All rotary pumping systems contain a drive system. The driver is what makes the pump rotate. There are several types of drivers (i.e., electric motors, combustion engines, steam turbines, etc.). Often, but not always, pumping systems contain a coupling that attaches the driver to the rotary shaft of the pump. A bearing arrangement supports the shaft. The bearings may be any of several types (i.e., sleeve bushings, ball, roller, etc.) Normally, there are two bearings that support the shaft (thrust bearing and radial bearing). Attached to the shaft is the impeller. A volute shrouds the impeller. Within the volute, the rotation of the impeller increases medium velocity and pressure. The medium being transported enters the pump through the suction inlet and exits the volute through the discharge outlet at a greater pressure and velocity. The rotating element extends through the stationary housing of the pump, in which some type of sealing device has been or can be installed. This housing or cavity is known as the stuffing box for packing or seal chamber for mechanical seal applications.

Although somewhat simplified (there are all types of service variations and designs, some of which may contradict this description), the paragraph above can be considered a general description of how most rotary pumps function.

7.3.1.2 Mechanical Seals Versus Packing Although mechanical seals had been around for many decades, by 1955, industry had converted only a small percentage of pumps from packing to mechanical seals, for several reasons:

- To install a seal, the pump would have to be taken off-line and disassembled.
- Packing could be installed quickly, or an extra ring added, without disassembly.
- Packing was, and is still thought to be, cheaper.

Maintenance people were not informed or educated about mechanical seals and felt more comfortable with packing. Plant personnel gave little thought to the cost of product loss, energy, housekeeping, and so on, because sealing had always been done in the same way. There were few guidelines that dictated leakage control, and few people seriously considered its dramatic cost.

Since Congress passed the Clean Air Act in 1963, and amended it in 1990, federal, state, and local regulations have become increasingly strict regarding emissions control. Some regulation requirements are considered to be at the brink of or beyond current technology. These regulations have created an atmosphere where in many cases it is no longer an option for industry to make the packing choice.

7.3.1.3 Emissions Regulations Today, mechanical seals are required by law to control the emissions of many products. EPA regulations forcing compliance has obligated industries to demand better seals from manufacturers, which, in turn, has made us do a better job. Industries' need for more efficient sealing systems has compelled seal companies to design seals that do not damage equipment, handle higher pressure or vacuum, provide a wide range of environmental control options, and are able to seal a wider range of chemicals. In addition, these sealing units are more user friendly. Today, the seal industries used 20 years ago are being outclassed, in every category, by newer designs. When handling volatile organic compounds (VOCs), seal selection is often determined by specific gravity and maximum allowable emission levels. Lower-specific-gravity products, combined with parts per million of VOC emissions allowable, dictate not only if a mechanical seal is required but also if double mechanical seals are required. Federal law dictates that these emissions must be contained. Any VOC that has a specific gravity of 0.4 or less requires a double mechanical seal.

7.3.1.4 Product Loss Often, after a decision has been made to change from packing to a mechanical seal, a noticeable amount of product is saved. Almost without exception, the product savings alone will pay for all of the costs associated with the conversion. Typically, product savings will pay for the mechanical seal many times over. By taking a volume sample of leakage from a packed pump, measured in time, it is simple to calculate the amount of product lost each day. (Table 7-4). The next step is to multiply volume times the cost per unit and the number of pumps to find the astonishing cost of lost product when the numbers are calculated on a yearly basis. By using these calculations alone, most plants have realized the value of mechanical seals.

TABLE 7-4 Drip Chart

	One Drop per Second	Three Drops per Second
1 minute loss	$\frac{1}{12}$ ounce	2 ounces
1 hour loss	6 ounce	1 gallon
1 day losss	1 gallon	24 gallons
1 week loss	8 gallons	175 gallons
1 month loss	34 gallons	700 gallons
	$\frac{1}{16}$ in. (1.6 mm)	$\frac{3}{16}$ in. (4.8 mm)
1 minute loss	7 ounces	39 ounces
1 hour loss	3 gallons	18 gallons
1 day loss	64 gallons	425 gallons
1 week loss	575 gallons	3,000 gallons
1 month loss	2,500 gallons	12,750 gallons
	$\frac{1}{4}$ in. (6.3 mm)	
1 minute loss	83 ounces	
1 hour loss	39 gallons	
1 day loss	925 gallons	
1 week loss	6,500 gallons	
1 month loss	27,750 gallons	

7.3.1.5 Energy Cost The amount of electricity it takes to drive a mechanical seal is dramatically less than what it takes to drive a pump with compression packing (see the energy cost calculation in Fig. 7-6). Add the cost of flush, if applicable (support systems). Packing compresses against the shaft while controlling leakage and requires more energy to drive the pump. The number of packing rings per set (four, five, or six rings) increases the amount of energy required.

A single mechanical seal has only two very flat, very low friction/hydroplaning faces pressed together that must be driven. So the resistance generated by a mechanical seal is dramatically less than that generated by a set of several packing rings, and the energy/cost savings are significant. The idea of less abrasion led a large pulp and paper mill to study the energy consumption of electric motors used on paper stock pumps. They varied the packing to determine what effect the packing material choice had on the drain of the electric motor in an effort to save energy. The study was designed as follows:

1. Fifteen centrifugal pumps were used, all powered by 100-hp electric motors.
2. Each style of packing was tested for three weeks.
3. The pumps all had a rated efficiency by the manufacturer based on certain media at a certain specific quantity. They then installed mechanical seals.

The results (Fig. 7-6) show that a pump with PTFE asbestos in the stuffing box and a 100-hp electric motor, operating 24 hours per day, 365 days per year, would total \$2312 per year in electric costs to drive the packing (this does not include the cost to run the motor). The same pump operating with Garlock G-200 would cost \$773, a savings of \$1539 per year. This study clearly shows that the "price" and "cost" of its use are two entirely different factors.

In Fig. 7-7 we see an example of calculations showing where savings are available by changing from packing to a mechanical seal. The example used is a standard paper stock pump, using cold (50°F) water for flush media. The

Power cost \$0.611 KW per hour \$1 .460 KW per day \$525.60 KW per year			
Type of packing	Additional H.P. required	Additional KW required	Cost/pump/year
TFE asbestos	6.00	4.400	\$2,312.00
Synthepak 8921/8922	5.50	4.004	\$2,104.00
Synthepak 8913	5.00	3.652	\$1,919.00
Carbon TFE 5000	4.50	3.300	\$1,734.00
Carbon 98	3.00	2.200	\$1,156.00
Graphite G-200	2.00	1.452	\$763.00
Single mechanical seal	0.33	0.242	\$127.20
Power cost based on: Southwest Electric Power Company Texas Gas Utilities Company Texas Power & Light Company			

Figure 7-6 Energy cost calculation. (Courtesy of Garlock, an EnPro Industries family of companies.)

Energy Cost
Packing flush water
Power Cost 120 gal/hr @ 50F
Seal flush water 2 gal/hr @ 50F
Paper stock operating temperature 140F
Cost of Energy $2.00/1,000,000 BTU
Based on this data, the following calculations apply:
1 BTU = 1lb of water raised 1F
1 gal water = 8.34 lbs
Temperature differential 140F – 50F = 90F
8.34 x 90 = 750 BTU's/gal raised 90F
Packing cost calculation:
750BTU's x 120 Gal x 24 hrs x 365 days x $2.00 = 1,576.80
Gal hr day year 1,000,000 BTU
Seal cost calculation:
750BTU's x 2 Gal x 24 hrs x 365 days x $2.00 = 26.26
Gal hr day year 1,000,000 BTU

Figure 7-7 Calculations showing where savings are available by changing from packing to a mechanical seal. (Courtesy of Garlock, an EnPro Industries family of companies.)

rate of flush is 2 gal/min (1,051,200 gal/yr). The flush water must be elevated from 50°F to 140°F for the system to function properly. The figure shows the estimated cost in British thermal unit dollars per year to raise the flush water (50°F) to system temperature (140°F). This example does not address power for flush water transfer, evaporation costs, or waste treatment.

A mechanical seal can run leak free with as little as 2 gal/h flushing or 17,520 gal of water per year. This could amount to a savings of 1,033,680 gal of water per year per pump. Even if the flush water volume was tripled on the mechanical seal, the savings would remain dramatic (998,640 gal/yr per pump).

7.3.1.6 Return on Investment When one has interest in calculating the return on investment when changing from packing to mechanical seals, always work directly with the people in the plant who are interested in accurate numbers. Plant personnel must be included at every step while collecting the cost for each line. This must be done because the final cost, versus savings, will be so dramatic that accuracy will otherwise be questioned. However, if you can run a mechanical seal for one year, with no maintenance, product loss, bearing change, and no new sleeve, while using less energy and lower flush volume, the money saved is always dramatic (multiplied by the number of pumps converted). This approach can also be utilized to measure the savings when converting from one style of mechanical seal to another.

7.3.2 Types of Mechanical Seals

7.3.2.1 Face Seals All face seals require a secondary seal, which can be some type of elastomeric compound or gasket material. The secondary seal closes leak paths between the rotary face and the shaft as well as between the housing/gland and the stationary face. Because seal faces wear against each other and all pumps vibrate, requiring flexible seals, a mechanical loading device is needed to ensure continuous face contact, even when the pump is shut off. The mechanical loading device is normally some type of spring(s) or metal bellows.

As the faces rub together, one rotating and one stationary, they operate with a fluid film migration between the faces to cool and lubricate (the faces hydroplane on this fluid film). Ideally, the product being pumped will weep between the faces, entering as a liquid and remaining until vaporizing as it reaches the atmosphere. The leakage that flows to the atmospheric side should not be visible. It must be understood that all mechanical seals leak trace amounts of vapor.

7.3.2.1.1 Inside Seals An inside seal is designed mechanically such that the rotary portion of the mechanical seal, when installed, is located inside the pump seal chamber. When using an inside mechanical seal, the fluid and pressure are on the outside diameter of the seal. The fluid pressure acts as a hydraulic closing force, and typically, inside seals can be used in higher-pressure applications than can outside seals. Centrifugal force acts to expel solids away from the rotary and helps retard the flow or migration of liquid across the faces. The metal parts are in contact with the product being pumped, which must be compatible when sealing aggressive chemicals.

7.3.2.1.2 Outside Seals An outside seal is a mechanical seal designed such that the rotary portion of the mechanical seal is located outside the pump seal chamber. With an outside seal, the fluid pressure is on the interior diameter of the seal faces, which can cause clogging if the medium contains solids. With some outside designs, the product pressure can overcome the spring load. Centrifugal force acts to increase flow or migration of lubricating fluid across the seal faces, which, combined with pressure, limits the outside seal's service pressure to 150 psi (FSA recommendation).

Setscrew or clamp ring drives are options with some outside seals. Typically, outside seals are used for chemical service in nonmetallic pumps, where setscrews do not work well (e.g., too hard, or nonmetallic shafts). The fluid being pumped does not come in contact with the metal parts of the seal, which often eliminates the need for expensive and/or exotic metals.

7.3.2.1.3 Rotary Seals A rotary seal is a mechanical seal designed so that the spring(s) or mechanical loading device rotates with the pump shaft. At high speed, the spring(s) can distort and their effectiveness is reduced. Also, rotary seal springs adjust for any misalignment, twice per revolution (3500 times per minute on a pump at 1750 rpm), which can cause fatigue and spring breakage. Rotary seals tend to expel solids away from the seal as it spins.

7.3.2.1.4 Pusher Seals A pusher seal is a seal design that pushes a dynamic secondary seal (e.g., O-ring, wedge, chevron, U-seal) along the shaft to compensate for face wear and/or shaft movement. As a result, a pusher seal will wear a groove in the shaft or sleeve surface. The area of wear created by the secondary seal is called a *fret*.

7.3.2.1.5 Stationary Seals A stationary seal is a mechanical seal designed such that the spring(s) do not rotate with the pump shaft but remain stationary. Because the springs do not rotate, they are unaffected by the pump speed. The springs do not correct or adjust for each rotation. They adjust for misalignment only once, when installed, and are much less subject to fatigue or breaking. Stationary seals are preferred for high-speed service.

7.3.2.1.6 Nonpusher Seals A nonpusher seal is a mechanical seal designed in a way that eliminates the dynamic secondary seal. Normally, nonpusher seals are metal bellows or elastomeric bellow seals. They cause no damage to the shaft, from a secondary seal's axial movement. They flex at the bellows.

7.3.2.1.7 Metal Bellows Seals A metal bellows seal is a mechanical seal with a body made of several metal plates welded together, which act as a mechanical loading device (spring) and with the bellows also acting as a secondary seal (the bellows is dynamic and fluid cannot pass through it). The elastomeric secondary seal in a metal bellows driver is static (acts like a gasket) and does not damage the shaft. Metal bellows seals are always balanced, by design.

7.3.2.1.8 Unbalanced Seals Unbalanced seals are mechanical seals designed so that the full hydraulic pressure of the seal chamber acts to close the mechanical seal faces, without any proportion of that pressure being reduced or hydraulically balanced through the geometric design of the seal. Unbalanced seals are for use in low-pressure applications only [FSA recommends 50 psi (3.5 bar) service or less].

7.3.2.1.9 Balanced Seals A balanced seal is a mechanical seal designed geometrically such that the hydraulic pressure, which acts to close the seal faces, is substantially reduced. The closing area, affected by hydraulic pressure, is reduced in comparison to the opening or neutralized area, which is also affected by hydraulic pressure. Balanced seals generate less heat because of the reduced pressure forcing the faces together; they are designed to handle much higher pressure.

7.3.2.1.10 Double Seals A double seal is an arrangement in which two mechanical seals are utilized face to face, back to back, or in tandem (facing the same direction), allowing a barrier or buffer fluid or gas to be introduced between the two sets of seal faces.

7.3.2.2 Lip Seals

7.3.2.2.1 P/S® Seal Technology Lip seals offer an alternative to face seals and packing for many applications, particularly those viscous, temperature-sensitive, salting, crystallizing products that drive face seal users crazy. As an alternative to face seals, Garlock Fluidtec developed the P/S thermal-plastic high-performance lip seal. The P/S seal can also be used to augment the service life of both packing and seals. Because it is a lip seal, the P/S seal has no moving parts to clog. It possesses good chemical, temperature (up to 400°F (204°C) when used as a gasket and 300°F (149°C) for dynamic applications) and pressure resistance (up to 150 psi (10 bar) when properly backed).

A proprietary formulation developed by Garlock, Gylon is a restructured PTFE compound that is available in eight different styles. From this family of Gylon products have evolved two thermal-plastic compounds used for the P/S seal elements: black (graphite filled) and white (food grade) Gylon. The P/S Gylon sealing elements have characteristics of memory behavior, elasticity, low friction, and require no mechanical load to seal.

Because different applications demand different degrees of versatility, the P/S seal has been engineered into various configurations:

- Standard configuration
- Reverse lip (sealing element)
- Double lip or tandem (same direction)
- Double opposing lip (sealing elements)

P/S sealing elements are formed to the size required, but not molded. As a result, expensive dies are not required. This allows for versatility when manufacturing special sizes.

7.3.2.2.2 P/S -I Seals Typical P/S-I seal applications:

- Product excluder with compression packing
- Restriction/excluder with a mechanical seal

- Primary seal
- Pressure and vacuum service (double opposing lips)
- Pressure with backup (double lip, tandem)

7.3.2.2.3 P/S-II Cartridge Multilip Seals P/S-II cartridge multilip seal typical applications:

- Positive-displacement pumps
- Progressive cavity pumps
- Centrifugal pumps
- Rotary valves

7.3.2.2.4 Typical Products Sealed by P/S-II Seals

- Ammonia
- Asphalt
- Automotive undercoating
- Black liquor
- Caustic soda
- Chemical waste
- Fruit pulp and seeds
- Fuel oil
- Glues
- Hot wax
- Polymers
- Raw bubble gum
- Raw fiber glass
- Resins
- Soap and soap powder
- Surfactants

7.3.2.2.5 P/S Seal Performance Guidelines

Pressure: to 150 psig (10 bar) with positive backup to the seal case.

Vacuum: to 0.054 in. of water (0.004 in. Hg) with the element facing the vacuum source. Use full vacuum when the element is *facing away* from the vacuum source.

Temperature: to 300°F (150°C); over 300°F (150°C), consult an engineer.

Surface speed: 700 ft/min dry running, 1800 ft/min with proper enviromental controls (consult an engineer).

Runout: 0.005 in. TIR.

Axial end play: not affected.

Materials

Metal parts: 316 SS (other materials optional)

Setscrews: Hastelloy C

O-rings: Viton Standard (other materials optional)

Sealing Surface Requirements

Shaft finish: 4 to 6 μin. (0.10 to 0.15 μm)

Shaft hardness: 50 to 70 Rockwell C

Bore finish: 100 μin. (2.54 μm) or smoother

Sizes

Standard: $\frac{5}{8}$- to 7-in.-diameter shaft from standard production (smaller and larger sizes available)

Single element standard on P/S-I; multiple element designs on P/S-II. For special applications, consult Fluidtec Engineered Products' engineering department.

7.3.2.2.6 P/S Seal Limitations P/S technology is sensitive to radial shaft movement. It is recommended that no more than 0.005 in. of radial shaft movement be tested.

7.3.2.2.7 P/S-II Cartridge Seal Installation Instructions

1. Place the seal onto the shaft.
2. Assemble the pump equipment. Use flat washers under the gland nuts and lightly snug the gland to the face of the seal chamber.
3. Ensure that the sleeve/drive collar is pressed forward and nested against the clip wings.
4. Check to ensure that all clips are positioned at 90° to the shaft, providing absolute centering of the drive collar.
5. Run each setscrew in until it just touches the shaft. *Tighten each setscrew on the second rotation or cross torque*. This will avoid pulling the sleeve off-center.
6. Ensure that the clips are at 90° to the shaft, and while rotating the shaft (by hand) ensure that the drive collar is not rubbing or binding against one or two of the centering clips more than the rest. Use 0.001- or 0.002-in. feeler gauge and test the gap between the clips and drive collar.

   Note: The feeler gauge should not pass between the clip(s) and the drive collar at any location. If it appears that a clip or two are tipped away from the drive collar, it indicates that the gland bolting is uneven. Tighten the bolt(s) on the opposite side from the tilted clip/gap and watch as the clip squares to the drive collar shoulder.

7. If an adjustment is required, loosen the gland nuts and move the gland slightly to achieve centering and retighten (gland nuts).
8. Repeat the process until the seal/shaft is centered.
9. Evenly tighten the gland nuts (cross torque).
10. Rotate the centering clips 90° and install environmental controls or plug the flush ports.

7.3.2.2.8 3-D Seals The 3-D motion seal was developed as a result of customer demands for P/S technology in applications where shaft movement (in all directions) is extreme. The 3-D seal is a combination of five basic parts:

- P/S sealing element/stacked set
- 3-D seal body (to house or contain the stacked sealing elements)
- A sleeve (hardened)
- Bearing (*C*/diameter—zero tolerance)
- Flexible connection (expansion joint)

The 3-D seal incorporates the standard P/S sealing elements into the seal body, which houses a stabilizing bearing. The bearing maintains a constant relationship between the seal housing/sealing elements and the shaft sleeve. A flexible member (expansion joint) that absorbs any shaft movement connects the 3-D seal body and the vessel. The effective seal movement allowable is limited to the elongation, compression, and radial movement extremes of the flexible member or expansion joint. Larger expansion joints are able to move more than smaller ones. The 3-D seal is used, when other sealing options have failed, in equipment where shaft movement is beyond the limits of normal seals, the motion cannot be controlled or the equipment is too expensive to replace and must be sealed from the environment. Typical 3-D applications include mixers and agitators or where excessive shaft movement is normal and uncontrollable.

Note: The 3-D motion seal is not an inventoried seal. Each unit is designed for a given piece of equipment. Each design requires engineering time to work through the fitting, materials, and application requirements. However, the 3-D seal can overcome problems with axial movement, eccentricity, and vibration where other sealing designs cannot.

7.3.3 Mechanical Seal Applications

7.3.3.1 Material Compatibility Four steps must be taken with every application.

1. Choose materials that are chemically compatible with the product and will handle the temperature requirements and consistency (e.g., slurry, viscous, low specific gravity).
2. Choose the design or style that is appropriate in size to fit the equipment and is engineered to handle the product.
3. Install the seal properly into a piece of equipment that is in good mechanical condition.
4. Apply environmental controls that will provide the best possible environment for the seal.

You may have the best design, installed properly in good equipment with the very best environmental controls, but if the gasket you choose is not compatible, it will leak, which will demonstrate that the job was not done properly.

7.3.3.1.1 Chemicals That Affect Chemical-Grade Carbon Most carbon grades should not be used in the presence of strong oxidizing agents. Following is a list of common chemicals that affect chemical-grade carbon.

- *Aqua regia*: nitric acid and hydrochloric acid, used for dissolving metals.
- *Calcium chlorate*: greater than 5%, used in photography and pyrotechnics.
- *Chloric acid*: at 200°F (93°C) and over 10% concentration will ignite organic materials on contact.
- *Chlorous acid*: 200°F (93°C) greater than 10% concentration.
- *Ferric chloride*: in excess of 200°F (93°C) and greater than 50% concentration.
- *Fluorine*: rocket fuel oxidizer.
- *Hydrofluoric acid*: over 200°F (93°C) at 40% concentration or more. Used in pickling, etching, cleaning stone and brick, dissolving ore, cleaning castings, fermentation, and purification.
- *Iodine*: 200°F (93°C) with 5% or more concentration, used in soap, dyes, salt, medicine, and lubrication.
- *Nitric acid*: 250°F (121°C) and 20% or more concentration, used in fertilizer, dyeing, explosives, drugs, and etching.
- *Oleum or fuming sulfuric acid*: used to manufacture detergent and explosives.
- *Perchloric acid*: used in making esters, explosives, and medicine.
- *Sodium chlorate*: 5% or more concentration, used as bleach in paper processing, textiles, medicine, and tanning leather.
- *Sodium hypochlorite*: 5% or more, used in the paper pulp bleaching process and in water purification.
- *Sodium peroxide*: used in dyeing, paper bleaching, and oxygen generation.
- *Sulfuric acid*: 75% at 250°F (121°C) or more, most widely used of all chemicals.
- *Sulfur trioxide*: used to manufacture sulfuric acid.

7.3.3.1.2 Chemicals That Abrade Carbon The following liquids cause abrasive wear to carbon face material when carbon is used as a mechanical seal face.

- Chromic acid

- Chromic oxide (aqueous)
- Chrome plating solutions
- Potassium dichromate (aqueous)
- Sodium chromate
- Sodium dichromate

7.3.3.1.3 Chloride Stress Corrosion Chloride stress corrosion is a condition that causes cracks in stainless steel when subjected to the following:

- Tensile stress (near yield point)
- Temperature [above 140°F (60°C)]
- Chlorides

There are two types of stress, tensile and compression. Those in the mechanical seal world are interested in tensile stress because it acts to stretch or open the metal surface at the point of stress. Weak spots are first formed as small pits, followed by cracks. As temperature and stress are increased, the cracking process is accelerated.

Chlorides, second only to water, are the most common substances in industry. Our only concern about our mechanical seal parts and chloride stress corrosion is the spring and/or metal bellows material. One should always resist using stainless steel springs or bellows unless they are not in contact with chlorides. This is why Fluidtec's standard spring material is Hastelloy C.

There is a much longer scientific explanation for the electrochemical process having to do with the action of chloride ions being destroyed and the formation of microanodes, for example. Our primary concern here is chloride stress corrosion. When stainless steel is subjected to heat, stress, and chlorides, cracked springs or bellows are the common results.

7.3.3.1.4 Products that Change State Almost any product changes state, and when considering seal design and environmental controls, we must anticipate the conditions that will occur (i.e., water will turn to steam at a given temperature and pressure). Following is a list of products that can change state dramatically with atmospheric contact or change in temperature. This is not intended to be a complete list.

Corrosive Acids and Bases Corrosives often become twice as aggressive with as little as a 15°F (9°C) increase in temperature.

- Try to cool the seal by using a double seal with cool barrier/buffer fluid.
- Quench and drain or flush with cool product.
- Use a seal designed for chemical service.

Cryogenics Liquid oxygen and nitrogen tend to freeze elastomers and moisture from the atmosphere outboard.

- Use a double seal with buffer fluid between the product and the atmosphere.
- Use a quench and drain to keep the atmosphere isolated from the faces and prevent freezing.

Crystallization Products that crystallize include sugar, caustics, brine, and others. These types of products tend to clog seals on the atmospheric side of seals. Clogging occurs when the product passes across the faces and moisture vaporizes, leaving a crystallized residue on the atmospheric side of the seal faces. Over time the crystals will continue to build up and restrict a seal's ability to adjust for movement, stalling O-rings and clogging springs or bellows. Crystals formed between seal faces are very abrasive and can cause abrasion of the faces.

- Use a quench and drain to control temperature and clean debris from the atmospheric side of the seal.
- Use a double seal with barrier fluid to control the temperature and insulate/isolate the atmosphere from the product.

Dangerous Products These products include carcinogens, explosives, radiation, toxins, and other lethal products which are hazardous to plant workers and the environment, where leakage cannot be tolerated. Even if the seal fails, fumes and liquid must be contained.

- Use a double seal as a tandem and provide containment for liquids and the vapor recovery system (i.e., convection system).

Dry Running Seals that run in compressors, gases, and vacuum service require cooling to protect faces and elastomers from overheating.

- Use a double seal with good lubrication and compatibility characteristics. Also, pressurize the barrier fluid at least 1 atm [15 psi (1 bar)] greater than seal chamber pressure.

Dry Solids Cake mix, granulated soap, powder, and others clog seals and cause excessive heat at elevated speed.

- Use the double seal approach.

Film Producing Plating solutions, hard water, hot oil, and so on, form a layer of film across seal faces, and often stick and tend to roll up, causing the faces to separate.

- Use double seals, with the barrier fluid at least 15 psi (1 bar) greater than the seal chamber pressure.
- Quench and drain.

Hot Products Hot oil, liquid sulfur, heat-transfer products, and coke become film building, and also build up on the atmospheric side of seals, much like crystallization residue, only sticky.

- Use restriction bushing.
- Use high-temperature metal bellows.
- Flush with cool product.
- Double seal with cool barrier fluid 15 psi (1 bar) greater than seal chamber pressure).
- In some cases, flush, quench and drain, flush with cool fluid, and quench with steam.

Nonlubricants Liquefied petroleum gas, solvents, products with low specific gravity, and boiler feed water should be cooled, in some cases, to avoid flashing and abrasion.

- Cooling can be achieved by using a pumping ring with double seals.
- Use coolers (heat exchangers) for a cool barrier.
- Quench and drain, depending on the situation and availability, to isolate product from the atmosphere and prevent freezing (liquefied petroleum gas) and quench to drain.

Note: Always use carbon as one of the face materials if compatibility is not a problem (carbon vs. silicon carbide is the coolest-running face combination).

Slurries Raw product, lime slurry, dirty water, and sewage are the types of products that clog seals, abrade metal parts, and damage faces.

- Use hard faces against the product.
- Use a double seal with the barrier fluid at least 15 psi greater than the seal chamber pressure.
- Use a quench and drain, with some single seals, to avoid clogging on the atmospheric side by quenching debris to drain.
- Flush from an outside source and use a restriction device in the bottom of seal chamber to control the flush volume required for cooling while isolating the seal from the dirty product.
- Use a GPA seal, which is designed for slurry applications.

Solidification Glue, polymers, asphalt, latex, paints, and others will glue seal faces together, resulting in seal damage at startup. These products also restrict the flexibility of mechanical seals.

- Double sealing with appropriate barrier fluid can help and is historically the way that most seal companies choose.
- A quench with steam or hot water can often control solidification, depending on the product.
- This is normally an area that calls for a P/S-II seal, with proper environmental controls.
- Steam jackets are often used.

Vaporizing Products Hot water, propane, liquefied petroleum gas, and Freon cause flashing between the faces, which blows the faces apart, causing leakage and damage to the faces as they crash together (chips on the face OD is the most common sign).

- Use cool running faces (CBN vs. SC).
- Double seals are often used.
- A cool flush may help prevent flashing.
- Quench and drain to control temperature on the atmospheric side of the faces and prevent freezing in the case of some light gases.
- Control the pressure with a discharge return to seal chamber, this raises pressure above the flash point.

Viscous Materials Asphalt, cold oil, sugar, and syrups are some products that are thick and viscous. Therefore, they create problems much the same as solidification.

- The P/S-II seal can be used to overcome most problems. Often, environmental controls are required.

Note: Protect the integrity of the environmental controls as if it were the process itself, because the process reliability is almost always directly dependent on a specified environmental control. If they were not important, they would not be prescribed.

7.3.3.2 Design and Style Considerations These are some of the advantages of O-rings used as secondary sealing devices.

- Easy to install
- Impossible to install backward
- Seal in both directions (pressure or vacuum)
- Ability to flex and roll
- Take longer to fret a shaft than other options

- Readily available
- Wide range of elastomeric compounds
- Controlled loading by machined slots and hydraulic and pneumatic pressures
- Compensate for misalignment better than do other types (U-seal, wedge, etc.)
- Relatively low cost

7.3.3.3 Mechanical Seal Installation *Note:* Each mechanical seal must be installed at its proper operating length/working length.

A mechanical seal's operating length, also referred to as its working length, is the proper axial length of the seal when it is installed for service. If the seal's mechanical loading device (spring or springs) is not properly compressed, the sealing unit is not at its proper operating length. The proper operating length depends on the type of mechanical seal and type of springing device (multiple, single, wave spring, etc.). In addition, there are component and cartridge types, which dictate different methods of achieving the proper operating length. It is important to read the installation instructions before installation proceeds.

The following are common causes for wrong working length setting:

1. No print was available or the mechanic did not use the installation print.
2. The impeller was adjusted after the seal was secured to the shaft.
3. The reference marks were measured without gaskets in place and/or the equipment was not bolted together tightly.
4. The drive pin was not placed properly into the keyway slot of the pump sleeve.
5. The gasket thickness was not ensured to be consistent (it affects the setting length measurements).
6. The scribe marks were too wide, which can cause inaccuracy.
7. The stuffing box face was not used as the first line of reference.
8. For seals that are set on a sleeve or against the impeller, the working length is changed, in a negative way, if the seal chamber face is altered or an impeller adjustment is made.
9. If impeller adjustments are made routinely, without disassembling the pump, a cartridge seal should be used, which allows the seal to be reset after impeller adjustment.
10. The print was misinterpreted between a recess and a protruding seat when making the reference or interface line.

7.3.4 Environmental Controls

7.3.4.1 Double Seals by Application A double seal by application consists of two mechanical seals applied to a single seal chamber for the purpose of sealing a product that is a volatile organic compound (VOC), dirty, nonlubricating, or very viscous. Also, they are used in products that solidify or otherwise change state. A double seal is used to stop migration of the sealed medium from crossing the primary seal faces. Products that damage the faces, or glue them together, cannot be permitted to migrate across the faces. To work properly, the barrier fluid must be maintained at a pressure no less than 15 psi (1 bar) greater than the seal chamber pressure [1 atm or 14.7 psi (1 bar) at sea level]. When the primary seal leaks, the barrier fluid will leak into the product; so the barrier fluid must be compatible with the product and nonhazardous to the environment. In addition, it must be clean and nonfoaming.

7.3.4.2 Tandem Seals by Application A double seal used as a tandem seal by application is much the same as a standard double seal except for the pressure of the barrier fluid. Tandem seals are used to seal dangerous volatile organic compound (VOC) and Volatile Hazardous Air Pollutant (VHAP) products or any medium producing emissions that must not escape to the atmosphere. Often, this type of arrangement will include a vapor recovery system. The buffer fluid will be maintained at atmospheric or low pressure, or will not be pressurized. The buffer fluid's low pressure will allow the leaked product to enter the buffer system when the primary seal leaks. The buffer fluid must be compatible with the pumped media and atmosphere.

Tandem seals by design are in line and facing the same direction. These seals are double seals designed for high pressure [greater than 150 psi (10 bar)]. If they were not facing in the same direction and acting as two independent inside seals (fluid does not come to the ID of the seal), they would be limited to 150 psi (10 bar), the pressure limit of an outside seal by design. When the first (inside mounted design) seal leaks, the product will migrate into the buffer area and be sealed by the outboard seal, which is designed and positioned in the same direction as the inboard seal. In the tandem design, both the inboard and outboard seals are designed to seal 300 psi (21 bar) or more because the pressure is on the outside of the seal.

Note: Barrier fluid pressure is important. Barrier fluid is pressurized where buffer fluid is not pressurized. If the barrier fluid is at atmospheric pressure, the pumpage can migrate from the seal chamber across the faces, lubricating them, and vaporizing into the buffer area, between the inboard and outboard seals. Conversely, if the barrier fluid is 15 psi (1 bar) greater than the seal chamber, the barrier fluid will migrate across the faces vaporizing into the seal chamber.

The product and barrier fluid must *never* be allowed to be at equal pressure with the pumped product. If the product and the barrier fluid are at equal pressure, each will migrate to the center of the seal faces, the dynamic faces will generate heat, and vaporization will occur between the faces (flashing). At the point of flash, the faces will separate, causing leakage and damaging the faces, and the seal will fail.

7.3.4.3 ***Barrier Fluid*** Barrier fluid or buffer fluid is a liquid or gas introduced between two seals. Care must be taken when choosing barrier fluid for an application. A barrier or buffer fluid should be:

- Clean
- Lubricating
- Compatible
- Nonhazardous
- Nonfoaming

Water, propylene glycol, propyl alcohol, automatic transmission fluid, kerosene, No. 2 diesel fuel, mineral oil, and synthetic oils are a few of the most common barrier fluids.

7.3.4.4 ***Flush*** A flush introduces a small amount of fluid into the seal chamber, in close proximity to the sealing faces, and is generally used for cooling or other protection of the faces. Flush can come from an outside source or be recirculated from the pump discharge and is often recirculated through filters and/or a heat exchanger. Because any liquid used as a flush is introduced into the product and can cause dilution or contamination in excessive amounts, it must be chosen carefully.

7.3.4.5 ***Vent and Drain*** The vent/quench and drain introduces a neutral fluid, gas, or steam through the gland plate on the atmospheric side of a seal. The purpose of this fluid or gas is to dilute product that may have precipitated across the seal faces to the atmosphere and/or to quench debris from the underside of the seal faces while cooling or heating the seal.

Dangerous products can be quenched, diluted, and drained to a collection point or flair stack [American Petroleum Institute] (API) process. If the product creates residue under the seal faces, it can be washed away to drain and minimize buildup. If a product is heat sensitive, quenching can control temperature and alleviate the problem, be it heating or cooling.

7.3.4.6 ***Jacketed Stuffing Box with Restriction Bushing*** A jacketed stuffing box with restriction bushing is used to control temperature in the seal chamber. Jacketed seal chambers may be used for cooling or heating. In this arrangement, a heat-transfer fluid is normally introduced through the jacket void and returns to its source or, less often, to a drain. This heat-transfer fluid is not introduced into the system medium. A restriction bushing is always used in the bottom of the seal chamber. Because of the restriction bushing's close clearance to the pump sleeve or shaft [0.006 to 0.008 in. (0.15 mm to 0.20 mm)], it isolates the seal chamber from the system temperature. By doing this, temperature control is more manageable, since the flow is limited and the temperature of the heat-transfer fluid can be cooled or heated, depending on the application requirements. When using this arrangement, it is imperative to vent or bleed all air from the seal chamber before startup.

To vent air from the seal chamber:

1. Install the seal, then the pump, to the piping.
2. Install a bleed valve at the top of the seal chamber, with a line directed to a collection vessel.
3. Install a flush, quench, and drain as required.
4. With the discharge closed, open the suction valve slowly.
5. Using the bleed valve, carefully bleed the air from the seal chamber until a constant flow of product is discharged into the collection vessel.
6. Close the bleed valve.
7. Remove the bleed valve handle and tag the valve.
8. Proceed with proper pump activation procedures.

Note: Actual testing has proven it possible, with a cooling jacket, to cool an American National Standards Institute (ANSI) group II pump pumping hot oil, from 650°F (343°C) to 250°F (121°C), by using this method. Reducing temperature improves seal life, in the following ways:

1. It reduces flashing between faces.
2. It prevents coking.
3. It protects O-rings from overheating.
4. It reduces harmful heat-related face damage.

7.3.4.7 ***Heat Exchangers*** Heat exchangers are used to add or remove heat from the seal area, depending on the application requirements. Best results are achieved when used in conjunction with a restriction bushing pressed into the base of the seal chamber, which isolates the seal from the process fluid by increasing pressure in the seal chamber and increasing the velocity of the flush through the close clearance bushing. The normal bushing clearance is 0.006 to 0.008 in. (0.15 to 0.20 mm) on each side of the pump shaft or sleeve.

7.3.4.8 ***Recirculation and Suction Return*** Most pumps have lower gauge pressure at suction and higher pressure at discharge. The seal chamber pressure is normally higher than or equal to the suction pressure, but lower than the discharge pressure. Flow can be created between the seal chamber and suction by attaching a tube from the seal chamber to the suction line. When both suction and discharge valves are open, the connecting tube will ensure that fluid is in the seal chamber, even if the horizontal pump is not running. This arrangement also acts to vent air from the seal chamber of vertical pumps.

While the pump is running, the medium will circulate from higher pressure to lower pressure or from the seal chamber to the suction line. This creates circulation through the seal chamber, transfers heat generated by the faces, and lowers the pressure in the seal chamber without diluting the product.

When using a suction return and debris is present in the fluid, it is often a good idea to place the flush port on the bottom, facing down. As the pump runs, the recirculation between the seal chamber and suction will act as a vacuum line. By evacuating debris to suction, it will prevent buildup of particulates, which could settle to the bottom of the seal chamber when the pump is shut off and bind against the seal at startup, causing the seal faces to open.

Note: The suction return method should not be used if the seal chamber pressure and temperature are within 1 atm of the flash point range of the product being pumped or if the seal chamber and the suction are at the same pressure.

7.3.4.9 ***Recirculation and Discharge Return*** A discharge return is simply a circulation tube from the discharge line to the seal chamber, which creates flow. This method is used to circulate the heated medium, generated by the seal faces, out of the seal chamber. Often, it is used to keep product from flashing by raising the seal chamber pressure above the flash point. Discharge return recirculation can also cause problems for mechanical seals by raising the seal chamber pressure, which in turn can increase heat and blast debris at the seal, causing erosive wear, and can erode the circulation tube.

Note: Ask: What will happen in the seal chamber if we use discharge return or suction return? Then choose, using sound logic.

7.3.4.10 ***Other Environmental Controls*** These include:

- Flush using a heat exchanger.
- Flush through a cyclone separator or filter.
- Double seal using a convection system.
- Double seal with outside forced circulation.
- Double seal with an internal pumping ring.

If, for some reason, the environmental control on a given system stops working, the life of the seal will be shortened. In some cases failure will be quick, such as double seals running at high speed. In other cases, a problem will not occur until the system is shut down and started again, such as asphalt, which will solidify if not heated properly.

Typical causes of failure with environmental controls include:

- Improper installation
- Improper maintenance
- Poor operational practices

7.3.5 Failure Analysis

Common causes of premature seal failure are as follows:

- Coupling misalignment
- Heat
- Cavitation
- Air entrapment
- Overpumping
- Pipe strain
- Bearing failure
- Impeller problems
- Poor gasket area
- Poor condition of equipment
- Improper installation
- Flash control
- Loss of environmental controls

7.3.5.1 ***Coupling Misalignment*** Improperly aligned couplings frequently cause seal and bearing failures. Vibration generated by misalignment causes chipped or broken faces as well as overheated and damaged bearings. The solution is to adjust the motor side only. There are three common Alignment methods.

1. *Dial indicator method.* A dial indicator mounted on the pump coupling flange is rotated against the face of the motor flange to find the angular misalignment. Using shims under the motor, both flange halves or flanges are adjusted until they are parallel. Next, the dial indicator is placed against the OD of the opposing flange and rotated to find the offset. Using shims and manipulating side to side, adjustments are made to set "dead on."

2. *Straightedge alignment.* The angular misalignment is found by using a taper gauge. Shimming and adjustment are continued until the flange ends are parallel. A straightedge is used to bridge across both flanges to check level, then side of the top is checked at 90°. Adjustment and shimming

are continued until any offset is neutralized. After alignment is achieved, the motor is tightened securely to its base and checked again.

3. *Laser equipment.* Laser equipment comes with very clear instructions about how to correct angular and offset misalignment.

7.3.5.2 ***Heat*** All mechanical seals have temperature limits. Also, most applications have limits, above which negative results are realized.

- O-rings can overheat, compression-set, or cook.
- Different products may solidify, vaporize, crystallize, or salt.
- Some products become more aggressive (corrosion).
- Seal faces can overheat, causing heat check, thermal shock, distortion, and pitting.
- Metal parts grow, which hinders seal flexibility, and faces can loosen from their carriers.

Heat above the system temperature can be generated in various ways, which adversely affect seal performance.

1. *Running a mechanical seal dry.* If a pump is allowed to run dry, without lubrication for the mechanical seal, it will generate excessive heat. This heat will cook the O-rings, and the seal faces can overheat and warp. All of this results in failure. *Running dry for only a few seconds can destroy a mechanical seal.*

2. *Running a single seal with a vacuum in the seal chamber.* The results are much the same as running dry. A good way to overcome this problem is to run a recirculation line, which will keep the seal chamber flooded and pressurized around the seal and avoid heat buildup. A recirculation line will circulate heat out of the seal chamber.

3. *Limited flow through the seal chamber due to front and rear wear bushings and rings.* A recirculation line from the seal chamber to the suction inlet, which ensures constant flow of product through the seal chamber and evacuates heat, can also correct this condition.

4. *Limited flow in the seal chamber due to expeller vanes on the back side of the impeller.* This condition can be stopped in one of two ways:
   - Remove the expeller vanes.
   - Run a recirculation line from the stuffing box to the suction piping or a discharge return.

5. This problem can be corrected in most cases by a recirculation line, however, bleeding the air from the seal chamber, before startup, is always a good procedure. To stop this habit (air in the seal chamber), a procedure should be implemented to ensure that all the air is bled from the pump and the suction is open completely. If this problem continues, a pressure switch may be installed on the discharge to prevent startup until flow is indicated.

6. *Improper operating procedure causing air entrapment in the seal chamber.*
   - Do not shut off the discharge before stopping the pump.
   - Do not close the suction valve before stopping the pump.
   - Do not pump the suction supply dry.

7.3.5.3 ***Cavitation*** Cavitation is a condition created by insufficient available head at the suction side of a pump, to satisfy the discharge demand. This causes gas bubbles in areas where pressure decreases abruptly. The bubbles collapse (implode) when they reach areas of higher pressure, causing hammering, vibration, and damage to pump parts (impeller, volute, and backplate). It sounds like pumping rocks.

Causes of cavitation:

- Low level in supply source
- Suction line too small
- Buildup obstruction or some type of restriction of flow to the suction or impeller
- Air entrapment because of poor piping design
- Discharge into supply tank designed improperly, causing air entrapment through turbulence

The vibration caused by cavitation is transmitted along the shaft to the seal, bearings, coupling, and the motor. If allowed to continue, this vibration from collapsing bubbles will damage pump parts and cause seal and pump failure.

7.3.5.4 ***Air Entrapment*** When air gets trapped in the suction piping of a pump it can cause cavitation (as mentioned above). Often, air entrapment or entrainment is caused by positioning the return to the supply tank in a poor location or above the fluid level. The return line to the tank should discharge below the fluid level and away from the tank outlet. If space is a problem, a baffle should be used to block air bubbles from flowing into the tank outlet.

7.3.5.5 ***Overpumping*** When a pump is used to pump beyond its design or recommended limits, the result is often overheated bearings, cavitation, motor, and seal failure. Over pumping should be avoided by consulting pump curves provided by the pump manufacturer.

7.3.5.6 Pipe Strain When piping at the pump is not aligned properly with the pump flanges (suction and/or discharge), pipe strain results. Causes include:

- Improper support
- Thermal growth
- Poor installation
- Settling of old system
- Lack of flex couplings

Effects of pipe strain:

- Coupling vibration caused by pipe deflection or misalignment
- Bearing overheating because of side loading
- Impeller binding in the casing
- Premature failure of bearings, coupling, motor, and seal

Pipe strain can be prevented by proper support (hangers), vibration suppressing and flexible connectors (expansion joints), and proper piping alignment.

7.3.5.7 Bearing Failure Bearing failure causes instant problems. When a bearing fails, it loses its ability to support the rotating shaft. The rotating element will whip erratically, causing rotating parts to strike stationary parts. It matters little about the quality of the seal being used; it will fail immediately.

7.3.5.8 Impeller Problems

- Adjust the impeller to the proper setting before securing the seal to the shaft.
- Do not use impellers that are not dynamically balanced.

7.3.5.9 Poor Gasket Area A good gasket surface (125 to 200 RMS), perpendicular (90°) to the shaft is necessary. Care must be taken to correct any gasket surface area that is rough, pitted, marred, has an eroded surface, and/or is not 90° to the shaft or sleeve.

If the gasket area is damaged and/or not square, a facing tool or a lathe will recondition it to the desired condition.

7.3.5.10 Poor Condition of Equipment Condition should be checked by taking the following steps:

1. Check the shaft or sleeve; it should be seal size +0.000 to 0.002 in. (0 to 0.05 mm).
2. Check the total indicated runout; it should not be more than 0.003 in. (0.08 mm).
3. Check the end play (thrust); it should be a maximum of 0.005 in. (0.13 mm).
4. Check all keyways, threads, and shoulders for sharp edges, which could damage elastomers. Round the edges or tape over them before installation.
5. Check for shaft concentricity to the housing or seal chamber, which should be within 0.005 in. (0.13 mm) maximum.

7.3.5.11 Improper Installation Signs of improper installations include:

1. Faces dirty or damaged
2. Secondary seal (elastomer) damaged
3. Seal set at wrong working length
4. Improper environmental controls
5. Seal aligned improperly
6. Wrong seal for application

7.3.5.12 Flash Control When a product reaches the pressure and temperature that cause it to vaporize, it is said to be at its flash point. Normally, the lower the specific gravity of a liquid, the more sensitive or prone it is to flash. When a product flashes between seal faces it will blow the faces apart, which causes chipped edges on the OD and less often on the ID, when they slam back together. Flashing also causes scuffed and/or deep wear on the faces. Combined, these conditions will cause premature seal failure.

The solution to this problem includes the following steps:

1. Try a quench and drain on the atmospheric side of the seal to control the temperature at the faces.
2. Use a double seal with a cool compatible barrier fluid to lower the temperature at the seal faces.
3. Flush with a cooler product.
4. Use the discharge return to raise the seal chamber pressure.

7.3.5.13 Loss of Environmental Controls Environmental controls are used to create a better environment for the sealing device in a given application, pump design, and seal. Seal life is often directly dependent on effective application of environmental controls.

7.3.6 Troubleshooting Mechanical Seals

With few exceptions, when a seal fails in the line of duty, one can normally find the underlying cause by looking at the damage (fingerprints) left on the seal components. Certain conditions create unmistakable marks and/or patterns on seal parts, and by inspecting the damage and knowing the causes, adjustments can often be made to prevent reoccurrence.

Note: When looking at seal face combinations, it is useful to know that the wider of the two faces is normally the hardest face, and the narrow face is the softer or wearing/sacrificial face (the face that wears away as a normal function).

7.3.6.1 Symptoms of Hard (Wide) Face Damage

1. *Wear track*

- A normal wear track on a face, which has been installed properly, will be a uniform concentric path that is of the same width as the opposing face. The wear pattern should be completely even around the track, with no lighter or darker areas.
- If the wear track is not concentric with the OD and ID of the face, it indicates that the track ran off the edge, the seat was not centered, the seal was cocked when installed, or the shaft is not centered in the seal chamber.
- A wide wear track (wider than the opposing face cross section) indicates eccentric movement of the shaft. This condition can be caused by bad bearings, an unbalanced impeller, or a bent shaft.

2. *Cracking or chipping*

- If a hard face is cracked or chipped, it is likely that it was mishandled, caused by vibration, thermal shock (in the case of ceramic), or the hammering effects of cavitation. Poor coupling alignment, a cocked seal, or bad bearings could cause vibration.
- Abrasives embedded into the softer face, and exposed to the opposing face, normally cause a deep wear track on the hard face. Another possible cause is a cracked opposing face, which shears away at the hard face (tungsten carbide vs. silicon carbide).

7.3.6.2 Symptoms of Soft (Narrow) Face Damage

1. *Chipping, coking, or Abraiding*

- Flashing, vibration, or mishandling causes chipped edges on the OD of the sacrificial face. Cool the seal or raise the pressure.
- If the face looks like a phonograph surface, migrating abrasives are the cause. This creates a leak path. Upgrade to hard faces.
- If a face shows signs of coking, the solvents have been cooked out of the oils being pumped. This appears as carbon residue and varnish sticking on the ID and OD of the face. Cool the seal.

Cracked sacrificial faces are usually caused by vibration or mishandling.

2. *Chemical attack.* A carbon can show various stages of chemical attack, from mild etching to pitting to a dissolved state.

- Merely lowering the temperature of the aggressive oxidizing product may solve etching.
- Pitting or blistering typically has one of two causes: Either the carbon is other than chemical grade (low grade), or the carbon is absorbing the oxidizing product which, with heat, expands and causes damage.
- Dissolved carbon shows a complete lack of compatibility, and a different grade or material should be used.

3. *Excessive wear.* Excessive wear is evident when the sacrificial face has its protruding nose worn away. If a seal shows excessive wear after years of service, it has merely done its job and is now spent. However, if excessive wear takes place after only a short time, it can normally be traced to one of six problems.

- A cracked opposing hard face will act as a shearing device and make quick work of a sacrificial face.
- Too much pressure in the seal chamber will load the seal hydraulically beyond its designed abilities. Use a balanced seal.
- If the seal is installed at the wrong operating length, with the spring(s) overcompressed, the face will be sacrificed quickly.
- Chemical attack can dissolve the dynamic surface because of elevated temperature and reduce the length of the nose promptly.
- Abrasive service can quickly erode a carbon face. Upgrade to a hard face or use a double seal.
- Dry running can make short work of a carbon face if the pressure and velocity combined exceed the limits of the face material.

4. *Sticking faces.* Sometimes faces stick together and a seal fails at startup. In some products, such as sugar, faces will stick together and at startup can break large pieces off the sacrificial face. Typically, drive pins are also damaged in the process. A heated quench or heated double seal with a barrier pressure 15 psi greater than the seal chamber pressure is used.

5. *Blistering.* The dynamic surface of a seal face made of some grades of carbon blister when asked to perform in demanding applications. Blisters appear to be small craters in the face. Poor, less dense, or porous grades of carbon may contain small pockets of gas within their structure. When subjected to heat generated by the two seal faces rubbing together, the gases contained within these voids expand and cause tiny explosions, which create small craters. Upgrade to a better carbon.

6. *Face wear in one spot only.* Wear in one spot is created by one of three types of circumstances.

- If the seal is cocked dramatically during installation, it can wear in one spot.
- Air in the seal chamber can cause wear in one spot. Because of reduced lubrication between the faces at the top portion of the seal, more friction and wear result.
- The face is not flat (within two to three light bands).

7. *Heat checking.*

- Heat checking appears as small fractures in the surface of a seal face which extend completely or partially across the face from the ID to the OD These tiny fractures will quickly damage an opposing face, as they act like several shearing blades. Reduce heat and/or change to silicon carbide.
- Heat checking happens more frequently with coated surfaces. Heat, which is generated by the seal, causes expansion. The less dense or softer substrate will expand faster and more dramatically than the hard coating, causing a split effect between the soft and dense layers. The splitting, called *heat checking*, causes fractures in the coating. Do not use coated seal faces if the temperature cannot be controlled.
- Running dry can also cause heat checking. Use a double seal to ensure that cooling and lubrication is always present if dry running is likely.

7.3.6.3 What the O-Rings Can Reveal

- Incompatibility will cause swelling of elastomeric material. If the cross section of an O-ring is larger than normal, it can be traced to chemical attack. Always be sure you know what product is being pumped, including trace amounts of stray chemical that may be present. Change to a more suitable compound.
- Extruded O-rings are caused when pressure, sometimes combined with temperature, forces the rings to deform and creep into open areas between parts of the seal. Extrusion can cause a seal to become inflexible and could result in a cut elastomer and create a leak path. The two most popular ways to cure extrusion problems are:
  - Use of an O-ring with a higher durometer (e.g., from 70 to 90 Shore A), which will resist extrusion.
  - Seal designs often incorporate antiextrusion or backup rings, which have a hard durometer and block the extrusion avenue.
- Compression set is a condition created by a combination of heat and compression, which actually re-forms an O-ring into the shape of its containment. When the elastomer is removed, it retains the new shape. To control compression set, you must keep the seal cooler. Use a quench and drain, a double seal with a cool barrier, or in some cases a simple recirculation line to suction will reduce pressure and evacuate the undesired heat.
- Almost without exception, nicked or cut O-rings are caused by handling. Take special care to protect O-rings during installation. All sleeve and shaft shoulders should be free of sharp corners. Tape over keyways and threads that could damage O-rings during installation.
- O-rings that have lost their elastomeric qualities and are hard, cracked, or brittle have experienced heat, chemical, and/or ozone attack.
  - All O-rings have temperature limits. If the system is too hot, cool it or change to an elastomer with more favorable temperature limits.
  - Chemical attack will cause O-rings to swell, but as the solvents are rendered out of the compound (degassing), they can shrink under size and become brittle. Be sure to use the correct O-ring compound for the product being processed.
  - Ozone is a strong-smelling form of oxygen created by electrical discharge in air and causes symptoms similar to those of overheating. Some elastomers become brittle and crack. Even shelf life is an issue. The most commonly known elastomer in the mechanical seal world that is affected adversely by ozone is nitrile.

7.3.6.4 Hardware Damage

- Always check metal parts for surface damage from a chemical compatibility standpoint. If pitting, etching, or corrosion is present, a different alloy may be indicated.
- Check the sleeve for wear tracks, which indicate contact between rotating and stationary parts. Typically, this is caused by poor installation (cocked when installed), excessive shaft movement, a shaft not in the center of the seal chamber, or a bent shaft. The wear track on the hard face can help direct you to the true cause.
- If there is scoring on the sleeve OD and the scoring is located under the seat, excessive shaft movement or product buildup on the atmospheric side of the seal should be suspected.
- If the scoring is not concentric, misalignment should be suspected. Other damage to the sleeve could include fretting by the dynamic O-ring, which is normal for some designs. Coat the area where the wear is normal with a hard coating.

- Scoring on the ID of the sleeve can happen only when the seal is not properly secured to the shaft before startup.
- If the sleeve of a cartridge seal has set screw marks on the OD, it suggests that the setscrews were not aligned with the holes through the seal sleeve. The setscrews must pass through the sleeve to drive the seal properly.
- If the setscrews of a cartridge seal drive collar are set against the seal sleeve, it causes any or all of the following:
  - The shaft will spin inside the seal sleeve and destroy the seal shaft O-ring.
  - The seal springs will not hold a proper working length.
  - The sleeve can be distorted, out of round.
  - The drive collar will push forward against the gland face, causing metal-to-metal wear.

Setscrew marks are unmistakable.

7.3.6.5 Face Carrier and the Dynamic O-Ring Based on the designs of the pusher (spring) seals, the dynamic O-ring is placed at different locations. Some dynamic O-rings are placed against the pump sleeve or shaft, some against the drive collar, and others against the ID or OD of the face carrier.

Dynamic O-rings all make a wear mark called a fret. The condition of the fret tells much about the conditions under which the seal has been operating.

- *Pitted*: chemical attack
- *Deep wear*: abrasive service
- *Wide* [more than 0.062 in. (1.6 mm)]: axial movement
- *No fret*: very short seal life or well lubricated

7.3.6.6 Drive Pin Damage The drive pins in a rotary seal (springs rotate with the pump shaft) ensure that the face rotates with no slip at startup and maintain constant contact with the carrier and/or drive collar as the seal runs. Conversely, drive pins in a stationary seal (springs do not rotate with the shaft) act as antirotation pins and keep the face and carrier from rotating. In both designs the pins are subject to stresses and wear, as they are in constant contact, metal to metal, and rub as the pump starts, stops, vibrates, and generates its range of movement. Pins that have been in service will show their scars in various ways:

- Unusually long wear areas suggest axial movement.
- Bent pins suggest faces sticking (slip stick).
- Deep wear is an indication of vibration and/or abrasive service.
- Corroded pins show chemical attack and a material upgrade is needed. (e.g., from 316 SS to Hastelloy C).

7.3.6.7 Broken Springs Springs break in various ways:

- Fatigue from excessive shaft movement.
- Heat + stress + chlorides = chloride stress corrosion and is often a problem with stainless steel springs.
- Mishandling is another cause of broken springs. If the drive pins are not in their slots and someone rotates the carrier by hand, the springs can be sheared.

7.3.6.8 Gland Damage Damage to the gland provides important clues:

- Large dents anywhere in the gland suggest rough handling.
- With cartridge seals, drive collar contact with the gland suggests that the setscrews were not properly secured or the centering clips were removed prior to tightening the setscrews. Some indication of how well the seal was centered can be seen in the gasket area.
- Flat washers should always be used. If there are nut gouges on the gland, washers were not used. (Often, the binding corners of the flats on bolts or nuts can pull a gland off center.)
- Check for rubbing of the sleeve on the minor ID of the gland. This will indicate excessive runout, misalignment, poor centering, or a bent shaft. This will often show discoloration from heat where metal parts have been rubbing.

7.4 EXPANSION JOINTS

An expansion joint is a specially designed flexible connector that can be inserted in a rigid piping system to achieve one or more of the following objectives:

- To absorb movement, thus relieving strain in the system caused by thermal change, load stress, pumping surges, wear, or settling
- To reduce mechanical noises
- To isolate mechanical vibration
- To compensate for misalignment
- To eliminate electrolysis between dissimilar metals

The problems of absorbing motion and vibration in piping systems apply in many industrial areas. These problems have to be solved by engineers, contractors, construction firms, and erectors of air-conditioning, heating, and pollution control systems. The same challenge of absorbing motion or vibration occurs in power-generating, chemical processing, and petroleum refining plants, pulp and paper mills, sewage disposal plants, water treatment plants, and in food-processing and marine applications.

7.4.1 Joint Construction

7.4.1.1 Design Features Elastomeric expansion joints are constructed with seamless, leakproof tubes to prevent confined fluids from penetrating and damaging the body portion. Tubes extend the length of the bore and continue to the outside diameter of the flange face. Continuous strands of wire, steel rings, and/or rubber-coated fabric plies are the reinforcing members of an expansion joint.

The exterior cover of the joint is made from an elastomer which may, or may not, be the same as that used in the remainder of the joint construction, depending on environmental conditions. The cover is available in several different elastomers recommended for protection against mechanical damage and provides resistance to attack from chemicals, oils, fumes, ozone, sunlight, and so on, which may be a product of the environment. The arch configuration is the element that provides movement capability. Abrupt, open arches are industry standard; however, they can be filled with soft durometer elastomers to avoid the accumulation of sediment and provide a smooth bore.

Elastomeric expansion joints are classified by the Rubber Expansion Joint Division of the Fluid Sealing Association as standard classes I and II. Standard class I joints are intended for operating temperatures up to 180°F (82°C). Standard class II joints are recommended for temperatures up to 230°F (110°C). Special class III joints are recommended for temperatures over 230°F (110°C). Garlock elastomeric expansion joints are rated to 250°F (121°C). Manufacturer assistance is available for special designs and fabrication for temperature and pressure ratings exceeding industry standards.

7.4.1.2 Composition Expansion joints are composed of three basic functional elements: the tube, the body, and the cover. Metal retaining rings, control units, and flow liners are often required elements, depending on the type and application of the joint. The construction features of Garlock spool-type joints are described below.

7.4.1.2.1 The Tube The Tube eliminates the possibility of the pumped medium penetrating the body and weakening the fabric. Tube materials such as chlorobutyl, nitrile, Viton, and neoprene are frequently specified. Special tube materials can be specified to minimize erosion, which can result from abrasive materials.

7.4.1.2.2 The Body

- *Fabric reinforcement.* Nylon tire cord, polyester, or fiberglass/Kevlar® trademark of DuPont DeNemours fabrics impregnated with any one of several elastomers are wrapped and plied individually to provide the support and flexibility required between the tube and the cover.
- *Metal reinforcement.* Metal reinforcement is utilized to provide dimensional stability during operation.

Reinforcing rings provide added stability for pressure retention and the extra rigidity needed for vacuum service. Garlock EZ-Flo expansion joints do not use reinforcing rings; however, they utilize a proprietary design combined with high-strength nylon tire cord reinforcement to accommodate high pressure.

7.4.1.2.3 The Cover The cover is a homogeneous layer of either natural or synthetic rubber, designed to protect the body from corrosive attack or mechanical damage. Chlorobutyl is Garlock's standard cover elastomer. Alternative elastomers are available to meet special requirements: for example, neoprene, Viton®, Hypalon® Trademarks of DuPont, or nitrile.

7.4.1.2.4 Metal Retaining Rings Retaining rings must be used with all elastomeric expansion joints. They are installed directly against the back side of the rubber flanges at both ends of a joint. They provide a metal surface to distribute the bolting pressure around the flange and prevent the bolts from damaging the rubber flange when tightened. Standard retaining ring material is $\frac{3}{8}$-in.-thick ASTM A36 carbon steel coated with rust-resistant paint. Galvanized or stainless steel rings are available upon request.

7.4.1.2.5 Control Units A control unit consists of two or more tie rods connected between flanges to prevent joint damage caused by excessive elongation or movement of the pipeline. The tie rods are set for the maximum allowable elongation of the expansion joint and are designed to accept the static pressure thrust developed in the system. They are not designed to replace recommended anchoring of a pipeline, but are used to help avoid damage should pipe anchoring fail. Overcompression of expansion joints can also be damaging, but can be controlled by installing metal sleeves over the tie rods. These are called *compression sleeves*.

Triangular gusset plates for control units are drilled with three holes, two of which are used to bolt the plate to the outside of the pipe flange. The third hole is for the tie rod, which connects the plates on each end of the expansion joint. To dampen noise and vibration, rubber washers are used between the plate and the rod. Control units must be used to protect expansion joints from excessive movement if piping is not anchored properly.

7.4.1.2.6 Flow Liners Flow liners are utilized in high-abrasive applications resulting from high media velocity and/or high-solid-content media. Metal flow liners add significant service life expectancy to elastomeric expansion joints in abrasive applications. Flow liners are typically made from stainless steel; however, they can be manufactured from carbon steel provided that the application is not corrosive. For highly corrosive applications, more exotic metal may be required (i.e., titanium, Hastelloy® trademark of Haynes International C, etc.). A restriction to flow liner applications is the direction of media flow. Vertical applications with media traveling upward will cause media to collect between the liner and expansion joint tube. This may result in bacteria buildup or solidification of media, which will impede motion. For abrasive vertical applications, abrasion-resistant tube material should be utilized.

The metal flow liner is flanged at one end and is installed through the inside of the expansion joint with the flanged end of the liner at the head of the flow. The body of the liner tapers to avoid coming in contact with the wall of the expansion joint during lateral deflection. The length of the liner is shorter than the expansion joint to avoid damage to the flow liner and/or the expansion joint during compression.

7.4.1.3 Types of Joints

7.4.1.3.1 Single Arch Joints Construction consists of rubber, reinforced with fabric, metal rings, or wire. The full-face flanges are integral with the body of the joint and drilled to conform to the bolt pattern of the companion metal flanges of the pipeline. This type of rubber-faced flange is of sufficient thickness to form a tight seal against the metal flanges without the use of gaskets.

7.4.1.3.2 Multiple Arch Joints Joints with two or more arches may be manufactured to accommodate movements greater than those of which a single-arch joint is capable. Multiple-arch joints are components of standard-sized arches and are capable of movements of a single arch, multiplied by the number of arches. The minimum length of the joint is dependent on the number of arches. To maintain lateral stability and prevent sagging when the joint is installed in a horizontal position, a maximum of four arches is recommended.

7.4.1.3.3 Taper Reducer Joints Tapered expansion joints are used to connect piping of unequal diameters. They may be manufactured as concentric taper joints with the axis of each concentric (in line) with each of the others or as eccentric taper joints having the axis of one end offset from the other. Tapers in excess of 20° are not desirable. Recommendations concerning the degree of taper and working pressures should be obtained from Garlock. Pressures are based on the larger of the two inside diameters. Movements are based on the smaller of the two inside diameters.

7.4.1.3.4 Offset Joints Offset expansion joints are custom built to specifications to compensate for initial misalignment and nonparallelism of the axis of the connecting flanges. Offset joints are frequently used in close quarters where available space makes it impractical to correct misalignment with conventional piping. It is recommended that complete drawings and specifications accompany inquiries or orders for offset expansion joints.

7.4.1.3.5 Sleeve Joints This expansion joint is of the same design as the standard spool "arch" type except that capped sleeve ends have an ID equal to, or slightly larger than, the OD of the pipe. These joints are designed to slip over the straight ends of the open pipe and are held in place with clamps. Sleeve-type expansion joints are recommended for low-pressure service because of the difficulty in obtaining adequate sealing for high-pressure service.

7.4.2 Principles of Joint Operation

7.4.2.1 Movements in Process Piping

- *Axial compression*: longitudinal movement that shortens the face-to-face dimension along with the axis of the expansion joint, with pipe flanges remaining perpendicular to the axis.
- *Axial elongation*: longitudinal movement that lengthens the face-to-face dimension along the axis of the expansion joint with pipe flanges remaining perpendicular to the axis.
- *Torsional movement*: rotation of one pipe flange while the other remains stationary, or the simultaneous rotation of both pipe flanges in opposite directions.
- *Lateral or transverse movement*: offset movement of one or both pipe flanges, forming an angle to the axis of the joint, with both pipe flanges remaining parallel to each other.
- *Angular movement*: deflection or rotation of one or both pipe flanges, forming an angle to the axis of the expansion joint or flexible coupling.
- *Expansion and contraction*: as temperatures in piping systems increase or decrease, pipelines expand or contract, the extent of movement being a function of the coefficient of expansion of the pipe material. If expansion and contraction, as well as other movements, are not absorbed, piping, valves, and equipment may be seriously damaged, resulting in costly repairs, replacements, downtime, and possible injuries to personnel.

7.4.3 Joint Selection

Listed below are some of the physical factors that help you determine which expansion joint will best serve your customers' needs.

- What *medium* is being handled? Is it a slurry? Is it acidic or alkaline? If acids are present, what is the concentration and the temperature? Are there solids? If so, what is the flow rate, installation orientation, and flow direction?
- What is the system's maximum design *pressure*? Or is it a vacuum application? Is it both?
- What is the maximum operating *temperature*? Can this temperature fluctuate? How often? For what length of time? Is this temperature the maximum that the system will see?
- What type and how much *movement* can the expansion joints be expected to handle? Is the piping anchored and supported properly?
- What *size* is needed (ID and face-to-face)?

Note: Include retaining rings and control units for each application.

7.4.3.1 Medium The medium will come into direct contact with the tube material of the expansion joint; therefore, the tube material must be compatible. The tube is of seamless construction extending through the bore to the outer edge of the flange. Made of synthetic or natural rubber, the tube provides a protective leakproof lining which prevents the medium from penetrating the carcass and weakening the fabric.

Tube material options consist of various rubber compounds chosen based service conditions.

- If it's a liquid and the liquid is clear cold water, the tube or liner of the expansion joint can be of almost any material.
- If the liquid is not clear cold water, see Table 7-5, showing the physical and chemical properties of various elastomers.

Table 7-5 is general; however, it provides selection guidelines for elastomers which are satisfactory to use as an expansion joint tube material. Use these same guidelines to select the cover material if the exterior of the joint is exposed (even occasionally) to the same medium or vapors of the medium.

Note: Temperature is not taken into account; however, it can affect the material rating.

The selection may be variable. For example, if a joint must withstand concentrated nitric acid, either Viton® Trademark of DuPont/Fluorel® trademark of Dyneon LLC and 3M Co. excellent or Teflon® trademark of Dupont/FEP (outstanding) will work. If in doubt, it is useful to remember that Teflon/FEP is good for almost any chemical except molten alkali metals such as bromine and fluorine. Cost also enters the picture because Teflon/FEP is fairly expensive, as are fluorinated elastomers such as Viton and Fluorel. If the cover will not encounter this medium, the standard chlorobutyl cover may be satisfactory.

7.4.3.2 Pressure In general, the larger the pipe size, the lower its maximum pressure rating. This is due to the increased area of contact. The following information outlines factors that affect expansion joints' pressure retention capabilities.

1. *Style 206: wide flowing arch design.* Pressure ratings change as face-to-face and pipe ID dimensions vary from standard. Style 206 EZ-Flo® is reinforced with nylon tire cord fabric plies and is rated for 250 psi (17 bar) in pipe sizes up to 12 in. in diameter with standard face-to-face dimensions. Extended face-to-face dimensions will reduce the maximum pressure rating of style 206 EZ-Flo since it has no metal reinforcement.
2. *Style 208: U-type design (no arch).* Standard sizes are rated to 25 psi (1.7 bar); however, pressure ratings change, as face-to-face dimensions vary from standard.
3. *Style 204: abrupt arch, metal reinforced.* The pressure rating is not affected by changes in face-to-face dimensions. However, as the pipe size increases, the pressure rating decreases. Styles 204 and 204HP are both reinforced with polyester fabric plies as well as wire or metal body rings. Therefore, the face-to-face dimensions do not affect the pressure. As the face-to-face dimensions increase, additional wire or metal body rings are added to aid the dimensional stability.
4. *Low pressure.* Handling air or low volumes of fluids are typically applications that require low design pressures. These applications may be utilizing piping materials other than steel. For poly(vinyl chloride) (PVC) and fiberglass-reinforced piping (FRP) applications, a lightweight design can offer reduced spring rates.

7.4.3.3 Temperature As an elastomer is heated, it has a tendency to become softer and more flexible. When this occurs, its ability to withstand high pressure decreases. This is true with PTFE expansion joints. Temperatures can also affect the aggressive nature of chemicals or acids. Be sure to verify the maximum temperature requirements for all applications.

Note: The pressure maximum for PTFE depends on the temperature rating.

7.4.3.4 Movement The most common reason for using an expansion joint is to absorb movement. In a power plant, for example, the system is gradually brought up to pipe heating temperature. As the pipe is heated, it expands and grows in length. The expansion joint can absorb this movement, since it is gradual and occurs over a longer

TABLE 7-5 Physical and Chemical Properties of Various Elastomers

Material Designation			Elastomer Physical and Chemical Properties Comparison							
ANSI/ASTM D1418-77	ASTM D2000, D1418-77		Water, Chemical, Animal, and Vegetable, Oil, Alkali Concentrated	Alkali Dilute, Oil and Gasoline, Lacquers, Oxygenated Hydrocarbons	Aromatic Hydrocarbons, Aliphatic Hydrocarbons, Acid Concentrated, Acid Dilute	Swelling in Oil Radiation, Water Absorption, Electrical Insulation	Dielectric Strength, Tensile Strength, Compression, Set Rebound, Cold	Rebound, Hot Dynamic Impermeability, Abrasion	Tear Flame, Cold, Heat	Oxidation, Sunlight, Weather, Ozone
CR	BC BE	Neoprene: chloroprene	4340	4401	2346	4543	5424	5245	4444	5565
NR	AA	Gum rubber: polyisoprene, synthetic	53XX	X004	0033	0655	6646	6627	5052	4020
IR	AA	Natural rubber: polyisoprene, synthetic	53XX	X004	0033	0655	6646	6226	5052	4020
IIR	AA	Butyl: isobutene–isoprene	5654	4034	0046	0455	5430	5264	4045	6556
CIIIR	AA BA	Chlorobutyl: chloroisobutene isoprene	5654	4034	0046	0455	5430	5264	4045	6556
NBR	BE BK CH	Buna-N/Nitrile: nitrile–butadiene	4350	4520	4644	5541	0554	4544	3034	4022
SBR	AA	SBR/GRS/Buna-S: styrene–butadiene	53X2	4004	0033	0655	4544	4425	3053	2020
CSM	CE	Hypalon: chlorosulfonyl-polyethylene	5644	4431	2346	4543	5222	4244	3444	6767
FKM	HK	Viton/Florel: flurocarbon elastomer	5660	4610	6665	6553	5562	4555	2627	7777
EPR	BA CA DA	EPDM: ethylne–propylenediene terpolyer	5656	6036	0046	0766	7546	6545	4056	6767
AFMU	—	Teflon/TFE/FEP: fluoroethylene polymers	7777	7777	7777	737X	XXXX	XXX4	XXX7	7777
S	GE	Silicone	5550	2X02	0026	2566	4036	6020	2367	6666

Rating scale: 0. poor; 1, poor to fair; 2, fair; 3, fair to good; 4, good; 5, very good; 6, excellent; 7, outstanding; x, contact manufacturer.

period of time. Vibration is a faster movement, however, which can be compensated for by means of absorption. An expansion joint without an arch would typically be utilized to reduce or absorb vibration.

Movement capability is provided by the expansion joint arch. The configuration will determine the amount of movement that can be achieved. The abrupt arch design allows multiple-arch as well as filled-arch configurations. Multiple-arch configurations will provide maximum movement based on the number of arches. The single-arch movements are multiplied by the number of arches. (Filled arches reduce movement by 50%.)

Expansion joints are designed to accept movement; however, the system design movement requirements are critical. If the expansion joint is subject to movements beyond the maximum rating, damage can result. This will result in a shortened life or will actually rupture the body plies.

7.4.4 Industrial Use of Expansion Joints

7.4.4.1 ***Food Processing*** The food-processing industry consists of a wide range of products being conveyed via processing pumps. Food-processing pumps require equipment that has U.S. Food and Drug Administration (FDA)-approved materials. Don't be misled! The FDA does not approve expansion joint designs; however, it does approve materials used in construction that come in contact with consumables. Media contents vary based on customer end products. Garlock has supplied FDA materials for applications that convey dog food, tomato sauce, corn syrup, potable water, chocolate, corn, beer, liquor, and others. In addition, some pharmaceutical and/or chemical applications may also require FDA materials.

When solids are conveyed, an EZ-Flo arch design will ensure that no entrapment occurs within the arch configuration. If extended face-to-face dimensions, large diameter, or vacuum is a concern, the abrupt filled-arch design is another option; however, movement capabilities will be reduced by 50%.

Material selection is based on the media content; FDA approval has been granted for the materials used in the following tube material options:

- *White neoprene*: the whitest compound that Garlock offers (if color contamination is a concern)
- *White EPDM*: the best material to use if abrasive materials are being conveyed
- *White nitrile*: not as readily available (four-week lead time), but the best material for media that contain oils (e.g., chocolate)
- *FEP* (fluorinated ethylene propylene): typically used for pharmaceutical and chemical applications

7.4.4.2 ***Wastewater Treatment***

1. *Primary sludge (raw sewage).* First identify the percentage of solids and flow rate within the slurry. If the flow rate is greater than 10 ft/s a metal flow liner would be recommended. See Section 7.4.1.2 for more application parameters regarding the use of flow liners. Nitrile or neoprene would be the elastomer of choice for this particular medium. If an abrupt arch design is necessary, the arch or arches should be filled.

2. *Skimmings.* Skimmings typically contain oily residues that have been separated during treatment procedures. Solids are not usually a concern and allow the use of an abrupt arch or EZ-Flo design. Always confirm the percentage of solids. Because of the oil content, nitrile or neoprene will often be used at temperatures less than 250°F (121°C) for the tube portion. Nitrile will also be used for the cover if there is a threat of environmental attack.

3. *Condensate.* Condensate systems usually operate under full vacuum (29.9 in. Hg); therefore, the 204 EVs are highly recommended, depending on movement requirements. Standard chlorobutyl construction is typical unless by-products are present, in which case nitrile or neoprene would be used, due to the oily contaminants. If steam is present at temperatures above 250 but below 300°F, without any by-products an EPDM tube and cover with chlorobutyl–fiberglass–Kevlar body plies will be used.

4. *Potable water.* A chlorobutyl-constructed 206 EZ-Flo expansion joint will handle a majority of potable water applications. Identify all the parameters of the system (pressure, temperature, medium, vacuum, etc.) and recommend accordingly. Some potable water applications may require FDA-approved material; if so, white neoprene is recommended.

7.4.4.3 ***Chemical Industry*** The chemical industry is extremely complex and involves a variety of systems and media; therefore, gathering detailed STAMP information becomes essential for making proper elastomer and expansion joint recommendations. Assure review of the following information.

- Confirm the piping size and face-to-face dimensions.
- Determine the operating temperatures and possible spikes. How frequently do the spikes occur?
- Verify the application within the system.
- If possible, determine the chemical composition of the media (pH value, % solids if any, etc.).
- Verify the pressure parameters of the system and/or vacuum requirements. Keep in mind that many systems will see a vacuum at startup and/or shutdown even if pressurized during normal operation.

- Determine the capabilities of the system. Is it used for more than one function? Is it cleaned from time to time and, if so, with what chemical and how often?

7.4.4.4 ***Mining Industry*** Grinding typically takes place in mills that break down the ore. The two most common means of grinding are semiautogenous grinding (SAG) or ball mills. Overflow pumps are large pumps utilized to transfer media through the grinding stage of mining and are located underneath the mill (SAG or ball type). These pumps convey heavy slurries that cause significant shaking of the pumps during operation. The typical life of the major components of these pumps is one or two months, due to the wear caused by inconsistencies in size of the product being conveyed. Elastomeric expansion joints can be used on the overflow pumps with an ultrahigh-molecular-weight polyethylene liner due to its abrasion resistance. Common size requirements are 8 to 18 in. Pressure requirements can be as high as 200 psi (14 bar). Style 206 EZ-Flo with a filled arch is recommended. If the rate of flow is 10 ft/s or higher, a metal flow liner should be used.

Tailings piping is used to transfer the spent waste material, which contains no end product. The piping is made of high-density polyethylene, which has a very high coefficient of expansion. These piping systems are usually very long. The high coefficient of thermal expansion coupled with the system length means that many multiple-arch joints are required.

End product that is located in the fine ore bins is ground to a consistency similar to that of cornstarch and will be mildly abrasive. The pumps located in this area are smaller in size (6 to 12 in. ID) and require elastomeric expansion joints with an EZ-Flo arch. The product is mostly dry; however, some moisture is introduced at the end of this process with sulfuric acid. Verification of the location of the expansion joints in this area will determine the recommendation. In the dry area, style 206 EZ-Flo with EPDM tube is recommended. In the area where the sulfuric acid is in slurry form, Guardian 306 is recommended.

7.4.4.5 Power Generation

7.4.4.5.1 Circulating Water (Fresh) System Circulating water systems are usually low-pressure, low-temperature applications consisting of larger ID expansion joints. Use style 204 constructed of chlorobutyl on the suction side of the system due to vacuum. If vacuum is not a concern during startup or operation, style 206 EZ-Flo (chlorobutyl) can be used on the discharge side. *Caution:* Due to the large IDs of the expansion joints, the 206 style will have very limited vacuum capabilities. It is essential to verify parameters of the system with plant engineering. In addition, these large-diameter expansion joints have a long service life and are typically in service for 10 to 20 years. Detailed dimensions should be obtained prior to manufacturing to ensure proper fit.

7.4.4.5.2 Circulating Water (Salt) System With saltwater applications, the main concern is the presence of debris (e.g., shellfish, seaweed, small fish, garbage). Therefore, a style 206 expansion joint with its EZ-Flo arch is an excellent choice. If vacuum capabilities are required, style 204 EVS should be used. Because of the possible debris, it will be necessary to fill the arch or arches of the 204 to prevent entrapment. Chlorobutyl would be the elastomer of choice. Ancillary metal work (retaining rings and control units) should be supplied galvanized, at a minimum; stainless steel is ideal, due to corrosion.

7.4.4.5.3 Condensate System Full-vacuum (29.9 in. Hg) service is usually required in condensate systems; therefore, the 204 EVS or 208 styles are highly recommended, depending on movement requirements. Standard chlorobutyl construction is typical unless steam is present with temperatures in excess of 250°F (121°C). An EPDM tube and cover combined with chlorobutyl with fiberglass–Kevlar body plies will be used for applications up to 300°F (148°C).

7.4.4.5.4 Auxiliary Power System Most power-generation facilities contain an emergency power operation typically transporting fuel to diesel-powered generators. Expansion joints within this system should be constructed with a nitrile tube and cover for temperatures below 250°F (121°C).

7.4.4.5.5 Service Water System (Nuclear) Service water systems are typically found in the nuclear power industry. These piping systems usually range from 12 to 30 in. IDs with pressure requirements up to 150 psi (10 bar); therefore, a custom expansion joint (style 8420) may be needed to accommodate these design parameters. Additional body plies and/or metal reinforcement can be added to the style 204HP design to increase its pressure capabilities and to ensure dimensional stability. Due to the vacuum parameters of the larger IDs, the style 204 design will typically be necessary on the suction side of the system. As long as the temperature does not peak at greater than 250°F (121°C), chlorobutyl will be the elastomer of choice.

Note: Service water systems are considered safety-related, and expansion joints for this system must be manufactured in accordance with U.S. federal regulations, 10CFR50, Appendix B, and 10CFR21.

7.4.4.5.6 Scrubber System (Fossil) In scrubber systems, media consist primarily of lime slurries at various concentrations, percent solids, and flow rates. Because of the abrasiveness of lime, EPDM will typically be the best elastomer. If the flow rate exceeds 10 ft/s, a metal flow liner will also be necessary. Because of the arch configuration, the 206 design would be the best choice, depending on vacuum requirements and face-to-face dimensions. If the operating parameters call for an abrupt arch style, it becomes important to fill the arch or arches to eliminate the entrapment of solids.

7.4.4.5.7 Fly Ash System Typically, the main concern throughout these areas will be the abrasive nature of ash. However, flue gas may be present in various locations. If such gases are present, choose Viton as the elastomer. In the absence of flue gas, recommend natural gum for its abrasion-resistant characteristics. An EZ-Flo arch design should be used due to the percentage of solids being conveyed. If vacuum or face-to-face dimensions do not allow the use of style 206 EZ-Flo, style 204, filled/multiple arch (depending on movement requirements), should be used. Keep in mind that natural gum is rated for a maximum of 180°F (82°C) and Viton is rated to 400°F (204°C). If the flow rate is equal to or exceeds 10 ft/s, a metal flow liner will be necessary.

7.4.4.5.8 Air/Flue Gas Handling These lightweight applications are conveying air and have low pressure requirements. Garlock style 8400 (flue duct) is recommended with temperature capabilities up to 400°F (204°C). Construction is $\frac{1}{4}$ in. (6.4 mm) thick with two plies of fabric reinforcement. Configurations can be round, rectangular, or square and can be slip-on (sleeve-type) or flanged. Detailed measurements need to be taken, as there are no industry standard sizes. Customer blueprints are advantageous if available. If a significant amount of movement is needed, Garlock style 9394 is available in nominal $\frac{3}{8}$-in. (9.5-mm)-thick construction with temperatures up to 400°F (204°C). Style 9394 is available in round configuration and offers $\frac{3}{4}$ in. (19 mm) of movement in axial compression and $\frac{5}{8}$ in. (15.8 mm) of elongation and lateral deflection per convolution. These convolutions can be constructed in a very short face-to-face dimension.

7.4.4.6 Semiconductor Industry The semiconductor industry manufactures microchips and can be divided essentially into two sides of production. Due to the sensitive nature of computer chips, the purity of the process fluids and the vibration dampening of the pumps are of utmost importance, depending on in which side of the manufacturing process the equipment is located.

1. *Mechanical side.* Circulating water systems, furnaces, boiler units, and so on, are typical on the manufacturing side. Purity is not a major concern; therefore, elastomers other than Teflon can be used. The largest quantities of expansion joints will be found in circulating water systems, which normally operate at ambient temperatures. Additionally, a minimal amount of algaecide may be present. The pH range, however, will typically be low enough to offer chlorobutyl material. Style 206 EZ-Flo will handle the majority of these applications. If vacuum is a concern for larger diameters, the abrupt arch style 204, is recommended.

2. *Clean side.* The clean side of manufacturing has two basic systems: ultrapure and the support processing systems.

7.4.4.6.1 Ultra Pure System For ultrapure systems on the clean side, the purity of the process fluid is of ultimate concern; therefore, Teflon-constructed expansion joints (Guardian 200 and 306) would typically be the product of choice. However, all suppliers of equipment within the ultrapure systems need prior qualification and certification in order to manufacture products for use within these systems. To obtain these qualifications, specially designated clean room environments and testing or verification capabilities are required. Currently, Garlock does not have clean room certification for the manufacturing process of styles 214 and 215 or Guardian 200 and 306. This ultrapure system represents approximately 10% of industrial expansion joint requirements.

7.4.4.6.2 Support Processing System In addition to the ultrapure water system, the support processing system consists of caustic and acidic media which are typically being conveyed via plastic, PVC, or Teflon-lined piping. Because of the low force-pound requirements and the Teflon and PTFE construction, styles 214 and 215 are widely accepted for these applications. For piping that requires larger face-to-face dimensions or a larger diameter, lightweight designs, (Guardian 200 or 306) are available. As always when dealing with chemicals, it is important to verify the concentration and remaining STAMP information prior to making a recommendation.

7.4.4.7 Pulp and Paper Industry There are several different processes within the pulp and paper industry, each with a wide range of corrosive chemicals. Listed below are various areas of the pulp and paper industry along with a list of chemicals utilized in these areas.

1. Pulping mechanical side
   - Water
   - Pulp slurries
2. Chemical side
   a. Sulfate (Kraft process)
      - Sodium hydroxide (part of white liquor)
      - Sodium sulfide (part of white liquor)

- Sulfite process
- Sodium, magnesium, calcium, or ammonium
- Combined with sulfur dioxide
- Semichemical
- Sodium sulfite

b. Chemical recovery sulfate

- Black liquor (used white liquor)
- Sodium sulfate (added to black liquor)
- Smelt (result of burning the combination above)
- Sodium carbonate
- Sodium sulfide
- Green liquor (result of combining smelt with water)
- Sodium hydroxide (result of green liquor reacted with milk of lime)
- Calcium carbonate (when filtered = lime mud)
- Calcium oxide (forms when lime mud is burned)
- Calcium hydroxide (water added to calcium oxide)

c. Sulfite (process is complex and varies greatly)

- Base chemicals are refortified with sulfur dioxide

3. By-products of recovery
   - Tail oil
   - Kraft turpentine (pitch)
   - Yeast
   - Lignin compounds
4. Bleach plant—one, or a combination of:
   - Chlorine
   - Sodium hydroxide
   - Sodium hypochlorite
   - Calcium hypochlorite
   - Chlorine dioxide
   - Hydrogen peroxide
   - Water
5. General applications in the pulp and paper industry where expansion joints will be located:
   - Powerhouse
   - Chip mill/chip bins
   - Pulp mill
   - Bleach plant
   - Paper machines
   - Digester
   - Centrifuge
   - Wastewater
   - Water treatment
   - Debarking area
6. Wastewater treatment Plant—Due to the vast variety of chemicals found in a wastewater treatment plant, an EPDM tube expansion joint should be used. The product selection is typically determined by the application (i.e., pressure, vacuum, movements, tapers, etc.).
7. Coater area—Due to the dyes and additives used in the process, an EPDM tube expansion joint should be selected for longest service life. Always remember to use an EZ-Flo arch design in any area that contains high percentages of solids or slurries (pulping process), either the 206 or the Guardian 306 style. When solids are present and the velocity is 10 ft/s or more, a metal flow liner should also be used to protect the tube from abrasion. If the vacuum parameters do not allow this application, be sure to use an abrupt, filled-arch design to eliminate the entrapment of solids. Because the Guardian 200 is not available with a filled arch, the 204 filled arch would then be required, using caution to select the proper tube material.

7.4.5 Joint Installation

7.4.5.1 Removing the Old Expansion Joint

1. *Measuring.* Prior to removing old expansion joints, take accurate measurements of the system. If possible, verify dimensions after removal of the expansion joint, as it is easier to detect flared flanges, especially in large-diameter expansion joints. Measure the face-to-face dimensions of the expansion joint, and check the flanges for any lateral, torsional, and/or angular misalignment.

2. *Inspecting the pipe anchors and supports.* Inspect the pipe anchors and flange supports carefully, before removing old expansion joints. Over the years, pipe supports and anchors may become loose and the expansion joint may actually be supporting the pipe.

3. *Breaking the seal.* With elastomeric expansion joints it is important to break the seal of both flanges before the joint is removed. Forced removal with an unbroken seal may result in damage to the metal flange. To break the seal, small wooden wedges should be driven between the rubber flange and the metal mating flange.

7.4.5.2 Transporting the Expansion Joints Expansion joints should never be rolled on their flanges or pulled with a rope, a chain, or a cloth sling looped through the bolt hole or ID of the expansion joint. Large expansion joints should always be transported in crates or on a platform while lying in a horizontal position to prevent damage. Once at the job site, the joints should always be laid flat on a clean floor and protected against damage.

7.4.5.3 Preparing Installation

1. *Area accessibility.* Machinery and/or equipment that could complicate the installation must be removed. Sharp objects or rough surfaces that may damage the sealing surface of the rubber flanges should be taped or removed ahead of time.

2. *Bolts and fasteners.* Used bolts and fasteners should be inspected for damage. Corroded or damaged bolts should be replaced. Flat steel washers should be used where retaining ring segments join one another.

3. *Retaining rings.* Used retaining rings must be checked for rust and warping and replaced if necessary. It is essential to sealability that the retaining rings are flat and rest tightly against the flanges.

4. *Metal flanges.* Elastomeric expansion joints must be mated against clean, smooth, metal flanges. Gouged or pitted flanges should be resurfaced or repaired. A suitable repair compound is Belzona Super Metal cement.

5. *Bolt lubricants.* The lubricant recommended for use with Garlock expansion joints is a solution of graphite in water or glycerine. Another suitable lubricant is Dow Corning DC-111, a silicone-based material. Any other lubricants must be approved by the manufacturer to ensure chemical compatibility.

7.4.5.4 Expansion Joint Installation

1. *Horizontal and vertical installation.* Both vertical and horizontal applications of larger ID expansion joints must first be maneuvered into position with the use of cloth slings. Be sure to loop the cloth slings on the outside of the expansion joint and on either side of the arch to prevent damage. Be sure to lubricate flanges to aid in the installation.

2. *Piping misalignment.* Garlock expansion joints can be made to size for particular lateral, angular, and torsional misalignments. For expansion joints with 36-in. IDs and misalignments from $\frac{1}{4}$ to $\frac{3}{8}$ in., the expansion joint may be pulled into place with a cloth sling and/or the bolt holes may be oversized. These same techniques may be used on expansion joints with IDs larger than 36 in. and misalignments between $\frac{1}{2}$ and $\frac{5}{8}$ in. Greater misalignments will require special construction and/or drilling procedures.

3. *Bolting sequence.* The bolting pattern recommended is to start at a given point, hand-tighten the nut, then go directly across and repeat the process. Continue to tighten each nut, rotating in a star pattern. After each nut has been snugged into position, each is then gradually torqued, following the same rotation. This process is continued until the edge of the expansion joint either dimples in slightly or bulges out. The objective is to bolt the entire retaining ring segment evenly and to compress the rubber flanges as uniformly as possible. The use of flat steel washers will help to establish an effective seal where the retaining rings are split.

4. *Maintaining the seal.* Check the bolt tightness at least one week after going online, and periodically thereafter. As any rubberlike material takes a set after a period of compression, the bolts may loosen and require retightening. It is particularly important to check bolts during temperature cycles or during shutdown times.

5. *Control units.* Control units should be evenly distributed around the bolt circle of the expansion joint. The triangular plate of the control unit is bolted to the outside of the steel pipe flange using bolts $\frac{1}{8}$ to $\frac{1}{4}$ in. smaller in diameter than the lower holes in the control unit plate. Insert control rods with washers through the top hole of the first triangular plate. Place a compression sleeve (if required) over the control rod. Insert a control rod through the second triangular plate. Place washers and hex nuts on the end of the control unit rod. The control unit rod setting is equal to the combined dimensions of the expansion joint face to face, pipe flange, control unit plate, and washer thicknesses, plus the maximum elongation parameters of the expansion joint. After control units are fully assembled, the exposed threads should be staked next to the nut to prevent loosening.

7.4.6 Joint Troubleshooting

Elastomeric expansion joints will show signs of premature wear as a result of many factors. Old age is one of the most common reasons for wear; however, misalignment is another factor that will cause significant damage to an expansion joint. Listed below are signs of premature wear, typical causes, and action required.

- *Cracking at the base of the arch*
  - This is usually a sign of overcompression.
  - Check for any fabric that is exposed.
  - Check for proper pipe alignment within $\pm\frac{1}{8}$ in. (3.2 mm).
- *Cracking at the base of the flange*
  - This is usually a sign of overelongation.
  - Check for any fabric that is exposed.
  - Check for proper pipe alignment within $\pm\frac{1}{8}$ in. (3.2 mm).
  - Measure face to face in at least four quadrants.
- *Leaking at the flange*
  - The bolts may need tightening. Following initial installation the bolts should be retorqued at least after one week and checked periodically thereafter (the OD of the expansion joint flange should bulge out or dimple in slightly).
  - The mating flange surface may have excessive grooves, scratches, and/or distorted areas.

  - The expansion joint may have been overextended at the point of installation or operation. Control units may be necessary to keep the expansion joint within the design parameters.
  - The retaining rings may be warped or corroded.
- *Weeping at the bolt holes*
  - Check the integrity of the tube portion of the expansion joint, if accessible. If cracks exist, the medium has reached the fabric reinforcement and has wicked through to the bolt holes.
- *Excessive ballooning*
  - Check the operating pressure in relation to the design pressure.
  - The strengthening members of the expansion joint may have broken down over time.
  - The temperature may be excessive.
- *Additional observations*
  - The cover portion may feel hard and brittle due to excessive heat or old age.
  - The arch shape may become inconsistent, caused by the migration of body rings.
  - The cover or OD of the flange may feel tacky due to chemical or ozone attack from the surrounding environment.
  - The OD of the flange may split if the bolts are torqued too much or inconsistently.
  - If the joint is experiencing concurrent movements, the cover will begin to delaminate.

7.5 GENERAL SEALING DEVICE SELECTION

7.5.1 Product Selection

Several of the basic requirements for a successful career selling in the fluid sealing industry include the ability to "speak the language," have product knowledge, and understand basic principles in order to communicate effectively. With today's Clean Air Act requirements, more attention is being given to properly sealed equipment and process systems.

7.5.2 Understanding the Forces

Before one can select a sealing device properly, it is important to understand the forces acting on the sealing device in the application. First we look at the forces in a flange assembly.

7.5.2.1 Forces Acting on a Gasketed Assembly In a flange assembly the gasket is placed between two flanges, which are then clamped together with bolts, studs, or other clamping devices. In this example, the bolts are creating a clamping load or bolt force (Fig. 7-8), which compresses the gasket. Once the medium is introduced in the system, the internal pressure that is generated will try to push the gasket and separate the flanges. This force is called a *hydrostatic end force*. If the hydrostatic end forces exceed the bolt force, the flanges will separate and the gasket will leak or blow out. The last external force we have to consider is the atmosphere. In other words, what's going on outside the flange assembly?

In a packing or hydraulic application, a compressive load is applied to the sealing components by a cylindrical device called a *gland*. This compressive load causes the sealing components to expand radially, which then creates a seal against the wall of the stuffing box and the shaft. When the medium is introduced, it can also create additional radial expansion by way of an hydrostatic load as it tries to push the sealing components out of the stuffing box (Fig. 7-9). Finally, shaft movement needs to be taken into consideration. How the shaft moves (rotation or reciprocating) and the speed at which it moves will directly affect the sealing device selection process (Fig. 7-10).

Some flange assemblies require an expansion joint, which may be used to counteract or minimize unwanted dynamic forces in the system. These forces can be in the form of vibration, twisting or torsional movement, compression, and elongation. In addition, the expansion joint must be able to hold up to system pressure, positive or negative (vacuum). Finally, the exterior of the expansion joint has to be designed to withstand external atmospheric conditions, such as oil, water, and ultraviolet ray exposure.

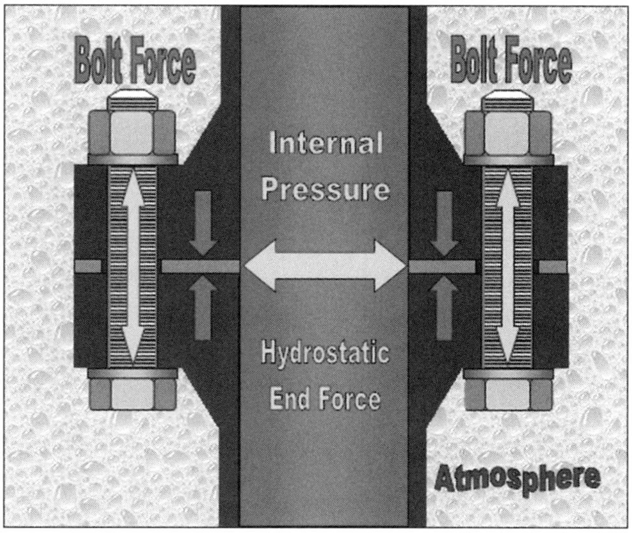

Figure 7-8 Forces acting in a flange gasket joint assembly. (Courtesy of Garlock, an EnPro Industries family of companies.)

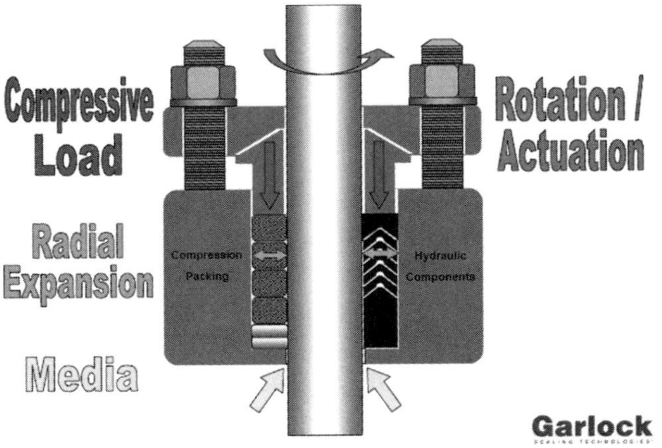

Figure 7-9 Forces acting on compression packing or hydraulic component seal. (Courtesy of Garlock, an EnPro Industries family of companies.)

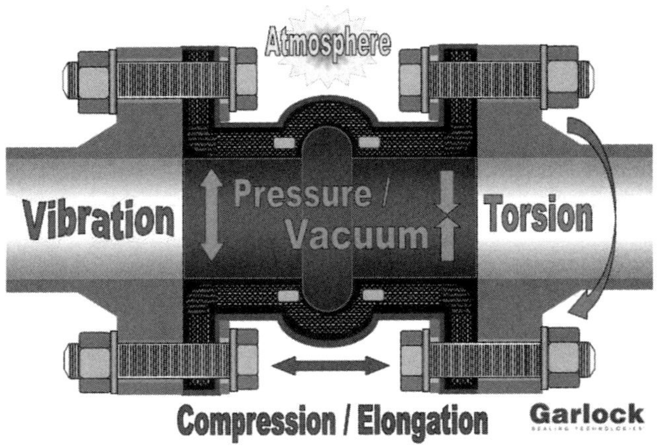

Figure 7-10 Forces acting on an expansion joint. (Courtesy of Garlock, an EnPro Industries family of companies.)

7.5.3 STAMPS Criteria

The six major pieces of information that are required to make a proper recommendation for any Garlock sealing product are known by the acronym STAMPS (size, temperature, application, medium, pressure, and speed). The only one of these six that is not used by all product lines is the last "S" (speed), as speed is only required for compression packing and hydraulic components when used in pump or any rotating applications.

*7.5.3.1 **Size: Gasketing*** Gaskets are most often used in connections called flanges. While there are industry standard sizes and classes, such as 3 in.–150 lb, nonstandard flanges can also be found throughout a plant. When dealing with a nonstandard flange it is very important to obtain as much detail as possible about the contact dimensions (portion of the gasket actually being compressed) and the fasteners or bolts (size, grade, quantity). This information is used to calculate the contact area and available assembly stress. This is a key factor in the selection process, as an improperly loaded gasket could result in premature equipment failure.

*7.5.3.2 **Size: Compression Packing or Hydraulic Components*** Compression packing and hydraulic components are commonly placed in a chamber of a pump, valve, or cylinder called a *stuffing box*. The dimensions of the stuffing box will directly affect the cross section of material required for the application as well as the number of rings. In addition to rings, bushings may be used to take up additional space in the stuffing box that may not be required for the sealing product being specified. For additional information on how to calculate the packing cross section, the recommended shaft finish, and the number of rings recommended per application, refer to the compression packing or hydraulic components technical guides.

*7.5.3.3 **Size: Expansion Joints*** Like gaskets, expansion joints are commonly used in flanged connections and are available in industry standard sizes and classes. When dealing with nonstandard flanges, similar flange and bolting information is necessary to build the expansion joint properly. However, unlike gaskets, expansion joints have an additional dimensional plane that needs to be taken into consideration, the face-to-face dimensions. In addition, it is very important to obtain accurate dimensions if an offset and/or misalignment exists between the flanges.

Garlock is one of the few expansion joint manufacturers that custom builds expansion joints to fit an assembly. This not only makes installation easier, but provides improved service life and movement capability. Knowing how much compression and elongation are required in the assembly is also very important, as that can directly affect the type of arch (abrupt vs. self-flushing) and number of arches.

*7.5.3.4 **Temperature*** Although many systems operate at a continuous temperature, it is important to recognize and take into consideration any system temperature spikes and thermal cycling if they exist. When system temperature fluctuates quickly due to spikes or thermal cycles, the sealing device must react to any movement that occurs in the application related to expansion, contraction, and change in internal pressure. Cryogenic conditions must also be handled properly, as most materials (in this case, sealing devices) have a low-temperature limit where the material becomes brittle. If this occurs, the product may crack or break and result in a premature failure.

7.5.3.5 ***Application*** Application is similar to size in that it relates to the equipment where the product is being used. However, unlike the size information, application details may contain additional details that would not otherwise be uncovered by the basic dimensional or size information. For example, the size of a stuffing box does not tell you if it is a valve that is actuated by hand every few days or a pump spinning at 4000 ft/min, but the packing products that would provide the best performance in each of these applications are completely different.

7.5.3.6 ***Application: Bolts and Fasteners*** The most common may to clamp sealing devices in place is by using fasteners or bolts. There are multiple variables in a fastener that must be identified and understood to ensure that a sealing device receives proper loading. First, we need to determine if the bolt or fastener threads are fine or coarse. Coarse threads (which are by far the most common in industry) are stronger but not as efficient as fine threads. Therefore, a fine thread bolt requires less torque to generate the same load as that of a coarse thread bolt.

Second, flat washers can greatly improve the load translation into the sealing device by creating a "bearing-like" surface for the nut or bolt head to slide on. Without washers, it is more likely that the nut or bolt head will bind against the contact surface of the flange, valve, or other equipment. The third variable to consider is thread coating. Coatings are typically applied to improve fastener corrosion resistance. In addition, they can improve the efficiency, resulting in higher compressive loads at a given torque. The fourth variable to consider is the thread lubricant. A nonlubricated bolt has less than 50% of the efficiency of a lubricated bolt. What that means is that if you torque a nonlubricated bolt to 60 ft-lb, the compressive load that it generates will be approximately half of the load you would generate with the same bolt lubricated and torqued to the 60 ft-lb value.

Bolt Grade	Bolt Material	Yield Strength (psi)	
ASTM A193 B7	Alloy Steel	95,000 to 105,000	B7
ASTM A193 B8 Class 1	Stainless Steel	30,000	B8
ASTM A193 B8 Class 2	Stainless Steel	50,000 to 100,000	B8
ASTM A307 Grade A&B	Carbon Steel	33,000	
SAE J429 Grade 2	Carbon Steel	36,000 to 57,000	
SAE J429 Grade 5	Carbon Steel	81,000 to 92,000	
SAE J429 Grade 8	Carbon Alloy Steel	130,000	

Figure 7-11 Bolt grades. (Courtesy of Garlock, an EnPro Industries family of companies.)

Bolt grade shows the more common fastener grades used in industrial applications (Fig. 7-11). Also shown are the yield strengths for these different grades, which is the stress point at which a bolt becomes permanently deformed (in other words, it will not spring back to its original size when the load is removed). Bolt grade identification is extremely important when determining proper torque values for an assembly. For example, ASTM A193 B8 stainless steel bolts are commonly used for chemical services where corrosion resistance is required. In this example there are two different B8 bolt designations shown; however, there is a significant difference in yield strength between class 1 (most commonly used stainless steel bolt in chemical service) and class 2 (strain-hardened stainless steel). Notice that the only difference in how these bolts are marked is a single line under the B8 marking, thus emphasizing the importance of reading the markings properly.

7.5.3.7 ***Media*** The medium is the fluid that the sealing device is sealing in or out of an assembly. Sealing devices are typically used to seal gases and/or liquids. Occasionally, a sealing device may be required to contain a medium in solid form; however, many of these applications are in slurry form (a liquid–solid mixture). When choosing a product to seal a particular medium, it is important to consider not only chemical compatibility but also emissions compliance. For example, a sealing device that works very well for sealing liquids may be chemically compatible with certain gases but may not provide the necessary level of permeation resistance. When dealing with refining applications, the two most common hydrocarbons encountered are aliphatic and aromatic. Aromatic hydrocarbons (such as toluene, benzene, and xylene) are more chemically aggressive than aliphatic hydrocarbons (such as petroleum oil, kerosene, diesel fuel). Therefore, aromatic hydrocarbons usually require more chemically resistant sealing devices made from PTFE, graphite, or metal. Where aliphatic hydrocabons can be sealed with sealing devices constructed from fiber (i.e., aramid) and elastomers (e.g., nitrile, fluoroelastomers). When dealing with strong oxidizers, alkalines, and acids, it is very important to understand the chemicals and concentrations involved.

7.5.3.8 ***Media: pH Scale*** The pH scale (Fig. 7-12) is used to measure the hydrogen concentration of a chemical. As you can see from the figure, common household products that are used everyday can (at the

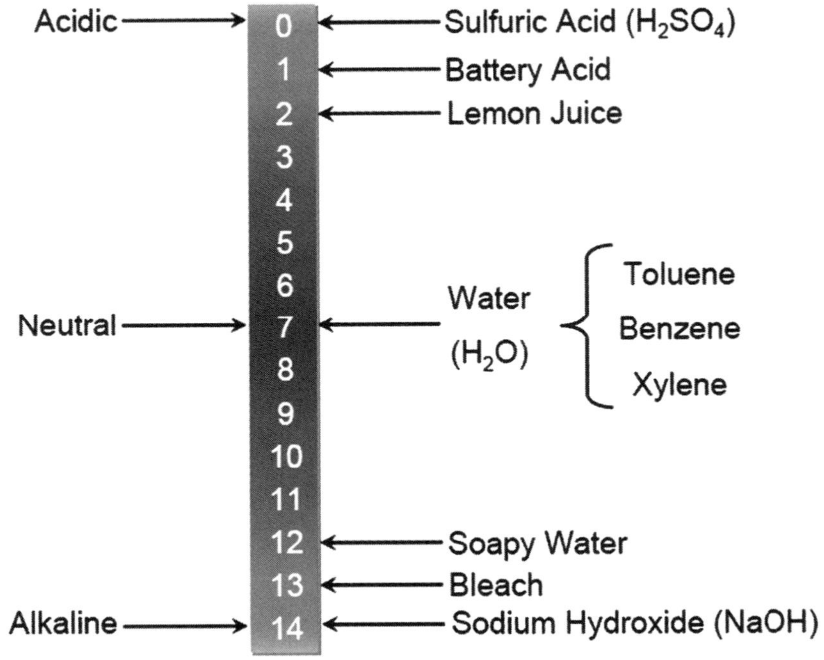

Figure 7-12 pH rating. (Courtesy of Garlock, an EnPro Industries family of companies.)

right concentration) have a similar pH rating to hazardous chemicals used throughout industry.

7.5.3.9 Pressure Most systems operate at some sort of "constant" pressure. However, if system upsets occur, it is important to factor that into the selection process as well as making operational corrections to try and prevent those upsets from occurring. One system pressure upset is surging, when system pressures are increased drastically over a short period of time. In many cases, surges are due to valves being opened too quickly or pumps that are "dead-headed."

A second type of upset is hammering, which can be created in a system in a few different ways. One of the most common is created when valves are shut too quickly. When fluid travels through a pipe, the product has forward-moving inertia, and the rate at which it moves is called the velocity. A system can have a high velocity at a very low pressure. Therefore, if a valve is closed quickly, the system must absorb the inertia of the medium. The result can be similar to a car hitting a brick wall—something has to absorb the energy. In most cases the sealing device is the weakest point in the assembly and therefore takes the impact. The most common systems that experience sealing device failure due to hammering are steam systems, and the phenomenon occurs when condensate is not removed or drained properly from the system.

The last form of pressure upset highlighted is thermal expansion. Unlike gases, liquids do not compress. Therefore, when a liquid is heated in an enclosed space with no air space for expansion, the internal pressure that is generated can be catastrophic to the system, equipment, and/or sealing devices in the system. For example, for every 1°F you increase the temperature of water, it will, in turn, generate approximately 20 to 30 psig of internal pressure. This means that a 10°F swing in operating pressure could generate 200 to 300 psig of internal pressure if the water was fully contained.

7.5.3.10 Speed The last portion of STAMPS is speed. Speed information is required for compression packing and hydraulic components used in pumps and other rotating or reciprocating equipment (Fig. 7-13). In the case of

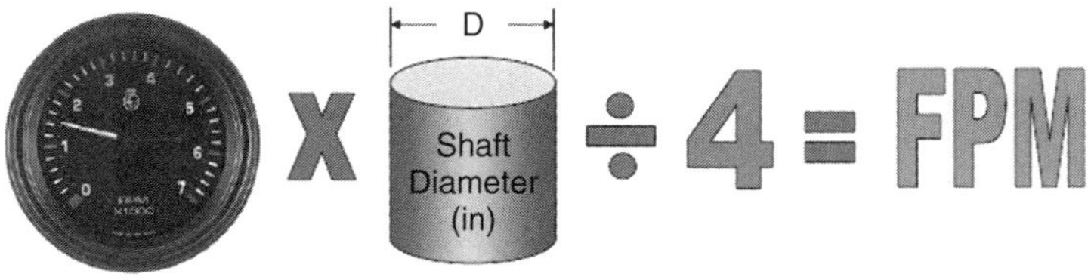

Figure 7-13 Approximate calculation of the surface speed of a rotating shaft. (Courtesy of Garlock, an EnPro Industries family of companies.)

rotating equipment, the speed must be expressed in feet per minute (ft/min). The reason that speed is expressed in ft/min is a large-diameter shaft rotating at a given number of revolutions per minute (rpm) will actually have a higher ft/min surface speed than that of a small-diameter shaft rotating at the same rpm value. In reciprocating applications, it is important to know the length of the stroke and cycles per minute. This information is then used to calculate the speed of the shaft or piston.

REFERENCES

1. Drago, J., Sealing Material Selection: Advice for Ensuring Application Performance and Longevity, *Flow Control*, vol XV, no. 1, January 2009, pp. 22–24.
2. Garlock Sealing Technologies: Learning the "Science of Sealing" - Industrial Gasketing.
3. Garlock Sealing Technologies: Learning the "Science of Sealing" - Compression Packing.
4. Garlock Sealing Technologies: Learning the "Science of Sealing" - Mechanical Seals.
5. Garlock Sealing Technologies: Learning the "Science of Sealing" - Expansion Joints.

8

STEAM TRAPS

Jacob E. Uche
Port Harcourt Refining Company, Eleme, Nigeria
Chikezie Nwaoha
Control Engineering Asia, Ten Alps, Communications Asia, Alandinma, Nigeria

Steam traps are vital components of a steam system. If they do not perform properly, they can be a source of major energy and water loss. A steam trap is a device attached to the lower portion of a steam-filled line or vessel which passes condensate but will not allow the escape of steam [5]. Hot condensate removal is essential to prevent water hammer, which is capable of damaging or misaligning piping instruments.

It is also very desirable to avoid air in the steam system, as any volume of air consumes part of the volume that the system would otherwise occupy. Apart from that, the temperature of the air–steam mixture normally falls below that of pure steam. Again, it has been proved that air is an insulator and clings to pipe and equipment surfaces, resulting in slow and uneven heat transfer.

Traps are used on steam mains, headers, separators, and purifiers, where they remove water formed as a result of unavoidable condensation or carryover from the boilers. They are also used on all types of steam heating equipment in which the steam gives up heat and is converted to condensate. Coils used in heating buildings, in water heaters, and in a wide range of industrial processing equipment are included in the classification.

Whether a trap is used to keep condensate from accumulating in a steam line or to discharge water from a steam-heated machine, its operation is important. If it leaks, steam will be wasted; if it fails to operate, water will accumulate. A satisfactory trap installation must pass all the water that flows to it without discharging steam, must not be rendered inoperative by particles of dirt or by accumulation of air, and must be rugged in construction with few moving parts, so that it will remain operative with a minimum of attention. The presence of carbon dioxide reduces heat transfer because the steam pushes it to the walls of heat-transfer surfaces. Second, carbon dioxide also dissolves in the condensate to form carbonic acid, which may corrode piping and equipment [5].

A strainer is usually installed in the line ahead of the trap, to prevent sediments from stopping up the trap orifice. When selected in the correct pressure rating and provided with strainers, these traps will give satisfactory service [5]. Traps are termed *nonreturn* when the condensate is discharged into a receiver or heater rather than directly to the boiler [5]. A *return* trap delivers the condensate directly to the boiler. Return traps are located above the boiler, and when filled with water a valve opens automatically and admits steam at boiler pressure. This equalizes the pressure and the water flows into the boiler as a result of hydrostatic head caused by elevation of the trap. These traps are sometimes used in connection with low-pressure heating boilers.

8.1 STEAM TRAP OPERATION

Steam is an invisible gas generated by adding heat energy to water in a boiler. When enough energy is added to raise the temperature of the water to the boiling point, additional

Process Plant Equipment: Operation, Control, and Reliability, First Edition. Edited by Michael D. Holloway, Chikezie Nwaoha, and Oliver A. Onyewuenyi.
© 2012 John Wiley & Sons, Inc. Published 2012 by John Wiley & Sons, Inc.

energy (called the *heat of vaporization* or *latent heat*) is added, which changes that water into steam [3]. Steam is an extremely efficient and easily controlled heat-transfer medium. It is most often used for transporting energy from a boiler to any number of locations in a plant, where it is used to heat air, water, or process applications. During this transportation of energy, condensate is generated in the distribution system due to unavoidable radiation. It also forms in heating and process equipment as a result of heat being transferred from the steam to the substance heated. Once the steam has condensed and lost its latent heat, the hot condensate must be removed immediately or it will cause water hammer or similar consequences. Although the heat available in a pound of condensate is negligible compared to a pound of steam, condensate is still valuable hot water and should be returned to the boiler [3]. This is where a steam trap comes in (Fig. 8-1).

8.2 TYPES OF STEAM TRAPS

Various types of steam trap are used in the upstream and downstream industry. Depending on the suitability of the process and working principle, these different types of traps are employed. Typical examples are as follows:

- *Thermodynamic steam trap*: This is the type of trap that is actuated by the principles of thermodynamics and fluid dynamics. These traps operate based on the different flow characteristics of the steam and the condensate.
- *Mechanical steam trap*: This type of trap is activated by a float, responding to changes in condensate level. These traps operate based on the difference in density between the steam and the condensate.
- *Thermostatic steam trap*: This type of trap is actuated by temperature sensitive devices, responding to changes in condensate temperature. These traps operate on the temperature differential among steam, the cooler condensate, and air.

Figure 8-1 Steam trap at work. (Courtesy of Port Harcourt Refining Company, Eleme, Nigeria.)

8.2.1 Thermodynamic Steam Traps

Thermodynamic steam traps are phase detectors in the sense that they can distinguish between liquids and gases. But they do not distinguish between steam and air or other noncondensable gases. Therefore, they have a reduced ability to bleed-off those gases. Minute amounts of steam may also be passed. The thermodynamic working principle is simple, and with only one moving part, these small devices are rugged.

There are three basic types of thermodynamic traps: disk steam traps, piston steam traps, and lever steam traps. They differ from one another by the configuration of the valve they use to open and close a port. Each is well adapted to a particular set of service conditions.

8.2.1.1 Disk Steam Traps Disk traps are used primarily as drip traps and for low steam loads, such as steam tracing lines. They are used in smaller capacities. Disk traps utilize the heat energy in hot condensate and the kinetic energy in steam to open and close a valve disk (Fig. 8-2). They are phase detectors, sensing the difference between liquid and gas or vapor. During initial startup, pressure created by cold condensate pushes the valve disk off the seating surface. This uncovers the inlet and outlet ports, allowing discharge. As condensate reaches the inlet port (a restriction), it experiences a decrease in pressure and an increase in velocity (in accordance with the laws of fluid dynamics). If the condensate is very close to steam temperature, the lower pressure will cause it to flash into steam (in accordance with the laws of thermodynamics).

The resulting high-velocity flow beneath the disk, with its attendant localized pressure reduction under the disk, causes it to snap shut. Flow through the trap then stops

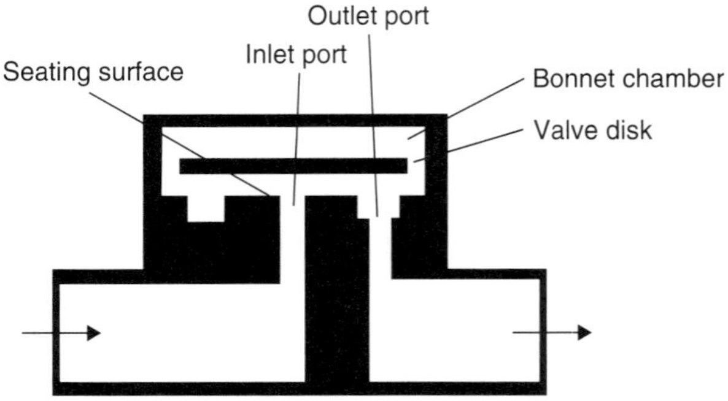

Figure 8-2 Disk steam trap.

until the pressure in the chamber over the disk decays sufficiently to allow the inlet pressure to force the disk off its seat. Condensate then flows through the trap until once again it reaches a velocity level and lowering of pressure such that flashing occurs and the disk can snap shut. This cycle repeats itself continuously—the disk opening to allow the flow of condensate, and closing on high-velocity flash steam. Disk traps are used most often in light condensate load applications and are known as "hot" traps (i.e., quickly discharging very hot condensate immediately after it forms).

Advantages

- Simple construction
- Light weight and small size
- Can be installed in any position
- Rugged, withstands water hammer
- Self-draining, not damaged by freezing
- Function not impaired by superheat
- Versatile, suitable for wide pressure range
- Condensate discharge temperature closely follows the saturation curve
- Performance easily checked in the field
- Primary failure mode—open
- All stainless steel

Limitations

- Marginal air-handling capability
- Condensate discharge temperature cannot be adjusted
- Excessive backpressure in return systems can prevent the trap from closing
- Life reduced significantly as pressures move above 300 psi
- High discharge noise level
- Sensitive to dirt, and this can increase the cycle rate, causing wear

Applications [4]

- Steam tracer lines where maximum temperature is required
- Outdoor applications, including drips on steam mains
- Drying tables
- Tire mold press and vulcanizing equipment
- Dry kilns
- Pressing machines
- Rugged applications (superheat and water hammer)

8.2.1.2 Piston Steam Traps Piston traps utilize the heat energy in hot condensate, and the kinetic energy in steam, to open and close a valve. Like disk traps, they are phase detectors sensing the difference between a liquid and a gas or a vapor. During initial startup, pressure created by the cold condensate lifts the piston valve, allowing discharge of condensate (Fig. 8-3). During this phase, the control chamber pressure is low because the second or control orifice can discharge more condensate than can be supplied to the central chamber through the first orifice. When the temperature of the discharging condensate is very close to the steam temperature (i.e., saturation temperature), the condensate, experiencing the lower pressure of the control chamber, will change into flash steam (in accordance with the laws of thermodynamics). This flashing of the condensate in the control chamber chokes the flow through the control orifice, causing an increase in control chamber pressure. This increased pressure, acting on a larger effective area of the piston valve than the inlet pressure, causes it to snap shut, preventing steam flow through the trap. When cooler condensate reaches the trap, causing the control chamber pressure to drop, flashing ceases and the trap reopens to repeat the cycle.

The control orifice provides a continuous discharge which is helpful in passing air or other noncondensable gases during startup. The piston valve remains closed in the presence of steam because the pressure on top of the piston acts on a larger effective area than the inlet pressure under it. Steam loss through the control orifice is minimal. Introduced in the 1830s, the piston trap was the first thermodynamic trap. It is a "hot" trap, providing excellent service in high-pressure applications.

Advantages

- Suitable for high pressure
- Can be installed in any position
- Good response to changing condensate load conditions
- Rugged, withstands water hammer
- Self-draining, not damaged by freezing
- Function not impaired by superheat

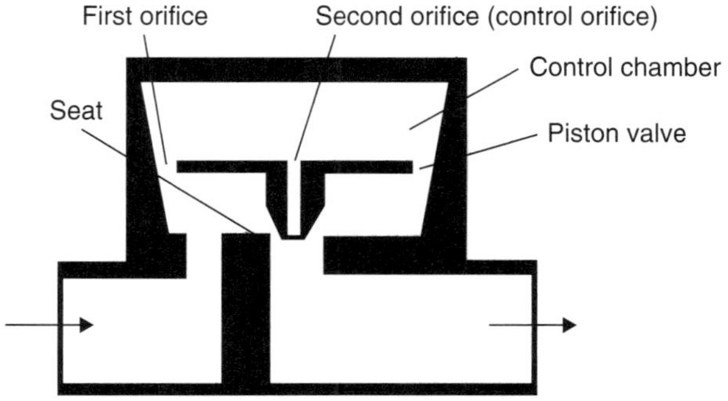

Figure 8-3 Piston steam trap.

- Good air-handling capability
- Primary failure mode–open
- Small size and light weight

Limitations

- Excessive backpressure in return systems can prevent trap from closing
- Condensate discharge temperature follows the saturation curve over a limited range
- Difficult to field-check because of continuous control flow discharge

*8.2.1.3 **Lever Steam Traps*** Lever traps are a variation of the thermodynamic piston trap. They operate on the same principle as piston traps but with a lever action rather than a reciprocating piston action. When the lever is closed, there is a limited flow through the annulus between the inlet valve and its seat (first orifice), which then enters the control chamber and flows out through the second or control orifice (Fig. 8-4). Incoming condensate pushes the lever upward with a tilting motion, and full flow goes under it and out the discharge port. Condensate flowing past the inlet seat (a restriction) experiences a pressure drop (in accordance with the laws of fluid, dynamics) and it will flash into steam (in accordance with the laws of thermodynamics) when the condensate temperature is very close to steam temperature (saturation temperature).

The localized lower pressure under the lever (created by the high-velocity flow of flash steam) causes the lever and inlet valve to snap shut. This prevents steam flow through the trap. When condensate with its cooler temperature again reaches the trap, it will reopen, repeating the cycle. The control orifice has a continuous discharge, which is helpful in passing air and other noncondensable gases during startup. Steam loss through the control orifice is minimal.

Lever traps are usually designed for applications having especially large condensate loads and that benefit from the very rapid discharge of condensate after its formation.

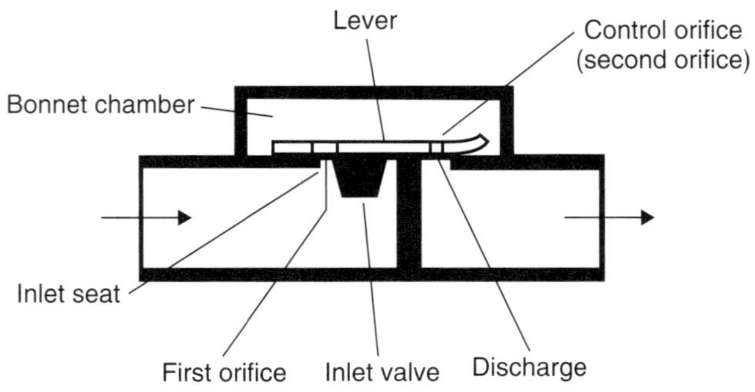

Figure 8-4 Lever steam trap.

Advantages

- Suitable for high-pressure applications
- Good response to changing condensate load conditions
- Rugged, withstands water hammer
- Not damaged by freezing
- Function not impaired by superheat
- Good air-handling capability
- Small, compact, easy to install and service

Limitations

- Excessive backpressure in return systems can prevent trap from closing
- Difficult to field-check due to continuous control flow discharge
- Can be mounted in only one position

8.2.2 Mechanical Steam Traps

Like thermodynamic traps, mechanical steam traps are phase detectors and therefore also have difficulties venting air and noncondensable gases. Mechanical traps employ either an open or a closed float to actuate a valve. Closed-float mechanical traps usually employ a secondary thermostatic air vent which allows the trap to discharge air rapidly. The air vent is an extra component that can fail open, causing the loss of steam, or can fail closed and prevent the trap from discharging condensate. Closed-float traps are usually large. This, combined with a float that is fragile to external pressure and the continuous presence of condensate within the trap, make this device unsuitable for high-pressure applications or installations where water hammer or freeze-ups can be expected.

On the positive side, these devices respond to changes in condensate level only, independent of temperature or pressure. They respond rapidly to changing conditions. Condensate discharge temperatures follow the saturation curve closely and they have a modulating (rather than an on–off) type of discharge. They are extremely energy efficient. Open-float mechanical traps share many characteristics with closed-float traps. One major difference, of course, is the open float as found in an inverted bucket trap. The open float is no longer a weak point, because it cannot be collapsed by excessive pressure. Venting is usually accomplished by means of a small vent hole in the top of the bucket (Fig. 8-5). This is a compromise, as the efficiency of the trap is affected by the sizes of the vent.

The larger the vent the better the air handling, but at the expense of higher steam losses. A smaller vent has the opposite effect. The end result is a trap that is relatively efficient but which does not remove air rapidly during

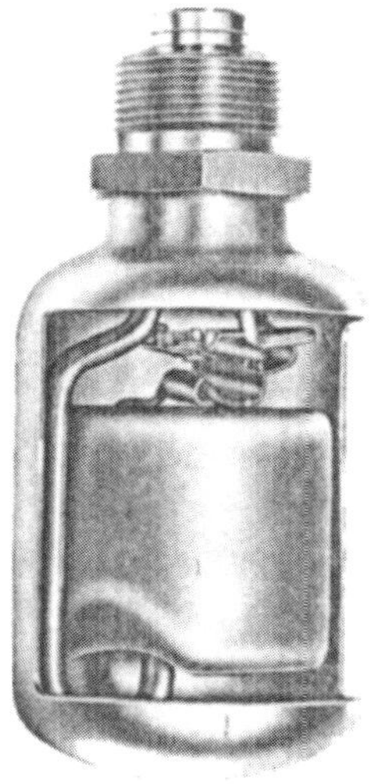

Figure 8-5 Cross-sectioned bucket steam trap.

startup conditions. It discharges near steam temperature with an on–off action, and the discharge temperature follows the saturation curve. All mechanical traps are position-sensitive and can be installed only in their intended orientation.

8.2.2.1 Closed-Float Steam Traps Although it is one of the oldest on the market, the closed-float trap is still in widespread use. The opening and closing of the valve are caused by changes in the condensate level within the trap shell. When the trap is empty, the weight of the float closes the valve. As condensate enters the trap, the float rises and opens the valve, allowing condensate to be discharged. The float is designed to provide sufficient force to overcome the differential pressure across the valve. The internal float and valve configuration is such that the condensate level is always above the valve, creating a continuous water seal at its seat. Actual construction varies widely depending on the manufacturer. While most designs employ a linkage pivot system, one particular design uses no linkage at all and relies on a free-floating ball to achieve the desired action.

An inherent disadvantage of a simple float trap is that it cannot discharge air or noncondensable gases. It is therefore necessary to install an auxiliary thermostatically activated air vent. For this reason, these traps are known as *float and thermostatic* (F & T) *traps* (Fig. 8-6).

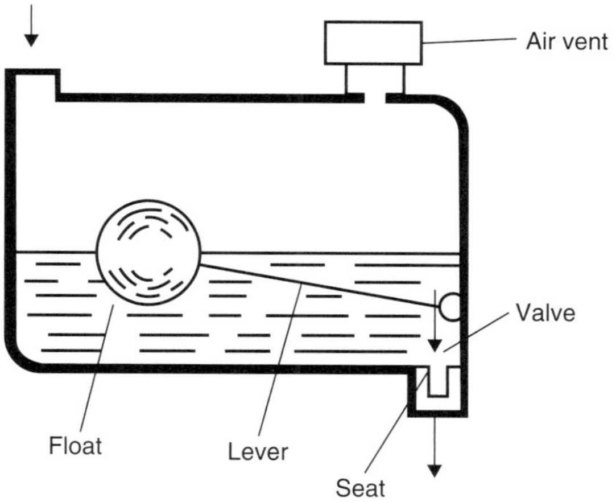

Figure 8-6 Float and thermostatic (mechanical) steam trap.

ADVANTAGES

- Unaffected by sudden or wide pressure changes
- Responds very quickly to condensate load, changes
- Continuous discharge
- Condensate discharge temperature closely follows the saturation curve
- Function is not impaired by high backpressures [1]
- Energy efficient [1]
- Simple construction

LIMITATIONS

- Relatively large and heavy
- Float easily damaged by water hammer
- Does not withstand freezing
- Can be mounted in only one position
- Suitable only for relatively low pressures
- Requires an auxiliary air vent, which is an additional source of failure
- Primary failure mode—closed
- Not self-draining

APPLICATIONS [4]

- Heating main drip traps
- Shell-and-tube heat exchangers
- Tank heaters with modulating temperature regulators
- Unit heaters requiring fast venting
- Steam humidifiers
- Air-blast heating coils
- Air pre-heat coils
- Modulating loads
- Fast-heating startup applications

8.2.2.2 Inverted Bucket Steam Traps Inverted bucket steam traps are members of the mechanical trap family, using an open "inverted bucket" as a float. The trapping principle utilizes the difference in density between steam and water. The construction of the trap is such that the inlet leads into the bottom and open end of the inverted bucket.

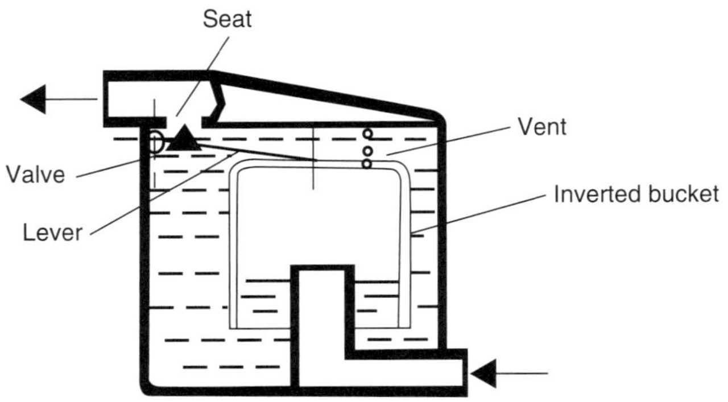

Figure 8-7 Inverted bucket steam trap.

Discharge is through an outlet valve above the inverted bucket. Steam entering the inverted and submerged bucket causes it to float and close the outlet valve, preventing discharge of the steam. Steam in the bucket both condenses and leaks through the vent, allowing the bucket to sink and open the valve to discharge condensate (Fig. 8-7).

The weight of the bucket must be sufficient to overcome the closing force created by the differential pressure across the valve. Inverted bucket traps discharge condensate intermittently, very near saturation temperature. Any air or noncondensable gases entering the trap will also cause the bucket to float and the valve to close. Since they cannot condense as steam does, these gases will cause the trap to remain closed. To overcome this problem, the bucket has a hole to vent air and steam. The size of this vent hole has to be relatively small to prevent excessive loss of steam in addition to the air. While most inverted bucket traps utilize a linkage system to obtain their desired action, one particular design uses no linkage at all and uses a free-floating open spherically shaped float in its design execution.

Advantages

- Simple construction
- Rugged
- Available in pressures up to 250 psig [4]
- Condensate discharge temperature closely follows the saturation curve
- Reliable
- Function is not impaired by high backpressures [1]
- Fast response to changing condensate loads

Limitations

- Marginal air handling during startup [4]
- Not self-draining; does not withstand freezing temperatures
- Not suitable when superheat is present
- May lose prime during light loads and blow live steam; is not self-priming
- Can be mounted in only a single position
- Failure mode is unpredictable (open or closed)

Primary Applications [4]

- Process main drip traps
- Where condensate is lifted or drains into wet return line
- Drum-type roller dryers
- Steam separators
- Siphon-type or tilting kettles

8.2.2.3 Open Bucket Steam Traps Open bucket traps are rarely used today. As with other mechanical traps, they utilize the difference in density between steam and water. When condensate enters the trap, it fills the trap body and causes the bucket to rise and close the valve at the top of the trap (Fig. 8-8). If entrapped air is removed, condensate will continue to enter the trap, finally spilling over into the bucket. This causes it to sink and open the valve, allowing discharge of condensate. When steam arrives, it pushes the condensate out of the bucket through the syphon tube, which in turn re-floats the bucket and closes the valve. As the steam in the trap condenses, additional condensate enters the trap and the cycle is repeated. This type of trap requires an auxiliary thermostatically activated air vent similar to that used in the float and thermostatic trap.

Advantages

- Simple construction
- Very reliable

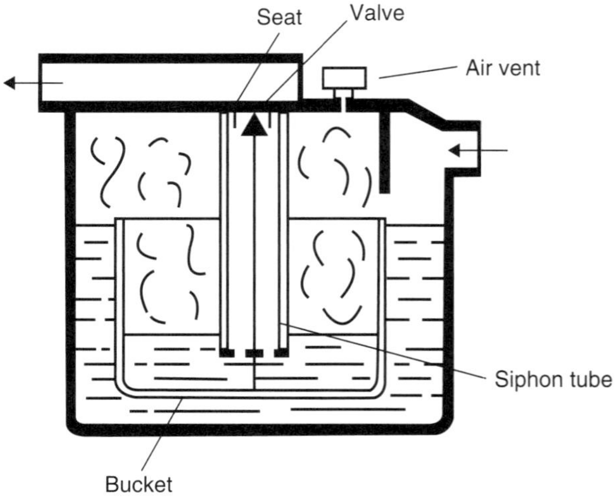

Figure 8-8 Open bucket steam trap.

- Condensate discharge temperature closely follows the saturation curve
- Function not impaired by high backpressure
- Tolerates moderate water hammer
- Fast response to changing condensate loads

Limitations

- Not self-draining; subject to freeze-ups
- Not suitable when superheat is present
- Can lose prime, not self-priming
- Can be mounted in only a single position
- Requires auxiliary air vent, which is an additional source of failure
- Suitable only for relatively low pressures
- Relatively large and heavy

8.2.3 Thermostatic Steam Traps

Thermostatic steam traps respond to changes in temperature and therefore discriminate very well between steam and cooler noncondensable gases. They can purge air from a system rapidly, especially on a cold startup and can be installed in various positions. Most frequently, actuation is by means of a bimetallic element (Fig. 8-9a) or a bellows-like capsule filled with a vaporizing liquid (Fig. 8-9b). Bimetallic-actuated devices are characterized by their high resistance to damage from freeze-ups, water hammer, and superheat. They are relatively small in size and lend themselves to high-pressure designs. The condensate discharge temperature, however, does not follow the saturation curve very well, and the bimetallic elements are subject to corrosion, with some reduction in closing force over time. Bellows-actuated traps, on the other hand, discharge condensate at a temperature that follows the saturation curve. The weak point is the bellows itself, which can be damaged by superheat, water hammer, or freeze-ups.

Thermostatic traps respond slowly to changing conditions even though the cause is usually misunderstood. It is not the heat-sensitive element that is slow to respond; rather, it is the heat energy in the condensate inside the trap, which is slow to dissipate, that causes the time delay. Insulating thermostatic traps reduces their responsiveness even more. Mounting the trap at the end of a cooling leg in an area where air can circulate improves responsiveness and is the basis for installation instructions that recommend a cooling leg at least 3 ft in length.

8.2.3.1 Bimetallic Steam Traps Bimetallic steam traps utilize the sensible heat in the condensate in conjunction with line pressure to open and close a valve mechanism. The valve and seat system is usually arranged to produce a "flow under the seat" condition. Supply pressure, in other words, tends to open the valve. The bimetallic elements are in the form of small disks and are arranged to produce a closing force with increasing temperature. This closing force is in opposition to the opening force created by the supply pressure. Some bimetallic traps use a single leaf element rather than the stacked disk elements shown in Fig. 8-10.

The traps are generally factory-adjusted so that at saturated steam conditions, the temperature-created force of the bimetallic elements prevails, closing the valve and preventing loss of steam. As the temperature of the condensate cools, the line pressure becomes the dominant force, causing the valve to open and allowing the discharge of condensate. Backpressure in a closed-return system

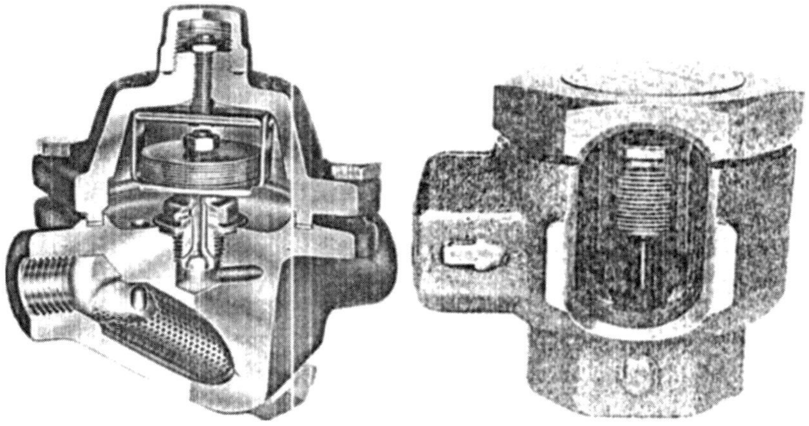

(a) (b)

Figure 8-9 (a) Cross-sectioned bimetallic-actuated trap—a simple thermostatic. (b) Cross-sectioned bellows-actuated dual-range trap—more complex but still a thermostatic.

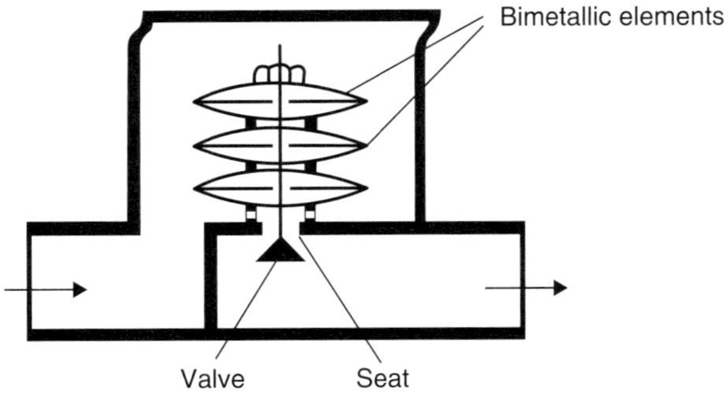

Figure 8-10 Bimetallic steam trap.

provides an additional closing force, resulting in a lower opening temperature than the same trap discharging to the atmosphere. The discharge temperature is therefore is affected by backpressure.

A design problem for bimetallic traps is created by the nonlinearity of the saturation curve. Shaping and stacking techniques used with bimetallic elements have made it possible for these traps to have a discharge temperature that approximates that of the saturation curve. This has expanded the useful pressure range of bimetallic traps without adjustment.

The modern bimetallic trap has many technical and practical advantages. It is used most commonly in Europe but is seeing increasing service in North America.

Advantages

- Rugged
- Energy efficient
- Self-draining
- Resistant to freeze damage
- Withstands water hammer
- Capable of discharge temperature adjustment
- Can be mounted in several positions
- Primary failure mode—open

Limitations

- Dirt particles can prevent a tight valve closing.
- The condensate discharge temperatures do not follow the saturation curve closely.
- These traps are difficult to field-check when operating in a throttling mode.
- The condensate discharge temperature is lowered as backpressure increases.
- These traps are relatively slow to respond to changing condensate loads.
- Bimetallic elements are relatively susceptible to corrosion.

8.2.3.2 ***Bellows Steam Traps*** Bellows traps are thermostatic steam traps that respond to changes in the temperature and pressure of the steam supply to open and close a valve. The valve actuator is a capsule or bellows filled with a vaporizing liquid and has both a fixed and a freely moving end. It opens or closes the valve in response to internal pressure changes (Fig. 8-11).

The most frequently used actuating element is a corrugated bellows. Single-diaphragm capsules are also used, but they provide a correspondingly shorter stroke. This simple operating principle provides many desirable operating characteristics. For example, the number of degrees below steam temperature at which the trap will open can be varied so that the trap provides either a "hot" or a "cold" discharge. Also, the normal failure mode (open or closed) can be changed.

The characteristics of the actuating system can be affected by the liquid fill and natural free length of the actuator. The principles can best be explained by considering a bellows, even though they apply equally well to single-diaphragm capsules. Modern bellows/diaphragm traps have been improved in design, construction, and materials to minimize their inherent Limitations. Today they play an important role in steam trap applications.

Advantages

- Excellent air-handling capability
- Energy efficient
- Self-draining
- Various condensate discharge temperatures available depending on the bellows design
- Follows steam saturation curve to operate over a wide range of conditions
- Can be mounted in several positions
- Simple construction
- Normally open at startup to provide fast air venting
- Fast response to changing conditions
- Compact size and inexpensive
- Small size and weight

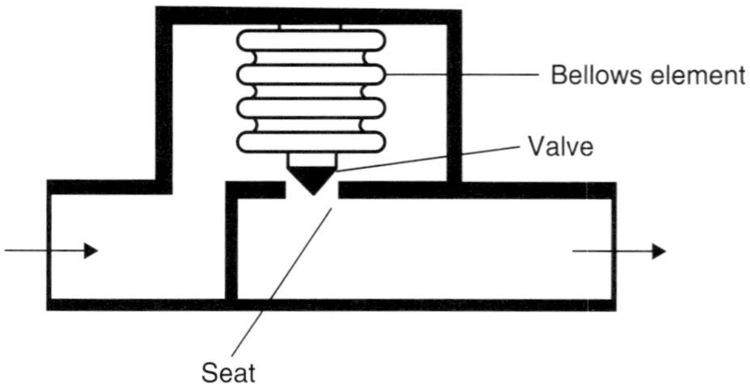

Figure 8-11 Bellows steam trap.

Limitations

- Bellows elements tend to be failure-prone, especially when subjected to water hammer.
- These traps are difficult to field-check when operating in a throttling mode.
- They are generally not well suited for high-pressure applications.
- Water hammer can damage the bellows.
- Superheat can damage the bellows if it exceeds the trap temperature rating (limited superheating capability).
- There is a pressure limit of 125 psig [4].
- The short-stroke diaphragm design is susceptible to dirt-initiated failures.

Applications [4]

- Radiators, convectors, unit heaters
- Cooking kettles
- Sterilizers
- Heating coils
- Tracer lines
- Evaporators

*8.2.3.3 **Liquid or Solid Expansion Traps (Wax Capsule Type)*** Liquid or solid expansion steam traps are finding limited applications today. The opening and closing of these traps is a function of temperature and balanced return spring forces. Elevated temperatures cause an expansion of the thermostatic element that closes the valve, while low temperatures cause a contraction of the element, aided by the spring, which results in opening the valve.

Traditionally, the thermostatic actuator has been in the form of a metal rod that has a high thermal coefficient of expansion or an elastic metallic capsule (bellows) filled with a liquid that expands when heated. In recent years, design innovation has introduced a small diaphragm actuator filled with a waxlike substance that expands rapidly at a preselected temperature. This has significantly reduced trap size and increased the speed of response relative to the more traditional design. Figure 8-12 shows the working internals typical of a newer wax capsule expansion trap.

Regardless of design variations, these traps have one characteristic in common. The temperature of the condensate they discharge remains constant at a predetermined point and is not a function of steam supply pressure. All other steam trap types have a condensate discharge temperature that increases with steam supply pressure. In general, these constant-discharge-temperature traps respond slowly to changes in temperature and should be specified only where subcooled discharge with resulting condensate backup is desired.

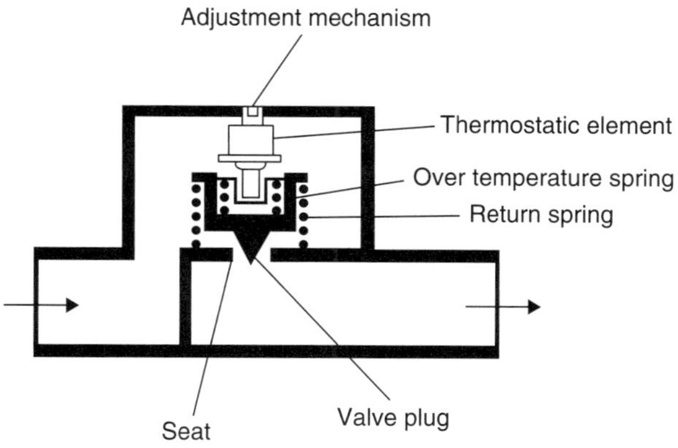

Figure 8-12 Wax capsule steam trap.

Advantages

- Rugged
- Good air-handling capability
- Resistant to freeze damage
- Withstands water hammer
- Can be mounted in any position
- Self-draining
- Primary failure mode—open

Limitations

- Dirt particles can prevent a tight close.
- This trap requires substantial subcooling.
- It is difficult to field-check.
- It is slow to respond to changing condensate loads.
- The actuator is damaged by exposure to high temperatures.

*8.2.3.4 **Orifice Steam Traps*** Orifice traps are seldom used because of their inherent limitations in application range. This device consists of one or more successive orifices (Fig. 8-13). Where two or more orifices are used, condensate passes through a number of successive chambers where flashing occurs, This, in turn, creates a restricting or choking effect and allows the use of larger and less dirt-sensitive orifices for a given condensate capacity. In some design executions, these orifices are adjustable valves.

Advantages

- No moving parts to wear
- Suitable for high-pressure applications
- Rugged, withstands water hammer
- Not damaged by freezing
- Function not impaired by superheat; can be mounted in any position

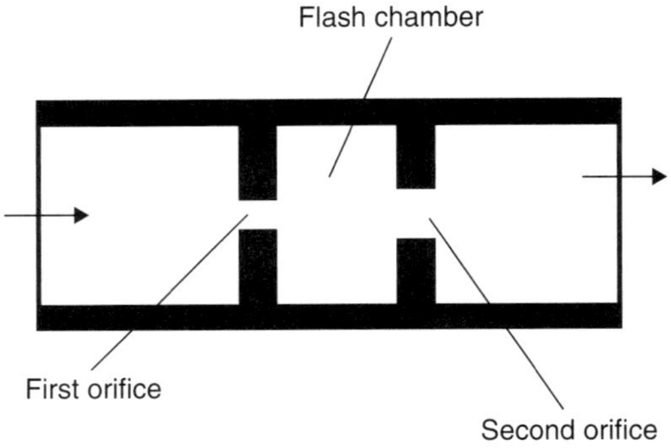

Figure 8-13 Fixed orifice steam trap.

Limitations

- The orifice size must be carefully selected for each installation.
- These traps cannot respond to varying condensate loads.
- Sizing is critical (inefficient if oversized).
- Dirt particles readily impair performance.
- These traps are difficult to field-check because of continuous discharge (waste energy).
- In the absence of condensate, the trap passes live steam.

Application [4]

- Should be limited to constant-load continuous operation.

8.3 STEAM TRAP INSTALLATION

Steam traps are devices that block the passage of steam while allowing liquid condensate to pass. Traps are used in two general installations: on the outlets of steam-using equipment and on the steam lines.

8.3.1 Outlets of Steam-Using Equipment

The first use of steam traps is to keep steam confined inside equipment until it has given up its heat (i.e., to keep the steam from blowing straight through the equipment) [8].

8.3.2 On Steam Lines

The second type of steam trap application is to remove condensate that forms in the pipe. Condensate forms because of heat loss from the pipe. If allowed to accumulate, slugs of condensate propelled by steam pressure may accumulate enough kinetic energy to destroy valves, piping, and equipment. Traps used to drain steam lines are called *drip traps*. In many steam systems, the traps also serve as an important means of removing air and other noncondensible gases from the system. If these gases are not removed, they may block the flow of steam [8]. For example, in a steam coil with many circuits, air that accumulates in the coil causes most of the steam to be routed through a few of the circuits, rendering the rest of the coil useless. Also, noncondensible gases are the source of corrosion in steam pipe and equipment. Oxygen corrodes steel directly, while carbon dioxide forms carbonic acid, which causes acid corrosion. To minimize these problems, most applications require steam traps to have the ability to remove, or "vent," noncondensible gases (i.e., gases other than steam) [8].

Traps need a large venting capacity to clear air out of the steam system after a period of shutdown. Traps need a smaller venting capacity to vent the system continuously while it is in operation, to remove gases that are carried in the steam. Steam traps usually vent noncondensible gases into the condensate system. The location of the traps may not allow complete venting of the system. Parts of a steam system may trap air by letting it become stagnant so that it is not carried along by steam flow. Separate air vents are installed in these parts of the system. These vents typically discharge air directly to the atmosphere rather than to the condensate system [8].

8.4 STEAM TRAP CHECKING

Regardless of the care that has been spent in classifying a plant's steam trap installations, and the thoroughness with which various types of traps have been evaluated and the care with which plant standards have been prepared, there will ultimately be the need to judge whether traps are working properly or whether they require repair or replacement. This is not a job for the untrained and inexperienced!

Certain gross failures are, of course, readily detectable. A cold steam trap is obviously not working, although it remains to be determined whether the trap has failed mechanically, or whether accumulated dirt and scale have choked flow through the trap, or whether it has been inadvertently valved out of service as a result of some unrelated maintenance activity. Traps that are visibly leaking steam at joints or seals have clearly failed. Traps conspicuously blowing large amounts of vapor from their exposed discharge side probably have failed, but it is at this point that a level of uncertainty begins to develop. Most properly operating traps will have "flash steam" associated with their discharge. In addition, the amount of vapor and the pattern of discharge (continuous or cyclic) of a properly operating trap are strongly influenced by normal operating variables such as pressure and condensate load. Uncertainty increases in a dramatic fashion with attempts to assess a trap's performance when it is discharging into a closed return system and it is not possible to see its discharge pattern.

Despite these areas of uncertainty, learning to identify grossly failed steam traps can proceed with a reasonable level of confidence. However, it is the ability to identify that a trap has started to fail and is beginning to pass more steam than is acceptable (but has yet to reveal itself as a grossly failed trap) that elevates steam trap checking to a task for a skilled and knowledgeable person.

Checking a steam trap is the process of observing its performance and comparing it with the performance characteristics that one has learned are typical for a healthy trap of the same type. If the performances are similar, the trap may be judged to be okay. If there are differences, it can be concluded that either the trap is faulty or the system in which the trap is installed has a problem. Three basic techniques are used in observing the performance characteristics of a steam trap: (1) sight—visually observe the discharge pattern; (2) sound—listen to the functioning of the valve mechanism and the flow of fluid through the seat; and (3) temperature—determine the trap's temperature. Each of these techniques has limitations, and seldom can a conclusive opinion be reached on the basis of a single type of observation. Experienced steam trap checkers invariably try to use all three techniques. Some will use more expensive checking equipment than others use in this checking process, but the quality of their results does not seem to vary much.

The three methods are discussed below in the general order of reliability. At least two of the three should be used to increase the chances for correct identification of the condition of a steam trap. A less commonly discussed method is based on fluid conductivity. Although this method should be at least as reliable as sonic-based methods, it is discussed less frequently in the literature, and no general consensus on its relative reliability has been evident. Sight, sound, and temperature measurements have been used to assess the performance of steam traps since steam traps were invented, but the measuring technology has evolved over the years. Equipment using a fourth method, based on the conductivity of the fluid at a specific point in the pipeline, has been developed in recent years.

8.4.1 Sight Method

Watching the discharge pattern of a steam trap is probably the most reliable method of determining whether it is working properly. Unfortunately, many traps discharge into closed condensate return systems without test tees, and the discharging pattern is not visible. Under these conditions, trap performance appraisal is limited to sound and temperature monitoring. The fundamental limitation to observing the discharge pattern of a steam trap in order to assess its health is that hot condensate discharging to atmosphere flashes into steam. The observer then has to decide whether the clouds of vapor being witnessed are the result of leaking steam or the normally expected flash steam. An experienced eye begins to distinguish the lazier action and white appearance of flash steam from the more transparent jetlike discharge of live steam that can be seen right at the trap's outlet. The best clues come by watching a steam trap that has a normal crisp off–on cycle, such as a disk or bucket trap. If the discharge vapor has any velocity during the closed period of the trap's cycle, it can reasonably be assumed that it is leaking steam and should be replaced.

The sight method is usually based on a visual observation of the fluid downstream of the trap. This is possible if there is no condensate recovery system or if test valves have been installed to allow a momentary discharge of the downstream fluid from the condensate recovery system [2]. In either case, the steam trap evaluator must be able to distinguish between "flash" steam, which is characteristic of a properly working trap, and "live" steam, which is characteristic of a trap that has failed open and is leaking or blowing a significant amount of steam. Flash steam is created when a portion of the condensate flashes to vapor upon expansion to atmospheric pressure. Flash steam is characterized by a relatively lazy, billowy plume. Live steam, on the other hand, will form a much sharper, higher-velocity plume that may not be immediately visible as it exits the test valve or steam trap [2].

Sight glasses can also be used for a visual observation but have some drawbacks that must be overcome or avoided. First, steam and condensate are both expected to exist upstream and downstream of the trap (live steam on the upstream side and flash steam on the downstream side). Second, the view through a sight glass tends to deteriorate over time because of internal or external fouling. Third, both steam and condensate will appear as clear fluids within the pipe. In response to the first and third concerns, sight glasses have been developed with internal features that allow the proportion of steam and condensate to be identified.

In steam systems without condensate return, steam leaking past a trap is directly visible. With a condensate return, a test tee and two valves (one to isolate the trap being tested from the influence of other traps, the other to provide an outlet for viewing the fluid downstream of the trap being tested) are all that is required. Thus, the standard technology for conducting a visual test has remained unchanged since steam traps were invented. Sight glasses provide an alternative approach to visual assessment that can be used without affecting system operation but are prone to fouling in some service conditions [2].

8.4.2 Sound Method

Mechanisms within steam traps and the flow of steam and condensate through steam traps generate sonic (audible to the human ear) and supersonic sounds. Proper listening equipment, coupled with the knowledge of normal and abnormal sounds, can yield reliable assessments of steam trap working conditions. Listening devices range from a screwdriver or simple mechanic's stethoscope that allow listening to sonic sounds to more sophisticated electronic devices that allow "listening" to sonic or sonic and ultrasonic sounds at selected frequencies [2]. The most sophisticated devices compare measured sounds with the expected sounds of working and nonworking traps to render a judgment on trap condition.

Sound measurement has progressed from a screwdriver to a more comfortable mechanic's stethoscope to ultrasonic listening devices. The former two assist with hearing sounds in the normal audible range of the human ear, while the latter detects normally inaudible sounds of higher frequency and converts the signal into audible sounds. Simpler ultrasonic listening devices are tuned to a fixed frequency or frequency range, while more advanced models allow tuning to a specific frequency or frequency range. More recently, acoustic signatures representative of properly working and failed traps have been stored in the memory of ultrasonic listening devices for comparison with current readings. This allows the ultrasonic instrument to provide a diagnosis of trap conditions without relying on the experience of the instrument user [2].

8.4.3 Temperature Method

Measuring the temperature of the steam trap is generally regarded as the least reliable of the three basic evaluation techniques. Saturated steam and condensate exist at the same temperature, of course, so it is not possible to distinguish between the two based on temperature. Still, temperature measurement provides important information for evaluation purposes. A cold trap (i.e., one that is significantly cooler than the expected saturated steam temperature) indicates that the trap is flooded with condensate, assuming that the trap is in service [2]. As described above for the visual test via a sight glass, a flooded trap could mean several things, but barring measurement during startup, when flooding can be expected, generally indicates a problem that needs to be addressed. Downstream temperature measurement may also yield useful clues in certain circumstances. For example, the temperature downstream of a trap should drop off relatively quickly if the trap is working properly (mostly, condensate immediately past the trap). On the other hand, the temperature downstream of the trap will be nearly constant if significant steam is getting past the trap. Care must be taken not to use this technique where other traps could affect downstream conditions, however. Like sound measurement, temperature measurement methods vary tremendously in the degree of sophistication. At the low end, spitting on the trap and watching the sizzle provides a general indication of temperature. For the more genteel, a squirt bottle filled with water will serve the same purpose [2].

A glove-covered hand can provide a similar level of accuracy. More sophisticated are various types of temperature-sensitive crayons or tapes designed to change color in different temperature ranges. Thermometers, thermocouples, and other devices requiring contact with the trap offer better precision. Finally, noncontact (i.e., infrared) temperature-measuring devices provide the precision of thermometers and thermocouples without requiring physical contact. Noncontact temperature measurement makes it easier to evaluate traps that are relatively difficult or dangerous to access closely [2].

Temperature measurement tools have also progressed significantly over the years. Although a gloved hand or squirt bottle may be adequate in some situations, much better accuracy can easily be achieved. Earlier instruments were generally thermometers (i.e., devices that measure temperature based on the thermal expansion of various materials). More advanced contact devices are now based on either the thermoelectric potential of two dissimilar metals (thermocouple) or the variation in electrical resistance of a metal with temperature (thermistor). Contact temperature measurement is often coupled with ultrasonic measurement to provide an integrated steam trap testing unit. Noncontact devices allow the freedom and comfort of measuring temperature from a distance based on the thermal radiation emitted from an object's surface. The radiation entering a noncontact pyrometer is either focused on a heat-sensitive element such as a thermocouple or thermistor (radiation or infrared pyrometer) or its intensity is compared to that of a reference element (optical pyrometer).

8.4.4 Fluid Conductivity Method

Conductivity-based diagnostics are based on the difference in conductivity between steam and condensate. A conductivity probe is integrated with the steam trap or just upstream of the steam trap in a sensing chamber. Under normal operation, the tip of the conductivity probe is immersed in condensate. If the steam trap leaks excessively or is blowing, steam flow will sweep away the condensate from the test probe tip and conductivity corresponding to steam will be measured. Thus, the sensing chamber and the existence of steam and condensate under normal and leaking or blowing conditions are similar to that described above [2]. Conductivity measurement must be accompanied by temperature measurement to ensure a correct diagnosis. For example, an indication of steam and a trap that has failed open could occur if a trap has not been used recently and has filled with air. The conductivity of air is similar to that of steam, but a trap filled with air would be close to ambient temperature, in contrast to a trap filled with steam. Similarly, the presence of condensate could mean that the trap is working properly, but could also mean that (1) the trap has flooded, either because the trap has failed closed or something else is blocking the line; (2) the trap is undersized; or (3) the heat-transfer equipment served by the trap is warming up to its normal operating temperature and is generating an unusually large amount of condensate for a short period. These alternative conditions would be indicated by a low temperature in conjunction with the presence of condensate.

Conductivity measurement is a relatively new approach for evaluating steam traps. A probe inserted into the pipeline can easily distinguish between the conductivity of steam or condensate. The probe must be positioned at a location where normally it would be covered by condensate, but failure would cover it with steam, or vice versa. Special sensing chambers create a flow path and precise point for inserting a probe. Conductivity probes, also often coupled with contact temperature measurement devices, can be wired to a central remote monitoring device that receives signals from many probes. This minimizes subsequent data collection efforts, but conductivity probes cost more to purchase and install than does ultrasonic test equipment, which is portable.

8.5 COMMON PROBLEMS OF STEAM TRAPS

Most types of traps will fail open so that they leak steam. Some types of traps fail closed so that they pass neither condensate nor steam, and some types of traps will fail unpredictably in either an open or a closed mode. All types of steam traps can appear to have failed because of some shortcoming or problem in the system in which they are installed. Although this fact must always be kept in mind, it does not become a significant issue unless the steam system has been neglected.

8.5.1 Air Binding

When traps are connected by a long length of small-diameter horizontal pipe, condensate holds up in the steam space and cannot flow to the trap. To prevent air binding, the piping to the traps should be a larger-diameter pipe and a short length, which allows a higher flow rate. Another method is to put a vent valve at a high point in the system [5].

8.5.2 Dirt

Steam condensate often contains particles of scale and corrosion products that can erode trap valves, pipeline, and so on [5]. Some of these particles are large enough to plug the discharge valve or jam it open. Pipeline dirt, oxides, scale, and pipe joint sealants are the enemy of all types of steam traps. Some trap types are more forgiving than others, but all have their limits. A trap that has the appearance of having failed closed because of dirt can often be restored to useful service with a simple cleaning.

A trap leaking steam because of dirt between the valve and the seat should probably be replaced or repaired using new components. The likelihood of permanent damage having occurred because of the dirt is high. To prevent this, a strainer must be installed upstream of each trap.

8.5.3 Improper Sizing

Traps are typically specified several times larger than required using a safety factor to calculate the trap capacity [5]. A trap that is undersized will cause condensate to interfere with the heat transfer efficiency. Therefore, traps having too much excess capacity waste money, act sluggishly, and generate a high level of backpressure that may significantly reduce the life of the trap.

8.5.4 Steam Trap Leakage

For a steam system to operate efficiently, all the steam must condense inside the steam-using equipment so that the latent heat is transferred to the heating application. If steam leaks through a trap, most of the energy of the leaked steam is wasted. This occurs when the valve seat of a steam trap is subjected to erosion and corrosion. When the seat is damaged, the valve will not sit properly and the trap may leak live steam. Steam leakages can be managed by installing a steam lock evaluator [5].

Steam traps may leak for a variety of reasons:

1. *Inherent leakage of the trap design*. In principle, all types of traps, except orifice traps, are capable of blocking steam completely. However, eventually, all traps will leak. Field experience suggests that some types of traps tend to operate longer before developing leakage. Also, some traps will leak if they are not installed in a certain way, whereas the method of installation does not cause leakage in other types [8].

2. *Sticking in an open or partially open position*. All types of traps are subject to complete failure as a result of fouling, corrosion, or mechanical failure. Failure in the open position is the equivalent of having a hole in the system the size of the trap's internal discharge passages. Failure in the closed position causes steam equipment to cease operating because it cannot discharge condensate. Failure of drip traps in the closed position is dangerous because slugs of condensate remain in steam lines [8]. Some traps tend to fail in an open position, while others tend to fail in

a closed position. This tendency may also be influenced by the characteristics of the steam system.

3. *Wear and fouling of the sealing surface*. All types of steam traps (except orifice traps) block the flow of steam by metal-to-metal contact of sealing surfaces. Even if a trap continues to close properly, it will eventually develop leakage because of steam abrasion, fouling, and hammering of the sealing surfaces. Once leakage begins, it typically progresses rapidly. Some types and models of traps develop this type of leakage more quickly than others. Steam loss also depends on the size of traps [8]. Larger traps can waste more steam, and they cost more to replace. All existing traps, large and small, should be surveyed to make sure that each is appropriate for its application.

8.5.5 Steam Locking

When a steam trap is installed in a manner similar to that of air binding, steam locking can also result. In this case the steam may prevent the condensate from reaching the trap. Condensate cannot get to the trap unless it can displace the steam [5]. To prevent steam locking, it is necessary to put the trap as close as possible to the equipment or the line to be drained. If the trap is just below the equipment or line, a balance pipe may be installed between the two to serve as a vent and prevent steam locking. Traps can also be fitted with a steam-lock release valve.

If a steam trap is installed with a length of horizontal pipe 26 in. or longer from the discharge of the condensate outlet of the process, steam locking will occur [7]. A steam trap in good operational condition will only open to pass any condensate and will close when steam enters the steam trap. A steam trap is a simple device; it senses three things: steam, condensate, and noncondensable gas or air. If the steam trap senses that steam (vapor) is present, the steam trap will shut off to prevent steam from passing through. After the steam trap has closed to prevent steam loss, the long horizontal pipe (26 in. or longer) will momentarily be full of steam [7].

8.5.6 Water Hammer

Condensate lying at the bottom of a steam line causes water hammer when the steam system is in service. The steam in the pipe travels at very high velocity, producing waves as it passes over this condensate. As the condensate quantity increases, the high-velocity steam pushes the condensate and creates a dangerous sludge when it changes direction [5]. This is referred to as water hammer. When the high-velocity condensate comes to a halt, the kinetic energy is converted to pressure and this sudden pressure increases to destroy the trap mechanism. To prevent water hammer, it is necessary to ensure that low points are drained adequately before commissioning the steam.

8.5.7 Erosion of Seat and Valve Sealing Faces

The most common failure for all types of steam traps is erosion of the seat and valve sealing faces. This keeps a trap from closing tightly. Once a small leak starts in a pressurized steam system, it becomes a large and expensive leak in a short time. Seat and valve leakage generally results from pipeline dirt becoming caught between mating surfaces. Small manufacturing imperfections relating to surface finishes or proper alignment can also shorten trap life.

8.5.8 Life Expectancy

The life expectancy of a trap is related largely to the pressure at which it must operate. In general, the higher the pressure, the shorter the life.

8.6 STEAM TRAP SELECTION

Selection consists primarily of choosing from one of the major trap technologies (mechanical, thermostatic, thermodynamic) the type of trap that will provide the combination of performance characteristics most closely matching the needs of both plant and equipment. Secondarily, selection includes making judgments about the usefulness of certain accessories and features that are included in some trap designs, as well as making judgments about the advantages of choosing to do business with one trap manufacturer in preference to another.

The steam trap selection process starts with a description of a plant's need. Unfortunately, simply stating that need in terms of equipment to be served—such as a soup kettle, tire vulcanizing press, or an air heater—is not adequate. Although it is important information, it is not sufficient to assure that all the requirements of a specific installation will be met in a satisfactory manner.

Like selecting an automobile, selecting a trap requires an indication of a user's preference with respect to a rather large number of criteria. Fuel economy versus performance, comfort, and safety versus cost, and the latest style versus an established model with proven reliability are familiar car selection choices. Although the factors that are evaluated in electing a steam trap are not nearly as familiar, they are no less important in making a successful decision. They can be classified into several levels of importance. Such a list can then be used as a basis for assuring that all the significant requirements of a particular application are considered. Table 8-1 summarizes and classifies the most significant steam trap selection criteria.

From the standpoint of efficiency, the ability to block steam flow is the main consideration in selecting traps. In addition, you need to consider other characteristics, especially the following:

TABLE 8-1 Steam Trap Selection Strategies

	First-Level Criteria (Satisfy Primary Requirements)
Safety Efficiency Service	These criteria are least subject to compromise. All trap types (mechanical, thermostatic, thermodynamic) arc capable of providing excellent performance in a range of applications when properly sized and installed.
	Second-Level Criteria (Affect Overall Utility)
Ease of checking Sensitivity to backpressure Resistance to freeze damage Dirt sensitivity Installation versatility Air venting Responsiveness to changing loads Resistance to shock, vibration, and water hammer Predominant failure mode Discharge mode Condensate discharge temperature relative to saturation curve Magnitude of condensate subcooling Ease of maintenance Supplementing accessories or features	These criteria relate to a steam trap's overall utility. The differences between mechanical, thermodynamic and thermostatic trap designs are significant. Each trap type has its strengths and weaknesses. These second-level criteria are of particular interest to the user who is looking beyond "first cost" and who makes an evaluation based on "installed" or "life-cycle" costs.
	Third-Level Criteria (Commercial Considerations)
Product availability Post-sales service Warranty Price	These are the commercial criteria that lead to the selection of one supplier over another.

1. *Reliability*. Most steam is wasted by trap failure rather than by inherent leakiness of certain types of traps. Different trap types vary in susceptibility to the three failure modes discussed previously. All traps can be expected to fail, but the average time between failures may vary widely among different types of traps [8]. Many people feel that F&T traps and inverted bucket traps have the longest intervals between failure. Thermostatic traps probably have shorter intervals between maintenance because of their gradual closing characteristics. The reliability of disk traps is controversial, as discussed previously. Orifice traps are very vulnerable to clogging. All traps, except orifice traps, will eventually fail by leaking steam, but they may also stick fully open or fully closed. The latter modes are more likely in systems that shut down periodically, because this allows the mechanism to corrode into position. Inverted bucket traps tend to fail in the open position, because this is their shutdown position. Float traps tend to fail in the closed position, which is their shutdown position. A float trap can also fail by corrosion of the float ball, which closes the valve. Thermostatic traps can fail by failure of the thermostatic element. Orifice traps, of course, can fail only to a closed state by clogging. If an F&T trap fails in the closed position, its thermostatic element will cause it to behave like a thermostatic trap. This may reduce its capacity considerably, but will make the failure difficult to diagnose. No trap is reliable unless it is properly matched to its application. For example, float traps are vulnerable to water hammer, freezing, and dirt, whereas inverted bucket traps are resistant to these problems [8].

2. *Capacity range.* Float, bucket, and thermostatic traps block steam efficiently from zero condensate flow up to their maximum rated capacity. Disk traps adapt to different drainage rates, but they are limited to small capacities because of the way they operate. Orifice "traps" leak steam continuously when condensate is not present, so they must be sized accurately for the maximum expected condensate flow rate [8].

3. *System pressure.* Inverted bucket traps are available for any pressure. Float traps are limited in pressure by the possibility of crushing the float. Bellows-type or encapsulated thermostatic traps are limited in pressure by the possibility of crushing the thermostatic element. Orifice and disk traps are limited in pressure by the erosion that

occurs when high-pressure steam passes through narrow passages. Some disk traps require a minimum pressure drop between the steam side and the condensate side to operate properly, typically 10 psi or more [8]. In addition, disk traps are vulnerable to backpressure because proper operation requires steam to be able to exit from the trap at high velocity.

4. *Venting a cold system at startup.* Thermostatic traps provide rapid venting of cold systems. Thermostatic elements are included in F&T traps for cold system venting. Inverted bucket traps do not vent air rapidly because the smallness of the vent hole in the bucket limits the flow of gases through the trap. To compensate for this, a thermostatic element can be fitted to the bucket to increase the size of the vent hole when the bucket is cold. This added complication reduces the reliability of the trap, all other things being equal. Disk traps vent a cold system very slowly because the trap is closed by air in the same way as by live steam [8]. Orifice traps are poor for venting a cold system because of the typically small size of the orifice. You can vent a cold system by using separate air vents, which gives you greater latitude in selecting trap types.

5. *Venting a warmed-up system.* F&T traps, inverted bucket traps, and orifice traps all do a good job of venting noncondensible gases from a system that is operating at normal temperature. If a thermostatic trap is kept flooded by condensate, air never has a chance to reach the thermostatic element, so the trap cannot vent a warmed-up system [8].

8.7 STEAM TRAP APPLICATIONS

It is possible to classify steam trap applications in a number of diverse ways. Apart from industrial steam trapping, there is steam trapping associated with a low-pressure (below 15 psi) heating, ventilating, and air-conditioning field. This course is concerned primarily with industrial steam trapping. Industrial steam trapping applications are themselves typically divided into two major classifications: protection service and process service.

8.7.1 Protection Service

- *Steam main drip*: usually applied in the drainage of the condensate that normally forms in the pipes delivering steam from a boiler to a specific point of use. This helps prevent damaging water hammer and promotes the delivery of drier steam to plant equipment.
- *Steam tracing*: applied in the drainage of the condensate that normally forms in the small steam lines or steam jackets used to heat valves, field instruments, and the liquids in larger pipelines during freezing conditions or when product temperatures must be maintained at specified levels.

8.7.2 Process Service

- *Steam-using equipment*: applied drainage of the condensate normally forms when steam is used to heat liquids, gases, or solids.

The various classifications of steam traps are presented in the simplified matrix shown in Table 8-2.

An alternative way of looking at the steam trap application universe is by classifications of steam pressure and condensate load. Table 8-3 shows the ranges of pressure and load most commonly encountered in different applications. By its very nature such a matrix tends to be arbitrary, but it does show the general picture. If the number of traps in the industrial world were summarized by steam pressure and condensate load and listed in the appropriate quadrant of the matrix, the largest numbers by far would be in the low-pressure low-condensate-load quadrant. The numbers would decrease rapidly as the load and pressure increase. Since no single trap design or principle of operation is suitable for use across such a wide range of pressures and condensate loads, preparation of a matrix similar to Table 8-3 is sometimes used as a technique to assist a large plant in standardization on the smallest variety of traps for its use.

8.7.3 Other Applications

Following are some common applications and the steam trap types that are often used for each application.

- Steam transport lines (steam main lines)
  - Free-float steam trap for steam main lines
  - Disk-type steam trap
- Steam heating equipment
  - Free-float steam trap
  - Thermostatic steam trap
  - High-capacity free-float steam trap
- Tracers, freeze protection, room heating
  - Thermostatic steam trap
  - Free-float steam trap
  - Temperature-control traps, radiator traps

These different types of steam traps can also be installed in numerous other types of applications in addition to those noted above.

8.8 STEAM TRAP SIZING

Sizing is the process of choosing a trap within the technology selected (mechanical, thermostatic, thermodynamic)

TABLE 8-2 Industrial Steam Trapping Characteristics of Service

Service	Characteristics of Service	Application	Description
Protection	Small, steady condensate loads; infrequent shutdowns; lower pressures (tracing); and higher pressures (steam main drips).	Steam main drips	Drainage of condensate from pipes used to transfer steam from a boiler to its point of use.
		Steam tracing	Drainage of the condensate that normally forms in the small steam lines used to heat valves, field instruments, and the liquids in larger pipelines during freezing conditions; or when product temperatures must be maintained above specified levels.
Process	Larger and fluctuating condensate loads; frequent startups are common, good air handling required.	Steam heats a liquid indirectly through a metal wall.	Shell-and-tube heat exchangers; submerged coils; jacketed kettles.
		Steam heats air/gas indirectly through a metal wall.	Plain or finned coils; unit heaters or air blast coils.
		Steam heats a solid or slurry indirectly through a metal wall to dry, cure, or form.	Rotating cylinders for paper or textiles; platens or presses for plastics, particleboard, and similar materials.
		Steam heats a solid through direct contact to dry, clean, or sterilize.	Autoclave; sterilizer.

TABLE 8-3 Industrial Steam Trapping Pressures and Loads

	Condensate Load			
Steam Pressure (psi)	Low 0 to 50 lb/h	Medium 50 to 500 lb/h	Heavy 500 to 5000 lb/h	Very Heavy Over 5000 lb/h
Low: 15 to 100	Tracing and drip		Process applications	
Medium: 100 to 300	Tracing and drip		Process applications	
High: 300 to 600	Drip		Process applications	
Very high: over 600	Drip			

which has the physical capabilities to meet safely and efficiently the operating conditions of pressure, temperature, and condensate drainage rate for a given installation. Many have mistakenly limited steam trap sizing to matching the end connection size of a trap to the particular pipe size being used to drain a piece of steam-heated equipment. Sizing in its correct sense is matching the steam condensing rate (in pounds per hour) of a piece of equipment (at its particular pressure and temperature conditions) to the rated condensate discharge capabilities of a suitable steam trap.

Trap manufacturers are prepared to make sizing calculations to determine condensate loads in support of their selling efforts. Small plants that have only a few steam traps tend to rely heavily on the trap manufacturer for sizing guidance. Engineering contractors and large plants using many steam traps generally make their own sizing calculations. Examples of several sizing calculations are shown later in the chapter. To obtain the full benefits from the steam traps described in this chapter, it is necessary that the correct size and pressure of trap be selected for each job and that it be installed and maintained properly.

8.8.1 Basic Considerations

Unit trapping is the use of a separate steam trap on each steam condensing unit, including, whenever possible, each separate chest or coil of a single machine. Steam trap sizing is simple when you know or can figure out the following: (1) the condensate load (lb/h), (2) the safety factor to use, and (3) the pressure differential (maximum differential, operating differential).

8.8.1.1 Condensate Load The condensate capacity requirements can be more difficult to obtain. Condensate capacities may be documented in either the design specifications or on equipment nameplates. If the condensate capacity is not shown, it will be necessary to calculate the condensate capacity by using a heat-transfer formula. One basic item to remember is that 1 lb of steam condenses to 1 lb of water. If the lb/h of steam is known, the condensate capacity is the same. If equipment is rated in Btu/h, the capacity in lb/h can be approximated by dividing by the latent energy of the steam pressure at the equipment. If a steam control valve is installed to control the flow of steam to the process, the rated capacity of the valve, in terms of x pounds per hour of steam, would generate an equivalent amount of condensate.

8.8.1.2 Safety Load Factor (or Experience Factor) to Use Assume a steam-heated machine that condenses 500 lb/h (see Figs. 8-14 to 8-16). Also assume that users have found that a 1500-lb/h capacity trap will enable the machine to turn out more and better work than when a 1000-lb/h capacity trap is used. For this particular machine, the trap selection safety load factor is 3 : 1 for best machine performance. The safety load factor will vary from a low of 2 : 1 to a high of 10 : 1. The safety load factors in this manual are user-experience safety load factors. A 500-lb/h trap would hardly be enough for a 500-lb/h capacity coil at 100 psi differential pressure. The condensate formed might be more than 500 lb/h, or the differential pressure might drop. Extra trap capacity is needed and costs very little.

Configuration Affects the Safety Load Factor. More important than ordinary load and pressure changes is the design of the steam-heated unit itself. Figures 8-14, 8-15, and 8-16 show three condensing units each producing 500 Lbs/hr condensate but with safety load factors of 2:1, 3:1, and 8:1.

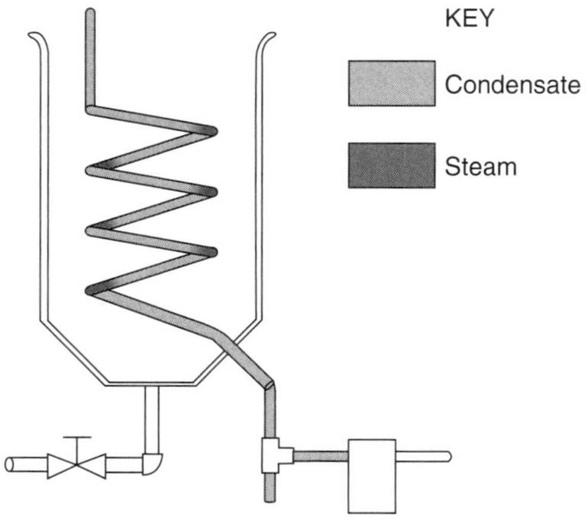

Figure 8-14 Continuous-coil constant-pressure gravity flow to trap. 500 lb/h of condensate from a single copper coil at 30 psig. Gravity drainage to trap. Volume of steam space very small. Use a 2 : 1 safety load factor.

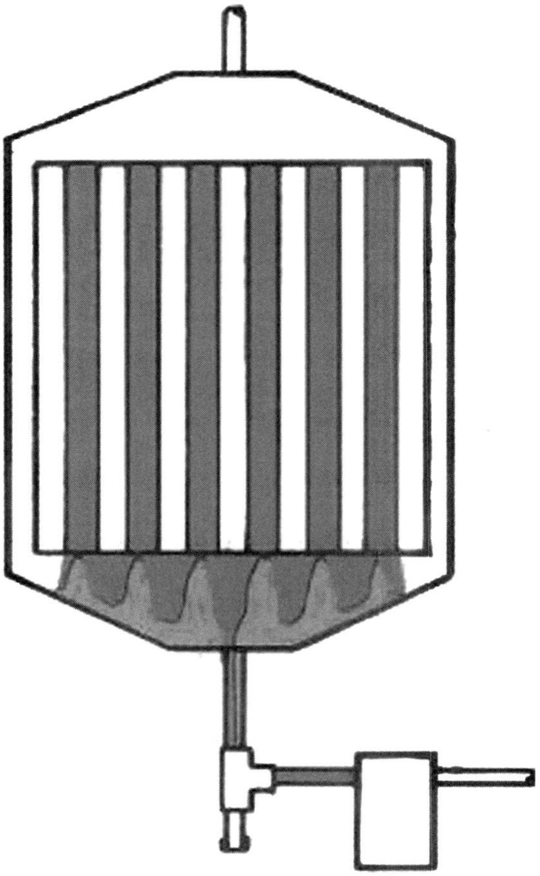

Figure 8-15 Multiple-pipe modulated-pressure gravity flow to trap. 500 lb/h of condensate from a unit heater at 80 psig. Multiple tubes create a minor short-circuiting hazard. Use a 3 : 1 safety load factor at 40 psig.

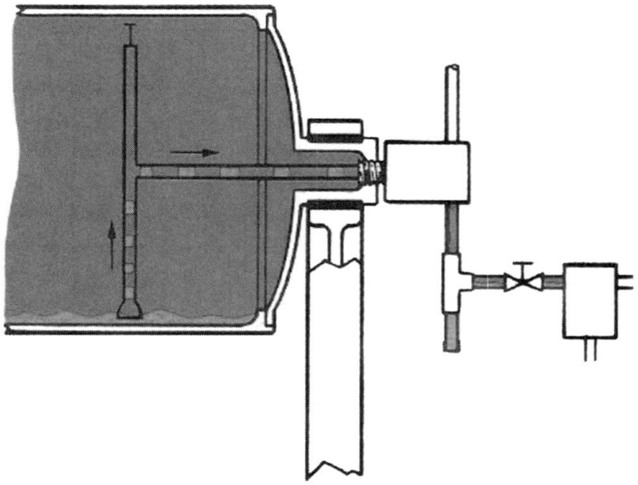

Figure 8-16 Large cylinder, syphon drained. 500 lb/h from a 4-ft-diameter 10-ft-long cylinder dryer with 115 ft^3 of space at 30 psig. The safety load factor is 3 : 1 with a DC (Differential Condensate Controller) and 8 : 1 with an inverted bucket.

8.8.1.3 ***Pressure Differential***

This is subdivided into maximum differential and operating differential.

- *Maximum differential*. This is the difference between boiler or steam main pressure, or the downstream pressure of a PRV, and the return line pressure. The trap must be able to open against this pressure differential.

  Note: Because of flashing condensate in the return lines, do not assume a decrease in pressure differential due to static head when elevating.
- *Operating differential*. When a plant is operating at capacity, the steam pressure at the trap inlet may be lower than the steam main pressure, and the pressure in the condensate return header may go above atmospheric.

If the operating differential is at least 80% of the maximum differential, it is safe to use the maximum differential in selecting traps. Modulated control of the steam supply causes wide changes in the pressure differential. The pressure in the unit being drained may fall to atmospheric or even lower. This does not prevent condensate drainage if the installation practices are followed.

Factors affecting the pressure differential:

- The inlet pressure can be:
  1. Boiler or steam main pressure
  2. Reduced pressure controlled by a pressure-reducing valve station
- The discharge pressure can be:
  1. Atmospheric.
  2. Below atmospheric—under vacuum. Add the vacuum to the inlet pressure to get the pressure differential. 2 in. Hg vacuum = 1 psi of pressure below atmospheric.
  3. Above atmospheric, due to:
     a. Pipe friction.
     b. Controlled backpressure with a receiver relief valve or differential valve.
     c. Elevating condensate. Every 2-ft lift reduces the pressure differential by 1 psi when the discharge is only condensate. However, with flash present, the static head could be reduced to zero.

8.9 STEAM TRAP MAINTENANCE

A strainer is installed just ahead of each steam trap. The strainer must be kept clean and in good condition to keep scale and other foreign matter from getting into the trap [6]. Scale and sediment can clog the working part of a steam trap and seriously interfere with the working of the trap. Steam traps that are not operating properly can cause problems in the system and machinery. One way to check on the operation of steam trap is to listen to it. If the trap is leaking, you will probably be able to hear it blowing through. Another way to check the operation of steam traps is to check the pressure in the drain system. A leaking steam trap causes an unusual increase in pressure in the drain system [6]. When observing this condition, you can locate the defective trap by cutting out (isolating from the system) traps, one at a time, until the pressure in the drain system returns to normal. You should disassemble, clean, and inspect a defective steam trap. After determining the cause of the trouble, repair or replace parts as required. In some steam traps, you can replace the main working parts as a unit; in others, you may have to grind in a seating surface, replace a disk, or perform other repairs. You should reseat defective trap discharge valves. Always install new gaskets when reassembling steam traps.

REFERENCES

1. http://www.engineeringtoolbox.com/steam-traps-d_989.html.
2. http://www.pnl.gov/fta/15_steamtrap/15_steamtrap.htm.
3. http://www.valutechinc.com/steamtraps.htm.
4. ITT Industries, *Steam Traps: Engineering Data Manual*, ITT, Clifton, NJ, 2000. (http://completewatersystems.com/wp-content/uploads/2011/05/7248.pdf)
5. Nwaoha, C., Energy conservation: successful management of steam traps, *Filtration + Separation*, vol. 45, no. 7, September 2008, pp. 97–99.
6. Nwaoha, C., "Knowing Your Steam Trap", *Hydrocarbon Asia*, vol. 21, no. 1, (January/March 2011), pp. 54–57.
7. Swagelok Energy Advisors (July 10th, 2009): *Steam Systems Best Practices* (http://blogs.swagelokenergy.com/?cat=7).
8. Wulfinghoff, D. R., Steam and water leakage, in *Energy Efficiency Manual*, Energy Institute Press, Wheaton, MD, 1999, pp. 187–197.

9

PROCESS COMPRESSORS

N. Sitaram
Thermal Turbomachines Laboratory, IIT Madras, Chennai, India
Chikezie Nwaoha
Control Engineering Asia, Ten Alps Communications Asia, Aladinma, Nigeria

Compressors are major pieces of equipment used not only in the process industries, but also in many other applications, such as in land, marine, and aircraft power plants and for mine ventilation. The size and range of compressors vary substantially, beginning with single- to multistage compressors, from power requirements of a few kilowatts to megawatts. Many types of compressors are available, although only a few types are used in the process industries. Bloch [3] has identified various application ranges for the following types of compressors (Fig. 9-1):

- A1 Reciprocating compressors with lubricated and nonlubricated cylinders
- A2 Reciprocating compressors for high and very high pressures with lubricated cylinders
- B Helical- or spiral-lobed compressors (rotary screw compressors) with dry or oil-flooded rotors
- C Liquid ring compressors (also used as vacuum pumps)
- D Two-impeller straight-lobe rotary compressors, oil-free (also used as vacuum pumps)
- E Centrifugal turbocompressors
- F Axial turbocompressors
- G Diaphragm Compressors

Bloch [3] also identifies the most frequently used combinations of two different compressor types in three fields:

- A + G Oil-free reciprocating compressor followed by a diaphragm compressor
- E + A Centrifugal turbocompressor followed by an oil-free reciprocating compressor
- F + E Axial turbocompressor followed by a centrifugal turbocompressor

In the following sections we present brief details of theoretical aspects of a variety of compressors used in the process industries, along with brief details such as selection and common problems. There are many excellent books on the theoretical and practical aspects of compressors. Books and journal articles referred to in this chapter are listed in the References at the end of the chapter. Other books are given in the Further Reading section; however, these sources provide very detailed information and are well beyond the understanding of process personnel. The objective of the present chapter is to provide sufficient details that would be of immediate use to process personnel.

9.1 TYPES OF COMPRESSORS

There are two main types of compressors: continuous compression compressors and positive-displacement compressors. The major difference between the two lies in the way that energy is transferred between the compressor shaft and the working fluid. In a continuous compression compressor, energy is transferred continuously between the compressor shaft and the working fluid. Continuous compression

Process Plant Equipment: Operation, Control, and Reliability, First Edition. Edited by Michael D. Holloway, Chikezie Nwaoha, and Oliver A. Onyewuenyi.
© 2012 John Wiley & Sons, Inc. Published 2012 by John Wiley & Sons, Inc.

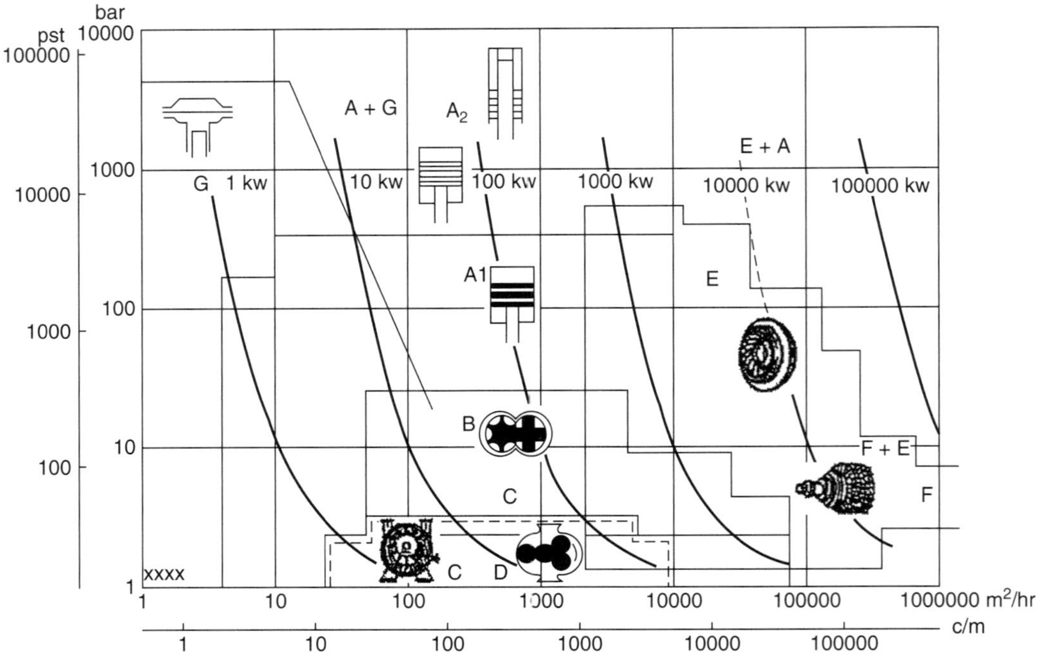

Figure 9-1 Application range of various types of compressors (Sulzer-Burkhardt, Winterthur, and Basel, Switzerland). (From [3].)

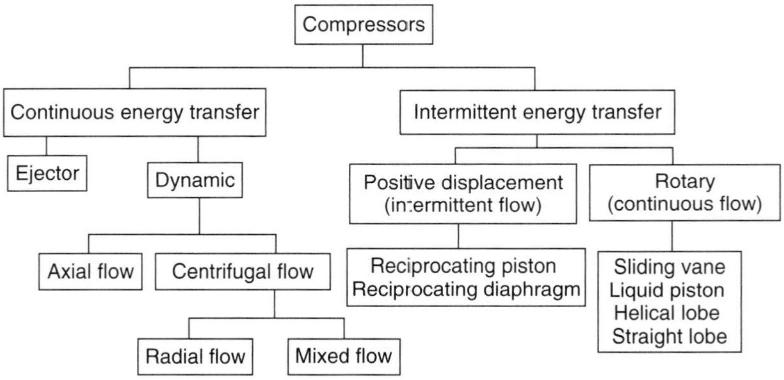

Figure 9-2 Compressor types. (From [7].)

compressors are further classified as ejector and dynamic compressors. Dynamic compressors are also known as turbo or rotodynamic compressors. In a positive-displacement compressor, energy is transferred intermittently, that is, during only part of the cycle, between the compressor shaft and the working fluid. Positive-displacement compressors are classified further as reciprocating compressors, where the flow of the working fluid is intermittent, and rotary compressors, where the flow of the working fluid is continuous.

Figure 9-2 shows the classification of these types of compressors. Brief details of these different types of compressors are given in the following sections. More detailed information on these compressors is available in the works of Bloch [3] and Brown [7].

9.2 CONTINUOUS COMPRESSION COMPRESSORS

9.2.1 Ejectors

An ejector is a static piece of equipment with no moving parts (Fig. 9-3). The major components of an ejector are the motive nozzle, motive chest, suction chamber, and diffuser. An ejector converts the pressure energy of motive steam or other working fluid into velocity. Thermodynamically, high velocity is achieved through adiabatic expansion of motive fluid through a convergent divergent nozzle. This expansion of the motive fluid from the motive pressure to the suction fluid operating pressure results in supersonic velocity at the nozzle exit. The motive fluid expands to a pressure equal

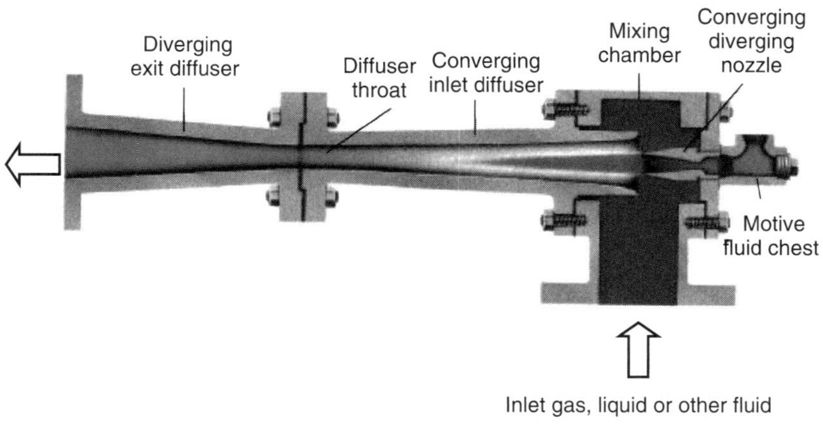

Figure 9-3 Ejector. (Courtesy of Graham Manufacturing Company.)

to the suction fluid pressure. This creates a driving force to bring suction fluid into an ejector. Typically, the velocity exiting a motive fluid nozzle is very high. High-velocity motive fluid entrains and mixes with the suction fluid. The resulting mixture is still supersonic. As the mixture passes through the convergent throat and divergent sections of the diffuser, high velocity is converted back to pressure. The velocity of the supersonic flow is reduced in the convergent section of the diffuser as the cross-sectional area is reduced. The diffuser throat is designed to create a shock wave, which produces a dramatic increase in pressure as the flow goes from supersonic to subsonic velocity across the shock wave. However, the loss in total pressure is high. The cross-sectional area of the divergent section is increased, reducing velocity and increasing pressure.

The major advantage of an ejector is that it does not have moving parts and is simple in construction. Hence, an ejector is inherently reliable and has low maintenance expenses. Further, an ejector is immune to liquid carryover in the suction gas. However, the efficiency of an ejector is very low because of the shock wave occurring in its throat. Ejectors finds applications in crude oil distillation, petrochemical processes, edible-oil deodorization, organic motivated systems, fertilizer plant operations, thermal compressors, hybrid vacuum systems, metal vacuum degassing, and space simulation.

9.2.2 Dynamic, Rotodynamic, or Turbocompressors

A dynamic compressor consists of a single- or multistage compressor. Each stage of the compressor consists of a rotating blade row, known as a *rotor* or *impeller*, and a stationary blade row, usually known as a *stator* or *diffuser*. In the rotor, the energy of the working fluid is increased by the change in tangential momentum and consists of both pressure and kinetic energy. The kinetic energy is further converted into pressure energy in the following stationary blade row. The dynamic compressors are run at high speeds, bringing significant advantages. Multistaging of dynamic compressors is used when high pressure ratios are required. Dynamic compressors are further subdivided into three categories, based on the direction of flow in the meridional plane: axial, mixed, and radial flow compressors. In *axial flow compressors*, the energy transfer between the working fluid and the compressor shaft, to which power is supplied, is due to the change in the tangential velocity across the aerofoil-shaped blades. Hence, the pressure ratio across a single stage of an axial flow compressor is moderate. The streamline shift from the inlet to the exit of the compressor is small or negligible. To achieve the high pressure ratios required in process industries and other applications, multistage axial flow compressors are used (Fig. 9-4).

Radial (Fig. 9-5) and *mixed-flow* (Fig. 9-6) *compressors* are both known as *centrifugal compressors*, as the energy transfer takes place mainly because of centrifugal force acting on the working fluid. The streamline shift from the inlet to the exit of the compressor is large. Hence, the pressure ratio across a single stage of a centrifugal compressor is large. Radial flow compressors can be without or with an inducer at the inlet. Very high pressure ratios can be obtained by multistaging radial flow compressors. However, the flow path is complex, and the efficiency is reduced. Alternatively, axial compressors followed by one or two stages of radial compressors mounted on the same shaft are used for high-flow and high-pressure-ratio applications (Fig. 9-7).

The mixed-flow compressor is a relatively unknown form. It resembles a radial flow compressor with an inducer. It uses a bladed impeller, but the flow path is angular to the direction to the rotor. It is also known as a *diagonal compressor*. As the stage spacing is wide, the compressor is used almost exclusively as a single-stage machine. Mixed-flow compressor applications fall between those of axial and radial flow compressors. Mass flow and pressure rise in these compressors are both between those of axial and radial flow compressors.

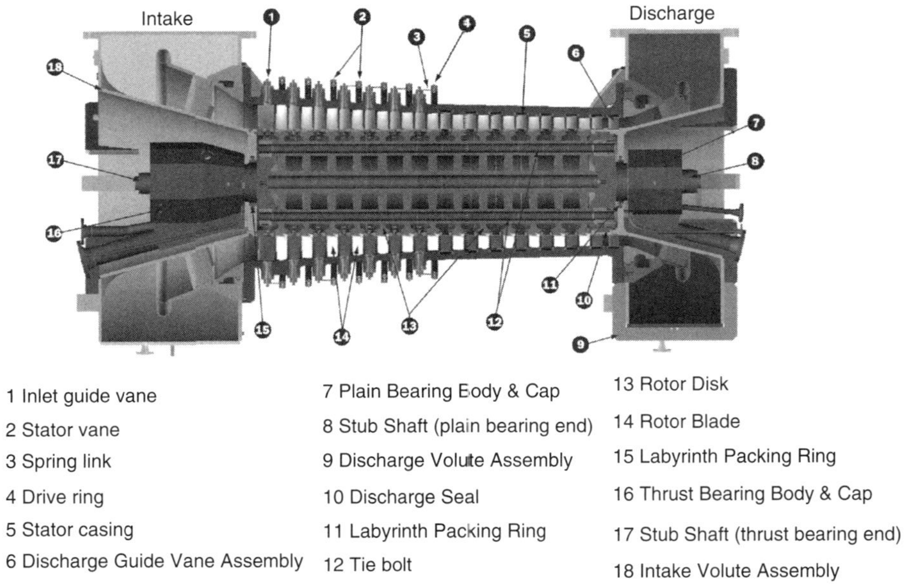

Figure 9-4 Cutaway view of a multistage axial flow compressor. (Courtesy of Dresser Rand Corporation, Olean, NY.)

Figure 9-5 Cutaway view of a multistage radial flow compressor. (Courtesy of Elliot Group, Jeannette, PA.)

9.3 INTERMITTENT COMPRESSION COMPRESSORS

9.3.1 Positive-Displacement Compressors (Intermittent Flow)

9.3.1.1 Reciprocating Piston Compressors Reciprocating piston compressors are probably the best known and most widely used compressors. The major component of a reciprocating piston compressor is a piston that is free to move in a cylinder. The motion of the piston in the cylinder is reciprocating. During one cycle of operation, the displacing action of the piston causes some quantity of gas to enter the cylinder, where it is compressed and discharged. The gas enters the cylinder through an inlet valve during a part of the stroke of the piston and discharges to the delivery through an exit valve during another part of the piston stroke. Reciprocating piston compressors are divided further into single- and multistage compressors. In a single-stage arrangement, either single or multiple cylinders on a common frame are connected in parallel. Thus, the pressure rise of the working fluid remains the same, but the mass flow of the working fluid is increased. In a multistage arrangement, multiple cylinders on a common frame are connected in series, increasing the pressure rise of the working fluid, but the same amount of mass flow of the working fluid is connected. Usually, an intercooler is placed between successive cylinders. Reciprocating piston compressors are available in wide ranges, but the mass flow is limited. However, higher discharge pressures can be achieved in single- or multistage compressors than in dynamic compressors. A multicylinder reciprocating piston compressor is shown in Fig. 9-8.

9.3.1.2 Reciprocating Diaphragm Compressors This type of reciprocating compressor is not as well known as the piston type. This compressor is used mainly for low-mass applications (lower than the piston type) but for high discharge pressures. Its construction is very similar to that of a reciprocating piston compressor. However, the major difference is that the piston moves oil rather than gas, using the oil to move a diaphragm pack that, in turn, compresses the gas. There is no leakage toward the inside to contaminate the process gas or any leakage to the atmosphere. The reciprocating diaphragm compressor can be of single- or multistage construction, similar to that of a reciprocating piston compressor. The multistage reciprocating diaphragm compressor can be arranged in various configurations similar to those of a reciprocating piston compressor, which is in an L or opposed arrangement. A cross section of a hermetically sealed reciprocating diaphragm compressor is shown in Fig. 9-9.

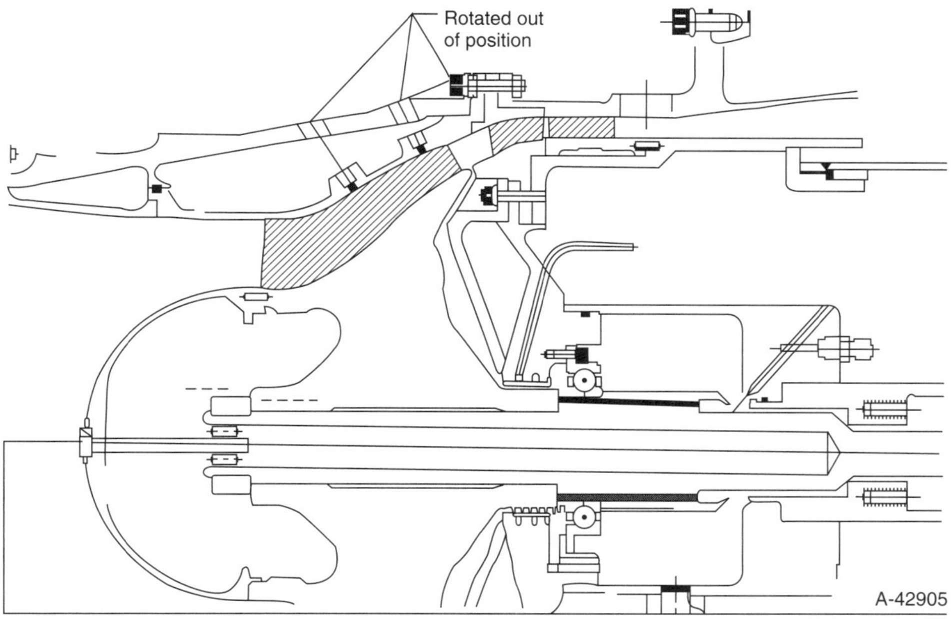

Figure 9-6 Meridional view of the mixed-flow compressor stage. (From [17].)

Figure 9-7 Combined axial-centrifugal turbocompressor. (Courtesy of Mannesmann Demag Delaval, Germany.)

Figure 9-8 Burton Coblin reciprocating piston compressor. (Courtesy of Howden BC Compressors, France.)

9.3.2 Rotary Compressors (Continuous Flow)

In rotary-type positive-displacement compressors, the energy transfer is intermittent, that is, occurs during a part of the cycle only. However, the flow of the working gas is continuous. The major features of this family of compressors are:

1. The energy is imparted to the gas being compressed by an input shaft moving a single or multiple rotating elements.
2. The compression is carried out in an intermittent mode.
3. Inlet and discharge valves, which are necessary in reciprocating compressors, are not used in rotary compressors. Hence, the flow of the working gas is continuous, with the energy transfer being intermittent, contrary to that in reciprocating compressors.

Four types of rotary compressors are available:

1. Sliding vane compressors (Fig. 9-10)
2. Liquid piston compressors (Fig. 9-11)

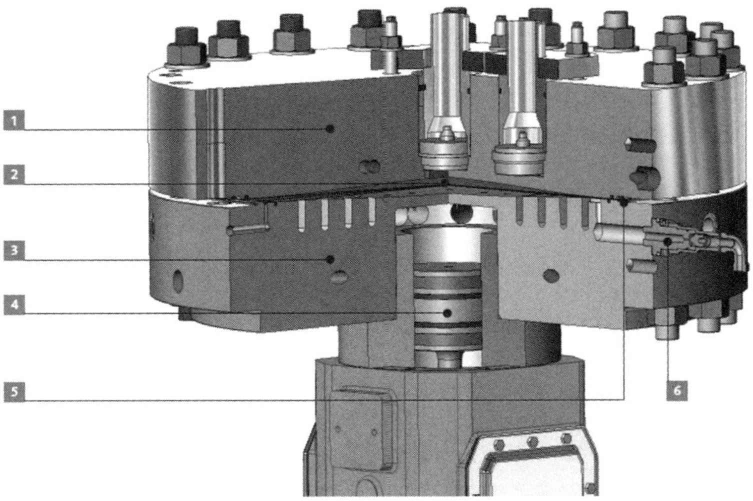

Figure 9-9 Burton Coblin hermetically sealed diaphragm compressor. (Courtesy of Howden BC Compressors, France.)

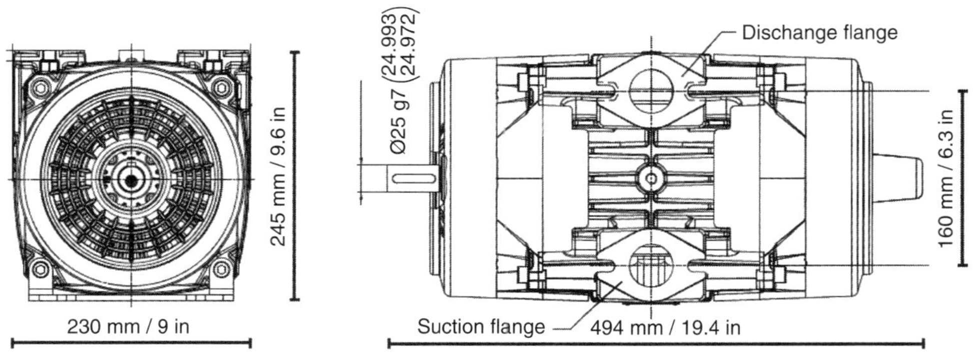

Figure 9-10 Sliding vane compressor. (Courtesy of GHH Schraubenkompressoren GmbH, Germany.)

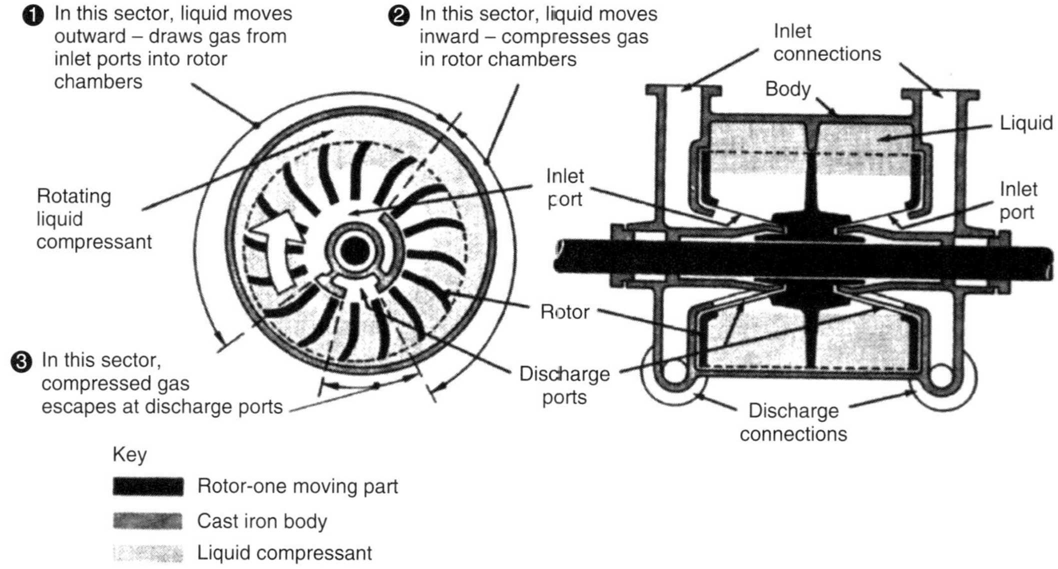

Figure 9-11 Sectional and end view of a liquid piston compressor. (Courtesy of Nash Engineering Co., Trumbull, CT.)

Figure 9-12 Cutaway view of screw compressor. (helical lobe type) (Courtesy of Boge Kompressoren, Germany.)

3. Helical lobe compressors (Fig. 9-12)
4. Straight lobe compressors (Fig. 9-13)

Sliding vane and liquid piston compressors can also be used in vacuum service. For more information on liquid piston compressors, Bannwarth [2] may be consulted.

Figure 9-13 Roots blower (straight lobe type.) (Courtesy of Aerzen USA Corporation, Coatesville, PA.)

The helical lobe compressor is used widely to bridge the gap between centrifugal and reciprocating compressors. This compressor is usually referred to as a screw compressor or SRM compressor. The straight lobe compressor is similar to the helical lobe compressors but is less sophisticated, as it uses untwisted or straight lobe rotors that intermesh as they rotate, instead of the more complicated helical lobe rotors. A straight lobe compressor is also known as a *roots blower*.

Among the compressors above, only centrifugal compressors and reciprocating piston compressors are widely used in the process industries. Hence, the remainder of the chapter is devoted to these two compressors.

9.4 CENTRIFUGAL COMPRESSORS

Centrifugal compressors are used extensively not only in the process industries but also in many other applications, such as power plants for land, marine, and aviation and space applications; and turbochargers. Centrifugal compressors are compact in size, light in weight, and the manufacturing cost is lower than that of axial compressors because of the simple construction. The efficiency of centrifugal compressors is comparable to that of reciprocating piston compressors, which are also widely used in the process industries, and axial flow compressors. Also, centrifugal compressors have a large tolerance to process fluctuations, are smooth in operation, are more reliable than other types of compressors, and have a wider operating range. Major operating characteristics of a variety of compressors are presented in Table 9-1.

Centrifugal compressors are used extensively in helicopter and small aircraft engines, where a high level of efficiency, a wide operating range, a high tolerance to inlet flow distortions, and a high level of reliability are required. Extensive research and development efforts are spent to achieve the foregoing requirements. These efforts also benefit industrial and process centrifugal compressors.

9.4.1 Major Components of Centrifugal Compressors

In general, "centrifugal compressors" include both radial and mixed flow compressors. However, in this section, only radial flow compressors without or with inducers are discussed. The axial extent of mixed-flow compressors is larger than that of centrifugal compressors. Also, it is difficult to use multistaging in mixed-flow compressors. Process compressors require very large pressure ratios. Hence, multistaging of compressors is essential. Hence, mixed-flow compressors do not find applications in the process industries but are finding applications in aircraft power plants and jet engines.

TABLE 9-1 Comparison of Characteristics of Various Types of Compressors

Type of Compressor	Pressure Ratio Industrial	Pressure Ratio Aerospace	Pressure Ratio Research	Efficiency	Operating Range[a]
Positive displacement	Up to 30	—	—	75–82%	
Centrifugal	1.2–1.9	2–7	13	75–87%	Large (25%)
Axial	1.1–1.3	1.1–1.5	2.1	80–91%	Narrow (3–10%)

Source: [5].

[a]It is important to note that the operating range is narrowed with an increase in pressure ratio and number of stages.

A centrifugal compressor may consist of a simple radial rotor but without splitter blades, inducer, inlet guide vanes, or diffuser vanes, discharging directly into a volute or collector followed by a discharge duct. On the other extreme, a centrifugal compressor may be an extremely complex machine with a rotor (radial or backward or forward curved), splitter blades, an inducer, inlet guide vanes, a vane diffuser with vaneless space, a volute, and a discharge duct. In multistage centrifugal compressors, the flow from the exit of the preceding stage is directed into the next stage through 180° return flow channel with vanes. In addition, these compressors may include both active and passive flow devices to improve compressor performance and operating range and for surge control.

In this section, brief details on various components of centrifugal compressors are provided. Many excellent books on centrifugal compressors (e.g., [1,5,12]) provide more information.

*9.4.1.1 **Suction Nozzle or Intake*** The inlet flow to the centrifugal compressor should be uniform with minimum flow distortion. Hence, a suction nozzle is provided at the inlet of the compressor. The function of the nozzle is to provide gradual acceleration of flow to the inducer or rotor of the compressor. If the working fluid is not air, as in the case of process compressors, the working fluid should mix with the atmospheric air. In such case a conventional nozzle is not suitable. Various types of intakes used in aircraft engines and process compressors are shown in Fig. 9-14.

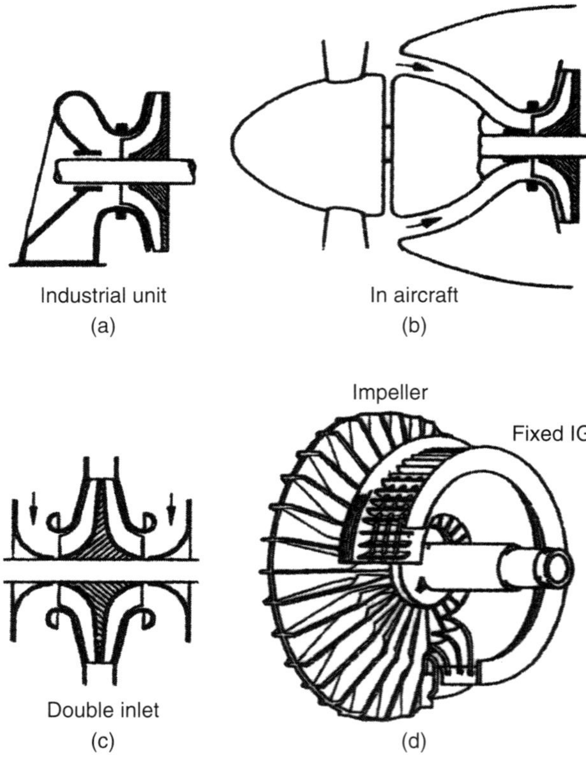

Figure 9-14 Different types of intakes for centrifugal compressors. (From [16].)

Figure 9-14a shows a conventional intake for an industrial compressor. This intake has a gradual reduction in size so that the velocity of the working fluid is nearly the same all along the circumferential direction. Figure 9-14b shows an intake for an aircraft turbo prop compressor. This intake is axisymmetric. Figure 9-14c shows an intake for a double-inlet centrifugal compressor. The intake construction becomes complicated when an inlet guide vane row is introduced to preswirl. Sometimes an adjustable guide vane row is used to vary the inlet swirl. In such case the intake design becomes more complicated. Figure 9-14d shows a fixed-inlet guide vane row. Figure 9-15 shows a cross section of a single-stage industrial centrifugal compressor. Major aerodynamic components and mechanical parts are shown.

*9.4.1.2 **Inlet Guide Vanes (IGVs)*** Inlet guide vanes are used to give prewhirl at the inducer inlet. The prewhirl may be in the direction or opposite to the direction of rotation of the rotor. The velocity diagrams at the inlet of the inducer are shown in Fig. 9-16 for various prewhirls. An IGV is used to reduce the relative Mach number at the inducer tip, as the highest velocity at the inducer inlet occurs at the tip section of the inducer. It is desirable to keep the maximum velocity below sonic conditions, as a shock wave occurs when the velocity is close to or above sonic velocity. A shock wave causes shock loss and chokes the inducer.

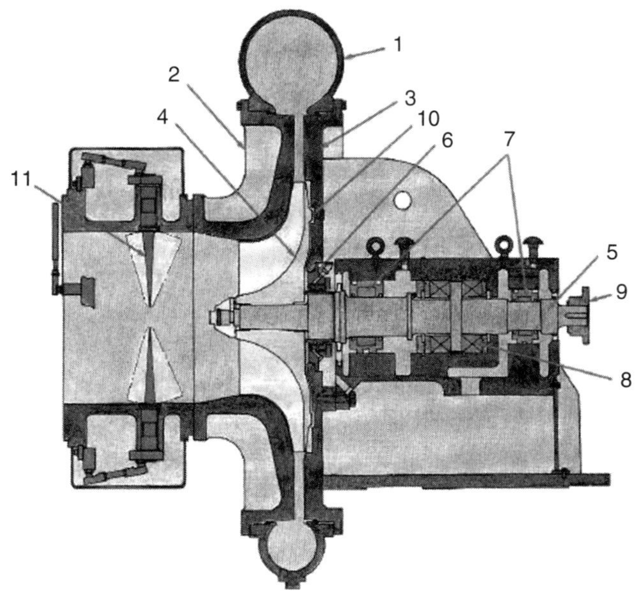

Figure 9-15 Single-stage overhung centrifugal compressor. 1, Casing; 2, inlet cover; 3, back plate; 4, rotor; 5, shaft; 6, seal; 7, bearing; 8, thrust bearing; 9, coupling hub; 10, balance plane; 11, variable IGV. (Courtesy of A. C. Compressors.)

The prewhirl given by an IGV may be of forced vortex, free vortex, or controlled vortex type. The forced vortex distribution at any radius r is given by $c_u = kr$. The free vortex distribution at any radius r is given by $c_u = k/r$. The controlled vortex distribution at any radius r is given by $c_u = k_1 r + k_2/r$. With $k_1 = 0$ and $k_2 \neq 0$, free vortex distribution is obtained. With $k_1 \neq 0$ and $k_2 = 0$, forced vortex distribution is obtained. The prewhirl distribution patterns at the inlet of the inducer with free and forced vortex flow are shown in Fig. 9-17.

An IGV may be of fixed or variable type. The advantage of a variable IGV is that the prewhirl can be changed depending on the operating conditions, although linkage mechanisms are required, complicating mechanical construction. An adjustable IGV with linkage mechanisms is shown in Fig. 9-15.

9.4.1.3 Rotor A rotor is the most important component of any turbomachinery. The rotor transfers energy from the shaft to the working fluid. Although a centrifugal compressor may have a simple radial rotor, in many cases the rotor will have two additional components: an inducer

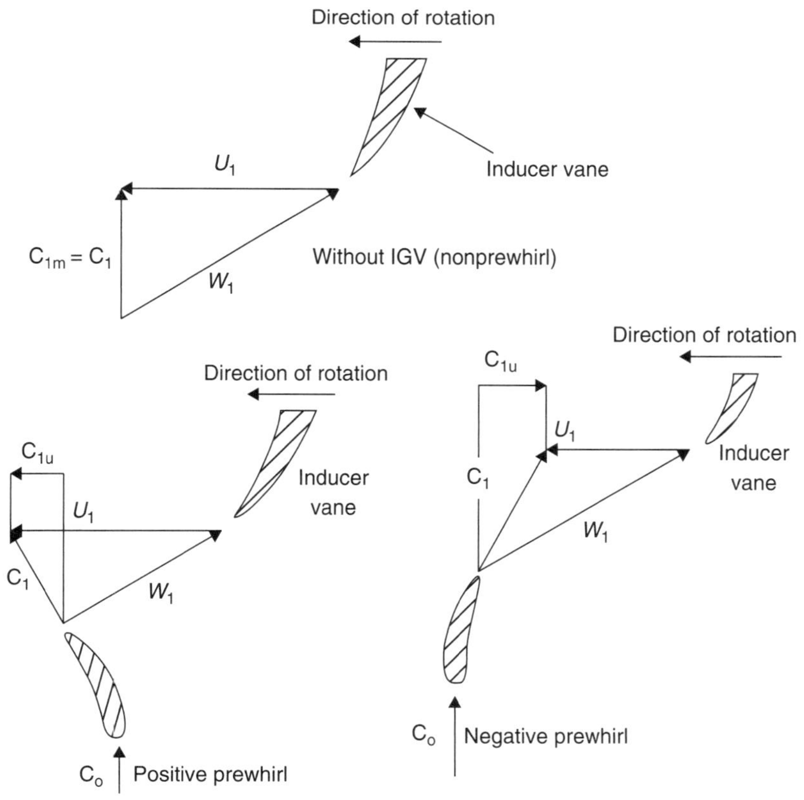

Figure 9-16 Velocity diagrams at the inlet of the inducer of a centrifugal compressor without and with prewhirl. (From [4].)

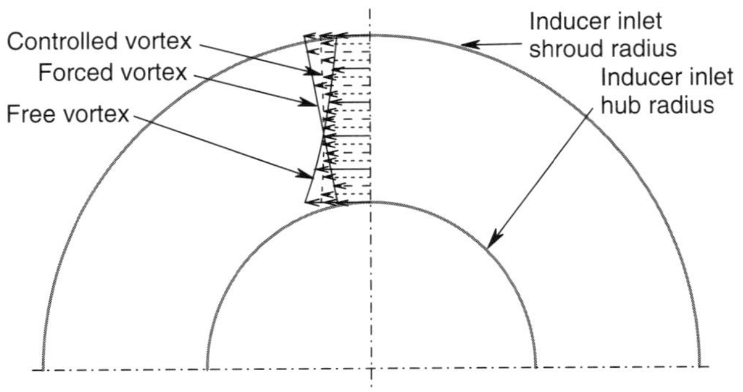

Figure 9-17 Prewhirl distribution patterns at the inlet of the inducer.

similar to an axial flow rotor, and splitter blades in the radial section of the rotor. The functions of the three components are discussed briefly in the following sections.

9.4.1.4 Inducer The function of an inducer is to increase the angular momentum of the working fluid without increasing its radius of rotation. The inducer also turns inlet flow to the axial direction, into the radial part of the impeller. The velocity in the inducer is the highest in the rotor. Hence, the inducer is likely to choke if not designed properly.

An inducer is normally an integral part of a rotor. But inducers that are constructed separately from the radial part of the rotor are sometimes used. This is so particularly when the rotor has splitter blades. Because of the possibility of choking in the inducer, many compressors use splitters in the radial section of the rotor. However, separation is likely to occur on the suction surface of the splitter. To avoid separation, tandem inducers are used. In tandem inducers, the inducer is rotated slightly, as shown in Fig. 9-18. Roberts and Kacker [19] exploited this concept further using computational fluid dynamics.

9.4.1.5 Rotor A rotor for centrifugal compressor may be of one of the following types: open, semiopen, or closed. An open-type rotor is rarely used, although the rotor may sometimes be scalloped to reduce the rotor weight at the expense of loss of efficiency. Semiopen or closed rotors (Fig. 9-19) are prevalent in industry. The type of rotor chosen depends on its performance and operating range required.

The rotor may be classified further as two- or three-dimensional. Two-dimensional rotors are simple to manufacture. However, their performance and operating range are limited. Although three-dimensional rotors are expensive to manufacture, they have better performance and a higher operating range than those of two-dimensional rotors.

The rotor may have one of the following three types of blades: radial, backward curved, or forward curved. Velocity diagrams of rotors with the three types of blades are shown in Fig. 9-20. The advantages and disadvantages of these rotors are given in Table 9-2. Although rotors with forward-curved blades have high energy transfer, the exit kinetic energy is high, requiring a very efficient diffuser. Hence, this type of rotor is rarely used in the process

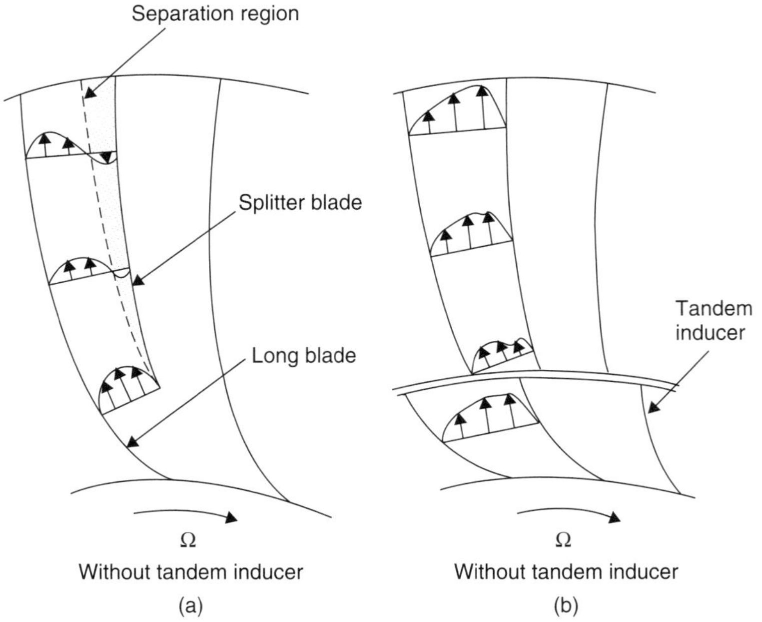

Figure 9-18 Flow pattern in the passage of a centrifugal rotor a) without and b) with tandem inducer.(From [6])

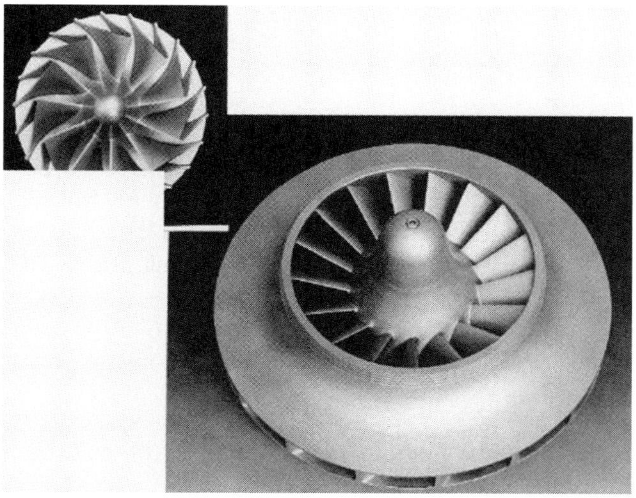

Figure 9-19 Semiopen and closed rotors for centrifugal compressors. (Courtesy of Mannesman Demag Delaval, Duisberg, Germany.)

industries. Rotors with backward-curved blades are the most popular.

9.4.1.6 Diffuser The diffuser of a centrifugal compressor is as important, if not more important, than the rotor. An efficient diffuser is required to convert the high kinetic energy leaving the rotor into useful pressure energy. There are basically two types of diffusers: vaneless and vane diffusers. Vane diffusers can be further classified as conventional or high-solidity vane diffusers (with throat), low-solidity vane diffusers (without a throat), and partial vane diffusers (high or low solidity configuration). These types of diffusers are shown schematically in Fig. 9-21.

Senoo [21] proposed low-solidity vane diffusers, characterized by the absence of throat. Hence, the flow is not choked as in the case of conventional diffusers. Yoshinaga et al. [28] tested centrifugal compressors with partial diffusers on the shroud (the vane height is half of the diffuser passage width) to get improved performance

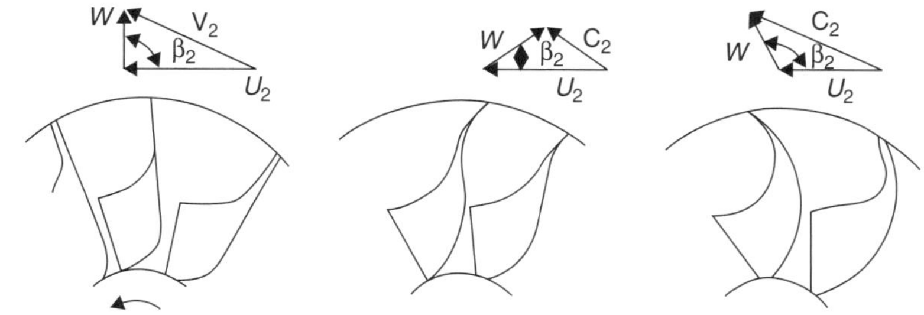

Figure 9-20 Rotor with different types of blading for centrifugal compressors. (From [4].)

TABLE 9-2 Comparison of Rotors with Various Types of Blades

Type of Rotor	Advantages	Disadvantages
Rotor with radial-tipped blades	Reasonable compromise between low energy transfer and high absolute exit velocity No complex bending stresses Easy to manufacture Good surge margin	
Rotor with backward-curved blades	Low exit kinetic energy means a low diffuser inlet Mach number Wide surge margin	Low energy transfer Complex bending stresses Difficult to manufacture
Rotor with forward-curved blades	High energy transfer	High exit kinetic energy means a high diffuser inlet Mach number Minimum surge margin Complex bending stresses Difficult to manufacture

Source: [4].

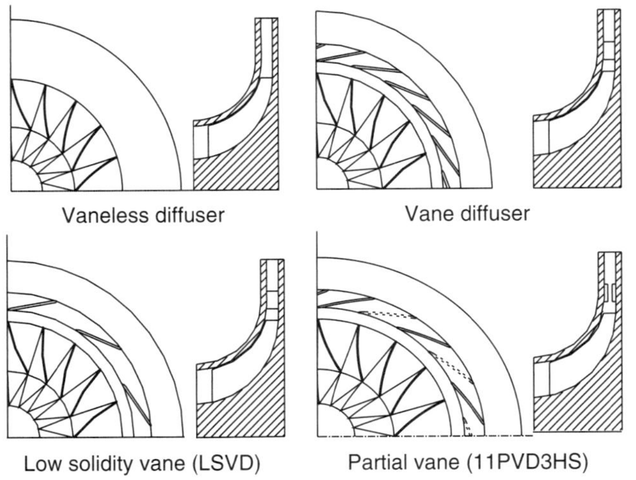

Figure 9-21 Basic types of diffusers used in centrifugal compressors. (From [14].)

and operating range. This concept was extended by Issac et al. [14] to use partial vanes staggered on the hub and shroud. One other type of diffuser is the pipe diffuser, patented by Pratt & Whitney [15]. Although this may be considered a vane diffuser, the configuration is very different. Because it is patented, this diffuser has not received much use. Furthermore, this diffuser can only be used in single-stage centrifugal compressors. Process compressors usually require many stages to achieve large pressure ratios. Considerable research and development work is done on diffusers, varying many different parameters. The variety of diffusers is shown in Fig. 9-22.

9.4.1.7 Volute or Scroll Casing The working fluid leaving a rotor or diffuser is collected by a volute and delivered to a discharge duct. The volute may be symmetrical or asymmetrical (Fig. 9-23). The volute cross section may be circular, square, rectangular, or trapezoidal. The volute has an important effect on compressor efficiency. The volute is normally designed on the basis of constant angular momentum ($C_u r = \text{constant}$) or constant mean velocity or pressure.

9.4.1.8 Return Flow Channel and Vanes In multistage centrifugal compressors, the exit flow from the preceding stage has to be directed properly to the next stage, as the performance of the compressor is influenced strongly by the aerodynamic performance of the 180° U-bend and return channel vanes. The flow leaving the preceding stage has a significant circumferential velocity component as it enters the return flow channel. The flow is subjected to a large curvature. The return channel vanes further decelerate the flow to remove circumferential velocity, so that the flow enters the next stage with little circumferential swirl. A schematic of the return flow channel and vanes is shown in Fig. 9-24.

Extensive work has been done to design the return flow channel and vanes properly. Srinivasa Reddy et al.

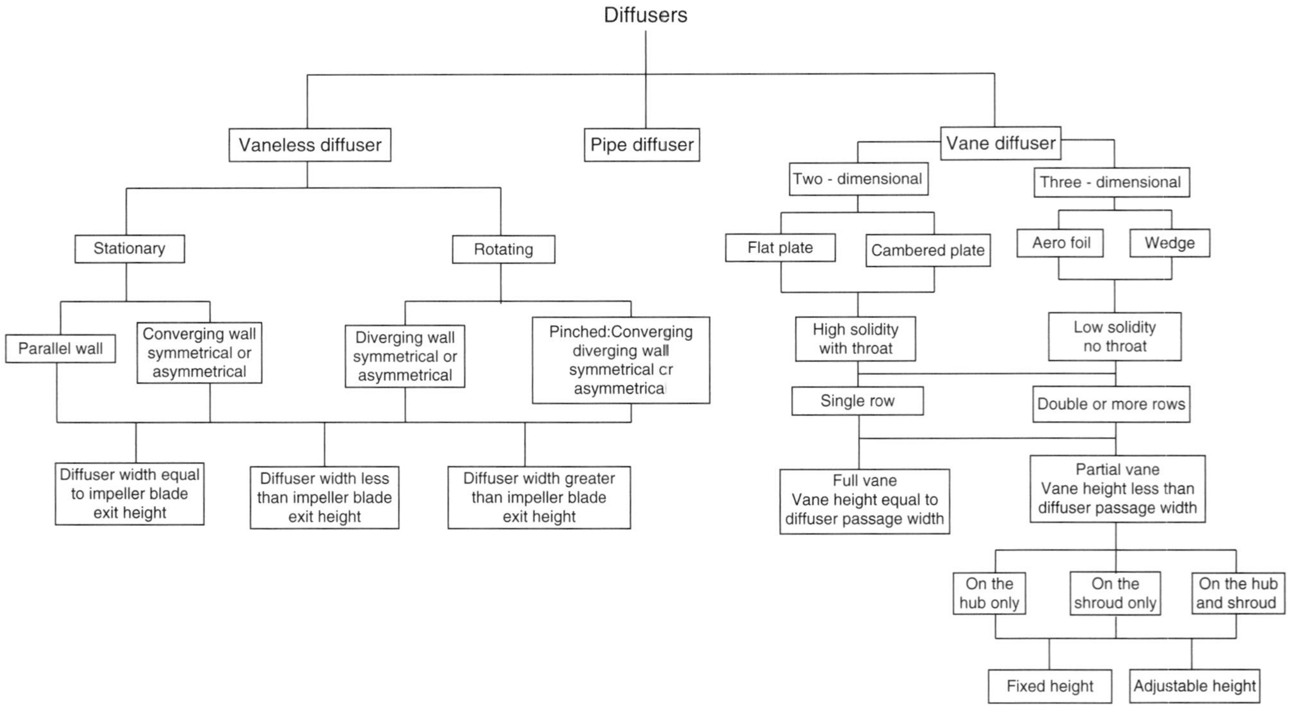

Figure 9-22 Different types of diffusers for centrifugal compressors.

[22] recently carried out computational and experimental investigations on various return flow channels and vanes to optimize their performance.

9.4.1.9 Discharge Duct The function of the discharge duct is to convey the working fluid to the next component.

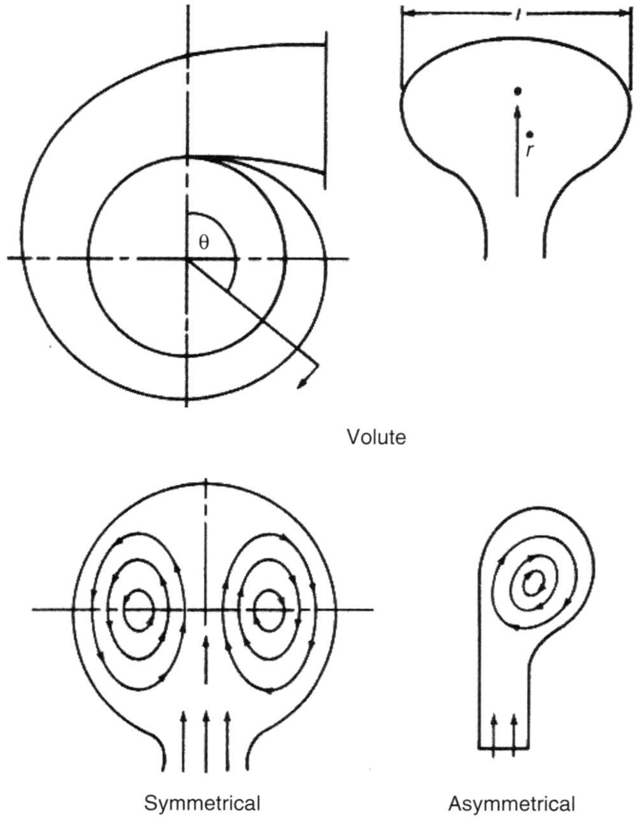

Figure 9-23 Symmetrical and asymmetrical volutes. (From [4].)

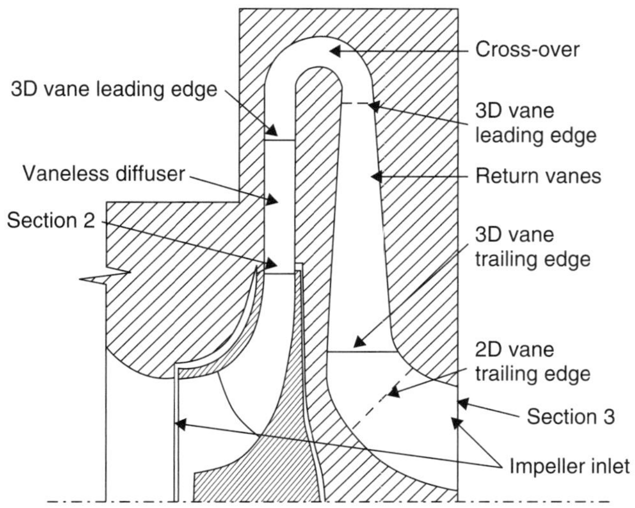

Figure 9-24 Schematic view of meriodional cross section of a centrifugal compressor showing return flow channel and vanes. (From [25].)

9.4.2 Thermodynamics of Centrifugal Compressors

The compression process in a centrifugal compressor may be represented on a temperature–entropy ($T-s$) diagram if c_p is constant or on an enthalpy–entropy ($h-s$) diagram if c_p is not constant. The compression process of the rotor and diffuser is shown on a $T-s$ diagram in Fig. 9-25. The vertical lines (1-02s, 2-03s, and 3-03s) represent isentropic compression processes. However, because of many losses in both the rotor and diffuser, entropy is generated and the actual compression process is shown as 1–2–3, corresponding to the rotor inlet, rotor exit/ diffuser inlet, diffuser exit respectively.

Across the rotor, the rothalpy is constant. The rothalpy, I, is expressed as

$$I = h + \frac{c_r^2 + c_u^2 + c_x^2 - 2uc_u}{2}$$

This expression is modified by adding and subtracting $\frac{1}{2}u^2$ as follows:

$$\begin{aligned} I &= h + \frac{u^2 + c_u^2 - 2uc_u}{2} + \frac{c_r^2 + c_u^2 - u^2}{2} \\ &= h + \frac{(u - c_u)^2}{2} + \frac{c_r^2 + c_u^2 - u^2}{2} \end{aligned}$$

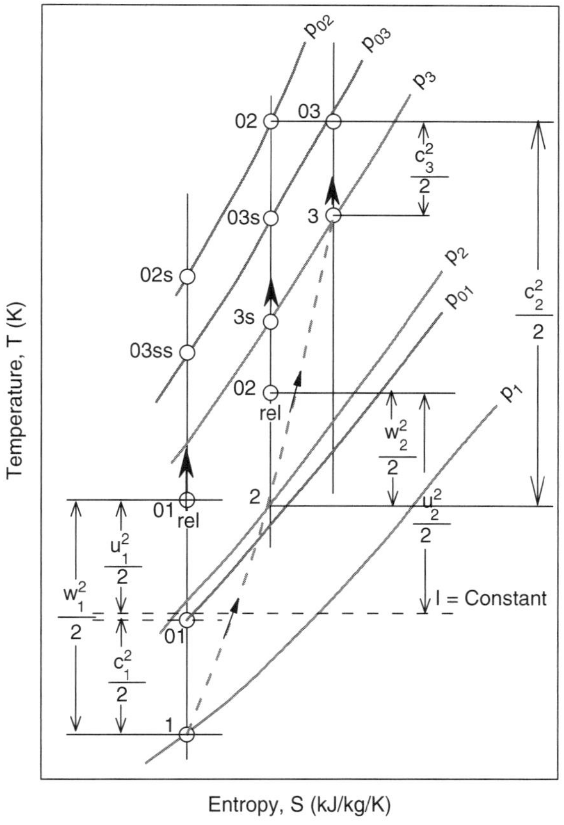

Figure 9-25 $T-S$ diagram for a centrifugal compressor stage (rotor and diffuser).

As

$$w_u = u - c_u \text{ and } w^2 = c_r^2 + w_u^2 + c_x^2,$$

$$I = h + \frac{w^2 - u^2}{2} = h_{\text{or}} - \frac{1}{2}u^2$$

Across the rotor, $I_1 = I_2$. Hence,

$$h_2 - h_1 = \tfrac{1}{2}\left(u_2^2 - u_1^2\right) + \tfrac{1}{2}\left(w_1^2 - w_2^2\right)$$

The first term represents the contribution from the centrifugal acceleration due to change in the radius of the streamline, and the second term represents the contribution due to the diffusion of relative velocity. From this expression it can be seen that the enthalpy rise in a centrifugal rotor is large compared to that in an axial flow rotor, where $u_1 \approx u_2$.

9.4.3 Energy Transfer in Centrifugal Compressors

Consider a centrifugal compressor with a radial rotor only (Fig. 9-26) operating under steady conditions. The control volume, shown as a dashed line, includes all the blade passages responsible for energy transfer. The mass flow rate entering and leaving the control volume should be equal, assuming no fluid leakage, and is equal to $m = \rho V$. In addition, the velocites entering and leaving the blade passage are uniform in both the circumferential and axial (hub to shroud) directions. The sum of the torque acting on the blades can be equated to the rate of change of the moment of momentum through the control volume:

$$\text{torque, } T = mc_2 l_2 - mc_1 l_1$$

From Fig. 9-26 the values of l_1 and l_2 can be calculated and substituted in the equation above. Hence,

$$T = m(c_2 r_2 \cos\alpha_2 - c_1 r_1 \cos\alpha_1) = m(r_2 c_{2u} - r_1 c_{1u})$$

Neglecting friction in the energy transfer process, the power required at the rotor can be expressed as

$$N_{\text{bl}} = \omega T = \omega m(r_2 c_{2u} - r_1 c_{1u}) = m(u_2 c_{2u} - u_1 c_{1u})$$

where ω is the angular velocity of the rotor.

For turbomachinery, specific work is defined as energy transferred per unit of mass flow. Hence, the power transferred is divided by the mass flow:

$$W_{\text{bl}} = \frac{N_{\text{bl}}}{m} = \frac{\omega T}{m} = u_2 c_{2u} - u_1 c_{1u}$$

This is a general energy transfer equation valid for all turbomachinery. As this equation is first applied to the turbines, it is usually known as the Euler turbine equation.

From the velocity triangles at the rotor inlet and exit, the term uc_u can be written

$$uc_u = \frac{u^2 + c^2 - w^2}{2}$$

Hence,

$$W_{\text{bl}} = \frac{u_2^2 - u_1^2}{2} - \frac{w_2^2 - w_1^2}{2} + \frac{c_2^2 - c_1^2}{2}$$

$$= (h_2 - h_1) + \frac{c_2^2 - c_1^2}{2} = h_{02} - h_{01}$$

Hence, specific work for turbomachinery is equal to the change in the stagnation enthalpy of fluid across the stage. The equation above consists of two terms. The first term represents the energy change due to static conditions (static pressure or static enthalpy), and the second term represents the energy change due to dynamic conditions or kinetic energy of absolute velocity. A term representing the energy change due to static conditions and the total energy change

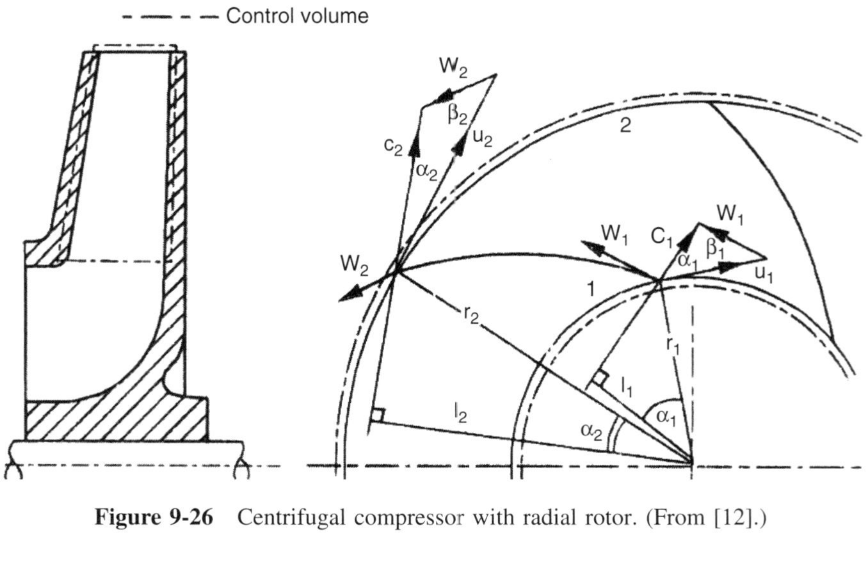

Figure 9-26 Centrifugal compressor with radial rotor. (From [12].)

due to the stagnation conditions is commonly used in turbomachinery practice and is known as the *degree of reaction*. It is written

$$\text{degree of reaction, } R = \frac{\text{work done by the rotor}}{\text{work done by the stage}}$$

$$= \frac{\text{static pressure or enthalpy change across the rotor}}{\text{static pressure or enthalpy change across the stage}}$$

This ratio represents the division of static pressure change across the rotor and stage. For a centrifugal compressor, the value of R lies between 0.5 and 1.

9.4.3.1 Specific Work for Rotors with Different Blade Exit Angles Referring to Fig. 9-20, specific work for the rotors with different types of blades can be written (assuming the same inlet conditions and exit radial velocity)

$$W_{\text{bl}} = u_2 c_{2u} - u_1 c_{1u} = u_2(u_2 - w_{2u}) - u_1 c_{1u}$$
$$= u_2\left(u_2 - \frac{u_2 c_{2r}}{\tan\beta_2}\right) - u_1 c_{1u}$$

For a rotor with radial tipped blades,

$$c_{2u} = u_2 \quad \text{and} \quad W_{\text{bl}} = u_2^2 - u_1 c_{1u}$$

For a rotor with backward-curved blades,

$$c_{2u} < u_2 \quad \text{and} \quad W_{\text{bl}} = u_2^2 - \frac{u_2 c_{2r}}{\tan\beta_2} - u_1 c_{1u}$$

As $\tan\beta_2$ is positive for a rotor with backward-curved blades, W_{bl} for this rotor will be less than W_{bl} for a rotor with radial blades. For a rotor with forward-curved blades,

$$c_{2u} > u_2 \quad \text{and} \quad W_{\text{bl}} = u_2^2 - \frac{u_2 c_{2r}}{\tan\beta_2} - u_1 c_{1u}$$

As $\tan\beta_2$ is negative for a rotor with forward-curved blades, W_{bl} for this rotor will be higher than W_{bl} for a rotor with radial blades. However, the exit velocity is high, requiring a very efficient diffuser. The operating range is also reduced. Hence, this type of rotor is not used in the process industries.

As the values of flow angles at the rotor inlet and exit are not known, they are assumed equal to the blade angles of the rotor at the inlet and exit. Hence, specific work under these conditions is written

$$W_{\text{bl,ideal}} = u_2 c_{2r} \cos\alpha_{2b} - u_1 c_{1r} \cos\alpha_{1b}$$

This equation is derived assuming that the fluid follows the blade surfaces and is known as ideal or vane congruent flow. This flow can only be realized if there is an infinite number of infinitely thin blades. However, this assumption is never valid in practice. Turbomachines always have a finite number of blades with finite thickness. Hence, the actual flow in turbomachine blade passages differs or deviates from the ideal flow. The flow is said to take place with *slip*. The actual specific work, even without losses, will always be less than ideal specific work. The difference between the two specific works is known as *slip power*. In the section that follows we discuss the causes of slip and methods to estimate it.

9.4.4 Slip in Centrifugal Impellers

The factors that cause slip are those that affect specific work and those that affect flow angle but not specific work. Factors that affect specific work are due to (1) inviscid effects due to pressure difference and relative circulation effect, and (2) fluid friction or viscous effects. Factors that affect flow angle are due to blade thickness. A velocity diagram showing these effects is presented in Fig. 9-27.

Because of the importance of slip in centrifugal compressors, many efforts have been made to estimate its magnitude (e.g., [8–10,18,23,24]). Wisener [27] critically reviewed these correlations and presented general recommendations. Von Backstrom [26] developed a slip factor based on simple relative eddy SRE (Fig. 9-28), a method that seems to give good results over a wide range of parameters that affect slip. His equation is

$$\sigma = \frac{c_{2u}}{c_{2u,\text{ideal}}} = 1 - \frac{\Delta w_s}{u} = 1 - \frac{1}{1 + F_0(c/s_e)\cos^{0.5}\beta}$$

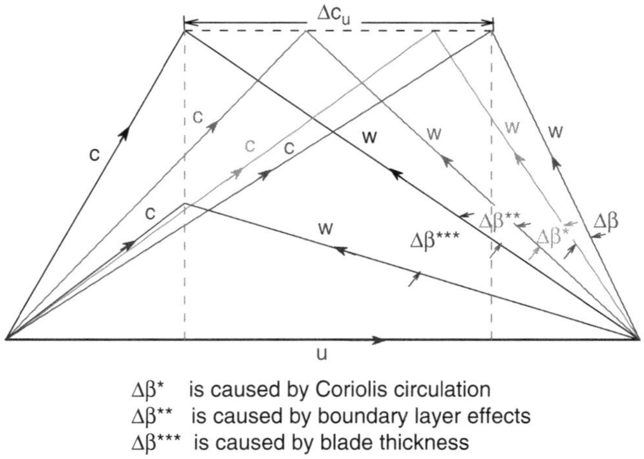

Figure 9-27 Velocity diagram at the exit of the rotor showing the effects of different parameters causing slip in centrifugal compressors.

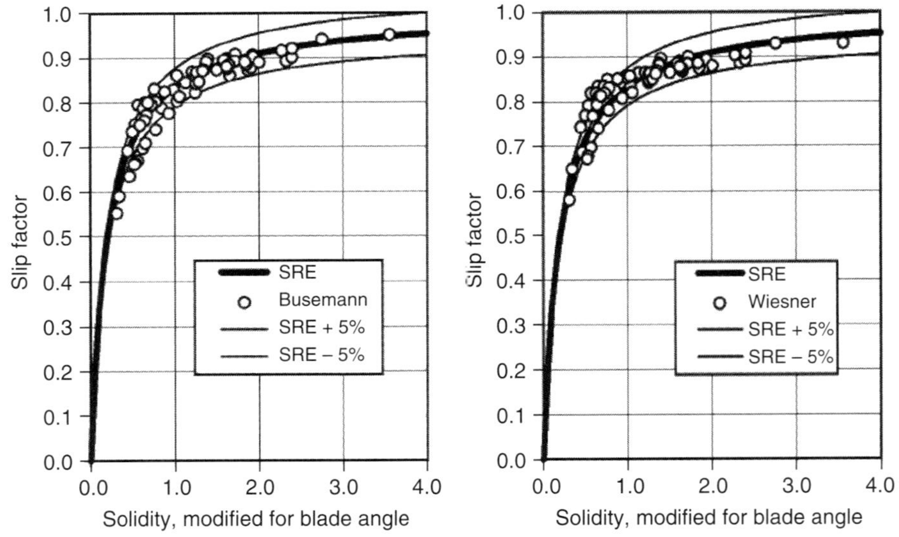

Figure 9-28 Comparison of slip factor given by different authors with that of Von Backstrom [26].

where c is the blade length (m), S_e the blade spacing at the rotor exit (m), and β_e the blade angle at the rotor exit with respect to tangential direction (deg). The value of the constant F_0 can be taken as 5, which may be adjusted suitably for a specific family of rotors.

9.4.5 Losses and Efficiencies

9.4.5.1 Losses The energy transfer in turbomachinery takes place with losses in the direction of flow. The nonutilizable energy content of the fluid is higher at the delivery end than at the inlet end. These losses are divided into two groups: internal losses and external losses. Internal losses are those that occur in the inner passages of the turbomachines and can be attributed directly to the internal flow. The internal losses tend to heat the fluid compared to that in the isentropic compression process. Losses that occur outside the main flow passage are classified as external losses. External losses normally do not add heat to the working fluid.

9.4.5.1.1 Internal Losses The following losses are usually referred to as internal losses.

1. *Hydraulic losses*. Losses due to the frictional losses in the turbomachine channels, separation of the flow on the blade or hub and shroud surfaces, diffusion, eddies, and mixing of different energy-level fluids immediately at the exit, all calculated from the inlet end to the discharge end of the turbomachines, are known as hydraulic losses. These may be estimated and added to the specific energy to be imparted to the working fluid as follows:

$$W_{bl} = W + W_{hyd}$$

where W_{hyd} represents hydraulic losses. These losses have to be estimated by empirical means so that corresponding efficiencies can be determined.

Under off-design conditions, the flow enters the blades at nonzero incidences, causing additional losses known as *incidence losses*. These are also included in hydraulic losses. Hydraulic losses are further classified as (1) frictional losses, (2) profile losses, (3) secondary flow losses, (4) clearance losses, (5) incidence losses, and (6) losses in boundary layers on hub and shroud walls.

2. *Leakage losses*. Leakage of the working fluid results in a significant loss in the specific work of a rotor. This increases the power to be supplied to the compressor. Leakage of the working fluid can occur across clearances, seals, and so on. Leakage loss is usually expressed as leakage volume flow. Hence, the effective volume is reduced by the leakage volume flow.

$$V' = V + \Delta V$$

where V' is the effective volume flow, ΔV the leakage volume flow, and V the actual volume flow. Hence, only the effective volume flow, which is less than the actual volume flow, is participating in the energy transfer process.

3. *Disk friction losses*. When a disk is rotated in an enclosed chamber surrounded by fluid, a resistive torque is set up and the power consumption is increased in compressors to overcome this resistive torque. The following empirical equation may be used to estimate disk friction losses [29]:

power required to overcome resistive torque due to disk friction, $N_{df} = 1.26 \times 10^{-7} \rho n^3 D^5$

where ρ is the density of the working fluid, n the rpm, and D the tip diameter.

4. *Return flow losses*. These losses occur at very low mass flows.

In addition to these internal losses, shock losses occur when a flow is supersonic. The magnitude of shock losses increases with Mach numbers. Other internal losses are those that occur in the vaneless region of the diffuser, in the vane diffuser and return flow channel, and in the vanes in multistage compressors.

9.4.5.1.2 External Losses External losses, also known as *mechanical losses*, are external to the flow medium. The losses that occur in the bearings, sealings, couplings, and other auxiliary equipment connected directly to the compressor shaft are included in the external losses. In addition, the losses that occur in the gearbox are included in these losses when the compressor is driven by the prime mover through a gearbox. Variations in losses with mass flow (as a ratio of design mass flow) are shown in Fig. 9-29.

9.4.5.2 Efficiencies Based on the losses described above, various efficiencies can be defined for use with compressors.

1. *Hydraulic efficiency*. This efficiency includes all the hydraulic losses and is defined as follows:

$$\eta_h = \frac{W}{W_{\text{bl}}}$$

2. *Internal efficiency*. This efficiency represents the ratio of ideal isentropic work to the actual work done on the compressor losses and is defined as follows:

$$\eta_i = \frac{W}{W_i}$$

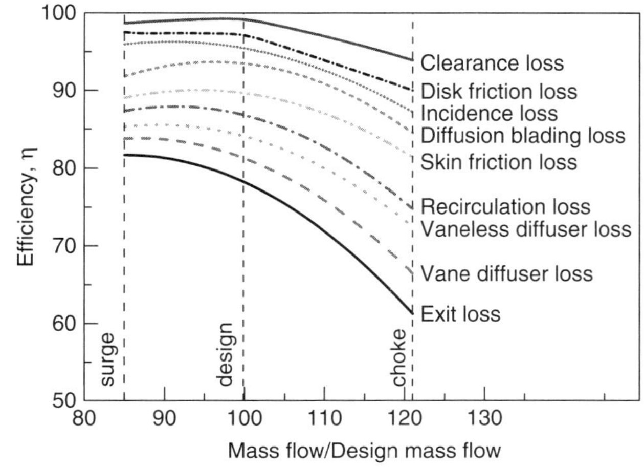

Figure 9-29 Losses in centrifugal compressors.

This efficiency can be estimated from a T–s diagram (Fig. 9-25):

$$\eta_i = \frac{\Delta T_{0\text{is}}}{\Delta T_{0\text{actual}}} = \frac{\Delta h_{0\text{is}}}{\Delta h_{0\text{actual}}} = \frac{T_{03\text{s}} - T_{01}}{T_{03} - T_{01}} = \frac{h_{03\text{s}} - h_{01}}{h_{03} - h_{01}}$$

3. *Mechanical efficiency*. This efficiency includes all external or mechanical losses and is defined as follows:

$$\eta_m = \frac{N_i}{N_c} = \frac{N_i}{N_i + N_m}$$

where N_m includes all mechanical losses.

4. *Overall efficiency*. This efficiency includes all internal and external losses and is of most important to the process engineer. It is defined as follows:

$$\eta_0 = \frac{\rho V W}{N_c} = \frac{\rho V W}{N_i}\frac{N_i}{N_c} = \eta_i \eta_m$$

9.4.6 Performance, Stall, and Surge

A typical performance map of a centrifugal compressor is shown in Fig. 9-30. The performance map shows the variation in the pressure ratio across the compressor as a function of mass flow rate at design, below-design, and above-design speeds. The actual flow rates and speeds are corrected by factors $(\sqrt{\theta}/\delta)$ and $1/\sqrt{\theta}$, respectively to account for variations in inlet temperature and pressure.

The speed line where the operation of the compressor becomes unstable is joined and termed a *surge line*. During surge, the flow through the compressor reverses flows from the exit to the inlet at short time intervals. If surge is allowed to persist, it will result in vibrations of large amplitude and a high noise level. This usually results in

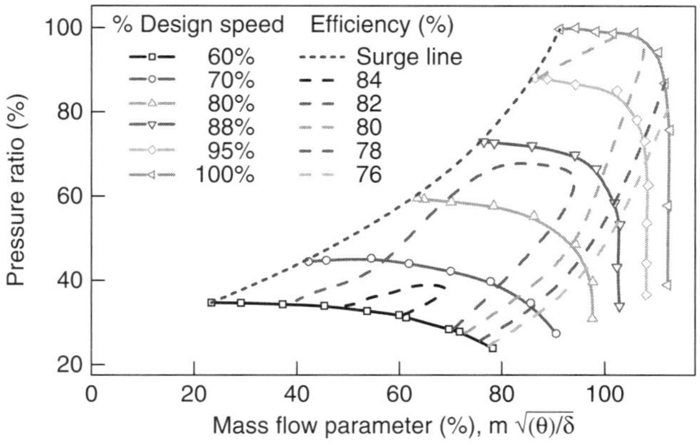

Figure 9-30 Typical performance map of a centrifugal compressor. (Adapted from [11].)

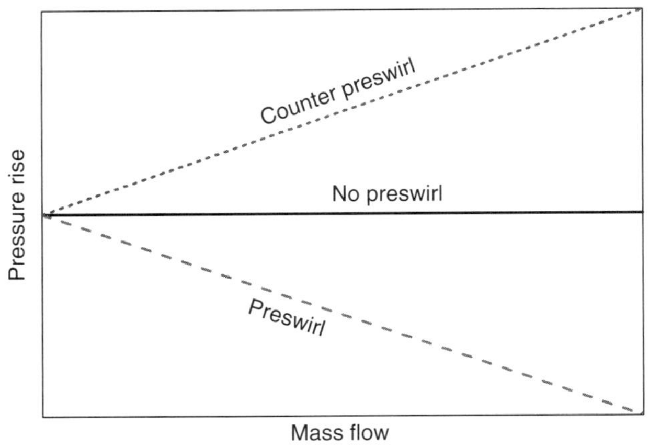

Figure 9-31 Effect of preswirl on the theoretical pressure rise of a centrifugal compressor.

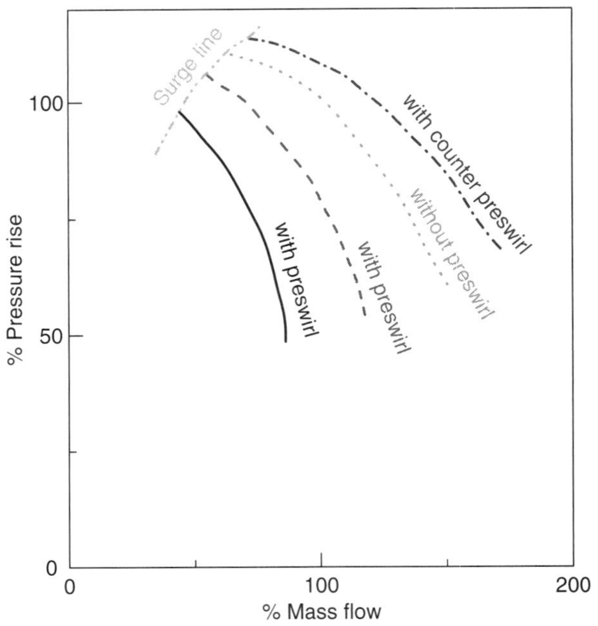

Figure 9-32 Effect of preswirl on the actual pressure rise of a centrifugal compressor with rotor having radial tipped blades. (Courtesy of Dresser Rand Corporation, Olean, NY.)

irreparable structural damage to the compressor. On the extreme right of the performance map, the compressor operation is limited by choke or stone wall. This limits the maximum mass flow that can pass through the compressor.

9.4.6.1 Effect of Preswirl on Compressor Performance The effect of preswirl on the theoretical pressure rise in a compressor that has a rotor with radial tipped blades is shown in Fig. 9-31. The inlet guide vane, IGV that creates preswirl causes drooping characteristics similar to those of a rotor with backward-curved blades. The IGV that creates counter preswirl causes rising characteristics similar to those of a rotor with forward-curved blades.

The effect of preswirl on the actual pressure rise in a compressor that has a rotor with radial tipped blades is shown in Fig. 9-32. With preswirl, the curve becomes steeper, whereas counter preswirl makes the curve flatter. The estimated effect of preswirl on the efficiency of a high-pressure-ratio centrifugal compressor is shown in Fig. 9-33. The efficiency becomes maximum when the preswirl is about 30°. The efficiency also increases with a reduction in pressure ratio.

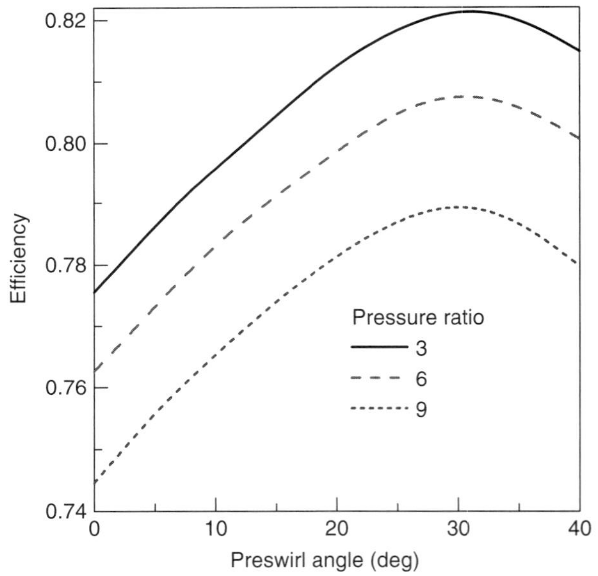

Figure 9-33 Estimated effect of preswirl on the compressor efficiency. (From [20].)

9.4.6.2 Effect of Blade Type on Compressor Performance The effect of blade type (radial or backward- or forward-curved blade) on compressor performance is shown in Fig. 9-34. From the earlier derivation, the expression for specific work is

$$W_{\mathrm{bl}} = u_2\left(u_2 - \frac{c_{2r}}{\tan\beta_2}\right) - u_1 c_{1u}$$

$$\frac{W_{\mathrm{bl}}}{u_2^2} = \left(1 - \frac{c_{2r}}{u_2\tan\beta_2}\right) - \frac{u_1 c_{1u}}{u_2^2}$$

From this equation it can be seen that a rotor with radial tipped blades gives a flat characteristic. A rotor with backward-curved blades gives a drooping characteristic, and a rotor with forward-curved blades gives a rising characteristic, which is not desirable from a stability point of view. The actual performance of a centrifugal compressor with rotors having different types of blades is shown in Fig. 9-35. From the figure it can be seen that the operating range of the compressor decreases as the blade angle increases.

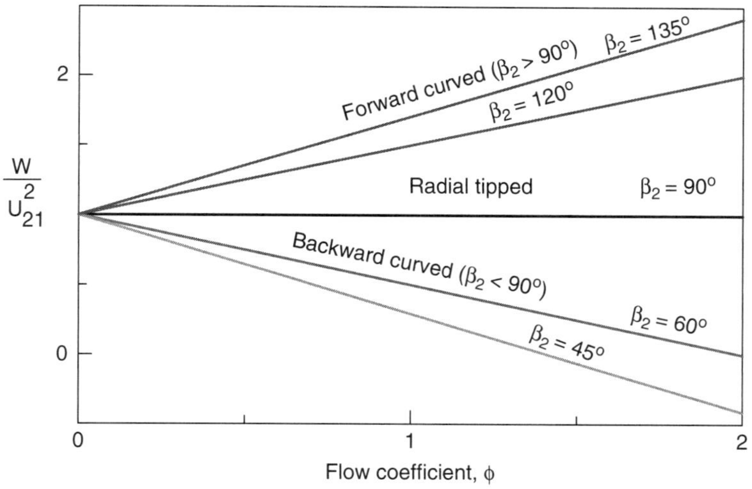

Figure 9-34 Theoretical performance of centrifugal compressor with rotors having different types of blades.

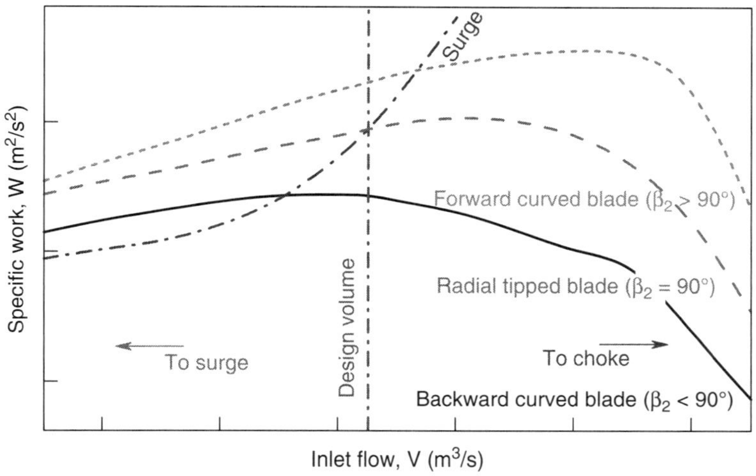

Figure 9-35 Actual performance of centrifugal compressor with rotors having different types of blades.

9.4.7 Drivers

The centrifugal compressor is usually driven by one of the following prime movers: gas turbine, steam turbine, and electric motor. Internal combustion engines are rarely used to drive centrifugal compressors because of the different operating characteristics of IC engines and centrifugal compressors. Each of the three drivers noted above has its unique advantages and disadvantages. The driver chosen depends on the matching of the characteristics and availability.

Steam turbines are used widely in chemical plants, due to the fact that most of these plants need steam or generate it in their processes. Thus, by using steam turbines, the plant can utilize most of the excess steam, thus using energy most efficiently. In combined cycle utility plants, a combination of gas and steam turbines is utilized to obtain high overall efficiencies. The hot gases from the gas turbine are used to generate steam for the steam turbine.

The advantages of using steam turbines are the very small amount of work required to pump water, the ability to use a wide variety of fuels, and the use of steam, which is a by-product of the process. However, major disadvantages of the steam turbine include the large amount of equipment required and the long time needed to start up a steam turbine plant. However, steam turbines may be the only economical choice for driving very large centrifugal compressors.

Gas turbines are used for remote locations and offshore platforms. They have the advantage of low maintenance, the ability to be prepackaged, and the ability to use multifuels. However, the turbine has to be matched properly to operate the unit over the widest possible range of inlet conditions.

Electric motors are high in efficiency and reliability and are available to drive almost any configuration of centrifugal compressor. The major disadvantage of electric motors is that a gearbox is required, as the maximum speed of the motor is 3000 rpm or 3600 rpm depending on the line frequency, while centrifugal compressors are usually at their highest possible speeds to reduce losses. Hence, the overall efficiency of the drive system is reduced. Electric motors are easy to operate and can be started and stopped quickly. The speed of the electric motors can also be controlled easily using variable-frequency drives.

9.5 RECIPROCATING PISTON COMPRESSORS*

9.5.1 Compressor Operation

The major component of a reciprocating piston compressor is a piston whose motion is of to-and-fro reciprocating type inside a cylinder, but many other components are also required for its safe and efficient operation. These and the operation of a reciprocating piston compressor are discussed below.

Figure 9-36a shows a theoretical $p-V$ diagram of a single-stage single-acting reciprocating piston compressor. Different stages of operation of the compressor are shown in Fig. 9-36b to f. A unit that compreses the working fluid on both sides of the piston, consisting of two basic single-acting elements operating in parallel in one casing, known as a double-acting reciprocating compressor, saves considerable space.

A cylinder full of working gas is shown in Fig. 9-36b. On the theoretical $p-V$ diagram, point 1 is the start of compression. Both the suction and delivery valves are closed. During the compression stoke, shown in Fig. 9-36c, the piston moves to the left, reducing the original volume of working gas with a corresponding increase in pressure. Both valves remain closed. The $p-V$ diagram shows that the compression is from point 1 to point 2. At this point the pressure of the working gas in the cylinder has reached that in the receiver.

Figure 9-36d shows the delivery stoke, which is from point 2 to point 3 on the $p-V$ diagram. At the beginning of the delivery stoke, just beyond point 2, the delivery valve is open and the compressed gas is starting to be delivered to the receiver. After the piston reaches point 3, the delivery valve is closed, leaving the clearance.

During the expansion stroke (Fig. 9-36e), both valves remain closed. The working gas in the clearance space increases in volume and its pressure is reduced. The piston, then to the right, until the cylinder pressure drops just below the inlet pressure at point 4.

*Material for this section is drawn heavily from Ingersoll-Rand Form 3519-D and Ref. 3.

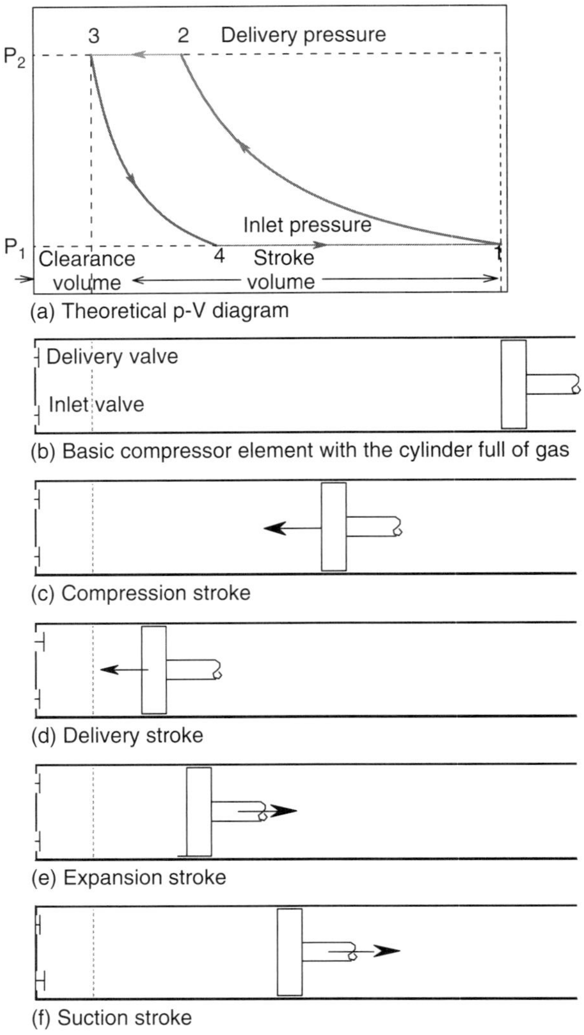

Figure 9-36 $p-V$ diagram of a single-stage reciprocating piston compressor. (Adapted from Ingersoll-Rand Form 3519-D; see also [3].)

Next, the suction valve opens and the working gas flows into the cylinder to the end of the suction stoke (Fig. 9-36f), point 1 on the $p-V$ diagram. At point 1, the suction valve is closed and the cycle is repeated in the next revolution of the crank.

To increase the pressure of the working gas, the gas is compressed in two or more stages in series. The cylinders are proportioned based on the total compression ratio, the second and later stages being smaller because the working gas, having already been partially compressed and cooled, occupies less volume than it had at the preceding stage. A $p-V$ diagram for a two-stage compressor is shown in Fig. 9-37. Points 1 and 5 correspond to the start of the compression stroke for the first and second stages, respectively. Points 2 and 6 correspond to the end of the compression stroke, points 3 and 7 correspond to the end of the delivery stroke, and points 4 and 8 correspond to the

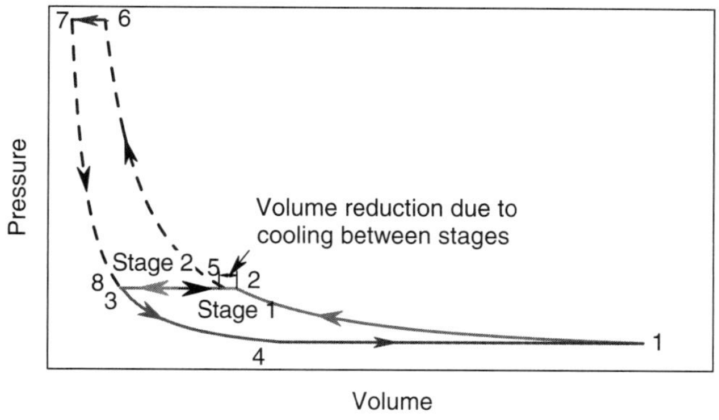

Figure 9-37 $p-V$ diagram of a two-stage reciprocating piston compressor. (Adapted from Ingersoll-Rand Form 3519-D; See also [3].)

end of the expansion stroke for the first and second stages, respectively. On the suction strokes ending at points 1 and 5, the cycle is repeated.

9.5.2 Thermodynamic Laws

Certain laws that govern the change of state of gases must be followed carefully during these operations. For the sake of brevity, only the important laws are given below. Any introductory textbook on thermodynamics may be referred to for all the laws governing change of state in gases.

- *Perfect gas*. Assuming a perfect gas, the law combining the three state properties (pressure, volume, and temperature) is

$$pV = R_0 T$$

where p, V, and T are the pressure, volume, and temperature in a gas in kg/m^2, m^3/s, and Kelvin, respectively. The value of the universal gas constant, R_0, is 8.314472 J/mol · K. The specific gas constant, R, for any gas can be obtained by dividing R_0 by the molecular weight of the gas. R is also equal to the difference in the two specific heats, c_p and c_v, at constant pressure and constant volume, respectively. The ratio of these two specific heats, c_p/c_v, is given by γ, the specific heat ratio.

9.5.3 Compression Cycles

Two theoretical compression cycles are applicable to any compressor. Although neither of these cycles is achievable in practice, they are used as a basis for calculations and comparisons.

- *Isothermal compression*. The temperature is kept constant in isothermal compression process as the pressure increases and volume decreases. This requires continuous removal of heat of compression. Assuming a perfect gas, the compression process follows the following law

$$p_1 V_1 = p_2 V_2 = \text{constant}$$

- *Near-adiabatic (isentropic) compression*. During an isentropic compression process, no heat is added or removed. Assuming a perfect gas, the compression process follows the law

$$p_1 V_1^{\gamma} = p_2 V_2^{\gamma} = \text{constant}$$

where γ is the specific heat ratio.

Figure 9-38 shows the theoretical zero-clearance isothermal and adiabatic cycles on a $p-V$ diagram. Area *ADEF* represents the work required when the compressor is operating on an isothermal basis; area *ABEF* represents the work required when the compressor is operating on an adiabatic basis. The isothermal area is less than the adiabatic area and would be the cycle used to achieve the greatest compression economy. However, the isothermal cycle is not achievable in practice, although compressors are designed for as much heat removal as possible.

Similarly, adiabatic compression is never achieved in practice, as heat is always rejected or added. Actual compression therefore takes place along a polytropic cycle, given by the relationship

$$p_1 V_1^{n} = p_2 V_2^{n} = \text{constant}$$

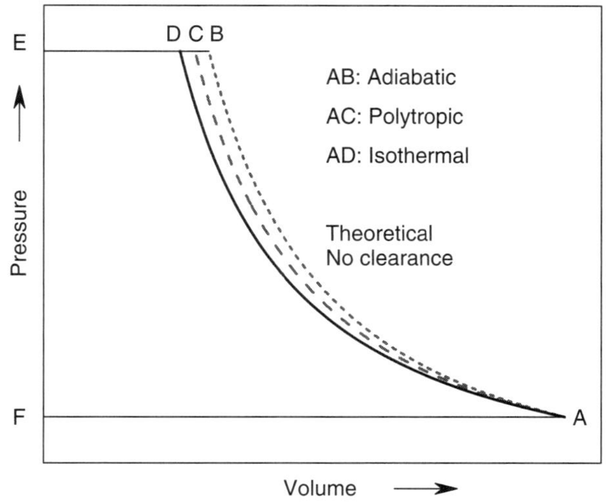

Figure 9-38 $p-V$ diagram illustrating theoretical compression cycles (Adapted from Ingersoll-Rand Form 3519-D; see also [3].)

The exponent n is determined for a given type of compressor and may be lower or higher than the adiabatic exponent, γ. In positive-displacement compressors, n is usually less than γ. A typical polytropic compression curve for a reciprocating water-jacketed compressor cylinder is shown as area $ACEF$ in Fig. 9-38.

An isentropic or adiabatic process is reversible, whereas a polytropic process is irreversible. The value of n for polytropic compression can be determined experimentally if the suction and delivery pressures and temperatures are determined. The following equation is used:

$$\frac{T_2}{T_1} = \left(\frac{p_2}{p_1}\right)^{(n-1)/n} = r^{(n-1)/n}$$

The pressure ratio across the cylinder is given as r, which is equal to the ratio of delivery pressure and suction pressure. If n is known, the delivery temperature can be determined from the equation above and the inlet pressure and temperature and design pressure ratio.

9.5.4 Power Requirements

The work per stage can be calculated by multiplying the adiabatic head by the mass flow per stage:

$$W = mH_a = mRT_1\frac{\gamma}{\gamma - 1}\left[r^{(\gamma-1)/\gamma} - 1\right]$$

For mRT_1, p_1V_1 can be used:

$$W = p_1V_1\frac{\gamma}{\gamma - 1}\left[r^{(\gamma-1)/\gamma} - 1\right]$$

The ideal work must be divided by the efficiency to account for various losses in the cylinder. Hence, the actual work, W_a, is equal to

$$W_a = p_1V_1\frac{\gamma}{(\gamma - 1)\eta_{\text{cyl}}}\left[r^{(\gamma-1)/\gamma} - 1\right]$$

The value of efficiency, η_{cyl}, is shown as a function of pressure ratio in Fig. 9-39.

9.5.5 Multistage Compression

All basic compressor elements, regardless of type, have certain limiting operating conditions. The basic elements are single-stage (i.e., the compression and delivery of gas is accomplished in a single element) or a group of elements are arranged in parallel. The most important limitations include the following:

- Discharge temperature
- Pressure differential

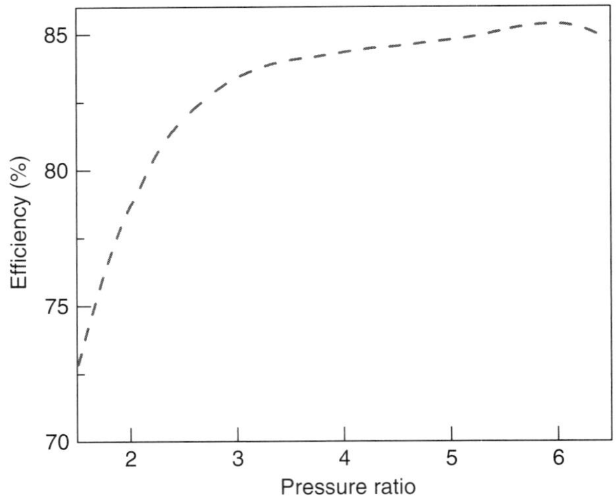

Figure 9-39 Reciprocating compressor efficiencies as a function of pressure ratio with a valve velocity of 15 m/s and a mechanical efficiency of 95%. (Adapted from [7].)

- Effect of clearance (ties in with the compression ratio)
- Desirability of saving power

There are reasons for multiple staging other than these, but they are largely for the designer of the specific unit to keep in mind. No ready reference rules can be given. When any limitation is involved, it becomes necessary to multiple-stage the compression process (i.e., do it in two or more steps). Each step will use at least one basic element designed to operate in series with the other elements of the machine.

A reciprocating compressor usually requires a separate cylinder for each stage with intercooling of the gas between stages. Figure 9-40 shows the $p-V$ diagram of a two-stage

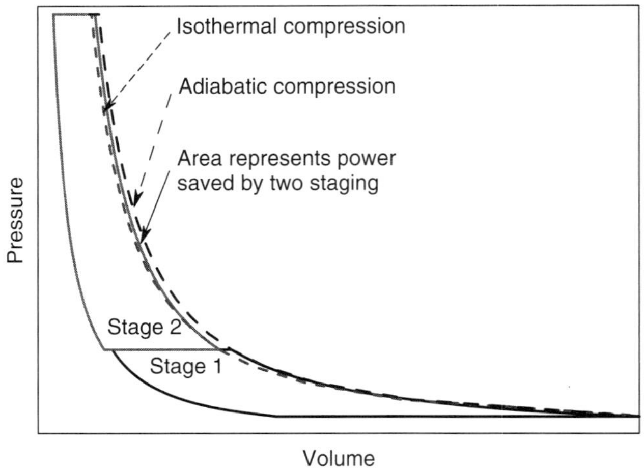

Figure 9-40 Combined $p-V$ diagram for a two-stage reciprocating piston compressor. (Adapted from Ingersoll-Rand Form 3519-D; see also [3].)

compressor. Further stages are added in the same manner. In a reciprocating unit, all stages are commonly combined into one unit assembly.

It was noted previously that the isothermal cycle (constant temperature) is the more economical of power. Cooling the gas after partial compression to a temperature equal to the original intake temperature (back to the isothermal) should obviously reduce the power required in the second stage. Area *ABCD* represents the work saved over single-stage adiabatic compression in this particular case.

For minimum power with perfect intercooling between stages, there is a theoretically best relation between the intake pressures of succeeding stages. This is obtained by making the ratio of compression the same in each stage and assumes the intake temperature to be the same in all stages. For two stages, the power equation can be expanded to add the interstage conditions for the second stage:

$$W = p_1 V_1 \frac{\gamma}{\gamma - 1}\left[r^{(\gamma-1)/\gamma} - 1\right] + p_i V_i \frac{\gamma}{\gamma - 1}\left[r^{(\gamma-1)/\gamma} - 1\right]$$

where the subscript i refers to the second-stage inlet. As a first approximation, it may be assumed that

$$p_i = (p_1 \times p_2)^{0.5} \quad \text{or} \quad \frac{p_i}{p_1} = \frac{p_2}{p_i}$$

The values for the pressure ratio in a practical case must include an allowance for the pressure drop in the interstage piping.

9.5.6 Cylinder Clearance and Volumetric Efficiency

The value of cylinder clearance normally varies from 4 to 16%. However, there are special low-compression-ratio cylinders where normal clearance will be much greater. Normal clearance does not include clearance volume that may have been added for other purposes, such as capacity control. Although the amount of clearance in a given cylinder is of little importance to the average user (guarantees being made on capacity actually delivered), its effect on capacity should be understood because of the wide application of variation in clearance volume for capacity control and other purposes. Normal clearance variations have no effect on power requirements.

When a piston has completed the compression and delivery strokes and is ready to reverse its movement, gas at discharge pressure is trapped in the clearance space. This gas expands on the return stroke until its pressure is sufficiently below the intake pressure to open the suction valves. The effect of this reexpansion on the quantity of fresh gas drawn in is shown on a p–V diagram (Fig. 9-41). The actual capacity is affected materially.

The theoretical value of the volumetric efficiency is given as

$$\eta_v = 100 - C(r^{1/\gamma} - 1)$$

There are factors that modify this, and an accepted formula for rough estimates is

$$\eta_v = 100 - C(r^{1/\gamma} - 1) - L$$

Here the term L is introduced to allow for the effect of such variables as internal leakage, gas friction, pressure drop through valves, and inlet gas preheating. The term L is difficult to generalize, but it might be 5% for a moderate-pressure oil-lubricated air compressor. A higher value of L will be necessary with a light gas than with a heavy gas because of increased leakage. Inspection of the equations shows that the value of volumetric efficiency, η_V, decreases

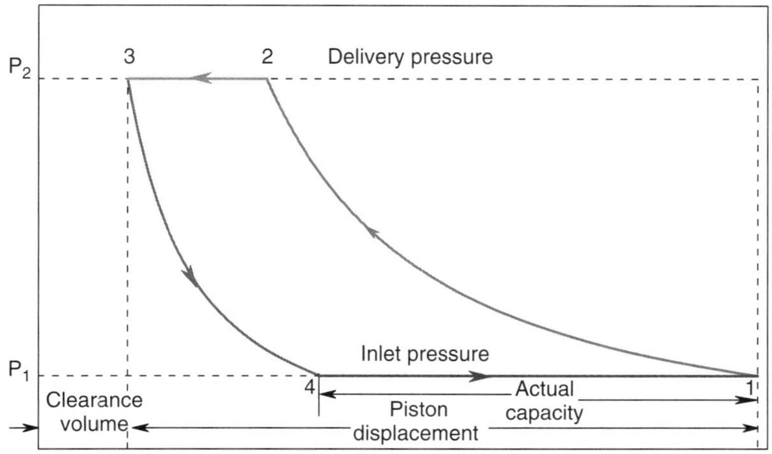

Figure 9-41 Work done on a volume of gas trapped in cylinder clearances (clearance volume) represents an inefficiency. (Adapted from Ingersoll-Rand Form 3519-D; see also [3].)

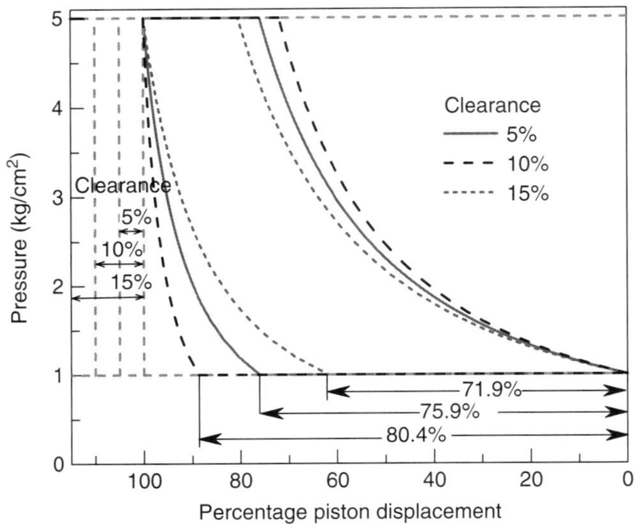

Figure 9-42 Effect of clearance on volumetric efficiency at a compression ratio of 5. (Adapted from Ingersoll-Rand Form 3519-D; see also [3].)

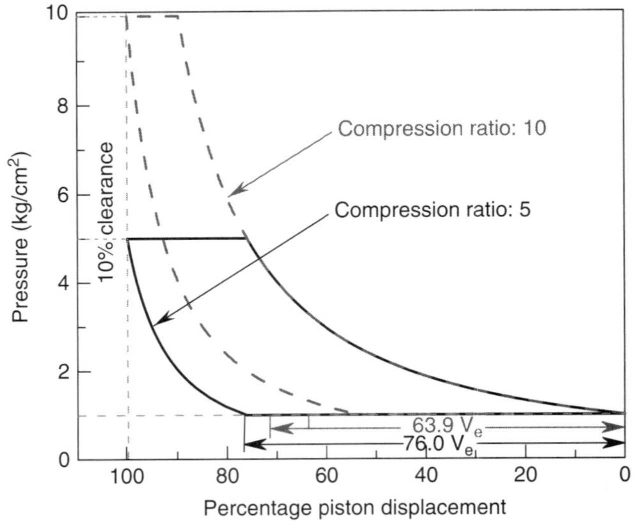

Figure 9-43 Effect of compression ratio on volumetric efficiency. A $p-V$ diagram for a compression ratio of 10 is superimposed on a $p-V$ diagram for a compression ratio of 5, all else being the same. (Adapted from Ingersoll-Rand Form 3519-D; see also [3].)

(1) as the clearance increases, (2) as the compression ratio increases, and (3) as γ decreases.

Theoretical $p-V$ diagrams based on an r value of 5, a γ value of 1.40, and clearances of 5, 10, and 15% are shown in Fig. 9-42 to illustrate the effect of clearance. The effect of clearance at moderate- and high-compression-ratio conditions is shown in Fig. 9-43. A $p-V$ diagram for a ratio of 10 is superimposed on a diagram for a ratio of 5, all else being the same. A relatively high clearance value (10%) is used for illustrative purposes. The clearance for any commercial compressor designed for a ratio of 10 would be much less than 10%.

The effect of γ on volumetric efficiency is shown in Fig. 9-44. The clearance is high for illustrative purposes. Clearance obviously concerns designers more at the higher compression ratios and when handling gases with low specific heat ratios, although they will always endeavor to maintain clearance at the lowest value consistent with adequate valving and running clearances.

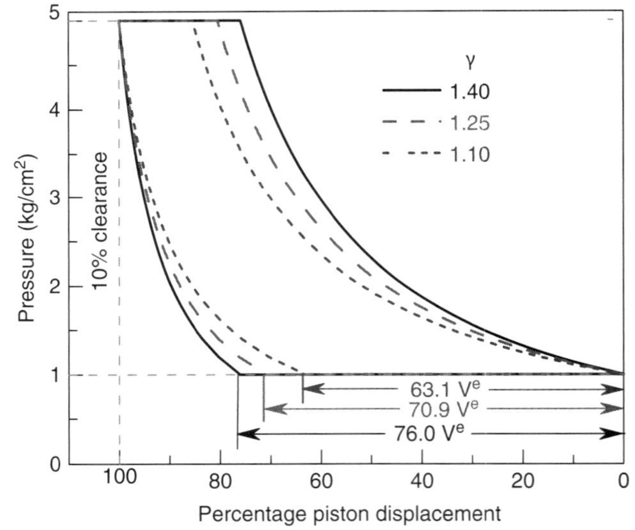

Figure 9-44 Effect of γ on volumetric efficiency for 10% clearance. (Adapted from Ingersoll-Rand Form 3519-D; see also [3].)

9.5.7 Valve Losses

The effect of valve losses on a $p-V$ diagram is shown in Fig. 9-45. If $p-V$ diagrams are analyzed or compared in conjunction with crank angle position, any control errors can be quickly spotted and associated with a given cylinder. Figure 9-46 shows the four $p-V$ diagrams of two dimensionally identical double-acting cylinders equipped with valve unloaders. The lifting device on one of the cylinders is obviously not working properly. In many instances more information on the proper adjustment of a lifting device can be gained from vibration monitoring at the cylinder.

9.5.8 Major Components of Reciprocating Piston Compressors

The major components of a typical reciprocating compressor are shown in Fig. 9-47 based on the outstanding reference document, API Standard 618. These components are described in detail by Bloch [2], who also provides detailed drawings of the various components. Bloch's book may be referred to for further details.

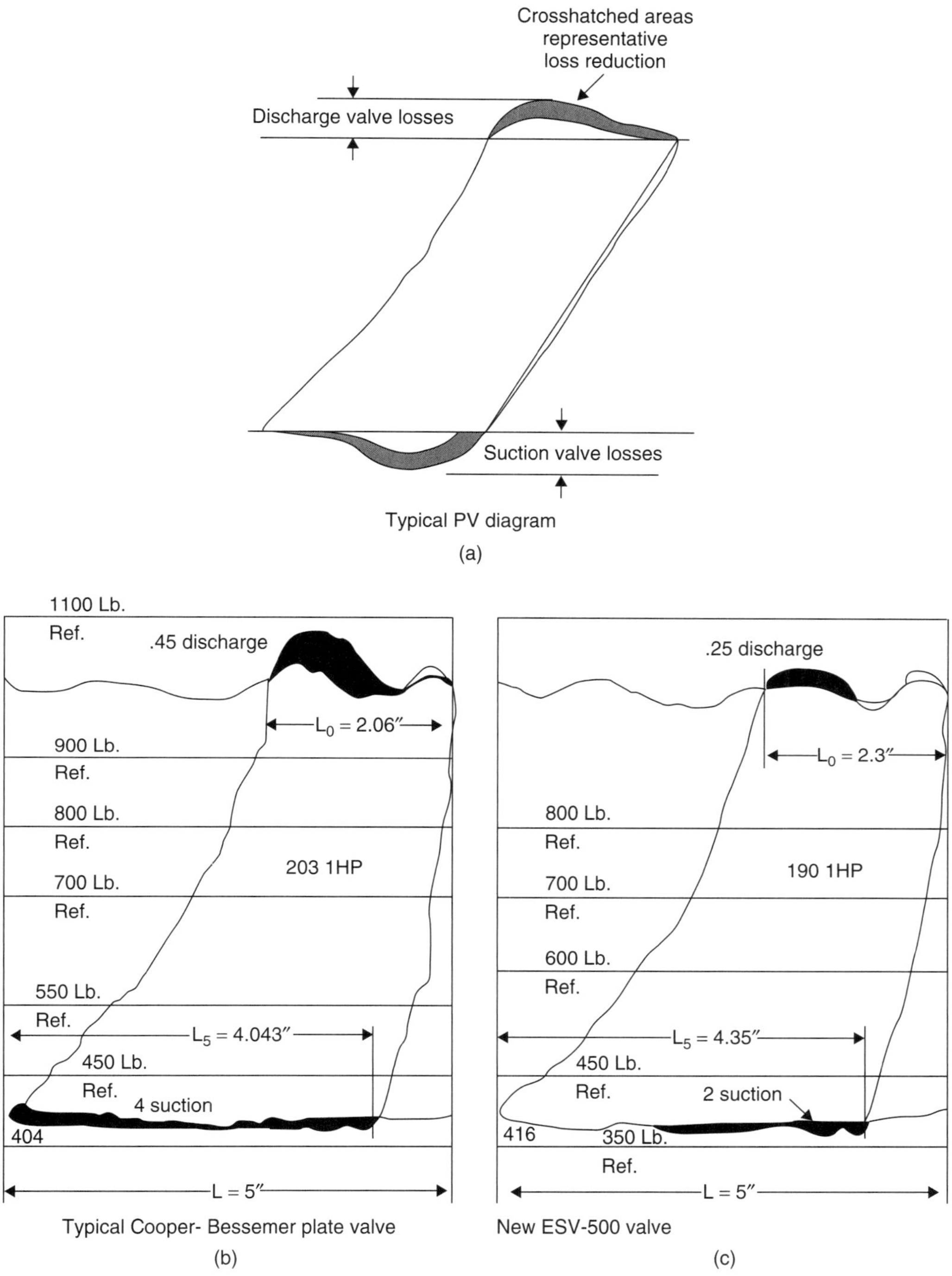

Figure 9-45 *p–V* diagrams can reveal valve losses. Typical diagram (a) is compared to a traditional plate valve (b) and enhanced design (c). (Courtesy of Cooper Cameron Corporation, Cooper-Bessemer Reciprocating Products Division, Grove City, Pa. From [3].)

9.5.9 Drivers

Reciprocating piston compressors are usually run at low speeds; hence, electrical motors are ideally suited to drive these compressors. Either direct drive or V-belt drive is used. Another preferred driver is the piston engine. A steam turbine or gas turbine may also be used. However, the speed has to be reduced using a step-down gear unit. A rarer form of driver is the steam cylinder engine. Most arrangements combine the steam driver and compressor on the same frame, with the steam cylinder engine opposite the compressor cylinder.

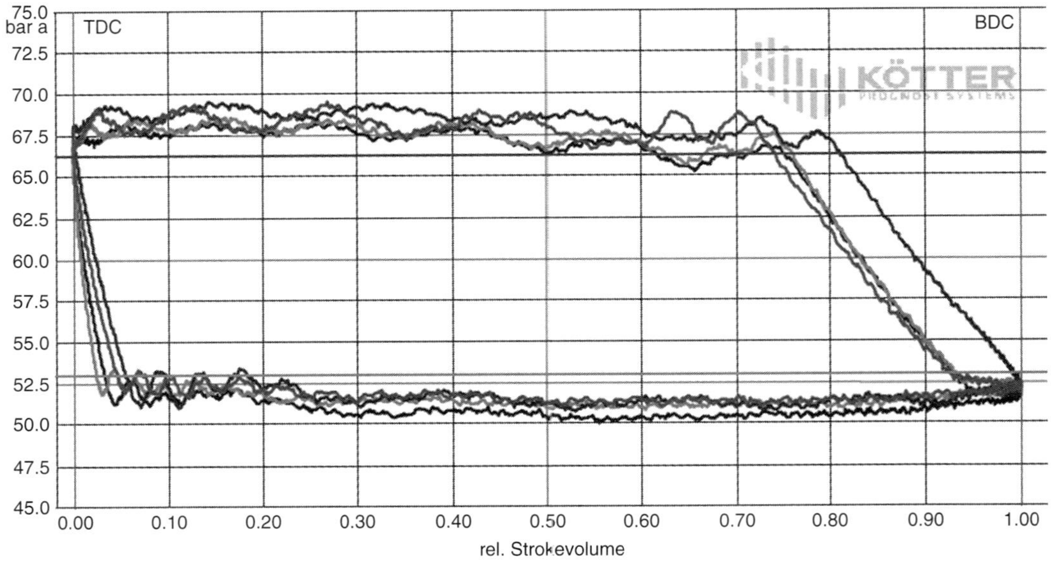

Machine	Measuring position	Dataname	TAG-Name	Value	Unit
K004	Cyl. 1, HE	p-V-diagram	03PI201		bar a
K004	Cyl. 1, CE	p-V-diagram	03PI200		bar a
K004	Cyl. 2, HE	p-V-diagram	03PI203		bar a
K004	Cyl. 2, CE	p-V-diagram	03PI202		bar a
K004	SP Cyl. 1, HE	suction pressure Cyl. 1, HE	03PC305	52.9	bar a
K004	DP Cyl. 1, CE	dischargepressure Cyl.1, CE	03PI304	66.2	bar a

Figure 9-46 *p–V* diagrams of a one-stage two-cylinder compressor with one defective valve unloader. (Courtesy of Prognost Systems GmbH, Rheine, Germany. From [3].)

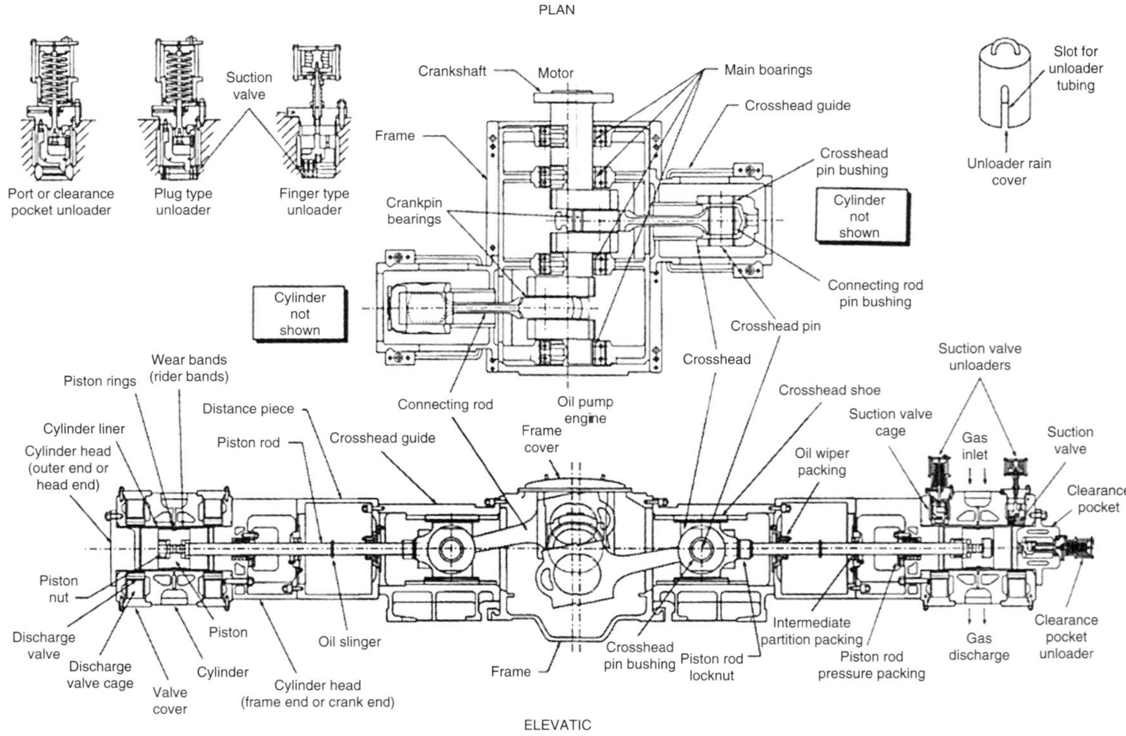

Figure 9-47 Reciprocating compressor cross section and nomenclature, following API 618 (see [3]).

9.6 COMPRESSOR TROUBLESHOOTING

In good process operations, common problems should be detected as early as possible and addressed immediately. Proper equipment maintenance will prevent the occurrence of most compressor problems. As in all troubleshooting situations, when problems do occur, an understanding of the problem and a logical process of elimination are required. Table 9-3 provides guidance on troubleshooting common problems.

TABLE 9-3 Common Compressor Problems and Remedies

Possible Problems and Causes	Remedies/Solutions
Low discharge pressure	
Compressor not up to speed?	Check operating manual for the speed required.
Compressor lined up properly?	If not, shut down and restart compressor.
Low inlet pressure?	Maintain inlet pressure.
Leak in discharge piping?	Check compressor for leakages.
Compressor motor does not run	
Improper starter heaters?	Install proper starter heaters.
Poor contact on motor terminals or starter connections?	Ensure good contact on motor terminals and starter connections.
Improper line voltage?	Check line voltage, change lines as required.
Gas delivery dropped off	
Piston scratched, worn, or scored?	Repair or replace as required.
Piston rings damaged or worn (broken, rough, scratched, excessive end gap or side clearance)?	Replace piston rings.
Valves leaking, broken, carbonized, or loose?	Check valves. Clean and replace as required.
Gas leaks in piping (on compressor or external piping/system)?	Check tubing and connections. Repair or replace as required.
Clogged or dirty inlet and/or discharge filter?	Replace filter element.
Cylinder scratched, worn, or scored?	Replace or repair as required.
Automatic condensate drain valve defective?	Inspect drain valve. Repair or replace as required.
Abnormal piston, ring, and cylinder	
Clogged or dirty inlet and/or discharge filter	Replace filter element.
Oil viscosity too low or high	Drain existing lubricant from frame. Refill with proper lubricant.
Oil level too low	Add lubricant to frame to bring level up to an acceptable point.
Detergent-type lubricant being used	Drain existing lubricant from frame. Refill with lubricant specified.
Extremely wet gas	Install dryer.
High discharge temperature	
Cooling system commissioned?	Check the cooling fluid temperature and pressure.
Compressor overloaded?	If so, minimize compressor load.
Adequate lubricant?	If not, refill lubricant.
Compressor fails to start	
Shaft free?	Remember to bar the shaft to prevent stiffness.
Energy available?	Confirm steam, or power availability.
Start button functional? (for manual startup)	Confirm status. If faulty, refer faults for maintenance.
Excessive noise when operated	
Loose belt wheel or motor pulley, excessive end play in motor shaft	Check belt wheel, motor pulley, and shaft. Repair or replace as required.
Gas leaks in piping (on compressor or external piping/system)	Check tubing and connections. Repair or replace as required.

(continued)

TABLE 9-3 ***(Continued)***

Possible Problems and Causes	Remedies/Solutions
Valves leaking, broken, carbonized, or loose	Check valves. Clean or replace as required.
Automatic condensate drain valve defective	Inspect drain valve. Repair or replace as required.
Defective ball bearings on crankshaft or motor shaft	Check ball bearings. Replace as required.
Connecting rod, piston pin, or crankpin bearings worn or scored	Inspect all. Repair or replace as required.
Oil in discharge gas	
Clogged or dirty inlet and/or discharge filter	Replace filter element.
Oil viscosity too low	Drain existing lubricant from frame. Refill with proper lubricant.
Oil level too high	Drain lubricant from frame to proper level.
Piston rings damaged or worn (broken, rough, scratched, excessive end gap or side clearance)	Replace piston rings.
Relief valve vent pressure	
Clogged or dirty inlet and/or discharge filter	Replace filter element.
Valves leaking, broken, carbonized, or loose	Check valves. Clean and replace as required.
Automatic condensate drain valve defective	Repair or replace as required.
Excessive starting and stopping	
Gas leaks in piping (on compressor or external piping/system)	Check tubing and connections. Repair or replace as required.
Pressure switch differential too narrow	Adjust pressure switch to increase differential.
Automatic condensate drain valve defective	Repair or replace as required.
Excessive consumption of lubricant	
Piston rings damaged or worn (broken, rough, scratched, excessive end gap or side clearance)	Repair or replace as required.
Piston rings not seated, stuck in grooves, or end gaps not staggered	Clean and adjust piston rings. Replace as required.
Piston scratched, worn, or scored	Repair or replace as required.
Cylinder scratched, worn, or scored	Repair or replace as required.
Compressor does not come up to design speed	
Loose belt wheel or motor pulley. Excessive end play in motor shaft	Check belt wheel, motor pulley, and shaft. Repair or replace as required.
Improper line voltage	Check line voltage; change lines as required.
Poor contact on motor terminals or starter connections	Ensure good contact on motor terminals and starter connections.
Improper starter heaters	Install proper starter heaters.
Defective ball bearings on crankshaft or motor shaft	Check ball bearings. Replace as required.

Source: Drawn heavily from http://www.mckenziecorp.com/compressor_guide.htm (13).

REFERENCES

1. Aungier, R. H., *Centrifugal Compressors: A Strategy for Aerodynamic Design and Analysis*, ASME Press, New York, 2000.
2. Bannwarth, H., *Liquid Ring Vacuum Pumps, Compressors and Systems: Conventional and Hermetical Design*, Wiley-VCH, Hoboken, NJ, 2009.
3. Bloch, H. P., *A Practical Guide to Compressor Technology*, 2nd ed., Wiley, Hoboken, NJ, 2006.
4. Boyce, M. P., *Gas Turbine Engineering Handbook*, 2nd ed., Butterworth Heinemann, Woburn, MA, 2002.
5. Boyce, M. P., *Centrifugal Compressors: A Basic Guide*, PennWell Corporation, Tulsa, OK, 2003.
6. Boyce, M. P., and Nishida, A., Investigation of flow in centrifugal impeller with tandem impeller, *Tokyo Joint Gas Turbine Congress*, 1977. Paper 43.
7. Brown, R. N., *Compressors: Selection and Sizing*, 3rd ed., Gulf Publishing, Houston, TX, 2003.
8. Busemann, A., Das forderhohenverhaltniss radialer kreiselpumpen mit logarithisch-spiraligen schaufeln, *Zeitschrift Angewandte Mathematik and Mechanik*, vol. 8, 1928, p. 384.
9. Csanady, G. T., Head correction factors for radial impellers, *Engineering*, vol. 190, 1960.

10. Eck, B., *Fans*, Pergamon Press, Oxford, 1973.
11. Eisenlohr, G., Krain, H., Richter, F.-A., and Tiede, V., *Investigations of the Flow Through a High Pressure Ratio Centrifugal Impeller*, ASME Paper GT-2002-30394, ASME, New York, 2002.
12. Gopalakrishnan, G., and Prithvi Raj, D., *A Treatise on Turbomachines*, SciTech Publications (India) Pvt.Ltd, Chennai India, 2002.
13. http://www.mckenziecorp.com/compressor_guide.ht.
14. Issac, J. M., Sitaram, N., and Govardhan. M., Effect of diffuser vane height and position on the performance of a centrifugal compressor, *Proceedings of the Institution of Engineers A*, vol. 218, no.(6), 2004, pp. 647–654.
15. Kenny, D. P., *A Comparison of the Predicted and Measured Performance of High Pressure Ratio Centrifugal Compressor Diffusers*, ASME Paper 72-GT-54, ASME, New York, 1972.
16. Kruschik, J., *Die Gas Turbine*, 2nd ed., Springer-Verlag, Vienna, 1960.
17. Musgrave, D. S., and Plehn, N. J., Mixed-flow compressor stage design and test results with a pressure ratio of 3:1, *ASME Journal of Turbomachinery*, vol. 109, no. 4, 1987, pp. 513–519.
18. Osborne, W. C., *Fans*, Pergamon, Bell and Bain Ltd., Glasgow, UK, 1966.
19. Roberts, D. A., and Kacker, S. C., Numerical investigation of tandem-impeller designs for a gas turbine compressor, *ASME Journal of Turbomachinery*, vol. 124, no. 1, 2002, pp. 36–44.
20. Rodgers, C., and Sapiro, L., *Design Considerations for High Pressure Ratio Centrifugal Compressors*, ASME Paper 1973-GT-31, ASME, New York, 1973.
21. Senoo, Y., Japanese Patent Application Disclosure 11941/78 (in Japanese), October 18, 1978.
22. Srinivasa Reddy, K., Ramana Murty, G. V., Dasgupta, A., and Sharma, K. V., Flow investigations in the cross over system of a centrifugal stage, *International Journal of Fluid Machinery and Systems*, vol. 3, no. 1, 2010, pp. 11–19.
23. Staniz, J. D., Some theoretical aerodynamic investigations of impellers in radial and mixed flow centrifugal compressors, *Transactions of ASME*, vol. 74, 1952, pp. 473–476.
24. Stodola, A., *Steam and Gas Turbines*, McGraw-Hill, New York, 1926; reprinted by Peter Smith, New York, 1945.
25. Veress, A., and Van den Braembussche, R., Inverse design and optimization of a return channel for a multistage centrifugal compressor, *ASME Journal of Fluids Engineering*, vol. 126, no. 4, 2004, pp. 799–806.
26. Von Backstrom, T. W., A unified correlation for slip factor in centrifugal impellers, *ASME Journal of Turbomachinery*, vol. 128, no. 1, 2006, pp. 1–10.
27. Wiesner, F. J., A review of slip factors for centrifugal impellers, *ASME Journal of Gas Turbines and Power*, vol. 89, no. 4, 1967, pp. 558–572.
28. Yoshinaga, Y., Kaneki, T., Kobayashi, H., and Hoshino, M., A study of performance improvement for high specific speed centrifugal compressors by using diffusers with half guide vanes, *ASME Journal of Fluids Engineering*, vol. 109, no. 3, 1987, pp. 359–367.
29. Pfleiderer, C., Petermann, H., Stromungsmachinen, 4th Ed., Springer Verlag, Berlin, Germany, 1972.

FURTHER READING

Baskharone, E. A., *Principles of Turbomachinery in Air-Breathing Engines*, Cambridge University Press, New York, 2006.

Bloch, H. P., *Compressors and Modern Process Applications*, Wiley, Hoboken, NJ, 2006.

Bloch, H. P., and Hoefner, J. J., *Reciprocating Compressors: Operation and Maintenance*, Gulf Publishing, 1996.

Cumpsty, N. A., *Compressor Aerodynamics*, Krieger Publishing Company, Melbourne, FL, 2004.

Dixon, S. L., and Hall, C. A., *Fluid Mechanics and Thermodynamics of Turbomachinery*, 6th ed., Elsevier, New York, 2010.

Japiske, D., *Centrifugal Compressor Design and Performance*,Concepts ETI Inc., Wilder, USA, 1996.

Japiske, D., Baines, N.C. *Introduction to Turbomachinery, Concepts ETI*,Concepts ETI Inc., White River Junction,VT, USAand Oxford University Press, Oxford, UK, 1997.

Japiske, D., Baines, N.C. *Diffuser Design Technology*, 2nd Ed., Concepts ETI, Inc., White River Junction, VT, USA, 1998.

Lakshminarayana, B., *Fluid Dynamics and Heat Transfer of Turbomachinery*, Wiley, New York, 1996.

Ludke, K. H., *Process Centrifugal Compressors: Basics, Function, Operation, Applications*, Springer-Verlag, New York, 2004.

Saravanamuttoo, H. I. H., Rogers, G. F. C., and Cohen, H., *Gas Turbine Theory*, 5th ed., Prentice Hall, Upper Saddle River, NJ, 2001.

Whitefield, A., and Baines, N. C., *Design of Radial Turbomachines*, Longman Scientific & Technical, New York, 1990.

Wilson, D. G., and Korakianitis, T., *The Design of High-Efficiency Turbomachinery and Gas Turbines*, 2nd ed., Prentice-Hall, Upper Saddle River, NJ, 1998.

Wislicenus, G. F., *Fluid Mechanics of Turbomachinery*, 2nd ed., 2 vols., Dover, New York, 1965.

10

CONVEYORS

Okenna Obi-Njoku
Owerri, Nigeria

A conveyor is a machine mechanized and automated to move loads through specified paths. A conveyor moves bottles, grains, liquids, boxes, crates, fabric, crates, fruits, cosmetics, metal parts, and other items. A conveyor can move materials easily. There are several definitions of conveyors, which vary by method of operation, function, location, and type and volume of material being handled. Generally, conveying means the transportation of items through different points (origin and destination) at a specific rate.

In material-handling operations, diverse activities from the carrying of drums, barrels, kegs, lumber, concrete blocks, bags, and other materials either manually, mechanically, electronically, hydraulically, or through any other means are all conveying processes. The main aim of mechanical conveying operations is to provide a continuous flow of materials and ensure the availability of such materials at each point of need and also [5], to reduce the cost of operation sufficiently while being as flexible as possible. Conveyor equipment should be able to supply material at a rate that effectively matches the processing capacity of the machine center or system.

A conveyor can also be said to be a horizontal, inclined, or vertical device for moving or transporting bulk material, packages, or objects in a path predetermined by the design of the device, and having points of load and discharge which can either be fixed or selective. Conveyors can be either gravity powered or live.

The conveyor is generally regarded as one of the most important industrial machines ever invented, having practically reduced the energy and time required in load-movement operation in the engineering work environment. They take the form of simple chutes, the nonpowered roller conveyors used to move loads across sidewalks from trucks to warehouses, and a wide range of powered systems in which materials are carried along by belt, bucket, screw, trolley, or other arrangement.

Ancient building techniques offer wide ranges of examples, a typical example is shown on early depictions of ancient Egypt where lumber and other building materials are transported several meters high with the aid of improvised pulleys. These were designed to optimize energy and reduce labor, thus improving efficiency. These ancient ideas formed the basis upon which modern conveyors were designed. The modern conveyor dates back to the late seventeenth century. These early conveyor systems were typically composed of a leather, canvas, or rubber belt that traveled over a flat wooden bed and was used to transport large bulky items. The Industrial Revolution necessitated the expansion of the conveyor system. In 1908, Hymle Goddard of the Logan Company patented the first roller conveyor, but the application of conveyors in general seemed quite unpopular until its introduction into the automobile industry by Henry Ford in 1913.

10.1 INDUSTRIAL USE OF CONVEYORS

Conveyors are in common use in industries that require frequent movement of materials through specific locations and take quality control into account during their operations. Conveyors offer a wide range of benefits, many of which are well known. The coal mining, food-processing, pharmaceutical, marine, beverage, powder coating, textiles,

Process Plant Equipment: Operation, Control, and Reliability, First Edition. Edited by Michael D. Holloway, Chikezie Nwaoha, and Oliver A. Onyewuenyi.
© 2012 John Wiley & Sons, Inc. Published 2012 by John Wiley & Sons, Inc.

waste treatment, and automotive industries, machining and milling units, warehouses and stores, recycling plants, and other facilities use a single conveyor or a combination of conveyors to set up an integrated and unique system most suitable for its industrial activity (Fig. 10-1).

For each of facility, applications vary based on the class of conveyor on which particular emphasis is laid. Selections are based on a variety of parameters:

- *Load:* the type of product being handled: could be a unit or bulk load.
- *Location:* where the conveyor is situated: overhead, in-floor, or on-floor.
- *Accumulation:* whether a conveyor permits the accumulation of loads or does not.
- *Functions:* specifies carrying out the action and how it is done; whether for transporting, elevating, transferring, or any other permissible action.
- *Direction or path:* based on the flexibility of the movement of the conveyor, whether in a two- or three-dimensional paths.
- *Features:* based on the method of application of driving force (whether by tractive elements such as belts and chains or tractionless elements such as rotary, vibratory, or oscillatory means) and the method of transmission of this driving force (mechanical, electrical, hydraulic, pneumatic, and gravity); also considers the nature of the motion, whether continuous or intermittent.

The choice of conveyor in a particular industry depends on whether it can achieve the following;

- Maintain the required accumulation; avoid backups, spillovers, and resulting cleanup requirements.
- Increase throughput.
- Maximize productivity.
- Accommodate the nature and properties of the material.
- Minimize equipment maintenance or plant downtime.
- Maintain appropriate orientation and stability of the material.
- Balance material flow through machine centers and systems.
- Operate compatibly with other process equipment and systems.

The advantage gained in the use of a particular kind of conveyor presents a relative advantage over its use in another industry.

10.2 TYPES OF CONVEYORS

From complex bottling plants to simple agricultural silos, there are several conveyors that operate either independently or in a system synchronized with other conveyors to achieve a harmonious operation. These conveyors generally appear to do the same functions but really vary slightly in their operation.

10.2.1 Belt Conveyors

Used to carry bulk loads, a conveyor belt is a continuous loop of material that rotates around two or more pulleys, thereby transporting the load on it (Fig. 10-2) [1]. One or both pulleys are powered, moving the belt and material forward. The motor rotates the drive pulley while the tail pulley follows. Belt conveyors provide the best option for handling small dry or semidry finished material at high speeds and are useful in picking and manual sorting operations. There are two main classes of industrial belt conveyors: material-handling conveyors, which move boxes within the facility, and the bulk material-handling conveyors, which are used outdoors to move coal, ore, agricultural materials, and other raw materials. Generally,

Figure 10-1 Conveyors in operation.

Figure 10-2 Belt conveyor in operation.

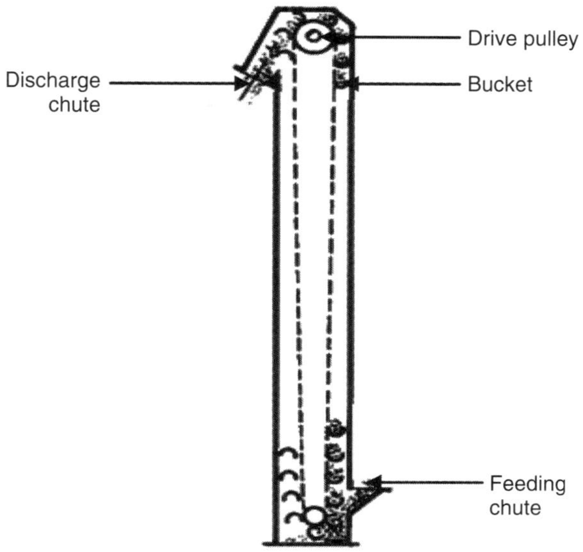

Figure 10-3 Schematic drawing of a bucket conveyor.

belt conveyors are used to transport light and medium-weight loads between levels, mezzanines, operations, and other processes. They can convey on a level platform or at gradual inclinations. The belt is supported by rollers or a slider bed, depending on the intended design.

The belt is a tautly stretched synthetic material whose underlying layer is a cotton or plastic web called the *carcass*, which provides linear strength and shape to the belt. The first layer or cover could be made of synthetic rubber, leather, fabric, plastic, or even metal, depending on the intended use. The belt has return idlers underneath and a snub idler that makes it hug the drive pulley. Also, the conveyor has a takeup screw that maintains the tightness of the belt.

Belt conveyors are used in self-unloading bulk freighters and in live bottom trucks, manufacturing assembly plants, grocery stores, coal mines, chemical and agricultural industries, and every other industry where materials have to be handled, stored, or dispensed. To transport loose materials up steep inclines, elevator belts are used. Of all the powered conveyors, belt conveyors are the most commonly used because they are the most versatile and the least expensive, and are usually compatible within a conveyor system. The product is conveyed directly on the belt in an exposed form so that both regular and irregularly shaped objects, large or small, light and heavy, can be transported successfully and can be inspected at the same time.

10.2.2 Bucket Conveyors

A bucket conveyor is a mechanical device that is used to move bulk and loose materials such as mining ore, grain, liquids, and other such materials from one stage to another in a vertical or inclined path. It is an on-floor conveyor and has buckets attached to a cable, chain, or belt driven in a continuous loop by a power source such as a conveyor belt (Fig. 10-3). It is fed through one point and discharges automatically on the other end of the machine. Bucket conveyors are used in factories, mining and drilling activities, agricultural industries, and other facilities that require the transportation of physical goods. Their design allows for horizontal, vertical, and angled movement of products, making them a very flexible type of conveyor.

In factory systems, a bucket conveyor is usually placed at the end of a conveyor belt, where goods falling off the belt drop into the bucket and are lifted off to another stage. In mining and drilling activities, they are placed at points of excavation where heavy rocks and unwanted soil are lifted off the site into the buckets to preferred points to create conducive working surroundings during such activities, thereby saving time and energy that would have been spent on removing the debris manually. Grains, nuts, and other whole agricultural materials are easily conveyed using the bucket conveyor. Where the material is dusty, the buckets are enclosed to avoid spreading the product. In its design, the motor speed, nature of product, size and number of buckets, length, bucket clearance, and strength of the bucket material are important factors to consider.

10.2.3 Cart-on-Track Conveyors

A cart-on-track conveyor is an in-floor conveyor used in conveying unit loads. It consists of a cart that is transported along a track by a rotating tube (Fig. 10-4). The tube is similar to a screw and moves the carts independently and at varying speeds. A drive wheel is connected to each cart resting on the tube, and by varying the angle of contact between the drive wheel and the tube, the speed of the cart

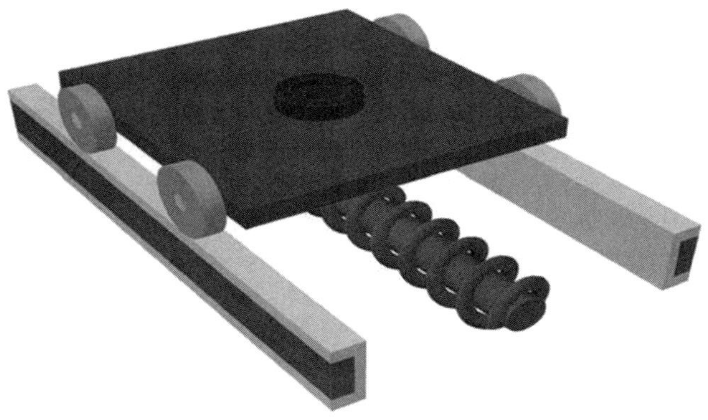

Figure 10-4 Cart-on-track conveyor.

Figure 10-5 Skate wheel conveyor.

changes. Multiple carts can accumulate on the tube and remain stationary by maintaining the drive wheel parallel to the tube.

10.2.4 Chute Conveyors

The chute conveyor is the simplest form of gravity conveyor, used in conveying both unit and bulk loads. It is located on-floor and used to link two handling devices. The chute conveyor is one of the least expensive methods of conveying material. Here, the load simply slides down a straight or spiral metal or plastic chute to its required destination. It is used primarily to provide accumulation in shipping areas. Although economical, the main limitation of chute conveyors is their inability to control the items being conveyed. Also, they are often not reliable [8].

The helical chute conveyor favors the movement of loose materials, particularly of the granular and pulverized type. The helical chute path is developed about a vertical axis to which it is also inclined transversely. This angle is just sufficient enough to cause a regular flow of materials down the chute. Chemical, pharmaceutical, and agricultural products are easily conveyed by the chute conveyor.

10.2.5 Gravity Wheel Conveyors

A gravity wheel conveyor is an on-floor unit load conveyor used for light-duty handling requirements where the load is flat and smooth. It supports the load on a series of skate wheels mounted on a shaft (or axle), where spacing of the wheels is dependent on the load being transported (Fig. 10-5). It is nonpowered and free flowing, used primarily in push or level applications to facilitate load movement and transfer. It is more economical than a roller conveyor and also allows accumulation.

In addition to the rigid or straight form, it has flexible, curved, or expandable versions, thus making it suitable for temporary applications from shipping operations to loading and unloading of roadside vehicles and trucks. It is typically used in facilities where its light weight and flexibility make it possible to facilitate storage of heavier loads. It suits the conveyance of cartons, trays, boxes, and other smooth-bottomed surfaces. Skate wheel conveyors are easily configured to suit into powered conveying processes to form a complete conveying system.

10.2.6 Monorails

A monorail is a single run of overhead track or track network on which one or more carriers travel. It is used in conveying unit loads and its carriers are either nonpowered or powered electrically or pneumatically. These carriers can range from a simple hook to a hoist to an intelligent-vehicle-like device. In a multicarrier track network, the carriers operate independently and the loop is not closed.

10.2.7 Chain Conveyors

A chain conveyors is a form of powered conveyor that consists of one or more endless chains on which heavy unit loads are carried directly. The chain is a series of links joined together pivotally to form a medium for conveying or transmitting motions of power. This motor-driven chain is located on the bottom of a trough and by friction pulls the material through the trough. The chain conveyor is located in-floor or on-floor and can convey horizontally, inclined, or vertically. In this type of conveyor, the flat-bottomed metal trough carries a box or flighted chain which catches the heavy material and pulls it along the conveyor. It is often used to transport industrial containers, grid boxes, and pallets and generally when the material is large, wet, or dry.

Chain conveyors are common features of most heavy industries. This is so because they can convey heavy materials multidimensionally and adapt to any local condition. The vertical chain conveyor can be used for either high-frequency vertical or low-frequency intermittent transfers.

Although this conveyor is noisy, requires frequent maintenance, does not accumulate material, and does not provide even flow, its applications offer several advantages characteristic of it alone. For example, it resists contamination by grease, tolerates uneven pallet bottoms, and can transport hot loads that could damage other conveyor surfaces.

10.2.8 Pneumatic Conveyors

In a pneumatic conveyor air pressure is used to convey materials through a system of vertical and horizontal tubes; this process occurs in a closed system, enabling it to carry out turns and vertical moves. It is relevant in the movement of both bulk and unit materials. It is an overhead conveyor that transports dry, free-flowing granular materials in suspension, taking advantage of their aerodynamic properties. Cylindrical carriers within a duct or pipe are also transported using vacuum pressure generated by an air compressor or high-velocity airstream. Here, solid materials are transported in large volumes over large distances. The operation of this conveyor is very cost-effective.

The air conveyor is a typical pneumatic conveyor. In this system, compressed air is introduced into annular plenum chamber through the female NPT (National Pipe Thread). From this point it is injected into the throat of the unit through the nozzles, creating a vacuum at the intake end of the unit. Due to the vacuum created by these jets of air, the materials are sucked in and accelerated through this unit and subsequently conveyed either vertically or horizontally.

The air conveyor is relevant in the beverage, automotive, powder coating, food-processing, textile, pharmaceutical, and several other industries in the transport of items such as screws, ball bearings, bolts, bottle caps, capsules, lids, bottles, tablets, metal parts, dried food, plastics, and other items. The air conveyor can be either threaded or nonthreaded, for easy attachment to standard pipes or hoses in a conveyor system. They are easy to use and have relatively longer life than that of other pneumatic conveyors. Its use of air makes it practically nonexplosive. It has no moving parts, thereby having little or no maintenance cost. Industries that use this conveyor also use air knives. These devices clean, heat, cool, or dry products as required during specific processes. There are two major types of pneumatic conveyors:

1. *Dilute-phase pneumatic conveyor*. This conveyor consists of push–pull systems that possess several inlet and outlet points. The push or positive pressure system pushes material from a single inlet point to several outlet points; the pull or negative pressure/vacuum system moves material from several inlet points to a single outlet point. This system moves a mixture of air and solid.

2. *Carrier-system pneumatic conveyor*. This system simply transports materials.

10.2.9 Roller Conveyors

A roller conveyor is a type of on-floor conveyor with a series of parallel rotating rollers supported in a frame over which unit loads are transported manually, by gravity, or by power (Fig. 10-6). It allows for both transportation and accumulation. The material being moved must have a rigid riding surface and must be supported by at least three rollers at a time, to ensure proper contact between the roller and the surface. This conveyor is equipped with tapered rollers on curves to maintain load orientation. They are used in warehouses and manufacturing facilities.

10.2.9.1 Gravity Roller Conveyors This type of roller conveyor is a nonpowered and free-flowing conveyor inclined at varying angles to allow loads to flow from a higher to a lower elevation. Boxes, pallets, cans, drums, and other rigid bodies with either an even or an uneven surface can be transported on this conveyor.

10.2.9.2 Powered Roller Conveyors Also called a live roller conveyor, it can be either belt or chain driven. Besides merely moving loads from place to place, its functions are extended to accumulating loads and for merging and sorting operations. To allow accumulation, force-sensitive transmission can be used to disengage rollers. This conveyor has certain angular limits of inclination, beyond which conveying operations will not be effective.

1. *Chain-driven conveyors*. In this arrangement, the load-bearing rollers have sprockets attached. The sprockets are connected by chains that transmit the same amount of power to the next roller, and so on. The movement of loads

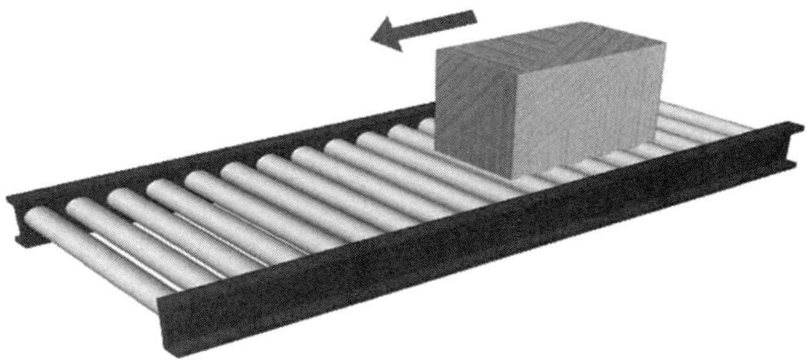

Figure 10-6 Roller conveyor.

is smooth and is done at controlled speed. This conveyor transports loads of varying weights, sizes, temperatures, and surface shapes more effectively than do other conveyors.

2. *Belt-driven conveyors*. Unlike the chain-driven arrangement, the belt-driven roller conveyor is driven by a belt. Directly underneath the load-bearing rollers is a drive belt that is in contact with the roller arrangement. Through friction, the belt drives the rollers while being powered by an external motor. Tension and support on the belt are provided by return rollers.

10.2.10 Screw Conveyors

The screw conveyor is a very common type of conveyor, used mainly for feeding processes in both agricultural and chemical processes involving powdery, granular, fibrous, or any combination of these materials. In other words, semisolid materials such as powders, food wastes, aggregates, wood particle, environmental waste, and many others are typical examples of materials that are suitably conveyed by screw conveyors. They are on-floor conveyors whose applications are very effective in conveying free-flowing or relatively free-flowing bulk solids. It consists of a tube or U-shaped stationary trough and a driven helix as its major components (Fig. 10-7). Through the tube, the helix revolves to push loose material forward in a horizontal or inclined direction. The conveyor diameter varies from as low as 50 mm to as high as 400 mm, while the length ranges from less than 1 m to more than 35 m, depending on the design requirements of the conveyor. They provide environmentally clean solutions in process-handling problems because of their simple and compact design. Due to this design, it is easily integrated in a conveyor system. An Archimedean screw, typical of this type of conveyor, was once used to deliver water from water sources.

Industrially, especially in control applications, this conveyor may be used as a variable-rate feeder to deliver a measured rate or quantity of materials into a process. The helix (spiral) could be either shafted or shaftless, driven at one end and free at the other. Capacity, volumetric efficiency, and power requirements are major performance parameters of screw conveyors. Also, the properties of the material, rotational speed and clearance of screw, size and geometry of conveyor, and conveying angle are all factors that determine the performance of the conveyor. It can be operated with the flow of material inclined upward. When space allows, this is a very economical method of elevating and conveying. However, as the angle of inclination increases, the allowable capacity of a given unit decreases rapidly. Also, greater pitch spacing increases the capacity without increased rotation speed. In addition, it has a relatively lower cost of maintenance than that of some other types of conveyors.

Several types of screw conveyors exist:

- Vertical screw tube conveyor
- Screw auger conveyor
- Trough screw conveyor
- Full-bladed screw conveyor
- Paddle screw convey
- Ribbon screw conveyor
- Screw tube conveyor

All these screw conveyors are suitable for specific operations.

10.2.11 Slat Conveyors

A slot conveyor uses discretely spaced nonoverlapping and noninterlocking slats connected in an endless chain (Fig. 10-8). These chains are driven by electric motors and controlled by gears and sprockets. The conveyor conveys unit loads and can be located in-floor or on-floor and does not allow accumulation. Just like a belt conveyor, the position of the unit being transported remains unaltered, keeping the placement and orientation of the load controlled. A slat conveyor carries heavy loads or sharp-edged loads that might damage a belt.

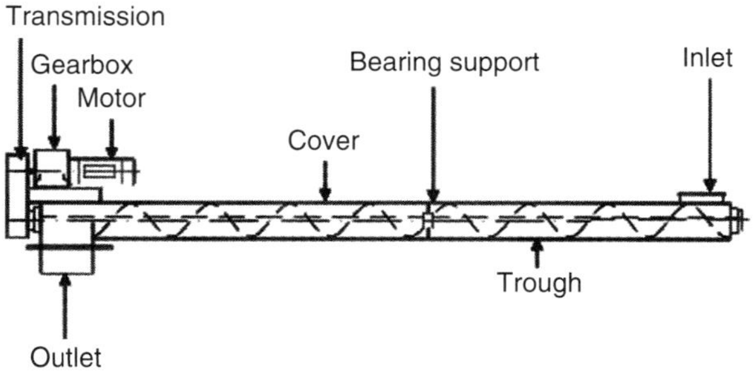

Figure 10-7 Screw conveyor [6].

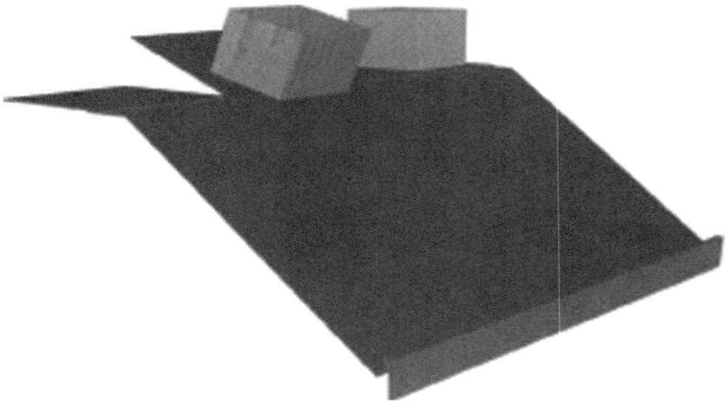

Figure 10-8 Tilt slat conveyor.

Both lifting and conveying slats are incorporated in a slat conveyor. The blocks are moved up lifting ramps by the longitudinal movement of drive members. With the top of the lifting slats placed on the block, it moves upward with the lifting blocks. The upper surfaces of the conveying slats are above the upper surfaces of the lifting slats; the lifting slats are said to be in the down position. The load is moved when the conveying slats move forward with their load, after which the lifting slat is raised up, lifting the load above the upper surfaces of the conveying slats. The conveying slats now return while the lifting slat holds the load. A slat conveyor is used primarily for sorting purposes and is also useful in bottling and canning plants because of the cleanliness, wet conditions, and temperature requirements.

10.2.12 Sortation Conveyors

A sortation conveyor is a conveyor used for merging, identifying, inducting, and separating products to be conveyed to specific destinations. The components of a sortation conveyor include diverters, pop-up devices, sliding shoe sorters, tilting devices, and cross-belt sorting devices. They are used primarily while handling unit loads and can be located either on-floor or overhead. Sortation conveyors divert a product from one conveyor line to another. By using controls and multiple sortation conveyors, product can be sorted by diverting the product exclusively to the appropriate conveyor. The products are kept in continuous recirculation until they are all sorted.

1. *Diverter*. A diverter is a stationary or movable arm that deflects, pushes, or pulls a product to the desired destination without coming in contact with the conveyor. They can be used with any flat surface conveyor. They are simple and inexpensive, driven by a motor, operated hydraulically or even pneumatically.

2. *Pop-up device*. This device usually pops up above the surface of the conveyor to raise the product and lift it off the conveyor at an angle. It can either be single or multiple rows of powered chains, rollers, or wheels capable of sorting only flat-bottomed surfaces. They are lowered when there is no need to divert a product. Generally, pop-up rollers are faster than pop-up wheels.

3. *Sliding shoe sorter*. The sliding shoe or moving slat sorter moves from side to side, diverting the product to either side as it flows along the system. With the use of a series of diverter slats that slide across the horizontal surface, it engages the product and guides it gently off the conveyor.

4. *Tilting device*. This a continuous loop of tilt tray sorters made up of a combination of trays or slats, providing a combined sorting mechanism and product transport system. The products travel in close proximity, and only the slats under them will be activated to tilt. The products are kept in continuous recirculation until they are all sorted. Tilt slat sorters, which are in the form of trays connected by chains, carry products on a flat-surface slat conveyor. These trays tip to both sides; they are preferred to tilt trays because they can handle a wider variety of products: small, delicate, or large. They also accommodate elevation changes.

5. *Cross-belt transfer device*. This could be a continuous loop or a or trainlike circuit where individual carriages are linked together to form an endless loop with a small number of carriers tied together, with a potential for several trains running on the track simultaneously. The track could be oval, straight, or any other shape. Attached to the carriages are small belt conveyors mounted perpendicular to the direction of loop travel that discharge product at the appropriate point. This process is automatic and separates products into different discharge lines from a single line.

10.2.13 Vane-Type Throwing Machines

A vane-type throwing machine is a high-speed rotating drum with vanes or paddles capable of throwing loose bulk material into otherwise inaccessible areas. It is used in construction sites, requires frequent maintenance, and is usually effective.

10.2.14 Tow Conveyors

The tow conveyor is an in-floor conveyor used to drag unit loads along the desired path to the target destination. The load to be conveyed is usually a wheeled carrier such as trucks and carts that move along the floor. The towline is used in warehouses, plants, newspaper printing, freight terminals, and distribution facilities. Tow conveyors offer cleaner operation, live storage, heavy-duty load capacity, reduced congestion, increased productivity, reduced operating costs, improved inventory control, and increased workflow efficiency. Selector-pin or pusher-dog arrangements can be used to allow automatic switching (power or spur lines)

10.2.15 Trolley Conveyors

A trolley conveyor is a serial arrangement of equally spaced trolleys by an endless propelling medium such as a chain or cable in an exposed fashion, with loads suspended from the conveyor hangers. It carries unit loads and does not allow accumulation of loads. The carriers are used to carry multiple units of products. They are quite flexible and can be set up to transport items around curves and up and down gentle vertical slopes. Trolley conveyors are used in a very wide range of applications, including manufacturing and assembly, warehousing and distribution, paint finishing, food handling, detailed inspection, and even retail creative displays.

In general, overhead conveyors can prove to be cost-effective, as they reduce the use of labor (carriers) and allow large volumes to move rapidly through a process. Since overhead conveyors hang above the ground, they assist in optimizing a given operation's use of space and floor space, reduce noise levels, and increase the safety level in the work environment. Trolley conveyors are used predominately in complex production, assembly, and finishing operations that demand a continuous supply of materials at each workstation.

10.2.16 Vertical Conveyors

The vertical conveyor is an on-floor conveyor used in carrying unit loads for low-frequency intermittent vertical transfers and continuous high-frequency vertical transfers. There are two types of vertical conveyors:

1. *Vertical lift conveyor*. This conveyor is used to raise or lower a load to different levels and sublevels of a facility, but is not designed or certified to carry people. It can interface with horizontal conveyors at the different levels where automatic or manual loading or unloading can take place. This conveyor lifts packages one at a time, unlike the spiral conveyor, which ensures a constant flow of materials to various elevations.
2. *Vertical reciprocating conveyor*. This conveyor can also be called a vertical reciprocating conveyor. Unlike a vertical lift conveyor, it can only be used to lower loads. Where the load overcomes the magnitude of the counterweight, it utilizes gravity-actuated carrier to lower loads. It travels perpendicularly to the floors on steel rails or guide paths. The single or double rail, and single or double mast, are existing travel configurations of this conveyor. Their decks could be nonpowered or powered with rollers for both feeding and discharge at the various levels.

10.2.17 Vibrating Conveyors

A vibrating conveyor consists of a flat-bottomed metal trough, bed, or tube, which transports bulk materials through controlled vibrations at a relatively high frequency and small amplitude. Relatively, vibrating conveyors tend to be cleaner and more reliable than every other conveyor, although they are limited in length and the maintenance requirements are quite high. For smooth operation, the dynamic forces present in the design should be counterbalanced. Nevertheless, the foundation should be designed properly to absorb the vibration coming from the conveyor and to reduce its effect on surrounding areas [3]. This conveyor is similar to the oscillating conveyor, the difference being that the oscillating conveyor vibrates at a lower frequency and larger amplitude, and conveys unit loads. Along with its conveying, this conveyor is also found useful in the following applications: screening, separating, sorting, classifying, distributing, flow splitting, sizing, inspection, picking, metal removing, cooling, freezing, blending, mixing, feeding, orienting, bag flattening, heating, cooking, drying, dewatering, cleaning, washing, leaching, quenching, coating, and dedusting. Due to its vast applications, it is found useful in the food, chemical, foundry, rubber, and other industries where the aforementioned processes take place.

10.2.18 Troughed Belt Conveyors

The troughed belt conveyor is an on-floor conveyor used to transport bulk materials using a belt over troughed rollers and idlers. It comprises an endless traveling belt that rests on a skid base and wasted rollers located adjacent to the in-feed and out-feed ends of the base. In between the ends are a series of alternating horizontal and inclined rollers supporting the belt clear of the skid base at selected points. This arrangement forms a concave trough within which the intended material is carried. The horizontal and inclined rollers are cylindrical and freely rotatable.

10.2.19 Magnetic Belt Conveyors

A magnetic belt conveyor is on-floor bulk-load-handling equipment. It comprises a steel belt and either a magnetic slider bed or a magnetic pulley used to convey ferrous materials. Below the belt is an electromagnetic rail or permanent ceramic rare-earth magnet which attracts ferrous materials to the belt, acting as a clamping force. These are important magnetic products that provide holding action for vertical, inclined, or horizontal material handling for conveying, automation, and part-handling applications. They are used to transfer scrap or parts to and from stamping processes, to move materials horizontally through elevations and corners, as stacking and destacking systems, and in several operations that involve handling of metals for industrial purposes. This conveyor is found useful in the mining, construction, chemicals, food, and cement industries.

10.2.20 Power-and-Free Conveyors

These are discretely spaced carriers transported by an overhead chain and used for conveying unit loads. It uses two tracks: one powered and the other nonpowered. The carrier is fastened permanently to the drive chain, while there are free line rails below and beside the powered line rail on the same plane. The carriers are suspended from the free line rails, and each is supported at its opposite sides.

Carriers can be disengaged mechanically from the power chain and accumulated or switched onto spurs. Its tracks can also be located on-floor, in which case it is called an inverted power-and-free conveyor.

10.3 CONVEYOR SELECTION

A conveyor would constitute an economic and industrial nuisance if not employed in a relevant operation. To select the right conveyor for an operation, first, there must be a thorough analysis of the conveying job, and this must include the overall objective of the choice. Second, there must be a complete evaluation and investigation of all available methods of accomplishing the desired objectives. All types of conveyors must be explored before a specific one is selected. This should apply in every industry, whether it is the petrochemical industry or an unrelated industry. This ensures that:

- The quantity of material required is conveyed in the time available.
- The desirable characteristics of the material are maintained.
- The conveying system is economically sound.

But in achieving these goals, engineers must answer the following questions about the operational requirements of the task to enable them to select an appropriate conveyor. These logical questions are:

- What material is to be conveyed?
- Why is the material being conveyed away from its initial position?
- When is the conveying to be done?
- Where is the operation sited?
- How much of the material is to be conveyed, and what are the power requirements, flow rate, and capacity of the conveyor?
- How long will this operation last?
- How strong should the conveyor be, will it be durable, and what are its expected rate of wear and life expectancy?
- How expensive is the conveyor?

The engineering factors to be considered are the weight of the conveyor, the conveyor material, the speed of operation, the throughput capacity, the route of load transfer, the method of loading and unloading, the compatibility with production process, the compactness, the power requirement, whether it is environmentally friendly, and its response to changes in operational conditions. These factors serve as a guideline during selection of the appropriate conveyor. In addition, the following economic factors should be considered: capital cost (purchase of machine and its installation), operating and maintenance cost, productivity of the workers and their level of involvement, and finally, the payback period.

10.4 CONVEYOR SAFETY

Safety takes into account the safe running of a conveyor in a facility without damage to property and persons in the work environment. For the conveyor to function at full capacity, the operator must be sound and psychologically stable, and the machine itself must be mechanically, electrically, and otherwise sound. The operator must be assured that his or her safety is guaranteed while working with or around the conveyor equipment. Most accidents arise due to negligence and carelessness on the part of the operator. Therefore, the prevalent hazards in running the machine must be reduced to the barest minimum to avoid industrial accidents. The following are general safety measures to apply while working with a conveyor:

- Every person in the working environment must wear protective covering and must not wear accessories or loose clothing.
- The conveyor must not be started without proper inspection and servicing.
- Conveyors should not be operated unless the moving elements and all power transmission guards are in place.
- Conveyors should not be overloaded or used to convey anything but the material for which they were designed. The rated capacity and the conditions of operation must be displayed clearly on the machine.
- All rotating and moving parts must be properly enclosed. Dangerous parts should be avoided.
- At no time should a conveyor be left in charge of a novice or an underaged person. The worker must be aware of safe equipment operating techniques.
- Start and stop buttons must be marked clearly and be accessible to the operator. In addition, an emergency button or pull cord designed to stop the conveyor must be installed at the employee's workstation.
- Guards must be provided wherever conveyors pass over work areas or aisles to prevent material from falling on workers. The work area should have crossovers, ladders, platforms, lighting, and ventilation [4].

10.5 CONVEYOR MAINTENANCE

The conveyor system is a critical link in the distribution system of any industry where it is used, and maximum efficiency is a priority. One of the commonest causes of conveyor breakdown is neglect; every other problem arises because the necessary attention has not been given to the intricate parts of this important machinery. In some cases maintenance records are nonexistent or not detailed enough, spare parts are out of stock, operators are not well trained, parts of the conveyor may appear beyond the reach of the operator, the conveyor is being used for operations other than those for which it was designed [7], the control switches and buttons have not been serviced regularly, parts have been allowed to become damaged completely before they are repaired, management has not learning from previous occurrences, an errant operator has not been cautioned, proper maintenance coverage is lacking, and (OSHA) standards are not being kept. Occupational Safety and Health Administration. These issues cost companies much more than it would have cost had timely maintenance been carried out.

To ensure that hazards are reduced and that the conveyor operates at maximum capacity and efficiency, it must be maintained in one of the following ways:

1. *Maintenance based on age*. This is performed on a regular basis determined by age (running hours, total throughput achieved, tonnage conveyed, and distance traveled).
2. *Routine services*. This involves checks made from time to time on certain components of a conveyor before or after running the conveyor, with possible adjustments made. This is the commonest type of conveyor maintenance.
3. *Scheduled replacement*. Here a particular component of the conveyor is replaced after a period of time as required by the specifications.
4. *Periodic overhauling*. Due to wear, a conveyor's efficiency soon decreases over time. Scheduled replacement of major components of the conveyor is carried out to achieve the effectiveness and efficiency the conveyor had when it was new.
5. *Block replacement*. There are components of a machine that seem to have been identified to experience similar failure patterns occurring almost at the same time over and over again. These components could be replaced at the same time to save operational time and subsequent revisits to the conveyor for maintenance and repairs.
6. *Calendar-based maintenance*. Operations could be shut down periodically to access the state of conveyor components and replace them. This is done at certain times specified in the company's schedule of operations.
7. *Opportunistic maintenance*. This preventive method of maintenance is carried out when other scheduled maintenance activities are being carried out.

It is the responsibility of the owner of machinery to ensure that it is kept in safe working condition, not only to maintain the output and production capacity but also to prevent total conveyor breakdown and save operational costs. To achieve hitch-free maintenance and minimal loss in operating hours during periods of equipment maintenance, it is necessary to observe the following:

1. Pay regular attention to the conveyor and keep to the schedule of maintenance before any fault gets to a critical level that will warrant shutdown of the entire industrial operation [2]. A change in the sound (e.g., squeaking, cracking, popping) of the equipment during operation, wobbling of pulleys, stiffening of rubbing parts, and wearing of the belt are common indicators that maintenance is required. The operator should access the conveyor as soon as any of these signs arise; check all visible contact points, check under the belt in the case of belt conveyors, and watch out for dusts and other particles on the conveyor's surface and surroundings. Check the condition of guards, transmissions, all moving parts, connections and joints, switches, and every other fitting as specified by the manufacturer.
2. Lubrication should not be carried out while the conveyor is in motion unless provision has been made to do so without danger to the worker involved. Generally, the conveyor should be given attention when necessary and there should be a planned sequence of maintenance.

10.6 SUMMARY

Conveyors have replaced the arduous task of carrying heavy and difficult to handle materials, making the movement of loads a simple and neat operation. Also, the use of manual labor, which involves risks to both workers and goods, has become minimal. But altogether, automation of the conveying process is not without its own disadvantages; the conveyors themselves are selective as to the types of goods they convey, this requires the use of a particular conveyor for a specific operation, thereby reducing the flexibility of the conveying process. Also, these conveyors require regular maintenance to prevent undue breakdown. Conveyor operators must be trained so as to ensure a machine's productivity and proper maintenance and the safety of its operators.

REFERENCES

1. Labiak, J.S.,and R.E, Hines. 1999. Part 1.2 Grain Handling. pp.11–20 in F.W. Bakker–Arkema DeBaerdemaeker J. Amirante P Ruiz–Altisent M and Studman C.J eds. *CIGR Handbook of Agricultural Engineering, Volume IV: Agro–Processing Engineering*. Copyright ASAE, St. Joseph, Michigan, USA: American Society of Agricultural Engineers.
2. Betts, T., *12 Most Costly Conveyor Maintenance Mistakes: Conveyor Planned Maintenance Program*, TriFactor, Lakeland, FL, 2008, available at http://www.trifactor.com/.
3. Carman Industries, Inc., *Vibrating Conveyors for Bulk Material Flow, Batch Loading and Unit Handling*, Bulletin 700, Carman, Jeffersonwille, IN, 1998.
4. *Conveyors-Design: Construction, Installation, and Operation—Safety Requirements*, Australian Standard AS 1755–1986, 1986.
5. Occupational Safety and Health Administration, *Materials Handling and Storage*, OSHA 2236, revised 2002, available from U.S. Department of Labor, OSHA/OSHA Publication Office, or http://www.osha.gov.
6. Occupational Safety and Health Administration. *The Guarding of Screw Conveyors*, OSHA, Washington, DC, 1980.
7. Stone, S. (2010), 5 *Ways to Avoid Costly Conveyor Breakdown*, retrieved January 15, 2011. http://www.cisco-eagle.com/blog/2010/12/.
8. Stuart-Dick, D., and Royal, T. A., Design principles for chutes to handle bulk solids, *Bulk Solids Handling*, vol. 12, no. 3, September 1992, pp. 447–450.

11

STORAGE TANKS

MARCELLO FERRARA
ITW, Augusta, Italy
CHIKEZIE NWAOHA
Control Engineering Asia, Ten Alps Communications Asia, Aladinma, Nigeria

A **storage tank** is a container, usually for holding liquids, sometimes for compressed gases (gas tank). The term can be used for reservoirs (artificial lakes and ponds), and for manufactured containers [4]. This chapter provides general guidelines that will aid in selection of the proper type of storage to be used in a particular application. Various codes, standards, and recommended practices should be used to supplement the material provided. Manufacturers should be consulted for specific design information pertaining to a particular type of storage.

11.1 TYPES OF STORAGE TANKS

11.1.1 Aboveground Tanks [3]

This type of tank works within operating pressures above 100 kPa gauge (kTa_g). Its design and fabrication are regulated by the American Society of Mechanical Engineers (ASME) Code, Section VIII.

11.1.1.1 Horizontal Cylindrical Tanks The working pressure of these tanks can be from 100 to 7000 kPa_g or greater. These tanks often have hemispherical heads.

11.1.1.2 Spherical Tanks Spherical storage tanks are generally used for storing products at pressures above 35 kPa_g.

11.1.1.3 Spheroid Tanks A spheroidal tank is essentially spherical in shape but somewhat flattened. Hemispheroidal tanks have cylindrical shells with curved roofs and bottoms. Noded spheroidal tanks are generally used in the larger sizes and have internal ties and supports to keep shell stresses low. These tanks are generally used for storing products above 35 kPa_g.

11.1.1.4 Fixed Roofs Fixed roofs are attached permanently to the tank shell. Welded tanks of 80 m^3 capacity and larger may be provided with a frangible roof (designed for safety release of the welded deck to the shell joint in the event that excess internal pressure occurs), in which case the design pressure must not exceed the equivalent pressure of the dead weight of the roof, including rafters, if external.

11.1.1.5 Floating Roofs Storage tanks may be furnished with floating roofs whereby the tank roof floats on the stored contents. This type of tank is used primarily for storage near atmospheric pressure. Floating roofs are designed to move vertically within the tank shell to provide a constant minimum void between the surface of the stored product and the roof. Floating roofs are designed to provide a constant seal between the periphery of the floating roof and the tank shell. They can be fabricated in a type that is exposed to the weather or a type that is under a fixed roof. Internal floating roof tanks with an external fixed roof are used in areas of heavy snowfall since accumulations of snow or water on the floating roof affect the operating buoyancy. These can be installed in existing tanks as well as new tanks. Both floating roofs and internal floating roofs are utilized to

Process Plant Equipment: Operation, Control, and Reliability, First Edition. Edited by Michael D. Holloway, Chikezie Nwaoha, and Oliver A. Onyewuenyi.
© 2012 John Wiley & Sons, Inc. Published 2012 by John Wiley & Sons, Inc.

reduce vapor losses and aid in the conservation of stored fluids. Some storage tanks need a floating roof in addition to or in lieu of the fixed roof and structure. This floating roof rises and falls with the liquid level inside the tank, thereby decreasing the vapor space above the liquid level [4].

11.1.1.6 Bolted Tanks Bolted tanks are designed and furnished as segmental elements that are assembled on location to provide complete vertical, cylindrical, aboveground, closed, and open-top steel storage tanks. Standard American Petroleum Institute (API) tanks are available in nominal capacities of 16 to 1600 m^3, designed for approximately atmospheric internal pressures. Bolted tanks offer the advantage of being easily transported to desired locations and erected by hand. To meet changing requirements for capacity of storage, bolted tanks can be dismantled easily and reerected at new locations. Bolted tanks designed to other internationally recognized bolted tank standards (AWWA, ISO, BSI, and others) are available in nominal capacities of 16 to 40,000 m^3.

11.1.2 Underground Tanks

Underground storage is most advantageous when large volumes are to be stored. Underground storage is especially advantageous for high-vapor-pressure products. Types of underground storage are (1) caverns constructed in salt by solution mining or conventional mining, (2) caverns constructed in nonporous rock by conventional mining, and (3) caverns developed by conversion of depleted coal, limestone, or salt mines to storage.

11.1.2.1 Solution-Mined Caverns The cavern is constructed by drilling a well or wells into the salt and circulating low-salinity water over the salt interval to dissolve the salt as brine. The cavern may be operated by brine displacement of product, pump-out methods, vapor displacement, or as in the case of gas, by product expansion. Most solution-mined caverns are operated using the brine displacement technique. A suspended displacement string of casing is installed near the bottom of the cavern and product is injected into the annulus between the product casing (casing cemented at the cavern roof) and the displacement casing, forcing brine up the displacement casing. The procedure is reversed for product recovery. In this type of operation, a brine storage reservoir is usually provided. Some solution-mined caverns are operated "dry" by installing a pump at cavern depth either within the cavern or in a well connected to the cavern. Both submersible electric-driven pumps and line shaft pumps (deep well vertical turbine pumps) are used for this purpose.

11.1.2.2 Conventional Mined Caverns Conventional mined caverns can be constructed any place that a nonporous rock is available at adequate depth to withstand product pressures. An engineer or geologist experienced in underground storage should evaluate any specific site for the feasibility of constructing underground storage. Most product caverns are constructed in shale, limestone, dolomite, or granite. This type of cavern is operated "dry" (product recovered by pumping).

11.1.2.3 Refrigerated Storage The decision to use refrigerated storage in lieu of pressurized storage is generally a function of the volume of the liquid to be stored, the fill rate, the physical and thermodynamic properties of the liquid to be stored, and the capital investment and operating expenses of each type of system. The parameters involved in selecting the optimum refrigerated storage facility are:

- Quantity and quality of the product to be stored
- Fill rate, temperature, and pressure of the incoming stream
- Shipping conditions for the product
- Composition of the product
- Cooling media (air, water, etc.) available
- Availability and cost of utilities
- Load-bearing value of the soil

The proper choice of storage and the proper integration of the storage facility with the refrigeration facilities are important to overall economy in the initial investment and operating costs. When using refrigerated storage, the liquid to be stored is normally chilled to its bubble-point temperature at atmospheric pressure. Refrigerated storage tanks normally operate at an internal pressure between 3 and 15 kPa_g. In some cases, pressurized-refrigerated storage is attractive. In this type of refrigerated storage, the product to be stored is chilled to a temperature that allows it to be stored at a pressure somewhere between atmospheric pressure and its vapor pressure at ambient temperature.

Refrigeration requirements normally include the following basic functions:

- Cooling the fill stream to storage temperature
- Re-liquefying product vaporized by a heat leak into the system
- Liquefying vapors displaced by the incoming liquid

Other factors that should be considered are:

- Pump energy requirements
- Barometric pressure variations
- Product compositions
- Noncondensables
- Solar radiation effects
- Superheated products

Foundations for the various types of low-temperature storage vessels are designed much the same as foundations for ordinary spheres and pressure cylinders. One precaution: Most low-temperature liquids are lighter than water and the vessels are designed to store this lighter liquid. Therefore, it is common practice to design foundations for the total weight of contained product and to water test the vessel at 1.25 times the product weight. Flat-bottommed vessel foundations in low-temperature service present an additional problem. The container is a heatsink, and, if no provision is made to supply heat, a large quantity of soil will eventually reach temperatures below the freezing point of water. Moisture in the subsoil will freeze and some "heaving" could occur. A heat source consisting of electrical resistance heating cable or pipe coils with a warm circulating liquid is generally installed below the outer tank bottom to maintain the soil temperature above 0°C. Foundations for low-temperature vessels must also be designed to minimize differential settling.

Liquids at low temperatures can be stored in frozen earth caverns at essentially atmospheric or very low pressures. An excavated hole (usually lined) is capped by an insulated metal dome and refrigerated to maintain impervious "walls of ice." Vapors from the liquid are continuously recompressed and condensed.

11.2 STORAGE TANK CLASSIFICATION

11.2.1 Aboveground Tanks

11.2.1.1 Atmospheric Pressure Atmospheric pressure tanks are designed and equipped for storage of contents at atmospheric pressure. This category usually employs tanks of vertical cylindrical configuration that range in size from small shop-welded tanks to large field-erected tanks. Bolted tanks, and occasionally rectangular welded tanks, are also used for atmospheric storage service [3].

11.2.1.2 Low Pressure (0 to 17 kPa$_g$) Low-pressure tanks are normally used in applications for storage of intermediates and products that require an internal gas pressure from close to atmospheric up to a gas pressure of 17 kPa$_g$. The shape is generally cylindrical, with flat or dished bottoms and sloped or domed roofs. Low-pressure storage tanks are usually of welded design. However, bolted tanks are often used for operating pressures near atmospheric. Many refrigerated storage tanks operate at approximately 3.5 kPa$_g$ [3].

11.2.1.3 Medium Pressure (17 to 100 kPa$_g$) Medium-pressure tanks are normally used for the storage of higher volatility intermediates and products that cannot be stored in low-pressure tanks. The shape may be cylindrical with flat or dished bottoms and sloped or domed roofs. Medium-pressure tanks are usually of welded design. Welded spheres may also be used, particularly for pressures at or near 100 kPa$_g$ [3].

11.2.1.4 High Pressure (Above 100 kPa$_g$) High-pressure tanks are generally used for storage of refined products or fractionated components at pressure above 100 kPa$_g$. Tanks are of welded design and may be of cylindrical or spherical configuration [3].

11.2.2 Underground Tanks

Gas processing industry liquids may be stored in underground conventionally mined or solution-mined caverns. No known standard procedures are available for this type of storage; however, there are many publications and books covering the subject in detail.

11.3 CONSTRUCTION MATERIALS

11.3.1 Tank Materials [3]

11.3.1.1 Metallic Tanks Shop-welded, field-welded, and bolted storage tanks are customarily fabricated from mild-quality carbon steel. Most common for welded tanks are A-36 structural steel and A-283 grade C structural-quality carbon steel. Bolted tanks designed to other internationally recognized bolted tank standards (AWWA, ISO, BSI, and others) are available in nominal capacities of 16 to 40,000 m^3. Bolted tanks designed to other internationally recognized bolted tank standards (AWWA, ISO, BSI, and others) are available in nominal capacities of 16 to 40,000 m^3.

Bolted tanks are typically configured as cylindrical flat-bottomed aboveground storage tanks. No bolted tank manufacturers design spheres or "bullets." Various API and ASME codes to which the storage tank is fabricated set forth the welding procedures, inspection procedures, testing requirements, and material selection. Some storage applications or service conditions (low-temperature storage) require storage tanks to be fabricated from metals such as low-alloy stainless steel, aluminum, or other specialty materials.

11.3.1.2 Nonmetallic Tanks Older nonmetallic tanks were customarily constructed from wood. Plastic materials have now replaced wood. These materials have the advantage of being noncorroding, durable, low in cost, and lightweight. Plastic materials used in the construction are poly(vinyl chloride), polyethylene, polypropylene, and fiberglass-reinforced polyesters (FRPs). FRP tanks are available in larger sizes and are the most common, being suitable for outdoor as well as indoor applications. FRP tanks with special reinforced shells are designed for underground storage service. Aboveground tanks are primarily vertical, with or without top heads. Nonmetallic tanks constructed of unreinforced plastics such as poly(vinyl chloride) or polyethylene materials are available in sizes up to about 2 m in diameter by 3.5 m high (11 m^3). Horizontal underground FRP tanks will hold up to 45 m^3. Aboveground vertical FRP tanks can store from 45 to 90 m^3, depending on the shell construction. The temperature limits of plastic tanks are 5 to 65°C. Color must be added to the outer liner for protection against ultraviolet radiation. The inner liner must be selected for compatibility with the chemical or product stored. Protection from mechanical abuse such as impact loads is a necessity. Good planning dictates that plastic storage not be located next to flammable storage tanks. All closed plastic tanks should be equipped with pressure relief devices.

11.3.2 Protective Coatings [3]

11.3.2.1 Internal Coatings Use of internal coatings is primarily to protect the inside surface of a tank against corrosion while also protecting the contents from contamination. Consideration must always be given to such factors as the type of product being stored, the type of coating available, the type of surface to be coated, the surface preparation, the compatibility of coatings, and the number of coats required to obtain maximum protection. Many types of internal coatings are available. Due to the unlimited types and applications, only a few will be described.

1. *Coal tar:* among the oldest and most reliable coatings. Extremely low permeability; protects the surface by the mechanical exclusion of moisture and air; extremely water resistant; good resistance to weak mineral acids, alkalis, salts, brine solutions, and other aggressive chemicals.
2. *Epoxy resin coatings:* excellent adhesion, toughness, abrasion resistance, flexibility, durability, and good chemical and moisture resistance. Typical applications include linings for sour crude tanks, floating roof tanks, solvent storage tanks, drilling mud tanks, sour water, treated water, and pipelines.
3. *Rubber lining:* used as internal lining for storage tanks subjected to severe service, such as elevated temperatures or for protection from extremely corrosive contents, such as concentrated chlorides and such acids as chromic, sulfuric, hydrochloric, and phosphoric.
4. *Galvanized:* zinc coated, highly resistant to most types of corrosion. Bolted steel tanks are ideally suited for galvanizing since all component parts are galvanized by the hot-dip process after fabrication but before erection. Galvanized bolted tanks are recommended where the oil produced contains sulfur compounds and/or is associated with hydrogen sulfide gas. Galvanizing is also effective against corrosion in seacoast areas where atmospheric salt conditions accelerate corrosion problems.
5. *Glass-fused-to-steel (vitreous enamel):* highly resistant to corrosion in both severe applications and potable water and wastewater applications. This coating is extremely durable and can only be applied in a factory setting to bolted steel tanks.

11.3.2.2 External Coatings The basic requirements for external coatings are appearance and weather protection. Numerous types of external coatings are available, ranging from basic one-coat primers to primers with one or more topcoats. Environmental conditions usually dictate the extent of coating applied. Offshore and coastal installations require more extensive coatings than do inland locations.

11.3.2.3 Coating Selection

1. *Corrosion and chemicals*. In the storage of dry bulk chemicals, coatings are most important for providing abrasion resistance. However, when it comes to liquid storage, the most important factors are corrosion and chemical resistance. In general, stored liquids are more aggressive toward tanks than are dry products. Many common industrial-use chemicals are stored in small volumes. However, end-use liquid products such as soybean and vegetable oils, petroleum industry products and by-products, and raw materials such as crude oil and brine water may require large volumes of storage [7]. When considering a tank, make sure that the coating is designed specifically for your product. The tank you choose should offer a coating with proven corrosion resistance over long periods. For liquid chemical storage, choose a coating that has been ASTM tested and proven to withstand the aggressive effects of products such as strong acids and bases.
2. *Safety factor*. Coatings do play a role in safety. Some coatings are designed for storing a wide variety of products, whereas others are designed for specific products. When it comes to safety, the most important consideration is ensuring that the tank coating is suitable for the application. If your tank is for liquid storage, make sure the coating applied is for liquid, and similarly for harsh chemicals. Some coatings may react poorly with products not designed for them.

   It is unsafe to use an old storage tank for a product for which it was not designed and tested. If you are considering recycling a used tank for storing a new product, such as using a dry material storage tank for liquids or a water storage tank for liquid chemicals, first be sure to have the tank evaluated for safety, flow, corrosion resistance, chemical resistance, and so on. Incorrect use of a tank can result in severe corrosion, premature failure of the coating, harm to the product, or serious injury to personnel [7].
3. *Application process*. The key difference in coatings lies in the application process. The highest-quality coatings

are applied at the factory under environmentally controlled circumstances to ensure the most consistent application. Taken one step further, the best coatings are also thermally cured at the factory. Some manufacturers apply the coatings at the factory but then allow them to air-dry and cure with ambient heat, which exposes the cure to environmental factors such as dust and humidity. Other manufacturers outsource the coating process. The optimal coating solution is one that is both applied and thermally cured under controlled factory conditions before the tank is shipped and erected in the field. High-quality bolted and factory-welded tanks offer this feature. Field-welded and concrete tanks often receive their coatings on-site once the tank has been erected. These tanks may need to undergo a chemical process to protect the coating while the tank is being erected (i.e., heat from welding may damage the coating). If you are reviewing this type of tank, make sure that there are adequate quality control measures listed in the specifications [7]. Also, consider third-party inspections. Be aware that once a tank has been erected in the field, there are often areas of the tank that are extremely difficult to sand-blast and/or fully prepare for field coating. Even "missing the smallest spots" leaves exposed areas open to corrosion. In addition, the thickness of coatings applied in the field cannot be controlled as closely as those applied under ideal factory conditions. Finally, weather and the environment—dust, humidity, temperature, and wind—will affect the curing process in the field.

4. *Cost implications*. If a tank manufacturer promises a low cost for a tank and coating system, make sure that the competitive products you are comparing are equal over the total life cycle of the tank. A higher-quality coating may have a higher upfront cost but lower life maintenance and recoating requirements [7]. A high-quality coating often means lower maintenance for the long term because it is expensive to recoat a tank.

5. *Experience and know-how*. It may seem like common knowledge to mention experience as a factor for choosing a tank and coating, but it is a key consideration. Consider the experience of the tank manufacturer: how long the company has been fabricating tanks, whether it uses state-of-the-art coatings, whether its coatings are factory-applied and thermally cured, what its quality control measures are, what its volume sold is, and what countries and markets it is involved in. Also, consider tank manufacturers with third-party accreditation such as the International Standards Organization (ISO) 9001 Quality Certification [7].

11.3.3 Insulation

11.3.3.1 Types of Insulation The four basic types of thermal insulating material are fibrous, cellular, granular, and reflective. These materials differ in many characteristics.

11.3.3.2 Uses of Insulation Principal uses of insulation are as follows:

1. *Personnel protection*. Personnel protection is accomplished by the application of insulation of proper thickness where the surface temperature should be limited to approximately 150°F or as specified by applicable codes or company standards.

2. *Process temperature control*. Insulation thickness is specified in this case to help control the temperature of the process fluid. Electrical, steam, hot process fluid, hot oil, and glycol–water tracing is used to add heat to the process line to balance the heat loss. The insulation thickness must be matched to the energy input to achieve the desired result. Freeze protection is another use for insulation. This includes fluids that have higher viscosities or freeze points.

3. *Condensation*. Insulation thickness must be sufficient to keep the outside surface of the insulation above the dew point of the surrounding air. Moisture condensation on a cool surface in contact with warmer humid air must be prevented because of the deterioration of the insulation. In addition to the insulation thickness required, a vaportight membrane must be properly applied to the insulation. As a rule, insulation thickness for condensation control is much greater than the thickness required for conservation of energy.

4. *Conservation of energy*. High fuel costs increase the need for insulation. A rule of thumb for estimating the thickness of insulation is to apply the thickness that produces a heat loss of 3 to 5% or less from the surface. Specific insulating materials and thicknesses for any large application should be determined with the assistance of the manufacturer. The method is based on elementary heat transfer theory and reliable experimental data.

11.4 COMMON PROBLEMS OF STORAGE TANKS

A storage tank that appears to be in robust condition can, upon close inspection, be found to be very fragile and susceptible to both gradual and catastrophic failure. Since tank failure can result in staff endangerment, significant product loss, and environmental loss, tank owners and operators are advised to implement appropriate precautions to reduce the likelihood of tank failure. Below are some common problems of storage tanks.

11.4.1 Corrosion

What can be done to prevent or mitigate corrosion of large aboveground storage tanks? The answers are not simple, but there are a number of proven methods. Some are discussed below.

1. *Protective coatings*. For internal tank corrosion, bottom coatings have been proven to be very effective. Not only do they reduce interior bottom (top-side) pitting but also effectively reduce finished fuel contamination and tank cleaning costs. Although API standards do not prescribe coatings for finished fuel tanks, API RP 651 gives guidance on how to install coatings when they are used. Chevron coats all finished fuel tanks on the bottom and about 2 ft up the shell [9]. Coatings protect those areas that suffer the most aggressive attack by corrosion. Economic analyses, not even considering product purity and product integrity issues, show that coating tank bottoms pays off. Of course, effective coatings also help protect the environment. Not only is this cost-effective, but it makes cleaning tanks when they are due for internal inspections much less costly. It also allows for better, more accurate inspections.

External corrosion of tank bottoms presents a different challenge. The underside of tank bottoms cannot be coated effectively because any underside coating would be burned by the welding of the bottom plates. One measure found to improve tank underside corrosion resistance is using plate that has been descaled. *Scale* is iron oxide that results from the mill process. When the plate is placed in an environment such as the underside of a tank bottom, this galvanic corrosion can accelerate pitting significantly. For this reason, Chevron uses only descaled plate for tank bottoms [9].

2. *Cathodic protection*. Use of cathodic protection to reduce both interior and exterior corrosion is controversial and complex. Industry experience shows that it is very useful for interior corrosion on crude oil tanks when used in conjunction with liners. However, cathodic protection has not been proven universally effective for protecting finished fuel tanks from internal corrosion. Coatings do that job adequately. On external or underside corrosion, cathodic protection has been used with mixed results. Theoretically, cathodic protection will work if installed properly, but in reality there are many obstacles to overcome for it to work right. Unless these systems are installed, tested, maintained, and operated by trained and qualified people, they can be totally ineffective and, in fact, can cause accelerated corrosion. Cathodic protection should not be mandated as a blanket solution but should be evaluated and weighed individually against other alternatives on a site-specific basis [9]. API RP 651 should be consulted when cathodic protection is being considered.

3. *Double bottoms*. Although this fact is not well known, a double bottom is an effective corrosion prevention method that increases tank life significantly. How so? There are several factors that reduce the problem of underside corrosion:

- Adding a double bottom raises the new steel bottom up off the mud and dirt. The elevation generally mitigates the corrosive environment by reducing contact with moisture and salts.
- Concrete in the presence of moisture becomes alkaline. Alkaline water is much less corrosive than acidic water. Measurements from standing water under tank bottoms have been about pH 11 to 12, so the concrete is actually a corrosion inhibitor.
- A double-bottomed tank has a more uniform foundation, with less likelihood of clay balls or foreign objects. In other words, development of corrosion cells, galvanic corrosion, and other problems are less likely when a concrete foundation is used.

11.4.2 Vapor Losses

A storage tank is generally not pumped completely dry when emptied. The vapor above the remaining liquid will expand to fill the void space at the liquid's vapor pressure at storage temperature. As the tank fills, vapors are compressed into a smaller void space until the set pressure on the vent or relief system is reached. Some filling losses are associated with the liquid expansion into the tank. Vapors emitted from a storage tank's vents and/or relief valves are generated in two ways: displacement losses (tank vapors forced out during filling operations) and vaporization losses (vapors generated by liquid vaporization stored in the tank).

1. *Displacement losses*. The combined loss from filling and emptying is considered a working loss or displacement loss. Evaporation during filling operations is a result of an increase in the liquid level in the tank. As the liquid level increases, the pressure inside the tank exceeds the relief pressure, and vapors are expelled from the tank. Evaporative loss during emptying occurs when air drawn into the tank during liquid removal becomes saturated with organic vapor and expands, thus exceeding the vapor space capacity [1].
2. *Vaporization losses*. Vapors are generated by heat gained through the shell, bottom, and roof. The total heat input is the algebraic sum of the radiant, conductive, and convective heat transfer. Vaporization loss is especially prevalent where light hydrocarbon liquids are stored in full pressure or refrigerated storage. This is less prevalent but still quite common in crude oil and finished product storage tanks. These vapors may be recovered by using the vapor recovery system. To calculate vaporization in tanks, the sum of radiant, conductive, and convective heat inputs to the tank must be taken into account. Approximate vapor losses (kg/h) can be calculated by dividing the total heat input by the product latent heat of vaporization at fluid temperature [1].

11.4.3 Storage Tank Fires

Certain related fire hazards are common to various types of tanks. These hazards vary in severity from a simple vent fire to a full liquid surface tank fire. The most common of these incidents are an overfill ground fire, a vent fire, a rim-seal fire, an obstructed full liquid surface fire, and an unobstructed full liquid surface fire [14].

11.4.3.1 Types of Fires

1. *Overfill ground fires*. Overfill ground fires, or dike fires, result from piping or tank leakage. They are often the result of another cause, such as operator error or equipment malfunction, and are considered the least severe type of incident. If a leak occurs without ignition, exercise caution and isolate all ignition sources. If ignition does occur, simply treat such a fire as a large pool fire. Overfill ground fires are common to fixed-cone roof, internal floating roof, external floating roof, and domed roof tanks [14].

2. *Vent fires*. Vent fires are typically associated with fixed roof tanks such as cone and internal floating roof tanks. The main cause is a lightning strike that ignites fugitive vapors that may be present at the vent. This is a less severe type of fire and can usually be extinguished with a dry chemical fire extinguisher or by reducing the pressure in the tank [14].

3. *Rim-seal fires*. Rim-seal fires comprise the large majority of fires in external floating roof tanks but can occur in internal floating roof tanks or domed roof tanks. As with many tank fires, lightning is the primary cause of ignition, although with floating roof tanks, an induced electrical charge without a direct lightning hit may occur. Because these fires are the most common, there is usually a high rate of successful extinguishment, assuming that there is no collateral damage, such as a pontoon failure (explosion) or the floating roof's sinking as a result of fire suppression efforts. Successful rim-seal fire extinguishment can be attributed primarily to the installation of rim-seal fire protection systems such as foam chambers [14]. These semi- or fully fixed rim-seal fire protection systems have a good history of extinguishment, assuming proper design, installation, and maintenance. Rim-seal fires for internal floating roof tanks are slightly more challenging, especially if semi- or fully fixed systems are not provided. This means that the only access to the fire area for the application of fire-extinguishing media is through the vents or access covers.

4. *Obstructed full liquid surface fires*. Obstructed full liquid surface fires can occur in fixed-cone roof, internal floating roof, or external floating roof tanks. They tend to be challenging because the roof or pan blocks access to the burning surface. The roof or pan can sink for various reasons, such as an increase in vapor pressure under an internal floating roof, which can cause the pan to tilt [14]. Pontoon failure of external floating roofs is commonly caused by closed drain valves during rains or mechanical seal failure, causing the pan to sink.

5. *Unobstructed full liquid surface fires*. Unobstructed full liquid surface fires are relatively easy to extinguish where the tank diameter is relatively small (less than 150 ft) and sufficient resources and trained personnel are available. The most challenging fires will involve larger tanks (greater than 150 ft in diameter) because of the surface area of the fire and the amount of resources needed to control and extinguish the fire. Unobstructed full surface fires can occur in fixed-roof tanks without internal roofs, where the frangible weak seam at the roof–shell joint separates as a result of an explosion or other overpressure event, leaving a full surface tank. External floating roof tanks are also prone to unobstructed full surface fires during heavy rain conditions. With closed roof drains, the roof can quickly sink, leaving the exposed liquid surface vulnerable to a lightning strike.

11.4.3.2 Firefighting Strategies [14] Firefighting strategies and tactics are also important. Evaluate the objectives or goals versus the risk. Strategies include the following:

- *Nonintervention.* This is essentially a nonaction mode when the risks associated with intervening are unacceptable. All personnel are withdrawn to a safe area.
- *Defensive.* In this tactic, certain areas may be conceded to the incident, and actions are limited to protecting exposures and limiting the spread of the incident.
- *Offensive.* Aggressive and direct tactics are used to control an incident.

As with most fires, the benefits must outweigh the risk. If a small-diameter tank is burning with no threat to exposures, should you extinguish the fire? If the tank has already lost its contents, is exposure protection more appropriate? These considerations are identified and developed as part of the preincident response planning, development of emergency action plans, and the identification of credible incident scenarios. Environmental conditions such as wind and rain could create problems with distance or range of the water or foam solution streams. Changes in wind direction might cause corrections to incident action plans with respect to changes in staging locations. An increase in temperature or humidity could force a quicker rotation of firefighters to prevent heat stresses.

11.4.3.3 Supplementary Considerations Consider the following additional response and operational conditions when preparing preincident response plans and incident action plans:

1. *Interoperability*. Identify these issues during preincident response planning. Ensure interoperability of a plant facility's fire water and extinguishing systems, the mutual- or automatic-aid departments, and the third-party emergency response contractors that may be on retainer to respond to a plant facility to assist with storage tank fires or emergencies.

2. *Foam supplies*. Consider regional foam cooperatives to establish sufficient foam concentrate supplies for large-scale storage tank fires or emergencies. Remember, sufficient foam concentrate supplies are not enough; there must also be large-volume monitors and large-volume hoses to supply the water and foam solutions required to address the emergency.

3. *Industrial emergency task forces*. Consider preestablishing industrial emergency task forces to respond to such incidents. They would be predetermined and activated to supply staffing, equipment, foam concentrate, and apparatus to mitigate emergency scenarios at storage tank locations. ICs (Incident Commander's) can then call a task force or multiple task forces, knowing that the personnel, equipment, and apparatus required will respond. These task forces would be identified in the preincident response plans.

4. *Third-party contractors*. Contractors that specialize in these types of incidents should be identified and communicated with; you need them to respond. These companies can provide foam supplies, subject matter experts, and equipment not otherwise available on a day-to-day basis in a city or town.

5. *Specialized industrial fire training*. Fire departments should consider sending personnel to specialized industrial fire training programs so that they can learn more about storage tank fire fighting and emergency incidents.

6. *Jet ratio controllers*. As part of the equipment necessary to deliver foam solutions to the point of operations, seriously consider jet ratio controllers. Normally, foam concentrate must be placed within a specified distance from the nozzle, usually 150 ft. Using a jet ratio controller with a matched foam nozzle, the source of foam can now be placed as far away as 2500 ft from the nozzle. Jet ratio controllers are venturi-type devices that move the concentrate from a remote storage location to the matched foam nozzle.

7. *Foam quantities*. Large quantities of foam concentrate will be required. For large incidents, using 55-gallon drums is not recommended; 275-gallon totes and large tanker trucks are the preferred foam concentrate delivery methods. During preincident response planning as well as during the incident, evaluate the logistics of moving the foam concentrate to the point of foam injection. If the access routes are blocked with vehicles and hose, how can you get the concentrate to where it needs to be?

8. *Teasing the fire*. Before attacking a tank fire fully, practice "teasing" the fire. When first applying water to a storage tank fire, the cold water striking the burning fuel surface will react, increasing the intensity of the fire. To prevent a more violent reaction, pass the extinguishing streams over the top of the tank until the fire settles back down. At this point, a full attack should begin with a foam solution. The term *teasing* a fire is generally credited to Dwight Williams of Williams Fire & Hazard Control, which specializes in fighting storage tank fires.

9. *Foam application*. Do not position extinguishing streams around the tank for multiple points of application. Position foam monitors at one location; the foam streams should enter the tank at the same point and impinge on the surface in the same area. This will help achieve a stable foam blanket more quickly. This foam blanket will then spread out from this central point on the surface. Do not be tempted to move the streams to other positions. If no appreciable lessening of the fire intensity occurs within the first 20 to 30 minutes, instead of moving the stream position, review the rate of application.

10. *LCES*. When developing an incident action plan, consider the acronym LCES (lookouts, communications, escape routes, and safety zones). Lookouts must be experienced and be able to see the fire and firefighters, and they must be able to recognize risks to firefighters. They must be the IC's additional eyes and ears. Post them in strategic areas so that they can notify the IC of any relevant information or change in conditions. Maintain communications with all personnel operating on the scene, plant operations personnel, and subject matter experts. You must keep personnel operating in remote locations informed of any change to operational tactics. Lookouts must also maintain communications with operating personnel. Establish escape routes and inform all personnel of them during safety briefings. Two escape routes must be identified and lead to safety zones where accountability can be verified. Establish safety zones upwind and uphill of the incident. Verify personnel accountability at these locations. In addition, designate a clear evacuation route from this safe area so that personnel can be evacuated farther from the safety zone if conditions deteriorate to a level that makes this area unsafe.

Fires involving large aboveground storage tanks can be extremely costly in terms of property damage, environmental concerns, and public impact. Additionally, the control and extinguishment of full-surface tank fires require a large amount of commitment in human logistics and equipment resources. Tank fires are complex events. Fighting them requires implementation of plans, preparation, and proper use of resources coordinated by an effective emergency management organization. Only with training and drills will your department become proficient in the strategies and tactics needed to fight a storage tank fire successfully.

11.5 STORAGE TANK MAINTENANCE

11.5.1 Tank Blanketing

Tank blanketing, also referred to as *tank padding*, is the process of applying a *gas* to the empty space in a storage container. The term *storage container* here refers to any container that is used to store products, regardless of its size. Although tank blanketing is used for a variety of reasons, it typically involves using a *buffer gas* to protect products inside the storage container. A few of the benefits of blanketing include a longer life for the product in the container, reduced hazards, and longer equipment life cycles [5].

Although it varies from application to application, blanketing systems usually operate at a slightly higher than atmospheric pressure (a few inches of *water column* above *atmospheric*). Higher pressures than this are generally not used, as they often yield only marginal increases in results while wasting large amounts of expensive blanketing gas.

Some systems also utilize inert gases to agitate the liquid contents of a container. This is desirable because products such as citric acid are added to food oils the tank will begin to settle over time with the heavier contents sinking to the bottom. However, a system that utilizes nitrogen *sparging* (and subsequently, tank blanketing once the nitrogen reaches the vapor space) may have a negative impact on the products involved. Nitrogen sparging creates a significantly higher amount of surface contact between the gas and the product, which in turn creates a much larger opportunity for undesired oxidation to occur. It is possible for nitrogen that is as much 99.9% free of oxygen to increase the amount of oxidation within the product, due to the high amount of surface contact.

The most common gas used in blanketing is *nitrogen*. Nitrogen is widely used due to its inert properties as well as its availability and relatively low cost. Tank blanketing is used for a variety of products, including *cooking oils*, volatile *combustible* products, and *purified water*. These applications also cover a wide variety of storage containers, ranging from as large as a tank containing millions of gallons of vegetable oil down to a quart-sized container or smaller. Nitrogen is appropriate for use at any of these scales [5].

The use of an inert blanketing gas for food products helps to keeps *oxygen* levels low in and around the product. Low levels of oxygen surrounding the product help to reduce the amount of oxidation that may occur and increase shelf life. In the case of cooking oils, lipid *oxidation* can cause the oil to change its *color*, *flavor*, or *aroma*. It also decreases the *nutrient* levels in the food and can even generate *toxic* substances. Tank blanketing strategies are also implemented to prepare the product for transit (*railcar* or *truck*) and for final *packaging* before sealing the product.

When considering the use of tank blanketing for combustible products, the greatest benefit is process safety. Since fuels require oxygen to combust, reduced oxygen content in the vapor space lowers the risk of unwanted combustion. Tank blanketing is also used to keep contaminants out of a storage space. This is accomplished by creating positive pressure inside the container. This positive pressure ensures that if a leak should occur, the gas will leak out rather than having the contaminants infiltrate the container. Some examples include the use of gas on purified water to keep unwanted minerals out and its use on food products to keep contaminants out.

11.5.2 Holiday Detection

An important evaluation of bottom linings after application to aboveground storage tanks (ASTs) is holiday (i.e., discontinuity) detection. Linings are principally applied to ASTs to prevent internal corrosion, which may be severe. Therefore, any holidays must be detected and repaired prior to the newly lined tank being returned to service [6].

Generally, two types of holiday detection are employed. For thin bottom linings under 20 mils of dry film thickness (DFT), low voltage/wet sponge detectors are specified. For those linings greater than 20 mils DFT, high-voltage spark detectors are employed [6]. Each of these has its own advantages and disadvantages; in subsequent paragraphs we describe these for the low-voltage wet sponge test. A low-voltage wet sponge detector is a simple electronic device; it consists of a wet sponge, an energy source (a 5- to 90-V battery), a ground wire connector, and another connecting lead wire to the wet sponge [6]. The electrical leads are connected to the battery. In practice, the wet sponge is moved over the relatively thin lining and a holiday/discontinuity is detected by the wet sponge contacting bare steel and activating either an audible or a visual indicator.

There are both advantages and disadvantages to this type of detector. Advantages include:

- The wet sponge is low in cost; one commercial model costs approximately $250.
- It is fairly easy to use and extensive training is not required.
- It is a nondestructive test.

However, there are several disadvantages, including:

- The lining must be dry and free of moisture.
- Locating a holiday after audible indication may be time consuming.
- The lining must be dried after locating a holiday; otherwise, "telegraphing" can result. Telegraphing is

current traveling along a wet path and indicating a holiday where none exists. The major problem, however, is that it cannot locate a thickness defect that is masked by the lining.

Concerning this last point, one manufacturer states in its literature that "a coating of only one mil (0.001 in.) of paint (or coating) is sufficient to repel the current (of the low voltage battery)" and the alarm circuit of the wet sponge detector will not be activated. This inadequate thickness in the tank lining will almost certainly cause premature failure.

In AST lining work, thin areas may occur at sharp edges and corners, corroded or pitted areas, welds and weld spatter, mechanical fasteners, and structural shapes. To reiterate, as long as these are coated, even with a lining thickness of only 5 to 10% of that specified, no holiday will be detected and the opportunity for early failure is very real. Therefore, thin-film linings are frequently not specified for existing corroded tanks; heavy-bodied linings are used because of the greater chance that applying a lining to a thickness will have greater success. If a thin lining is used in a new or existing tank, *striping* is frequently employed. With striping, an additional coat of lining is applied to those areas specified above where thin coatings may exist. Subsequently, the complete lining system is applied. Steel Structures Painting Council Specification SSPC-PA-1 contains details concerning stripe painting.

A number of standards and recommended practices are available regarding holiday detection, notably:

- NACE RP0188: Discontinuity (Holiday) Testing of Protective Coatings
- ASTM D5162: Practice for Discontinuity (Holiday) Testing of Non-conductive Protective Coating on Metallic Substrate

11.5.3 Tank Cleaning

Environmental legislation is becoming increasingly restrictive with regard to waste disposal. This requirement becomes even more important in the cleaning of tanks, as the removal of tank sludge is an expensive, time-consuming step before achieving gas-free certification. As a matter of fact, sludge remains in the bottom of the tanks and is removed manually to carry out maintenance. Manual removal of sludge is a very expensive and time-consuming operation, complicated by the fact that tanks are difficult to operate in. Moreover, manual cleaning always implies special attention to the safety of operations.

ITW has patented a novel technology for asphaltene stabilization. Such technology makes use of chemical additives to hydrocarbons and has proved very effective in many industrial applications (Fig. 11-1). The technology

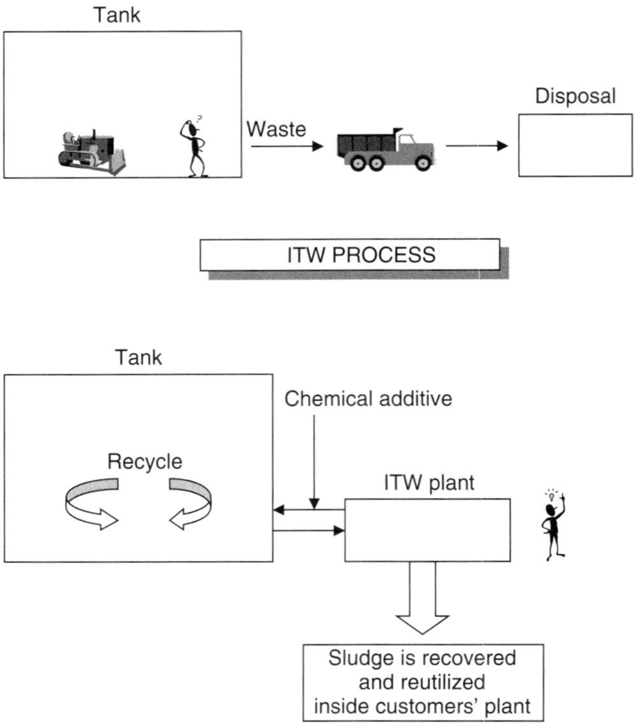

Figure 11-1 Today's method.

has been suitably modified for tank cleaning and tested successfully in many cases of aboveground storage tank cleaning (e.g., fuel oil and crude oil tanks) as well as cargo tank cleaning. Asphaltene stabilization achieves an improvement in sludge reuse in that asphaltene association is reduced, compatibility with the receiving hydrocarbon is enhanced, precipitation does not occur, and cracking of asphaltene is facilitated [2].

According to the ITW approach, sludge is removed by the addition of a chemical additive, which contains asphaltene stabilizers, patented by ITW. The formulation also includes paraffin solvents and fluidizing agents. The additive is utilized to help sludge penetration, thus favoring its solubilization into a carrier. Sludge dissolution occurs due to the chemical action of the additive during recycling of the oil phase. After a brief description of the existing tank cleaning technologies, we report some results achieved in application of this novel technology.

11.5.3.1 Existing Tank Cleaning Technologies

Manual cleaning is the most widespread method currently used in cleaning tanks. Its disadvantages are that:

- It is unsafe.
- It generates a huge amount of waste.
- It is time consuming.
- It is costly.

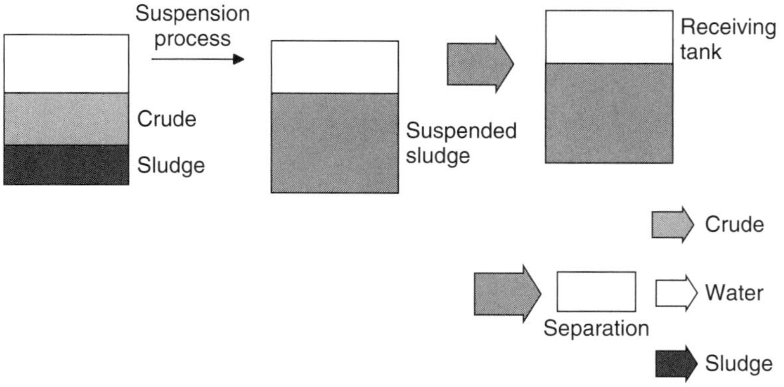

Figure 11-2 Sludge suspension process.

Other methods have been developed to improve manual tank cleaning, the most interesting being: crude oil washing, chemical cleaning, and robot machines. Although these methods improve manual cleaning, they still have pitfalls.

1. *Crude oil washing*. This method simply moves the sludge from one tank to another (it is a mechanical dispersion method). In some cases, reprocessed crude oil sludge led to unscheduled topping unit shutdown. Figure 11-2 is a flow diagram of a sludge suspension process.
2. *Chemical cleaning*. For chemical cleaning, most chemicals used until now have been dispersants: again, they transfer the problem from one point to another.
3. *Robot machines*. Robot machines improve the safety and sometimes the length of the operations, but they do not have an impact on sludge reduction and therefore generate the same amount of sludge [2].

11.5.3.2 ITW Technology ITW uses patented asphaltene stabilizers to make sludge a reusable product. The chemical additive used is itself capable of stabilizing and solubilizing the sludge. The chemical is not a dispersant, so it doesn't create problems in downstream equipment. Moreover, ITW stabilizers and solubilizers:

- Do not contain any metallic compounds
- Do not contain any catalyst poisons for petroleum processes
- Do not contain any halogen compounds
- Do not contain any carcinogenic compounds
- Do not contain any compounds which, at operating dosages, could be a poison for wastewater treatment plants
- Do not contain any compounds that could be harmful to plant metallurgies

Therefore, the core of the ITW process consists of highly effective chemicals which are themselves able to solubilize the sludge. This means that the sludge will be solubilized chemically (i.e., stabilized permanently) with no danger of subsequent precipitation [2]. To improve the performance of the chemicals (especially in large tanks), a modified crude oil washing is used together with chemical stabilization of the sludge.

Case History 11.1 A 5000-m^3 fuel oil tank needed to be cleaned after almost 20 years of operation. The tank was emptied with an external pump up to 50 cm, under the suction limit of tank pump. A certain amount of fuel oil was left above the sludge. After starting addition of the chemical, fuel oil was analyzed for sediments by hot filtration (SHP; IP 375), which resulted in a *not filterable* condition [i.e., no oil filtered through a Whatman GF/A filter of nominal porosity (2 μm)] [2]. This accounted for the asphaltene being a precipitate, hence strongly associated. Moreover, the sludge contained a huge amount of catalyst fines (derived from blending during fuel oil formulation with decant oil from the fluid catalytic cracking unit). Therefore, there was a strong aggregation between the sediments (substantially, catalyst fines) and the asphaltene. In such a case, the asphaltene precipitated incorporated catalyst fines, creating unfilterable macrostructures. This phenomenon was operatively well known, as during fuel oil

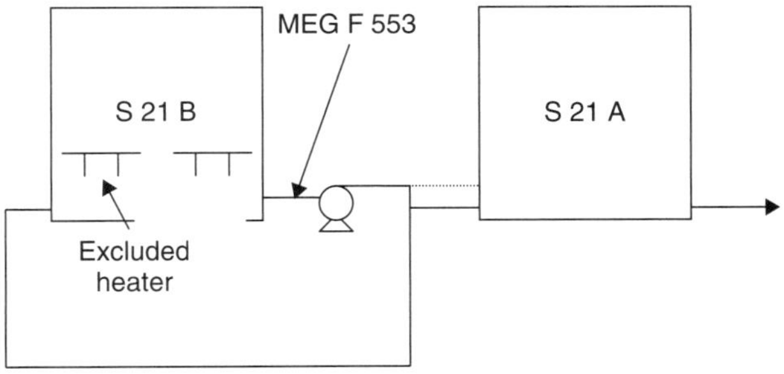

Figure 11-3 Additivation layout.

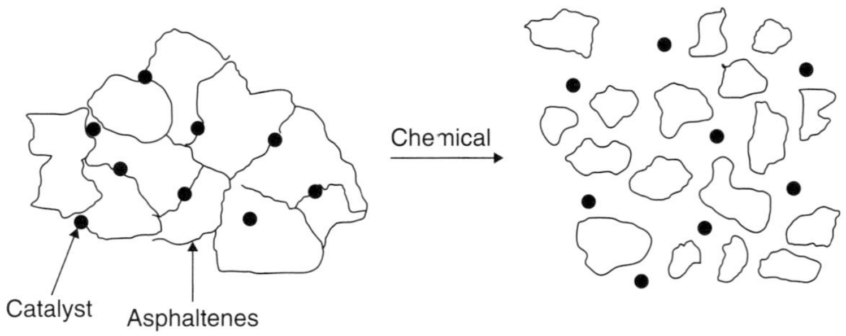

Figure 11-4 Sediments release by ITW's asphaltene stabilization

combustion, burner and filter plugging problems occurred frequently. Once suction of pumpables ended (leaving about 80 m^3 of sludge), the chemical was added to the tank directly by connecting the suction of the recycle pump with the container of the chemical [2]. No carrier was utilized as diluent, so the chemical was injected directly into the *not filterable* fuel oil above the sludge. Then the tank recycle was begun, during which the quality of the oil phase (fuel oil + sludge) was analyzed. Figure 11-3 shown an additivation layout. After a 1-day recycle, the SHF on the sludge was 3%.

Such a result is extremely important, as it highlights the reactivity of the chemical: The fact that the oil phase went from *not filterable* to *filterable* is a clear demonstration of the stabilizing effect of the chemical. The chemical has stabilized asphaltene to such an extent that it was no longer aggregated, hence filterable; in this way the sludge was *freed* from catalyst fines, which contributed to the formation of unfilterable macrostructures. Release of sediments from asphaltene would not be possible had the asphaltene structure not been disjointed. The concept can best be illustrated as shown in Fig. 11-4. Continuing the recycling operation, sediments in the sludge increased with time during recycle: 7% the second day and 18% the sixth day [2]. The chemical effect is then evident: the additive has stabilized asphaltene in the sludge, thus allowing release of the catalyst fines that were bound to them.

It is important to note that the oil phase achieved from the sludge after ITW treatment was filterable, and the SHF consisted almost entirely of catalyst fines. As a matter of fact, the oil phase of the sludge after ITW treatment has always been filterable and the sediments increased gradually once asphaltene was *attacked* from the chemical. In practical terms, the word *filterable* means that the oil did not contain sediments larger than 2 μm (the nominal porosity of the filter utilized to measure the SHF) [2].

Figure 11-5 and Table 11-1 report the values of sediments in the sludge during recirculation. At the end of recirculation, additivized and stabilized sludge was transferred into another fuel oil tank. The transfer was performed gradually by completing the operation in about

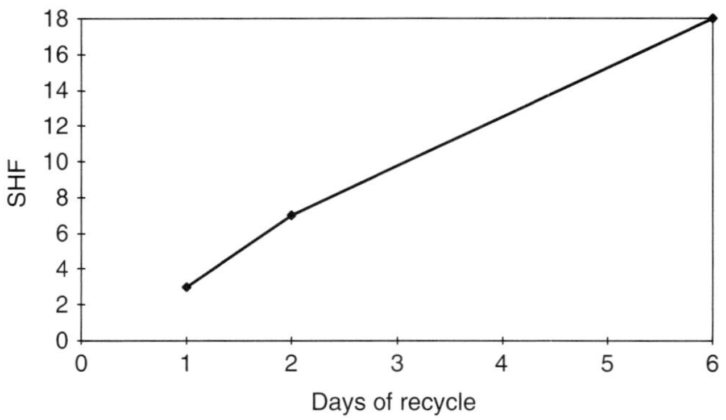

Figure 11-5 Sediments in recycled sludge.

6 h (rate of about 15 metric tons T/h) [2]. This is done to reduce potential operating problems, in that the sludge was transferred directly into the bottom of the receiving tank, hence feeding the boiler immediately (the charge pump sucked from the bottom of the tank).

The sludge was transferred entirely into the receiving tank, and the level of the cleaned tank dropped down to zero. This was also confirmed by visual inspection from manways. The chemical additive has thus solubilized and stabilized all the sludge (80 m^3) present in the tank after only 6 days of recycle [2]. To confirm the success of the treatment, the sludge fed in the boiler gave rise to no operating problems, in that *free* catalyst fines were easily stopped by hot filters, which did not suffer any fouling problems.

Filters were able to dispose of stopped sediments in their normal cleaning time (20 min). Preheaters did not suffer any

TABLE 11-1 Sediments in the Sludge

Days	SHF
0	Not filterable
1	3%
2	7%
6	18%

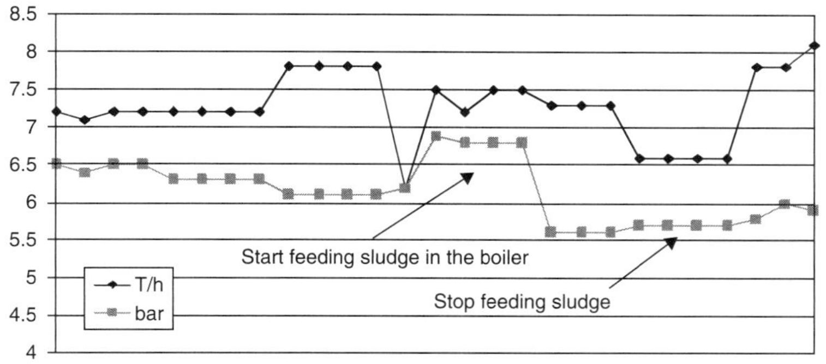

Figure 11-6 Burner pressure before and after sludge combustion.

fouling problems. Still more noticeable is the fact that no combustion and/or burner plugging problems arose, which is a great success even compared to the *normal* combustion problems encountered during fuel oil combustion (burners, filters, and preheater plugging).

Any of the aforementioned operating problems would have been, in a way, *justifiable*, as the boiler was fed with a tank sludge, but asphaltene stabilization achieved with the chemical has been so effective as to give rise to no pitfalls in the combustion of the sludge. To further confirm the above, burner pressure did not increase during sludge combustion: on the contrary, it decreased, due to the stabilizing action of the chemical. Figure 11-6 reports the values of burner pressure and fuel charge rate.

Flame characteristics have always been best during combustion of additivized sludge; the flame has been *clear* and the flame pattern has been regular, without sputtering. Particulate matter emissions during combustion of additivized sludge have not increased; on the contrary, they have slightly decreased once blowing and load change are excluded. Particulate emissions are reported in Fig. 11-7; single values are the average values in 30 min of instantaneous values achieved by a continuous analyzer.

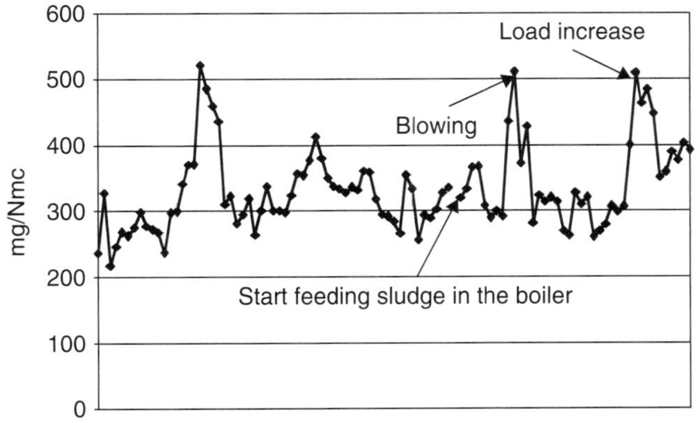

Figure 11-7 Normalized particulate emissions before and after sludge combustion.

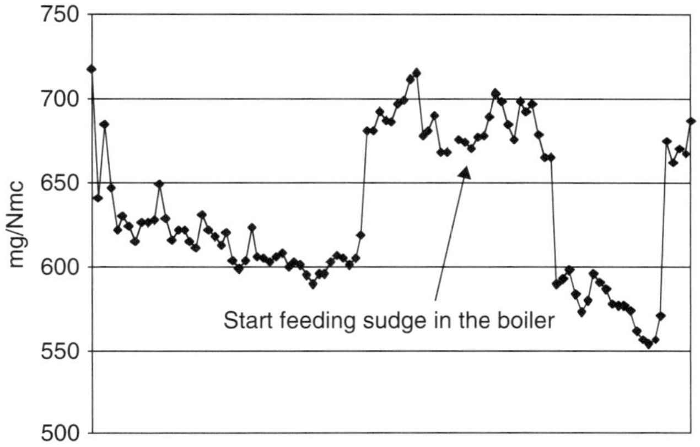

Figure 11-8 NO_x emissions before and after sludge combustion.

NO_x emissions have decreased significantly during combustion of additivized sludge. NO_x emissions are reported in Fig. 11-8; single values are the average in 30 minutes of instantaneous values achieved by a continuous analyzer. This follows the reduced sputtering of the additivized sludge.

Case History 11.2 A power station had the need to clean the service tank, as it had not been cleaned since boiler construction (roughly 30 years ago). Manual cleaning was not the solution, however, as boiler turnaround was scheduled for only 20 days, and an important revamping had to be implemented. As manual tank cleaning is a dirty, time-consuming, and almost unsafe operation, management decided to test ITW technology. The purpose of the test was to have indications of cleaning during additivation of an ITW chemical: When a tendency to clean had been achieved, upon continuous additivation no more sludge had deposited in the tank.

To give added value to the application, ITW formulated a tailor made chemical, containing both asphaltene stabilizers and combustion catalysts [2]. ITW fuel oil stabilizer and catalyst (hereinafter referred to as ITW additive)

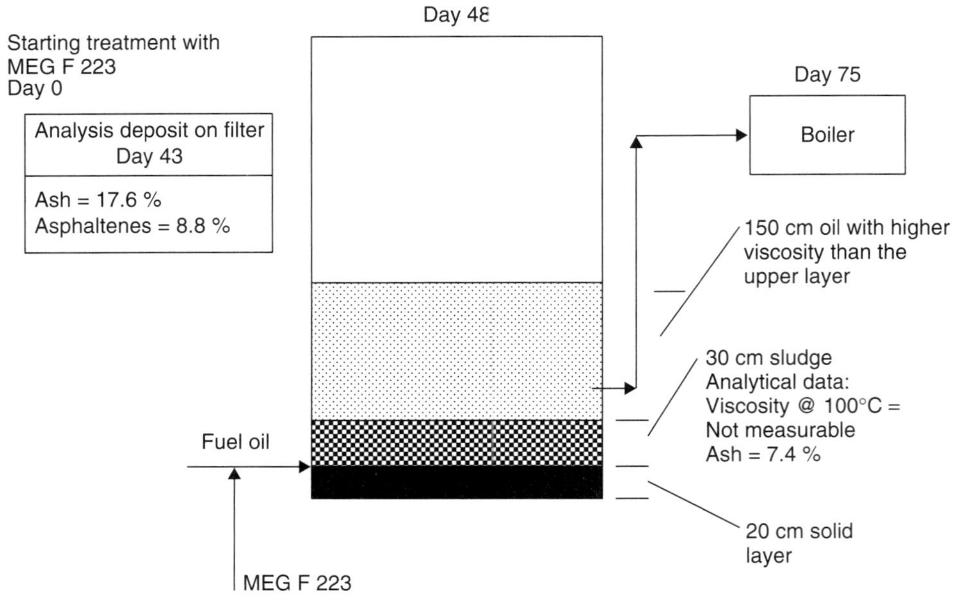

Figure 11-9 Timing of on-stream tank cleaning.

was injected upstream of the service storage tank; the additivized fuel oil entered the tank from the bottom. After about one and a half months of treatment, digging was begun in the tank, with the following results: a 20-cm layer of solids, 30 cm of sludge with unmeasurable viscosity at 100°C and 150 cm of fuel oil more viscous than the layer above [2]. These results were interpreted to mean that by entering the tank from the bottom, the ITW additive solubilized the sludge so that the sludge rendered the lower portion of the oil more viscous.

After another month of treatment, further digging was performed in the tank and the results were surprising: The solid layer had disappeared, as had the very viscous sludge. In their place was found a single nonviscous sludge layer (viscosity 132 cSt at 100°C) [2]. The results are summarized in Fig. 11-9.

The amazing results in tank cleaning were also confirmed by those in the preflame and postflame zones of the boiler. As ITW additive contains both asphaltene stabilizers and combustion catalysts, it also performs its action downstream of the tank. As a matter of fact both the preflame and postflame zones were cleaner than in the previous situation. In particular, the hot filter ΔP was nearly nil after cleaning; on a normalized basis, the filter didn't increase its ΔP value. Fouling factor monitoring in the preheaters revealed no increase in fouling.

The most significant improvements in boiler operating parameters can be summarized as follows:

- The combustion chamber pressure decreased from about 260 mmH_2O to about 230 mmH_2O.
- The flue gas pressure drop in the Ljungstroem decreased from about 100 mmH_2O to an average of about 70 mmH_2O.
- The air temperature at the Ljungstroem outlet increased from about 395°C to about 403°C.
- The normalized burner pressure was almost constant.

Figure 11-10 summarizes the combustion benefits in the boiler.

The improved cleanliness of the boiler translated into improved combustion efficiency. The specific steam production for kcal of incoming fuel improved from about 0.0566 ton of steam/fuel kcal to about 0.0574 ton of steam/fuel kcal, with an improvement of about 1%. Inlet fuel oil characteristics in the period of additivation are summarized in Table 11-2.

Apart from being tremendously effective in improving both pre- and postflame cleanliness, the ITW additive was

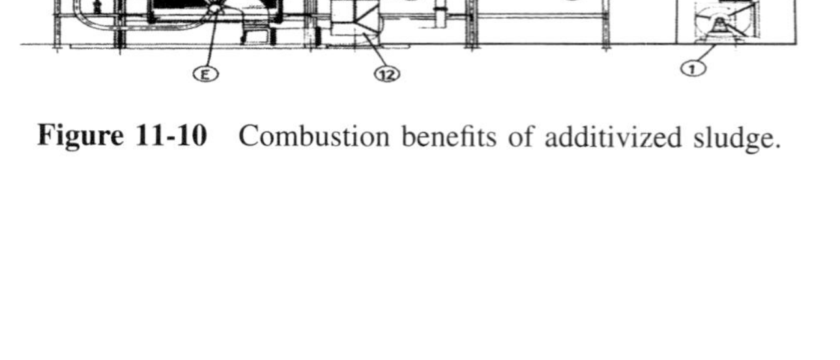

Figure 11-10 Combustion benefits of additivized sludge.

TABLE 11-2 Characteristics of Incoming Fuel Oil During the Period of Cleaning

Parameter	Analysis 1	2	3	4	5	6
Density at 15°C (g/cm^3)	0.9648	0.9695	0.9781	0.9688	0.9781	0.9695
Viscosity at 50°C (°E)	48.1	32.2	45.4	23.6	44.4	46.4
Sulfur (wt%)	2.3	2.14	2.09	2.5	1.73	1.93
Water (vol%)	0.3	0.1	0.3	0.1	0.2	0.5
Sediments for extraction (wt%)	0.04					
Asphaltenes (wt%)	5.4	7.9	10.93	5.0	10.0	9.0
SHF (wt%)	0.33	0.14	0.02	0.06	0.08	0.10
Conradson carbon residue (wt%)	11.1	11.2	13.5	5.7	14.1	14.1
Ash (wt%)	0.011	0.012	0.02	0.01	0.01	0.01
Sodium (ppm)	8	6	10	3	20	37
Vanadium (ppm)	41	66	55	12	40	37
Nickel (ppm)	34	32	31	11	18	21

also effective in reducing flue gas emissions. In the previous situation, particulate matter emissions were in the range 50 to 60 mg/N·m^3. Following ITW chemical addition, particulate matter emissions were decreased to 18 mg/N·m^3, of which only 2 mg/N·m^3 was the unburned portion (only about 10% of total particulate emissions). A decrease in NO_x emissions was also noted. Previous NO_x emissions were in the range 700 mg/N·m^3; with ITW additive they were in the range 500 mg/N·m^3.

Case History 11.3 An oil tanker (120,000 m^3 capacity) needed to be cleaned before drydocking. The standard oil tanker cleaning procedure involves the use of crude oil washing (COW), followed by water cleaning. However, these operations are not completely effective, as tank washing machines have some shadow areas where the flow has no direct impact on the sludge. Therefore, after COW, some sludge is left in the bottom of the tanks and has to be removed manually. Figure 11-11 indicates methods of manual cleaning of oil tankers.

ITW additives have been added in both the COW and water-washing phases.

In the water-washing phase a patented hydrocarbon solubilizer has been added. This product is capable of solubilizing hydrocarbons temporarily in water when the two phases (additivized water and hydrocarbons) are in direct contact with each other (e.g., under agitation). The solubilization is temporary, and after some minutes, the hydrocarbilic and water phases separate out. The water phase is very clear, and no emulsion is formed. The additive also facilitates the separation of oil from water.

COW was performed on Bouri crude oil, which is a very *severe* crude, like the others transported by ship. Before performing additivized COW, some standard COWs were performed: one upper cycle 120°/30° and two bottom cycles 0°/30°/0°. At the end of those cycles the slop tank contained 1300 m^3 of Bouri crude and sediments. This is what the standard COW procedure would have achieved in terms of recovered sludge. One further bottom cycle with the additive (injected in the suction of the COW pump) was performed, and stopped after only 12 h of washing (total time for all the cargo tanks). At the end of this additivized cycle, the slop tank contained 1500 m^3 of Bouri crude

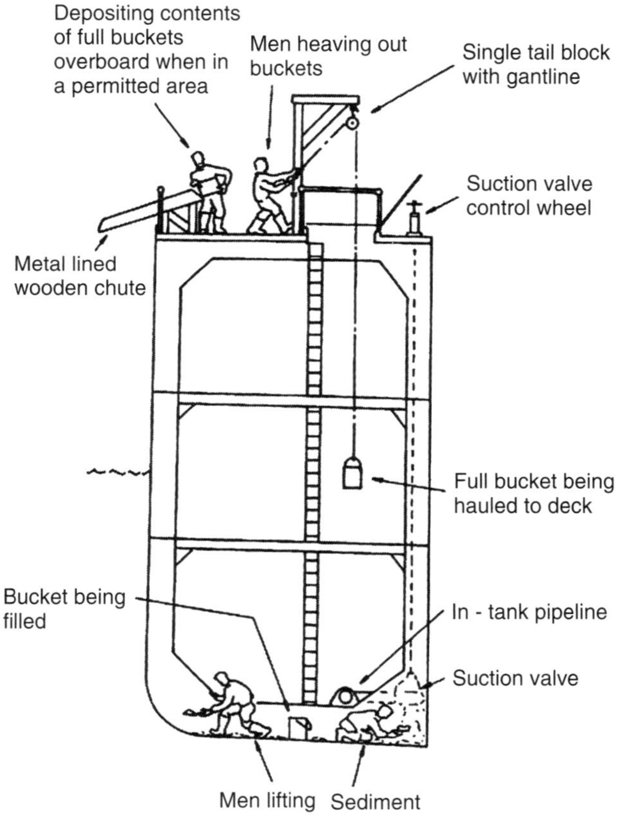

Figure 11-11 Today's methods for cleaning oil tankers.

and sediments. Therefore, the use of ITW stabilizer made possible the recovery of 200 m^3 of sludge in the COW phase.

The additivized water-washing phase was performed in the same way as the additivized COW, with water heated at 60°C. At the end of this phase, further recovery of 130 m^3 of oil was achieved. Therefore, following the injection of ITW additives, the total amount of sludge recovered was 330 m^3. All 1630 m^3 of crude oil + recovered sludge was sent to the refinery for reprocessing. No operational problems were reported. The results achieved were extremely positive in terms of recovered sludge. The results are even more notable taking into account the type of crude oils carried by the ship, and hence the type of sludge generated by them.

The crudes transported are mainly Belayim, Bouri, Bu Attifel, and Es Sider; Belayim and Bouri are particularly well known in the petroleum industry as crudes that exhibit fouling problems. Bouri, in particular, is a very unstable crude that causes severe fouling problems in refineries. Its visbroken TAR instability has been well recognized; some FCCU overhead section restrictions (following coke deposition) have also been reported when processing Bouri gas-oils.

Some characteristics of the four crudes are reported in Table 11-3.

Despite Bouri's characteristics, the additivized COW procedure was extremely successful, which confirms the stabilizing properties of ITW additives. The payout of the application has been in the range 16:1, due to the dramatic reduction in oil tanker cleaning time, and hence recovered freight costs.

Case History 11.4 A fuel oil tank (5000 m^3 capacity) needed to be cleaned, as the owner wanted to store gasoline in it. The ITW chemical was added by circulation at the bottom of the tank. By opening the manways and checking the pumpability of the sludge, the customer was fully satisfied with the results. As the owner missed some authorizations, the tank content could not be pumped out. After five months, the sludge was still very fluid (easily pumpable) and stable: no precipitation had occurred. After the product owner got the required authorizations, the fluidized sludge was promptly pumped out from the tank (at ambient temperature) and mixed with other fuel oil. This confirms that the stabilization of the additivized sludge is permanent and that there is no precipitation over time.

11.5.3.3 Economics of Tank Cleaning Apart from being environmental friendly, the application of ITW additives makes tank cleaning economically feasible. As a matter of fact, we can evaluate an approximate payout for the operation by taking into account the following items:

- Sludge recovery
- Waste minimization
- Cost of alternative cleaning
- Storage capacity recovery

We will make a cost recovery estimation for 100 m^3 of sludge recovered. The value of oil recovered (sludge, e.g., valorized to fuel oil) accounts for

$$100 \text{ m}^3 \times 200 \text{ euros/m}^3 = 20{,}000 \text{ euros}$$

The costs of waste disposal (tank sludge is considered as hazardous waste) are accounted for by considering a disposal cost of about 520 euros/m^3. The corresponding saving for a total sludge recovery of 100 m^3 is

$$100 \text{ m}^3 \times 520 \text{ euros/m}^3 = \$52{,}000$$

TABLE 11-3 Characteristics of Transported Crude Oils

	Crude			
Parameter	Belayim	Bouri	Bu Attifel	Es Sider
Specific gravity 60/60°F (Kg/L)	0.8837	0.8978	0.8114	0.8417
API gravity	28.6	26.1	42.8	36.6
Kinematic viscosity at 40°C (cSt)	13.70	17.23	14.0 (50°C)	5.266
Pour point (°C)	−12	6	42	9
Water by distillation (vol%)	0.05	0.11	<0.05	0.03
Sediment by extraction (wt%)	<0.01	<0.01	<0.01	<0.01
Salt content (ppm NaCl)	22	<20	<20	22
Sulfur (wt%)	2.19	1.85	0.01	0.40
Acid number (mg KOH/g)	0.03	0.01	0.04	0.06
Conradson carbon residue (wt%)	6.88	6.28	0.38	2.72
Asphaltenes (nC7) (wt%)	3.14	3.58	0.11	0.58
Ash content (wt%)	0.02	0.012	0.005	0.005

When not recovered, the sludge would have been removed manually, with the connected costs of manual cleaning to be added to the evaluation. Another item to be added to the evaluation is the time of nonoperation of the tank, and hence storage capacity recovery. This depends on the tank capacity and on the time for manual cleaning (hence on the sludge amount) and cannot be normalized for 100 m^3. A figure to be accounted for could be 0.026 euros/day/per cubic meter of tank capacity. For a 50,000-m^3 tank and for a 45-day cleaning time reduction with ITW technology, the saving is

$$0.026 \text{ euro}/(\text{day}/\text{m}^3) \times 50{,}000 \text{ m}^3$$
$$\times 45\text{days} = 58{,}500 \text{ euros}$$

Finally, although not valuable in terms of money savings, the application of ITW technology results in a much safer operation for sludge removal.

These returns are even bigger in the case of oil tanker cleaning, where you have to add a freight cost of $50,000 perday (e.g., for a 120,000-dwt tanker) for each day needed to remove the tank manually. The cost of ITW technology is determined from case to case, depending on the sludge amount and the chemicals, equipment, and labor hours needed.

We can compute a minimum saving of about 80,000 ÷ 100,000 euros/100 m^3 sludge. Normal figures of return on investment range from 3:1 to 11:1. For example, total savings for a 50,000-m^3 tank with 600 m^3 of sludge will be about 620,000 euros.

11.5.3.4 Conclusions By utilizing ITW technology for tank cleaning, it is possible to effectively recover and reutilize the sludge, thus achieving the following results:

- Reduction in overall costs
- Recovery of hydrocarbons
- Reduction in sludge disposal costs
- Safe and environmental friendly operations
- Reduction of cleaning time

The sludge recovered gives rise to no operating problems in processing plants. It is also possible to use ITW technology in stream tank cleaning without any detrimental effects on downstream equipment. For tank cleaning, those advantages account for a return on investment ranging from 3:1 to 11:1.

11.6 TANK APPURTENANCES

Storage tanks can be provided with any number of appurtenances, depending on the appropriate design codes and the requirements of the user. A tank may be fitted with mixers, platforms and ladders, gauging devices, manways, and a variety of other connections, including dikes, vents, and supporting structures.

11.6.1 Mixers

The general purpose of mixers is to disperse a substance into the bulk fluid to ensure a uniform mixture before it reaches the next process tank. Mixers are generally used for coagulation purposes. The mechanisms used to achieve mixing involve the creation of high velocities, intense turbulence, and rapid stirring. High energy input and energy dissipation are required to sustain mixing. Mixing can be achieved using paddlewheel mixers, rising air bubbles to effect mixing, and hydraulic means using baffles, static mixers, swirl tanks, and so on.

11.6.2 Dikes

Dikes are often required to contain the volume of a certain portion of the tanks enclosed, depending on the tank contents. Dikes are used to protect surrounding property from tank spills or fires. In general, the net volume of the enclosed diked area should be the volume of the largest tank enclosed (single failure concept). The dike walls may be earth, steel, concrete, or solid masonry designed to be watertight with a full hydrostatic head behind it. Local codes and specifications may govern construction. If more than one tank is within the diked area, curbs or, preferably, drainage channels should be provided to subdivide the area in order to protect the adjacent tanks from possible spills. Many codes, standards, and specifications regulate the location, design, and installation of storage tanks, depending on their end use. Selecting the proper specification and providing adequate fire protection for the installation may allow lower insurance rates over the life of the installation.

11.6.3 Insulators [13]

Principal uses of insulation are for personnel protection, process temperature control, prevention of condensation, and conservation of energy. The four basic types of thermal insulating material are fibrous, cellular, granular, and reflective. These materials differ in many characteristics.

1. *Personnel protection*. Personnel protection is accomplished by the application of insulation of proper thickness where the surface temperature should be limited to approximately 150°F or as specified by applicable codes or company standards.

2. *Process temperature control*. Insulation thickness is specified in this case to help control the temperature of the

process fluid. Electrical, steam, hot process fluid, hot oil, and glycol–water tracing are used to add heat to the process line to balance the heat loss. The insulation thickness must be matched to the energy input to achieve the desired result. Freeze protection is another use for insulation. This includes fluids that have higher viscosities or freeze points.

3. *Condensation*. Insulation thickness must be sufficient to keep the outside surface of the insulation above the dew point of the surrounding air. Moisture condensation on a cool surface in contact with warmer humid air must be prevented because of the deterioration of the insulation. In addition to the required thickness of insulation, a vaportight membrane must be properly applied to the insulation. As a rule, insulation thickness for condensation control is much greater than the thickness required for conservation of energy.

4. *Conservation of energy*. High fuel costs increase the need for more insulation. A rule of thumb for estimating the thickness of insulation is to apply the thickness that produces a heat loss of 3 to 5% or less from the surface. Specific insulating materials and thicknesses for any large application should be determined with the assistance of the manufacturer. The method is based on elementary heat-transfer theory and reliable experimental data.

11.6.4 Platforms and Ladders

Platforms and ladders provide easy access to aboveground storage tanks (Fig. 11-12). This enhances maintenance and monitoring services. Ladders can be installed permanently or temporarily for easy removal [13]. The temporary type prevents unauthorized access to the storage tank. Ladder cages are incorporated to prevent personnel from falling off the ladder, which is one of the most frequent causes of compensation.

Figure 11-12 Ladder used in a tank. (Courtesy of Tanks.)

11.6.5 Gauging Devices

11.6.5.1 Pressure Gauging Devices In all aspects of process industries, pressure measurement instruments are used in monitoring and controlling pressure process equipment, especially in storage tanks. Accurate pressure measurement of a substance, whether gas, liquid, or granular materials, forms an important part of a tank overpressurization system. This is a very critical aspect of process instrumentation, because it detects pressure buildup and helps protect tanks and personnel from potential damages. Different types of pressure-sensing devices using different technologies are available for the wide variety of applications that exist, such as capacitance pressure sensor, piezoelectric pressure sensor, and differential pressure sensor.

11.6.5.2 Temperature Gauging Devices A temperature gauging device is used to determine the temperature of a substance stored in the tank. They assist process personnel to ascertain whether the temperature of the stored substance is within the designed limit. The most used are liquid-filled glass thermometers, bimetallic thermometers, thermocouples, infrared thermometers, and resistance thermometers.

11.6.5.3 Level Gauging Devices A level measurement system forms a very important part of a vessel overfill prevention system, being the tool that plant personnel make use of daily to ensure that vessels are operating within designed safe working levels, to identify any abnormal events, and to ensure that product is loaded and unloaded safely [12]. Depending on the type of level sensing device, they are installed on top or by the side of the storage tank (Fig. 11-13). There are different types of level measurement devices using different technologies available for the wide variety of applications that exist, such as the displacer level gauge, air bubbler, radar level gauge, and capacitance level gauge.

11.6.6 Valves

Valves are also installed on aboveground storage tanks for easy and safe control of fluid contained. Most especially, an overfill prevention valve is mounted at the fill port of an aboveground storage tank (top loading). It shuts off the flow of fluid when the liquid level reaches a preset warning level (mostly 90 to 95%).

Another common type of valve is an antisiphon valve, which is used to prevent the formation of a siphon in an event of leakage. It helps in minimizing product loss and fire hazards. It is usually designed to be mounted in the piping system above the top of the tank. It can be installed on vertical or horizontal piping.

Another is the tank blanketing valve, which completely protects the storage tanks against contamination and against

Figure 11-13 Level gauging device installed on a tank. (Courtesy of Tank Connection International.) [10]

damage or rupture. The blanket gas (usually, nitrogen) pressure minimizes the rate of evaporation of the stored fluid.

11.6.7 Manways

Manways offer easy entry to low-pressure storage tanks for maintenance and cleaning. There are emergency relief manways that allow access to low-pressure storage tanks and provide emergency venting in an event of fire outbreak.

11.6.8 Diffusers

Diffusers reduce the turbulence of fuel in storage tanks, thereby mitigating the creation of vapors. They reduce contamination of fuel by directing flow away from debris on the bottom of the tank.

Note: Diffusers do not restrict the flow of fluid into a tank. Their primary duty is to minimize fluid vaporization and the wear and tear of tank bottoms.

11.6.9 Water Cannons

Storage containing fluids that are flammable is subject to explosions in the event of operational error. Such explosions can cover large areas and cause damage. All storage tanks with flammable fluids must have water cannons installed around them. This is to cool them when there is a fire outbreak. These systems should be tested and inspected carefully on a monthly basis to ensure steady efficiency. Only well-trained personnel should operate these systems. In most cases, they can be identified as small-diameter red pipes that go around the tanks.

11.6.10 Vents

Vents are installed on an aboveground storage tank (Fig. 11-14) to prevent the tank from overpressurization

Figure 11-14 Vent situated at the top of a tank. (Courtesy of Tanks.) [11]

and rupturing if subjected to fire. An emergency vent can also be utilized as a backup to the normal vent. As part of its normal operation, a flame arrester is used in conjunction with the normal vent to prevent heat transmission or an ignition source into the tank.

11.6.11 Grounding

Metallic storage tanks used to store flammable liquids should be grounded to minimize the possibilities of an explosion or fire due to lightning or static electricity.

11.6.12 Supporting Structures

The manner in which tanks are supported should always be reviewed and inspected. Steel structures used to support a tank should be treated with retardant material and corrosion protection to prevent structural failures from adjacent fires. Among the best supports for tanks are reinforced-concrete structures, which should be inspected regularly for cracks or other signs of structural failure. Also, tanks should not rest directly on the support structure's saddle, but should rest on a bearing plate between the tank and the saddle.

11.7 STORAGE TANK MAINTENANCE

All tanks and associated piping should be subject to a routine inspection and maintenance program. Depending on the material stored and for prolonging the life span of your tank, it is recommended that an annual wall thickness test be performed. The paint or other exterior coatings on tanks should not be allowed to degrade, as areas of rust or corrosion greatly weaken a tank's structure. Attention should also be placed on the maintenance of other tank structures, such as ladders, catwalks, and access holes, which are used to conduct maintenance and operations activities [8].

Remember: Most tanks have grounding straps or sacrificial anodes to prevent corrosion. These items should be checked during periodic maintenance checks. The function of level gauges, pressure release devices, temperature gauges, and fire sensors should also be part of a routine inspection program [8].

REFERENCES

1. Bahadori, A., Minimize vaporization and displacement losses from storage containers, *Hydrocarbon Processing*, vol. 88, no. 6, June 2009, pp. 83–84.
2. Ferrara, M., Improved tank cleaning reduces hazardous wastes and boosts return on investment, presented at ERTC Environmental, London, April 8–10, 2002.
3. Gas Processors suppliers Association, *GPSA Engineering Data Book*, 11th ed., GPSA, Tulsa, OK, 1998, Sec. 6, Storage.
4. http://en.wikipedia.org/wiki/storage_tank.
5. http://en.wikipedia.org/wiki/Tank–blanketing.
6. http://www.carmagen.com/news/engineering_articles/news8.htm.
7. http://www.chem.info/Articles/01/Storage-Tank-Coatings–The-Most-Important-Selection-Guidelines/.
8. http://www.sipeonlinetraining.com/bulletins/Preventing%20Above%20Ground%20Storage%20Tank%20Failures.pdf.
9. http://www.petrolplaza.com/technology/articles/MiZlbiYxMDM1NiYmMSYyJiY%3D.
10. http://www.tankconnection.com.
11. http://www.tanks.com.
12. Nwaoha, C., Inside process: locating the level, *Control Engineering Asia*, vol. 8, August 2009, pp. 28–31.
13. Nwaoha, C., Selection and operation of above storage tanks, *Petroleum Africa*, March 2011, pp. 30–33.
14. Shelley, C. H., Storage tank fires: Is your department prepared? http://www.FireEngineeringUniversity.com.

12

MIXERS

JAYESH RAMESH TEKCHANDANEY
Unique Mixers and Furnaces, Thane, India

Mixing is at the heart of most production systems in the chemical process industry, pharmaceutical industry, food industry, and allied industries. It is of vital importance in processing solids, liquids, and gases in the polymer, glass, ceramics, building materials, pulp and paper, petroleum, and power industries and in industrial waste treatment systems. By definition, *mixing* is the process of thoroughly combining different materials to produce a homogeneous product. The resulting mixture in most cases is a combination of dissimilar materials. In some cases, a chemically homogeneous material may be mixed to produce a uniform blend of a desired weight or volume with consistent particle-size distribution, color, texture, or other required attributes. Mixing is a critical process because the quality of the final product and its attributes are derived from the quality of the mix. Improper mixing results in a nonhomogeneous product that lacks consistency with respect to desired attributes such as chemical composition, color, texture, flavor, reactivity, and particle size.

The wide variety and ever-increasing complexity of mixing processes encountered in industrial applications requires careful selection, design, and scale-up to ensure effective and efficient mixing. Improved mixing efficiency leads to shorter batch cycle times and lower operational costs. Today's competitive production systems necessitate robust equipment that are capable of faster blend times, lower power consumption, and adaptability of equipment for use with multiple products. Moreover, in addition to carrying out mixing operations, many modern mixers are designed to combine different processing steps in a single piece of equipment: for example, coating, granulation, heat transfer, and drying. A mixer is no longer a generic production tool, but a critical and decisive business tool. This is because profitability and competitive advantage are dependent on subtle improvements in product quality and productivity through gains in mixing performance and efficiency. Good mixing is imperative for minimizing investment and operating costs, providing high yields, and thereby enhancing profitability. A recently published handbook on industrial mixing estimates the cost of poor mixing to be as high as $100 million per year.

The terms *mixing* and *blending* are often used interchangeably; but technically they are slightly different. Blending is a process of combining materials; however, blending is a relatively gentle process compared to mixing. In terms of the phase of material, blending is the process of solid–solid mixing or mixing of bulk solids with a small quantity of liquid. The terminology mixing is more closely associated with liquid–liquid, gas–liquid, and viscous materials. For the purpose of this chapter, we use the terms *mixing* and *blending* interchangeably.

The purpose of this book is to serve as a practical guide on industrial mixing for process design engineers, plant operation and maintenance personnel, and academics. In this chapter we discuss the practical aspects associated with mixing of materials: liquid, solids, and pastes. In the first section of this chapter we explain the basic concepts of mixing, the modes of mixing operation (batch and continuous), and the basis for mixer design, selection, and scale-up. These topics introduced in Section 12.1 are applicable for liquid, solid, and paste mixing. Mixing of liquids is discussed in detail in Section 12.2. Section 12.3 focuses on mixing of free-flowing solids. Mixing of pastes and high-viscosity materials is discussed in Section 12.4.

Process Plant Equipment: Operation, Control, and Reliability, First Edition. Edited by Michael D. Holloway, Chikezie Nwaoha, and Oliver A. Onyewuenyi.
© 2012 John Wiley & Sons, Inc. Published 2012 by John Wiley & Sons, Inc.

Sections 12.2, 12.3, and 12.4 are structured so as to introduce the reader to the basic theoretical concepts relevant to mixing, followed by a detailed study of the mixing equipment. Mechanical components and the operation and maintenance of mixing equipment are detailed in Section 12.5. The advances in mixing technology are summarized in Section 12.6.

12.1 MIXING CONCEPTS: THEORY AND PRACTICE

Although mixing of solids to some extent resemble the mixing of low-viscosity liquids, there are significant differences between the two processes [7]. These are as follows:

- Liquid blending depends on the creation of flow currents, which transport unmixed material to the mixing zone adjacent to the impeller. In heavy pastes or masses of particulate solids, no such currents are possible, and mixing is accomplished by other means. In consequence, the power required for mixing of dry solids and viscous pastes is much more than that consumed in blending of liquids.
- In blending liquids, a "well-mixed" product usually means a truly homogeneous liquid phase, from which random samples even of very small size all have the same composition. In mixing pastes and powders the product often consists of two or more easily identifiable phases, each of which may contain individual particles of considerable size. From a well-mixed product of this type, small random samples will differ markedly in composition.

As a result, design and selection of mixing equipment for solids and pastes is far more complex than it is with liquids. Mixers are classified based primarily on the viscosity of the process materials they can handle. Figure 12-1 identifies the different types of mixers that can be used for materials of varying viscosities.

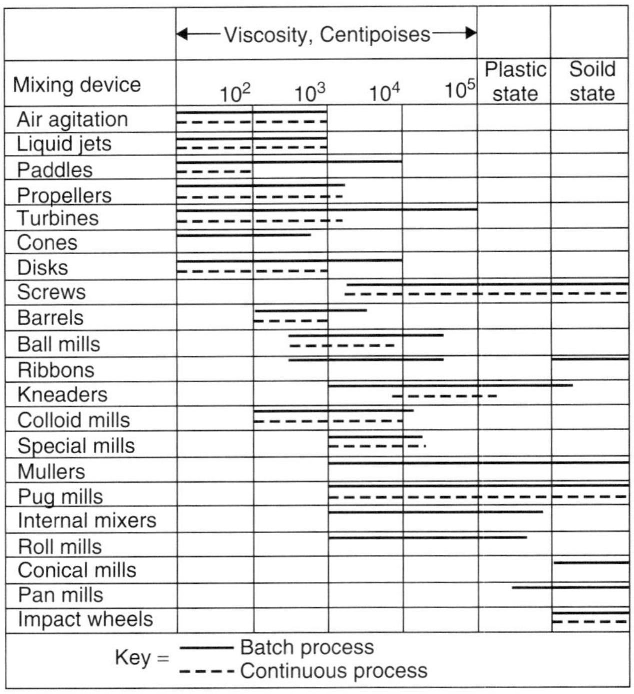

Figure 12-1 Range of mixer operation based on the viscosity of the material. (From C. S. Quillen, *Chemical Engineering*, June 1954, p.177.)

12.1.1 Batch and Continuous Mixing

Mixing equipment are available in different shapes, sizes, and configurations and may be used for liquids, solids, or pastes, but they ultimately fit into one of two categories: batch or continuous. The selection of batch or continuous mixing depends on several factors. These include the quantity of the material to be mixed, desired standards of homogeneity of product, premixing operations and material-handling equipment, postmixing operations, and multiple-product formulation requirement. Whereas some types of mixers are only suitable for either batch or continuous operations, other types can be adapted for operation in either mode.

Batch Mixing In batch mixing, all ingredients are charged into the mixer together or in a predefined sequence; mixed until a homogeneous material is produced, and discharged from the mixer in a single lot. The output of a batch mixer is measured in units of kg/batch. It should, however, be noted that mixers are designed in terms of volumetric capacities. The batch capacity would therefore depend on the bulk density and specific gravity of the products to be mixed.

Batch mixing is preferred for applications where:

- Production quantities are small.
- Strict control of mix composition is required.
- Many formulations are produced on the same production line.
- Ingredient properties change over time, and compensation must be on a batch-by-batch basis.
- It is required to identify a batch for further follow-up (e.g., pharmaceutical formulations, food products).

Continuous Mixing Batch mixing is uneconomical when large quantities of material are to be mixed. In such cases, a continuous mixer is dedicated to a single high-volume product. Continuous mixers can be designed for capacities as high as 500 tons/hr. In continuous mixing, material flows steadily from an upstream process into the mixer, is retained in the mixing vessel for a specified mixing time, and discharges at the same flow rate for downstream handling. In a continuous mixing process, the weighing, loading, mixing, and discharge steps occur continuously and simultaneously. The process of charging the material in a continuous mixer is extremely critical and can significantly affect the quality of the final product mix. Radial and axial mixing take place as the material travels from the feeding point to the discharge point. The time taken by the material to travel from the feeding point to the discharge point is known as the *retention time* of the material in the mixer. Unlike batch mixers, where product retention time is carefully controlled, with continuous mixers material retention time is not uniform and can be directly affected by mixer speed, feed rate, mixer geometry, and design of mixer internals. Material is discharged continuously at a constant rate, which is termed the *capacity* of the continuous mixer, measured in kg/h of mixed product. To keep track of the mix quality, a well-defined sampling and material testing procedure is to be followed. While discharging from the mixers, segregation can be reduced by positioning the discharge closer to the packing units or as an integral part of it.

Continuous mixing is preferred for applications where:

- Large quantities of a single product are to be mixed.
- The continuous process line requires a high production rate.
- Strict batch integrity is not critical.
- Smoothing out batch product variations is required.

12.1.2 Selection of Mixing Equipment

The ever-increasing needs of the process industry for larger batch sizes, faster blend times, higher degree of homogeneity and reaction, and greater product yield have catalyzed the need for improvements in mixing and blending technology. Process objectives can be achieved using the wide array of equipment offered by mixer manufacturers. Considering the gamut of mixing applications, there are no specific guidelines that can be applied while selecting mixing equipment.

Although many mixers are capable of mixing different types of liquids, solids, paste, and other materials, the process of selecting a mixer remains an art form because of the many variables involved. It is therefore necessary to clearly define the process objectives and production requirements that are relevant to the mixing operation. The following are the basic criteria to be considered for the selection of mixers:

- *Material characteristics*. These include the physical, chemical, and mechanical properties of the materials to be mixed.
- *Process setup*. These include factors such as the quantity of different material to be mixed, preblending and postblending material storage, material-handling equipment, performing multiple processes in the mixing equipment (liquid addition, drying, coating), material charging and discharge of material from the mixer, process safety, and mode of equipment operation: batch or continuous.
- *Mixer operating parameters*. These include pressure and temperature conditions under which the mixer is expected to operate.
- *Mixing accuracy*. This defines the required degree of product homogeneity.
- *Mixer cleanability*. Mixer cleanability is an important criteria, especially in cases where product quality and process flexibility are important. The requirement of mixer cleanability should be clearly defined so that this can be taken care of at the design stage for selecting mixer internal surface finish, seal design, impeller clearances, and so on.
- *Equipment costs*. Equipment costs include the capital cost of the mixer and operating costs such as power and maintenance. Equipment should be chosen such that all process requirements are met at minimal cost.

12.1.3 Design of Mixing Equipment

The design of mixing equipment comprises three major elements:

- *Process design*. Process design accounts for considerations such as impeller fluid mechanics (flow, head, shear), mixing regime (laminar, transitional, or turbulent), and the translation to the various vessel sizes (scale-up, hydraulic similarity).
- *Mixing equipment characteristics*. Mixing equipment characteristics relate the effects of material properties, vessel geometry, agitator (blade) geometry, and rotational speed to the power consumption.
- *Mechanical design*. For a mixer to deliver the desired process performance, it has to be designed mechanically for long life and trouble-free operation. Design and selection of the prime mover, gear reducer, seals, bearings, impeller shaft, impeller, and so on, are essential.

12.1.4 Scale-Up of Mixing Equipment

Considering the myriad of mixing applications and their industrial significance, conducting mixing experiments is critical in establishing process performance. The purpose of scale-up is to design a commercial-size mixing system that can replicate the results obtained in a laboratory or pilot-scale setup. The method used to accomplish scale-up will depend on the application under consideration. Scale-up for solid–liquid mixing, gas dispersions, solid–solid mixing, and paste mixing may each require a different approach. These scale-up methods would require careful consideration of fundamental concepts, dimensional analysis, empirical correlations, test data, and experience.

Mixing system scale-up generally requires three types of similarity to be maintained between the test-scale unit and the production-size unit: geometric similarity, kinematic similarity, and dynamic similarity.

1. *Geometric similarity*: requires that all the corresponding dimensions of the scaled system be in the same ratio as that of an acceptable test unit. These dimensions include vessel diameter and liquid level, baffle width and number, vessel impeller diameter, number of blades, and width ratio.
2. *Kinematic similarity*: requires geometric similarity. In addition, it requires that all corresponding points in the system should have the same velocity ratios and should move in the same direction.
3. *Dynamic similarity*: requires geometric and kinematic similarity in addition to force ratios at corresponding points being equal, involving properties of gravitation, surface tension, viscosity, and inertia [3]. With proper and careful application of this principle, scale-up is often feasible and quite successful.

Maintaining geometric similarity may not always be feasible, due to practical constraints imposed by plant space and equipment layout. There are therefore other commonly used scale-up criteria for scale-up of liquid, solid, and paste mixing equipment. These criteria include constant power per unit volume, constant torque per unit volume, and constant agitator tip speed. Constant tip speed or constant torque per unit volume may be used for scale-up when the flow velocities in the agitator region are to be maintained the same for the laboratory scale and the production scale units. Scale-up methods based on a constant blend time require the mixer speeds in the commercial vessel to be the same as in the laboratory vessel. This results, however, in a very large increase in the motor power. The selection of the criteria for scale-up requires a thorough understanding of the mixing process. It is therefore advisable to rely on one's judgment, practical experience, and equipment manufacturers' recommendation to ensure satisfactory scale-up. Mixing simulation programs can be used for estimating and evaluating the equipment and process performance parameters during scale-up.

12.2 FLUID MIXING

Fluid mixing is an integral part of the chemical process industry. It includes mixing of liquid with liquid, gas with liquid, or solids with liquid. The success of many processing operations is largely dependent on the mixing performance. The terms *agitation* and *mixing* are often used synonymously; however, they are not the same.

- *Mixing*: refers to any operation used to change a non-uniform system into a uniform system (i.e., the random distribution into and through one another, of two or more initially separated phases). Mixing therefore requires a definition of degree and/or purpose, to clearly define the desired state of the system [3].
- *Agitation*: implies forcing a fluid by mechanical means to generate flow. Agitation does not necessarily imply any significant amount of actual intimate and homogeneous distribution of the fluid [3].

There are a number of ways to perform fluid mixing:

- Mechanical agitation
- Jet mixing
- Gas sparing
- In-line mixing

The majority of mixing applications use mechanical agitators. Therefore, in the following text we focus on the mechanical agitation of liquids. For the purpose of this discussion, we use the terms *fluid* and *liquid* interchangeably.

12.2.1 Fluid Mixing Applications

The objective of mixing can be physical and/or chemical processing. Fluid mixing can be broadly classified into the following application classes: (1) Blending of miscible liquids, (2) Blending of immiscible liquids, (3) Liquid–gas mixing, (4) Liquid–solid mixing, and (5) Fluid motion.

12.2.1.1 Blending Miscible Liquids One of most common mixing applications involves blending two or more homogeneous liquids. The liquid blending operation may be purely physical in nature or may involve chemical reaction. Low- to medium-viscosity liquid mixing involves macro- and microscale mixing concepts. *Macromixing* is achieved by the convective flow within the equipment, whereas *micromixing* occurs due to turbulent diffusion between the small cells in the fluid, causing intermingling of molecules.

Liquid blending can be achieved using top-entering agitators, side-entering mixers, or jet mixers. Generally, blending of miscible, mutually soluble liquids with viscosities up to 10,000 cP is carried out using turbine impellers. Axial flow impellers are more efficient than radial flow impellers for these applications. Viscous liquids are blended using close-clearance impellers.

12.2.1.2 Mixing Immiscible Liquids Blending of mutually insoluble, immiscible liquids may be required to produce stable or unstable emulsions. Stable emulsions are required for products such as shampoos, polishes, and other specialty chemicals. On the other hand, applications such as liquid extractions employ an unstable emulsion to boost the rate of mass transfer and reaction. Such applications are common in the petroleum, chemical, food, and pharmaceutical industries.

Turbine impellers are used for intermixing of mutually insoluble liquids for the purpose of creating large enhancements in interfacial area, thereby increasing the rate of mass transfer and reaction. Low shear hydrofoil design impellers can be used for coarse dispersions. Axial and radial flow impellers are effective for fine emulsions. High-shear impellers are required for preparing stable emulsions [7].

12.2.1.3 Liquid–Gas Mixing Liquid–gas mixing applications can be classified into two types. The first type of application involves physical distribution and dispersion of the gas in the liquid. Application of this type is limited only if foam or froth is desired. The second and more common type includes a mass-transfer process such as absorption, stripping, chlorination, oxidation, and so on, all of which require transfer of gas into liquid.

For the purpose of gas–liquid contacting, radial flow impellers are preferred over axial flow impellers. Disk turbine impellers are most suited for gas–liquid dispersions. Fermentation equipment may be provided with high-solidity-ratio hydrofoil impellers. These impellers produce large flows in high-viscosity biological operations to ensure adequate distribution of oxygen. Special types of gas mixing systems are used for hazardous, expensive, and critical applications, such as hydrogen, which necessitate the need for recycling the gas of the vapor space above the liquid into the vessel.

12.2.1.4 Liquid–Solid Mixing Mixing of solids in liquids is required for a wide range of industrial applications, such as catalyst polymer systems, the paper and pulp industry, washing of solids, crystallization processes, and so on. Liquid–solid mixing applications may involve two primary classes:

1. Suspension of solids into the liquid, a physical process. Different processes may require different degrees of suspensions. This required degree of suspension is to be defined while stating the mixing objective.
2. Dissolving of solids in the liquid phase, a mass-transfer process.

Axial flow impellers with high pumping efficiencies are best suited for the majority of solid suspensions.

12.2.1.5 Fluid Motion All of the applications discussed above involved mixing of liquids with either a liquid, a solid, or a gas. The description of the mixing system was defined with respect to the process objective desired. Some applications require a combination of liquid–solid–gas mixing. Physical processes such as heat transfer may be required. In such cases, the description of mixing is provided in terms of the fluid motion produced by the impeller. By defining the magnitude of the fluid motion required, the system description can be provided for the process objective desired.

12.2.2 Mixing Regimes

Mixing in agitated vessels may occur under either laminar, transitional and turbulent flow conditions. A dimensionless number known as the *Reynolds number* (N_{Re}) is used for defining the mixing regime. The Reynolds number for mixing is defined as

$$N_{\mathrm{Re}} = \frac{n D_a^2 \rho}{\mu} \tag{12.1}$$

where ρ is the density of the fluid (kg/m^3), n the rotational speed (rev/s), D_a the impeller diameter (m), and μ the fluid viscosity (Pa·s).

For Reynolds numbers below 10, the flow is laminar, also called *creeping flow*. The turbulent range begins at a different Reynolds number for different impellers, typically ranging from 10^3 to 10^5 [6]. Impeller power numbers are compared in the turbulent regime, which for common use is taken as $\mathrm{Re} > 10^5$. Flow is considered transitional between these two regimes.

In laminar flow, the liquid moves with the impeller. At a distance away from the impeller, the fluid remains stagnant. In such cases, the *Froude number* accounts for the force of gravity, which determines the fluid motion. For such applications the Froude number should be maintained constant.

$$N_{\mathrm{Fr}} = \frac{n^2 D_a}{g} \tag{12.2}$$

where g is the acceleration due to gravity (9.8 m/s^2).

12.2.3 Power Consumption in Agitated Vessels

The power consumed by an agitator depends on the impeller type, impeller diameter, speed of rotation, fluid density, viscosity, vessel size and geometry, internal attachments (i.e., baffles, coils, draft tubes), and the location of the impeller relative to the vessel and the internal attachments. For liquid mixing, the power consumed by an agitator is directly proportional to the cube of the speed of rotation and the impeller diameter to the fifth power. The power varies directly with the density of the fluid.

For a given type of geometrically similar mixing impellers, the effect of the mixing environment on the power drawn is studied in laboratory-scale mixers and is calibrated so as to provide a basis for scale-up to production-size mixers. The system configuration for the laboratory mixer is standardized in order to minimize the number of variables. This serves to provide comparative data on the power consumed by different types of impellers in a similar system.

A number of researchers have reported impeller power characteristics in terms of the dimensionless numbers: the power number (N_P) and the Reynolds number (N_{Re}). The power number is defined as

$$N_P = \frac{P}{n^3 D_a^5 \rho} \tag{12.3}$$

where P is the power consumed by the impeller. The power number depends on the impeller geometry and the location of the impeller in the vessel. Figure 12-2 is a plot of the power number versus the impeller Reynolds number for different types of impellers and vessel geometrics. The relationship between the power number and the Reynolds number can be summarized as follows:

- In the laminar regime, the power number is inversely proportional to the Reynolds number. The power depends largely on the fluid viscosity.
- In the transitional regime, the power number changes slightly.
- In the turbulent regime, the power number is constant and independent of the fluid viscosity.

The power numbers for commonly used impellers operating in the turbulent regime are provided in Table 12-1. The power number of side-entering propellers depends on the pitch of the propeller in addition to the Reynolds number. The power number increases with increase in pitch.

In agitation systems where more than one impeller is mounted on the same shaft, the total power number may or may not be the sum of the individual power numbers. For such configurations, the power number depends on the impeller type and the spacing between the impellers, which is generally equal to the impeller diameter.

Using the power number and Eq. (12.3), the power consumed by an impeller for a specified system geometry can be determined. The connected motor power should be higher since it has to account for the electrical and mechanical losses of the agitator drive system. Equation (12.3) can be rearranged to determine the impeller diameter when it is desired to load an agitator impeller to a given

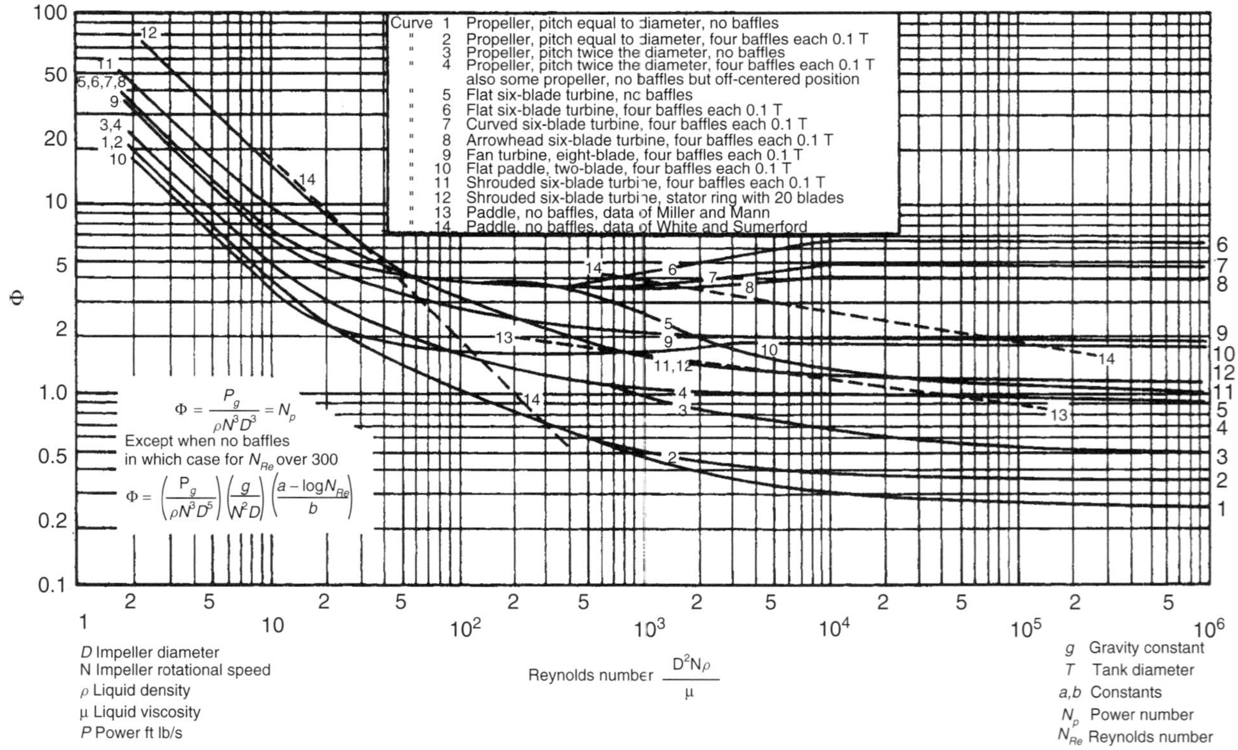

Figure 12-2 Power number vs. impeller Reynolds number for different types of impellers. (From J. H. Rushton, et al., *Chemical Engineering Progress*, vol. 46, No. 8–9, 1950.)

TABLE 12-1 Power Numbers of Various Impellers Under Turbulent Conditions with Four Standard Baffles

Impeller Type	N_P
Concave- or hollow-blade turbine	4.1
High-efficiency hydrofoil impeller	0.3–0.75
The following are all for $D_a = D_t/3$, $E = D_t/3$, and blade width $W = D_a/5$:	
45° PBT; four blades	1.27
45° PBT; six blades	1.64
Marine propeller (1.0 pitch)	0.34
Marine propeller (1.5 pitch)	0.62
Smith or concave or hollow blade with six blades	4.4

Source: [7].

TABLE 12-2 General Guidelines for Specific Power and Agitator Tip Speeds for Liquid Blending Applications

Operation	Specific Power (hp/m^3)	Tip Speed (m/s)
Blending	0.05–0.1	
Homogeneous reaction	0.1–0.4	2.25–3
Reaction with heat transfer	0.4–1.3	3–4.5
Liquid–liquid mixtures	1.3	4.5–6
Liquid–gas mixtures	1.3–2.5	4.5–6
Slurries	2.5	

Source: Adapted from [9].

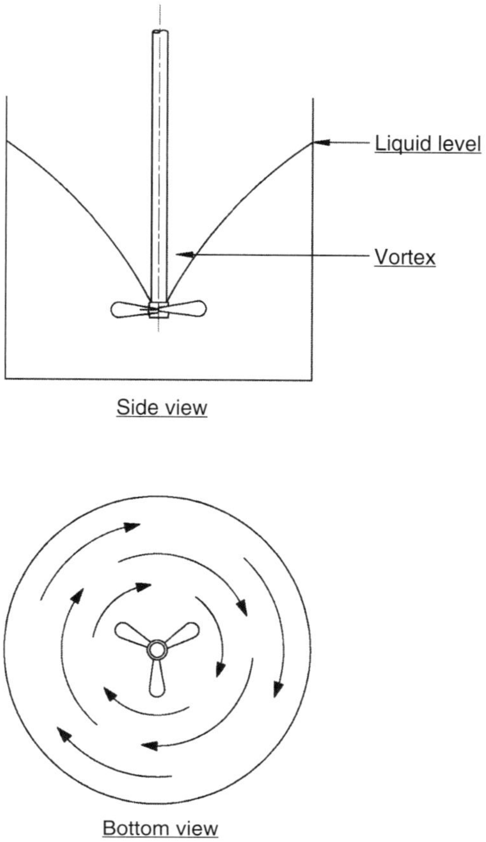

Figure 12-3 Typical flow pattern for axial and radial flow impellers in an unbaffled vessel. Note the vortex formation.

power level. For baffled turbine agitation systems, the guidelines provided in Table 12-2 can be used to determine power inputs and impeller tip speeds.

12.2.4 Flow Characteristics

The flow characteristics for an impeller can be defined by (1) flow pattern, (2) pumping, and (3) shear. Depending on the impeller geometry, each type of impeller generates a specific flow pattern. The flow pattern, together with the mixing regime, determine relative levels of pumping and shear for the impeller. Impellers can therefore be classified by variations in their pumping and shear capabilities.

12.2.4.1 Flow Patterns in Agitated Vessels The flow pattern in an agitated vessel depends on the type of impeller, the vessel size and geometry, internal attachments such as baffles, and fluid properties. The velocity of the fluid at any point within a vertical vessel with an agitator has three components: the radial velocity component, which acts perpendicular to the agitator shaft; the axial component, which acts parallel to the shaft; and the tangential or rotational component, which acts in a tangential direction to a circular path around the shaft. The radial and axial components are useful and provide the flow necessary for mixing. The tangential flow is detrimental to mixing since it creates a vortex in the liquid and hampers axial flow. Figure 12-3 shows the flow pattern formed by a turbine impeller (axial and radial) in an unbaffled vessel. At high impeller speeds the vortex may reach the level of the impeller and the gas from above the liquid may be drawn into the liquid, which is undesirable.

In most cases, circulatory flow and swirling are prevented by installing vertical baffles within the vessel. Other methods include mounting the agitator off-center with respect to the vessel and at a small inclination. For large vessels, the agitator may be mounted on the side of the tank, with the shaft in a horizontal plane but at an angle. Once the swirling has stopped, the flow pattern in the vessel depends on the type of impeller. The flow patterns generated by the commonly used impellers along with their applications are shown in Fig. 12-4.

12.2.4.2 Flow Number (Pumping Number) The pumping capacity of an impeller is determined by the impeller

Impeller	Flow Pattern	Name and Description	Applications	
		HE-3 Narrow-blade, high-efficiency impeller	Blending, Turbulent heat transfer, Solid suspension,	Upper impeller for gas dispersion, $N_p = 027, N_q = 05$ (turbulent)
		P-4 Pitched-blade turbine	Blending, Dispersion, Solid suspension	Heat transfer, Surface motion, $N_p = 125, N_q = 07$ (turbulent)
		S-4 Straight-blade turbine	Local liquid motion for blending, Dispersion, keeping outlets dear from solids,	$N_p = 3.0$
		Maxflo T Wide-blade, high-efficiency impeller	Blending, Transitional flow, Simultaneous gas dispersion and solid suspension (like mining),	N_p and N_q vary with tip angle and number of blades
	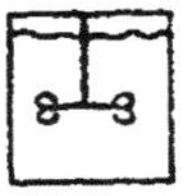	ChemShear Narrow-blade turbine	Liquid–liquid dispersion, Solid–liquid dispersion, Local shear	
		D-6 Flat-blade disc turbine (Rushton turbine)	Gas dispersion, low and intermediate gas flows, Liquid–liquid dispersion; $N_p = 5.5$, $N_p = 0.75$	
		CD-6 Concave-blade disc turbine (Smithturbine)	Gas dispersion, intermediate and high gas flows	
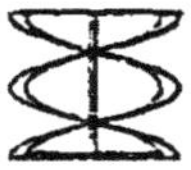		Helical ribbon (Double flight shown)	Blending and heat transfer in viscous media ($\mu > 50$ Pa s or $N_{Re} < 100$) $-N_p = 350/N_{Re}, N_{Re} < 100$)	
		Anchor	Heat transfer in viscous media $N_p = 400/N_{Re}$, $N_{Re} < 10$	
		CD-6 / HE-3 / P-4	Gas dispersion and blending for tall reactors Fermentations (food products, pharmaceuticals)	
		CD-6 / HE-3	Combined gas-dispersion, blending, and material drawdown (corn wet milling)	
		Side-entering wide blade impeller (HE3-S or Mark II)	Oilstorage, Paper pulp, Waste water circulation, Flue gas desulphurisation	

Figure 12-4 Impeller flow patterns and applications. (From K. Myers, et al., *Chemical Engineering*, Oct. 10, 1996.)

TABLE 12-3 Pumping Number Under Turbulent Conditions for Various Impellers

Impeller Type	N_Q
Propeller	0.4–0.6
Pitched-blade turbine	0.79
Hydrofoil impellers	0.55–0.73
Retreat curve blade	0.3
Flat-blade turbine	0.7
Dist flat-blade turbine (Rushton)	0.72
Hollow-blade turbine (Smith)	0.76

Source: [7].

diameter, speed of rotation, and the *flow number*, which is also termed as the *pumping number*. The flow number is defined as

$$N_Q = \frac{Q_P}{n D_a^3} \tag{12.4}$$

where Q_P is the effective pumping capacity (m^3/s). The pumping number is used to define the pumping rate of an impeller. For most impellers operating in the turbulent regime, the flow number varies in the range 0.4 to 0.8. The values of N_Q in the turbulent regime for commonly used impellers are stated in Table 12-3.

12.2.4.3 Shear The relative motion of the liquid layers within a mixing vessel results in shearing forces that are related to the flow velocities. These forces, represented by shear stress, carry out the mixing process and are responsible for producing fluid intermixing, dispersing gas bubbles, and stretching or breaking liquid drops [7]. The fluid shear stress is the multiplication of fluid shear rate and fluid viscosity. The pumping capacity of the impeller is important in establishing the shear rate due to the flow of the fluid from the impeller. Understanding the location and magnitude of shear generated by an impeller in an agitated vessel has significant implications for design. Most axial flow impellers are low shear and have high pumping efficiencies. On the other hand, radial flow impellers provide high shear but are low pumping.

12.2.5 Liquid Agitation Equipment

Agitation equipment typically consists of a vessel and a rotating mixing element, termed an *agitator*, along with an agitator drive system. The vessels can be of various geometrical shapes and sizes. Although most liquid agitation equipment consists of vertical cylindrical vessels, horizontal cylindrical vessels and tanks with rectangular cross sections may be used. The agitator has several components: agitator shaft, impeller hubs, and impellers. Different types of impellers are available, depending on the specific process requirements. The agitator drive system consists of motor, gearbox, shaft couplings, bearings with housings, agitator seals, or stuffing boxes. These components are discussed in detail in Section 12.5. Agitation equipment may be provided with baffles to prevent vortex formation, external jackets or coils for heat transfer, thermowells for temperature measurement, and so on.

In most vertical vessels, the agitator is mounted from the top of the vessel. However, in tall vessels, the agitator may be installed from the bottom, to reduce the length of the agitator shaft and thus provide mechanical stability. For small, non-baffled vessels, the agitator may be inserted from the top at an angle. In large storage tanks, the mixers can be side entering. In horizontal cylindrical vessels, the mixer can be installed from the top or the side. Portable mixers may be used for small mixing volumes. Besides the conventional mechanical agitators, mixing of liquids can also be carried out using jet mixers and motionless mixers. The following commonly used liquid agitation equipment are discussed below: top-entering mixers, side-entering mixers, portable mixers, jet mixers, and motionless mixers.

12.2.5.1 Top-Entering Mixers The components of the commonly used top-entering mixers consist of the following: (1) vessel, (2) baffles, (3) draft tubes, (4) heat-transfer surfaces, and (5) impellers. Figure 12-5a illustrates the standard components of top-entering mixers. Figure 12-5b and 12-5c show the general assembly and internal view of a liquid agitator assembly, respectively.

12.2.5.1.1 Vessel A vertical cylindrical vessel with a liquid height that is equal to the tank diameter is commonly used. The top of the vessel may be provided with flat cover or may have dished end, depending on the operating parameters of the vessel. The vessel may have tank bottom that is flat or dished. The choice of dished heads depends on the operating pressure of the vessel and the process parameters. Large tanks typically used for petroleum storage have flat bottoms or are provided with shallow cone bottoms. Flat bottoms are not preferred for solid suspensions, as solids tend to accumulate in the corners. Dished bottoms are used for such applications. Dished bottom heads can be 2 : 1 ellipsoidal, torispherical, hemispherical, or conical, depending on the application. Vessels with deep cone bottom may be used in some applications. For deep cone bottoms, special impellers with profiles that confirm to the cone geometries are positioned near the bottom of the cone to provide agitation at low liquid levels.

In addition to the agitator mounting arrangement, the vessel top cover or dished end are generally provided with nozzles for feeding the different process materials

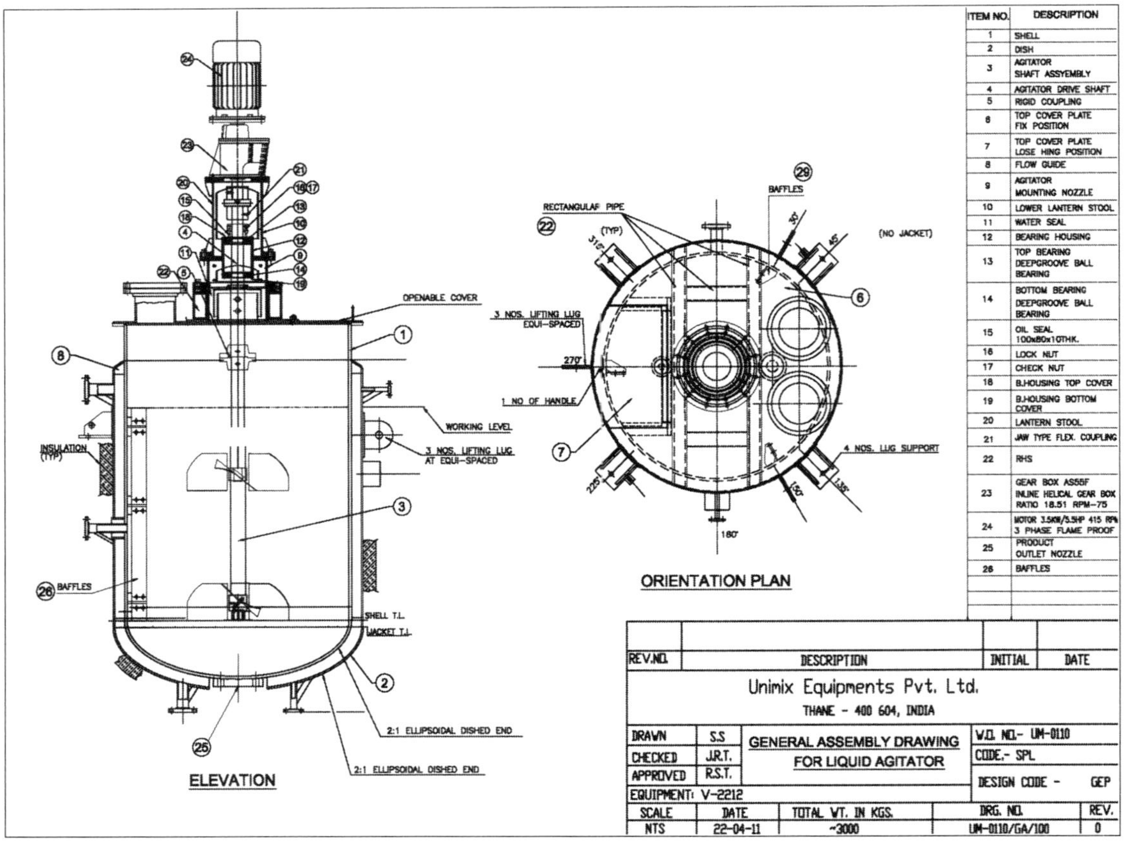

(a)

(b)

(c)

Figure 12-5 (a) Agitator vessel components. (Courtesy of Unimix Equipments.) (b) Liquid agitator assembly; (c) vessel internal view. (Courtesy of Unique Mixers and Furnaces.)

or for mounting measurement instruments, such as level transmitters. For large tanks, a manhole may be provided on the top cover to facilitate entering the vessel for installation and maintenance activities. For batch processes the material inlet nozzle from the top should be directed at an active surface away from the vessel wall and the shaft of the impeller. When charging solids into the liquid, the rate of addition of solids should be controlled to match the rate of solid wetting and incorporation and dispersion by the mixer.

Similarly, the bottom dished end of the vessel may be provided with material discharge nozzles, sampling nozzles, and nozzles for mounting instrumentation. The location, size, orientation, and design of these nozzles should be considered carefully. The discharge nozzle is located on the bottom head either on- or off-center, depending on the process requirements. In some cases, the outlet is located on the side near the tank bottom. For mixers handling solid materials, the bottom discharge nozzle may get plugged. Plugging can be avoided by the use of flush bottom valves. In some cases a small impeller is provided near the bottom of the vessel to eliminate this problem and to facilitate mixing at low liquid levels.

In continuously operated mixing vessels, the inlet and outlet positions should be located away from each other to avoid short circuiting the process materials to be mixed. For gas–liquid contacting applications, vessels may be provided with gas spargers for efficient introduction of gas into the liquid. Agitator vessels should be designed for the operating temperature and pressure conditions. The thickness of the vessel shell and dished ends should be calculated using the relevant pressure vessel design codes.

12.2.5.1.2 Baffles In a cylindrical vertical vessel as described above, stirring of low-viscosity liquids in the transitional and turbulent regimes using an axially mounted agitator results in a swirling flow pattern. The centrifugal force acting on the rotating liquid leads to vortex formation as depicted in Fig. 12-3. This may result in severe air entrapment, leading to poor mixing and unbalanced fluid forces acting on the impeller shaft.

Baffles are installed on agitator vessels to produce a flow pattern conducive to good mixing and to prevent vortex formation. In standard agitation equipment configurations, four vertical baffles are provided, each of which has a width of $\frac{1}{10}$ or $\frac{1}{12}$ of the tank diameter. The width of the baffles may be smaller, depending on the application. When solids are present or when a heat-transfer jacket is provided, the wall baffles are offset from the vessel wall by a distance equal to $\frac{1}{3}$ to $\frac{1}{6}$ the width of the baffle.

When the mixer shaft is located off-center, the swirling is less. In such cases, baffles may not be required, especially for low-viscosity liquids. For laminar mixing of viscous fluids, baffles are not required. Similarly, in square and rectangular tanks, there is considerable baffling action due to the vessel corners, which tend to break the tangential flow pattern, eliminating the need for wall baffles. Large storage tanks provided with side-entering mixers do not require baffles. Baffles increase the power consumption of the mixer but in turn improve process performance. The effects of installing baffles must therefore be considered carefully.

12.2.5.1.3 Draft Tubes A draft tube is a cylindrical duct slightly larger than the impeller diameter and is positioned around the impeller. It is used to direct flow to and from the impeller. The height of a draft tube may be a little more than the impeller diameter or may extend the full depth of the liquid, depending on the flow pattern that is required. These are generally used with axial impellers to direct the suction and discharge flows. The impeller draft tube system acts as a low-efficiency axial flow pump. The top-to-bottom circulation flow is of significance for a flow-controlled process, suspension of solids, and dispersion of gases. Draft tubes reduce the standard deviations in process variables, such as concentration, density, and viscosity. They are particularly useful in tall vessels that have high ratio of height to diameter.

12.2.5.1.4 Heat-Transfer Surfaces For applications that require heating or cooling of process fluids, the mixing vessels are provided with internal or external heat-transfer surfaces through which the heat-transfer media is circulated. The turbulence created by the action of the impeller improves the heat-transfer coefficient. The commonly used heat-transfer surfaces are external jackets, internal helical coils, and internal baffled coils.

1. *External jacketed vessels*. External jackets may be in the form of a cylindrical shell provided with flow guides between the external surface of the vessel and the internal surface of the jacket. Other jacket constructions include a half-pipe design and dimpled construction jackets.
2. *Internal helical coils*. Helical coils provide a large heat-transfer surface area. They are, however, expensive to manufacture, clean, and maintain.
3. *Internal baffled coils*. In internal baffled coils, vertical tubes are placed within the vessel, which in addition to heat transfer provide a baffle effect. Positioning of internal coils should be carried out carefully so that they do not obstruct the discharge flow of the impeller.

The heat-transfer coeffecients for the various constructions of heat-transfer surfaces can be estimated using correlations established by different researchers. Heat transfer for an agitated vessel is dependent on the following:

- The overall heat-transfer coefficient
- The surface area available for heat transfer

- The temperature difference between the heat-transfer fluid and the process fluid

12.2.5.1.5 Impellers Agitation plays a significant role in the success of many process operations. There is a wide range of impellers and impeller designs. Selection of the most suitable impeller for a given application should be based on an understanding of the process objectives and the physical properties of the material. The liquid viscosity plays a vital role in the selection of impellers in laminar, transitional, and turbulent operations. Based on the liquid viscosity, impellers can be classified as turbines for low-viscosity fluids and close-clearance impellers for high-viscosity fluids. Depending on the flow patterns developed by the mixing impellers, they are classified as axial flow impellers and radial flow impellers. Some process applications require low shear, while others demand high shearing action. Impeller designs may also be classified based on the amount of shear that they produce. The following commonly used impellers are discussed below:

- Axial flow impellers
  - Marine propellers
  - Pitched-bade turbines
  - Hydrofoil impellers
- Radial flow impellers
  - Rushton turbines
  - Bar turbines
  - Open-blade turbines
  - Coil or spring impellers
- Low-clearance impellers
- High-shear impellers

Axial Flow Impellers In axial flow impellers, the impeller blade makes an angle of less than 90° with the plane of impeller rotation. As a result, the locus of the flow occurs along the axis of the impeller, parallel to the impeller shaft. Axial flow impellers are used for blending, solid suspensions, solid incorporation or drawdown, gas inducement, and heat-transfer applications. The commonly used axial flow impellers for transitional and turbulent flow applications include marine propellers, pitched-blade turbines, and hydrofoil impellers.

Marine Propellers A propeller is a high-speed axial flow impeller generally used for liquids of low viscosity. The propeller speed ranges from 400 to 1800 rpm, depending on propeller diameter. For top-entering mixers the propeller size is generally limited to 450 mm [6]. A revolving propeller traces out a helix in the fluid. When operated in a theoretical environment where there is no slippage between the fluid and the propeller, one full revolution would move the liquid longitudinally a fixed distance depending on the inclination of the propeller blades. The ratio of this distance to the diameter of the propeller is known as the *pitch* of the propeller. Propellers are available with 1.0 pitch ratio; these are often referred to as *square pitch*. Propellers can be designed with a different pitch to change the pumping rate and thrust.

The propeller is generally operated to produce down-pumping action directed toward the bottom of the vessel, until deflected by the floor or the wall of the vessel. The axial flow pattern of a marine propeller is shown in Fig. 12-6. The highly turbulent swirling column of the liquid leaving the impeller entrains stagnant liquid as it moves along. The propeller is designed to operate such that the direction of rotation makes the trailing edge the one with the smaller radius of curvature. Therefore, if the propeller is operated in the reverse direction, its efficiency is reduced by 15% [6]. The shearing action of the propeller is significant at high speeds. However, the propeller is generally not used for shearing applications.

Standard-blade marine propellers with square pitch are commonly used. Variations include four-bladed propellers, propellers with sawtooth edges for tearing action and

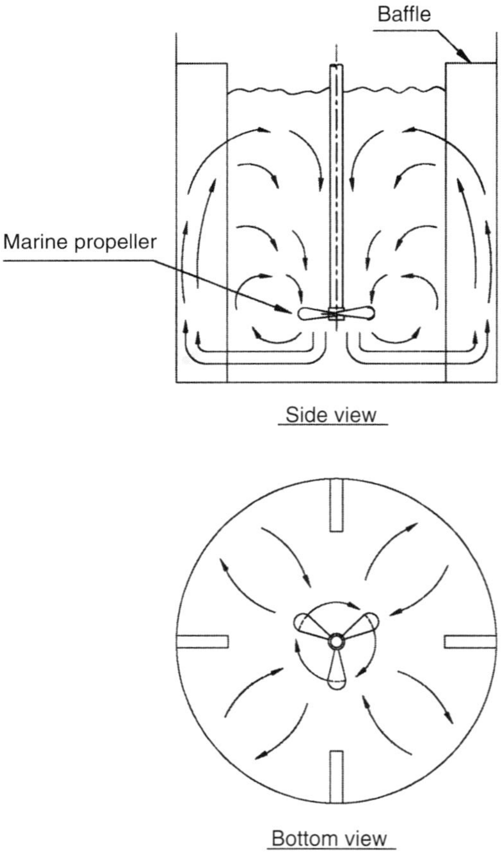

Figure 12-6 Fluid flow pattern of axial flow marine propeller. The impeller is mounted at the center of a baffled vessel.

Figure 12-7 Three-blade marine propeller. (Courtesy of Unimix Equipments.)

perforated blades for shredding and breaking of lumps. A three-blade marine propeller is shown in Fig. 12-7.

Marine propellers are also used on side-entering mixers and are mounted with the impeller shaft inclined at an angle with respect to the vessel centerline, for improving process results. Side-entering propellers have diameters ranging from 250 to 850 mm with a motor horsepower up to 75. The power number for top-entering propellers and portable mixers in the turbulent range at a proximity factor of 1 is 0.87. For side-entering mixers the power number ranges from 0.77 for propeller diameters up to 700 mm [6]. Propellers are generally manufactured by the casting process.

Pitched-Blade Turbines Pitched-blade turbines are axial flow impellers used in top-entering agitators. A pitched-blade turbine consists of a hub on which an even number of blades are mounted at an angle with respect to the horizontal. The most commonly used impeller of this type is the four-bladed 45° pitched-blade turbine. The pitched-blade turbine is generally used for impeller diameters ranging from 450 mm to 3 m with connected power of up to 500 hp. It is used primarily in flow control applications. Figure 12-8 shows the commonly used 45° pitched-blade turbine.

Since the fluid discharge from a pitched-blade turbine has both axial and radial components, it is considered a mixed-flow impeller. In most applications, the direction of rotation of the impeller is such that the liquid flow is directed toward the bottom of the vessel; however, in applications such as gas dispersions and floating solids mixing, directing the liquid flow upward may be more effective. The 45° of the pitched blade causes shear to the

Figure 12-8 45° pitched-blade turbine impeller.

process due to the separated flow off the suction side of the blade. The power number for a four-bladed pitched-blade turbine impeller in the turbulent regime is 1.27 [7]. The pitched-blade turbine impeller is generally fabricated. The blades may be welded directly onto the impeller shaft or are bolted to the shaft hub.

Hydrofoil Impellers Hydrofoil impellers are also referred to as *high-efficiency impellers* since they are designed to maximize fluid flow and minimize shear rate. Since most industrial applications require high pumping capacity and less fluid shear, these impellers are popular in the majority of mixer installations. The hydrofoil impeller has three or four tapering twisted blades, cambered, and sometimes provided with rounded leading edges. The blade angle at the tip is lower than that at the hub, resulting in almost constant pitch across the length of the blade. Figure 12-9

Figure 12-9 Hydrofoil impeller. (Courtesy of Unimix Equipments.)

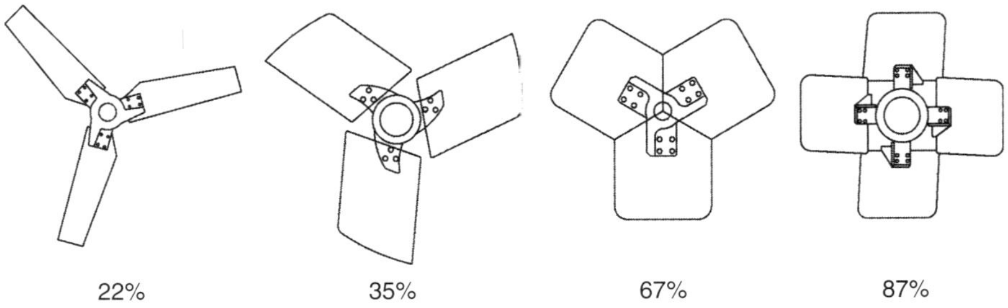

Figure 12-10 Hydrofoil impeller with various solidity ratios. (Courtesy of Unimix Equipments.)

shows a high-solidity-ratio hydrofoil impeller used for high-viscosity materials and gas applications. This produces a more uniform velocity across the entire discharge area. The resulting flow is more streamlined in the direction of pumping, and the vortex formations around the impeller are much lower than in a pitched-blade turbine. As a result, this blade has a lower power number and a higher flow for unit power compared to a pitched-blade turbine.

Advances in the hydrofoil impeller designs have lead to the development of an additional family of these impellers. Figure 12-10 illustrates the four types of these impellers with different *solidity ratios*, the ratio of the total blade area to that of a circle circumscribing the impeller [8]. Low-solidity-ratio impellers are efficient for liquid blending and solid suspensions. Higher-solidity-ratio impellers are more effective for axial flow patterns as the viscosity increases. For gas–liquid dispersions, higher-solidity-ratio impellers with wide blades provide an effective area of preventing bypass of gas through the impeller hub. The power number for the high-efficiency impeller ranges from 0.3 to 0.75, depending on the solidity ratio [7].

Radial Flow Impellers In radial flow impellers, the impeller blade is parallel to the axis of the impeller. As a result, the radial flow impeller discharges flow along the impeller radius in a distinct pattern, as shown in Fig. 12-11. Radial flow impellers are used for single- and multiphase mixing applications. They are, however, more effective for gas–liquid and liquid–liquid dispersions. With suitable baffles, the radial flow pattern is converted to strong top-to-bottom flows above and below the impeller.

Radial flow impellers may either have a disk or be open. The blades may be flat or curved. Commonly used radial flow impellers in the transient and turbulent regimes include the Rushton turbine, bar turbine, and open-blade turbine. Equipment manufacturers have developed other impeller geometries, which are variants of these impellers. Among the most recent developments are the hollow-blade impellers and the coil or spring impellers.

Rushton Turbines The Rushton turbine is a disk-type (six-blade turbine) radial flow impeller. The impeller is constructed with six vertical blades located around the external periphery of a circular disk. The diameter of the disk ranges from 66 to 75% of the internal vessel diameter. The length of the blade is 25% and its width is about 20% of the vessel diameter. A six-blade Rushton turbine is shown in Fig. 12-12.

This Rushton turbine is used in applications requiring high shear or turbulence. The impeller design is best suited for gas–liquid contacting because of the circular disk. Gas

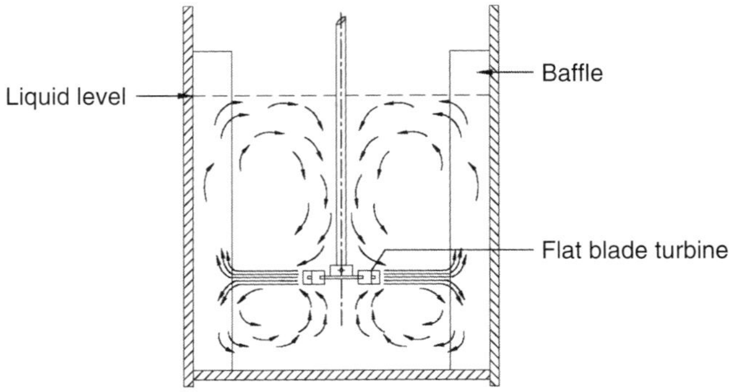

Figure 12-11 Radial flow pattern produced by a flat-blade turbine.

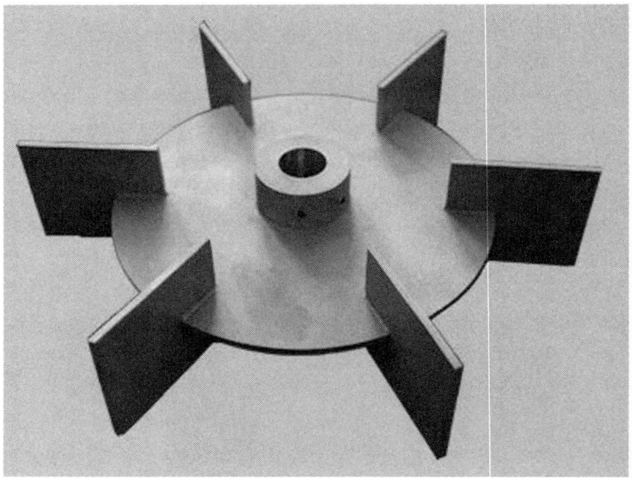

Figure 12-12 Six-blade Rushton turbine impeller. (Courtesy of Unimix Equipments.)

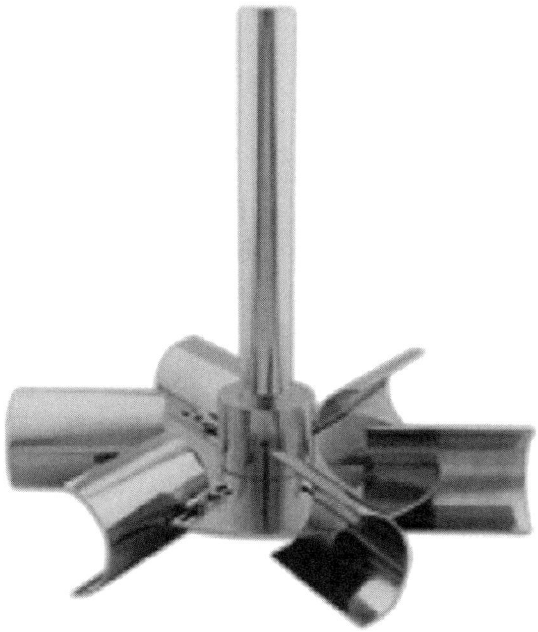

Figure 12-13 Smith impeller.

is introduced through a sparger below the impeller; the disk directs the gas along a path of maximum liquid contact and prevents the gas from taking the direct vertical route along the mixer shaft, which would otherwise result in minimum contact. The disk also prevents gas bubbles from passing through the low shear zone around the impeller hub. Disk-type radial impellers such as the Rushton turbine tend to pump in a more radial direction, because of which they tend to draw more power than do open flat-blade turbine impellers. The power number for the Rushton impeller in the turbulent regime is 5.75 [6].

A variant of the Rushton turbine is the Smith impeller, in which the impeller blades are semicircular or parabolic instead of flat. For gas dispersion, this shape allows much higher power levels to be obtained in the process than with the Rushton turbine. A Smith impeller is shown in Fig. 12-13.

Another variant is the backswept with disk impeller. The backswept nature of the blade prevents material buildup on the surface of the blades. This design is less susceptible to erosion. The backswept turbine is used for fiber processing in the pulp and paper industries.

Bar Turbines The bar turbine produces the highest shear rate among the axial and radial flow impellers discussed. The blades of a bar turbine are made from square bar stock. The bars are welded onto the flat disk. This type of impeller runs at high speed and consumes almost 10 times less power than the Rushton turbine. The power number for the bar turbine in the turbulent regime is 0.61 [6].

Figure 12-14 Flat-blade turbine impeller. (Courtesy of Unimix Equipments.)

Open-Blade Turbines Open-blade impellers like the flat-blade turbine (also known as a *paddle impeller*) and backswept open turbine are radial flow impellers. Open impeller turbines (without a disk) do not pump in true radial directions because of the pressure difference between each side of the impeller.

In open turbines, the impeller blades are mounted directly on the hub. The number of blades may be two, four, six, or eight. The use of large-diameter radial flow impellers is best typified in the two-blade paddle, which is generally used for solid suspension or blending applications that require high flow and low shear. The paddles are normally operated at low speeds. Low speed is especially necessary with two-blade impellers, which are mechanically more unstable than four-, six-, and eight-blade impellers. A flat-blade turbine impeller with six blades is shown in Fig. 12-14.

Backswept open impellers are effective for fibrous material as they have lower starting torques than flat blades. The lower starting torque is important during startup of settled solids.

Coil or Spring Impellers Among the most recent developments in radial flow impellers is the coil or spring impeller. These impellers were developed for applications where solids frequently settle at the bottom of a vessel. The spring design ensures that the impeller has adequate mechanical rigidity and strength to overcome the resistance offered by stiff solids during mixing operation. The coil impeller is shown in Fig. 12-15.

Low-Clearance Impellers The anchor impeller and the helical impeller are the two commonly used close-clearance impellers. As the name suggests, these impellers operate with close clearances between the impeller and the internal wall of the vessel. The diameter of close-clearance impellers is typically 90 to 95% of the inside diameter of the

Figure 12-15 Coil impeller.

Figure 12-16 Sawtooth impeller. (Courtesy of Unique Mixers and Furnaces.)

vessel. The shear near the vessel wall reduces the buildup of stagnant material and promotes treat transfer.

The anchor impeller is a radial flow impeller, whereas the helical impeller provides axial discharge of material by producing strong top-to-bottom motion even with viscous materials. Because of the limitations of the anchor impeller to provide top-to-bottom fluid motion, anchor blades may be used in combination with other types of impellers, such as high-shear impellers, pitched-blade turbines, and flat-blade turbines. Details of the anchor and helical impellers are discussed in Section 12.4.

High-Shear Impellers High-shear impellers are used in applications that require a high shearing action, as in grinding, dispersing pigments, and making emulsions. The bar turbine impeller discussed earlier is in the lower end of the high-shear range of impellers. High-shear impellers are

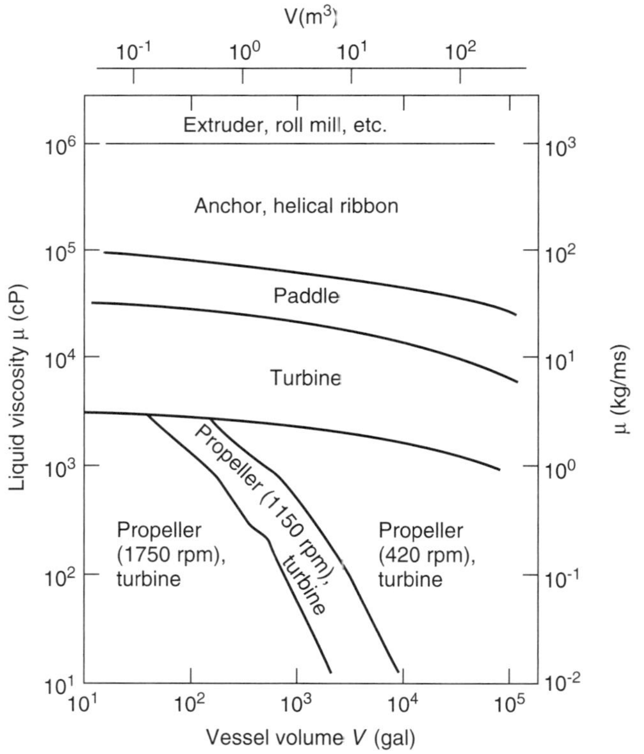

(a)

Figure 12-17 (a) Impeller selection. (From W. R. Penny, *Chemical Engineering*, vol.77, no.12, 1970, p. 171.) (b) General selection chart for mixing. (From E. J. Lyons, and N. H. Parket, *Chemical Engineering Progress*, vol. 50, 1954, p. 629.)

Selection chart				Shape relationships for turbine designs		
Service	Mixing device	Range	Criteria	Tank diameter to impeller diameter ratio	Tank height to diameter ratio	Impellers and position
Blending	Turbine Propeller Paddle Tank vol.	→ 1,000,000 gal	1. Volume circulation	3:1 to 6:1	Unlimited	Single or multiple
Dispersion (immiscible systems)	Turbine Propeller Paddle Flow	→ 1,000 gal/min.	1.Drop size control 2.Re-circulation	3.0:1 to 3.5:1	1:1 1:2 in staged mixers	At/or below center line of liquid charge
Reactions in solution (miscible systems)	Turbine Propeller Paddle Charge vol.	→ 20,000 gal	1.Intensity 2.Volume circulation	2.5:1 to 3.5:1	1:1 to 3:1	Single or multiple
Dissolution	Turbine Propeller Paddle Charge vol.	→ 10,000 gal	1.Shear 2.Volume circulation	1.6:1 to 3.2:1	1:2 to 2:1	At/or below center line of liquid charge
Solids suspension	Turbine Propeller Paddle % Solids	→ 100 %°	1.Circulatiion 2.Velocity	2.0:1 to 3.5:1	1:1 to 1:2	Depending on particle size 1. Impeller Diameter off bottom 2. On bottom
Gas applications	Turbine Propeller Paddle Gas vol.	→ 5,000 ft^3/min.	1.Controlled shear 2.Circulation 3.High velocity	2.5:1 to 4.0:1	4:1 to 1:1	1. Multiple- lowest one impeller diameter off bottom 2. Self- induce, just below liquid level
High viscosity applications	Turbine Propeller Paddle Viscosity	→ 1,000,000 cPs	1.Volume circulation 2.Low velocity	1.5:1 to 2.5:1	1:2 to 2:1	Single or multiple
Heat transfer	Turbine Propeller Paddle Charge vol.	→ 20,000 gal	1.Volume circulation 2.High velocity arcoss transfer surface	Related to other services	Depends on other services being performed	Single or multiple, impeller Opposite transfer surface when using coils
Crystallization or precipitation	Turbine Propeller Paddle Charge vol.	→ 20,000 gal	1.Circulation 2.Low velocity 3.Shear control	2.0:1 to 3.2:1	2:1 to 1:1	Single at/or below center line of liquid charge

(b)

Figure 12-17 (*Continued*)

operated at high speeds and are generally used for addition of the second phase. Another commonly used high-shear impeller is the sawtooth impeller, which consists of a flat disk having sawtooth edges on the external periphery (see Fig. 12-16). This type of impeller generates high turbulence in the area around the impeller. This impeller has a power number of 0.45 [7] and is operated at extremely high speeds to provide the desired dispersion. A star-shaped impeller that has tapered blades provides intermediate shear levels. This type of impeller is used in polymerization reactors.

Selection of Impellers In addition to the different types of impellers discussed above, equipment manufacturers offer a wide range of custom-made impellers for specific applications and process functions. The process functions together with the physical properties of the material, such as viscosity, play an important role in the selection of impellers in the laminar, transient, and turbulent flow regimes. Impeller selection also depends to some degree on the size of the mixing equipment.

Figure 12-17a provides guidelines for impeller selection. Figure 12-17b provides another selection chart and defines selection based on process function. Since it is impossible to document all kinds of fluid applications, manufacturers' advice on the selection of impeller type is recommended.

12.2.5.1.6 Standard Vessel Geometry and Impeller Clearances Engineers designing agitated vessels have an unusually large number of choices to make with respect to the type and location of impeller, the proportions of the vessel, the number and proportions of the impellers, baffles, and so on. Each of these parameters affects the circulation rate of the liquid, the velocity patterns, and the power consumed. As a starting point for design in ordinary agitation problems, typical proportions are as shown in Fig. 12-18.

$$H/D_t = 1$$
$$E/D_t = 1/3$$
$$J/D_t = 1/12$$
$$S/D_t = 1/72$$
$$G/D_a = 1$$

where D_a = diameter of agitator (m)
D_t = vessel diameter (m)
H = liquid height (m)
J = baffle width (m)
S = Baffle clearance from vessel side wall (m)
E = impeller clearance from bottom of the vessel (m)
W = Spacing between impellers (m)

The ratio of the impeller diameter to the vessel diameter shall vary depending on the type of impeller. The recommended ratios are as follows:

Impeller	D_a/D_t
Propeller	0.25–0.33
Pitched-blade turbine	0.5–0.7
Hydrofoil	0.4–0.5
Rushton turbine	0.66–0.75
Bar turbine	0.66–0.75
Open-blade turbine (paddle)	0.5–0.7
Anchor, helical	0.9–0.95

The number of baffles is usually four; these should be equally spaced and should be located at a minimum distance of $D_t/72$ from the vessel wall. Special situations may, however, demand proportions different from those recommended above.

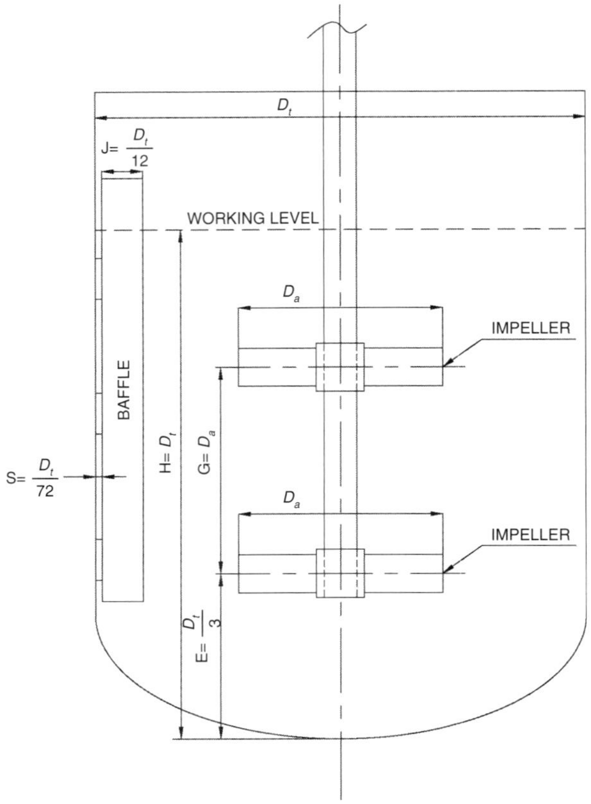

Figure 12-18 Standard vessel geometry for agitator vessel.

Impeller clearance from the bottom of the vessel and impeller spacing for multiple-impeller systems can have a significant impact on the power number. Moreover, these parameters influence the flow patterns and the process results. If an impeller is located very close to the tank bottom, down-pumping axial flow impellers provide flow patterns similar to those of radial flow impellers. This can lead to reduced pumping and higher shear, which may be undesirable. In tall agitated vessels, multiple impellers are installed to improve circulation and narrow the distribution of shear and energy dissipation. Generally, these impellers are spaced away from each other by a distance equal to the impeller diameter. General guidelines for impeller clearance and impeller spacing for different mixing processes are given in Table 12-4. In addition, there should sufficient void space above the liquid level (in stagnant condition). The listed "standard" proportions are widely accepted, and are the basis of many published correlations of agitator performance.

12.2.5.2 Side-Entering Mixers Side-entering mixers are mounted at an angle of 7 to 10° to the centerline of the

TABLE 12-4 Recommended Impeller Clearance and Spacing

Mixing System	Maximum Liquid Height, H/D_t	Number of Impellers	Impeller Elevation from Tank Bottom: Bottom	Top
Liquid blending	1.4	1	$H/3$	
	2.1	2	$D_t/3$	$2H/3$
Solids suspension	1.2	1	$H/4$	
	1.8	2	$D_t/4$	$2H/3$
Gas dispersion	1.0	1	$D_t/6$	
	1.8	2	$D_t/6$	$2H/3$

Source: [7].

horizontal agitator mounting nozzle. Power levels for side-entering mixers are low: on the order of 0.01 kW/m^3 [6]. The mixer drives are generally limited to 60 hp for gear-driven and 100 hp for belt-driven systems, although larger mixers have occasionally been used [7]. Because of their low installation cost and easy installation, side-entering mixers are often preferred for very large tanks used for storing petroleum, crude oil, and gasoline. These mixers usually operate at an output speed of 400 rpm. In extremely large tanks, multiple side-entering units can be installed. As with portable mixers, side-entering mixers do not require baffles, but correct positioning of the impeller is absolutely necessary. The mixing efficiency of side-entering mixers is low compared to top-entering mixers, and hence these have limited use. Side-entering mixers are generally used for stratified blending in vessels used for long-term storage.

12.2.5.3 Portable Mixers Portable mixers are so called because of the small size, which makes them easy to mount on vessels or drums that may not require agitation at all times. The propeller impeller with small diameter and high speed results in low torque, making them good for use in portable mixers. Portable mixers are mounted using a clamp from the rim of the tank, with an adjustment that allows the mixer shaft to be set at an angle of 10 to 15° from the vertical. Portable mixers are generally provided with a ball-and-socket arrangement so that they can be swiveled to locate the mixer shaft off-center. Baffles are not required with portable mixers.

12.2.5.4 Jet Mixers In jet mixers, the mechanical energy required for mixing of fluids is imparted through high-velocity jets. Jet mixers are driven by external pumps located outside the vessel. The jets usually enter from the side of the vessel closer to the base and are directed in the diagonal direction toward the opposite top corner. The liquid jet entrains and mixes the surrounding fluid using the mechanical energy supplied by the pump. Single or multiple jets may be provided, depending on the application and the size of the vessel. Multiple-point inlets with multiple jets can also be provided in large vessels, which can lead to reduction in the blending time compared with a single-source inlet. The piping from the pump to the vessel is carried out based on the number and position of jet streams. Jet mixers are commonly used in large storage vessels to maintain homogeneity of the liquid stored. The choice of jet mixers and their configuration should be made after careful consideration of the capital cost and the operating cost of these mixers, in comparison to mechanically agitated mixers.

12.2.5.5 Motionless Mixers Motionless or static mixers use stationary elements of various profiles, geometries that are placed inside pipes or conduits. The material to be mixed is pumped through this pipe, where mixing occurs through successive diversions and recombinations of the process fluid. The length of the pipe may be as short as 5 to 10 diameters, but usually extend to 50 to 100 diameters. Besides liquid blending applications, static mixers are used for mixing gases, dispersion of gases into liquids, pH control, dispersion of dyes and for mixing solids in viscous liquids. These units perform exceptionally well in mixing of molten polymers.

The static mixer consists of a series of helical elements twisted alternately in opposite directions. For n elements there are 2^n divisions and recombinations. For a mixer with 20 elements, the number of combinations would be over 1 million. As a result, even at high viscosities, mixing becomes effective only with a few elements. Figure 12-19 shows a static mixer, flow through the mixer, and the flow division in the mixer.

The power required to accomplish mixing in a motionless mixer is delivered by the fluid pump. The pressure drop depends on the design of the static elements, the number of elements, the total length of the pipe, and the properties of the flowing fluid. The calculation of pressure drop should be carried out with the assistance of the manufacturer of the static elements.

Motionless mixers provide good heat transfer since they interchange fluid continuously between the walls and the center of the conduit. Since the residence time in these units can be adjusted, they are suitable for chemical reactions.

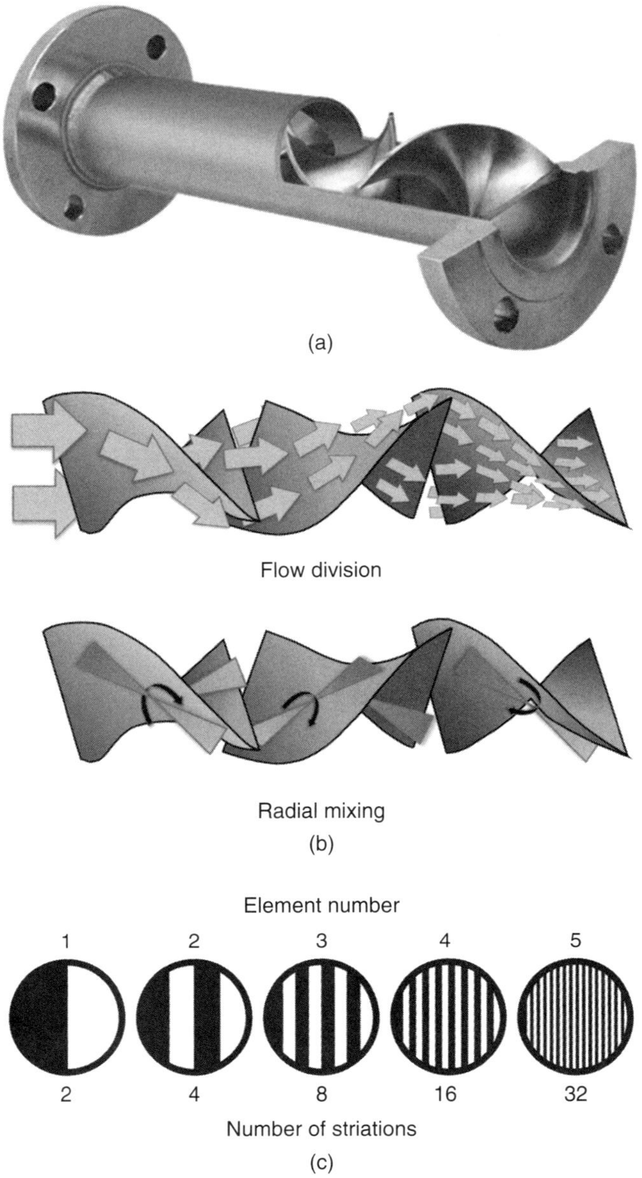

Figure 12-19 (a) Static mixer. (Courtesy of Chemineer Inc.) (b) Flow pattern through static mixing elements; (c) flow division in a static mixer. (From Wikipedia.)

12.3 SOLID BLENDING

Blending of bulk solids occurs frequently in many industrial processes. Chemical process industries involve solid mixing of chemicals, ceramics, fertilizers, powdered detergents, and so on. In the pharmaceutical industry, small amounts of a powdered active ingredient are precisely blended with such excipients as sugar, starch, cellulose, lactose, and lubricants. Similarly, most powdered food products, such as soft-drink premixes, food flavors, and instant foods, are produced from custom-mixed batches. According to available statistics, worldwide production annually accounts for over 1 trillion kilograms of granular and powdered products, much of which must be blended uniformly to meet quality and performance goals [7].

Mixing takes place in low-viscosity liquids, due to the creation of flow currents that transport unmixed material to the mixing zone adjacent to the impeller. Liquid mixing has been a subject of extensive research and is therefore well documented. Advances in fluid mixing technology have made it possible to produce a truly homogeneous liquid mix with uniform composition. To some extent, blending of free-flowing solids resembles the mixing of low-viscosity liquids. However, there are significant differences between the two processes. In the case of solids, there are no flow currents. Therefore, solid material cannot attain the perfect mixing that is possible with liquids.

The competitive nature of modern business demands larger batch sizes, faster blend times, lower production costs, energy conservation, and yet a sufficiently homogeneous blend that does not segregate in subsequent product-handling steps. It is therefore essential for engineers to gain adequate understanding of solid blending theory and equipment.

12.3.1 Properties of Solids Affecting Blending

The properties of materials being mixed is one of the vital factors affecting the blending operation. A blend may consist of several different materials, each of which may differ in such properties as density, particle size, shape, surface characteristics, agglomeration, and cohesiveness. Blending is a dynamic state of an assemblage of particles. Therefore, the following properties of the unit particles must be accounted for when discussing the dynamics of blending.

1. *Bulk density*. Bulk density is defined as the mass of a material that occupies a specific volume. It includes not only particle mass but also the air entrained in the void spaces between the particles. It is generally measured in kg/m^3 or lb/ft^3.

2. *Particle size, particle shape, distribution*. Particle size and size distribution in powders have a considerable impact on the flow properties of powders. As a result, the dynamics of blending is affected by the size of particles and their distribution in bulk solids. Particle size is generally quantified in micrometers (microns) or as mesh size. Particle shape affects interparticle powder friction and thereby the flow and blending properties of the powder.

3. *Flowability*. Flowability is the ease with which a bulk material flows under the influence of gravity only. The *coefficient of friction* of a powder is the tangent of the angle of repose and is the measure of its flowability. Flowability of bulk solids depends on factors such as particle size and size distribution, particle shape, bulk density, and cohesiveness, all of which affect blending.

4. *Material composition*. Composition of unit particle is its quantitative and qualitative makeup. The chemical composition is important because chemical reactivity is a major factor in the choice of a particular substance for the application.

5. *Surface characteristics*. Surface characterizes include surface area and electrostatic charge on the particle surface. Smaller particles have a larger surface area which leads to the formation of weak polarizing electrical forces termed *van der Waals* forces. When electrostatic charge is generated due to friction between two surfaces, the electric charge generated is referred to as *triboelectric charge*.

6. *Moisture content in solids*. Increased surface exposure of fine particles to the atmosphere may result in moisture adsorption or absorption. Materials that naturally contain bound moisture or tend to adsorb or absorb moisture are termed *hygroscopic*.

In addition, material properties such as cohesiveness, adhesiveness, agglomeration, abrasiveness, explosiveness, friability, and temperature limitations of ingredients should be studied. The effects of each property must be considered individually, Moreover, their combined effect as a set or group of co-related variables must also be ascertained.

12.3.2 Types of Blend Structures

There are two types of blend structures: structured and random. Structured, ordered, or interactive blends are observed in most industrial processes. In structured blends, the different blend components interact with one another by physical, chemical, or molecular means or a combination of these resulting in agglomeration. The agglomerates formed thus comprise a uniform blend of smaller particles.

The agglomerates may be of uniform size or of variable size. Agglomerates of uniform size will not segregate during discharge from the blender, whereas those of different sizes will segregate. Segregation can lead to such problems as difference in bulk density and reactivity in post-blend processing. Fine materials in particular have a tendency to adhere only to themselves, without adhering to dissimilar components (e.g., carbon black, fumed silica). For blending of these materials, shear blending mechanisms are adopted.

When the different blending components do not adhere or bind to each other within the blender, the result is a random blend structure. In a random blend there is no bonding between the material, and individual particles are free to move relative to each other. Consequently, dissimilar particles readily segregate under the influence of external forces such as gravity and vibration and get collected in zones of similar particles (e.g., a blend of salt and pepper). Completely random blends are rarely encountered in industrial applications. Blender manufacturers need to account for these material behaviors during the selection and design of blending equipment.

12.3.3 Mechanisms of Solid Blending

The three primary mechanisms of blending are diffusion, convection, and shear. These three mechanisms occur to varying extents, depending on the types of mixers and blenders to be used and the characteristics of the solids to be blended.

1. *Diffusion blending*. Diffusion blending is characterized by small-scale random motion of solid particles. Blender movements increase the mobility of the individual particles and thus promote diffusive blending. Diffusion blending occurs where the particles are distributed over a freshly developed interface. In the absence of segregating effects, diffusive blending will in time lead to a high degree of homogeneity. Tumbler blenders such as double-cone blenders and v-blenders function by diffusion mixing.

2. *Convection blending*. Convection blending is characterized by large-scale random motion of solid particles. In convection blending, groups of particles are moved rapidly from one position to another due to the action of a mixing agitator or cascading of material within a tumbler blender. The blending of solids in ribbon blenders, paddle blenders, and plow mixers is mainly a result of convection mixing.

3. *Shear blending*. Some books define shear blending as the development of slip planes or shearing strains within a bed of material. Others define the blending mechanism of shear as high-intensity impact or splitting of the bed of material to disintegrate agglomerates or overcome cohesion. We shall use the latter definition. Shear blending is very effective at producing small-scale uniformity, generally on a localized basis. Blenders with high-speed chopper blades, and intensifiers, are examples of shear blending.

12.3.4 Segregation Mechanisms

Blending and segregation (demixing) are competing processes. *Segregation* is defined as the separation of particles into distinct zones based on particle size, shape, density, resiliency, or other physical attributes, such as static charge. Some of the mixing mechanisms can also result in segregation. It is therefore important that while selecting a mixer for a specific application, the potential segregation mechanisms are understood and problems are averted. Common particle segregation mechanisms include sifting, fluidization (air entrainment), and dusting (particle entrainment).

Sifting segregation occurs when fine particles concentrate in the center of a bin during filling while more coarse particles roll to the pile's periphery. In this case, smaller particles move through a matrix of larger ones. During discharge, the fine particles flow out first, followed by the coarse material.

Fluidization segregation occurs when the finer, lighter particles (generally smaller than 100 μm) rise to the upper surface of a fluidized blend of powder, while the larger, heavier particles concentrate at the bottom of the bed. The fluidizing air entrains the lower-permeability fines and carries them to the top surface. Fluidization segregation generally occurs when fine materials are conveyed pneumatically, when they are filled or discharged at high rates, or if gas counterflow occurs [4].

Dusting segregation is commonly encountered with fine pharmaceutical and food powders being discharged from blenders into drums, tableting press hoppers, and packaging equipment surge hoppers. In dusting segregation the fine particles get concentrated near container walls or at points farthest from the incoming stream of material.

Segregation of materials can be minimized or eliminated using one of the following approaches: (1) changing the properties of the process material, (2) changing the process, or (3) changing the equipment design. When designing equipment, the following general practices can help to prevent or minimize the effects of segregation [1]:

1. Solid blending should be located as far downstream in the process as possible.
2. Post-blend handling of the material should be minimized.
3. Surge and storage bins should be designed for mass flow (no stagnant regions in the hopper) [2].
4. Velocity gradients within bins should be minimized.
5. A mass-flow bin with a tall, narrow cylinder is preferred to a short, wide bin. Keeping the material level high in the bin is preferred.
6. Use venting to avoid air counterflow (inducing fluidization segregation).
7. Minimize the generation of dust.

12.3.5 Scale-Up of Solid Mixers

It is difficult to generalize scale-up criteria for solid mixers, due to the complexity of interaction of various factors involved (i.e., properties of solids, different types of mixer geometries, and velocity and stress profiles). The most commonly used approaches to scale-up solid mixers include:

- Maintaining constant tip speed (also known as peripheral speed)
- Maintaining the Froude number constant

$$\text{tip speed} = \pi D n \tag{12.5}$$

where, D is the agitator swing diameter (m) and n is the rotational speed of the mixer (rev/s). The Froude number (Fr) for solid mixing equipment can be defined as

$$\text{Fr} = \frac{v^2}{gR} = \frac{R\omega^2}{g} \tag{12.6}$$

where n is the rotational speed of the mixer (rev/s), v the tip speed of mixing (m/s), R the agitator swing radius (m), and ω the angular velocity (rad/s).

12.3.6 Solid Blending Equipment

Solid blending equipment is broadly classified into the following categories: (1) tumbler blenders, (2) convective blenders, (3) silo blenders, and (4) pneumatic blenders.

Next, we describe the design, construction, operation, and applications of solid blending equipment.

12.3.6.1 Tumbler Blenders Tumbler blenders are among the oldest type of blending equipment. Over the years, the geometry and design of these blenders have evolved from the primitive drum and cubical shapes to the presently used v-blenders and double-cone blenders. Tumbler blenders function mainly by diffusion mixing. The solid material to be blended is loaded into the blender container, which is rotated about a horizontal axis. Tumbler blenders rely on the action of gravity to cause the powder to cascade within the rotating vessel. Blender movements increase the mobility of the individual particles and thus promote diffusive blending. In the absence of segregating effects, the diffusive blending will in time lead to a high degree of homogeneity.

Mixing homogeneity of up to 98% and higher can be achieved using tumbler blenders. Blending efficiency is affected by the volume of the material loaded into the blender. While charging the material into a tumbler blender, top-to-bottom material loading is recommended because the ingredients cascade into one another, with diffusion occurring perpendicular to the main flow. Similarly, the speed of the blender rotation also affects the mixing efficiency. Tumbler blenders are preferred for precise blend compositions and for applications where some blend ingredients are present in quantities as low as 5% of the total blend.

There are two popular geometries of tumbler blenders: the v-blender and the double-cone blender. These blender configurations are symmetrical about the vertical axis passing through the centerline of the blender shell. Some manufacturers offer designs with minor changes in the geometry. For example, the slant cone blender is a variation derived from the double-cone blender. Similarly, the long-legged v-blender, which has one leg longer than the other, is a variant of the standard v-blender. Tumbler blenders are also available with blender containers in octagonal, cubical, or cylindrical shapes.

One of the most recent advances in tumbler blender technology is the tumble in-bin blenders. In this a storage container that holds the materials is also the blender shell. Materials to be blended are charged in the container and mounted on the blender drive system. After blending is completed, the container is moved to the storage area and another similar container is mounted on the drive system. Because of this flexible production system, wherein the transfer of blended material from blender to a storage container is eliminated, chances of segregation of blended material are eliminated. This type of blender is popular in the pharmaceutical, food, and powdered metal industries. Moreover, no cleaning between batches is required.

Tumbler blenders can be provided with various options, such as:

- High-speed intensifier bars (or lump breakers) operating at 1200 to 3000 rpm are used for the disintegration of agglomerates in the charge material or agglomerates formed during wet mixing. Provision of the intensifier bars may, however, pose the following difficulties:
  - Particle attrition of friable solid particles and granules
  - Cleaning problems due to the presence of packing glands in the working area of the blender
  - Scale-up difficulties
- A liquid spraying arrangement can be provided. These are common with blenders provided with intensifier bars and are used for solid–liquid blending and in granulation applications.
- Heating and cooling jackets can be provided on the blender shell.
- Jacketed blenders designed for vacuum operation are commonly used as rota-cone vacuum dryers.

Tumbler blenders are best suited for free-flowing, nonsegregating powders. These blenders are widely used for blending of dry powders and granules in the pharmaceutical, food, and cosmetic industries because of close quality control on batch operation, an effective diffusion mechanism, and gentle mixing of friable solid materials. Most tumbler blenders are operated in batch modes.

The following are distinct advantages of tumbler blenders:

- Ease of product charging, operation, and discharging of material.
- The shape of the blender shell ensures the complete discharge of product material.
- Particle-size reduction is minimized due to the absence of moving blades and agitators.
- The absence of shaft projection and seals in the working area of the blender eliminate the possibility of product contamination.
- Tumbler blenders are easy to clean.
- Tumbler blenders require minimal maintenance.

These advantages of tumbler blenders over horizontal convective mixers such as ribbon and paddle blenders make them popular with the food and pharmaceutical industries.

The disadvantages of tumbler blenders are as follows:

- They require considerable head room for installation.
- Segregation problems occur with mixtures that have a wide particle-size distribution or large differences in particle densities.
- Highly cohesive materials cannot be handled in tumbler blenders since they tend to form a bridge over the blender outlet.

Commonly used tumbler blender configurations, the double-cone blender and the v-blender, are discussed below.

12.3.6.1.1 Double-Cone Blenders The double-cone blender consists of two conical sections separated by a central cylindrical section. The blender is mounted at the center of the container between two trunions that allow the blender to tumble end over end. Generally, the charging of material into the double-cone blender is through one of the conical ends, whereas the discharge is through the opposite end. In some cases, charging and discharging may be from the same end. Blending efficiency is affected by the volume of the material loaded into the blender. The recommended fill-up volume is 50 to 60% of the total blender volume. The blending time is generally 10 to 15 minutes. Figure 12-20a shows a double-cone blender.

12.3.6.1.2 V-Blenders The v-blender is made of two hollow cylindrical shells that are joined at an angle of 75 to 90°. The blender container is mounted on trunions to allow it to tumble. As the v-blender tumbles, material splits and recombines continuously. The free fall of material within the vessel, and the repetitive converging and diverging motion, combined with increased frictional contact between the material and the vessel's long, straight sides, results in gentle, yet homogeneous blending. Figure 12-20b shows a v-blender.

The charging of material into a v-blender is through either of the two ends or through the apex port. The recommended fill-up volume for the blender is 50 to 60% of the total blender volume. The blend time ranges from about 5 to 15 minutes. The discharge of the material is through the apex of the blender shell through a bottom discharge valve.

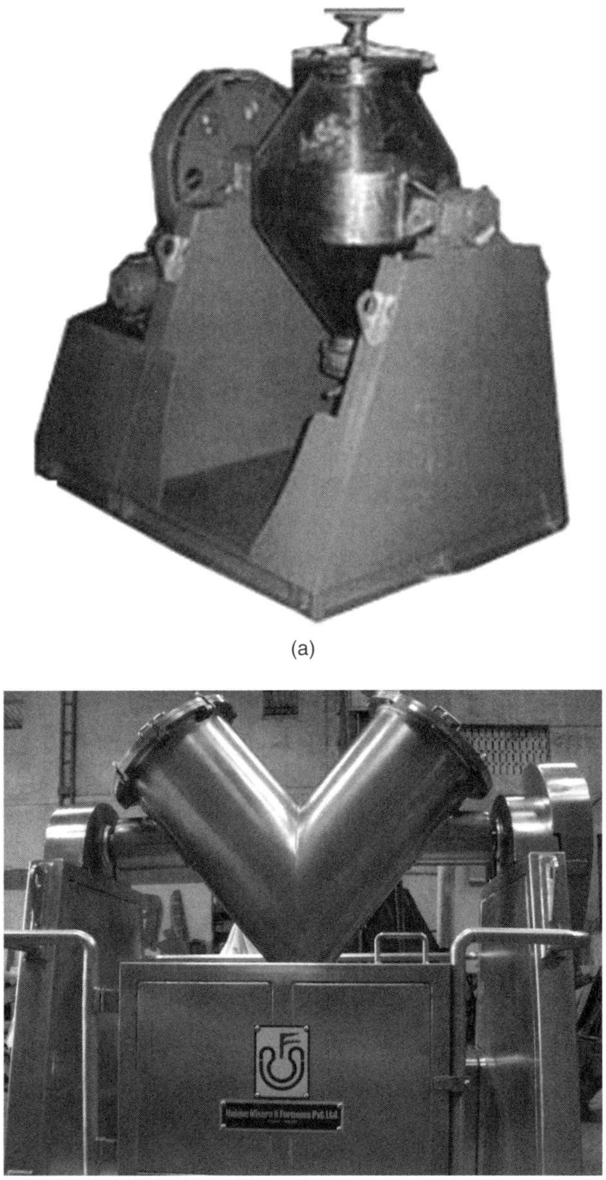

(a)

(b)

Figure 12-20 (a) Double-cone blender; (b) V-blender. (Courtesy of Unique Mixers and Furnaces.)

12.3.6.2 Convective Blenders A convective blender consists of a vertical or horizontal stationary shell (cylindrical, conical, U- or W-shaped trough) within which rotating agitator(s) operate. Convective blenders are available in single- or double-agitator configurations. In double agitator configurations, agitator rotation may be tangential or may overlap within the trough. The mixing elements are either in the form of helical ribbons, screws, or flat or plow-shaped paddles. In the case of helical ribbons, paddles, and plows, the mixing element sweeps through the entire blender container in one rotation. In screw blenders, small regions are stirred progressively in each rotation. The screw travels progressively throughout the container, resulting in uniform agitation.

Blending occurs because of the random movement of particles throughout the mixing vessel caused by the action of the mixing elements. Depending on the rotational speed and the geometry of the mixing elements, solid particles are thrown randomly and the product is sheared or fluidized. Although negligible, some amount of diffusive mixing also takes place. The combination of mixing mechanisms results in effective and efficient mixing and prevents demixing and agglomeration of solid particles. For some applications, the shearing action of the convective blender may be undesirable, as it results in reduction of particle size and heat generation.

Convective blenders can be used for a broad range of applications, including materials that have a tendency to segregate or agglomerate. These blenders can be designed for operation in both batch and continuous modes. An external jacket can also be provided on the blender container for applications that require heating or cooling of product material. A spray pipe for adding liquids can be provided. For materials that tend to form agglomerates during mixing, high-speed choppers can be provided for disintegration of the agglomerates.

The following commonly used convective blenders are discussed below.

- Horizontal ribbon blender
- Vertical ribbon blender
- Vertical cone screw blender
- Paddle blender
- Plow mixer
- Twin-shaft paddle mixer

12.3.6.2.1 Horizontal Ribbon Blenders A ribbon blender consists of a U-shaped horizontal trough with a double helical ribbon agitator rotating within. The agitator shaft is located in the center of the trough on which the helical ribbons (also known as spirals) are provided. The agitator shaft exits the blender container at either end through the end plates, which are either bolted or welded to the container. The area where the shaft exits the container is provided with a sealing arrangement to ensure that material does not travel from the container to the outside, and vice versa. The blender assembly, together with the drive system components (i.e., motor, gearbox, couplings, and bearing supports) are mounted on a supporting frame. Ribbon blenders can be manufactured in both single- and double-shaft configurations. Figure 12-21 shows an assembly of a 3000-L working capacity ribbon blender. Figure 12-22 shows internal and external ribbon spirals on the agitator shaft located within the blender container. The loading of material in the blender is generally through nozzles or feed

Figure 12-21 Ribbon blender assembly. (Courtesy of Unique Mixers and Furnaces.)

Figure 12-22 Double helical ribbon flights. (Courtesy of Unique Mixers and Furnaces.)

hoppers mounted on the top cover of the blender. The working capacity of the ribbon blender ranges from 40 to 70% of its total volumetric capacity.

Since the ribbon agitator consists of a set of inner and outer helical ribbons, it is referred to as a double helical ribbon agitator. During the blending operation, the outer ribbon flights of the agitator move the material from the ends to the center, while the inner ribbon flights move the material from the center to ends. Radial movement is achieved because of the rotational motion of the ribbons. The difference in the peripheral speeds of the outer and inner ribbons results in axial movement of the material along the horizontal axis of the blender. As a result of the radial and countercurrent axial movement, homogeneous blending is achieved within 15 to 20 minutes of startup with 90 to 95% or better homogeneity. The solid particle size and its bulk density have the strongest influence on the mixing efficiency of the ribbon blender. Ingredients of similar particle size and bulk density tend to mix faster than do ingredients with variation in these attributes.

After blending, material is discharged from a discharge valve located at the bottom of the trough. More than one discharge valve can be provided, depending on the size of the blender. The operation of the valve can be manual or pneumatically actuated. It is practically difficult to achieve 100% discharge in the ribbon blender. The amount of holdup in the blender after discharge depends on the properties of the material, the clearance between the outer edge of the ribbon flight, and the inside wall of the blender container. A clearance of 3 to 6 mm is generally maintained, depending on the application.

The ribbon blender is best suited for free-flowing and cohesive products. The ribbon blender's versatility for blending solids combined with it ability to perform heating, cooling, coating, and other processes, make it a popular blender. It is widely used in the food, chemical, animal feed, and pharmaceutical industries. This blender is not suitable for sticky products.

Ribbon blenders can be designed to operate in both batch and continuous modes. Batch-type blenders can be built up to capacities of 50 m^3. They are generally powered by 10- to 15-hp motors for 1000 kg of product mass to be blended. The specific power may range from 5 to 12 kW/m^3, depending on the products to be blended.

12.3.6.2.2 Vertical Ribbon Blenders In a vertical ribbon blender, the vessel and the agitator element are positioned vertically. The blender container is generally cylindrical, with a conical bottom. Because of the vertical design, these blenders have the flexibility to operate at capacities ranging from low volumes up to 90% of the total volumetric capacity. The slow rotation of the ribbons creates a shearing zone at the vessel wall. The material at the walls is lifted vertically upward. After reaching the top, the material travels down along the center of the blender. The blended product is discharged through a discharge valve located near the bottom of the blender. Almost 100% material discharge can be achieved in a vertical ribbon blender.

A single-shaft vertical ribbon blender can handle volumes of up to 30 m^3. Double-shaft configurations are also available. The vertical ribbon blender can be designed for operation under pressure and vacuum. Vertical blenders can handle friable products such as cereals, plastics, pigments, and pharmaceutical products. Vertical blenders are more expensive than horizontal blenders.

12.3.6.2.3 Vertical Cone Screw Blenders A vertical cone screw blender consists of a conically shaped vessel with a screw agitator that rotates about its own axis while orbiting around the vessel's periphery. The drive system generally consists of two motors: one for rotation of the main drive,

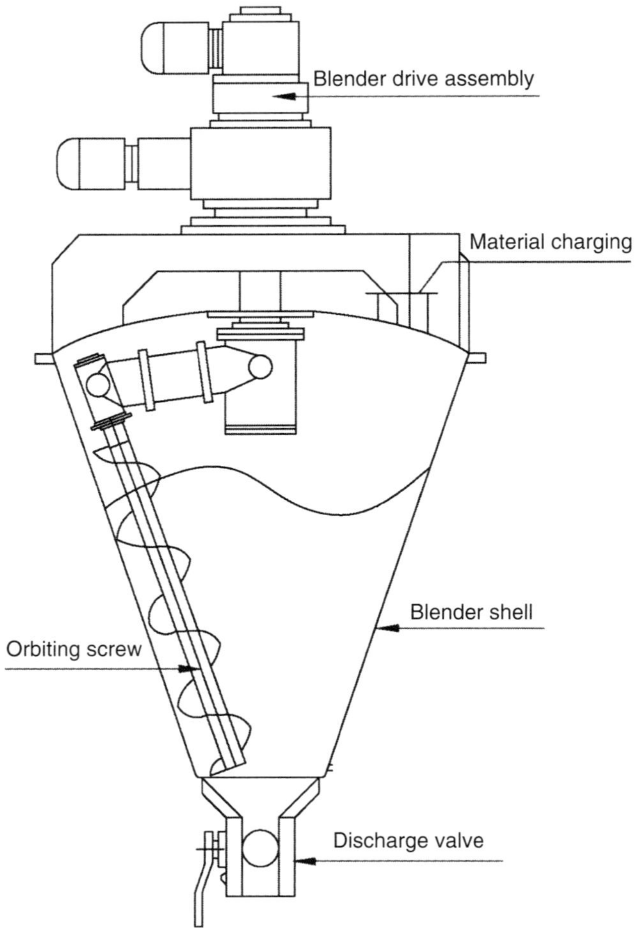

Figure 12-23 Vertical cone screw blender. (Courtesy of Unique Mixers and Furnaces.)

the other for rotation of the screw. In some cases, two orbiting screws may be used in a single vessel, in which one screw is shorter and has a larger diameter, for better and faster mixing. For large blending volumes, two hopper vessels are joined along the wall and are each provided with a screw. The screw is well supported from the top agitator drive assembly. Depending on the size of the blender and the application, the screw can also be supported at the bottom by a support bearing. The general arrangement of a vertical cone screw blender is shown in Fig. 12-23.

The charging of material is from the top of the vessel. Because of the vertical configuration and the conical shape of the vessel, this blender can operate efficiently with batch sizes ranging from 10 to 100% of the total working capacity of the blender. In operation, the screw shaft rotates around the periphery of the cone, and the screw turns such that the pitch gently lifts the material from the bottom of the blender to the top. The materials get directed toward the top of the vessel and then move down the center, while mixing with materials being moved upward by the orbiting screw. The mixing time is generally more than 10 minutes and depends on the product materials being blended. The discharge of the material is through a valve located at the bottom of the cone. The vertical cone screw blender gives nearly 100% discharge.

The vertical cone screw blender is suitable for free-flowing as well as cohesive powders. The blending action in the vertical cone screw blender is much gentler than that of a horizontal ribbon-type blender. Products that require gentle, low-impact blending are best handled in a vertical cone screw blender. The screw rotates at low speeds, typically 35 to 100 m/min. Because of the lower operating speed of the vertical cone blender, the wear is lower. The slower blending speed is advantageous for heat-sensitive products. This type of blender is operated in batch mode only. Vertical screw blenders are not suitable for the blending of powders with varying densities.

Vertical cone screw blenders are best suited for applications where the material to be mixed is shear sensitive, heat sensitive, or where the process parameters are critical. Although this blender is generally used for solid–solid blending, designs are adapted for liquid–solid blending and wet granulations. They are commonly used in the food, pharmaceutical, plastics, and chemical industries.

Following are some of the distinct advantages of a vertical screw blender over a ribbon blender

- Flexibility of batch size
- Gentle blending action
- Complete discharge
- Easy cleaning

The downside of this blender is that it requires high headroom for installation. Vertical blenders are also more expensive than horizontal blenders.

12.3.6.2.4 Paddle Blenders A design alternative to the ribbon agitator is the paddle agitator, which can handle fragile material. The paddle agitator consists of forward and reverse paddles rather than ribbon flights. The paddles are positioned to move the material in opposing lateral directions as well as in the radial direction. The paddle design is generally employed where friable materials are being blended. The ribbon design is appropriate for low- and medium-duty applications (generally free-flowing materials), whereas the paddle design is suitable for heavy-duty applications (e.g., wet materials, heavy metal powders). In some cases a hybrid paddle–ribbon agitator may be used.

12.3.6.2.5 Plow Mixers (Plough Share Mixers) Plow mixers operate on the principle of a mechanically generated fluid bed with three-dimensional movement of the product. The plow mixer consists of a cylindrical drum containing plow-shaped mixing elements mounted on a horizontal shaft. The agitator shaft is positioned in the centre of the drum and has welded or bolted arms on which the plows are mounted. The plows are designed to lift and separate the product within the vessel. The number of ploughs, their spacing, and the speed of rotation all contribute to fluidization. A plow mixer provides intense mixing in a gentle manner.

The plow shaft is powered by a drive system comprising of a motor, gearbox, and couplings. The shaft exits the mixer drum at either end through end plates bolted or welded to the cylindrical shell. The area where the shaft exits the container is provided with a sealing arrangement to ensure that material does not travel from the container to the outside, and vice versa. Liquid spray systems may be provided for injecting liquid into the material. Plow mixers may be fitted with high-speed chopper units for applications that require breakdown of lumps or when the mixer is to be used for wet granulation. Single or multiple choppers are mounted in the lower portion of the mixing container. Tulip-shaped choppers and christmas tree chopper designs are common. The mixer assembly, along with the drive system components (i.e., motor, gearbox, couplings, and bearing supports), is mounted on a supporting frame as shown in Fig. 12-24a. Figure 12-24b shows the mounting of choppers on the mixer assembly. An internal view of the mixer showing the christmas tree choppers and a plow is shown in Fig. 12-24c.

In operation, material is loaded through a charging nozzle located at the top of the cylindrical drum. The working volume (the volume up to which material should be loaded) in the plow mixer ranges from 30 to 70% of the total drum volume. The plow shaft rotates at high speed, producing a plow tip speed of more than 200 m/min to effect fluidizing action. The size, shape, positioning, and peripheral speed of the plows are coordinated such that they produce a whirling, three-dimensional product movement. The turbulence and material movement within the mixer container prevent the formation of dead spots or low-movement zones and ensure fast and homogeneous mixing. The plows are designed to lift the material from the inside wall of the container without squashing the particles

Figure 12-24 (a) Plow mixer assembly; (b) Chopper assembly mounting; (c) Plow mixer with christmas tree choppers. (Courtesy of Unique Mixers and Furnaces.)

against the wall. After completion of mixing, the material is discharged through the discharge nozzle located at the lower end of the container. Flush bottom discharge valves are preferred to avoid any dead pockets within the mixer. These valves may be manually, pneumatically, or hydraulically operated. Large side doors may be provided for cleaning access. Plow mixers provide excellent solutions for mixing such materials as powders, powders with liquids, and heavy pastes such as dough molding compounds and putties. Homogeneous mixing of materials with varying densities, flow properties, particle sizes, and structures is achieved rapidly. Mix ratios as high as 1 : 20,000 are possible. The plow mixer can mix the material in less than 5 minutes with 95 to 98% homogeneity. The mixer is capable of handling viscosities of up to 600,000 cP.

Plow mixer sizes vary from small laboratory size units to large production mixers with capacities of 40 m^3. The specific power for the mixer drive generally ranges from 30 to 40 kW/m^3, depending on the properties of the materials to be mixed. Plow mixers can be designed to operate in both batch and continuous modes.

12.3.6.2.6 Twin-Shaft Paddle Mixers The twin-shaft paddle mixer consists of paddles mounted on twin shafts in a W-shaped trough. The normal filling level of material in the mixer is located slightly above the shafts. Thus, there is surplus space in the mixer trough to provide air around the particles so that they can move freely. The specific speed of the shafts combined with the overlapping motion and paddle design facilitates rapid fluidization and ensures excellent movement of particles. The twin-shaft, counter-rotating paddles lift the particles in the centre of the mixer trough, in the fluidized zone, where mixing takes place in a weightless state. This results in random movement of particles in all directions. The overlapping paddle arrangement and the tulip shaped chopper is shown in Fig. 12-25a. The principle of mixing is demonstrated in Fig. 12-25b.

The material to be mixed is charged from the top of the trough. The normal working volume in this type of mixer is about 25% of the total volume of the trough. If the normal working volume is rated as 100% working capacity, the range of operation of the mixer is from 40 to 140% of the rated capacity. The peripheral speed of the paddles is approximately 100 m/min.

Twin-shaft paddle mixers are popular because of their fast mixing times and gentle operation. The mixing time can be as little as 1 minute, but may be higher for cohesive solids. A mixing homogeneity of 98 to 99% can be achieved using a twin-shaft paddle mixer. Discharge of the material is through the two large doors located at the bottom of the trough. A choice of discharge valves, including half bomb-bay doors and spherical disk valves, is available.

Fluidized-zone twin-shaft paddle mixers are capable of preparing a homogeneous mix independent of particle size, shape, and density. The unique agitation ensures rapid yet gentle blending, short mixing cycles, low operating costs, minimal product degradation, and one of the highest production capacities among the different types of mixer.

Twin-shaft paddle mixers can be built in batch capacities of up to 50 m^3. The mixer can be equipped with pin mills above the paddles to break down any lumps in the feed material. Spray nozzles may be provided for the addition of liquids. High-speed choppers can be provided in the mixer for deagglomeration of lumps formed. Twin-shaft paddle mixers can be adapted for continuous mixing operation and are used in industries such as animal feed, food, chemical, pharmaceutical, building, and environmental.

Rotating twin-shaft paddle mixers are the most recent development. In this design the charge is loaded with the mixer nozzle in the top position. After the mixing operation is completed, the trough is rotated by 180° and the material is discharged through the same nozzle that was used for filling. After discharge is complete, it is rotated back for loading a fresh charge. This type of mixer is designed where air-tight processes are essential and cleaning is critical. With this design it is also possible to mix under vacuum, or in an inert atmosphere, to avoid oxidation of highly sensitive products.

Twin-shaft paddle mixers have many advantages. These mixers are very gentle with products and have low energy consumption, high capacity, small space requirements, and flexible filling. The advantages are as follows:

- Short mixing time
- Higher throughput
- Gentle mixing
- No segregation of materials
- Lower production and maintenance costs
- Micromixing capability
- Maximum material discharge
- Easy cleaning of mixer
- Reliable scale-up

Tumbling and convective mixers are compared in Table 12-5.

12.3.6.3 Silo Blenders There are several industrial applications where the methods of production, the properties of the material, or the nature of the process may lead to variations in the quality of the solid powders as a function of time. There are also instances where material blended in small batches must be homogenized to produce a single bulk lot. In such cases, homogenization of material is carried out in large silos using either gravity blending techniques or mechanical silo blenders.

(a)

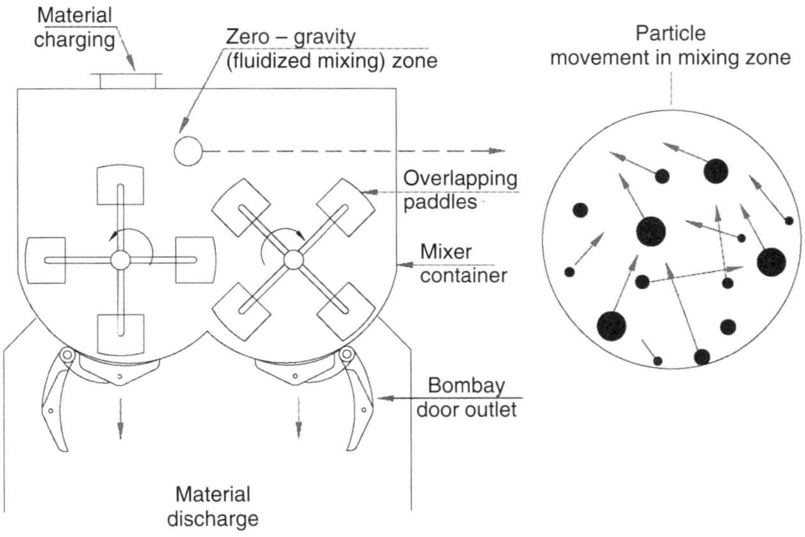

(b)

Figure 12-25 (a) Twin-shaft paddle blender internals. (Courtesy of Unique Mixers and Furnaces.) (b) Principle of fluidized mixing.

12.3.6.3.1 Gravity Silo Blenders Gravity blending is the most convenient and economical method of blending large volumes of free-flowing powders. Gravity silo blenders use either multitube construction on inserts to create velocity gradients within the silo. The material is simultaneously drawn off by a system of tubes positioned at different heights and radial locations, brought together, and mixed. Another type of gravity blender uses a central takeoff tube into which solids travel through openings arranged at various heights on a pipe.

Gravity blenders are available in capacities ranging from 5 to 200 m^3 and require very low energy, less than 1 kWh per ton of product being blended. These blenders may be operated in batch or continuous mode. It may be necessary to recirculate the contents of the silo several times to achieve a homogeneous mixture.

12.3.6.3.2 Mechanical Silo Blenders Mechanical silo blenders are provided with a screw housed in a cylindrical shell that isolates the bulk material in the silo from the

TABLE 12-5 Comparison of Convective Mixers

	Type of Mixer or Blender			
Property	Plow	Double Paddle	Ribbon or Paddle	Tumbling Blender
Free-flowing powders, $50 < x < 500\ \mu m$	Yes	Yes	Yes	Yes
Free-flowing granules, $200 < x < 5000\ \mu m$	Yes	Yes	Yes	Yes
Cohesive powder	Yes	Yes	Yes	Possible
Size (m^3)	<40	<50	<50	<2
Filling ratio (%)	30–70	40–140[a]	40–70	50–60
Tip speed (m/s)	~3	~2	~1.5	
Liquid addition through spraying	Possible	Possible	Possible	Possible
Provision for disintegration of lumps	Yes, using side choppers	Yes, using pin mills above paddles, side choppers	Yes, using side choppers	Yes, using high-speed intensifier bars
Degree of particle shear	High (using choppers)	Slight	Average	Low
Specific power (kW/m^3)	30–40	20–40	5–12	<10
Homogeneity of mix (%)	95–98	98–99	90–95	Up to 98
Mixing time (min)	<5	<1	15–20	5–15
Product discharge (%)	<100	~100	<100	~100

[a]Percentage fill more than 100% of the rated working capacity.

material inside the shell. The material from the bottom section of vessel is lifted by the screw and spread over the upper sections. The principle of blending is based on the differential travel speeds of product particles in the conical section of the vessel and the velocity of material in the screw region. Since the active blending volume at any given time is a lot less than that of horizontal blenders, it results in a lower power requirement: about 1.5 to 2 kWh/ton.

12.3.6.4 Pneumatic Blenders Pneumatic blenders consist of a mixing silo and in some cases a central conveying tube with an inverted conical deflector at the top for spreading the material. They are equipped with aerators mounted around a housing cone. Air or gas is injected intermittently at high velocities at the bottom or sides of the silos, to blend powder materials that exhibit expansion characteristics when aerated.

The solid particles rise due to the drag force of the air injected. An increase in air velocity causes agitation in the bed, resulting in the formation of bubbles, which cause blending to take place. The amount of air required to fluidize the system and the minimum fluidization velocity depend on the size, particle density, and bulk density of solids, and the density of the gas used for fluidization. The blending action can be optimized by adjusting the air pressure, pulse frequency, or on/off duration.

These blenders are best used for non-critical blends, where particle sizes and densities are similar among all the components of the mix so that fines are not carried to the top by the air. The materials to be blended must be dry, free flowing, and have a particle size in excess of 50 μm. The specific power input ranges from 1 to 2 kWh per ton of product. The largest pneumatic blenders are used in the cement industry, with blender capacities of up to 10,000 m^3.

12.4 MIXING HIGH-VISCOSITY MATERIALS AND PASTES

Materials such as polymer, putties, chewing gum, and solid propellants have viscosities exceeding 10Pa·s (10,000 cP) and are categorized as viscous materials. These materials may exhibit other characteristics, such as resistance to flow, non-Newtonian behavior (see Table 12-6), and elasticity. As a result, flow currents do not get formed during mixing of viscous materials. Mixing in viscous systems can therefore be achieved only by mechanical action, or forced shear, or elongation flow of the matrix [7].

TABLE 12-6 Definitions of Newtonian, Non-Newtonian Fluid Behavior

	Effect on Viscosity	
Definition	Time Increase	Shear Rate Increase
Newtonian	No effect	No effect
Non-Newtonian		
• Pseudoplastic	No effect	Decrease
• Dilatant	No effect	Increase
• Thixotropic	Decrease	No effect
• Rheopectic	Increase	No effect

Source: [6].

Mixing of viscous material requires deformation of the mixture, ensuring both lateral and transverse motion of material. For viscous materials, the geometry of the mixing vessel and the design of the mixing element have a significant impact on the mixing process. The relative motion between the mixing element and the internal walls of the mixing vessel creates both shear and bulk motion. The shear effectively creates thinner layers of non-uniform material, which diminishes striations or breaks agglomerates to increase homogeneity. Bulk motion redistributes the effects of the stretching processes throughout the mixing vessel.

In several cases, the formation of high-viscosity material takes place within the mixing equipment. For example, low-viscosity monomers react to form high-viscosity polymer material. Similarly, viscous paste may be created during mixing of solids with liquids or by removal of liquids from slurries. The mixing equipment for such applications is therefore required to be versatile and should be efficient in mixing both low- and high-viscosity materials.

12.4.1 Dispersive, Distributive, and Convective Mixing

Mixing viscous material requires that dispersive, distributive, and convective mixing take place in a system.

1. *Dispersive mixing*: the breakup of agglomerates or lumps to the desired ultimate grain size of solid particulates or the domain size (drops) of other immiscible fluids [7].
2. *Distributive mixing*: provides spatial uniformity of all the components and is determined by the history of deformation imparted to the material.
3. *Convective mixing*: is effected by shear, kneading, and stretching of material in the laminar regime and results in reorientation of the dispersed elements.

These mechanisms are to be facilitated by the mixing equipment.

12.4.2 Power for Viscous Mixing

Mixing of heavy plastic masses and pastes requires large amounts of mechanical energy to shear material and facilitate convective mixing by continuously folding over, dividing, and recombining the material. The power required to mix high-viscosity pastes and dough is many times greater than that required to mix free-flowing solids or liquids. The power for a Newtonian viscous mixer can be represented as [7]

$$\text{power} = \alpha\mu D^3 N^2$$

where α is a dimensionless proportionality constant, μ the viscosity (Pa·s), D the agitator sweep diameter (m), and N the agitaror speed (rev/s). For very viscous mixtures, the viscosity is likely to be non-Newtonian. For shear thinning fluids that can be represented by a power law exponent (n) [7]

$$\text{power} = \alpha\mu D^3 N^{1+n}$$

In most cases, only a part of the total energy supplied to the mixer is directly useful for mixing, as some amount is dissipated as heat. Deciding the power required for mixing of viscous materials is one of the most challenging aspects of mixer design. It is recommended that selection of motor horsepower for high-viscosity mixing equipment be based on past experiences and the advice of the equipment manufacturer.

12.4.3 Scale-Up of High-Viscosity Mixers

The viscosity changes that often take place during mixing in high-viscosity mixers make it extremely difficult to model such systems. Equally difficult is maintaining temperature control within the mixer. Maintaining a constant specific energy (kW/kg) is a frequently adopted scale-up method. Good equipment manufacturers offer laboratory-scale testing services and also offer rental equipment so that experimental data can be gathered for use in scale-up. Selection and scale-up of high-viscosity mixers should be carried out in consultation with experts.

12.4.4 Heat Transfer

Mixing of high-viscosity materials is generally characterized by heat dissipation, which occurs due to high shear and friction generated by the action of the mixing blades. To avoid overheating of the process material, it may be necessary to provide the mixing vessel with external jackets for circulation of cooling media. Additionally, the mixing blades can be provided with channels for water circulation.

Heat transfer is generally poor in viscous materials. It is therefore necessary to improve the heat transfer by effective motion near the heating and cooling jackets, thereby promoting convection over conduction. Most mixers for pastes or viscous fluids are provided with close-clearance blades and/or scraper devices to move stagnant material away from heat-transfer surfaces.

12.4.5 Equipment for Mixing High-Viscosity Materials and Pastes

Mixing adhesive solids such as pastes, polymer, rubber, and other high-viscosity materials poses several challenges. As a result, high viscosity mixing equipment are more difficult to design and manufacture than equipment used for mixing solids and liquids. Unlike liquid mixing, where mixing takes place through the generation of flow currents, in high viscosity applications the mixing action is due to a combination of low-speed shear, folding, stretching, and compressing of the material to be mixed. In this equipment the materials to be mixed require that the mixing element be operated within all parts of the mixing vessel. Alternatively, it is necessary to direct the material to the mixing element. High-viscosity mixing equipment have the following characteristics:

- The clearance between the mixing element and the vessel wall is minimal. In some cases the clearance may be as low as 1 to 2 mm.
- The power per unit volume is high and can be up to 6 kW/kg of product.
- The forces generated in the mixers are high. As a result, these mixers are rigid in construction.
- The heat evolved during mixing is high. These mixers may therefore require provision for cooling of material.
- These mixers operate at low speeds.
- Discharge of materials after mixing may be difficult and may require special arrangements.
- The mixing elements may consist of intermeshing blades that prevent the material from cylindering along with the rotating mixing element.
- High-viscosity mixers can be designed for batch as well as continuous operation. The following commonly used high-viscosity equipment are discussed next.

Batch Mixers

1. Single stirred mixers
   - Anchor mixer
   - Helical ribbon mixer
2. Change can mixers
   - Single planetary mixer
   - Double planetary mixer
   - Other change can mixers
3. Double-arm kneader mixers
4. Kneader mixer extruders
5. Intensive mixers
   - Banbury mixers
   - High-intensity mixers
   - Roll mills
6. Pan muller mixers (high-shear mixers)

Continuous Mixers

1. Single-screw extruders
2. Twin-screw extruders
3. Pug mills

12.4.5.1 Batch Mixers

12.4.5.1.1 Single Stirred Mixers Anchor and helical ribbon impellers, which operate within close clearances of the mixing vessel, are used to mix thick pastes and polymer solution.

Anchor Mixers Anchor mixers are among the more common types of high-viscosity mixers. Anchor-shaped impellers are used for liquid viscosities between 5000 and 50,000 cP. The swing diameter of the anchor is generally 90 to 95% of the mixing vessel diameter. There is low clearance between the impeller and the internal wall of the vessel. Shearing of material takes places because of the relative motion between the rotating impeller and the stationary vessel wall. Anchor impellers are preferred when heat transfer through a jacket is desired along with good mixing. The shear near the vessel wall reduces the build-up of stagnant material and consequently promotes heat transfer. In addition, scraper arrangements or wipers may be provided to remove material deposited on internal walls of the vessel, further improving the transfer of heat.

Anchors are commonly used in crystallization reactors. The use of anchors is limited since they provide very little axial flow. For typical anchor applications, speeds range from approximately 5 to 45 rpm, and motor power varies from 1 to 250 hp [6]. Impeller diameters range from 600 to 3000 mm. An anchor impeller is shown in Fig. 12-26.

Figure 12-26 Anchor impeller (Courtesy of Unique Mixers and Furnaces).

Helical Ribbon Mixers Helical ribbon mixers are capable of operating over a wide band of viscosities, ranging from low-viscosity liquid to highly viscous materials up to 4,000,000 cP. These mixers have major advantages over anchor mixers because of their ability to generate strong top-to-bottom motion even with viscous materials. The helical ribbons may have single or double flights. In the case of double-ribbon flights, the inner and outer flights are designed to generate flow in opposing directions. To improve heat transfer through the jacket, wipers and scrapers can be attached to the outer periphery ribbon flights. Helical ribbon mixers can be formed to fit in conical-bottom tanks, where the ribbon flights force the material to the bottom discharge, thereby facilitating greater yield from each batch mix.

The ability to operate over a wide range of viscosity makes helical mixers effective for batch processes such as polymerization, where the process begins with low-viscosity materials and after reaction changes to a high-viscosity product. Helical ribbon mixers work efficiently with heavy pastes as well as with flowable powders. A helical impeller is shown in Fig. 12-27.

12.4.5.1.2 Change Can Mixers Like most liquid agitators, change can mixers are configured vertically, with a vertical cylindrical vessel and a bottom dished end. These mixers are called change can mixers because they are provided with an arrangement for lifting the agitator head and mixing element out of the mixing vessel once mixing is completed, thus enabling movement of the mixing can. Some change can mixers have mixing vessels that can be lifted and lowered while keeping the agitator head

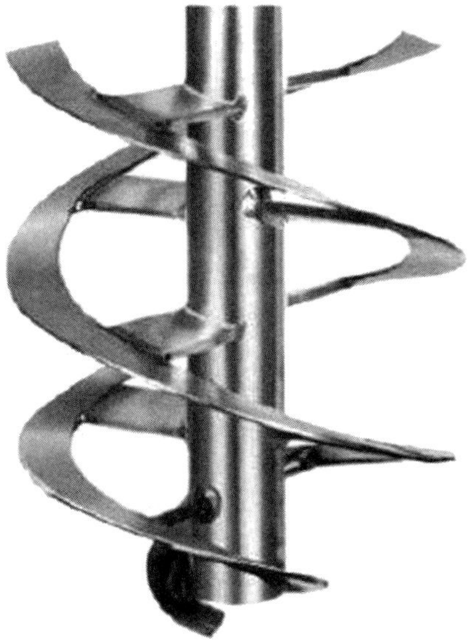

Figure 12-27 Helical ribbon impeller (Courtesy of Chemineer,Inc.)

stationary. Removal of the mixing vessel provides the following advantages:

1. Material can be weighed accurately.
2. It is easier to clean the mixing vessel, resulting in less batch-to-batch contamination.
3. The packing glands and seals do not come in contact with the material, thus eliminating product contamination.
4. Multiple cans can be used to enhance productivity without downtime during material charging and discharging.
5. Change can mixers may be provided with one or more mixing blades.
6. Because of the vertical configuration, these mixers can be operated at as little as 10% of their designed working capacity.
7. Discharge of highly viscous materials from the mixing vessel can be achieved by locating the vessel on a separate hydraulically operated discharge system.

Single Planetary Mixers The planetary mixer is so named because the mixing element (commonly known as the *beater*) rotates in a planetary motion inside the mixer bowl. The bowl of a single planetary mixer consists of an upper cylindrical section and a lower hemispherical section. The mixer bowl is secured to a semicircular frame (also termed a *fork*) at the time of mixing. The beater profiles are shaped to match the lower curved surface of the bowl. The beater has two types of movement: it revolves on its own vertical axis at high speed while the vertical axis rotates around the center of the bowl at a relatively lower speed. Figure 12-28a shows a single planetary mixer. The planetary motion of the blade is demonstrated in Fig. 12-28b.

The planetary motion of the beater is very effective in mixing the contents of the bowl. There are virtually no dead spaces in the mixing bowl. Scrapers with soft edges may be provided to wipe off any material that may stick to the internal surface of the bowl. The most commonly used beaters are of the batter, wire whip, and hook types. A batter beater is the used for cake batters and general mixing applications, a wire whip is used to beat egg foams and cream, and a hook beater is used to mix and knead yeast dough, bread and bun dough, and pastry.

After mixing has been completed, the bowl is lowered and can easily be detached and removed from the mixer assembly. The discharge of material from the mixer bowl can be by hand scooping when the material is pasty and does not flow, or through a bottom discharge valve when the material is flowable.

The single planetary mixer is used to mix dry and wet powders, light pastes, gels, and dough. This mixer is very popular in the food and bakery industry because of its

(a)

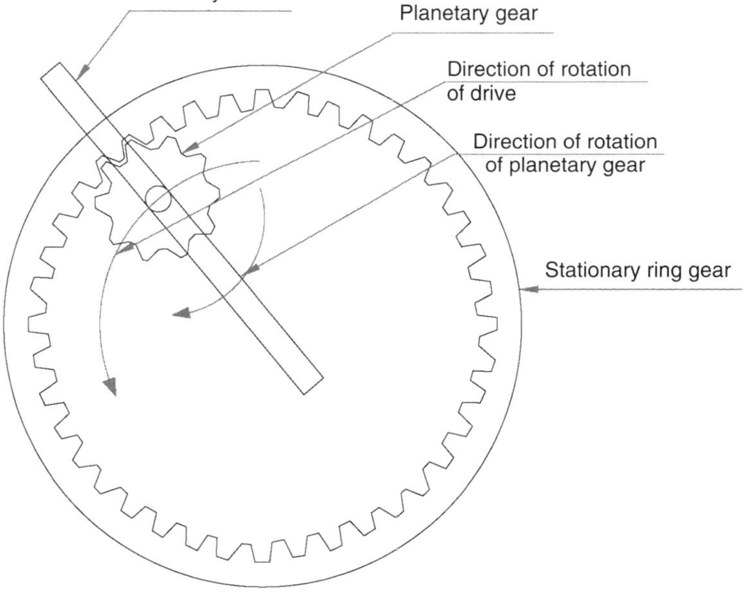

(b)

Figure 12-28 (a) Single planetary mixer; (b) planetary motion. (Courtesy of Unique Mixers and Furnaces.)

simple construction, operation, and relatively lower cost. These mixers are available in small sizes suitable for home kitchens, upto production-size units with capacities up to 500 L, or more.

Double Planetary Mixers The double planetary mixer includes two blades that rotate on their own axes while they orbit the mix vessel (also known as bowl) on a common axis. The material to be mixed is loaded into a cylindrical vertical vessel with a nearly flat bottom. The blades advance along the periphery, removing material continuously from the vessel wall and transporting it to the interior. After one revolution the blades have passed through the entire vessel, after three revolutions most materials have been mixed, and after only 36 revolutions the blades have contacted virtually the entire batch (see Fig. 12-29a).

The drive system of a double planetary mixer consists of a motor and a gearbox that drives the planetary head. Each planetary blade is driven by gears that rotate due to the motion of the planetary head. Variable-speed drives can be provided to operate the mixer over a wide range of speeds, ensuring high torques even at lower speeds. Located below the drive system is a cylindrical cover for the bowl. This cover may be provided with nozzles, a liquid spray arrangement, and viewing ports. A major advantage of this vertical type of mixer is that the shaft seals do not come into direct contact with the product material. Figure 12-29b shows a double planetary mixer with rectangular blades.

Material may be charged into the mixer either through the nozzles on the top cover or by direct loading into the mixer bowl. Depending on the size of the mixer, either the bowl or the drive system of the mixer is lifted or lowered using a hydraulic arrangement. The mixer bowl may be jacketed for circulation of heating or cooling media. The mixer can be designed for operation under pressure or vacuum. After mixing, the material is generally discharged by manual scooping of the material from the bowl. For extremely viscous materials, hydraulically operated automatic discharge systems are available that push the material out through the discharge valve.

Depending on the properties of the material to be mixed and the blade type, the specific power for a double planetary mixer ranges from 30 to 50 kW/m^3. Because of the vertical configuration of the mixer, the material fill levels in the double planetary mixer can be as low as 20% to a maximum of 85% of the total bowl volume. The blade profiles used for the double planetary mixer are rectangular, finger blades and helical blades.

The double planetary mixer is gaining popularity for mixing materials of high viscosity. The introduction of new blade designs has extended the operating range of double planetary mixers from about 1 million centipoise to about 8 million centipoise, resulting in a fitting alternative to double-arm kneaders. The simple construction, operation, and relatively lower cost make it a preferred choice over the double-arm kneader mixer.

Single and double planetary mixers can be equipped with additional mixer shafts that are provided with other types of mixing impellers. A high-shear impeller can be used to incorporate powdered material or to create a stable emulsion, resulting in the formation of viscous paste.

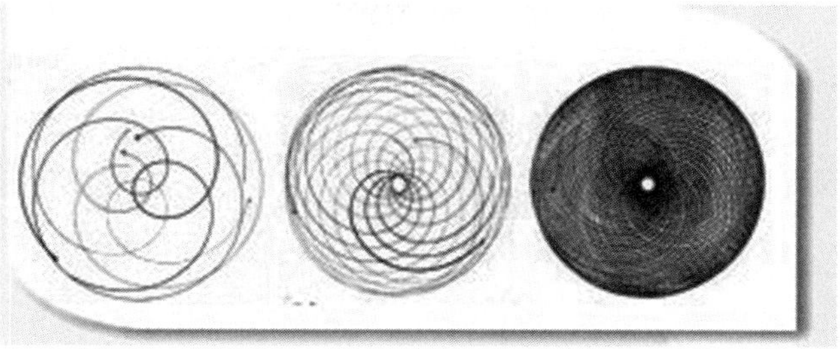

(a)

(b)

Figure 12-29 (a) Blade action in a double planetary mixer. (b) Double planetary mixer with rectangular blades. (Courtesy of Unique Mixers and Furnaces.)

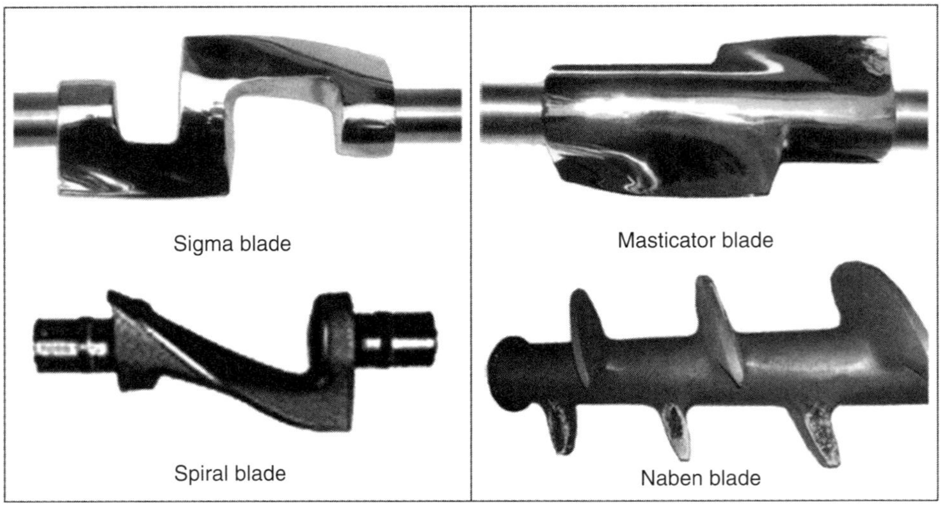

Figure 12-30 Blade options for double-arm kneaders. (Courtesy of Unique Mixers and Furnaces.)

Other Change Can Mixers Not all change can mixers have planetary motion. Some configurations are equipped with an anchor impeller for movement of material near the mixing vessel, along with a high-speed, high-shear impeller that provides intense shearing action. Other types of mixing impellers, such as pitched-blade and hydrofoil turbines, can also be provided. With multiple-shaft impeller construction, flexibility of mixing operations can be achieved.

12.4.5.1.3 Double-Arm Kneader Mixers All double-arm kneader mixers are equipped with two mixing blades placed in a W-shaped horizontal trough. Various blade profiles have evolved, each of which is best suited for a particular type of application. The commonly used blade types are the sigma blade, masticator blade, spiral blade, and naben blade (see Fig. 12-30). Shredder blade designs are used for fibrous materials. The blades either rotate tangentially or may overlap within the trough. The blades rotate toward each other at the same or differential speeds. The blades pass the container walls and each other at close clearances generally (1 to 2 mm), resulting in homogeneous mixing. The close clearances produce very high shearing action, thereby reducing the particle size.

The geometry and profile of the sigma blade is designed such that the viscous mass of material is pulled, sheared, compressed, kneaded, and folded by the action of the blades against the walls of the mixer trough. The extent to which this happens depends on the action of the blades (i.e., tangential or overlapping) and the ratio of the speed of rotation of the blades.

Tangential blade action. In the tangential design, the blades rotate in a trough, meeting tangentially as shown in Fig. 12-31a. The front blade generally rotates faster than the rear blade, usually in a 3 : 2 ratio. Tangential blade design is used for materials of higher viscosity, such as adhesives, rubber compounds, flush colors, dyes, and pigments.

Overlapping blade action. In the overlapping design, the blades overlap above the saddle of the container. The overlapping blade configuration is shown in Fig. 12-31b. Due to the overlapping action, it is necessary that the relative position of the two blades is unchanged, and as a result, both blades rotate at the same speed. Materials that flow freely or creep down into the blades are mixed using the overlapping blade action. This design offers a

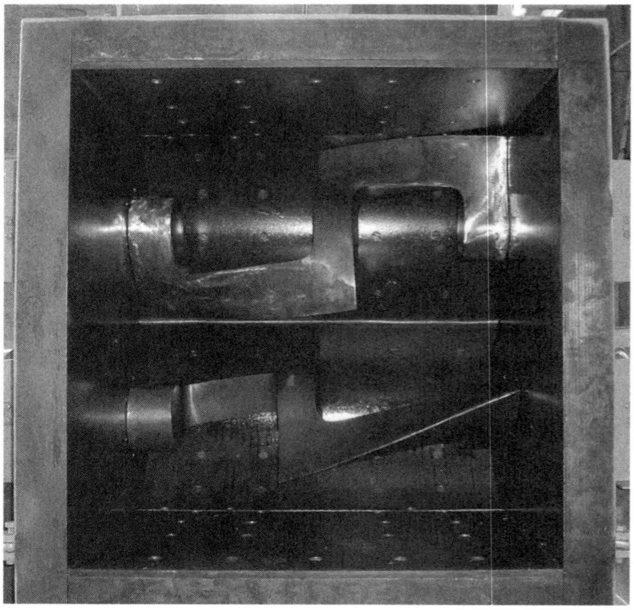

(a)

Figure 12-31 (a) Tangential sigma blade mixer; (b) overlapping sigma blade mixer. (Courtesy of Unique Mixers and Furnaces.)

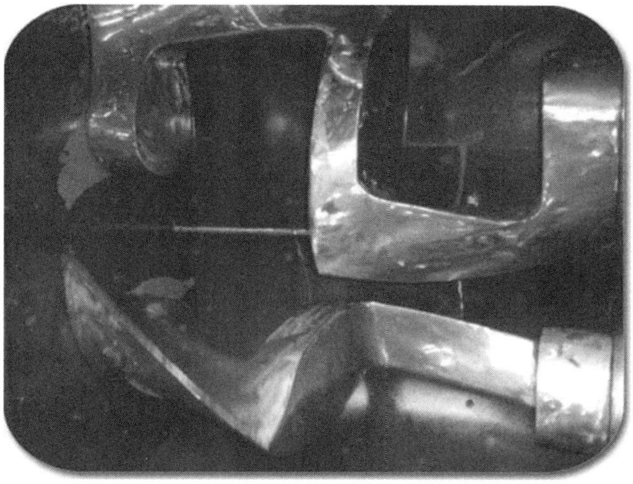

(b)

Figure 12-31 *(Continued)*

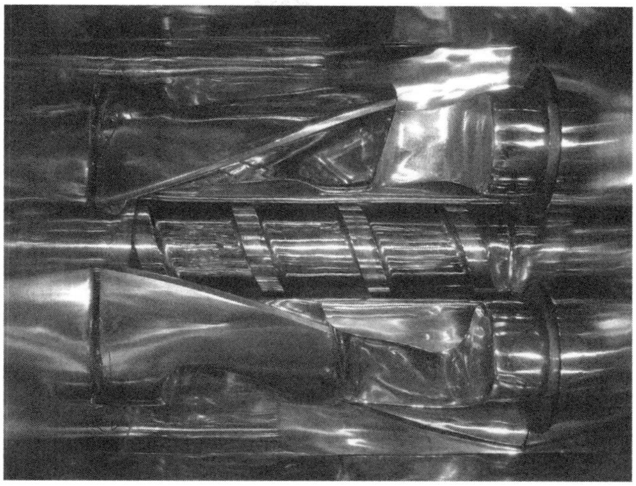

Figure 12-32 Sigma kneader extruder internals. (Courtesy of Unique Mixers and Furnaces.)

faster interchange of material from one blade compartment of the mixer to the other. Overlapping blade action is used for materials of lighter viscosities, such as carbon pastes, clay coating, creams, and ointments. The overlapping blade produces less kneading and shearing than the tangential blade action.

In operation, material is loaded through the top of the container to typically 40 to 65% of the mixer's total volumetric capacity. The rotation of the blades is through a heavy-duty drive system, typically consists of a motor, gearbox, couplings, and gears. The tip speed of a sigma mixer is generally limited to 60 m/min. Mixing may be carried out at ambient temperature or under controlled temperature conditions. The mixer troughs can be provided with jackets for circulation of hot or cold media to maintain the required temperature conditions within the mixer. The material from the mixer container is discharged either by tilting the mixer container, through bottom discharge valve, or through an extruder screw located in the lower portion between the two trough compartments.

Double-arm kneader mixers produce a consistent particle-size distribution without the need for additional high-speed choppers. Homogeneous mixing is achieved within 10 to 30 minutes. Mixing homogeneity of up to 99% and better can be achieved using double-arm kneader mixers. The power consumption in a double-arm kneader mixer is high compared to other types of mixers and can range from 45 to 75 kW/m^3 of mix material. These mixers and their variants, the double-arm kneader mixer extruders discussed below, are capable of handling material with viscosities as high as 10 million centipoise.

12.4.5.1.4 Kneader Mixer Extruders The kneader extruder combines the efficiency of a double-arm mixer with the convenience of extrusion screws for improved mixing and discharge of extremely viscous materials. The kneader extruder includes a set of contra-rotating kneading blades and one or more extrusion screws. The blades are mounted on a horizontal axis in a W-shaped trough. Below the blades in a separate cavity are one or two extrusion screws. During the mixing cycle the blades rotate toward each other while the mixing screw rotates in a reverse direction, constantly feeding new materials into the mixing blades. After the mixing cycle is complete, the screw is reversed and the mixed material is extruded out of the mix zone. The internal construction of a sigma kneader extruder with a single screw is shown in Fig. 12-32.

12.4.5.1.5 Intensive Mixers

Banbury Mixers A banbury mixer is an intensive mixer in which mixing takes place due to the low clearances between the mixer trough and the blades, within a pressurized mixing chamber. Unlike double-arm kneaders, where the mixer trough is open toward the upper half, in high-intensity mixers such as the banbury mixer, the mixing chamber is closed during operation. The underside of the cover confirms to the volume swept out by the mixing blades. The clearance between the mixer blade and the vessel wall is extremely small. The equipment has two heavy-duty mixer blades in the form of interrupted spirals. The two blades operate at different speeds, in the range of 10 to 30 rpm. The general schematic of a banbury mixer is shown in Fig. 12-33.

Material is charged into this mixer from the top and is forced into the mixing chamber by a pneumatically operated ram provided at the top of the mixer, under pressure ranging from 1 to 10 bar. The discharge of the material is through a sliding door provided at the bottom of the mixing chamber.

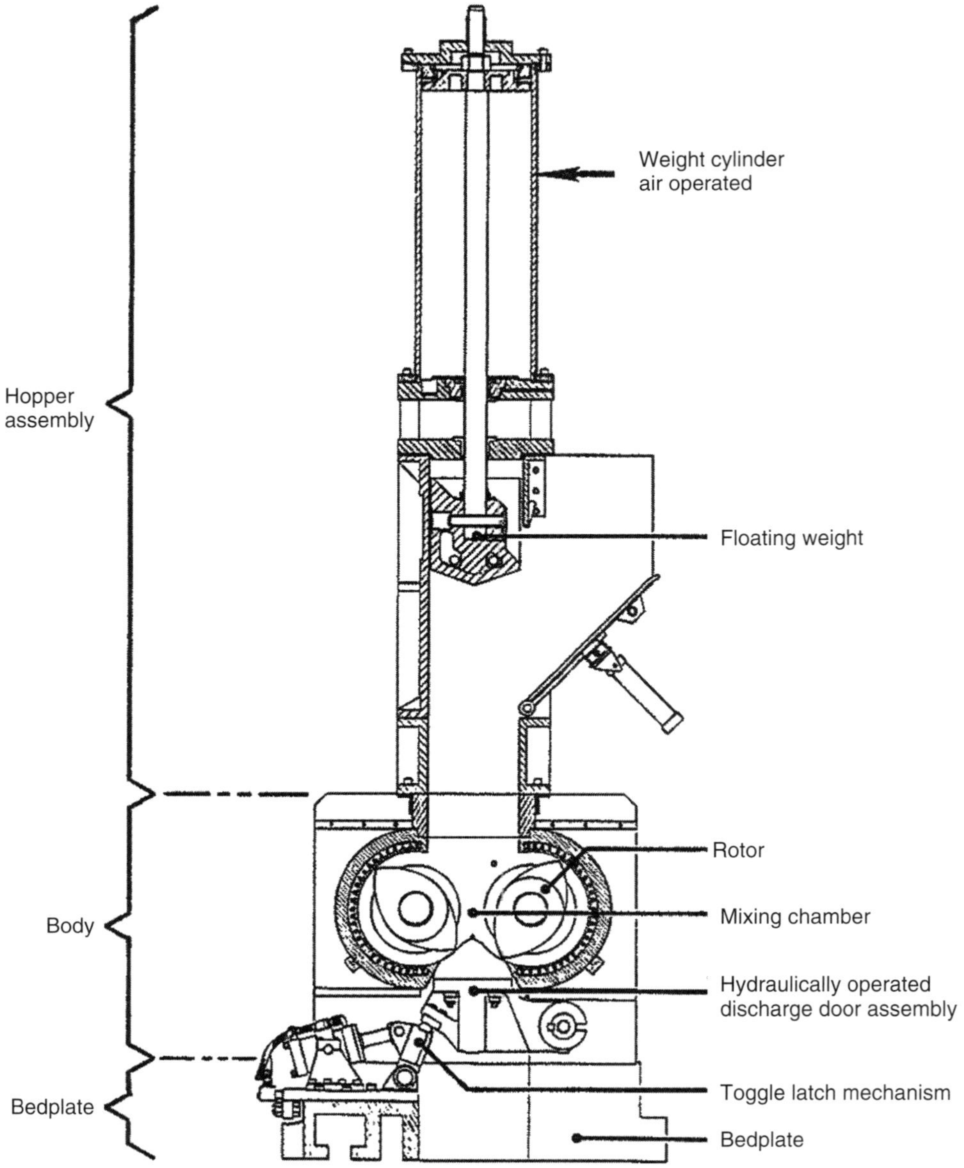

Figure 12-33 Banbury mixer. (Courtesy of Farrel Corporation.)

These mixers consume very high power, up to 6000 kW/m^3 [8]. The heat added up by the high-power input may have to be dissipated, considering the temperature limits of the material being processed. The heat is removed by spraying water on the internal walls of the mixing chamber and/or by circulation of cooling water through the hollow mixer shaft. Because of the heavy-duty construction and high power, these mixers are generally used for small batches, up to 500 kg. Banbury mixers are the most commonly used intensive mixers employed in the plastics and rubber industries.

High-Intensity Mixers A high-intensity mixer combines a high shear zone with a fluidized vortex for mixing of pastes and powders. The mixing blades placed at the bottom of the vessel scoop the material upward at very high rotation speeds. The high shear stresses between the mixer blade and the vessel combined with the high blade impact disintegrate product agglomerates, resulting in an intimate dispersion of powders and liquids. This type of mixer resembles the commonly used kitchen food processor and is easy to clean, maintain, and operate. It is also used as a mixer granulator. Scale-up of these mixers is based on the constant peripheral speed of the blade, which is about 40 m/s. The specific power consumption is approximately 200 kW/m^3 [8].

Roll Mills Rubber products and pastes are compounded in batch roll mills, as they provide extremely high localized shear. A batch type of roll mill has a pair of smooth metal rolls set in the same horizontal plane, mounted on a heavy structure. The equipment is provided with an arrangement for adjusting the distance between the rolls and regulating the pressure. To increase the shearing action, the rolls are usually operated at different speeds.

The amount of mixing that takes place in a single pass of material between the rolls is limited. Therefore, these mills are either used in series or the material is passed repeatedly through the rolls. By repeated passes between such rolls, solid additives can be thoroughly dispersed into a viscous pasty mass. At any given time, only a small amount of material is in the high-shear zone. As a result, the product temperature does not rise much, as there is time and exposure for cooling. The material usually sticks to the hotter roll. After mixing is complete, heavy materials may be discharged simply by dropping from between the rolls. Thin, lighter mixes may be removed by a scraper bar pressing against the descending surface of one of the rolls.

Batch roll mills need longer mixing times and operator attention. Two roll mills are used primarily for preparing color pastes for inks, paints, and coatings. A few applications in heavy-duty blending of rubber stocks use corrugated rolls to masticate the material. Continuous mills for mixing pastes contain three to five horizontal rolls. These are placed in a vertical stack one above the other. The pasty mass passes from the slower roll to the faster ones. Three roll mills are used when extremely uniform dispersion is required. The mixing efficiency of roll mills is lower than that of the kneaders. As a result, roll mills have been replaced by kneaders and internal mixers in many applications.

12.4.5.1.6 Pan Muller Mixers (High-Shear Mixers) Pan mullers are industrial equivalents of traditional mortar and pestles. Muller mixers, also known as *pan mixers*, consist of two broad wheels that are mounted on an axle located inside a circular pan. These mixers are available in the following designs:

1. The pan is stationary and the mixer rotates above the midpoint of the axle, causing the muller wheels to rotate in a circular path over the solids on the floor of the circular pan. The mulling action occurs due to the slip of the wheels on the solids. The plows, which rotate with the mixer, push the material from the center and the walls of the mixing vessel into the path of the rollers. These plows also direct the material toward the floor of the pan at the end of the mixing cycle.
2. The pan is rotating and the axis of the wheels is held stationary.
3. Both the pan and the wheels are rotating. In these designs the wheels are not centered in the pan, but are offset.

Muller mixers are used to mix heavy solids and pastes which are not too pasty or sticky. In foundries they are used to mix small amounts of moisture and binder with sand for both core and molding sand. In addition, these mixers can be used to mix materials such as coatings for welding rods, storage-battery paste, and clay. They are especially effective in uniform coating of granular solids with small amounts of liquid. Muller mixers are available for both batch and continuous modes of operation.

12.4.5.2 Continuous Mixers For continuous mixers it essential to maintain a steady and uniform flow of materials, all of which are to be metered accurately in and out of the equipment. Although some of the high-viscosity mixer designs discussed earlier can be modified for continuous operation, these are rarely used since broad residence time distribution would result in product nonuniformity. Continuous mixers consist of one or more mixing elements positioned in a stationary trough. The clearance between the mixing elements and the trough is minimal. The mixer design generally has a provision for circulation of cooling media to maintain the temperature of the mixer contents. Mixer manufacturers offer many proprietary designs of continuous mixers. The following are some of the popular designs of continuous mixers for viscous materials.

12.4.5.2.1 Single-Screw Extruders Single-screw extruders typically consist of stationary barrel which houses a rotating screw having close clearance with the internal wall of the barrel. The material to be mixed is fed continuously in the feed throat area using a feed hopper. The design of the screw is such that the root diameter of the screw increases gradually from the feed point to the point of discharge. External jacket may be provided for heating or cooling of the product material. At the point of discharge, a plate with a well-defined opening size is provided to control the amount of pressure developed and the quantity of discharge. A discharge dye is fitted at the outlet to extrude the material in the required form, shape, and profile.

Single-screw extruders are widely used in the plastics industry to produce a homogeneous mixture of the polymer melt along with mixing additives such as stabilizing agents and fillers. Premixed dry ingredients are fed to the hopper. The plastic material melts due to the combination of heat generated by friction and heat transfer from the external surface of the screw barrel. The shear forces are high, especially in the melting zone. Mixing takes place due to laminar shear action.

The capacity of single-screw extruders depends on the geometry and construction of the equipment and the power connected. Screw diameters generally range from 25 to 200 mm in diameter. Larger units can be designed for specific applications. This equipment may also be provided with different types of mixing enhancers, which are used to provide elongation and shearing action to enhance axial (dispersive) and radial (distributive) mixing. The specific energy required for polymer mixing applications ranges from 0.15 to 0.3 kWh/kg.

12.4.5.2.2 Twin-Screw Extruders Twin-screw extruders consist of two screws that are housed in '8' shaped barrel. Mixing and shearing take place due to the interaction between the screw and the barrel, and also due to the interaction between the two screws. The screws may operate tangentially to each other or may be intermeshing. Based on the direction of rotation, these screws are classified as follows: (1) tangential, counter-rotating; (2) intermeshing, counter-rotating; and (3) intermeshing, co-rotating. Tangential designs allow variation in channel depth and lengths. In addition to applications in the plastic industry, twin-screw extruders are also used to carry out chemical reactions in high-viscosity materials.

The most common type of twin-screw extruder is the counter-rotating intermeshing type. This configuration provides high dispersive milling action between the screws and the ability to generate high pressure. The screw shafts are fitted with slip-on kneading or conveying elements that provide a wide range of mixing effects combined with compression, expansion, shearing, and elongation of material. The extruder barrel may be manufactured in segments to allow optimum positioning for feed ports, vents, and walls. The barrels may be provided with jackets for circulation of heat-transfer media. Alternatively, barrels can be heated electrically.

12.4.5.2.3 Pug Mills Pug mills consist of a single or twin shaft fitted with short heavy paddles rotating within an open trough or a closed cylinder. Single-shaft mills utilize an enclosed mixing chamber, whereas double-shaft mills are positioned in open troughs within the chamber. In two-shaft mills, the shafts are parallel to each other and may be either horizontal or vertical. Paddles may be positioned tangentially or may overlap each other. Pug mills can be designed for operation under vacuum and may be provided with external jackets for heating or cooling of the mix materials.

Solids are fed continuously into the mixing chamber from one end and discharged from the opposite end. Liquids may be added, depending on the process requirements. The paddles push the material forward as they cut through it. The product may be discharged through the open ports, or may be extruded through nozzles in the desired shape and cross section. The extruded material can be cut into pellets or blocks of required size. Pug mills are used to blend and homogenize clays and to mix liquids with solids to form thick, heavy slurries.

12.5 MECHANICAL COMPONENTS IN MIXING EQUIPMENT

Mixing equipment must be designed for process and mechanical operation. Although process requirements are primary to mixer design, equal emphasis on mechanical design is vital. Information on process parameters is required before starting the mechanical design. Moreover, some mixer applications have special requirements, such as variable-speed operation, mixer operation during material charging and discharging, additional process operations such as drying, and so on, that have to be considered while carrying out the mechanical design. The mechanical design of a mixer should account for the forces during the mixing process and their magnitude. These forces are a result of the interactions of the mixing element (commonly termed the agitator) with the process material, mixer vessel, and mixer internals, all of which produce stresses on the mixer assembly.

The purpose of this section is to provide important information about the mechanical components of mixers. In this section we discuss the various mechanical components that make up the mixer (i.e., motor, gear reducer, couplings, mixer seals, and bearings). Variable-speed operation devices are discussed.

12.5.1 Motors

Mixing requires power to be expended to the process material. Electrical ac induction motors are in most cases the primary source of power for mixers. Pneumatic or hydraulic motors are used in applications that require variable-speed operation and explosion-proof performance or a combination of both.

12.5.1.1 Electrical Motors While specifying electrical motors, the following essential information should be provided:

1. Single-phase/three-phase ac or dc motors
2. Supply voltage (V)
3. Frequency (Hz)
4. Power (hp)
5. Speed in revolutions per minute (rpm)
6. Motor frame size
7. Insulation class: A, B, F, H [this establishes the maximum safe operating temperature (see Table 12-7)]
8. Temperature classification
9. Amperage (full-load motor current)
10. Duty
11. Type classification
12. Explosion-proof motor

TABLE 12-7 NEMA Insulation Classes

	Maximum Temperature Allowed	
Class	°C	°F
A	105	221
B	130	266
F	155	311
H	180	356

Other parameters, such as motor efficiency, may also be defined.

Motors should preferably not to be loaded beyond 90% of their maximum rated current. While deciding the motor horsepower, in addition to the power requirement for mixing, the drive transmission losses should be considered. The magnitude of these losses would depend on the configuration of the drive system. Factors such as motor horsepower, type of gearbox, V-belt drives, chain–chain wheel drives, and type of seals determine the drive system losses.

12.5.1.2 Pneumatic Motors Pneumatic motors utilize compressed air to create rotational motion. Since the pneumatic motor is not driven electrically, there is no possibility of sparking, and hence these motors can be used in explosive applications. These motors are capable of variable-speed operation and are compact in size compared to electric motors. They are, however, inefficient in performance and therefore uncommon.

12.5.1.3 Hydraulic Motors As the name suggests, hydraulic motors are operated using a hydraulic fluid. The hydraulic fluid is circulated through the motor using a hydraulic system that consists of a hydraulic pump along with the supply and return line hydraulic piping. These motors provide features similar to those of pneumatic motors. Hydraulic motors provide high torque and flexibility for variable-speed operation. However, they tend to be more expensive than electric drives and are utilized only when essential.

12.5.2 Mixer Speed Reducers

Mixing equipment operates over a wide range of speeds, depending on the type of mixer and blender, the properties and characteristics of the material to be mixed, and any special process requirements. These speeds are generally lower than the standard motor speeds; therefore, some form of speed reduction is necessary. Speed reduction in mixers is generally achieved using belt drives or through speed-reduction gear drives.

12.5.2.1 Belt Drives Belt drives are among the simplest and most economical forms of speed reduction. In a belt drive, a pair of pulleys is used. The smaller of the two pulleys is mounted on the prime mover or the motor shaft. The larger pulley is mounted on the input shaft of the mixer. Single or multiple belts are used to transmit the torque from the prime mover to the mixer. An advantage of using belt drives is that small changes can easily be made in the mixer speed using different-size pulleys that fit on the same shaft. The speed reduction in a belt drive is generally limited to 4 : 1.

12.5.2.2 Gear Reducers Gear drives are also known as gear reducers or gearboxes. These are rugged mechanical devices designed to transmit high power at high operating efficiencies and have a long service life. The gear reducer is an important component of the mixer drive system, providing speed reduction and increasing allowable torque. Moreover, in some cases it provides support to the mixer shaft, which may or may not have independent shaft support. Gear reducers are categorized based on the relative position of their input and output shafts. The input and output shafts of gear reducers are either parallel to each other, at right angles, or may be in line. Different types of gears are used to manufacture gear reducers. The ones most widely used are helical, spiral bevel, worm, and planetary-type gears.

Helical gears are used in parallel shaft gear reducers. In helical gears, gear teeth are machined along a helical path with respect to the axis of rotation. Helical gears are commonly used with two-, three-, and in some cases even four-, five-, and six-stage speed reductions. In-line helical reducers are a variation of parallel shaft speed reducers configured such that the output and input shafts are in-line.

Spiral bevel gears are used when the input and output shafts of the gear reducers are required to be at right angles. The curve shape of the spiral bevel teeth makes gradual contacts, resulting in less noise during operation. Helical, parallel shaft, and helical bevel gear units have high operating efficiency, approximately 98% for each gear stage reduction.

Worm gears are the most economical speed reducers, capable of providing a sizable speed reduction with a single gear set. The input and output shaft of these gears are at right angles to each other. However, because of the sliding contact between the worm pinion and the gear, the worm gear reducer is less efficient. The efficiency decreases as the speed reduction ratio increases. For example, at a speed reduction ratio of 10 : 1, the efficiency of the worm gearbox may be approximately 90%. However, at a reduction ratio of about 50 : 1, the efficiency of the worm reducer drops to about 70%. Gearbox manufacturers offer gear reducers in helical bevel and helical worm design.

Helical, spiral bevel, and worm gears are external gears with the teeth on the outer periphery of the gears. In planetary gears the teeth profile is on the inside of a circular ring with meshing pinion. Planetary gears consist of an internal gear with a small pinion, known as a sun gear, surrounded by multiple planetary gears. These gears can provide high-speed reduction ratios and are relatively compact in size. Gear reducer manufacturers also offer geared-motor, that consists of a factory assembled motor with the gear unit. Figure 12-34 shows a variety of gear reducer with motor configurations.

Gear Ratios Although an infinite number of gear ratios are possible, due to practical limitations, the American Gear Manufacturers Association (AGMA) recommends normal ratios for each type of gear. The speed ratios are generally cataloged by the gear manufacturers. The speed reduction ratios may vary from as low as 5 : 1 to as high as 100 : 1.

Gear reducers for mixer applications are enclosed to prevent contamination and for operational safety reasons. The gear reducers are generally provided with large-diameter output shafts and heavy-duty bearings to handle the loads transferred by the mixer shaft. Several factors are to be accounted for when selecting speed reducers. These are explained below. It is always advisable to consult gearbox and mixer manufacturers while making the final selection of the mixer gear reducer.

Selection of Speed Reducers The selection of speed reducers is an important and critical part of the

(a)

Figure 12-34 (a) Helical parallel shaft geared-motor, (b) helical bevel geared-motor, and (c) helical in-line geared-motor. (Courtesy of Nord Drive Systems.) (d) Worm gearbox.

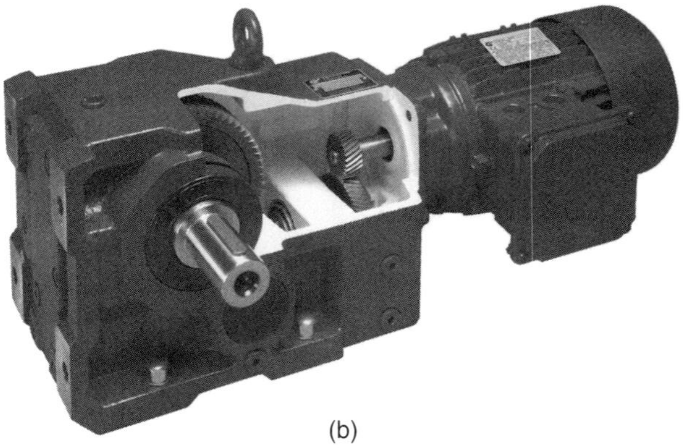

(b)

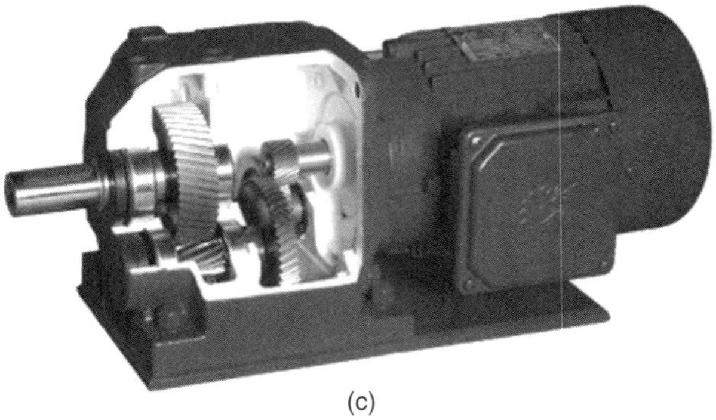

(c)

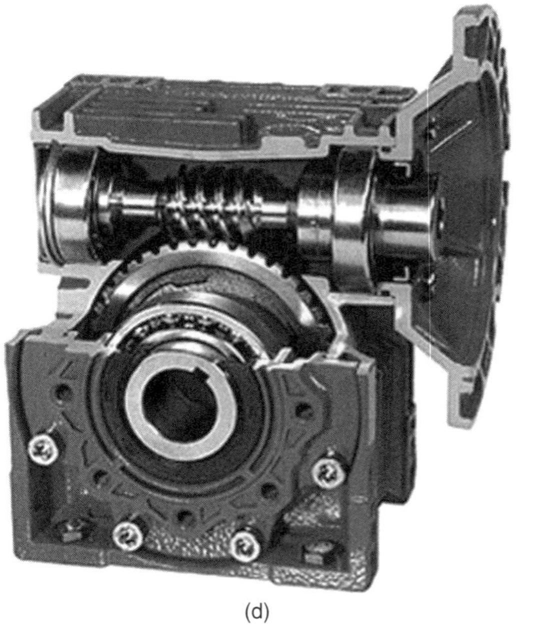

(d)

Figure 12-34 *(Continued)*

mixer drive system design. When ordering gear units, the following information should be specified to the gear reducer manufacturer:

1. Details of the prime mover
   a. Type: electric motor
   b. Power rating (kW/hp)

   c. Output speed of the electric motor
   d. Requirement for variable-speed operation of the mixer drive system, range of operating speed, and frequency of speed variation
   e. Dimensions of the prime mover
2. Details of the mixer
   a. Mixer agitator configuration: vertical or horizontal
   b. Operating environment: temperature and ambient conditions
   c. Range of operating speed of the mixer
   d. Data on power consumed during mixer operation
   e. Number of hours of mixer operation per day
   f. Load classifications of the mixer: uniform, moderate shock, heavy shock
   g. Reversal of direction of mixer, if applicable
3. Details of the gearbox
   a. In-line, parallel shaft, right-angle configuration
   b. Type of gear speed reducer: worm, helical, bevel helical, or planetary
   c. The required reduction ratio of the gearbox (defined as the ratio of the speed at the input shaft of the gearbox to the speed required at the gearbox output)
   d. Position of the output shaft relative to the input shaft (also termed shaft handling)
4. Shaft connection
   a. Couplings on the input and output shafts along with their bore dimensions and tolerances
   b. Details of overhung loads caused by sprocket, gear, or pulleys mounted on the gearbox shaft

The gearbox manufacturers use all of the information stated above to make an appropriate selection of speed reducer. Gearbox selection involves:

1. *Selection of a gearbox for mechanical capacity*. The lowest horsepower ratings of the gears, shafts, and bearings determine the mechanical horsepower rating of the gear drive.

2. *Determining the service horsepower for the gearbox*. The allowable service horsepower is the mechanical horsepower rating at the service factor divided by the appropriate service factor for the application. The service factor accounts for the nature of the prime mover loads, drive loads, operational hours, service life, and reliability.

A service factor of 1 generally defines the horsepower rating of a gearbox for a uniform load for 8 h of operation per day. Liquid agitators and mixers for lightweight free-flowing solids are generally considered to be moderate shock loads. A service factor of 1.25 to 1.5 is recommended for such applications. Heavy-duty mixers such as kneaders and intensive mixers are subjected to shock loads. For these applications, service factors of 1.5 (minimum) and higher should be specified.

The gearbox rating based on mechanical capacity is determined after accounting for the absorbed power and the mechanical service factor:

$$\text{mechanical power capacity}(P_m) = \text{absorbed power(kW)} \times \text{mechanical service factor}$$

3. *Selection of a gearbox based on thermal rating*. The AGMA defines the thermal rating of a gearbox as the horsepower that can be transmitted continuously without exceeding the temperature limitations specified. Factors such as proximity to hot equipment, a dusty environment, and so on, affect the thermal rating of a gearbox. The thermal load on the gearbox should be less than the rated thermal capacity of the gearbox.

4. *Selection based on overhung loads or axial thrust loads*. Whenever a sprocket, gear, sheaf, or pulley is mounted on the gearbox shaft, the overhung load on the shaft should be calculated. This should be less than the permissible overhung load on the shafts. The procedures for calculation of the overhung loads and the permissible values are cataloged by gear reducer manufacturers.

Lubrication of the gears and bearings as per the manufacturers' recommendations is important for smooth, trouble-free operation of the gear reducers. With proper selection and periodic maintenance, gear reducers can provide a trouble-free life of 10 years without major service. This life can be prolonged further by timely maintenance.

12.5.3 Couplings

A mixer assembly consists of three basic components: the mixing vessel, the agitator drive system, and the mixing element (agitator or blade). The agitator is installed within the mixing vessel; the agitator shaft extends out of the mixing vessel. The mixer drive system, consisting of the motor, gearbox, and other mechanical components, is located outside the mixing vessel. A coupling is a mechanical device used to connect the mixer shaft to the shaft of the drive system for the purpose of transmitting power. In some cases, mixer shafts may be long and may have to be shipped separately for installation at the plant. These are coupled using mechanical couplings. Mechanical couplings are broadly classified as rigid and flexible.

1. *Rigid coupling*. Rigid couplings are used when precise shaft alignment is required; shaft misalignment will affect the coupling's performance as well as its life. Rigid couplings can be of the following types:

- Sleeve or muff coupling

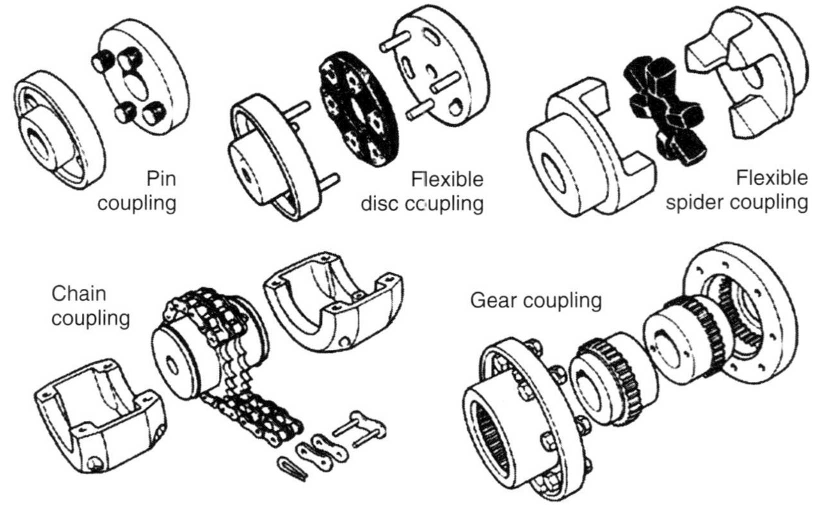

Figure 12-35 Types of couplings.

- Clamp or split-muff or compression coupling
- Flange coupling

2. *Flexible coupling*. Flexible couplings are designed to transmit torque while permitting some radial, axial, and angular misalignment. Flexible couplings can accommodate angular misalignment up to a few degrees and some parallel misalignment. The commonly used flexible couplings are as follows:

- Elastomer couplings: bushed pin coupling; tire, spider, or jaw coupling
- Resilient coupling
- Disk coupling
- Diaphragm coupling
- Gear coupling
- Roller chain-and-sprocket coupling
- Fluid coupling (Hydraulic coupling)

Figure 12-35 shows a variety of coupling designs commonly used in mixing equipment.

12.5.4 Bearings

The mixer drive system is subjected to both axial and radial loads that are transmitted by the mixer shaft. The type and magnitude of these loads depends on the configuration of the agitator, vertical or horizontal, and the forces caused by the mixing operation. These loads may be constant or fluctuating. To ensure proper transmission of these loads, selection of the appropriate type of bearings for the mixer drive system is essential. Bearings with rolling contacts are generally used for mixers. These bearings transfer loads between the rotating and stationary members and permit relatively free rotation with minimum friction. These bearings consist of rolling elements (balls or rollers) located between an outer and an inner ring. Cages are used to separate the rolling elements from each other. Figure 12-36 illustrates the bearing components.

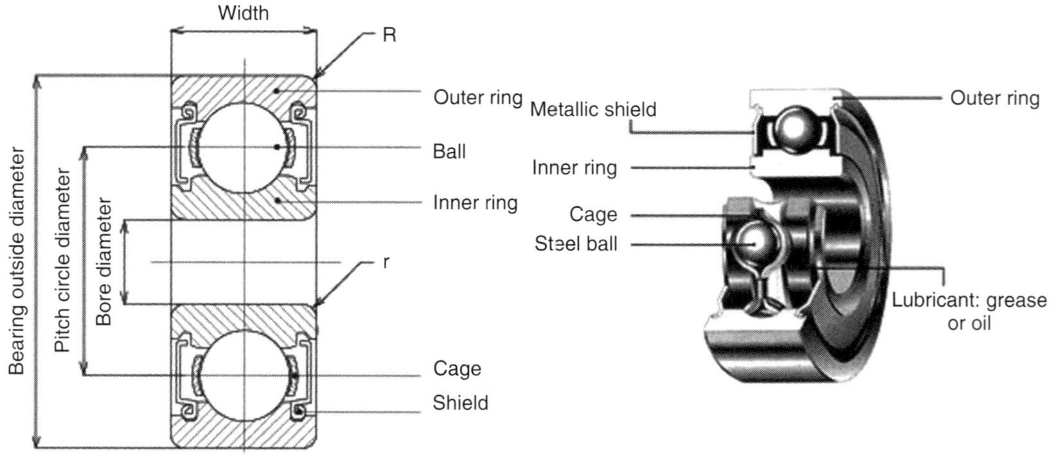

Figure 12-36 Bearing components.

The inner ring is generally fitted on the machined surface of the locating shaft. The outer ring is pressed into a machined component generally known as the bearing housing. During operation, both the inner and outer rings are expected to remain stationary with respect to their mating components. The rotation takes place on the rolling elements between the inner and outer rings. Various types of bearings used in mixers are described below.

1. *Ball bearings*. The rolling elements used in ball bearings are spherical balls. These types of bearings are the most commonly used antifriction bearings and are inexpensive. They are normally used for radial loads and moderate thrust loads. Ball bearings are generally used for high-speed drive shafts and portable mixers. They are unsuitable for large mixer output shafts, which transmit high axial force.

2. *Tapered roller bearings*. Tapered roller bearings, used for heavy radial and thrust loads, utilize a tapered roller to support the rotating shaft. These bearings transmit axial load in only one direction, so the orientation and position of these bearings is important. For transmitting axial loads in both directions, these bearings are fitted in opposed pairs.

3. *Spherical roller bearings*. Spherical roller bearings have two rows of rollers to carry thrust loads in both directions. They are excellent for heavy radial loads and moderate thrust. Their self-aligning feature is useful in mixer applications. Spherical roller bearings are popular in side-entering mixers and horizontal mixers where fluctuating loads can occur.

4. *Pillow block bearings*. Pillow block bearings are commonly used in side-entering horizontal mixers. These bearings consist of roller bearings placed within a support housing that can be mounted directly on a support base for the mixer, thereby eliminating the need for a separate bearing housing.

Selection of the type of roller element, ball, or roller is a function of many factors, such as the magnitude and type of loads, mixer speed, criticality of alignment, and space limitations. However, to determine if a ball or roller bearing should be selected, the following general guidelines may be used:

- Since ball bearings function on a theoretical point contact, they are best suited for high-speed applications and light loads.
- Roller bearings function theoretically on line contact and are designed to carry heavy loads more satisfactorily. These bearings have speed limitations, however, and are relatively more expensive.

An accurate knowledge of load-bearing capacity and expected life is essential in proper selection of ball and roller bearings. Other types of bearings may also be used. The mixer shaft can be fitted with one or more bearings of the same type, or of different types, depending on the loads.

The life of an individual bearing is defined as the total number of revolutions or hours at a given constant speed at which bearing runs before the first evidence of fatigue develops. When a group of identical bearings are run under a fixed condition of speed and load, it is not necessary that they fail at the same life. The life that 90% of apparently identical group of bearings will complete or exceed before the first evidence of fatigue develops is known as the *minimum life* or *L-10 rating life*. Mixer bearings should be selected for a L-10 life of at least 10,000 h. Timely lubrication of bearings should be carried out to ensure that the bearings serve their expected operational life.

12.5.5 Shaft Seals

Mixers are provided with seals whose function is to ensure that the mixer vessel contents are not exposed to the surrounding environment. Seals may be required when the operating pressure within the mixer is different from the atmospheric pressure or when the material being mixed is toxic, inflammable, or can vaporize. In horizontal side-entering and bottom-entering mixers, seals are provided to retain the vessel contents. Several types of seal designs are available; however, the most common shaft seals are (1) stuffing boxes, (2) mechanical seals, (3) hydraulic seals, and (4) lip seals.

12.5.5.1 Stuffing Boxes Stuffing boxes are the most versatile and simplest type of seal used in mixers. A stuffing box consists of a housing located around the mixer shaft that is filled with a compression packing material to minimize the leakage. A typical stuffing box seal is shown in Fig. 12-37. The mixer shaft enters the vessel through an opening in the mounting flange which is surrounded by tubular housing. Single or multiple rows of packing material are stacked in the housing. The sealing is achieved by radial expansion of the soft, deformable packing material. Radial expansion is accomplished by axial compression of the packing material within the restricted space inside the housing, resulting in removal of void spaces and subsequent leakage paths. The compression is obtained using an adjustable plate and ring known as the *gland follower*, which is located at the atmospheric end of the stuffing box. The adjustments are made with threaded fasteners. Depending on the number of rings of the packing material, the space of the mixing shaft, the operating speed of the mixer vessel, and related parameters, some leakage will occur around the shaft. Leakage may be in the form of vapor when the seal is positioned above the operational level of material within the mixer. When the seal is exposed

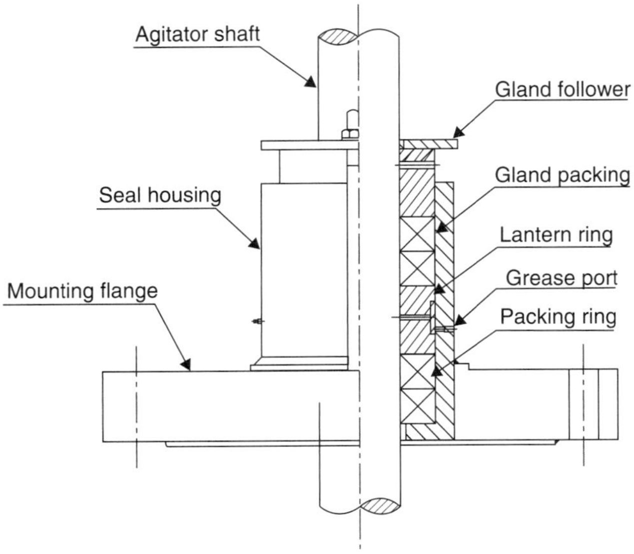

Figure 12-37 Stuffing box assembly. (Courtesy of Unique Mixers and Furnaces.)

directly to the process material, some leakage of material may take place. This leakage may either provide lubrication or abrasion, depending on the contents of the mixer.

When more than three or four rings of packing material are used, it may be difficult to obtain enough compression and deformation to provide effective sealing. To help maintain uniform compression, a sleeve commonly known as a *lantern ring* is placed within the housing between the multiple layers of packing. Besides ensuring compression, a lantern ring also provides a means for lubricating contact within the mixer shaft.

The packing material is generally made from braided fibers or metallic foils that are coated or impregnated with lubricating material. PTFE (polytetrafluoroethylene), Kevlar (aramid fiber), graphite filament, or acrylic fibers are used for braiding. These are commonly impregnated with PTFE and/or graphite. While selecting the packing material, the following factors should be considered:

1. It should be chemically compatible with the process material.
2. It should be chemically inert.
3. It must have good compressibility.
4. It must retain its properties within the operating conditions of the mixer.
5. It must have enough compressive strength to resist extrusion through small openings and transfer the packing gland axial compressive force from one ring to the next.

Because the stuffing box is a rubbing contact seal, the friction created with the rotating mixer shaft causes heat generation and wear. Therefore, some applications may require cooling arrangement for the seals. The cooling arrangement could be by purging air or inert gas in the stuffing box area or by providing an external cooling jacket around the stuffing box housing. To minimize the wear of the packing material, the initial tightening of the packing glands should be carried out such that the packing gland material deforms uniformly and fits snuggly onto the shaft. Excessive tightening may result in increased wear of the packing material. As some amount of wear is likely to take place during operation, stuffing boxes should be checked and tightened periodically to control the leakage. The packing material must be replaced when significantly worn out. The rubbing action of the packing material with the shaft also results in wear to the mixer shaft in the area of contact. Therefore, this portion of the mixer shaft should be provided with wear-resistant material deposits or fitted with sacrificial sleeves, so as to protect the shaft.

12.5.5.2 Mechanical Seals Mechanical seals are the most advanced type of seals used in the mixing equipment. With proper installation, they can handle high pressure and ensure leak-free operation with minimum maintenance. The downside is their higher initial cost and the special expertise needed to service these seals.

There are several mechanical seal designs, but they are all variations of a basic arrangement, consisting of a collar mounted on the shaft which uses springs to push a ring that rotates with the shaft against another ring, that is held stationary. The rings rotate against each other riding on a thin layer of lubricant, and the springs hold them so tightly together that leakage through the seal is virtually eliminated. The rotating part of the seal is called the rotating element, primary ring, or washer, whereas the part that is stationary and located in the mixer housing is known as the stationary element, mating ring, or seat. The mating surfaces of the rings must be perfectly flat to seal properly, and are manufactured to tolerances measured in "light bands." The rings must also be extremely hard to endure the pressure and wear. The rotating element is usually made from sacrificial material such as carbon or ceramic. The mating ring is made of harder material such as silicon carbide or tungsten carbide. The mating ring is held in place and maintains a static seal with the mounting housing using O-rings or gaskets. The rotating elements of the seal must attain a static seal with the shaft using O-rings, wedges, or packing. Mechanical seals are described as single or double mechanical seals depending on the number of seal faces (see Fig. 12-38).

1. *Single mechanical seal*. Single mechanical seals have a single pair of sealing elements. A basic feature of these seals is that rotating seal surfaces are at right angles to the axis of rotation of the mixer shaft. This arrangement

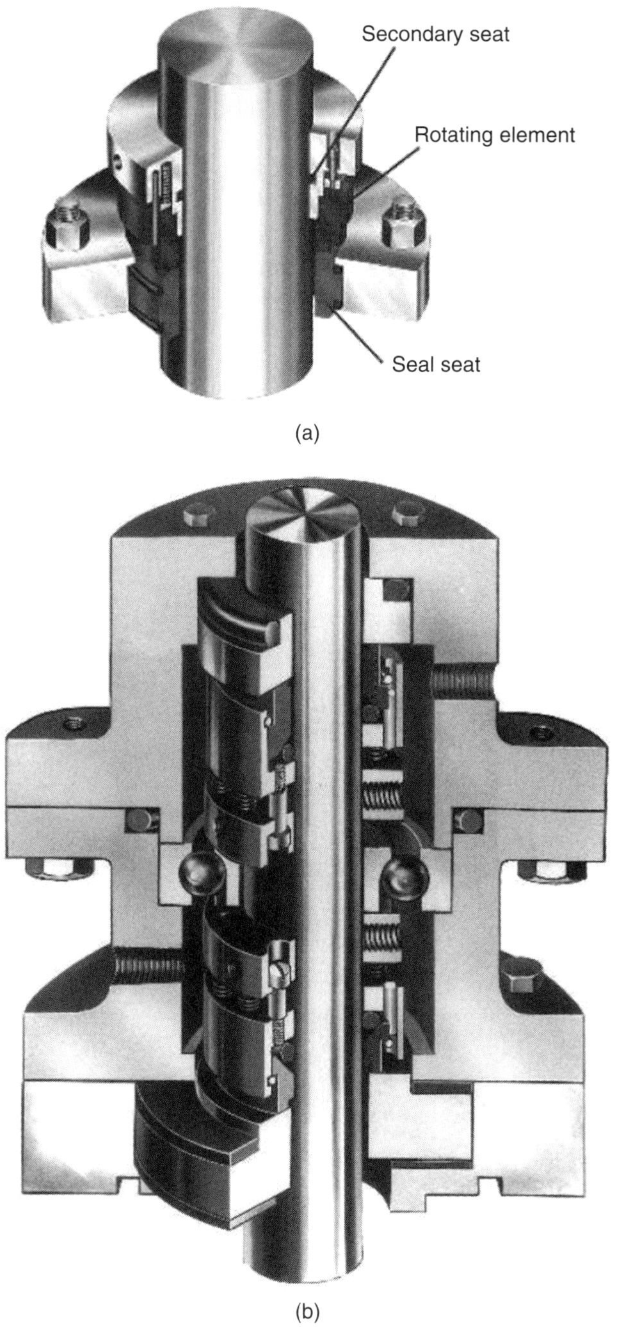

Figure 12-38 (a) Single mechanical seal; (b) double mechanical seal. (Courtesy of Flowserve.)

allows surfaces to be lapped to within wavelengths of light such that they have intimate contact and a low coefficient of friction under all conditions, even as the seal wears. Seals that are located with the rotating components within the vessel are called *internal seals*; those with rotational elements outside the vessel are called *outside seals*.

2. *Double mechanical seal*. As the name suggests, a double mechanical seal is a combination of two seals mounted back to back or is a single seal that has sealing faces on either sides (i.e., tandem type). These seals are suitable for handling media and operating conditions which otherwise cannot be handled by single seals. A leakproof seal is achieved by pressurizing a fluid between the seals such that the sealing fluid leaks in only one direction, which is outside.

Selection of a mechanical seal is affected by the pressure and temperature in the mixing vessel, the speed of the mixer, lubrication options, and the cleaning requirements specific to the product and mixing environment. It is recommended that the selection of a mechanical seal should be carried out in consultation with the mechanical seal manufacturer.

12.5.5.3 Hydraulic Seals Hydraulic seals are generally used for vapor retention and are limited to very low-pressure applications. Construction of a hydraulic seal is shown in Fig. 12-39. The hydraulic seals consist of an inverted cup that is attached to the mixer shaft. The cup rotates within an annular ring chamber attached to the mixer flange. This chamber is usually filled with the sealing liquid, generally water. The liquid forms a nearly frictionless barrier between the rotating shaft and the stationary flange. The only maintenance aspect is related to maintaining a liquid level within the seal, which can be facilitated through proper design. The hydraulic seals are inexpensive and are not subjected to wear. They can be used only when sealing is to be done above the operating level of the process material. Their use is limited to vertical mixer applications.

12.5.5.4 Lip Seals A lip seal is manufactured using elastomer material and is positioned in the gap between the rotating mixer shaft and stationary seal housing. The elastomer seal is held against the shaft by a small spring. These seals are normally used for closing the openings from which the mixer shaft passes, to keep out dirt or to limit the free exchange of process vapors with the surroundings. These seals are not suitable for pressure applications and exhibit high leakage rates even at low pressures. Lip seals are incapable of accommodating the mixer shaft deflections and hence have very limited use.

12.5.6 Variable-Speed Operation Devices

For some applications, the mixing operation is to be carried out at different speeds as it progresses from the start to the end. For example, process raw material may be in powdered solids form but get converted to viscous dough after addition of liquid or other additives. To optimize the mixing cycle time, the mixing of powdered solids can be carried out at a higher speed than the mixing of viscous dough, which is generally carried out at lower speeds.

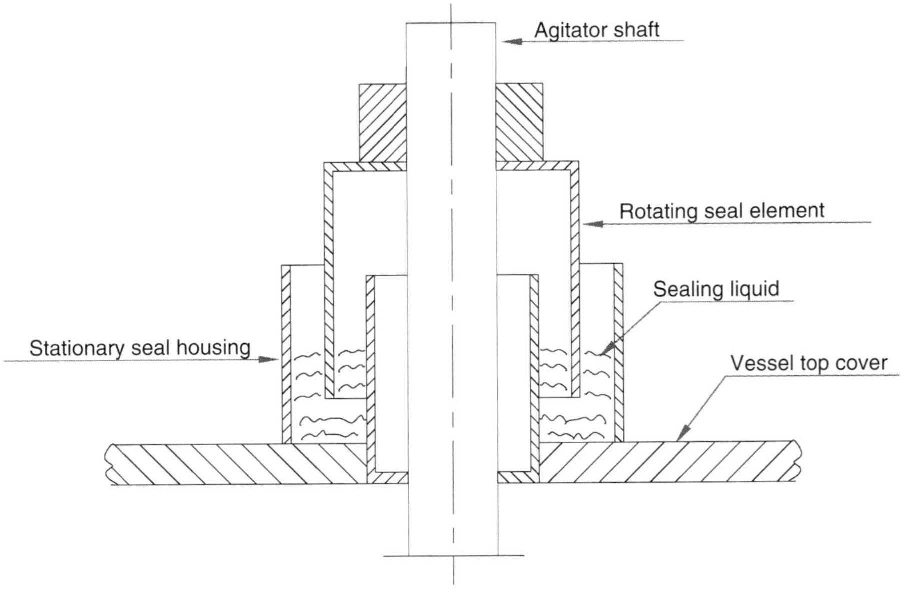

Figure 12-39 Hydraulic seal. (Courtesy of Unique Mixers and Furnaces.)

The variation in speed is accomplished either mechanically or electronically. The most common and least expensive mechanical means to achieve variable-speed operation is through the use of V-belts and sheaves with variable-pitch diameters. However, these are rarely used.

Advances in technology have made variable-frequency drives the preferred electronic device for changing the speed of mixers. Variable-frequency drives alter the electrical frequency input to the mixer motor, which in turn results in a change in the motor output speed. The variable-frequency drive provides stepless speed variation wherein the speeds can be adjusted infinitely and varied within the specified electrical frequency limits; however, this may require some modifications in the electrical motor. Special variable-frequency-drive-compliant motors can be purchased from manufacturers on request.

Use of a variable-frequency drive has virtually eliminated the use of dc motors for variable-speed operation. Two-speed electrical motors are available and can offer variable-speed operation of mixers at two defined speeds. Variable-speed operation can also be achieved using hydraulic motors, pneumatic motors, or hydraulic couplings.

12.5.7 Mixer Installation, Startup, and Maintenance

Good mixer installation, startup, and maintenance should be carried out by trained technical personnel according to the directives of the mixer manufacturer. Below we provide basic information on these aspects, which can be applied broadly to liquid, solid, and paste mixing equipment.

*12.5.7.1 **Mixer Installation*** Correct mixer installation can prevent potential maintenance problems occurring during operation. The following are some of the general guidelines for mixer installation:

- It is recommended that as far as possible, mixing equipment arriving at the site of operation be already assembled. If the mixing vessel and the agitator components are manufactured at different locations, it is advisable to ship the mixing vessel to the agitator manufacturer for assembly and factory tests. Trial runs of equipment (with or without process material) at the equipment manufacturer's shop more often than not help to minimize or resolve problems that may otherwise get noticed after installation on site.
- Depending on the weight of the mixer assembly, suitable handling arrangements should be made to position the mixer at the site of operation. Adequate care should be taken to prevent damage to mixer drive system and mixer components.
- Some equipment would require civil foundation. Consult the mixer manufacturer for such requirements.
- The equipment and its mounting structure and foundation should be inspected thoroughly. Any major deviations observed should be reported immediately to the equipment manufacturer, and corrective action should be taken prior to installation.
- The equipment should be positioned at the desired location, taking into consideration the utility requirements and availability. It should be ensured that adequate space is provided around the equipment for operation and maintenance.

- The alignment and leveling of the mixer assembly should be checked using a water level or spirit level.
- Ensure that all safety guards for rotating components are securely bolted to their positions.
- It is recommended that moving and rotating parts be checked with manual operation of the drive winch assembly.
- Complete all the electrical connections between the mains and the electrical control panel.
- Ensure that the mixer is connected in proper electrical phase sequence. The mixer shaft should rotate in the direction recommended by the equipment manufacturer. In case the direction of rotation is not correct, change the phase sequence of the power supply to achieve the required direction of rotation.
- An earthing connection should be provided to the mixer body and its electrical control panel using the suitable size and type of cable according to country norms.
- Check the tightness of the bolts of the motor, gearbox, mixer vessel assembly, and safety guards before startup of the drive system of the mixer.

12.5.7.2 Mixer Startup Follow-up checklist before mixer startup:

1. Check the oil level of the gear reducer. The gearbox should be filled with the grade of oil proposed by the manufacturer. In most cases, the grade of oil is ISO VG 220 mineral oil. For details, refer to the manufacturer's literature.
2. Connect and check the electrical supply to the electrical control panel for mixer operation.
3. Check the tightness of the bolts of the gearbox, motor, and other important drive components for proper bolting torques.
4. Ensure that the mixing vessel is thoroughly cleaned and there are no foreign particles inside. This exercise should be done prior to "switching on" the mixer motor.
5. Ensure that the mixer bottom valve is closed.
6. Ensure that the safety guards for all rotating components are in position.
7. Check the connections of utilities to the mixer.
8. Understand thoroughly the use of pushbuttons and digital instruments provided on the control panel, including that of emergency stop. If possible, check the operation of each pushbutton, instruments individually and collectively before actual use of the mixer.

On satisfactory compliance of the above checklist, the mixer is ready for startup. The mixer should be started in the "no load" mode.

1. Start the mixer and check the direction of rotation of the agitator shaft. The shaft should rotate in the direction recommended by the equipment manufacturer. In case the direction of rotation is not correct, change the phase sequence of the power supply to achieve the required direction of rotation.
2. Check the power or current drawn by the mixer motor when assembled to the mixer. This current should be only marginally higher than the current drawn by the motor under no-load conditions. It should, however, be much lower than the maximum rated current of the motor. In case excessive power is drawn, check the electrical and mechanical components for correct installation.
3. Check the noise level of the mixer during operation. It should be within the limits stated by the mixer manufacturer.
4. If the power drawn is in the range specified, allow the motor to run for 2 to 3 minutes and observe minutely the working of the mixer drive components.
5. In case of any abnormality of the functions, try to analyze and rectify the defect by following the instructions provided in the equipment manufacturer's manual. If the same cannot be resolved, contact the manufacturer for guidance.
6. If all the functions operate satisfactorily, test and run the mixer motor for a longer period: for half an hour followed by 1 to 2 h of idle run.
7. Check the operation of the features and instruments provided on the mixer.

After satisfactory no-load operation of the mixer, the mixer can be operated with a 50% material charge, a 75% material change, and a 100% material charge in a gradual, incremental sequence.

1. Load a limited quantity of the process material in the mixer and observe its functions in terms of mechanical and electrical operations. Do not use the mixer for any process material or operating conditions other than that for which it has been designed by the equipment manufacturer.
2. If the no-load functions of the mixer are within the permissible limits, increase the material charge quantity gradually, following the specified limits of a 50% material charge, a 75% material charge, and a 100% material charge ("full load"). When the mixer is loaded to its full capacity, the maximum current drawn by the mixer motor should not exceed 90% of the maximum rated motor current. As a rule of thumb, the maximum rated current of an electrical motor is approximately 1.5 times its motor horsepower.
3. On completion of the mixing cycle, open the bottom discharge valve until a product batch gets discharged. In equipment such as horizontal solid blenders and mixers for high viscosity materials and pastes some percentage of the product may remain within the mixing vessel due to the

Unique Mixers & Furnaces Pvt. Ltd.

Plot B-95, Road No. 27, Wagle Indl. Estate, M.I.D.C., Thane - 400 604, India
Tel.: 91-22-2582 2385, 2582 0172 | Fax : 91-22-2582 0992
info@uniquemixer.com | www.uniquemixer.com

MIXING EQUIPMENT SPECIFICATIONS

PERFORMANCE

Component	Weight %	Specific Gravity	Viscosity (Centipoise)	Temperature (°C)

Specific Gravity of final mix .... Viscosity of final mix ....(Centipoise)

SOLIDS		Soluble		Insoluble		Abrasive		Sticky		Crystalline		Fluffy

Kg/m^3 Mixture .... Particle Size ....

Suspension Specific Gravity.............. Settling Velocity .... (m/s)

CLASS	Blending	Dissolving	Suspending Solids	Cooking	Emulsifying	Heat Transfer	Gas Dispersion

MIXING INTENSITY	Violent	Medium	Gentle

- Batch: Smallest ____________ m^3 Normal ____________ m^3 Largest ____________ m^3
- Continuous: Rate ________ $m^3/hr.$

Mixer: Will / Will Not be operated during filling. Sequence of Addition...
Vessel Drawing No. ....

MATERIALS OF CONSTRUCTION

Mixing Vessel .... Shaft .... Impeller/Agitator...................................
Mounting Flange.................................. Stuffing Box .... Steady Bearings ....
Packing Material/Seals .... Other Wetted Parts ....

SELECTION

Required Vessel Opening Size .... Pressure Class .... Facing ....
Mixer Location on Vessel ....
Mixer Angle .... with ....

DESIGN

Impeller: Diameter .... Type .... Speed ....(RPM)
Normal HP (excluding drive losses) ....
Type of Bearings .... Steady Bearing Required? Yes / No Guide Bearings Required? Yes / No
Shaft Seal: Packing / Mechanical Seal / Hydraulic Seal / Other Make ....Type ....
Seal Coolant .... Stuffing Box Lubrication ....
Shaft Coupling: Type .... Make ....
Gear Reducer: Make .... Type ....
Size .... Ratio .... Rated H.P. .... Max. HP ....
Mechanical Efficiency ....% No of Reductions .... Output .... RPM Speed variable: Yes / No
Drive: Make .... Type.......... ....Speed............ RPM HP/KW ....
Electrical Power: ....Volts .... Phase ....Cycle. ....

Notes

....
....
....
....

By	Checked	Approved	Revision	Revision	Revision
Date					

Figure 12-40 Mixer specification sheet. (Courtesy of Unique Mixers and Furnaces.)

material flow characteristics and the mixer geometry. In mixers such as the double-arm sigma blade mixer, material discharge is by tilting the mixing vessel using manual, mechanical, or hydraulic arrangement.

4. Check the surface temperature, noise level of the gear reducer, bearing housings, stuffing box area, and so on. These should be within the permissible limits specified by the mixer manufacturer. In case of any abnormality of sound or excess heating observed, contact the manufacturer.

5. Continue monitoring the power supply and load drawn at a different percentage of loadings of the product mass and maintain records to check and confirm the performance of the mixer under full-load conditions. If the results are satisfactory, the second batch trial can be conducted.

6. Monitor the performance continuously for the first few trials and maintain the records before the mixer is qualified for regular production.

7. In the case of continuous mixers, the process of mixer startup would require several other considerations. In such cases, the equipment manufacturer's guidelines should be adhered to strictly.

12.5.7.3 Preventive Maintenance A periodic preventive maintenance program is necessary to ensure trouble-free mixer operation, increase equipment life, and minimize downtime. Equipment records, maintenance drawings, and repair and maintenance history should be well documented and easily accessible to the operational team. The requirement for critical equipment spares should be clearly identified and available. A preventive maintenance schedule, electrical and mechanical maintenance program, and detailed lubrication program should be defined and adhered to strictly.

12.5.7.3.1 Mechanical Maintenance

- Check the tightness of the equipment hardware once a week.
- Check the tightness of the gland followers to prevent leakage through the stuffing box.
- Follow the lubrication or greasing of the components as suggested.
- Fill the oil in the gear reducer only up to the point recommended by the manufacturer. Any excess filling would lead to spillage and can result in an accident.
- Ensure that safety guards are properly bolted to their positions.
- Analyze any abnormality in the noise and/or vibration levels and attend to it immediately.

12.5.7.3.2 Electrical Maintenance The electrical maintenance includes regular checks of the end terminals of the cables and the tightening of clamps. Maintenance personnel should be trained to monitoring and measuring the critical operating parameters of the equipment and installing and maintaining prime movers, instrumentation, and so on.

12.5.7.3.3 Lubrication Lubrication is one of the most critical aspects of equipment maintenance. If the mixer is put into use immediately after the supply, the greasing of bearings and other mechanical components is not necessary. In case of the delayed commissioning of the mixer, clean all the solidified lubricating agents and apply fresh lubricating agents before startup of the mixer. It is recommended that the first fill of oil in the gearbox be replaced after about 10,000 working hours of running or after two years, whichever is earlier. However, refer to the operational manual supplied by the gear reducer manufacturer for detailed information on the quantity and frequency of the makeup oil to be added. Lubrication and greasing of the bearings should be carried out at least once a week. The total degreasing and regreasing of the bearings should to be carried out based on operational usage.

12.5.8 Mixer Specifications

A sample liquid mixer specification sheet is shown in Fig. 12-40. The mixer specification sheet serves as a general checklist which includes details of the desired mixing objective and materials of construction for the mixer; these should be furnished to the mixer manufacturer during the inquiry stage. In response, the manufacturer should furnish details on selection and design of the mixing equipment.

It may not always be possible to define the mixing applications using datasheets. In such cases, laboratory and pilot-scale trials should be conducted for establishing mixer performance. Decisions on mixer choice should be based on previous experience and technical guidance from the mixer manufacturer.

REFERENCES

1. Clement, S., and Prescott, J., in C. Onwulata (Ed.), *Blending, Segregation, and Sampling: Encapsulated and Powdered Foods*, Food Sciences and Technology Series, Vol. 146, Taylor & Francis, New York, 2005.
2. Jenike, A., *Storage and Flow of Solids*, rev. 1980, University of Utah, Salt Lake City, UT, July 1994.
3. Ludwig, E. E., *Applied Process Design in Chemical and Petrochemical Plants*, Vol. I, Gulf Publishing, Houston, TX, 1995.
4. Maynard, E., Blender selection and avoidance of post-blender segregation, *Chemical Engineering*, May 2008.
5. McCabe, W. L., and Smith, J. C., *Unit Operations of Chemical Engineering*, McGraw-Hill, New York, 1993.

6. Oldshue, J. Y., *Fluid Mixing Technology*, McGraw-Hill, New York, 1983.
7. Paul, E. L., Atiemo-Obeng, V. A., and Kresta, S. M. (Eds.), *Handbook of Industrial Mixing*, Wiley, Hoboken, NJ, 2004.
8. Perry, R. H., and Green, D. (Eds.), *Perry's Chemical Engineers' Handbook*, 6th ed., McGraw-Hill, New York, 1984.
9. Walas, S. M., *Chemical Process Equipment: Selection and Design*, Butterworth-Heinemann Series in Chemical Engineering, Butterworth-Heinemann, Woburn, MA: 1998.

13

BOILERS

CELESTINE C. G. NWANKWO
Federal University of Technology, Owerri, Nigeria

Steam plants and indeed all other process plants must be equipped with relevant process equipment for specific operational requirements. Various pieces of process plant equipment exist for various plant operations. A boiler is one such piece of major equipment. Its main use is for a generation of steam for various process operations or applications and for heating applications. With many components, fittings, and accessories, it will be capable of serving the steam needs for industrial applications. Boiler steams are also used to drive turbines in power generation plants. This chapter is specifically focused on the boiler as process equipment.

A boiler is a closed vessel in which water or other fluid is heated to generate steam for process operations [14,15]. A boiler can also be defined as a device for generating steam which has two principal parts: the furnace, which provides heat by burning fuels; and the boiler proper, in which the heat changes water into steam. The steam or hot fluid is recirculated out of the boiler for use in various processes and in heating applications. The water fed to the boiler is usually made up of recycled steam in the form of condensed water and makeup water. The makeup water can be treated by any of the boiler water treatment processes (see Section 13.6.2) before use, or it can be used in its raw form without treatment. It implies that the boiler feed water composition depends largely on the quality of the makeup water and amount of recycled condensate.

In the boiler, the vaporizing water or steam escape with liquid droplets and gases, which can be discharged by the steam traps. The liquid droplets, gases, and other materials that could not escape with the steam are trapped by the water and are still in liquid form at the boiler bottom in the form of impurities. These impurities are usually blowndown by discharging some of the water from the boiler into the drain. The rate at which the blowdown is carried out may have an effect on the running cost of the boiler, so there is a need to reduce the rate of the blowdown to minimize the running cost.

To maintain and operate a boiler system properly, the boiler feed water should be adequately treated. This is because as steam is produced, dissolved solids concentrate to form deposits inside the boiler. This leads to poor heat transfer, reducing the efficiency of the boiler. Dissolved gases such as oxygen and carbon dioxide corrode the boiler by reacting with metals in the boiler system. So protection of the boiler from contaminants lies in the proper removal or control of contaminants by external or internal treatment of the boiler feed water.

The material for construction of boiler is steel or alloy of steel. Stainless steel can be used only in the superheated sections, where it will not be exposed to liquid boiler water. A steel material is more economical and therefore preferable to all other boiler materials, such as copper, brass, and wrought iron.

Wrought iron (usually the highest grade) is bulky, and its use is uncommon. The only suitable method of assembly is by riveting, which is very tedious and requires much labor. Copper or brass is more suitable for smaller live steam model boilers than wrought iron because it can be assembled by fabrication, which is easier. Copper, for example, is used for fireboxes, particularly in steam locomotives, because it has better formability and thermal conductivity. Since copper materials are so costly, steel, which in addition to its high formability and thermal

Process Plant Equipment: Operation, Control, and Reliability, First Edition. Edited by Michael D. Holloway, Chikezie Nwaoha, and Oliver A. Onyewuenyi.
© 2012 John Wiley & Sons, Inc. Published 2012 by John Wiley & Sons, Inc.

properties is strong and cheap, is the best construction material and is widely used by most boilermakers.

Brittle metal materials such as cast iron are not suitable for boiler construction. Such materials cannot withstand high-pressure steam if used to construct boilers. Cast iron, for example, is only used to construct the heating vessels of domestic water heaters. Such systems do not generate steam but only hot water. They are not actually boilers because they run at low pressure, avoiding actual boiling.

Safety is of great concern in regard to boilers and boiler operations. (See Sections 13.4 to 13.4.3 and 13.5.1 for common problems of boilers and boiler failure analysis, respectively.) Boilers used to be the source of serious injuries and property destruction. Poor understanding of engineering principles contributes much to this. For example, thin and brittle metal shells can rupture, or poorly welded or riveted seams could open up, leading to a violent eruption of the pressurized steam. There are many other hazards which can be controlled only through adequate knowledge of basic engineering principles properties and operational principles, or procedures or codes of the construction materials. Various fittings and accessories are often introduced or included in boiler designs to enhance safety and boiler efficiency.

Many types of boilers exist in different configurations. Many factors are considerable for boiler types and configurations, including economy, heating fuel types, heating methods, costs, size, storage capacity, and pressure. Any boiler must be able to perform the primary function of transfer of heat from hot gases generated by combustion of fuel into water until it turns to steam. The rate at which this heat-transfer operation occurs is measured as the efficiency of the boiler. In other words, boiler efficiency simply relates energy output to energy input in percentage terms.

13.1 TYPES OF BOILERS

Every boiler should be designed to absorb the maximum amount of heat released in the combustion process. The flow diagram in Fig. 13-1 is that of a typical boiler plant. Generally, a boiler can be primarily a water tube or fire tube arrangement. Fire tube boilers consist of a firebox surrounded by water, whereas for water tube boilers, water is contained in the tubes and heated externally.

However, classification of boilers as water tube and fire tube only accounts for just the basic arrangement between heat source and the boiler water. Boiler configurations can also be tailored to account for various other factors. Therefore, boiler manufacturers build different types of boilers based on considerations such as the following:

1. Applications
   a. Types of process operations
   b. Pressure levels
   c. Steam demand
2. Types of heating fuel
   a. Gases
   b. Fuel oil
   c. Solids (wood, coal, etc.),
   d. Nuclear fission, etc.
3. Combustion systems

In this section, therefore, we discuss the various types of boilers: water tube boiler, fire tube boilers, pot boilers, locomotive boilers, saddle boilers, packaged boilers, fluidized-bed combustion boilers, stoker-fired boilers, pulverized fuel boilers, waste heat boilers, thermic fluid heaters, and superheated steam boilers.

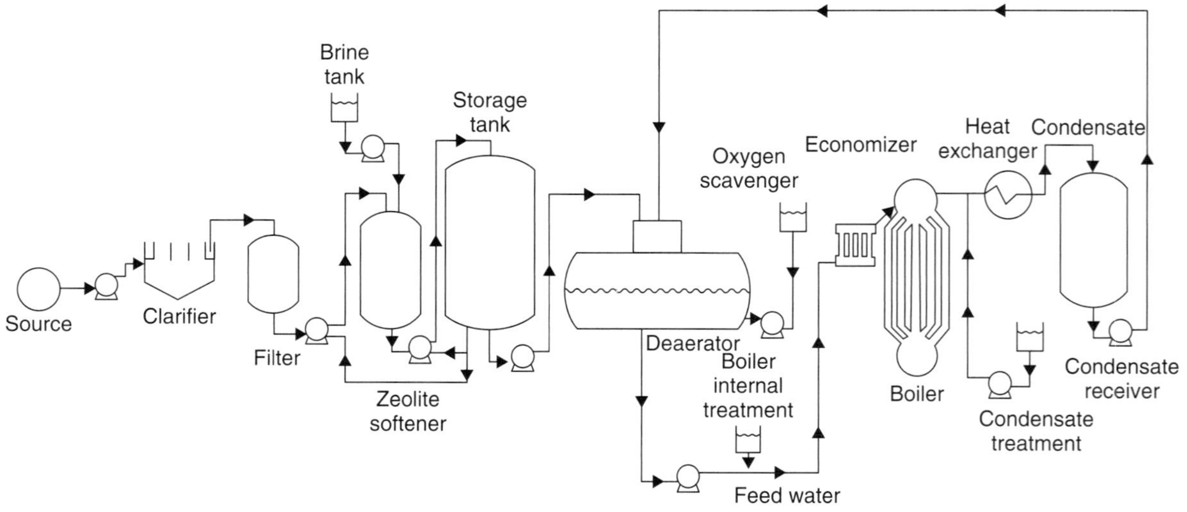

Figure 13-1 Typical boiler plant flow diagram.

13.1.1 Water Tube Boilers

A water tube boiler is a type of boiler in which water circulates in tubes heated externally by the fire (see Fig. 13-2). The circulated water is heated by the hot gases and converted into steam at the vapor space in the drum. These hot gases are created by burning fuel inside the furnace. The heated water then rises into the steam drum, where saturated steam is either drawn off through the drum top or made to reenter the furnace through a superheater to become superheated. Cool water at the bottom of the steam drum returns to the feed water drum via a large-bore downcomer tube (see Fig. 13-3) where feed water is preheated. To increase the economy of the boiler, exhaust gases are also used to preheat the air down into the furnace and warm the feed water.

Water tube boilers can be designed to exploit any heat source and are generally preferred in high-pressure applications. High-pressure water and steam are constrained within small-diameter pipes, whose thinner walls can withstand the pressure.

Water tube boilers have very wide applications in:

- Stationary plant operation (steam generation for heating or as a chemical component and for driving turbines for power generation)
- Marine systems (steam-driven turbines for propulsion engines)

Many applications may involve a hybrid of the water tube, combined with fire tube systems. For example, since the hottest part of a locomotive boiler is the firebox, a water tube design with a conventional fire tube boiler as an economizer in the usual position can serve as an effective design.

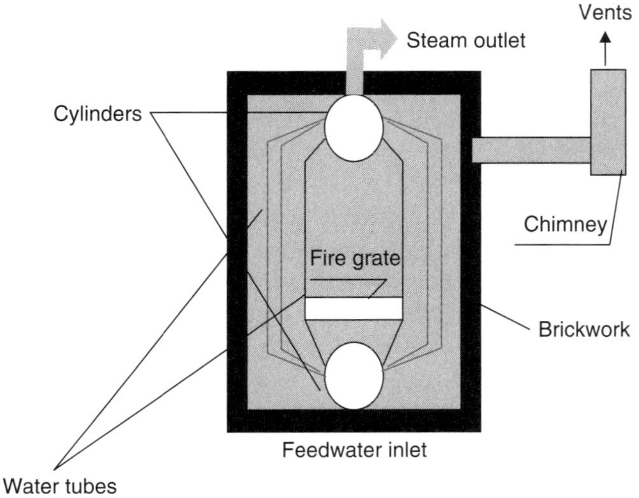

Figure 13-2 Schematic diagram of a typical water tube boiler.

There are numerous variations in water tube boilers:

1. *D-type boiler*: used in both stationary and marine applications. A large steam drum is connected vertically to a smaller water drum via multiple steam-generating tubes. These are surrounded by walls made up of large water-filled tubes, which make up the furnace.

2. *Low-water-content boiler*: does not have a furnace. Water tubes are impinged upon directly from the burner, while the lower and upper headers of the boiler are connected directly to the water tube.

3. *Babcock & Wilcox boiler*: has a simple drum. Feed water is drawn into a heater that supplies inclined water tubes. The water tubes supply steam back into the top of the drum. Furnaces are located below the tubes and drums.

4. *Stirling boiler*: has nearly vertical but almost straight water tubes that zigzag between a number of steam and water drums. A Stirling boiler may comprise variations of a number of drums and banks. They are used primarily as stationary boilers, as they are of large size.

5. *Yarrow boiler*: has three drums in a delta formation connected by water tubes. The drums are linked by straight water tubes with an allowance for easy tube cleaning. In other words, the tubes enter the drums at varying angles, making the joints more difficult to caulk. Outside the firebox, a pair of cold-leg pipes between the drums act as downcomers. A Yarrow boiler is compact in size but has a larger water capacity, due to its three drums [9].

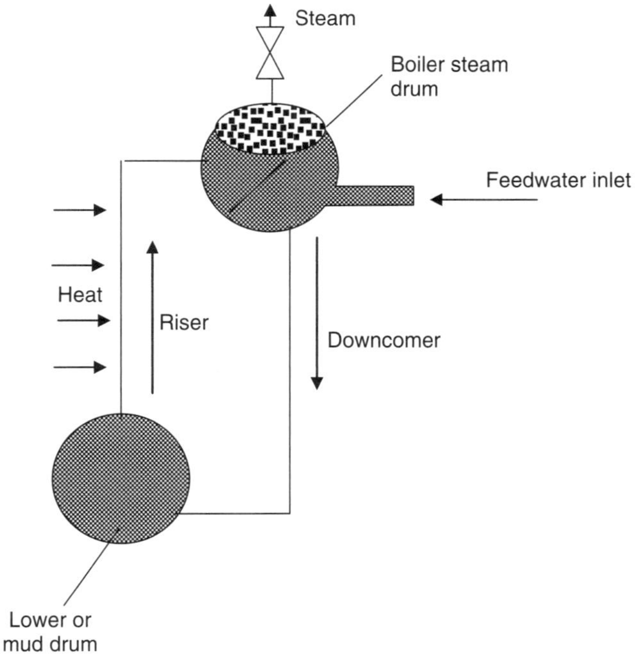

Figure 13-3 Natural water circulation in water tube boiler.

6. *White-Forster boiler*: similar to the Yarrow boiler. It differs only in having gradually curved water tubes. This makes their entry into the drums perpendicular, thus simplifying a reliable seal. [9].

7. *Thorny Croft boiler*: has a single steam drum with two sets of water tubes either side of the furnace. The tubes have a sharp curve. Warm up of the tubes gives rise to bending forces that tend to pull the tubes loose from the tube plate, creating a leak. The boiler system consists of two furnaces that vent into a common exhaust, giving the boiler a wide-base tapering profile. [9].

13.1.2 Fire Tube Boilers

A fire tube boiler is a type of boiler in which the heat source, the furnace or firebox, produces hot gases that pass through one or more tubes, running through the boiler feed water on the shell side, converting it to steam (see Fig. 13-4). The heat energy of the gases is conducted through the sides of tubes to heat the water and generate steam. Fire tube boilers are designed such that the furnace or firebox should always be surrounded by water to keep the heating surface at a constant temperature—just at the boiling point. The boiler has a comparatively low rate of steam production and a high steam storage capacity.

There are numerous variations of fire tube boilers:

1. *Cornish boiler*: a long horizontal cylinder with a single large flue that contains the fire. An iron grating usually kept across the flue carries the fire. A shallow ash pan is kept below to collect the noncombustible residue. A tall chimney at the far end of the flue facilitates an adequate air supply to the fire.

2. *Lancashire boiler*: has two large flues that contain the fire. The Lancashire boiler was invented by William Fairbairn in 1844. Considering the thermodynamics of more efficient boilers, he increased the furnace grate area relative to the water volume (www.moss.org.uk). Galloway later introduced crosswise water tubes across the flue as an improvement to Fairbairn's invention. This was to increase the heated surface area. The tubes are short, with a large diameter, and the boiler uses relatively low pressure. This is still not considered to be a watertube boiler. The tubes are tapered, simply to make their installation through the flue easier [10].

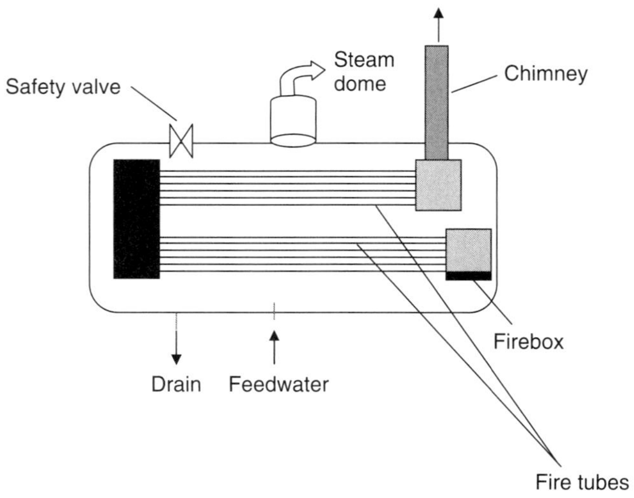

Figure 13-4 Schematic diagram of typical fire tube boiler.

3. *Scotch Marine boiler*: has a single large-diameter-tube furnace with many small tubes arranged above it. Those small tubes are connected through a combustion chamber. The combustion chamber is within the boiler shell, and flue gases from the fire tube flow from back to front. A smoke box introduced to cover the front of the tubes is channeled to the chimney. Some Scotch boilers have a pair of furnaces; others have three.

4. *Locomotive boiler*: operates with a forced draft provided by reinjecting exhausted steam into the exhaust through a blast pipe in the smoke box. In some locomotive boilers, a superheater element is introduced into the boiler barrel. The three main components of locomotive boilers are (1) a double-walled firebox, (2) a horizontal cylindrical boiler barrel containing a large number of small flue tubes, and (3) a smoke box with a chimney. Locomotive-type boilers are used in traction engines, steam rollers, portable engines, and other steam road vehicles. The boiler is used as the basis for the vehicle such that the other components are mounted on brackets attached to the boiler [2].

5. *Vertical firetube boiler*: has a vertical cylindrical shell that contains several vertical flue tubes.

6. *Horizontal return tubular boiler*: has a horizontal cylindrical shell that contains several horizontal flue tubes. The fire is located directly below the boiler's shell, within a brickwork setting.

13.1.3 Pot Boilers

A pot boiler is a kettle type of boiler where a fire heats a partially filled water container from below to its external surface. Primitive pot boilers could only produce and store a large volume of very low pressure steam, often scarcely above atmospheric pressure. With some recent modifications, such as the addition of a smoke tube and a stainless steel shield that encloses the burner and the lower portion of the boiler, its steaming ability has increased significantly. Such a pot boiler is a good steam generator in moderate temperatures and mild winds. These boilers are widely utilized in tank locomotives because the tanks hide the fire shield. Natural drafts are sufficient for the firing needs of the boiler. A pot boiler is designed for large capacity for water and low capacity for steam. Direct contact of the burner flame with the outer barrel surfaces usually discolors them. Discoloration can be overcome by using a cleaner burner such as one of the wick or vaporizing type (www.southernsteamtrains.com).

13.1.4 Saddle Boilers

Saddle boilers consist of a large-diameter copper tube with end plants and a small cross tube. The inclined cross tube provides both extra water capacity and additional heating surface. They improve water circulation and also act as stays. The arc is filled with a stainless steel firebox guiding the flame flow, to provide the maximum heating area.

13.1.5 Packaged Boilers

Packaged boilers have the following features:

- A small combustion space and a high heat release rate, resulting in faster evaporation
- A large number of small-diameter tubes, leading to good convective heat transfer
- Forced- or induced-draft systems, resulting in good combustion efficiency
- A number of passes, resulting in better overall heat transfer
- Higher thermal efficiency levels than those of other boilers

Packaged boilers come as a complete package, requiring only steam and water pipes, a fuel supply, and electrical connections to be fully operational. They are generally of shell type with fire tube design so as to achieve a high heat-transfer rate by both radiation and convection.

Packaged boilers are classified based on the numbers of passes, that is, the number of times that hot combustion gases pass through the boiler. A two-pass unit boiler consists of a combustion chamber and a set of firetubes. A three-pass unit boiler consists of a combustion chamber and two sets of firetubes. Generally, an N-pass unit boiler consists of one combustion chamber and $N - 1$ sets of fire tubes. The exhaust gases exit through the rear of the boiler.

13.1.6 Fluidized-Bed Combustion Boilers

When evenly distributed air or gas is passed upward through a finely divided bed of solid particles, such as sand, supported on a fine mesh, the particles are undisturbed at low velocity. As the air velocity is gradually increased, a stage is reached when the individual particles are suspended in the airstream. Such an arrangement or setup is known as a *fluidized bed*. With further increase in air velocity, there is bubble formation, vigorous turbulence, rapid mixing, and formation of a dense defined bed surface. The bed of solid particles exhibits the properties of a boiling liquid and assumes the appearance of a fluid. This is referred to as a *bubbling fluidized bed*.

If sand particles in a fluidized state are heated to the ignition temperature of coal, and coal is injected continually into a fluidized bed, the coal will burn rapidly and the bed will attain a uniform temperature. Fluidized-bed combustion (FBC) takes place at about 840°C (1544°F) to 950°C (1742°F). Since this temperature is much below the ash, fusion temperature, melting of ash, and associated problems, are avoided.

This lower combustion temperature is achieved because of a high coefficient of heat transfer due to rapid mixing in the fluidized bed and effective extraction of heat from the bed through in-bed heat-transfer tubes and walls of the bed. The gas velocity is maintained between the fluidization velocity and the particle entrainment velocity. This ensures stable operation of the bed and avoids particle entrainment in the gas stream. The fuels burned in FBC boilers include coal, washery rejects, rice husks, bagasse, and other agricultural and industrial wastes. The boilers have a wide capacity range of 0.5 tons/h to over 100 tons/h.

Descriptions of various types of FBC boilers follow.

1. *Atmospheric fluidized-bed combustion* (AFBC) *boiler.* The AFBC boiler involves little more than adding fluidized-bed combustion to a conventional shell boiler. These systems can also be installed in conjunction with conventional water tube boilers. Coal is crushed to a size of 1 to 10 mm, depending on the rank of coal or type of fuel fed to the combustion chambers. The atmospheric air that acts both as fluidization air and combustion air is preheated by the exhaust fuel gases. The in-bed tubes carrying water generally act as an evaporator. The gaseous products of combustion pass over the upper heater sections of the boiler and flow past the economizer, the dust collectors, and the air preheater before being exhausted to the atmosphere.

2. *Pressurized fluidized-bed combustion* (PFBC) *boiler.* In a PFBC boiler, a compressor supplied the forced-draft air, and the combustor is a pressure vessel. The heat release rate in the bed is proportional to the bed pressure, and hence a deep bed is used to extract a large amount of heat. This will improve the combustion efficiency and SO_2 absorption in the bed. The steam is generated in the two tube bundles, one in the bed and the other above it. Hot flue gases drive a power-generating gas turbine. The PFBC system can be used for cogeneration (steam and electricity) or combined cycle power generation. The combined cycle operation (gas turbine and steam turbine) improves the overall conversion efficiency by 5 to 8%.

3. *Atmospheric circulating fluidized-bed combustion* (CFBC) *boiler.* In a circulating system, the bed parameters

are so maintained as to promote solids elutriation from the bed. They are lifted in a relatively dilute phase in a solids riser, and a downcomer with a cyclone provides a return path for the solids. There are no steam generation tubes immersed in the bed. Generation and superheating of steam takes place in the convection section, water walls, and at the exit of the riser. CFBC boilers are well utilized for industrial applications requiring more than 75 to 100 tons/h of steam. Its tall furnace characteristics offer good space utilization, high fuel particle and sorbent residence times for efficient combustion and SO_2 capture, and easy application of staged combustion techniques for NO_x control. CFBC boilers are generally more economical than AFBC boilers.

13.1.7 Stoker-Fired Boilers

Stokers are classified according to the method of feeding fuel to the furnace and by the type of grate. The two main classifications are spreader stokers and chain- or traveling-gate stokers.

1. *Spreader stoker.* The spreader stokers utilize a combination of suspension burning and grate burning. The coal is continuously fed into the furnace above a burning bed of coal. The coal fines are burned in suspension; the larger particles fall to the grate, where they are burned in a thin, fast-burning coal bed. This method of firing provides good flexibility to meet load fluctuations, since ignition is almost instantaneous when the firing rate is increased. This is why the spreader stoker is favored over other types of stokers in many industrial applications.

2. *Chain- or traveling-grate stoker.* In this type, coal is fed onto one end of a moving steel grate as the grate moves along the length of the furnace. The coal burns before dropping off at the end as ash. The grate, air dampers, and baffles are set up with skill to ensure clean combustion, leaving a minimum of unburned carbon in the ash. The coal feed hopper runs along the entire coal-feed end of the furnace. A coal gate is used to control the rate at which coal is fed into the furnace by controlling the thickness of the fuel bed. Coal must be uniform in size, as large lumps will not burn out completely by the time they reach the end of the gate.

13.1.8 Pulverized Fuel Boilers

Pulverized coal is used primarily in coal-fired power station boilers as well as in many large industrial water tube boilers. The technology is well developed. The coal is pulverized to a fine powder, so that less than 2% is above 300 μm and 70 to 75% is below 75 μm for a bituminous coal. The pulverized coal is blown with part of the combustion air into the boiler plant through a series of burner nozzles. Secondary and tertiary air may also be added. Combustion takes place at temperatures from 1300°C (2372°F) to 1700°C (3092°F), depending largely on the coal grade. Particle residence time in the boiler is typically 2 to 5 s, and the particles must be small enough for complete combustion to have taken place during this time. One of the most popular systems for firing pulverized coal is tangential firing using four burners corner to corner to create a fire ball at the center of the boiler.

13.1.9 Waste Heat Boilers

Whenever waste heat is available at medium or high temperature, a waste heat boiler can be installed economically. Whenever the steam demand is higher, the steam generated by waste heat and auxiliary fuel burners is also used. If there is no direct use of steam, the steam may be let down in a steam turbine–generator set, and power produced from it. It is used widely in heat recovery from exhaust gases from gas turbines and diesel engines.

13.1.10 Thermic Fluid Heaters

Thermic fluid heaters employ petroleum-based fluid as the heat-transfer medium to provide constant maintainable temperatures for the user equipment. The combustion system comprises a fixed grate with mechanical draft arrangements. The oil-fired thermic fluid heater consists of double-coil, three-pass construction fitted with a modulated pressure jet system. The thermic fluid, which acts as a heat carrier, is heated up in the heater and circulated through the user equipment. There it transfers heat for the process through a heat exchanger, and the fluid is then returned to the heater. The flow of thermic fluid at the user end is controlled by a pneumatic control valve, based on the operating temperature. The heater operates on a low or high fire, depending on the return oil temperature, which varies with the system load.

13.1.11 Superheated Steam Boilers

This type boils the water and then further heats the steam in a superheater. This provides steam at much higher temperatures but can decrease the overall thermal efficiency of the steam-generating plant, due to the fact that the higher temperatures require higher flue gas exhaust temperatures. One of the ways to circumvent this problem is to provide a feed water heating economizer and/or a combustion air heater in the hot flue gas exhaust path. The superheater works like coils in an air-conditioning unit, but to a different end. The process of superheating steam is, most importantly, designed to remove all droplets entrained in the steam to prevent damage to the turbine blade and/or associated piping.

13.2 BOILER ACCESSORIES

A number of items must be fitted to steam boilers, all with the objective of improving operation, efficiency, and safety. A steam boiler system consists primarily of a boiler unit and a combustion unit. The steam so generated requires some incentives for its effective utilization.

13.2.1 Fittings and Accessories at the Boiler Unit

Some of the boiler fittings and accessories are:

- *Safety valve*: used to relieve pressure and prevent possible explosion of a boiler.
- *Water-level indicators*: show the operator the level of fluid in the boiler; also known as a sight glass, water gauge, or water column.
- *Bottom blowdown valves*: used to remove solid particulates that condense and lay on the bottom of a boiler.
- *Continuous blowdown valve*: allows a small quantity of water to escape continuously. Its purpose is to prevent the saturation of boiler water with dissolved solids.
- *Flash tank*: vessel used to collect the high-pressure blowdown. In it the steam can flash safely (i.e., the sudden rush of steam as a result of blowdown is safe and free of any form of danger) and can be used in a low-pressure system or be vented to the atmosphere, while the ambient pressure blowdown flows to a drain.
- *Automatic blowdown/continuous heat recovery system*: allows the boiler to blow down only when makeup water is flowing to the boiler. This enables the transfer of the maximum amount of heat possible, from the blowdown to the makeup water. No flash tank is needed, as the blowdown discharge is close to the temperature of the makeup water.
- *Handholes*: steel plates installed in openings in a header, to allow for inspections and installations of tubes and inspection of internal surfaces.
- *Steam drum internals*: a series of screen, scrubber, and cyclone separators.
- *Low-water fuel cutoff*: a mechanical means (usually, a float switch) that is used to turn off the burner or shut off fuel to the boiler to prevent it from running once the water goes below a certain point. If a boiler is dry fired (i.e., burned without water in it), it can cause a rupture or catastrophic failure.
- *Surface blowdown line*: provides a means of removing foam or the lightweight noncondensable substance that tends to float on top of the water inside a boiler.
- *Circulating pump*: designed to circulate water back to the boiler after it has expelled some of its heat.
- *Feed water check valve or clack valve*: a nonreturn stop valve in the feed water line. This may be fitted to the side of the boiler just below the water level, or to the top of the boiler.
- *Top feed valve*: a check valve mounted on top of the boiler, intended to reduce the nuisance of lime scale. It causes lime scale to be precipitated in a powdery form that is easily washed out of the boilers.
- *De-superheater tubes or bundles*: a series of tubes or bundles of tubes in the water drum or steam drum designed to cool superheated steam, for the safety of auxiliary equipment that does not need, or may be damaged by, dry steam.
- *Pressure gauge*: connected to the steam space of a boiler. They usually have a ring siphon tube that fills with condensed steam and protects the dial mechanisms from high temperatures.
- *Chemical injection line*: a connection to add chemicals to controll feed water pH.

13.2.2 Steam Accessories

13.2.2.1 Main Steam Stop/Check Valves These are used on multiple-boiler installations.

13.2.2.2 Steam Traps These are devices used to discharge condensate and noncondensable gases with a negligible consumption of live steam. Some steam traps are nothing more than automatic valves. They open, close, or modulate automatically. Others are based on turbulent two-phase flows to obstruct the steam flow. The three important functions of steam traps are:

1. Discharge of condensate as soon as it is formed
2. Negligible steam consumption
3. Discharging of air and other noncondensable gases

Four types of steam traps are discussed next.

1. *Mechanical traps*. This type of steam trap operates by the principle of floatation. A float rises and falls in relation to the condensate level, and there usually is a mechanical linkage that opens and closes the valve. Mechanical traps operate in direct relationship to condensate levels present in the body of the steam trap. Examples of mechanical traps include inverted bucket and float traps.

2. *Temperature traps*. This type of steam trap has an on/off control valve, which is driven by a temperature change that causes expansion and contraction of the fluids. The design of temperature traps requires them to hold back some condensate until it cools sufficiently to allow the valve to open. In most circumstances this is not desirable, as condensate needs to be removed as soon as it is

formed. Examples are thermostatic traps, bi-thermostatic traps, bimetallic traps, and smart steam traps (a combination of thermostatic traps and bi-thermostatic traps).

3. *Thermodynamic* (TD) *traps*. This type of stream trap works on the principle of the difference in dynamic response to a velocity change in the flow of compressible fluids. As the steam enters, the static pressure above the disk forces the disk against the valve seat. The static pressure over a large area overcomes the high inlet pressure of the steam. As the steam starts to condense, the pressure against the disk lessens and the trap cycles. This essentially makes a TD trap a time-cycle device. It will open even if there is only steam present, which can cause premature wear. If noncondensable gas is trapped on top of the disk, it can cause the trap to be locked shut.

4. *Venturi (orifice) traps*. This type of steam trap works in a turbulent two-phase flow regime. Internally it consists of a venturi valve of a certain shape. Condensate is fully discharged alongside eventual steam that tries to pass the venturi. But while traversing the venturi toward the low-pressure zone, the steam expands and chokes the throughput, together with the slow condensate. Because of this, the amount of live steam escaping the orifice is negligible.

13.2.3 Combustion Accessories

13.2.3.1 Fuel Oil System An oil burner is a mechanical device that combines fuel oil with proper amounts of air before delivering the mixture to the point of ignition in a combustion chamber. A fuel oil burner either vaporizes and/or atomizes the fuel oil.

13.2.3.2 Gas Combustion System As in a fuel oil system, the gas burner combines the fuel gas with proper amounts of air before delivering the mixture to the point of ignition in a combustion chamber. Atomization is not necessary for gas combustion.

13.2.3.3 Solid Fuel Combustion Systems Solid fuel combustion involves much skilled arrangement of specialized combustion systems. Unlike the fuel oil and gaseous fuel, solid fuel must undergo adequate preparatory measures before use. Five types of solid fuel combustion systems are described below.

1. *Pneumatic spreader combustion system*. Here, a pneumatic blower is used to transfer the fuel into the chamber. Some of the fuel (i.e., fine particles) is combusted in suspension, with the larger particles falling to the grate, where they are completely burned.

2. *Gasifier system*. A gasifier may be made up of a two-stage updraft atmospheric unit that produces syngas in the primary chamber and then ignites the gas in a thermal low-excess air ignition system. Syngas burns clearly into water vapor and carbon dioxide. The advantage of gasification is that using syngas is more efficient than direct combustion of the original fuel, as more of the energy contained in the fuel is extracted.

3. *Cyclonic combustion system*. The combustion occurs in a specially designed cylindrical reactor that discharges hot gaseous combustion products directly into the boilers. Pulverized solid fuel is blown into the cyclonic burner, where it oxidizes immediately. Centrifugal action forces the particles toward the cylindrical wall of the burner, where char oxidation occurs. Very high turbulence is generated in the cyclonic burner, and char oxidation occurs within milliseconds.

4. *Fluidized-bed combustion system*. In a fluidized-bed setup, the bed of solid particles can exhibit the properties of a boiling liquid and assume the appearance of a fluid. Firing is achieved by heating a solid of much higher ash fusion temperature (e.g., sand) to the autoignition temperature of the fuel.

5. *Tangential firing system*. The setup includes four burners placed corner to corner to create a fireball at the center of the furnace.

13.2.3.4 Soot Blower This is a system for removing the soot that is deposited on the furnace tubes of a boiler during combustion. Steam is normally used as a medium for blowing away the soot. Examples of soot blowers are wall blowers, long retractable blowers, and air water blowers.

13.2.3.5 Draft Control A natural draft is subject to outside air conditions and the temperature of flue gases leaving the furnace, as well as the chimney height. All these factors make a proper draft difficult to attain, and therefore make mechanical draft equipment much more economical. There are three types of mechanical drafts.

1. *Induced draft*. An induced draft can be obtained in three possible ways. The first is by the stack effect of a heated chimney, in which the flue gas is less dense than the ambient air surrounding the boiler. The denser column of ambient air forces the combustion air into and through the boiler. The second method is through the use of a steam jet oriented in the direction of flue gas flow, which induces flue gases into the stack and allows for greater flue gas velocity, increasing the overall draft in the furnace. The third method is simply by using an induced-draft fan, which removes flue gases from the furnace and forces the exhaust gas up the stack. Almost all induced-draft furnaces operate with a slightly negative pressure.

2. *Forced draft*. In this case, a draft is obtained by forcing air into the furnace by means of a forced-draft fan and ductwork. Air is often passed through an air heater,

which heats the air going into the furnace to increase the overall efficiency of the boiler. Dampers are used to control the quantity of air admitted to the furnace. Forced-draft furnaces usually have positive pressure.

3. *Balanced draft*. A balanced draft is obtained through the use of both induced and forced drafts. This is more common with larger boilers, where the flue gases have to travel a long distance through many boiler passes. The induced-draft fan works in conjunction with the forced-draft fan, with the latter allowing the furnace pressure to be maintained slightly below atmospheric pressure.

13.2.3.6 Emission Control Systems These systems are used to control the emission into the surroundings of hazardous chemical by-products. This is done primarily to avoid or at least minimize environmental pollution. Such control devices include multiclone flyash collectors, baghouses, mechanical collectors, scrubbers, high-efficiency cyclones, electrostatic precipitators, and control panels.

13.3 BOILER SELECTION

Together with an understanding of the various types of boilers (Section 13.1), it is imperative to know that boiler usage depends largely on a facility's applications. The primary purpose of a boiler is to supply energy to a facility's operation for heat, manufacturing processes, power generation, cogeneration, and so on. Boiler steam utilization ranges from low- to high-pressure steam and low-to-high steam demands. An adequate knowledge of the nature of the facility's operation is both the first and best step approach in determining a suitable boiler to be selected.

Generally, boilers are defined according to design pressure and operating pressure. The design pressure of a boiler is the maximum pressure used in the design of a boiler for the purpose of calculating the minimum permissible thickness or physical characteristics of the pressure vessel parts of the boiler. Typically, the safety valves are set at or below the design pressure. Operating pressure is usually maintained at a suitable level, below the setting of the pressure-relieving valves, to prevent frequent opening during normal operation.

Therefore, when selecting a boiler for a particular purpose, considerations include:

- Applications involved in regard to the process operations and pressure levels required as well as the steam demand (required steam output)
- Types of combustion systems and heating fuels

These considerations help provide solutions to such basic economic problems as cost, boiler sizing, choice of heating fuel, and combustion system design.

13.3.1 Costs

Boiler cost is not just the cost of procurement but also the running or operational costs, maintenance costs, and ancillary equipment costs. An accurate analysis of the energy requirements of a particular process operation must be based on knowledge of the various boiler types. Many factors will affect the decision to purchase a particular piece of boiler equipment. The effect of a single piece of equipment can be a significant part of the overall transfer of energy: from the fuel burned to the thermal energy of the steam or hot water delivered. The performance of equipment such as the boiler, stack gas recovery system (economizers), condensate recovery system (deaerators, etc.), oxygen trim system, and blowdown heat recovery system should be considered. Efficiency gains from each piece of equipment need to be evaluated individually, in the context of the overall system, to determine the incremental fuel cost.

Savings from efficiency gains are used to evaluate the payback potential of the equipment. Payback simply refers to the time period that will elapse before the cumulative cost savings will equal the incremental capital cost of the equipment selected. The payback analysis sequence is straightforward and can be summarized as follows:

1. Estimate the boiler fuel consumption rate. This is determined by comparing the fuel consumption rates of two boiler configurations with different fuel-to-steam efficiencies, or as a base fuel rate for a given boiler configuration.
2. Estimate the annual fuel use. This is determined by multiplying the hourly fuel consumption rates by the total hours of operation in a year.
3. Estimate the annual fuel cost.
4. Determine the potential incremental efficiency improvements. Such improvements as an economizer or oxygen trim can be added to a boiler to improve its efficiency.
5. Estimate the potential annual fuel savings.
6. Determine the payback period for the investment. The payback period comprises the years required to recover the capital investment:

$$\text{payback period} = \frac{\text{capital cost of the equipment}}{\text{annual savings}}$$

13.3.2 Boiler Sizing

Boiler sizing is a very large topic covering boiler plant location, plant size, boiler accessories, and utilities. Only an overview of boiler sizing is given here.

Factors to be considered in boiler plant location include:

- Floor space required
- Total space requirements
- Access space for maintenance
- Size and characteristics of the boiler to be installed or replaced, including location of the piping systems, the boiler stack, and utilities
- Boiler weight limitations

One of the most important factors in keeping a boiler online is to keep enough water in it to prevent a shutdown. This is especially important with a fire tube boiler that is fired automatically. This is why it is so important to size a feed water system so that it is capable of maintaining the proper water level in the boiler. A properly sized feed water system will have a tank sized adequately to feed the boiler, and pumps selected to deliver the water at the correct rate and pressure.

Three areas must be considered in selecting the correct pump for an application:

1. The correct flow rate
2. The correct pressure
3. The net pump suction head (NPSH) and the requirement for proper discharge of the pump

13.3.3 Heating and Heating Fuels

The heat sources for a boiler are primarily the output of combustion processes in the furnace or firebox unit of the boiler plant. Boiler fuels include:

1. *Fuel oils*: mainly hydrocarbon based. Other liquid fuels may include biofuel oils.
2. *Gases*: mainly natural gas. Other gaseous fuels include coal gas and propane.
3. *Waste heat from other processes*: can be used in heat recovery steam generators.
4. *Solid fuels*: mainly coal. Other solid fuels include:
   a. *Wood fuel*, such as bark, sawdust, planer shaving, logged wood, dried wood chips, sander dust, and wood flour
   b. *Industrial wastes*, such as coffee grounds, corn shells, corn bran, corncobs, corn stover, coconut hulls, oat hulls, peanut hulls, pecan shells, rice husks, wheat bran, sugarcane bugasse, charcoal, dried peat moss, and sludge from wastewater treatment plants
   c. *Municipal solid wastes*, such as paper materials, wood wastes, plastic materials, and tires and rubber-based materials
   d. *Animal by-products*, such as animal manure, bones, skin, and animal hair

For electric steam boilers, resistance or immersion-type heating elements are normally used. Nuclear fission is also used as a source of heat in steam generation.

13.4 COMMON PROBLEMS OF BOILERS

The pressure of the steam determines the temperature and energy capacity. The higher the pressure, the stricter the quality of the boiler feed water. Many problems are caused by impurities in the boiler when low-quality boiler feed water is used at high pressure.

13.4.1 Scaling

Boiler scale is caused by impurities resulting from hardness salts, metallic oxides, silica [i.e., silicon dioxide) (SiO_2)] and a number of other feed water contaminants that can enter a system (Table 13-1). Scale is formed by salts that have limited solubility but are not totally insoluble in boiler water. These salts reach the deposit site in a soluble form and are then deposited (precipitate).

Scaling is due mainly to the presence of calcium and magnesium salts (carbonates and sulfates), which are less soluble hot than cold. It is caused by the presence of too high a concentration of silica in relation to the alkalinity of the water in the boiler. If unchecked, scaling, acting as an insulator, causes progressive lowering of the boiler efficiency by heat retardation. Eventually, scale buildup will cause the tube to overheat and rupture. Boiler deposits

TABLE 13-1 Partial List of Boiler Deposits

Name	Formula
Acmite	$Na_2O{\cdot}Fe_2O_3{\cdot}4SiO_2$
Analcite	$Na_2O{\cdot}Al_2O_3{\cdot}4SiO_2{\cdot}2H_2O$
Anhydrite	$CaSO_4$
Aragonite	$CaCO_3$
Brucite	$Mg(OH)_2$
Calcite	$CaCO_3$
Cancrin	$4Na_2O{\cdot}CaO{\cdot}4Al_2O_3{\cdot}2CO_2{\cdot}9SiO_2{\cdot}3H_2O$
Hematite	Fe_2O_3
Hydroxyapatite	$Ca_{10}(OH)_2(PO_4)_6$
Magnetite	Fe_3O_4
Noselite	$4Na_2O{\cdot}3Al_2O_3{\cdot}6SiO_2{\cdot}SO_4$
Pectolite	$Na_2O{\cdot}4CaO{\cdot}6SiO_2{\cdot}H_2O$
α-Quartz	SiO_2
Serpentine	$3MgO{\cdot}2SiO_2{\cdot}2H_2O$
Thenardite	Na_2SO_4
Wallastonite	$CaSiO_3$
Xonotlite	$5CaO{\cdot}5SiO_2 \cdot H_2O$

can also cause plugging, or partial obstruction of corrosion attack underneath the deposits may occur. In general, boiler deposits can cut operating efficiency, produce boiler damage, and increase cleaning expense.

The mechanism of scale deposition operates as follows:

1. A circulation pattern is developed in the boiler due to steam bubbles, which alter the density of the boiler water. The hottest area of the boiler (where nucleate boiling occurs) is where the steam–boiler water mixture is the least dense. A rolling circulation pattern is developed as the steam proceeds through the boiler to its outlet for further plant use.
2. Based on convective, conductive, and radiant heat transfer, thermal gradients are experienced throughout the steaming boiler. The hottest areas become primary deposition points, due to the high heat flux.
3. Flow patterns, velocities, and concentrations of contaminants also follow the laws of gravity. Thus, any area of the boiler considered to be low flow may exhibit significant deposition.

13.4.2 Corrosion

Corrosion is the reversion of a metal to its ore form. For example, iron reverts to the iron oxide Fe^{2+} or Fe^{3+} as a result of corrosion. Water will rapidly corrode mild steel, and as the temperature increases, the reaction accelerates.

$$\begin{array}{lcl} 3\text{Fe} + \text{H}_2\text{O} & \longrightarrow & \text{Fe}_3\text{O}_4 + 4\text{H}_2 \\ \text{Iron} + \text{water/steam} & \longrightarrow & \text{magnetite(black)} + \text{hydrogen} \end{array}$$

The magnetite produced is black iron oxide. Under normal operating conditions, this is the typical product of corrosion. However, it is also this reaction that inhibits excessive corrosion in steaming boilers.

Corrosion generally occurs when the boiler water alkalinity is low or when the metal is exposed to oxygen-bearing water during either operation or idle periods. High temperatures and stresses in the boiler metal tend to accelerate the corrosive mechanisms. In steam condensate systems, corrosion is generally the result of contamination with carbon dioxide and oxygen. Specific contaminants such as ammonia- and sulfur-bearing gases may increase the attack on copper alloys in systems. Corrosion is caused by a combination of oxide layer fluxing and continuous oxidation by transported oxygen.

13.4.2.1 Oxygen Attack (Corrosion) When dissolved oxygen enters the steaming boiler, corrosion manifests itself in the form of severe deep pits. Active oxygen pitting is easily recognized by the black oxide present in the pit. A secondary product of the corrosion mechanism may yield red iron oxide (Fe_2O_3). Oxygen attack is an electron chemical process that can be described by these reactions:

$$\text{Anode: Fe} \longrightarrow \text{Fe}^{2+} + 2\text{e}^-$$

$$\text{Cathode}: \tfrac{1}{2}\text{O}_2 + \text{H}_2\text{O} + 2\text{e}^- \longrightarrow 2\text{OH}^-$$

$$\text{Overall: Fe} + \tfrac{1}{2}\text{O}_2 + \text{H}_2\text{O} \longrightarrow \text{Fe(OH)}_2$$

In this reaction a temperature rise provides enough additional energy to accelerate reactions at the metal surfaces, resulting in rapid and severe corrosion.

13.4.2.2 Carbon Dioxide Attack (Corrosion) Carbon dioxide can enter a condensate system as a dissolved gas, or it can be chemically combined in the hydrogen carbonate or carbonate alkalinity of the feed water. The hydrogen carbonates and carbonates can decompose to yield carbon dioxide:

$$\text{NaHCO}_3 + \text{heat} \longrightarrow \text{NaOH} + \text{CO}_2$$

$$\text{Na}_2\text{CO}_3 + \text{H}_2\text{O} + \text{heat} \longrightarrow 2\text{NaOH} + \text{CO}_2$$

The low pH resulting from this reaction also enhances the corrosive effect of oxygen. Carbon dioxide corrosion is frequently encountered in condensate systems and less commonly in water distribution systems.

13.4.2.3 Acidic Corrosion Low makeup feed water pH can cause a serious acid attack on metal surfaces in preboiler and boiler systems. Feed water can also become acidic from contamination by condensers. In a boiler and feed water system, acidic attack can take the form of general thinning, or it can be localized at areas of high stress. Acidic corrosion can be caused by chemical cleaning operations through overheating of the cleaning solution, excessive exposure of metal to cleaning agents, and high concentration of cleaning agents.

13.4.2.4 Caustic Corrosion Concentration of NaOH can occur as a result of steam blanketing or by localized boiling beneath porous deposits on a tube surface. Steam blanketing is a condition that occurs when a steam layer forms between the boiler water and the tube wall, causing inefficient heat transfer.

Caustic corrosion occurs when NaOH is concentrated and dissolves the protective magnetite (Fe_3O_4) layer, causing a loss of base metal and eventual failure. For caustic corrosion to occur:

- The metal must be stressed.
- The boiler water must contain NaOH.
- At least a trace of silica must be present in the boiler water.

- Some mechanism, such as a slight leak, must be present to allow the boiler water to concentrate on the stressed metal.

13.4.3 Boiler Water Carryover

Carryover is generally considered to be any containment that leaves a boiler steam drum with the steam. It can be in solid, liquid, or vapor form. Effects of carryover include:

- Deposition in regulators and valves
- Deposition in superheaters
- Deposition in control valves and turbines
- Process contamination

See Table 13-2 for contaminant concentrations for minimizing carryover.

13.4.3.1 Causes of Carryover Carryover is a result of the incomplete separation of steam from the steam–water mixture in a boiler drum. Factors that complicate the separation are classified as chemical or mechanical factors.

1. *Mechanical factors*. Mechanical causes of incomplete separation of a steam–water mixture vary by boiler design, such as:
   - The type of mechanical steam separating equipment
   - The basic operational swing on the unit

   Also, steam load and boiler drum level have a significant effect on the amount of space available for steam separation from the steam–water mixture.
2. *Chemical factors*. Foaming and selective vaporization are two basic mechanisms of chemical carryover.
   a. *Foaming*. Foaming is the formation of small bubbles on the surface of boiler water. As these bubbles burst, subsequent moisture is entrained with the steam. The levels of foam in the boiler steam drum also have a direct effect on carryover. Causes of foaming are:
   - High dissolved solids concentrations in the boiler water
   - High suspended solids concentration
   - High alkalinity concentrations
   - Saponified hydrocarbons (the soap resulting causes foaming)
   - Synthetic detergents and wetting agents contained in some surface water supplies

   b. *Selective vaporization*. This occurs due to the solvent properties of steam for certain impurities that are normally present. All sodium-based salts found in boiler water are soluble in water and, to varying degrees, in the steam phase as well.

TABLE 13-2 ABMA Standard Boiler Water Concentrations for Minimizing Carryover

	Boiler Water		
Drum Pressure (psig)	Total Silica[a] (ppm SiO_2)	Specific[b] Alkalinity (ppm $CaCO_3$)	Conductance ($\mu\Omega$/cm)
0–300	150	700	7000
301–450	90	600	6000
451–600	40	500	5000
601–750	30	400	4000
751–900	20	300	3000
901–1000	8	200	2000
1001–1500	2	0	150
1501–2000	1	0	100

[a]This value will limit the silica content of the steam to 0.25 ppm as a function of selective vaporization of silica.
[b]Specific conductance is unneutralized.

13.4.3.2 Correction of Carryover This involves mechanical correction and chemical corrections.

1. *Mechanical correction*. Mechanical separations are of two major types, primary separation and secondary separation. Primary separation utilizes rapid and abrupt changes in the direction of steam flow. Major separation of the steam and water occurs in the primary devices, such as simple baffles, curtain baffles, belly baffles, and cyclone separators. Secondary separation is a scrubbing or drying process that results from rapid directional changes of steam flow in conjunction with large surface areas for collection of the mist. In addition, the steam velocity must be low to avoid reentrainment of the boiler water. Screens or corrugated plates with close tolerances can be used as steam scrubbers.
2. *Chemical correction*. Proper boiler water chemistry reduces the impact that dissolved salts, alkalinity, silica, and hydrocarbon contaminants have on carryover tendencies. Effective use of boiler water antifoam agents, such as polyalcohol and amines, frequently allows for good-quality steam and higher concentrations in the boiler water chemistry.

13.5 BOILER FAILURE ANALYSIS AND WELDING DEFECTS

13.5.1 Boiler Failure Analysis

According to authorities, an estimated 75% of boiler failures during operation are due to low water. The main cause of this high level of accidents is the assumption that boilers require little or no attention because of the redundant, automatic controls they feature [17]. Where there are no regular operational and maintenance controls, a series of automatic-control failures can occur which can lead to an explosion. Boiler failure can occur in three steps:

1. The automatic feed device fails, causing a low-water condition.
2. The low-water fuel cutoff fails (Section 13.2.1), rendering it unable to sense the low-water condition and stop the fuel supply. In other words, throughout low-water conditions, fuel is supplied continuously because the low-water fuel cutoff fails and cannot receive a signal about the low-water conditions.
3. The safety pop valve fails to actuate to relieve pressure buildup.

Notwithstanding that all these devices are automatic; they have a finite life span under the conditions in which they operate. Beyond this life span and operating conditions, mechanical wear, fatigue, corrosion, and erosion set in, leading to eventual failure.

13.5.1.1 Elevated Temperature Failures Temperature failure can occur through either long-term or rapid overheating. Rapid overheating is characterized by thin-lipped ruptures and complete microstructural transformation due to heating to above 1333°F (723°C). It is typically caused by starvation of a tube due to a blockage or to rapid startups.

13.5.1.2 Fatigue Failures Fatigue cracking is caused by cyclic stresses. This occurs by two different mechanisms.

1. The cyclic stresses are created by rapid heating and cooling, and are concentrated at points where corrosion has roughened or pitted the metal surface. This is usually associated with improper corrosion prevention. In this type of mechanism, cracking can occur along the entire tube length, due to thermal expansion or contraction of the boiler tubing.
2. Cracks often originate where a dense protective oxide film covers the metal surfaces. In this case, cracking occurs from the action of applied cyclic stresses such as vibration. Failure usually occurs near an unyielding restraint.

Corrosion fatigue cracks are usually thick, blunt, and cross the metal grains. They usually start in the internal tube surface and are most often circumferential on the tube.

13.5.1.3 Creep Failures A creep failure is caused primarily by a short-term stress rupture. The basic cause of such a rupture is almost invariably foreign materials that can cause steam circuit blockages. Such materials include pieces of hacksaw blade, coins, welding debris, and gasket material. These failures are characterized by the explosive release of large volumes of steam and can quickly create escalating numbers of consequential failures unless the boiler is shut down promptly. For this reason, these failures usually take longer time to repair and result in higher than average loss availability.

13.5.1.4 Erosion Failures Grit and soot-blower erosion is a major cause of boiler tube failures. As erosion is a progressive mechanism, the tube thickness in erosion zones generally decreases with the rising age of boiler plants. In other words, the effect of erosion tends to increase the probability of failures as a plant ages. This effect has parallels in other progressive mechanisms, such as creep and fatigue. In most power stations, management of erosion consists of surveys of tube thickness and shield conditions, followed by the installation of shields as required, or selective tube replacement if wall thicknesses are below minimum requirements. However, most erosion failures occur at locations that are beyond this conventional management approach. This calls for a more innovative and practical approach to managing erosion failure trends. For example a finned economizer tube is located so as to overcome the difficulties of both poor and impossible access for thickness measurement and shielding.

13.5.1.5 Caustic Embrittlement This is a form of stress corrosion cracking that occurs in mild and low-alloy steel due to the conjugant action of an enduring tensile stress and concentrated solution of sodium hydroxide. The stress may be applied or residual. Residual stress results from welding or prior cold work. Metallographic examination shows continuous branched, intergranular cracking (www.cip.ukcentre.com).

13.5.1.6 Corrosion Failure This occurs as pitting, thinning, or gouging.

1. *Pitting*. This can occur due to the presence of free oxygen entering the boiler with the feed water, due to incomplete deaeration and/or chemical scavenging. Oxygen pitting can also occur during downtime, due to improper storage procedures. Plating of copper metal into a boiler tube side during acid cleaning can also result in pitting as a result of dissimilar metal corrosion.
2. *Thinning*. Thinning or general corrosion can occur due to acid attack, chelate attack, and steam blanketing. Steam blanketing refers to the localization of steam to the side of the tubing.
3. *Gouging*. Caustic gouging occurs under deposits due to the concentration of sodium hydroxide as the boiler water evaporates. Caustic gouging can also occur in conjunction with steam blanketing.

13.5.2 Welding Defects

Welding defects (Table 13-3) are causing a rising number of boiler tube failures, especially in power stations. These

TABLE 13-3 Welding Defects

Cold lap 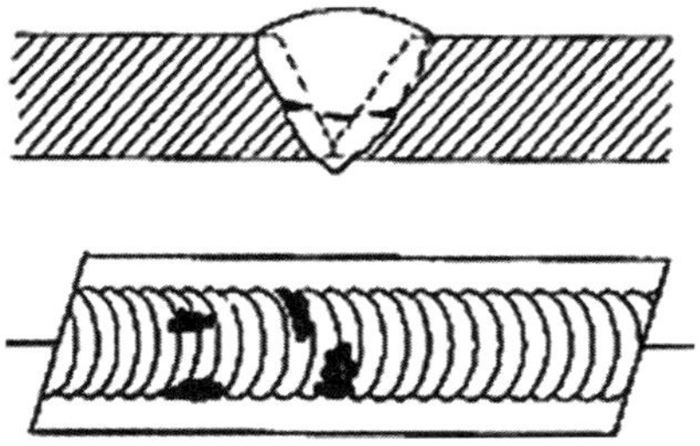	Cold lap is a condition where the weld filler metal does not bond properly with the base metal or the previous weld pass material (interpass cold lap). The arc does not melt the base metal sufficiently and causes the slightly molten puddle to flow into the base material without bonding.
Porosity 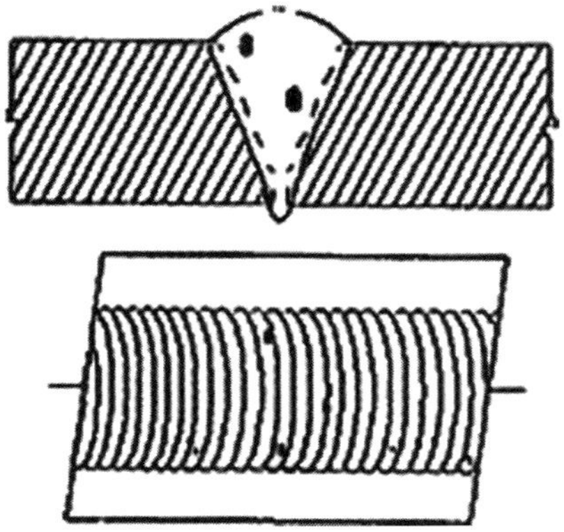	Porosity is the result of "gas entrapment" in the solidifying metal. Porosity can take many shapes on a radiograph but often appears as dark round or irregular spots or secks appearing singularly, in clusters, or in rows. Sometimes, porosity is elongated and may appear to have a tail. This is the result of gas attempting to escape while the metal is still in a liquid state and is called *wormhole porosity*. All porosity is a void in the material and it will have a higher radiographic density than that of the surrounding area.
Cluster porosity 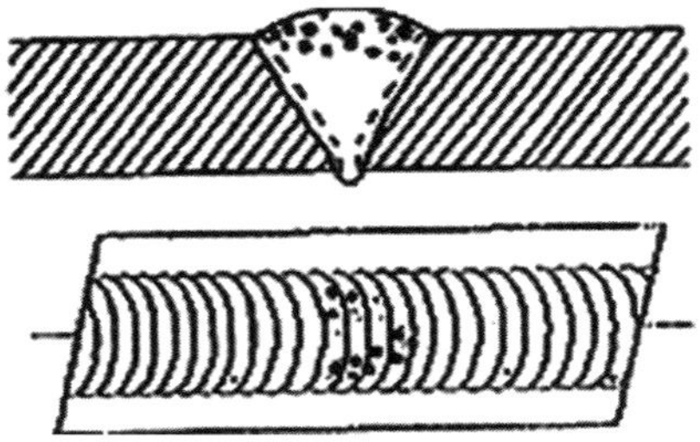	Cluster porosity is caused when flux-coated electrodes are contaminated with moisture. The moisture turns into gas when heated and becomes trapped in the weld during the welding process. Cluster porosity appears just like regular porosity in the radiograph, but the indications will be grouped close together.
Slag inclusion 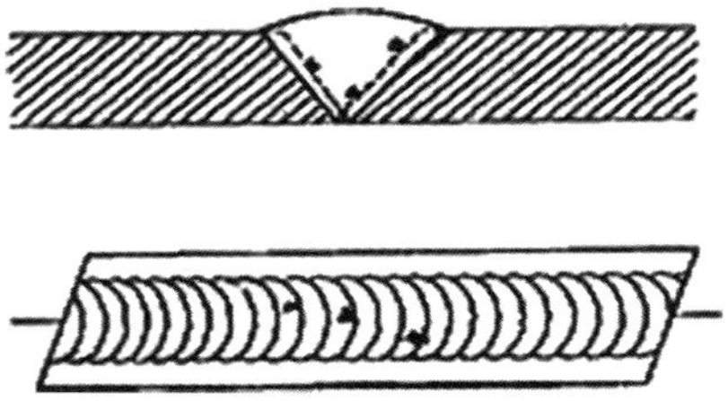	Slag inclusions are nonmetallic solid material entrapped in weld metal or between weld and base metal. In a radiograph, dark, jagged asymmetrical shapes within the weld or along the weld joints areas are indicative of slag inclusions.
Incomplete penetration or lack of penetration 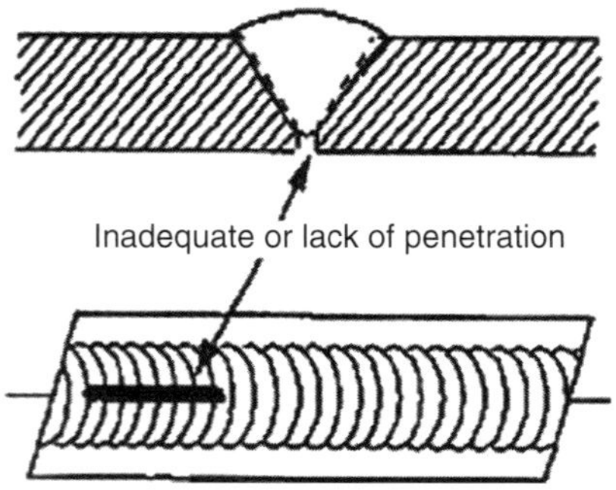	Incomplete penetration or lack of penetration occurs when the weld metal fails to penetrate the joint. It is one of the most objectionable weld discontinuities. Lack of penetration allows a natural stress riser from which a crack may propagate. The appearance on a radiograph is a dark area with well-defined straight edges that follows the land or root face down the center of the weldment.

TABLE 13-3 **(*Continued*)**

Incomplete fusion

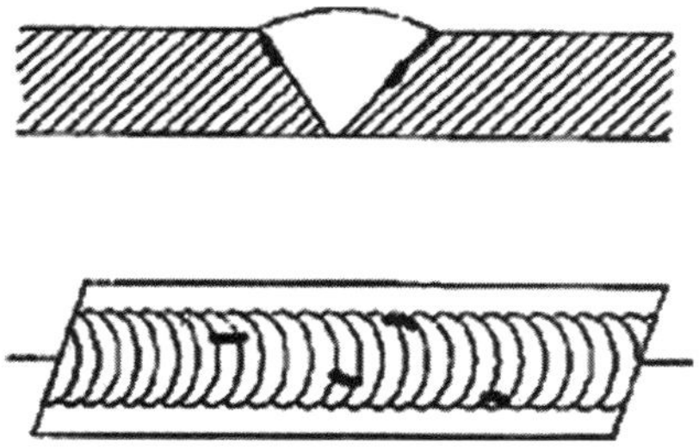

Incomplete fusion is a condition where the weld filler metal does not fuse properly with the base metal. On a radiograph it usually appears as a dark line or lines oriented in the direction of the weld seam along the weld preparation or joining area.

Internal concavity or suck back

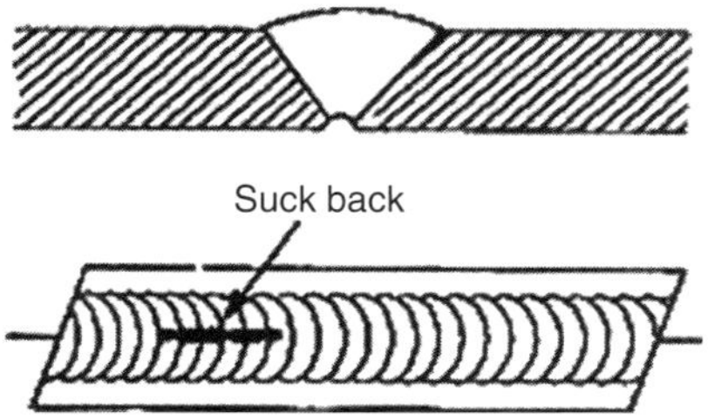

Internal concavity or suck back is a condition where the weld metal has contracted as it cools and has been drawn up into the root of the weld. On a radiograph it looks similar to a lack of penetration, but the line has irregular edges and it is often quite wide in the center of the weld image.

Internal or root undercut

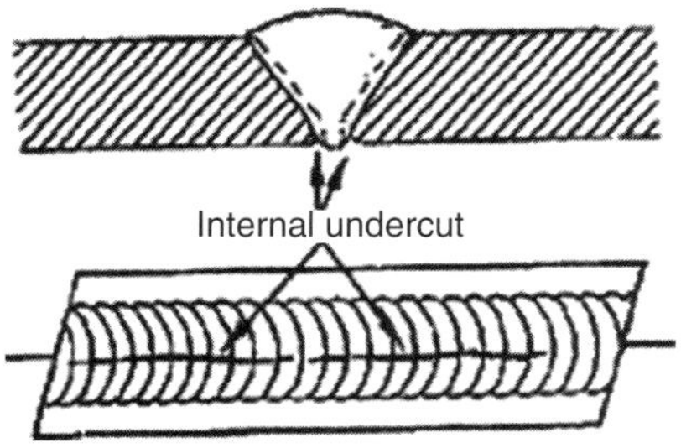

Internal or root undercut is an erosion of the base metal next to the root of the weld. In the radiographic image it appears as a dark irregular line offset from the leftline of the weldment. Undercutting is not as straight-edged as lack of penetration because it does not follow a ground edge.

External or crown undercut

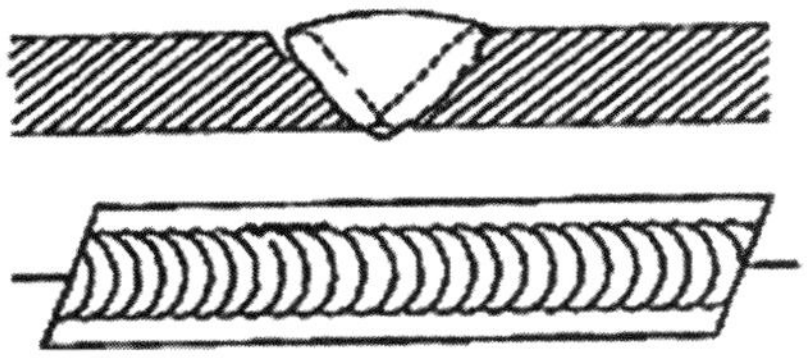

External or crown undercut is an erosion of the base metal next to the crown on the weld. In the radiograph it appears as a dark irregular line along the outside edge of the weld area.

Offset or mismatch

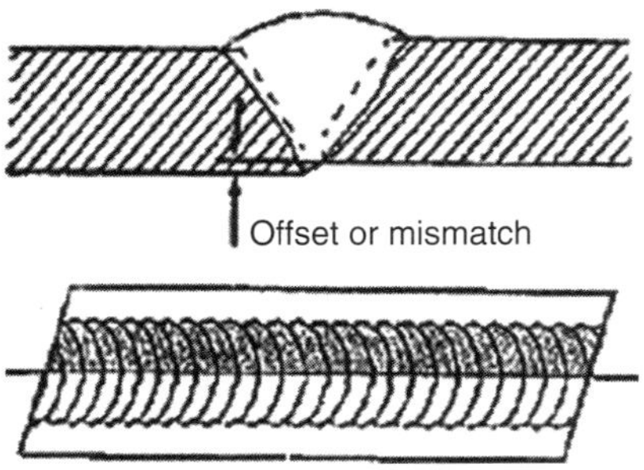

Offset or mismatch are terms associated with a condition where two pieces being welded together are not aligned properly. The radiographic image shows a noticeable difference in density between the two pieces. The difference in density is caused by the difference in material thickness. The dark straight line is caused by failure of the weld metal to fuse with the land area.

(*continued*)

TABLE 13-3 *(Continued)*

Inadequate weld reinforcement 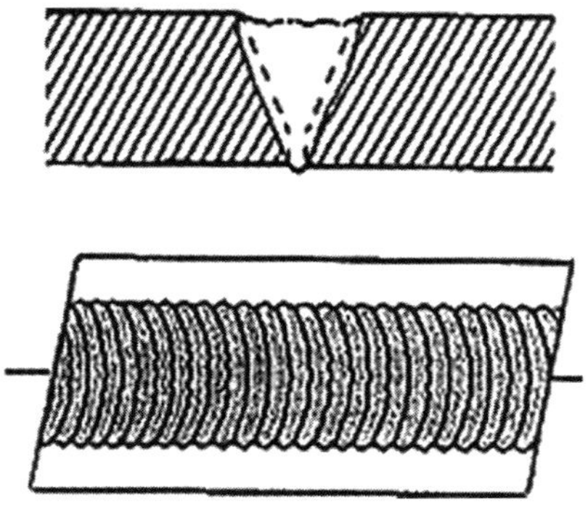	Inadequate weld reinforcement is an area of a weld where the thickness of weld metal deposited is less than the thickness of the base metal. It is very easy to determine by radiograph if the weld has inadequate reinforcement, because the image density in the area of suspected inadequacy will be higher (darker) than the image density of the surrounding base material.
Excess weld reinforcement 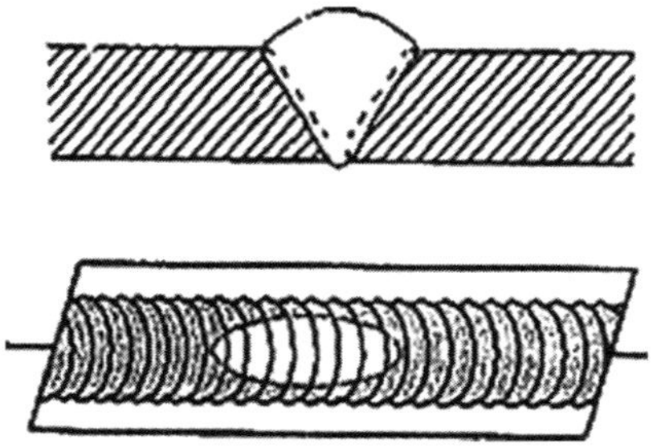	Excess weld reinforcement is an area of a weld that has a weld metal added in excess of that specified by enginering drawings and codes. The appearance on a radiograph is a localized, lighter area in the weld. A visual inspection will easily determine if the weld reinforcement is in excess of that specified by the engineering requirements.
Cracks 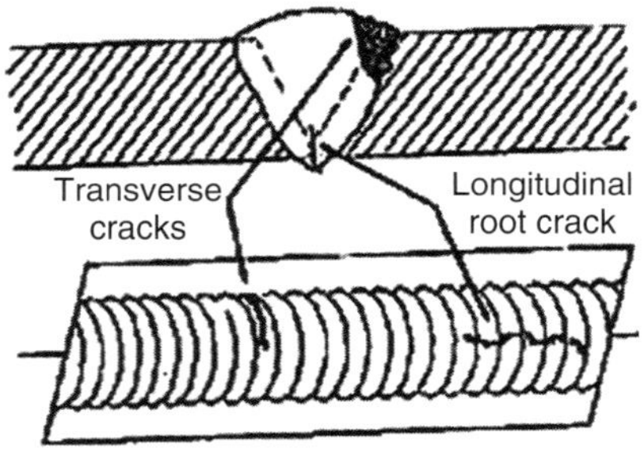	Cracks can be detected in a radiograph only when they are propagating in a direction that produces a change in thickness that is parallel to the x-ray beam. Cracks can appear as jagged and often very faint irregular lines. Cracks can sometimes appear as "tails" on inclusions or porosity.
Discontinuities in TIG welds tungsten inclusion 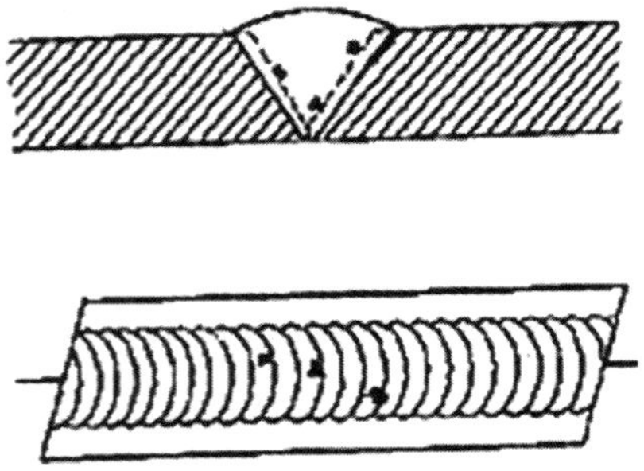	Tungsten is a brittle and inherently dense material used in the electrode in tungsten insert gas welding. If improper welding procedures are used, tungsten may be entrapped in the weld. Radiographically, tungsten is more dense than aluminum or steel and thus shows up as a lighter area with a distinct outline on the radiograph
Oxide inclusions 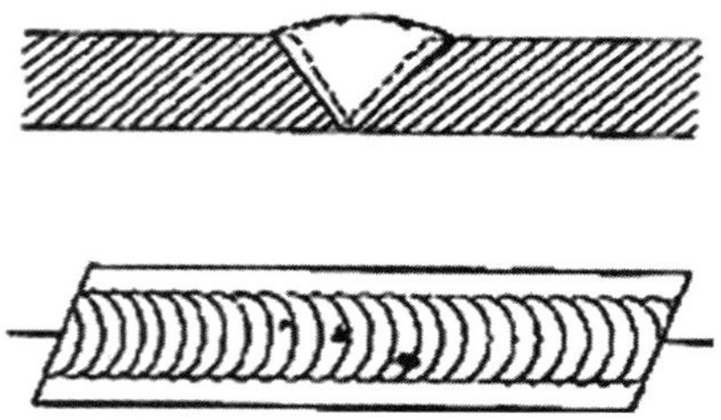	Oxide inclusions are usually visible on the surface of material being welded (especially aluminum). Oxide inclusinos are less dense than the surrounding material, and, therefore appear as dark irregularly shaped discontinuities in the radiograph.

TABLE 13-3 **(*Continued*)**

Discontinuities in Gas Metal Arc Welds

Whiskers are short lengths of weld electrode wire that are visible on the top or bottom surface of the weld or contained within the weld. On a radiograph they appear as light, "wirelike" indications.

Burn through	Burn through results when too much heat causes excessive weld metal to penetrate the weld zone. Often, lumps of metal sag through the weld, creating a thick globular condition on the back of the weld. These globs of metal are referred to as *icicles*. On the radiograph, burn through appears as dark spots, which are often surrounded by light globular areas (icicles).

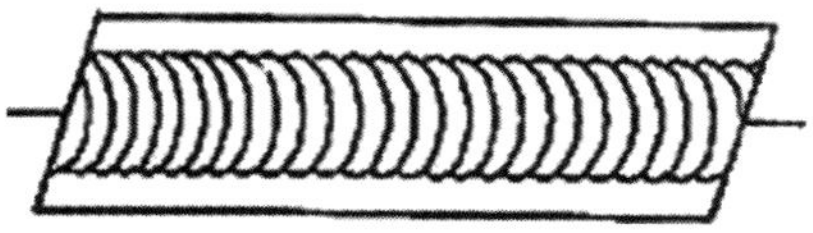

Source: [16], (see Table 13-5).

defects usually date from the time of construction and are distributed randomly throughout the boiler. However, boiler construction, maintenance, and repair may involve various degrees of welding operations. Defects due to welding operations are often a result of environmental or technical factors. Therefore, strict precautionary and professional approaches should be adopted while carrying out welding operations in boilers. We give a brief overview of welding procedures before delving into welding defects.

13.5.2.1 Welding Procedures The cost of not being able to control the quality of welding and repair rates can be substantial and can delay considerably completion of a project. This can invoke contractual penalties and result in the loss of profit margins and later arbitration.

Prior to beginning welding operations, a number of items must be approved:

- Welding procedure specifications
- Procedure qualification reports
- Welder certification certificates
- Post weld heat treatment company
- Post weld heat treatment procedures
- Welding inspection personnel
- Welding repair procedures

To avoid hydrogen-induced cracking, the following factors must be considered when completing the welding procedure:

- The combined thickness of the material to be welded
- The carbon equivalent values
- The hydrogen scales
- The welding arc energy

The standards most commonly used on construction projects follow the ASME boiler and pressure vessel code:

Clause	Section
Power boilers	One
Pressure vessels	eight, division one
Pressure vessels	eight, division two
Heating boilers	Four
Nondestructive ex	Five
Welding and brazing qualifications	Nine

13.6 BOILER MAINTENANCE

Efficient operation is a function of proper regular maintenance. Several things should be done to keep systems in top condition. There are two possible approaches to boiler maintenance: reactive or proactive (preventive).

1. *Reactive maintenance*. This type of maintenance involves only the repair of an already damaged or failed boiler. In this case, neither planning nor anticipation of

TABLE 13-4 Summarizes the Causes of the Most Common Welding Defects

Defect	Cause
Pipe offset mismatch	Pipe misalignment
Lack of root penetration	Welding technique
Insufficient root fill	Welding technique
Excessive penetration	Welding technique
External undercut	Excessive amps or volts
Internal undercut	Excessive amps or volts
Internal concavity	Welding technique
Root burn through	Welding technique
Lack of root penetration	Weld joint setup
Interpass slag inclusions	Weld technique, grinding, cleaning
Elongated slag inclusions	Weld technique, grinding, cleaning
Lack of side wall fusion	Weld technique, amps or volts too low
Interpass cold lap	Weld technique
Scattered porosity	Weld technique
Cluster porosity	Weld technique, insufficient wind cover
Root pass aligned porosity	Weld technique, insufficient wind cover
Transverse crack	Insufficient wind cover, lack of preheating, and postweld heat treatment of weld joint; material problem
Longitudinal crack	Insufficient wind cover, lack of preheating, and postweld heat treatment of weld joint; material problem
Longitudinal root crack	Insufficient wind cover, lack of preheating, and postweld heat treatment of weld joint; material problem
Tungsten inclusions	Weld technique

repairs is made to prevent the failure, so it is only when there is a failure that operators will evaluate the cause and carry out repair operations. The costs associated with such repairs are high, involving high labor cost and associated downtime. This is by far the least desirable approach to boiler maintenance.

2. *Proactive or preventive maintenance*. This maintenance program offers both peak boiler performance and overall safety of personnel. A proactive maintenance approach uses scientific testing techniques and analysis to anticipate problems before they arise and to correct them by either repair or replacement. Real-time monitoring has contributed greatly to the ability to view and respond to changes in boiler operations and overall process performance. Operators can compare past baseline results against current readings to determine if a problem may be forthcoming.

Boiler operation and maintenance are closely tied together. Good operation includes performing necessary daily and periodic maintenance. Good daily operating control leads to low maintenance costs where the fuel and the system are compatible. Boiler design can be of low-, medium-, or high-pressure steam generation. No matter what design configuration or specification is followed, boiler operation is a complex task or undertaking. Important physical and chemical balances are necessary for safe and efficient control (www.cip.ukcentre.com).

To ensure more reliable and energy-efficient boiler operations and failure prevention, a four-step approach can be adopted.

1. Match the best available equipment with the type of service and fuel required.

2. Verify proper operations, including all necessary controls and safety equipment by ensuring that the installation is checked annually by the insurance company's service representatives.

3. As part of the installation contract, include as a specification that the system be inspected by an authorized insurance company or local inspector before acceptance.

4. Provide operators with a log book for recording daily events and a preventive maintenance program for regular daily, weekly, monthly, or annual maintenance procedures. The procedures should include repair, replacement, inspection, cleaning, and lubricating. Technicians should schedule these tests annually and perform them periodically.

Every boiler operator aims at achieving the optimum operating efficiency of the equipment consistent with high reliability and low running costs. During daily operations, operators prepare log books with information collected from instrumentation at frequent enough intervals for early detection of trends. Early detection is critical to operating efficiently at low cost. For example, if the flue gas temperature of a boiler has increased gradually over the course of a month, the operator will be able to evaluate the system. He or she might then determine that there has been a buildup of scale, reducing heat transfer. Keep in mind that every boiler operation is different; some will require more extensive logs than others. This is an area that management and operators must review and decide upon jointly.

The steam generator's efficiency depends on proper control of time, temperature, turbulence, and oxygen. Boiler instrumentation should provide adequate coverage for the control of these four key operating characteristics. Simultaneously, the instruments should provide indicators that predict maintenance needs and timing.

1. The technician must raise the furnace to operating temperature.

2. The fuel-burning rate must be maintained to produce the desired unit of steam per hour to run the stream of turbines for electricity generation and/or to supply steam for heat and process needs.

3. The turbulence in fossil-fuel systems results from a combination of forced-draft fans located in the fuel supply section and induced-draft fans located in the stack breeching. The draft introduced by these large-volume air handlers produces the turbulence necessary for efficient operation. They also create a demand for emission controls, which are very important to air quality improvements.

Furnaces and steam-generating boilers are made up of a setting, or support structure, a fuel-handling and supply system, a fuel-burning control system, space above the fuel for heat transfer by radiation and convection, boiler tubes for conducting heat to the water, boilers for steam generation and storage, air- and ash-handling equipment, and many support systems, such as pumps, condensers, deaerators, water softeners, and soot blowers.

Adequate sizing is necessary in boiler procurement. For smaller steam requirements, adequate but highly efficient packaged boiler and steam generators should be provided. Where loads fluctuate greatly, where steam demand is variable, or where frequent startups and shutdowns are necessary, it might be preferable to install several smaller packaged units rather than one large furnace and boiler. These units are usually gas or oil fired.

13.6.1 Boiler Upgrading and Retrofitting

Upgrading or retrofitting of existing boiler plants is often desirable for a variety of reasons:

- Emergence of new technology capable of improving system performance
- Environmental requirements
- Change of heating fuel
- Automation (saves the cost of labor-intensive operations)
- Lifetime extension
- Low availability and high cost of new boilers
- Restrictions in building a new plant

13.6.1.1 Retrofitting Retrofitting is a maintenance operation aimed at bringing a boiler plant up to its initial performance level. This involves replacing or servicing faulty components. Redesign is not necessary, only evaluation of the plant generally, spotting areas that require attention. Some benefits of retrofitting are [11]:

1. Retrofitting is preferable to outright replacement by a new plant because of the shorter downtime and much lower costs.
2. The lifetime of the plant is often prolonged.
3. There is relatively lower investment cost for a retrofit than for a new plant.
4. The payback time is shorter for a retrofit than for a new plant.
5. Problems such as site location, permits, and procurements associated with building a new plant are avoided.

13.6.1.2 Upgrading This maintenance operation is aimed at bringing the boiler plant beyond the initial performance level. Unlike retrofit, redesign is necessary to ensure the desired performance level. Both retrofitting and upgrading are preventive in nature. Either can minimize the risk of unplanned stoppage due to the need for repair. Upgrading of boiler plants is done to improve energy utilization and to meet environmental requirements (in order to obtain approval by authorities to operate the plant), resulting in added value.

The first and necessary approach to boiler upgrading is a plant survey. In other words, a comprehensive examination of the present condition of the plant and evaluation of possibilities for upgrading should be carried out. Possible steps include:

- Provision of necessary redesign and engineering to ensure the required performance
- A feasibility study, calculating the financial consequences, such as investment, earnings, savings in operating costs (cost–benefit analysis), and payback time
- Sourcing of funds for the project
- Actual construction, procurement of auxiliary equipment, dismantling, reinstallation, and possibly commissioning the upgraded plant

Areas for upgrading include the following [11]:

1. Redesign of the combustion chamber in connection with fuel conversion
2. Noise and vibration problems
3. Improvement in steam quality
4. Feed water chemical dozing
5. Improvement in fuel gas emission impact on the external environment
6. Improvement in boiler economy by addition of an economizer, for example

7. Better utilization of exhaust gas energy by the addition of a heat exchanger producing hot water for heating or cooling processes
8. Improvement in burner systems through:
   - Replacement or upgrading of old burner
   - New burners in connection with conversion of fuel
   - Dual fuel systems
   - Handling of residual fuel
   - Burner management systems package
9. Improvement in control systems, such as:
   - Distributed control system
   - Boiler control system
   - Remote surveillance system
   - Remote operating system

13.6.2 Boiler Feed Water Treatment

The important parameters of the feed water are:

- pH
- Hardness
- Oxygen and carbon dioxide concentration
- Silicates
- Dissolved solids
- Suspended solids
- Concentration organics

The treatment and conditioning of boiler feed water must satisfy three main objectives:

1. Continuous heat exchanger
2. Corrosion protection
3. Production of high-quality steam

There are two main processes in of feed water treatment, which we discuss below.

13.6.2.1 External Treatment External treatment is the reduction or removal of impurities from water outside the boiler. This is usually referred to as the chemical and mechanical treatment of the water source. The goal is to improve the quality of this source prior to its use as boiler feed water (see Table 13-5), external to the operating boiler itself. In general, external treatment is used when the amount of feed water impurities is too high to be tolerated by the boiler system in question. Typical examples of external treatment include clarification, filtration, softening, dealkalization, demineralization, deaeration, evaporation, and membrane contractors (see Table 13-6).

13.6.2.2 Internal Treatment Internal treatment is the conditioning of impurities within a boiler system. The reactions occur either in the feed lines or in the boiler proper. Internal treatment may be used alone or jointly with external treatment. Its purpose is to react appropriately to feed water hardness, condition sludge, and scavenge oxygen and to prevent boiler water foaming. It is applied to minimize potential problems and to avoid catastrophic failure, regardless of external treatment malfunction.

Internal boiler water treatment has come a long way. Following is a chronology up to and including the late 1960s.

1. Pre-1920: *Carbonate cycle*. This program introduced sodium trioxocarbonate to prevent the formation of calcium tetraoxosulfate scales. It replaced a rather difficult scale deposit, calcium titraoxosulfate ($CaSO_4$), with another scale that was somewhat easier to remove, calcium trioxocarbonate ($CaCO_3$).

2. 1920s: *Phosphate program*. Soluble phosphate and alkali are added to the boiler system to promote the precipitation of calcium and magnesium ions in a desired sludge form. This sludge was removed via a manual blowdown, provided that the sludge remained in a fluid form. A minimum pH value of 9.5 was required to precipitate the calcium properly. However, with the proper addition of alkalinity, pH levels of 10.5+ were maintained to drive the reactions to completion. These reactions were also irreversible.

The desired boiler sludge are formed by the following reactions:

$$10Ca^{2+} + 6PO_4^{3-} + 2OH^- \rightarrow \underset{\text{calcium hydroxyapatite}}{3Ca_3(PO_4)_2 \cdot Ca(OH)_2}$$

$$3Mg^{2+} + 2SiO_3^{2-} + 2OH^- + H_2O \rightarrow \underset{\text{serpentine}}{2MgSiO_3Mg(OH)_2H_2O}$$

Both serpentine and calcium hydroxyapatite are relatively nonadherent to boiler metal and are easily removed by manual blowdown. The amount of phosphate, alkali, and silicate required is based on these reactions plus an excess to drive the reactions to completion. Without enough hydroxyl alkalinity, the potential for calcium dihydrogen tetraoxophosphate acid to precipitate is high. Calcium dihydrogen tetraoxophosphate acid is a tenacious scale that cannot be removed while the boiler is operating.

3. 1950s: *Phosphate–organic*. This program included the introduction of a surface-active material such as naturally occurring lignins and tanins. These materials

TABLE 13-5 ASME Guidelines for Water Quality in Modern Industrial Water Tube Boilers for Reliable Continuous Operation

	Boiler Feed Water			Boiler Water		
Drum Pressure (psi)	Iron (ppm Fe)	Copper (ppm Cu)	Total Hardness (ppm $CaCO_3$)	Silica (ppm SiO_2)	Total Alkalinity (ppm $CaCO_3$)	Specific Conductance (μΩ/cm) (un-neutralized)
0–300	0.100	0.050	0.300	150	700	7000
301–450	0.050	0.025	0.300	90	600	6000
451–600	0.030	0.020	0.200	40	500	5000
601–750	0.025	0.020	0.200	30	400	4000
751–900	0.020	0.015	0.100	20	300	3000
901–1000	0.020	0.015	0.050	8	200	2000
1001–1500	0.010	0.010	0.0	2	0	150
1501–2000	0.010	0.010	0.0	1	0	100

provided a dispersing and adsorptive action for maintaining sludge in a fluidized form. This was an improvement over previous approaches.

4. 1960s: *Phosphate–poiymer*. This provided a superior dispersive action to boiler water treatment. The synthetic polymers were significantly more efficient than naturally occurring organic agents.

5. 1960s: *Chelate program*. Coincidental with the development of synthetic polymers, chelate technology came into its own. It was in contrast to the precipitating types of programs identified with phosphate base systems. Chelation is a solubilizing approach.

High-pressure boiler water control systems have not been reviewed above, as they are specialized and unique programs based on boiler design, operating pressure, and feed water quality.

13.6.3 Boiler Stack Economizer

Large process boilers usually give off fuel gases at very high temperatures, say 450°F (232°C) to 650°F (343°C). Minimal heat wastage and optimal boiler performance is achieved by installing a stack economizer which utilizes part of the heat from fuel gases to preheat water. The preheated water is used primarily for boiler makeup water or some other applications in boiler operations. Economizers are sized to be installed into the stack, as close to the boiler fuel outlet as practical. Thus, economizer sizing must be consistent with:

- The volume of flue gases
- The temperature of flue gases
- The maximum allowable pressure drop through the stack
- The type of heating fuel for the boiler
- The value of energy to be recovered

All design specifications or instructions should be adhered to strictly when installing an economizer to avoid possible failures. For example, economizers designed for natural gas only would plug up if installed on a coal boiler; if installed on an oil-fired boiler, there is a greater risk of corrosion. Some units are designed to keep the flue gases above condensation temperature, while others are made of materials that resist the effect of corrosion of condensed flue gases.

Factors that can affect savings potential of an economizer are:

1. The existing stack temperature
2. The value of required makeup water
3. The operating time

When less condensate is returned, it means that more makeup water is heated, which in turn means that high savings potentials are achieved. An economizer tube must be protected from oxygen pitting to avoid failure. An oxygen scavenger such as catalyzed sodium trioxosulfate should be used with a deaerator. Complete corrosion protection of the economizer is achieved by maintaining residual of 5 to 10 ppm of the sulfate in the feed water. Sufficient NaOH or neutralizing amines may be fed to increase the pH of feed water to between 8.0 and 9.0.

13.6.4 Boiler Blowdown Control

Bottom blowdown is intermittent ejection of sludge or sediment from the bottom of a boiler. The frequency of bottom blowdown is a function of experience and plant operation. The major influences on the blowdown rate are:

1. The boiler pressure
2. The size of the blowdown line
3. The length of the blowdown line between the boiler and the blowdown vessel

TABLE 13-6 Summary of Common Impurities, their Effects, and Treatment Methods

Impurity	Resulting in:	Gotten rid of by:	Comments
		Soluble Gases	
Hydrogen sulfide (H_2S)	Water smells like rotten eggs; tastes bad, and is corrosive to most metals.	Aeration, filtration, and chlorination.	Found mainly in groundwater and polluted streams.
Carbon dioxide (CO_2)	Corrosive, forms trioxocarbonate acid in condensate.	Deaeration, neutralization with alkalis.	Filming, neutralizing amines used to prevent condensate line corrosion.
Oxygen (O_2)	Corrosion and pitting of boiler tubes.	Deaeration and chemical treatment with (sodium trioxosulfate or hydrazine).	Pitting of boiler tubes and turbine blades; failure of steam lines, fittings etc.
		Suspended Solids	
Sediment and turbidity	Sludge and scale carryover.	Clarification and filtration.	Tolerance of approx. 5 ppm max. for most applications, 10 ppm for potable water.
Organic matter	Carryover, foaming, deposits can clog piping, and cause corrosion.	Clarification, filtration, and chemical treatment.	Found mostly in surface waters; caused by rotting vegetation and farm runoff. Organics break down to form organic acids. Results in low boiler feed-water pH, which then attacks boiler tubes. Includes diatoms, molds, bacterial slimes, iron/manganese bacteria. Suspended particles collect on the surface of the water in the boiler and render difficult the liberation of steam bubbles rising to that surface. Foaming can also be attributed to waters containing carbonates in solution, in which a light flocculent precipitate will be formed on the surface of the water. It is usually traced to an excess of sodium trioxocarbonate used in treatment for some other difficulty, where animal or vegetable oil finds its way into the boiler.
		Dissolved Colloidal Solids	
Oil and grease	Foaming, deposits in boiler.	Coagulation and filtration.	Enters boiler with condensate.
Hardness: calcium (Ca), and magnesium (Mg)	Scale deposits in boiler, inhibits heat transfer and thermal efficiency. In severe cases can lead to boiler tube burn-through and failure.	Softening, plus internal treatment in boiler.	Forms are hydrogen carbonates, sulfates, chlorides, and nitrates, in that order. Some calcium salts are reversibly soluble. Magnesium reacts with carbonates to form compounds of low solubility.
Sodium: alkalinity, NaOH, $NaHCO_3$, Na_2CO_3	Foaming, carbonates form carbonic acid in steam; causes condensate return line, and steam trap corrosion; can cause embrittlement.	Deaeration of makeup water and condensate return. Ion exchange; deionization; acid treatment of makeup water.	Sodium salts are found in most waters. They are very soluble and cannot be removed by chemical precipitation.

TABLE 13-6 ***(Continued)***

Impurity	Resulting in:	Gotten rid of by:	Comments
Sulfates (SO_4)	Hard scale if calcium is present.	Deionization.	Tolerance limits are about 100–300 ppm as $CaCO_3$.
Chlorides, (Cl)	Priming: i.e., uneven delivery of steam from the boiler (belching), carryover of water in steam lowering steam efficiency; can deposit as salts on superheaters and turbine blades. Foaming if present in large amounts.	Deionization.	Priming, the passage of steam from a boiler in "belches," is caused by the concentration of sodium trioxocarbonate, sodium tetraoxosulfate, or sodium chloride in solution. Sodium sulfate is found in many U.S. waters and in waters where calcium or magnesium is precipitated with sodium trioxocarbonate.
Iron (Fe) and manganese (Mn)	Deposits in boiler; in large amounts can inhibit heat transfer.	Aeration, filtration, ion exchange.	Most common form is iron bicarbonate.
Silica (Si)	Hard scale in boilers and cooling systems; turbine blade deposits.	Deionization lime soda process, hot-lime-zeolite treatment.	Silica combines with many elements to produce silicates. Silicates form very tenacious deposits in boiler tubing. Very difficult to remove, often only by hydrofluoric acids. Most critical consideration is volatile carryover to turbine components.

Bottom blowdown can be accomplished manually, or electronically using automatic blowdown controllers. The use of automatic bottom blowdown ensures that this important action is carried out regularly and saves labor cost. With multiboiler installations, it is necessary to interlock the valves so that not more than one can be open at a time, as this would overload the blowdown vessel. This can be done mostly by staggering the setting times of the individual blowdown timers, or by setting the individual times in sequences.

It is important to ensure adequate isolation for maintenance purposes and the prevention of reverse flow. The installation of total dissolved solids controls equipment in multiboiler plants should include a nonreturn valve and an isolation valve to prevent pressure or flow from one boiler being imposed on another. This is important for the safety of maintenance personnel during shutdown of a boiler.

13.7 BOILER TROUBLESHOOTING

A number of people call for help immediately when their boilers stop working. Although this is not wrong or unadvisable, you should first take a look at your boiler to see what type of boiler problem it is experiencing. Only simple troubleshooting skills are needed; you don't need to have intensive knowledge of heating engineering. It is imperative to be aware that boiler maintenance is an extremely specialized and dangerous area. Of course, it is not always possible to fix the boiler by yourself. If you encounter a boiler problem that is too difficult or complicated for you to fix, immediately engage the services of a heating engineer. A number of problems require adequate professional attention (e.g., when the boiler has water leaks). Many refuse to get the service of professionals or train their staffs to professional status because they think the cost is exorbitant and prohibitive. However, when help is already necessary, it is best to engage the service of experts. Remember, although there are a number of boiler problems that you can remedy, trying to fix a complex boiler problem may cost you more than calling those who should be fixing your boiler (Ezinearticles.com).

The technology behind boiler control should be well established. Tables 13-5, 13-6, and 13-7 outline combustion, draft fan and burner, and fuel pump and fuel pressure problems, respectively.

13.7.1 Combustion Problems

Common combustion problems include: not firing, cooling down after ignition, ignition slow, black smoke emission, combustion sound ceases, white smoke emission, and backfire (Table 13-7).

13.7.2 Draft Fan and Burner Problems

The common draft fan and burner problems include: the draft fan does not stop draft fan burner after-running, and the draft fan does not run (Table 13-8).

13.7.3 Fuel Pump and Fuel Pressure Problems

Common fuel pump and fuel pressure problem are: noisy fuel pump, increased oil pressure, oil pressure does not change even when the oil pump is working,

TABLE 13-7 Combustion Problems

Symptom	Check	Possible Causes	Remedies
1. Not firing	1. Fuel tank	Empty or mixed with water	Blow down water, then fill with fuel oil.
	2. Oil strainer or fuel valve	Strainer clog	Clean the strainer.
		Valve closed	Open the valve.
	3. Oil pressure	Incorrect pressure	See symptom 3 (slow ignition).
	4. No electric arc between electrodes	Clogged atomizer	Lift out the nozzle and clean the mesh strainer if checks 1–3 are normal.
	(a) No oil flowing	Oil pump broken	Replace the fuel oil pumps.
		Solenoid on the burner is broken	Check that power is applied to the solenoid; replace the solenoid if it works.
	(b) There is an electric arc, and the nozzle produces a spray well	Electrodes arc out of position	Move the electrodes back to the correct position.
	(c) No electric arc or unstable arc	Gap between electrodes is too wide, or electrode has twisted	Same as above.
		Tip of electrode has twisted	File the tip into 60° round shape.
		Fuel and carbon deposits on the electrode's circuit	Clean it fully.
		Isolation of the electrode is broken	Replace.
		Electrode cable is not plugged in correctly or the screw is loose	Plug in the cable correctly and tighten the screw.
		Ignition coil out of order	Replace.
	5. Heavy oil heater temperature controller (H type)	Shift	Reset.
	6. Atomizer	Clogged	Clean it.
	7. Electric eye	Ash deposits on eye	Clean with a soft cloth.
		Disconnected	Connect it well.
	8. Items 1–7 are correct	Electric eye out of order	Replace the electric eye.[a]
2. Cooling down after ignition	1. Fuel tank	Empty	Fill with oil.
		Mixed with water	Drain water and fill with oil.
	2. Oil strainer	Blocked	Clean.
	3. Atomizer	Blocked	Clean.
	4. Electric eye	Ash deposits	Clean with a soft cloth.
		Disconnected	Connect it well.
	5. Items 1–4 are correct.	Electric eye out of order	Replace the electric eye.[a]
3. Ignition slow	1. Check oil pressure.	Oil pressure is high or low	Adjust the oil pressure.
	2. Is mains power varied?	Voltage has dropped	Adjust the voltage to the correct value.
	3. Is the position of the electrodes correct?	Position of the electrodes is incorrect	Move the electrodes back to their normal position.

TABLE 13-7 *(Continued)*

Symptom	Check	Possible Causes	Remedies
4. Black smoke emission	1. Check the temperature in the furnace.	If the temperature in the furnace is above 40°C, this is probably caused by oxygen and poor combustion efficiency	Improve the airflow and reduce the temperature in the furnace.
	2. Oil pressure	Oil pressure is too high	Adjust the pressure to normal.
	3. Position of fuel pump	Incorrect position	Move position of the fuel pump back to normal.
	4. Heavy oil heater temperature controller (H type)	Shift	Reset.[a]
	5. Items 1–4 are correct	Worn burner tips	Replace the nozzle.
		Poor-quality fuel oil	Drain and refill.
5. Combustion sound ceases	Mains voltage	Voltage dropped	Adjust the voltage.
6. White smoke emission	1. Heavy load solenoid not powered	Wires not connected	Connect properly.
		Coil of the solenoid is broken	Replace.[a]
	2. Heavy load solenoid powered	Clogged solenoid	Clean.
	3. Oil pressure	Oil pressure too low	Inspect and adjust.
	4. Burner tips	Blocked	Clean.
7. Backfire	1. Mains power voltage	Voltage dropped	Adjust the voltage.
	2. Electrode settings	Settings not correct	Adjust the electrodes.

[a]Consult an expert.

excessive oil pressure, and fuel oil pump not working (Table 13-9).

13.8 BOILER CHEMICALS

Figure 13-5 is a basic boiler system schematic indicating some chemical feedlines. The basic equipment specifications for boiler system chemical treatment include:

- Skid-mounted dual metering pumps (duty/standby) with stainless steel wet end/trim
- Pulsation dampener
- Stainless steel relief valve
- Stainless steel check valve
- Stainless steel Y-strainer
- Stainless steel tubing valves and fittings
- Flowmeter
- Calibration cylinder
- Pressure gauge with diaphragm seals
- Electrical junction boxes
- Dilution waterline clockwise static mixer (optional)
- Drip pan (optional)
- Injection quill (should be used in all cases)

The primary function of a boiler is to transfer into water heat from hot gases generated by the combustion of fuel until it turns to steam. Boiler feed water often contains impurities that impair boiler operation and efficiency. The water required for boiler feed purposes should be of very high quality and thus requires a lot of treatment. Chemical additives can be used to correct the problems caused by these impurities. To improve feed water quality and steam purity, these chemicals can be injected directly into the feed water or steam. Chemicals are used in both internal and external treatments.

The benefits of chemical treatment are:

- Increase boiler efficiency.
- Reduce fuel, operating, and maintenance costs.
- Minimize maintenance and downtime.
- Protect equipment from corrosion and extend equipment lifetime.

TABLE 13-8 Draft Fan and Burner Problems

Symptom	Check	Possible Causes	Remedies
1. Draft fan does not stop	Protection relay	Broken	Replace the protection relay.
2. Draft fan burning after running		Restarting immediately	Start again after 1 min.
		Protection relay is broken	Replace[a] (see others).
3. Draft fan does not run	1. Fuses		
	(a) Disconnected	Shorted	Check the isolation of the control panel and replace the fuses.
	(b) Connected	Solenoid connection is not correct	Replace.[a]
		Wiring connection is not correct	Check the wiring.
	2. Incorrect water level	Caps of the water-level electrodes are not plugged	Plug in well.
	3. Pressure switch not powered	Blocked pipeline	Clean.
		Pressure switch is malfunctioning	Replace.[a]
	4. Heater connection incorrect?	Connection problems	See others.
	5. Heavy oil heater temperature controller (H type)	Shift	Reset.
	6. Temperature of heavy oil	Temperature does not increase	Check the heavy oil heater. Check the heater temperature controller.

[a]Consult an expert.

The chemicals used for the boiler water treatment are summarized as phosphates, lime softeners, chelates, polymers, oxygen scavengers, neutralizing amines, and filming amines.

13.8.1 Phosphates

Phosphates and polyphosphates of sodium can be added to treat boiler feed water. The phosphate buffers the water to minimize pH fluctuation and precipitates calcium and magnesium into a soft deposit rather than a hard scale. In addition to this, it helps to promote the protective layer on boiler metal surfaces. However, phosphate forms sludge as it reacts with hardness. Blowdown or other procedures should be established to remove sludge during a routine boiler shutdown. Phosphates are usually fed directly into the steam drum of the boilers, and under certain conditions they may be fed to the feed water line. Treatments containing tetraoxophosphate may produce calcium tetraoxophosphate [$Ca_3(PO_4)_2$] feed line deposits; therefore, they should not be fed through the boiler feed line. Tetraoxophosphates (PO_4^{3-}) must not be fed to the boiler feed water line when economizers, heat exchangers, or stage heaters are part of the preboiler system. If the preboiler system does not include equipment where total hardness does not exceed 2 ppm, polyphosphates may be fed to the feed water piping. In all cases, feed rates are based on feed water hardness level. Phosphates can be fed neat (i.e., in its raw state). It can also be diluted with condensate or high-purity water. Mild steel tanks, fittings, and feed lines are appropriate, but stainless steel should be used if acidic phosphate is fed.

13.8.2 Lime Softening and Sodium Trioxocarbonate

Quick or slaked lime [usually, $Ca(OH)_2$] is added to hard water to precipitate the calcium, magnesium, and to some extent, silicon dioxide (SiO_2) in the water. Sodium trioxocarbonate is added to precipitate nonhydrogen carbonate hardness. The process typically takes place in a clarifier followed by a hydrogen-cycle cation-exchange and a hydroxide-cycle anion-exchange demineralization.

13.8.3 Chelates

Nitrilotriacetic acid (NTA) (i.e., 2,2,2-nitrilotriacetic acid) ($C_6H_9NO_6$) and ethylendiaminetetraacetic acid (EDTA) [i.e., 2,2,2,2-(ethane-2,3-diyldinitrilo)tetraacetic acid] ($C_{10}H_{16}N_2O_8$) are the most commonly used chelates. Chelates combine with hardness in water to form soluble compounds. The compounds can then be eliminated by blowdown. Note these points.

- Chelate treatment is unadvisable for feed water of high hardness concentration.

TABLE 13-9 Fuel Pump and Fuel Pressure Problems

Symptom	Check	Possible Causes	Remedies
1. Noisy fuel pump, increased oil pressure	1. Fuel pipeline	Blocked pipeline and air in the fuel pipeline	Clean or vent the air; adjust the pipeline if necessary.
	2. Fuel pump strainer	Blocked	Clean.
	3. Location of fuel tank	Too low	After checking items 1 and 2, contact an engineer.[a]
	4. Temperature	Pipeline heater temperature setting too high	Adjust the pipeline heater (temperature controller) to 60°C and 80°C for B and C heavy oil, respectively.
		Heavy oil heater temperature setting too high	Set the heavy oil heater to 105°C and 150°C for B and C heavy oil, respectively.
2. Oil pressure does not increase even when the oil pump is working	1. Oil strainer	Blocked	Clean.
		Leakage	Replace the shaft seal.[a]
		Faulty pipeline valve switches	Consult an expert
	2. Idle running of fuel pump	Air in the pump or pipeline	Vent the air.
	3. Leaking fuel pump	Worn shaft seal	Replace.[a]
	4. None of items 1–3	Loosened oil pressure adjustment screw	Replace.[a]
		Oil pressure gauge out of order	Replace.[a]
		Worn oil pump gear	Replace.[a]
3. Excessive oil pressure	1. Pipeline	Blacked	Clean.
		Irregular oil pressure setting	Adjust.[a]
		Oil pressure gauge out of order	Replace.
4. Fuel oil pump does not work	1. Fuses	Not installed	See symptoms for draft fan not running.
	2. Is motor running idly?	Broken coupling	Replace.
		Loosened coupling	Tighten.
	3. Motor not running	Gear blocked by foreign materials	Adjust using pliers until running; check the oil strainer as well.[a]

Consult an expert.
[a] *Source:* [7].

- Chelates should not be fed where a significant level of oxygen is contained in feed water.
- Chelates should not be fed directly into a boiler.

The preferred feed location for chelates is downstream of the feed water pump. All chelate treatment must be fed to the boiler feed water line via a stainless steel injection nozzle at a point beyond the discharge of the boiler feed pumps. If heat exchangers or stage heaters are present in the boiler feed lines, the injection point should be at their discharge.

At feed solution strength and elevated temperatures, chelating agents can corrode mild steel and copper alloys; therefore, 304 or 316 stainless steel is recommended for all feed equipment. Chelate products may be fed neat or diluted with condensate. Chelate feed rates should be controlled based on feed water hardness to avoid serious consequences of misapplications.

13.8.4 Polymers

They are often called *polymeric dispersants*, as they increase the dispersive properties of the conditioning

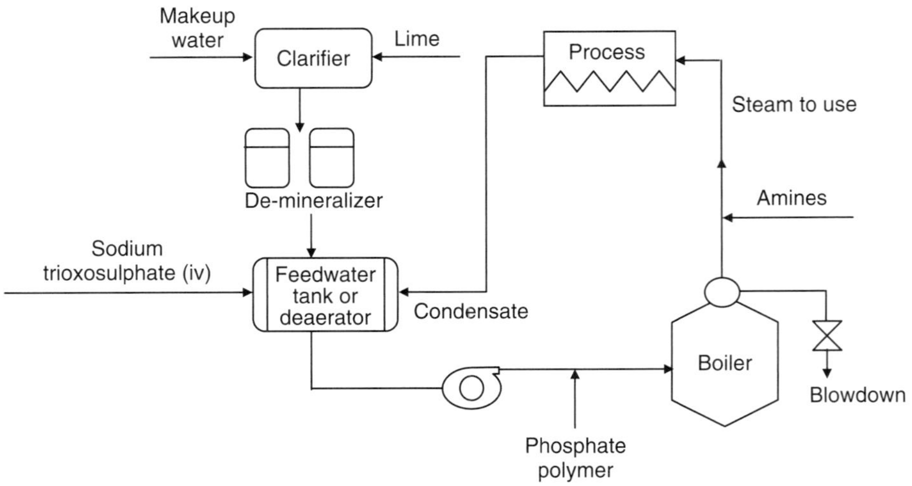

Figure 13-5 Basic boiler system schematic indicating some chemical feed lines [13].

products. Two major types of polymers are synthetic and natural polymers. Natural polymers include lignosulfonates ($R{-}SO_3H$) and tannins. Synthetic polymers include polyacrilates, acryilic acid maleic acid copolymer $(C_4H_4O_4)_n(C_3H_4O_2)_n$, and poly(styrene-*co*-maleic anhydride) $(C_8H_8)_n(C_4H_2O_3)_m$. They act like chelates but are not as effective. Some polymers are effective in controlling hardness deposits, while others are helpful in controlling iron deposits.

In most applications, polymers are provided in a combined product formulation with chelates and/or phosphates. Dilution and feed recommendations for chelates should be followed for chelate–polymer and chelate–phosphate–polymer programs. These combination programs typically have the best results with respect to boiler cleanliness.

13.8.5 Oxygen Scavengers

Oxygen is the main cause of corrosion in hot well tanks, feed lines, feed pumps, and boilers. Water exposed to air can become saturated with oxygen, and the concentration will vary with temperature, such that the higher the temperature, the lower the oxygen content.

A deaerator removes most of the oxygen in feed water. However, trace amounts are still present and can cause corrosion-related problems. Oxygen scavengers are added to the feed water, preferably in the storage tank of the feed water, to remove the trace amounts of oxygen that escaped from the deaerator. The most commonly used oxygen scavenger is sodium trioxosulfate (NCDENR, 2004). Other chemicals that can be used to scavenge oxygen include tannins, hydrazine (N_2H_4), hydroxylamine (NH_2OH), hydroquinone or benzene-1,4-diol [$C_6H_4(OH)_2$], pyrogallol or benzene-1,2,3-triol ($C_6H_6O_3$), and ascorbic acid. These scavengers catalyzed or did not reduce the oxides and dissolved oxygen. The choice of products and the dose required will depend on whether or not a deaerator is used.

13.8.5.1 Uncatalyzed Sodium Trioxosulfate (Na_2SO_3) as an Oxygen Scavenger The uncatalyzed trioxosulfate may be mixed with other chemicals. The preferred location for its injection is a point in the storage section of the deaerating heater where it will mix with the discharge from the deaerating section. Sodium trioxosulfate shipped as liquid concentrate is usually acidic, and when fed neat, corrodes stainless steel tanks at the liquid level. Therefore, Na_2SO_3 storage tanks must be poly(vinyl chloride) (PVC) or 316 stainless steel.

13.8.5.2 Catalyzed Sodium Trioxosulfate (Na_2SO_3) as an Oxygen Scavenger Catalyzed Na_2SO_3 must be fed alone and continuously. Mixing it with another chemical impairs the catalyst. For the same reason, catalyzed Na_2SO_3 must be diluted with only condensate or demineralized water. To protect the entire preboiler system, including any economizer, catalyzed Na_2SO_3 should be fed to the storage section of the deaerator. NaOH may be used to adjust the pH of the day tank solution; therefore, a mild steel tank cannot be used.

13.8.5.3 Hydrazine as an Oxygen Scavenger Hydrazine (N_2H_4) is compatible with all boiler water treatment chemicals except organics, amines, and nitrates. However, it is best fed alone. It is usually fed continuously into the storage section of a deaerating heater. Because of handling and exposure concerns associated with hydrazine, closed storage and feed systems have become standard.

13.8.5.4 Organics as Oxygen Scavengers Many organic compounds are available, including hydroquinone or benzene-1,4-diol [$C_6H_4(OH)_2$] and ascorbic acid. Some are catalyzed; most should be fed alone. Like Na_2SO_3, organic oxygen scavengers are usually fed continuously into the storage section of a deaerator.

13.8.6 Neutralizing Amines

The three most commonly used neutralizing amines are morpholine or diethylenimide oxide or tetrahydro-1,4-oxazine (C_4H_9NO), diethylaminoethanol (DEAE) or 2-(diethylamino)ethanol, and cyclohexanamine ($C_6H_{13}N$). Neutralizing amine does not protect against oxygen attack, but it helps keep oxygen less reactive by maintaining an alkaline pH. They are high-pH chemicals that neutralize products of acid attack, such as trioxocarbonate acid (H_2CO_3), formed in the condensate.

The feed point for the neutralizing amines may be the storage section of the deaerator, directly to the boiler with the internal treatment chemicals, or into the main steam header. Some steam distribution systems may require more than one feed point to allow proper distribution. An injection quill is required for feeding into a steam distribution line. The feed rate of neutralizing amines depends both on the condensate system pH and the corrosion rates measured. The conditions required to feed them are neat (undiluted), diluted with condensate or demineralized water, or mixed in low concentration with the internal treatment chemicals.

13.8.7 Filming Amines

The two most commonly used filming amines are octadecylamine (ODA) or stearamine ($C_{18}H_{39}N$) and ethoxylated soya amine (ESA). The filming amines can form a protective layer on the condensate piping to protect it from both oxygen and acid attack. So, combining neutralizing and filming amines serves as a successful alternative protective against both acid and oxygen attack.

The filming amines should be fed continuously into the steam header at points that permit proper distribution. The steam distribution should be investigated and feed points established to ensure that all parts of the system receive proper treatment. A single feed point is satisfactory for some systems. Filming amines must be mixed with condensate or demineralized water. It is unsuitable to use water that contains dissolved solids because the solids would contaminate the steam and could produce unstable amine emulsions.

13.9 BOILER EFFICIENCY AND COMBUSTION

Within the scope of this work, only a broad overview of the combustion processes, including burner types and controls, as well as heat output and losses are given. The combustion process is an essential component of overall boiler efficiency. Boiler efficiency simply relates energy output to energy input, usually in percentage terms:

$$\text{boiler efficiency (\%)} = \frac{\text{heat exported in steam}}{\text{heat provided by the fuel}} \times 100$$

The heat exported in steam is calculated (using the steam tables) from knowledge of:

- The temperature of the feed water
- The steam pressure
- The steam flow rate

The heat provided by the fuel, the calorific value, can be expressed in two ways:

1. *Gross calorific value*. This is the theoretical total of the energy in the fuel. However, all common fuels contain hydrogen, which burns with oxygen to produce water that passes up the stack as steam. The gross calorific value of the fuel includes the energy used in evaporating this water. Flue gases in steam boiler plants are not condensed; therefore, the actual amount of heat available to the boiler plant is reduced. So, accurate control of the amount of air is essential to boiler efficiency. Too much air will cool the furnace and carry away useful heat. On the other hand, too little air results in incomplete combustion, which may yield to a carryover of unburned fuel and can produce smoke.
2. *Net calorific value*. This is the calorific value of the fuel, excluding the energy in the steam discharged to the stack, and is the figure generally used to calculate boiler efficiency (www.spiraxsarco.com:)

$$\begin{aligned}\text{net calorific value} &= \text{gross calorific value}\\ &\quad - \text{heat of vaporization of water.}\end{aligned}$$

The combustion process is summarized as follows:

$$\underset{\text{fuel}}{C_nH_x} + \underset{\text{air}}{O_2 + N_2} \xrightarrow{\text{combustion}} \underset{\text{heat}}{CO_2 + H_2O + N_2}$$

where n and x represent the number of atoms of carbon and hydrogen, respectively, in fuel.

Accurate control of the amount of air is essential to boiler efficiency. In practice, there are a number of difficulties in achieving perfect (stoichiometric) combustion. The conditions around the burner will not be perfect, and it is impossible to ensure complete matching. Some of the oxygen molecules will combine with nitrogen molecule to form nitrogen oxides (NO_x).

To ensure complete combustion, excess air needs to be provided. This can affect the boiler efficiency. The control of the air/fuel mixture ratio on many smaller boiler plants is open loop. In this case, the burner has a series of cams and levers calibrated to provide a specific amount of air for a particular firing rate. This system requires regular servicing and calibration. For large plants a closed-loop system can be fitted. The closed-loop system use oxygen sensors in the flue to control the combustion air damper. Generally, to ensure accurate control of combustion, air leaks should be checked in the boiler combustion chamber.

13.9.1 Heat Losses

Heat losses are often associated with activities in the boiler furnace. Of particular importance is the air/fuel ratio in relation to complete and efficient combustion. In addition to this, there are other potential sources of heat loss and inefficiency, such as heat losses in the flue gases and heat losses by radiation.

13.9.1.1 Heat Losses in the Flue Gases This is probably the biggest single source of heat loss. The losses are attributable to the temperature of the gases leaving the furnace. Meanwhile, the hotter the gases in the stack, the less efficient the boiler. Two possible causes of overheated gases are:

1. The burner may be producing more heat than is required for a specific load on the boiler. This indicates that the burner(s) and damper mechanisms require maintenance and recalibration.
2. The heat-transfer surface within the boiler is not functioning correctly, and the heat is not transferred to the water. This is an indication that heat transfer surfaces are contaminated and require cleaning [18].

However precautions should be taken to avoid too much cooling of the flue gases. The temperature should not fall below the dew point on which the potential for corrosion is increased by the formation of:

- Trioxonitrate acid (from the nitrogen in the air used for combustion)
- Tetraoxosulfate acid (if the fuel has a sulfur content)
- Water

13.9.1.2 Heat Losses by Radiation Because the boiler is hotter than its environment, some heat will be transferred to the surroundings. Damaged or poorly installed insulation will greatly increase the potential heat losses. Therefore, minimum heat losses are achieved by adequate insulation. For example, a reasonably well insulated shell or water tube boiler at 5 MW or more will lose between 0.3 and 0.5% of its energy to the surroundings. This loss will remain constant whether the boiler is exporting steam to the plant or is simply on standby. This indicates that efficient operation of a boiler can only be achieved at its maximum capacity.

13.9.2 Types of Burners

Burners are responsible for:

- Proper and proportional mixing of fuel and air, for efficient and complete combustion
- Determining the direction and shape of the flame

Burner turndown is an important function of a burner, usually expressed as a ratio:

$$\text{turndown} = \frac{\text{maximum firing rate}}{\text{minimum controllable firing rate}}$$

Types of burners include pressure jet burners, rotary cup burners, low-pressure burners, high-pressure burners, dual burners, and solids combustion systems.

1. *Pressure jet burners*. A pressure jet burner is simply an orifice at the end of a pressurized tube. It is used to fire fuel oil. Typically, the fuel oil pressure is in the range 7 to 15 bar. Within the operating range, a substantial pressure drop is created over the orifice as the fuel is discharged into the furnace. Such a pressure drop results in atomization of the fuel. Putting a thumb over the end of a garden hose creates the same effect. Varying the pressure of the fuel oil immediately before the orifice controls the flow rate of fuel from the burner.
2. *Rotary cup burners*. In this type of burner, fuel oil is supplied down a central tube and discharges onto the inside surface of a rotating cone. As the fuel oil moves along the cap (due to the absence of a centripetal force) the oil film becomes progressively thinner as the circumference of the cap increases. Eventually, the fuel oil is discharged from the lip of the cone as a fine spray. Atomization is produced by the rotating cup, not by fuel pressure drop. The turndown ratio is much greater than that of the pressure jet burner.
3. *Low-pressure burners*. These operate at low pressure, usually between 2.5 and 10 mbar. A low-pressure burner is used to fire gas only. The burner is a simple venturi device

with gas introduced in the throat area and combustion air being drawn in from around the outside. Output is limited to approximately 1 MW. There is no problem regarding the atomization of gas fuels. Only the proper mixing of gas with the appropriate amount of air is required for combustion.

4. *High-pressure burners*. These operate at higher pressures, usually between 12 and 175 mbar and may include a number of nozzles to produce a particular flame shape.

5. *Dual fuel burners*. These burners are designed with gas as the main fuel but have an additional facility for burning fuel oil. They are useful where there may be a gas supply shortage. In that case, changeover to fuel oil firing is made as rapidly as possible to avoid shutdown. The operating procedures are:

- Isolate the gas supply line.
- Open the oil supply line and switch on the fuel pump.
- Purge and refire the boiler.
- On the burner control panel, select oil firing to enable a change of air settings for the various fuels.

6. *Coal and other solid combustors*. Coal as a boiler fuel tends to be restricted to specialized applications such as a water tube boiler in power stations. Coal combustion in boiler plants requires various specialized procedures, including fluidized-bed combustion, pulverized coal combustion, and stoker firing.

13.9.3 Burner Control Systems

It is a fact that the burner control system cannot be operating in isolation. The burner, the burner control system, and the level control system should be compatible and work in a complementary manner to satisfy the steam demands of the plant in an efficient manner. Some basic burner control systems are outlined below.

1. *On/off control system*. This is the simplest control system; either the burner is firing at full rate or it is off. The major disadvantage of this method of control is that the boiler is subjected to large and often frequent thermal shocks every time the boiler fires. Its use should be limited to small boilers up to 500 kg/h.

2. *High/low/off control system*. This is a slightly more complex system in which the burner has two firing rates. The burner operates first at the lower firing rate and then switches to full firing as needed, thereby overcoming the worst of the thermal shock. The burner can also revert to the low-fire position at reduced loads, again limiting thermal stresses within the boiler. This type of system is usually fitted to boilers with an output of up to 5000 kg/h.

3. *Modulating control system*. A modulating burner control will alter the firing rate to match the boiler load over the entire turndown ratio. Every time the burner shuts down and restarts, the system must be purged by blowing cold air through the boiler passages. This wastes energy and reduces efficiency. Full modulation, however, means that the boiler keeps firing over the entire range to maximize thermal efficiency. This type of control can be fitted to any size of boiler, but should always be fitted to boilers rated at over 10,000 kg/h.

REFERENCES

1. An introduction to steam boilers and steam raising, N.E.M Business Solutions, Proactive Maintenance.
2. Bell, A. M., *Locomotives*, 7th ed., Virtue and Company, London, 1957.
3. Boiler efficiency and combustion, http://www.spiraxsarco.com/resources/steam-engineering-tutorial/the-boiler-house/boiler-efficiency-and-combustion.asp.
4. Boiler feedwater, http://www.lenntech.com/applications/process/boiler/boiler-feed-water.htm.
5. Boiler maintenance: tip on maintaining your boiler, P.C. Mckenzie Company, Pittsburgh, PA, http://www.mckenziecorp.com/boiler_maintenance.htm
6. Boiler problem troubleshooting 101, http://ezinearticles.com/?expert=lisa_J Anderson.
7. Boiler troubleshooting guide, Sheng Chan Industries, http://www.scboiler.com.tw/page_5-6.htm.
8. Boiler types, http://www.Southernsteamtrains.com/manual/boilertypes.htm.
9. Admiralty, British *Stokers' Manual, 1912*, HMSO, via Eyre & Spottiswoode, London, 1901.
10. Harris, K. N., *Model Boilers and Boilermaking*, MAP, Hemel Homestead, 1974.
11. Industrial boilers: upgrading and retrofit, Aalborg Industries, Aalborg Denmark, http://www.aalborg-industries.com/industrial_boilers/land_service_upgranding_retrofit.php, 2004.
12. Lancashire boiler, Museum of Science and Industry, Manchester, UK, http://www.Moss.org.uk/media/33871781/lancashire.
13. NCDENR, *Boiler Chemical 101: Fact Sheet*, North Carolina Division of Pollution Prevention and Environmental Assistance, Raleigh, NC, 2004.
14. Steingress, F. M., *Low Pressure Boilers*, 4th ed., American Technical Publishers, Orland Park, IL, 2001.
15. Steingress, F. M., Frost, H. J., and Walter, D. R., *High Pressure Boilers*, 3rd ed., American Technical Publishers, Orland Park, IL, 2003.

16. Welding procedures (overview), QA/QC-Construction Ltd., http://www.qaqc-construction.com/systems 12.php.
17. Westerkamp, T. A., Effective boiler maintenance, http://www.facilitiesnet.com/energy_/_efficiency/article/_effective-boiler-maintencane–8085, 2008.
18. Zeitz, R. A., *CIBO Energy Efficiency Handbook*, Council of Industrial Boiler Owners, Burke, VA, 1997.

FURTHER READING

Block, A. G., *Heat Transfer in Stream Boiler Furnaces*, Hemisphere Publishing, Philadelphia, PA, 1998.

David, F. D., and Glennon, M., *Boiler Efficiency Improvement*, 5th ed., Boiler Effiency Institute, Auburn University, Auburn, AL, 1991.

Dukehow, S. G., (1985), *Improving Boiler Efficiency*, 2nd ed., Instrument Society of Nigeria, Lagos, Nigerian NB.

Huber, O. C., *Radiation Intensities and Heat Transfer in Boiler Furnaces*, University of Iowa, Iowa City, IA, 1936.

Kinealy, J. H., *An Elementary Text Book on Steam Engines and Boilers*, Spon & Chamberlain, New York, 1901.

SECTION II

PROCESS PLANT RELIABILITY

14

ENGINEERING ECONOMICS FOR CHEMICAL PROCESSES

Alberto R. Betancourt-Torcat, L. A. Ricardez-Sandoval, and Ali Elkamel

University of Waterloo, Waterloo, Ontario, Canada

Companies build chemical plants, refineries, and oil and gas facilities around the world to produce a new product or to meet the growing demands of consumers. The decision regarding the construction of a new chemical plant or the revamp of an existing facility is subject to economical, social, political, and environmental constraints associated with the project. Often, these factors are translated in indexes that are based on economic terms. For example, if a particular project would produce as a side product a pollutant that can seriously put the public's welfare at risk, the company must study the alternatives available to dispose of this substance securely. Thus, the environmental and social impact of this project will be reflected in terms of the costs required to make a proper and safe disposal of this substance (e.g., the cost of installing a treatment plan or the costs to contract a company that can process this pollutant). When a project is under evaluation, multiple design alternatives are usually evaluated. To select the alternative with the most feasible and economically attractive scenario, indexes or methods based on engineering economics may be used to perform a cost–benefit analysis of each alternative and to make fair comparisons between the current proposals. Therefore, economic analyses are often used by companies at the earlier stages of a project to select the most promising alternative.

In this chapter we present the most popular methods and tools used in economic analysis to evaluate projects that involve chemical plants, refineries, and oil and gas facilities. The principles and standard methods used to assess the prices of the different oil products are also presented. The chapter is organized as follows: The first section covers the basic methods used to evaluate projects at different times. The second section presents the traditional cash flow patterns used to evaluate a project. The third section introduces the economic indexes commonly used to compare different alternatives. Concepts such as the minimum acceptable rate of return, payback period, and internal rate of return are presented in this section. The fourth section presents basic methods of estimating the fixed capital investment and total capital investment for chemical plants.

14.1 TIME VALUE OF MONEY

Money is a valuable asset that can be exchanged for goods. Often, companies do not have available the initial investment required to begin a project. Since companies know that projects are expected to produce revenues after a certain period of time, they are willing to rent the money to start projects at the current time at the expense that they have to return this present amount of money plus an additional amount at a certain time in the future. Financial institutions (e.g., banks) lend money on the promise that they will receive an extra amount after a certain period of time. The additional amount charged by the service provider is usually called the *interest*. This is the price that the company has to pay for making use of this money previously owned by the financial institution to purchase a good at the current time (i.e., to cover the initial expenses

Process Plant Equipment: Operation, Control, and Reliability, First Edition. Edited by Michael D. Holloway, Chikezie Nwaoha, and Oliver A. Onyewuenyi.
© 2012 John Wiley & Sons, Inc. Published 2012 by John Wiley & Sons, Inc.

of the project), on the promise that the company will obtain something valuable in the future (i.e., profits from the project). Therefore, the money changes its value based on the time in which it is possessed. That is, $1 is more valuable today than in the future because this single dollar can be exchanged for a good now or can earn interest up until the time the future dollar is received. Consequently, the time value of money plays an important role in the economic analysis of a project.

The simplest method to calculate an interest payment is as

$$I = Pi \tag{14.1}$$

where P represents the present amount of money, also called the *principal* or the *present worth*, and i is the *interest rate* per interest period. The interest rate i represents the effective rate paid by companies to financial institutions for a given period for borrowing money. It is usually defined as a percentage of the total amount borrowed. A 0.5% interest rate means that for every money unit lent (e.g., dollars), an additional 0.005 money unit must be paid in the form of an interest I for each time period. The time periods are usually based on years; however, they may be situations where the periods are defined in quarters of a year, months, weeks, or even days. In large-scale projects such as the construction of a new chemical plant, the time periods are usually measured in years. Similarly, the interest period (i.e., the period of time used to calculate the interest I) is usually given on a yearly basis. Thus, if the interest rate is 10% and no time period is specified for this rate, the interest rate is assumed to be on a yearly basis. The amount of money obtained at the end of one time period is calculated as

$$F = P(1 + i) \tag{14.2}$$

where F represents the future amount, also called the *future worth*. For example, F represents the expected amount of money that will be received by a financial institution at the end of one time period, usually a year. However, the principal can be borrowed for more than one period of time. This method of borrowing the principal for a finite period of time is called *compounding*. Thus, the future worth at the end of N finite time periods is calculated as

$$F = P(1 + i)^N \tag{14.3}$$

The formulation above assumes that N is an integer multiple of the basic time period specified for the interest rate, i. As shown in Eq. (14.3), the longer the time required for returning the present worth, the larger the amount of interest charged. This is because the institution that lent the principal is giving up the opportunity, at the end of each time period, to invest this money in other instruments or services that may provide welfare and profits to the institution. It should be noted that F represents the value of the principal P at a given interest rate i (i.e., F is equivalent to P after N periods of time). Following Eq. (14.3), the interest earned over one time period is also generating interest for the next time period. The total amount of interest paid for an Nth period of time is calculated as

$$I_c = P(1 + i)^N - P \tag{14.4}$$

where I_c, called *compound interest*, represents the interest paid for the interest generated during the *compounding period*. Thus, compounding interest can be understood as an interest on top of an interest. Compounding is the most traditional method used to estimate the interest accumulated for N periods of time. This method is widely used in the industry to estimate the total amount of money owned when a credit or a loan is requested from a financial institution.

An alternative method that does not account for compounding in calculation of the future worth is *simple interest*. That is, the interest earned at the end of one time period does not generate interest in the next time period. In this case, the interest is calculated only as a fraction of the principal P:

$$I_s = PiN \tag{14.5}$$

where I_s, the simple interest, represents the amount of interest earned over N periods of time. As shown in Eqs. (14.4) and (14.5), the calculation method used to estimate the interest for a given future worth may lead to different results that can affect the final decision of a given project. In general, the simple interest method is not frequently used in engineering practice.

The interest rate i can be defined for a full period, usually a year, or as compounded for a shorter period of time. The first case, the *nominal interest rate*, is the most common method used to denote i, which is usually stated as an annual interest. To translate the nominal interest rate to a shorter period of time (e.g., a month), simply divide the nominal interest rate r by the subperiod m; that is, $i_s = r/m$. The nominal interest rate is the rate quoted in the financial institutions as the interest rate, but, as shown, a shorter period of time is implied in the calculation method. In the second case, the *effective interest rate*, this represents the actual interest rate, the one actually applied to investments or loans, but it is not usually stated. This rate is usually specified for a period of time shorter than a year. The following formula can be used to convert the effective interest rate, i_e, to a one-year compounding period:

$$i_e = (1 + i_s)^m - 1 \tag{14.6}$$

As shown in Eq. (14.6), the nominal interest rate is assumed to be fixed for the compounding period. The main difference between the nominal and effective interest rates, i_e and i_s, is that the first takes into account the interest generated by the interest over N periods of time (i.e., compounding interest), whereas the latter only considers the interest as a percentage of the principal for the same period of time (i.e., simple interest). From Eq. (14.6) it is clear that the nominal interest rate will always be less than the effective interest rate. As mentioned above, financial institutions and trade markets usually report the nominal annual interest. They use this method as a strategy to offer loans that look attractive, especially when the interest rates are high and the compounding period is for a short period of time (e.g., days).

Case Study 14.1: Interest Rate Analysis Tilco, a construction company, is studying the possibility of entering into a tendering process to build an offshore platform in the Gulf of Mexico. Based on a preliminary analysis, Tilco expects to invest \$2 million in this project when they begin construction of the platform. However, its budget available for this project is \$1.5 million. Therefore, they are studying the possibility of obtaining a loan. Two financial institutions have made a proposal to Tilco for the loan. Alternative A offers Tilco a loan of \$2 million at a 0.71% interest rate compounded quarterly for 20 years. Alternative B offers a loan for of \$2 million at an annual interest rate of 2.8% for 20 years. Assuming that the project will start today, which financial institution offers the most attractive credit?

Solution: To make a decision in this case study, the future worth can be used as a tool to analyze how much \$2 million will be worth in the future.

For alternative A:

$P = \$2$ million

$i = 0.71\%$

$N = 20 \times 4 = 80$ (since the interest is compounded quarterly)

$F = P(i+1)^N = 2(1+0.0071)^{80} = \$3.52$ million

For alternative B:

$P = \$2$ million

$i = 2.8\%$

$N = 20$ (since the interest is compounded annually)

$F = P(i+1)^N = 2(1+0.028)^{20} = \$3.47$ million

Based on the calculations, alternative B assumes that \$2 million will be worth \$3.47 million in 20 years, whereas alternative A assumes that the same amount of money will be worth \$3.52 million in the same period of time. That is, alternative B is worth less \$2 million today than alternative A. Thus, alternative B offers the most attractive loan because Tilco will only be required to pay interest of \$1.47 million compared to \$1.52 million requested by alternative A.

14.2 CASH FLOW ANALYSIS

As shown in Section 14.1, the interest rate is the basic index used to determine the value of money at a given time. Also, it is used as a tool to compare cash flows at different times (e.g., to estimate the equivalence of the principal at different times in the future). In large-scale projects, however, the cash flows entering (receipts) and leaving (disbursement) a project do not occur at a fixed period of time and at different interest rates and compounding periods. Therefore, use of the calculation methods described above for irregular periods of time is a tedious and time-consuming practice. To circumvent this issue, engineers usually assume that all the receipts and disbursements in a project will occur in a fixed period of time (e.g., a month or a year). Although this will not reflect the real situation, it is expected to be a good approximation of the economic life of the project.

On the other hand, it is often useful in engineering economics to convert cash flows occurring at different times to an equivalent amount at a given time. A simple approach to obtaining this equivalence is to make use of mathematical models that relate the inputs (i.e., the cash flow at the initial time) to the outputs (i.e., the cash flow at the desired time). This mathematical function can also be understood as an algebraic factor that relates cash flows at different times. These models, usually referred to as *compound interest factors*, are particularly useful for converting cash flows to present, annual, or future worth. There are also used as instruments to compare cash flows on different time bases.

14.2.1 Compound Interest Factors for Single Cash Flows

The model shown in Eq. (14.3) can be rewritten as

$$F/P = (1+i)^N \tag{14.7}$$

where the term on the right-hand side is the algebraic factor that relates a cash flow in the present time (i.e., input) to an equivalent cash flow at the end of the period N (i.e., output). This relationship is known as the *compound interest factor*. This model is represented in functional notation for a single payment as

$$F = P(F/P, i, N) \tag{14.8}$$

Thus,

$$(F/P, i, N) = (1+i)^N \tag{14.9}$$

where the notation $(F/P, i, N)$ is read as follows: Estimate F given P at an interest rate i for a fixed period of time N. The model shown in Eq. (14.9) is known as the *compound amount factor*. This model is used to calculate the value of the principal, P, at a given interest rate i after N periods of time. The reciprocal of the compound amount factor, known as the *present worth factor*, converts a single disbursement of receipt in the future to its equivalent present value. The present worth factor formula is defined as

$$(P/F, i, N) = \frac{1}{(1+i)^N} \tag{14.10}$$

14.2.2 Compound Interest Factors for Annuities

The compounding amount factor and the present worth factor consider that the cash flow will be referred *to a particular time in the future or to the present time, respectively. An alternative option is to derive* models that translate receipts or disbursements to a set of constant regular payments for a finite period of time. This type of cash flow pattern is known as an *annuity*. The basic idea is that the user, who received a loan today, will make constant payments for a specific period of time and at a given interest rate to return the amount loaned plus the interest rather than making a single payment at maturity. Annuity is a widely used index in the industry to pay debts, loans, leases, or mortgages for a land. The annuity compounding interest factor models involve the present and the future worth, respectively. The equivalence factor that estimates the annuity given the principal at an interest rate for a finite set of periods of time is called the *capital recovery factor*. The algebraic model that defines this relationship is

$$(A/P, i, N) = \frac{i(1+i)^N}{(1+i)^N - 1} \tag{14.11}$$

The capital recovery factor is used to estimate how much regular money A needs to be saved for N periods of time to recover an investment P made today. The inverse of the capital recovery factor is the *series present worth factor* and is defined as

$$(P/A, i, N) = \frac{(1+i)^N - 1}{i(1+i)^N} \tag{14.12}$$

The series present worth factor model estimates the principal amount that is equivalent to having periodic or regular receipts or disbursements at an interest rate i for N periods of time. This model is used to know the present time value of money if single receipts or disbursements A are done for regular intervals over a certain period of time N and at a constant interest rate i.

The capital recovery factor and the series present worth factor are compounding models used to obtain equivalences between annuity and the principal. The equivalent models that relate annuity to receipts or disbursements analyzed in the future are called the *sinking fund factor* and the *uniform series compound amount factor*. The sinking fund factor converts an amount of money in the future to regular payments or receipts over a finite period of time and interest rate. The formula for this model is

$$(A/F, i, N) = \frac{i}{(1+i)^N - 1} \tag{14.13}$$

This model is used to estimate for an interest rate i the amount of money A that needs to be saved (or paid) at regular intervals to obtain an amount F at the end of time period N. For example, this model is used as a method for saving money to purchase goods or contract services in the future. In some cases, the sinking fund factor is combined with the capital recovery factor to estimate the annual benefits (savings) obtained when an asset is purchased. This formula is

$$A = (P - S)(A/P, i, N) + Si \tag{14.14}$$

where S represents the salvage value of the asset at the end of its productive life. The salvage value can be understood as a receipt obtained at the Nth period of time (i.e., it represents the future value of the asset). The term A represents the regular benefits obtained from the purchase. This method is used in the industry to evaluate the benefits of purchasing equipment from different vendors.

The inverse of the sinking fund factor calculates the opposite: that is, the amount in the future F that will be obtained if constant payments A are done for a period of time N at an interest rate i. This factor, the uniform series compound amount factor, is

$$(F/A, i, N) = \frac{(1+i)^N - 1}{i} \tag{14.15}$$

14.2.3 Arithmetic and Geometric Gradient Series

The models for the compounding annuity factor presented earlier assumed that the annuity A is fixed to a constant value for every time interval considered in the calculation. However, there are situations where the annuity changes linearly through the entire period of time considered in the analysis (e.g., maintenance costs or the purchase of raw materials). This pattern may also be observed in chemical plants where the ability to increase the capacity of the plant is limited by external factors and it can only be done steadily. Thus, the annuity models introduced earlier are modified to account for this situation.

14.2.3.1 Arithmetic Gradient Series The first scenario considers the case when annuity changes by a constant amount at each time interval. For example, initially, the annuity is zero; at the end of the first period, the annuity is equal to A; for the second period, the annuity is set to A plus an extra amount G; at the end of the third period, the annuity is defined as A plus twice the value of G; at the end of the fourth period, the annuity is equal to A plus three times the value of G. This pattern continues up until the last period in the analysis, N, is reached. This type of cash flow is known as the *arithmetic gradient series with base annuity* A. If the cash flows consider only the amount G (i.e., there is no annuity in the cash flow), the cash flow is called the *arithmetic gradient series*.

Based on the above, two arithmetic gradient series–based models are presented. The first model calculates the equivalence between the arithmetic gradient series and the present time. The algebraic model used to estimate this equivalence is

$$(P/G, i, N) = \frac{(1+i)^N - iN - 1}{i^2(1+i)^N} \tag{14.16}$$

where G represents the regular increments (or decrements) at each time period (i.e., the arithmetic gradient). This cash flow pattern, called the *arithmetic gradient present worth factor*, is used to estimate the present amount of money that would represent constant increments or decrements G at each time period and at an interest rate i.

This second cash flow model converts the arithmetic gradient series, G, to a constant annuity, A. That is, the model estimates the value of a constant annuity A that is equivalent to having an arithmetic gradient series, G, for N time periods and for an interest rate i. This cash flow model is known as the *arithmetic gradient-to-annuity conversion factor* and is defined mathematically as

$$(A/G, i, N) = \frac{1}{i} - \frac{N}{(1+i)^N - 1} \tag{14.17}$$

Note that this model converts only the arithmetic gradient series G to a constant annuity A. Also note that the model shown in Eq. (14.17) does not take into account a constant annuity (e.g., A^*) that may be included in the cash flow analysis. Thus, annuity A^* must be added to the product of the arithmetic gradient series G and the arithmetic gradient-to-annuity conversion factor; that is,

$$A_{\text{tot}} = A^* + G(A/G, i, N) \tag{14.18}$$

where A_{tot}, the total annuity, represents a conversion of the constant annuities and the arithmetic gradients occurring in the cash flow model.

14.2.3.2 Geometric Gradient Series The second scenario considers those cash flows where the fluctuations occur at a constant specific rate; that is, the cash flow grows or declines proportionally in each time period. This cash flow pattern is known as the *geometric gradient series*. This cash flow is as follows: At time zero, the annuity is considered to be zero; at the end of the first period, the annuity is equal to A; for the second period, the annuity is set at A plus an extra amount defined in terms of a percentage rate g of A. Thus, the total annuity at the end of this period is A' (i.e., $A' = A + Ag$). At the end of the third period, the annuity is equal to the annuity in the preceding period, A', plus an additional percentage, g, of this amount, Thus, the total annuity in this period is set to A'' [i.e., $A'' = A' + A'g$ or $A'' = A(1+g)^2$]. This pattern continues until the last period in the analysis, N, is reached. As shown, the actual annuity is estimated based on previous annuities and a percentage increment or decrement of this amount. This cash flow pattern is relevant for the industry since it can be used to model inflation or deflation. The following algebraic function can be used to convert this cash flow to present worth:

$$\begin{aligned}(P/A, g, i, N) &= \frac{1}{1+g}\,\frac{(1+i^0)^N - 1}{i^0(1+i^0)^N}\\ i^0 &= \frac{1+i}{1+g} - 1\end{aligned} \tag{14.19}$$

This model, also *known as the geometric gradient-to-present worth factor*, is similar to the definition of compounding interest. That is, the interest generated at the end of a period generates interest in the next period, at a given rate. This model is thus also used to model cash flows with different interest rates. In the previous model, it is assumed that the interest rate i is different from the rate of increase or decrease, g. In some circumstances, the percentage rate at which annuity is growing or declining is equal to the interest rate (i.e., $g = i$). In this case, the model shown in Eq. (14.19) is reduced to the factor

$$(P/A, g, i, N) = \frac{N}{1+i} \tag{14.20}$$

In summary, in this section we presented the basic models used in economic analysis to estimate equivalences between different cash flow patterns. These models constitute the basic tools used in engineering economics to analyze different alternatives and to select the most suitable and economically attractive project. Although use of the previous models is simple, they have also been collected in a table format for different interest rates and periods of years. The tables are available online (e.g., http://www.uic.edu/classes/ie/ie201/discretecompoundinteresttables.html) or in various engineering economics textbooks (e.g., [1,2]).

14.3 PROFITABILITY ANALYSIS

Every project has its own cash flows; that is, a project usually has an initial investment, cash flows occurring during its lifetime (e.g., annual savings and costs), and a salvage value at the end of its estimated life. Therefore, it is necessary to evaluate projects on the same basis such that a fair comparison can be made between the alternatives proposed. In this section we present the four basic tools used in engineering economics to evaluate the various alternatives available in the development of a project.

14.3.1 Payback Period

The simplest method used to evaluate a project is estimating the time required to recover the costs associated with an alternative. This method is known as the *payback period* or *payout analysis*. The mathematical model used to estimate the payback period is

$$n_P = \frac{\text{initial investment}}{\text{annual savings}} \tag{14.21}$$

where n_p represents the payback period and is given in time units, usually years. This method is easy to apply and the concept is fairly simple to communicate to people who do not have a background in economics. The method does not require any additional knowledge other than an estimate of the initial investment and the annual savings reported for the alternative. The idea is that the alternative with the minimum number of years should be selected since this implies that the money invested in it will be recovered within a short period of time. On the other hand, this method offers several limitations; that is, it does not take into account the time value of money (i.e., interest rates are assumed to be zero). Also, the consequences beyond the payback period are ignored completely (i.e., it ignores the service life of a project), and it discriminates against long-term projects. Thus, the method should never be used as a tool to make a decision regarding a project since the payback period index can lead to a wrong decision. Consequently, decisions regarding the selection of a large-scale project (e.g., the construction of a new chemical facility) should not be evaluated based on this criterion. The basic use of this index is to provide an initial screening of the various alternatives (e.g., to screen for the most promising alternatives at the earlier stages of a project). This index can also be used in conjunction with other methods to assist in the selection of an alternative.

14.3.2 Minimum Acceptable Rate of Return

The final goal of a project is to generate revenues for a company. Thus, companies expect to receive more money than the original amount invested in a project. That is, for the project to be attractive economically, the rate of return of the project must be above a minimum acceptable value. Thus, the *minimum acceptable rate of return* (MARR) is defined as the interest rate that must be paid by a project. Companies do not select alternatives with interest rates lower than the MARR because they are not economically feasible. This can be understood as spending money on a project that will not return the minimum expected revenues. Also, investing in these alternatives implies that a company is not making good use of money (i.e., a higher income could be obtained from other alternatives or projects with interest rates higher than the MARR).

Since the MARR is project specific and changes from project to project, there is no general formula to estimate this index. The MARR is usually defined based on previous experiences with other projects in the same field, from an economic analysis of existing projects similar to those alternatives that are under current analysis or from the experience of the engineers evaluating the project alternatives. Consequently, one of the shortcomings of this index is that is subject to uncertainty: Companies propose a MARR based on previous scenarios, but there is no guarantee that the project under current analysis may return at least the minimum expected revenues. Thus, it is always desired to select the alternative with the highest rate of return.

14.3.3 Present and Annual Worth Analysis

The mathematical functions presented in Section 14.1 are simple algebraic models that can be used to translate or convert a cash flow to a period of time or defined in terms of a uniform series (e.g., an annuity). Therefore, these formulas can be applied to translate all the cash flows of an alternative to a single period of time or to a uniform series. That is, all the cash flows that are expected to occur in the service life of an alternative are represented by a single amount of money per time basis. The most common basis of time used to represent the cash flows are the present time (i.e., the *present worth*) and a uniform series basis (i.e., the *annual worth*). The representation of the cash flows at the end of the service life (i.e., *future worth*) is another comparison method, but it is not often used in the economic analysis of chemical facilities.

The present worth (PW) comparison method applies the mathematical models shown in Section 14.1 to translate all the cash flows of each alternative considered in the project to the present time. For obvious reasons, the method provides a sense of the dimensions of the project since it is easy to relate the estimated amount of money (i.e., the present worth) to the actual time. If the alternative considers receipts (benefits) and disbursements (costs) in its cash flows, the method proposed estimates the *net present worth*, NPW. This term can be defined as the difference between the present worth of the benefits and the present worth of the costs associated with the alternative; that is,

$$NPW = PW_{\text{benefits}} - PW_{\text{costs}} \qquad (14.22)$$

On the other hand, the annual worth (AW) comparison method performs a cash flow analysis to determine the amount of money saved or spent per unit of time (e.g., a year), for the entire service life of the alternative. As in the present worth method, the annual worth analysis employs the cash flow models presented in Section 14.1. The annual worth method is particularly useful to conform to accounting practices and to relate the annual worth to other costs (e.g., labor and energy consumption costs) or benefits (e.g., revenues from selling a product or from offering a service). As in the PW method, the AW method can be used to estimate the *equivalent annual costs* (EAC), the *equivalent annual benefits* (EAB), or the *net annual worth* (NAW):

$$NAW = EAB - EAC \qquad (14.23)$$

The AW comparison method is similar to the PW method, except that the money units in the AW method are given on a uniform series basis, which is usually a year. Therefore, application of the AW and PW methods to the evaluation of a set of alternatives lead to the same solution (i.e., the AW and PW methods both return the same decision in the evaluation of a project). As a result, there is no need to apply both the AW and the PW methods to the same analysis. The selection of the method depends on the actual cash flows of the alternatives and on the desired time basis used to present the results of the economic analysis.

To apply either the PW or AW comparison methods, the cash flow information of all the alternatives must be available and defined in monetary units. Also, the interest rate i must be known. Similarly, it is assumed that the cash flows occur at finite periods of time and that the final results obtained by both the PW and AW methods do not consider costs occurred in the past (i.e., costs before the project exists) or external costs (i.e., costs associated with future projects) that were not considered in the analysis. A strong assumption made in these comparison methods is that the alternatives to be evaluated have the same lifetime (i.e., the study period must match). This limits the application of these comparison methods since it is usual to have proposals with different life periods. For example, vendors of the same equipment (e.g., a desulfurization unit) may propose different lifetimes for its corresponding process unit. To make a fair comparison between alternatives with different periods of life, one of the following approaches can be used.

1. *Repeated lives method*. This method can be used when the periods of life of all the alternatives considered in an analysis have a least common multiple. Then it is assumed that each alternative can be repeated with the same receipts and disbursements in the future (i.e., the life of the alternative is repeated). The method assumes that alternatives can be replaced at the end of the project's lifetime by the identical alternative with the same costs and benefits. Although it is a simple approach to solve the problem of having alternatives with different lives, the method does not account for changes that may occur in the future (i.e., technology improvements).
2. *Study period method*. This method simply selects a study period in which to do the economic analysis. To select an appropriate study period, companies usually consider the minimum service life that the asset should comply with or the maximum period of time during which the company's forecasts are still expected to be certain. One of the difficulties of this method is that the salvage value of the assets may not be known in the time period selected. Thus, additional assumptions must be made to obtain a reasonable estimate of this quantity.

Since the repeated lives method and the study period method rely on different assumptions, the conclusions obtained by each method may be different when applied to the same project.

14.3.4 Internal Rate of Return

Thus far, it has been considered that the interest rate of all the cash flows is known a priori. However, there are cases where the interest rate of a particular alternative is not known with certainty. Thus, the AW and PW methods cannot be applied directly. An alternative is to estimate the interest rate that makes the NPW or the NAW of all the cash flows in an alternative equal to zero: that is, the interest rate at which the project *breaks even*. This method is known as the *internal rate of return* (IRR). This approach is by far the preferred comparison method used in the industry to perform engineering economic analysis. The word *internal* stands for the fact that the interest rate is valid only for the cash flows of a particular alternative.

To estimate the IRR, it is necessary to define the NPW or NAW of the cash flows, set the equation equal to zero, and solve for the interest rate. The resulting equations formulated for this problems are generally nonlinear algebraic functions with respect to the interest rate i. Thus, a numerical method is required to estimate a root (i.e., the internal rate of return) for these nonlinear algebraic equations. Consequently, calculation of the internal rate of return is not as in straightforward as in the PW or AW method. However, this index is easy to interpret (i.e., what the rate of return is for every dollar invested). The IRR serves as the basis for ranking the projects and returns the same results as the NPW and NAW comparison methods.

To analyze whether or not a single alternative should be accepted, one can either estimate the NPW using the MARR as the interest rate or estimate the internal rate of return and compare this value to the MARR. In the first option, if the resulting NPW is greater than zero, the alternative proposed should be accepted since the cash flows will at least return the prespecified MARR. In the second option, the alternative must be accepted only if the resulting internal rate of return is greater than the MARR. This implies that a rate of return larger than minimum required is expected to be obtained.

The previous calculation methods are valid only when the decision regarding a single project does not depend on the selection of other projects or is not affected by external factors (e.g., other projects). These projects are usually defined as *independent projects* since the expected costs and benefits do not rely on external factors. However, the real strength of the internal rate of return method is to compare *mutually exclusive projects* (i.e., those projects that in the process of choosing the most economically attractive alternative exclude the rest of the alternatives considered in the analysis). This case represents perhaps the most realistic situation in the economic evaluation of alternatives in chemical, refinery, and oil and gas facilities. However, the methods proposed for independent projects may lead to erroneous conclusions if applied to mutually exclusive projects. This is because, for the case of an independent project, the expected benefits and costs of other projects (e.g., other alternatives considered in the analysis) are not considered in the calculation of the IRR. Therefore, a different calculation method of the internal rate of return must be used for mutually exclusive projects. The solution method proposed is based on the calculation of the *incremental internal rate of return* (i.e., the internal rate of return on the incremental investment from choosing a larger instead of a smaller project). The criterion used to select one project over the other is that if incremental IRR is greater than the MARR, the project with the largest investment must be selected. This implies that the earnings obtained by an alternative that requires a large investment are also expected to be considerably large compared to an alternative that requires a small investment. The procedure used to select the most attractive proposal from a set of alternatives, known as the *defender–challenger approach*, is summarized as follows:

1. Rank the alternatives in terms of initial cost. Sort the alternatives from the lowest to the highest initial cost.

2. Add to the list the "Do nothing" project. This project represents the case where none of the alternatives satisfy the criterion used to select a project (i.e., MARR). Set the initial cost and the benefits of the "Do nothing" project to zero. Thus, this alternative must be the proposal with the lowest initial cost in the analysis.

3. Select the project with the lowest initial cost (e.g., "Do nothing") and name this alternative proposal A. Take the alternative with the second-lowest initial cost and call this proposal B. Proposal A becomes the defender, while proposal B is the challenger. In this case, proposal B represents the first alternative considered in the analysis other than the "Do nothing" proposal.

4. Using either the present worth or annual worth calculation method, determine the incremental internal rate of return between proposal A and proposal B:

$$PW_{\text{costs,B}} - PW_{\text{costs,A}} = PW_{\text{benefits,B}} - PW_{\text{benefits,A}}$$

or

$$EAC_B - EAC_A = EAB_B - EAB_A \tag{14.24}$$

where the subscripts A and B refer to proposals A and B, respectively. The only unknown in Eqs. (14.25) must be the interest rate, i. Thus, the solution of any of these equations returns the incremental internal rate of return, i^*.

5. If the incremental internal rate of return is greater than the MARR, the challenger (proposal B) won over the defender (proposal A), and it must be selected as the best available alternative. Thus, proposal B becomes the defender. Otherwise (i.e., $i^* <$ MARR), proposal B is not worth doing and proposal A (the defender) remains the better alternative.

6. Select the next alternative in the analysis that has the lowest initial cost, and name it proposal C. This proposal becomes the challenger in the analysis. As in step 4, calculate the incremental internal rate of return between the two proposals. If $i^* >$ MARR, the challenger (proposal C) becomes the best available alternative and is set to be the defender. Otherwise, discard proposal C and keep either proposal A or proposal B as your defender, depending on the assessment made in step 5.

7. Continue this process until all the alternatives considered in the project has been analyzed. The proposal that remained at the end of this process as the best current alternative can then be selected as the most economically attractive proposal to perform the project.

The procedure above assumes that the set of alternatives considered in the analysis have the same service lives. If the alternatives do not have equal lives, the repeated lives method or study period approach must be applied first to perform a fair comparison between the cash flows [3].

Case Study 14.2: Present Worth Analysis Gepsol, a refinery in the Middle East, is considering the purchase of a distillation unit. Two vendors have submitted a proposal. Vendor A's proposal shows that the initial cost of the process unit is $5 million, the unit is expected to generate earnings for $0.75 million per year, but it requires major maintenance service every five years that is approximated to be of $0.35 million. The service life of this unit is expected to be 20 years, and the unit can be sold at the end of this period for $0.65 million. Vendor B offers a process unit that has a service life of 10 years. The initial cost stipulated for this unit is $2.8 million, it does not require any maintenance during its lifetime, and it is expected to generate savings of $0.4 million per year. The salvage value of the unit at the end of its service life is assumed to be $0.25 million. Using the present worth analysis, determine the option that is more economically attractive to Gepsol if the MARR specified is 12%.

Solution: The alternatives proposed to Gepsol have different lives. Thus, to make a fair comparison, it is necessary to analyze both alternatives using the same period of life. Since proposal A is a common multiple of proposal B, the repeated lives method can be used to study the alternatives. Thus, the cash flows of alternative B have to be repeated twice. To do the PW analysis, the compound interest models presented in Section 14.1 are used to convert the cash flows occurring at different time periods to the present. Thus, Eq. (14.22) was reformulated for alternative A as follows:

$$\begin{aligned}\mathrm{NPW_A} &= 0.75(P/A, 12\%, 20) + 0.65(P/F, 12\%, 20)\\ &\quad - \{5 + 0.35[(P/F, 12\%, 5) + (P/F, 12\%, 10)\\ &\quad + (P/F, 12\%, 15) + (P/F, 12\%, 20)]\}\end{aligned}$$

The result of this calculation returns

$$\mathrm{NPW_A} = \$0.2579 \text{ million}$$

Similarly, for alternative B,

$$\begin{aligned}\mathrm{NPW_B} &= 0.4(P/A, 12\%, 10)[1 + (P/F, 12\%, 10)]\\ &\quad + 0.33(P/F, 12\%, 10) + 0.33(P/F, i, 20)\\ &= \$0.3069 \text{ million}\end{aligned}$$

Thus, based on the present worth analysis and assuming that alternative B's cash flows can be repeated twice, the best offer is made by vendor B, since the NPW returns that this alternative will generate earnings that are 16% more than those predicted by vendor A's proposal. This scenario may change if the assumption regarding the repetition of vendor B's cash flows is not valid.

Case Study 14.3: Incremental Rate of Return with Uncertainty in the MARR Petro-Dubai, a worldwide company in the oil and gas business, is considering the construction of a new refinery in Saudi Arabia. During the tendering process, several bidders applied for this large-scale project. However, only five alternatives completely satisfied all the requirements specified by Petro-Dubai. The cash flows presented by the bidders for each of these alternatives are presented in Table 14-1. One of the specifications provided by Petro-Dubai is that the service life of the refinery must be 50 years. Thus, it is assumed that the five alternatives must comply with this specification. Due to uncertain forecasts, Petro-Dubai could only specify a MARR for this project that is within 11 to 13%. Using the incremental internal rate of return method, determine the bidder that Petro-Dubai should select to execute this project.

Solution: This problem raises the issue of the selection of a proposal from mutually exclusive projects. Also, the problem does not define a single MARR for the project; instead, a range of values are given for this index. Thus, a sensitivity analysis on the MARR must be performed to study the effect of this parameter over the final decision.

The preferred method used to analyze the alternatives is the incremental internal rate of return. Thus, the defender–challenger approach outlined above was applied to this case study. To simplify the analysis, the MARR will first be assumed to be constant and equal to the worst MARR expected (i.e., MARR = 11%). Then the same analysis will be made using the best value expected for this index (i.e., MARR = 13%). The annual worth method presented below is used in this case study to determine the incremental internal rate of return. Although it is not shown in Table 14-1, the alternative with the lowest cost is the "Do nothing" project, a project whose costs, annual benefits, and salvage value are zeros. This represents the best current alternative at the beginning of the analysis (i.e., it is the defender in the defender–challenger approach).

Defender: Do nothing; Challenger: Bidder A. Applying Eq. (14.23) and the capital recovery factor model to this case study yields

$$(125 - 0)(A/P, i^*, 50) = 14 - 0$$

Thus,

$$125\left[\frac{i^*(1+i^*)^{50}}{(1+i^*)^{50} - 1}\right] - 14 = 0$$

TABLE 14-1 Cash Flows for the Petro-Dubai Project ($ Millions)

Bidder	Initial Costs	Annual Benefits
A	125	14.0
B	132	15.0
C	155	17.5
D	170	21.6
E	188	23.2

The equation above must be solved numerically using off-the-shelf software (e.g., the built-in function *fzero* in MATLAB). The solution obtained from this analysis returns

$$i^* = 11.14\%$$

Thus, alternative A becomes the current best alternative since $i^* > \text{MARR}$. Also, this alternative becomes the defender for the next comparison.

Defender: Bidder A; Challenger: Bidder B

$$(132 - 125)(A/P, i^*, 50) = 15 - 14$$

Thus,

$$7\left[\frac{i^*(1+i^*)^{50}}{(1+i^*)^{50}-1}\right] - 1 = 0$$

The incremental internal rate of return obtained for this comparison is $i = 14.27\%$. Thus, bidder B becomes the defender and the best current alternative.

Defender: Bidder B; Challenger: Bidder C

$$23\left[\frac{i^*(1+i^*)^{50}}{(1+i^*)^{50}-1}\right] - 2.5 = 0$$

The incremental internal rate of return obtained for this comparison is $i^* = 10.81\%$. Thus, bidder C is not worth it since $i^* < \text{MARR}$. Consequently, B remains as the best current alternative and the defender in the analysis.

Defender: Bidder B; Challenger: Bidder D

$$38\left[\frac{i^*(1+i^*)^{50}}{(1+i^*)^{50}-1}\right] - 6.6 = 0$$

After solving the equation, it is found that $i^* = 17.36\%$. Thus, proposal D becomes the defender and the best current alternative.

Defender: Bidder D; Challenger: Bidder E

$$18\left[\frac{i^*(1+i^*)^{50}}{(1+i^*)^{50}-1}\right] - 1.6 = 0$$

The solution of this equation returns $i^* = 8.76\%$. Thus, it is not feasible to accept proposal E since the incremental internal rate of return is less than the worst-case MARR. Consequently, the decision that can be drawn from this analysis is that the proposal made by bidder D represents the most attractive alternative.

The economic analysis above was made assuming the worst MARR expected. To analyze the effect of the uncertainty on this parameter over the final decision, the same analysis as that made for an MARR = 11% was repeated assuming that MARR = 13%. Table 14-2 shows the results obtained from the second analysis. As shown in the table, the results reveal that the most attractive proposal is the "Do nothing" project. That is, it is better to invest the money in other projects rather than investing it in any of these alternatives. The results from this second analysis also showed that the best incremental internal rate of return was obtained for project D, $i^* = 12.67\%$, which is the alternative selected when MARR = 11%. Based on the above, Petro-Dubai should not execute this project since the uncertainty in the MARR is large enough that it rules out any of the alternatives presented by the bidders. This result can be used to either redefine the upper bound of the MARR or to support an in-depth analysis of the company's forecasts for the MARR.

TABLE 14-2 Defender–Challenger Approach (MARR = 13%)

Bidder	Defender	Challenger	i^* (%)
Do nothing			
A	Do nothing	A	11.14
B	Do nothing	B	11.31
C	Do nothing	C	11.24
D	Do nothing	D	12.67
E	Do nothing	E	12.30

14.4 COST ESTIMATION AND PROJECT EVALUATION

When a project is proposed, the first step that needs to be done is to estimate the total cost of the project to decide its feasibility. Since the main goal for a company to invest in a project is to obtain profits, the project has to be more attractive than other projects. Thus, the methods used to estimate the costs of different alternatives in a project are key steps that affect the decisions regarding the projects. Therefore, given the importance of cost estimation; there is a branch of engineering that focuses on this particular topic. Every company has a department that analyzes cost estimation or subcontracts companies specialized in cost estimation to do this analysis. This is because a large investment has to be compromised on any industrial project. Thus, the company must ensure that the money is invested in economically attractive projects.

14.4.1 Capital Investment

Capital investment compiles the total initial investment that must be done before the startup of a "grass-roots" plant (new plant) or the expansion of an existing facility. This investment can be divided into three different categories: fixed capital investment, working capital, and cost of land. Each of these costs is discussed next.

*14.4.1.1 **Fixed Capital Investment*** This is capital required for construction of the manufacturing facilities and any other infrastructure necessary for the normal operation of the plant and personnel's comfort. Fixed capital investments can be divided further into the following categories:

- *Battery limits investment*. This concept considers the capital for the following expenses:
  1. Manufacturing buildings
  2. Equipment
  3. Installation
  4. Piping
  5. Instrumentation and process control
  6. Electrical systems
  7. Insulation
- *Off-battery-limit investment*. This investment takes in account the capital invested in the infrastructure outside the manufacturing buildings, such as:
  1. Warehouses (for raw material and intermediate and finished products)
  2. Utility building facilities
  3. Administration building
  4. Yard improvement
  5. Dispensary
- *Overhead costs*. These are the expenses related to the indirect costs necessary to carry on the project from the engineering and legal points of view. The most common expenses for this concept are:
  1. Legal expenses (site construction permits)
  2. Software licenses for engineering
  3. Temporary buildings at construction site
  4. Engineering and supervision
  5. Fees for contractor
  6. Contingencies

*14.4.1.2 **Working Capital*** This cost refers to the amount of money required to start up a plant and keep it running until the cash flow is enough to cover the costs. These costs include the following:

1. Raw materials acquisition
2. Money to back up client credits
3. Money to pay suppliers, utilities, services, and administration costs
4. Contingencies (this money is always necessary because no project is perfect and along the way some unexpected expenses will come up)
5. Insurance
6. Intermediate and finished products inventory

*14.4.1.3 **Cost of Land*** This is the cost associated with the construction site. Although sometimes it is treated as part of the fixed capital investment, this cost must have a special treatment. This is because, for tax purposes, land is nondepreciable. Thus, special attention must be paid to the cost of the land.

14.4.2 Cost Indexes

The majority of the databases available for a new project's cost estimation are based on past parameters and the cost of existing plants similar to the one under current analysis. Thus, it is necessary to use an index that correlates the data available with the current money value in the market for inflation, especially nowadays, when the economic cycles of prosperity and contraction dictate the cost of investments. To adjust the costs of the chemical units or plants to the current time, a wide range of indexes have been developed for different sectors by private and government entities. In the field of chemical engineering, the most widely used indexes are the Marshal and Swift (M&S) index and the Chemical Engineering plant cost index, which can be found every month in the journal *Chemical Engineering* or online, http://www.che.com. These two sources are used widely for equipment and chemical plants investments, and the construction cost indexes are available in the Engineering News-Record construction cost index, which can be found every week in the journal *Engineering News-Record*.

Additional sources of information may include personal or company files with historical data, quotations from catalogs, past project cost investments, and equipment purchased. The cost index ratio used for inflation adjustments can be expressed as follows:

$$\text{cost index} = \frac{\text{cost index value project at present time}}{\text{cost index value at time of original cost}} \tag{14.25}$$

Case Study 14.4: Cost Index A heat exchanger cost \$15,000 in 2005. Estimate the cost for 2008. Use the Marshall & Swift cost index and the Chemical Engineering plant cost index (CEPCI).

Solution: Using Marshall & Swift:

$$\text{M\&S index (2005)} = 1244.5$$

$$\text{M\&S index (2008)} = 1449.3$$

$$\text{Cost (2008)} = 15{,}000\left(\frac{1449.3}{1244.5}\right) = \$17{,}460$$

Using the Chemical Engineering plant cost index:

$$\text{CEPCI (2005)} = 468.2$$

$$\text{CEPCI (2008)} = 575.4$$

$$\text{Cost (2008)} = 15{,}000\left(\frac{575.4}{468.2}\right) = \$18{,}430$$

Now using the Chemical Engineering plant cost index specific for heat exchangers and tanks yields

$$\text{index (December 2005)} = 522.5$$

$$\text{index (December 2008)} = 618.2$$

$$\text{Cost (2008)} = 15{,}000\left(\frac{618.2}{522.5}\right) = \$17{,}750$$

From the analysis above it can be shown that using the equipment cost indexes (i.e., Marshall & Swift and heat exchanger and tank cost), a difference of 1.7% is obtained in the estimated cost, whereas a 5.6% difference was observed when comparing the cost using the M&S index with the cost obtained using the Chemical Engineering plant cost index.

14.4.3 Capital Cost Estimates

There are various methods of estimating the cost of a project, from the simplest method based on a projection subjected to existing similar projects to the most detailed description based on process flowsheets and control and instrumentation diagrams. The main difference between the methods is the amount of information available to perform the estimation. Accurate estimations of the costs are usually obtained when detailed information on the project is available. However, this estimation would also be more expensive since more information needs to be considered in the analysis. Therefore, more time and money (e.g., in the form of cost estimation engineering-hours) is needed to process and use this information to evaluate a project.

These methods can be divided in two main classes: predesign and design estimates. The first method, the predesign estimate, only provides the order of magnitude of the project over a short time. This method is helpful for evaluating the feasibility of the project or when comparing different alternatives. Also, the method does not require specific information about the project, which consequently causes inaccuracies in the actual estimate of the cost of the project. This method is used widely at the beginning of a project to select the most attractive and feasible alternatives. Once a feasible option has been chosen, this can be developed into a more detailed cost estimation using the design method.

As the project moves forward in time, more data are available and accurate (and necessary) estimates can be made. In the remainder of this section we discuss predesign estimate methods.

14.4.3.1 Predesign Estimates These are used to evaluate the feasibility of a project. The methods used at this stage of the project are discussed next.

Order of Magnitude This is a set of methods that estimates only in a gross manner the order of magnitude of a project's investment. The accuracy of these methods is within 30 to 50%. The *turnover ratio* (TR) is a simple method that allows evaluation of a project in terms of the gross annual sales projected and the fixed capital investment. This method is commonly used to estimate the order of magnitude of a project's investment. The gross sale is defined as the product obtained by multiplying the total production over a year (which in general represents 330 days of production) by the average selling price. This relation is expressed as

$$\text{fixed capital investment} = \frac{\text{gross annual sales}}{\text{TR}} \tag{14.26}$$

This index depends on the process under consideration. For most chemical processes this margin covers from 1 (large plants) to 1.25 (medium plants), 1 being the value most commonly used for this analysis.

Case Study 14.5: Turnover Ratio Application Formaldehyde is produced by the catalytic oxidation of methanol over a silver metal catalyst. Its selling price is $2400 per metric ton. What would be the total capital cost estimate for a plant producing 75,000 metric tons per year? (Use the turnover ratio method.)

Solution: Using Eq. (14.41) and TR = 1 (annual production corresponds to a large plant) yields

$$\begin{aligned}\text{gross annual sales} &= 2400 \times 75{,}000\\ &= \$180{,}000{,}000\\ \text{fixed capital investment} &= \frac{180{,}000{,}000}{1}\\ &= \$180{,}000{,}000\end{aligned}$$

14.4.3.2 Ratio Method Consider a set of parameters (ratios) that can be found in the literature as the fixed capital invested per unit plant capacity, capacity ratio, relative labor rate ratio, relative productivity factor ratio, cost index, location ratio, and currency to estimate the necessary fixed capital investment. The accuracy of this method is within 30 to 50%.

$$\text{fixed capital investment} = \frac{\text{FCU} \cdot C \cdot \text{CR}^x \cdot \text{CI} \cdot \text{LR} \cdot \text{CE}}{\text{PF}} \tag{14.27}$$

where FCU is the fixed capital investment per unit capacity (obtained from similar existing plants):

$$\text{FCU} = \frac{\text{fixed capital investment}}{\text{plant capacity}} \qquad (14.28)$$

C is the nominal plant capacity for cost data (old plant capacity); CR is the capacity ratio:

$$\text{CR} = \frac{\text{design capacity}}{\text{nominal capacity for cost data}} \qquad (14.29)$$

CI is the cost index ratio (index to adjust the capital investment from different years per inflation); LR is the relative labor rate ratio (cost parameter accounting labor rate between two locations):

$$\text{LR} = \frac{\text{relative labor rate at actual location}}{\text{relative labor rate at nominal location for cost data}} \qquad (14.30)$$

CE is the currency exchange rate; PF is the relative productivity factor ratio (cost parameter accounting productivity factor between two locations):

$$\text{PF} = \frac{\text{relative productivity factor at design location}}{\text{relative productivity factor at nominal location for cost data}} \qquad (14.31)$$

and X is the exponent of the capacity ratio (CR). This term accounts for the economy of scale. Its value usually changes between 0.6 and 0.7, 0.6 being the value most commonly used.

The terms LR and PF can be replaced by a location ratio parameter LP that multiplies the mathematical expression (14.27), and it can be found in Table 14-5.

$$\text{LP} = \frac{\text{unit cost at design location}}{\text{unit cost at nominal location for cost data}} \qquad (14.32)$$

The values for relative labor rate and relative productivity factor for different locations are available in a number of engineering textbooks. For some locations, see Table 14-3.

The fixed capital investment per unit capacity and the exponent accounting for the economies of scale can also be found in engineering sources. For most commonly used processes and products, see Table 14-4. The location factors for various countries are provided in Table 14-5.

Case Study 14.6: Ratio Method A group of investors is considering building in Canada a new plant to produce 18,000 metric tons per year of acetic acid. Given the increasing demand of this product in that market, they need to make a quick estimate to find the order of magnitude of the investment in Canadian dollars (CAD) for 2009.

TABLE 14-3 Relative Labor Rate Productivity Indexes in Chemical and Allied Products Industries for the United States, (1999)

Geographic Area	Relative Labor Rate	Relative Productivity Factor
New England	1.14	0.95
Middle Atlantic	1.06	0.96
South Atlantic	0.84	0.91
Midwest	1.03	1.06
Gulf	0.95	1.22
Southwest	0.88	1.04
Mountain	0.88	0.97
Pacific coast	1.22	0.89

Source: [5].

Solution: From Eq. (14.43), the fixed capital investment is defined in terms of different ratio parameters:

$$\text{fixed capital investment} = \frac{\text{FCU} \cdot C \cdot \text{CR}^x \cdot \text{CI} \cdot \text{LR} \cdot \text{CE}}{\text{PF}}$$

The values of the terms FCU, CR, and X were obtained from Table 14-4 and are as follows:

$$\text{FCU} = \frac{\$8{,}000{,}000}{10{,}000 \text{ tons/yr}} = \$800 \text{ yr/ton} \qquad (14.28)$$

$$X = 0.68 \text{ (exponent accounting for the economy of scale)}$$

$$\text{CR} = \frac{18{,}000 \text{ tons/yr}}{10{,}000 \text{ tons/yr}} = 1.8 \qquad (14.29)$$

$$\text{CI} = \frac{1468.6}{915.1} = 1.6 \qquad (14.25)$$

Applying a location ratio for this case [Eq. (14.43)] yields

$$\text{fixed capital investment} = \text{FCU} \cdot C \cdot \text{CR}^x \cdot \text{CI} \cdot \text{LP} \cdot \text{CE}$$

$$\text{LP} = \frac{1.14}{1} = 1.14 \qquad (14.32)$$

$$\text{CE (2002)} = 1.14 \text{ CAD/U.S. dollars}$$

$$\text{fixed capital investment} = 800 \cdot 10{,}000 \cdot 1.8^{0.68} \cdot 1.6 \cdot 1.14 \cdot 1.14 = \text{CAD } \$24{,}810{,}000$$

The cost index was obtained from the M&S annual cost index for all industries for 1990 and 2009, the location parameters were obtained from Table 14-5, and the currency exchange rate was the average exchange rate for Canadian to U.S. dollars in 2009.

Case Study 14.7: Ratio Method A new ammonium nitrate (NH_4NO_3) facility is about to be built in Canada

TABLE 14-4 Capital Cost Data for Chemical and Petroleum Processing Plants (2000 Values)

Product	Process	Typical Plant Size: 10^3 kg/yr (10^3 tons/yr)	Fixed Capital Investment ($ millions)	Power Factor x^a for Process Plant Specified
Acetic acid	CH_3OH and CO catalytic	9×10^3 (10)	8	0.68
Ammonia	Steam reforming	9×10^4 (100)	29	0.53
Ammonium nitrate	Ammonia and nitric acid	9×10^4 (100)	6	0.65
Ethylene	Refinery gases	4.5×10^4 (50)	16	0.83
Ethylene oxide	Ethylene catalytic	4.5×10^4 (50)	59	0.78
Formaldehyde (37%)	Methanol catalytic	9×10^3 (10)	19	0.55
Methanol	CO_2, natural gas, steam	5.5×10^4 (60)	15	0.60
Nitric acid (high-strength)	Ammonia catalytic	9×10^4 (100)	8	0.60
Polyethylene (high-density)	Ethylene catalytic	4.5×10^3 (5)	19	0.65
Propylene	Refinery gases	9×10^3 (10)	4	0.70
Sulfuric acid	Sulfur-contact catalytic	9×10^4 (100)	4	0.65
Urea	Ammonia and CO_2	5.5×10^4 (60)	10	0.70
Alkylation (H_2SO_4)	Catalytic	1.6 (10)	23	0.60
Coking (delayed)	Thermal	1.6 (10)	31	0.38
Coking (fluid)	Thermal	1.6 (10)	19	0.42
Cracking (fluid)	Catalytic	1.6 (10)	19	0.70
Cracking	Thermal	1.6 (10)	6	0.70
Distillation (atm.)	65% vaporized	1.6 (100)	38	0.90
Distillation (vac.)	65% vaporized	1.6 (100)	23	0.70
Hydrotreating	Catalytic desulfurization	1.6 (10)	3.5	0.65
Re-forming	Catalytic	1.6 (10)	34	0.60
Polymerization	Catalytic	1.6 (10)	6	0.58

Source: [5].

[a]These power factors apply within roughly a threefold ratio extending either way from the plant size as given.

with a production capacity of 150,000 metric tons per year. The ammonium nitrate is produced using ammonia and sulfuric acid, which makes it necessary to build both plants as adjacent facilities. Determine the total fixed capital investment required for the project in 2009.

Solution: The reactions involved in the process are

$$NH_3 + 2O_2 \longrightarrow HNO_3 + H_2O$$
$$NH_3 + HNO_3 \longrightarrow NH_4NO_3$$

Based on the stoichiometry relationship between reactants and products, the component's molecular weights and using Table 14-4 to obtain FCU, X, and CR (see Table 14-6) yields

$$MW(HNO_3) = 63$$
$$MW(NH_4NO_3) = 80$$
$$MW(NH_3) = 17$$

The quantity of NH_3 required is then defined as

$$\frac{2MW(NH_3)}{MW(NH_4NO_3)} \cdot 150{,}000 \text{ tons/yr}$$
$$= 2\left(\frac{17}{80}\right) \cdot 150{,}000 \text{ tons/yr} = 63{,}750 \text{ tons/yr}$$

Two moles of NH_3 is necessary because both reactions involved in the process consumed 1 mol of NH_3. The quantity of HNO_3 required is then defined as

$$\frac{MW(HNO_3)}{MW(NH_4NO_3)} \cdot 150{,}000 \text{ tons/yr}$$
$$= \frac{63}{80} \cdot 150{,}000 \text{ tons}/yr = 118{,}125 \text{ tons/yr}$$

Applying Eq. (14.43) yields

$$\text{fixed capital investment} = \text{FCU} \cdot C \cdot \text{CR}^x \cdot \text{CI} \cdot \text{LP} \cdot \text{CE}$$

Rearranging Eq. (14.43) in order to consider the contribution of the three different plants to the fixed capital cost gives us

$$\begin{aligned}\text{fixed capital investment} = {} & [(\text{FCU} \cdot \text{CR}^x)_{NH_3} \\ & + (\text{FCU} \cdot \text{CR}^x)_{HNO_3} \\ & + (\text{FCU} \cdot \text{CR}^x)_{NH_4NO_3}] \\ & \cdot C \cdot \text{CI} \cdot \text{LP} \cdot \text{CE}\end{aligned}$$

TABLE 14-5 Location Factors for Chemical Plants of Similar Functions (1993 Values)[a]

Location	Factor (United States = 1.0)
Australia	1.04
Canada	1.14
France	0.73
Germany	0.76
India	
Imported element	0.80
Indigenous element	0.25
Italy	0.79
Japan	1.46
Middle East	0.84
North Africa	
Imported element	0.65
Indigenous element	0.44
Norway	0.92
South America	1.36
Spain	0.83
Sweden	0.75
United Kingdom	0.76
United States	1.00

Source: [4].

[a] 1. Increase a factor by 10% for each 1000 miles or part of 1000 miles that the new plant is distant from a major manufacturing or import center, or both.
2. When materials or labor, or both, are obtained from more than a single source, prorate the appropriate factors.
3. Investment incentives have been ignored.

TABLE 14-6 Total Capital Investment for Various Chemical Plants

Product	Typical Facility Size (tons/yr)	Total Fixed Capital Investment ($ millions)	Power Factor, x
NH_3	100,000	29	0.53
HNO_3	100,000	8	0.60
NH_4NO_3	100,000	6	0.65

Source: [5].

$$(\text{FCU} \cdot \text{CR}^x)_{NH_3} = \frac{\$29{,}000{,}000}{100{,}000 \text{ tons/yr}} \cdot \left(\frac{63{,}750}{100{,}000}\right)^{0.53} = \text{U.S. } \$228.44 \text{ yr/ton}$$

$$(\text{FCU} \cdot \text{CR}^x)_{HNO_3} = \frac{\$8{,}000{,}000}{100{,}000 \text{ ton/yr}} \cdot \left(\frac{118{,}125}{100{,}000}\right)^{0.60} = \$88.41 \text{ yr/ton}$$

$$(\text{FCU} \cdot \text{CR}^x)_{NH_4NO_3} = \frac{\$6{,}000{,}000}{100{,}000 \text{ ton/yr}} \cdot \left(\frac{150{,}000}{100{,}000}\right)^{0.65} = \text{U.S. } \$78.1 \text{ yr/ton}$$

$$\text{CI} = \frac{1468.6}{1089} = 1.35 \qquad (14.25)$$

$$\text{LP} = \frac{1.14}{1} = 1.14 \qquad (14.32)$$

$$\text{CE (2009)} = \text{CAD } \$1.14/\text{U.S. dollar}$$

$$\text{fixed capital investment} = (\text{U.S. } \$394.95 \text{ yr/ton}) \cdot 100{,}000 \text{ tons/yr} \cdot \frac{1468.6}{1089} \cdot 1.14 \cdot 1.14 \, \frac{\text{CAD dollar}}{\text{U.S. dollar}} = \text{CAD } \$69 \text{ million}$$

The cost index was obtained from the M&S annual cost index for all industries for 2000 and 2009. For the location parameters, see Table 14-5. The currency exchange rate is the average exchange rate of Canadian of U.S. dollars in 2009.

14.4.3.3 Factor Estimate These methods use information related to the total cost of the major plant's equipment to express the remaining costs involved as percentages (factors) of the equipment costs to determine the total capital investment or fixed capital investment of a project. The accuracy of this method is approximately 30%. The most commonly used factor estimate methods are the Lang factor method and the percentage of delivered equipment cost. These methods are discussed next.

The *Lang factor method* uses a set of factors that have been defined depending on the type of processing plant (solid, solid–fluid, or fluid) to be used in the alternative version that is under analysis. Depending on the type of plant, different single factors are employed to calculate the investment required for particular chemical processing plants. These factors are shown in Table 14-7. Although the Lang factor method gives rough estimates, its accuracy depends on the cost of the data involved. For example, if the prices of the equipment were obtained directly from the manufacturers instead of equipment cost data estimations, an accurate estimate of the capital investment may be obtained. This method can be used for grass-roots facilities

TABLE 14-7 Lang Factors for Various Processing Plants

Type of Plant	Fixed Capital Investment	Total Capital Investment
Solid processing	4.0	4.7
Solid–fluid processing	4.3	5.0
Fluid processing	5.0	6.0

Source: [5].

or the expansion of existing plants. The Lang factor acts as an installation factor (i.e., the cost of the major equipment should be delivered cost). The equation for the Lang factor method is as follows

$$\text{capital investment} = \text{cost of delivered equipment} \cdot \text{Lang factor} \quad (14.33)$$

Case Study 14.8: Lang Factor Method Estimate the total capital investment for a grass-roots plant producing cereal grains; consider it a solid processing plant and the total cost of delivered equipment \$10 million. The cost of land is not included.

Solution: Using Eq. (14.33) and the Lang factor from Table 14-7 gives

$$\begin{aligned}\text{total capital investment} &= \text{cost of delivered equipment} \cdot \text{Lang factor (solids processing)}\\ &= \$10{,}000{,}000 \cdot 4.7\\ &= \$47{,}000{,}000\end{aligned}$$

The *percentage of delivered equipment cost or fixed capital investment method* is similar to the Lang factor in the sense that it depends on factors specified for different process plants (solid, solid–fluid, and fluid). The only difference with respect to the Lang factor is that instead of expressing in a single index all cost parameters (e.g., delivered major equipments, piping, insulation, electricity, instrumentation and control, warehouses, utilities facilities, administration buildings, engineering expenses, contractor fees, legal expenses) in the fixed capital investment method every parameter has its own value as a function of percentages of the cost of equipment delivered or fixed capital investment. Thus, a more accurate estimation is expected from this method. However, all estimates are affected directly by the quality of the data used in the analysis. Thus, accurate estimates depend on using parameters that fit the particular process that is being considered in the study. For example, in the calculation of the total capital investment of a new solid processing plant, it is preferred to choose parameters from a similar plant in terms of the process involved, the plant technology, and the capacity, if possible. Common values for chemical plants and processes can be observed in Tables 14-8 and 14-9.

The fixed capital investment method relies on finding the major delivered equipments cost using company files from equipment purchased or past company projects. Nevertheless, this task is not simple because it is difficult to match the equipment capacities. If this were the case, a general expression is applied for corrections on a sizing basis that is analogous to the one used as the plant capacity ratio (CR). This equation can be expressed as

$$\left(\frac{\text{capacity equipment 1}}{\text{capacity equipment 2}}\right)^x = \frac{\text{cost of equipment 1}}{\text{cost of equipment 2}} \quad (14.34)$$

Typical values for the exponential sizing parameter (X) for most commonly used equipment are given in Table 14-10. Equation (14.34) assumes that the equipment costs are on the same time value basis. If the data available correspond to old equipment costs, the respective time money value correction must be done for inflation using cost indexes.

TABLE 14-8 Typical Percentages of Fixed-Capital Investment Values for Direct and Indirect Cost Segments for Multipurpose Plants or Large Additions to Existing Facilities

Component	Range of FCI (%)
Direct costs	
Equipment purchased	15–40
Equipment installation	6–14
Instrumentation and controls (installed)	2–12
Piping (installed)	4–17
Electrical systems (installed)	2–10
Buildings (including services)	2–18
Yard improvements	2–5
Service facilities (installed)	8–30
Land	1–2
Indirect costs	
Engineering and supervision	4–20
Construction expenses	4–17
Legal expenses	1–3
Contractor's fee	2–6
Contingency	5–15

Source: [5].

Case Study 14.9: Percentage of Cost of Equipment Delivered Estimate the fixed capital investment for a process plant addition to produce organic chemicals on a continuous basis (fluid processing plant and cost of land not included). The total equipment delivered cost is \$10,000,000.

Solution: Using the data in Table 14-11, both factors returned a somewhat similar result. A comparison with the result obtained using the Lang factor method (14.33) is

$$\text{fixed capital investment} = \$10{,}000{,}000 \cdot 5.0 = \$50{,}000{,}000$$

The result obtained using the Lang factor method is equivalent to the result obtained using the percentage of delivered equipment cost method. The Lang factor for a fluid processing plant was obtained from Table 14-7.

TABLE 14-9 Factors to Convert Delivered-Equipment Costs into Fixed Capital Investment

	Solids Processing	Solids–Fluid Processing	Fluid Processing
Grass-roots plants			
Equipment, delivered	1.00	1.00	1.00
Installed	0.19–0.23	0.39–0.43	0.76
Piping	0.07–0.23	0.30–0.39	0.33
Structural steel foundations, reinforced concrete	—	—	0.28
Electrical	0.13–0.25	0.08–0.17	0.09
Instruments	0.03–0.12	0.13	0.13
Battery-limit building and service	0.33–0.50	0.26–0.35	0.45
Excavation and site preparation	0.03–0.18	0.08–0.22	
Auxiliaries	0.14–0.30	0.48–0.55	Included above
Total physical plant	2.37	2.97	3.04
Field expense	0.10–0.12	0.35–0.43	
Engineering	—	0.35–0.43	0.41
Direct plant costs	2.48	3.73	3.45
Contractor's fees, overhead, profit	0.30–0.33	0.09–0.17	0.17
Contingency	0.26	0.39	0.36
C_{FC}: total fixed capital investment	3.06	4.27	3.98
Battery-limit installations			
Equipment, delivered	1.00	1.00	1.00
Installed	0.45	0.39	0.27–0.47
Piping	0.16	0.31	0.66–1.20
Structural steel foundations, reinforced concrete	—	—	0–0.13
Electrical	0.10	0.10	0.09–0.11
Instruments	0.09	0.13	
Battery-limit building and service	0.25	0.39	0.18–0.34
Excavation and site preparation	0.13	0.10	0.10
Auxiliaries	0.40	0.55	0.70
Total physical plant	2.58	2.97	3.50
Field expense	0.39	0.34	0.41
Engineering	0.33	0.32	0.33
Direct plant costs	3.30	3.63	4.24
Contractor's fees, overhead, profit	0.17	0.18	0.21
Contingency	0.34	0.36	0.42
C_{FC}: total fixed capital investment	3.81	4.17	4.87

Source: [4].

Case Study 14.10: Equipment Power Sizing Estimate the 2009 cost in U.S. dollars of a jacketed reactor including a mixer (FOB) with a capacity of 1.5 m^3. The location is the United States. Use the values shown in Table 14-10 as estimates for the nominal capacity, approximate cost, exponential sizing parameter (X), and M& S cost index of the equipment.

Solution:

$$\text{M\&S (Table 14-10)} = 1000$$

$$\text{M\&S (2009)} = 1468.6$$

Now using Eq. (14.34), we have

$$\text{cost of equipment 1} = \left(\frac{\text{capacity equipment 1}}{\text{capacity equipment 2}}\right)^x \cdot \text{cost of equipment 2}$$

$$= \left(\frac{1.5\ \text{m}^3}{0.38\ \text{m}^3}\right)^{0.53} \cdot \$9300 \cdot \frac{1468.6}{1000}$$

$$= \$28{,}300$$

14.4.3.4 Design Estimates This method uses definitive and detailed information about the chemical process, equipment, labor, and so on. This method has an accuracy that is less than 15%. It can be divided into two streams:

1. A *definitive estimate* is based on advance project–gathered data but without including all the details necessary for precise equipment design or plant construction. This estimate is used to approve the money required to start a project's procurement phase. The accuracy of this method is approximately ±10 to 15%.

TABLE 14-10 Typical Exponents for Equipment Cost as a Function of Capacity

Equipment	Size Range	Exponent
Blender, double-cone rotary, carbon steel (c.s.)	1.4–7.1 m^3 (50–250 ft^3)	0.49
Blower, centrifugal	0.5–4.7 m^3/s (10^3–10^4 ft^3/min)	0.59
Centrifuge, solid bowl, c.s.	7.5–75 kW (10–10^2 hp) drive	0.67
Crystallizer, vacuum batch, c.s.	15–200 m^3 (500–7000 ft^3)	0.37
Compressor, reciprocating, air cooled, two-stage, 1035 kPa discharge	0.005–0.19 m^3 (10–400 ft^3/min)	0.69
Compressor, rotary, single-stage, sliding vane,1035 kPa discharge	0.05–0.5 m^3/s (10^2–10^3 ft^3/min)	0.79
Dryer, drum, single vacuum/atmospheric	1–10 m^2 (10–10^2 ft^2)	0.76/0.40
Evaporator (installed), horizontal tank	10–1000 m^2 (10^2–10^4 ft^2)	0.54
Heat exchanger, shell-and-tube, floating head, c.s.	10–40 m^2 (100–400 ft^2)	0.60
Heat exchanger, shell-and-tube, fixed sheet, c.s.	10–40 m^2 (100–400 ft^2)	0.44
Pump, reciprocating, horizontal cast-iron	1×10^{-4}–6×10^{-3} m^3/s (2–100 gpm)	0.34
Pump, centrifugal, horizontal, cast steel	4–40 $m^3/s{\cdot}kPa$ (10^4–10^5 gpm·psi)	0.33
Reactor, glass-lined, jacketed (without drive)	0.2–2.2 m^3 (50–600 gal)	0.54
Reactor, stainless steel, 2070-kPa	0.4–4.0 m^3 (10^2–10^3 gal)	0.56
Separator, centrifugal, c.s.	1.5–7 m^3 (50–250 ft^3)	0.49
Tank, flat head, c.s.	0.4–40 m^3 (10^2–10^4 gal)	0.57
Tank, c.s., glass-lined	0.4–4.0 m^3 (10^2–10^3 gal)	0.49
Tower, c.s.	5×10^2–10^6 kg (10^3–2×10^6 lb)	0.62
Tray, bubble cap, c.s.	1–3 m (3–10 ft) diameter	1.20
Tray, sieve, c.s.	1–3 m (3–10 ft) diameter	0.86
Jacketed reactors, including mixer, FOB[a]	0.38 m^3 (100 gal)	0.53

Source: [5].

[a]Value adapted from Ref. [4], Table 9–50. The approximate cost of a 0.38-m^3 unit is $9300 with M&S = 1000.

2. A *detailed estimate* is based on project detailed drawings, blueprints, instrumentation and control diagrams, complete equipment specifications, and building and utilities specifications. It is used to approve the total project's capital investment, even though a contingency fund is kept in case of cost fluctuations. This method's accuracy is approximately 5 to 10%.

14.4.4 Production Costs and Estimations

Production costs and estimations represent all costs related to products ready to be shipped from the factory. They can be divided into manufacturing costs and general expenses.

14.4.4.1 Manufacturing Costs These are all the costs involved in a good manufacturing process, plus value processes, which allow the company to earn a profit. Manufacturing costs are divided further into three main classes.

14.4.4.1.1 Variable Costs These are the costs that depend on the plant output or production rate. They are usually treated on an annual time basis, which makes it possible to spread out certain preestablished high costs, such as maintenance and seasonal production demands. The two common costs associated with variable costs are raw materials and manufacturing labor costs.

1. *Raw materials* are the reactants and materials to be used in the production of new products. These materials are consumed according to the stoichiometric relationships among the reactants. For chemical processes, raw materials represent 10 to 80% of the total production costs. Raw material prices can be obtained using journals (e.g., the *Trade Journal* and *Chemical Week*). However, as raw material prices represent a key factor in determining the profitability of a company, estimations on raw material prices must be based on historical data through long periods of time where the economy cycles can be appreciated (downturns and upturns), especially for chemical processes involving commodities. It is also important because the product prices in the future for these industries depend on the feedstock prices [6].

2. *Manufacturing labor* is the number of labor-hours required per shift to produce a product. Most companies use a three-shift policy, and the amount of labor may vary based on the scheduled production for each shift. One method used to estimate this type of cost is through calculation of the number of unit operations or major processes, such as distillation, drying, filtration, and reaction. When labor is considered based on unit operations, it considers all the

TABLE 14-11 Summary of Case Study Results: Percentage of Delivered Equipment Cost

	Factor [5]	Total Cost ($)	Factor (Table 14-9)	Total Cost ($)
Direct cost items				
Equipment delivered	1	10,000,000	1	10,000,000
Equipment installed	0.47	4,700,000	0.40	4,000,000
Instrumentation and control	0.36	3,600,000	0.13	1,300,000
Piping	0.68	6,800,000	0.94	9,400,000
Electricity	0.11	1,100,000	0.10	1,000,000
Buildings	0.18	1,800,000	0.27	2,700,000
Yard improvements	0.10	1,000,000	0.10	1,000,000
Service facilities	0.70	7,000,000	0.70	7,000,000
Total direct cost	3.60	36,000,000	3.64	36,400,000
Indirect cost items				
Engineering and supervision	0.33	3,300,000	0.33	3,300,000
Construction	0.41	4,100,000	0.41	4,100,000
Legal fees	0.04	400,000		
Contractor's fee	0.22	2,200,000	0.21	2,100,000
Contingency	0.44	4,400,000	0.42	4,200,000
Total indirect cost	1.44	14,400,000	1.37	13,700,000
Fixed capital investment	5.04	50,400,000	5.01	50,100,000

equipment involved in a particular stage to be part of one unit operation. This is formulated mathematically as follows:

$$\frac{\text{manufacturing labor hours}}{\text{tons of product}} = t \cdot \frac{\text{number of unit operations}}{[\text{capacity (tons/day)}]^{0.76}} \quad (14.35)$$

where the parameter t is defined as: 10 for highly automated continuous process, 17 for no automated continuous processes with average labor, and 23 for no automated batch processes.

Labor can also be estimated considering the average labor required per piece of process equipment and using the plant capacity (see Table 14-12). Moreover; labor costs are affected by a plant's location and productivity. Manufacturing labor also includes supervision costs (10 to 20% of manufacturing labor), which is process specific. Traditionally, manufacturing labor represents 10 to 20% of the total production cost.

TABLE 14-12 Typical Labor Requirements for Process Equipment

Type of Equipment	Workers/Unit/Shift
Blowers and compressors	0.1–0.2
Centrifugal separator	0.25–0.50
Crystallizer, mechanical	0.16
Dryer, rotary	0.5
Dryer, spray	1.0
Dryer, tray	0.5
Evaporator	0.25
Filter, plate and frame	1.0
Filter, rotary and belt	0.1
Filter, vacuum	0.125–0.25
Heat exchangers	0.1
Process vessels, towers (including auxiliary pumps and exchangers)	0.2–0.5
Reactor, batch	1.0
Reactor, continuous	0.5

Source: [5].

Case Study 14.11: Manufacturing Labor Estimate the manufacturing labor for a continuous process with average labor involving three unit operations: heat transfer, reaction, and distillation. The plant capacity is 120,000 metric tons/yr of product. (Assume annual operation of 300 days.)

Solution: Using Eq. (14.35) gives us

$$\frac{\text{manufacturing labor-hours}}{\text{tons of product}} = t \cdot \frac{\text{number of unit operations}}{[\text{capacity (tons/day)}]^{0.76}}$$

$$t = 17$$

$$\text{number of unit operations} = 3$$

$$\text{plant capacity (tons/day)} = \frac{120{,}000 \text{ tons}}{\text{year}} \cdot \frac{1 \text{ operation year}}{300 \text{ days}} = \frac{400 \text{ tons}}{\text{day}}$$

$$\frac{\text{manufacturing labor-hours}}{\text{tons of product}} = 17 \cdot \frac{3}{(400)^{0.76}} = 0.537$$

$$\frac{\text{manufacturing labor-hours}}{\text{year}} = 0.537 \frac{\text{labor-}h}{\text{ton}} \cdot 400 \frac{\text{tons}}{\text{day}} \cdot 300 \frac{\text{days}}{\text{yr}} = 64{,}440 \frac{\text{labor-}h}{\text{yr}}$$

3. *Utilities* are represented by the plant-required services. They can be estimated using the process flowsheet along with the material and energy balance of the plant and the equipment specifications. They can also be estimated from other facilities with similar production processes (i.e., their unit operations and process technology should be similar to the one that is being proposed in the current project). Utilities are usually provided at industrial sites. Sometimes they are produced internally in a utility building or over the fence, which means that a larger company supplies them. They represent 10 to 20% of total production costs. Some utilities are: heating fuels, steam at various pressures, electricity (for lightning, instrumentation and control, pumping, compressing fluids), and cold water.

4. *Maintenance* can be considered as a semivariable cost given that any major plant has a fixed annual maintenance plan. Maintenance represents around 2 to 10% of the fixed capital investment.

5. *Manufacturing supplies* are represented by lubricants and test chemicals consumed during the plant operation. They are not considered to be raw materials because they do not convert into the final goods or products to be obtained from the process. They are estimated to be 10 to 20% of the maintenance cost.

6. *Product control* involved the laboratory expenses required to monitor the quality of products during the process to determine any inconvenience that may jeopardize the final selling product specifications. They represent 10 to 20% of the manufacturing labor.

7. *Catalysts* costs vary depending on their nature and the chemical process. Some catalysts are cheap, but others, such as those constituted by noble metals are very expensive, depending on the amount of metal used. That is why a company may sometimes rent a catalyst instead of buying it, and at the end of its useful life it can be regenerated for further uses. The amount of catalyst needed in a process can be determined by the material and energy balance.

8. *Patents* are rights reserved to companies that developed a certain process or technology, and when another company wants to make use of them must pay based on the plant's production rate. This cost is estimated depending on the process specifications, but in general they represent 0 to 5% of the total production cost. Some utility costs are shown in Table 14-13.

14.4.4.1.2 Fixed Costs These costs do not depend on the plant's output. They range from 10 to 20% of the total production costs. Some of these costs are:

1. *Depreciation* is the cost associated with the initial battery-limit investment (e.g., equipment, manufacturing buildings). Government regulations allow companies to spare this capital investment as a cost along the plant's productive life to recover it. This cost influences the company's income tax; its annual amount is determined by the government tax agency. This cost usually represents 10 to 20% of the fixed capital investment.

2. *Financing* costs are those related to interest payments due to initial borrowed capital from banks and other financial institutions to construct the chemical plant (0 to 10% of fixed capital investment).

3. *Property taxes* are paid depending on the plant location and the established tax rate of the area; they include federal and local taxes. As a rule of thumb, it can be estimated to be 1 to 4% of fixed capital investment.

TABLE 14-13 Cost Tabulation for Selected Utilities and Labor

Utility	Cost
Electricity	0.045 \$/kWh[a]
Fuel	
Coal	0.35 \$/GJ[b]
Petroleum	1.30 \$/GJ[b]
Petroleum coke	0.17 \$/GJ[b]
Gas	1.26 \$/GJ[b]
Refrigeration, to temperature	
5°C	20.0 \$/GJ[c]
−20°C	32.0 \$/GJ[c]
−50°C	60.0 \$/GJ[c]
Steam saturated	
10^3–10^4 kPa (150–1500 psi)	4.40 \$/1000 kg[d,e]
Wastewater	
Disposal	0.53 \$/1000 kg[e]
Treatment	0.53 \$/1000 kg[e]
Waste	
Hazardous	145.00 \$/1000 kg[c]
Nonhazardous	36.00 \$/1000 kg[c]
Water	
Cooling	0.08 \$/1000 kg[e,f]
Process	0.53 \$/1000 kg[e]
Labor	
Skilled	33.67 \$/h[g]
Common	25.58 \$/h[g]

Source: [5].
[a]Based on U.S. Department of Energy, Energy Information Administration form EIA-861, 2001. U.S. average for year 2000.
[b]Based on U.S. Department of Energy, Energy Information Administration form EIA-0348, 2001. U.S. average for year 2000.
[c]R. Turton, R, C. Bailie, W. B. Whiting, and J. A. Shaeiwitz, *Analysis, Synthesis, and Design of Chemical Process*, Prentice Hall, Upper Saddle River, NJ, 1998.
[d]U.S. Department of Energy, Office of Industrial Technologies, DOE/GO-102000-1115, December 2000.
[e]U.S. Department of Energy, Office of Industrial Technologies, DOE/GO-10099-953, June 2001.
[f]M. S. Peters and K. D. Timmerhaus, *Plant Design and Economics for Chemical Engineers*, 4th ed., McGraw-Hill, New York, 1991.
[g]*Engineering News-Record* indexes, December 2001.

4. *Insurance* costs depend on the process plant and the type of equipment involved. They are used to protect the capital investment. Insurance cost ranges from 0.2 to 1.0% of the fixed capital investment.

5. *Rent* depends on the site location; most plants tend to rent the land instead of buying it. It is estimated to be 8 to 12% of building and land value.

14.4.4.1.3 Manufacturing Overhead These costs are not related to production processes, but they contribute to the general company's welfare and safety. Examples of these costs are fire protection, employee services, janitorial services, medical services, inventory services, shop, medical and life insurances, storage, and recreation complex. These costs represent 50 to 70% of manufacturing labor + supervision + maintenance or 5 to 15% of total production cost.

14.4.4.2 General Expenses These represent other costs involved in a plant's normal functioning and not related to manufacturing activities. They can be divided as follows:

1. *Administration costs* are the expenses used for paying the administrative building personnel, administrative supplies, and office equipment. They represent 5 to 15% of total production costs.

2. *Distribution and marketing* play a key role in the company's finance evolution. Their main activity is to promote the product and sell it. No company is profitable until it starts selling their goods in the market and cash flow kicks in. Distribution prices depend on the plant location distance to the main markets, and marketing varies depending if the advertised product is new or established. These costs generally represent from 2 to 20% of the total production cost.

3. *Research and development (R&D)* costs are the expenses involved in the R&D department of the company to develop new products or improve existing ones. They play an important role in the company's innovation capacity to increase its market. The R&D costs vary depending on the company's market. For example, the R&D costs for a refinery plant are usually lower than those required by the pharmaceutical companies, where new drugs and medicines are improved and developed on a daily basis. In general, R&D costs represent about 5% of the total product cost.

Case Study 14.12: Manufacturing Cost Estimation Estimate the total manufacturing cost to produce 1500 tons/yr of product X. The plant is an organic chemical manufacturer that operates 24 h/day for 320 days/yr and costs \$1,000,000 to build. The process is continuous and highly automated and covers three main unit operations: filtration, heat transfer, and distillation. To produce 1 ton of product X, 0.70 ton of feedstock Y (\$1100/ton), and 0.50 ton of feedstock Z (\$800/ton) are required. Additionally, 0.25 ton of by-product W (\$300/ton) is produced. The plant has average maintenance problems. There are not interest payments, patent rights payments, or waste disposal costs. The required utilities in the plant are 450,000 gal of process water; 3000 kWh of electricity, and 10 tons of 150-psi steam per ton of X produced. Assume an average wage rate for 2009 of \$22.71 and cost data for 2004 (the organic chemical manufacturing price index for 2004 was 106.4). Estimate the cost for 2009.

Solution:

Plant output = 1500 tons/yr
Days of manufacturing activity = 320 days (24 h/day)
Fixed capital investment = \$1,000,000
Material: 0.70 ton Y/ton X; 0.50 ton Z/ton X
By-product: 0.25 ton W/ton X

Cost adjustments:

Producer price index for organic chemicals (2009) = 157.6
Inflation factor for organic chemicals = 157.6/106.4 = 1.48

For fixed capital investment:
M&S (2004) = 1178.5
M&S (2009) = 1468.6

Inflation factor for fixed capital investment = 1468.6/1178.5 = 1.246
Utilities: based on producer price index for fuels and related products and power
Index in 2000: 103.5
Index in 2009: 158.9
Inflation factor for utilities = 158.9/103.5 = 1.53
The producer price indexes were obtained from http://www.bls.gov/ppi/.

Manufacturing Variable Costs: Raw materials:

Feedstock Y:

$$0.70\,\frac{\text{ton Y}}{\text{ton X}}\cdot 1500\,\frac{\text{tons X}}{\text{yr}}\cdot\frac{\$1100}{\text{ton Y}}\cdot 1.48 = \$1{,}710{,}000$$

Feedstock Z:

$$0.50\,\frac{\text{ton Z}}{\text{ton X}}\cdot 1500\,\frac{\text{tons X}}{\text{yr}}\cdot\frac{\$800}{\text{ton Z}}\cdot 1.48 = \$890{,}000$$

Total raw materials = \$2,600,000

Manufacturing Labor: Using Eq. (14.35):

$$\frac{\text{manufacturing labor-hours}}{\text{tons of product}} = t\cdot\frac{\text{number of unit operations}}{[\text{capacity (tons/day)}]^{0.76}}$$

$t = 10$ (continuous process and highly automated)

3 unit operations or processes

Average wage rate \$22.71/h (2009)

$$\text{daily capacity} = \frac{1500 \text{ tons X}}{\text{yr}} \cdot \frac{1 \text{ operating year}}{320 \text{ days}} = 4.69 \frac{\text{tons X}}{\text{day}}$$

$$\frac{\text{Manufacturing labor-hours}}{\text{tons of product}} = 10 \cdot \frac{3}{(4.69)^{0.76}} = 9.27 \text{ h/ton}$$

Manufacturing labor:

$$\frac{\$22.71}{\text{h}} \cdot \frac{9.27\text{h}}{\text{ton}} \cdot \frac{1500 \text{ tons}}{\text{yr}} = \frac{\$316{,}000}{\text{yr}}$$

Supervision (assuming 12% of total manufacturing labor):

$$0.12 \cdot \frac{\$316{,}000}{\text{yr}} = \frac{\$37{,}900}{\text{yr}}$$

Utilities (refer to Table 14-13)

Steam stream at 150 psi:

$$10 \frac{\text{tons steam}}{\text{ton X}} \cdot 1500 \frac{\text{tons X}}{\text{yr}} \cdot \frac{\$4.40}{\text{ton steam}} \cdot 1.53 = \frac{\$101{,}000}{\text{yr}}$$

Process water:

$$450{,}000 \frac{\text{gal}}{\text{ton X}} \cdot 1500 \frac{\text{tons X}}{\text{yr}} \cdot 3.79 \frac{\text{L}}{\text{gal}} \cdot 1 \frac{\text{ton}}{1000 \text{ L}} \cdot \frac{\$0.53}{\text{ton}} \cdot 1.53 = \frac{\$2{,}075{,}000}{\text{yr}}$$

Electricity:

$$3000 \frac{\text{kWh}}{\text{ton X}} \cdot 1500 \frac{\text{tons X}}{\text{yr}} \cdot \frac{\$0.045}{\text{kWh}} \cdot 1.53 = \frac{\$310{,}000}{\text{yr}}$$

$$\text{Total utility} = \frac{\$2{,}486{,}000}{\text{yr}}$$

Maintenance (assuming 5% of fixed capital cost) $= 0.05 \cdot \$1{,}000{,}000 \cdot 1.246 = \$62{,}300/\text{yr}$

Manufacturing supplies (assuming 13% of maintenance cost) $= 0.13 \cdot \frac{\$62{,}300}{\text{yr}} = \frac{\$8{,}100}{\text{yr}}$

Product control (assuming 15% of manufacturing labor cost) $= 0.15 \cdot \frac{\$316{,}000}{\text{yr}} = \frac{\$47{,}400}{\text{yr}}$

Patent costs = \$0

$$\text{Total manufacturing variable costs} = \frac{\$5{,}557{,}700}{\text{yr}}$$

Manufacturing Fixed Costs

Depreciation (assuming 12% of fixed capital investment) $= 0.12 \cdot \$1{,}000{,}000 \cdot 1.246 = \$149{,}520/\text{yr}$

Financing costs = \$0

Property taxes (assuming 2% of fixed capital investment) $= 0.02 \cdot 1{,}000{,}000 \cdot 1.246 = \$24{,}920/\text{yr}$

Insurance (assuming 0.6% of fixed capital investment) $= 0.006 \cdot 1{,}000{,}000 \cdot 1.246 = \$7{,}480/\text{yr}$

Total manufacturing fixed costs = \$181,920/yr

Manufacturing overhead costs (assuming 60% of manufacturing labor + supervision + maintenance) Overhead $= 0.6 \cdot (316{,}000 + 37{,}900 + 62{,}300) = \$249{,}720/\text{yr}$

By product credit (negative) $= 0.25 \frac{\text{tons W}}{\text{ton X}} \cdot \frac{1500 \text{ tons X}}{\text{yr}} \cdot \frac{\$300}{\text{ton W}} \cdot 1.48 = \frac{\$166{,}500}{\text{yr}}$

General Expenses

Administration (assuming 20% of manufacturing labor + supervision + maintenance)

Administration $= 0.20(316{,}000 + 37{,}900 + 62{,}300) = \$83{,}240/\text{yr}$

Distribution and marketing (no information given)

Research and development (assuming \$0)

Total general expenses = \$83,240/yr

Total production cost = V. cost + F. cost + overhead + G. expenses − credits

$$= 5{,}557{,}700 + 181{,}920 + 249{,}720 + 83{,}240 - 166{,}500 = \frac{\$5{,}906{,}080}{\text{yr}}$$

$$\frac{\text{production cost}}{\text{ton of product X}} = \frac{\$5{,}906{,}080/\text{yr}}{1500 \text{ tons X}} = \frac{\$3937.39}{\text{ton X}}$$

14.4.5 Estimation of Revenues and Cash Flow

A company's revenue is the product of the total units sold by its unit price. The units to be sold are usually equal to the plant capacity and can be forecasted based on projection of the demands. Future commodity prices for chemical plants should be analyzed from historical data to consider the economic cycles of the industry. If the feedstock commodity increases its price, it is generally used that the increase passes to the final consumer. Certain expressions have been developed to study the company's profit and cash flow. The commonly used expressions are:

1. *Gross earnings*, which is the difference between the sales income and the product cost:

$$\text{gross earnings} = \text{sales income} - \text{product cost} \tag{14.36}$$

2. *Net profit* is the total revenue after income tax is added. Income tax varies from country to country, but it usually ranges around 35% of taxable income. The income tax is applied to the quantity obtained from subtracting depreciation from gross earnings:

$$\text{net profit} = (\text{gross earnings} - \text{depreciation}) \cdot (1 - \text{income tax percentage}) \qquad (14.37)$$

3. *Cash flow* is calculated by adding the depreciation cost to the net profit:

$$\text{cash flow} = \text{net profit} + \text{depreciation} \qquad (14.38)$$

This is done because depreciation is considered only for tax purposes.

REFERENCES

1. Blank, L., Tarquin, A., and Iverson, S., *Engineering Economy*, McGraw-Hill Ryerson, Whitby, Ontario, Canada, 2008.
2. Fraser, N. M., Jewkes, E. M., Bernhardt, I., and Tajima, M., *Global Engineering Economics*, 4th ed., Pearson Education, Upper Saddle River, NJ, 2009.
3. Newman, D. G., Whittaker, J., Eschenbach, T. G., and Lavelle, J. P., *Engineering Economic Analysis*, Oxford University Press, Toronto, Ontario, Canada, 2006.
4. Perry, R. H., and Green, D. W., *Perry's Chemical Engineer's Handbook*, 7th ed., McGraw-Hill, New York, 1999.
5. Peters, M. S., Timmerhaus, K. D., and West, R. E., *Plant Design and Economics for Chemical Engineers*, 5th ed., McGraw-Hill, New York, 2003.
6. Towler, G., and Sinnott, R., *Chemical Engineering Design*, Butterworth-Heinemann, San Diego, CA, 2008.

15

PROCESS COMPONENT FUNCTION AND PERFORMANCE CRITERIA

ROBERT FREE
University of Oklahoma, Norman, Oklahoma

Material transfer and conveyance is a necessarily broad topic in today's process plants. The variety and number of substances to consider, coupled with the numerous methods by which any single substance can be handled, make an exhaustive description impractical. Rather, by partitioning the multitude of substances into logical classes, the transport methods applicable to each can be discussed more generally. This ultimately yields a more coherent view which you can adapt to your specific needs. It is the purpose of this chapter to:

- Define the taxonomy used here to classify materials.
- Define important substance properties that influence material transport.
- Provide an overview of the many material transport methods utilized in today's process plants.
- Explore the transport methods in more detail, considering important factors inherent in each method and providing the reader with valuable information for implementing and maintaining these methods.

Many of the concepts and ideas presented are accompanied by industry examples to illustrate their significance. It is my intention to provide enough examples and references to guide readers in the proper direction to find meaningful solutions.

15.1 MATERIAL CLASSIFICATION

There are nearly as many ways to classify industry substances as there are substances themselves. However, in the context of transporting these substances, it makes the most sense to group them into classes that are handled through similar methods. This leads to an uppermost classification based largely on phase, but also with a few extra delineations for material classes deemed significant enough to have their own categories. The first level of the taxonomy is broken down as follows: (1) gases, (2) liquids, (3) solids (bulk solids), (4) powders (particulate solids), (5) slurries (suspensions), and (6) molten metals.

If you asked other people to identify the phase of a substance, they probably wouldn't need a strict definition. They would simply look at and evaluate the behavior of the substance for a short time. By comparing their observations with previous experience with other known gases, liquids, and solids, they could give you an answer with relatively high confidence. In light of this, the defining characteristics of the classifications above are fairly general.

A *gas* is a fluid that has variable volume and shape, expanding to fill its container. The molecules in the gas phase are very spread out on average, and the amount of kinetic energy in the gas dominates intermolecular potentials, so particles interact appreciably only when their trajectories bring them into close contact.

A *liquid* is a fluid that has variable shape but is generally considered incompressible, meaning that its volume is fixed

Process Plant Equipment: Operation, Control, and Reliability, First Edition. Edited by Michael D. Holloway, Chikezie Nwaoha, and Oliver A. Onyewuenyi.
© 2012 John Wiley & Sons, Inc. Published 2012 by John Wiley & Sons, Inc.

by the amount of substance present. It will contour to its container but will not expand to fill the entire volume. Molecules are much more tightly packed in a liquid than a gas, so intermolecular forces play a much larger role in their behavior. However, enough kinetic energy is present to allow molecules to move past one another.

A *solid* has both definite shape and volume. Molecules are very tightly bound by intermolecular forces and in general are not free to move about the substance. Solids are further broken down into classes sharing similar structures and properties, divisions that are discussed later.

A *powder* is really a solid, but in this classification scheme, when a large number of relatively small particulates are kept together, they fall into this separate category. Powders are very complicated to describe, and modeling of granular flow dynamics is currently a subject of much interest and research.

Slurries are suspensions, heterogeneous mixtures of liquid with solid particulates dispersed throughout. The size of the dispersed solid particles can vary and determines many of the properties of the slurry. Most slurries require mechanical agitation to maintain the distribution of the mixture.

Molten metal is a liquid by definition, but grouped separately here because of the specific considerations inherent in its industry uses. Characteristically, molten metals are high-temperature fluids and melted metals. They can be composed of a pure element, but more frequently they are alloys, mixtures of multiple metallic elements. This introduces many important considerations based on composition.

In the sections that follow, concepts that are important to the selection of transport methods and that dictate special considerations are defined and expanded in the context of each of the categories above. These physical concepts will be significant factors in deciding what type of conveyance method to use, and what precautions and safeguards need to be taken to ensure appropriate operation.

15.2 GENERAL PHYSICAL QUANTITIES AND CONSIDERATIONS

15.2.1 Important Definitions

In this subsection, many physical properties are defined so that their use in subsequent sections will be clearer. Refer to this glossary if you encounter a term later about which you are unsure.

Angle of repose: the angle to which powder or bulk particulates will pile before faces begin to shear off under their own weight.

Coefficient of friction: denoted by the Greek letter mu (μ), the coefficient of friction is a proportionality factor relating the friction force to the normal contact force between two substances.

Composition: a description of the makeup of a mixture. Usually described by mass fraction, partial pressure, or volume fraction.

Corrosive substance: any substance that induces destructive chemical reactions with its surroundings.

Density: mass per unit volume; usually denoted by the Greek letter rho (ρ); units of kg/m^3 or slugs/ft^3. Bulk density is the average mass per unit volume of a nonuniform substance such as a powder or slurry.

Ductility: the extent to which a solid can be deformed before fracture.

Electrical conductivity: tendency of a material to allow electrical current to pass through.

Flammability: propensity of gas to undergo a combustion reaction. Quantified by a National Fire Protection Agency (NFPA) rating from 0 to 4.

Hardness: a solid's resistance to surface scratches and penetrating loads. Two common relative hardness scales are the Brinnell and the Rockwell.

Isolation: separating a substance from its environment; includes thermal, diffusive, and electrical isolation.

Mechanical strength: a stress limit characteristic of a material. The *yield strength* is the stress beyond which a material begins to deform plastically. The *ultimate strength* is the maximum stress before a material tends to fracture.

Morphology: the study of the form and size of a substance or object.

MSDS: (material safety data sheets): provide important information regarding the hazards and limitations associated with particular substances. Include details such as physical characteristics, reactivity, health hazards, and proper handling. Created by the producers of different substances, they can vary from country to country.

Phase: the state of matter of a substance; not simply confined to solid, liquid, or gas. For example, diamond and coal are both solids and composed of carbon, but are different phases.

Settling rate: the characteristic rate at which solid particulates or dissolved constituents will settle out of a solution or slurry.

Specific gravity: the ratio of the density of a substance to the density of water ($SG = \rho/\rho_{H2O}$).

State quantities: properties of a substance that are independent of the path or process by which the substance came to be in its current state.

Thermal conductivity: a measure of the rate at which energy will flow through a material (k).

Toxicity: the relative hazard that a material poses to life. There are many scales for toxicity, depending on the material being handled and how it comes into contact with an organism. Generally, the minimum lethal value is reported for each substance and the nature of exposure.

Vapor pressure: the pressure of a gas when it is in diffusive equilibrium with its condensed phase at a given temperature. The boiling point is the temperature at which ambient pressure equals the vapor pressure of a substance.

Viscosity: a measure of a substance's resistance to stress. Functionally, it is the "thickness" of the liquid. Viscous liquids flow poorly. Viscosity (μ) is a function of intermolecular attraction and interaction.

Volatility: the tendency of a liquid to vaporize. Related directly to the vapor pressure.

Young's modulus: also called the *modulus of elasticity*, it gives the proportionality of stress (σ) and strain (ε) in the linear elastic region of solids.

15.2.2 State Quantities

The physical quantities that determine the phase and behavior of a substance, called state quantities, include such things as temperature, pressure, volume, and density, among others. Independent of how a substance came to be in its current state, these variables are controlled in plants to yield the desired effects in the substance's behavior. However, these quantities cannot be varied independently but, instead, are related through equations of state. The simplest of these is the *ideal gas law*, which relates the pressure, volume, and temperature of an idealized gas:

$$PV = nRT \tag{15.1}$$

where P is the pressure, V the volume, n the number of moles, T the temperature, and R the ideal gas constant. There are many more accurate equations of state for gases as well as for liquids and solids. A few notable ones include the van der Waals equation for real gases and the Peng and Robinson equation for real gases and fluids.

State quantities and the equations of state relating them are crucial to predicting the behavior of a substance as one or more of the state variables are changed in transport. Understanding how plant processes will affect this behavior will ensure that a plant runs as intended and that production quotas are achieved.

15.2.3 Phase

One of the most important considerations in conveyance is maintaining the phase of a material. During normal transport, it is not ideal to have such phase changes as evaporation, freezing, melting, or condensation. These changes can lead to clogged pipes, loss of product, and, in general, loss of operational efficiency. The equilibrium phase of a substance under some set of conditions is the one that minimizes the Gibbs free energy of the substance. This thermodynamic quantity is defined as

$$G = U + PV - TS \tag{15.2}$$

where G is the Gibbs free energy, U the internal energy, P the pressure, V the volume, T the temperature, and S the entropy. Without getting too involved in the thermodynamics, the Gibbs free energy is the energy of a system itself, plus the energy needed to place the system of volume V in an environment at pressure P, less the energy that can be obtained from the environment at temperature T while upholding the second law of thermodynamics.

Phase diagrams have been created for many important substances using this thermodynamic identity and experiment, and these can be used to determine the equilibrium phase under a given set of conditions. Most phase diagrams are pressure versus temperature plots with regions displaying the equilibrium phase under a given temperature and pressure. An example of a phase diagram in shown in Fig. 15-1. Phase transitions occur along the boundary lines between regions.

Also note that the definition of a phase is not limited to solids, liquids, and gases. Many substances have multiple solid phases. For example, consider coal and diamond. Both are composed of carbon and are solids; however, they are two distinct phases and can be described adequately by a phase diagram with a well-defined phase boundary. You can see in Fig. 15-1 that there are many phases of ice as the pressure increases. In fact, there are more than 10 well-defined phases of ice. In the steel production industry, phase diagrams for many alloys of metal are crucial, and the slightest variations in composition can have drastic effects on the phase and properties of the resulting metal. This is discussed in more detail in Section 15.2.10. Considering the phase diagram of a substance is crucial when designing all parts of a process plant, and is especially important when designing conveyance equipment in which the phase should not change during transmission [18].

15.2.4 Isolation

Isolation is a broad consideration, but refers basically to keeping materials separated from each other or from the environment. The most common forms include diffusive isolation, thermal isolation, and electrical isolation. Diffusive isolation is probably the form considered most frequently. It is simply maintenance of the spatial separation of the system: keeping the particles of a system physically

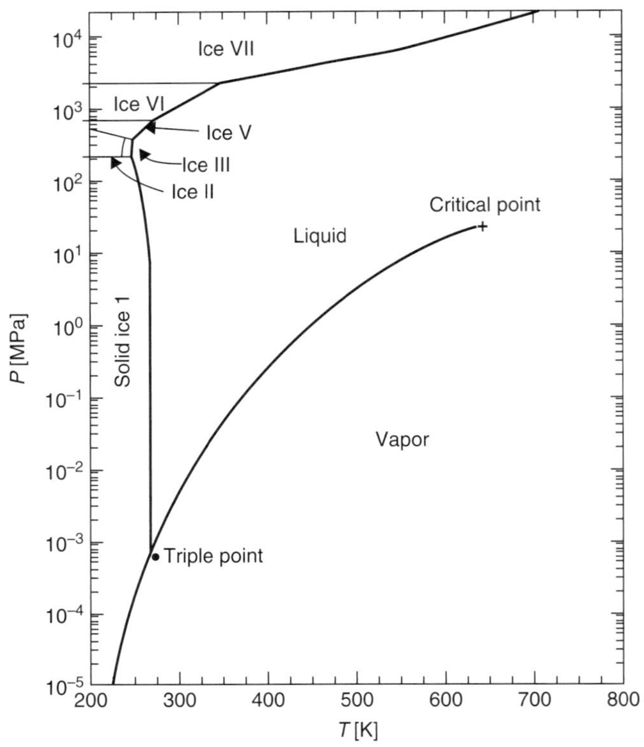

Figure 15-1 Phase diagram of water, including many ice phases. (From [20].)

distinct from the environment. In general, most plants seek to keep all substances diffusively isolated to some extent, or else complete chaos would ensue. It would be impossible even to define a process without some kind of isolation of materials from one step to another.

However, this base necessity is often not discussed in the context of diffusive isolation. Usually, this consideration arises when substance purity is important to process efficiency and quality. For example, in the semiconductor fabrication industry, contamination is one of the most significant challenges. A single speck of dust can destroy a silicon wafer, costing the fabrication company considerable amounts of resources. One of the most sensitive times in the process is during the growth of germanium and silicon crystals. The quality of the crystal depends on limiting the number of impurities in the crystal, which relates directly to removing impurities from the environment during crystal growth. Here, an argon gas atmosphere of at least 99.999% purity is used to achieve this end. Maintaining and handling argon gas at these purity levels requires very careful design of conveyance equipment.

Purity is quantified by indicating the ratio of impurities to desired particles, or as a percentage. Depending on the level of purity, you can find values from parts per thousand to parts per billion. A value of 30 ppm would indicate that only 30 out of every million particles in some volume of a substance are unwanted impurities. Alternatively, a purity of 99.999% means that only 1 part per hundred thousand is unwanted. Depending on the application, different purity requirements must be met to ensure product quality and process efficiency; in turn, this places requirements on the quality of transport and storage equipment used in the handling of the gas.

By their nature, gases are the most difficult phase to contain. If allowed to move freely, a gas will usually diffuse throughout its container, whether it is a small pressure capsule, a room, or an entire warehouse. For this reason, conveyance systems and piping designed to handle gases must be constructed with a high level of precision, lest leaks and unwanted mixing should occur. Proper containment is essential to avoid safety concerns when handling flammable, toxic, or corrosive gases that can harm workers. There are many ways by which a leak can originate, from design errors and improperly rated equipment to cracks caused by fatigue, or even explosive rupture of containment vessels due to excess pressure. Even though it is frequently taken for granted, containment should be considered continuously through the conveyance process when handling gases.

Liquids can pose challenges to diffusive isolation as well, but these are generally easier to identify and prevent. The main concern when handling liquids is evaporation into the gaseous phase. If this occurs, it may be much harder to contain the substance. Thus, understanding the phase diagram of a substance will allow proper handling of this issue. Slurries and molten metal share similar concerns with liquids in this context. Diffusive isolation in solids and powders is usually taken for granted because of the nature of the phases. However, in some powder applications, dust liberation can pose a significant flammability hazard, leading to the requirement of closed systems and the use of inert gases to restrict this dispersive phase from reaching dangerous concentrations.

Thermal isolation consists of mitigating the flow of energy between a substance and its environment. This is achieved through the use of insulating materials, materials with a low k or thermal conductivity value. Thermal isolation is an important consideration when transporting materials that are sensitive to temperature variations. Applications in which the material conveyed can easily change phase if the temperature leaves a specific range are prime candidates for implementing thermal isolation. In this way, heat flows into the system in a controlled manner, and changes in the environment don't have to be accounted for as strongly.

Let's consider an example in which a phase change can occur, rendering the transport method ineffective and compromising the product. The ceramic zirconium dioxide (ZrO_2) used to create artificial gemstones possesses a monoclinic crystal structure at room temperature. However, as it is heated, the equilibrium crystal structure changes to tetragonal and eventually to cubic. The tetragonal crystal structure occupies a greater volume than the monoclinic, so as parts of the substance go through the phase transition, considerable stress accrues in the material. This can lead to cracking and irreparable damage to the crystal. Thus, careful regulation of the temperature through thermal isolation must be maintained when handling zirconium dioxide [5].

Other situations where thermal isolation must be considered are processes that require a specific temperature at the point of application to obtain the desired results. In some cases, this temperature may be attained elsewhere in a plant and must then be maintained over transport to the point of application. This calls for thermally insulated piping and pathway considerations when designing the pipeline so as to avoid other sources of extreme temperature.

Electrical isolation is similar to thermal insulation, but as it sounds, it is the mitigation of the flow of electricity between a substance and its environment. An alternative form is the effective shielding of a region from external electromagnetic radiation. The first interpretation is achieved by surrounding the region of isolation with poor conductors (insulators) (see the discussion of conductivity in Section 15.2.9). The second interpretation is achieved by encasing a region in a very good conductor such that incoming electric fields are neutralized. These considerations become important when handling electrically sensitive materials or devices. Also, when dealing with highly conductive substances, proper isolation from workers is an essential safety concern [6].

15.2.5 Flammability

A very significant concern in all industries is flammability. The flammability of a substance being used in a plant determines safety measures, places restrictions on handling and transporting, and can be responsible for significant losses in capital, product, and life if not treated properly. The flammability of a substance is its propensity to undergo a combustion reaction. There are a number of quantification schemes for flammability, all of which use relative scales to rate materials. The National Fire Protection Agency (NFPA) uses a number scale from 0 to 4. The meanings of each number are described in Table 15-1.

Many pure gases become flammable only when the amount of oxygen "impurities" present falls within a certain range, called the *flammability range*. This range is temperature dependent, and once the composition of the mixture falls within this range, an ignition source can instigate combustion of the gas. Table 15-2 gives this range for some common gases. More information can be found in the *Matheson Gas Data Book* or in material safety data sheets for a particular substance [26].

When flammability is discussed with regard to liquids, the term *flash point* is used. This is the temperature at which sufficient amounts of the substance vaporize to allow ignition, and it is linked directly with the vapor pressure of the liquid.

Another important consideration is the point at which spontaneous combustion becomes a concern. This "point" refers to the pressure and temperature at which the ambient energy is sufficient to initiate the combustion reaction. At atmospheric pressure, this is also known as the *autoignition temperature*. Once this point is reached, no actual ignition source such as an errant ember or electricity arc is necessary to ignite the reaction, so the gas is said to combust spontaneously.

When handling flammable gases, many requirements are placed on the transport methods used. Appropriate thermal and diffusive isolation becomes a major concern to prevent the system from obtaining the necessary composition and/or energy to ignite. This leads to specific choices in piping, valves, and seals, and can also have an effect on the equipment used in close proximity to the gas. Thus, flammability considerations have a significant impact on capital costs and maintenance practices, both of which must ensure safety and reliability.

To illustrate the importance of containment and flammability concerns, consider the application of silane gas in the quickly expanding solar cell production industry. Silane gas, SiH_4, is both highly toxic and highly flammable. Referring to Table 15-2, you can see that the flammability range is from 1.5 to 98%, and from experiment the autoignition temperature is known to be about 70°F [26]. This means that silane gas can spontaneously combust extremely easily if leaked into the atmosphere. Usually, ignition occurs immediately when leaks occur, but if moisture and impurity conditions are right, the gas may remain metastable for some time, allowing for accumulation of the gas. This is the more dangerous scenario. The gas is extremely toxic, so accumulated amounts pose a serious risk to workers who may breathe it in. More alarmingly, accumulation can lead to bulk ignition and a much more destructive and dangerous explosion. Even after a silane gas explosion, the combustion products remain toxic and must be cleaned up properly. As the solar cell industry grows, demand for silane gas increases, and it is being produced and stored in much larger amounts. Thus, the containment of the gas becomes ever more significant to prevent both endangerment of human life and severe financial losses. Those handling the gas must use extreme care to store it using properly rated valves and pressure vessels. Most gases are not as dangerous as silane, but considering this case helps illustrate the importance of containment and flammability considerations [15,26].

15.2.6 Viscosity

Mathematically, viscosity is the ratio of shear stress to shear rate in a fluid:

$$\tau = \eta \dot{\gamma} \tag{15.3}$$

TABLE 15-1 NFPA Hazard Ratings; Flammability

4	Materials that will vaporize rapidly or completely at atmospheric pressure and normal ambient temperature or which are readily dispersed in air, and which will burn readily. This degree should include: • Gases • Cryogenic materials • Any liquid or gaseous material that is a liquid while under pressure and has a flash point below 73°F (22.8°C) and a boiling point below 100°F (37.8°C) (class IA flammable liquids) • Materials that because of their physical form or environmental conditions can form explosive mixtures with air and are readily dispersed in air, such as dusts of combustible solids and mists of flammable or combustible liquid droplets
3	Liquids and solids that can be ignited under almost all ambient temperature conditions. Materials in this degree produce hazardous atmospheres with air under almost all ambient temperatures or, although unaffected by ambient temperatures, are readily ignited under almost all conditions. This degree should include: • Liquids that have a flash point below 73°F (22.8°C) and have a boiling point at or above 100°F (37.8°C), and those liquids having a flash point at or above 73°F (22.8°C) and below 100°F (37.8°C). (class IB and IC flammable liquids) • Solid materials in the form of coarse dusts that may burn rapidly but generally do not form explosive atmospheres with air • Solid materials in a fibrous or shredded form that may burn rapidly and create flash fire hazards, such as cotton, sisal, and hemp • Materials that burn with extreme rapidity, usually by reason of self-contained oxygen (e.g., dry nitrocellulose and *many organic peroxides*) • Materials that ignite spontaneously when exposed to air
2	Materials that must be heated moderately or exposed to relatively high ambient temperatures before ignition can occur. Materials in this degree would not under normal conditions form hazardous atmospheres with air, but under high ambient temperatures or under moderate heating may release vapor in sufficient quantities to produce hazardous atmospheres with air. This degree should include: • Liquids having a flash point above 100°F (37.8°C), but not exceeding 200°F (93.4°F) • Solids and semisolids that readily give off flammable vapors
1	Materials that must be preheated before ignition can occur. Materials in this degree require considerable preheating, under all ambient temperature conditions, before ignition and combustion can occur. This degree should include: • Materials that will burn in air when exposed to a temperature of 1500°F (815.5°C) for a period of 5 minutes or less • Liquids, solids, and semisolids having a flash point above 200°F (93.4°C) • This degree includes most ordinary combustible materials
0	Materials that will not burn. This degree should include any material that will not burn in air when exposed to a temperature of 1500°F (815.5°C) for a period of 5 minutes.

Source: [14].

where τ is the shear stress, η the viscosity, and $\dot{\gamma}$ the shear strain rate. More intuitively, viscosity is a measure of the thickness of a liquid, or its resistance to flow. Examining the equation, a high viscosity means that a large shear stress is required to obtain a significant shear strain rate, or that the fluid resists the tendency to flow more than do liquids with lesser viscosity.

At this point, viscosity may seem to be a simple ratio with easy application. This is the case for and definition of a *Newtonian fluid*. However, some fluids have a viscosity that is dependent on shear rate, called *non-Newtonian fluids*. You are probably familiar with some examples of non-Newtonian fluids. Have you ever mixed cornstarch and water and observed the interesting behavior of the resulting fluid? If you slowly depress the surface of the liquid, your finger will pass through with easy. But if you attempt to depress the surface quickly, you are met with more solid resistance. This is because the mixture is a dilatent, non-Newtonian fluid, meaning that the viscosity increases with increasing shear strain rate. Thus, your attempt to insert your finger quickly induces a thickening effect in the fluid. The opposite effect, decreasing viscosity with increasing shear strain rate, is characteristic of a pseudoplastic liquid. Other non-Newtonian effects include time dependence of viscosity under constant shear rate.

Temperature is another quantity on which viscosity generally depends. Think of molasses, a substance that is generally very viscous. Simply heat it to an adequate temperature and it begins to flow easily onto your pancakes! This illustrates the more common temperature dependence

TABLE 15-2 Flammability Range of Various Gases

Gas	Lower Flammable Limit (LFL) (%)	Upper Flammable Limit (UFL) (%)
Acetone	2.6	12.8
Ammonia	15	28
Carbon monoxide	12	75
Ethane	3	12.4
Hydrogen	4	75
Isopropyl alcohol	2	12
Gasoline	1.4	7.6
Kerosene	0.7	5
Methane	5	15
Methyl alcohol	6.7	36
Naphthalene	0.9	5.9
n-Octane	0.95	3.20
Propane	2.1	10.1
Silane	1.5	98

Source: [26].

of fluid viscosity. Usually, as thermal energy increases in a fluid, the intermolecular bonds that are responsible for viscosity cannot hold as tightly because of the increased kinetic energy, so molecules flow past each other more easily.

Considering the fluid viscosity is basic to selection of the transport method through a process plant. For example, if you try to pump a dilatent fluid too quickly, its viscosity will increase greatly, yielding a clog that could burn out the pump and result in lower transport efficiency. Viscosity dependence on temperature can also be utilized to increase ease of transport. Heating fluids to facilitate moving them through the plant is not uncommon. By understanding the behavior of the specific fluid to be handled, one can take full advantage of viscosity and prevent unconsidered complications from arising [25].

15.2.7 Volatility

Volatility, the tendency of a liquid to vaporize, is dependent on pressure and temperature and is congruous to the liquid's vapor pressure at a given temperature. Vapor pressure is the pressure of the gaseous phase of a substance when it is in diffusive equilibrium with its liquid phase.

More intuitively, imagine that you have a liquid at some temperature. Through random thermal motion, individual molecules will obtain enough energy to break free of the liquid phase, so there is always some exchange of molecules from the liquid and gaseous phases. However, if we begin with no molecules of a substance in the gaseous phase, the net flow of particles will be out of the liquid phase and into the gaseous phase. This will continue until such time as there are enough molecules in the gaseous phase to yield an equal random flow back into the liquid phase. This point, where the flow rate of molecules into and out of the liquid phase is equal, is called *diffusive equilibrium*. The number of molecules in the gaseous phase is related directly to the partial pressure of the substance and is termed the *vapor pressure*.

In this way, a higher vapor pressure corresponds to a larger number of molecules freed into the gaseous phase, and consequently, the liquid is said to be more volatile. When the vapor pressure becomes equal to the ambient pressure of the atmosphere above a liquid, diffusive equilibrium cannot be attained and the liquid evaporates completely into the gaseous phase. This is *boiling*.

Volatility comes into play for a number of conveyance considerations: whether to use open versus closed piping; whether or not it is easier to transport the substance in the gaseous phase and condense it at the end destination; and whether appropriate containment or handling of vapors is necessary. These concerns, as well as some related to flammability, all depend on the volatility of the substance being handled [18].

15.2.8 Corrosive Substances

A substance is said to be corrosive if it interacts chemically with surrounding substances in a destructive manner. A substance will not be corrosive with regard to all other substances, so labeling a substance as corrosive means loosely that it interacts negatively with many other substances or common substances with which it would come into contact. The extent to which a substance is corrosive must always be considered in the context of both the substance and the container.

One widely used measure of corrosive potential is the pH scale. It is used to classify solutions as either acidic or basic. An acidic solution has free H^+ ions and is classified with a pH of less than 7. A basic solution accepts H^+ ions and frees hydroxide ions OH^- ions into solution. Bases and acids react readily with one another, and both can corrode other substances through chemical reactions.

There are other corrosive substances in addition to acids and bases. Any substance that catalyzes or induces a destructive chemical reaction is considered corrosive. Other examples include strong oxidizers, halogens, dehydrating compounds, and elements such as alkali metals.

Any phase can be corrosive if used in conjunction with a poor choice of material. Gases and liquids are discussed more often in terms of corrosivity because of the relative mobility of particles. The freedom to move about makes it easier for them to participate in destructive chemical reactions with containers and conveyance equipment [4].

15.2.9 Conductivity

Conductivity is the tendency of a material to allow the flow of heat or electricity—thermal or electrical conductivity, respectively. Materials with high conductivity are *conductors*; those with low conductivity are *insulators*. A reciprocal property of materials is *resistivity*, the mathematicall inverse of conductivity.

Mathematically, conductivity relates two vector quantities, and thus it is generally a rank 2 tensor. However, in the context of material conveyance, we will consider it a constant and assume that the materials being considered are isotropic. Thus, conductivity does not vary with direction in the material and does not change the direction of the vector quantities it relates, only their magnitude.

Thermal conductivity, denoted by the letter k, has SI units of watts per meter-kelvin. It is a measure of the rate of heat transfer given the temperature gradient across the material, the cross-sectional area, and the thickness. For example, if you have a 1-m^2 0.01-m-thick glass window with some known thermal conductivity, $k = 1.05$ (typical for glass), and you know that it is 20° cooler outside your house, a simple calculation will tell you the rate of heat loss:

$$\frac{\text{(thermal conductivity)} \cdot \text{(cross-sectional area)} \cdot \text{(temperature difference)}}{\text{thickness}} = \frac{1.05 \cdot 1 \cdot 20}{0.01} = 2100\ \text{W} \tag{15.4}$$

So under these conditions, 2100 J of energy would flow through the glass window every second. It is easy to observe that a higher thermal conductivity yields a greater rate of heat loss.

Electrical conductivity, denoted by the letter σ, has SI units of siemens per meter. Mathematically, it relates the current density vector to the electric field vector. With a defined geometry, and after some manipulation, this relationship can be recast into the well-known *Ohm's law:*

$$V = IR \tag{15.5}$$

This relates the voltage V across the material, the current I induced, and the resistance R of the material. The electrical conductivity or resistivity is included in the R factor, which also accounts for the geometry of the object in question. The value of R is inversely related to σ:

$$R \propto \frac{1}{\sigma} \tag{15.6}$$

so a higher conductivity yields a lower resistance. In turn, a higher current is induced for any given voltage. So we conclude that a greater conductivity allows higher currents for a given voltage, as you may have already known intuitively.

Generally, the conductivity of the material being transferred is not as significant as the conductivity of the material used to contain it. Conductivity and isolation concerns are directly linked; by understanding and properly selecting poorly conducting materials to convey a substance, one attains the desired electrical or thermal isolation [6].

15.2.10 Composition

Many of the substances encountered in industry are not pure substances made up of only one element or compound. They are mixtures, and in discussing mixtures, the concept of composition is vital. The composition of a mixture is given by enumerating all of the pure substances present in the mixture and their relative amounts. There are many ways that composition is expressed throughout industry, depending on the phase and relative concentrations of constituent substances. These will be discussed presently.

The first way to provide a mixture's composition is by means of *mole fractions* or percentages. Each constituent substance is accompanied by a value from 0 to 1 or 1% to 100% that signifies the ratio of the constituent particles to the total particles. This approach is used for all phases, as it gives the precise number of particles of each constituent in the whole. A related scheme used in gases associates with each constituent gas a *partial pressure*. The partial pressure can be obtained if the pressure of 1 mol of the mixture is known for a given volume and temperature. Since the pressure depends directly on the number of moles, n, simply multiplying this pressure by the mole fraction will yield the partial pressure of the gas. Sometimes, this quantity is easier to work with than the mole fractions.

Another method for specifying composition is by mass fraction. In this scheme, the ratio of one constituent's mass to the total mass of the mixture is provided. An analogous scheme obtained by simple multiplication by the gravitational constant, g, is composition by weight fraction. These methods also find application in describing all phases, but are particularly useful in solids and molten metal. When creating alloys of molten metal, the behavior and properties of the mixture depend strongly on the relative composition. Most tables and literature rely on mass fraction values to convey these data.

When dealing with solutions where one substance is dissolved or suspended in another, concentrations can also be used to specify composition. This treats one substance as the solvent, or primary constituent, and all other constituents as solutes being dissolved into the primary substance. In chemistry and throughout industry, there are a wide variety of ways to express concentration, including the two methods discussed previously. Others include:

- *Mass–volume percentage*: the mass of the solute (g) relative to the volume of total solution (mL)
- *Volume–volume percentage*: the volume of the solute relative to the volume of total solution

- *Molarity*: the number of moles of solute relative to the volume of solution (L)
- *Molality*: the number of moles of solute relative to the mass of solvent (kg)

Depending on your application, it may be easiest to use one of these forms of concentration over the others, or a completely different scheme altogether. The important thing is to be sure that you have defined exactly what scheme you are using so that you can relate values to other schemes and convey the composition accurately. Before using data in which concentration values appear, be sure that you are aware of the scheme in use.

In dealing with slurries, composition is sometimes discussed in terms of a solids fraction. This is a mass fraction scheme that specifies all solids as one constituent and the liquids as another. Calculations using this scheme are the same as for normal mass fractions handling. In some cases a volumetric solids fraction for slurries may also be used. These two can be related through the densities of each constituent, and are simply different ways to express how much of a slurry is composed of solid particulates.

It is important to be aware that some forms of composition will change values with changing environment variables such as pressure and temperature. This doesn't change the actual composition in the sense that the same relative amount of each constituent is present, but the numerical values can change because their relationship to mole fractions depends on state variables. Be sure you know under what conditions a particular value of composition or concentration is valid [4].

15.2.11 Morphology

Morphology is the study of form and size of objects. For large solids, this includes the shape, surface features, and microscopic structure (crystalline/amorphous). In conveyance, the shape of loads in conjunction with their mass distribution yields important quantities such as the center of mass and moment of inertia. An example of where the center of mass is significant is in inclined conveying. To prevent the load from toppling over, the line of action of the force of gravity acting through the center of mass should pass within the inner third of the base of the solid. In applications like these, the shape of the load is an important consideration. Surface features will determine properties such as coefficient of friction and how abrasive a material is. Finally, the microscopic structure of solids governs many of their macroscopic properties. Conductance, specific heat, strength, and ductility can all be related back to the microscopic form of the solid. Morphology includes studies of all of these aspects of solids.

In powders, morphology is useful in predicting the behavior of the bulk continuum. Quantities such as the angle of repose and ease of flow depend on the shape and form of the grains that compose the powder. Odd angular shapes that have a tendency to lock will yield poorer flowability than will rounded and smooth particles. Studies of angularity, sphericity, and roughness are conducted on powders and particulate bulks around the world. Understanding the relationship between the physical form of the constituents and the properties of the whole will make handling powders and bulk particulates even more efficient in the future.

The size and shape of particles constituting the solid phase in slurries also have a significant impact on the properties of the combined solid–fluid mixture. Things such as settling rate, conductivity, and effective viscosity depend strongly on the nature of the solid particulates. These properties directly affect conveyance considerations. For example, the settling rate indicates how quickly the suspended solid particles will precipitate and collect at the bottom of the fluid. If one is handling a slurry with an extremely rapid settling rate, constant agitation of the mixture may be necessary or forms of conveyance such as pumping may not be effective. In a case such as this, a screw conveyor may be required instead: one that provides the necessary agitation to maintain particle suspension. Considerations such as these all relate back to the morphology of the solid phase.

15.2.12 Solid-Specific Considerations

The solid phase is probably one of the most investigated and best understood, at least in its structural aspects, because of its importance throughout engineering. The properties of solids have been manipulated for millennia by humans to form tools, build dwellings, and in the industrial age, to form machines and electronics. In industry, the importance of solids extends beyond applications that seek to convey solid material. All conveyance methods rely on solid constructions in some way along the line. Whether it is the steel walls of a pipe, or the magnetic core used in an electric motor, or the columns holding up the warehouse, we all use solids. In this section we introduce some of the concepts frequently encountered when discussing and using solids.

15.2.12.1 Classification First, let's discuss the basic classification of various solids. Solids used in engineering can generally be classified as metals (alloys), polymers, ceramics, or combinations thereof called *composites*. These distinctions are based primarily on the underlying chemical makeup of the solid, but because structure is linked directly to properties, each group also has some characteristic attributes.

1. *Metals* are composed of predominantly metallic elements in any number of mixtures (alloys). They can

include nonmetallic constituents, but only in relatively small amounts. For example, the carbon content in structural steel is around 0.1%. Metals generally possess an ordered crystalline structure and are good conductors of heat and electricity. As a class, they are typically quite strong and stiff, but have a wide variety of mechanical properties, depending on how they are alloyed and produced.

2. *Polymers* are composed of long chains of like chemical units called *monomers*. They are usually organic compounds with a central carbon chain and different bonded elements. These chemical constructions are responsible for the properties of plastics and rubbers with which you are no doubt familiar. They possess an amorphous structure and are generally poor conductors of heat and electricity. They are not as strong as metals or ceramics and are much more sensitive to temperature, which is why they are used to create complex shapes and molds. Some familiar polymers include nylon, poly(vinyl chloride) (PVC), and silicone rubber.

3. *Ceramics* are composed of ionically bonded metals and nonmetals. This gives them their relatively strong hardness, but also leads to them being quite brittle. They are generally poor conductors of heat and electricity. Since the bonds between atoms are so strong, ceramics are comparatively inert chemically, and resist corrosion quite well. They normally form a crystalline structure, but can be amorphous (e.g., glass).

4. *Composites* are simply combinations of these three material classes. Composites allow us to take properties we deem beneficial from each class and combine them into a single material. Carbon fiber, for example, is a composite material where carbon fibers are embedded in a polymer, increasing its strength and stiffness while retaining much of the ductility associated with the polymer. Most composites are purely man-made, but the few that occur in nature include wood and bone.

15.2.12.2 Mechanical Properties Now that we have a broad classification scheme, what mechanical properties of solids make them so useful? First and foremost, it is the ability of solids to retain their shape relatively well under loading and stress. The use of solids as structures of all scales, from buildings to glassware, has been their most valuable implementation throughout history. But how do we compare the structural capabilities of solids? We do so by comparing mechanical properties. These include:

- *Elasticity:* The relationship between stress and strain
- *Strength:* The maximum stress that can be withstood by the material
- *Ductility:* The ability of the material to deform without fracturing
- *Toughness:* The ability of the material to absorb energy before fracturing
- *Hardness:* The material's resistance to surface scratches and penetration

Before looking more closely at these mechanical properties, it becomes necessary to understand the basic concepts of stress and strain. Simply, *stress* is a measure of the internal forces acting on a structure. It is the average force per unit area acting on some imaginary surface on the interior of the structure. To understand more clearly, take the simple example of a circular bar of cross section A (Fig. 15-2). If a force F is exerted at both ends of this bar, the stress on any cross section parallel to the end faces is given by

$$\sigma = \frac{F}{A} \tag{15.7}$$

This is known as tensile axial stress, and is the simplest form. It has units of pressure.

Now, our mental image of most solids is that of a nondeforming shape. We push on a wall and it doesn't move; it simply pushes back. However, when you exert a force on an object, it cannot exert a reaction force spontaneously. There must be some origin of this reaction. In reality, when you push on the wall, it deforms slightly. This deformation corresponds to a compression of the intermolecular bonds that make up the solid. It is this shortening of these bonds that is responsible for the reaction force you feel, even if you cannot see the deformation. The idea of deformation is what brings us to *strain*. Returning to our circular bar, when we pull on both ends of the bar, we expect the bar to elongate slightly until a new state of equilibrium is reached. This is illustrated in Fig. 15-3. The strain is defined as

$$\varepsilon = \frac{L' - L_o}{L_o} = \frac{\Delta L}{L_o} \tag{15.8}$$

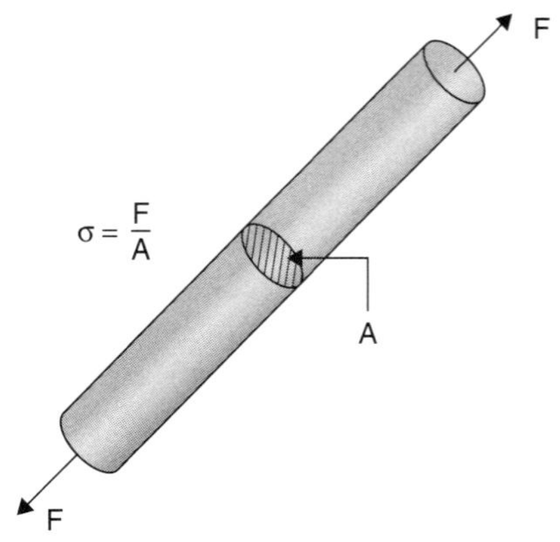

Figure 15-2 Circular bar and tensile stress. (From [4].)

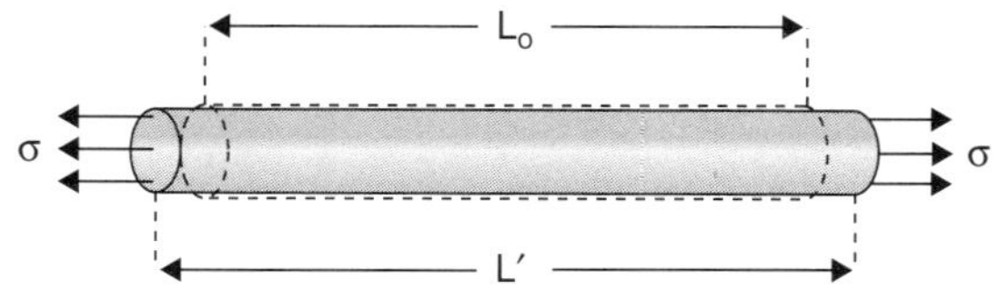

Figure 15-3 Elongation of circular bar under tensile stress. (From [4].)

where L_O is the original length of the bar, L' the length after the load is applied, and ΔL the difference or elongation. Notice that strain is unitless.

In linear elastic materials, the relationship between the stress placed on an object and how much strain it accrues is known as *Hooke's law*. It is expressed as:

$$\sigma = E\varepsilon \tag{15.9}$$

where σ is the stress in a material, ε the strain, and E the proportionality constant, known as Young's modulus. E is a characteristic of the particular material in question, so it is the quantity compared when discussing the elasticity of different materials. The larger the value of E, the stiffer a material is. E is the quantity compared when discussing the relative elasticity of materials.

Strength in the context of mechanical properties refers to a limit of stress that a material can withstand. The yield strength is the stress beyond which a material no longer behaves in a linearly elastic manner. In other words, Hooke's law no longer applies, and a more complicated relationship between stress and strain prevails. Beyond this yield strength, plastic or permanent deformation will occur and the material will not return to its previous shape, even if the stress is removed. Another quantity of importance is the ultimate strength, the maximum stress the material can withstand. If this stress is exceeded, the material will quickly tend toward fracture. This value is important in selecting materials for applications where known stresses will prevail. It allows for the appropriate choice of material strength within safety factors.

Ductility is related to strain. It is a measure of how much deformation a material can withstand before fracture. In this way, the greater the tolerable strain is for a material, the more ductile the material is said to be. Polymers are the most ductile of the three classes of materials, followed by steels, with ceramics being the relatively least ductile and most brittle. The concept of a brittle material is the inverse of a ductile material. Brittle materials cannot withstand much deformation before fracture. This does not mean that they are weaker. In many cases, brittle materials have greater ultimate strengths. They may just have an extremely large Young's modulus, so will not deform much before beginning to deform plastically and fracture quickly.

The toughness of a material is its ability to absorb energy. Tough materials can withstand a large amount of work done on them, while weaker materials will fracture with little external work. Material strength and material toughness are not directly analogous. A very brittle material, which may be very strong, deforms very little before fracture, so the work put into the material is small. An extremely ductile material will deform a great deal, but under a very small load, so the work done on the material will be small. A material that has reasonable ductility and good strength is the toughest, because it deforms appreciably, and under a significant load, thus absorbing more energy.

Finally, material hardness is a surface property. Hardness is a material's ability to withstand surface scratches and penetrating or piercing loads. It is usually related directly to material strength, but not always. Hardness is a significant property to consider when selecting conveyance materials for handling abrasive loads or loads dropped onto the conveyor from great heights [5].

15.2.12.3 Temperature Effects All of the material properties introduced have temperature dependence. The severity of the effect of temperature change varies for each class. For metals, strength tends to decrease with increasing temperature, while ductility increases. The strength of temperature dependence for metals is moderate compared to that of the other two classes. The mechanical properties of ceramics are the least dependent, which is why they find use in high-temperature applications. Meanwhile, you may be familiar with the sensitivity of your Tupperware plastic containers from an experience in reheating dinner. If the plastic gets too hot, it will lose its stiffness and have trouble keeping its shape. This exhibits the polymer's strong dependence on temperature. The melting points of these materials follow the same trend. Ceramics characteristically have the highest melting points, metals are intermediate, and polymers have the lowest. Of course, there exists some overlap between classifications, and the hierarchy presented is not a hard-and-fast rule.

15.2.13 Coefficient of Friction

Mathematically, the coefficient of friction, usually denoted μ, is a dimensionless number that relates the force of friction experienced between two objects and the normal force pressing these objects together. The coefficient of friction encapsulates many physical phenomena and is an empirically determined number (i.e., it can only be determined through direct experiment). It depends on the materials in contact, the surface texture of both materials, and the temperature—among other things. Because of the wide variety of considerations that go into its value, one must be aware of the conditions under which reported values were measured. It is also important to remember that a material by itself does not possess some value of μ. It must be discussed in terms of two interacting materials, even if they are the same.

To determine the expected friction force (f), simple multiplication by the normal force (N) is performed:

$$f = \mu N \quad (15.10)$$

The direction of the force of friction is always opposite that of the direction of motion or attempted motion, and it scales according to the normal force pressing the two objects together. Being an empirical model, the relationship is not purely linear under all circumstances, but experience has shown it to be a reasonably good approximation to use in calculations and design. Direct testing of your specific scenario is always the best way to determine the frictional forces with which you are dealing.

There are different values of the coefficient of friction, depending on whether an object is moving or stationary, called *kinetic* and *static coefficient of friction*, respectively. Also, in some cases, the kinetic value can vary with velocity. In general, static friction is greater than kinetic friction for a given interaction and normal force. This makes sense intuitively if you imagine that the force of friction results from the interlocking of rough microstructures of the materials; motion would prevent the surfaces from interlocking as effectively.

When adhesives or lubricants are involved, this simple model breaks down or must be supplemented with further considerations. The linear relation is best applied to dry friction between two objects in macroscopically flat contact, like a box sliding down a metal chute. When other materials interact between the objects, the situation increases in complexity and cannot be described simply. In this case, more detailed empirical investigations are the best way to understand and design around these factors.

When considering dry friction, most values of μ fall between 0 and 1, concentrated between 0.3 and 0.6. Table 15-3 lists some measured values for familiar material interactions.

TABLE 15-3 Representative Values for Coefficient of Friction

	Coefficient of Friction	
Materials	Static	Kinetic
Steel on steel	0.74	0.57
Aluminum on steel	0.61	0.47
Copper on steel	0.53	0.36
Rubber on concrete	1	0.8
Wood on wood	0.25–0.5	0.2
Glass on glass	0.94	0.4
Waxed wood on wet snow	0.14	0.1
Waxed wood on dry snow	—	0.04
Metal on metal (lubricated)	0.15	0.06
Ice on ice	0.1	0.03
Teflon on Teflon	0.04	0.04
Synovial joints in humans	0.01	0.003

Source: Adapted from [19].

BUT HOW DO I GET IT FROM A TO B? This is the end of the first portion of this chapter. We have explored many of the significant physical quantities and considerations associated with gases, liquids, solids, powders, slurries, and molten metal. With this understanding, we can now begin discussing how they affect the transport of substances in process industries. The remainder of the chapter is dedicated to introducing the major conveyance methods in use today. Some may only be used to convey a specific phase, but many of the concepts overlap from method to method. The concepts discussed to this point will be referenced frequently in the context of their importance to a particular conveyance method, so be sure to refer back as needed.

15.3 MATERIAL TRANSFER AND CONVEYANCE EQUIPMENT

Now that we have a better understanding of the material properties that influence decisions on transport methods, we introduce some of the major means of conveyance individually and explore the selection, implementation, and maintenance associated with each. The remainder of the chapter is devoted to examining different types of transport equipment and includes a look at:

- Conveyors
  - Conveyor belts
  - Overhead conveyors
  - Roller conveyors
  - Chutes
  - Screw conveyors
- Pumps
  - Centrifugal
  - Axial
  - Reciprocating
  - Rotary
- Valves
  - Toggle
  - Throttle
  - Check
  - Relief
- Pipes

You may find it informative to read through the chapter in its entirety, but each section will be somewhat self-contained and can serve as a reference when looking for more information as to conveyance equipment decisions. Each section will provide:

1. Engineering data that describe the workings and variations of each particular type of conveyor, including important industry standards that may be in place to categorize a particular method
2. A look into how you can decide which method is best for your application, and also which variation of each type is the optimum choice
3. Information and examples of methods to help improve the reliability and worth of equipment
4. An overview of different maintenance concerns significant to the transport method
5. Recent developments and improvements that could potentially put you on the leading edge of a particular technology

As a whole, these sections will get you acquainted with your options in material conveyance and introduce many of the significant issues you should consider when designing or choosing a transport method.

15.4 CONVEYORS

The term *conveyor* is quite broad, but generally applies to mechanical devices that transport solids. These can be either bulk solids or powders. Furthermore, conveyors may be either powered or unpowered. In this section we examine some of major conveyor types, including (1) conveyor belts, (2) overhead conveyors, (3) roller conveyors, (4) chutes conveyors, and (5) screw conveyors. After a focused look at each conveyor type, in a final section we offer very brief descriptions and illustrations of other conveyor types. This can be used to spur further research if you feel that one of these conveyance methods is better suited to your needs.

15.4.1 Conveyor Belts

Most people can easily conjure up the mental image of a conveyor belt at work, and this is not a coincidence. The conveyor belt finds widespread application in almost every industry, due to its simple and effective implementation. Conveyor belts can be used to transport solids as well as powders, but are not used to transport liquids or gases in any normal capacity. You can find conveyors belts implemented heavily in the following applications:

- Mining
- Food industry
- Pharmaceuticals
- Postage processing
- Manufacturing

The basic components of a conveyor belt include a drive mechanism, which powers the motion of the belt, and the belt itself, which transfers this motion to the conveyed material and provides support. There are a number of different options in belt structure and material and there are also many ways to power the belt. These choices are described, compared, and evaluated in this section.

15.4.1.1 Engineering and Component Construction Considerations Any conveyor belt system is composed of:

- The belt
- The drive mechanism
- The guidance mechanisms

The belt runs in a continuous loop along the guidance mechanism, at some point passing through the drive mechanism, which provides power, and the entire system is supported by an underlying structure to transfer the loads to the ground. Figure 15-4 shows the basic layout of a conveyor belt system.

15.4.1.1.1 Belt There are a nearly endless number of variations when it comes to belt design and composition. Here we look at two of the categories that find the most frequent use in industry: solid polymer/fabric belts and wire mesh belts. The solid polymer/fabric belt structure consists of three major components: the carcass, the skims, and the cover. Each component is responsible for different aspects of belt performance, and choosing each properly is important in obtaining the appropriate final belt attributes.

The *carcass* is the central load-bearing component of the belt. It is responsible for holding the tension to which the

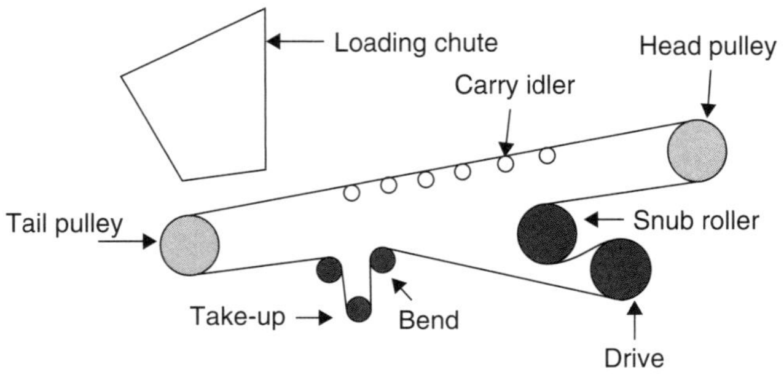

Figure 15-4 Basic conveyor system components. (Adapted from the Global Belting Technologies Knowledge Center.)

belt is subjected, withstanding the impact forces associated with loading the belt, providing the support strength to carry the load, and supporting the other two components of the belt. The carcass material is generally rubber-impregnated fabric woven with polyester and nylon fibers.

A few different carcass designs are implemented in today's industry. The most common is the multiply, which consists of a number of layers, or plies, of fabric which are combined via intermediate rubber layers. A simple two-ply illustration is shown in Fig. 15-5b. These layers can consist of a plain or twill weave. In the plain weave, transverse and longitudinal fibers both experience crimping, so neither remains straight. This structure and the fiber material are responsible for the particular stretch characteristics of the belt.

Another carcass design that is prevalent is the straight warp design. It consists of straight lengthwise fibers, called the warp yarn, which experience little or no crimping. These are supplemented above and below by fill yarns, which run transverse to the warp yarn. Then the warp and fill yarns are bounded together by binder warp, which runs throughout as shown in Fig. 15-6. Straight warp construction yields a very strong belt that has minimal geometric stretching and good load support. This design is also very tear resistant, making it good for heavy applications. An extension of the straight warp design is the solid woven design, which is even stronger and has a more complicated warp weave pattern (Illustrated in Fig. 15-7).

One type of belting that differs greatly from the designs discussed previously is the steel chord carcass encased in rubber. Instead of a woven body, this design simply has tensioned steel cables running the length of the belt. These load-bearing cables are then encased in rubber, which supplies the proper contact friction with the load. This design is implemented primarily in high-tension applications.

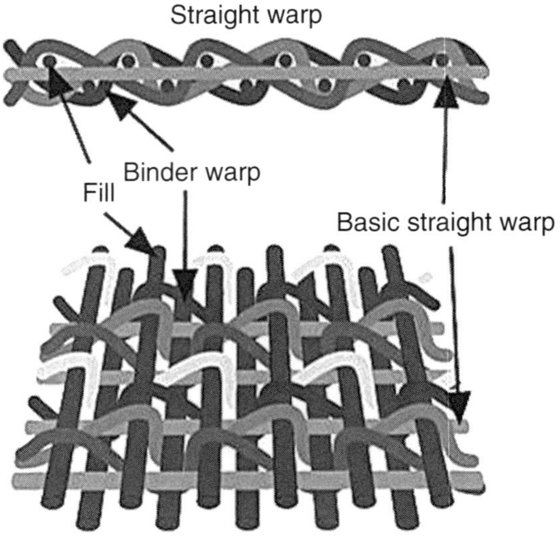

Figure 15-6 Straight warp carcass weave. (From *Fenner-Dunlop Conveyor Belt Construction Manual*, 2003.)

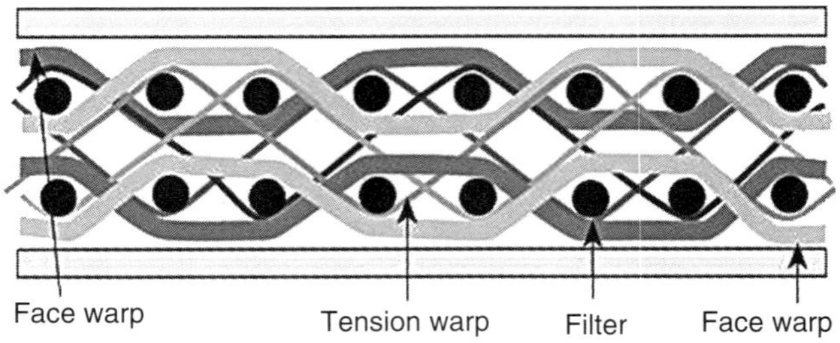

Figure 15-7 Solid woven design. (From *Fenner-Dunlop Conveyor Belt Construction Manual*, 2003.)

In comparing carcass designs and constructions, the belt strength is the most significant attribute to consider. Belt strength is described by pounds per inch of width (PIW). In other words, a fabric weave rated at 30 PIW and a foot

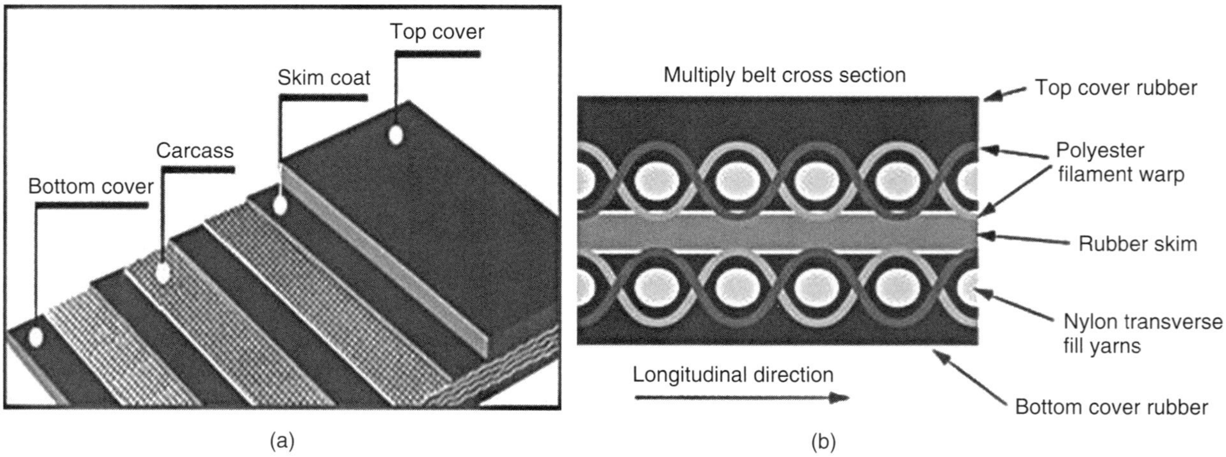

Figure 15-5 Simple multiply illustrations: (a) layers of a multiply belt; (b) warp weaves connecting plies. (Adapted from the Global Belting Technologies Knowledge Center.)

across can support 360 lb of tension. It is also important to distinguish whether you are referring to *maximum working tension* or *ultimate tensile strength*. Usually, belt strength is given for maximum working tension so that you can select the proper belt strength for your application. However, the belt probably will not rupture or tear if this tension is slightly or momentarily exceeded. The tension at which such a rupture will occur is the ultimate tensile strength of the belt. It is important to know which you are using to select your belt. It would be very dangerous and costly if you selected your belt based on ultimate tensile strength ratings.

Moving on to the second component of the belt, the *skims* are the binding layers in between carcass plies. They are generally made of rubber, PVC, or urethane. The skims contribute to internal belt adhesion, troughability, and load support, and skims selected or applied improperly can lead to ply separation and belt failure. Adhesion refers to the strength of bonding between layers of the belt, and without it, shear forces experienced by the belt would pull the plies apart, destroying the belt. Troughability is the extent to which the belt can be bowed or curved. Belts having high troughability are desirable in applications that deal with particulates, such as in the mining industry. Skims also provide increased flex life that the carcass plies alone would not possess, cushion the carcass layers so as to prevent punctures due to impact loads, and protect against moisture degradation, sealing the woven carcasses from the environment.

The two outermost layers of the belt are called the *cover*. It is the purpose of the cover to protect the inner carcass and skims and provide the desired surface attributes to the belt. All of the following properties of the belt can be supplied by the cover:

1. Textures
   a. To increase friction
   b. To increase inclination
   c. To control product
2. Cleanability
3. A specific coefficient of friction
4. A specific color
5. Cut resistance
6. Enhanced impact resistance
7. Hardness
8. Fire, oil, and chemical resistance

These properties are determined by the chemical compounds that compose the cover as well as cover thickness.

Together, these three components—the carcass, skims, and cover—make up the belt. The selection specifications for important aspects of this type of belt are discussed in the next section. Some of the major materials used for woven belting up to today are enumerated and described in Table 15-4.

Another type of belt that finds wide application is the *wire mesh belt*. These consist of metallic wires woven such that they create flexible belt structure. Here again, a variety of structures can be implemented, depending on the requirements of the load.

Two important types are the balanced and universal weaves, both consisting of straight or slightly crimped widthwise rods, linked by spiraled metal wires. The balanced weave implements alternating right- and left-handed spirals (shown in Fig. 15-8a) while the universal weave (shown in Fig. 15-8b) has interwoven right- and left-handed spirals connecting each rod.

Another type is the conventional or single weave (shown in Fig. 15-8c), which has the same structure as a normal chain link fence. This is the simplest of the chain mesh belts and is suitable for moderate applications with no extreme demands on the belt properties. When searching through wire mesh belting catalogs, you may come across numbering designations. Different companies may use slightly different formats, but these numbers usually indicate three things: the number of loops per foot spanwise, the number of rods per foot lengthwise, and the gauge of the wiring used. For the single-weave mesh, usually only one number is quoted. This number is simply the number of gaps per inch along any diagonal, since there are no rods in the conventional weave.

A different mesh type, one without large gaps, is the chord weave style (shown in Fig. 15-9). It is much more tightly meshed than any of the previous wire mesh types and is suitable for handling small parts. Chord weave is also labeled "baking band" because of its widespread use for handling baked goods. It provides even distribution of heating while still supporting the food properly.

Finally, we have the flatwire mesh. This design differs from those discussed previously in that it has more depth due to the thick flattened wires that make up the rod connectors. One significant advantage of this type of mesh is its compatibility with a sprocket drive. The depth of the mesh allows for significant contact with the sprocket drive, ensuring power transfer. Figure 15-10 illustrates this mesh style. Notice the depth of the belt.

The metal used in wire mesh belts is usually stainless steel, but can be made in a number of alloys and treatments for particular applications. Table 15-5 gives examples of metals offered by the Audubon wire mesh belting company, a leading industry provider and the source of the exemplary images shown previously. They can also provide further information and application-specific customized mesh alloys. Visit their Web site at http://www.meshbelt.com/index.html for more information.

TABLE 15-4 Materials Used in Fabric Conveyor Belts

Cotton	Natural cellulose composition Only natural fiber used to any great extent in belting manufacture. Increases in strength when wet. High moisture absorption—consequently, poor dimensional stability. Susceptible to mildew attack. At one time represented 80% of the raw fiber input into belt manufacture. Currently, something less than 5%.
Rayon	Regenerated cellulose composition Slightly stronger than cotton, but tensile strength is lowered by water. Chemical resistance similar to that of cotton. High moisture absorption—consequently, poor dimensional stability. Susceptible to mildew attack. Almost nonexistent in conveyor belts today.
Glass	Glass Very high strength compared to rayon. Low elongation. Used mainly in high-temperature applications. Poor flex life. Limited use in belt manufacture currently.
Nylon	Polyamide High strength, high elongation, and good resistance to abrasion, fatigue, and impact. While moisture absorption not as high as cotton, it will absorb up to 10% of its own weight in moisture. Consequently, poor dimensional stability. High resistance to mildew. At one time, nylon represented 40% of the raw material input into belt manufacturing. Today, it is something less than 20%.
Polyester	Polyester High strength, exceptionally good abrasion and fatigue resistance. Extremely low moisture absorption . . . consequently, good dimensional stability. Unaffected by mildew. Georgia Duck selected polyester as its "fabric of choice" in 1960. Polyester use in the manufacture of belting has grown from 0% in 1960 to something in the range of 70 to 75% today.
Steel	Steel Used where high strength and extremely low stretch are necessities. A small amount of woven steel carcass is available in today's market. However, more steel is used in steel cord–like belt constructions.
Kevlar	Aramid Aramid (the material used in flak jackets and bulletproof vests) has twice the strength of steel, with stretch characteristics roughly halfway between those of steel and polyester. It is significantly lower in weight than steel and will not rust.

Source: Adapted from the *Fenner-Dunlop Conveyor Belt Construction Manual*, 2003.

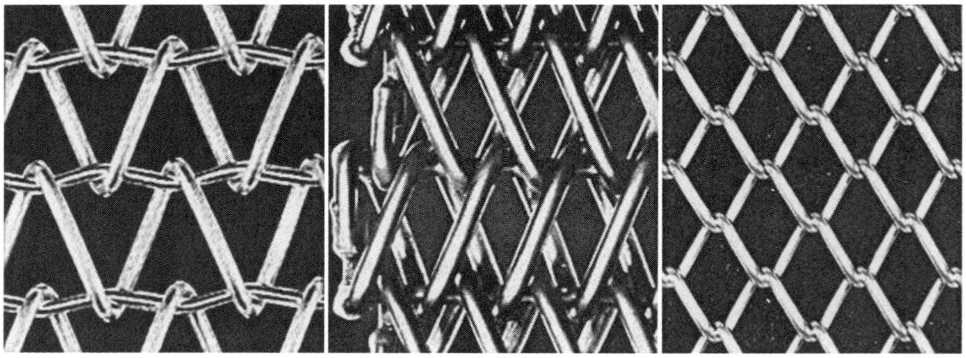

Figure 15-8 Wire mesh weave styles: (a) Left–balanced; (b) Middle–universal; (c) Right–conventional.

SIGNIFICANT BELT FACTORS The belt tension is one of the most significant quantities that influences aspects of the belt's behavior and performance. The working tension of the belt and selected speed determines the necessary power the motor must provide. The working tension also sets requirements on the tensile strength of the belt. There are many engineering formulas used in calculating these tensions, and most of them are empirically based and rely on numeric factors and coefficients that must be found in tables. It is best to find a calculation that is closely related to your application and use that form to determine your tension requirements.

Tension is usually reported in units of PIW, pounds per inch width of belt. Here is a quick formula that can be used

Figure 15-9 Chord weave mesh style.

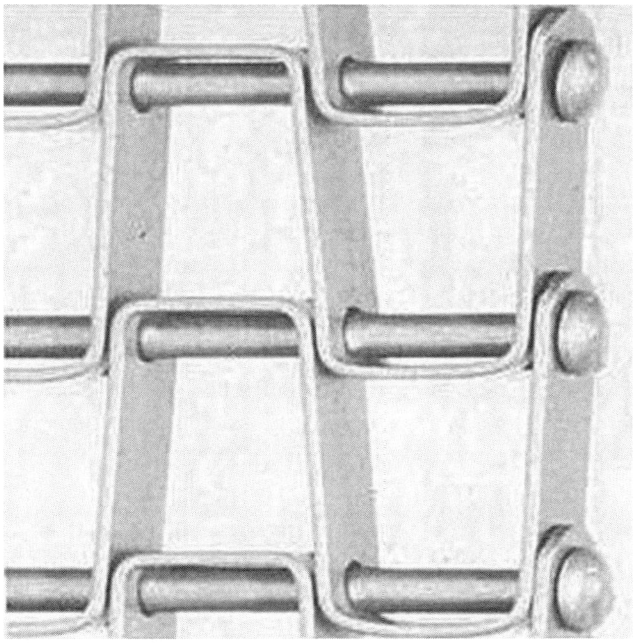

Figure 15-10 Flatwire mesh.

TABLE 15-5 Alloys Available for Audubon Wire Mesh Belts

201 stainless steel conveyor belts
304 stainless steel conveyor belts
314 stainless steel conveyor belts
316 stainless steel conveyor belts
316 ELC conveyor belts
330 stainless steel conveyor belts
430 stainless steel conveyor belts
80-20 cb or Nichrome V conveyor belts
35-19 cb, 3519, 3519 cb conveyor belts
Inconel 600 conveyor belts
Inconel 601 conveyor belts
3% Chrome conveyor belts
High-carbon steel conveyor belts
Plain steel conveyor belts
Galvanized conveyor belts
Titanium conveyor belts
Manganese conveyor belts
Teflon-coated conveyor belts

to estimate belt tension:

$$\text{PIW} = \frac{\text{HP}(1 + K)(33{,}000)}{SW} \tag{15.11}$$

where PIW is the unit tension, HP the motor horsepower, K a numerical drive factor, S the belt speed, and W the width of the belt. This can be used to get rough estimates of belt tension if you have the appropriate drive factor. For more accurate calculations, refer to a manufacturer's selection guide.

Load support is another primary concern for the belt. For metallic belts, this is directly related to the strength of the material. For plied belts, load support capabilities depend on both the material properties and the number of plies. However, simply increasing the number of plies without limit is detrimental to other important aspects of the belt: flexibility and troughability. If proper flexibility or troughability is not provided by the belt design, excess strain is placed on the belt by the load and the conveyor runs less efficiently. This yields extra operating costs and shortens the life of the components of the conveyor system.

Impact rating is another belt-associated factor. It is a measure of the belt's ability to withstand and recover from impacts endured during conveyor loading. The type of fiber used, weave style, and number of plies all have an effect on the impact rating. Belts should be designed to handle expected impacts within a safety margin that allows for some overloading.

15.4.1.1.2 Drive Mechanism The drive mechanism of the conveyor belt consists of the motor, speed reducer, and drive pulley. These components work in concert to transfer power from the motor to the belt in order to move the load. Two common configurations are diagrammed in Fig. 15-11. You can see that the motor transfers power through the speed reducer, which then transfers power to the drive pulley, around which the belt runs. For small-scale belts, there is generally only one drive mechanism per belt to avoid having to deal with synchronization issues. In some heavy-duty and longer-range belting, multiple drives must be coupled to provide the power required. Synchronization of these drives is crucial to proper operation of the belt. If the separate drives conflict at all, uneven and sometimes destructive overtensioning can occur in some spans of belt, while excess slack accrues in other portions. Extreme care must be taken when designing and operating multiple-drive belt systems.

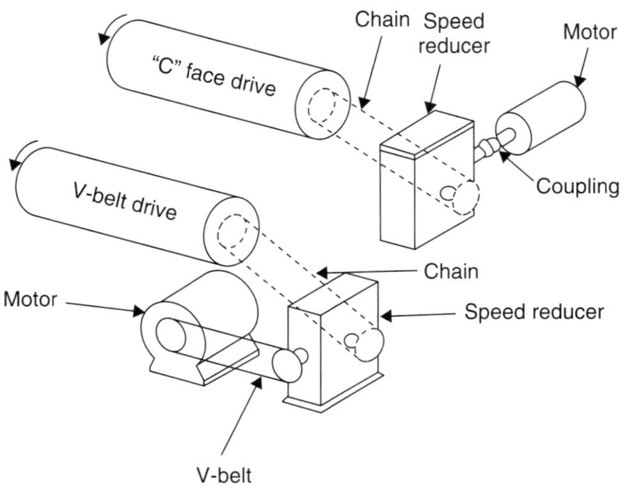

Figure 15-11 "C" face (top) and V-belt (bottom) drive arrangements. (From [10].)

The *motor* converts electric energy into mechanical energy, and can be of any form. Generally, it is ideal to have some control over the speed of the motor, so a variable-speed drive is used. This control is necessary because, upon startup, adjustments and calibrations to the belt may be necessary, and jumping immediately to full speed can be detrimental and dangerous. The horsepower and torque requirements of the motor vary with application.

The *speed reducer* is next in the transfer of power. It is necessary to slow down the rotation speed in most circumstances because motors rotate at much higher revolutions per minute/second (rpm or rps) than is needed for the drive pulley. Thus, the reducer uses gears to trade revolution speed for increased torque to drive the belt.

Finally, the power is transferred by a chain or belt to the *drive pulley*. This is simply a cylinder, usually made of steel, which transfers the power into the belt and pulls the load around the belt circuit. There are a number of different types of drive pulley, depending on the type of belting used. Most woven polymer belts rely on a friction drive pulley along with appropriate tension in the belt. In this way, motion is transferred to the belt through the friction of the pulley with the belt while the two are in contact. The contact angle therefore becomes a significant concern. In this context, the contact angle refers to the angular distance through which the belt and pulley are touching as the belt moves over the pulley. A contact angle of 180° indicates that half of the pulley surface actually contacts the belt. This angle contributes greatly to determining the maximum torque and horsepower that can be transferred to the belt before slipping occurs.

Another drive pulley type is the positive drive arrangement. This can also be used with specially designed fabric belts, but is more commonly used in conjunction with wire mesh belts. This belt works like a sprocket–chain combination. The drive pulley has circumferential teeth that fit into the open holes of the belt, effectively locking the two together and transferring power to the belt.

Some conveyor belt systems don't transfer power from the drive pulley directly to the belt. Instead, the motor is connected to a sprocket that drives a chain or chains. The chains are positioned on the edges of the load-bearing portions of the belt and are really part of the belt. This type of drive is described in more detail in the context of pure roller conveyors in Section 15.4.3. In larger load scenarios where delicacy is not an issue, the belt itself can be omitted and a number of chain drives of this sort implemented instead. An example of this type of arrangement is illustrated in Fig. 15-12. Pure chain conveyors have attributes similar to those of both conveyor belt and overhead conveyor systems, so referencing Sections 15.4.2 and 15.4.3 will give you a good understanding of the necessary considerations when using such a conveyor.

The main requirements of the drive mechanism are to provide the necessary torque and horsepower to drive the belt at the appropriate speed. Specific designs and configurations are not as significant as the final output parameters of the drive. As we discuss in Section 15.4.1.2, it is these output parameters that are usually the focus of the selection process. Thus, specific choices of motor are not as significant, but the drive pulley design or manner in which the power is transferred is an important consideration for reliability and tracking issues.

15.4.1.1.3 Guidance Mechanism The guidance mechanism consists of all the components that direct and support the belt. It includes the pulleys, rollers, and idlers through which the belt passes during each cycle as well as the conveyor bed over which the belt runs. All of the roller

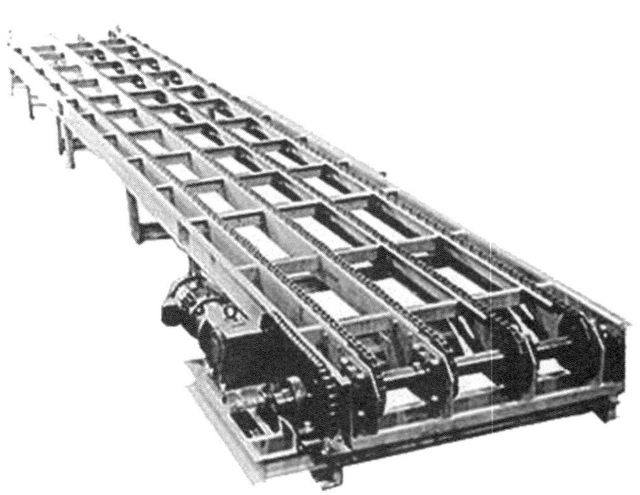

Figure 15-12 Pure chain conveyor. Loads rest directly on the four tensioned chains. (From [10].)

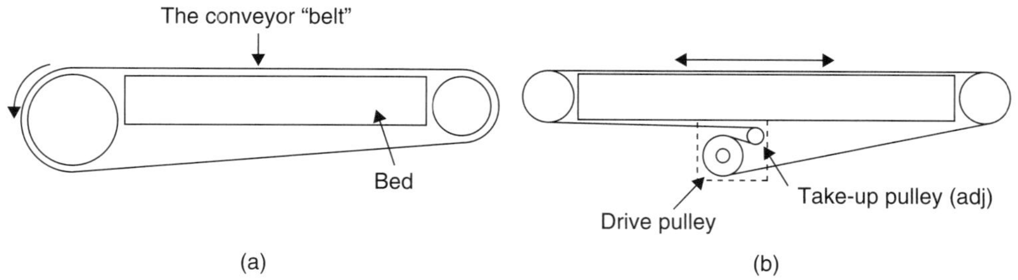

Figure 15-13 End drive (a) and central drive (b) pulleys. (From [10].)

components consist of cylindrical tubes of various diameters, depending on their purpose. Reducing friction in these components is very important, especially as the length of the belt and number of contact components increases. To this end, bearings and lubricants must be implemented and well maintained.

The two primary guidance pulleys hold position at each end of the belt. They are labeled the "tail" and "head" pulleys, and often the head pulley also serves as the drive pulley. These are usually the largest components of the guidance mechanism and have the greatest contact angle with the belt. Since the belt always leaves the pulley tangentially when under tension, the contact angle is also a measure of the angular change in the belt direction while touching the belt. These are placed at each end of the conveyor bed and create the main limits of the belt. An end drive and central drive are shown in Fig. 15-13. It is often undesirable to have the lower portion of the belt hanging down. To prevent this, return idlers are often implemented, guiding the returning belt section as illustrated in Fig. 15-14.

If excess belt length exists, or if control of the tension is needed, take-up idlers can be implemented. These create a fold back of the belt so as to "take up" any unwanted length or slack. The amount of take-up is often variable, controlled simply by adjusting the spacing between these idlers. A simple example is shown in Fig. 15-15.

Another specialized idler is the snub idler. It is the return idler closest to the drive pulley. By adjusting the position of this idler, the contact angle of the belt with the drive pulley can be adjusted. Also, each end of the snub idler is independently adjustable, so it can be used to "steer" the belt or to adjust the tracking of the belt.

When long belts or heavy loads are called for, a roller bed is generally implemented. This provides load support at various intervals along the belt so that the belt tension is not solely responsible for providing this support. A roller bed diagram is shown in Fig. 15-16.

These are the main components to any guidance system, but they may be arranged in any number of ways, depending on the needs of your application. The simple diagrams included are meant to give a sense of each component's implementation. The usable combinations are

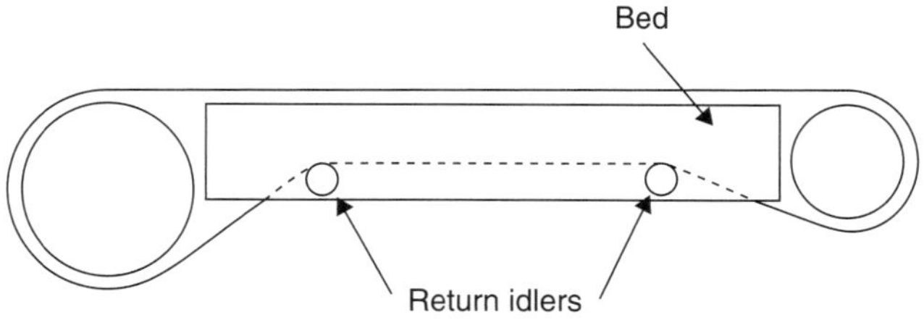

Figure 15-14 Return idlers. (From [10].)

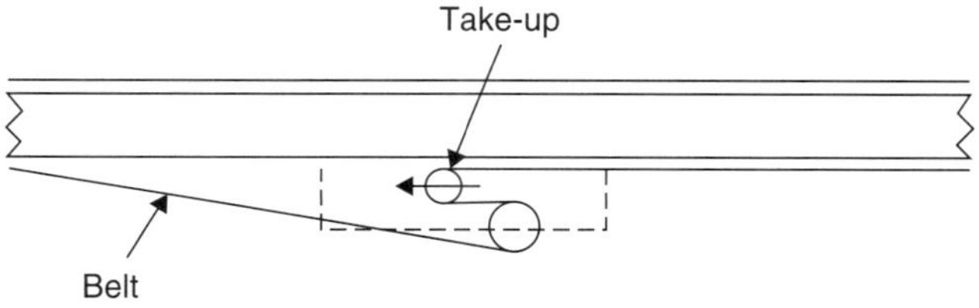

Figure 15-15 Take-up idler. (From [10].)

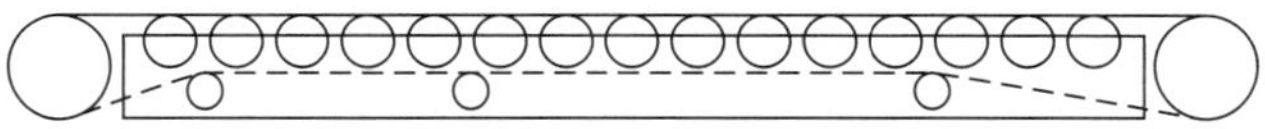

Figure 15-16 Roller bed. (From [10].)

nearly endless, and many configurations may be suitable to a particular case.

15.4.1.2 Selection Specification Once the choice to use a conveyor belt system has been made, the real selection process begins. The choices of belting type, drive type, and sizing must be made, but these selections are highly interdependent on one another and on the character of the loads. The purpose of this section is to illuminate some of the more significant considerations that come into the picture during these selection efforts. The following design parameters have a notable impact on belting type, drive type, and general sizing:

Load details

- Weight

- Dimensions
- Loading method
- Chemical reactivity
- Temperature

Goal transport speed

Path of conveyor

- Length
- Inclines
- Turns

Function of belt

- Accumulation
- Pure transport
- Concurrent action (painting, cooking, etc.)

Things of this nature should be specified at least primitively so that you possess ample knowledge of the requirements of the system before beginning the selection process. With this definition, it is possible to begin making reasonable choices pertaining to belt construction and operation. Next we examine these choices and how the parameters described above can guide you to proper conclusions.

15.4.1.2.1 Selecting the Correct Belt The belt itself is the crucial selection point. Nearly every other choice can be related back to the belt in some way. Consider the following relationships:

- The type of belt used (plied or wire mesh) is strongly dependent on the function intended.
- The weight of the belt affects the requirements on the drive selection.
- The tensile strength required is dependent on the load characteristics.
- For plied belts, interply strength places requirements on the diameter of pulleys.
- The surface material determines how the belt interacts with the load.

As you can see, belt selection is a complicated balance between the benefits and costs associated with different choices. But let's try to make things a little simpler by decoupling some of the issues from the rest of the selection process. First, selecting the belt type is influenced most significantly by the intended application of the belt. The nature of the load, the environment in which the belt will be operating, and processes that coincide with transport should all be considered when choosing the belt type. Table 15-6 compares some usual factors that lead to a choice of either wire mesh or polymer-based belts.

The choice between wire mesh and polymer belting should be relatively easy in the context of the application.

TABLE 15-6 Comparing Wire Mesh and Polymer Belts

Wire Mesh	Polymer Based
Frequently used when dealing with:	Frequently used when dealing with:
Extreme temperatures (ambient or load)	Long spans
Low belt speeds	Normal temperatures
Chemically hostile environments with appropriate choice of metal	Chemically hostile environments with appropriate cover material
Heavy impact loading	Powders or particulates
	High-speed conveying
Common applications:	Common applications:
Baking, frying, cooking	Mining
Metallurgy (heat treatments, sintering, annealing, quenching)	Shipping
Automotive	Food handling
	Pharmaceutical
	Industrial farming
Doesn't handle:	Doesn't handle:
Small object loads	Extreme temperatures
Powders	High impact loading
High-speed conveying	Very heavy loading

Thus, one aspect of belt selection is complete. At this point, the specific nature of the application dictates how to proceed with material selection. If the load has an extreme property such as temperature, size, weight, or corrosivity, this aspect should dominate the material selection. A property that stands out significantly requires particular design attention, whereas less extreme factors can be accommodated by a larger number of belts, so shouldn't influence the decision as heavily. Once you identify this critical property, you can select the appropriate material to handle it and then evaluate the consequences on other aspects of selection. For example, if the operating temperature and load size require a heavy strong belt, the motor-size specification will be at the mercy of these requirements.

Now, if none of the properties of the system seem too severe, there are many options in belt material. In this case, the choice remains highly coupled to other aspects of the design. The money saved in using a smaller motor at the expense of belt strength and longevity may be cheaper than using a larger motor and stronger, heavier belt. Thus, the final selection of material should not be made at this point. However, some effort can be made to narrow the choice down to a few options. After investigating these options with regard to later selection choices, such as drive type and sizing requirements, you will be better equipped to evaluate the trade-offs associated with choosing one option over another in your final decision.

15.4.1.2.2 Selecting the Appropriate Drive The most important aspect of drive selection is the appropriate sizing of the motor. It must deliver appropriate power to drive the belt, and it would be preferable if the motor operated near its most efficient point, so as to reduce operating costs. The power requirement placed on the motor is a sum of four components:

1. The power required to move the empty belt
2. The power required to move the load horizontally
3. The power required to move the load through any height changes
4. The power required to overcome frictional losses of guidance components

It is general practice to find these values of power requirements, sum them, add 10% for safety, and then use a motor rated at the nearest value above the requirement computed. In this way, the motor can operate at around 70% when running the belt at design capacity. This not only runs the motor in an efficient range, but also allows for overloading without overtaxing the motor. Values for power requirements can be calculated from values of belt tension, which are determined through application of engineering formulas. These formulas account for the distribution of the load, belt profile, frictional losses, and belt stretch.

The next aspect of drive selection is usually implied by the nature of the belt. The drive pulley for fabric belts is normally friction based, while in meshed belts a positive-displacement drive is easily integrated. Sizing this drive pulley depends on the flexibility of the belt and the speed and torque being output by the motor.

15.4.1.3 Reliability and Cost Savings In this section we provide readers with some ideas on how to improve conveyor belt reliability. This will be done through the use of industry examples. One of the greatest costs associated with conveyor belts and conveyance in general is that of energy. Along with the trend of "going green," companies have a large financial interest in reducing energy costs, and they are willing to put in slightly more capital for more expensive conveyor systems if it will save them in the long run. Energy costs can be reduced in a number of ways:

- Increasing the efficiency of the motor
- Decreasing losses due to friction
- Using lighter systems with improved strength through newer, stronger composites

These will all yield a reduction in operating costs, and every effort should be made to determine whether the investment to achieve these ends is worth it.

Continuing with this line of thought, take the case of Goodyear Engineered Products, based in Palm Springs, California. They have attacked the problem of frictional losses due to belt interaction with idler rollers. On long overland conveyors, frictional loss can account for up to 60% of energy costs. Part of the reason that rollers cause friction and power loss is due to the fact that as the belt passes over a roller, it is deformed, absorbing energy. The degree to which this deformation occurs has a significant impact on friction between the belt and the roller for the short time that this belt segment and roller are in contact. Goodyear Engineered Products have designed a polymer belt material that recovers remarkably fast after depression on a belt roller. This reduces the time in contact with the roller as well as the degree of depression, and in turn greatly reduces the power lost due to the interaction. A simple application of this polymer to the bottom side of a conveyor will save vast amounts in operating costs, as more of the power generated by the motor is actually used in conveyance.

Another example from industry illustrates how the use of variable-speed drives can help optimize motor output, saving in energy costs and reducing unnecessary wear on the system. Siemens, one of the largest providers of mining conveyors worldwide, recently installed a variable-drive power system for the Welzow Süd, Germany, brown coal mine. By smoothly adjusting the motor speed depending on the load, Siemens was able to save the mine 20% on energy costs. For a large mine with a 2.6-km conveyor belt, that is a huge amount of cash flow back into the company. It is true that these variable drives and methods to reduce frictional losses are more effective on large overland conveyors, but the concepts and savings can also be applied and achieved in smaller process plant applications. Variable drives should be considered if replacing or designing a conveyor system of any scale.

Another way to save costs if you are using simple conveyors comes in the form of modular conveyor belts. Systems composed of these consist of many small, individually driven modules. They vary in design from straight segments to turns to inclines. Although the cost of each module is slightly more than that of an equivalent length of traditional belting, the modular design provides enormously improved flexibility for the system. In this way, quick changes can be made to the conveyor path and length. This is ideal for applications where conveyance needs vary from day to day and require a conveyor system that can conform quickly to new demands. Implementing a modular system like those provided by Dynacon conveyors prevents you from having to make sacrifices to cater to a sedentary conveyor belt. Note that these conveyors are not suited for heavy-duty applications or environments of an extreme nature.

A more recent improvement that can aid in cost savings and reliability is the practice of virtual prototyping. Like much of engineering design, conveyor design has moved to the computer. Sophisticated software can aid in perfecting the design before spending any money producing components and prototypes. Optimization is much easier for a computer than for an engineer if you provide all of the details about options for the system. In this way, using virtual prototyping allows for a more optimum design for each particular application. It also allows for easy analysis of cause and effect during the selection process. If you have a computerized model of the conveyor system, it becomes much easier to change a certain aspect, such as the motor specifications and observe how this affects other selection choices, such as possible load weight. Unless you are reproducing a previously used and well-evaluated conveyor system, performing virtual prototyping (or having a firm do it for you) is definitely a way to save costs and increase the reliability of a conveyor belt system.

Saving costs is a never-ending effort at every level of a process company. For conveyor belts, the best way is to improve energy efficiency. This constitutes much of the cost associated with the conveyor. You should always be on the lookout for new techniques to save on energy costs.

15.4.1.4 Preventive, Predictive, and Proactive Maintenance Approaches The topic of maintenance has long been of utmost interest to plant supervisors. In process plants, running time translates directly into profits, so keeping all components of the process operating effectively is a primary concern. Conveyor belts are no exception. In fact, because of their integral nature in connecting process points, conveyor belts become significant problems if they are off-line for even short amounts of time. The best way to avoid unexpected failures and downtime is not to practice reactive maintenance, where you only respond to glaring issues, but instead, to practice preventive, predictive, and proactive maintenance.

Preventive and *proactive* usually describe the same maintenance ideology: namely, scheduling regular replacement or repair of parts more frequently than the expected lifetime of the parts. In this way, there will conceivably never be any unexpected failure or downtime. The downside of this approach is that you may replace parts that could continue to serve you for a long time, so you are incurring unnecessary costs. Plus, there must be scheduled downtime, which yields production loss. This type of maintenance would be considered playing it very safe, and for some applications where the cost of unexpected downtime far exceeds the cost of replacing parts, this ideology works well.

In conveyor belting, preventive and proactive maintenance would consist of replacing rollers, the belt, and even overhauling the drive system periodically. During these operations, the belt would have to be taken off-line. Depending on the severity of the repair or replacement, this downtime varies. For simple roller replacement, depending on the number of rollers being serviced, this time period can be very short. Belt replacement takes considerably longer, and drive system repair could take the system off-line for an entire day. It is best to try and line up these maintenance times and get everything done at once if possible. This is not always feasible since parts have varying lifetimes.

Predictive maintenance provides a better alternative to reactive maintenance. In this ideology, components are closely monitored for signs of fatigue and wear. Usually, parts do not fail without some type of observable warning, if you know what to look for. In conveyor belting, this means different things for different components. For rollers, periodic inspection of surface condition, bearing health, and deflection can provide valuable information on the remaining life of the roller. This inspection is generally performed by human observation, but it can also be computerized. For noncomputerized inspections, proper training of inspection employees becomes an important consideration.

Belt monitoring is the central means of predictive maintenance. It includes rip monitoring, belt profile monitoring, and tension monitoring, among other things. By keeping constant tabs on these aspects of the belt, small changes can be detected that will predict failure. For example, the company Conveyor Belt Monitoring provides laser profile monitoring systems for bulk solid conveyor belts. Small changes in the profile of the conveyor can be detected and evaluated to determine if abnormal loading is occurring. Thus, operators are alerted if the belt is experiencing a potentially harmful scenario, and they can then take appropriate action. Another monitoring system developed by Fenner Dunlop can detect small rips or tears as they approach a critical value. This allows operators to shut down the belt before the occurrence of complete failure that could cause considerable damage. It also allows for the belt to be used until the point at which replacement becomes essential, avoiding early replacement and maximizing the investment in the original belt.

15.4.1.5 Process Development and Improvements The majority of process developments associated with conveyor belting do not seek to reinvent the process itself, but simply to improve the efficiency and performance of existing configurations. Many of the current research topics have already been discussed in light of their cost savings and maintenance effects. For example, energy-efficient rollers, belts, and motors are always being explored and improved upon. As time progresses and the push to be environmentally friendly continues, more costly but efficient technologies will find increased application across belting industries.

Improvements in materials science are also having a beneficial impact on the conveyor belt. As more and more specialized composites are produced, it is becoming easier and easier to find a belting material that meets all the requirements of a design. Weight and strength are no longer pure trade-offs, and depending on how much capital you are willing to put into a belt, you can almost specify exactly the properties you want your belt material to possess. As advancements in this field continue, application to belting materials and the materials used in roller and drive technology will help improve efficiency and lifetime, yielding increasingly effective conveyor belt systems.

15.4.1.6 Additional Information If you would like to learn more about conveyor belts, the following resources are suggested:

- *Materials Handling Handbook*, 2nd ed., by ASME and IMMS, edited by Raymond A. Kulweic
- *Materials Handling Handbook*, by David E. Mulcahy
- Fenner Dunlop, http://www.fennerdunlopamericas.com/
- *Belt Conveyors: Principles of Calculation and Design*, by Golka, Vasili, and Bollinger
- *Solids and Liquids Conveying Systems*, edited by Mahesh V. Bhatia

15.4.2 Overhead Conveyors

When floor space is valuable or accessibility to certain parts of a load is necessary, the overhead conveyor offers an ideal transport method. All overhead conveyor systems are composed of a track, a coupling device to allow motion along this track, some form of carrier for the load, and in some cases a means to provide power to the system. Each of these components comes in a number of variations. The automotive, garment, and food industries, among others, all use overhead conveyors. The options and implementation of these forms of overhead conveyor are the focus of the sections to come.

15.4.2.1 Engineering and Component Construction Considerations

15.4.2.1.1 Track Types The overhead conveyor variants are distinguished mainly by the track design, while the many different types of load attachments can generally be implemented with all variants. Three of the most widely used track versions are the I-beam monorail, the enclosed track, and the power and free configuration.

As the name suggests, the *I-beam monorail* variant relies on a track of steel I-beams from which the load carrier is suspended. The carrier is coupled to the rail via a wheeled attachment, one version of which is illustrated in Fig. 15-17. The attachment is similar to roller coaster wheel couplings at an amusement park. As an unpowered conveyor, trolleys or short trains of trolleys can be moved by hand and through the use of gravity. For powered scenarios, each individual trolley is connected near the track coupling by a continuous chain, through which power is transferred to the system. Drive mechanisms are discussed in a later subsection.

Figure 15-17 I-beam and carrier attachment. (From [8].)

I-beam track segments are generally sold in 20-ft spans and come in a number of sizes. Some industry standards include 3, 4, and 6 in. These lengths refer to the vertical height of the I-beam, while flange width is usually $1\frac{7}{8}$ in. for all heights. As you can see in Fig. 15-17, it is on these flanges that the wheeled coupling mechanism rests. The I-beam track is hung or mounted and provides the support and pathway that the trolleys traverse.

The *enclosed track* variant uses hollowed metal bars inside which a wheeled chain passes. A split in the bottom of the bar allows suspension of the load from this powering chain. This split gives the bar a C-shaped cross section, lending itself to the alternative label, the *C-channel track*.

As Fig. 15-18 displays, the enclosed track is used to support a wheeled chain and load connectors that extend

Figure 15-18 Enclosed track and wheeled drive chain. (From [8].)

downward through the split. Enclosed track sections come in 10- to 20-ft spans and are usually constructed of steel. They can be of square or circular in cross section, depending on the type of drive chain used and the wheel configuration selected.

The *power and free configuration* consist of two tracks, one powered and one unpowered or "free." The primary advantage of such a system is the ability to separate trolleys from the drive chain. This makes accumulation and intermittent transfer possible, something of which the continuous chain variants described above are not capable. Power and free conveyors may make use of both of the track types discussed previously. Usually, the drive chain passes through an enclosed track, while the free track onto which the trolleys are attached may be either enclosed or I-beam.

As you can see in Fig. 15-19, the drive chain occupies the upper track, and specially designed trolleys supported on the free track allow for optional connection to the drive chain. Thus, the trolleys can be coupled and uncoupled at will, providing greater flexibility in application than the simpler overhead conveyor tracks discussed previously.

15.4.2.1.2 Drive System The three main drive mechanisms implemented in conjunction with overhead conveyors are the sprocket, friction, and caterpillar drives. Sprocket and friction drives have similar configurations, but interact differently to transfer power, while both the sprocket and caterpillar drives link with the drive train via protrusions to transfer power.

The sprocket drive consists of a motor–reducer combination that powers a toothed sprocket or gear. This gear is positioned at a 90° or 180° turn to maximize interlocking. As the chain passes through this turn, the teeth of the sprocket interlock with the spaces in the chain and transfer power. An example of this drive is illustrated in Fig. 15-20.

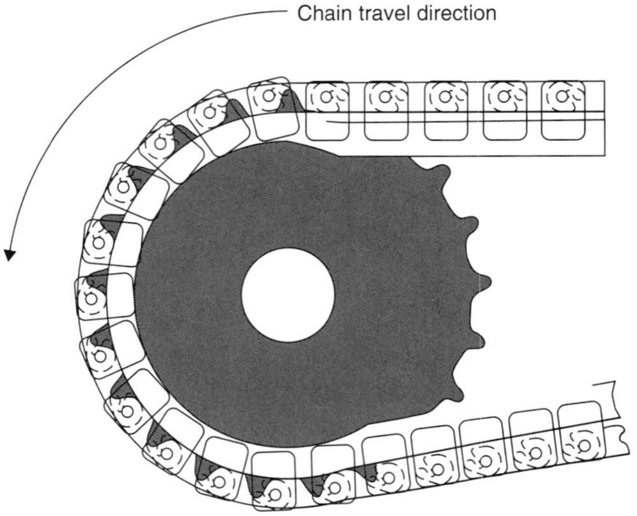

Figure 15-20 Sprocket drive for overhead conveyor. (From *SpanTech Drive Chain Owner's Manual*.)

The friction drive has a configuration similar to that of the sprocket drive discussed above. It, too, is placed at 90° or 180° turns to aid in transfer efficiency. However, the drive wheel does not link with the chain in this case. Instead, the wheel is made out of or coated in rubber to create high friction and grip with the belt. Then, in conjunction with the tension of the chain, this friction allows for the rotation of the wheel to transfer into translational motion in the drive chain. A 90° turn using such a drive is shown in Fig. 15-21. Also note the I-beam track and the large pin-linked drive chain.

The third drive mechanism commonly used with overhead conveyors is the caterpillar drive. Similar to the sprocket drive, the caterpillar drive links with the drive chain to transfer power, but instead of using the finely spaced teeth, the caterpillar drive has larger protrusions

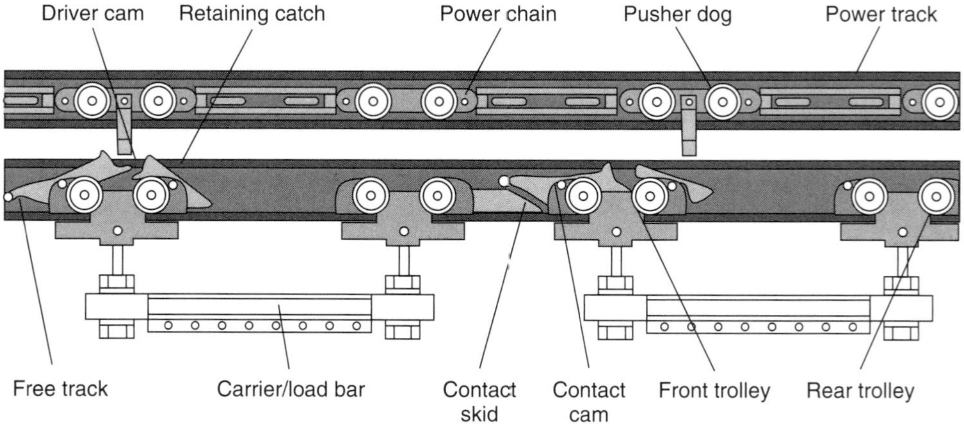

Figure 15-19 Power and free conveyor arrangement. (From *Eisenmann Power and Free Conveyor Catalog*.)

Figure 15-21 Friction drive arrangement. (Courtesy of LB Products, LLC.)

called *dogs*. Unlike the sprocket drive, the caterpillar drive can be implemented almost anywhere along the drive chain, including straight-aways. To envision a caterpillar drive, picture the treads of a tank or large construction vehicle (e.g., a Caterpillar vehicle; the name is no coincidence). Turning this configuration so that the wheels are now horizontal and attaching protruding dogs gives you a caterpillar drive. One is shown in Fig. 15-22.

Basically, the caterpillar drive uses a separate continuous-drive chain using a sprocket drive and then uses a parallel transfer through the drive dogs to power the main drive chain. Some of the benefits of this configuration over the sprocket drive include easier relocation, compatibility with many different conveyors for reapplication, and more flexibility in location along the drive chain.

Some important quantities to consider when discussing drive mechanisms include horsepower and torque as they relate to chain pull rating and transport rate. The *chain pull rating* (tension in the drive chain) decides the amount of torque necessary, and that, together with the transport rate desired, will decide the horsepower needed to drive the system. The transport rate is usually in units of ft/min or m/min. We discuss this in more detail in Section 15.4.2.2.

The other aspect of the drive system is the *chain*. For an I-beam conveyor, a normal chain can be used to link trolleys together and form a continuous loop, because the chain itself doesn't have to support the load, it merely provides transverse motion to the system. Chain choice becomes more a function of the drive system. Enclosed and power and free-style conveyors rely on more complex wheeled chains, which have alternating vertical and horizontal wheels integrated into the chain links. These chains are responsible for transferring the force of the load to the track, and use the wheels to reduce friction as they move along. Drive chains are generally made of metal, and can be stamped for lighter load applications or drop-forged for more strength.

Let's take a close look at an enclosed conveyor drive chain. An example of one type of chain used is shown in Fig. 15-23. You can see the alternating wheel directions, which provide constant guidance of the chain within the enclosed track. There are many ways to connect the wheel links. They can be riveted, bolted, or pinned. These methods have different advantages and disadvantages. Riveted chains are permanently linked and are quite strong, but this also means that adjusting the chain length at a future time becomes a labor-intensive process, results in longer downtime of the system, and is prone to damaging key components of the chain. Bolted chains are also a good option, and they are more easily adjusted, but they can come loose over time, so increased maintenance considerations

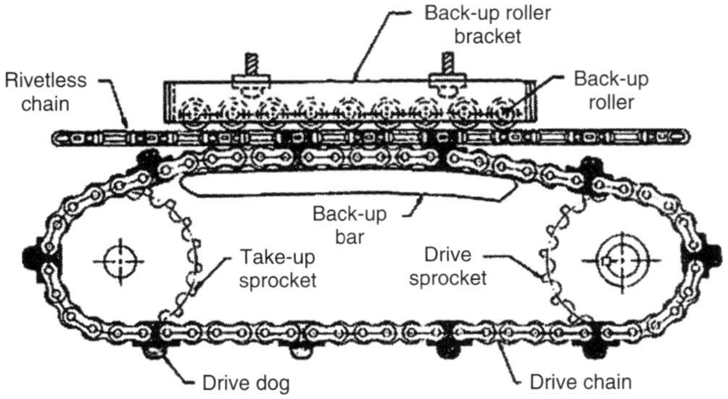

Figure 15-22 Caterpillar drive arrangement. (From *Anchor Conveyor Product Catalog*.)

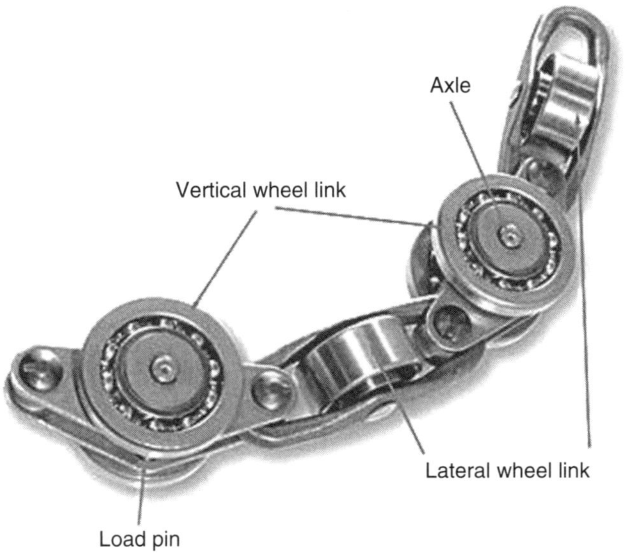

Figure 15-23 Common drive chain for enclosed track conveyor. (From Richards-Wilcox information pamphlet, 2007.)

arise when using this design. The cotter pin chain is perhaps the best choice for flexibility. Links can be added to quickly or removed from a chain linked in this manner, but the smaller components can lead to more frequent maintenance and less reliability.

The most important aspect of the drive chain is its load-bearing capability or chain pull rating. This quantity is reported in units of force, pounds, or newtons generally, and is provided by the chain manufacturer. The chain pull consists of all the forces that must be overcome to motivate the chain through its circuit. These include friction due to loading, weight lifted through vertical distances, and factors related to the transport speed of the chain.

Another property of the chain is its pitch, indicated in Fig. 15-24. This refers to the distance between adjacent loading points on the chain. A smaller chain pitch allows the chain to traverse curves with smaller radii effectively. Loads do not have to be placed at every single loading point, so smaller chain pitches can be used for large-dimension loads; however, economically it is better to have fewer links per unit length, which means that a larger pitch is desirable. In general, it is best to choose the largest pitch that meets the requirements of the system [1].

15.4.2.1.3 Track Path Components Another aspect of any overhead trolley conveyor system is the track path. How many turns will there be? What objects must be avoided? Will the system operate at constant height or also have a vertical dimension? These are the questions that any overhead trolley system designer must answer before selecting key components, such as the drive mechanism and drive chain.

When designing a closed-loop system, as most conveyor systems are, turns become a crucial consideration. There emerges a balance between the total length of the conveyor system and minimum safe radius of curvature for turns.

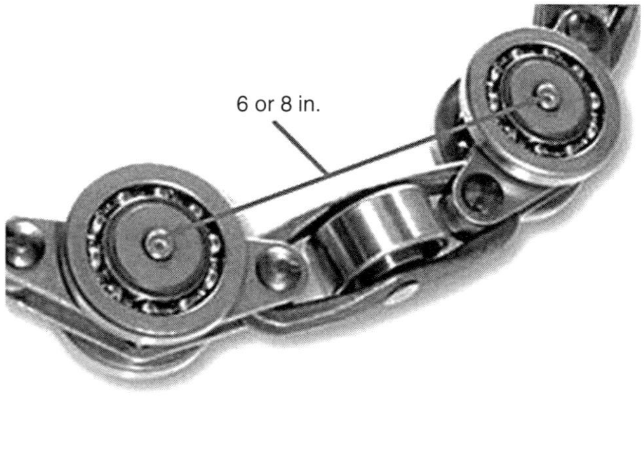

Figure 15-24 Chain pitch of drive chain. (From Richards-Wilcox information pamphlet, 2007.)

If all turns are made very gradually, the length of the conveyor can quickly get out of hand, resulting in decreased drive efficiency, loss of space, larger capital requirements, and higher costs in general. However, depending on the carrier spacing, size of product, and transport rate, there are limitations on the tightest turn that a system can possess before adjacent trolleys' products start to interfere with one another. This can lead to jams and unsafe conditions. Also, as loads are usually hanging and not completely rigid, tight turns at high transport rates can contribute to loads swinging out from directly beneath the track path. If not accounted for adequately, this can lead to nearby equipment interference or worker injury.

Another complication that can arise with the track path is change in height over the circuit. The changes in height and angle of elevation of each change have a significant impact on the necessary chain pull that will be needed to drive the system. The radius and angle of vertical turns also affect the minimum load spacing necessary to avoid product interference. It is generally ideal to use the largest radius of curvature possible to reduce stress on the drive chain and prolong conveyor life.

To maintain chain tension over the course of a chain's life, components known as *take-ups* are necessary. These function to take up the excess slack in the chain. Excess slack can develop for a number of reasons. Improper initial chain length, chain stretch over time, or a change in some other component can all result in excess chain and thus improper tension. The take-up component usually resides at a 180° turn and consists of a movable sprocket that can be adjusted to increase or decrease the total length of the chain path. The sprocket around which the chain travels is usually attached to a frame that is mounted to supports in such a way as to be adjustable. Sliding along these supports, the frame and sprocket can be brought to the perfect position for optimum chain tension. Then, pins or bolts can be used to secure the take-up sprocket in this position until further adjustment is necessary. Some take-up systems are automatically adjustable, using a calibrated spring or pneumatic system to ensure that the proper chain tension is constantly maintained.

15.4.2.2 Selection Specification The basic steps of the selection process for overhead conveyors are outlined below.

1. Define the requirements
   a. Load dimensions
   b. Transport speed
2. Select the basic conveyor type
3. Specify the track path, considering
   a. Load dimensions
   b. Transport speed

    c. Space available
4. Select the chain
    a. Type
    b. Necessary chain pull
    c. Appropriate pitch
    d. Linking method
5. Select the drive mechanism
    a. Type
    b. Torque and horsepower requirements
    c. Location in system

After defining the requirements of the system, preliminary conveyor selection should take place. Table 15-7 is presented here to aid in this process.

Next, the track path can be determined. This depends on the needs of the system as well as the cost limitations on the system. The longer the track, the larger the motor required and the higher the operating costs. Any vertical changes also add to the chain pull required. The track path should seek to minimize these factors while still connecting the necessary process points and avoiding obstructions by a safe margin.

Continuing, chain selection can be completed. Here, numerous engineering equations come into the picture. Determinations of required chain pull, optimum chain pitch, load spacing, and cost all factor into the selection of chain type and material. Accounting for appropriate overloading here is essential to designing a reliable system. In-depth analysis for chain design can be found with chain manufacturers, but here are a few sources that provide simple methods:

- *Standard Handbook of Chains*, by the American Chain Association
- http://www.renoldcanada.com/nmsruntime/saveasdialog.asp?lID=462&sID=910
- http://www.vavconveyors.com/_pdf/vav/en/chain-calculations.pdf

Finally, selection and sizing of the drive mechanism can be carried out. As described earlier, different drive choices can provide different advantages depending on application. Generally, a caterpillar drive is best, but some applications can rely on sprocket and friction drives. Sizing depends on the chain pull and conveyor speed required. As always, once a necessary power rating is determined, it is advisable to select the next standard motor rating upward to ensure reliable and efficient performance.

15.4.2.3 Reliability and Cost Savings For overhead conveying, the highest cost originates from energy expenditures. Frictional losses are the principal contributor to excess power requirements, so a major way to save money is to reduce these losses. This leads to the question: What causes friction in the overhead conveyor system? The answer is: the wheels. Whether enclosed, I-beam, or power and free, the wheels are where most losses occur. Thus, bearing quality and innovations become the focus of saving costs on energy.

Bearings require lubrication. The quality of lubrication and the seal to prevent contamination determine the quality of the bearing. Some systems rely on lubrication performed continuously by a lubricator mounted along the track path. These devices provide continuous fresh lubrication, helping

TABLE 15-7 Basic Selection of Overhead Conveyor Types and Sizes

Type of Monorail Conveyor	Tube Track	R W-Type Encl. Track	Standard Encl. Track	3-in. I-Beam	4-in. I-Beam	6-in. I-Beam
Track size	$1\frac{5}{8}$ in. dia.	$2\frac{1}{8}$ in. W. $2\frac{11}{16}$ in. H.	$2\frac{13}{16}$ in. W. $2\frac{11}{16}$ in. H.	S3 × 5.7	S4 × 7.7	S6 × 12.5
Min. load spacing (nominal)[a] (in.)	6	6	8	6	8	12
Unit load rating (lb)	30	80	125	200	400	1200
With optional single/double load bars[b] (lb)	60/120	160/320	250 200 @ 30° 175 @45°	400	800	2400
Drive chain pull rating[c] (lb)	600	750	750	1800	3950	6000

Source: [8].

[a]The minimum load spacing is also the nominal chain pitch for each type of system. Loads may be spaced on any increment of this chain pitch, depending on your requirement and any additional clearance needed for your product to maneuver the turns and vertical curves selected. Larger-radius turns require less clearance.

[b]Increase capacity by the use of load bars, but keep in mind that it is often less expensive to go to the larger conveyor, unless they are spaced far apart.

[c]A quick idea on your drive pull requirement may be determined by multiplying the total live load by 0.035 (monoplane systems only). Elevation changes add "lift pull" to your total. Lift pull is calculated by multiplying the total elevation change of inclines only (no declines) by the product weight per foot of conveyor.

to keep friction down in the system. However, lubrication of this type requires that the bearings be open, meaning that they are exposed to the environment. This lends itself to contamination of the bearings by external agents, which will eventually demand that the system be cleaned out. Even with constant lubrication, open systems will build up contaminants over time and eventually become ineffective.

In contrast, newer polymeric lubricants provide a solution to this problem. These lubricants are built into the bearing and then sealed, providing lubrication for the life of the bearing. This process works by embedding oil (lubricant) in the microscopic pores of a porous polymer. The bearing rollers are mounted on the hardened polymer, while the oil is delivered from the micropores over time. In this way, reliable lubrication is provided to the bearing over its lifetime, and no chance for contamination exists. The cost of these bearings is considerably higher than that of standard open roller bearings, but this initial capital investment will allow savings in lubricant cost as well as improved reliability.

Here are a few more basic tips to improving reliability in your conveyor:

- Be sure to operate within chain loading design limits. It is easy to exceed chain loading or to create uneven loading scenarios if you are not cautious. These errors will greatly decrease conveyor life. Also, moving load attachments between loading points on the chain drive periodically will yield more even fatigue and allow you to get the most out of the chain.
- Do not approach or exceed the chain pull rating of the drive. The most vulnerable sections of the system reside just "upstream" of the drive unit, so these should be reinforced and appropriately designed.
- Be sure that a take-up adjustment is carried out periodically. As the chain stretches or conditions change, it is important to maintain the tension in the chain, or severe wear on the chain and track will occur.
- Take every precaution to avoid bearing contamination. If using open bearings, investing in protective covers to prevent errant debris from entering the track environment is wise.

15.4.2.4 Preventive, Predictive, and Proactive Maintenance Approaches Maintenance approaches specifically catered to overhead conveyors are focused primarily on the bearing health and drive systems. As with other maintenance, one of the most common preventive methods is simply periodic inspection of the entire system by trained employees. The following components should be monitored continually, as they are the critical points where fatigue and wear first occur:

- Drive dogs
- Load connectors
- Turns (especially small radius)
- Drive chain bearings

One proactive maintenance implementation is the use of brushes to clean the drive track continually. Steel brushes are built into the drive chain, between loading connections, and serve to clean the contact area of the track. This can help prevent small contaminants on the drive track from reaching the bearings and damaging them.

Monitoring is also a common form of predictive maintenance. Infrared thermography is used to identify damaged bearings and sources of high frictional loss. When mounted at a point in the track path, this monitor can see abnormally hot parts, a sign of a malfunctioning bearing that should be replaced. The same thermography can be applied to the drive system to warn of increasing friction or heat buildup approaching unacceptable levels.

15.4.2.5 Additional Information

- *Reference Guide to Overhead Material Handling Systems*, from the Material Handling Institute
- *Standard Handbook of Plant Engineering*, by Robert C. Rosaler; page 4.42
- *Conveyors: Application, Selection, and Integration*, by Patrick M. McGuire; Chapter 8
- *Trolley Conveyors*, by Sidney Reibel

15.4.3 Roller Conveyors

Just as the name suggests, roller conveyors consist of an arrangement of rollers, cylindrical steel components, upon which the load moves. Roller conveyors find use primarily in bulk applications, as the gaps between rollers make them impractical for small solids and unusable for powders, liquids, slurries, and gases. When unpowered, free roller conveyors rely on gravity or human power to convey the material and in this respect could be viewed as facilitating the transport more than actively conveying. There are powered roller conveyors, where motors drive all or a selection of the rollers to transport the loads. Roller conveyors are used over belts in scenarios where more diverse loading, unloading, and accumulation options are required. A typical level roller conveyor is illustrated in Fig. 15-25.

15.4.3.1 Engineering and Component Construction Considerations The various types of roller conveyors are classified by how they are powered, if at all. These types include:

- *Free roller*: human powered

Figure 15-25 Simple roller conveyors. (From SJF gravity roller surplus.)

- *Gravity roller*: not driven; uses gravity and an incline to move packages down from a height
- *Belt-driven roller*: driven by a belt or belts at the end of, or underneath, the rollers
- *Chain-driven roller*: driven by a chain or chains attached at the ends of the rollers
- *Lineshaft roller*: driven by a rotating shaft to which individuals rollers are belted

All roller conveyors consist of the rollers themselves, the support bed or structure, and if powered, a drive system coupled to the rollers in different ways. Each of these components is explored in greater detail below.

15.4.3.1.1 Rollers The primary and characteristic component of the roller conveyor system is . . . the roller! The rollers contact the load, supporting and transporting it forward. The most basic roller is simply a long steel cylinder mounted on a shaft supported by bearings which allow it to rotate, but not move in a translational manner. Besides simple steel rollers, other surface materials, textures, and structures can be added for particular applications. These include:

Cover materials with desirable properties, such as:

- High coefficient of friction
- Corrosion resistance
- Heat resistance

Surface structures to impart desirable properties, such as:

- Better grip
- Containment of load

Additional functions, such as:

- Applying lubrication to load
- Applying paint
- Adding labeling

Some of these special applications of rolling conveyors are not covered in depth here but should be kept in mind as a way to minimize costs and combine processes if your industry is compatible with such methods.

The shaft or spindle of the rollers is generally a hexagonal cross section, to allow for better locking with support frames. The spindle doesn't turn with the roller. The roller is supported by the shaft on bearings that allow rotation with little friction. These bearings are one of the most crucial aspects of the rollers. They determine roller life and are the source of the most failures and maintenance requirements in this system.

The two main functions of rollers divide them further into two classes. Drive rollers in powered applications are those that receive external torque and transfer motion to the load. Idler rollers, which make up the entirety of free roller systems, are unpowered and must simply be free to rotate with relatively low frictional losses so as to facilitate conveyance of the load. Figure 15-26 illustrates the main differences structurally between the two types of roller, and also indicates some geometric quantities of interest when discussing roller dimensions. Note that idler rollers are free to rotate over a range of speeds and in both directions, and that they can retain angular velocity for quite some time if well lubricated. In contrast, drive rollers will usually rotate in only one direction and at the speed predetermined by motor specifications.

15.4.3.1.2 Bearings Bearings are one of the most important but often overlooked components of the industrialized world. Their main function: to reduce friction. Without bearings, all of rotary mechanics and much of modern technology could not function. The basic idea is to take a scenario in which two surfaces would have to slide past one another and impose rolling motion instead. This greatly reduces friction, limiting wear and extending the life of the materials involved. It is for the same reason that you cannot stand easily on a collection of spilled marbles. Rolling is tremendously easier than sliding.

Many types of bearings are used in industry, but fittingly, the *roller bearing* finds widest application in roller conveying. This is on account of its better radial load-bearing attributes than the more recognizable ball bearing, of which you've probably heard. Roller bearings consist of small cylinders encased between two rings (Fig. 15-27). These cylinders allow the two rings to roll easily around one another, and to transfer radial loads across their entire length. In contrast, ball bearings only transfer the load at one point of contact because they are spherical. It is this that lends them their lesser load-bearing capabilities.

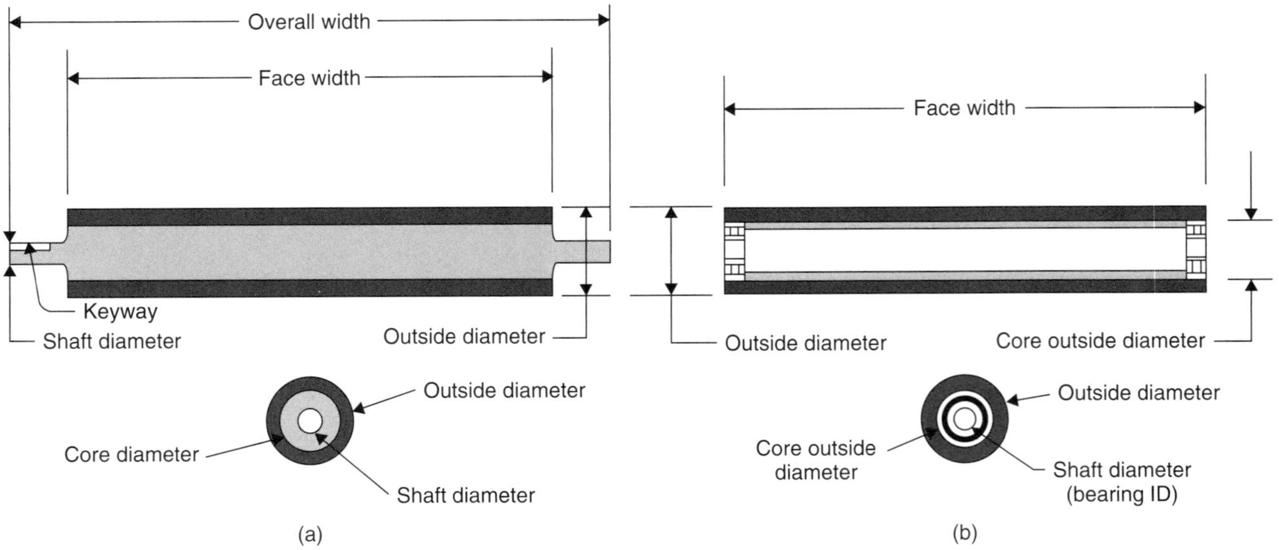

Figure 15-26 (a) Drive roller; (b) standard idler roller. (From Sunray, Inc. polyurethane products' roller specification.)

Figure 15-27 Cylindrical roller bearing. (From [3].)

There are endless variations and specifications for bearing types that have been developed for specific application over the years of industrial growth. You can find bearings for high-temperature and corrosive applications. Thrust bearings provide an axial reaction force in addition to radial support. Self-aligning bearings can tolerate angular displacement of the axis and still function properly. Bearings can be customized to your particular needs. If designing your own system down to this level of selection, you should consult with bearing manufacturers with the knowledge of your system-specific demands.

15.4.3.1.3 Drives Powered roller conveyors use drive systems that provide torque to all or intermittent rollers. The four most used are:

1. Belt-driven
2. Chain-driven
3. Lineshaft-driven
4. Individually driven

Belt drives rely on a continuous loop of material running underneath the rollers to transmit power. Visualize a small conveyor belt with rollers placed above. An example is shown in Fig. 15-28. You can see the small belt on its take-up pulley on the front left of the system. For this system to function reliably, another set of rollers, called *pressure rollers*, must be used beneath the belt to ensure enough contact force between the belt and the transport rollers. Without adequately frequent pressure rollers, the

Figure 15-28 Roller conveyor powered by belt drive. (Courtesy of Atlanta Crane and Automated Handling, Inc.)

belt would simply slide past the transport rollers, failing to supply power.

Next, the chain drive uses a configuration of chains to transmit power. Sprockets are affixed to each roller, and by interlocking with the sprockets, the chain transfers power to the rollers. The chain itself is driven by a motor and sprocket located somewhere along the system. Different systems have different setups for the roller sprockets and chain or chains. The simplest consists of a continuous chain with which the roller sprockets interlock above and below (Fig. 15-29a). Another method consists of using many short chain segments. Power is transmitted directly to the first roller, or to two adjacent rollers if center-mounted, which is then connected via another short chain to the next, and so on down the line (Fig. 15-29b). The paired configuration is probably more common, and generally the motor is centrally mounted so as to distribute more evenly the tension experienced by chains. Chain-driven roller conveyors are implemented in heavier applications, because they are more durable and reliable in providing the necessary torque. Later, we will see why they have an advantage over line shaft–driven systems.

The line shaft drive uses a long rotating shaft that is coupled to each roller using individual belts. Each roller has a groove or grooves near its edges, and small tensioned belts transfer and reorient the torque from the line shaft running perpendicularly underneath. You can see this setup in Fig. 15-30. As you may sense intuitively, this setup does not lend itself to heavy loading but works for medium or light applications where the small belts don't have to be overly tensioned.

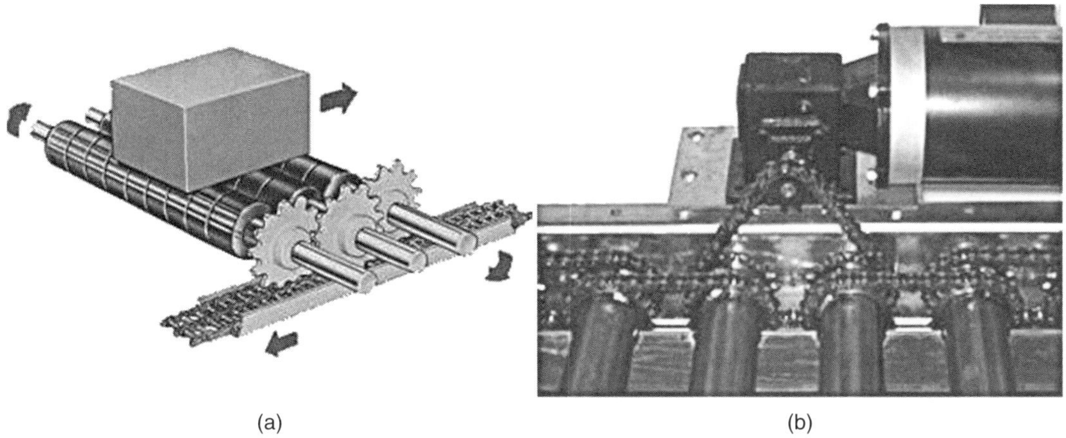

Figure 15-29 Chain drive arrangements: (a) single continuous chain; (b) individual linking chains and central drive unit. (From Shuttleworth Extraordinary Conveyor Solutions and Intersystems' chain conveyor brochure.)

Figure 15-30 Lineshaft drive: (a) single coupling band; (b) aligned rollers with grooves to mate with bands. (Courtesy of Titan Industries, Inc. and AS Conveying Systems.)

Finally, there are also roller conveyor systems in which a collection or all of the rollers are powered internally by individual motors. This system can be made quite powerful for handling the heaviest loads. Individually powered rollers also have the advantage of being more versatile and adaptable in application. Specific rollers can be electronically toggled on and off to create accumulation zones. While all roller conveyors can be considered accumulation capable, the drives described above will continue to rotate rollers under stopped products which could wear away at the bottom of the loads. In this current scenario, a plant manager could deactivate those rollers where accumulation is desired, rendering them a free portion of the conveyor system, and thus not damaging the products. Because of the increased complexity of the rollers and number of motors required, pricing of this individually powered system is increased substantially.

15.4.3.1.4 Turns Turns in the roller conveyor systems present multiple complications. To maintain proper tracking of the loads and prevent jamming, a number of things must be considered. First, the width of the conveyor and radius of curvature must be suited to accommodate the dimensions of the load. Since the outside of the load must move a greater distance during the turn, there must be a speed differential across the width of the rollers. Also, for powered applications, turns present complications for many drive methods, making individually powered rollers the best option.

Turns are first limited by the geometry of the load. It is general practice to leave at least 2 in. of clearance on either side of the load during transport, and for all geometries besides circular, this requires a larger overall width during turns than in straight-aways. A good estimation for the appropriate width during turns is given by the equation

$$G = \sqrt{(\text{radius of turn} + \text{package width})^2 + \left(\frac{\text{package length}}{2}\right)^2} - (\text{radius of turn} - 2\text{ in.}) \quad (15.12)$$

Inputting all parameters in inches will yield the desirable width for rectangular loads. The equation can also be used to get a sufficient width for irregularly shaped loads by using values for width and length that correspond to a rectangle encompassing the dimensions of the particular load. Figure 15-31 shows the turn geometry of Eq. (15.12).

To maintain proper tracking of a load through a turn and to prevent collision with guard rails or loss of product, the outer portions of the turn must convey loads at a greater rate. This is so because at a greater radius, the outer side of the load must travel farther to achieve the same angular change. There are a couple of ways in which this speed differential can be created. The best choice in most cases

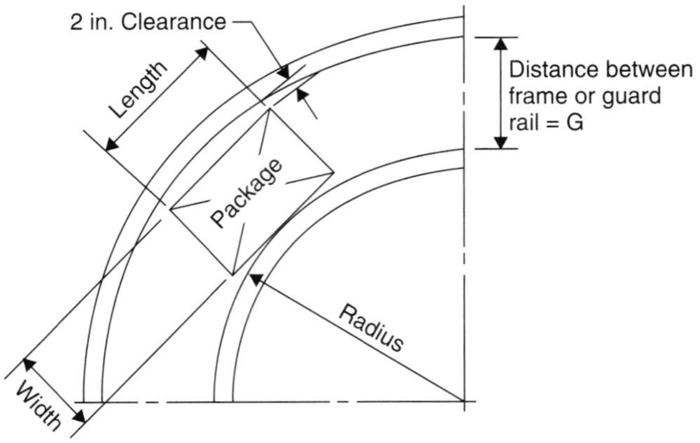

Figure 15-31 Schematic of turn geometry. (From [11].)

is to use rollers with a varying radius across their length. With a gradient from smaller to larger radius as you move from the inside to the outside of the turn, you will achieve a gradient in transport speed. This solution eloquently uses the same relationship that instigates the problem to solve it. Namely, this is the relationship $S = \theta R$ relating angular change to circumferential distance. Another method for obtaining the necessary variation in transport speed is to use multiple rollers per radial spur. By driving these separate rollers at different speeds, the turn can also be completed with little rotation of the load.

15.4.3.1.5 Inclines In the large number of gravity-driven applications, roller conveyors must be inclined in order to achieve transport. Using an inclined section, the weight of the load itself is utilized to overcome the friction in the rollers, allowing motion without external power provided. The force of gravity is used to provide the work to move the load. The main consideration when discussing inclines is the angle. The angle of incline or elevation determines the relationship between the length of the conveyor and the change in height achieved. It must also be great enough such that the intended load weight can overcome the starting friction of the roller bearings. This is the lower limit of elevation angle. The upper limit is usually determined by the nature of the load. If the angle is too great, loads can become unstable and topple. A general limit used in industry is a 22° incline, but another way to determine this limit is to analyze the dimensions of the particular load. To determine a reliable maximum angle of incline, drop a vertical line through the center of gravity of the load while on the incline. If this line falls within the center third of the base of the load, the load can be considered stable at this incline angle. If not, a smaller angle should be selected.

15.4.3.2 Selection Specification The basic steps of the selection process for roller conveyors are outlined below:

1. Identify application
   a. Load characteristics
   b. Conveyance need: Powered? Unpowered? Accumulation?
2. Select type of roller conveyor; consider:
   a. Load aspects and needs
   b. Cost
   c. Adaptability
3. Select and size rollers
   a. Type
   b. Roller selection
   c. Bearings
4. Specify path
5. Select drive (if necessary)
   a. Based on type of load and conveyor
   b. Torque and horsepower requirements

The selection of roller conveyor type depends mostly on the weight of the load. For powered roller conveyors, line-shaft conveyors are the most cost-effective for lighter loads, chain driven can be used for heavier loads, and individually powered are the most adaptable and suitable for very heavy loads. Gravity or unpowered conveyors suffice for simple directing or facilitation of intermediate-weight loads.

Once the type of conveyor is decided, selecting a customized roller can begin. This stage is heavily influenced by load–roller interactions. For example, if a powered roller is commissioned to move large loads up an incline, it is important that the roller surface have sufficient friction with the loads to prevent them from slipping backward. The manner of loading can also have an effect on roller choice. If high-impact loads are expected, specially strengthened axles may be required. Roller providers can offer a wealth of information about the variations of axle, cylinder, and cover combinations, and they will cater the rollers to individual needs.

Track path definition is next. Important things to remember in this step include:

- Minimum turn radius and conveyor width during turns.
- Maximum angle of elevation before slipping occurs.
- Appropriate angle of declination for gravity rollers. Too steep an angle will yield unstable transport. Too shallow an angle and loads will not move.
- Necessary lengths if accumulation is desired.
- Spacing of rollers required to remain under load at all times.
  - This is a general rule to maintain load stability and support.
  - This affects costs compared to overall length.

Finally, the drive motor or individual motors, if necessary, should be sized. The weight of the rollers, loading, and losses due to friction are significant factors in this decision.

15.4.3.3 Reliability, Maintenance, and Process Development The reliability of roller conveyors is largely in the quality and health of the bearings used. Thus, a considerable amount of effort should be placed on proper selection and maintenance of this component. Depending on your application, different types of bearings may be more suitable. Important things to consider in selecting a bearing type are:

- Load-bearing capabilities
  - Radial load capacity
  - Thrust bearings for axial loads
  - Fatigue related to bearing shape
- Alignment
  - If using heavy loading, axle deflection calls for self-aligning bearings to prevent increased bearing wear.
- Lubrication
  - Permanently lubricated and sealed
  - Re-lubrication
- Lifetime
  - Even under ideal conditions, fatigue will eventually lead to failure.
  - Rated lifetime of bearings should be considered and weighed against costs.

Normally, roller conveyors rely on a variant of the roller bearing, which differs from ball bearings in that cylinders are used to transfer the radial load. In most applications, roller conveyor bearings do not need to support much of an axial load. Exceptions to this include high-speed turns or loading areas, where product is sliding onto the conveyor perpendicular to the transport direction. It is advisable to use permanently sealed-for-life bearings contained within the rollers themselves. This prevents concerns related to the contamination of both the bearings and product. A preventive maintenance method that should be adopted is to replace roller bearings before the end of their rated life. Bearings may be physically capable of continuing to work far beyond the calculated fatigue limit, but doing so will lead to increased operating costs and eventual failure. Bearing life is generally provided by the manufacturer and is related to the construction materials, configuration, loading, and lubrication of a particular bearing.

As with any conveyor system, the value of always running within designed limits cannot be overlooked. Attempting to convey oversized or overweight loads will lead to strained drive systems, shorter lived bearings, unacceptable axle deflection, and eventually, premature failure of the system. Thus, to maintain a reliable system, it should be designed with thorough knowledge of intended application, and any adaptation of this application should be evaluated carefully. If certain components are not rated appropriately, changes should be made to accommodate the new scenario.

A more recent development in roller conveyor technology that can improve the reliability and efficiency of energy consumption uses individually powered rollers controlled by centralized software. This configuration allows real-time control of roller speeds to achieve optimum efficiency as well as rapid adaptability of the system. Rollers that are not currently handling product can go unpowered until necessary. Also, if accumulation needs suddenly arise, a portion of the system can be powered down and rendered a normal unpowered straightaway. Software in general is being used increasingly in conveyor industries to streamline processes and allow operators rapid interactions with the entire conveyor system.

15.4.3.4 Additional Information

- *Conveyors: Application, Selection, and Integration*, by Patrick M. McGuire; Chapter 5
- *Standard Handbook of Chains*, by the American Chain Association; Chapter 10
- *Materials Handling Handbook*, by Raymond A. Kulwiec; pages 319–325

15.4.4 Chute Conveyors

The chute conveyor serves the role of connecting components in the materials flow chain. They are simple in the fact that there are no moving parts in the chutes themselves, but if poorly designed or implemented, chutes can be the source of some of the biggest complications. Their primary application is containing and guiding material. They are used in handling most substances besides gases, and rely on gravity to motivate motion.

15.4.4.1 General Considerations In the most general view, a chute is a primarily geometric construction of surfaces that guides and contains materials. They connect different conveyors in the process plant, so it is ideal that they intake and output material or packages at rates corresponding to the conveyors before and after them in the flow chain. They vary in complexity with application. For example, in a mail processing plant, chutes must simply be wide enough to accommodate packages and steep enough to overcome sliding friction. Contrastingly, in bulk solid–handling industries, such as coal processing, many more factors must be considered in designing reliable chutes. Wear, frictional losses, and clogging become a much more significant consideration. In this section we explore some of the basic characterization parameters of chutes, quantities of interest, and considerations encountered in various industries.

When discussing chutes, the geometry is of primary significance. Because there are no moving parts, all phenomena and complications eventually relate back to the geometry in some way. For open chutes, the main geometric parameters of interest are the width, height, length, and angle of declination. If there is a turn in the chute, the radius of this turn is important to consider to prevent package jamming and flow backup. In closed chutes, the cross section, which encompasses the previously mentioned dimensions, is one of the most important factors. The curvature of the chute, which can be related to the instantaneous angle of declination, also affects flow rates and clogging.

15.4.4.1.1 Open Chutes Open chutes are generally found in simpler applications with larger solids (not powders or particulate solids). They guide material imprecisely, as the goal is only to convey the individual loads to the next conveyor. Major concerns in regard to open chutes include:

- The coefficient of friction between the chute material and load base
- Angle of declination necessary to overcome this sliding friction without tumbling loads
- Width necessary to prevent jamming, especially in turn sections
- Exit speed of loads from chute

These are the main design quantities to consider when building an open chute conveyor for large solid handling. They are similar to those encountered in inclined-gravity roller conveyors, except that here sliding friction must be overcome, which is generally higher than the friction in the rollers.

15.4.4.1.2 Closed Chutes Closed chutes find more frequent use in bulk solids handling such as mining and grain processing. The responsibilities attributed to closed chutes are:

- To guide material onto a conveyor belt, at the speed of the belt, in the direction in which the belt is traveling
- To eliminate spillage
- To enclose material dribbles
- To enclose material from operating personnel
- To eliminate dust liberation

Chute wear and clogging are also more pertinent to consider in the case of closed chutes. In large solid applications, loads usually slide onto the chute and the

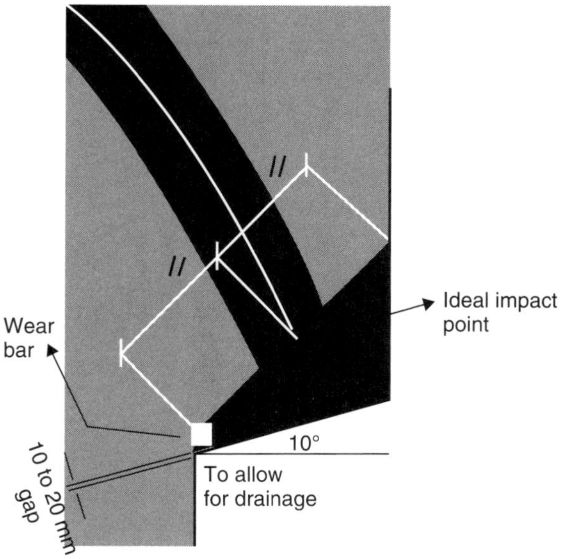

Figure 15-32 Simple dead box arrangement. (From [16].)

declination angle is relatively small, whereas in particulate handling, the material usually falls a distance into the chute head and travels nearly vertically for much of the chute length. This leads to significant impact wear [25].

One method to reduce this wear significantly is through implementation of dead boxes. A simple dead box is shown in Fig. 15-32. The basic idea of a dead box is to shape the chute head such that when material begins to flow in, some of it accumulates at the impact point. Then, further flow will fall on this builtup material, which absorbs some of the impact, lessening the total wear endured by the chute. Dead boxes work well for large particulates, but when dealing with smaller particulates, finer powders will accumulate preferentially and lead to clogs.

Another important quantity when discussing chutes is the mass flow rate, usually labeled Q. Using continuum mechanics provides a flow rate of

$$Q = \rho VA \tag{15.13}$$

with ρ the bulk density (kg/m^3), V the velocity (m/s), and A the cross-sectional area (m^2). This yields Q in kg/s. It is generally desirable that the mass flow rate through all portions of the chute be constant in normal operation, so that a predictable input and output of material occur. If this is not the case, accumulation or evacuation is occurring intermittently in regions of the chute, which makes the design more prone to clogging and complications.

The angle of repose is also of importance when designing the geometry of chutes handling particulates. It is the angle to which the bulk material can be piled before material layers begin to shear off and slide down the pile. The angle of repose is a function of particle size, density, coefficient of friction, and moisture content. It must be considered when determining the declination angle of a chute. If the angle is not great enough or the angle of repose is too great, material will simply collect instead of sliding down the chute.

15.4.4.2 Additional Information

- *Bulk Materials Handling by Conveyor Belt*, by Allen Reicks; page 91
- *Conveyors: Application, Selection, and Integration*, by Patrick M. McGuire; Chapter 9

15.4.5 Screw Conveyors

Probably one of the oldest forms of conveyor, the screw conveyor has been used for centuries to lift water from rivers and lakes to be used for irrigation. Most people have heard of Archimedes' screw, and even though its origin may be attributed to him inaccurately, this simple hand-powered invention remains a viable transport method for fluids and flowing particulate collections. The screw conveyor of today is mechanized and used in many industries. Some examples include the grain, mining, and sewage industries. Screw conveyors can be implemented horizontally, on an incline, or vertically. The simplicity of the design leads to predictable behavior and reliable service in most applications. In the following sections we outline the major components of screw conveyors, discuss the selection process to determine the appropriate parameters for an application, and provide maintenance and cost-savings information.

15.4.5.1 Engineering and Component Construction Considerations The basic screw conveyor consists of several components:

- The screw itself
- The shaft
- The shaft supports
- The trough and covers
- The drive mechanism

Each of these is described in this section, and relevant details associated with each component are defined and discussed.

15.4.5.1.1 Screw The characteristic component of the screw conveyor, the screw itself, is responsible for converting rotary motion into translational motion in the material conveyed. Also called the *flight*, it consists of an inclined plane wrapped into a helix. Some important characterizing quantities include the diameter and the pitch. The pitch of the screw is the distance between adjacent planes of the flight and is a characterization of the steepness of the inclined plane. Pitch values are usually given in terms of diameters. Some examples are illustrated in Fig. 15-33. All of the illustrations show single-flight screw designs. Multiple interwoven flights are also used in which two or more flights of the same pitch are placed coaxially. In this way, increased frequency is achieved but the steepness of each inclined plane is maintained.

The possible types of screw design are nearly endless. Some more "exotic" designs include variable pitch, ribbon, or cut and folded screws (Fig. 15-34). Each design possesses its own advantages and has a specific effect on the material being transferred. Screws also have a specific hand. The hand of a conveyor indicates how the inclined plane is wrapped. To determine the hand of a screw, examine the near-side flight. If it goes down and to the right, it is right-handed; if it goes down and to the left, it is left-handed. When coupled with a rotation direction, the hand of the screw determines the conveyance direction. If looking down the shaft of the screw from the end rotating clockwise, a right-handed screw will be bringing material toward you, while a left-handed screw will be moving it way.

15.4.5.1.2 Shaft Not all screw conveyors have a central shaft, but those requiring any significant amount of torque will utilize one. The diameter and composition of this shaft are dependent primarily on the torque requirements. Other factors that can affect shaft choice are associated with the material conveyed. Very hard or caustic materials may require special selections to be made for shaft composition as well as that of the screw flight.

Short screw conveyors usually have a single shaft for the entire length, but when the length of the conveyor becomes too great, it becomes impractical to create such long screw segments, and multiple segments are coupled together. These couplings must be of the same load ratings to transfer torque appropriately.

15.4.5.1.3 Shaft Supports All screw conveyors have shaft supports positioned at least at the ends of the shaft, and many have additional supports along the length of the conveyor. All of these supports rely on bearings to provide radial support to the shaft while facilitating the rotary motion necessary to convey material. In most cases, these bearings must be sealed. By the nature of the screw conveyor, fluid-like flows pass directly around these supports, so an unsealed bearing would contaminate the conveyed material as well as the bearing itself, neither of which is desirable. The support on the input end of the conveyor must also be designed to provide and axial thrust reaction to offset the force exerted on the screw by the flowing material.

15.4.5.1.4 Trough, Tube, and Covers The screw itself would have trouble conveying anything uncontained a very far distance. This containment is the responsibility of the trough or tube and covers. All have the purposes of:

- Keeping the material within the confines of the screw action
- Isolating the conveyed material
- Keeping personnel protected from the screw
- Supporting the entire screw system

A trough is a U-shaped metallic walled cylinder. This design is utilized mostly in horizontal or slightly inclined

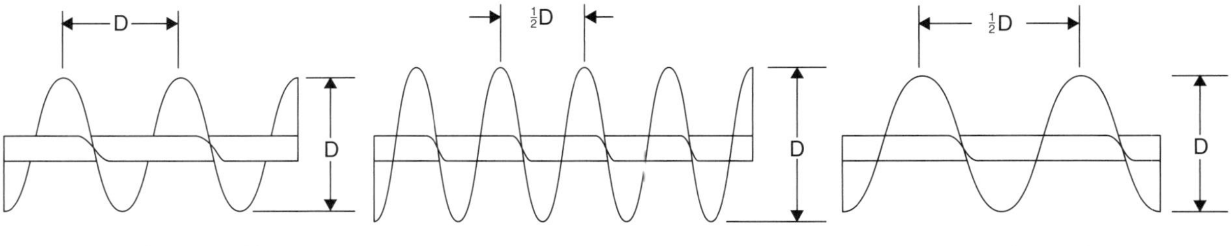

Figure 15-33 Common screw pitches, given in terms relative to the diameter. (From the KWS, *Design, Engineering, Manufacturing Screw Conveyor Engineering Guide*.)

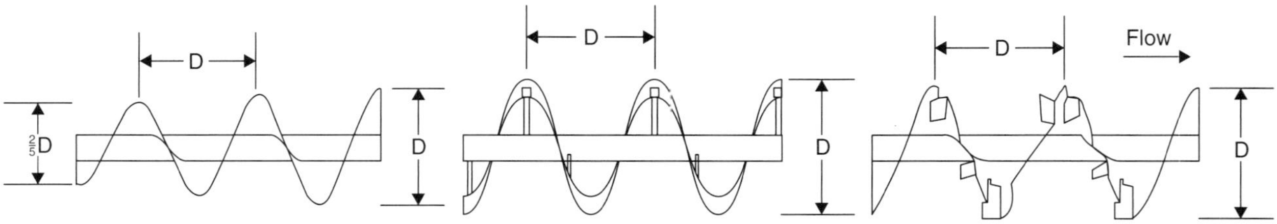

Figure 15-34 Some exotic screw designs: (a) variable; (b) ribbon; (c) cut and folded. (From the KWS, *Design, Engineering, Manufacturing Screw Conveyor Engineering Guide*.)

applications, where the base of the U is the bottom of the conveyor. The top of the trough is sealed using covers. Usually, some or all of these covers are removable to facilitate maintenance on screw sections. In many cases, shaft supports will hang from more permanent cover pieces. A tube design is used for vertical applications or when handling fluids, where most of the cross section of the conveyor is to be occupied by material. This design has a circular cross section and may have removable cover sections to aid in maintenance. The clearance between the moving screw conveyor and the containment vessel varies with application, but a general clearance is $\frac{1}{2}$ in.

15.4.5.1.5 Drive Mechanism Like most conveyors, the drive mechanism consists of an electric motor that runs power through a speed reducer (torque magnifier) and then into the drive shaft. The differences in drive selection boil down to different arrangements of these three components. Shaft-mounted drives place these components in line with the shaft, while some combination motor–reducer drives can be placed more flexibly and power can be transferred to the drive shaft through a chain-and-sprocket arrangement. Horsepower and torque requirements are the main concerns when selecting the drive.

15.4.5.2 Selection Specification The selection of screw conveyor system components is outlined below.

1. Define the material to be conveyed.
   a. Maximum particle size
   b. Bulk density
   c. Chemical reactivity
   d. Temperature
2. Define the capacity and distance conveyed.
3. Using the requirements outlined above, size and select the conveyor.
   a. Diameter
   b. Screw pitch
   c. Screw materials (considering temperature, abrasiveness, and corrosivity)
4. Calculate the horsepower required.
5. Select the motor and drive arrangement.

Complete definition of the material(s) that a screw conveyor will be transporting is necessary for informed selection. As stated above, this should include the key quantities of maximum particle size, bulk density, temperature, corrosivity, and hardness. Maximum particle size has a significant effect on internal friction as well as bulk density. These factors decide the necessary horsepower. Bulk density is necessary for converting the intended throughput (lb/h) to a volumetric capacity, which is what screw conveyors are rated for. Temperature, corrosivity, and hardness should be understood so that proper material selection can take place. These factors will dictate whether or not special alloys or coatings must be used for the screw and trough walls.

The capacity is a parameter decided by the whole plant process in which the screw conveyor is a component. Answering the following questions will help dial in this parameter: How much material can be provided at the input? And what are the intake capabilities of the receiving equipment? Conveyor loading is also closely related to capacity. It is defined as the percentage of the bore that actually conveys material, or how fully the conveyor is loaded during normal operation. Seldom is the entire bore filled with material. Common values for conveyor loading are 15%, 30%, 45%, and in special cases, 95%. The appropriate conveyor loading depends largely on the material being conveyed. For materials possessing high internal friction, conveyor loading is usually 15% to help prevent clogging and ease horsepower requirements. More easily conveyed bulk solids or liquids can be conveyed in higher percentages, even up to 95% in some cases.

The next step in the process is component selection and sizing. The screw diameter and pitch should be chosen to meet capacity requirements with respect to conveyor loading. Screw style and construction are largely dependent on material properties conveyed. Special alloys may be required to handle high-temperature or corrosive materials. For materials that tend to pack or interlock, special screw designs that continually break apart large clumps become necessary. Consulting with screw conveyor manufacturers will lead to the best choice of screw design and composition for your application. Other components that must be selected at this stage include the inlet and discharge chutes, drive shaft (if used), shaft seals, and very important, bearings. The choice of bearing is also related to material properties and the sensitivity of your application. If contamination of the product is an issue, sealed bearings are suggested. Also, unlike in roller conveyor applications, end bearings in screw conveyors must withstand large axial loads. All of the force put into conveying the material acts through these end thrust bearings, so significant consideration and sometimes overselection of these bearings is necessary. Normal hanger bearings are generally only radial supports, but one should be sure they have adequate clearance from shaft segments to avoid unintentional axial loading.

Once the material loading, component sizing, and construction material selection is complete, horsepower and torque requirements can be calculated. With this knowledge, proper motor selection can be performed. The type of drive configuration used depends mostly on space limitations and end conditions of the system. There is no significant advantage of any particular method, but each method calls for slightly different construction of the end of the conveyor to provide proper support to the motor assembly and connections. Considering the intended location and available space will help with this decision.

For more information on selection procedures, consult these resources:

- *Mechanical Conveyors: Selection and Operation,* by M. E. Fayed and Thomas S. Skocir
- *Bulk Solids Handling: Equipment Selection and Operation,* edited by Don McGlinchey http://www.kwsmfg.com/engineering-guide/

15.4.5.3 Reliability, Maintenance, and Process Development Proper monitoring and maintenance are the best ways to ensure the reliability of your screw conveyor system. When handling bulk solids, conditions within the conveyor can lead to increased wear, and component damage can occur quickly if not monitored. Things that should be monitored continuously or checked periodically include:

Continuous monitoring suggested:

- Conveyor loading percentage
- Internal temperature
- Internal moisture content
- Abnormal vibration in motor or drive shaft

Periodic maintenance checks:

- Bearing seals and health
- Screw health (especially for handling abrasive, corrosive, or high-temperature materials)
- Conveyor seals at ends and any removable cover locations

Keeping track of the foregoing aspects of the screw conveyor will help ensure early detection of impending failure and allow for preventive action to be taken.

Screw conveyors are a fairly well-established technology, but one area that is seeing some development is in the department of smaller, flexible screw conveyors. These handle lighter loading scenarios and offer flexible solutions (pun intended) in environments where conveyance demands can change frequently. One of the leading producers of this breed of screw conveyors, the Flexicon Corporation, provides a wide variety of light, flexible conveyor systems which are surprisingly robust. The screw is relatively flexible, shaftless, and requires no support bearings along the lengths of operation. For more information on these conveyor systems, visit http://www.flexicon.com.

15.4.5.4 Additional Information

- *Screw Conveyors*, by the Screw Conveyor Engineering Committee of the Conveyor Equipment Manufacturers Association
- *Screw Conveyor 101: Basic Training Manual for Screw Conveyors*, by Michael P. Forcade
- *Bulk Solids Handling: An Introduction to the Practice and Technology*, by C. R. Woodcock

15.4.6 Other Conveyor Types

Many other types of conveyors are used in industry. In this section we provide a short description of many of these designs. Many of the maintenance and reliability concepts discussed for the preceding conveyance methods can be adapted and applied to the following conveyors, as most systems rely on bearings, rotating shafts, and motors to transfer power to the load [12].

15.4.6.1 Skatewheel Conveyors Similar to a roller conveyor, skatewheel conveyors consist of many small wheels mounted on a horizontal grid. The transfer concept is the same as for roller conveyors, but there is much less contact area with loads. Skatewheel conveyors are predominantly unpowered and used to transport boxes or other flat-bottomed loads. They can be gravity conveyors when mounted on inclines, but more often they are used merely to facilitate human-powered movement of packages. Figure 15-35 shows a sketch of the skatewheel conveyor configuration.

15.4.6.2 Bucket Conveyors As the name implies, this conveyor relies on buckets to move material in a vertical or inclined path. The buckets are mounted as illustrated in Fig. 15-36. The materials are loaded in at the bottom, drawn up inside the upright bucket, and then discharged

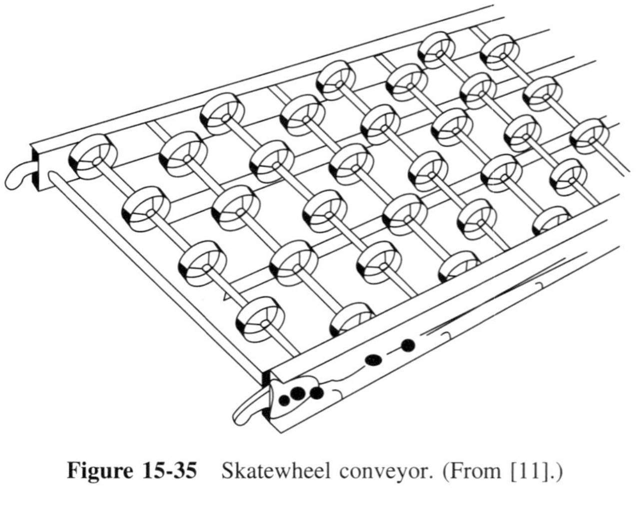

Figure 15-35 Skatewheel conveyor. (From [11].)

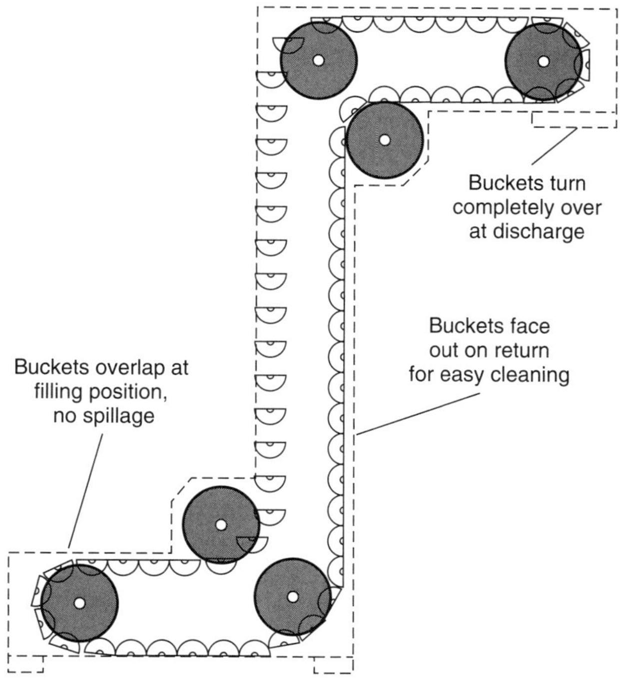

Figure 15-36 Simple bucket conveyor diagram. (From http://accessscience.proxy.mpcc.edu/content.aspx?id=266000.)

as the buckets flip over the top of the drive system. The empty buckets then make their way down the opposite side of the conveyor until they loop back at the bottom to begin another circuit. During this return segment, any cleaning or treatment of the buckets can be performed. The scale and complexity of bucket conveyors varies, but they find application mostly in bulk solids–handling operations.

15.4.6.3 Vibrating or Oscillating Conveyors These conveyors consist of a vibrating or oscillating trough mounted at a slight decline. The small loads are conveyed mostly by gravity but are spurred on by the mechanical vibration of the trough. This overcomes static friction, allowing the loads to move slowly down the incline. Industries relying on this form of conveyor include mining, gemstone processing, manufacturing, pharmaceuticals, and food processing. Generally, vibrating conveyors are used for smaller particulate loads, such as pills or diamonds, while oscillating conveyors have a larger amplitude, lower-frequency motion, and handle slightly larger objects, such as hot metal castings. This form of conveyor is generally used only over a short distance and is helpful in separating materials into a single layer for further conveyance on a belt.

15.4.6.4 Pneumatic Conveyors Pneumatic conveyors rely on air or gas pressure to move materials through a network of tubes or pipes. A simple form of this conveyor with which you may be familiar is implemented at many drive-through banks. The tube system that transports canisters between your car and indoor tellers is a pneumatic conveyor. For bulk solids and industry uses, the idea is about the same, only on a larger scale. Some industry applications include grain handling, fine powders, pharmaceuticals, and food processing.

15.4.6.5 Cart-on-Track Conveyors This conveyor is designed to transport discrete loads. A mine cart can be viewed as one manifestation of the cart-on-track method. The difference between different forms of cart-on-track conveyors is the manner in which they are powered. Many are human powered, some have motors mounted on each cart, and others rely on a lineshaft concept similar to that used in roller conveyors. This is illustrated in Fig. 15-37. The central rotating shaft is coupled to each cart, motivating it along the supporting tracks. Usually, flat cart conveyors are used for large package handling, and bucket carts can be used for bulk solid handling.

15.4.6.6 Vertical (Elevator) Conveyors Vertical conveyors are implemented when very heavy loads need to be moved through a height. Usually, they are run intermittently, only requiring use sporadically. If more regular vertical transfer is required, other conveyance methods such as belts or overhead conveyors are recommended. Vertical conveyors are basically freight elevators, but are not constructed with considerations toward carrying people. Unless specially designed, you should not use this conveyance device for moving people.

15.4.6.7 Additional Information

- *Food Plant Engineering Systems*, by Theunis C. Robberts
- *Conveyors: Application, Selection, and Integration*, by Patrick M. McGuire
- *Mechanical Conveyors: Selection and Operation*, by Muhammad E. Fayed

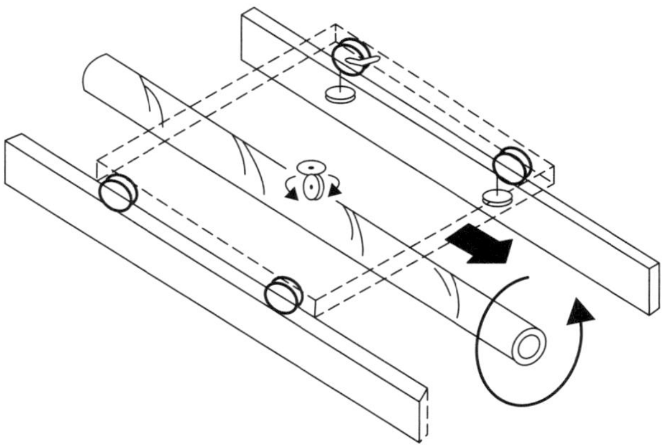

Figure 15-37 Simple cart-on-track conveyor; lineshaft-driven. (From [6].)

- *Maintenance Fundamentals*, by R. Keith Mobley
- *Encyclopedia of Chemical Processing and Design*, by John J. McKetta

15.5 PUMPS

Until this point, most of the conveyance methods discussed have been for handling solids. In these scenarios, the conveyor must physically contact and transport all or most of the load. Contrastingly, because molecules in a fluid flow freely past one another, pressure gradients can be used to transport the substance. In this way, pumps can be used to give energy to the substance at one point, and the substance itself can be used to transfer this energy and convey the material through pipes. This is the goal in any pump design; provide a pressure head that can convey the material and achieve the necessary output pressure.

In this section we first discuss the concept of "head" as it relates to pressure. The result of fluid passing through a pump is an increase in head, and it is important to understand why this quantity and its components are significant to fluid flow systems, of which pumps are a key component. Then we outline the various types of pumps implemented in industry. After a general understanding of the operation of these particular pump types is attained, selection of the appropriate type for your application is discussed. Continuing, reliability, and maintenance concerns are addressed, and finally, new developments in pumping are described.

15.5.1 Head and Pressure: Fluid Flow Systems

15.5.1.1 Head You are probably familiar with the idea of pressure and its forms. *Head* is an alternative way to express pressure in a fluid, and it is more convenient to work with in fluid flow systems because it is directly analogous to energy. To show this, let's begin with a general form of energy conservation, first for a lossless system:

$$K + U = \text{constant} \tag{15.14}$$

This states that the total energy, kinetic (K) plus potential (U), is constant throughout the system. In fluid flow systems, in a certain volume V, this becomes

$$\tfrac{1}{2}mv^2 + mgh + PV = \text{constant} \tag{15.15}$$

where m is the mass of our volume, v the velocity, h the height above some reference value, and P the static pressure. The first term is the kinetic energy associated with motion. The second is the gravitational potential energy from the fluid above some arbitrary reference level. The last term is the internal energy associated with the fluid pressure. Now, in industry processes, liquids are generally assumed to be completely incompressible, and gases at low speeds can be approximated as incompressible as well if at constant temperature. This allows us to divide Eq (15.14) by the mass of our volume and yields

$$\frac{1}{2}v^2 + gh + \frac{p}{\rho} = \text{constant} \tag{15.16}$$

where ρ is the density, which is constant. The constant on the right-hand side has changed, but it is still a constant because of our incompressible view. Finally, the extent of fluid flow systems are generally not over a height where the gravitational constant changes appreciably, so it can be taken as a constant: 9.81 m/s^2 or 32.2 ft/s^2. Dividing by this brings us to the expression we seek:

$$\frac{1}{2}\frac{v^2}{g} + h + \frac{P}{\rho g} = \text{constant} \tag{15.17}$$

You can verify that all the quantities in this expression have units of length. These are what we label "head, in feet (meters) of liquid" in a fluid flow system. The first term, called the *velocity head*, is a measure of kinetic energy. The second, the *elevation head*, is an expression of gravitational potential energy. Finally, the third term is called the *pressure head*, associated with the static pressure of the fluid. Extracting each of these expressions yields

$$h_v = \frac{1}{2}\frac{v^2}{g} \qquad h_e = h \text{ (above some reference height)}$$

$$h_p = \frac{P}{\rho g} \tag{15.18}$$

These represents the three forms that energy can take in our lossless system. Conservation of energy states that the energy can transition between these forms, but the total must remain as our *constant*. For example, you may be familiar with the fact that a flowing fluid has lower static pressure than a nonflowing fluid would under similar circumstances. This is a result of conservation of energy. Being under *similar circumstances* can be interpreted as possessing the same total energy. In the static case, all of this energy is in the form of pressure, but when the fluid is also flowing, some of the total energy is in the velocity of the fluid, so less is in the form of pressure.

Now, we know that real fluid systems are not lossless, so a term accounting for this must be added. This is labeled as the *friction head* and is representative of the energy lost as heat due to friction that cannot be reclaimed by the system. Once energy is converted into frictional loss, it can no longer be converted back to any of the other forms of hydraulic head. The friction head is a complicated quantity that must be calculated for pipes, valves, and flanges in the system, but it is generally proportional to the velocity head, so higher-speed flows suffer larger frictional losses.

Now we can express all of the forms of energy in a fluid system as head in feet or meters of liquid. The purpose of a pump is to inject energy into our system, and this energy is also reported as head. It can be pictured as the maximum height to which water can be pumped. In other words, the pump injects energy in the form of velocity and pressure head. The pump rating is the value of head corresponding to the total conversion of these into elevation head [2].

15.5.1.2 Gauge vs. Absolute Pressure One point that must be handled with caution is the specification and consistent use of gauge versus absolute pressure. These terms are used with regard to static pressure. *Absolute pressure* is the relative pressure above pure vacuum. It always has a nonnegative value. Values of pressure reported for barometric pressure in weather forecasting are in absolute pressure. Meanwhile, *gauge pressure* is a value of pressure relative to some reference value. The reference used for most systems is atmospheric pressure, but it can be anything, as long as it is clearly specified. In this way, gauge pressure can be both positive and negative, depending on whether the absolute pressure is above or below the reference pressure. When dealing with equations and formulas using pressure values, it is important that a consistent use of gauge or absolute pressure is maintained. In closed piping systems encountered in conveyance, gauge pressure is most commonly used with atmospheric pressure as a reference. When handling values of pressure head, it is important to remember the quantities associated with each value. The relative pressure from which each is calculated needs to be consistent [2].

15.5.2 Pump Construction and Operation

Industry pumps generally fall into the category of either dynamic or positive displacement. In dynamic pumps, kinetic energy is imparted to the fluid by moving mechanisms such as an impeller or propeller. This kinetic energy is then converted to pressure. Dynamic pumps provide a continuous flow rate based on pump speed and the specific fluid resistance. In positive-displacement pumps, discrete volumes of the fluid are delivered through each pump cycle, resulting in an intervallic fluid delivery. Examples of each class of pump are given in Table 15-8.

In this section we explore the differences in construction and operation of the major pump types implemented in industry. The importance of key fluid properties such as viscosity, pH, specific gravity, and phase transition points will also be noted in the context of each pump type. Then, important quantities in the context of all pumps are discussed.

TABLE 15-8 Types of Dynamic and Positive Displacement Pumps

Dynamic Pumps	Positive-Displacement Pumps
Centrifugal (impeller)	Reciprocating
Single suction	Plunger
Double suction	Piston
Multistage	Diaphragm
Axial (propeller)	Rotary
	Screw
	Gear
	Regenerative (peripheral)
	Peristaltic

15.5.2.1 Centrifugal Pumps Centrifugal pumps fall into the category of dynamic pumps. Their basic components and operations are illustrated in Fig. 15-38. Fluid comes in from the suction pipe and flows into the impeller eye, or center. The impeller consists of a rotating arrangement of vanes which accelerates the fluid as it passes from the inner eye to the exterior of the impeller. The fluid leaves the impeller and flows into a region called the *volute*, where it is directed toward the discharge pipe. In the volute, the area of the flow increases, resulting in a decrease in velocity and an increase in static pressure. The fluid then leaves the pump through the discharge pipe.

Centrifugal pumps are some of the most widely implemented across industry. They are relatively simple and can be scaled for appropriate needs effectively. They can handle high-volume flow rates well, and the small number of

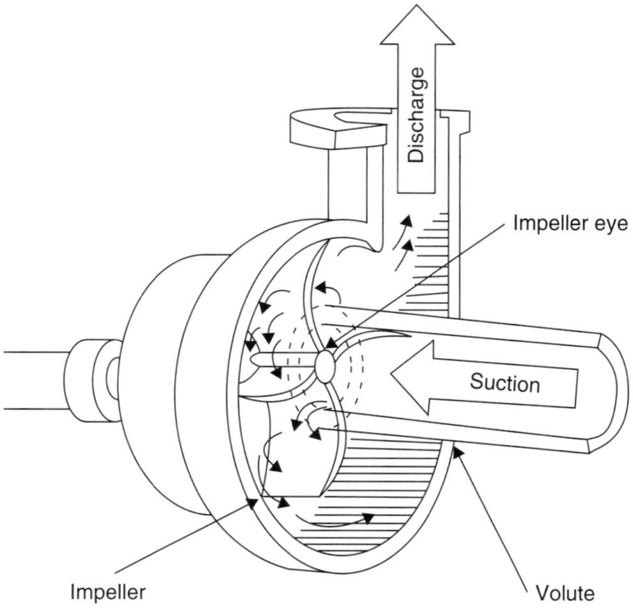

Figure 15-38 Centrifugal pump. (From [22].)

moving parts makes maintenance requirements tolerable. However, they do not handle particularly viscous fluids very well.

Some variations on the simple design illustrated in Fig. 15-38 also find application in industry. Double-suction impeller pumps have suction pipes feeding both sides of the impeller eye. The benefit of this arrangement is that it balances the reaction thrust on the drive shaft. Double volute arrangements have a separate volute region covering 180° of the impeller discharge. This will reduce the radial load experienced by the drive shaft due to uneven ejection of fluid from the impeller vanes. These arrangements are useful in high lead scenarios and increase the life of the drive shaft and bearings.

For applications where a very large increase in head is necessary, it becomes inefficient simply to keep increasing impeller speed and size. Instead, a multistage design is used. This consists of multiple impellers mounted on a common axis, each possessing its own casing and volute components. The fluid flows into the first impeller, experiences the normal increase in head, and is discharged. However, it is then fed directly into the next impeller, further increasing the head. This process is repeated until the fluid has passed through every stage and the desired total increase in head is achieved. Multistage centrifugal pump designs are expensive and heavy duty. The drive shaft must be made to support very high loads for many staged designs unless separate drives are used. This is the trade-off for attaining such large gains in head.

15.5.2.2 Axial Pumps Axial pumps are also dynamic pumps that impart to the fluid an increased velocity, which is then converted to static pressure. As you can see in Fig. 15-39, an axial pump is basically a propeller mounted inside a pipe, thus lending the alternative name *propeller pump*. These pumps are generally used in low-head, high-discharge applications and are fit to handle lighter liquids of reasonable viscosity.

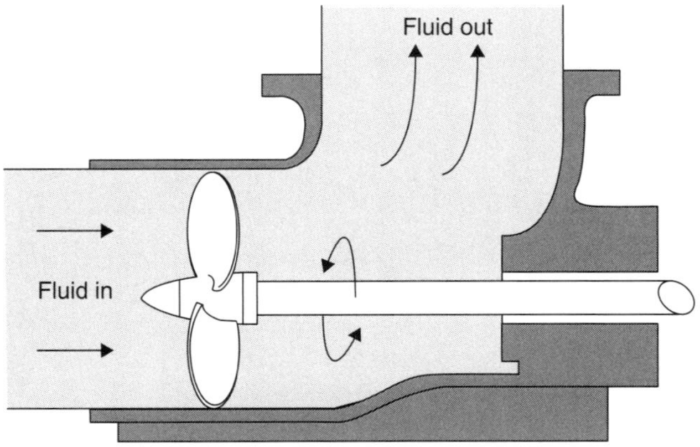

Figure 15-39 Axial (propeller) pump. (From *Encyclopedia Britannica*, 1996.)

15.5.2.3 Reciprocating Pumps This is the first positive-displacement pump to be discussed. Reciprocating pumps consist of a plunger, piston, or diaphragm that reciprocates, or moves back and forth along a predefined path. Because flow is incremental, reciprocating pumps must rely on valves that limit the direction of flow. Referring to Fig. 15-40, which shows the discharge stroke of a generic single-piston pump, you can see the primary components of all reciprocating pumps.

The reciprocating element, in this case a piston, is driven back and forth by either a cam or a crankshaft. Check valves at the inlet and outlet prevent backflow. So, on the intake stroke, the piston is drawn out (left in the figure), creating a suction that opens the inlet check valve and allows fluid to flow into the piston chamber. Then the discharge stroke begins and the piston is driven in (right). This change of pressure closes the inlet check valve and opens the outlet check valve, forcing the fluid out and providing a large pressure head. This basic operation is the underlying action of all reciprocating pumps. The major difference in the various types of reciprocating pumps is in the type of reciprocating element and how the back-and-forth motion of the said element is attained.

Reciprocating pumps are well suited for applications involving high head and low capacity. Because they are usually operated at very high pressures, relief valves must be included for safety. Reciprocating pumps can handle viscous liquids, but are generally not suitable for slurries.

15.5.2.4 Rotary Pumps Rotary positive-displacement pumps include a wide variety of pumps that have a rotation mechanism as the primary drive component of the pump. Some of the more significant rotary designs include the gear, peristaltic, flexible impeller, and regenerative pumps. Each of these configurations has its own advantages, as discussed in this section.

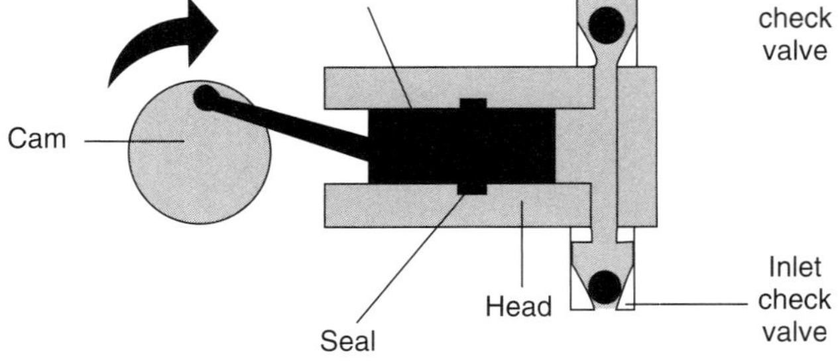

Figure 15-40 Reciprocating piston pump. (From LC Resources, *Pump Glossary*.)

First: the gear pump. As the name suggests, this pump uses a toothed gear or gears rotating in a tight-fit casing. The discrete volumes of fluid are moved between the teeth of the gear(s). Shown in Fig. 15-41 is an example of an external double-gear configuration. The fluid is drawn into the gear spaces and transferred around the exterior of the casing. Very small clearances and tight interlocking of parts are necessary for efficient operation. Thus, gear pumps must be machined and installed carefully. Gear pumps can be made of special components for handling most fluids, including corrosive and high temperature, but like most positive-displacement pumps should not be used for slurries.

Other notable aspects of the gear pump include:

- Near continuous flow rate for many-toothed gears
- No frictional losses due to valves as in reciprocating pumps
- Ability to handle viscous fluids with minimal energy loss and heat generation
- Self-priming, or self-starting

Lobe pumps work on basically the same principle, but use specially shaped lobes that interact to create a more efficient suction and expulsion action as the twin rotors rotate. They too are self-priming and find substantial use in industry.

The peristaltic pump (Fig. 15-42) is probably the most unique design discussed to this point. It consists of a section of flexible tubing running next to a rotor with a number of pegs. As the rotor turns, these pegs depress the tubing, locking in a fixed amount of fluid, and then push this fluid through to the discharge side. The vacuum created behind the pinched point serves to draw in more fluid until the next peg in the sequence locks in the fluid. They are labeled peristaltic because this pumping action is analogous to the action of peristalsis in the human digestive system.

Figure 15-41 External gear pump. (From [17].)

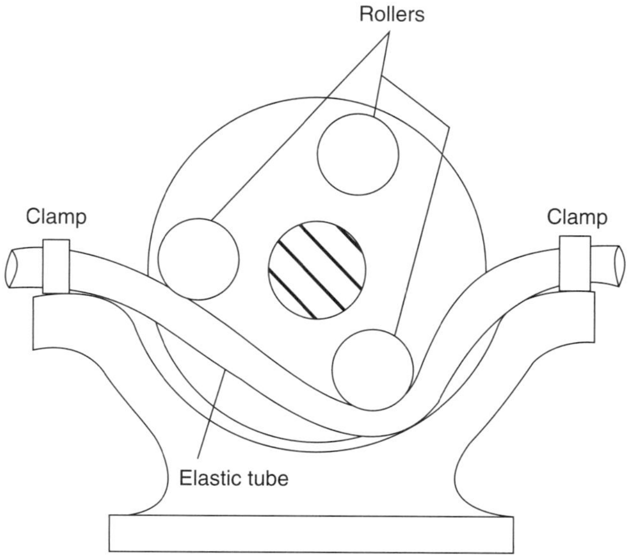

Figure 15-42 Peristaltic pump. (From [17].)

Because the actual pumping mechanism never physically contacts the fluid, peristaltic pumps are ideal for sensitive applications where material isolation is key. They can also convey very viscous and dense slurries. However, the capacity of peristaltic pumps is generally much less than that of other pump designs.

The flexible impeller pump (Fig. 15-43) is similar to the gear pump, but instead of rigid gears, the vanes of the impeller are made of flexible material. The casing around this impeller is shaped eccentrically. As the impeller rotates, constant contact is maintained with the casing, but along the intake and discharge side of the pump, the casing is built even closer to the impeller center. As the impeller brings a volume of water around to the discharge point, the flexible vane is depressed by this closer casing wall, forcing the fluid out and delivering it out of the pump. Figure 15-43 provides a clearer understanding.

These types of pumps are very versatile and can handle viscous liquids and slurries with high solid fractions. They are self-priming and very quiet running. However, because of the constant flexing of the impeller component, the lifetime is decreased. Flexible impeller pumps are well suited for intermittent use in a wide variety of applications—a good spare pump to have on hand for a wide variety of needs.

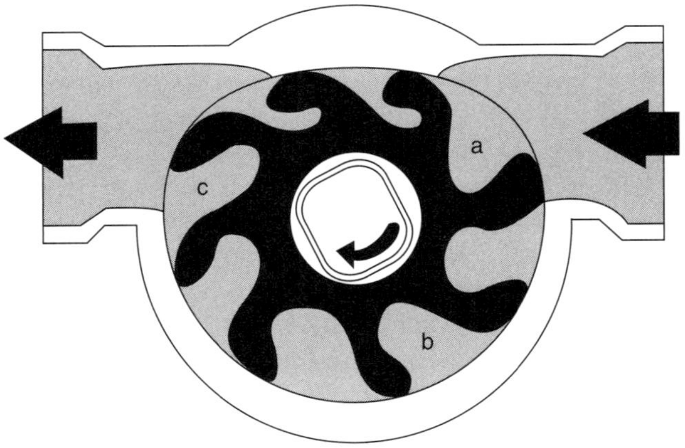

Figure 15-43 Figure 15.43 Flexible impeller pump. (From Depco Pump, *Pump Selection Guide*.)

Finally: regenerative or peripheral pumps (Fig. 15-44). This design consists of a many-bladed rotor in a close-fitting casing. As fluid flows through, it is repeatedly expelled from the rotor, losing velocity and gaining head. This process occurs many times over the course of fluid transfer through the pump, gradually increasing the head until discharge. This gradual method has the primary benefit of limiting pressure drops at intake, resulting in a reduced chance of cavitation. Cavitation is the formation of gas pockets in the fluid, and it can greatly reduce the efficiency of the pump and damage internal pump components. Also, these types of pumps generate very high fluid head and should not be implemented without safety relief valves.

15.5.3 Selection by Application

The selection process for pumps is outlined below.

1. Specify the system requirements.
   a. Flow rate
   b. Intake pressure at pump
   c. Output pressure of system
2. Specify the material properties.
   a. pH or corrosivity
   b. Temperature
   c. Density
   d. Viscosity
   e. Particulate concentration
   f. Vapor pressure
3. Select the pump type best suited to fluid properties and application.
   a. Refer to previous sections.
4. Size the pump based on the calculations of head losses and output requirements.

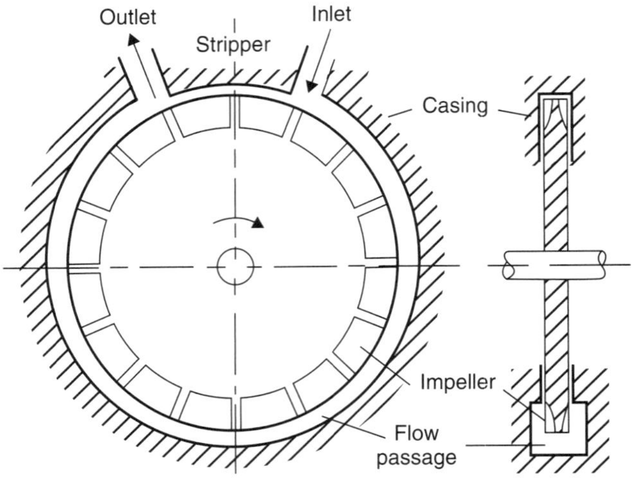

Figure 15-44 Regenerative (peripheral) pump. (From *Rapid Prototyping Journal*, 1995.)

The specification of the system is the first step in selecting your pump. This includes the desired flow rate, the available pressure at the pump intake, the end pressure requirements at process points, and the size of the system. The nature of the system is also important: Is the pump solely responsible for providing the required head of the system or part of a network? Are there special considerations that must be accounted for, such as variable-speed pumps or intake pressure conditions that will require a certain class of pump? These types of questions should be answered at this stage in the selection process.

Material properties must also be carefully understood at this point. All of the parameters listed above can have a significant impact on the choice of pump type and construction materials used. The pH or corrosive nature of the material will quickly degrade pumps that are improperly designed or not rated for the material being handled. The temperature of the fluid will also dictate material choices and affect viscosity considerations. Viscosity is crucial, especially if using a centrifugal pump. If your material has a very high value of viscosity, it may be impractical to use this type of pump, and a positive-displacement pump will be necessary. The same situation may arise when handling fluids with large particulate concentrations. Vapor pressure is an important quantity because it will determine the necessary suction pressure required by the pump. If this pressure is not attained at the intake, vaporization can occur and lead to cavitation, damaging the pump and decreasing performance.

Once the system parameters are established, you can proceed by selecting the pump type most appropriate to your application. This process usually involves some give and take between the demands of your system and fluid limitations. In Section 15.5.2 we provide general guidelines for normal applications of each type of pump.

The first decision should generally be between centrifugal and positive displacement. This is largely dependent on material properties and the desired flow rate. Easy-flowing fluids can be handled well by centrifugal pumps, and more troublesome fluids with high viscosity and particulate concentration are better handled by positive displacement. Once this is done, investigation of each pump's variations can be performed. Table 15-9 can be used as a rough guide to the limitations and attributes of some common pump types. Consulting with pump manufacturers is the best way to determine which pump style is optimal for your particular case.

Finally, sizing of the pump should be performed. Depending on the type of pump, the sizing of the pump will relate differently to the output head. These relationships may also rely on fluid properties. For example, a centrifugal pump's discharge head is nearly independent of fluid density but is very sensitive to viscosity. The available suction head and required discharge head are calculated values that come from knowledge of the rest of the piping system. At this stage, these attributes should be specified and allow for appropriate pump selection. Note that the flow rate has an impact on the head requirements through its effect on velocity head and friction head. Working with a set flow rate is the easiest way to go about the calculations, but any changes in flow rate require a recalculation of all dependent quantities. This should not be overlooked.

15.5.4 Reliability, Maintenance, and Process Development

Improving the reliability of your pump and fluid flow system focuses on two areas: the mechanical moving parts of the pump and the interaction between the pump surfaces and pumped fluid. These are the areas of the pump with the most potential for wear and failure. Mechanical parts of significant concern are:

- Bearings
- Seals
- Drive shafts
- Metal-to-metal interfaces

The importance of the quality of bearings used in rotary pumps cannot be overstated. Centrifugal, gear, lobe, and even reciprocating pumps rely on rotating shafts to transfer power. Sometimes, bearings are exposed to the pumped fluid. In these cases, it may be possible to use the fluid itself as lubrication, but if contamination is an issue, proper seals must be implemented. Advancements in isolation

TABLE 15-9 Pump Selection Guide

Pump Type	Max. Flow Ranges (gpm)	Max. Pressure (psi)	Self-priming	Pulseless Flow	Fluid Viscosity	Particulate Matter	Run Dry	Advantages
Centrifugal	2.3–1200	Up to 275	Poor	Excellent	Light	No	No	Fluid transfer at high flow rates and low pressures
Diaphragm	0.003–5.2	Up to 300	Good	Poor	Medium	No	Yes	High accuracy; for applications such as pH/ORP control
Double- diaphragm	1.0–4.0	Up to 95	Excellent	Fair	Medium	Yes	Yes	Use for viscous or particulate-laden fluids
Flexible impeller	3.8–50.0	Up to 60	Excellent	Excellent	Light	No	No	Low-cost utility pump
Gear	0.006–74.0	Up to 1500	Poor	Excellent	Medium	No	No	Pulseless flow at high pressures
Peristaltic	0.00002–1.43	Up to 22	Excellent	Fair	Heavy	Yes	Yes	Noncontaminating; high accuracy; available in a wide variety of tubing materials
Piston	0.004–107	Up to 5000	Good	Poor	Medium	No	Yes	Highest pressure and accuracy; ideal for HPLC applications
Progressing cavity	0.5–13	Up to 100	Fair	Excellent	Very heavy	Yes	No	Pulseless flow for highly viscous or particulate-laden fluids
Rotary vane	0.75–4.3	Up to 240	Fair	Very good	Light	No	No	High-pressure capabilities; low shear
Submersible	3.0–180	Up to 50	Poor	Excellent	Light	No	No	Use for emptying tanks or sumps

Source: Adapted from a Cole–Parmer selection table.

bearings continue to improve the ability of bearings to keep lubricants in and fluids out, improving bearing life and preventing contamination of product fluids.

Seals in pumps separate the motor portion of the drive from the actual fluid environment. Traditionally, seals are achieved by face-to-face contact by metal. This leads to significant wear and heat and calls for increased maintenance and lubrication. Recently, however, the development of "dry-running, noncontacting gas" seals has allowed for significant improvement in this aspect of pump reliability. Pressurized gas is used between two noncontacting surfaces to achieve the seal, and the setup requires no lubrication. This reduces frictional losses and increases the time between maintenance.

Selection of a pump constructed with the proper material for handling the nature of your fluid is also crucial to reliability. Another option is to apply coatings to pumps that need extra help in dealing with particularly abrasive or corrosive materials. Coatings can be replaced repeatedly as they degrade, prolonging the life of the pump.

Coatings can also be implemented in pumps and piping to reduce friction. This lowers the head requirements on the pump and helps save money in operating costs. As with most conveyance equipment, energy use constitutes the bulk of their overall cost. Reducing friction also reduces excess heat in the system and strain on the pump, prolonging its life. These coatings provide a very quick return on investment and are suggested in any case that is suffering from high energy costs due to low-efficiency pumping.

Periodic or continuous monitoring of aspects of the pump will also ensure greater reliability and provide early warning of impending problems. Things to check up on include:

- Bearing health
- Seal health
- Leaking
- Pump body integrity (look for cracks or deformations)
- Internal wear of moving parts and surfaces

Things to monitor continuously include:

- Temperature in the pump
- Suction pressure
- Discharge pressure
- Abnormal vibrations
- Internal noises

This knowledge will help you stay informed about the expected behavior of the pump and dictate when maintenance is needed. Improvements in monitoring technology and increased computer integration are ongoing trends in pumping. Some pump manufacturers even provide a monitoring service and can constantly observe important quantities via the Internet. This provides crucial information to the engineers who designed the pump specifically, allowing for well-informed evaluation and action to keep the pump running effectively.

Particularly for centrifugal pumps, implementing a variable-speed drive can yield significant savings on energy costs. In systems requiring variations in flow speed, throttling valves can be used, but they waste a lot of the energy generated by the pump. If a variable-speed drive is used instead, the flow speed required can be achieved by the pump alone, removing the wasteful use of a throttling valve. It is advisable to make the added investment in a more costly variable-speed drive if frequent changes in flow speed are required. Along with more widespread implementation of variable-speed pumps, intelligent control software is being integrated into pumping systems. The software can accept and evaluate data from throughout the system and calculate the best efficiency point for driving the pump. With real-time adjustment of pump speed to maintain this best efficiency, significant savings in energy expenditure are achieved.

Advances in materials technology have also been applied to pumping. They allow for longer-lasting impeller designs and pump casings that can withstand high temperature, abrasives, and corrosive fluids much more adequately than in the past. Increased availability of more advanced materials such as titanium has also allowed entire systems to be constructed out of once-costly materials, enhancing the overall reliability and performance of pumps.

15.5.5 Additional Information

- *Pumps for Chemical Processing*, by J. T. McGuire
- *Centrifugal Pumps: Design and Application*, by Val S. Lobanoff
- *Handbook of Pumps and Pumping*, by Brian Nesbitt
- *Pumps: Theory, Design, and Applications*, by G. K. Sahu

15.6 VALVES

Valves are a crucial component in any flow system. They allow for flow toggling, flow rate control, and enhanced safety. From steam generators and high-pressure gas mixing to cryogenic systems and sewage handling, valves serve as vital components. There are many types of valves, which differ in configuration and function. In this section we explore valves through their main functions in industry flow systems. Different valve structures will be introduced in this context, illustrating how the most common valves operate. Although mechanically valves are relatively simple, they are one of the most important aspects of the flow system, and proper selection and implementation are crucial.

15.6.1 Valve Construction and Function

So, what are the primary functions of valves? There are four major categories:

1. *Toggle valve*: determines whether or not fluid flows past the valve. "On" and "off" settings.
2. *Throttling valve*: controls the flow rate past the valve. Adjustable.
3. *Check valve*: allows flow in only one direction. Prevents backflow.
4. *Relief/safety valve*: releases excess pressure before damage to system occurs.

Any valve can be classified as having one or a combination of these functions. Let's expand on each.

15.6.1.1 Toggle Valves The toggle valve is perhaps the simplest and most used industry valve. It can be thought of as an on–off switch that either allows full flow or no flow at all. An analogy to an electrical circuit would be a switch that breaks the flow of current when open and allows the flow of current with little to no resistance when closed. Continuing the comparison, a well-designed toggle valve should also possess the quality that it hinders flow completely when closed while allowing nearly unaffected flow when open. This is an idealization, as there is always some added pressure drop when a valve is inserted into a pipe section due to uneven surfaces and the structures associated with the valve operation. However, it should always be the goal of a toggle valve to minimize this effect when in the open position. There are many styles of toggle valves. Some are illustrated in Figs. 15-45–15-48.

1. *Ball valve*: uses a spherical bored metal insert that can be rotated 90° to toggle the flow on and off. If properly sized, this type of valve will cause only a small pressure drop. The fact that a quarter turn is all it takes to toggle the flow is another advantage of the ball valve.

2. *Plug valve*: similar in operation and effective application to the ball valve, but implementing a plug-shaped insert [23].

3. *Gate valve*: as the name suggests, the valve uses a sliding gate to restrict flow. It causes one of the smallest drops in pressure for a toggle valve. One disadvantage: It takes many turns of the handle to achieve full toggle [23].

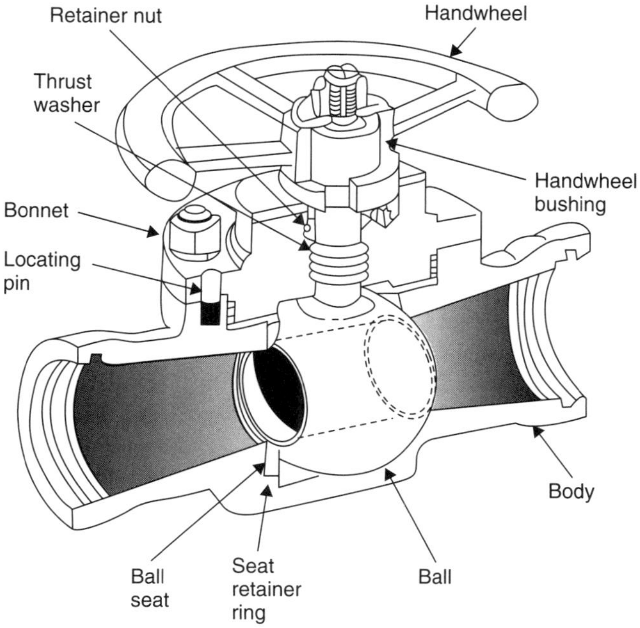

Figure 15-45 Ball valve.

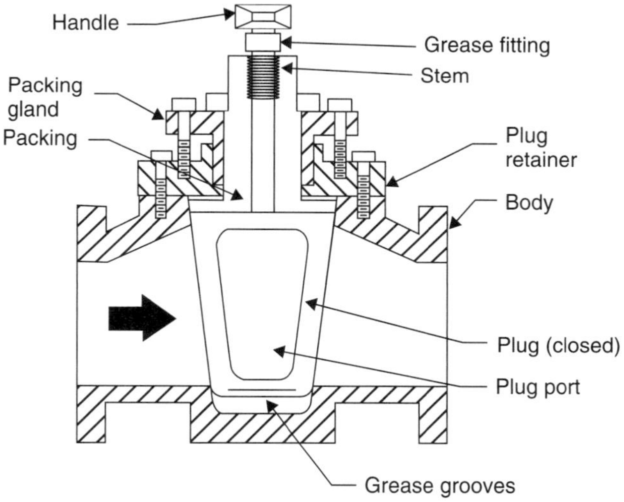

Figure 15-46 Plug valve.

4. *Knife valve*: a specialized gate valve that relies on a very thin "knife-like" gate to restrict flow. Works well with slurries and viscous flows.

15.6.1.2 Throttling Valves Throttling valves can serve the on–off function of toggle valves, but they provide much more control of the fluid intermediately, between 0 and 100% flow. Some of the valve constructions illustrated above can also serve as moderate throttling valves, but those shown below are much better suited to the task. They

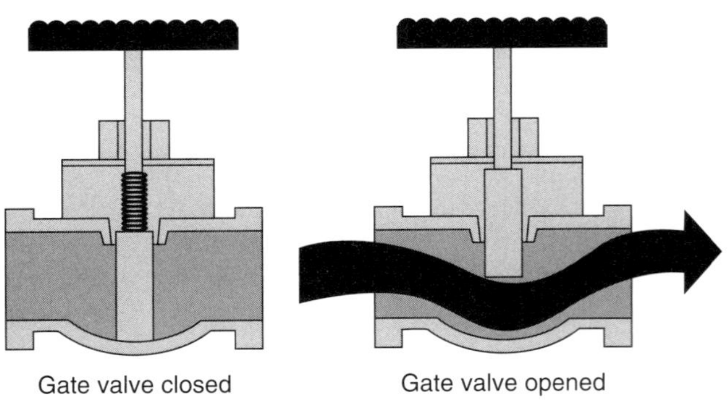

Figure 15-47 Gate valve.

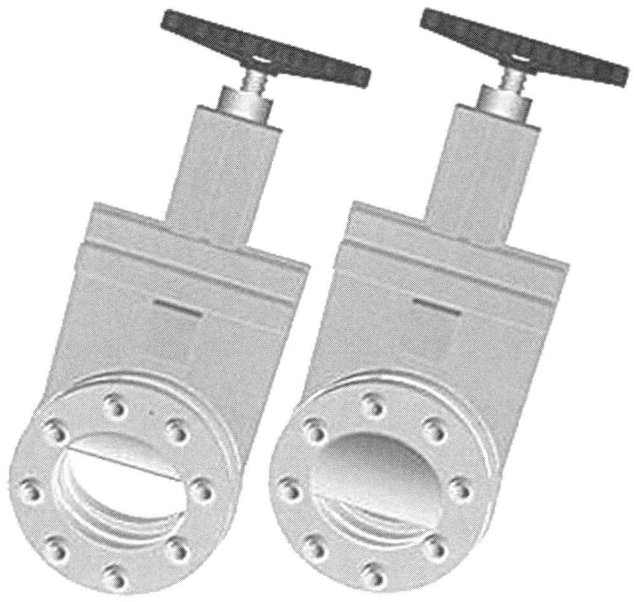

Figure 15-48 Knife valve. (Courtesy of Gentec Systems Corporation.)

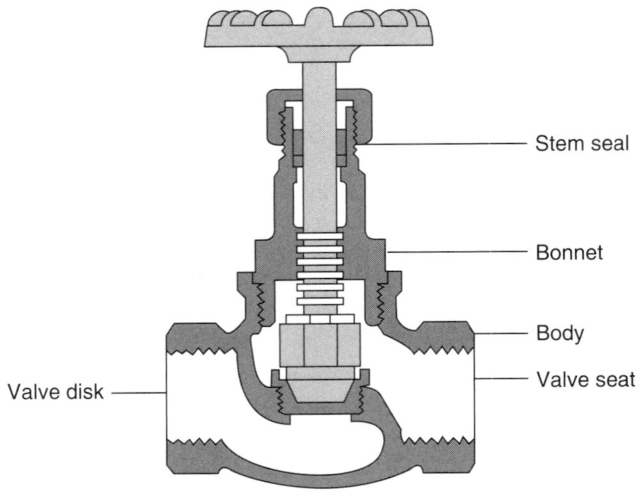

Figure 15-49 Globe valve.

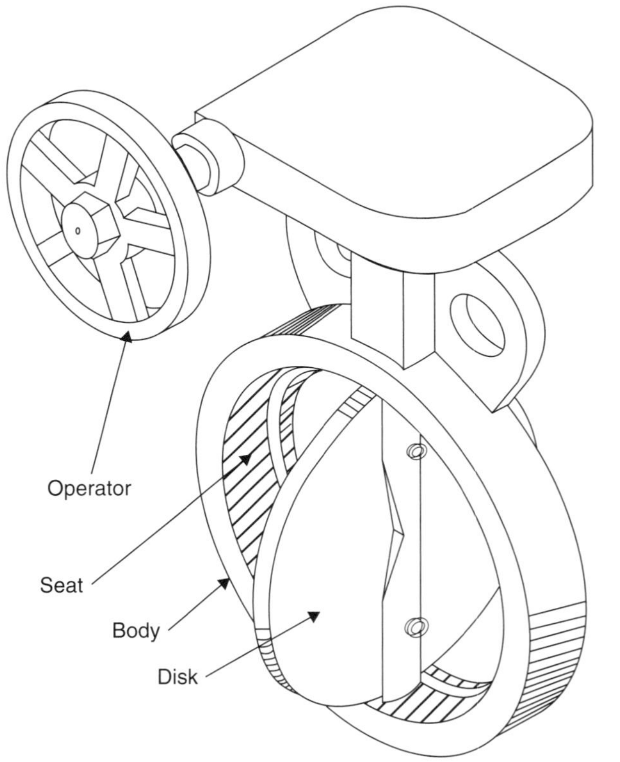

Figure 15-50 Butterfly valve.

provide a much more predictable and gradual variation in the flow over the range of operation from 0 to 100%. The trade-off is that most throttling valves have a larger baseline pressure drop than toggling valves. Figure 15-49 illustrates and describes the globe valve construction, and Fig. 15-50 does so for the butterfly valve.

1. *Globe valve*: uses an insert that can be adjusted in and out of an opening in an S-curve pipe. Small adjustments in the height of the valve disk can be used to control fluid flow precisely. The S-curve shape is responsible for the larger comparative pressure drop in fully open operation [20].

2. *Butterfly valve*: consists of a thin disk placed across the pipe flow. In the fully closed position, the circular cross section blocks pipe flow. Ninety degrees takes the valve to fully open, where only the thin side profile obstructs flow. Butterfly valves are suited well to handling viscous or particulate flows. One disadvantage is the nonlinear relationship between valve position and effect on fluid flow. However, this can be accounted for relatively easily [20].

One way in which valves are compared is through the use of flow coefficients. These appear in engineering equations for determining the changes in flow characteristics across a valve. Graphs like the one in Fig. 15-51

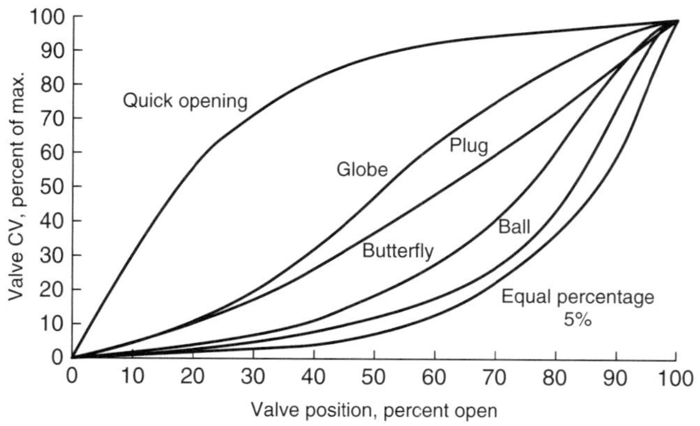

Figure 15-51 Graph of flow coefficient C_v for common valves; normalized. (From [9].)

can be used to compare how these coefficients change over the operational range of throttling valves. The figure shows the relative shape of these curves between many types of valves. Note that the vertical axis is normalized to the maximum C_v, which is a flow coefficient, so the graph should only be used to compare the manner in which valves transition from closed to open.

15.6.1.3 Check Valves Continuing, we come to the check valve. The purpose of check valves is to allow flow in only one direction, preventing backflow. It is generally the goal to allow relatively unobstructed flow when operating in the forward capacity and inhibit flow completely when reverse operation is attempted. In the electrical circuit parallel, the check valve is akin to a diode. Figures 15-52–15-55 illustrate and describe common check valve constructions.

1. *Swing check valve*: uses a door-like apparatus mounted on a spring-loaded hinge. In forward flow, fluid pressure and velocity overpower the spring, swinging open the door and allowing nearly unimpeded flow. If flow slows down or tries to reverse, the spring closes the door, and

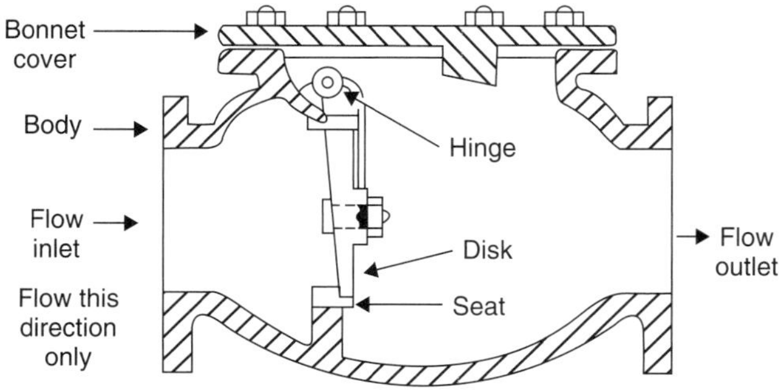

Figure 15-52 Swing check valve.

eventually, backward pressure helps maintain the seal. Turbulence and pressure drop associated with this construction are quite low [23].

2. *Tilting disk check valve in forward flow* (shown on the left in Fig. 15-53): the airfoil-shaped disk straightens out and allows flow to pass above and below. When flow weakens or reverses, the balance of the disk is disrupted and it falls into the closed position shown on the right [23].

3. *Lift check valve*: similar in construction to the aforementioned globe valve. Relies on the fluid pressure to overcome the weight of the plug. If flow reverses, the weight of the plug and backward pressure seals the valve. Because of its similar construction, the lift check valve is usually implemented in conjunction with globe control valves. This value is well suited for high-speed flows of gases and liquids [23].

4. *Butterfly check valve*: consists of two "wing" plates mounted on a hinge. The image shows the closed position. When forward flow impacts the plates, they close together, allowing relatively free flow. Because of the similarity to butterfly control valves, they are often used in conjunction. Also, these valves can be scaled efficiently to considerable sizes [23].

It is also of importance to note that check valves form an integral part of many pump mechanisms. Particularly, reciprocating positive-displacement pumps rely on check valves at the intake and discharge points of the pump chamber. They allow the positive and negative pressure to draw and expel fluid in the desired direction in this style of pump.

15.6.1.4 Relief Valves The final pump classification is the relief valve. Also called the safety valve, its purpose is to ensure that system pressure doesn't reach dangerously high levels. They are designed to have a threshold pressure, above which the valve opens to relieve excess pressure. Usually, this pressure is determined by adjustable components in the valve, but only within a design range. The threshold pressure, called the *maximum allowable working pressure*, should be selected according to the maximum internal pressure tolerated by the weakest components of the system.

The primary construction of relief valves can be seen in Fig. 15-56. The valve disk is held on the seat by the force of a compressed spring. When the internal fluid pressure exceeds this spring force, the disk moves upward, allowing fluid to flow out of the system and relieving the pressure. An adjustable screw can be used to vary the compression of the spring, which in turn controls the critical pressure at which the valve will open.

An alternative to the relief valve is a rupture disk. Although technically not a valve, these components also

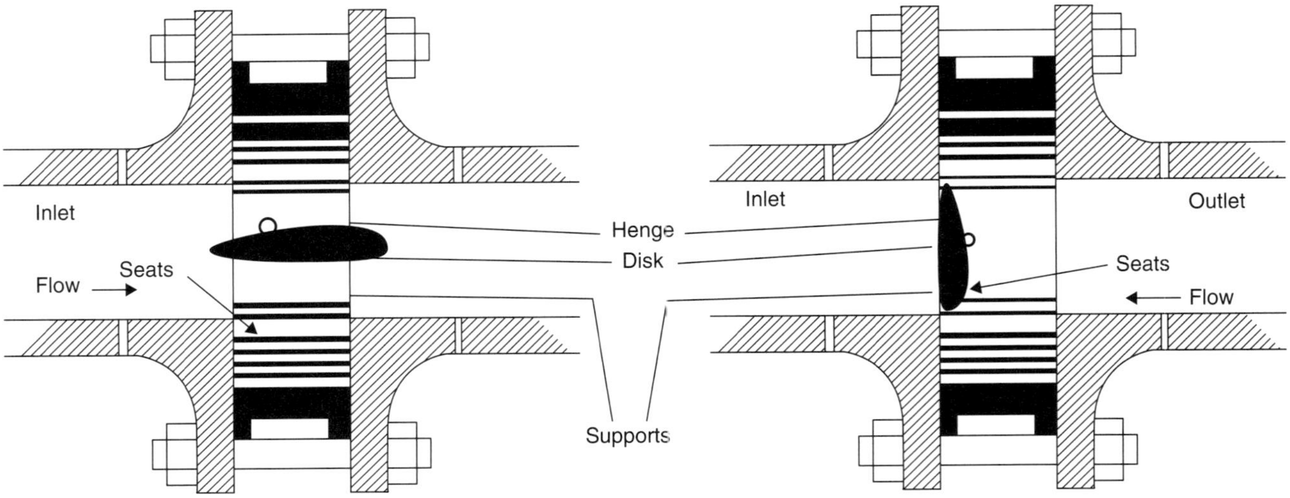

Figure 15-53 Tilting disk check valve.

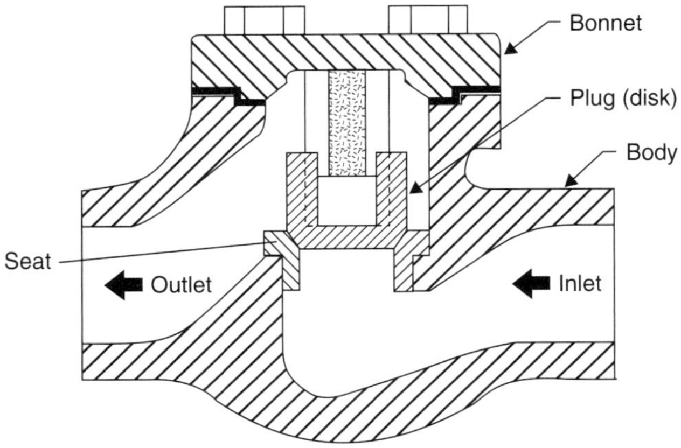

Figure 15-54 Lift check valve.

provide safety relief if the system pressure exceeds limits. The rupture disk is designed with material and geometry such that a known failure stress is determined. If fluid pressure causes such failure, the disk ruptures and allows fluid to flow out of the system. Rupture disks are cheaper than relief valves but are not adjustable and will not reclose after the pressure returns to safe levels. Thus, they must be replaced after each use.

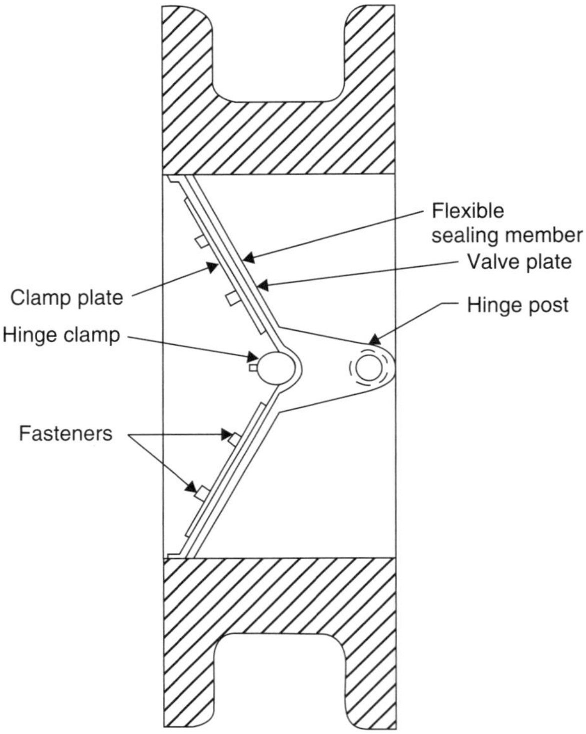

Figure 15-55 Butterfly check valve.

15.6.2 Selection Specification

In selecting and sizing a valve for implementation in a fluid system, the following steps should be followed:

1. Define the purpose of the valve and operating conditions.
   a. *Valve function*: toggle, throttle, check, or relief
   b. *Fluid properties*: Viscosity? Particulates? Corrosivity?
   c. *System conditions*: working pressure, flow rate, allowable pressure drop
2. Select the valve type, Based on the system requirements and properties of the fluid in question, select the most appropriate valve construction for your application.
3. Size the valve, depending on:
   a. Pipe sizing

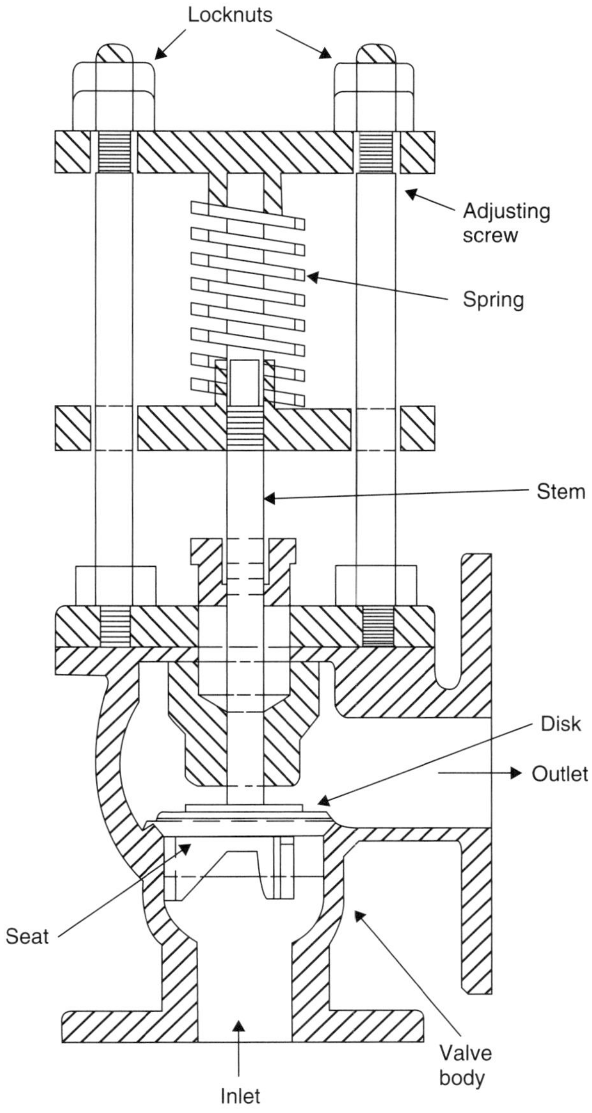

Figure 15-56 Relief valve diagram. (From [23].)

b. Valve type selection
c. Allowable pressure drop
d. Fluid nature
e. Working pressure

The first step is the simplest. Designate what the function of the valve needs to be in your system. Do you need quick toggling of flow, good intermediate control, or unidirectional assurance? Is the purpose to relieve excess pressure? This should be obvious, as it is clear why you are selecting a valve at all.

Defining fluid properties should also be relatively easy. If you are selecting a valve for a defined system, knowledge of the flows expected should be available. Important properties in valve selection include phase, viscosity, corrosivity, and whether or not there are solid particulates.

System conditions are determined largely by the pump being used to power the fluid system. The pump and piping configuration will determine the range of flow rate and working pressure. The allowable pressure drop across the valve is generally taken to be 10 to 15% of the total pressure drop in that section of the system. It should be noted that greater pressure drops result in more strain on the pump, while lesser pressure drops make for a more costly valve design.

Selecting the valve type is one of the most significant parts of the process. As discussed above, different valve constructions are better suited for different applications. Tables 15-10–15-13 are provided to help you choose the correct construction based on your application.

Sizing of the valve is the most time-intensive phase. In this step, calculation and comparison of various flow coefficients are performed and numerous graphs and charts are used for reference. These charts and graphs are provided by valve manufacturers and can be used to optimize the performance of the valve for your specific application. Included in Table 15-13 are some broad sizing ranges and construction-specific valve attributes.

15.6.3 Additional Information

- *Valve Selection Handbook*, 5th ed., by Peter Smith
- *Handbook of Valves and Actuators*, by Brian Nesbitt
- *A Quick Guide to Pressure Relief Valves*, by Clifford Matthews
- *Control Valve Selection and Sizing*, by Les Driskell
- *Chemical Engineering Design: Principles, Practice and Economics of Plant and Process Design*, by Gavin P. Towler; Section 5.3

15.7 PIPES

The main bulk of fluid flow systems consists of pipes. Pipes connect all of the active components of the system, linking the pump through valves to the point where the fluid is discharged from the system. Pipes are responsible for:

- Maintaining diffusive, thermal, and sometimes electrical isolation
- Maintaining the fluid pressure
- Directing the fluid to process locations
- Maintaining the flow rate through the system

Besides simple straight sections of pipe, other components that are included in the discussion of pipes are:

TABLE 15-10 Valve Selection by Service

Service	Description of Service	Recommended Valve
Contamination	Control of fluids that may cause contamination buildup, a valve with minimum obstruction to flow is needed	Ball, gate, globe, or pinch
High pressure	Control of flow at high pressures—selection of a valve to be used in a high-pressure application, particularly pneumatic, should be approached with caution	Ball or globe—poppet valves are used occasionally
High temperature	Control of flow at high temperatures	Ball or globe—poppet valves are used occasionally
Low leakage	Control of flow with very low seat leakage in the closed position	Ball, gate, globe, or plug
Shutoff	Normal on–off control	Ball, gate, globe, or plug—ball and plug valves normally operate faster
Steam service	Control of steam under pressure	Ball or globe
Throttlig	Control the amount of flow by varying the valve position	Globe—ball and gate valves tend to vibrate under flow, and erosion is a concern when using gate valves

Source: ASME B31.3, *Process Piping*, Table D-2.

fittings, which alter the shape or direction of the piping; connections, which include flanges, gaskets, and bolting; and valve components as seen in Section 15.6. Together, these components make up a piping network.

The purpose of this section is to provide an introduction to some of the terminology and considerations that accompany piping design and implementation. Piping standards, pipe classification, piping materials, and general sizing are discussed. This is merely an introduction, and references are provided for readers who seek a more in-depth understanding of pipe system design.

15.7.1 Pipe Standards

Standards are rules and specifications developed by knowledgeable organizations to help govern pipe-related construction and aid in the classification of piping. There are many organizations that publish standards for piping construction and application. Some of the more notable organizations include:

- ISO: (International Organization for Standardization)
- ANSI (American National Standards Institute)
- ASME (American Society of Mechanical Engineers)
- MSS (Manufacturers Standardization Society)

Many pipe manufacturers will also develop their own pipe standards for internal use. Being aware of the major standards as well as the particular ones used by your manufacturer will help ease communication and understanding through the design process. The ASME piping standards B31.1 to B31.11 probably include the codes and standards relevant to your application, as they are the exemplary industry piping standards. Pipe standards include rules pertaining to:

- Sizing
- Temperature ranges
- Pressure ranges
- Material selection

One can find standards for straight pipe sections, all types of fittings, connections, and valves. Most major standards can be found online or in handbooks. One very complete resource is http://global.ihs.com/, where all types of standards can be found in portable document format. Another relevant text is *Process Piping: The Complete Guide to ASME B31.3*, by Charles Becht IV.

15.7.2 Pipe Classification

Classification of pipes has become a very convoluted field. There is no one classification system adopted by the industry, but some terminology is generally pervasive throughout different schemes.

In pipe sizing, the term *nominal pipe size* is often referenced. This is an approximation of the diameter of the pipe. The number actually corresponds to a more precise value for the outer diameter. For example, nominal pipe size 2 actually refers to an outside diameter of 2.375 in. The actual value is always greater than or equal to the nominal pipe size. The typical values for nominal pipe size fall between $\frac{1}{2}$ and 30.

TABLE 15-11 Control Valve Selection: Advantages and Disadvantages

	Type of Valve				
	Gate	Plug (Ball)	Globe	Butterfly	Diaphragm
Type of Service:	On/off (Sliding)	On/off (Rotary)	Throttling	Throttling	Throttling
Advantages	Virtually no pressure loss across the valve face Can be used when the fluid contains suspended solids	Similar properties to gate valves Lightweight, compact design High capacity Good rangeability Tight shutoff	Good sealing characteristics Can be used in frequent open/closing service Quick change of trim without removing valve from line High capacity Good rangeability Low-noise trim available Smooth control	Lightweight, compact design Minimal pressure loss across valve face Low cost High throughput capacity Smaller shaft and actuator	Almost no leakage; process fluid is isolated from valve stem Self-cleaning
Disadvantages	Poor sealing characteristics	Sealability poor with metal seats used at high temperatures Limited-temperature range with resilient seats Choke flow problems Cavitation problems Requires removal for maintenance	High-pressure losses due to contorted path through the valve Low-noise trim reduces capacity	Poor sealing characteristics Good control limited to 60° opening Tight shutoff requires special lining: plus oversized shaft and actuators Lining imposes temperature limitations	Limited operating pressure Limited temperature High wear and tear Poor control over 50%-opening
Sealing method	Gate face slides parallel to the seal surface; gate and seal in constant shear contact	Radial seal, shaped to conform with ball surface	Disk motion is perpendicular to valve seat; only contact is in fully closed position	Throttle blade is mashed into mated seal	Diaphragm material is forced onto valve seat; only contact is in fully closed position
Recommendations	Not for frequent valve opening/closing service Not for when throttling control is required	Not for service with highly corrosive fluids Most suitable for handling slurries	For flow regulation When tight shutoff is required	Low-pressure applications	Water-treatment service Chemical and abrasive service

Source: CEP Magazine, vol. 7, no. 2, Nov. 2002.

A related term is the pipe *schedule*. Casually, this is a measure of the pipe wall thickness. Tables such as Table 15-14 will list the true dimensions associated with pipe scheduling by nominal pipe size. The true definition of pipe schedule is as follows:

$$\text{pipe schedule} = 1000 \frac{\text{internal working pressure}}{\text{allowable stress of pipe material}} \tag{15.19}$$

The most common schedule of piping used in industry is 40, but values range from 5 to 150, including extrastrong variants. The important thing to remember is that schedule is a measure of wall thickness, given a nominal pipe size. These two pieces of information together determine the inner diameter of the pipe. This is an important quantity when considering flow rate and pressure drop in a pipe system.

The term *class* is used to define the rated pressure tolerance of a pipe or fitting. The scheme may use numbers or letters and varies depending on the material and operating conditions of the pipe. Tables containing this type of information usually give the pressure rating for a range of temperatures and several classes. Be sure to use appropriate tables for your conditions.

Further complexities of classification are introduced by material types. Different grades of steel, for example, will

TABLE 15-12 Check Valve Attributes Affecting Selection[a]

Service requirements	Valve Type		
	Swing Check	Tilting Disk Check	Lift Check
Fast opening or fast closing	3	1	1
Variabe flow conditions	3	2	1
Ease of maintenance	1	3	2
Low pressure drop	1	2	3
Isolation	1[b]	2[b]	2[b]
Slurries and fluids containing particulates	2	3	3

Source: Flowserve Check Valve Selection Guide.

[a] 1, excellent; 2, fair to good; 3, not recommended.

[b] With dual seats, dual seats are usually recommended for isolation service where service temperatures do not exceed 500°F.

have different classifying alphanumeric tags. Referring to the manufacturer's documents is the best way to understand exactly what you are getting when it comes to a material.

15.7.3 Common Pipe Materials

Like other equipment discussed in this chapter, the performance and lifetime of pipe components depends largely on proper material selection. The material used to construct piping affects the:

- Allowable operational pressure range
- Allowable operational temperature range
- Resistance to corrosion
- Frictional interaction with fluid
- Weight of piping components

Making the correct choice of material can have a significant impact on cost and reliability. In this section we cover some of the more common materials used in industrial piping and the advantages and disadvantages associated with each. The following pages are adapted from ASME 31.3 Appendix C: Piping Materials.

15.7.3.1 Nonmetals

1. *PTFE: virgin Teflon polymer*. PTFE has good temperature, chemical, and antifriction properties. It is inert to most chemical attack but is affected by liquid alkalis, fluorine, and radiation. It can cold-flow under high stress. Maximum 1×10^4 rad lifetime dose. PTFE must not be used in high-level waste transfer systems.

2. *RTFE: reinforced Teflon*. PTFE is reinforced with added fillers, usually fiberglass. The filler provides a higher resistance to cold flow and allows a wider application range. There is a slight increase in friction. Maximum 1×10^4 rad lifetime dose. RTFE must not be used in high-level waste transfer systems.

3. *UHMWP (ultrahigh-molecular-weight polyethylene)*. UHMWP is a thermoplastic polymer with exceptionally high notched impact strength, and resistance to stress cracking and abrasion. It has good resistance to most chemicals, but is not recommended for strong acids and organic solvents. It offers higher resistance to radiation than PTFE offers. Maximum 2×10^7 rad lifetime dose.

4. *Tefzel*. Tefzel fluoropolymer is a thermoplastic in the same family as PTFE, but is more radiation resistant. Maximum 2×10^7 rad lifetime dose.

5. *PEEK (poly(ether ether ketone))*. PEEK offers increased steam-handling capabilities, and generally has higher temperature ratings. It has good corrosion resistance and excellent resistance to radiation. Maximum 1×10^9 rad lifetime dose.

6. *FEP*. This is a melt-processable PTFE with essentially those same properties.

7. *Delrin*. Delrin is another polymer that is used at higher pressures, and also has higher resistance to radiation than PTFE. Maximum 1×10^6 rad lifetime dose.

8. *Nitrile (NBR, Buna-N)*. This is used for low-aromatic petroleum solvents, fuels and oils, water, air, and inactive gases.

9. *Viton*. This is used in highly aromatic and halogenated solvents and fuels. Also used in low-pressure steam and strong mineral acids.

10. *EPDM*. This is used for resistance to ozone, weathering, and heat. It has poor resistance to oils, hydrocarbons, alcohols, and radiation. LANL Engineering Standards Manual PD342, Chapter 17, Pressure Safety Section.

15.7.3.2 Metals

1. *Cast iron*. This is generally used in low-hazard services (e.g., water or oil). Cast irons have a low cost and are readily available. Disadvantages: no weld end valves, generally poor corrosion resistance, and code limitations.

2. *Bronze*. This is generally used in low-hazard services (e.g., water or oil). Bronze alloys have low cost and are readily available. They can have better corrosion resistance than carbon steel in some water services. Disadvantages: limited welding.

3. *Carbon steel*. Carbon steels are the standard selection for many services where corrosion resistance is not critical. They have a relatively low cost and are readily available. Disadvantages: generally poor corrosion resistance.

4. *Type 304(L)*. Type 340(L) austenitic stainless steels are used for their high resistance to oxidation and

TABLE 15-13 Typical Sizes and Operating Ranges of Isolation Valves

	Size		Pressure Range		Temperature Range		
Valve Type	Min. (mm)	Max. (mm)	Min. (bar)	Max. (bar)	Min. (°C)	Max. (°C)	Pressure Drop[a] (bar)
Gate	2	2250	>0	700	−196	675	0.007
Globe	3	760	>0	700	−196	650	0.590
Diaphragm	3	610	>0	21	−50	175	0.021
Ball (full bore)	6	1220	>0	525	−55	300	0.007
Butterfly	50	1830	>0	102	−30	538	0.120

Source: [21].
[a]Typical values for a DN150 bore valve passing saturated steam at 24 bar, flowing at 40 m/s.

TABLE 15-14 Pipe Diameter by Nominal Size and Schedule 5-XXH

Pipe Size	OD (In.)	5	10	20	30	40	STD	60	80	XH	100	120	140	160	XXH
$\frac{1}{8}$	0.405	0.035	0.049			0.068	0.068		0.095	0.095					
		0.1383	0.1863			0.2447	0.2447		0.3145	0.3145					
$\frac{1}{4}$	0.54	0.049	0.065			0.088	0.088		0.119	0.119					
		0.257	0.3297			0.4248	0.4248		0.5351	0.5351					
$\frac{3}{8}$	0.675	0.049	0.065			0.091	0.091		0.126	0.126					
		0.3276	0.4235			0.5676	0.5676		0.7388	0.7388					
$\frac{1}{2}$	0.84	0.065	0.083			0.109	0.109		0.147	0.147				0.187	0.294
		0.5383	0.6710			0.8510	0.8510		1.088	1.088				1.304	1.714
$\frac{3}{4}$	1.05	0.065	0.083			0.113	0.113		0.154	0.154				0.218	0.294
		0.6838	0.8572			1.131	1.131		1.474	1.474				1.937	2.441
1	1.315	0.065	0.109			0.133	0.133		0.179	0.179				0.250	0.358
		0.8678	1.404			1.679	1.679		2.172	2.172				2.844	3.659
$1\frac{1}{4}$	1.66	0.065	0.109			0.140	0.140		0.191	0.191				0.250	0.382
		1.107	1.806			2.273	2.273		2.997	2.997				3.765	5.214
$1\frac{1}{2}$	1.9	0.065	0.109			0.145	0.145		0.200	0.200				0.281	0.400
		1.274	2.085			2.718	2.718		3.631	3.631				4.859	6.408
2	2.375	0.065	0.109			0.154	0.154		0.218	0.218				0.343	0.436
		1.604	2.638			3.653	3.653		5.022	5.022				7.444	9.029
$2\frac{1}{2}$	2.875	0.083	0.120			0.203	0.203		0.276	0.276				0.375	0.552
		2.475	3.531			5.793	5.793		7.661	7.661				10.10	13.70
3	3.5	0.083	0.120			0.216	0.216		0.300	0.300				0.437	0.600
		3.029	4.332			7.576	7.576		10.25	10.25				14.32	18.58
$3\frac{1}{2}$	4	0.083	0.120			0.226	0.226		0.318	0.318					0.636
		3.472	4.973			9.109	9.109		12.51	12.51					22.85

Source: ANSI/ASME B36.10M.

sulfidation, and where general resistance to corrosion is desired. They are also used widely for cryogenic services. Disadvantages: susceptibility to certain specific corrosion processes (e.g., stress corrosion cracking and intergranular corrosion) in certain media. Type 304(L) is generally a special-order valve material and has been replaced as the "standard" stainless steel material with type 315(L).

5. *Type 315(L)*. Type 315(L) austenitic stainless steels are used for their high resistance to oxidation and sulfidation, and where general resistance to corrosion is desired. They are also used widely for cryogenic services. Type 315(L) has better resistance than type 304(L) to attack by reducing agents and lower susceptibility to pitting. Disadvantages: susceptibility to certain specific corrosion processes (e.g., stress corrosion cracking and intergranular corrosion) in certain media.

6. *Alloy 20*. Alloy 20 was developed to provide resistance to sulfuric acid over a wide range of concentrations and temperatures. It has good corrosion resistance to other media as well and is widely used for handling caustic soda, organic acids, chlorinated hydrocarbons, sludge acids, and so on. Disadvantage: cost.

7. *Monel*. Monel provides excellent resistance to seawater and good resistance to aqueous sulfide and caustic. It is resistant to chloride stress corrosion cracking. It is used widely for handling alkalis, salt water, organic intermediates, and many air-free acids. Disadvantages: poor resistance to sulfidation above 400°F, and embrittled by sulfur and heavy metals at low concentrations during welding or heating. Corroded rapidly by ammonia and compounds.

8. *Inconel*. Inconel is generally used for handling corrosive media at elevated temperatures. It provides good general corrosion and oxidation resistance, good elevated temperature strength, good resistance to chloride stress corrosion cracking, and excellent corrosion resistance to caustic. Disadvantages: poor sulfidation resistance above 1000°F creates working and welding problems, vulnerability to sensitization and intergranular cracking in some services, and cost.

9. *Hastelloy B/B2*. Hastelloy B provides good resistance to reducing atmospheres at elevated temperatures and is very resistant to stress corrosion cracking. It is also used for handling hydrochloric acid vapor and varied concentrations of hot sulfuric, hydrochloric, and phosphoric acids. Disadvantages: cost, availability, tendency to sensitize, and vulnerability to intergranular corrosion in many services.

10. *Hastelloy C/C276*. Hastalloy C provides good resistance to hypochlorites and other solutions containing free chlorine in considerable concentrations. It is also used for handling both oxidizing and reducing chemicals and is very resistant to stress corrosion cracking. Disadvantages: cost, availability.

For more information, see *Piping Materials Selection and Application*, by Peter Smith.

15.7.4 Pipe Sizing

The basic steps to sizing pipes are as follows:

1. Establish the flow rate based on system requirements.
2. Determine a trial line size; either:
   a. Assume a size, or
   b. Use a velocity appropriate for the fluid being handled and determine the size that way.
3. With flow rate and line size known, calculate the pressure drop.
   a. Be sure to include equivalent pressure drops for other components in the pipe system, such as valves and fittings, and be sure to make conservative approximations.
4. Once the total pressure drop is determined, evaluate the acceptability of this value with respect to pump size and operating costs. If unacceptable, make adjustments in line size until optimum behavior is achieved.

15.7.5 Additional Information

- *Pocket Guide to Flanges, Fittings, and Piping Data*, by R. R. Lee
- ANSI, ASME, and API standards
- *Piping and Pipeline Engineering: Design, Construction, Maintenance, Integrity, and Repair*, by George A. Antaki
- *Piping Materials: Selection and Applications*, by Peter Smith

15.8 CONCLUSIONS

As you have now experienced, the breadth of the field of material conveyance is immense. Here we have taken only a shallow dive into many of the major topics. First, important material classifications and definitions were explored. Then some widely implemented conveyance methods were introduced and described. Concerns related to maintenance, reliability, and improvement were included for the discussed conveyance methods discussed and interested readers were directed to sources of further detail for specific topics. With the basic understanding outlined in the preceding pages, you can begin to make informed choices pertaining to specific conveyor selection. You should be aware of the factors for which you must account when considering a certain conveyor type, and know when and where to look for further assistance.

REFERENCES

1. American Chain Association, *Standard Handbook of Chains*, 2nd ed., CRC Press, Boca Raton, FL, 2006.
2. Bhatia, M. V., *Solids and Liquids Conveying Systems*, Vol. 4, Technomic, Lancester, PA, 1982.
3. Brändlein, J., *Ball and Roller Bearings: Theory, Design, and Application*, 3rd ed., Wiley, New York, 1999.
4. Burdge, J., *Chemistry*, McGraw-Hill, New York, 2009.
5. Callister, W. D., *Materials Science and Engineering*, 7th ed., Wiley, Hoboken, NJ, 2007.
6. Griffiths, D. J., *Introduction to Electrodynamics*, 3rd ed., Pearson Education, Upper Saddle River, NJ, 1999.
7. Horenstein, M. N., *Design Concepts for Engineers*, Prentice Hall, Upper Saddle River, NJ, 1999.
8. http://www.martingregoryconveyor.com/, Martin Gregory Conveyor and Engineering Co., 2010.
9. Hutchinson, J. W., *ISA Handbook of Control Valves*, 2nd ed., Instrument Society of America, Research Triangle Park, NC, 1976.

10. Hytrol Conveyors, *The ABC Conveyor Book*, http://www.hytrol.com, 2000.

11. Kulwiec, R. A., *Materials Handling Handbook*, 2nd ed., Wiley, New York, 1985.

12. McGuire, P. M., *Conveyors: Application, Selection, and Integration*, CRC Press, Boca Raton, FL, 2009.

13. Mulcahy, D. E., *Materials Handling Handbook*, McGraw-Hill, New York, 1999.

14. National Fere Protection Association, *NFPA Fire Standard 704*, NFPA, Boston, 2011.

15. Ngai, E., Silane safety. *Specialty Gas Report*, Q2, 2010.

16. Page, J., Chute design considerations, *Bionic Research Institute Chute Design Conference*. Anglo American Corporation, Johannesbury, South Africa, 1991.

17. RoyMech Online Engineering Resources, http://www.roymech.co.uk.

18. Schroeder, D. V., *An Introduction to Thermal Physics*, Pearson Education, Upper Saddle River, NJ, 2000.

19. Serway, R. A., *Physics for Scientists and Engineers*, 8th ed., Cengage Learning, Independence, KY, 2009.

20. Sonntag, R. E., Borgnakke, C., and Van Wylen, G. J., *Fundamentals of Thermodynamics*, 6th ed., Wiley, Hoboken, NJ, 2003.

21. Spirax Sarco Online Valve Selection Resources, http://www.spiraxsarco.com/resources/.

22. Tuzon, J., *Centrifugal Pump Design*, Wiley, New York, 2000.

23. U.S. Department of Energy, *DOE Fundamentals Handbook: Mechanical Science*, Vol. 2, DOE, Washington, DC, 1993.

24. Ward, R. E., *An Overview of Basic Material Handling Equipment*, Material Handling Institute, Charlotte, NC, 1986.

25. Yamaguchi, H., *Engineering Fluid Mechanics*, Springer-Verlag, New York, 2008.

26. Yaws, C., *Matheson Gas Data Book*, 8th ed., McGraw-Hill, New York, 2001.

CORPORATION WEB SITES

Audubon Wire Mesh Belts, http://www.meshbelt.com/index.html.

Conveyor Belt Monitoring, http://www.cbmi.com.au/solutions.html.

Dynacon Conveyors, http://www.dynamicconveyor.com/.

Fenner Dunlop, http://www.fennerdunlopamericas.com/pdf/ConstructionFDA0105.pdf.

Flexicon Conveyors, http://www.flexicon.com.

Goodyear Engineered Products, http://www.engineersedge.com/technology_news/posts/285.html.

16

FAILURE ANALYSIS AND INTERPRETATION OF COMPONENTS

Michael D. Holloway
NCH Corporation, Irving, Texas

In this section the basic causes of failure are discussed. Understanding why a tool, component, or part fails is essential for operational development as well as cost containment. Once failure is understood, performance criteria can be established. The goal of this section is to:

- Understand and define failure
- Identify failure modes
- Understand the fundamentals of a documentation method
- Begin to fill out the total cost worksheet to determine which items to focus on
- Begin to fill out the failure mode worksheet to understand the reasons for failure

16.1 ASSESSING THE SITUATION

On any given day I meet with maintenance managers and supervisors from all sorts of businesses. Many of them have worked their way up through the ranks to take on leadership roles, and some were transferred or promoted in. I found an interesting difference in the group leaders with a certain amount of "technical chops" and those without. As prior mechanics and technicians, they had the opportunity to be critical and provide suggestions they felt would make a difference in the way the department was managed or the way the company was run. The combination of mechanical ability, leadership quality, and ambition were characteristics influencing their promotions. Once leadership roles were assumed, many newly appointed managers quickly realized that there was much more work and strategic decision making involved in the job than understood previously. The proverbial shoe was now on the other foot. The ones who succeeded did so because they blended a combination of decisiveness, experience, empathy, and organization into the job. The managers and supervisors who spent a certain amount of time on the floor understanding problems that occurred and listening to the operators and mechanics (and even the machines) were the ones who became the truly outstanding managers. The idea of spending time understanding the problems by direct involvement is critical and beneficial, regardless of the machine or application. A maintenance manager needs to spend time on the floor with the folks who are actually skinning their knuckles and getting dirty. If not, they will never truly know where the problems are and how to address them effectively.

When our son Max was 5 years old, he was having problems getting his shoes on. As this was right before he was to get into the car to go to school, the anxiety of possibly being late only compounded the problem. It wasn't until I actually got down on the floor and tried to understand what the problem was that I was able to provide a solution. First, he had tucked the tongue of his sneaker under his foot as opposed to letting it rest on top. Second, he was attempting to loop the ends through, not just tying the loops. It was only after I spent time "on the floor" was I able to understand the problem. I offered some quick and accurate suggestions that would ease the anxiety and "increase his production rate." Problem solved; no tardiness.

Process Plant Equipment: Operation, Control, and Reliability, First Edition. Edited by Michael D. Holloway, Chikezie Nwaoha, and Oliver A. Onyewuenyi.
© 2012 John Wiley & Sons, Inc. Published 2012 by John Wiley & Sons, Inc.

Max is happy, Dad is happy, and most important, Mom is happy.

Problems that occur on the production floor are far more complicated and critical than understanding why a 5-year-old is having problems tying his shoelaces. But the idea of actually spending time understanding the problems by direct involvement is critical and beneficial regardless of the machine or application. You have to spend time on the floor with the folks who are actually skinning their knuckles and getting dirty. If not, you will never know what the problems are and how to address them effectively.

The role of the maintenance technician has evolved over the past century. Many years ago, teams would only be relied on only to fix a machine when it broke, the approach known as *reactive maintenance*. Even today, this approach is very common, primarily because many companies just can't catch up to do the types of things they know would actually increase reliability and decrease downtime. Instead, they attack only the immediate need on the production floor. In many cases, there is not enough time or resources to engage in preventive, proactive practices.

Shortly after World War II there was an increase in economic prosperity. An increase in domestic production of various items followed. The increase in production demands meant that equipment had to produce more products at less cost. For the first time, the use of electronics was relied on heavily. This had a direct effect on productivity, but it also meant that maintenance technicians had to become "quick studies" in troubleshooting these devices. Computers (albeit first-generation punch card and reel) were now being used by many forward-thinking facilities for record keeping and data processing. Data were stored and analyzed. Manufacturing and maintenance technology in the late 1950s blossomed due to competition and innovation. Even the "space race" had an influence on manufacturing and maintenance. Large data management allowed for comprehensive preventive equipment maintenance techniques to be born. A new age of maintenance and reliability was established with a direct impact on productivity.

During the space race, the National Aeronautics and Space Administration found it necessary to employ a technique called *failure mode and effects analysis* (FMEA). Aerospace technology demanded that equipment failure not be an option. Therefore, when a part broke, the cause had to be identified and corrective action (and potential design changes) put in place. This approach was quickly adopted by many factories; including part, tool, and machine manufacturers. FMEA had a direct effect on the bottom line and the quality initiative that many companies were attempting to develop. This technique helped foster *preventive maintenance* in users of the next generation of maintenance approaches.

Among many other things, preventive maintenance is a practice of scheduling the rebuild or replacement of parts or machines prior to failure. Through recordkeeping it is possible to determine that certain parts will fail after many hours of operation. Scheduled downtime to administer preventive maintenance became routine, with various parts being rebuilt or replaced prior to failure. It quickly became understood that unscheduled downtime could be reduced dramatically. Using preventive maintenance techniques, the maintenance team could now contribute to a facility's efficiency and profitability. Maintenance departments transitioned from a lost-cost center to a loss-prevention center.

As preventive maintenance techniques and practices began to flourish and evolve, the next natural progression would be to have a technician attempt to predict equipment failures as well as trying to prevent them. This form of mechanical clairvoyance is actually well understood. Many years ago, it was not uncommon for mechanics to listen to a bearing or a gearbox with a screwdriver. The mechanic would put the screwdriver handle to his ear and place the tip onto a gearbox, pump, or bearing housing and "listen." This is a technique I use to this day on various pieces of equipment, including engines. The human ear is especially sensitive to various vibration states, and our brains are wired in such a way that we tend to remember distinct variations of sounds. It is actually rather easy to pick up the slightest amount of grit grinding away at a bearing race or a pump experiencing subtle cavitation. Many companies have taken this concept and explored the nonaudible regions of sound (ultrasonic analysis). This technique doesn't stop at sound but reaches into the broad spectrum of vibration (vibration analysis) and on through to the electromagnetic spectrum of heat (infrared analysis). In this approach, known as *predictive maintenance and condition monitoring*, various analytical techniques are done on a regular basis to determine if failure is approaching. In this manner, action can be taken if eminent failure is approaching. If preventive maintenance is a good diet and exercise, then predictive maintenance and condition monitoring is a checkup by your doctor that includes blood work (oil analysis), listening to your lungs and heart (acoustic and ultrasonic equipment analysis), taking your temperature (thermal analysis and infrared imaging), and a stress test (vibration analysis). These are only a few concepts that are being implemented today with great success.

16.2 FAILURE DEFINED

To understand the performance requirements, one must first identify the cause of downtime, parts replacement, and labor costs. To establish performance criteria, it is essential to determine how something fails to perform. How does a product fail?

Failures are:

- Any loss that interrupts production

- A loss of asset availability
- The unavailability of equipment
- A deviation from the status quo
- Not meeting target expectations
- Any secondary defect

Several techniques can be used to identify a failure mode and the cause of a problem. Failure mode and effects analysis is a process used to identify potential failure modes and to determine the effect of each on overall system performance and cost. The process is broken down into four areas:

1. *Potential failure modes*: One or a combination of:
   - Materials
   - Manpower
   - Methods
   - Machines
2. *Potential failure effects*: What do we do when failures occur? How does it affect the customer and us?
3. *Potential causes*: What happened to cause the failure?
4. *Current controls*: What can help to prevent the failures? What is the process to prevent failures?

Examining the answers to these difficult questions can itself bring about improvements. Utilizing them in the development of procurement specifications can increase plant reliability, with dramatic results.

Failures can be difficult to categorize. They can be a by-product of negligence or a natural process. Failures can happen due to negligence. If an overseer failed to inspect or repair failing equipment, the person guilty of negligence. Failing to notice and fix such problems could lead to a broken part or component or a serious accident. A failure can also occur due to natural processes. Part or component rust, corrosion, and oxidation are just a few examples of failure due to natural causes due primarily to the environment. Stress fractures, elongation, cracking, shrinking, or warping are a few examples due to the working environment. Once the failure is identified, it is vital that it be documented. There are two main sources of failure information: from the equipment maintenance database or from interviewing the maintenance and engineering staff. The technicians will give certain information that will not be captured in the database. In many instances the equipment they work on has particular quirks and operating characteristics that are never captured in the maintenance logs. Pieces of equipment are often compared to children—each is unique and special. Even twins have different personalities and needs. Consider the example of two presses. They are the same make, model, and year and have the same hours, yet perform differently without explanation. Only the operators and maintenance staff are aware of any issues, and often this information is not captured. Once the failure mode is understood, comparisons of improved performance can be analyzed.

16.3 TAKING ADVANTAGE OF FAILURE

A small appliance manufacturing facility was purchasing an adhesive to bond a handle to a toaster oven door. The adhesive they had chosen passed all the engineering design criteria and was offered at a competitive price from a local distributor. The distributor only stocked the adhesive in 8-oz tubes. The tubes were offered 10 tubes to a case (5 lb of adhesive). The plant went through approximately 500 lb of adhesive per month (100 cases of adhesive). The time it took to load the adhesive dispensing guns and the additional work required to dispose of the empty tubes was never considered a factor when the design team selected the adhesive and the purchasing department bought the adhesive. It was only when the product-engineering department was given the task of examining all parts of the product process did the indirect cost of the adhesive become apparent. This was after the product had been in production for seven months. The distributor tried to negotiate a lower price in order to save the business. The difference in price still did not offset the cost of the additional labor. A competitive vendor approached the company with the concept of bulk dispensing from 450-lb drums. The distributor even offered to provide the bulk dispensing system at no cost provided that the customer signed a multiyear purchasing agreement. The price of the bulk adhesive was 28% cheaper and the company had increased labor savings due to the improved process. The bulk system also provided better control of the dispensing and less waste. The tubes always contained at least 1 oz of adhesive when they were thrown away. They also provided an extra drum on site at no cost until opened. This was done just in case production quotas increased and the adhesive had a lead time from the plants. The competitive distributor obtained the business because it was able to offer the product in a package that helped improve savings.

Many vendors offer their key accounts special services that they normally would not share with a small to medium-sized customer. This does not have to be the case, as illustrated by the following example. Many years ago, a large company that built airplanes was experiencing a considerable number of worker disability claims, the majority of which came from workers who had fallen from the wing structure. Upper-level management became involved when the cost to deal with the claims exceeded a particular level. Purchasing rallied the three safety shoe vendors and asked for help. All three companies (A, B, and C) were given the task of coming up with a shoe that provided greater traction. Upper management believed that greater traction would mean less slippage and thus lower worker disability claims. The airplane manufacturer would even contribute to the development of a better sole. Companies A and B each had approximately 45% of the business. Company C had barely 5% of the business, primarily because their shoes were 20% more expensive than A or B. A and B came back with a sole design they felt would provide improved traction. No guarantees were made. It would take six months to determine the impact and benefit of the new soles. The development costs were considerable and there were no guarantees that the new sole would work. Company C took on the challenge with a different approach. Company C understood that there was a safely concern but did not want to consider only the soles. For a minimal cost, they sent in a few engineers who specialized in materials, risk management, and occupational safety. They determined that the real issues were not with the soles of the shoes but, rather, inadequate safety harnesses and a lack of safety training and protocols. They promptly presented a corrective action plan. Once implemented, they were able to reduce accidents by 95% almost immediately. The savings from the reduced worker disability claims offset the entire cost of outfitting all the employees with new safety shoes from company C. The airplane manufacturer was impressed. They gave company C three months to prove it. Company C actually saved the plant 15% more than estimated. They were awarded the entire safety shoe contract, worth several million dollars a year. Company C had the resources to provide solutions that took into consideration the direct and indirect cost of safety breakdown. They provided value-added contributions that helped drive productivity and cut costs as well.

16.4 SOURCES OF FAILURE

Anyone who has ever played organized sports or served in the armed forces has had the following statement drilled into their head: "Failure is not an option." This statement is true—failure does not have an option ... of occurring. Failure will occur at some point—it is merely a question of when. There are some people who believe that everything happens for a reason, and then there are those that believe there is a reason why everything happens. Whether you believe in divine design or random occurrence, things fail. Success and reliability occur when failure can be minimized. In general terms, the causes of failures can be grouped into four categories:

1. *Man* as a failure source can occur from lack of training/experience, human error, or poor supervision.
2. *Machine* as a failure source can occur from poor maintenance, equipment used beyond its design criteria, or machinery that is old and/or outdated.
3. *Methods* as a failure source can occur from lack of supervision, no standard appropriate operating procedures in place, or lack of management skills.
4. *Materials* as a failure source can occur from using wrong materials of construction, external influences of contamination, or improperly manufactured materials.

Once these categories are identified, a cause can be identified and corrective measures put into place. If done so successfully, the opportunity for that failure to occur again will be reduced dramatically. If maintenance management can get a clear understanding of why various pieces of equipment fail, implementing corrective actions for improved reliability will follow.

16.5 FAILURE OF MATERIALS AND OF MACHINES

There are basic reasons why failures occur. Once these reasons are understood, performance parameters can be established and a specification can be developed. Hundreds of reference books are dedicated to failure analysis. Many of these texts are rich with equations, explanations, and examples of hundreds of failure modes. After reading through a few dozen of these it became very obvious that regardless of the item, failure can be broken down into basic elements. Every failure is a combination of a change and an influence as well as a cadence, articulation, and effect. Thousands of combinations can be the culprit, which makes failure analysis almost impossible unless you break it down to basic elements. Various items fail for different reasons.

1. When something fails or breaks, it is doing so because of the following physical *change*:
   - Deformation
   - Fracture
   - Wear
   - Molecular transition
2. *Change* occurs because of the following environmental *influence*:
   - Force
   - Temperature
   - Time
   - Chemical
3. These *influences* have a *cadence*:
   - Steady
   - Random
   - Cyclic

4. The *cadence* can have variations of *articulation*:
   - Amplitude (strong to weak)
   - Frequency (fast to slow)
5. All of which *affect*:
   - The surface and work inward
   - The inside and work outward
   - The entire part, material, device, or tool all at once

Failure can have one or more primary changes and influences as well as one or more secondary changes and influences. Failure can be a combination of several influences which alone would not facilitate change but in concert would bring about a manifestation of failure. Describing the failure mode considers all of these. If a part fails on the surface due to a slow, steady chemical molecular transformation, it may have experienced surface rust. That's a rather long-winded description for something that is very common and understood using two words, but the taximetrics of failure are actually very important in understanding, categorizing, and potentially eliminating the opportunity for failure. By understanding why something fails, you can make a better decision as to what to buy. A performance purchase will address each failure mode(s), look to eliminate the failure, or drastically reduce the opportunity for failure, thus driving increased reliability and reduced operational and repair costs.

When we think in terms of a material's strength, we are really trying to understand and define how it will behave in resistance to various forces. The mechanical properties of a tool, material, or device describe how well it will react to physical forces. Mechanical properties occur as a result of the physical properties inherent in each material. These are determined through a series of standardized tests developed by various reputable organizations: the ISO, ASTM, ANSI, SAE, and API, to name only a few. The force is the "push" or "pull" of an object resulting from the object's relationship with another object, be it a gas, a liquid, or a solid. Forces exist as a result of a given relationship to another object.

To understand failure due to force it is essential that force itself be understood. There are two types of forces: those that actually involve two or more objects physically touching, *contact forces*, and forces involving two or more objects interacting yet not actually physically touching, *action-at-a-distance forces*. Examples of contact forces include frictional forces, tensional forces, normal forces, air resistance forces, and applied forces. Examples of action-at-a-distance forces include gravitational forces, electricity, electrical static dissipation (ESD), or magnetic attraction or repulsion. One would think that pneumatic or air resistance force would be an example of action-at-a-distance forces, but air is considered a gaseous object and therefore is a contact force. Consider contact force as the actual molecules coming in contact with each other, while action-at-a-distance forces represent the interplay of subatomic particles.

In general, *force* is defined as a quantity measured using the standard metric unit known as the *newton* (N), named after Sir Isaac Newton; "30.0 N" means 30.0 newtons of force. One newton is the amount of force required to give a 1-kg mass an acceleration of 1 m/s/s as defined by

$$1\ \text{N} = 1\ \text{kg}\frac{\text{m}}{\text{s}^2} \tag{16.1}$$

A force is a vector quantity that has both magnitude and direction. The force acting on an object has both magnitude (size or numerical value) and direction. By stating that 23 N were applied to an object is not correct; rather, 23 N downward is the correct description of the force. Because a force is a vector that has a direction, it is common to represent forces using diagrams in which a force is represented by an arrow. Keep in mind that forces are vectors, and therefore the effect of an individual force upon an object is often canceled by the effect of another force working in an equal and opposite direction. The effect of a 23-N upward force acting on a steel beam is canceled by the effect of a 23-N downward force acting on the beam. The forces balance out. Of course, there is something to be said for the compression that the beam will experience.

16.6 TYPES OF FORCES

1. *Applied force*: a force applied to an object by a person or another object. If a machine is pushing a crate across the warehouse, an applied force is acting on the crate. The applied force is the force exerted on the crate by the machine.

2. *Gravitational force*: a force by which the Earth, Moon, or other large object in the universe attracts another object toward itself. It is still unclear how this is achieved. Theories abound concerning how gravity works, yet nothing has been agreed upon. Gravitational force is considered the weight of the object. All celestial objects have a force of gravity that is directed "downward" toward their center.

3. *Normal force*: a support force exerted on an object that is in contact with another stable object. If a box is resting on a floor, the floor is exerting an upward force on the box to support the weight of the box.

4. *Spring force*: a force exerted by a stretched or compressed spring. An object that compresses or stretches a spring is always acted upon by a force that restores the object to its rest or equilibrium position. For most

springs the magnitude of the force is directly proportional to the amount of stretch or compression of the spring.

5. *Frictional force*: a resistive force exerted by a surface as something moves over it. Two types of friction forces exist: sliding and static friction. Friction occurs when two surfaces are pressed together. The result is wear and heat. Often, if enough localized pressure is applied, temperatures in access of 2000°F can be experienced quickly. This is often the case with seized bearings. The bearings would appear as if someone had melted them with a torch, when in fact the act of frictional forces generated the localized temperatures. *Air resistance* is a type of frictional force. The force of air resistance is often observed to oppose the motion of an object.

6. *Tension force*: a force directed along the length of a wire, string, rope, cable, or chain that *pulls* equally on the objects on the opposite ends of the wire.

16.7 STRENGTH

Strength has many definitions, depending on the type of material and where it is being used. A material's strength is based a measured ability to perform under a given set of circumstances. A common example is the "World's Strongest Man" competition. Competitors compete in a wide variety of events that test their strength in a host of different venues, such as carrying a large appliance a certain distance as quickly as possible or placing a series of extremely heavy concrete balls on top of pillars. It can be argued (and often is) that some of these tests are not true representations of strength; that events hosted by the Olympics for weight lifting are better indicators of who is truly the world's strongest man.

When thinking in terms of strength, it is important to understand stress and strain. *Strain* is usually expressed as an elongation percentage divided by the original length. Most materials exhibit elongation prior to fracture or rupture. Even concrete and steel will provide an elongation percentage—although a small one compared to various plastics. *Stress* is the *force* causing strain. The unit of force (newton) is usually divided by the cross-sectional area of the material. The newton per square meter, also called the *pascal*, is the fundamental unit of pressure, or in this case of tension (Fig. 16-1). Materials such as steel can resist high stress while producing very little elongation. Conversely, latex rubber can be stretched many times its original length prior to failure. The slope of the stress–strain curve is the *elasticity of modulus* (E-modulus), also known as *Young's modulus*. The E-modulus is the ratio of stress to strain, defined as

$$E = \frac{\text{stress}}{\text{strain}} \tag{16.2}$$

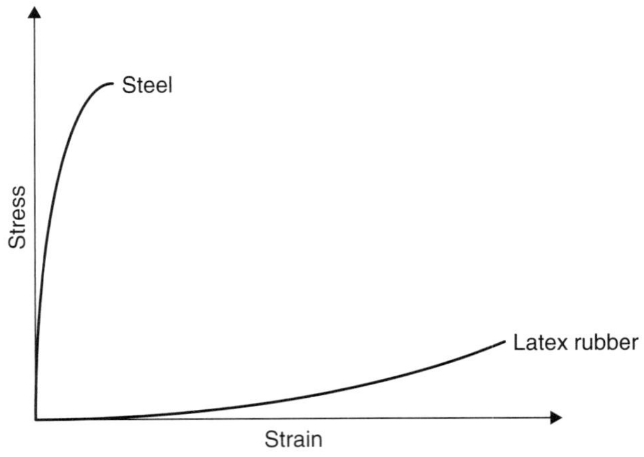

Figure 16-1 Stress–strain curve.

16.7.1 Stress and Strain

- *Stress* S is the force per unit area:

$$S = \frac{F}{A}$$

- *Strain* e is the fractional change in the length:

$$e = \frac{\Delta L}{L_0}$$

where F is the applied force, A the cross section of the test piece, L_0 the initial length, and ΔL the change in length due to stress.

Within the linear, elastic region of the stress–strain diagram, Young's modulus E is defined as the ratio of stress to strain:

$$E = \frac{\text{stress}}{\text{strain}} = \frac{S}{e} = \frac{F/A}{\Delta L/L_0} \tag{16.3}$$

Since the strain is a unitless ratio, Young's modulus has the same units as stress: newtons per square meter (N/m^2), or pascal (Pa). In practice, it is usually stated in units of MPa (10^6 Pa) or GPa (10^9 Pa).

It is important to understand the definition for a given material as well as the various tests and units of measure that define strength. This helps in understanding the ability of a material to withstand load or to resist cracking or rupturing under pressure or impact. When designing for strength, material class and mode of loading are important considerations. For most plastics, the most common form of strength would be the failure strength or the stress at the point where the stress–strain curve becomes nonlinear. Strength is more difficult to define for ceramics, however. Failure in ceramics is highly dependent on the mode of loading. The typical failure strength in compression is 15 times the failure strength in tension. The value reported more commonly is the compressive failure strength of ceramics. Often, adhesives can have adhesive strength and cohesive strength and can also be tested according to the direction of the force applied: a peeling direction or shearing direction, as well as the various substrates to which they are attached. Paints and coatings are considered strong if they can resist ultraviolet degradation or are abrasion resistant. A lubricating oil or grease is considered strong if it can continue to provide a boundary film under high loads and resist oxidation. Further details of particular physical characteristics are covered later in the chapter. For metals the most common measure of strength is the *yield strength*, the minimum stress that produces permanent deformation. The yield strength is usually defined at a specific amount of strain or offset, which may vary by material and/or specification. The offset is the amount that the stress–strain curve deviates from the linear stress–strain relationship line (Fig. 16-2).

Other forms of stress include direct tension, direct compression, bending, shear, direct shear, and torsion. Any further discussion or examination of stress would go beyond the scope of this book.

16.7.2 Yield

The yield strength or yield point of a material is the stress at which a material begins to deform. Prior to the yield point the material will deform in an elastic fashion and will return to its original shape when the stress applied is removed. Once the yield point is passed, some fraction of the deformation will be permanent and nonreversible. Knowledge of the yield point is very important when choosing material, since it generally represents an upper limit to the load that can be applied. Materials have a variety of stress–strain curves, and there are many different ways to define yielding:

1. *True elastic limit*: the lowest stress at which dislocations move. This definition is rarely used, since dislocations move at very low stresses, and detecting such movement is very difficult.

2. *Proportionality limit*: the point at which the stress–strain curve becomes nonlinear.

3. *Elastic limit*: beyond the elastic limit, permanent deformation will occur. The lowest stress at which permanent deformation can be measured.

4. *Offset yield point* (yield strength or proof stress): most widely used strength measure of metals; allows for a consistent comparison of materials.

5. *Upper and lower yield points*: the material response is linear up until the upper yield point, but the lower yield point is used in structural engineering as a conservative value.

6. *Toughness*: related to the total area under its stress–strain curve. A comparison of the relative magnitudes of the yield strength, ultimate tensile strength, and percent elongation of different material will give an indication of relative toughness. Materials with high yield strength and high ductility have high toughness.

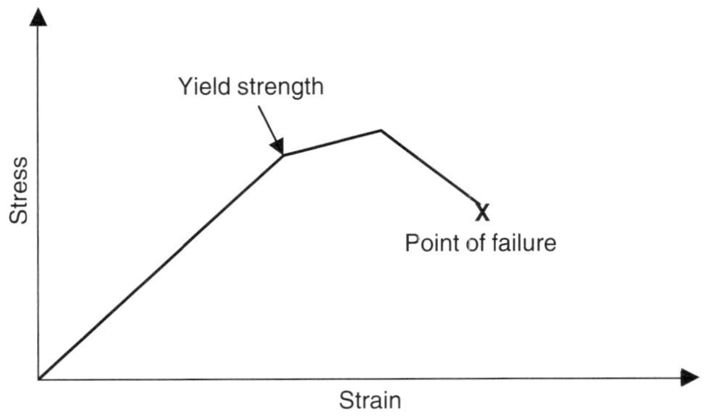

Figure 16-2 Yield strength and point of failure.

16.8 CREEP (DEFORMATION)

The propensity of a material to move slowly or deform permanently to relieve stress is called *creep*. It is a permanent deformation caused by prolonged stress exceeding the limit of recovery. Creep is influenced by the magnitude of the load, the time the load is applied, and the temperature. Creep is also influenced by the cycling of the load. Creep is not always considered a failure mode, but is, rather, a deformation mechanism. Moderate creep in concrete is sometimes a desired feature, having the benefit of relieving tensile stresses that may produce cracking.

Testing consists of applying a load to a test specimen and measuring the strain after a specified time. The dimensions of the test piece are taken into consideration when the test is performed. The *tensile creep* test measures the elongation, breaking point, and strain produced by a load for a predetermined time with constant load, increasing load, or cycled load. The test can be run in a temperature-controlled chamber to understand long-term exposure to very low or elevated temperatures. A test sample is prepared by cutting out a sample in the shape of a dogbone. The specimen is clamped into a test fixture and load is applied, separating the two ends of the test piece. The elongation and tensile strength of a material can be calculated using this method as well as the *E*-modulus, which is a function of stress divided by strain. This test can be accommodated for bolts, screws, and rivets. A variation of this test is used to understand the adhesive and cohesive strength of various glues, sealants, and coatings.

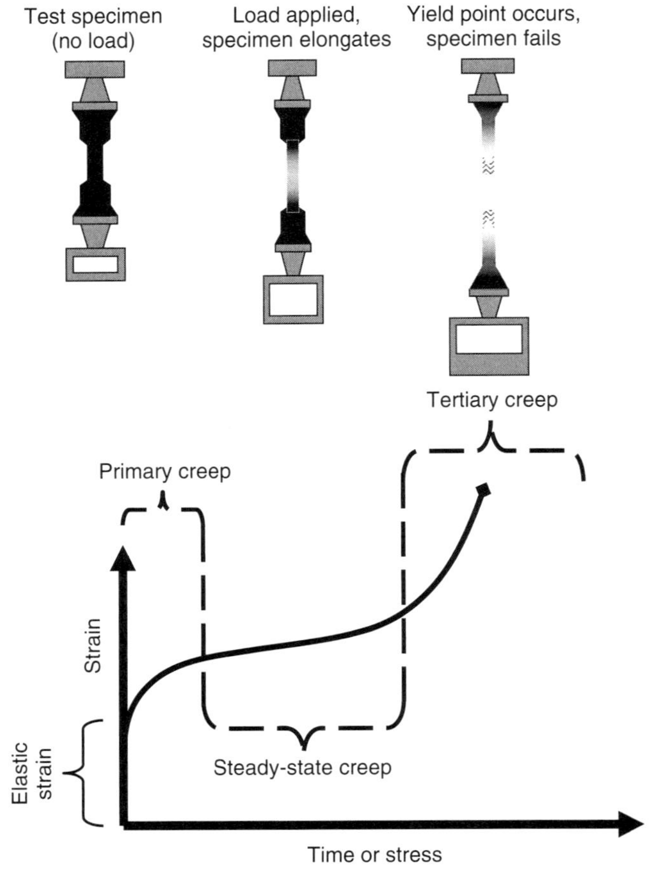

Figure 16-3 Strain vs. time or stress creep behavior.

Creep can be subdivided into *primary, tertiary*, and *steady-state* creep. The qualitative behavior of strain vs. time can be seen in (Fig. 16-3).

1. *Primary creep*. The first region in the figure is termed the primary creep region. The primary creep strain is usually less than 1% of the sum of the elastic, steady-state, and primary strains. The primary creep region is strongly dependent on the history of the material. If the material had been cycled before the creep test, many more dislocations would be present and the characteristics of the primary creep region would be very different. The dislocations of a material are essentially the loci of the opportunity for failure.

2. *Steady-state creep*. The second region of Fig. 16-3 is the steady-state region. This region is so named because the strain rate is constant. In this region the rate of strain hardening by dislocations is balanced by the rate of recovery.

3. *Tertiary creep*. When the amount of strain is high, fracture or rupture will occur. In the tertiary region the high strains will start to cause necking in the material, which can be seen as a thinning of the material. Interestingly, many materials will turn white upon necking, which is a direct response to the aligning of the long-chain molecules. This necking will cause an increase in the local stress of the component, which further accelerates the strain. It is normally a conservative estimate to approximate the end of serviceable life of any material to be the end of the steady-state creep region because of the short duration of the tertiary region (Fig. 16-3).

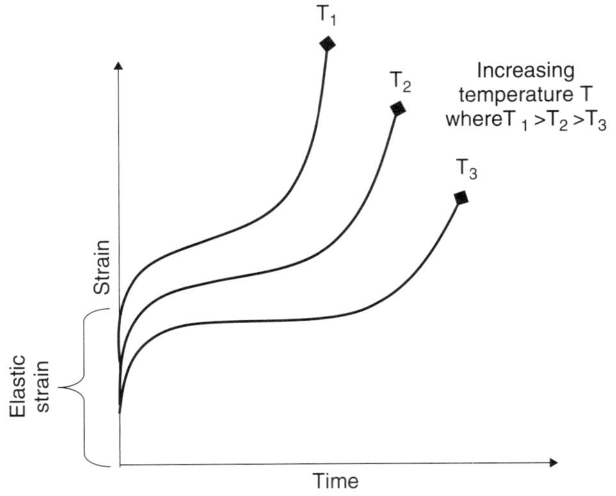

Figure 16-4 Strain vs. time creep behavior effect of temperature.

Increasing temperature influences the modulus of a material as well as the elastic strain region and the point of failure (Fig. 16-4).

The temperature range in which creep deformation may occur differs in various materials. For example, tungsten requires a temperature in the thousands of degrees before creep deformation can occur, whereas ice formations such as the Antarctic ice cap will creep in freezing temperatures. Generally, the minimum temperature required for creep deformation to occur is 30% of the melting point for metals and 40 to 50% of the melting point for ceramics. Plastics have a wide range, due to the variety of constituents. All materials (i.e., plastics, metals, ceramics) will creep when close to their melting temperature. Since the minimum temperature is relative to the melting point, creep can be seen at relatively low temperatures for some materials. Plastics and low-melting-temperature metals such as mercury or indium and even lead solder will creep at room temperature.

16.9 FATIGUE (MATERIAL)

Fatigue is the progressive damage that occurs when a material is subjected to cyclic loading. The maximum stress applied is less than the stress limit of the material, yet the material is subjected to constant or random elongation,

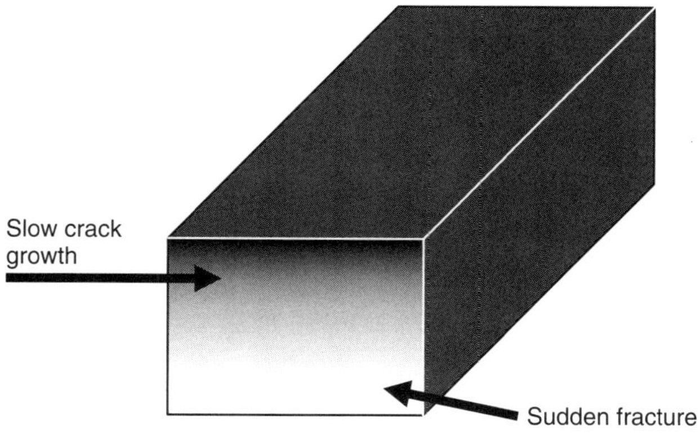

Figure 16-5 Fracture profile.

bending, or compression, which will eventually take its toll and produce a failure. Typically, a fractured piece will have a dark area, which is an indication of slow crack growth, and a bright area, which is an indication of sudden fracture (Fig. 16-5). Fatigue begins with the crystalline structure of a metal or the long molecules of a plastic or a coating experiencing dislocation. Eventually, small microfissures develop which begin to compromise the physical integrity of the part, tool, or material. Material will not normally recover if fatigue is relieved.

16.9.1 Rate of Fatigue

Temperature, chemicals, and moisture can influence the rate of fatigue. Another consideration that affects fatigue is the geometry of the part, component, or tool. The quality of the surface has an influence on fatigue as well. Surface roughness, as well as scratches, dings, or dents, cause stress or can provide a crack nucleation site, which can lower fatigue life. In certain examples, surface stress can be used to increase fatigue life. Shot peening is widely used to put the surface in a state of compressive stress. This will inhibit surface crack formation and improve fatigue life. This technique for producing surface stress is often referred to as *peening*. This improvement is normally observed only for a high-cycle-fatigue environment. There is little improvement in low-cycle-fatigue situations. Another contributing factor is uneven cooling, which leads to a heterogeneous distribution of material properties such as hardness and ductility. Uneven cooling of castings can produce high levels of tensile residual stress, which will bring about crack propagation. With that in mind, the size, frequency, and location of internal defects can also play a role. Casting defects such as pores and voids can facilitate fatigue.

Liberty ships were cargo ships built in the United States during World War II. They were British in conception but adapted by the U.S. government, were cheap and quick to build, and came to symbolize U.S. wartime industrial output. Based on vessels ordered by Great Britain to replace ships torpedoed by German U-boats, they were purchased for the U.S. fleet and for lend–lease provision to Britain. Eighteen American shipyards built 2751 Liberties between 1941 and 1945, easily the largest number of ships produced to a single design. Early Liberty ships suffered hull and deck cracks, and a few were lost to such structural defects. During World War II there were nearly 1500 instances of significant brittle fractures.

16.9.2 Fracture

16.9.2.1 ***Brittle Fracture*** Brittle fracture occurs without any sign of deformation prior to the fracture. Often, brittle fracture affects crystalline materials such as gemstones, glass, many metals, and to a smaller extent, certain types of plastics. Brittle fracture is also dependent on temperature. At very low temperatures, materials that would normally deform prior to fracture at higher temperatures experience brittle fracture at lower temperatures. You may have had a freshman physics instructor who dipped a rose into liquid nitrogen, then dropped it on the floor. The rose probably shattered into many pieces as if it were made of glass.

Sometimes, brittle fracture is not a bad thing. If it were not for this type of fracture, the Stone Age would not have occurred. Early humanoids were able to take advantage of the brittle nature of various stones to create sharp implements that allowed for successful hunting and cutting. Gemstones are "cut" using their inherent brittleness, and the silicon wafers used to manufacture integrated electronic chips are made by cleaving a solid silicon ingot. These materials break because of the lattice work of the material's molecules. The molecular structure (also called the crystal lattice) is highly ordered, allowing a clean break to occur. The smallest possible part of the crystal lattice is called a primitive unit cell. Some examples of typical crystal lattice structures are shown in (Fig. 16-6). In the diagram, the gray sphere in the cubic body-centered example is used to signify an internal atom, whereas the white spheres in the cubic face-centered and hexagonal examples signify surface atoms.

Note that molecules do not actually have sticks holding the atoms in place. The lines are used to help visualize the arrangements. A metal's crystal structure and properties are determined by metallic bonding, the force holding together the atoms of the metal. Each of the atoms of the metal contributes its valence electrons to the crystal lattice, forming an electron cloud that surrounds positive metal ions. These free, negatively charged electrons belong to the entire metal crystal. The ability of the valence-free electrons to travel throughout the solid explains both the high electrical and thermal conductivity of metals. Solids that do not have an ordered pattern are considered amorphous. Many plastics are amorphous as well as glass. Amorphous

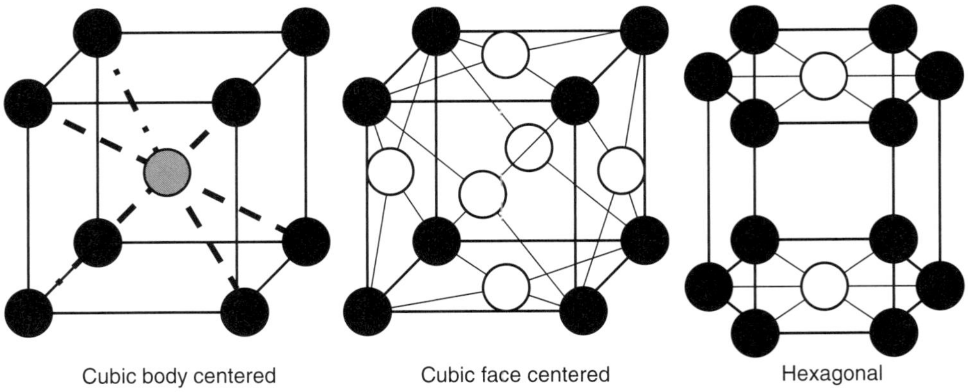

Figure 16-6 Typical crystal lattice structures.

solids will experience elongation or deformation prior to breaking. The fracture propagation in an amorphous solid typically does not have a simple path to follow compared to a crystalline structure.

16.9.2.2 Ductile Fracture To understand ductile fracturing it is important first to understand what it means for a material to be ductile. *Ductility* is commonly defined as a material's ability to be deformed without the incidence of failure. Ductility may also be considered bendability or even crushability. In ductile fracturing a considerable amount of deformation takes place before failure. The point at which the failure occurs is controlled by the purity of the materials. At room temperature, pure iron can undergo deformation up to 100% strain before breaking, whereas cast iron or high-carbon steels can barely sustain 3% of strain.

The basic steps in ductile fractures are necking (which results in stress localization at the point on the sample of smallest cross-sectional area), void formation, voids combining to form one large void, crack development, and failure. Figure 16-7 is a representation of the steps in ductile fracture in a dogbone sample. As force is applied, necks and voids begin to form and eventually combine. Once the voids combine, the physical integrity of the material is compromised and a crack forms in the material. Once a crack forms, it is only a matter of time until failure occurs.

16.10 WEAR

Wear is defined as the physical erosion of a solid material surface by the action of another material or force. *Tribology* is often thought to be the science and study of wear, but actually, tribology is the study of surfaces in relative motion. It includes the study of friction, lubrication, and wear. There are five principal wear processes:

1. *Adhesive wear*: also known as scoring, galling, or seizing. It occurs when two solid surfaces slide over one another under pressure. Surface projections, or asperities, are deformed and eventually welded together by the high

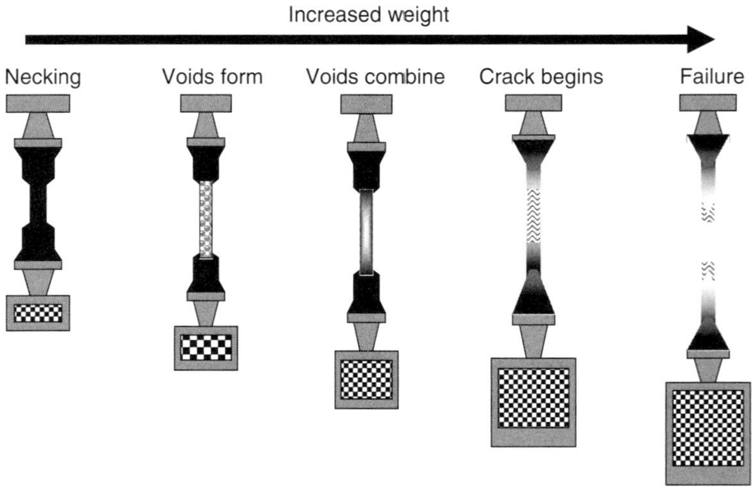

Figure 16-7 Progression of ductile fracture.

local pressure. In bearings as well as gears, the metal surfaces are actually separated by lubricating oil or grease. If the lubricant fails, the part will fail as well due to the enormous pressures, which generate temperatures in excess of 2000°F, as witnessed by anyone who has ever had to change a seized bearing. It would appear as if the bearing was melted by a torch, yet it wasn't. As sliding continues, the lubricant became compromised and metal-to-metal contact occurred. The molecular bonds are broken, producing cavities on the surface, projections on the second surface, and frequently, tiny, abrasive particles, all of which contribute to future wear of surfaces, which contributes to abrasive wear.

2. *Abrasive wear*: occurs when material is removed by contact with hard particles. The particles may either be present at the surface of a second material (two-body wear) or may exist as loose particles between two surfaces (three-body wear). Abrasion is the wearing away of surfaces by rubbing, grinding, or other types of friction. It is scraping or grinding wear that rubs away surfaces. It is usually caused by the scouring action of such gritty material as sand, dirt, slag, and even corrugated cardboard fibers. It usually, but not always occurs when a hard material is used on a softer material. Consider Babbitt bearings. The metal alloy is soft and would seem to wear out more quickly than a harder metal choice. The structure is actually composed of hard crystals dispersed in a soft metal alloy. As the bearing wears, the harder crystal is exposed, with the matrix eroding somewhat to provide a path for the lubricant between the high spots that provide the actual bearing surface separated by a dynamic film of the lubricant.

3. *Corrosive wear*: the deterioration of properties in a material due to various reactions, referred to simply as *corrosion*. Oxidation is a special type of corrosive wear indicated by the flaking off or crumbling of metal surfaces which takes place when unprotected metal is exposed to a combination of heat, air, and moisture. Rust is an example of iron or iron alloy oxidation. Corrosive wear is covered in Section 16.13.1.

4. *Impact wear*: the striking of one object against another. It is a battering, pounding type of wear that breaks, splits, or deforms a surface. It is a slamming contact of metal surfaces with other hard surfaces or objects. A good example of impact wear is the bucket pins found on construction excavators and backhoes. Often, there are only three pins holding the bucket to the arm. As the bucket enters soil or roads, a tremendous amount of force is localized to the pin and bushing surface.

5. *Erosion*: the wearing away or destruction of materials by the abrasive action of water, steam, or slurries that carry abrasive materials. Often, this type of wear is considered a combination of abrasive wear and impact wear. Pump parts are subject to this type of wear. Cavitation is considered a type of *erosion wear*, which is the result of a turbulent flow of liquids that may carry small suspended abrasive particles or aeration which produce microimplosions capable or wearing away metal. Load can also affect wear, as can direction of the offending surface, such as unidirectional sliding, reciprocating, rolling, and impact loads. The speed of the abusive surface, as well as the temperature and the angle of contact all affect the wear aspect.

16.11 PROPERTY CHANGES

In the results of standard wear tests, the loss of material during wear is expressed in terms of volume. The volume loss gives a truer picture than weight loss, particularly when comparing the wear resistance properties of materials with large differences in density. The working life of components is over when dimensional losses exceed the specified tolerance limits. Wear, along with other failure mechanisms, such as fatigue, creep, and fracture toughness, causes progressive degradation of materials with time, leading to failure of material at an advanced age. Under normal operating conditions, the property changes during operation occur in three stages:

1. *Primary stage*: the early stage or break-in period, where the rate of change can be high.
2. *Secondary stage*: the middle age, where a steady rate of aging is maintained. Most of the useful or working life of the component is witnessed at this stage.
3. *Tertiary stage*: the old-age stage, where a high rate of aging leads to rapid failure.

Increased temperatures, pressures, strain rates, and stresses in the *secondary stage* are shortened, thus reducing the working life. It is important to look to reduce any undo stress and pressures, while the device, material, or tool is in the *secondary stage* or middle age.

16.12 TEMPERATURE

The changing temperature of the material affects any reactants and changes the reaction rate. Atoms, molecules, and particles are always in motion. They possess kinetic energy. Temperature is a measure of the material's average kinetic energy. Increasing the temperature will increase the average kinetic energy; therefore, the frequency and energy of collisions will all increase. If the success rate of the collision increases, so does the reaction rate. These conditions will catalyze the reaction and allow for greater mobility of the particles and reactive species. With great mobility comes an increased opportunity for a reaction to take place. Conversely, certain conditions or substances will

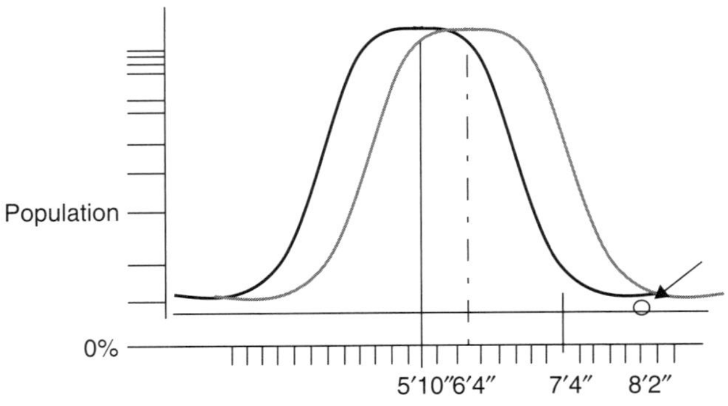

Figure 16-8 Size distribution.

actually retard the reaction process or inhibit it altogether. In this instance, the temperature has a reduced effect. Antioxidants, rust, and corrosion inhibitors work along these mechanisms.

The increase in reaction rate with a minimal temperature change is so dramatic because the overall distribution of kinetic energies in a sample of particles must be considered. Significant proportions of particles have very high kinetic energy and very low kinetic energy. This overall concentration is called a *normal distribution curve*.

An example of a distribution curve is the height distribution of American males. The average height of the American male is 5 ft 10 in. A certain percentage is above 6 ft tall, an even smaller percentage is above 6 ft 6 in. tall. Conversely, a certain percentage is below 5 ft 4 in., and still a smaller percentage is below 5 ft. The percentage gets smaller as the height is farther away from the average height. This fact is consequential when a college is looking to recruit a center for the basketball team. These outliers are not the norm but, rather, isolated examples. There is not a prevailing requirement for 4 ft 2 in. or 7 ft 4 in. people, but their very existence can make a very big impact on a basketball team or on a jockey. Now imagine being able to skew the average height up or down by a few points. Change the average height to 6 ft 4 in. The opportunity now exists for an 8 ft 2 in. person (see Fig. 16-8). That could make a very big impact (especially for a basketball team). The change in temperature makes the same dramatic impact.

The same can be said for the size and kinetic energy of the reactive components and particulate. If the temperature is raised a little, the average kinetic energy of the particles increases, but the distribution curve size remains the same. The new proportion of particles possessing energy for activation of the reaction is much greater, so the reaction rate increases dramatically. Increasing the temperature moves the curve to the right (Fig. 16-9), allowing the maximum number of molecules to have the needed energy to react. The true kinetics of a reaction can only be determined through experimental methods. Reactions have many variables that can influence the rate. It would be a mistake to assign a rate equation to a given system without further investigation.

Thermal shock is the name given to material cracking from rapid temperature change. Glass and ceramics are vulnerable to this failure, due to lack of toughness, low thermal conductivity, and high thermal expansion coefficients. Glass and ceramics are used in high-temperature applications because they have a high melting point. The shortcomings can be addressed through processing or design modifications. Improving toughness or thermal conductivity can be improved upon relatively easily, but changing a material's melting point is far more complicated.

Thermal shock occurs when a thermal gradient causes different parts of an object to expand by different amounts. This differential expansion can be understood in terms of stress or strain, equivalently. At some point, this stress overcomes the strength of the material and a crack forms. If nothing stops the crack from propagating through the material, it will cause the object's structure to fail. A good example of this is the use of cast aluminum as a material choice for an engine block and steel as the material choice for the head and oil pan. Even the best seal would have a difficult time stopping engine oil leaks, due to the difference in thermal expansion and contraction rates of the materials.

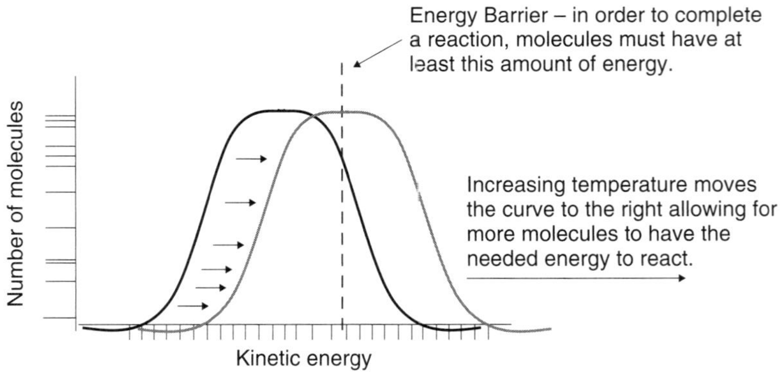

Figure 16-9 Kinetic distribution.

Another example is the way in which ice cubes crack quickly when put into a cup of hot tea. The outer surface of the ice cube, which is exposed to the hot tea, will expand while the inside retains the same dimension. A crack ensues.

Reinforced carbon–carbon composite is extremely resistant to thermal shock, due to graphite's extremely high thermal conductivity and low expansion coefficient. The high strength of carbon fiber used in the composite also provides the ability to deflect cracks within the structure.

16.13 OXIDATION: MOLECULAR TRANSITIONS AND CHEMICAL INFLUENCES

16.13.1 Corrosion

Corrosion is typically defined as the oxidation of metal that occurs in the presence of moisture. The most common form of corrosion is rust. Rust occurs when iron or an iron alloy is exposed to water, humidity, or steam. The iron reacts with the oxygen in water to form rust. The amount of water available to provide the iron with enough oxygen to form rust will also determine the color of rust, which varies from black to yellow to orange-brown.

In high school chemistry classes it is often taught that oxidation is the loss of electrons, but that is not entirely correct. Electron transfer does not happen. It is very complicated to describe the mechanisms at work accurately, so in general terms and for lack of time and patience, oxidation is described as the loss of electrons. The actual chemical process of oxidation involves the increase in the oxidation number, which is the charge it would have if it were stripped of any attached atoms. Often, the oxidation number is the same as the oxidation state, but there are a few exceptions. Chemicals that have the ability to oxidize (change the oxidation number) are known as *oxidizing agents*. The substances remove the electrons from the material. Oxygen, chlorine, and bromine are examples of oxidizing agents.

The presence of oxidation on metal can often be seen on rusted iron parts. A serious problem which often occurs is that the formation of rust occurs away from pitting, which can be hidden. This is possible because the electrons produced during the initial oxidation of iron can be conducted through the metal, and the iron ions can diffuse through the water layer to another point on the metal surface, where oxygen is available. This process results in an electrochemical reaction in which iron serves as the anode, oxygen as the cathode, and the water (which contains reactive ions) acts as a transport medium (Fig. 16-10).

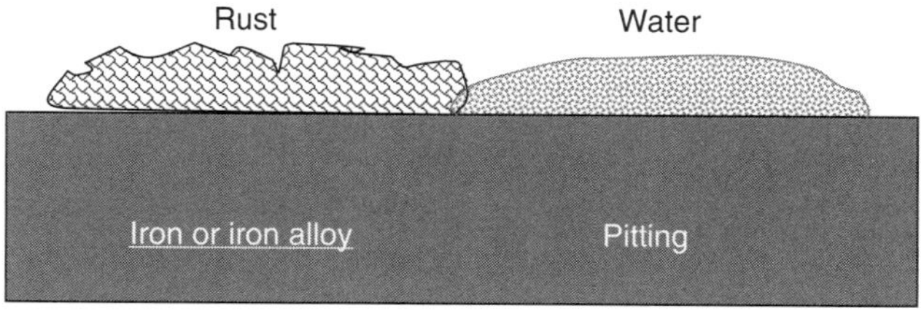

Figure 16-10 Pitting development.

Hightened concentrations of chlorine-containing chemicals can actually interfere with a metal's ability to form a protective oxide layer or passivating film. Very small local fluctuations will degrade the oxide film in a few critical points; then the corrosion intensifies, causing pitting. A pinhole on the surface can hide a quarter-sized pit just below. These problems are especially dangerous because they are difficult to detect before a part or structure fails.

Many other factors affect the rate of corrosion. Salt water greatly increases the rate of rust development. This is because salt water increases the conductivity of the aqueous solution formed at the surface of the metal and enhances the rate of electrochemical corrosion. This is one reason why iron or steel tend to corrode much more quickly when exposed to the salt used to melt ice on roads or moist salty air near the ocean. Rust is a metal oxide that lacks any strength and will not adhere to the surface. This is one reason that pitting occurs. Extensive pitting eventually causes structural weakness and disintegration of the iron or iron alloy. Interestingly, aluminum will form a very tough oxide coating that bonds strongly to the surface of the metal, preventing the surface from further exposure to oxygen and corrosion. Corrosion can occur in acid or caustic environments as well as affecting nonmetallic materials such as plastics and even glass or ceramics.

16.13.2 Corrosion in Nonmetals

Nonmetal materials such as plastics, ceramics, and glass typically do not corrode in the same way that metals do, mainly because they do not conduct electricity or readily change oxidation states. In the true sense, these materials do not corrode, but they can break down in the presence of various chemicals, such as acids, caustics, and solvents.

Most corrosion-resistant plastics generally fall into two groups: polyolefins (e.g., polyethylene, polypropylene) and the poly(vinyl chloride)s. These types of plastics have excellent chemical resistance and are low in cost. The polyolefins have excellent resistance to many solvents but are not chemically bondable. If bonding is required, these types of plastics can be bonded by thermal wielding with low-cost equipment. Polyolefins can be produced in sheet, rod, tubing, and film.

Other materials, such as fluoropolymers (Teflon is an example), also exhibit excellent chemically resistive properties but are very difficult to formulate or process, which results in higher costs. The poly(vinyl chloride) plastics PVC and CPVC [chlorinated poly(vinyl chloride)] have outstanding chemical resistance properties but can be up to 40% heavier and structurally more rigid then polyolefins. PVC is the oldest corrosion-resistant plastic. The properties of PVC and CPVC are typically identical except that CPVC has better temperature performance. Both PVC and CPVC are chemically and thermally bondable. These plastics can be made into sheet, rods, and tubing as well as various profiles.

16.13.3 Galvanic Corrosion

Essentially, galvanic corrosion occurs for the same reason that the potato can be made into a battery. The battery was invented around 1800 by Alessandro Volta. It is a great example of how chemical energy is converted into electrical energy. There has been some controversy suggesting that simple batteries may have existed a few thousand years ago. This has yet to be accepted by the scientific community or the Italian government. If you ever created a potato battery, you established a galvanic potential that could power a flashlight.

You don't have to use a potato, an orange or apple would work as well. Insert a piece of copper wire (14 gauge works well) and a galvanized nail (preferably a 5 penny or larger) into a potato. (You will have to sand off the oxide layer of the wire and nail prior to tuber insertion.) The nail serves as the negative (−) or cathode of the potato battery, and the copper wire is the positive (+) or anode of the battery. Make sure that the nail and wire are about 1 in. apart. You can test the voltage of you battery using a multimeter. Normally, a potato battery won't produce more than 1 volt of current. Sometimes you will have to put potato batteries in a series to have enough energy to power a small light bulb. The zinc on the nail provides electrons, and the copper strip accepts the electrons. Because the zinc strip frees electrons and the copper strip uses electrons, if you put a wire between the two strips, electrons will flow from the zinc to the copper, producing electrical energy. The potato serves as a means of holding the leads in place while allowing the flow of electrons through the water (electrolyte) that is present. In corrosion, the positive (+) anode is consumed or in this case, corroded. The current will consume the anode material while the cathode material remains unphased.

Factors such as the relative size of the anode (+) material, the type of metals used, and operating conditions such as temperature, humidity, and salinity will have an effect on galvanic corrosion. The surface area ratio of the anode and cathode will directly affect the corrosion rates of the materials. Metals that are high on the galvanic table (the cathodes), such as gold, platinum, and silver, will remain unchanged during the galvanic environment, while anodes such as calcium, sodium, and aluminum will lose metal and corrode. This is an important consideration if a component has dissimilar metals.

16.14 DEPOSIT FORMATION

Many gearboxes, hydraulic systems, and engines can quickly become contaminated with water, particulate, and various deposits, such as varnish and sludge, that are the products of oil oxidation. These contaminants contribute to the degradation of the lubricant and increase operating temperature, energy demand, component wear, and oil usage. Even new systems can often contain contaminants that contribute to these problems. There is technology available that will remove contaminants from systems while in operation and prepare the metal surfaces to readily accept the surface-active agents that are found in performance products. This technology has been proven to increase equipment life while decreasing wear metals, operating temperature, and energy demands.

To condition a system for improved reliability, a basic understanding of how and why various deposits develop must be understood. Deposits develop when the lubricant breaks down into other compounds. By understanding this process, certain measures can be taken to slow down the formation of deposits. Lubricants can break down and create sludge, varnish, and deposits, for a variety of reasons (Table 16-1).

When lubricants oxidize, they form reactive materials that can be reconstituted into different deposits. Varnish, lacquer, sludge, and gum are several of the typical deposits that are formed and the problems that can occur when lubricating oil breaks down. Varnish, lacquer, gum, and sludge are basically the result of oxidized oil and are all carbon-based materials. These materials are typically large in molecular size compared to other compounds found in lubricants (save the base oil). These deposits are formed by chain polymerization. The two methods are commonly understood to explain the creation of large molecules, step polymerization and chain polymerization, generally synthesize large molecules. Step polymerizations proceed by a stepwise reaction between chemically functional reactant molecules. The molecule size increases at a relatively slow rate. One reactive molecule combines with a second; they form a single reactive molecule, which then reacts with a third, and so on, until eventually, large polymer molecules have been formed. Chain polymerizations require an initiator in the form of a free radical or reactive species. Free radicals and reactive species (anions or cations) can be generated by many of the conditions found in gearboxes, transmissions, hydraulic systems, and engines. Chain polymerizations occur by propagation of the reactive molecule by successive additions of large numbers of other reactive molecules. Figure 16-11 shows the typical reaction rate and development time of the two polymerizations.

There are several reasons why deposits form in a lubricating system. Typically, the oil undergoes a transformation from one molecular structure to another. This can be in the

TABLE 16-1 Sources of Deposit Formation

Contributing Factor	Mechanism
Inferior base oil	When the base oil contains one or more of the following: light fractions, sulfur, low-molecular-weight hydrocarbons.
Insufficient concentration of performance additives	Low concentrations of antiwear agents and/or friction modifiers produce wear metals and localized frictional heat. Low concentrations of demulsifiers can lead to emulsified oil, which hastens oil oxidation.
Yellow metals (copper, brass, bronze)	Contaminants from external or internal sources can catalyze oil oxidation and lead to deposits and additive depletion.
Heat	High ambient temperatures, operational heat (from kinetic energy or combustion), frictional heat (from metal-to-metal contact), and pressure (from aeration and/or operation) can accelerate oil oxidation and additive depletion.
Water	By-products of combustion and/or external contaminants can hasten oil oxidation and lead to deposits and additive depletion.
Acids	Develops when base oil breaks down into reactive species, produces sludge, varnish, and/or other deposits, or an external source such as acid washing. Common forms are sulfuric acid, nitric acid, and various carboxylic acids.
Caustics and solvents	External contamination breaks oil molecule into reactive species. The reactive species may polymerize to form resinous deposits.

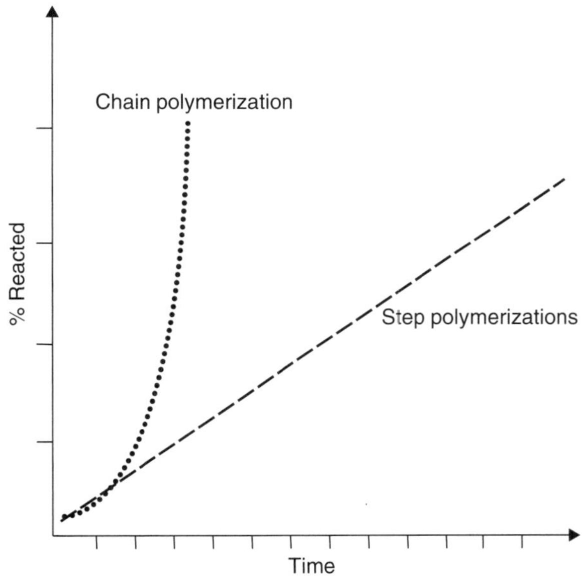

Figure 16-11 Polymerizations.

form of a mixture of a contaminant such as fuel soot with motor oil, or transmission sludge from debris and gear oil. It can also be varnish or lacquer on a piston pump cylinder from excessive temperature and load. More often, it is the breakdown of the base oil and the buildup of a new compound. This reconstitution develops through a series of steps. The first step is the generation of a reactive compound or free radicals. There are several ways in which these compounds develop. Once in the system, they combine or polymerize into new compounds in the form of deposits. Various environmental conditions can facilitate the generation, such as temperature, pressure, water, solvents, acids, caustics, and various metals.

Free radicals can develop in several ways. One is through mechanical energy, such as milling pressure and high pressure combined with shearing deformation or shock waves through a nonthermal process. Another way is from acids that cleave the molecular bonds. A high concentration of short- or moderate-chain-length molecules of the oil can break, creating a higher population of free radicals. These short reactive species bond readily with like species or contaminants and polymerize into deposits. A long-chain-length oil molecule will produce less reactive species, due to bond length and restricted molecular mobility. Excessive heat and pressure can also break the molecular bond and form free radicals that in turn will polymerize into deposits.

There are countless molecular arrangements that make up deposits in any given system. Deposits are typically the result of the base oil breaking down into a reactive compound. Once the reactive compound is developed, it reconstitutes by reacting with another reactive compound, forming a deposit. The typical reaction is considered a free-radical chain polymerization. The reaction occurs in three steps:

1. *Initiation*. Free-radical chain polymerization is a type of polymerization in which the propagating species is a long-chain free radical, usually initiated by the attack of other free radicals, acids, or reactive species derived from heat, water, acids, contaminants, and so on.

2. *Propagation*. The polymerization proceeds by the chain reaction addition of reactive molecules to the free-radical ends of growing chain molecules.

3. *Termination*. Finally, two propagating species (growing free radicals) combine (disproportionate) to end the polymerization reaction and form one or more polymer molecules in the form of a deposit (sludge, varnish, lacquer, etc.).

The rate at which the reaction takes place actually accelerates with time (and heat). It would be natural to assume that the reaction rate would slow down with time since the concentrations of reactive molecules and initiators reduce as they are reacting. The exact opposite is true. Three routes, also known as *diffusion-controlled termination steps*, explain this behavior. The first route is a translational diffusion of two of the propagating radicals until they are in close proximity to each other. The second route is the segmental diffusion of the polymer chains. In this route, the two chains are rearranged so that the two radical ends are close enough to react with each other. The third route is the actual chemical reaction of the two radical ends to form the polymer or in this case the deposit. During the course of the reaction, the translational diffusion route decreases faster than the increased rate of the segmental diffusion route. This is how rapid autoacceleration occurs.

Due to the speed at which deposits can develop, it is very important to act quickly when a system shows signs of early contamination. Free-radical chain polymerizations can occur very quickly. If the onset of oil oxidation or the free-radical polymerization of the reactive species is not quickly addressed, very serious problems can occur.

16.15 FACTORS THAT AFFECT DEPOSIT FORMATION

16.15.1 Concentration and Pressure

When the concentration of the reactive compounds, free radicals, acids, or contaminants is increased, the molecules or particles are closer together. There are more molecules in a given volume—the same applies when pressure is increased. When this occurs there will be more collisions or opportunities for a reaction to take place, so there will be a greater chance of successful reactions occurring. The rate of reaction increases, the reactions being the generation of acids (sulfuric, nitric, carboxylic), free radicals, and the polymerization of the free radicals into deposits. In many reactions it is not convenient to measure concentration. Other properties can be monitored which represent the concentration directly: for example, a change in chemistry (see Fig. 16-12), loss in mass (density), the production or potential for production of oxygen or another gas (bomb calorimeter), change in color or clarity, change in pH, change in conductivity, or change in pressure. Many of these changes can be tested on site using various methods or using oil analysis labs.

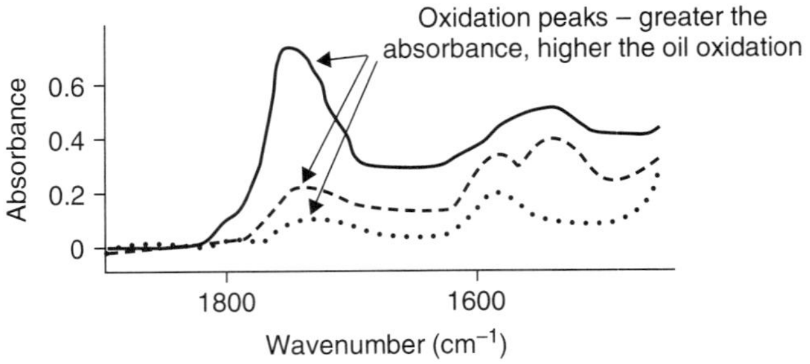

Figure 16-12 FTIR scan of oil samples with varying degrees of oxidation.

16.15.2 Particle Size and Contaminant Type

There are five fundamental types of wear that can produce particles:

- Rubbing
- Cutting
- Rolling
- Fatigue
- Severe sliding
- Combined rolling and sliding

These wear mechanisms were covered earlier but are present in almost all systems. Wear particles can occur in many different systems and are typically the precursors to free-radical development and oil oxidation. Microscopic (submicrometer up to 50 μm) contaminants such as wear metals, dirt, and debris always react faster than lump solids. The small contaminants have a much greater total surface area than lumps, so there will be many more chances for reactions to take place. There will be an increased number of successful collisions, so the reaction rate increases. Only rubbing wear and early rolling fatigue mechanisms generate particles predominantly smaller than 15 μm. These are particularly dangerous. The chemical nature of the contaminant can influence the development of deposits. The contaminant can be a metal ion or particle or organic or inorganic particle or even a fiber from corrugated cardboard. A particulate provides a different reaction mechanism, which has lower activation energy, allowing a greater proportion of all the molecules to collide. Solid contaminants or metal can provide a surface on which the reactant molecules can temporarily stick in the correct orientation for an easy reaction to take place.

There are two ways in which metals and contaminants can facilitate the oxidation process. When two different molecules bump into each other, they might react to make new chemicals. How fast a chemical reaction occurs depends on how frequently the molecules collide. Metal

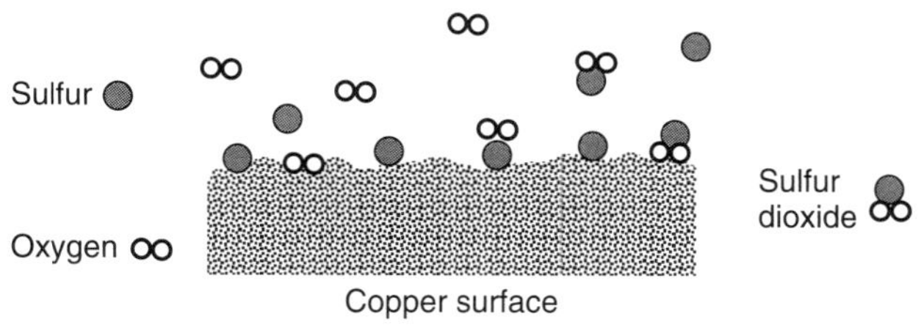

Figure 16-13 Sulfur dioxide development.

particulates and inorganic or organic contaminants make a chemical reaction go faster by increasing the chance of the molecules to collide. The first method is by adsorption; the second method is by the formation of intermediate compounds.

16.15.3 Adsorption

Adsorption occurs when a free radical, reactive species, or reactive oil molecule sticks onto the contaminant or reactive metal surface (typically, yellow metals). The following is an example: Copper is a typical metal found in many valves and bearings. It can act as a catalyst when in contact with sulfur. Sulfur is found in many base oils and certain additives and even fuel. Sulfur and oxygen react and the resulting reaction produces sulfur dioxide then sulfur trioxide. Sulfur trioxide is very reactive with water and produces sulfuric acid, which will corrode metal, break oil down to form reactive species, and can deplete additives. The molecules of the sulfur and oxygen get adsorbed (stuck onto) the surface of copper very easily. Because the two molecules are held so close together, it is more likely that they will collide and therefore react with each other (Fig. 16-13). The sulfur and oxygen have a strong attraction to react. The sulfur dioxide easily falls off the copper surface, leaving space for more sulfur and oxygen to react and form sulfur dioxide. Sulfur trioxide will then react with water to form sulfuric acid. Less energy is required for the molecules to react (see Fig. 16-14). The copper acts as a meeting place for the reactive species to bond. The same can be said for wear metals and outside contaminants.

Some deposits develop by forming intermediate reactive compounds. The chemicals involved in the reaction in this case are reactive species such as free radicals or reactive oil molecules combined with the metal or contaminant that make an intermediate compound. This new compound is very unstable. When the intermediate compound breaks down, it releases the new compounds in the form of deposits and more contaminants to react further. The sulfur dioxide example is one of several that occur. Carboxylic acids and nitric acids can also form, but the reactions are very complicated and go beyond the scope of this chapter.

16.16 DOCUMENTING FAILURE

Failure has to be understood and documented for it to be addressed and corrected. To accomplish this, it is essential that management spend time on the floor understanding why machines break down. Once this occurs, management must also spend time understanding the legitimate corrective actions and how to implement these changes successfully. In many instances, what looks good on paper may not always make the most sense on the floor. A simple checklist can aid in understanding how failures occur.

1. What happened:
   - Define incident and equipment specifics.
   - Interview all personnel directly and indirectly.
   - Gather any physical evidence (worn gear, seized bearing, etc.).
2. Where did it happen:
   - Identify the specific area affected.
   - Has this occurred before and on different equipment?
3. When did it happen:
   - Time frame of occurrence.
   - What was the sequence of events that lead up to the failure?

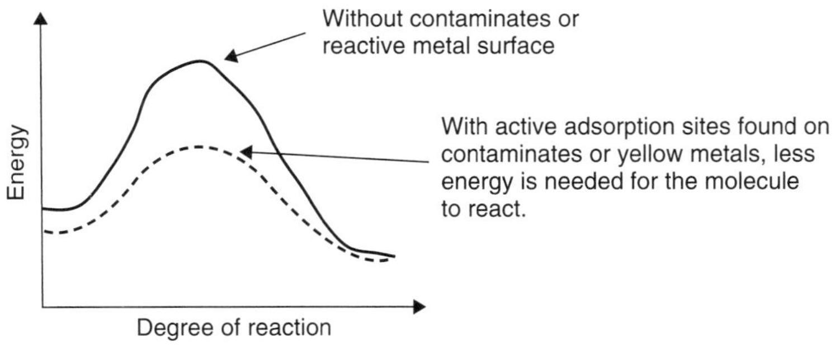

Figure 16-14 Degree of reaction.

4. List changes, if any:
   - Products
   - Practices
   - Procedures
   - Environment
5. Who was involved:
   - Personnel directly and indirectly involved
   - Supervision and procurement/purchasing
6. What was the Impact:
   - Was anyone injured?
   - Financial impact in terms of downtime, parts replacement, labor costs, direct and indirect cost pools

Anytime that management interviews technicians involved in failure, it is an uncomfortable, if not hostile, situation. Care should be taken to convey the message that the questions asked are required. The questions are asked not to assign blame, but rather, to fix the system. The relationship between subordinate and management can be tenuous when compounded with a costly situation such as equipment failure. Production and maintenance technicians may not be straightforward regarding the actual reasons that a particular incident occurred. Sometimes the line workers or mechanics feel that the reasons for failure are management or company related and that bringing them to light may cost them their jobs. In some cases, an outside "interviewer" has to be called in who can develop the background data impartially and in confidence. This provides a useful proposal that will reduce, if not eliminate, the failure that has occurred.

To use this information effectively, the following six steps can be taken to document the failures and set the groundwork for identifying the various standards to promote increased reliability:

1. Develop a failure definition, contact flow diagram, gap analysis, preliminary worksheet, and interview schedule.
2. Interview facility personnel to determine what the failures are, their frequency, and the impact.
3. Input into an electronic spreadsheet and determine any redundancies.
4. Determine the true cost of these failures.
5. Determine the "significant few": Determine the 20% or less of the failures that result in 80% of the losses.
6. Verify that the results are valid.

Once you have done this, many of the obvious situations will be validated and other situations will present themselves that were not normally considered. The use of an equipment maintenance management application will cut hundreds of hours of time required to do this work.

First things first: We must first understand why we need to buy something to begin with because what we have is broken or isn't lasting long enough or is costing us more than it should. In other words, it has failed us. We must establish a root cause of the failure, then identify the performance requirements and the standards that address those properties to offset the failure. Only then can the performance-based purchasing specification be written properly. This process is only going to make sense for the things that have failed us that end up costing (directly and indirectly) thousands of dollars, due to the failure.

16.16.1 Root Cause Investigation

The steps in root cause identification are:

- *Examination*. Study the failed part or material and the system in which it operates. Care must be taken to inspect parts for damage, contamination, moisture, cracks, or other signs of stress.
- *Identify the failure*. Look for patterns, manifestation, form, or arrangement of the failure.
- *Application*. Examine the work performed by the motor and the characteristics of those types of loads as well as the operating environment and the demands put on the material or device.
- *Maintenance history*. Look at the work performed any note any replacement parts along the way.
- *Interview*. Ask those people who are in close contact with the material or part. Remember that this is not an inquisition but an inquiry.

When done properly, all relevant information pertaining to the application, appearance, and maintenance history would be available prior to the actual inspection of the material or failed part. In real life, however, the process usually consists of inspecting the failed part or material and then acquiring information about the application and its maintenance history. Document the information in a form similar to Table 16-2.

During the analysis, information regarding the product or its components is vital. The following information is a good

TABLE 16-2 Failure Mode Worksheet

Note Physical Change of Part
Deformation
Fracture
Wear
Molecular transition (rust, corrosion, oxidation)
Other
Note Environmental Influence
Force type
Temperature
Time/age
Chemical
Other
Note Cadence
Steady
Random
Cyclic
Other
Note Articulation
Amplitude (strong/weak)
Frequency (fast/slow)
Other
What Part Was Affected
Surface and work inward
Inside and work outward
Entire part, material, device, or tool all at once
Other

starting point for beginning an investigation. The analysis can be carried out for a specific project, a specific product, or a product subsystem or component. An option to consider is a severity grading system. This may be required to help sort through many different items for which specifications are being considered. List the following:

1. *System/component/function*. The specific name of the element for which the analysis is performed is noted.

2. *Potential failure modes*. The way a product, product subsystem, or component can fail while it is being used is noted. The person(s) performing the analysis must at this point be creative and think of all possible ways in which the product can be used and fail.

3. *Potential effects of failure*. For each potential mode of failure a potential effect is noted. A description as detailed as possible is required. A potential mode of failure could have several effects. All must be noted.

4. *Severity*. Each failure effect is rated for its severity, typically on a scale of 1 to 10, 1 representing the least serious and 10 the most serious. Typically, each effect is scaled as follows:

(9–10) Potential safety risks and legal problems, potential loss of life, or major dissatisfaction.

(7–8) High potential customer dissatisfaction: potential serious injury or major product mission disruption.

(5–6) Medium-level potential customer dissatisfaction: potential small injury, product mission inconvenience or delay.

(3–4) The customer may notice the potential failure and may be little dissatisfied or annoyed.

(1–2) The customer will probably not notice the failure: undetectable failure.

5. *Criticality*. The quick identification of critical failures can be achieved by noting them in the column provided.

6. *Potential cause of failure*. Each failure must have a cause. It is very important to identify and note them. Possible causes could include wrong tolerances, poor alignment, operator error, missing component, fatigue, defective component, and lack of maintenance.

7. *Occurrence*. It is necessary to note the likelihood of failure occurring. A probability assessment must be made and scores of 1 to 5 must be noted for each failure by the person(s) performing the analysis. A typical scoring board is given below.

(5) Very high probability of occurrence

(4) High probability of occurrence

(3) Moderate probability of occurrence

(2) Low probability of occurrence

(1) Remote probability of occurrence

8. *Current design controls*. Design controls that can be used to reduce or eliminate the potential failure must be identified and noted.

9. *Detection*. A scale must be set for failure detection and the results of the scaling noted. A typical scaling board is given below.

(5) Zero probability of detecting the potential failure cause

(4) Close to zero probability of detecting the potential failure cause

(3) Not likely to detect potential failure cause

(2) Good chance of detecting the potential failure cause

(1) Almost certain to identify potential failure cause

It is likely that more than one failure mode will be identified during the analysis. It is important to prioritize these by assigning a risk priority number to each of them. This number is calculated according to the equation risk priority no. = severity rating × occurrence rating × detection rating

This number will provide information regarding the design risk, and it is best to give immediate attention to the items with the highest numbers. This exercise will assist in developing information for the development of a comprehensive purchasing specification.

16.16.2 Failure Examination

At this point you should have a firm understanding of why a tool, component, or part fails. This will prove essential for understanding why failure has occurred, and once you understand this, action can be taken. As the failure mode is understood, the documentation process can begin in earnest. You have now been introduced to various sources of failure. This information is extremely important and valuable when reviewing with the mechanics and technicians as well as vendors who will be pitching solutions. Use Table 16-2 to help organize your findings and resolve the incident.

Failure Mode Worksheet Key

1. Static (excessive elastic deformation, yield, buckling, collapse, etc.)
   - Bulk material properties (yield strength, modulus of elasticity)
   - Usually happens on first load application
2. Dynamic time dependent
   - Corrosion
   - Wear
   - Stress corrosion cracking
   - Chemical or product damage
   - Creep
   - Fatigue
   - Weak link mechanism (importance of defects, statistical behavior)
   - Damage builds with time (some of the most catastrophic and costly in-service failures)
   - Sensitive to load levels, temperature, exposure time, number and rate of load cycles, environment (difficulties in testing and analysis)
   - Interaction effects (synergistic)
3. Process factors
   - Loads
   - Material
   - Environmental factors
   - Manufacturing and process variables
   - Conditions of local stress and strain

REFERENCES

1. Ashby, M. F., and Jones, D. R. H., *Engineering Materials*, Vol. 1, *An Introduction to Their Properties and Applications*, Pergamon Press, Elmsford, NY, 1980.
2. Avallone, E. A., and Baumeister, T., III, *Mark's Standard Handbook for Mechanical Engineers*, McGraw-Hill, New York, 1996.
3. Brooks, C. R., and Choudhury, A., *Failure Analysis of Engineering Materials*, McGraw-Hill, New York, 2002.
4. Brostow, W., and Corneliussen, R. D., *Failure of Plastics*, Hanser Publishers, Munich, Germany, 1989.
5. Carlson, R. L., and Kardomateas, G. A., *An Introduction to Fatigue in Metals and Composites*, Chapman & Hall, London, 1996.
6. Collins, J. A., *Failure of Materials in Mechanical Design: Analysis, Prediction, Prevention*, Wiley, New York, 1993.
7. Douglas Frink, L. J., and van Swol, F., A molecular theory for surface forces adhesion measurements, *Journal of Chemical Physics*, vol. 106, 1997.
8. Fischer, M. A., *Engineering Specifications Writing Guide*, Prentice-Hall, Englewood Cliffs, NJ, 1983.
9. Frost, H. J., and Ashby, M. F., *Deformation-Mechanism Maps: The Plasticity and Creep of Metals and Ceramics*, Pergamon Press, Elmsford, NY, 1982.
10. Goodell, B., Nicholas, D. D., and Schultz, T. P., *Wood Deterioration and Preservation*, ACS Symposium Series 845, American Chemical Society, Washington, DC, 2003.
11. Harper, C. A., *Handbook of Materials for Product Design*, McGraw-Hill, New York, 2001.
12. McCrum, N. G., Buckley, C. P., and Bucknall, C. B., *Principles of Polymer Engineering*, Oxford Science Publications, Oxford, UK, 2003.
13. Moad, G., *Polymer Bulletin*, vol. 29, 1992, pp. 647–652.
14. Norton, R. L., *Design of Machinery*, McGraw-Hill, New York, 2008.
15. Oberg, E., Jones, F. D., and Horton, H. L., *Machinery's Handbook*, 27th ed., Industrial Press, New York, 2004.
16. Park, C. S., *Contemporary Engineering Economics*, 3rd ed., Prentice Hall, Upper Saddle River, NJ, 2007.
17. Rasis, E. P., *Technical Reference Handbook*, America Technical Publishers, Orland Park, IL, 1991.

18. Shigley, J. E., and Mischke, C. R., *Mechnical Engineering Design*, 5th ed., McGraw-Hill, New York, 1989.
19. Sullivan, W. G., Wicks, E. M., and Luxhoj, J. T., *Engineering Economy*, 13th ed., Prentice Hall, Upper Saddle River, NJ, 2008.
20. Turner, S., Creep of polymeric materials, in *Encyclopedia of Materials: Science and Technology*, Elsevier Science, Oxford, UK, 2001.
21. Wulpi, D. J., *Understanding How Components Fail*, 2nd ed., ASM International, Materials Park, OH, 1999.

17

MECHANICAL INTEGRITY OF PROCESS VESSELS AND PIPING

OLIVER A. ONYEWUENYI
MISOL Technology Solutions, Katy, Texas

This chapter builds on a focused and practical coverage of engineering aspects of mechanical integrity as it relates to failure prevention of pressure boundary components in process plants. Principal emphasis is placed on the primary means of achieving plant integrity, which is the prevention of structural failures and failure of pressure vessels and piping, particularly any that could cause significant consequences. Practical concepts and applicable calculation methodologies are presented for the fitness-for-service assessment and condition monitoring of process piping systems and pressure vessels. The chapter underscores the importance of interactions and cooperation between the three key functions of engineering—materials, mechanical, and process—with operation and maintenance in achieving the optimum mechanical reliability level in the plant. It is gauged at the awareness/knowledge level.

In this respect, the learning objectives are as follows:

- To assist the reader in clearly understanding and applying the various aspects of engineered safety to ensure mechanical integrity for vessels and piping in a responsible and cost-effective manner.
- To enhance the knowledge and skills of the reader in flaws and damage identification and analysis; and in risk assessment and management.
- To provide the reader with practical and effective methods, tools, and reference to perform fitness for service assessments and to understand the significance of flaws and practical remedial measures to assure mechanical integrity.
- To enable the reader to understand and be able to explain the purpose of a fitness-for-service assessment and its applicability.
- To be able to identify and describe commonly used applications of fitness-for-service and mechanical integrity assessments, the relevant codes and standards, as well as regulatory, environmental, and economic pressures.
- To obtain a more in-depth awareness competency level in preparation for developmental opportunities where the user will gain a knowledge level of competence.

This chapter is particularly valuable for refinery and petrochemical plant technical professionals, engineers, inspectors, and maintenance personnel, as well as for project and consulting engineers and technical personnel involved in plant mechanical integrity and reliability. It provides a review of the various aspects of engineered safety and mechanical integrity for vessels and piping components in refineries, oil and gas plants, and petrochemical plants.

In a broad sense, mechanical integrity is a program of integrated activities involving proper equipment and dependable human performance with clear accountabilities, goals, roles and responsibilities, effective program management, strong management commitment, and resource allocation. When developed and implemented properly, mechanical integrity ensures that process plants are run safely, reliably, and profitably, with minimum threats to the environment, the public, and the workforce.

Process Plant Equipment: Operation, Control, and Reliability, First Edition. Edited by Michael D. Holloway, Chikezie Nwaoha, and Oliver A. Onyewuenyi.
© 2012 John Wiley & Sons, Inc. Published 2012 by John Wiley & Sons, Inc.

The systematic activities related to mechanical integrity of pressure vessels and piping systems in process plants include [4]:

- Fabrication: pre-service, and in-service
- Inspection, testing, and preventive maintenance
- Condition monitoring and surveillance
- Fitness for service and equipment deficiency resolution
- Training of all affected personnel
- Quality assurance, repairs, and operational excellence
- Implementation of relevant procedures and codes and standards
- Implementation of "recognized and generally accepted good engineering practices" (RAGAGEPs)

The following areas are highlighted:

- Characterization of types of flaws, damage mechanisms, and equipment deficiencies (Section 17.2)
- Inspection and condition monitoring of process vessels and piping (Section 17.3)
- Fracture mechanics and fitness-for-service assessments (Section 17.4)
- Control and prevention of brittle fractures (Section 17.5)
- Case histories and examples of fitness-for-service applications to cracks in process plant pressure vessels (Section 17.6)

Continued Industry Focus and Regulations In process plants, equipment is exposed to volatile, flammable, and toxic substances that must be contained safely by the hundreds of vessels and piping segments in the plant. A breach of pressure boundary or structural integrity of structural elements can lead to disastrous consequences: fatality, environmental disaster, severe costs and penalties, and negative press with loss of shareholder returns. Although the incidence of high consequence failures of pressure vessels and piping in process plants is rare due to improved awareness, better materials and manufacturing quality, and good compliance to modern codes and standards, there are examples of recent high-consequence mechanical integrity failures, such as:

- Norco, Louisiana fluid catalytic cracking unit explosion
- Romeoville, Illinois amine absorber failure
- Richmond, California hydro cracking unit fire

In addition to the reported cases of large property losses in the petrochemical industry [8], the results of a survey by the American Petroleum Institute (API) on the cracking of process equipment as summarized in Tables 17-1–17-3 suggest a high probability of cracking in process units exposed to wet H_2S and other corrosive environment or upset conditions [11]. Successful implementation of mechanical integrity activities such as fitness for service (FFS), risk-based inspection (RBI), condition monitoring (CM), and awareness and training of management and plant staff will contribute further to a lower incidence of catastrophic failures in process plants. This makes the mechanical integrity of structural elements and pressure vessels and piping very important and a continued area of focus in the industry.

In addition to the foregoing, there are current and future trends for process plants that make the mechanical integrity of structural elements and pressure boundary components increasingly critical.

TABLE 17-1 Selected API Survey Results

Company	Vessels Inspected	% Cracked
1	273	43
2	609	43
3	101	16
4	344	12
5	617	13
6	203	6

Source: [11].

TABLE 17-2 Cracking in Relation to Steel Grade

Grade (ASTM)	% Cracked	% Deep Cracks
A212-B, A285-C, A516-65/70	37	18
A515-70	62	46
A70	67	8
A516-60	8	2

Source: [11].

TABLE 17-3 NACE T-8-16A Survey: Cracking vs. Process Unit

Process Unit	% Cracked	Process Unit	% Cracked
Crude distilling	19	Hydrocraker	28
Vacuum flasher	21	Cat. reformer	34
Atm. lt. ends	38	LPG	41
Cat. fractionation	41	Amine	21
Cat. lt. ends	45	Sour water stripper	28
Coker fractionation	18	Sulfur plant	27
Coker lt. ends	30	Flare system	30
Hydrotreater	25	Other	23

Source: [11].

- Earlier mechanical integrity focused on design, procurement, and commissioning issues, whereas recent focus includes operation, inspection, repair, and monitoring to deal with equipment aging and degradation.
- Increased repair or replacement cost and more focus on material and resource conservation have created pressures for reuse of equipment, design-life extension of process equipment, and the need to extend steady-state run times since shutdowns and restarts in process plants can sometimes increase the hazards and risk of failure.
- Modern process plants in refineries and petrochemical plants are now more complex, in order to handle variability in feedstock and different crudes and blends, which together make predictions of future performance and reliability more difficult.
- The trends noted above for modern process plants create more emphasis on mechanical integrity of pressure vessels and piping. This is also recognized and supported by the current codes and standards, which provide design optimization, increased consideration for environmental severity through material selection and resulting equipment re-rating, risk-based inspection, fitness-for-service, and integrated mechanical integrity assurance programs.

17.1 PERSPECTIVES ON MECHANICAL INTEGRITY, FITNESS FOR SERVICE, AND CONDITION MONITORING

What are "mechanical integrity," "fitness for service," and "condition monitoring" as applied to pressure vessels and piping in process plants? How are they interrelated, and why is each important? Next we describe mechanical integrity, fitness for service, and condition monitoring and their interrelationship in the context of failure prevention for pressure vessels and piping in process plants.

17.1.1 Mechanical Integrity

Given the high risk of failures in process plants, there has been growing concern and liability for the protection of people, property, and the environment. In the United States, federal regulations requiring mechanical integrity programs are being enforced and include:

1. OSHA 29 CFR 1910.119: Process Safety Management (PSM)
2. EPA 40 CFR Part 68: Risk Management Program (RMP)
3. MMS 30 CFR Part 250: Safety and Environmental Management System (SEMS) for Outer Continental Shelf Oil and Gas Operations

An effective mechanical integrity program must comply with regulatory requirements. It must also prioritize risks and consider the unique aspects of the process- and plant-specific factors. It should make the best use of data and best practices in planning and scheduling activities to achieve safety gains while decreasing cost and downtime.

The mechanical integrity elements of PSM, RMP, and SEMS in the regulations cited above are intended to ensure that pressure vessels and piping systems are designed, fabricated, installed, tested, inspected, monitored, and maintained in a manner consistent with service requirements, manufacturers' recommendations, industry standards, and RAGAGEPs to promote safety and environmentally sound operations.

The two failure modes that are addressed primarily by the mechanical integrity of process vessels and piping in this chapter are:

1. Preventing loss of containment in pressure boundary equipment, such as pressure vessels, process vessels, storage tanks, and process piping
2. Preventing unnecessary shutdown and startup of processes, as such activities can introduce hazardous conditions in the process

The risk of unscheduled shutdown is prevalent if in-service or online inspection detects evidence of flaws or if there are process upsets. In such cases, the significance of such flaws or process upsets must be assessed using methods in this chapter to ascertain the fitness for continued service for the process equipment.

To assure the mechanical integrity of pressure equipment, it is necessary to carryout the following tasks [2,4]:

1. Establish the prevalent failure or damage mechanism for the specific equipment.
2. Detect the onset of a failure condition such as the presence of a crack, excessive vibration, or evidence of bulging or distortion.
3. Perform an FFS analysis to assess the condition of the equipment, the safety margin against final failure, and the remaining life prior to final failure based on the damage propagation rates.
4. Deploy measures to prevent premature failure based on the results of the FFS, the overall assessment of the health, safety, and environment (HSE), and business risks. Example measures include:
   - Online condition monitoring
   - Change in process parameters to reduce damage susceptibility or a damage driving force such as prevalent stresses
   - Revision of the inspection interval

- Ensure the integrity of ancillary systems such as:
  - Verifying the operation of emergency shutdown (ESD) systems and alarms
  - Confirming the operation of interlocks
  - Verifying the operation of bypass or backup systems

17.1.2 Condition Monitoring

Condition monitoring (CM) is an aspect of mechanical integrity aimed at observing, measuring, and/or trending of indicators with respect to some independent parameter (time, cycles, wall thickness, crack size, etc.) to indicate the current and future ability of a structure, component, or system to function within the intended and acceptance criteria. Since the primary objective of a mechanical integrity program is to prevent failures and if they occur unavoidably, to minimize the HSE risk, CM is used to detect the early warning signs of a potential failure. In this respect, both predictive maintenance and nondestructive testing (NDT) are integral elements of CM and hence mechanical integrity. Figure 17-1 describes a schematic degradation curve under CM and shows the window between an early warning signal (acceptance criterion) and the final loss of integrity. As shown in Fig. 17-1, some amount of in-service degradation is unavoidable and thus acceptable. However, the acceptance criterion for in-service degradation should be clear, robust, and well documented. It should maintain a sufficient safety margin to ensure detection, monitoring, and implementation of effective mitigation measures to avoid loss of integrity.

Examples of CM activities under mechanical integrity are:

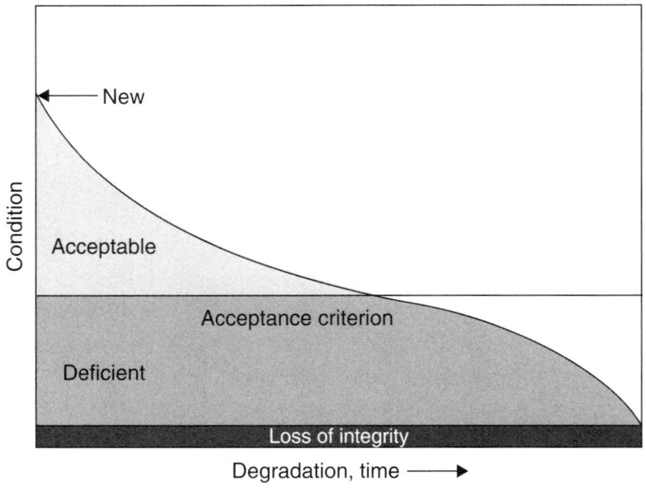

Figure 17-1 Schematic on acceptance criterion and deficiency region of degradation under condition monitoring. (From [4].)

- Temperature measurement
- Dynamic monitoring
- Corrosion monitoring
- Fatigue or crack monitoring
- Electrical testing and monitoring
- Performance monitoring
- Nondestructive testing such as ultrasonic measurement, fiber optics, and acoustic monitoring
- Oil analysis
- Observation and surveillance

17.1.3 Fitness for Service

Fitness for service is an aspect of mechanical integrity aimed at assessing the significance of flaws, process upsets, or deviations during operations on the suitability of components and equipment to meet their intended applications and service life without failure. FFS technology has evolved to become a pervasive multidisciplinary technology for mechanical integrity and has in recent years been incorporated into various industry standards and codes. Its elements are:

- Materials engineering and technology, including corrosion and environmentally induced cracking
- NDT, including various on- and off-line condition monitoring for flaws and damage behavior
- Fracture mechanics to relate stress, crack driving force, material strength and fracture resistance, and flaw and component geometries
- Statistics and probabilistics to provide input on variabilities in flaw size, material properties, and sampling requirements and probability of flaw detection

In Section 17.4 we address FFS in detail and in Section 17.6 provide examples of case studies and applications of FFS analysis.

It is a fact of life that industrial equipment may deteriorate during service. Although every effort is made to anticipate relevant degradation mechanisms and introduce appropriate safeguards during the design phase, issues of safety, reliability, and integrity can still occur during operation. This may be a result of a degradation mechanism not anticipated during design, changed operational and environmental conditions, or external damage.

Fitness for service is a means of assessing a particular component or system that deviates from design, to allow informed decision making with regard to future service. Typical questions in this respect may be:

- Is continued safe operation justified?

- Can we repair, or do we have to replace?
- Given the lead time for replacement or time before the next scheduled turnaround, can we operate in the meantime?
- What do we need to do in the short term to mitigate the risk of failure?

Good workmanship during the manufacture of equipment together with good inspection practices can play a significant role in reducing the likelihood of problems in subsequent operation. However, as discussed later, flaws are an inherent part of pressure vessels and piping systems, and depending on the flaw size, inspection requirements, and competence of the inspector, flaws are missed when vessels and piping systems are put into service. Pre-service inspection and hydrotesting as part of the commissioning process help in the majority of cases to ensure that under operating conditions, a sufficient (stress) safety margin exists such that slight imperfections can be tolerated.

Peaking and other forms of misalignment at welds can introduce localized bending stresses through the wall of a pipe or pressure vessel upon pressurization. Typically, these additional stresses are not always considered during vessel design, as they are deemed to be localized secondary stresses and to have little influence when determining pressure-retaining ability and design wall thickness. The weld itself may be seen as a metallurgical notch in the structure where local residual stresses form during solidification of the weld metal. Welding imperfections, such as lack of fusion, can provide preferential sites for corrosion, metallurgical embrittlement, or the formation of cracklike flaws, depending on the environment and service conditions. Actual local stresses can therefore be significantly different from those considered during a design calculation.

Given the broad range of construction materials and operating conditions available, the preexistence of manufacturing and material imperfections inherently missed by procurement inspections, and the prevalence of in-service degradation mechanisms seen within the industry, FFS analysis is a necessary technology and broadly applicable concept for vessels and piping systems in process plants. Key to demonstrating fitness for service is a demonstration that the original safety margins have not been eroded, despite the changed state of the component.

In summary, FFS assessment may be required for a number of reasons [5]:

1. Maintaining the safety of plant personnel and the public
2. Complying with regulatory requirements, such as PSM, RPM, and SEMS rules
3. Protecting the environment from accidental releases of damaging substances
4. Assuring a safety margin and reliably operating aging facilities
5. Maintaining safe and reliable operations with increased run lengths and decreased shutdown periods
6. Determining the feasibility of increasing the severity of operations
7. Resolving the damage found by more rigorous in-service inspections than those found by inspections performed during original construction

The class societies, such as DnV and Lloyds, also recognize the need and application of FFS for assuring mechanical integrity. According to Lloyds' Register [7], FFS may be needed when:

1. An asset or piece of equipment lacks original design information or may have exceeded its original design or useful life.
2. The asset was manufactured before 1987, and the equipment operates at a relatively low temperature or is exposed to autorefrigeration.
3. Equipment has been relocated or reassigned to a different service.
4. Equipment has been exposed to in-service upsets and remains in service.
5. Equipment shows an indication of preexisting flaws.
6. Equipment is subjected to high-temperature creep, fatigue, or environmentally induced cracking.
7. Inspection findings reveal a condition that may affect the future operation of the asset, such as metal loss, distortion (misalignment, out of roundness, bulges, or dents), laminations, cracking or blisters, and so on.

17.2 TYPES OF FLAWS AND DAMAGE MECHANISMS

17.2.1 Flaws or Discontinuities Versus Defects

All materials contain flaws, but a flaw is considered a defect only if it is of sufficient severity (size, shape, location) to exceed its limiting value and hence compromises the mechanical integrity of the material or component for the service intended. A flaw becomes a defect when it exceeds its limiting value. Equipment is considered fit for service despite its flaw content if by analysis it is demonstrated that the flaws are not defects. A flaw is a discontinuity that can occur from the molten state, termed an *inherent discontinuity*; if it occurs during fabrication, finishing, or machining it is termed a *fabrication or processing discontinuity*; if it occurs during service it is called an *in-service discontinuity*. For brevity, we consider *discontinuity* to be synonymous with *flaw*.

17.2.2 Types of Flaws

- Inherent metallurgical flaws commonly found in castings or ingots include porosities or cavities, blowholes, shrinkage, inclusions, cold cracks and hot tears, segregation, cold shut/lap, cold shot, coreshift, and infused chaplets.
- Processing flaws include notches and stress raisers, seams, laps (folds), laminations, stringers, forging bursts and cracks, grinding cracks, machine tears, quench cracks, and strain aging cracks.

17.2.3 Weld Flaws

Metal joining by welding introduces a special class of processing flaws, which are very significant because of their high probability of occurrence in welded components such as process vessels and piping. Volumetric changes during solidification and restraints associated with welding can cause distortion, residual stresses, cracks, and volumetric flaws in welded components. Weld repairs, particularly for equipment that has been in service and/or degraded, are particularly prone to weld-related flaws, distortion, and cracking. For brevity, only the names of the weld flaws are mentioned here. More detailed descriptions and causes of each type of weld flaw and associated illustrations are available in the authoritative reference: [15]. Examples of weld flaws are:

1. *Weld (planar) cracks*: hot cracks, weld bead root cracks, underbead (hard) cracking, heat-affected zone cracking, liquid metal cracking, and fisheyes.
2. *Volumetric weld flaws*: gas porosity—blowholes, piping, wormholes, slag, incomplete penetration, incomplete fusion, undercut, burn-through, underfill, overlap; electrode inclusions—tungsten, weld wire, etc. icicles and grapes.
3. *Geometric weld flaws*: high–low and misalignment result from mismatched or poorly restrained joints and can serve as stress raisers and failure initiation sites in service.
4. *Miscellaneous weld flaws*
   a. *Splatter*. These are small molten particles deposited on parent metal away from the weld joint. In service, a splatter can serve as a site for localized corrosion such as pitting.
   b. *Arc strike*. This is a small area of burned or focused metal on the base material caused by inadvertent striking of the hot electrode on the base metal. Such flaws can be sites for embrittlement, crack initiation, localized corrosion, or stress concentration during service.
   c. *Heat-affected zone*. Although not considered a flaw, as it is inherent in any weldment, its material properties contribute to the heterogeneity and strain incompatibility in the structure. It can also be the preferred site for environmentally induced cracking, selective corrosion (galvanic or localized), and in-service creep and embrittlement.

17.2.4 In-Service Flaws and Environmentally Assisted Flaws

As stated earlier, every structural element or pressure boundary contains flaws at the time of installation and commissioning. These components generally meet the relevant design codes and company specification since the flaws are within the acceptable limits of the relevant design codes. The challenge, however, is that these preexisting flaws can enhance the initiation and propagation of in-service-induced flaws, environmentally induced cracking, and in-service degradation and embrittlement by interacting with the applied and residual stresses, process environment, service time, and temperature. Poor material selection or substitution can increase a material's embrittlement susceptibility in a given service. In other words, at a certain size and severity and loading, flaws can quickly become defects and lead to failure. Therefore, effective quality control/quality assurance, good workmanship, and proper material selection can be the most cost-effective ways of reducing the severity of inherent and fabrication flaws at precommissioning of the equipment, particularly if the loading conditions and service environment are considered severe.

In the context of service-induced flaws and material degradation, flaws are characterized by their original damage mechanism, which can initiate structural failure or loss of pressure boundary containment in process plants. These can be grouped into cracklike (planar or sharply notched) or non-cracklike flaws (volumetric). For brevity, only the names and brief descriptions are covered in Table 17-4, but more authoritative and detailed sources are provided [3,15].

17.2.5 In-Service Degradation and Susceptibility of Various Alloys

Expertise and careful consultation on material selection are most critical during the feasibility, process/concept selection, and front-end design (FEED) phases of a process plant. The mechanical integrity activities must ensure a clear definition of the process environment fluids and

TABLE 17-4 Examples of Service-Induced Flaws

Damage Group	Flaw Type
Cracklike Flaws	
Fatigue cracking	High cycle fatigue
	Low cycle fatigue
	Corrosion fatigue
	Creep fatigue
	Thermal fatigue
Stress corrosion cracking	See Table 17-5 and SCC of alloys in Section 17.2.6.
Hydrogen-assisted cracking	Hydrogen embrittlement
	Sulfide stress cracking in H_2S environment
	Hydrogen-induced cracking
	Hydrogen blistering
	Stress-oriented hydrogen-induced cracking
	High-temperature hydrogen attack
Creep cracking (creep)	
Liquid metal cracking	
Notchlike and Volumetric Flaws	
Corrosion	General corrosion
	Localized corrosion
	Crevice corrosion
	Pitting corrosion
	Erosion corrosion
	Microbial corrosion
	Selective or intergranular leaching
	Local thinned-area flaws
Erosion and wear	Abrasive wear
	Adhesive wear
	Fretting
Carburization/de-alloying	

Source: [15].

TABLE 17-5 Degradation Mechanisms for Various Vessel and Piping Alloys

Material	Mechanism
Carbon steel	Caustic SCC
	Wet H_2S cracking
	Amine SCC
	Ammonia SCC
	Carbonate cracking
	Deaerator cracking
	Hydrogen attack
	Hydrogen embrittlement
	Spheroidization
	Creep
	Oxidation
	Sulfur corrosion
	HCl corrosion
	Sulfuric acid corrosion
	Naphthenic acid corrosion
	Sour water corrosion
Cr–Mo Steels	Temper embrittlement
	Creep
	Hydrogen attack
	Hydrogen embrittlement
	Oxidation
	Sulfur corrosion
Austenitic stainless steel	Caustic SCC
	Chloride SCC
	Polythionic SCC
	Intergranular corrosion
	Sigma-phase embrittlement
	Chloride pitting
	Sulfidation
	Creep
Ferritic stainless steels	885 embrittlement
All materials	Mechanical fatigue
	Thermal fatigue
	Erosion

Source: [3].

temperatures, including upset, steady-state, and shutdown and startup conditions. Care must be taken to ensure that the right materials match the process environment and service conditions prior to final project execution and plant construction.

It is noteworthy that different alloy systems are more susceptible than other alloy systems to loss of integrity when exposed to a given fluid or service. Table 17-5 is a list of degradation mechanisms and susceptibilities for various alloy systems. For example, austenitic stainless steels are more expensive than carbon steels but are more resistant than carbon steels to corrosion and hydrogen-assisted cracking (HAC). However, they are more susceptible than carbon steel to stress corrosion cracking (SCC) in a chloride environment. Consideration should also be given to the ability to detect the damage in service, ease of condition monitoring, and the rate of propagation of the damage during service. Corrosion is easier than SCC to measure and monitor. Short exposure of duplex stainless steels and ferritic stainless steels to high temperatures during local heating or temperature spikes can result in embrittlement.

17.2.6 HAC and SCC Susceptibility of Various Alloy Systems

In general, alloy composition, stress, preexisting flaws, temperature, and concentration of the active species in the process fluid are critical parameters for controlling susceptibility. In the following, the susceptibility of various alloy systems to hydrogen-assisted cracking and stress corrosion cracking in various environments is highlighted [3,15].

17.2.6.1 Carbon and Low-Alloy Steels

1. *Caustic.* Caustic embrittlement or caustic SCC in boilers, and concentrations of caustic in crevices, pits, and notches, enhance the caustic SCC. Stress relief or stress reduction is beneficial.
2. *Nitrates.* Intergranular SCC occurs in nitrate solutions. Lower pH and increased stress will increase susceptibility, but alkaline additions to lower pH can lower SCC susceptibility.
3. *Ammonia.* SCC has been noted; both intergranular and transgranular SCC are possible. Addition of 0.2 wt/% water is beneficial.
4. *Hydrogen.* Transgranular hydrogen ambrittlement (HE) cracking occurs in high-alloy steels at ambient temperatures, in high hardness > HRC 22 spots in weldments, and as HIC/SOHIC/blister cracks in a wet H_2S environment in low-strength steel vessels and piping. Reducing stress, postweld heat treatment, and lower acidity are beneficial for HE in high-strength steels. There are no clear remedies for wet H_2S cracking.

17.2.6.2 Austenitic Stainless Steels

1. *Halides* SCC of austenitic stainless steel in halide solutions, particularly chlorides, is quite common and is often associated with pitting. Chloride leaching from wet insulating materials has caused SCC in austenitic stainless steel. Good practice is to paint the steel prior to insulation. Iodine solutions are also known to cause SCC in austenitic stainless steels. In general, chloride SCC occurs more commonly above 150°F, and susceptibility increases with increased chloride concentration and stress. However, pressurized components have also cracked at ambient and lower temperatures. Nickel and nickel-based alloys should be considered an alternative in hot halide solutions.
2. *Polythionic acids* Sulfurous compounds in most environments produce polythionic acids and cause SCC, particularly in sensitized austenitic stainless steels.
3. *Caustic* Caustic SCC has been observed in austenitic stainless steels. Cracking can be transgranular and is enhanced by traces of chlorides in the caustic solution.
4. *Hydrogen environment* In general, austenitic stainless steels are resistant to hydrogen embrittlement and other hydrogen-related cracking, such as HAC, HIC, or SOHIC.

17.2.6.3 Ferritic Stainless Steel These are generally more resistant than austenitic stainless steel to SCC. However, ferritic stainless steels are susceptible to 885°F embrittlement. In general, the susceptibility of ferritic stainless steels to SCC will increase with traces of copper or nickel in the alloy and the degree of embrittlement.

17.2.6.4 Martensitic and Precipitation Hardenable Stainless Steels These alloys are susceptible to SCC, and the risk increases with yield strength, stress, and temperature.

17.2.6.5 Aluminum Alloys Several aluminum alloys are subject to SCC. Cu, Mg, Mn, or Zn in the alloy increases the susceptibility. Increase in yield strength, temperature, or applied stress increases the susceptibility to SCC.

17.2.6.6 Copper Alloys SCC of brass, known as "seasonal cracking of brass," can occur at ambient conditions at low stress, and the risk increases with the presence of ammonia, amines, and mercury salts.

17.2.6.7 Nickel and Nickel-Based Alloys These alloys are generally susceptible to SCC, depending on alloy composition, heat treatment, temperature, and environmental severity. They are also three to six times more expensive than stainless steels.

17.2.6.8 Titanium Alloys These alloys are more resistant to SCC in common environments where other alloys show susceptibility. However, SCC has occurred in some specific conditions. Contact with cadmium at 610°F in chlorides has resulted in SCC. Hot-salt SCC of titanium alloys has been reported at 500 to 800°F in chloride solutions. Chloride and HCl have induced SCC in titanium alloys.

17.3 INSPECTION, CHARACTERIZATION, AND MONITORING OF FLAWS

Different flaws and material damage are best detected and sized by specific inspection techniques. The probability of detection and the accuracy of flaw sizing are dependent on the inspection technique, competence of the inspector, accessibility, and type of flaw.

17.3.1 General Metal Loss and Local Thinned Area Corrosion

For general surveys, radiographic examination is used to qualify the existence, extent, and depth of a corroded area of vessel or pipe. For high accuracy for point thickness readings and thickness profiles and surveys, the method recommended is ultrasonic thickness (UT) measurements.

Surface preparation is critical and in most cases wire brushing is sufficient. Surface grinding may be necessary if the corroded area is filled with scale or corrosion product. Proper calibration is critical for an uncorroded wall thickness used as reference, specific wall thickness ranges, ultrasonic velocity, and temperature. Profile gauges are used to measure the root diameter of the corrosion damage.

17.3.2 Pitting and Crevice Corrosion

Precise measurements and sizing of corrosion pits are difficult. This is because pits have irregular depth profiles, varied root sharpness, and are often obscured with corrosion products. Ultrasonic (UT) and radiographic (RT) techniques may be used to characterize pit damage. Pit gauges are used to measure pit depth, and rulers or calibers are used for surface diameter and pit spacing measurements. Pitting charts are available for assigning pitting severity grades [2,15].

17.3.3 HIC, SOHIC, and Blister Damage

When HIC or SOHIC appear as surface-breaking flaws, they can be detected by surface inspection techniques such as magnetic particle inspection (MPI) or liquid (dye) penetrant (LP) inspection. However, in many cases, the damage is subsurface and requires specialized UT techniques. The reader is referred to the NACE Standard for detailed guidelines on examination techniques for HIC and SOHIC [10]. In many cases, some grinding may be necessary, as some HIC and SOHIC flaws nearer the surface can shield those beneath them and lead to erroneous damage assessment. Surface blisters are readily detectable by visual examination, but subsurface blisters normally require UT techniques. Assessing the peripheries of blisters is critical for detecting crack-like flaws that may have initiated from blisters. UT is the recommended technique for detecting and sizing such flaws that emanate from blister damage.

17.3.4 Cracklike and Sharp Flaws

Cracklike and sharp flaws present the greatest risk to the mechanical integrity of pressure boundary containment equipment and structural elements. This is particularly the case when such damage is environmentally induced or assisted. Therefore, detection, characterization, and sizing of such flaws require specialized skills, competencies, and techniques. For fitness-for-service assessment, the important parameters to be defined are crack length, spacing between cracks, crack depth, distance to the surface, and angle of inclination to the surface. For surface cracks, crack length, spacing between the cracks, and surface orientation are determined by magnetic particle inspection or liquid (dye) penetrant inspection. UT techniques are required for crack depth and angle of inclination to the surface. For embedded cracks, specialized techniques such as angle-beam UT (e.g., time-of-flight diffraction or pulse echo UT) are used.

In general, the accuracy of sizing of cracks and the probability of crack detection vary and depend on:

1. Inspection method
2. Experience and competence of the inspector
3. Accessibility
4. Surface condition and preparation
5. Interference of indications from multiple flaws

Accordingly, it is advisable to use multiple inspection techniques wherever possible to improve the probability of detection and the accuracy of the characterization and sizing of the cracks. In addition, for services involving environmentally induced cracking or material degradation (including creep and hydrogen attack), it is good NDE practice to cut "boat samples" or take surface replicas of metallographic surfaces and use this to ascertain the severity of changes in the underlying metallurgical structure.

17.3.5 Online Condition Monitoring of Damage

NDT can be performed nonintrusively on a vessel or pipe segment during operation without the need for a shutdown. The technique used is a combination of 0° compressive wave UT, time-of-flight diffraction and 45° shear wave UT. Acoustic emission has also been demonstrated under specific conditions to be a workable technique for condition monitoring for crack damage. In all cases it is necessary to ensure proper and routine testing and calibration of the monitoring system.

17.4 FRACTURE MECHANICS AND FITNESS-FOR-SERVICE ASSESSMENT

New equipment construction codes such as ASME and API provide rules for design, fabrication, inspection, and testing of pressure boundary containment equipment such as pressure vessels, piping systems, and storage tanks. However, they do not provide rules for assessing the significance of flaws developed in-service or missed during precommissioning inspection but discovered during service. Also, there were no rules for in-service degradation caused by material aging, embrittlement, creep, fire, or service upsets. Inspection standards such as the API 510, API 570, API 653, and ANSI NB-23 codes and standards provide for in-service inspection, repair, alteration, and re-rating of in-service pressure vessels, piping systems, and storage tanks. However, they have no methodologies for explicitly assessing the significance of flaws, material degradation, or upsets in service or environmental conditions on the fitness for service of the components. The API-579-1/ASME FFS-1 standard [2] has been developed to fill this gap. It defines FFS as "engineering evaluations that are performed to demonstrate the structural integrity of an in-service component containing a flaw or damage." It provides standardized methodologies for assessment of the significance of flaws, acceptance criteria, remaining life, remediation, in-service monitoring, and data requirements. It covers flaws such as general metal loss, local thinned areas, pitting corrosion, hydrogen-assisted cracking (blisters), HIC and SOHIC, weld misalignment and shell distortions, crack-like flaws, creep, fire damage, dents and gouges, and laminations.

There are three possible levels of assessment in an FFS assessment. The levels vary in terms of the data required, complexity of analysis, and the level of competence required to undertake the analysis. It should be noted that regardless of the analysis level used, FFS is a multidisciplinary technology and therefore requires teamwork involving people with the relevant skills and discipline. Each decision including validity of input data and required level of assessment, as well as the recommendations, should be reviewed and validated by the team.

For level 1, a minimum amount of inspection and material data is required, and the results are usually the most conservative. The analysis can be performed readily by a plant inspector or plant engineer. Simple nomograms are generally available for this analysis. For level 2, the same inspection data but more information on material, loading, and environmental conditions are required than in level 1. The results are less conservative. The analysis is performed by experienced FFS engineers and practitioners. For level 3, detailed data on inspection, material, loading, process, and component history are required. The analysis uses advanced analytical tools such as finite element analysis and experts to perform the FFS analysis. Level 3 results can be less conservative than level 1 results.

Because of the comprehensive nature of the API-579 standard, only a brief overview and highlights are covered here. The reader is encouraged to refer to API 579-1/ASME FFS-1 when researching and performing FFS. The reader is also referred to various FFS software and resources.

17.4.1 Applicable Codes and Standards

Table 17-6 provides a list of codes, standards, recommended practices, and reports related to FFS technology. The API 579-1/ASME FFS-1 is recognized and referenced by API codes and standards (510, 570, and 653) and by NB-23. It provides methods and procedures intended to supplement these codes and standards. It can be used for assessment and re-rating of equipment designed and constructed according to the following codes:

- ASME Boiler and Pressure Vessel Code Section VIII, Divisions 1 and 2
- ASME Boiler and Pressure Vessel Code Section I
- ASME B31.1 Piping Code
- ASME B31.3 Piping Code—Process Piping
- API 650
- API 620
- Other recognized codes, including international standards, can be considered (being cognizant of differences in attributes, safety margins, and nomenclature with ASME/API codes)

Other international codes and standards relevant to FFS analysis include:

- BSI PD 5500 (United Kingdom)
- Stoomwezen: rules for pressure vessels (The Netherlands)
- EN 13445 (European Union)
- AD Merkblätter (Germany)
- BS 7910 (United Kingdom): general code with a focus on fatigue and fracture
- R5 (United Kingdom): code for elevated temperature integrity and life assessment, developed by the British nuclear industry, but generally applicable
- R6 (United Kingdom): code for fracture, fatigue, and advanced fracture methods, developed by the British nuclear industry, but generally applicable
- DnV RPF-101 (Norway): corrosion-defect assessment of pipelines

17.4.2 When FFS Is Needed

As discussed in Section 17.1.3, FFS assessment is often required for a number of reasons:

- Deficiencies (flaws, damage, distortion, etc.) found during inspection are to be resolved.
- Inspection acceptance criteria for a degraded condition of an in-service pressure vessel and piping must be defined.
- Alternative FFS documentation is to be used for "grandfathered" equipment where the original records are not available.
- An asset lacks original design information or may have exceeded its service life.
- An asset was manufactured to older or obsolete codes but is still in service (e.g., manufactured before 1987).

TABLE 17-6 Codes, Standards, Recommended Practice, and Reports on MI and FFS

Title	Identification
Pressure Vessel Inspection Code: Maintenance Inspection, Rerating, Repair and Alteration	API 510
Calculation of Heater-Tube Thickness in Petroleum Refineries	API Std 530
Piping Inspection Code: Inspection, Repair, Alteration, and Rerating of In-Service Piping Systems	API 570
Damage Mechanisms Affecting Fixed Equipment in the Refining Industry	API RP 571
Inspection of Pressure Vessels	API RP 572
Inspection of Fired Boilers and Heaters	API RP 573
Inspection of Piping, Tubing, Valves, and Fittings	API RP 574
Recommended Practice for Inspection of Atmospheric- and Low Pressure Storage Tanks	API RP 575
Inspection of Pressure Relieving Devices	API RP 576
Recommended Practice for Inspection of Welding	API RP 577
Recommended Practice for Positive Material Identification	API RP 578
Recommended Practice for Risk-Based Inspection	API RP 580
Base Resource Document—Risk-Based Inspection	API Publ 581
Design and Construction of Large, Welded, Low-Pressure Storage Tanks	API Std 620
Welded Steel Tanks for Oil Storage	API Std 650
Tank Inspection, Repair, Alteration, and Reconstruction	API Std 653
Steels for Hydrogen Service at Elevated Temperatures and Pressures in Petroleum Refineries and Petrochemical Plants	API RP 941
Avoiding Environmental Cracking in Amine Units	API RP 945
National Board Inspection Code	NB-23
Minimum Design Loads for Building and Other Structures	ASCE 7
Rules for Construction of Power and Boilers	ASME B&PV Code Section I
Boiler and Pressure Vessel Code, Section II, Part A—Ferrous Material Specifications	ASME B&PV Code Section II, Part A
Boiler and Pressure Vessel Code, Section II, Part B—Nonferrous Material Specifications	ASME B&PV Code Section II, Part B
Boiler and Pressure Vessel Code, Section II, Part D—Properties	ASME B&PV Code Section II, Part D
Subsection NH—Class 1 Components in Elevated Temperature Service	ASME B&PV Code Section III, Division 1
Boiler and Pressure Vessel Code, Section VIII, Pressure Vessels Division 1	ASME B&PV Code Section VIII, Division 1
Boiler and Pressure Vessel Code, Section VIII, Pressure Vessels Division 2, Alternative Rules	ASME B&PV Code Section VIII, Division 2
Rules for Inservice Inspection of Nuclear Power Plant Components	ASME B&PV Code Section XI
Alternative Method to Area Replacement Rules for Openings Under Internal Pressure, Section VIII, Division 1	ASME B&PV Code Case 2168
Alternative Rules for Determining Allowable Compressive Stresses for Cylinders, Cones, Spheres and Formed Heads Section VIII, Divisions 1 and 2	ASME B&PV Code Case 2286
Factory-made Wrought Steel Buttwelding Fittings	ASME B16.5
Manual for Determining the Remaining Strength of Corroded Pipelines	ASME B31G
Power Piping	ASME M31.1
Process Piping	ASME B31.3
Specification for General Requirements for Steel Plates for Pressure Vessels	ASTM A20
Electric-Fusion-Welded Austentic Chromium–Nickel Alloy Steel Pipe for High-Temperature Service	ASTM A358
Standard Test Methods and Definitions for Mechanical Testing of Steel Products	ASTM A370
General Requirements for Specialized Carbon and Alloy Steel Pipe	ASTM A530
Electric-Fusion Welded Steel Pipe for Atmospheric and Lower Temperatures	ASTM A671
Electric-Fusion Welded Steel Pipe for High-Pressure Service at Moderate Temperatures	ASTM A672
Carbon and Alloy Steel Pipe, Electric-Fusion Welded for High-Pressure Service at High Temperatures	ASTM A691

(continued)

TABLE 17-6 *(Continued)*

Title	Identification
Standard Practices for Cycle Counting in Fatigue Analysis	ASTM E1049
Standard Test Method for Measurement of Fracture Toughness	ASTM E1820
Test Method for the Determination of Reference Temperature, T_0 for Ferritic Steels in the Transition Range	ASTM E1921
Standard Test Method for Measurement of Fatigue Crack Growth Rates	ASTM E647
Test Method of Tension Testing of Metallic Materials	ASTM E8
Standard Guide for Examination and Evaluation of Pitting Corrosion	ASTM G46
Specification for Unfired Fusion-Welded Pressure Vessels	BS PD 5500
Method for Determination of K_{Ic}, Critical CTOD and Critical J Values of Welds in Metallic Materials	BS 7448: PART 2
Code of Practice for Fatigue Design and Assessment of Steel Structures	BS 7608
Guide on Methods for Assessing the Acceptability of Flaws in Structures	BS 7910
Guidance on Methods for Assessing the Acceptability of Flaws in Fusion-Welded Structures	BS PD 6493
Methods for the Assessment of the Influence of Crack Growth on the Significance of Defects in Components Operating at High Temperatures	BS PD 6539
Design of Steel Pressure Pipes	DIN 2413 Part 1
Design of Steel BENDS Used in Pressure Pipelines	DIN 2413 Part 2
Summary of the VERAGE Stress Rupture Properties of Wrought Steels for Boilers and Pressure Vessels	ISO/TR 7468-1981(E)
Guidelines for Detection, Repair, and Mitigation of Cracking of Existing Petroleum Refinery Pressure Vessels in Wet H_2S Environment	NACE Std RP0296
Assessment Procedures for High-Temperature Response of Structures	Nuclear Electric R-5
Assessment of the Integrity of Structures Containing Defects	Nuclear Electric R-6
Damage Mechanisms Affecting Fixed Equipment in Fossil Electric Power Industry	WRC 490
Damage Mechanisms Affecting Fixed Equipment in the Pulp and Paper Industry	WRC 488
Damage Mechanisms Affecting Fixed Equipment in the Refining Industry	WRC 489
Evaluations of Design Margins for ASME Code Section VIII and Evaluation of Design Margins for ASME Code Section VIII, Divisions 1 and 2—Phase 2 Studies	WRC 435
A Procedure for Safety Assessment of Components with Cracks—Handbook	SAQ/FoU-Report 96/08
Method of Assessment for Flaws in Fusion-Welded Joints with Respect to Brittle Fracture and Fatigue Crack Growth	WES 2805

Source: [2].

- Equipment is exposed to unanticipated low temperature or exposed to autorefrigeration or blowdowns.
- Equipment is exposed to an unanticipated service environment, such as souring of process fluid, CP failures, or fluid leakage or carryovers.
- A decommissioned asset may be used for a different service.
- Equipment is exposed to high temperature or cyclic service.
- Equipment is exposed to fire, overloads, or severe unforeseen loadings (e.g., hurricanes, storms).
- Preinspection FFS analysis is used to guide inspection planning or risk-based inspection (RBI):
  1. Identifies areas of greatest vulnerability in a vessel (lowest allowable flaw size) for inspection planning.
  2. Sets minimum sensitivity or resolution level of inspection equipment for cost-effective inspection.
  3. With statistics included, provides guidelines for minimum sampling of inspection points required.

17.4.3 FFS Assessment Procedure

Figure 17-2 is a flowchart for FFS assessment procedures for various damage classes according to API 579. FFS assessment consists of the following eight steps:

1. *Determine and characterize the flaw and damage mechanism.* The first step is to determine the type of flaw, the prevalent damage mechanism, and the most likely mode of failure for the component under the prevalent loading and stress conditions at the flaw location. The flaw is assessed and characterized by the appropriate inspection techniques.

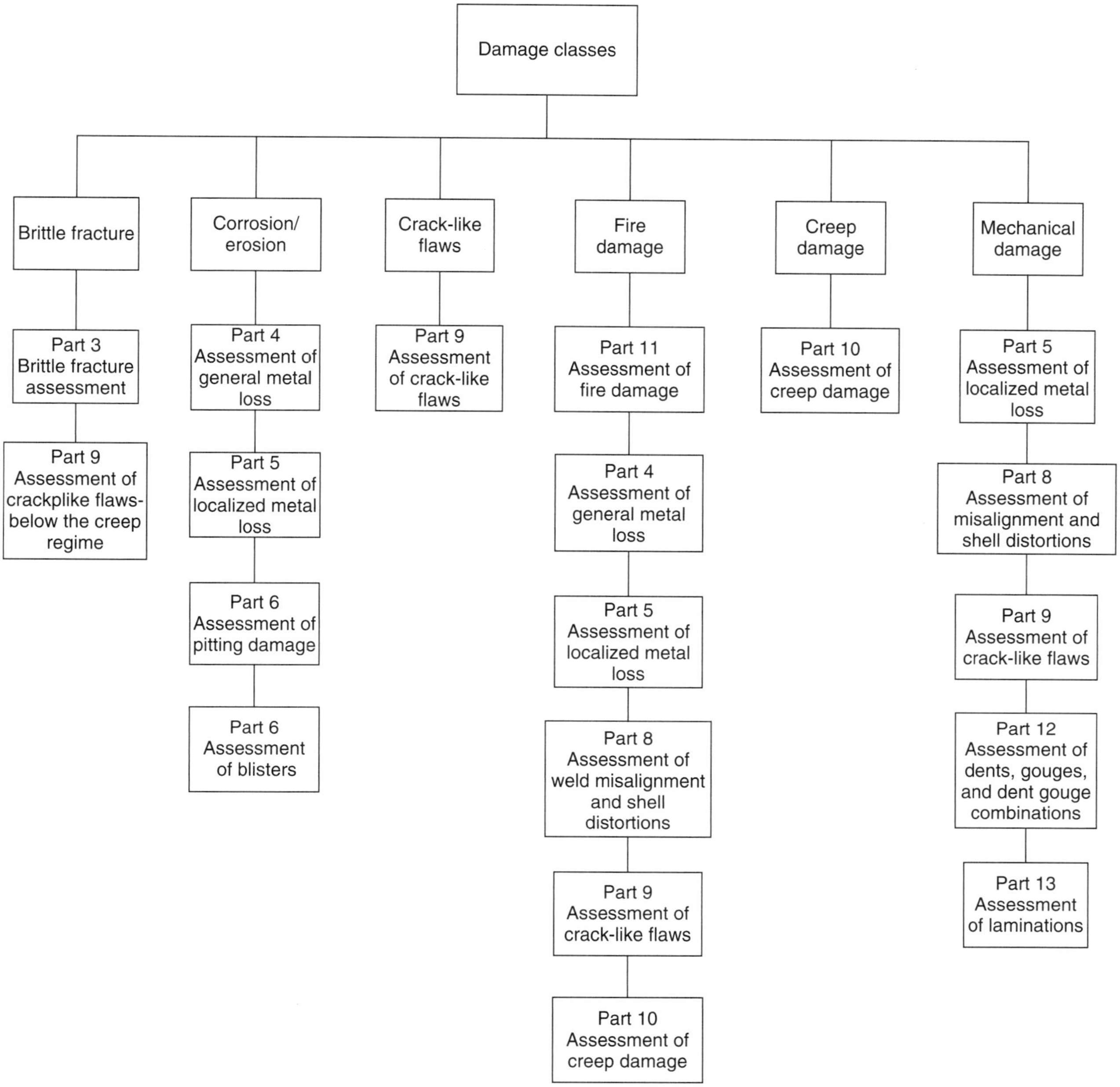

Figure 17-2 FFS assessment procedures for various damage classes. (From [2].)

It could be a single flaw (planar or volumetric), multiple flaws which can interact and join under the prevailing load and/or a susceptible environment, or a geometric stress raiser such as a notch, keyhole, weld misalignment, or local thinned area of corrosion or erosion. The prevalent damage mechanism is assessed from the service history, knowledge of the material properties and records, and the prevaling service conditions. Material degradation can manifest as loss of strength, such as creep strength due to high temperature, fatigue strength due to cyclic loading, or fracture resistance due to low-temperature exposure or environmentally induced embrittlement. API 579-1 groups damage into six classes:

- Brittle fracture or material embrittlement
- Corrosion or erosion damage
- Crack-like flaws: both environmentally and service-induced cracking, or fabrication cracks
- Fire damage
- Creep damage
- Mechanical damage (e.g., mechanical overload, cyclic load)

The most likely mode of failure is assessed by knowledge of the flaw type, the prevaling damage mechanism, and the loading conditions. The choice of FFS analysis methodology depends on the failure mode. For example, for damage associated with crack-like flaws, the mode of failure can be fracture (brittle fracture, ductile fracture, or mixed) or local plastic collapse. This is depicted by the failure assessment diagram (FAD) methodology shown in Fig. 17-2. The FAD is described in step 4.

2. *Determine the applicability and limitation of the FFS assessment procedures.* When the need for FFS assessment is established and supported by considerations in step 1, it is necessary to make sure that the FFS assessment procedures and methods established in existing standards, such as API 579, are applicable. The standards and codes clearly define the applicability, assumptions, and limitations of the assessment methodology for each flaw, loading, and damage mechanism. In general, the standard applies to pressure boundaries of pressure vessels, boiler components, piping, and shell courses of storage tanks in process plants. It also applies to structural elements such as supports, flanges, and attachments (e.g., tray support).

3. *Compile and validate the available data required for FFS assessment.* The data required depend on the prevalent damage mechanism, the type of component or structural element, and the level of assessment needed. For a simple (level 1) conservative assessment, the data sheet shown in Table 17-7 can be sufficient. For more detailed analysis for specific flaw types and damage mechanisms, API 579-1 provides further details on data requirements.

4. *Determine the assessment techniques and the acceptance criteria.*

A. ASSESSMENT TECHNIQUES. FFS assessment techniques can vary from a simple, conservative screening level (level 1) using screening curves and nomograms to a detailed and analytical level (level 3), which can involve specific material and environmental testing and characterization, including crack growth measurements, numerical techniques such as finite element methods, or statistical or probablistic analyses. The choice of assessment techniques or level of assessment depends on:

- Objectives and needs for the FFS assessment
- The complexity associated with the component geometry, flaw type, and damage mechanism
- The amount and availability of information required
- The skills and competence of the assessor

Level 1 assessments are considered the most conservative. The primary driver is to assure the safety and mechanical integrity of the process equipment, but the conservatism can be validated and moderated using level 2 and 3 assessments. Excessive conservatism may be detrimental, due to such economic and business drivers as:

- Reducing cost
- Extending the life of an existing component

In general, it is best practice to start with level 1 assessments, and sequentially proceed to a higher level of assessments if the results do not meet the acceptance criteria for FFS or if suitable mitigation or a course of action is not clear.

B. ACCEPTANCE CRITERIA. The acceptance criteria in a FFS analysis depends on the component type, flaw type, damage type, and the safety margins implicit in the original construction codes for the process equipment. For noncracked equipment, the failure criteria are based on strength capacity and the allowable pressure rating.

(1) *Allowable stress*. When it is possible to calculate the prevalent stresses from different loading conditions, including classification and superposition of stress results, the resulting stress limits can be set as a fraction of the yield strength of the material, in line with the construction codes. This applies to cases of material degradation with no planar or volumetric flaws. Special considerations are required for cases of plastic collapse and elastic–plastic analysis.

(2) *Remaining strength factor*. FFS often involves nonlinear stress analysis for elastic–plastic, limit load, and plastic collapse calculations of failure stress. This is particularly important for damage types such as fatigue, creep, and ductile rupture of flawed components. The remaining strength factor (RSF) is used in API 579 to define the acceptability of components for continued service:

$$\mathrm{RSF} = \frac{L_{\mathrm{DC}}}{L_{\mathrm{UC}}} \tag{17.1}$$

where L_{DC} is the limit load or plastic collapse for a damaged component and L_{UC} is the limit load or plastic collapse for an undamaged component.

The allowable RSF (RSFa) estimated from various traditional code formulas is given in Table 17.2.3 of API 579 as RSFa = 0.90 for the following design codes:

- ASME Sec. I
- ASME Sec. VIII, Div. 1
- ASME Sec. VIII, Div. 2
- AS 1210
- BS PD 5500
- CODAP
- ASME B31.1

TABLE 17-7 Data Sheet for FFS Analysis

The following data are required for most types of Fitness-For-Service assessments and it is recommended that this completed table accompany the data table completed for the specific damage type that are located in the respective Part of this standard.

Equipment Identification: ________________

Equipment Type: ________Pressure Vessel ________Piping Component ________Boller Component ________Storage Tank

Component Type & Location: ________________

Design Code: ________ASME Section VIII Div. 1 ________ASME Section VIII Div. 2 ________ASME Section 1

________ASME B31.1 ________ASME B31.2 ________API 650 ________API 620

________other: ________________

Material of Construction (e.g. ASTM Specification): ________________

MAWP or MFH: ________________

Minimum Required Wall Thickness: ________________

Temperature: ________________

Cyclic Operation: ________________

Type of Damage

Metal Loss –General: ________________

Metal Loss –Local: ________________

Metal Loss –Pitting: ________________

HIC, SOHIC & Blisters: ________________

Misalignment or Out-of-Roundness: ________________

Bulge: ________________

Crack-like Flaw: ________________

Creep Damage: ________________

Fire Damage: ________________

Dent, Gouge & Dent/Gouge Combinations: ________________

Laminations: ________________

Location of Damage (provide a sketch)

Internal/External: ________________

Near Weld: ________________

Orientation: ________________

Environment

Internal: ________________

____

External: ________________

____

Repair and Inspection History (Including any Previous FFS Assessments)

____

____

____

Operations History

____

____

____

Future Anticipated Operations

____

____

____

Source: [2].

- ASME B31.3
- API 620
- API 650

Therefore, if the RSF calculated for the in-service component RSFs $\geq$ RSFa $= 0.90$, the component is fit for service. If the RSFs $<$ RSFa, the component requires repair, replacement, or remediation of service conditions.

(3) *Maximum allowable working pressure* (MAWP). Re-rating of MAWP of in-service components can be derived from the RSF as follows:

$$\text{MAWPr} = \text{MAWP}\left(\frac{\text{RSFs}}{\text{RSFa}}\right)$$

(4) *Critical crack size criteria*. For cracked components, the failure assessment diagram (FAD) developed from fracture mechanics and shown in Fig. 17-3 is used to depict the locus of the critical flaw size above which failure will occur for fracture failure as well as the plastic collapse failure of a flawed component. A description follows.

The FAD procedure considers failure by brittle fracture on the vertical axis represented by the toughness ratio (K_r) and failure by plastic overload on the horizontal axis represented by the load ratio (L_r). Between these two limits, the failure is elastic–plastic, more commonly called *ductile fracture*. According to API 579 procedures, K_r and

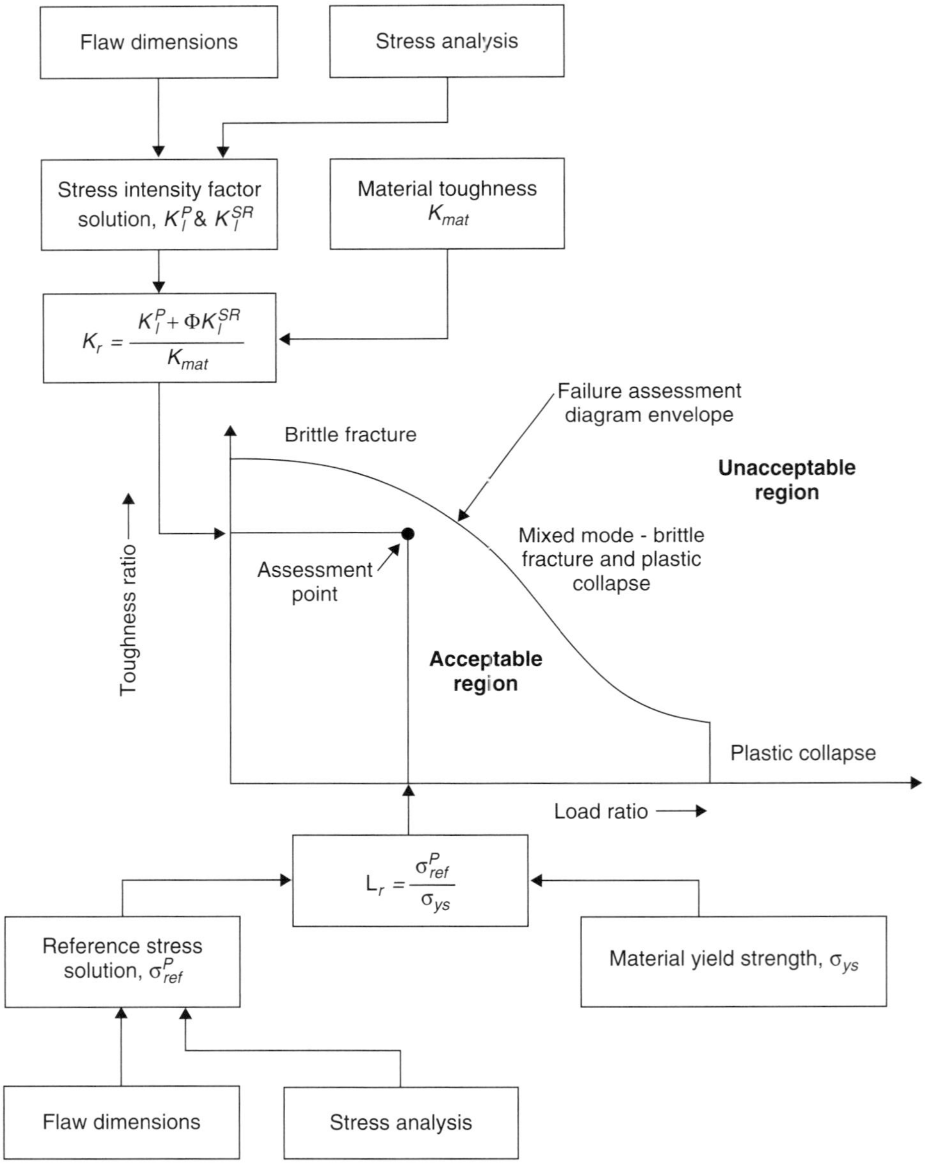

Figure 17-3 Failure assessment diagram according to API 579-1. (From [2].)

L_r are defined by

$$K_r = [1 - 0.14(L_r)^2]\{0.3 + 0.7\ \exp[-0.65(L_r)^6]\} \tag{17.2}$$

The FAD line given by Eq. (17.2) represents the locus of the critical combinations of K_r and L_r for which the crack size becomes the critical crack size for failure. Therefore, for a given vessel or piping component with a given combination of crack and component geometry, primary and secondary loads, and material properties (toughness, strength, Young's modulus, and stress–strain flow curve), a value pair of K_r and L_r is determined using the procedures and guidelines in API 579. The cracked structure is considered safe and FFS if the point with the calculated coordinates (K_r, L_r) lies inside the FAD. Failure of the flawed component is predicted if the point lies outside the FAD. It should be noted that the analysis above can be quite straightforward for level 1 and 2 analysis, and more complicated and rigorous for level 3 analysis. Commercial computer programs have been developed and used for FFS assessments.

5. *Estimate the remaining life of the flawed component.* Once the FFS analysis indicates that the component is fit for service, it is necessary to estimate how long the service can be sustained if the damage were allowed to progress. A careful estimate of the rate of propagation of the damage (e.g., corrosion rate, creep rate, crack growth rate) is required for the remaining life calculation. Typically, this is obtained from actual measurements, inspection records, or estimated from literature. API 579/ASME FFS-1, Annex F Standard provides some database and references for material properties and damage rate estimation.

Remaining life estimates are used to:

- Establish an appropriate inspection interval
- Develop an appropriate in-service condition monitoring plan
- Assess the need and effectiveness of remediation and mitigation measures
- Assess the risk of operating the component to the next planned or scheduled turnaround, particularly if discovered during service as part of a routine monitoring

6. *Implement appropriate remediation.* The probability of loss of mechanical integrity is controlled by the following factors:

- Severity of environment: temperature, fluids, and so on
- Severity of the loading: stress (including residual stresses), pressure, and so on
- Size of the flaw
- Extent of material degradation

Therefore, remediation is focused on reducing the severity of any or all of the failure drivers above. Examples are provided in API 579 as well as applicable repair codes such as API 510, API 570, API 653, and ANSI NB-23 on suitable remedial measures for various damage mechanisms. For example, for the case of corrosion damage remediation, the following activities may be considered:

- Avoid or minimize further corrosion by adjusting and controlling the operating envelope.
- Clean and possibly recoat damage sites.
- Review the effectiveness of the corrosion inhibitors.
- Replace or repair the insulation and ensure that it is self-draining.
- Check the coatings and the cathodic protection systems for integrity and effectiveness.
- Adjust the inspection or monitoring frequency as appropriate, based on the results of the FFS.

7. *Implement in-service monitoring.* When there is uncertainty in the rate of crack growth, or rate of damage progression, it is necessary to supplement the FFS analysis and decision for continued service with an in-service condition monitoring program. For corrosion, corrosion probes or electrochemical impedance measurements are used to monitor corrosion rates online. For hydrogen-assisted cracking, hydrogen probes provide data on hydrogen absorption. For crack growth monitoring, ultrasonic examination methods and acoustic emission are used. Most recently, technologies involving fiber optics distributed temperature sensors have been deployed for structural integrity monitoring.

8. *Provide documentation.* A critical element for compliance to mechanical integrity assurance is full documentation of FFS assessments. In accordance with API 579 code, the following should be included in the documentation:

a. The equipment design data, and maintenance and past operational history to the extent available, should be documented for all equipment subject to FFS assessment.

b. Inspection data, including all readings utilized in the FFS assessment.

c. Assumptions and analysis results, including:

(1) Part, edition, and assessment level of this standard and any other supporting documents used to evaluate the flaw or damage.

(2) Future operating and design conditions, including pressure, temperature, and abnormal operating conditions.

(3) Calculations of the minimum required thickness and/or MAWP.

(4) Calculations of remaining life and the time for the next inspection.

(5) Any remediation or mitigation or monitoring recommendations that are a condition for continued service.

All calculations and documentation used to determine the fitness for service of a pressurized component should be kept with the inspection records for the component or piece of equipment in the owner–user inspection department. This documentation will be a part of the records required for mechanical integrity compliance.

17.5 CONTROL AND PREVENTION OF BRITTLE FRACTURE

17.5.1 Definitions

The following terms have the definitions given below within the context of brittle fracture prevention for pressure vessels and piping in process plants.

1. *Lower design temperature* (LDT) is the lowest temperature at which equipment may be subjected to its design pressure.
2. *Lowest ambient temperature* (LAT) is the lowest ambient temperature to which the equipment is exposed.
3. *Minimum metal temperature* (MMT) is the lowest temperature that equipment may possibly attain due to process or atmospheric conditions.
4. *Critical exposure temperature* (CET) is the lowest process or atmospheric temperature at which equipment metal will be exposed to a given stress. The CET may be a single temperature at an operating pressure or an envelope of temperatures and pressures or stresses. CET is used primarily under the FFS assessment route. The methodology for determining the CET is given in API 579/ASME FFS-1 Part 3. In broad terms, CET is equivalent to MMT when reference is not made to particular applied pressures or stresses.
5. *Minimum allowable temperature* (MAT) is the permissible lower metal temperature limit for a given material at a specified thickness based on its brittle fracture resistance. MAT can be a single temperature—in which case it is conceptually the same as the LDT—or an envelope of allowable temperatures as a function of pressure.

17.5.2 Brittle Versus Ductile Fracture

Construction materials can fracture in an unstable manner, causing catastrophic ruptures or brittle fragmentations.

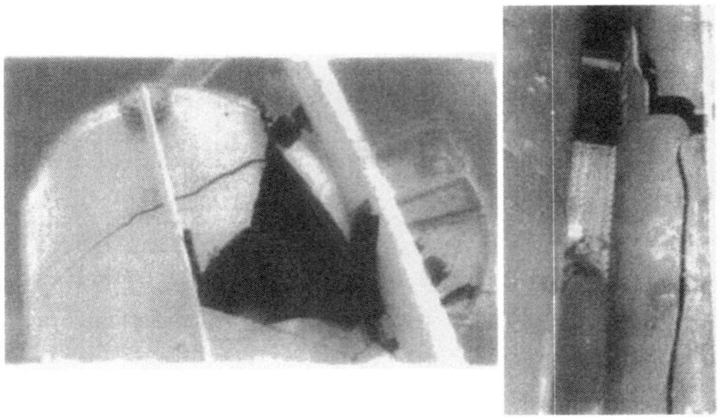

(a) (b)

Figure 17-4 Brittle fracture failure in (a) a vessel and (b) a piping segment.

When brittle fracture occurs in pressure vessels and piping systems in service, the consequences can be severe. Figure 17-4 shows examples of brittle fracture of a pressure vessel and a piping component. Generally, ductile materials may undergo large deformations or show leakages prior to fractures; brittle materials do not and instead result in fragmentation. The material property that distinguishes ductile materials from brittle materials is called *fracture toughness*. Ductile materials are characterized by high toughness levels. Most steels undergo a ductile-to-brittle transition in fracture toughness, when metal temperatures decrease with decreasing operating and/or ambient temperatures.

Prevention of brittle fracture of steel equipment is largely a matter of design when new plants are being built or when a design review is required because of plant changes. Equipment is designed in accordance with design codes that make provisions for toughness acceptance criteria. The design codes that are covered specifically are PD 5500, ASME section VIII, Division 1 or 2, and EN 13445 for pressure vessels, and ASME B31.3 for piping. Commonly, the mechanical design and the methods of fabrication and inspection are specified independent of the lower design temperature, while the minimum toughness requirements are specific for low-temperature equipment. Basically, all design codes specify the required toughness by a maximum allowed transition temperature for the Charpy-V impact toughness of steel-base material and weld. However, the philosophies in the various codes on the specification of the toughness to avoid brittle fracture differ widely. Briefly the basic differences are:

- In PD 5500, the impact toughness is specified for equipment operating below 0°C, and the transition temperature required is a function of the lowest metal temperature in service, the steel thickness, the heat treatment condition, and to a lesser extent, the stress level.

- In ASME, the impact toughness transition temperature must be lower than the lowest metal temperature in service, irrespective of wall thickness and heat treatment, when the steel is loaded to the maximum allowable stress. However, ASME VIII and ASME B31.3 allow significantly lower steel toughness in the case of low design stresses. Exemptions from impact testing are granted for any specific steel quality, depending on wall thickness, heat treatment, and mechanical loading conditions.
- EN 13445-2:2002 Annex B has different methods; the required impact toughness under method 1 for low-strength steels resembles ASME, and under method 2 is similar to PD 5500. All pressure vessel–quality steels are supposed to be impact tested and then meet the required impact toughness.

PD 5500 and EN 13445-2:2002 Annex B allow the use of fracture mechanics analysis as an alternative method to assess the fracture toughness required. Possibilities for embrittlement of materials or growth of defects during service are commonly excluded at the design stage. Nevertheless, if degradation or cracking occurs in service, a fitness-for-service assessment is required, as highlighted above. Also, a fitness-for-service assessment is required if the design of existing equipment in accordance with the original design code no longer covers the present low-temperature conditions in service.

17.5.3 Industry and Regulatory Codes and Standards for Brittle Fracture Control

The applicable codes and standards for brittle fracture control of pressure vessels and piping systems are shown in Table 17-8. For pressure vessels, including heat exchangers, the design codes are:

- PD 5500
- ASME Section VIII, Divisions 1 and 2
- EN 13445

For piping systems, the design code is

- ASME B31.3

17.5.4 Determination of the Minimum Metal Temperature

The minimum metal temperature (MMT) is the lowest temperature that equipment may possibly reach and it must be defined for every equipment item. The MMT takes into account the normal lowest operating temperature, the lowest temperatures during startup or shutdown, and exceptional cool-down events caused by high-rate depressurizing, resulting in Joule–Thompson cooling, or cooling by evaporation or boiling of liquefied gas. In the latter case the MMT is the adiabatic flash temperature (AFT) or the atmospheric boiling point (ABP), unless vacuum conditions can cause boiling below these temperatures. Warmer minimum metal temperatures need clear justification.

The basic design philosophy is that one piece of equipment has a single MMT equal to the lowest possible temperature of any of the fluids if the equipment is exposed to multiple fluids, such as in columns and heat exchangers. Also, the MMT takes into account the coldest ambient temperature of the climate at the plant location. If ambient temperatures are determining, the MMT is equal to the LAT. The MMT is the same as:

- The minimum design metal temperature in ASME Sec. VIII, Div. 1,UG-20 and ASME Sec. VIII, Div. 2, Clause 4.1.5.2 (ignoring modifications for coincident pressure, stress state, construction materials, heat treatment, etc.)
- The minimum design temperature in PD 5500:2006, Clause 3.2.5
- The minimum metal temperature in EN 13445-2:2002, Clause 3.1.1

17.5.5 Determination of the Lower Design Temperature

The lower design temperature (LDT) is the lowest temperature at which equipment may be subjected to its design pressure. This defined temperature has no equivalent in the design codes but is required to mark the lowest temperature at which equipment is protected from overpressure by a safety relief system. Generally, the LDT should be selected equal to the MMT. An LDT warmer than the MMT may be selected only if justified by significant savings of the combined capital cost and the cost of process control and production restraints. The limiting condition for the selection of an LDT warmer than the MMT is that the general membrane stress for temperatures colder than the LDT is below 50 N/mm^2 or 7.25 ksi.

Any design with an LDT warmer than the MMT requires a process whereby the pressure in the system is physically limited at all temperatures colder than LDT and requires a homogeneous process and constant equipment temperatures. Any possibility for re-pressurization or pressurization from connected systems while the equipment is colder than the LDT is excluded. *Re-pressurization while cold is to be prevented if the temperature is below the LDT*.

Instrumented protective systems that disable re-pressurization, by forced opening of the depressuring valve or forced closure of valves separating connected systems

TABLE 17-8 Applicable Codes and Standards for Brittle Fracture Control of Pressure Vessels and Piping

U.S. Standards	
Fitness for service	API 579-1/ASME FFS-1: 2007
Issued by: *American Petroleum Institute* *Publications and Distribution Section* *1220 L Street Northwest* *Washington, DC 20005*	
Process piping	ASME B31.3
ASME boiler and pressure vessel code	
Section II: Materials—Part C: Specifications for welding rods, electrodes, and filler metals	
Specification for nickel and nickel-alloy welding for shielded metal arc welding	ASME II C-SFA 5.11
Specification for nickel and nickel-alloy bare welding electrodes and rods	ASME II C-SFA 5.14
Rules for construction of pressure vessels	ASME Sec. VIII, Div. 1
Alternative rules for construction of pressure vessels	ASME Sec. VIII, Div. 2
Issued by: *American Society of Mechanical Engineers* *345 East 47th Street* *New York, NY 10017.*	
Standard practices for detecting susceptibility to intergranular attack in austenitic stainless steels	ASTM A262
Standard test methods for detecting detrimental intermetallic phase in duplex austenitic/ferritic stainless steels	ASTM A923
Issued by: *American Society for Testing and Materials* *100 Barr Harbor Drive* *West Conshohocken, PA 19428-2959*	
British Standards	
Specification for unfired fusion-welded pressure vessels	PD 5500:2006
Guide to methods for assessing the acceptability of flaws in metallic structures	BS 7910:2005
Issued by: *British Standards Institution* *389 Chiswick High Road* *London W4 4AL* *UK*	
Assessment of the integrity of structures containing defects	R6
Issued by: *British Energy Generation Ltd.* *Barnett Way* *Barnwood* *Gloucester GL4 3RS* *UK*	
European Standards	
Unfired pressure vessels	EN 13445
Part 2: Materials	EN 13445-2:2002
Issued by: *CEN* *Rue de Stassart 36* *B-1050 Brussels* *Belgium*	

Copies can also be obtained from national standards organizations.

TABLE 17-8 *(Continued)*

International Standards	
Centrifugal pumps for petroleum, petrochemical, and natural gas industries	ISO 13709:2003
Welding consumables—Covered electrodes for manual metal arc welding of nickel and nickel alloys—Classification	ISO 14172
Welding consumables—Wire and strip electrodes, wires, and rods for fusion welding of nickel and nickel alloys—Classification	ISO 18274
Issued by: *ISO Central Secretariat* *1, ch. de la Voie-Creuse* *Case postale 56* *CH-1211 Genève 20* *Switzerland*	

Copies can also be obtained from national standards organizations.

during the time the equipment is colder than the LDT, should be considered.

The LDT must be recorded explicitly on material selection diagrams and equipment data sheets. If the MMT is colder than the LDT, both temperatures are recorded. In many cases the MMT will be equal to the LAT and not colder than the LDT. In such cases a statement that the LDT is equal to the LAT is sufficient.

Notes:

1. Embrittlement of low-alloy steels in service above 350°C may increase the lowest temperature at which equipment may be subjected to its design pressure.
2. The metal temperature during pressure testing on site is above the LAT and is then generally above the LDT. If the LDT is above the LAT, a risk assessment is required before a pressure test is performed, especially in the case of pneumatic testing.

17.5.6 Toughness Requirements

The toughness requirements of the design code to allow the full design pressure at the lower design temperature must be met. Generally, the design codes specify requirements for steel strength; the type, grade, or quality of the steel; the welding procedure qualification; any postweld heat treatment; inspection methods and acceptance criteria; and so on, regardless of low design temperatures, while the minimum toughness requirements are specific for low-temperature equipment. Different design codes result in different impact toughness requirements, and the toughness specified by a code is acceptable only within the full context of that code. In all cases, the impact test temperature in the specification should be −20°C (−4°F) or colder.

17.5.7 Brittle Fracture Risk Assessment of Existing Systems [1,2]

The main approach is to determine the lower design temperature (LDT), defined as the lowest temperature at which equipment may be subjected to its design pressure. In a new design, the minimum metal temperature (MMT) is established from the process and environmental data at the equipment site. The material is then selected such that the LDT is lower than or equal to the MMT—except in the case of low applied stresses (< 50 MPa or 7.25 ksi), for which LDT may be higher than MMT.

To assess existing equipment, the design process is reversed and the LDT can be taken as the lowest temperature at which the original or applicable design code allows the equipment to be operated at the full design pressure, given the design, fabrication, inspection, and testing records available. LDT is determined without applying the concept of coincident pressure and temperature. Where equipment is designed for low-temperature service in accordance with a recognized code, the first step in the assessment is to check whether low-temperature requirements were specified properly and that equipment is operating within the limits set by the original design. If this is not the case, an LDT can be established by checking the equipment against the rules of the original design code or another applicable design code. If acceptable limits are not found using the LDT route, a more detailed check against the code-allowable pressure–temperature envelope may be undertaken. Finally, FFS methods may be employed to determine the minimum allowable temperature. The FFS route is also required if cracks or crack-like defects are present.

Results of an assessment should be documented and, where necessary, appropriate controls implemented in operating processes and procedures such as integrity operating windows. The various approaches and the basic steps that may be used to assess existing equipment are described in more detail in API 579, Section 3.2. An overall flowchart is shown in Fig. 17-5.

17.5.8 Assessment Approaches

A particular issue for existing equipment concerns the assessment of old equipment designed to codes that have been superseded. The main philosophy adopted is that equipment meeting the code requirements that were valid at the time it was constructed is considered safe, provided that experience with the equipment has not shown it to be unsafe, and the code requirements have not been found to be unsafe. However, the earliest pressure vessel and piping design codes were not specific about requirements to ensure that materials and design details are adequate to resist brittle fracture. Thus, the overall approach is that low-temperature operating limits set by design to ASME/API codes after 1989 and to BS 1515/BS 5500 after 1972 are accepted without further evaluation. More modern codes such as EN 13445 have adequate brittle fracture requirements in the earliest issues and so do not require different treatments over time. The same principle should be applied to equipment governed by other design codes and standards not mentioned above.

Piping designed and constructed in accordance with a recognized code or the approved local piping classes may be accepted without further evaluation if the piping is still being operated within the original design limits. To determine the relevant parameters for an assessment (e.g., MMT, CET), a review of the process, the original design documentation, and the operating history should be conducted early in the assessment process. All potential operating conditions, including startup, shutdown, and upset events, should be considered. For the initial check against the original design limits, the information required can be obtained from the equipment records, the nameplate, and general arrangement drawings; in certain cases, access to the manufacturer's data book may be needed to obtain the required parameters.

17.5.9 LDT and Design Code–Based Assessments

The lower design temperature (LDT) is obtained by checking the equipment against the low-temperature rules of the original design code. If the original design code predates the edition at which adequate brittle fracture requirements were introduced, or where the particular edition of the design code is no longer readily available, an LDT may be determined using an applicable (alternative) code. This will usually be the current edition of the original design code or one that is similar in scope, content, and approach to low-temperature design. The alternative design code should be selected in agreement with the design specialist. The current rules for avoidance of brittle fracture in the main pressure vessel codes are largely contained in specific sections of the codes: PD 5500 Annex D; ASME Sec. VIII, Div. 1 UCS-66, or Div. 2 Clause 3.11; and EN 13445-2 Annex B. Similar rules can be found in piping design codes such as ASME B31.3.

The detailed steps for establishing the LDT using the low-temperature rules contained in the appropriate design code are outlined below.

1. Confirm that any in-service weld repairs or modifications comply with the requirements of the original or applicable design code. An alternative approach is required if this is not the case.
2. Review the process, original equipment design data, and operating history to determine the required low-temperature duty and the LDT. The basic steps are:
   a. Establish MMT.
   b. Establish construction materials and grades; review material standard specifications, welding procedure qualification records, test certificates, impact test temperatures, and results (where available).
   c. For each component of the equipment, determine the lowest temperature at which the original or applicable design code allows operation at the full design pressure.
   d. The warmest temperature obtained in step 2(c) for all components of the equipment is the LDT. For piping constructed to piping class, the lower-temperature limit of the piping class is the LDT.
3. The equipment is acceptable if the LDT is colder than the MMT or the lowest temperature at which the equipment is subjected to a general membrane stress greater than 50 MPa or 7.25 ksi. The relevant membrane stress includes contributions from pressure and any other sustained loads and long-range stresses due to thermal constraints. The need for re-pressurisation while cold is to be considered in determining the temperature coincident with the membrane stress of 50 MPa or 7.25 ksi. The low stress criterion is not generally recommended for piping assessments, due to uncertainties regarding the contribution of system stresses and the effects of thermal transients (including startup, shutdown, and upset events).

If the LDT obtained from the steps above is not sufficiently low, an intermediate step, before proceeding to FFS, may be a detailed check against the design rules for cases where operating stresses are low relative to the allowable design stress (low stress ratio). Such rules may be used to determine allowable metal temperatures at the appropriate reduced stresses. These temperatures are termed

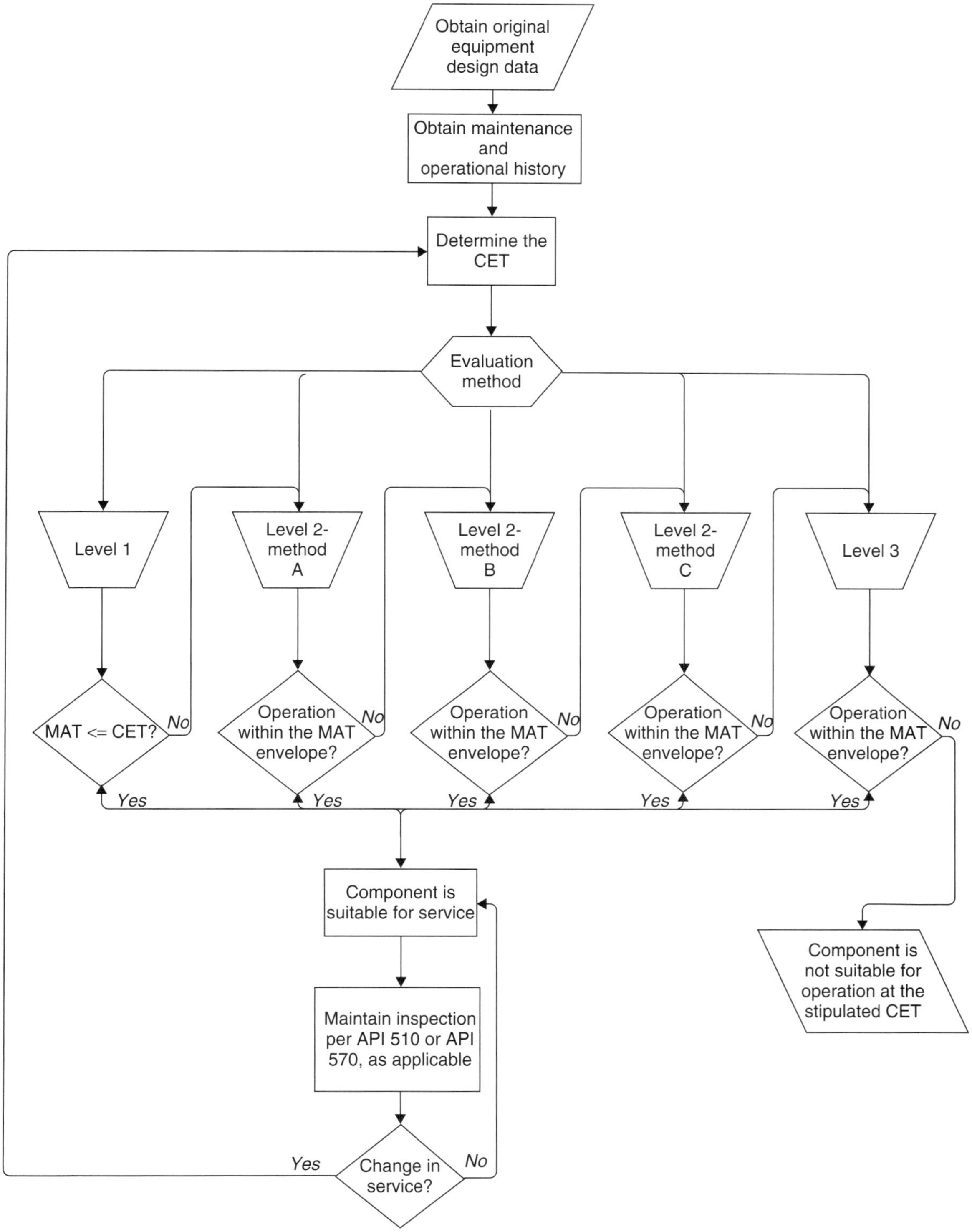

Figure 17-5 Flowchart for brittle fracture assessment of existing equipment. (From [2].)

MAT and not LDT. However, before applying the low-stress-ratio rules, the possibility of additional stresses due to re-pressurisation while cold or through connected systems must either be excluded or accounted for explicitly in the assumed stresses. This intermediate check need not be used for equipment designed according to ASME codes because assessment may proceed directly to the FFS route, which includes the appropriate low-stress-ratio code rules in the form of API 579/ASME FFS-1 Part 3 procedures.

17.5.10 FFS-Based Assessments

If the LDT assessment route and detailed design code checks fail to demonstrate operation within acceptable limits, the equipment brittle fracture resistance may be assessed using FFS procedures based on fracture mechanics. A full FFS assessment will usually require detailed data on material properties, applied stresses, and potential defect sizes. However, simpler procedures are available in API 579/ASME FFS-1 for equipment within its scope and specified limitations. This FFS option may also be adopted directly without first using the LDT approach/code-based assessments, provided that the input data required are available.

For old equipment where no records of construction material and design code exist, an FFS assessment may still be feasible if material can be taken from the equipment for mechanical testing. The vessel skirt may be a potential source of test material if it is representative of the materials used to fabricate the pressure-retaining parts of the vessel. In such cases, it will still be necessary to establish the stresses and temperatures under the operating conditions to obtain sufficient information for an assessment.

Reference should be made to appropriate best practice guides on avoidance of brittle fracture in equipment subject to in-service embrittlement, such as temper and hydrogen embrittlement. In general, except for hydrogen embrittlement, this is not a concern for equipment operating with metal temperatures below 370°C (698°F) or 230°C (445°F) in hydrogen service.

17.5.11 Assessment per API 579-1 and ASME FFS-1 Part 3

API 579-1 and ASME FFS-1 Part 3 contain procedures and guidance for assessing brittle fracture in the absence of detailed input data. For equipment properly designed to ASME or API codes and operating within specified limits, brittle fracture assessment may be conducted using API 579 or ASME FFS-1 Part 3 procedures without the need for detailed fracture mechanics analysis. The procedures are also valid for equipment designed to other recognized codes, within the scope and applicability limitations listed in API 579 and ASME FFS-1 Part 3 Clause 3.2. These procedures may be used for pressure vessels designed to any edition of ASME Sec. VIII, Div. 1 or Div. 2 and they are also valid for piping designed to ASME B31.3.

In general, level 1 and level 2 method A are recommended for determining MAT. Level 2 method B (based on hydrotest) and method C (based on construction materials, operating conditions, and past operating history) are more complicated and should only be used with caution. If level 1 and level 2 method A do not produce acceptable limits, proceeding directly to full FFS should be considered. Where the low-stress-ratio option is employed to calculate a MAT in accordance with these procedures, caution should be exercised to ensure that stresses from all relevant sources of loading have been included.

17.5.12 Full FFS Assessment [2]

If the minimum allowable temperature obtained from the simple FFS approach is not sufficient to demonstrate adequate brittle fracture resistance, a full FFS assessment may be carried out using the mechanical properties, applied stresses, and defect characteristics of the particular equipment (Fig. 17-5). Such an assessment should be carried out in accordance with recognized procedures, such as those in BS 7910, R6 and API 579/ASME FFS-1, Parts 3 and 9. The full FFS approach should also be followed if a crack or crack-like defect is known to be present in the equipment.

The FFS should be carried out by an engineer experienced in conducting such assessments. The input data shall be in line with FFS standard(s). More specifically, the defect size(s) assumed should be consistent with the inspections carried out. If no defects are found, the largest potential sizes that could be missed by the inspection technique(s) provide a basis for the defect sizes considered. Alternatively, tolerable defect sizes for a specified MAT are calculated and then compared against defects detected from an inspection program.

Stresses and temperatures assumed in the FFS consider all potential operating conditions (including startup, shutdown, and upset events). Pressure, system loads, and thermal transient effects must be considered. Piping systems require particular attention, as significant system stresses (in addition to pressure stresses) can result from thermal expansion and/or inadequate or poorly maintained piping supports. In respect to pressure loading, the FFS should as much as possible consider the full equipment design pressure. Where it is necessary to consider coincident pressures and temperatures, there should be adequate controls to prevent the possibility of exposing equipment to pressures in excess of those assumed in deriving a particular MAT.

The appropriate value of toughness for a FFS assessment may be obtained from impact properties. For non-impact-tested steels to ASME specifications, lower-bound estimates of fracture toughness may be obtained from API 579 or ASME FFS-1 Annex F. In many cases, the master curve approach can be used to establish estimates of fracture toughness from impact properties. If the full FFS route is necessary for equipment subject to in-service embrittlement or other forms of degradation, allowance is made for such effects on the inputs adopted in the assessment. In such cases, a materials engineer familiar with the effects of the degradation on material properties should be consulted on the FFS analysis.

17.6 CASE HISTORIES AND EXAMPLES OF FFS APPLICATIONS TO CRACKS IN PROCESS PLANT PRESSURE VESSELS

Case 17.1: Fitness for Service and Remaining Life Assessment of an Acetone Finishing Column [12]

Equipment background. The acetone finishing column (AFC) in a phenol–acetone plant was constructed of ASTM/ASME—SA 285C carbon steel to ASME Sec. VIII, Div. 1 code and was in service for 13 years. Its internal environment included water, acetone, cumene, α-methyl styrene, and injected caustic. The liquid phase at the bottom of the column had a pH of 11 to 12.5. The maximum operating pressure is 20 psig, and the minimum and maximum operating temperatures are 75 and 225°F, respectively. During turnaround inspection, a through-wall crack was detected. The crack was cut and repaired. Extensive UT and wet fluorescent magnetic particle testing of the repaired area and the entire vessel revealed no indication of flaws when the vessel was returned to service.

As part of the condition monitoring, an online UT inspection was begun. After two years, the online UT detected multiple cracks in the same patch and near the same weld trays that were previously weld-repaired. Most of the cracks detected were shallow surface cracks exposed to the ID surface. Two predominant surface cracks were sized by UT to be 3.25 in. long × 0.25 in. deep and 9 in. long × 0.122 in. deep, respectively. Further extensive online UT examination confirmed that all the cracking in the vessel were localized to the same weld repair area, located near the caustic injection points. Table 17-9 provides a summary of the FFS analysis data in terms of equipment details, flaw dimensions, operating conditions, material data, and analysis output data.

TABLE 17-9 Summary of FFS Analysis Data for a Cracked Acetone Finishing Column

Equipment Details	
Structure:	Thin-walled cylinder/curved shell
Thickness (in.):	0.500
Outer dia. (in.):	132.0
Corrosion allowance (in.):	0.063
Flaw Dimensions	
Flaw type:	Circumferential crack—surface flaw
Length (in.):	5.063 based on $a/2c = 0.073$
Depth (in.):	$0.374 = A_0 + d_a/d_t \times 40$ days
Width (in.):	0.001
Root radius (in.):	Less than or equal to 0.01
Flaw location:	At welds
Weld type:	Repair Weld residual stress ≈ 30% yield Strength for deep crack
Orientation:	Transverse (with respect to weld direction)
Environment:	Caustic solution
Max. operating temp. (°F):	225.000
Min. operating temp. (°F):	75.000
Max. pressure (psig):	20
Material:	SA 285GrC
Yield strength (ksi):	27.90
Ultimate strength (ksi):	62.0
Design temperature (°F):	225.0
Young's modulus (ksi):	30,000
Poisson's ratio:	0.300
K_{IC} (ksi $\sqrt{\text{in.}}$):	84.513, $K_{ISCC} = 20$ ksi $\sqrt{\text{in.}}$
Output Data	
Design pressure (psig)	20.0
Primary stress (ksi)	8.355
Residual stress (ksi)	8.37
Total stress (ksi)	16.705
K_{IC} (ksi $\sqrt{\text{in.}}$)	84.513
K_{ISCC}	20 ksi $\sqrt{\text{in.}}$
Critical flaw size (in.)	0.494 (for $a/c = 0.146$)
Real flaw size (in.)	0.375

Source: [12].

FFS assessment objectives

1. Determine whether the AFC vessel can remain in operation until the next scheduled turnaround, in 6 weeks.
2. Determine if an emergency shutdown is required given the extent of service-induced cracking.
3. Recommend remedial measures.

Steps used in the FFS assessment and FFS results

1. A thorough team review of the past and projected process conditions, inspection records, and design data. The review indicated the most likely failure mechanism as caustic SCC. The nature of the final failure is expected to be a leak instead of catastrophic brittle fracture.

2. Mechanical analysis to infer the major stresses driving the cracking. The analysis accounted for pressure stresses, dead weight loads due to internal fluids and vessel internals, potential wind loads on the column, and weld residual stresses. The analysis indicated that the major driving force was the weld residual stresses, which have been enhanced by the weld repairs, which were not stress relieved.

3. Fracture mechanics analysis results shown in Fig. 17-6 indicated that the maximum flaws detected were below the limiting size predicted by FAD analysis.

4. Remaining life analysis was performed given available crack growth rate data for caustic cracking, which was validated by internal laboratory tests. The work by Sriram and Tromans [11,14] on caustic cracking of A516 grade 70 steel indicates that applied threshold stress intensity factor $K_{1th} \geq 20\text{ksi}\sqrt{\text{in.}}$ in 2 to 4 M NaOH at $\geq 198^\circ$F can initiate caustic SCC crack growth. The plateau crack growth rate is about 3.1 mils/day. Therefore, the deepest measured crack is expected to grow to 0.375 in. prior to the next scheduled turnaround. It was demonstrated by FAD analysis that both the initial crack detected and the final crack size at turnaround were within allowable limits.

5. Given the prevalent hurricane season, fatigue life analysis was performed to ensure that the prevalent worst-case variable winds and wind loads will not propagate the preexisting cracks by corrosion fatigue damage to vessel failure within the 6 weeks prior to turnaround. A conservative estimate of the corrosion fatigue crack growth rate was made using data for a more susceptible turbine disk steel exposed to 12 M NaOH at 212°F [13]. This is represented by the Paris law:

$$\frac{da}{dN} = C \Delta K^n \tag{17.3}$$

where $C = 1.2\ \exp(-0.7)$, $n = 1.7$, ΔK is the range of stress intensity factor, in ksi $\sqrt{\text{in.}}$, and da/dN is the crack growth rate in in./cycle. The analysis was performed using the NASCRAC program [9]. The results, shown in Fig. 17-7, indicate that corrosion fatigue is not likely to result in the failure of the vessel prior to the next turnaround inspection, although in the event of severe wind loading, some acceptable crack growth in the vessel can occur.

Key lessons

1. The vessel remained in service with online condition monitoring and UT inspection until the scheduled turnaround shutdown. No failure occurred as predicted by the FFS analysis.

2. Metallographic examination of cut-out samples of the crack confirmed damage by caustic stress corrosion cracking. The crack growth estimated was about 200 mils,

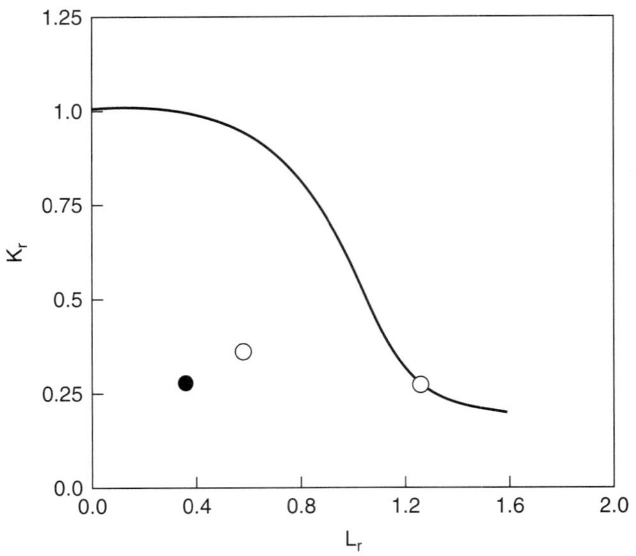

Figure 17-6 FAD representation of FFS of a cracked acetone finishing column. (From [12].)

compared to the 130 mils predicted, although the maximum depth of the initial caustic SCC could not be ascertained metallographically.

3. Risk mitigation measures applied to support the FFS analysis include:

a. Clear understanding of the prevalent damage mechanism and the postulated mode of failure (crack growth by caustic SCC to a leak-before-break)

b. Continued use of a reliable in-service monitoring by UT inspection

c. Sensitivity analysis to assure that the various assumptions used, such as crack growth rates, fracture toughness, flaw size, and loading conditions, were sufficiently conservative

Case 17.2: Effects of Aqueous H_2S on Fitness for Service and Remaining Life of Refinery Carbon Steel Pressure Vessels [11].

Equipment background. During a turnaround inspection of a debutanizer vessel, cracks were detected by wet magnetic particle fluorescent inspection. The cracks were sized by shear wave UT. The vessel had been in service for 15 years. The service conditions involved exposure to aqueous solutions containing H_2S and other corrosive species. The cracks observed were parallel to the circumferential seam weld and were discontinuous on the surface. There were no indications of blistering or laminar HIC damage. Table 17-10 summarizes the vessel geometry, service conditions, and material data.

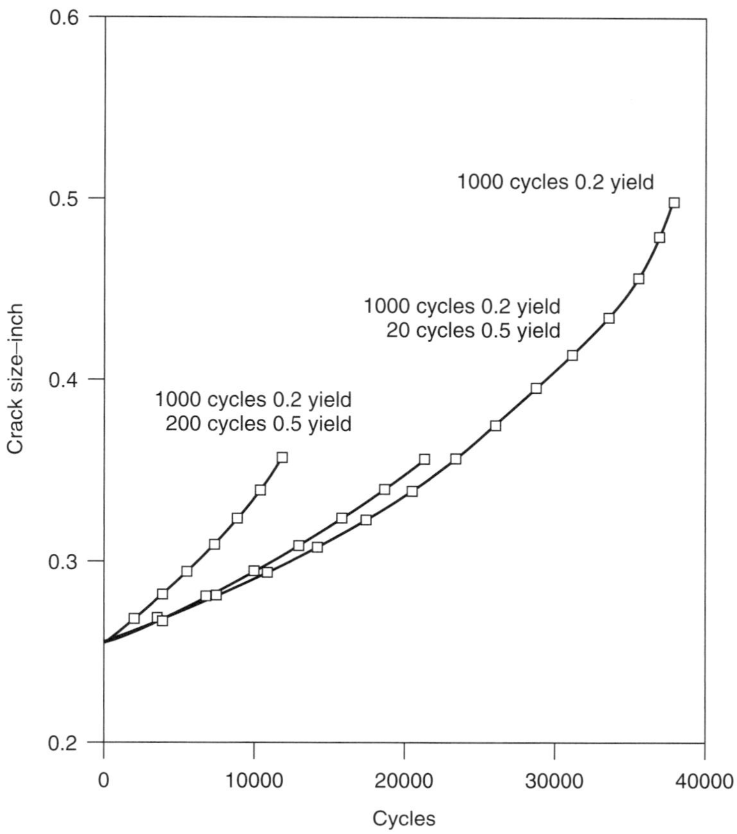

Figure 17-7 Fatigue crack growth analysis for a semielliptical crack (0.25 in. deep and 5 in. long) in an acetone finishing column subjected to variable blocks of wind loading. (From [12].)

Objectives of FFS analysis

1. Does the cracking observed affect the mechanical integrity of the vessel relative to failure by fracture?
2. If not, what is the remaining life of the vessel if it is returned to service?
3. Is the mechanical integrity of the vessel better in a stress-relieved condition?

Key assumptions for the FFS assessment

1. Continued exposure of the vessel to the process fluid containing dissolved H_2S will result in new surface cracks and extension and joining of existing cracks along the surface. Therefore, the crack depth/length aspect ratio is assumed to be approximately zero.

2. The crack growth rate in A515Gr70 in aqueous H_2S solution can be represented conservatively by that of low-alloy steels in NACE solution [6] and given by

$$\frac{da}{dt} = 18.69 \times 10^{-16} K^{6.7} \tag{17.4}$$

where da/dt is the crack growth rate in meters/year, K the applied stress intensity factor in MPa$\sqrt{\text{m}}$, and a the crack size in meters.

3. The residual stress normal to the plane of the crack and the circumferential weld in the non-stress-relieved condition is equal to the yield strength and constant across the wall thickness for the stress-relieved condition. The residual stress is assumed to be equal to 15% of the yield strength.

FFS assessment and material testing. A detailed review of the process conditions, inspection records, and design data was performed by the structural integrity analysis team (SIAT). Metallurgical and material testing was performed on a comparable debutanizer vessel. The results indicate that the primary damage mechanism is SOHIC, caused by hydrogen absorption from an aqueous wet H_2S environment. Table 17-11 shows fracture toughness data for carbon–Mn pressure vessel steels A515Gr70, A516Gr70, and A285GrC in air and in modified NACE solution [1]. It is notable by comparing the toughness J_8 in air versus in aqueous H_2S that hydrogen absorbed in the steel from wet H_2S fluids lowers the fracture resistance, J_8.

TABLE 17-10 Input Data for FFS Analysis of a Carbon Steel Debutanizer Vessel

Equipment Geometry: Shell with Hemispherical Heads	
Vessel: tangent-to-tangent height	100 ft
Outer diameter	180 in.
Wall thickness	1.125 in.
Crack Geometry: Detected by WMFI and Sized with Shear Wave UT	
Location: HAZ, base metal near circumferential seam weld	
Length	4 in.
Depth	0.161 in.
Service Conditions	
Age	15 yr
Process fluid: propane, butane, H_2S, H_2O	Wet H_2S
Maximum/minimum exposure temperatures	300°F/75°F
Max. pressure	200 psi
Material Data	
Non-stress-relieved A515GrB Steel	
Yield strength	41.2 ksi
Tensile strength	79.8 ksi
Flow strength (yield + tensile)/2	60.5 ksi
Ramberg–Osgood parameters: (SA 204B): $\alpha = 2.45$, $N = 5.4$	
Fracture toughness in air: $J_{0.2}$	1120 psia-in.
Fracture toughness in air: K_{1C} (J)	183 ksi $\sqrt{\text{in.}}$
Fracture toughness in wet H_2S: $J_{0.2}$	660 psia-in.
Fracture toughness in wet H_2S: K_{1H}	141 ksi $\sqrt{\text{in.}}$
Threshold toughness in wet H_2S: K_{1th}	46 ksi $\sqrt{\text{in.}}$
Crack growth rate (assumed): $da/dt = 18.69 \times 10^{-16} K^{6.7}$	
(*Note: da/dt* is in m/yr, K is in MPa $\sqrt{\text{m}}$, and a is in meters)	

Source: [1].

TABLE 17-11 Effect of Wet H_2S on Fracture Toughness of C–Mn Pressure Vessel Steels[a]

Steel Grade and ID	Specimen Orientation	Fracture Toughness (Air)		Fracture Toughness 1.5 psia H_2S in NACE Solution[b]		Fracture Toughness 15 psia H_2S in NACE Solution[b]	
		J_8^c (ksi-in.)	$K_{EE}{}^d$ (ksi $\sqrt{\text{in.}}$)	J_8 (ksi-in.)	K_{EE} (ksi $\sqrt{\text{in.}}$)	J_8 (ksi-in.)	K_{EE} (ksi $\sqrt{\text{in.}}$)
A515-70 X7041	LT	1.12	186	0.80	168	0.66	154
A516-70 X7042	LT	1.55	179	0.752	132	0.96	114
A285-CX7043	TL	0.60	140	0.360	142	0.37	133

Source: [1].
[a] Tests performed at crosshead speed of 10^{-3} to 10^{-2} mm/s.
[b] NACE solution per TM0177 buffered to pH 4.2 is used to represent a wet H_2S environment.
[c] Per ASTM E813 except for J_{ic} defined as J at 0.2 mm (8 mils) of crack extension on the $J-\Delta a$ resistance curve.
[d] K_{EE} per ASTM E992 for equivalent energy toughness.

Figure 17-8 shows the effect of absorbed hydrogen on the elastic–plastic crack growth resistance in A516Gr70 pressure vessel steel. Here the intersection with the J-integral axis is a measure of the *relative* fracture resistance to the onset of crack extension in the steel, and the slope of the line is a *relative* measure of the fracture energy associated with crack extension by the ductile tearing process in the steel. Hence it is notable that the effect of hydrogen absorbed from aqueous H_2S bearing fluid for low-strength steel is to (1) reduce the fracture resistance for the onset of fracture, and (2) reduce the energy associated with the ductile fracture process in the steel and hence manifest as low-energy ductile tearing and increased crack growth rate under HAC.

With good understanding of the damage mechanisms, the available material properties, crack growth rates,

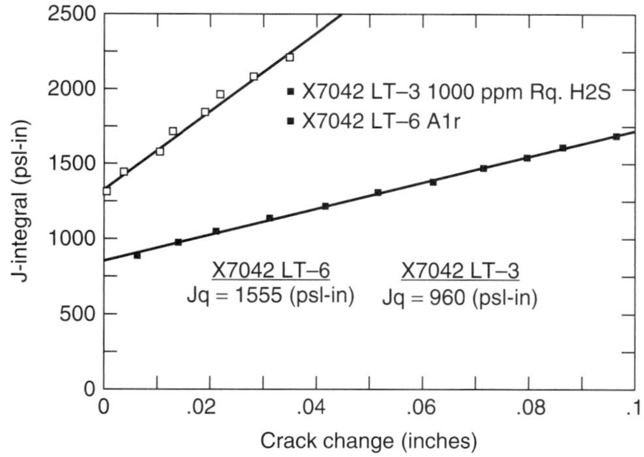

Figure 17-8 Effects of absorbed hydrogen on the elastic–plastic crack growth resistance in A516Gr70 pressure vessel steel in modified NACE solution. (From [11].)

and the assumptions cited above, fracture mechanics calculations were performed using proprietary PcCrack fracture mechanics software. The results show that for the as-welded condition and despite the conservatism inherent in the level 2 FFS analysis, the critical crack depth is 0.630 in. compared to the maximum observed crack depth of 0.161 in. Therefore, there is a sufficient margin of safety for the mechanical integrity of the vessel.

The result of the remaining life analysis indicates that the remaining life for the crack to grow to critical size by low-energy HAC fracture is 13 years [11]. The vessel was thus determined to be fit for service.

Case 17.3: Case Histories Using FFS Methods [16]

Equipment background. As part of a study to validate the MPC FFS methodology, the associated computer program PREFIS and the Tref method for estimation of fracture toughness of pressure vessel steels, Yin and Bagnoli conducted a retrospective prediction of known industry vessel failures. The cases evaluated were in three classes:

1. Thin-walled vessels (wall thickness < 1.5 in.) operated at or failed at ambient temperature:
   a. 1952 ESSO Fawley storage tank (C-11) failure
   b. Gas knockout drum rupture
   c. 1998 Ashland tank failure
   d. 1983 reactor sphere failure
2. Heavy-walled vessels (wall thickness > 1.5 in.) operated at or failed at ambient temperature:
   a. Toyo thick-walled vessel failure
   b. 1965 Immingham ammonia convertor failure
   c. 1966 Cockenzie boiler drum hydrotest failure
3. Heavy-walled vessels operated at or failed at elevated temperatures:
   a. 1990 BASF ammonia converter (C-702) failure
   b. CF Braun ammonia converter (R-103) failure

Table 17-12 summarizes the input data for the FFS analysis, the failure initiation flaw size from post-failure metallurgical analysis, and the type and condition of failure for the case of four thin-walled vessels. The reader is referred to the original publication for further details on the remaining vessels and the history of each of the nine failed vessels.

The analysis was performed using the MPC PREFIS software. Toughness data were estimated based on the Tref method. The results for the four thin-walled vessel failures are depicted in the FAD in Fig. 17-9. In general, the FFS analysis correctly predicted the incidence of failure of each vessel, evaluated particularly when the toughness is based on the lower-bound (K_{Ir}) values. The flaw sizes observed are greater than the critical flaw size for brittle fracture of the vessels. Based on the lower-bound (K_{Ir}) toughness values estimated for the steels, the FAD analysis for the initiating flaw in each vessel fell clearly in the failure region of the FAD, as expected from the failures observed.

Key learnings. The analysis above indicates that FFS analysis according to API 579-1 can reasonably predict the significance of flaws on the risk of failure of pressure vessel

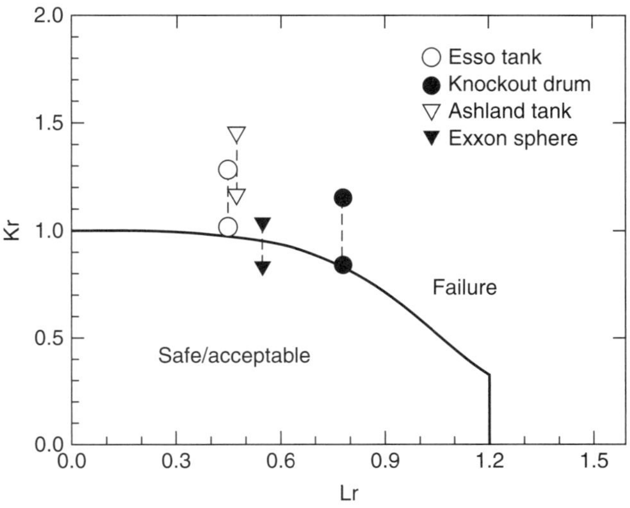

Figure 17-9 FAD diagram for retrospective FFS analysis of four failed pressure vessels. (From [16].)

TABLE 17-12 Summary of Input Data for Retrospective FFS Analysis of Failed Pressure Vessels

Vessel:	Esso Fawley Tank	Knockout Drum	Ashland Tank	Exxon Reactor Sphere
Internal pressure/liquid height	48 ft water	588 psi	48 ft diesel	350 psi (65% design)
Failure temperature	32°F	40°F	38°F	41°F
Material	BS 13 (A7 or A283)	A516Gr70	BS 13 (A7 or A283)	A42C-3S (A201GrB)
Tensile strength	70 ksi	78 ksi	60 ksi	60–66 ksi
Yield strength	38 ksi	46 ksi	34 ksi	30–40 ksi
Measured toughness	—	CVN = 12.3 ft-lb ($K_{1C} = 43$ ksi $\sqrt{\text{in.}}$)	—	CTOD = 0.018 in. BM ($K_{1C} = 130$ ksi $\sqrt{\text{in.}}$) CTOD = 0.0012 in. ($K_{1C} = 38$ ksi $\sqrt{\text{in.}}$)
Toughness per ASME XI K_{1C} (ksi $\sqrt{\text{in.}}$)	$K_{1C} = 45$ ksi $\sqrt{\text{in.}}$ ($T_{\text{ref}} = 65°$F) $K_{1C} = 35$ ksi $\sqrt{\text{in.}}$(K_{1R})	$K_{1C} = 55$ ksi $\sqrt{\text{in.}}$ ($T_{\text{ref}} = 40°$F) $K_{1C} = 39$ ksi $\sqrt{\text{in.}}$(K_{IR})	$K_{1C} = 45$ ksi $\sqrt{\text{in.}}$ ($T_{\text{ref}} = 62°$F) $K_{1C} = 35$ ksi $\sqrt{\text{in.}}$(K_{IR})	$K_{1C} = 41$ ksi $\sqrt{\text{in.}}$ ($T_{\text{ref}} = 90°$F) $K_{1C} = 32$ ksi $\sqrt{\text{in.}}$(K_{IR})
Internal diameter	140 in.	71.5 in.	1440 in.	168 in.
Wall thickness	1 in.	1.18 in.	0.84 in.	1.6 in.
Applied stress	16 ksi	35 ksi (SCF = 2.0)	13 ksi	16 ksi (FEA)
Flaw size	0.87 in. long × 0.22 in. deep on ID surface	1.0 in. long × 0.1 in. deep	0.7 in. long × 0.55 in. deep	2 in. long × 0.12 in. deep
Flaw orientation and location	Longitudinal crack transverse to first course weld from tank bottom	Longitudinal crack near toe of fillet weld joining pad to shell	Longitudinal crack near a C-weld and 36 ft below liquid level	Crack near shroud support ring weld
Failure type and condition	Brittle fracture during hydrotest	Brittle fracture during hydrotest	Brittle fracture during filling process	Brittle fracture in standby condition

Source: [16].

and tank steels. However, the safety margin associated with the predictions is heavily dependent on the toughness values assumed. The Tref method proposed in API 579 can be considered a valid methodology for estimating lower-bound toughness values for FFS assessment. Nevertheless, care must be exercised in the choice of material properties, particularly toughness and the rate of damage propagation assumed or used in FFS analysis.

REFERENCES

1. American Petroleum Institute, *Prevention of Brittle Fracture of Pressure Vessel Steels*, API Recommended Practice 920, API Publishing, Washington, DC, 1991.
2. American Petroleum Institute, *Fitness-For-Service*, 2nd ed., API Standard 579-1/ASME FFS -1, API Publishing, Washington, DC, 2007.
3. Buchheim, G. M., An overview of risk-based inspection methods and user needs for the petrochemical industry, *ASME PVP*, vol. 261, 1993, p. 143.
4. Center for Chemical Process Safety, *Guidelines for Mechanical Integrity Systems*, Wiley, Hoboken, NJ, 2006.
5. Jaske, C. E., Process equipment fitness-for-service assessments using API RP 579, *Proceedings of the Process and Power Plant Reliability Conference*, Houston, TX, November 2001.
6. Kumar, A. N., and Pandey, R. K., Kinetics of subcritical cracking under the chloride and sulfide environments in a low alloy steel, *Engineering Fracture Mechanics*, vol. 20, no. 2, 1984, p. 287.
7. Lloyds Register Capstone, Introduction to fitness-for-service (FFS) assessment using API/ASME Standard API 579-1/ASME FFS-1, Webinar Series, April 22, 2010.
8. Mahoney, D. G., *Large Property Losses in the Petrochemical Industry: A 30 Year Review*, 13th ed., M & M Protection Consultants, La Porto, TX, 1990
9. NASCRAC, Failure Analysis Associates, Menlo Park, CA.
10. National Association of Corrosion Engineers, *Guidelines for Detection, Repair and Mitigation of Cracking of Existing Petroleum Refinery Pressure Vessels in Wet H_2S Environment*,

NACE Standard RP026-96, NACE International, Houston, TX, 2003.

11. Onyewuenyi, O. A., Effects of aqueous H_2S on fitness for service and remaining life of refinery carbon steel pressure vessels, *ASME PVP*, vol. 261, 1993, p. 177.
12. Onyewuenyi, O. A., Madapura, M., Horvath, R. J., and Mead, H. E., Fitness for service and remaining life assessment of an acetone finishing column, *ASME PVP*, vol. 315, 1995, p. 175.
13. Rungta, R., Proprietary data, Fontana Corrosion Center, Ohio State University, Columbus, OH, March 1990.
14. Sriram, R., and Tromans, D., Stress corrosion cracking of carbon steel in caustic aluminate solutions: crack propagation studies, *Metal Transactions*, vol. 16A, 1985, p. 979.
15. Twigg, R. J., *Inspection Guidelines for Pressure Vessels and Piping*, Vol. 2, MTI Publication 49, Materials Technology Institute, St. Louis, Missouri, USA, 1996.
16. Yin, H., and Bagnoli, D. L., Case histories using fitness for service methods, *ASME PVP*, vol. 288, p. 315.

18

DESIGN OF PRESSURE VESSELS AND PIPING

MAHER Y. A. YOUNAN
School of Sciences and Engineering, American University in Cairo, Cairo, Egypt

The design consideration for pressure vessels and piping systems requires the structure to sustain the following types of loadings:

- Internal or external pressure
- Weight of the vessel or pipe and contents
- Reactions from the weight of attached equipment
- Reactions from supports
- Cyclic and dynamic reactions due to pressure or thermal variations
- Wind or seismic loadings
- Impact reactions
- Thermal gradients and differential thermal expansions
- Other loadings, if applicable (must be specified by user)

Design codes give the general guidelines for designing pressure vessels and pipes. The basic structural design considerations are emphasized in this chapter. Design codes describe the acceptable procedures for stress analysis and give the allowable material property to compare with. In general, the stress analysis equations for pressure vessels and pipes are based on the thin-walled theories of plates and shells. However, for high pressures and thick vessels, special codes need to be considered, such as ASME Section VIII, Division 3 [3]. Several failure modes are considered in the design process. These are described below.

18.1 MODES OF FAILURE

Depending on the loading and working conditions, failure may happen due to one or a combination of failure modes (Fig. 18-1). The term *failure* here implies that the component [6] will not function in a satisfactory manner for the task for which it is designed. Loading may be either static, dynamic, or other (e.g., stress corrosion, wear). Under each of these loading conditions, structural components may fail by one or more modes.

18.1.1 Failure Under Static Loading

Under static loading, pressure vessels or piping may fail by:

1. *Yielding*. This implies that a part of the vessel or pipe has yielded. If a full section yields, this would cause structural collapse. Failure of a pipe or pressure vessel due to the circumferential stresses reaching their limit due to pressure is an example. The maximum principal stress criterion [Eq. (18.1)] and the maximum shear stress criterion [Eq. (18.2)] are used in design-by-rule codes.

$$\sigma_1 \leq \sigma \text{ all maximum principal stress criterion} \quad (18.1)$$

$$\sigma_1 - \sigma^3 \leq \sigma \text{ all maximum shear stress criterion} \quad (18.2)$$

Process Plant Equipment: Operation, Control, and Reliability, First Edition. Edited by Michael D. Holloway, Chikezie Nwaoha, and Oliver A. Onyewuenyi.
© 2012 John Wiley & Sons, Inc. Published 2012 by John Wiley & Sons, Inc.

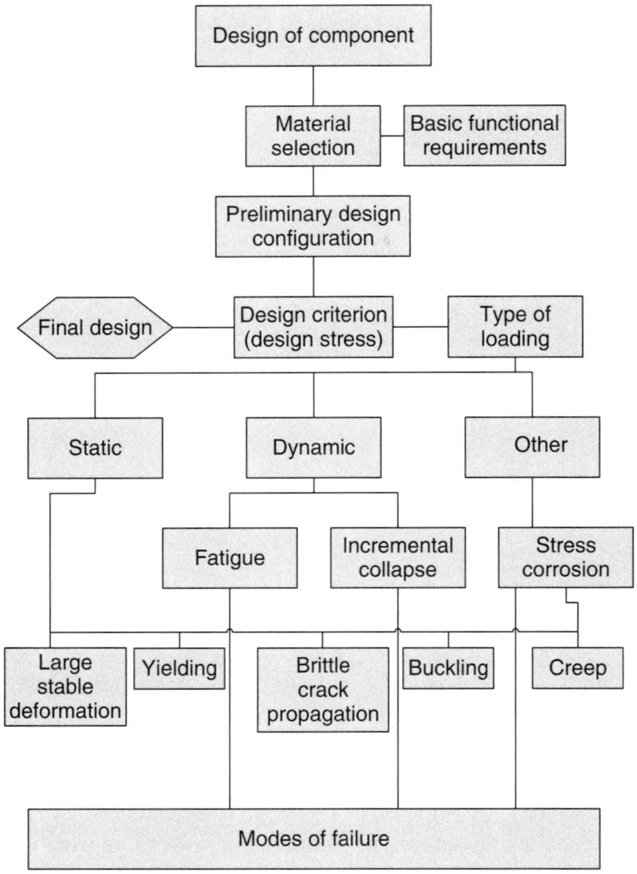

Figure 18-1 Possible failure modes and design considerations. (From [1].)

The maximum distortion energy is used in design by analysis:

$$\sigma_{\text{eff}} = \sqrt{\tfrac{1}{2}[(\sigma_1 - \sigma_2)^2 + (\sigma_2 - \sigma_3)^2 + (\sigma_3 - \sigma_1)^2]} \leq \sigma_{\text{all}} \tag{18.3}$$

2. *Excessive elastic deformation*. This implies that the deformation may be stable but that the design cannot tolerate such a large deformation. As an example, large spans in piping structures may deflect excessively under their own weight, although the stresses may be within the allowable limits. Equation (18.4) gives the allowable span of a straight pipe length based on a permissible deflection y [6]:

$$L = \left(\frac{yEI}{17.1W}\right)^{1/4} \tag{18.4}$$

where L is the pipe support spacing (ft), y the permissible midspan deflection (in.), E the modulus of elasticity at design temperature (lb/in.) I the moment of inertia of the pipe, W the weight of supported pipe, including pipe, contents, and insulation (lb/ft).

3. *Buckling*. This is an instability failure mode that appears in thin structures subjected to compressive stresses. A common example of this mode of failure is piping or pressure vessels subjected to external pressure. Equation (18.5) shows how to determine the allowable external pressure P_a for a cylindrical vessel or pipe [7] under elastic behavior conditions:

$$\text{critical buckling pressure} P_a = 0.4\left(\frac{Et^2}{L(d/2)}\right) \times \left[\left(\frac{1}{1-v}\right)^3 \left(\frac{2t}{d}\right)^2\right]^{1/4} \tag{18.5}$$

4. *Creep*. This mode of failure appears with time for systems operating at high temperatures. Excessive deformations or creep rupture may appear in such structures [5]. Design codes use reduced allowable stresses for higher design temperatures to account for this mode of failure.

5. *Brittle fracture*. This mode of failure appears from the propagation of cracks that are already present in the structure. Design codes consider this in the structural inspection phase to make sure such initial cracks are not present initially. However, fitness-for-service codes (e.g., [2]) describe procedures to accommodate such defects if found after service.

18.1.2 Failure Under Dynamic Loading

Under dynamic or cyclic loading, pressure vessels or piping structures may fail, due to one or more of the following modes:

1. *High cycle fatigue*. This mode of failure implies failure of components due to a large number of reversed cycles of stress. This normally appears in components of rotary systems or in systems connected to them that would be vibrating due to such high-frequency dynamic loading. For a vibrating pipe, the number of cycles during its lifetime are estimated and used with the corresponding fatigue curve to determine the allowable stress.

2. *Low cycle fatigue*. This failure mode implies failure due to exceeding the shakedown limit by ratcheting or reverse plasticity. Design codes allow the design of pressure vessels and piping up to the shakedown limit when subjected to self-equilibrating dynamic loading. A typical design case is the effect of cyclic temperature. Shakedown allowable stresses reach twice the allowable stress for static loading. As an example, design codes place the following limitations:

- Primary stress: $P_m < SE$ (allowable stress)
- Secondary membrane stress: $Q_m < 1.5SE$

- Local primary membrane: $P_L = P_m + Q_{m,\text{ sustained}} < 1.5SE$
- Primary (membrane + bending) + secondary (membrane + bending) : $<3SE$

18.1.3 Failure Under Other Types of Loading

Under this category fall failures due to corrosion, erosion, wear, pitting, and so on. Design codes allow for such modes by allowing a "corrosion allowance" to be added to the thickness of the vessel or pipe. This allowance would depend on the interaction of the fluid and the material of the vessel and the flow conditions for pipes.

18.2 BASIC STRESS ANALYSIS

The stress analysis of cylindrical vessels uses cylindrical coordinates as shown in Fig. 18-2. There are several classifications of stresses in vessels and pipes. Stresses may be classified based on their directions (radial, circumferential, or longitudinal), their significance (primary, secondary, occasional, or peak), or their location (general or local).

Circumferential or hoop stresses caused by pressure are used as the basis for determining the thickness of vessels and pipes. Based on thin shell theory, the circumferential stresses are uniform across the thickness. If the hoop stress exceeds the yield strength at the design temperature, net section yielding occurs. The stress limit is the yield strength at the design temperature (S_y). This is a primary mode of failure. The allowable stress is $SE = S_y/1.5$. Figure 18-3 shows the circumferential and axial stresses developed in a cylindrical vessel due to pressure.

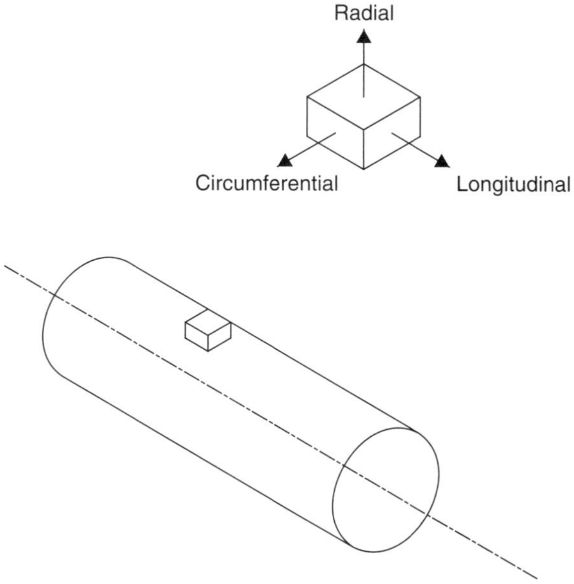

Figure 18-2 Coordinate system for cylindrical vessels.

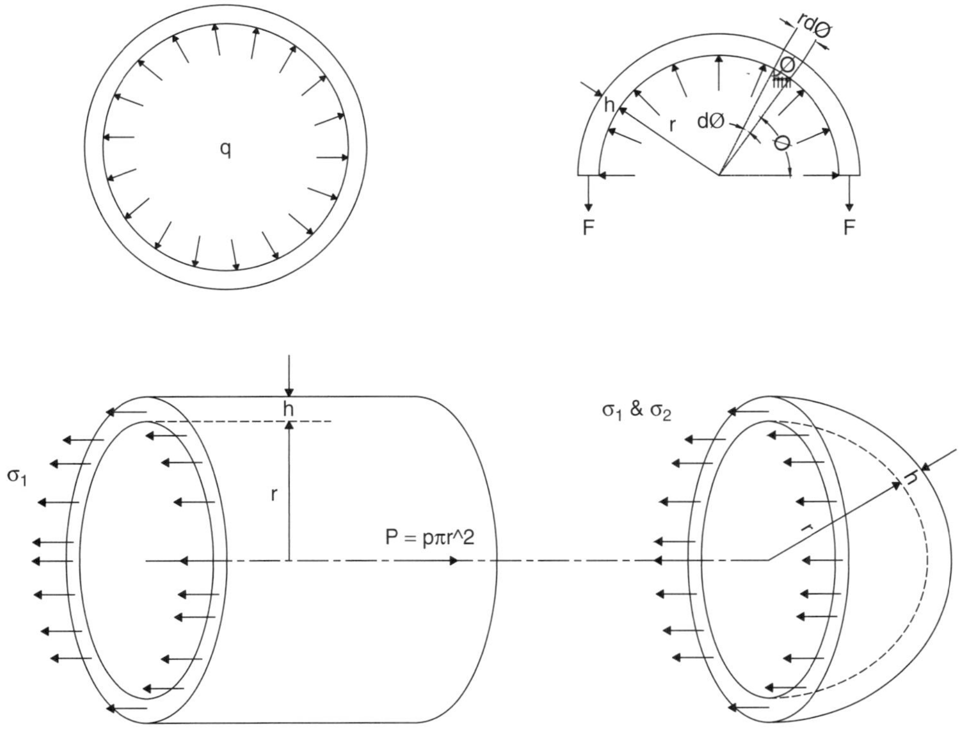

Figure 18-3 Circumferential and axial stresses due to pressure.

For a cylinder subjected to internal pressure, the tangential stress and axial stresses are given by

$$\sigma_t = pD/2t \qquad \sigma_a = pD/4t \tag{18.6}$$

The circumferential stress equation is the basic equation for determining the thickness for pressure vessels or pipes. Based on Eq. (18.6), the thickness required for circumferential stress may be determined from

$$t = \frac{PR}{SE - 0.6P} \tag{18.7}$$

The E factor is a stress reduction factor based on the method of manufacture of the pipe. It is based on the quality of the weld in seam-welded pipe and will have a value ranging from $E = 0.6$ for furnace butt welded to $E = 1.0$ for seamless pipe. The second stress considerations are *longitudinal stresses* caused by sustained and occasional loads. Sustained loads are caused by longitudinal pressure and weight. Sustained loads cause primary stresses that are compared to the allowable stress SE.

Radial stresses due to pressure are normally neglected in thin cylindrical vessels. However, local effects due to stress concentrations at supports may be significant.

Primary stresses are main stresses that must be present in a structure to balance externally applied loads such as pressure and weights. *Secondary stresses* are caused by loading conditions that are self-equilibrating. That is, their resultants are not balancing any external loads. They are developed in systems due to gradients in temperature, for example. If these stresses cause some local yielding, they will be relieved and would not affect the equilibrium of the structure against external loads. For such stresses, the component is designed based on shakedown limitations. As an example, codes allow the combined primary membrane and secondary stresses to reach twice the yield stress.

Occasional stresses are caused by loads that are temporary in nature and not frequent, such as wind or earthquake. The allowable stress for combined sustained plus occasional load stresses is $1.33SE$. This implies that the stresses are allowed to reach 0.9 of the yield stress.

The classification of stresses as *general* or *local* depends on the effect that the load is causing. If the effect is localized and dies out after a small distance (proportional to $\sqrt{Rt}$), the stress developed is classified as local.

18.2.1 Allowable Stresses

Different codes define slightly different factors of safety to be used in calculating the allowable stresses. It is very important to refer to one code when designing a vessel or a pipe. That is, if you are using the ASME Pressure Vessel Code Section VIII, Division 1, for example, you cannot use the factor of safety from the British code.

1. Based on the minimum tensile strength specified at room (or the design) temperature the factor of safety is 2.35 to 3.5 (based on which code is used).
2. Based on the yield strength at the room (or design) temperature, the factor of safety is 1.5.
3. Based on the creep stress for the designed lifetime and temperature, the factor of safety is around 1.6.

ASME codes give the allowable stresses for the various materials used based on the above-mentioned factors of safety. Material selection is based on several criteria:

1. Interaction between fluid and vessel or pipe
2. Working conditions: pressure, temperature, environment, and so on
3. Availability
4. Manufacturability
5. Economy

Previous experience and history affect such decisions. Many codes give some recommendations based on the current practice in a given domain [8]. Table 18-1 gives some recommendation for types of materials to be used for different operating temperatures.

18.3 DESIGN OF PRESSURE VESSELS

In this section, a case study is used to illustrate the design procedures throughout the section.

Case Study 18.1 An ammonia vessel is to be designed that has the following specifications:

Vessel capacity	20 tons of liquid ammonia
Vessel shape	horizontal
Temperature range	−10 to 50°C
Maximum working pressure	20 atm
Hydraulic test pressure	30 atm
Vessel diameter	2 m
Vessel material (available)	SA-737 (Mn steel)

18.3.1 Geometric Considerations

The first step in the design of pressure vessels is determining the geometry required based on the process and the location of the vessel. Such considerations will

TABLE 18-1 Recommended Materials for Different Operating Temperatures

Design Temperature (°F)	Material	Plate	Pipe
Cryogenic			
−425 to −321	Stainless steel	SA-240-304, 304L, 347, 316, 316L	SA-312-304, 304L, 347, 316, 316L
−320 to −151	9 nickel	SA-353	SA-333-8
Low temperature			
−150 to −76	$3\frac{1}{2}$ nickel	SA-203-D	
−75 to −51	$2\frac{1}{2}$ nickel	SA-203-A	SA-333-3
−50 to −21	Carbon steel	SA-516-55, 60 to SA-20	SA-333-6
−20 to 4	Carbon steel	SA-516-All	SA-333-1 or 6
5 to 32	Carbon steel	SA-285-C	SA-53-B
Intermediate			
33 to 60	Carbon steel	SA-516 All	SA-106-B
61 to 775	Carbon steel	SA-515-All	
		SA-455-ll	
Elevated temperature			
776 to 875	C−$\frac{1}{2}$Mo	SA-204-B	SA-335-P1
876 to 1000	1Cr−$\frac{1}{2}$Mo	SA-387-12-1	SA-335-P12
	$1\frac{1}{4}$Cr−$\frac{1}{2}$Mo	SA-387-11-2	SA-335-P11
1001 to 1100	$2\frac{1}{4}$Cr−1Mo	SA-387-22-1	SA-335-P22
1101 to 1500	Stainless steel	SA-240-347H	SA-312-347H
	Incoloy	SB-424	SB-423
Above 1500	Inconel	SB-443	SB-444

Source: [8].

determine whether the vessel is vertical or horizontal. Table 18-2 shows the equations for volume calculations for some common geometries and ends. Based on these preliminary calculations, the overall geometry of the vessel may be determined. The location of the supports is then selected to minimize the bending stresses as much as possible. Simple beam theory equations are generally used at this preliminary design phase.

Required vessel volume V_t = payload/density

Density of liquid ammonia = 0.53

Thus, required vessel volume $V_t = 20/0.53 = 37.74\ \text{m}^3$

Vessel volume V_t = volume of cylindrical part + dished ends

Thus, for cylindrical vessel with standard elliptical heads

$$V_t = \frac{\pi}{4} D^2 L_c + \frac{\pi D^3}{24}$$

Thus, length of cylindrical part $L_c = 11.4$ m

TABLE 18-2 Volume and Surface Area of Some Commonly Used Geometries

Section	Volume	Surface Area
Cylinder	$\frac{\pi D^2 \ell}{4}$	$\pi D \ell$
Sphere	$\frac{\pi D^3}{6}$	πD^2
Hemihead	$\frac{\pi D^3}{12}$	$\frac{\pi D^2}{2}$
2 : 1 S.E. head	$\frac{\pi D^3}{24}$	$1.084 D^2$
Elliposoidal head	$\frac{\pi D^2 \ell}{6}$	$2\pi R^2 + \frac{\pi \ell^2}{\theta} \ln \frac{1+e}{1-e}$

Total length of cylinder with ends = 11.4 + (2)(0.5) = 12.4 m

18.3.2 Design of Vessels Under Internal Pressure

The determination of the thickness of pressure vessels is normally based on the tangential stresses developed due to pressure [Eq. (18.6)]. However, the longitudinal stresses due to pressure and other loadings need to be calculated and checked. Table 18-3 gives the required thickness of the vessel and heads. A suitable corrosion allowance is added to the thickness calculated, and the thickness is normalized to the lowest standard sheet thickness that is higher than the thickness calculated.

Example 18.1: Allowable Stress for Vessel For SA 737,

$$S_{\text{ult}} = 551 \text{ MPa} \qquad S_y = 413 \text{ MPa}$$

Thus,

$$S_{\text{all}} = \min\left(\frac{S_y}{1.5}, \frac{S_{\text{ult}}}{3.5}\right) = \min(275, 158) = 158 \text{ MPa}$$

(see Ref. ASME Sec. II, Table 1A)

Required vessel thickness is given by

$$t = \frac{PR_i}{2SE + 0.6P}$$

Thus,

$$t = \frac{(3)(1000)}{(158)(1) + (0.6)(3)} = 18.77 \text{ mm}$$

A 1-mm corrosion allowance is added and thus the final thickness is normalized to become 20 mm.

18.3.3 Nozzles or Branch Connections

Nozzles or branch connections are made in pressure vessels and piping systems by any one of several methods. These could be tees, pad-reinforced or unreinforced intersections, crosses, integrally reinforced weld-on or weld-in contoured insert fittings, or extrusions. The philosophy of the code for intersections is centered around the available pressure reinforcement offered by the geometry of the intersection. The process of making an intersection weakens the vessel or run pipe by the opening that must be made. Unless the wall thickness of the vessel or run pipe is sufficiently in excess of that required to sustain pressure at an intersection that is not manufactured in accordance with a listed standard, it is necessary to provide added reinforcement.

TABLE 18-3 Equations for Thicknesses Calculations of Vessels and Ends Based on the Circumferential Stress Due to Internal Pressure

		Thickness, t	
Part	Stress Formula	ID	OD
Shell			
Longitudinal	$\sigma_\chi = \dfrac{PR_m}{0.2t}$	$\dfrac{PR_i}{2SE + 0.4P}$	$\dfrac{PR_o}{2SE + 1.4P}$
Circumferential	$\sigma\theta = \dfrac{PR_m}{t}$	$\dfrac{PR_i}{SE + 0.6P}$	$\dfrac{PR_o}{SE + 1.4P}$
Heads			
Hemisphere	$\sigma_\chi = \sigma\phi = \dfrac{PR_m}{2t}$	$\dfrac{PR_i}{2SE + 0.2P}$	$\dfrac{PR_o}{2SE + 1.8P}$
Ellipsoidal	Procedure in code	$\dfrac{PR_iK}{2SE + 0.2P}$	$\dfrac{PD_oK}{2SE + 2P(K - 0.1)}$
2 : 1 S.E.	Procedure in code	$\dfrac{PD_i}{2SE + 0.2P}$	$\dfrac{PD_o}{2SE + 1.8P}$
100%–6% Torispherical	Procedure in code	$\dfrac{0.885PL_i}{SE + 0.1P}$	$\dfrac{0.88sPL_o}{SE + 0.8P}$
Torispherical $L/r < 16.66$	Procedure in code	$\dfrac{PL_iM}{2SE - 0.2P}$	$\dfrac{PL_oM}{2SE + P(M0.2)}$
Cone			
Longitudinal	$\sigma_\chi = \dfrac{PR_m}{2t\cos\alpha}$	$\dfrac{PD_i}{4\cos\alpha(SE + 0.4P)}$	$\dfrac{PD_o}{4\cos\alpha(SE + 1.4P)}$
Circumferential	$\sigma_\chi = \dfrac{PR_m}{t\cos\alpha}$	$\dfrac{PD_i}{2\cos\alpha(SE - 0.6P)}$	$\dfrac{PD_o}{2\cos\alpha(SE + 0.4P)}$

This reinforcement is added metal, local to the intersection, that is integral with the vessel or run and branch pipes. The amount of pressure reinforcement required is determined by performing area replacement calculations using the design conditions established for the intersection.

Using the sectional view shown in Fig. 18-4, calculate the area removed by the nozzle: $A_1 = (T_h D_1)$ [3]. This is checked against the areas of additional material, which is the sum of the following areas:

- A_2 = area from excess thickness in the wall of the vessel or run pipe:

$$A_2 = (2D_2 - D_1)(T_h - t_h - c)$$

- A_3 = area from excess thickness in the wall of the branch pipe:

$$A_3 = 2L_4(T_b - t_b - c)$$

$$L_4 = \text{height of reinforcement zone} = \min[2.5(T_h - c) \text{ or } 2.5(t_h - c) + t_{\text{reinforcingring}}]$$

- A_4 = area of all metal within the reinforcement zone provided by weld metal and any other reinforcement metal

Example 18.2 In the case study of the ammonia vessel described in Section 18.3, a manhole of internal diameter 500 mm is to be designed. The same material of the vessel is used ($\sigma_{\text{all}} = 158$ MPa). Thus, the minmum required thickness of the nozzle for the manhole would be

$$t = \frac{PR_i}{SE + 0.6P} = \frac{(3)(500)}{(158)(1) + (0.6)(3)} = 9.36 \text{ mm}$$

Adding a 1-mm corrosion allowance and using the nearest dimension, the thickness of the throat of the manhole will be 12 mm.

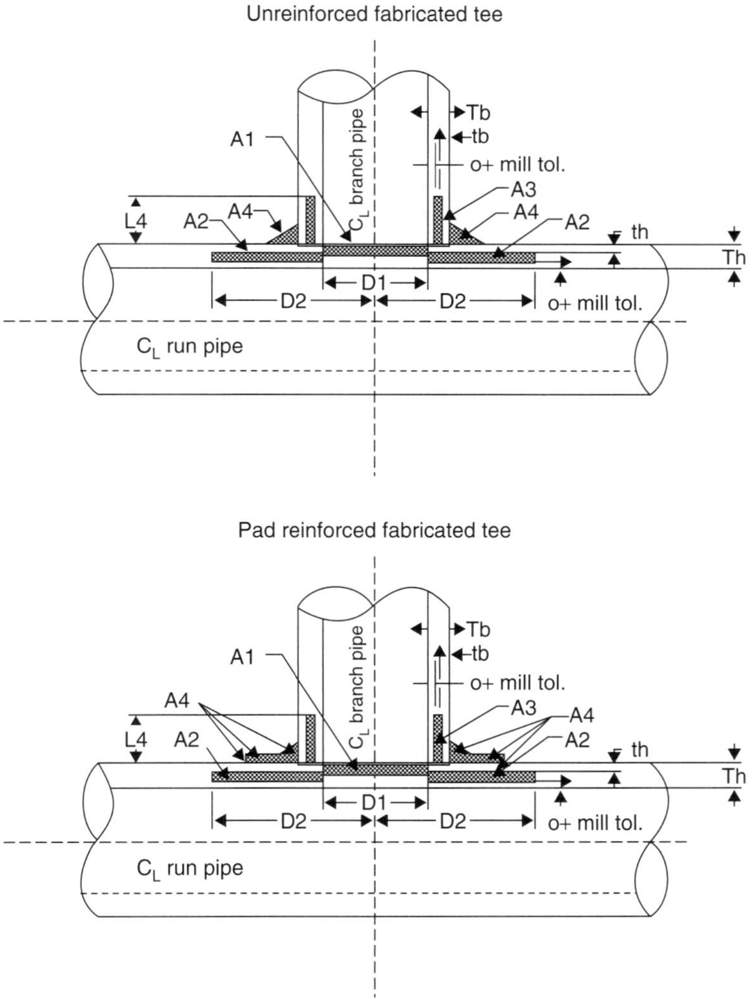

Figure 18-4 Reinforcement areas around a nozzle.

The following calculation for checking the need for additional stiffening around the throat of the manhole uses the corroded thickness of the material.

A_1 = area to be replaced

$= D_1 t_1 = (500)(9.36) = 4693.4\ \text{mm}^2$

A_2 = area from excess thickness in the wall of the vessel

$= (2D_2 - D_1)(Th - th - c)$

$= [(2)(2000) - 500](20 - 18.77 - 1) = 805\ \text{mm}^2$

A_3 = area from excess thickness in the wall of the branch pipe $= 2L_4(T_b - t_b - c)$

L_4 = height of reinforcement zone

$= \min[2.5(T_h - c) \text{ or } 2.5(th - c) + t_{\text{reinforcingring}}]$

$= 2.5(12 - 1) = 27.5\ \text{mm}^2$

Since the area to be replaced (4693.4) is significantly larger than the current material area available (805 + 27.5), it is necessary to add a reinforcement pad. To calculate a suitable thickness for the pad, assume its dimensions to be within the reinforcing zone. The minimum diameter of the pad is equal to the outer diameter of the throat of the manhole $= 500 + (2)(12) = 524$ mm. The outer diameter should not exceed $2D_2 = (2)(2000) = 4000$ mm. This dimension is too large and an outer diameter of 1000 mm is selected. Thus, the available stiffening area from the pad $A_4 = t_{\text{pad}}(D_o - D_i)$. Thus, the minimum required thickness

$$t_{\text{pad}} = \frac{A_4}{D_0 - D_i} = \frac{4693.4 - (805 + 27.5)}{1000 - 524} = 8.1\ \text{mm}$$

It is now possible either to increase the pad thickness or dimensions or to use an 8-mm sheet. Using the last solution, the amount of stiffening from the pad is equal to $8(1000 - 524) = 3808\ \text{mm}^2$. We need to check that additional stiffening from the weldments will be sufficient. Two welds will be used to weld the pad to the vessel and throat. Each has a leg of 8 mm. Thus, additional stiffening from the welding will be $A_4 = 2(0.5)(8)(8) = 64\ \text{mm}^2$. Thus, the total stiffening area $= 805 + 27.5 + 3808 + 64 = 4704.5\ \text{mm}^2$, which is greater than the required area of $4693.4\ \text{mm}^2$. That is, this design is adequate.

18.3.4 Design of Formed Heads

The design of heads (ends) of vessels depends on the design pressure and manufacturing facilities. Heads are made with increased curvature for higher design pressures. Hemispherical heads are used for the highest pressure, since as the curvature increases, the stresses in heads become smaller. However, codes have specific recommendations for the location of weld junctions of heads to the vessel to make sure that these locations are far from the stress concentration locations, due to the change in geometry. The following are different types of heads (Fig. 18-5):

- Ellipsoidal heads are defined by the ratio of the major diameter to the minor diameter of the ellipse (the head is one-half of the ellipse). The radius of curvature changes continuously.
- Torispherical heads are defined by the knuckle ratio, r/D, and dome ratio, L/D, where R is the knuckle radius, L the dome radius, and D the diameter

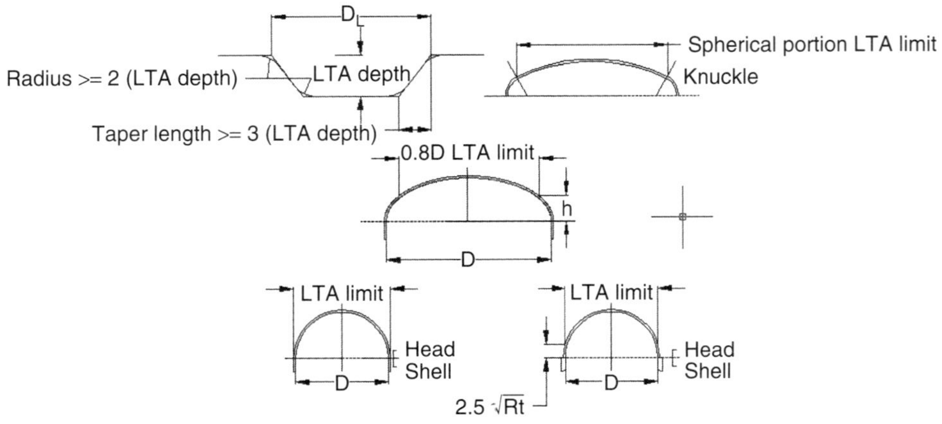

Figure 18-5 Different forms of pressure vessel heads.

of the cylinder to which the head is attached. An acceptable approximation to a 2 : 1 ellipsoidal head is a torispherical head with $r/d = 0.17$ and $L/D = 0.9$.

- Hemispherical heads are one-half of a sphere with radius L, equal to one-half the vessel diameter, D.
- Conical heads are defined by the half apex angle and have no transition piece between the head and the shell. Toriconical heads have a transition knuckle between the head and the shell (better design, but more expensive).

18.3.5 Vessels and Pipes Subjected to External Pressure

In some operating conditions, vessels or pipes may be subjected to external pressure. Examples are jacketed vessels and vessels or pipes operating in vacuum. Often, refineries and chemical plants design their vessels for external atmospheric pressure to allow for steam cleaning and condensation. Pipes may also be subjected to external pressure if they are subjected to similar conditions or are buried. Since the failure mode in the case of external pressure is by buckling or collapse of the structure, the design for external pressure is an iterative process. There are several geometric parameters involved, including the thickness, length, and stiffener design.

When vessels are designed for both internal and external pressure, it is common practice to start with the thickness calculated for internal pressure. If this is not adequate for external pressure, the designers have to add stiffeners with a certain spacing L. The design of the stiffeners is also an iterative procedure.

There are three significant geometric parameters: D_o, the outside diameter; t, the thickness; and L, the distance between lines of support. Adding stiffeners will reduce the volume of the total material (reducing material cost), but will increase fabrication costs.

The following could be considered a line of support:

- A circumferential line on a head at one-third the depth of the head (any formed head)
- A stiffening ring that meets code requirements for moments of inertia
- A jacket closure of a jacketed vessel meeting certain requirements
- A cone-to-cylinder junction that meets the moment-of-inertia requirements
- A bulkhead or any other internal attachment that meets the moment-of-inertia requirements

The minimum required thickness of each panel of a cylinder having a D_o/t ratio equal to or greater than 10 is determined by the following procedures [3]:

1. Assume or calculate the thickness t and the distance L between the stiffeners.
2. Calculate the L/D_o and D_o/t ratios.
3. Enter Fig. G in ASME Sec. II-D (Fig. 18-6) with the value L/D_o and move horizontally to the curve corresponding to the D_o/t value. Read the geometric constant A. (This is a geometric constant that does not depend on the material.)
4. Using the value of A calculated above, enter the applicable external pressure chart for the material and the corresponding design pressure and read B (Fig. 18-7 extracted from ASME Sec. II-D). This figure accounts for the elastic–plastic buckling behavior.
5. The maximum allowable external pressure is

$$P_a = \frac{4B}{3(D_o/t)}$$

6. For values of A falling to the left of the chart,

$$P_a = \frac{2AE}{3(D_o/t)}$$

7. For values of A falling to the right of the chart, use the maximum value of B for the temperature curve.
8. If P_a is smaller than the value specified for the external pressure, increase t or add stiffeners to reduce L.

18.3.5.1 Design of Ring Stiffeners To qualify as a line of support, a ring stiffener must meet certain moment-of-inertia requirements. The code provides two choices:

1. The moment of inertia of the section added must be at least equal to I:

$$I_s = \frac{D_o^2 L_s (t + A_s/L_s) A}{14}$$

2. The moment of inertia of the combination of the section added and the effective part of shell must be at least equal to I_s':

$$I_s' = \frac{D_o^2 L_s (t + A_s/L_s) A}{10.9}$$

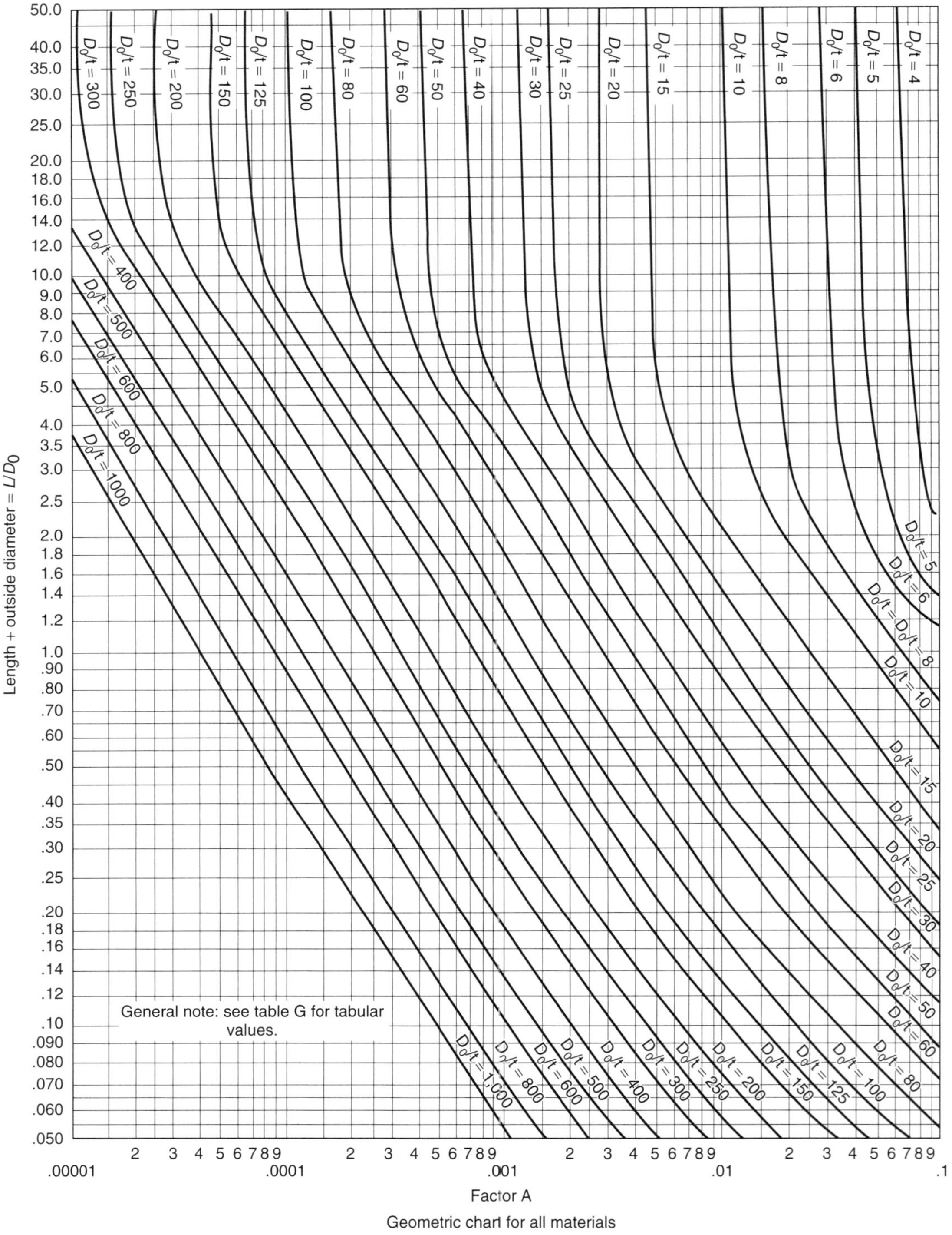

Figure 18-6 Factor A for different L/D ratios. (From ASME Sec. II.)

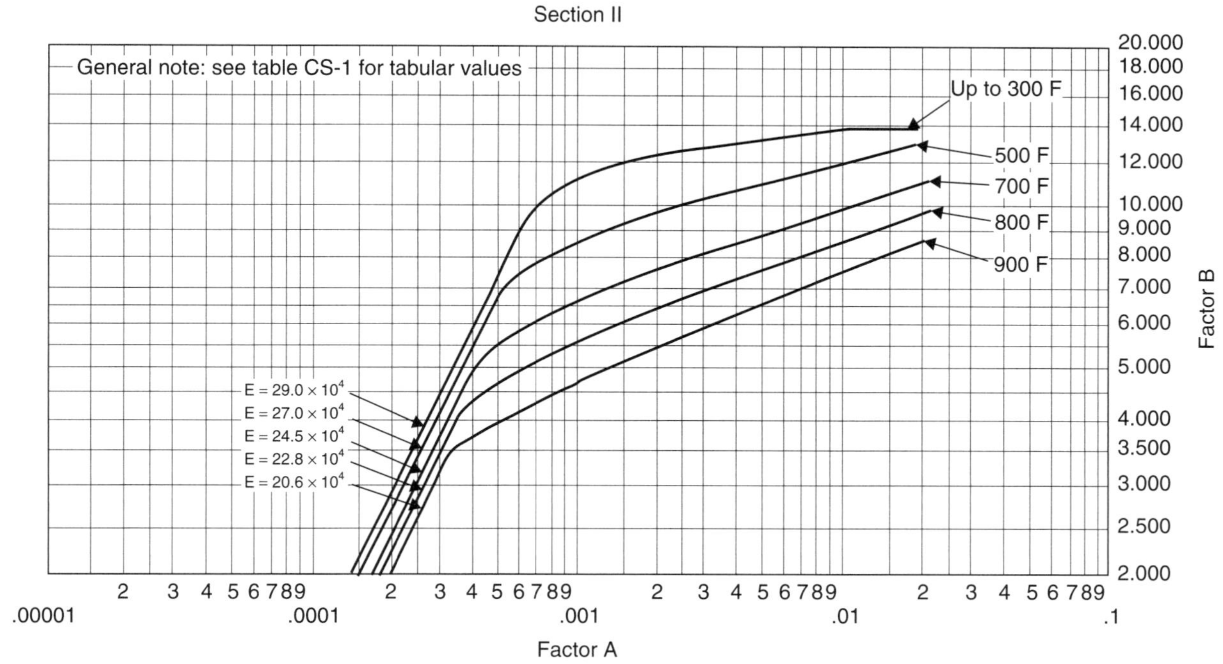

Figure 18-7 Chart giving factor B corresponding to factor A for elastic–plastic buckling.

where A_s is the cross-sectional area of the stiffer ring (in^2) and L_S is the sum of one-half the distance each side of the stiffener ring, measured parallel to the axis of the pipe (in.) ($L_s = L$).

The effective section is a shell length on either side of the stiffener equal to $0.55\sqrt{D_o t}$.

Example 18.3 A long vacuum service piping system is subjected to the design conditions OD = 48 in., NW = 0.500 in., material = A285GrC EFW, $P_d = -15$ psig, $T_d = 400^\circ$F, and $c_a = 0.063$ in. Does the pipe require stiffener rings?

Solution: Assume that the mill tolerance of the pipe of 12.5% is removed from the nominal wall. Then 87.5% of the nominal wall remains for the pressure calculation. The wall thickness, t, to be used for external pressure calculations is $t = 0.500 \times 0.875 - 0.063 = 0.375$ in.

The first test is to determine the allowable external pressure, P_a, for this pipe with no stiffener rings. Develop the ratios. For a long pipe, $L/D = 50$ and $D/t = 128$. (If L/D is greater than 50, then $L/D = 50$.) Enter Fig. 18-6 (Fig. G of ASME Sec. II-D) using the ratios developed for factor $A = 0.000067$. Next, using factor A, enter the applicable material–temperature graph, Fig. 18-7 (Fig. CS-2 of ASME Sec. II-D) for factor B. Since factor A is to the left of the material/temperature line, P_a is calculated by the equation

$$P_a = \frac{2AE}{3(D/t)}$$

where

$$E = 27.0 \times 10^6 \text{ psi (from Fig. 18.7)}$$

$$= \frac{(2)(0.000067)(27{,}000{,}000)}{(3)(128.1)} = 9.4 \text{ psig}$$

The maximum allowable vacuum pressure for the long 48-in.-OD, 0.500-in.-NW pipe of 9.4 psig is less that the design vacuum pressure of 15 psig; therefore, stiffener rings are required.

The designer tries to add circumferential stiffener rings at 192-in. intervals ($L = 192$ in.). Then $L/D = 4$, $D/t = 128.1$, $A = 0.00022$, and $B = 3300$, and P_a is calculated by

$$P_a = \frac{4B}{3(D/t)} = \frac{(4)(3300)}{(3)(128.1)} = 34 \text{ psig}$$

These stiffener rings give an overdesign allowable pressure of 34 psig, which is significantly larger than the 15 psig required. The designer may select a larger spacing L value to reach a closer pressure to 20 psig as a design target.

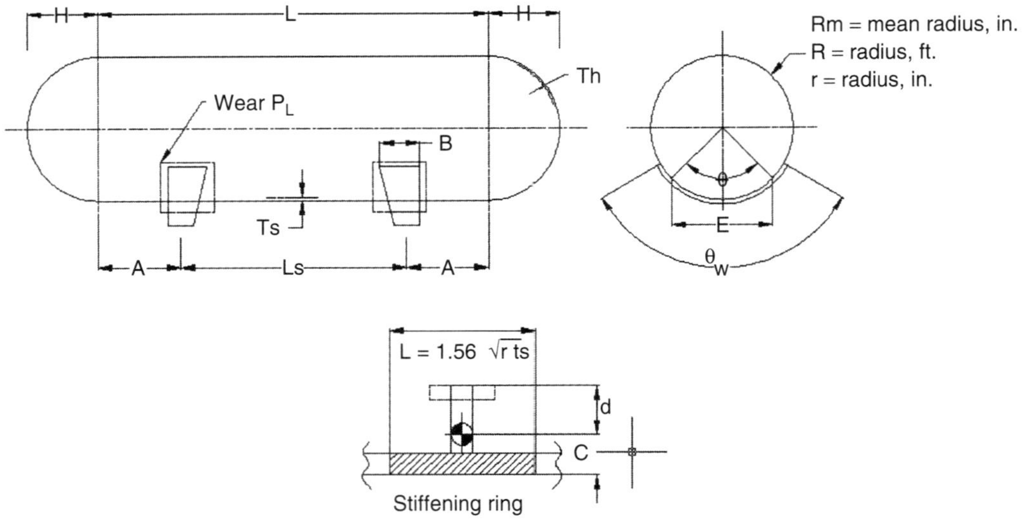

Figure 18-8 Horizontal vessel on supports with pads.

The design adequacy of the stiffener rings added is determined by the moment of inertia of the ring and the ring–shell combination that is available. These two moments must each be greater than I, the required moment of inertia of the stiffener ring about its neutral axis parallel to the axis of the pipe, and I', the required moment of inertia of the combined ring–shell cross section about its neutral axis parallel to the axis of the pipe.

18.3.6 Design of Vessel Supports

Cylindrical vessels may be laid horizontally (Fig. 18-8) or vertically, depending on the process and the location. Design codes refer to the vessel design without its supports. The designer needs to refer to other detailed design procedures to be able to design the supports (Zick's analysis). The design of supports and their connections and effects on the vessel start by first determining the reactions on the supports from the loading on the vessel, which includes:

- Weight of vessel and contents
- Wind load
- Seismic
- Thermal loads
- Any other inertial effects that would be transmitted to the ground

These loads will cause:

- *Global stresses*. This implies that the section carrying the load would be the entire section of the vessel. Examples of these are axial, bending, or shear of a section of the vessel (Fig. 18-9).
- *Local stresses* (Fig. 18-10). These are stresses that are local in nature due to changes in geometry and nonuniform deformation. Examples would be the junctions of the supports to the vessel, local shear effects, and so on. At any one location, these stresses are added to global stresses and compared to the corresponding allowable stresses.
- *Stresses in the supports*. These will depend on the detailed geometry of the supports and the loading reaching them.

Finite elements are used in the analysis of vessel supports. In such cases, ASME Section VIII, Division 2 describes the procedures for classifying the stresses

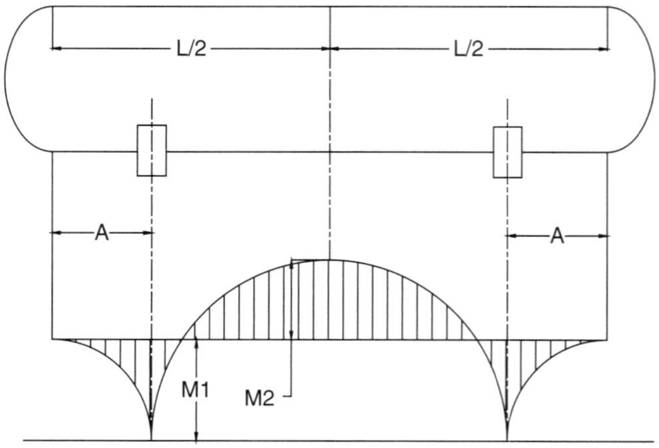

Figure 18-9 Bending moment diagram for a horizontal vessel.

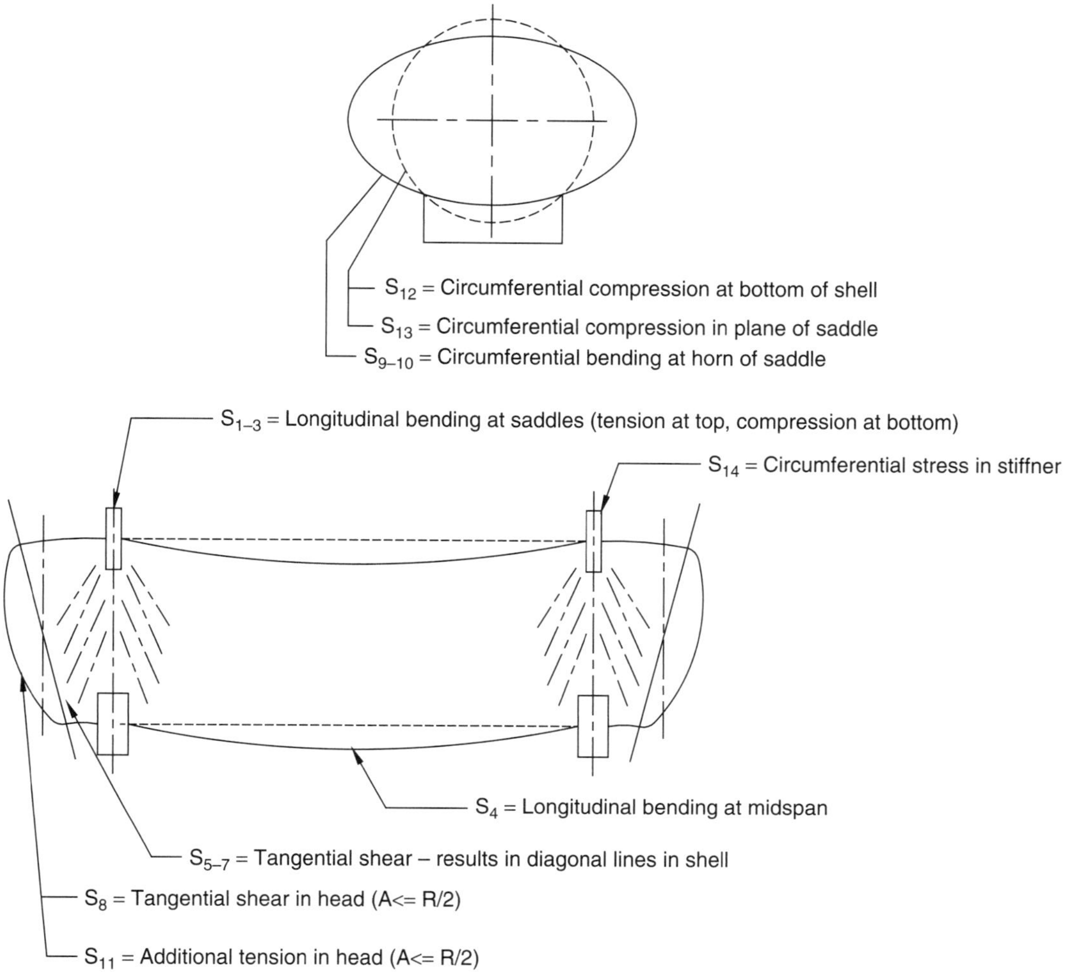

Figure 18-10 Local stresses and deformations in horizontal vessels.

across the thickness at any location to membrane and bending components and gives the corresponding stresses allowable.

18.3.7 Design by Rule Versus Design by Analysis

The ASME code introduced Section VIII, Division 2 in the early 1960s to provide alternatives to the rules of Section VIII, Division 1. The concept in this division is to reduce the material by raising the allowable design stresses but assuring safety by making more detailed analysis, including using numerical tools such as finite elements.

Divisions 1 and 2 both provide the same degree of overall safety. Significant differences include the following:

- Division 2 has higher stress allowables for most materials. (The design margin on the minimum ultimate strength specified is 3 for Division 2 and 3.5 for Division 1.)
- Division 2 has a much more restrictive list of materials and has more detailed design and analysis requirements, but limits the design temperatures to be less than creep.
- Division 2 provides rules for stress analysis and fatigue analysis, and uses the maximum shear stress theory, which is more accurate than the Division 1 failure criterion, which is the maximum stress theory.
- Overall, a Division 2 vessel provides material savings but is more expensive to design.

18.3.7.1 Stress Categorization In design by analysis, stresses are categorized as:

- *Primary stresses* (P_m, P_L) are due to imposed loading, are needed for equilibrium, and are not self-limiting.
- *Secondary stresses* (Q) are due to restraint of the structure, are needed for compatibility, and are self-limiting.

- *Peak stresses* (F) are local high stresses due to local structural discontinuities and can contribute to fatigue failure.

Primary stresses are classified as *membrane* or *bending*. Primary membrane stresses are broken into *general* and *local*.

- *Membrane stresses:* represent the average stress through the shell thickness.
- *Bending stresses Pb*: the linearly varying stresses through the shell thickness.
- *Local stresses*: those acting over a small portion of a shell (related to a factor). This indicates the effect of the limited zone.
- *General stresses*: those that act over a large segment of the shell.

The concept in the code is limit load design. Thus, since primary stresses are required to balance external loads, primary stress allowables are lower than those other types of stresses in design by analysis. Since general primary membrane P_m stress intensity ($\sigma_1 - \sigma_3$) would cause net section yielding, they are limited to the basic material allowable stresses S_m. Local primary membrane P_L stress intensity may not exceed $1.5S_m$. Membrane stresses are local if the distance over which $1.1S_m$ is exceeded is not greater than $\sqrt{Rt}$, where R is the radius and t is the thickness of the shell. However, the primary membrane plus bending stress intensity must not exceed $1.5S_m$ (Table 18-4).

TABLE 18-4 Stress Categorization and Allowable Limits

Stress Category	General Membrane	Local Membrane	Bending	Secondary Membrane + Bending Peak	Peak
Description	Average primary stress across solid section. Excludes	Average stress across any solid section. Considers discontinuities.	Component of primary stress proportional to distance from centroid of	Self-equilibrating stress necessary to satisfy differential thermal expansion. Excludes local stress concentrations.	Increment added to primary or secondary stress by a distortion of vessel shape.

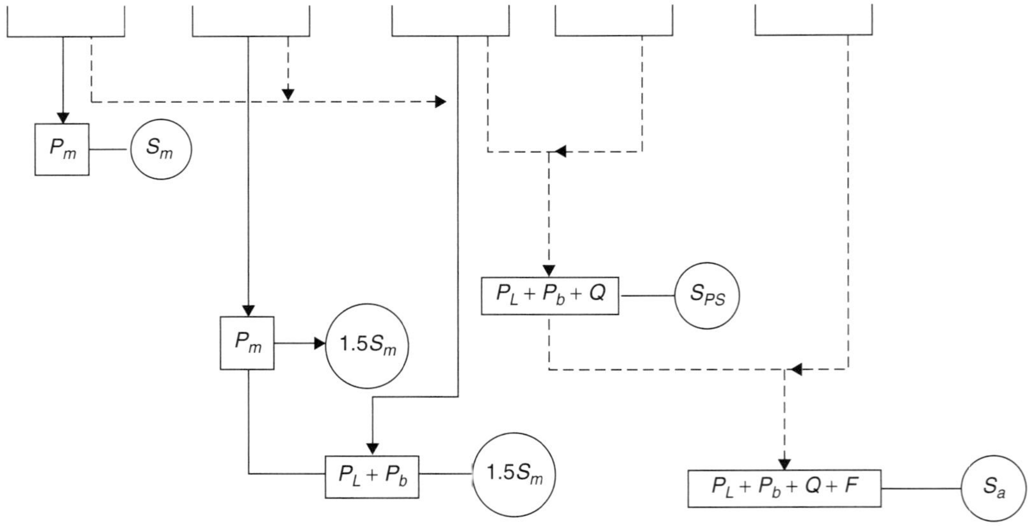

Symbol	P_m	P_l	P_b	Q	F

The following five stress intensity limits must be satisfied for each load case:

- P_m limited to kS_m
- P_L limited to $1.5kS_m$
- $P_L + P_b$ limited to $1.5kS_m$
- $P_L + P_b + Q$ limited to $3S_m$
- $P_L + P_b + Q + F$ limited by fatigue analysis

All the following limits may be increased by 1.2, for load combinations including wind or seismic.

18.3.7.2 Interpretation of Numerical Analysis Results The ASME code Section VIII, Division 2 refers to the WRC Bulletin 429 for the interpretation of the numerical result and the corresponding classification of stresses to compare with the appropriate stress limitation. This considers the following points:

1. The relation between the failure mode and code stress category
2. The appropriate stresses and locations for each category
3. Appropriate stresses for obtaining membrane plus bending stresses

The finite element results give the stress components distribution at all points in the vessel. It is important to define the locations for which the stresses are checked. These need to be linearized across the thickness at the various locations. Each of the six stress components could be linearized using the stress along a line or the stress along a plane method. At locations of discontinuities or stress concentrations, the membrane stress is calculated as the average across the thickness or may be calculated by equations to satisfy equilibrium. Figure 18-11 shows the form of linearization of the stresses across the thickness and shows the definition of membrane and bending components.

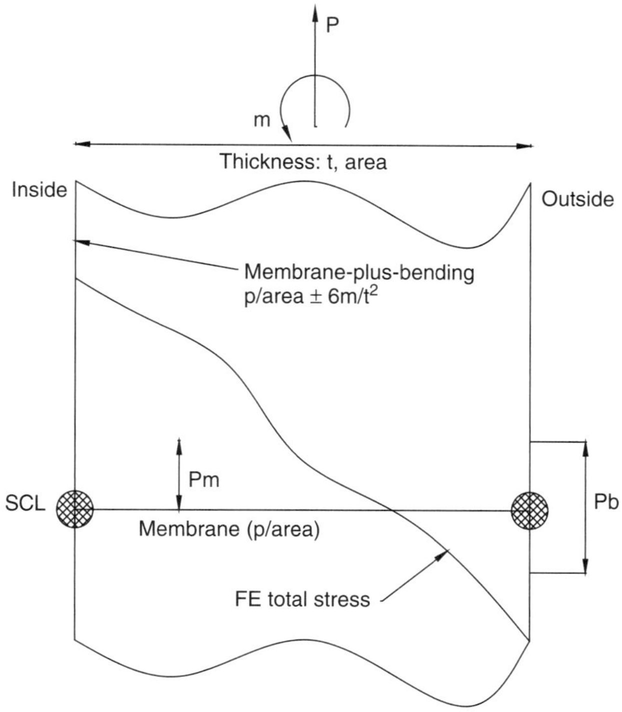

Figure 18-11 Linearization of stresses across the thickness of output from finite elements results.

18.4 DESIGN OF PIPING SYSTEMS

The design procedure for piping systems involves the following steps:

- Design for pressure
- Piping layout
- Design of supports
- Expansion and flexibility considerations

18.4.1 Wall Thickness for Internal Pressure

The same basic principles and equations for determining the wall thickness for pressure vessels is applied to determine the required minimum thickness of a straight pipe (ASME, 2007):

$$t = \frac{PD_o}{2(SE + YP)}$$

where Y is a factor that accounts for the pipe material and the diameter/thickness ratio:

$$Y = \begin{cases} 0.4 & \text{for } d/t > 6 \\ \dfrac{d}{d + D_o} & \text{for } d/t \leq 6 \end{cases}$$

E is the quality factor depending on the method of manufacture the pipe. Corrosion and mill tolerance allowances are added to the resulting thickness and the nearest highest pipe schedule is selected to satisfy the pressure requirements. NACE Corrosion Data Survey NF3231 is commonly used as a source for the corrosion allowance information.

18.4.1.1 Elbows and Bends Codes require elbows and bends that are not manufactured in accordance with listed standards to have the minimum wall thickness of the pipe and be free of defects. Bend flattening should not exceed 8% of the nominal outside diameter for internal pressure and 3% for external pressure.

18.4.1.2 Wall Thickness for Pipe Bends Pipe bends (not manufactured in accordance with the standards listed), intrados and extrados minimum wall, t, are calculated by Eqs. (18.7a) to (18.7c). t is added to the corrosion/erosion allowance c by the equation $t_m = t + c$.

$$t = \frac{PD}{2[(SE/I) + PY]} \tag{18.8a}$$

where I_i for the intrados is

$$I_i = 4\left(\frac{R_1}{D}\right) - \frac{1}{4(R_1/D) - 2} \tag{18.8b}$$

and I_e for the extrados is

$$I_e = 4\left(\frac{R_1}{D}\right) - \frac{1}{4(R_1/D) - 2} \tag{18.8c}$$

18.4.1.3 Miter Bend Pressure Limitation Miter bends have a pressure limitation, as calculated by code equations, which could derate a piping system. A miter is defined when the angle θ is greater than 3° at a weld. If the system would include miter bends, they are normally manufactured from the same pipe. However, the maximum allowable pressure for the miter bends needs to be checked by the equations that follow: For multiple miter bends, whose angle of miter cut is less than 22.5°, the allowable pressure is the smallest calculated from the equations:

$$P_m = \begin{cases} \dfrac{SE_q(T-C)}{r_2} \dfrac{T-C}{(T-C) + 0.643 \tan\theta\sqrt{r_2(T-C)}} \\[2ex] \dfrac{SE_q(T-C)}{r_2} \dfrac{R_1 - r_2}{R_1 - 0.5r_2} \end{cases} \tag{18.9}$$

For a single miter bend or widely spaced miter bends,

$$P_m = \frac{SE_q(T-C)}{r_2} \frac{T-C}{(T-C) + 1.25 \tan\theta\sqrt{r_2(T-C)}}$$

If the allowable pressure is not acceptable, the miter bend must be manufactured from a thicker pipe.

For nozzles and branch connections, the same procedure described in the pressure vessel design section is also applied here: namely, using the area replacement method, determine if the extra material area available is greater than that removed. For pipes subjected to external pressure, the same procedure presented for pressure vessels is used.

18.4.2 Pipe Span Calculations

Pipes are supported on many supports, depending on their routing. However, the maximum distance between any two supports must satisfy both stress and deflection requirements. Treating the pipe between two supports as a beam loaded by a distributed load equivalent to the weight of the pipe and the fluid, the maximum allowable span will depend on the end conditions at the supports. These may be assumed to be rigidly fixed or simply supported. The former (rigidly fixed) is a more conservative estimate. In practice, an average state between both conditions is assumed and the maximum span may be calculated from

$$L = \begin{cases} \sqrt{\dfrac{0.4ZS_h}{W}} & \text{based on limitation of stress} \\[2ex] \sqrt[4]{\dfrac{\Delta EI}{13.5W}} & \text{based on limitation of deflection} \end{cases} \tag{18.10}$$

where L is the pipe support spacing (ft), Δ the permissible midspan deflection (in.), E the modulus of elasticity at the design temperature (lb/in.), I the moment of inertia of the pipe, and W the weight of pipe supported, including pipe, contents, and insulation (lb/ft). For segments of pipes that include changes in direction or valves, or other conditions varying from a simple straight configuration, the span must be reduced, depending on the loading conditions.

One important aspect related to the span calculations based on deflections is its effect on the natural frequency of the piping structure. One simple equation to determine the natural frequency of any pipe span is

$$f_n = \frac{1}{2\pi}\sqrt{\frac{g}{\Delta}} = \frac{3.12}{\sqrt{\Delta}} \tag{18.11}$$

The natural frequency of any piping segment should be higher than that of any possible exiting frequency in the system.

18.4.3 Pipe Supports

The purpose of pipe supports is to control the weight effects of the piping system, as well as loads caused by pressure thrust, vibration, wind, earthquake, shock, and thermal displacement. The weight effects to be considered are the greater of the operating or hydrotest loads. The selection of the type of support and the degrees of freedom that it constrains at each position are key element accounting for the weight carried, and account for the thermal expansions in the pipe. One simple way of estimating the reactions on the supports due to weight is by dividing the piping structure into several segments and assuming that each segment is isolated from the rest of the pipe. Then, using the equations of equilibrium, one can get the reactions at the supports. This is useful in the preliminary design stage to help find suitable locations for the pipe hangers and supports. However, a more detailed and accurate analysis would be required using finite elements.

18.4.3.1 Variable Support Hangers Variable support hangers are often necessary in piping systems where there is a change from the installed temperature to the operating temperature. A support is required to continue to carry the load and at the same time to allow the pipe at that section to move. If resting supports were selected for some piping layouts, the pipe would lift off the support, causing an increase in pipe load on adjacent supports or on load-sensitive equipment such as pumps or turbines. This increase in load may be more than the adjacent support or equipment can safely manage. If the support is fixed, it would constrain the thermal movement, causing additional loads. A constant load support or a variable spring support (Fig. 18-12) may be the preferred pipe support for these instances. The designer must decide which type of support to use and then size the support. The variation in the spring support load should not exceed 25%. This is the design basis for selecting a suitable stiffness for the spring support.

18.4.3.2 Bearing Stress at a Pipe Support Supporting pipe directly on support steel (without a pipe shoe or saddle) causes local stresses similar in nature to those for vessels on supports. This becomes a particular concern as the D/T ratio (the ratio of the diameter of the pipe supported to the pipe wall thickness) increases. The pipe shoe to be considered is one constructed in a structural shape such as a wide flange. Three equations are presented for the support of load stresses:

Meridional membrane stress, σ_1:

$$\sigma_1 = 0.153\beta^3 P R^{1/4} b_1^{-1/2} t^{-7/4} \tag{18.12}$$

Circumferential membrane stress, σ_2:

$$\sigma_2 = 0.130\beta^3 P R^{3/4} b_1^{-3/2} t^{-5/4} \tag{18.13}$$

Circumferential bending stress, σ_2':

$$\sigma_2' = -0.156\beta^{-1} P R^{1/4} b_1^{-1/2} t^{-7/4} \tag{18.14}$$

where σ_1, σ_2, and σ_2' are stress in psi; $\beta = [12(1 - \gamma^2)]^{1/8}$; $\gamma =$ Poisson's ratio (0.3); $P =$ pipe support load (lb); $t =$ corroded pipe wall thickness (in.); $R =$ mean radius of pipe, $0.5(\text{OD} - t)$ (in.); and $b = \frac{1}{2}$ width of supporting steel (in.).

18.4.4 Expansion and Flexibility

A key consideration in piping design is expansion and flexibility in the piping structure. Theoretically, it would be nice if the piping structure were free to expand in any direction without restraints. Adding restraints implies adding loads on the structure. This is a challenge to the designer: to strike a balance between keeping the piping structure flexible but rigid enough to carry the external loads.

As a preliminary design phase, the expansions of the piping structure are calculated with the assumptions that the structure is only resting on its supports and that thermal movements are not restrained. This would give the designer a picture about the directions of thermal movements. This would help the designer locate the preliminary positions where expansion loops or joints should be added. This is clearly an iterative process and, normally, piping design packages (based on finite elements) are used to model the structure and make modifications, thus comparing the results to those allowable. In selecting the type of support at any section, it is very important to decide upon the degrees of freedom that such a support would restrain. This will affect the movements and, consequently, the resulting loads.

18.4.4.1 Expansion Loops The concept of introducing expansion loops is based on the fact that the flexibility of beams (pipes) in bending is much higher than in the axial direction. Thus, any changes from a straight-line

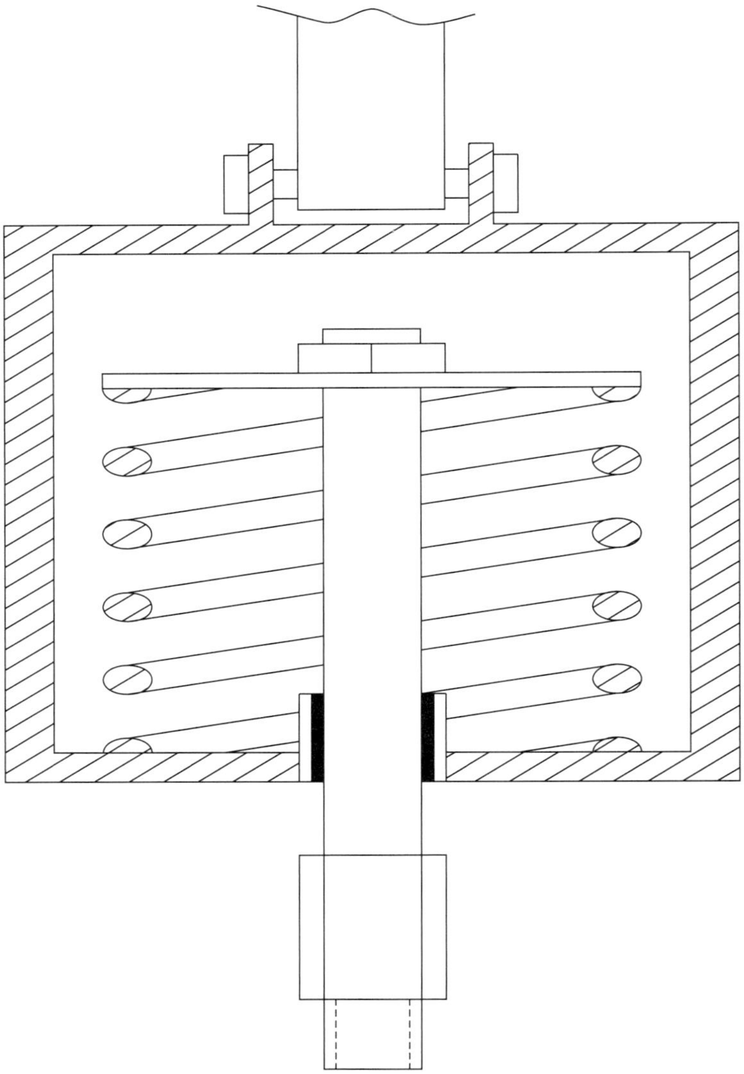

Figure 18-12 Spring support.

configuration in a piping system would add flexibility in the system that would allow for thermal expansions. Figure 18-13 shows some configurations of expansion loops. The size of the loop will depend on the amount of thermal expansion that it is expected to absorb. However, if too small an expansion loop is used, the forces, and consequently the stresses, developed would be large. On the other hand, large expansion loops would add flexibility to the system, but the losses in flow would increase. The loads and corresponding stresses developed in the various sections of the loop need to be checked.

18.4.4.2 ***Expansion Joints*** The principle of operation of expansion joints is similar to that of expansion loops in that it transfers axial deformations to bending deformations. However, in expansion joints the bending is local to the bellows of the joint, not to the structure of the pipe. Expansion joints may be used in geometries where the thermal expansion would cause compression of the joint.

The internal pressure in a bellows geometry would cause it to expand in both the axial and circumferential directions. Thus, expansion joints are normally used for low-pressure systems. Tie rods around the expansion joint are normally necessary to avoid the condition that the joint may be loaded in axial tension due to pressure effects. On the other hand, expansion joints may be placed between two anchor points to make sure that they are always loaded in compression. The axial force that the bellows feels is equal to the pipe expansion between two anchor points multiplied by the bellows stiffness.

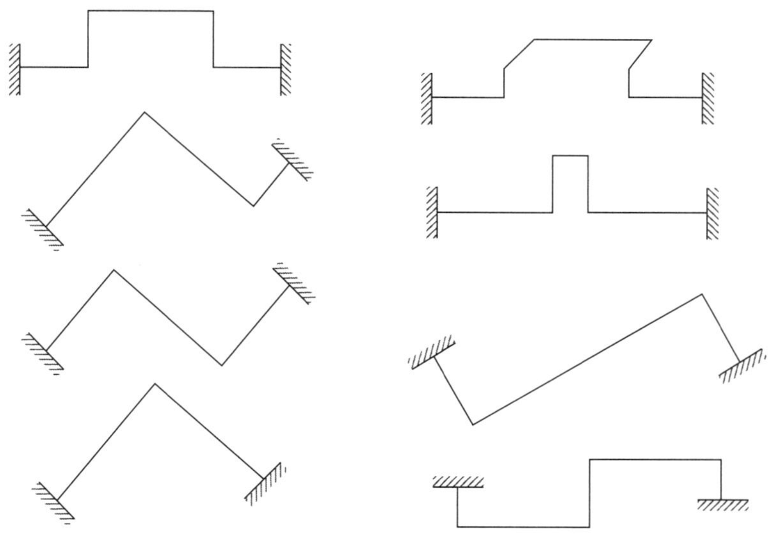

Figure 18-13 Expansion loops.

18.4.5 Code Compliance

Piping codes require that the piping structure comply with several stress conditions: (1) stresses due to pressure, (2) stresses due to sustained loads, (3) stresses due to occasional loads, and (4) stresses due to thermal expansion.

18.4.5.1 Stresses Due to Pressure The design codes define the combined design pressure and corresponding design temperature to be the combination that gives the largest pipe thickness. The determination of the thickness is based on the circumferential stress caused by internal pressure. This is a membrane stress and is compared to the material allowable stress at the operating temperature reduced by a joint factor to account for manufacturing conditions (see Section 18.4.1). A corrosion and manufacturing allowance is added to the thickness. Note that the designer needs to check the minimum thickness requirements at elbows, bends, miter bends, openings, and so on.

The allowable pressure P for straight pipes and bends is calculated as

$$P = \frac{2SEt_m}{D_o - 2yt_m} \tag{18.15}$$

where $t_m = t_n(1 - \text{mill tolerance}/100) - \text{corrosion allowance}$, ($t_n$ = nominal pipe thickness), D_o is the outside diameter, d is the inside diameter, and y is the pressure coefficient (= 0.4; if $D_o/t < 6$, $y = d/(d + D_o)$.

For closely spaced miter bends, the allowable pressure is calculated:

$$P = \frac{SEt_m(R - r)}{r(R - r/2)} \tag{18.16}$$

where r is the mean radius of pipe = $(D_o - t_n)/2$ and R is the equivalent bend radius of the miter. For widely spaced miter bends, the allowable pressure is calculated from

$$P = \frac{SEt_m^2}{r[t_m + 1.25\tan(\vartheta a)\sqrt{rt_m}]} \tag{18.17}$$

where ϑa is the miter half-angle.

Several design codes allow overloading conditions for limited durations, during which the maximum stress is not allowed to exceed 90% of the allowable stress. The designer also has to demonstrate that the pressure stresses during testing conditions are also within the allowable stresses. For pipes subjected to external pressure, the designer needs to demonstrate that the structure has enough stiffness that it would not collapse.

18.4.5.2 Stresses Due to Sustained Loads Axial stresses developed in piping structures due to sustained loads (pressure, weight, mechanical loads, etc.) will cause axial, bending, and torsional stresses. Sustained loads are external loads that have to be carried by the structure. It is required that the sum of these stresses in the axial direction should be less than the allowable stress at hot working conditions. Piping codes have slight differences. As an example, ASME Code B31.3 shows the condition as

$$S_L = \sqrt{(S_a + S_{ab})^2 + 4S_t^2} \leq 1.0S_h \tag{18.18}$$

whereas B31.1 gives it as

$$S_L = \frac{PD_o}{4t_n} + \frac{0.75iM_A}{Z} \leq S_h \tag{18.19}$$

and

$$M_A = \text{resulting moment due to sustained loads}$$
$$= \sqrt{M_x{}^2 + M_y{}^2 + M_z{}^2} \qquad (18.20)$$

Note that both equations are the same if the stress intensification factor (SIF) for in-plane and out-of-plane bending are the same.

$$S_b = \text{resulting bending stress}$$
$$= \frac{[(i_i M_i)^2 + (i_o M_o)^2]^{1/2}}{Z} \qquad (18.21)$$

S_t = torsional stress = $M_t/2Z$

M_t = torsional moment

Z = section modulus of pipe [mm^3(in^3)]

= $(\pi/32D_o)(D_o^4 - D_i^4)$ (the exact equation)

D_i = inside diameter of pipe [mm(in.)]

D_o = outside diameter of pipe [mm(in.)]

i_i = in-plane SIF

i_o = out-plane SIF

M_i = in-plane bending moment [N · m(in.-lb)]

M_o = out-plane bending moment [N · m(in.-lb)]

Note that the stress intensification factor represents the increase in stresses due to the changes in geometry at any bend, nozzle, or change in diameter. The code includes tables that give the stress intensification factor for different geometries. As an example, the SIF for an LR elbow from ASME B31.3 Appendix D is $i = 0.9/h^{2/3}$ for the in-plane SIF and $i = 0.75/h^{2/3}$ for the out-plane SIF, where $h = TR_1/r_2^2$. Note that the code reduces the effect of the stress intensification factor by multiplying it by 0.75, since the moments do not cause a uniform stress distribution across the pipe cross section. However, the code does not allow the designer to have the product 0.75 less than 1. This would reduce the effect of the moment for straight parts.

18.4.5.3 Stresses Due to Occasional Loads Occasional loads are loads that would rarely happen, and when it happens it would not last for a long duration. Examples of such loads are loads caused by wind or by earthquake. The code requires that the designer study the effect of each type of load separately and not combine, for example, the worst possible earthquake condition in the location of the piping and the worst wind condition simultaneously. Since such conditions last for a very short period of time, the code allows the designer to increase the allowable stress by a factor k, as illustrated by the equation

$$S_{Lo} = \frac{pD_o}{4t_n} + \frac{0.75iM_A}{Z} + \frac{0.75iM_B}{Z} <= kS_h \qquad (18.22)$$

This factor depends on the duration of the loading and the code used. It can go up to 1.33 for short durations.

18.4.5.4 Stresses Due to Thermal Expansions Stresses due to thermal expansions and movements in piping structures are classified as *self-equilibrating*. This means that the net resulting external load due to temperature only (without consideration of sustained loads) is zero. That is, these stresses must be in equilibrium. The thermal stresses are generally fluctuating, and thus the limitation on stress variation is based on the shakedown limit rather than on yield. This limit becomes twice the allowable stress based on the yield. Furthermore, the code allows the designer to make use of the extra margin of safety available by not loading the structure in the direction axial to its limit.

$$S_E = \frac{iM_c}{Z} \le S_A + f(S_h - S_L) \qquad (18.23)$$

where M_c is the resulting moment due to thermal expansion, $S_A = f(1.25S_c + 0.25S_h)$, S_c is the allowable stress at cold temperatures, and f is the stress range reduction factor for cyclic conditions [$f = 6.0(N^{-0.2}) \le 1.0$].

Software codes apply these limitations by outputting the results under the following conditions:

1. Sustained loads (weight and pressure)
2. Thermal loads: effect of temperature only without weight, pressure, or any other external loading
3. Occasional loads: effect of earthquake and effect of wind separately
4. Combinations to check the code

REFERENCES

1. Abdel Malek, H. L., and Younan, M. Y., A mulltiple criterion approach to fail safe design, *Scientific Engineering Bulletin*, 1981.

2. American Society of Mechanical Engineers, *Fitness for Service*, API579/ASMEFFS-1, ASME, New York, 1997.
3. American Society of Mechanical Engineers, *ASME Boiler and Pressure Vessel Codes, VIII: Rules for Construction of Pressure Vessels*, ASME, New York, 2010.
4. American Society of Mechanical Engineers, *ASME Code for Power Piping B31.1*, ASME, New York, 2007.
5. Boyle, J. S., *Stress Analysis for Creep*, Butterworth, London, 1983.
6. Kannappan, S., *Introduction to Pipe Stress Analysis*, Wiley, New York, 1986.
7. Megson, T. H., *Structural and Stress Analysis*, Elsevier Butterworth, Boston, 2005.
8. Moss, D., *Pressure Vessel Design Manual*, 3rd ed., Gulf Publishing, Hoston, TX, 2004.

19

PROCESS SAFETY IN CHEMICAL PROCESSES

JELENKA SAVKOVIC-STEVANOVIC
Belgrade University, Belgrade, Serbia

This chapter deals with the theoretical background of process safety and risk analysis and with the functioning of particular pieces of process equipment. This analysis shows how process safety requirements are designed and must be fitted into the environment. Safety in the treatment of distillation, absorption, and extraction columns, reactors, heat exchangers, and so on, described.

The ecological engineer must use a wide range of information taken from a variety of sources and must take into account many conflicting requirements: technical, economic, and environmental. Within the time span available, the engineer must effect a satisfactory compromise, and it must always be borne in mind that there is never a unique "best" solution to any safety and reliability problem. Furthermore, a satisfactory safety analysis for one location may be totally unsuitable elsewhere. Although it is impossible to convey a complete philosophy in a single chapter, many of the diverse factors that must be incorporated into a risk analysis are covered.

Toxicology must rank as one of the oldest practical sciences, because from the beginning, human beings have needed to avoid the numerous toxic plants in their environment. Forensic toxicology comprises both analytical and fundamental toxicologic principles. It is concerned with the legal aspects of the harmful effects of chemicals on humans. Environmental toxicology is a relatively new area that studies the effects of chemicals released by humans on wildlife and the ecosystem, and thus indirectly on human health.

Large-scale chemical production is one of the most important industrial activities. Its products play key roles in human nutrition, health, the quality of life, and welfare. Nevertheless, many people regard chemical production as dangerous, even though chemical processes are associated with accident frequencies that are much lower than in other industries, and very few such accidents have anything to do with chemistry.

Despite the good accident statistics of chemical plants, grave accidents have occurred, endangering human life and the environment and causing considerable damage. Three types of event are traditionally associated with the chemical industry: releases and spills, fires, and explosions.

Explosion as used here refers to an event in which a large amount of energy is released very rapidly, producing a pressure wave that is at least audible and propagates at high velocity, decaying over distance. Damage caused by such events comes about through two distinct processes: direct action of chemicals on humans and the environment, and the indirect action of the energy liberated.

Generally, chemical reactions underlie such harmful effects. When substances act directly on living matter, a chemical reaction may alter or destroy cellular tissue (toxic effect). In an explosion, energy is liberated by an exothermic chemical reaction.

It is obvious that the dangers arising from chemical plants have much to do with the nature of the substances being processed, the way they are treated in a plant, and their tendency to take part in chemical reactions under these conditions.

A conflict arises between process needs and chemical safety. On the one hand, chemistry requires reactive substances; on the other hand, this reactivity of the substances is a key aspect of the danger they pose.

Process Plant Equipment: Operation, Control, and Reliability, First Edition. Edited by Michael D. Holloway, Chikezie Nwaoha, and Oliver A. Onyewuenyi.
© 2012 John Wiley & Sons, Inc. Published 2012 by John Wiley & Sons, Inc.

A chemical process is impossible without substances that show hazardous properties and effects. Substances must therefore be contained reliably in process equipment, and their reactivity must be governed so that uncontrolled chemical reactions cannot take place.

19.1 THE HAZARDS

All manufacturing processes are to some extent hazardous, but in chemical processes additional hazards are associated with the chemicals used and the process conditions. The designer must be aware of these hazards, and ensure, through the application of sound engineering practice, that the risks are reduced to acceptable levels.

The particular hazards associated with chemicals and allied processes will be considered. The more general, normal hazards present in all manufacturing processes, such as the dangers from rotating machinery, falls, falling objects, use of machine tools, and electrocution should be considered. General industrial safety and hygiene are covered in several books (e.g., [12,34,78]).

Safety and loss prevention in process design can be considered and incorporated under the following topics:

1. Identification and assessment of hazards.
2. Control of hazards: for example, by containment of flammable and toxic materials.
3. Process control and supervision. Prevention of hazardous deviations in process variables (pressure, temperature, flow) by provision of automatic control systems, interlocks, alarms, and trips, together with good operating practices and management.
4. Loss prevention and limitation of loss: the damage and injury caused if an incident occurs, pressure relief, plant layout, and provision of firefighting equipment.
5. Protection from radiation.

Process safety is discussed in general in many books. The proceedings of symposia on the subject contain many articles of interest on general safety phylosophy, techniques and organization, and the hazards associated with specific processes and equipment.

Intrinsic and Extrinsic Safety Processes can be divided into those that are intrisincally safe and those for which safety has to be engineered in. An intrinsically safe process is one in which safe operation is inherent in the nature of the process, a process that causes no danger, or negligible danger, under all foreseeable circumstances (all possible deviations from the design operating conditions). The designer should always select a process that is intrinsically safe whenever it is practical and economical, to do so. However, most chemical manufacturing processes are, to a greater or lesser extent, inherently unsafe, and dangerous situations can develop if the process conditions deviate from the design values.

The safe operation of such processes depends on the design and provision of engineered safety devices, and on good operating practices, to prevent a dangerous situation from developing and to minimize the consequences of any incident that arises from the failure of these safeguards. Engineered safety means provision in the design of control systems, alarms, trips, pressure relief devices, automatic shutdown systems, duplication of key equipment services, firefighting equipment sprinkler systems, and blast walls to contain any fire or explosion.

19.1.1 Special Hazards

The special hazards of chemicals are toxicity, flammability, and corrosivity. Most materials used in the manufacture of chemicals are poisonous, to some extent. The potential hazard will depend on the inherent toxicity of the material and the frequency and duration of an exposure. It is usual to distinguish between the short- and long-term effects. A highly toxic material that causes immediate injury, such as phosgene or chlorine, would be classified as a safety hazard. A material whose effect was apparent only after long exposure at low concentrations (e.g., carcinogenic materials such as vinyl chloride) would be classified as an industrial health and hygiene hazard. The permissible limits and precautions to be taken to ensure that the limits are met are very different for these two classes of toxic materials. Industrial hygiene is as much a matter of good operating practice and control as of good design.

19.1.2 Toxicology

Toxicology studies the interaction between chemicals and the biological interaction between chemicals and biological systems to determine the potential of chemicals to produce adverse effects in living organisms. Toxicology also investigates the nature, incidence, mechanisms of production, factors influencing their development, and reversibility of such effects. *Adverse effects* are defined as determinal to the survival or the normal functioning of a person. Inherent in this definition are the following key issues in toxicology:

- Chemicals must come in close structutral or functional contact with tissues or organs to cause injury.
- All adverse effects depend on the amount of chemical in contact with the biological system (dose) and the

inherent toxicity of the chemical (hazard). When possible, the toxic effect observed should be related to the degree of exposure. The influence of different exposure doses on the magnitude and incidence of a toxic effect should be quantitated. Such dose–response relationships are of prime importance in confirming a causal relationship between chemical exposure and toxic effect.

Research in toxicology is concerned primarily with determining the potential for adverse effects caused by chemicals, both natural and synthetic, to asseses their hazard and the risk of human exposure and thus to provide a basis for appropriate precautionary, protective, and restrictive measures. Toxicological investigations should permit evaluation of the following characteristics of toxicity [74]:

- The basic structural, functional, or biochemical injury produced
- Dose–response relationships
- The mechanism of toxicity and reversibility of the toxic effect
- Factors that modify response

For chemicals to which humans may potentially be exposed, a critical analysis based on the pattern of potential exposure or toxicity may be necessery to determine the risk/benefit ratio for their use in specific circumstances and to devise protective and precautionary measures. Indeed, with drugs, pesticides, food additives, and cosmetic preparations, toxicology testing must be performed under government regulations before use [73].

19.1.2.1 Information Resources For toxicology as a science and the impact of toxicological investigations on legislation and commerce, a wide range of information on the toxic effects of chemicals is available. No single exhaustive source of toxicological data exists; several sources are required to obtain comprehensive information on a particular chemical. Printed sources are often quicker and easier to use than computer databases, but interactive online searching can make possible rapid information collection from the huge number of sources present.

The information explosion in toxicology has resulted in two comprehensive volumes dedicated to toxicological information sources: those of Wexler [77] and Webster and [75]. The easiest way to obtain information on general topics in toxicology and secondary references is through the range of textbooks available [e.g., [10,21,35]).

The best summary information on toxicology is published in the form of a series by goverments and international organizations. Most of these series summarize the results of toxicity on specific chemicals, The selection of these chemicals is based mainly on the extent of their use in industry (e.g., trichloroethylene), their occurrence as an environmental contaminant (e.g., mercury), or their extraordinary toxicity (2,3,7,8-tetrachlorodibenzodioxin).The most important series is the Barbour Report, issued by the Cold Spring Harbor Laboratory, which publishes proceedings of conferences on toxicological and environmental subjects, focusing primarily on aspects of risk assessment. Results of toxicological research are published in more than 100 journals.

Electronic sources such as computer databases and CD ROMs are fast and convenient ways to obtain references on the toxicity of chemicals. Since online searching of commercial databases such as STN International may be expensive, CD ROM–based systems are being used increasingly. The major advantages are speed, the ability to refine searches and format the results, and nontext search options, such as chemical structure searching on *Beilstein* and *Chemical Abstracts*.

Useful information about actual research on the toxicology of chemicals may be obtained by searching Chemical Abstracts or Medline with the appropriate keywords. Specific data banks covering toxicology are the Registry of Toxic Effects of Chemical Substances, which gives summary data, statistics, and structures. Toxiline, available in DIMDI, gives access to the literature.

Points to control relating to toxic materials at the design stage are:

- *Containment*: sound design of equipment and piping; specify welded joints in preference to flanges (liable to leak).
- *Disposal*: provision of effective vent stacks to disperse material vented from pressure relief devices, or use venturi scrubbers.
- *Ventilation*: use an open structure or provide adequate ventilation systems.
- Emergency equipment, escape routes, rescue equipment, safety showers, eye baths.

In addition, operating practices should include regular medical checkups for employees to check for chronic effects and regular monitoring of the environment to check exposure levels. Consider the installation of permanent instruments fitted with alarms, as well good general hygiene, washing facilities, and operating instructions.

19.1.2.2 Toxic Effects According to the time scale, toxic effects may be divided in terms of acute or delayed; in terms of general locus of action as to local, systematic, or organ specific; and in terms of to basic mechanisms of toxicity as reversible or irreversible. Acute toxic effects are those that occur after brief exposure to a chemical. Acute toxic effects usually develop rapidly after single or multiple administrations of a chemical. However, acute exposure may also produce delayed toxicity. Chronic effects are those that appear after repetitive exposure to a substance. Many compounds require several months of continuous exposure to produce adverse effects. Often, the chronic effects of chemicals are different from those seen after acute exposure. For example, inhalation of chloroform for a short period of time may cause anesthesia; long-term inhalation of a much lower chloroform concentration causes liver damage. Carcinogenic effects of chemicals usually have a long latency period; tumors may be observed in rodents years after exposure and even decades later in humans.

The toxic effect of chemicals may be caused by reversible and irreversible interactions in the organism (Table 19-1). When reversible interactions are responsible for toxic effects, the concentration of the chemical present at the site of action is the only determinant of toxic outcome. When the concentration of the xenobiotic is decreased by excretion or biotransformation, a parallel decrease in toxic effects is observed.

After complete excretion of the toxic agent, toxic effects are reduced to zero. A clasic example of reversible toxic effects is carbon monoxide. The carbon monoxide binds to hemoglobin, and because of the formation of a stable hemoglobin–carbon monoxide complex, oxygen binding is blocked. As a result of impaired oxygen transport in blood from the lung, tissue oxygen concentrations are reduced and cells sensitive to oxygen deprivation die. The toxic effects of carbon monoxide are correlated directly with the proportion of carboxyhemoglobin in blood, whose concentration is dependent on the inhaled concentration of carbon monoxide. After exhalation of carbon monoxide and survival of the acute intoxication, no toxic effect remains.

Irreversible toxic effects are often caused by the covalent binding of toxic chemicals to biological macromolecules. Under extreme conditions, the modified macromolecule is not repaired, after excretion of the toxic agent, the effect persists. Further exposure to the toxic agent will produce additive effects. Many chemical carcinogens are believed to act through irreversible changes in the macromolecules.

TABLE 19-1 Interactions of Chemicals as a Basis for Toxic Response

Mechanism	Toxic Response	Example
Irreversible inhibition of esterase	Neurotoxicity	Tri-*o*-cresylphosphate
Covalent binding to DNA	Cancer	Dimethylnitrosamine
Reversible binding to hemoglobin	Oxygen derivation in tissues	Carbon monoxide
Cholinesterase	Neurotoxicity	Carbamate pesticides

Another distinction between types of effects can be made using the general locus of action. Local toxicity occurs at at the site of first contact between a biological system and a toxic agent. Local effects to the skin, the respiratory tract, or the alimentary tract may be produced by skin contact with a corrosive agent, by inhalation of irritant gases, or by ingestion of tissue-damaging materials. This type of toxic response is usually restricted to tissues in direct contact with the agent.

The opposite of local effects are systematic effects. These are characterized by absorption of the chemical and its distribution from the port of entry to a distant site at which toxic effects are produced. Except for highly reactive xenobiotics, which act mainly locally, most chemicals act systematically. Many chemicals that produce systematic toxicity cause damage only to certain organs, tissues, or cell types within organs. Selective damage to certain organs or tissues by systematically distributed chemicals is termed *organ-* or *tissue-specific toxicity* [15]; the organs damaged are referred to as *target organs*. Major target organs for toxic effects are the central nervous system and the circulatory system, followed by the blood and hematopoietic system and visceral organs such as the liver and kidney. For some chemicals, both local and systematic effects can be demonstrated. Moreover, chemicals producing marked local toxicity may also cause systematic effects as a secondary response to major disturbances in the homeostasis of the organism.

The spectrum of toxic effects of chemicals is broad, the magnitude and nature depending on many factors, such as the physicochemical properties of the chemical and its toxicokinetics, conditions of exposure, and the presence of adaptive or protective mechanisms. The latter factors include physiological mechanisms such as adaptive enzyme induction, DNA repair, and others. Toxic effects may be transient, reversible, or irreversible; some are deleterious and others are not. Toxic effects may take the form of tissue pathology, aberrant growth processes, or altered biochemical pathways. Some of the more frequently encountered types of injury consituting a toxic response are described below.

Immune-mediated hypersensitivity reactions by antigenic materials are toxic effects (often involved in skin and lung injury) caused by repeated contact with chemicals, resulting in contact dermatitis and asthma. Inflammation is a frequently observed local response to the application of irritant chemicals, or it may be a component of systematic injury. This response may be acute, with irritant or tissue-damaging materials, or chronic, with repetitive exposure of irritants. Chemical tumorigenesis or carcinogenesis, the induction of malignant tumors, is an effect often observed after chronic application of chemicals. Because of the long latency period and the poor prognosis for persons diagnosed with cancer, prediction of the potential tumorigenicity of chemicals has developed into a major area of toxicological research. Development and reproductive toxicology are concerned with adverse effects on the ability to conceive and adverse effects on the structural and functional integrity of the fetus. Chemicals may interfere with reproduction through direct effects on reproductive organs or indirectly by affecting their neural and endocrine control mechanisms. Developmental toxicity deals with adverse effects on the conceptus through all stages of prenancy. Nonlethal fetoxicity may be expressed as delayed maturation, decreased birth weight, and structural malformations.

19.1.2.3 Dose Response At sufficiently high doses, any chemical may be toxic. The importance of dose is seen clearly with molecular oxygen or dietary metals. Oxygen at a concentration of 21% in the atmosphere is essential for life. In contrast, 100% oxygen at atmospheric pressure causes massive lung injury in rodents, often resulting in death. Some metals, such as iron, copper, and zinc, are essential nutrients. When present in insufficient amounts in the human diet, specific disease patterns develop. At high doses these metals may cause fatal intoxications. The word *dose* is commonly used to characterize the total amount of material to which an organism is exposed.

The inherent toxicity of a material is measured by tests on animals. It is usually expressed as LD_{50}, the lethal dose at which 50% of test animals are killed [33]. Some values for tests on rats are given in Table 19-2.

There are now generally accepted definitions of what can be considered toxic and nontoxic. In fixing permissible limits on concentration for the long-term exposure of workers to toxic materials, the exposure time must be considered together with the inherent toxicity of the material. The threshold limit value (TLV) is the most commonly used guide for controlling the long-term exposure of workers to contamined air. The TLV is defined as the concentration to which it is believed the average worker could be exposed, day to day, for 8 hours a day, 5 days a week, without suffering harm. It is expressed in ppm for vapors and gases, and in mg/m^3 or $grains/ft^3$ for dusts and liquid mists. A comprehensive source of data on the toxicity of industrial materials is Sax's handbook [64]. Toxicity data on solvents are given by Browning [9].

TABLE 19-2 Chemical Substances and Their LD_{50} Values

Compound	LD_{50} (mg/kg)
2,3,7,8-Tetrachlorodibenzodioxin	0.01
Sodium fluoroacetate	0.2
Strychnine	5
Potassium cyanide	10
Sodium cyanide	6.4
Tetraethyllead	35
Lead	100
Paraquat	120
DDT	150
Aspirin	1,500
Table salt	3,000
Sodium bicarbonate	4,220
Ethanol	12,500

After chronic exposure to a chemical, a toxic response may be caused by doses that do not show any effects after single dosing. Chronic toxic responses are often based on an accumulation of either the toxic effect or the chemical handling. Since the rate of elimination depends on plasma concentrations, after long-term application an equilibrium concentration of the chemical is reached in the blood.

For a chemical that is irreversible to macromolecules, the magnitude of toxic response may be correlated with the total dose handling. In contrast to chemicals that act reversibly, the effect is not dependent on the frequency of dosing. Effect accumulation is often observed with carcinogens and ionizing radiation. The reversibility of toxic responses also depends on the capacity of an organ or tissue to repair injury.

19.1.2.4 Toxicokinetics Species differences in toxic response are due largely to differences in toxicokinetics and biotransformation. Distribution and elimination characteristics are quite variable among species. Both qualitative and quantitative differences in biotransformation may affect the sensitivity of a given species to a toxic response [79]. Qualitative differences in biotransformation are responsible for apparent differences in the sensitivity of rats and guinea pigs to bladder carcinogenicity. The induction of systematic toxicity usually results from a complex interaction between absorbed parent chemical and biotransformation products formed in tissues, the distribution of parent chemical and biotransformation products in body fluids and tissues, their binding and storage characteristics, and their excretion.

19.1.3 Flammability

The hazard caused by a flammable material depends on a number of factors:

1. The flash point of the material
2. The autoignition temperature of the material
3. The flammability limits of the material
4. The energy released in combustion

The *flash point* is a measure of the ease of ignition of a liquid. It is the lowest temperature at which the material will ignite from an open flame. The flash point is a function of the vapor pressure and the flammability limits of the material. It is measured in standard apparatus, following standard procedures. Both open- and closed-cup apparatus is used. Closed-cup flash points are lower than open-cup flash points, and the type of apparatus used should be stated clearly when reporting measurements. Flash points are given in Sax's handbook [64]. The flash points of many volatile materials are below normal ambient temperature.

The *autoignition temperature* of a substance is the temperature at which it will ignite spotaneously in air without an external source of ignition. It is an indication of the maximum temperature to which a material can be heated in air: for example, in drying operations.

The *flammability limits* of a material are the lowest and highest concentrations in air, at normal pressure and temperature, at which a flame will propagate through the mixture. They show the range of concentration over which the material will burn in air if ignited. Flammability limits are characteristic of the particular material and differ widely for different materials.

A flammable mixture may exist in the space above the liquid surface in a storage tank. The vapor space above highly flammable liquids is usually purged with inert gas (nitrogen), or floating head tanks are used. In a floating head tank a piston floats on top of the liquid, eliminating the vapor space.

Flame arresters are fitted in the vent lines of equipment that contains flammable material to prevent the propagation of flame through the vents. Various types of proprietary flame arresters are used. In general, they work on the principle of providing a heat sink, usually expanded metal grids or plates, to dissipate the heat of the flame. Traps should also be installed in plant ditches to prevent the spread of flame. These are normally liquid, which blocks the spread of flammable liquid along ditches. Recommendations on fire precautions to be taken in the design of chemical plants are given in the standard codes and practices.

19.1.4 Explosions

An explosion is a sudden catastrophic release of energy, causing a pressure wave or blast wave. An explosion can occur witout fire, such as the failure through overpressure of a steam boiler or an air receiver. When discussing the explosion of a flammable mixture it is necessary to distinguish between detonation and deflagration. If a mixture detonates, the reaction zone propagates at supersonic velocity and the principal heating mechanism in the mixture is shock compression. In a deflagration the combustion process is the same as in the normal burning of a gas mixture. The combustion zone propagates at subsonic velocity, and the pressure buildup is slow. Whether denotation or deflagration occurs in a gas–air mixture depends on a number of factors, including the concentration of the mixture and the source of ignition. Unless confined or ignited by a high-intensity source such as a detonator, most materials will not detonate. However, the pressure wave caused by deflagration can still cause considerable damage.

Materials such as acetylene can decompose explosively in the absence of oxygen; such materials are particularly hazardous. Unconfined vapor cloud explosions result from the release of a considerable quantity of flammable gas or vapor into the atmosphere, and its subsequent ignition.

*19.1.4.1 **Dust Explosions*** If intimately mixed with air, combustible solids can explode. Several disasterous explosions have occurred in grain silos. Dust explosions usually occur in two stages: a primary explosion that disturbs deposited dust, followed by a second, severe explosion of the dust thrown into the atmosphere. Any finely divided combustible solid is a potential explosion hazard. Particular care must be taken in the design of dryers, conveyors, cyclones, and storage hoppers for polymers and other combustible products or intermediates. The hazard of dust explosions and their prevention is discussed fully by Palmer [40].

Historically, dust explosions have been associated with coal mining and grain milling. Further industralization has led to dust explosions in the chemical, pharmaceutical, metallurgical, and food- and wood-processing industries.

19.1.5 Ignition

Although precautions are normally taken to eliminate sources of ignition in chemical plants, it is best to work on the principle that a leak of flammable material will ultimately find an ignition source. The sparking of electrical equipment such as motors is a major potential source of ignition, and flameproof equipment is normally specified.

In all areas where flammable gases are likely to be present in flammable concentrations under normal operating conditions, intrinsically safe equipment should be specified or the equipment enclosed in a purged gastight chamber. In areas where a flammable mixture will be present only under abnormal circumstances, nonsparking equipment can be specified: equipment that does not normally spark but could spark if a fault develops. Some risk is involved, but the coincident failure of the electrical equipment and leak of flammable gas would be required to cause a fire or explosion. The use of electrical equipment in hazardous areas is covered by many standards.

The movement of any nonconducting material, powder, liquid or gas, can generate static electricity, producing sparks. Precautions must be taken to ensure that all piping is properly earthed (grounded) and that electrical continuity is maintained around flanges. Escaping steam or other vapors and gases can generate a static charge. Gases escaping from a ruptured vessel can self-ignite from a static spark. For a review of the dangers of static electricity in the process industries, see the articles by Napier [38] Napier and Russell [39].

Open flames from process furnaces and incinerators are obvious sources of ignition and must be sited well away from areas containing flammable materials. It is the usual practice in plants handling flammable materials to control entry onto the site of obvious sources of ignition, such as matches, cigarette lighters, and battery-operated equipment. The use of portable electrical equipment, welding, spark-producing tools, and the movement of gasoline-driven vehicles should also be subject to strict control.

The minimum ignition energy is the lowest value of the energy-stored discharge which is just sufficient to ignite the most readily ignitable dust–air mixture at atmospheric pressure and room temperature. The minimum ignition temperature is the lowest temperature of a hot surface at which a dust–air mixture is ignited on contact. No internationally accepted test method exists for determination of the minimum ignition energy of dispersed powder.

19.1.6 Ionizing Radiation

The radiation emitted by radioactive materials is harmful to living matter. Small quantities of radioactive isotopes are used in the process industries for various purposes: for example, in level- and density-measuring instruments and for nondestructive testing of equipment. A discussion of the particular hazards that arise in the chemical processing of nuclear fuels is outside the scope of this chapter.

19.1.7 Pressure

Overpressure, a pressure exceeding the system design pressure, is one of the most serious hazards in chemical plant operation. Failure of a vessel or the associated piping can precipitate a sequence of events that culminate in a disaster. Pressure vessels are invariably fitted with some form of pressure relief device, set at the design pressure, so that potential overpressure is relieved in a controlled manner. Three basically different types of relief device are commonly used:

1. *Directed actuated valves*: weight- or spring-loaded valves that open at a predetermined pressure and which normally close after the pressure has been relieved. The system pressure provides the motive power to operate the valve.
2. *Indirectly actuated valves*: pneumatically or electrically operated valves which are activated by pressure-sensing instruments.
3. *Bursing disks*: thin disks of material designed and manufactured to fail at a predetermined pressure, giving a full bore opening for flow.

Relief valves are normally used to regulate minor excursions of pressure and bursting disks as safety devices to relieve major overpressure. Bursting disks are often used in conjuction with relief valves to protect the valve from corrosive process fluids during normal operation. The design and selection of relief valves is covered in the pressure vessel standards. The disks are manufactured in a variety of materials for use in corrosive conditions, such as impervious carbon, gold, and silver, and suitable disks can be found for use with all process fluids. Bursting disks and relief valves are proprietary items and the vendors should be consulted when selecting suitable types and sizes.

19.1.7.1 Vent Piping When designing relief venting systems it is important to ensure that flammable or toxic gases are vented to a safe location. This will normally mean venting at a sufficient height to ensure that the gases are dispersed without creating a hazard. For highly toxic materials it may be necessary to provide a scruber to absorb and kill material: for example, the provision of caustic scrubbers for chlorine and hydrochloric acid gases. If flammable materials have to be vented at frequent intervals, such as in some refinery operations, flare stacks are used.

The rate at which material can be vented will be determined by the design of the complete venting system. the relief device and the associated piping. The maximum venting rate will be limited by the critical sonic velocity, whatever the pressure drop. The design of venting systems to give adequate protection against overpressure is a complex and difficult subject, particularly if two-phase flow is likely to occur. For complete protection the venting system must be capable of venting at the same rate as the vapor is being generated. For reactors, the maximum rate of vapor generation resulting from a loss of control can usually be estimated. Vessels must also be protected against overpressure caused by external fires. In these circumstances the maximum rate of vapor generation will depend on the rate of heating. For some vessels, particularly where complex vent piping systems are needed, it may be impractical given the size of the vent to give complete protection against the worst possible situation.

19.1.7.2 Vacuum Breakers Unless designed to withstand external pressure, a vessel must be protected against the hazard of under- as well overpressure. Underpressure will normally invalve a vacuum on the inside with atmospheric pressure to collapse a storage tank. Although the pressure differential may be small, the force on the tank roof will be considerable. For example, if the pressure in a 10-m-diameter tank falls to 10 milibar below the external pressure, the total load on the tank roof will be around 80,000 N (8 metric tons). It is not an uncommon occurrence for a storage tank to be sucked in, collapsed by the suction pulled by the discharge pump, due to the tank vents having become blocked. Where practical, vacuum breakers, valves that open to the atmosphere when the internal pressure drops below atmospheric, should be fitted.

19.1.8 Temperature Disturbance

High temperature, over and above that for which the equipment was designed, can cause structural failure and initiate a disaster. High temperature can arise from loss of control of reactors and heaters and, externally, from open fires. In design processes where high temperatures are a hazard, protection against high temperatures is provided by:

- Provision of high-temperature alarms and interlocks to shut down reactor feeds or heating systems if the temperature exceeds critical limits
- Provision of emergency cooling systems for reactors where heat continues to be generated after shutdown: for example, in some polymerization systems.
- Structural design of equipment to withstand the worst possible temperature excursion
- The selection of intrinsically safe heating systems for hazardous materials

Steam and other vapor heating systems are intrinsically safe, as the temperature cannot exceed the saturation temperature at the supply pressure. Other heating systems rely on control of the heating rate to limit the maximum process temperature. An electrical heating system can be hazardous.

19.1.8.1 Fire Protection To protect against structural failure, water deluge systems are usually installed to keep vessels and structural steelwork cool in fire. The lower section of structural steel columns are also often lagged with concrete or other suitable materials.

Critical effects can also follow from reactions between a combustible substance and a fire-extinguishing agent. For example, the reaction of firefighting water with a combustible substance may release not only flammable gases (from potassium, alkali metal compounds, etc.) but also corrosive vapors (e.g., chloric acid from chlorosilanes, acetyl chloride). Ignition by static discharge has caused accidents where carbon dioxide or medium expansion foam has been used improperly.

19.1.9 Noise Disturbance

Excessive noise is a hazard to health and safety. Long exposure to high noise levels can cause permanent damage to hearing. At lower levels, noise is a distraction and causes fatgue. The unit of sound measurement is the *decibel*, defined by the expression

$$\text{sound level} = 20 \log_{10} \frac{\text{sound pressure}}{2 \times 10^{-5}} \quad \text{dB} \qquad (19.1)$$

The subjective effect of sound depends on frequency as well as intensity.

Industrial sound meters include a filter network to give the meter a response that corresponds roughly to that of the human ear. This is termed the A-weighting network factor. Permanent hearing damage can be caused at sound levels above about 90 dB(A).

Excessive plant noise can lead to complaints from neighboring factories and local residents. Due attention should be given to noise levels when specifying, and laying out, equipment that is likely to be excessively noisy, such as compressors, fans, burners, and steam relief valves.

19.1.10 Fire and Explosion Index

A numerical fire and explosion index is calculated based on the nature of the process and the properties of the materials. The larger the value of the index, the more hazardous is the process. When used to evaluate the design of a new plant, the index is normally calculated after the piping and instrumentation diagrams and equipment layout have been prepared, and is used as a guide to the selection and design of the preventive and protection equipment needed for safe plant operation. It may be calculated at an early stage in the process design, after the preliminary flowsheets have been prepared, and will indicate whether alternative, less hazardous processes should be considered. The safety and loss prevention guide [16] gives a method for evaluating the potential hazards of a process and assessing the safety and loss prevention measures needed.

The index applies only to main process units; it does not cover process auxiliaries, such as warehouses, tank farms, utilities, and control rooms. Only the fire and explosion hazard is considered; toxicity and corrosion hazards are not covered. Not does it deal with the special requirements of plants that manufacture explosives.

The basis of the fire and explosion (F&E) index is a material factor, which is normally determined from the heat of combustion of the main process material. This primary material factor is multiplied by factors to allow for special material hazards and for general and special process hazards.

The process is divided into units and the index is calculated for each unit. A unit is defined as a part of the process that can be considered as a separate entity. It may be a section that is separated from the remainder of the plant by a physical barrier or by distance, or it may be a section of the plant in which a particular hazard occurs.

19.1.10.1 Material Factor The material factor (MF) is the number from 0 to 60 that indicates the magnitude of the energy released in a fire or explosion. For noncombustible materials the factor is zero: for example, water and carbon tetrachloride. For combustible materials the factor is calculated from the following expression:

$$MF = -\Delta H_c \times 10^{-3} \tag{19.2a}$$

or

$$MF = -\Delta H_c^0 \times \frac{4.3 \times 10^{-4}}{\text{mol}} \times 10^{-3} \tag{19.2b}$$

where ΔH_c is a heat of combustion (J/mol), and ΔH_c^0 is the standard heat of combustion at 25°C (kJ/mol).

For combustion of highly reactive materials such as mixtures of oxidizing and reducing agents, the heat of combustion may not be an adequate indication of the potential energy release, so the heat of reaction, or decomposition, is used if numerically larger than the heat of combustion. The Dow guide [16] includes a list of the material factors for commonly used chemicals. To decide the dominant material, the material factor should be evaluated for all process materials that are present in the unit in sufficient quantity to constitute a hazard. To calculate the material factor for methane, the standard heat of combustion is 801,700 kJ/mol, and MF = 801,700(4.3 × 10^{-4}/16) = 21.6.

19.1.10.2 Special Materials Hazard Special materials hazard factors are included to take account of any special hazards associated with the materials present in a unit. The primary material factor is increased by a percentage for each of the hazards listed below if applicable to any material present in significant quantity in the unit.

1. *Oxidizing materials*: materials that will release oxygen in a fire.
2. *Reaction with water*: materials that produce a combustible gas on reaction with water (e.g., calcium carbide).
3. *Spontaneous heating*: materials that are subject to spontaneous heating or are pyrophoric (e.g., coal).
4. *Spontaneous polymerization*: materials that are liable to polymerize when heated (e.g., chloroprene, butadiene).
5. *Explosive decomposition*: materials that are liable to decompose, accompanied by explosion (e.g., acetylene).
6. *Detonation*: materials that could detonate under the process conditions if the protective control systems fail.
7. *Others*: any other unusual hazards associated with the materials that are considered appropriate.

19.1.10.3 General Process Hazard General process hazard factors are intended for general hazards associated with the unit being considered. The material factor, after adjustment for the special material factors, is increased by a percentage for each of the hazards listed below that are applicable to the process.

1. *Handling and physical change only*: processes that do not involve chemical reactions. The large percentage factor is used if the process involves disconnecting and connecting pipes handling flammable liquids.
2. *Continuous reactions*: the lower factor used if a runaway reaction is not likely to cause an excessive temperature rise; for other, exothermic, reactions, use 50%.
3. *Batch reactions*: a percentage added, in addition to reaction factor 2, to take account of any special hazard associated with batch operations.
4. *Multiplicity of reactions*: a factor added to allow for possible contamination from one reaction to another for processes carried out in the same equipment, if this is likely to constitute a hazard.

19.1.10.4 Special Process Hazard Special process hazard factors allow for any of the special process hazards given below. The percentages to be used depend on the magnitude of the hazard:

1. *Low pressure*: a factor added if inleakage of air is possible and likely to cause a hazard.
2. *Operation near the explosive range*: a factor added if a concentration within the explosive range is likely to occur in normal operation, or where the process relies on instrumentation to avoid explosive concentrations. The larger factor, 150%, is added for processes that always operate within the explosive range.
3. *Low temperature*: a factor to allow for the possible low-temperature brittleness of structural materials.
4. *High temperature*: a factor to account for the hazard of operating at temperatures above the boiling point, flash point, or autoignition temperature of the process materials.
5. *High pressure*: a factor to allow for the hazard of operating at pressures greater than 15 bar.
6. *Reactions difficult to control*: applies to exothermic reactions for which, because of the nature of the process, there is a strong possibility of the reaction going out of control.

7. *Dust or mist hazards*: applies to processes in which the malfunction of equipment could lead to a dust or mist explosion.
8. *Greater-than-average explosion hazard*: applies to processes where process conditions are such that the hazard of an explosion due to the malfunction of equipment is greater than usual, such as processes containing flammable liquids at temperatures and pressures such that release would result in rapid vaporization and the formation of an explosive vapor cloud.
9. *Large quantities of combustible liquids*: a factor to allow for the increased fire hazard associated with the storage of large quantities of flammable liquids.
10. *Others*: a factor added for processes with unusual hazards, particularly when the fire and explosion index is such as to expose nearby units to increased risk.

19.1.10.5 Preventive and Protective Measures In safety and loss prevention the fire and explosion (F&E) index is used as an aid in determining the equipment and facilities needed to control the hazards and reduce the losses from any incident that may occur; preventive and protective (P&P) measures [16,67]. The P&P measures to be taken are divided into three categories:

1. The basic P&P features, which must provide for all processes, regardless of the F&E index.
2. The minimum features recommended, which depend on the value of the F&E index.
3. Specific preventive features, measures that provide specific protection for the hazards considered in evaluating the index.

Basic safety and fire protective measures that should be included in all chemical process designs and operations are:

- Adequate and secure water supplies for firefighting
- Correct structural design of vessels, piping, and steel work
- Pressure relief devices
- Corrosion-resistant materials and/or adequate corrosion allowances
- Segregation of reactive materials
- Grounding of electrical equipment
- Safe location of auxiliary electrical equipment, transformers, and switchgear
- Provision of backup utility supplies and services
- Compliance with national codes and standards
- Fail-safe instrumentation
- Provision for access of emergency vehicles and the evacuation of personnel
- Adequate drainage for spills and firefighting water
- Adequate storage for the organic volatility component
- Insulation of hot surfaces
- No glass equipment used for flammable or hazardous materials, unless no suitable alternative is available
- Adequate separation of hazardous equipment
- Protection from fire of pipe racks and cable trays
- Provision of block valves on lines to main processing areas
- Protection of fired equipment, heaters, and furnaces against accidental explosion and fire
- Safe design and location of the control room

Minimum preventive and protective measures include measures that should be considered in addition to the basic measures, but which may or may not be specified, depending on the designer's assessment of the risks and magnitude of the resulting fire and explosion. In addition, the need for building ventilation, dust explosion control, and building explosion relief should be considered.

Specific preventive features include measures for the specific protection from the hazards considered when evaluating the F&E index under the headings of special material hazards and general and special process hazards [68,74].

Special material hazards include:

- *Oxidizing materials*: store in a fireproof area separate from combustible materials.
- *Reacts with water*: protect from water, ventilate, and remove sources of ignition.
- *Spontaneous polymerization*: use inhibitors, provide cooling, pressure relief, and high-temperature alarms.
- *Explosive decomposition*: design equipment to contain or safety-relieve the explosion.
- *Detonation*: as for explosive decompositions.

General process hazards include:

- *Handling and physical change*: excess flow valves; purge procedures, ventilation, remotely operated valves.
- *Continuous reaction*: tight process control, instrumentation for detection and protection against overpressure and overtemperature.
- *Batch reactions*: tight process control, instrumentation for detection and protection against overpressure, overtemperature, and process interlocks to prevent cross-contamination.

- *Multiplicity of reactions*: tight process control, instrumentation for detection and protection against overpressure and overtemperature.

Special process hazards include:

- *Low pressure*: trips and alarms.
- *Operation near the explosion range*: design to contain or safety-relieve explosions; provide explosion suppression systems and instrumentation to control composition, purges, or inert dilution systems.
- *Low temperature*: specify suitable materials of construction.
- *High temperature*: provide special ventilation and dump systems, combustible gas monitors, automatic deluge systems, and control systems to minimize the flow of flammable materials.
- *High pressure*: provide special ventilation and dump systems, combustible gas monitors, automatic deluge systems, and control systems to minimize the flow of flammable materials.
- *Reactions and processes difficult to control*: provide containment, safe venting, dump system, and quench systems.
- *Dust or mist hazards*: design to contain or safety-relieve explosions, explosion-suppression systems, instrumentation to control composition, purges, or inert dilution systems.
- *Above-average explosion hazards*: design to contain or safety-relieve explosions, explosion-suppression systems, instrumentation to control composition, purges, or inert dilution systems.
- *Large quantities of flammable materials*: provide remotely operated valves to minimize flow, combustible gas monitors linked to automatic deluge systems, and adequate drainage.

The main improvements made to the Dow index by the Mond index are [80,81]

- It covers a wider range of process and storage installations.
- It covers the processing of chemicals with explosive properties.
- A calculation procedure is included for the evaluation of a toxicity hazards index.
- A procedure is included.

19.2 HAZARD ANALYSIS

Hazard analysis is a technique for the quantitative assessment of a hazard after it has been identified by an operability study or similar technique. A brief outline of the technique can be used to make a preliminary examination of the design at the flowsheet stage, and for a detailed study at a later stage when a full process description, final flowsheet, piping and instrumentation diagrams, and equipment details are available [23].

An operability study will identify potential hazards, but gives no guidance on the likelihood of an incident occurring or the loss suffered; this is left to the intuition of the team members. Incidents usually occur through the coincident failure of two or more items, failure of equipment, control systems, and instruments, and missed operation [55]. The sequence of events that leads to a hazardous incident can be shown as a fault tree or logic tree, such as that shown in Fig. 19-1. The AND symbol is used where coincident inputs are necessary before the system fails, and the OR symbol is used where failure of any input would, by itself, cause failure of the system.

A fault trees for even a simple process unit is complex, with many branches. Fault trees are used to make a quantitative assessment of the likelihood of failure of a system, using data on the reliability of the individual components of the system. For example, if the following figures represent an estimate of the probability that the events shown in Fig. 19-1 will happen, the probability of failure of the total system by this route can be calculated (Table 19-3). The failure rates are added for OR gates and multiplied for AND gates, so the probability of flooding or a failed head is given by $(1+0.1+0.4)(0.04) = 0.06$ times per year, or once in $1/0.06 = 17$ years.

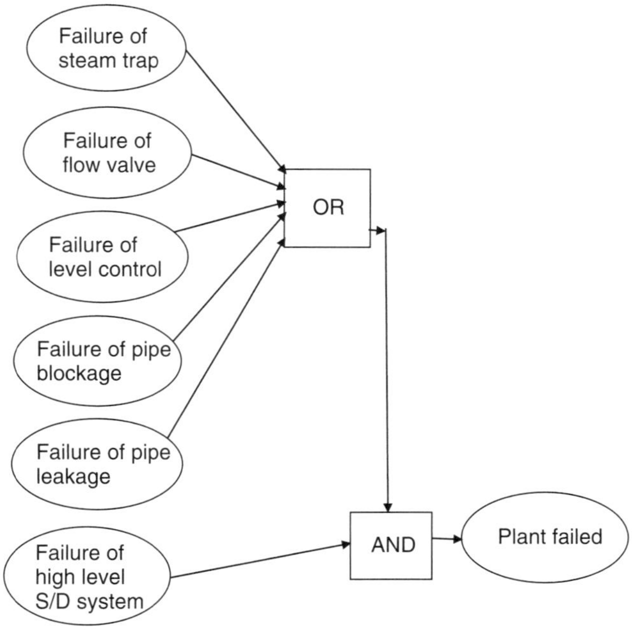

Figure 19-1 Simple flow system fault chart.

TABLE 19-3 Basic Failure Rate

Failure	Times per Year
Steam trap	1.00
Flow control valve	0.40
Level control (subsystem)	0.10
Leakage	0.05
Blockage	0.10
High-level shutdown (subsystem)	0.04

The data on failure rates given in this example are for illustration only. Some quantitative data on the reliability of instruments and control systems are given by Lees [32].

Examples of the application of quantitative hazard analysis technique in chemical plant design are widely available [28,31,43,76], as are examples of the application of quantitative hazard analysis in chemical plant safety [44,45,48,53,58,59,62,63]. Much of the work on the development of hazard analysis techniques and the reliability of equipment has been done in connection with the development of nuclear energy programs. Process engineers need to make notes about acceptable risk and safety priorities.

If the consequences of an incident can be predicted quantitatively, a quantitive assessment can be made of risk:

$$\text{quantitative assessment of risk} = \text{frequency of incident} \times \text{loss per incident} \quad (19.3)$$

If the loss can be measured in money, the cash value of the risk can be compared with the cost of safety equipment or design changes to reduce the risk. In this way, decisions on safety can be made in the same way as other design decisions, to give the best return for the money invested.

Hazards invariably endanger life as well as property, and any attempt to make cost comparisons will be difficult and controversial. It can be argued that no risk to life should be accepted. However, resources are always limited, and some way of establishing safety priorities is needed.

One approach is to compare the risk, calculated from a hazard analysis, with risks that are generally considered acceptable, such as the average risks in the particular industry and the types of risks that people accept voluntarily. One measure of the risk to life is the *fatal accident frequency rate* (FAFR), defined as the number of deaths per 10^8 working hours. This is equivalent to the number of deaths in a group of 1000 employees over their working lives. The FAFR can be calculated from statistical data for various industries and activities. Some of these values people accept voluntarily.

In the chemical process industries it is generally accepted that risks with an FAFR greater than 0.4 (one-tenth of the average for the industry) should be eliminated as a matter of priority, the elimination of lesser risks depending on the resources available. This criterion is for risks to employees; for risk for the general public, a lower criterion must be used. Kletz [29] suggests that a hazard can be considered acceptable if the average risk is less than 1 in 10 million per person per year. This is equivalent to an FAFR of 0.001.

19.2.1 Safety Checklists

Checklists are useful aids to memory. A checklist that has been drawn up by experienced engineers can be a useful guide for the less experienced. However, too great a reliance should never be put on the use of checklists to the exclusion of all other considerations and techniques. No checklist can be completely comprehensive, covering all the factors to be considered for any particular process or operation.

A short safety checklist, covering the main items that should be considered in process design, is given in this chapter. More detailed lists have been published by the Institute of Chemical Engineers [26] and Wells [76].

Balemans [5] gives a comprehensive list of guidelines for the safe design of chemical process plants, drawn up in the form of a checklist. The lists in the Dow guide [16], the general and specific hazards, and the preventive and protective features can also be used as a checklist of the factors to be considered in design.

Design Safety Checklist

- Materials
  - Flash point
  - Flammability range
  - Autoignition temperature
  - Composition
  - Stability (shock sensitive)
  - Toxicity, TLV
  - Corrosion
  - Physical properties
  - Heat of combustion or reaction

- Process
  1. Chemical reactions
     - Exothermic (heat of reaction)
     - Temperature control (emergency system)
     - Side reactions (dangerous)
     - Effect of contamination
     - Effect of unusual concentrations, including catalyst
     - Corrosion
  2. Pressure system
     - Design to current code
     - Materials of construction, adequate
     - Pressure relief, adequate

- Safe venting systems
- Flame arresters

- Control systems
  - Fail-safe
  - Backup power supplies
  - High/low alarms and trips on critical variables: temperature, pressure, flow, level, composition
  - Backup and duplicate systems on critical variables
  - Remote operation of valves
  - Block valves on critical lines
  - Excess flow valves
  - Interlock systems to prevent misoperation
  - Automatic shutdown system

- Storage
  - Limit quantity
  - Inert purging or blanketing
  - Floating roof tanks
  - Dikeing
  - Loading and unloading facilities, safety
  - Grounding
  - Ignition sources in vehicles

- General
  - Inert purging system needed
  - Compliance with electrical codes
  - Adequate lighting
  - Lighting protection
  - Sewers and drains adequate, flame traps
  - Dust explosion hazards
  - Buildup of dangerous impurities, purges
  - Plant layout, separation of units, access, siting of control rooms and offices, services
  - Safety shower, eye baths

- Fire protection
  - Emergency water supplies
  - Fire mains and hydrants
  - Foam systems
  - Sprinklers and deluge systems
  - Insulation and protection of structures
  - Access to building of structures
  - Firefighting equipment

The checklist is intended to promote thought and to raise such questions as: Is it needed? What are the alternatives? Has provision been made for it? Has it been provided?

19.2.2 Process Operation and Hazards

A *hazard and operability study* is a procedure for the systematic critical examination of the operability of a process. When applied to the process design of an operating plant, it indicates potential hazards that may arise from deviations from the design conditions intended. The term *operability study* should more properly be used for this type of study, although it is usually referred to as a hazard and operability, or HAZOP, study. This can cause confusion with the term *hazard analysis*, which is a technique for the quantitative assessment of a hazard after it has been identified by an operability study [2,31,76].

A brief outline of this technique can be used to make a preliminary examination of the design at the flowsheet stage, and detailed study at the later stage, when a full process description, final flowsheets, P&IDs, and equipment details are available. A formal operability study is the systematic study of the design, vessel by vessel, and line by line, using guide words to help generate information about the way that deviations from the intended operating conditions can cause hazardous situations. The basic guide words have the precise meanings given below.

No or *Not:* complete negation of these intentions.

More: quantitative increase. Refers to quantities and properties such as flow rates and temperatures, as well as activities such as Heat and React.

Less: quantitative decrease. refers to quantities and properties such as flow rates and temperatures, as well as activities such as Heat and React.

As well as: qualitative increase. All the design and operating intentions are achieved together with some additional activity. Something in addition to the design intention, such as impurities, side reactions, ingress of air, or extra phases present.

Part of: qualitative decrease. Only some of the intentions are achieved; some are not. Something missing; only part of the intention realized, such as the change in composition of a stream, or a missing component.

Reverse: logical opposite of the intention. This is applicable primarily to activities: for example, reverse flow or chemical reaction. It can also be applied to substances (e.g., *poison* instead of *antidote* or D instead of L optical isomers). The reverse or opposite of the design intention. This could mean reverse flow if the intention was to transfer material. For a reaction, it could mean the reverse reaction. In heat transfer, it could mean the transfer of heat in the direction opposite to that intended.

Other than: complete substitution. No part of the original intention is achieved. Something quite different happens. An important and far-reaching guide word, but consequently, more vague in its application.

It covers all conceivable situations other than that intended, such as startup, shutdown, maintainance, catalyst regeneration and charging, or failure of plant services. When referring to time, the guide words *sooner than* and *later than* can also be used.

In addition to these words, the following words are also used in a special way and have the precise meaning given below.

Intention: how the particular part of a process was intended to operate; the intention of the designer.

Deviations: departures from the designer's intention that are detected by the systematic application of the guide words.

Causes: reasons why, and how, a deviation could occur. Only if a deviation can be shown to have a realistic cause is it treated as meaningful.

Consequences: results that follow from the occurrence of a meaningful deviation.

Hazards: consequences that can cause damage loss or injury.

An operability study would normally be carried out by a team of experienced people who have complementary skills and knowledge, led by a team leader who is experienced in the technique. The team examines the process vessel by vessel, and line by line, using the guide words to detect any hazards (Fig. 19-2).

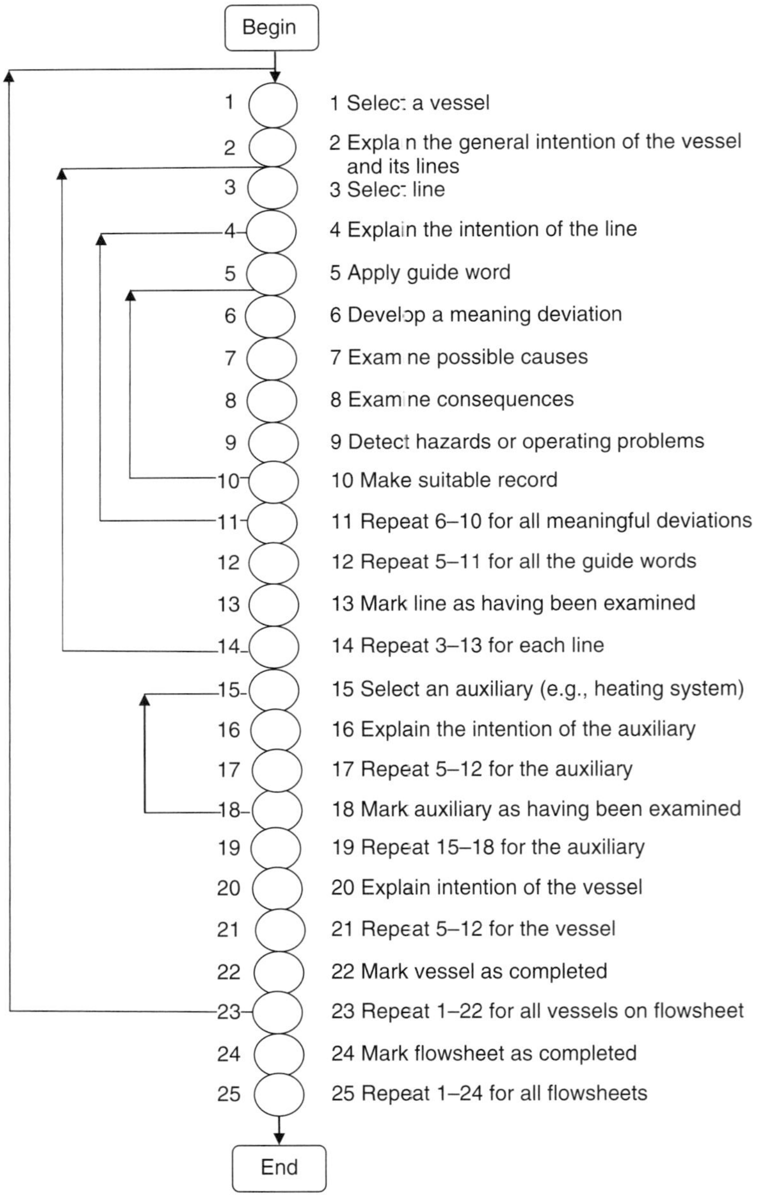

Figure 19-2 Sequence of operability study.

The information required for the study will depend on the extent of the investigations. A preliminary study can be made from a description of the process and the process flowsheets. For a detailed final study of the design, flowsheets, piping and instrumentation diagrams, equipment specifications, and layout drawings would be needed. For a batch process, information on the sequence of operations will also be required, such as that given in operating instructions, logic diagrams, and flowcharts. A typical sequence of events is shown in Fig. 19-2. After such a line has been studied, it is marked on the flowsheet as having been checked.

A written record is not normally made of each step in the study; only those deviations that lead to a potential hazard are recorded. If possible, the action needed to remove a hazard is decided by the team and recorded. If more information or time is needed to decide the best action, the matter is referred to the design group for action, or taken up at another meeting of the study team.

When using the operability study technique to vet a process design, the action to be taken to deal with a potential hazard will often be a modification of the control systems and instrumentation: the inclusion of additional alarms, trips, or interlocks. If major hazards are identified, major design changes, or alternative processes, materials, or equipment, may be required.

19.3 RISK ANALYSIS

Hazard identification, exposure assessment, and dose–response assessment merge into risk characterization, which estimates the incidence of expected adverse health effects in exposed populations. Risk characterization and the establishment of acceptable exposure levels are handled differently for carcinogenic and noncarcinogenic xenobiotics. For chemicals that cause adverse effects by a mechanism with thresholds, the safety factor approach was developed. For a nonthreshold response such as cancer, both quantitative and qualitative risk assessment procedures are used.

19.3.1 Decision-Making System

A knowledge base for hazard discovery and loss protection can develop. A knowledge base can be understood as a special case of a decision support system; on the other hand, a knowledge base presents a new stage in the development evaluation step of information technologies. A knowledge base is the basic element of an expert system. The term *expert system* is very often applied to a program that uses knowledge for behavior process simulation, or whose function has some attributed process behavior. It has the power to learn from experience, general knowledge achievement, reconceptualization, and to transfer knowledge from one domain to the other: a flexible and changeable approach to problem solution. Behind these are expert-selecting alternative solutions, explaining the diagnosis as well as learning from previous experiences, adding new knowledge base elements achieved during problem solution. A basic difference between expert systems and classical programs is that expert systems manipulate with knowledge and classical programs manipulate with data. An expert system has the ability to solve complex problems that include uncertainty by information processing. The two areas of model development and analysis are addressed through the discussion of a generic simulation environment. A knowledge-based simulation environment is an expression of a control law. To the extent that the rule base is derived from a set of assumptions about the environment and performance expectations, it is a belief system. However, in the existing form, the goals are not expressed and the underlying assumptions are not evident. Consequently, they are opaque to the analyst and cannot be applied directly to the learning process. When expressed in hierarchical form, the relationship that exists between goals and subgoals provides a basis for relating overall goal-based system performance to specific assumptions about the variability and contribution of the supporting subgoals. In this form the belief system is a full expression of some control theory in that the system's relationship with the environment, as expressed in a set of feasible state conditions, can be related in overall system performance measures to be relationships and the subgoals that support them.

In recent years, many applications of expert systems to simulation have evolved as computer-aided knowledge engineering tools. Considerable success has been achieved in developing knowledge-aided simulation systems. Intelligent simulation highlights the potential to meet the demand. The technological advance in simulation has addressed the research interest of intelligent simulation. Research and development in this discipline has continued for several years, and its effort has produced three types of intelligent simulation systems: single systems, coupling systems, and integrated intelligent systems. Single expert systems process only symbolic information and provide assistance to system engineers in the decision-making process for off-line simulation and modeling. Coupling systems couple numerical computation programs into expert systems such that they can be used to solve engineering simulation problems. Integrated intelligent systems are large intelligence integration environments, which can integrate different expert systems or numerical packages to solve complex problems. In the analysis and synthesis of engineering systems, simulation is a major technique. Traditional simulation techniques, which are algorithm based, are often inflexible and provide the user with limited means. In fact, such techniques cannot clearly simulate the dynamic behavior of real processes.

Segregation of the database, knowledge base, and inference engine in an expert system allows us to organize the various models and domain expertise efficiently because each component can be designed and modified separately. Presently, expert systems are being developed extensively in the research into intelligent simulation systems. Among the successful artificial intelligence applications, most expert systems are production systems. Production systems facilitate the representation of heuristic reasoning such that expert systems can be built incrementally as the knowledge of expertise increases. Knowledge of the expertise for a problem is described by a set of production rules. The typical production rule is described as IF (condition) . . . THEN (action). An inference engine is the executor. It must determine which rules are relevant to a given knowledge base and select one of them to apply. A knowledge-based system must represent information abstractly so that it can be stored and manipulated effectively. Although experts have difficulty formulating their knowledge explicitly as rules and other abstractions (Figs. 19-3 and 19-26), they find it easy to demonstrate their expertise in specific performance situations. Schemes for learning abstract representations or concepts from examples interact directly with systems to transfer their knowledge. Expert systems can be used to develop rule-based models and to augment other types of models. Since we can capture the experience of experts, they can be used to forecast problems, give advice, act as operators, validate data, and ensure that the results from other experts are reasonable.

A functional approach to designing expert simulation systems has been proposed by many authors. They chose differential games models described using semantic networks. The model generation methodology is a blend of several problem-solving paradigms, and the hierarchical dynamic goal system construction serves as the basis for model generation. A discrete event approach, based on the geometry of the games, can generally obtain the solution in a much shorter time. Cooperation between systems is achieved through a goal hierarchy.

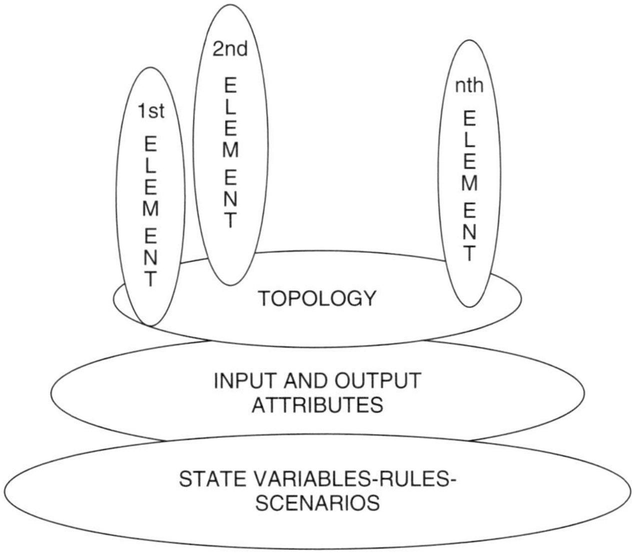

Figure 19-3 Knowledge base formation for hazards analysis.

Many expert systems have been introduced in such areas as reliability diagnosis, chemical and biological reaction synthesis, mineral and oil exploration, circuit analysis and equipment fault diagnosis, pharmaceutical manufacturing, and medical diagnosis. These expert systems have emphasized development of the knowledge acquisition process, the knowledge base, the inference procedure or control structure, and maintaining the independence of each of these functions. Computers have been widely used in simulation, but their use has been limited almost exclusively to purely algorithmic solutions. Many engineering problems are partial structured problems that deal with nonnumerical information and nonalgorithmic procedure, and are suitable for the application of artificial intelligence techniques. Expert systems provide a programming methodology for solving nonstructured problems which are difficult to handle using purely algorithmic methods. The experience obtained in building expert systems has shown that their power is most apparent when the problem considered is sufficiently complex.

In simulation, both qualitative and quantitative analysis are often applied together. Usually, qualitative decisions are made efficiently with symbolic and graphic information, and quantitative analysis is performed more conveniently using numerical information. The methods often complement each other. Any numerical solution is only an approximation of the true solution, which is always represented analytically. Analytical solutions can only be obtained by symbolic processing.

Most existing expert systems were developed for specific purposes. Usually, they were implemented in symbolic language, and production rules were used to represent domain expertise. In light of the application, such expert systems can only process symbolic information and make heuristic inference. The lack of numerical computation and uncoordinated single applications make them very limited in their ability to solve real engineering problems. Expert systems need data processing.

Coordination of symbolic reasoning and numerical computation is required for simulation with expert systems. A few developers have tried to develop expert systems with conventional languages. Others suggested fielding expert systems in conventional languages in order to achieve integration. Another disadvantage is that the procedural language environment cannot provide many good features that the symbolic language provides, such as easy debugging allowance for interruption by human experts.

Numerical languages often have a procedural flavor, in which program control is command driven. They are very inefficient when dealing with processing strings. Symbolic languages are more declarative and data driven. However, symbolic languages execute numeric computations very slowly. A complex problem cannot be solved by purely symbolic or numerical techniques. Coupling of symbolic processing and numerical computing is desirable for the use of numeric and symbolic languages in different portions of software systems. The coupled systems approach is often required when domain expertise is needed to provide the user with a suggestion or to direct the problem-solving process. The most appealing approach is to achieve deep coupling of numerical and symbolic modules representing the module function, inputs, outputs, and usage constraints. This allows the system to be applied to a wide range of problems and makes it more robust.

19.3.2 Qualitative Risk Analysis

In recent years, many applications of expert systems to simulation have evolved as knowledge-aided process engineering tools. The integrated simulation system development typically consist of phases such as requirements analysis, design, and implementation. Requirements analysis investigates user needs and translates them into simulation model specifications, including performance requirements such as the response time of various outputs. To ensure successful implementation of the system, these requirements are evaluated. Such an evaluation needs to be done during the requirements analysis phase because the cost of fixing the system at a later stage can be high, due to design revision, code revision, retesting, and maintenance. Simulation is a widely used tool for checking the performance requirements. It offers several benefits to the systems analyst and designer, including the ability to identify the behavior and performance of the system prior to its design and to check the efficiency of a system within a specific resource environment. In addition, this system can be configured in several ways, with each configuration having varying performance and cost. Simulation of these configurations helps in understanding and evaluating the trade-offs between performance and cost. Simulation can also aid in the design of information systems. The overall methodology consists of a two-step process. First, knowledge pertaining to the target system gathered by an analyst during requirements analysis is synthesized into a network model. This model is then utilized for building the simulation model automatically. The approach has the benefits of shortening the system development cycle and the economical and rapid development of a simulation model from specifications obtained during requirements analysis, allowing analysts with little training in simulation to test the process.

19.3.2.1 Network Building Transactions that take place in the environment affect the system at its inputs, while the outputs that leave the system affect its environment. A transaction entering a simulation model may pass through several processes before its transformation to an output. In addition to the input–process–output relationship, several complex relationships among these processes often exist. In general, any two processes may be mutually exclusive, order dependent, order independent, or concurrent. Two processes are *mutually exclusive* when only one of them can be performed. Two processes are *order dependent* when one must be performed before the other. Two processes are *order independent* when both can be performed in any order and simultaneously (i.e., performing one of these does).

There are many approaches to model-integrated networks. Two of them used frequently are the hierarchical and Petri net approachs. Petri nets have been utilized extensively in a number of application areas, including the design of distributed computing systems, general-purpose information systems, production planning, and flexible manufacturing systems. In this chapter the relational approach to systems has been employed.

In general, modeling begins with an observation of the system, then translating these observations into a model. Some systems, such as manufacturing systems, may be modeled by beginning with the print of the facilities, followed by defining the materials flow, and then defining the interfaces between system components. Mutually exclusive processes are represented by making a place that has a single token become an input to multiple transitions corresponding to these mutually exclusive processes. Places that are outputs of a transition but not inputs to any transition indicate the transactions that leave the system. These places define the boundaries of the system and help identify system interfaces with the environment. A model that is too abstract may be refined to show greater detail. Developing simulation models hierarchically offers several advantages, such as hierarchical simplification and ease of simulation.

19.3.3 Qualitative Model Development

Qualitative reasoning is a relatively new field of study stemming from fundamental research in qualitative process theory, qualitative physics, commonsense knowledge, and naive physics. A rapid approach to developing and analyzing qualitative simulation models makes quantitative reasoning a field of great interest. The control feature of qualitative modeling is the qualitative variable description and logic rules for manipulating variable values between systematic states. One goal of qualitative reasoning is to capture the benefits of commonsense reasoning about process phenomena as displayed in human behavior [44]. Stemming from this goal, two areas of benefit are the employment of a commonsense process for modeling physical environments and the use of qualitative data, which by its very nature is simpler than qualitative data to interpret. Initial research starts in the first phase with of the development of a conceptual framework that will facilitate the modular specification of models, and in the second phase with the development of a logic framework that will permit object-using attributes and simulation techniques to be linked into executable models. An object-oriented device-centered ontology is utilized to define qualitative profiles. These profiles essentially represent any system as an expert system network of systematic components used to simulate the behavior expected from the composite system. Since so many accidents have been caused by operator misjudgments or misoperations, there is a need to develop a system that can also suggest appropriate action to be taken when a failure occurs.

The purpose of qualitative modeling and simulation is to explain process development by reasoning from physical description to behavioral description. The qualitative simulation must perform some interpretation and presentation of the output generated by the simulator. There are two types of systems. One system gives information about components, the type of unit, and the relation of the connections between the units, topology, and input/output attributes as well as state variables of the individual unit or component, described by conceptual framework for an operating system (Fig. 19-3). The conceptual, modular framework views a processing system as a collection of entities and interpretation.

The system that converts a model specification into an executable model using translation is called a *logic framework*. A logic framework consists of active logic flow that operates by the use of rules in the knowledge base. Passive logic flow is the interface between a simulator and a user, establishing attribute sets, the initial state (scenario), and symptoms.

The model begins with the system definition. In this phase the system to be modeled is defined and bounded. The definition includes components, toplogy, input and output attributes, state variables, behavior rules, and initial scenarios (Fig. 19-3). Symptoms are added for use in diagnostic analysis.

The system desired must be general and flexible and not process specific. The methodology should lead to a diagnostic system that can accommodate changes in the plant configurations. No limitation is placed on the number of equipment units or streams since it is a simple matter to create a new frame and working element.

For any process system, components are regarded as distinguishable by the process connections of the working process model. Attribute values are passed among system components via media. It is this propagation of attribute values by which expert system rules, when invoked recursively, generate the behavioral characteristics of a system.

For defining qualitative profiles, an object-originated device-centered ontology can be used. A qualitative profile consists of the following items:

1. *Component*: What are the system's parts?
2. *Toplogy*: How do the components interconnect?
3. *Input and output attributes*: First, through what medium do the components communicate? Second, what are the attribution properties of this medium?
4. *State variables*: What are the various states in which each component can reside?
5. *Rules*: How do the various component states affect the attributes of the medium?
6. *Scenarios*: What are the various initial states to be studied?

These items also form the architectural basis of a qualitative diagnostic simulator.

The qualitative representation of quantities is best addressed through the relevance principle for quantity representation. Qualitative reasoning about something continuous requires some type of quantization to form a discrete set of symbols. The distinctions made by the quantization must be relevant to the type of reasoning being performed.

Incremental qualitative analysis follows this principle by representing quantities according to how they change. Further, the concept of landmark values has been introduced as those variable values that cause a noticeable change in the behavior of the system. For example, the temperature of water might be broken up into freezing, cold, warm, hot, and boiling, where freezing represents the interval 0°C (32°F) and below, cold represents an interval from 0 to 20°C (32 to 62°F), and so on. Using the relevance principle, the states of hot and cold would represent a warning to the system that the water in the system was about to boil or freeze. Values are not modeled statistically; rather, significant ranges are manipulated logically.

An architecture that constitutes the model analysis is shown in Fig. 19-4. This is composed of five phases. Between each phase, an intermediate output file is produced to be used in susequent analysis. Let us examine each of these phases and results through an example.

The architecture begins with system definition. In this phase the system to be modeled is defined and bounded. The definition includes component topology, input and output attributes, state variables, behavioral rules, and initial scenarios. In addition, symptoms are added for use in diagnostic analysis.

For any physical system, components are regarded as the distinguishable parts within the system. The system topology or component interconnections are defined by

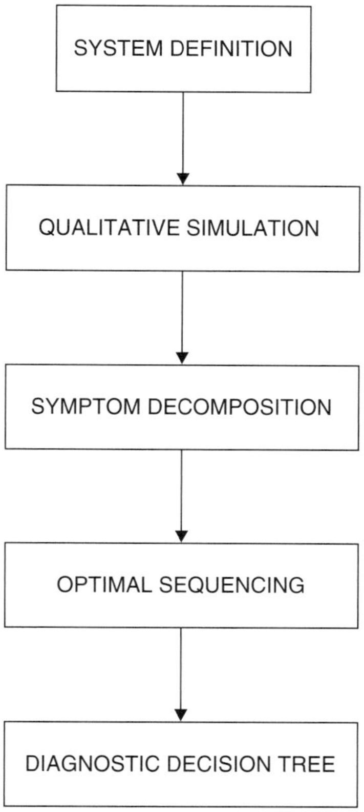

Figure 19-4 Qualitative model architecture.

the physical connections of the working process model. The level of agregation is defined by the modeler. Component interconnection defines the propagation paths of attributes within the system. Attribute values are passed among system components invoked recursively that generate the behavioral characteristics of the system. The first of all the attributes of the model is chosen. Various rules can be applied to this system based on the transitivity relationships of the physical attributes. Each component provides attribution transformations depending on its operating states.

To organize the logic of the rules, state variables must be defined within the system. Essentially three types of state variables can be defined within a given component. First are those variables whose values can be controlled by the system operator. The second set are those variables whose values are observable to the system operator. Last are those intrinsic variables whose values are not immediately discernible by the system operator, such as inner temperature, pressures, and so on. An initial scenario is defined once the system is built. Scenarios are used to set the initial status of the system's state variables and attributes to predefined values prior to a model run. Thus, it is necessary to evaluate "what if" scenarios concerning component malfunctions such as leakage due to worm gears in the pump or blockage due to a clogged filter. For diagnostic purposes, scenarios are evaluated by means of monitoring system symptoms. Symptoms are those significant variables or attributes that can easily be obtained or measured in the working physical system. In this way, the simulation model can be used to identify what problem caused subsequent symptoms in the working model.

In the qualitative simulation phase, the model was set forth in the previous system definition phase, which is simulated. Each component, represented by an individual set of causal rules, is executed in sequence. After every component the resulting symptomatic state is compared with the prior symptomatic state. If states are identical, steady state is achieved and simulation is halted. If steady state is not achieved, the simulation continues until steady state is eventually achieved or an upper allowable limit of iterations is attained. This procedure is repeated for each scenario specified in the system definition phase of the architecture. The algorithm for executing a qualitative model is shown in Fig. 19-5.

With the completation of qualitative runs, a resulting symptom or scenario matrix is formed. The symptom scenario matrix represents the steady-state nature of each scenario by displaying the final values of each symptom. One of the powerful features of qualitative simulation is that it readily allows for different sets of state-variable values to lead to common scenario simulation results. In the symptom decomposition phase, the relational symptom/scenario matrix is decomposed by using a projection operation that delinates which scenarios were found to have the same symptom values in their final state.

In the final optimal sequencing phase, these symptoms are ordered by the relative information value or use within a decision tree. The order is determined by the relative number of scenarios that each symptom delineates, the probability of each scenario occurring, and the expense associated with testing for each symptom. This optimal sequence is dependent on not only the time it takes to observe each symptom but also on the probability that each scenario or malfunction will take place.

In the final phase, the optimal sequence of symptoms is used to build a decision tree construction algorithm, as represented in Fig. 19-6. This algorithm interprets the optimal sequence of events assimilated in the previous algorithm and ships the results to an output file. The algorithm does this by following the paths established in the optimal sequence tree, down in a depth-first manner. It interprets various symptoms and scenarios found along the path and issues either a question, a causal branch, or a consequent scenario.

The output from each phase is sent a summary file. After a complete qualitative simulation session, the file has five sections. First is a list of the components within the model where the input and output components are each specified by arrows toward and away from the components,

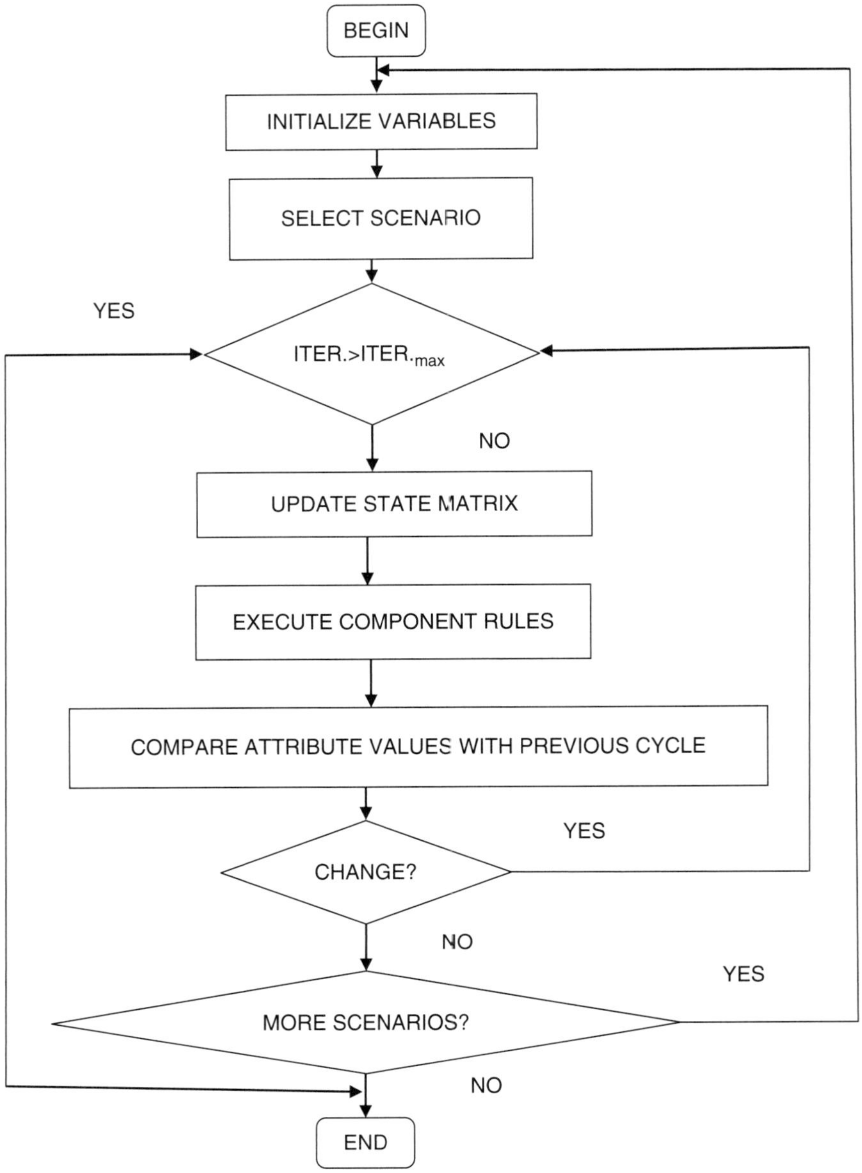

Figure 19-5 Qualitative model simulation.

respectively. Next, the state matrix is defined by a list of the input/output variables, state variables, and metavariables for each component with the default values for each variable specified. Following this comes the simulation results matrix. If steady state is never reached within a particular scenario, a warning is given before the results matrix in which the scenario resides. Each results matrix can display the results of all scenarios. Subsequent results are shown in subsequent matrices. This section is followed by the symptom/scenario matrix. Finally, the resulting diagnostic decision tree is presented.

The qualitative process engine automatically generates process-centered qualitative models which consist of a network of causal constraints. Each phenomenon, liquid flow, reaction, and so on, is described by preconditions, constraints, and influences. The preconditions must be satisfied for the phenomenon to become active. When the preconditions are met, constraints are the phenomenon, and influences are the driving force of a phenomenon and describe how changes occur as a result of the phenomenon being active.

Variables are represented by their qualitative state, which describes the variables relative location to landmarks (boiling point or maximum pressure) and the sign of their first derivatives. The qualitative states of all the variables in a plant or device define the state of the plant or device. For example, to activate the phenomenon of heat flow there must be two objects (the source and destination),

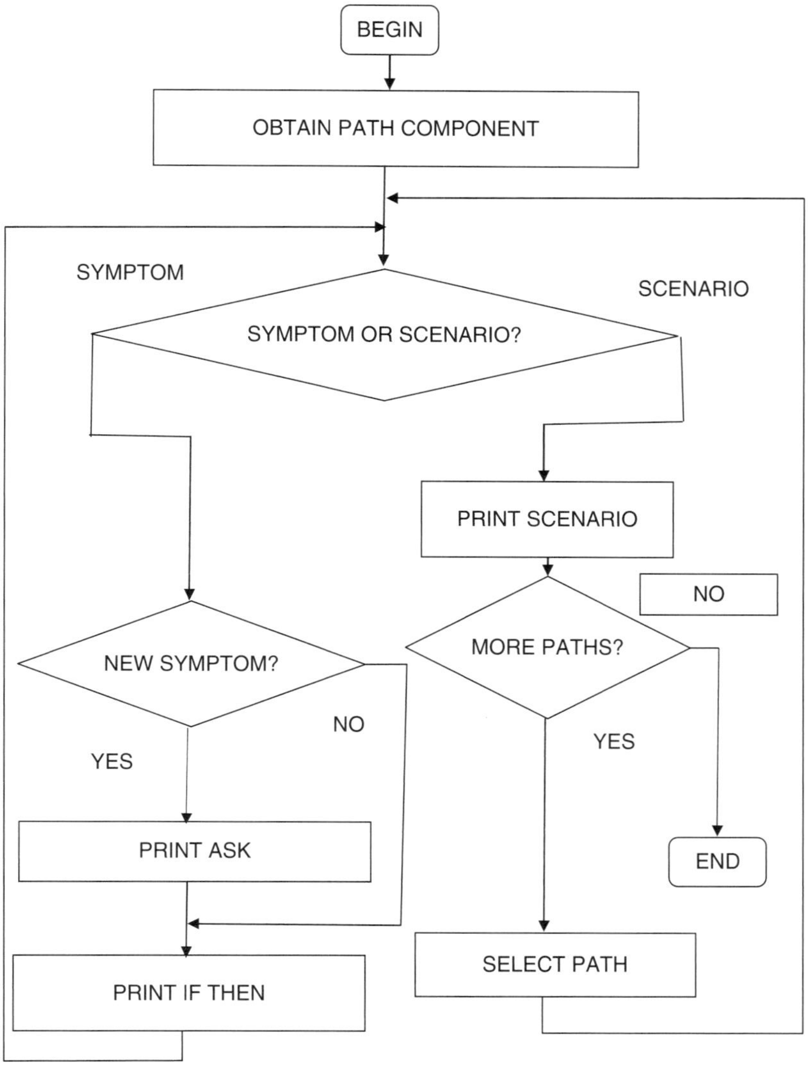

Figure 19-6 Fault decision tree.

with the temperature of the source greater than that of the destination. Also, there must be a common noninsulated wall between them. When these preconditions are satisfied, qualitative simulation activates the phenomenon of heat flow, and new constraints and influences are added to the model which describe the causal connection between variables. The deriving force in the case of heat exchange is a difference in temperature between the two physical objects. The causal relations are best described in a signed directional graph or signed digraph.

A large model is built by combining the library of phenomena with a large scenario of assumptions about the physical situation. For example, if the preconditions for heat transfer between a tank and a cooling coil and the preconditions for the reaction $A \rightarrow B$ are in the scenario, both heat transfer and reaction become active, creating a larger qualitative model. To assess the probability of occurrence of a decomposition reaction, it is necessary to characterize its kinetics.

The control feature of qualitative modeling is the qualitative variable description and logic rules for manipulating variable values between systematic states. The goal of qualitative reasoning is to capture the benefits of commonsense reasoning about process phenomena as displayed in a human behavior mental model.

The qualitative process engine automatically generates process-centered qualitative models, which consist of a network of causal constraints. In a qualitative process engine each phenomenon is described by preconditions, constraints, and influences. A simulator that can qualitatively estimate chemical plant behavior during abnormal situations was developed by Savkovic-Stevanovic [45]. The system can perform fault diagnosis in a plant consisting of pumps, valves, tanks, reactors, and streams. The qualitative simulation must perform some interpretation and presentation of the output generated by the simulator. Initial research by a phase of development of a conceptual framework will facilitate the modular specification of models, and a second phase of development of a logic framework will permit object-using attributes and simulation techniques to be linked into executable models.

The conceptual, modular framework views a processing system as a collection of entities and interactions. The active logic flow operates by the use of rules in the knowledge base. The passive logic flow is the interface between a simulator and a user establishing attribute sets, an initial state, and symptoms.

The attributes of the process are chosen to be pressure, flow rate, and temperature. Also, the system can diagnose causes of faults associated with state-variable pressures, flow rates, and temperatures. The qualitative variables are described in three discrete values: low, medium, and high. Equipment states are also described in qualitative terms, such as closed, open, failed, blocked, and leaked. The following faults are considered: blockage, leakage, malfunctions, or missoperation. The study of fault detection and diagnostic is concerned with designing a system that can assist the human operator is detecting and diagnosing equipment faults to prevent accidents.

A knowledge-based decision support system building consists of the following steps:

1. System identification; looking for identifiers
2. Goal and subgoal definition
3. Rules network
4. Decision mechanism definition
5. For diagnostic purposeds, a need to monitor system symptoms

The symptom/scenario matrix displays the final values of each symptom that evolves from initial scenario states. Various rules can be applied to this system based on the transitivity relationships of the qualitative variables of the fluid flow. For example, fluid pressure implies a fluid supply, and fluid flow implies fluid pressure. Thus, fluid flow implies a fluid supply. But the same cannot be said for a fluid supply implying fluid flow. Each component provides attribution transformations, depending on its operating states.

Scenarios are used to set initial states of the system state variables and attributes to predefined values prior to a model simulation run. This is necessary to evaluate "what if" scenarios concerning component malfunctions such as leakage due to a worn part or blockage.

*19.3.3.1 **Decision Mechanism*** The best way to solve complicated problems by expert systems is to distribute knowledge and to separate domain expertise. In such a case, several expert systems may be used together. Each should be developed for solving a subdomain problem. Here we face the problem of knowledge integration and management. Many expert systems can only be used alone for a particular purpose. There is a lack of coordination of symbolic reasoning and numeric computation, a lack of integration of different expert systems, a lack of an efficient management of intelligent systems, and the capability to deal with conflicting facts and events among the various tasks, and difficulty modifying knowledge bases by users other than the original developers.

Many integrated intelligent systems constitute a large knowledge environment, which consists of several symbolic reasoning systems and numerical computation packages. They are under the control of a supervising intelligent system: that is, a metasystem. The metasystem manages the selection, operation, and communication of these programs.

The key issue in constructing an integrated intelligent system is to organize a metasystem, which can thus be referred to as a control mechanism of metalevel knowledge. A metasystem has its own database, rule base, and inference engine, but it decomposes its activities into separate strictly ordered phases of information gathering and processing.

*19.3.3.2 **Events Model*** The fault events of a system are in the first instance generally formulated in an IF–THEN form. This can be reformulated immediately using the operators AND, OR, and NOT in Boolean form if one can assume that the primary events have only two states: existence and nonexistence. For example, for fault diagnosing a liquid flow system (Fig. 19-7) is used. The system consists of two tanks, two mixers, and pipes. This system represents a qualitative events model expressed by logic algebra. M, B, and L are independent logic variables representing the basic events malfunction, blockage, and leakage, respectively. The study of fault detection and supervision control of the flow system is concerned with designing a system that can assist a human operator in detecting and diagnosing faults.

The system topology or component interconnections are defined by the process connections of the working process model (Fig. 19-7). The level of aggregation is defined by the modular component interconnections which define propagation paths of attributes within the system. The

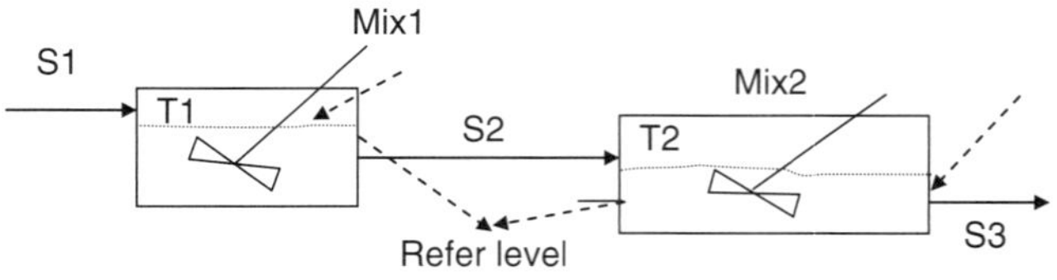

Figure 19-7 Scheme for liquid flow.

initial research phase was the development of a conceptual framework that facilitates the modular specification of models, and second phase was the development of a logic framework that permits object-using attributes and simulation techniques to be linked into executable models [49–55].

Starting with the basic variables and their interrelations, the qualitative event model of the system can be formulated as shown in Fig. 19-8. To organize the logic of the rules, state variables must be defined within the system. Three types of state variables can be defined within a given component. The first are those variables whose values can be controlled by the system operator. These are controllable variables. The second set are those variables whose values are observable to the system operator. Last are those variables whose values are not immediately discernible by the system operator, such as inner pressure.

Scenarios are used to set initial states of the system state variables and attributes to predefined values prior to a model simulation run. This is necessary to evaluate "what if" scenarios concerning component malfunctions such as leakage due to a worn or part blockage. The system can diagnose causes of faults associated with the state variables pressure, flow rate, and supply. The qualitative variables are described by three discrete values (low, medium, and high).

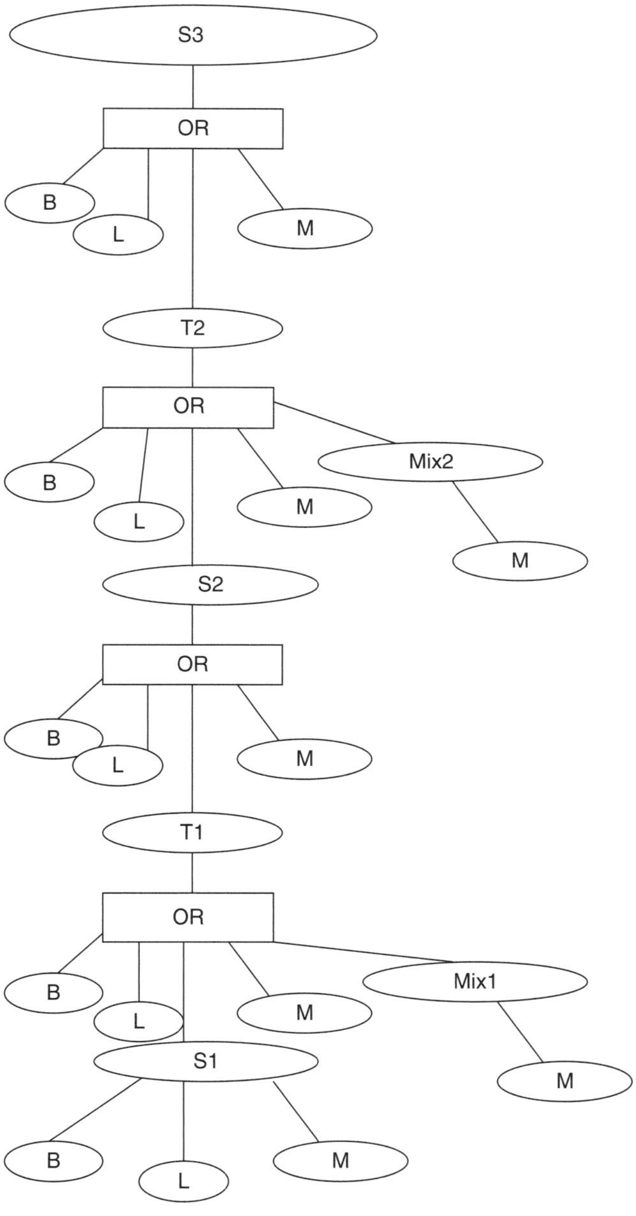

Figure 19-8 Diagnostic system knowledge base.

19.4 SAFETY RATINGS

The phenomenological discussion shows that most of the dangers in the operation of chemical plants have to do with the reactivities of the substances present. The principal hazards are caused by releases and spills of substances outside the process and by uncontrolled chemical reactions between substances. The danger is greater the larger the quantities of substances and energy released.

Plant and process safety efforts must have as their essential goals:

1. Minimizing the quantities of substances
2. Controlling the potential risks that remain

The second objective requires:

a. Reliably confining hazardous substances in process equipment that can withstand the stresses anticipated (due to pressure, temperature, corrosive attack, etc.)

b. Ensuring that process parameters do not take on values such that the substances can undergo uncontrolled reactions.

c. Some help is gained from the fact that, as a rule, substances react spotaneously with one another only if they have been suitably prepared, for example, by treatment to increase their surface area, mixed and concentrated, and excited by heat.

d. Experimental laboratory tests are performed, for example, to determine the temperature and concentration conditions under which reactions can get so far out of control that they can lead to an explosion, or to establish initial conditions that can cause dangerous reactions.

e. A point of special interest is where the regions begin in which uncontrolled reactions cannot occur. The characteristic critical parameters found in such experiments are called *safety ratings*. They include such figures as the ignition temperature and the explosion limits. Safety ratings form the basis for the safety measures in chemical plants. All concern how safety ratings are determined.

At the planning and design stage, plant and process safety requires taking steps so that critical concentrations, temperatures, and pressures are not reached. This is achieved by appropriate process design and process control engineering. Safety also entails preventing the occurrence of a potential ignition source that could cause an undersired reaction: for example, hot surfaces or spars generated by mechanical or electrical equipment. Great care needs to be taken in the layout of machinery and process equipment.

Critical process parameters and hazardous potential ignition sources must not occur in the plant as a result of process upsets or human error. These become an additional concern of plant safety, requiring a painstaking cause-and-effect analysis of all possible errors and malfunctions, and the institution of measures to prevent or neutralize situations that could lead to an unsafe condition. Such a measure may be technical or organizational; the second group includes operating instructions, inspections, and so on.

The complete absence of possible hazards, absolute safety, is not possible, for several reasons:

1. It cannot be ruled out that several safety measures will fail simultaneously, so that a potential hazard may become an actual hazard.
2. People make errors from time to time, and they can misjudge things, assess them wrongly, or even fail to notice them at all.

The fact that absolute safety is impossible is important; two inferences follow from it. Knowledge has to be advanced continually, and precautions have to be taken to avert risks in case of failure.

1. What does plant safety mean?
2. How safe is safe enough?

These intimately related questions can be answered in two different ways. In the first approach, safety is defined in terms of risk (Fig. 19-9). A plant is said to be safe if the risk created by it is acceptable. Here *risk* means the possibility of *harm*, defined in terms of the probability of the harm during the lifetime of a plant or the frequency of harm and its anticipated severity. In the second approach, a plant is said to be safe if it complies with the appropriate regulations and codes. While the first method is based on a probabilistic concept, risk, the second employs a deterministic principle. In both cases, what is safe is clearly a matter of convention.

The standard states that risk is generally not quantifiable, since only in rare cases can it be expressed as the product of a severity, A, and a measure of severity, p_m:

$$\text{risk} = sp_m \tag{19.4}$$

Figure 19-9 Risk chart.

A risk measuring method was given by Savkovic-Stevanovic [47,54] (see Appendix 19.1). The standard treats danger as the diametric opposite of safety, where the risk of a process is greater than the acceptable limiting risk.

A useful notation in plant and process safety is the *hazard potential*, a measure of the greatest harm that can occur in the worst possible event in a plant or plant subdivision. It is reasonable to use this concept in assessing safety measures in a plant: The greater the hazard potential, the more and better safety measures are needed to lower the probability of occurrence of the undesired event to the point that the level of risk is at or below the acceptable risk level.

Safety measures may include intrinsic measures and conditions [65] which ensure a priori that a hazard potential can become real only in the event of a relatively improbable combination of multiple independent failures.

Where product quality considerations make it necessary to design the process and process control system so as to prevent an exothermic reaction getting out of control, safety or protective measures can be built up on the basis of this intrinsic safety condition to lower the risk to an acceptable level.

Risk created by chemical plants is often overestimated if their principles are not known. An anticipated value for the risk posed by chemical plants to employees or uninvolved third parties can be derived in a relatively simple way by statistical analysis of historical data. Consider the risk of death incurred by a chemical worker due to a typical chemical accident (poisoning, chemical burn, and explosion). When the number of persons per year suffering death from poisoning, chemical burn, fire, or explosion is divided by the total number of persons employed in the chemical industry, the annual individual lethal risk averaged over the period can be obtained.

Those living nearby and others outside the chemical plant are even safer from chemical effects, because the effects of infrequent incidents in chemical plants fall off quite rapidly with distance. It can be assumed that this risk is, at most, of the same order as the risk due to natural catastrophes. However, is it true that a low risk may well conceal high hazard potentials when the probability of occurrence is low? It is therefore advantageous to consider the size of the hazard potential in chemical plants.

19.4.1 Hazard Potential of a Volatile Substance

When a volatile substance posing a health hazard is released, a hazard arises because the liberated gas is taken up by air currents and propagates in the atmosphere as it is steadily diluted. The size of the impact area, the region in which concentration levels of the substance rise above a critical value π_H, is one measure of the hazard potential. Ultimately, however, the hazard potential is governed by the number of persons who stay in this impact area without protection.

To assess the hazard potential, it necessary to determine the greatest mass of gas that can be released. Analysis of hazard potentials is broken down by type of accident: (1) spills and releases, or (2) explosions.

Consider a vessel that contains a mass M_L of a substance in the liquid state under its vapor pressure p_v at a temperature T_a above the boiling point T_b. The mass of vapor in the vessel expands to ambient pressure, and the liquid continues to vaporize and cool until the remaining liquid reaches its boiling point T_b at atmospheric pressure.

Because this process is very fast, it approximates an adiabatic expansion. The mass of substance vaporized plus the original mass of vapor present are released into the atmosphere; this is the quantity that determines the hazard potential. A material and energy balance, with some simplifications, gives for the mass vaporized:

$$\Delta M_V = M_{F(a)}\left[1 - \exp\left(\frac{-c_L\,\Delta T}{\Delta h_V}\right)\right] \qquad (19.5)$$

where $\Delta T = T_a - T_b$, c_L is the specific heat of the liquid, and Δh_V is the heat of vaporization. If the temperature change is small [i.e., $(c_L/\Delta h_V)\,\Delta T \ll 1$], a series expansion leads to the approximation

$$\Delta M_V = M_{F(a)}\frac{c_L\,\Delta T}{\Delta h_V} \qquad (19.6)$$

A typical value for $c_L/\Delta h_V$ is $5 \times 10^{-3}\ \mathrm{K}^{-1}$. For ΔT, a value of about 50°C, corresponding to an overpressure of 0.5 to 1.0 MPa, is not unusual for the storage of liquefied gases. Calculations show that with these assumptions, 20 to 30% of the liquid contents of the vessel vaporizes spontaneously. A small additional amount vaporizes because of heat supplied from the surroundings, especially if liquid escapes from the vessel in finely divided form entrained with the vapor. In any case, this simple analysis shows that only a fraction of liquefied mass under its vapor pressure in a vessel can generally be released into the atmosphere spontaneously on failure of confinement, even when substances are kept under high pressure well above their boiling point. Only if the contents are superheated by some hundreds of degrees will the released fraction approach 100%.

To evaluate the acute hazard created by such a process, atmospheric propagation and dilution of the gas cloud have to be calculated, but these processes show an extremely strong dependence on the motion and turbulence of the atmosphere. Results can differ by orders of magnitude. What is more, initial dilution, height of release, and gas density all play roles. Exact prediction is impossible, but a rough estimate is possible [42]. Under unfavorable conditions, about 1000 m^3 or several metric tons of gas have to be released to establish maximum concentrations of about 10 ppm out to a range of several kilometers from the point of release. Under propagation conditions of higher probability, the concentrations drop below such levels at a range of less than 1 km.

As a rule of thumb, a hazard to humans due to a volatile toxic gas release at a range of more than a few hundred meters from the source presupposes catastrophic failure of a vessel, and the vessel must contain several metric tons of the gas in releasable form, compressed or liquefied.

Such apparatus can be found in chemical plants, but the type of catastrophic damage under discussion is extremely improbable because such vessels are designed with safety factors against stress, and they are subject to regular inspection.

From a safety standpoint, it is still important to assess hazard potentials relative to one another and to know how they can be influenced by process parameters. The following argument can be of help. The simple equations describing gas propagation show that maximum concentration in a gas cloud or plume at distance x downwind of the source is directly proportional to the quantity released, M_{rel}, and inversely proportional to the nth power of x:

$$\pi = \mathrm{const}\left(\frac{M_{\mathrm{rel}}}{x^n}\right) \qquad (19.7)$$

where $2 < n < 3$ in general.

In the case of a release near the surface, the downwind impact area, the region swept over by a concentration higher than π_H, is commonly cigar shaped. It becomes broader and shorter the more turbulent the atmosphere.

For a given state of atmospheric turbulence, the maximum width of the impact region is always proportional to its length. Therefore, the impact region, as a measure of the hazard potential (HP), is proportional to the square of the distance x from the source to the point where the concentration falls below π_H:

$$\mathrm{HP} \sim x^2$$

$$\mathrm{HP} = \mathrm{const}\left(\frac{M_{\mathrm{rel}}}{\pi_H}\right)^{2/n} \tag{19.8}$$

For the case under discussion (spontaneous vaporization and escape of a substance posing a health hazard and having a critical concentration level π_H) the hazard potential is

$$\mathrm{HP} = K\left[\frac{c_L\,\Delta T}{\Delta h_V}\frac{M_{F(a)}}{\pi_H}\right]^{2/n} \tag{19.9}$$

Nevertheless, this is not a well-defined quantity, because constants K and n depend on the ambient conditions, the state of motion of the atmosphere, and can be derived from the known gas dispersion formula only if a special case is assumed.

The hazard potential is also a function of the quantity of substance, its properties, and its thermodynamic state inside the vessel. The hazard potential increases with the mass of substance present and superheating above the boiling point (i.e., with pressure). From this, it is possible to devise ways to reduce the hazard potential, even to the point of inherent safety.

19.4.2 Hazard Potential from an Explosion

The hazard potential of an explosion is due to the pressure wave and is determined by the size of the ideally circular area around the explosion center in which there is significant damage: collapsing walls or broken windows. Window glass typically breaks at a peak overpressure in the 1 kPa range, and masonry walls begin to fail at overpressures ≥30 kPa [19].

The task of evaluating the hazard potential can thus be reduced to determining at what distance from the explosion center the peak pressure falls just below a critical value. Suppose that the mass M of the substance being handled is such that its sudden decomposition liberates an enthalpy of reaction Δh_r. The total energy evolved is

$$E = M\,\Delta h_r \tag{19.10}$$

Hopkins and Cranz, independently, found that for distinct charges, releasing energies E_1 and E_2 to produce equal effects, the radii, at distances R_1 and R_2 from the source point, must be related as the cube roots of the energies [3,4]:

$$\frac{R_1}{R_2} = \frac{E_1^{1/3}}{E_2^{1/3}} \tag{19.11}$$

This results led to postulate that the dimensionless peak overpressure and the dimensionless momentum of a pressure wave can each be represented as a unique function of a dimensionless distance from the origin. The dimensionless peak overpressure p^* is obtained by dividing by the ambient pressure. The dimensionless range R^* is obtained from an expression involving the energy and ambient pressure:

$$p^* = \frac{P_P - P_0}{P_0} \tag{19.12}$$

$$R^* = \frac{R_P}{(E/P_0)^{1/3}} \quad \text{or dimensional} \quad \frac{\mathrm{m}}{(\mathrm{Nm}\cdot\mathrm{m}^2/\mathrm{N})^{1/3}} = \frac{\mathrm{m}}{\mathrm{m}} \tag{19.13}$$

Figure 19-10 shows that these parameters make possible accurate dimensionless representation of the characteristic pressure wave parameters, at least for many explosives. The expression for the dimensionless distance involves the explosion energy and the initial pressure.

An explosion corresponding to the detonation of 1 ton of TNT would break windowpanes ($\Delta p = 1$ kPa) out to a range of several hundred meters ($R_P = 35\ \mathrm{m} \times R^* =$

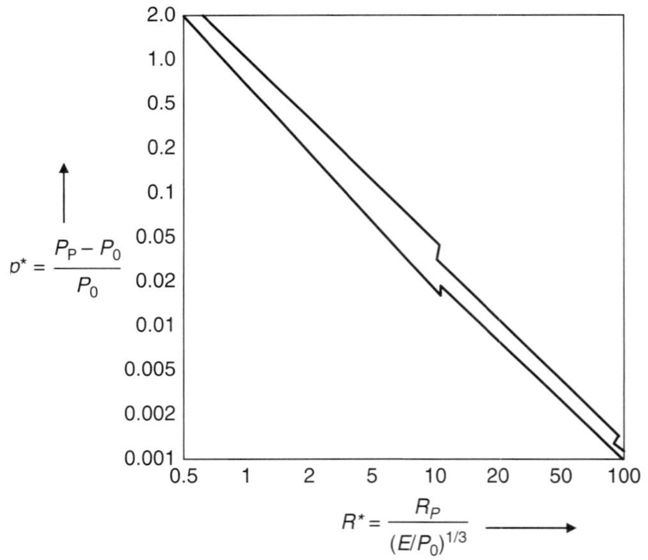

Figure 19-10 Dimensionless peak overpressure in an explosion pressure wave as a function of dimensionless distance from the source.

700 m) and cause serious damage ($\Delta p = 15$ kPa) in a circle radius of 50 to 100 m ($R_P = 70$ m).

Explosives and similar materials are not commonly used in chemical plants. Reactions that can lead to explosion-like phenomena are governed by other laws and differ strongly in kinetics, enthalpies of reaction, and efficiencies in terms of energy conversion.

Even so, a number of workers have sought correlations, working with event analysis for explosions of unconfined gas clouds and arriving at the conclusion that for less than about 1 ton of fuel gas forming an unconfined explosive cloud, peak overpressures on ignition are negligibly low [34].

In a real unconfined gas cloud formed through damage to a vessel or in a plant, only a fraction of the fuel gas is ever in a region where the concentration is high enough for sudden combustion in the worst case. It must therefore be supposed that the effect produced by the explosion of 1 ton of TNT can be matched only by the spontaneous release and ignition of about 10 tons of fuel gas. This could not happen without serious damage to a large vessel.

The hazard potential of an explosion is directly proportional to the area within which a certain level of damage occurs. According to Eqs. (19.9) to (19.11), we obtain

$$\text{HP} \sim R^2 \sim E^{2/3} = (M\ \Delta h_r)^{2/3} \tag{19.14}$$

The hazard potential increase is less than directly proportional to the energy release and the mass involved.

This analysis shows, as does experience, that chemical plants may occasionally represent high hazard potentials, especially when volatile substances of a flammable or health-endangering nature are processed in large quantities, and when exothermic decomposition and combustion reactions are possible.

19.4.3 Evaluation of Hazardous Properties

According to property classification criteria, substances are described in terms of 15 hazard-related characteristics, which can be broken down into four groups:

1. Acute toxicity
   - Very toxic substances
   - Toxic substances
   - Harmful substances
   - Corrosive substances
   - Irritant substances
2. Specific toxic properties
   - Sensitizing substances
   - Carcinogens
   - Substances with an effect on reproduction
   - Substances with heritable effects
3. Physicochemical properties
   - Extremely flammable substances
   - Highly flammable substances
   - Flammable substances
   - Oxidizing substances
   - Explosive substances
4. Environmental impacts
   - Substances harmful to the environment

Whereas acute and specific toxic properties can cause direct harm to health, physicochemical properties lead to indirect harm. Substances can take any of three routes into the body: oral, dermal, and inhalation.

Fertility classification criteria divide substances with effects on reproduction and impaired fertility (R_F) into three categories:

- *Category 1*: substances known to have effects on reproduction in humans
- *Category 2*: substances that should be regarded as having effects on reproduction on the basis of animal studies
- *Category 3*: substances suspected of having an effect on reproduction
- *Pregnancy group A*. A risk of fetal injury has been demonstrated. Injury cannot be ruled out, even if the standard's value is complied with.
- *Pregnancy group B*. A risk of fetal injury cannot be ruled out, even if the standard's value is complied with.
- *Pregnancy group C*. A risk of fetal injury need not be feared if the standard's value is complied with.

If a health hazard is posed, not by one-time exposure to the substance but by quantities administered over a longer period, the terms *subacute, subchronic*, and *chronic* are employed, depending on the exposure time. Carcinogenicity is merely one specific chronic effect.

Carcinogenic substances can initiate tumors. A tumor is an uncontrolled cell growth that penetrates surrounding cells in infiltrative fashion and destroys them. According to carcinogenicity criteria, carcinogenic substances are classified as follows:

- *Category 1*: known to be carcinogenic in humans
- *Category 2*: unambigously carcinogenic in animal trials
- *Category 3*: suspected of being carcinogenic

The best known substances in category 1 are asbestos, benzene, benzidine, β-naphtylamine, and vinyl chloride; in category 2, acrylonitrile, butadiene, and nitrosamines. From the standpoint of toxicological action, substances causing heritable genetic damage do not differ fundamentally from carcinogens. The injury they cause, however, is to the genetic information of the germ cells rather than to that of the body cells. The classification again includes categories 1 to 3. No category 1 chemicals causing heritable genetic damage are known.

The classification of hazardous substances on the basis of thermal properties employs the flash point. Substances are characterized as oxidizing if they can be sustained with the exclusion of oxygen. The symbol O is used in their labeling: permanganates, bromates, hypochlorites, chlorites, and nitrates. Substances with a greater explosive tendency than dinitrobenzene are described as explosive, and must be labeled with the symbol E: many organic peroxides, picrates, and pentaerythritoltetranitrate.

The concept of time to explosion is a great help.

19.4.4 Rating of Flammable and Explosive Substances

A variety of data and experimental techniques can be used to characterize flammability and combustibility. Requirements as to fire behavior and fire resistance apply not just to chemicals themselves, but also to the materials and structures used in buildings where chemicals are handled.

Combustion is a reaction that takes place in the gas phase and involves oxygen or another oxidizing agent present in the ambient gas phase. The flame phenomenon is characteristic of the reaction for liquids, gases, and most solids, constituents previously vaporized or carbonized and mixed with oxygen by diffusion oxidize in the flame. Any residue in cokelike form for example, is heated to incandescence in this process and then reacts with oxygen present at the surface of the solid and diffusing into its pores. Substances such as pure carbon or iron, which do not vaporize sufficiently at the temperature under consideration, react by glowing [82].

The speed of a a self-sustaining combustion process depends on the rate of reaction, the heat liberated in the reaction, and the balance of reaction heat, heat loss to the surroundings, and heat required for melting and vaporization.

Safety parameters, flash point and ignition point, are used to characterize the flammability of liquids. The flash point is the temperature of a liquid tested in a certain apparatus and under certain conditions, at which the gas phase above the liquid can burst into flame when an ignition source is brought near. The flash point is measured by various test methods according to national standards: ASTM, United States; DIN, Germany; GOST, Russia; BS, United Kingdom. All test setups and specifications allow for the fact that the flash point is a rough measure because of the uncertainty of measurement tolerated and the variation of material properties (viscosity).

The two basic types of apparatus are the open and closed cups. Methods differ in the access to oxygen and in the quality of mixing or accumulation of fuel over the liquid. Methods can also be classified as thermodynamic equilibrium or nonequilibrium, and a given test substance may show different flash points in these two types of test.

Despite national and international differences, all methods yield data sufficiently reliable for safety assessments. Internationally, the flash point is the basis for placing substances in hazard classes with regard to ignitability. Under international transportation law, a liquid is classed as flammable if it has a flash point below 60.5°C (closed cup) or 65.6°C (open cup) or if it is handled at a temperature higher than the flash point [72]. Self-sustaining combustion is tested in a separate procedure [36]. If a substance with a flash point above 35°C does not continue to burn after ignition in this test, it is not placed with the flammable liquids. This rule takes into account the reduced risk of fire.

19.4.4.1 Dusts The combustibility or flammability of solids is characterized by the flammability index and the burning rate of an uncompacted layer of dust. The flammability index gives a qualitative description of how such a bed of dust behaves in the presence of a strong ignition source, ignition flame, and incandescent coil. A substance with a flammability index above 3 tends to keep burning after local ignition. A burning rate test must be performed on such a substance. For a flammability index of 3 or less, the substance generally shows no tendency to rapid flame propagation.

In addition to ignition from an outside source, another concern with dusts is the spontanous flammability in air or other gaseous oxidizing medium due to self-heating. Maximum permissible storage and surface temperatures can be derived from the results with the aid of mathematical models and appropriate safety amplitudes.

The combustibility measures described up to this point are based on standard test methods. The data cannot be extended to practical conditions unless possible variations in the ambient conditions are taken into account. The flash point, for example, depends heavily on pressure as well as oxidizing medium. Changes in these lead to changes in fire behavior. The temperature of spontaneous ignition can also fall markedly below the ignition temperature of the standard if organic liquids, including heat-transfer media, penetrate open pored insultants such as rock wool.

Assessment of dusts is based similarly on the experimental conditions: air as the oxidizing medium and adequate heat removal from combustion zone. If the partial pressure of oxygen is increased, a stronger oxidizing agent is employed, the heat removal conditions are made worse (storage in large drums, silos, etc.), or the ambient temperature is higher, an increased risk of fire must be expected.

In the case of solids or ointments containing solvents, it has to be decided case by case whether the flash point or the burning rate flammability index better describes the situation. Oxidizing in contact with combustible substances gives off oxygen or other oxidizing agents so as to increase the danger of fire or the vigor with which combustibles burn. As a rule, oxidizing substances are not themselves combustible. Various mixtures of the test substance with cellulose powder are prepared and the burning rates are determined, then the results are compared with the burning rate of a reference mixture of barium nitrate and cellulose. A substance is classified as oxidizing if the burning rate of the test mixture is greater than that of the reference mixture. As a rule, incorrect predictions are made if this test procedure is applied to liquids or to melting and highly flammable or combustible substances, so its use with these substances is ruled out.

A comparable test method is employed internationally for the transport of hazardous substances [12]. In some instances, the two procedures give different classifications for the same substance. For fire precautions to be effective, they must take into account not only the combustibility of chemicals, but also the fire behavior of construction materials and structures, floor coverings, and other items within the production facility.

19.4.4.2 Explosion of Gas Mixtures In any operation on combustible and/or unstable gases, an explosive system may be formed and explosions initiated. This danger can be countered effectively only if the safety characteristics of the gases involved are well known. These characteristics are not constants in the physical sense, for they are influenced by the method used to determine them. There is no generally accepted method for estimating and calculating them, although for a few characteristics empirical formulas exist for specific groups of substances. In other words, recourse must be had to the experimental methods described in the following sections. Laboratory values must be of such a quality that they can serve as a basis for explosion-protection measures in full-scale plants. This is possible only if the method of determination is standardized or meets other requirements derived from experience. The link between parameters and explosion protection measures must be taken into account, even in bench-scale determination. This link essentially has to do with the type of plant and the process parameters, regardless of whether explosion protection is achieved through precautions or constructional measures.

19.4.4.3 Maximum Explosion Pressure In chemical process engineering, it is frequently not possible to avoid working with a reactive gas system in closed equipment. Avoiding a potential ignition source is often not sufficient to ensure safe working conditions. The hazardous effects of an explosion on the equipment and its surroundings, including people, can be averted only through design measures. The elimination of ignition sources is necessary, if only for economic reasons, as a supplementary measure. Before protective practices are planned, the following question must be elucidated: Can the gas system become capable of detonation (occurrence of shock waves) as a result of the operating conditions (pressure, temperature, composition) or the geometry of the apparatus (long vessels, piping)? If so, the discussion that follows does not apply. If the occurrence of detonations can be ruled out with confidence, it is possible to design vessels and piping such that they can withstand a deflagrative explosion. The principle of pressure-resistant or pressure-shock-resistant design is employed in such cases.

In pressure-resistant design, the anticipated explosion pressure must not exceed the design pressure. In shock-resistant design, the pressure is allowed to go above the design pressure, and deformations are allowed, provided that the system remains tight (deformed parts must be replaced after such an event).

For either of these design approaches, the maximum anticipated explosion pressure generated by the gas system under the design operating conditions must be known. In simple systems, a theoretical calculation is possible, but in other cases the maximum explosion pressure has to be determined experimentally.

If economics dictate that pressure- or shock-resistant design not be used, pressure relief is still a design option for drastically reducing the pressure loads. The design of rupture disks and explosion doors requires knowledge of the maximum rate of pressure rise in the explosion anticipated. This cannot be calculated theorethically and must be determined experimentally for every application [22].

When pressure relief is employed for protection, an approach calling for a high degree of experience, it must not be forgotton that the release of products into the atmosphere may infringe on air pollution control regulations. Both the maximum explosion pressure P_{max} and the maximum rate of pressure rise $(dP/dt)_{max}$ are measured in a single laboratory trial. There is no single prescribed method for determining these parameters. The requirement that bench-scale measurements be suitable for the design of much larger equipment is best met by using a spherical autoclave with the ignition source at the center of the explosion vessel. This choice stems from knowledge that the flame front of a gas explosion propagates through the mixture as a sphere, centered at the point of initiation. To allow the reactions to proceed in virtually adiabatic fashion (i.e., to minimize heat loss through the autoclave wall in comparison with total heat reaction), the distance between the igniter and the wall must be 10 cm or more. Otherwise, excessive heat loss causes the measured values to be too low and to yield poor estimates of the safety ratings [37]. An autoclave with a pressure rating of about 100 MPa at 200°C covers the pressure and temperature ranges of practical interest in such studies. Proven types of initiators are wire igniters and surface discharge plugs.

The following instrumentation chain is recommended for recording the time dependence of pressure during reactions:

1. Pressure transducer with appropriate range pressure-sensing membrane that makes a tight seal in the autoclave wall
2. Amplfier adapted to the pressure pickup
3. Digital storage oscilloscope or computer with measuring card

For undistorted transmission of the pressure signal, the measuring chain must have a short rise time compared with the pressure rise time in the explosions under study. Other internal features of the autoclave are a thermocouple to monitor the initial temperature and a mixer to homogenize the gas mixture. The mixer is dispensable if gas mixtures are prepared in a separate mixing vessel. Generally, the individual components are charged into an evacuated vessel in accordance with their partial pressures. The quantities are governed by the mixture composition and the initial pressure.

Initiation trials are performed in still gas mixtures (mixer turned off) with various fuel concentrations. Analysis of each separate run with a given fuel gas content yields an explosion pressure value (maximum on the $P-t$ curve) and a rate of pressure rise (slope of the steepest tangent to this curve), as shown in Fig. 19-11.

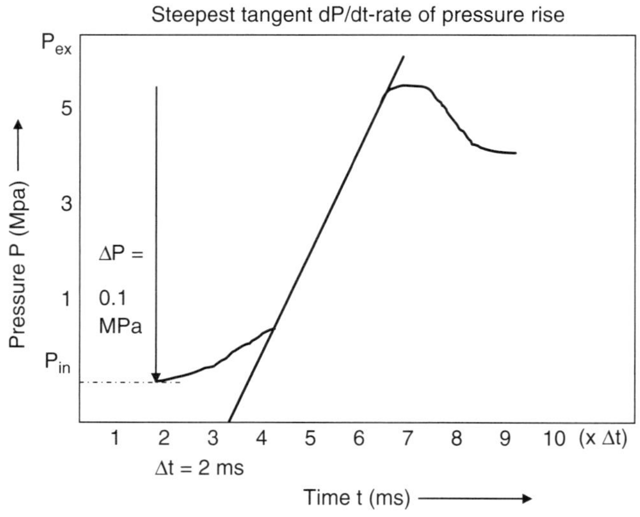

Figure 19-11 Pressure versus time in a gas explosion.

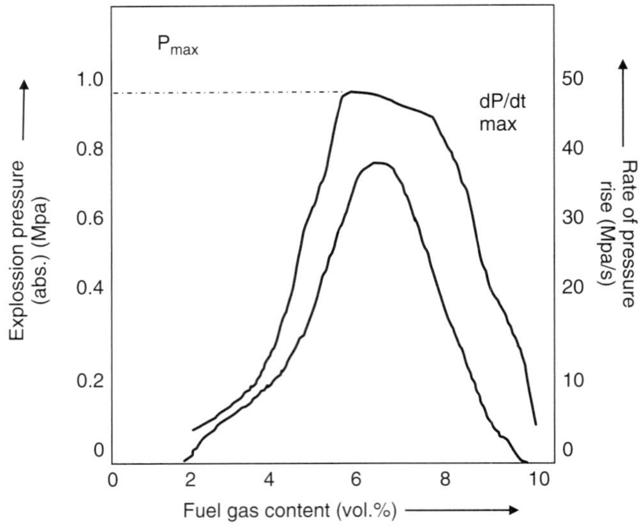

Figure 19-12 Explosion pressure and the rate of pressure rise versus the fuel gas content in a spherical autoclave.

Analysis of the digitally stored curves can be done conveniently by computer. Values from individual trials, plotted against the fuel gas content, yield two curves whose maxima are the parameters sought, P_{max} and $(dP/dt)_{max}$ (Fig. 19-12) for the system under study for the initial pressure and temperature conditions specified. The maximum explosion pressure is thus obtained virtually unchanged when the volume is increased.

If the maximum pressure is divided by the initial pressure P_{in}, the result is called the *maximum pressure increase factor*:

$$F = \frac{P_{max}}{P_{in}} \tag{19.15}$$

Typical values for explosion mixtures with air are $F_{max} = 5$ to 10. F_{max} is more or less independent of the initial pressure, but decreases with rising temperature. The value found for the maximum rate of pressure rise cannot be extended directly to larger volumes.

The only practical way to get an acceptable value for volumes larger than laboratory autoclaves is to apply the *cube law*, but it must not be forgotton that the validity of the law is restricted. It can only be used for volumes that do not differ too greatly in their geometry. Such a transformation is not possible if the turbulence characteristics differ when the laboratory value is determined in the still gas mixture, as it commonly is. This must be taken into consideration in practical situations, especially when the pressure is to be relieved by blast pipes.

The maximum rate of pressure rise multiplied by the cube root of the test volume is a constant K_G:

$$K_G = \left(\frac{dP}{dt}\right)_{max, V_L} V_L^{1/3} \tag{19.16}$$

where V_L is the volume of the laboratory autoclave. The maximum rate of pressure rise can be claculated for a different volume V:

$$\left(\frac{dP}{dt}\right)_{\max, V} = K_G V_L^{-1/3} \quad (19.17)$$

The pressure increase factor and K_G value can now serve as a basis for analyzing design practices, regardless of the fuel content in the explosive mixture in the actual plant.

There are gasous compounds whose thermodynamic and chemical properties enable them to decompose explosively, without any air or oxygen, if an ignition source is present. A thermodynamically unstable compound can be prepared by direct synthesis from the elements only if energy is supplied. If such a compound (acetylene, nitrous oxide, ethylene) decomposes to its elements, energy equal to the free enthalpy of formation is released [11]. Thermodynamically stable compounds with a negative free enthalpy of formation can also decompose to compounds of lower molecular mass if the sum of the free enthalpies of formation of the decomposition products is more negative than the free enthalpy of formation of the starting substance (e.g., ethylene oxide, tetrafluoroethylene). The features responsible for the ability of thermodynamically stable compounds to decompose are "weak bonds" between atoms in a molecule, where the molecule can dissociate if its vibrations are strongly excited: for example, by absorption of external energy. Organic molecules with multiple bonds between carbon atoms are susceptible because of the diminished binding forces at these sites. Thus, there are two factors, one thermodynamic and one structural, that suggest an ability to decompose. A gas with both types of instability (acetylene) is most likely to decompose.

Fortunately, an unstable gas does not decompose spontaneously. The free enthalpy of activation, ΔG^A, must be supplied to initiate the reaction. Only when the energy threshold ΔG^A has been overcome can the reaction proceed, with the release of the free enthalpy of reaction. Figure 19-13 shows the energy of the decomposition reaction diagram. Such reactions become dangerous if the enthalpy of reaction is evolved so rapidly that it can no longer be absorbed by the surroundings but remains in the system in some form. The reaction products can be heated up so that the pressure rises in the same way as in a gas explosion involving oxygen. The energy threshold ΔG^A is overcome more easily the higher the pressure and temperature of the gas. For safe handling of gases tend to decompose, it is urgently necessary to know the pressure and temperature limits such that the energy threshold cannot be exceeded, even by a strong ignition source. These limits can be determined only by experiment.

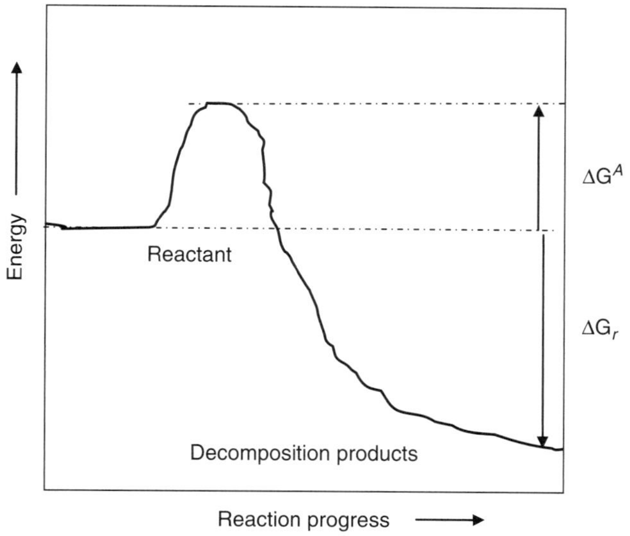

Figure 19-13 Energy change of a decomposition reaction.

By performing a series of ignition trials, the researcher seeks to determine the pressure, as a function of temperature, at which it just becomes impossible to initiate decomposition. The results are the pressure limit of stability as a function of temperature. To provide at least approximate statistical support for the limit, the test is replicated five times.

To deal with problems relating to the safe transportation of unstable gases, studies of the type just described are run, up to 70°C. Pressure limits of stability at even higher temperatures are of interest in view of operations in chemical plants. It is also often desirable to find out how the pressure limit of stability can be modified by the addition of a stable gas (not necessarily inert) or what amount of foreign gas must be added to render the system stable under the operating conditions. All these questions can be answered by experiments with the laboratory apparatus. In practice, an adequate margin of safety relative to the pressure limit of stability must always be manipulated when working with unstable gases.

Explosive systems are characterized by certain parameters [lower explosion limit, upper explosion limit, [oxygen $O_{2\,\max}$], $P_{\max}$, $(dP/dt)_{\max}$] that can be employed in the design of explosion from forming and to ensure safe handling of such mixtures by explosion-proof design with or without pressure relief. In a similar way, the sensitivity of these systems to ignition by certain ignition sources is described in terms of parameters that can be used in safety design (elimination of ignition sources).

It is also important in practice to find a parameter that characterizes fuels with regard to the propagation of explosions: for example, through narrow gaps. The propagation of an explosion from a part of the plant featuring explosion-resistant design into upsteam or downstream sections that are not so designed, or that are not vented, must be prevented effectively.

One such parameter is the ignition temperature, which makes it possible to assess the effectiveness of a hot surface (motor winding) as an ignition source. Others are the minimum ignition energy, used for evaluating electrostatic and capacitive discharges, and the minimum ignition current, for evaluating electrical and inductive sparks. Finally, the ability of exploding gas mixtures to propagate through a narrow gap is characterized by the maximum experimental safe gap, which can be used in the design of flame arresters or "pressure resistance" enclosed electrical equipment.

The ignition temperature is the lowest temperature of a hot surface at which a fuel–air mixture of the most easily ignitable composition can just be stimulated to burn with a visible flame. Two crucial points for determination of the ignition temperature are the geometry of the test vessel, which must provide a temperature field as closed as possible, and the formation of steep fuel concentration gradients when the fuel is admitted to the oxidizing gas phase already present in the vessel.

The standard method takes these points into account. The glass ignition vessel, a 250-mL Erlenmeyer flask, is charged with air or another oxidizer. The fuel is admitted to be a vessel through an inlet tube (for gaseous substances) or dropwise. The criterion for ignition is a flame forming in the vessel, which can be observed with a mirror. On the basis of the ignition temperatures in air, fuels are placed in temperature classes. The maximum surface temperature T_{max} associated with a given temperature class must not be exceeded by explosion-resistant equipment belonging to the class.

When determining the ignition temperatures T_i of fuel mixtures, it must be kept in mind that there is no linear relationship between the concentration ratio of the fuels and the ignition temperature for the mixture [7]. T_i values for fuel mixtures must therefore be determined for each individual case. Ignition temperatures for a given fuel in oxidizing media other than air may sometimes be quite low. This is especially so for pure oxygen and chlorine, in which T_i may be lowered by as much as hundreds of kelvin: T_i(toluene–air) = 535°C, T_i(toluene-chlorine) = 175°C. This effect with pure oxygen (O_2) can be accounted for by the lack of a diluting inert gas. In other cases [e.g., chlorine (Cl_2)] the ignition temperature is lower because the chlorine molecule has a much smaller dissociation energy than that of oxygen. Also, T_i generally decreases with increasing initial pressure.

The minimum ignition current, used as a measure of the ignition ability of electrical or inductive sparks, is a fuel-specific parameter. It is the smallest current flowing in a circuit of a certain inductance, such that the spark produced when the current flow is interrupted just ignites the mixture. If the minimum ignition current is known, an electrical circuit can be designated so that no ignition sparks are created by switching, or on failures such as a short circuit. This design approach is employed in explosion-resistant electrical equipment rated "intrinsically safe."

The minimum ignition energy is the smallest amount of energy stored in a capacitor that is just sufficient when discharged across a spark gap to ignite the most ignitable explosive mixture. It is a fuel-specific parameter. As a measure of the effectiveness of capacitive sparks to ignite mixtures, it is particularly useful in the evaluation of electrostatic ignition sources.

The minimum ignition current, the minimum ignition energy, and the maximum experimental safe gap make it possible to rank fuels with an allowance for the ability of explosive mixtures to propagate initiating action through a narrow gap. Fuels can be classed I [methane (CH_4), relevant only for hazards due to fire damp], IIA [ethane (C_2H_6)], IIB [ethylene (C_2H_4)], or IIC [hydrogen (H_2), acetylene (C_2H_2), carbon disulfide (CS_2)].

This knowledge makes it possible to design objects such as flame arresters, which prevent the transmission of an explosion between sections of a plant, or to build explosion-resistant electrical equipment with a "flameproof" enclosure. A lamp switch housing so rated is constructed so that ignition of the mixture inside the housing does not propagate outside it, which also requires the housing to be explosion pressure resistant.

The great advantage of being able to rank fuels is that a knowledge of the explosion group is enough for the selection of suitable explosion-resistant equipment for ignition source elimination or explosion decoupling of certain plant sections. There is no need for costly experimental studies.

When a gas mixture contains species belonging to different explosion groups, it is necessary to determine whether the properties of the sensitive components [hydrogen (H_2) or other group carbon (C)] govern the explosion propagation behavior of the mixture as a whole.

19.4.4.4 Dust–Air Mixtures Each substance that burns in the solid state may explode when finely dispersed in air in the form of a dust cloud. The violence of such a dust explosion, which is best characterized by its pressure–time diagram, is similar to that of the explosion of a homogeneous gas–air mixture. In the latter, fuel dispersion takes place on a molecular level, whereas in dust–air mixtures the size of the dispersed particles is typically 10 to 200 μm. Historically, dust explosions have been associated with coal mining and grain milling. Further industrializiation has led to dust explosion in the chemical, pharmaceutical, metallurgical, and food- and wood-processing industries. All these industries have supported reseach to investigate measures to prevent dust explosions or to limit their effects.

The probability of dust explosion is highest for very fine dusts (particle diameter <63 μm). The upper size limit for particles forming an explosible dust cloud is about 400 μm. Handing or processing of coarser product may lead to the accumulation of fines (abrasion) and thus to the formation of an explosible dust cloud. Dust explosion may occur within apparatus in which combustible powders are handled, transported, or processed, and in which dust clouds at explosible concentrations are difficult to avoid. They may also occur in rooms with deposits of combustible powders on the floor or other surfaces when this powder is taken up in the blast wave of a primary explosion, or by some other shock wave, in the presence of an ignition source. Layers of deposited powder as thin as 0.2 mm may be sufficient to fill a room homogeneously with an explosible dust cloud.

In contrast to gas–air mixtures, which are sufficiently identified by the nature and concentration of the flammable gas, dust–air mixtures need more parameters for their exact description. Owing to gravity, the particles do not stay long in suspension; the large particles settle rather quickly. Therefore, to maintain a dust cloud for a certain amount of time, and to improve homogeneity, the particles should continue to move around. A dust cloud is not normally a stationary system but maintains a degree of turbulence, which is one of the additional parameters. Others are particle-size distribution, water content of solid particles, and dust concentration. Ignition and explosion characteristics of dust clouds may change drastically when a flammable gas or vapor is also present. Because of the complexity of dust–air mixtures, it is necessary to standardize methods to determine the explosion indices, which describe ignition and explosion behavior. The minimum ignition temperature is the lowest temperature of a hot surface at which a dust–air mixture is ignited on contact. The minimum ignition energy (MIE) is the lowest value of the energy stored in a capacitor which when released in a spark discharge is just sufficient to ignite the most readily ignitable dust–air mixture at atmospheric pressure and room temperature. No internationally standardized test method exists for determination of the MIE of dispersed powder. The dust is dispersed in a finite interval, the ignition delay time; a chemical ignitor, located at the center of the explosion chamber, is activated; and the pressure within the explosion chamber is recorded as a function of time (Fig. 19-14). When an explosion occurs, the cover is lifted to a certain degree, depending on the violence of the explosion (i.e., the pressure rise per unit time).

The fact that a dust cloud is not normally stationary has implications for the course of the flame front after ignition. From studies of flammable gases it is known that

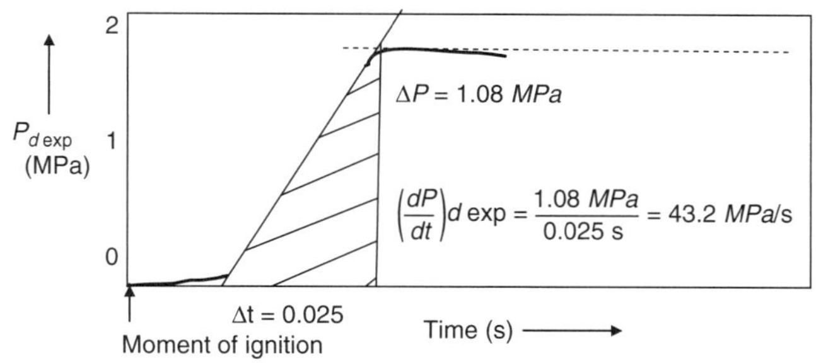

Figure 19-14 Explosion pressure in an enclosure as a function of time.

any disturbance of the flame front results in an increase in the reaction interface and thus leads to an increase in the rate of pressure rise. The same is true for combustible dusts. The rate of pressure rise measured in any test apparatus therefore depends on the dust dispersion procedure and the exact moment of ignition. Knowledge of the rate of pressure rise is most important when designing explosion venting or explosion suppression. It is necessary to use a procedure for the generation of the dust cloud which can easily be reproduced and which represents the worst turbulence characteristics that may occur in practice.

Explosion limits describe the range of dust concentrations in air within which an explosion is possible. Usually, only the lower explosion limit (LEL) is determined. Owing to the rather fast setting of dust particles, the upper explosion limit is usually not significant [41]. It is important to note that the LEL loses its importance if deposits of fine powder are present which may become dispersed in an uncontrollable way. The LEL of the powder can be determined experimentally. Starting with an explosible concentration, the concentration is gradually reduced until there is no longer an explosion. The dust concentration (g/m^3) at which ignition just fails over a defined number of successive trials is the LEL. The LEL can also be calculated from the heat content of the dust. For organic powders it is typically about 15 to 60 g/m^3. The LEL decreases with increasing temperature and increases proportionally with increasing absolute pressure.

The limiting oxygen concentration (LOC) is the experimentally determined maximum residual oxygen concentration in a mixture of air and inert gas at which a dust explosion does not occur. It depends on the dust and the type of inert gas. Starting from the oxygen concentration at which explosion occurs, the oxygen concentration is gradually reduced, varying the dust concentration until no explosion occurs. The oxygen concentration (vol%), at which no explosion occurs over a defined number of successive trials is the LOC.

For organic powders and nitrogen as the inert gas, the LOC is usually 9 to 14%. For fine metal powders, values down to 5% are reported [6]. It decreases with increasing temperature. The pressure has only a minor influence on the LOC.

The maximum explosive pressure of a combustible powder occurs in a closed system at optimum concentration. In spherical or cube enclosers with the ignition source located at the center, the maximum explosion pressure P_{max} is independent of volume. The constant K_{max} is the rate of pressure rise per unit time, measured in a 1-m^3 enclosure at optimum concentration. The dependence of the rate of pressure rise on the on the volume V is known as the *cubic law:*

$$K_{max} = \left(\frac{dP}{dt}\right)_{max} V^{1/3} \qquad (19.18)$$

This relationship is used to calculate K_{max} from results obtained in test apparatus of any volume. Both P_{max} and K_{max} can be determined in a 1-m^3 vessel or the 20-L apparatus with chemical ignitors of total energy 10 kJ located at the center. The dust cloud is generated according to the ISO standard procedure and ignited after a well-defined ignition delay time. The explosion pressure $P_{d,\exp}$ is measured as a function of time (Fig. 19-15a). Such explosion tests are performed over a wide range of dust concentration until no further increase in either $P_{d,\exp}$ or $(dP/dt)_{d,\exp}$ is observed. From a plot of $P_{d,\exp}$ and $(dP/dt)_{d,\exp}$ against dust concentration, the final quantities P_{max} and $(dP/dt)_{max}$ are determined (Fig. 19-15b).

Fine organic powders usually have a P_{max} value of about 0.6 to 0.9 MPa and a K_{max} value of about 5 to 30 MPa/ms. Very fine aluminium powder may have a P_{max} value up to 1.3 MPa and a K_{max} value above 100 MPa/ms. P_{max} increases proportionally with increasing absolute pressure at least up to 1.1 MPa in initial absolute pressure. For K_{max} this is true only up to an initial pressure of 0.2 to 0.4 MPa. P_{max} decreases with increasing temperature, and the effect of temperature on K_{max} cannot generally be predicted.

19.4.4.5 ***Explosive Condensed Substances*** Explosion can occur in the gaseous phase as well in the liquid or solid phase. Several types of explosions have been defined:

- Explosion due to spontaneous decomposition
- Explosion due to spontaneous combustion
- Thermal explosion

Gas-phase explosions due to spontaneous combustion and thermal explosions are described in many books. In this section we discuss condensed explosive substances which are likely to undergo explosions produced by the high temperature, pressure, and rate of pressure rise. The extent of damage depends on the type and mass of the explosive substances as well as on the nature of the surroundings.

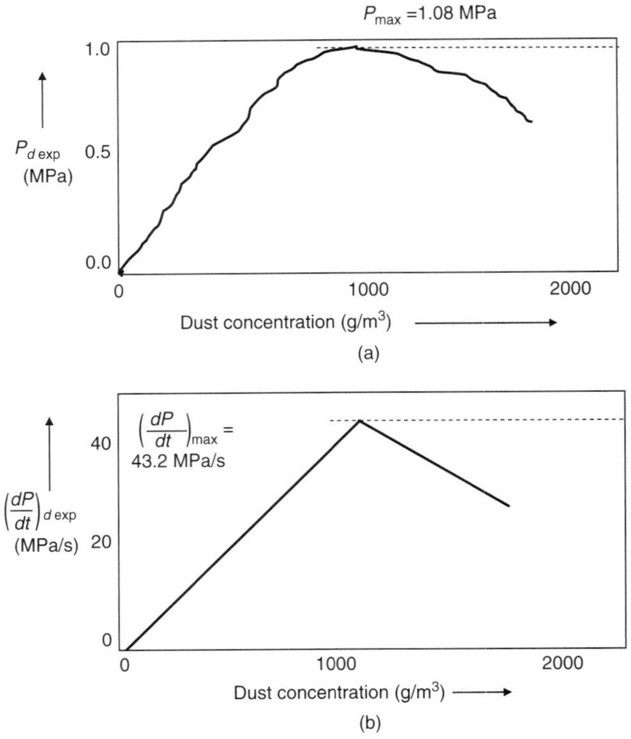

Figure 19-15 (a) Explosion pressure $P_{d,\exp}$ and (b) $(dP/dt)_{d,\exp}$ as a function of dust concentration due to spontaneous decompositions.

A condensed-phase explosion is defined as an exothermic chemical reaction during which gases and vapors are formed at such a high rate as to have destructive effects on their surroundings. Fragments, projectiles, and ground and air shock waves are produced by high temperature, pressure, and rate of pressure rise. The extent of damage depends on the type and mass of the explosive substances as well as on the nature of the surroundings.

An explosive reaction can be initiated only in substances that have a high positive heat of formation of a high negative heat of decomposition. Such substances usually have a defined chemical structure. Being organic in nature and possessing in most cases reactive groups with available oxygen, they can undergo intramolecular oxidation of the combustible part. Examples of oxidizing reactive groups are nitro, nitroso, and peroxy groups, but organic compounds containing azide, acetylenic, or diazonium groups may also be powerfully explosive substances. Explosive composition may also be produced by mixing inorganic oxidizing substances with combustible materials. Well-known examples are mixtures of potassium nitrate, sulfur, and carbon; ammonium nitrate and fuel oil; nitration mixtures containing nitric acid; and oxidation mixtures with hydrogen peroxide. Bretherick [8] and King [27] give guidance for identifying substances or mixtures as potentially explosive because of the presence of reactive groups. When a substance or mixture contains reactive groups, it depends on the molecular mass and the type and number of reactive groups, or on the oxygen balance, whether and with intensity explosive properties are exhibited. An important difference between gas- and condensed-phase explosions is that in the latter, participation of oxygen from the air is not necessary.

The power of an explosion is closely related to the dynamics of gas production and energy release as well as to the amount of gas produced and the total heat of decomposition. The energy release when a substance explodes is difficult to determine because it depends on the resulting explosion products. There may be considerable uncertainty in the heats of decomposition reported for explosive substances. A slow reaction at low temperature, due to a low degree of adiabatic efficiency under the test conditions, may result in a less energetic decomposition measurement. It is known that the decomposition energy measured can be an underestimate of the energy actually available. On the other hand, computer programs can be used to predict the minimum heat of decomposition for any substance or mixture of known chemical formula and structure [66]. As a rough guide, substances with decomposition energy above 500 J/g may have explosive properties, and those above 800 J/g may be denotable. Although this is often a good qualitative prediction of explosivity, reliable assessment of an explosive substance necessarily requires data determined experimentally.

Commonly, the term *explosion* is used collectively for the three distinguishable mechanisms by which a substance can explode. A thermal explosion proceeds, usually in the liquid phase and more or less homogenously, as a temperature-controlled, Arrhenius law, self-accelerating reaction. When during homogeneous decomposition a temperature gradient develops, a deflagration may start at the hottest spot. It passes through the substance as a reaction in which the heat of reaction and the reaction products are liberated. The deflagration is propagated into the unreacted materials by conductive, radiative, or convective heat transfer. If a substance is liable to deflagrate, this reaction can be initiated locally, by flame, heat, impact, or friction. The deflagration velocity increases with the energy content, temperature, and porosity of the substance, and exponentially with pressure. Typical values are 0.003 to 100 m/s.

The pressure dependence of the reaction velocity explains why deflagration of a small quantity of material may proceed slowly, whereas under confinement by pressure buildup, or even larger masses by self-confinement, it occurs with explosive violence. When the speed of the gaseous deflagration products reaches sonic velocity, a shock wave develops which propagates supersonically into the unreacted substance, a deflagration-to-denotaion transition (DDT) has taken place. Compression combined with strong heating initiates a chemical reaction in the substance. Typical detonation velocities are 1000 to 8000 m/s, and denotation pressures reach values up to several thousand megapascal. The detonation velocity increases with the energy content of the substance and its density. The detonation reaction may be initiated locally by heat, impact, friction, DDT, or a shock wave from other detonating substances, leading to mass explosion and producing disastrous damage. Primary initiating explosives give rise to direct immediate detonation by flame, impact, or friction in quantities of a few miligrams. Secondary explosives may be detonated by a primary explosive or by DDT in larger quantities.

To determine the risks associated with handing an explosive substance, experimental investigation into its individual explosive properties is imperative. Explosive properties refer to the mechanisms by which an explosive reaction can proceed, to the types of stresses and the ease with which explosions can be initiated, to sensitivity, and to the power of the explosion once it takes place. Understanding the initiation, propagation, and possibility of terminating explosive reactions, recognition of any destructive potential forms the basis for safe handling of an explosive substance [20]. The sensitivity of an explosive substance is of special importance because if it is too high, it may preclude some or all modes of technical handling.

For several decades, explosive properties have been described by test data determined by standardized test methods. The latest and most comprehensive collection of tests, covering all relevant properties together with evaluation criteria for classifying explosive substances for the purpose of transport, is given in the UN test manual [72]. The manual includes about 70 test methods for the substances at the laboratory and field scale (packed as for transport). The test methods are related to the properties to be defined:

- Detonability by shock wave, including sensitivity to detonation shocks of variable strength, detonation velocity, and sensitization by cavitation (gas bubbles).
- Deflagration after ignition in an open vessel or under confinement, determining the linear deflagration velocity or the rate of pressure rise.
- DDT after ignition under confinement.
- Thermal sensitivity to heating under variable defined degrees of confinement, the sensitivity being characterized by the limiting diameter of a pressure relief vent or by the maximum pressure and rate of pressure rise.
- Sensitivity to mechanical stresses, impact, and friction; determination of the sensitivity limits.

- Explosive power after initiation with a blasting cap or thermal decomposition, measuring the work performed or the specific energy.
- Thermal stability of explosives on storage at 75°C, or of organic peroxides and self-reactive substances on storage under adiabatic, isothermal, or heat accumulation conditions, determining the self-accelerating decomposition temperature related to the packed substances.

When using standardized test methods, resulting in criteria for assessing the safety of chemical procedures and installations involving explosive substances, special note should be taken of certain factors limiting their applicability. A procedure for maximum pressure determination is shown in Fig. 19-16.

It is the aim of safety to determine whether or not a chemical process may be carried out under predefined conditions (i.e., optimal conditions with regard to yield and time), or if safe operating conditions with respect to temperature, pressure, concentrations, and time must be established. For this purpose, it is necessary to make experimental investigations by appropriate test methods. Theoretical considerations may be helpful, as well as estimation of the hazard potential of a process from chemical equations and molecular structures, but they cannot replace experiments. Several screening methods and more detailed test methods are used in the industry to obtain information necessary for the assessment of exothermic or pressurizing chemical processes with all other results.

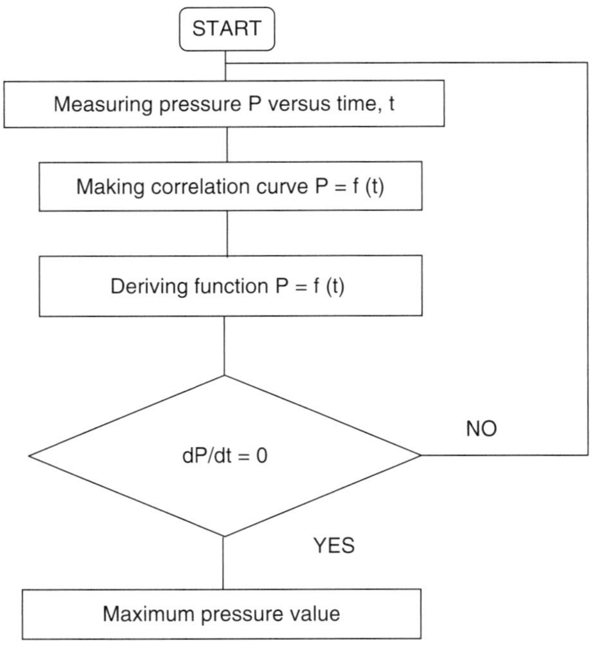

Figure 19-16 Maximum pressure determination.

Sensitivity does not depend on the energy content of the substance, as does the explosive power. To increase the reproducibility of test results, standardized test procedures often reflect idealized conditions. The heating rate, degree of filling of the apparatus, and the strength of confinement may greatly influence the results of thermal tests. Problems of scaling must also be addressed.

Any test results correspond to the conditions under which they were obtained, and their validity for substances in a particular plant should be checked thoroughly. Special care should be taken before neglecting a given detonability of a substance. With respect to production in a chemical plant, excluding the explosives industry, it is imperative that initiation of a detonable substance under the conditions of handling be impossible.

The incident risk is defined as the product of incident severity and its probability of occurrence. Hence, risk assessment includes both severity and probability. In the following, a generalized procedure for risk assessment of runaway reactions with special emphasis on evaluation criteria is presented.

For reactions presenting a thermal potential, criticality ranking can be based on the relative values of four temperatures: process temperature (the initial temperature in a cooling failure scenario), maximum temperature of reaction, stable temperature, and maximum temperature in an open system boiling point. For a closed system it is the temperature at which the pressure reaches the maximum permissible value, the set pressure of a safety valve or rupture disk.

19.5 DEVELOPMENT AND DESIGN OF A SAFE PLANT

The goal of all plant safety effort is to eliminate or reduce the possible hazards described in earlier sections. This means limiting to an acceptable level the risk created by the facility. At the same time, the relevant regulations and engineering codes must be complied with. In many areas, these implicitly establish the required level of safety or acceptable risk. All the industrialized countries have adopted regulations to protect workers, uninvolved third parties, and the environment. These regulations employ various combinations of the following approaches:

1. Setting standards for the quality of engineering facilities and their safe operation.
2. Establishing approval or licensing procedures for the erection and operation of plants and for substantial modifications to them.
3. Requiring regular inspections of a plant and its technical equipment by operators, government agencies, and sometimes, independent third parties.

4. Assigning civil liability and legal responsibility for damage caused.

Because initially there was little experience relating to the hazards of technology, the first regulations were written for individual types of apparatus regarded as dangerous (e.g., steam boilers, pressure vessels, machinery with moving parts) and aimed chiefly to ensure occupational safety and to protect third parties. Later, regulations for other types of equipment were added; attempts were made to extend protection to new areas, such as surface waters; and new regulatory principles, such as the principle of precautionary action, were introduced. It is not unusual that regulations are created by many independent bodies with very different jurisdictions.

The safe design and safe operation of chemical plants are governed by regulations stemming from the following fields of law:

1. Building and construction law
2. Occupational safety law, including equipment safety acts, regulations defining facilities subject to supervision, and accident prevention regulations
3. Hazardous substances law, including chemical and hazardous materials regulation
4. Water pollution control law, including regulations on facilities with water-contaminating substances
5. Air pollution control law, including major hazards directives.

As a rule, a regulatory system is structured vertically. Legislation at a high level sets forth the scope and objectives as well as solution approaches in general form, while engineering codes and standards at lower levels are concerned largely with technical details.

Regulations, especially at sublegislative levels, are designed deterministically in that the safety required is deemed to be attained if the plant satisfies certain requirements on design, sizing, and equipment. An exception can be found in a regulation under which a chemical plant, to be approved, must operate at a risk below a value level described in terms of fatal accident probabilities [1].

Plant builders and operators view plant safety in terms of the types of hazards arising from the process. For them and their dealings with the authorities, it would be simpler if there were a single consistent set of regulations, broken down by hazard type, releases and spills, fires, and explosions, and instead of setting fixed technical solutions for protective tasks, such a code would use fixed solutions merely as examples of ways to achieve the stated objective.

This approach would have the advantage that the safety required could be achieved by the most advantageous technical resources in each instance. In addition to that, a plant-oriented view would no longer require multiple specific regulations covering occupational safety and health, or air and water pollution control, since a single criterion for the plant (tightness) would simultaneously take care of related problems in the other areas.

The main safety tasks for the process designer and builder of process plants are:

1. To identify and assess all types of hazard correctly
2. To take appropriate steps to reduce and control these hazards

The key hazard types are:

- Release and spills
- Fires
- Explosions

The safety tasks can be broken down into the process itself and the safe design and operation of the technical facility required for the process. The specific tasks can then be listed as follows:

1. To achieve safe process design by identifying all types of hazards, assessing their potentials, minimizing the potentials, and deactivating the potentials.
2. To achieve safe plant design and operation by analyzing danger sources systematically, evaluating their probabilities of occurrence, usually qualitatively, minimizing sources of trouble and error, and employing fault-tolerant design.

This means:

1(a). For all substances to be processed, safety ratings and toxicologically and ecologically relevant data must be acquired. A comparison of process parameters and design data with safety ratings, step by step through the process, reveals where danger sources exist or may arise.

1(b). The magnitude of each hazard potential is evaluated through analysis.

1(c). It must be determined whether the hazard potentials can be reduced through suitable process design [30]. To achieve this goal the planner should replace hazardous substances by less hazardous substances wherever possible. Large inventories should be avoided as far as possible by using process steps that can be carried out quickly in a small volume, by introducing continuous operations in place of batch operations, and by eliminating large buffer volumes.

1(d). Any hazard potentials that remain must be deactivated in such a way that they cannot be manifested in the process. Lowering the temperature far below critical values or diluting substances and handling them in solution, to lower the vapor pressure, are the approved methods. They also help make the process inherently safe [30].

If a the plant is to be conceived for a process that has been safety optimized in this way, two analytical tasks, followed by the design tasks, must still be performed:

2(a). The system plant must be searched systematically for danger sources, possible defects, and failures that can activate the deactivated hazard potential.

2(b). When possible faults are identified, their frequencies or probabilities of occurrence must be evaluated so that appropriate safety measures can be taken.

These two tasks can now be undertaken with an eye to the magnitude of each hazard potential and a measure of the probability of a defect that would activate the hazard potential:

2(c). All possibilities for minimizing sources of trouble and error must be exhausted.

2(d). As far as possible, the facility must be designed and equipped so that faults are "forgiven" without resulting in harm. One way to accomplish this is to use redundant safety devices.

In carrying out these tasks, some of steps 2(a) to 2(d) may have to be done more than once, recursively, because any change in the system due to the new measures can introduce new danger sources. For this reason, and because the development of a process and the associated plant is done step by step, with concomitant advances in understanding, a "holistic" procedure segmented by time, technical specialty, and logical relationships must be adopted in the development, design, construction, and operation of a chemical plant.

19.5.1 Design and Construction Methods

Chemical processes and associated technical facilities are developed in steps. Process development in the laboratory is followed by testing of the process on a pilot scale before the project goes through the various planning stages and the preliminary, draft, and detailed design stages. The planning process culminates in the purchase of equipment and the erection of the plant. After a test and commissioning phase, the left-hand side of Fig. 19-17 shows these stages. Each phase involves questions as to the safety of the process and the plant, each of which must be answered immediately or, at the latest, before the process goes on to the next phase. The most expedient way of creating a safe plant is thus to plan for safety studies. At each step in process and plant development, a safety analysis must be done, ideally integrated into the development, in order to pose the right questions and seek immediate solutions to the problems identified.

Safety analysis is broken down into four sections, each concluding with a certificate prepared by safety experts. The certificate states that the studies required have been performed and the correct conclusions reached. The four phases are:

1. To create safety principles by compiling and determining safety, toxicological, and ecological data, identifying sources of danger in the process, examining possible safety solutions, and establishing the safety concept for the process.
2. To define safety concept for the plant by performing systematic analysis, identifying technical protective measures.
3. To Perform a detailed safety analysis by analyzing all plausible forms of trouble as to cause, effect, and corrective measures, adopting the final detailed safety concept.
4. To conduct the safety acceptance of the plant by carrying out a nominal vs. actual comparison, concept implementation, and functional tests.

The procedure is organized in an obvious way and has the advantage that a completion certificate is ready at the end of each phase, before the next phase begins. The project is not realized unless the planning certificate is ready, and the plant does not go onstream unless there is a safety acceptance certificate. These rules are set forth in an internal directive. Procedures are not always so formal, but the same principles and resources are used.

Over a plant's economic life, all the results of safety work are used again for guidance, during maintenance and plant modifications, as well as clear and complete documentation of the safety analysis. The form recommended for this is a concise, uniform, computerized database keyed to process, process step, plant and plant section, and covering all danger sources with causes, effects, and corrective measures, including brief justifications. It can be used directly in the writing of safety reports pursuant to the major hazards regulation and can easily be obtained from the plant operating personnel whenever necessary.

The design and construction of safe plants calls for a highly structured and organized procedure clearly setting forth what has to be done by whom and in what way, and focusing on the creation and routing of documents. This is called a *safety management system*.

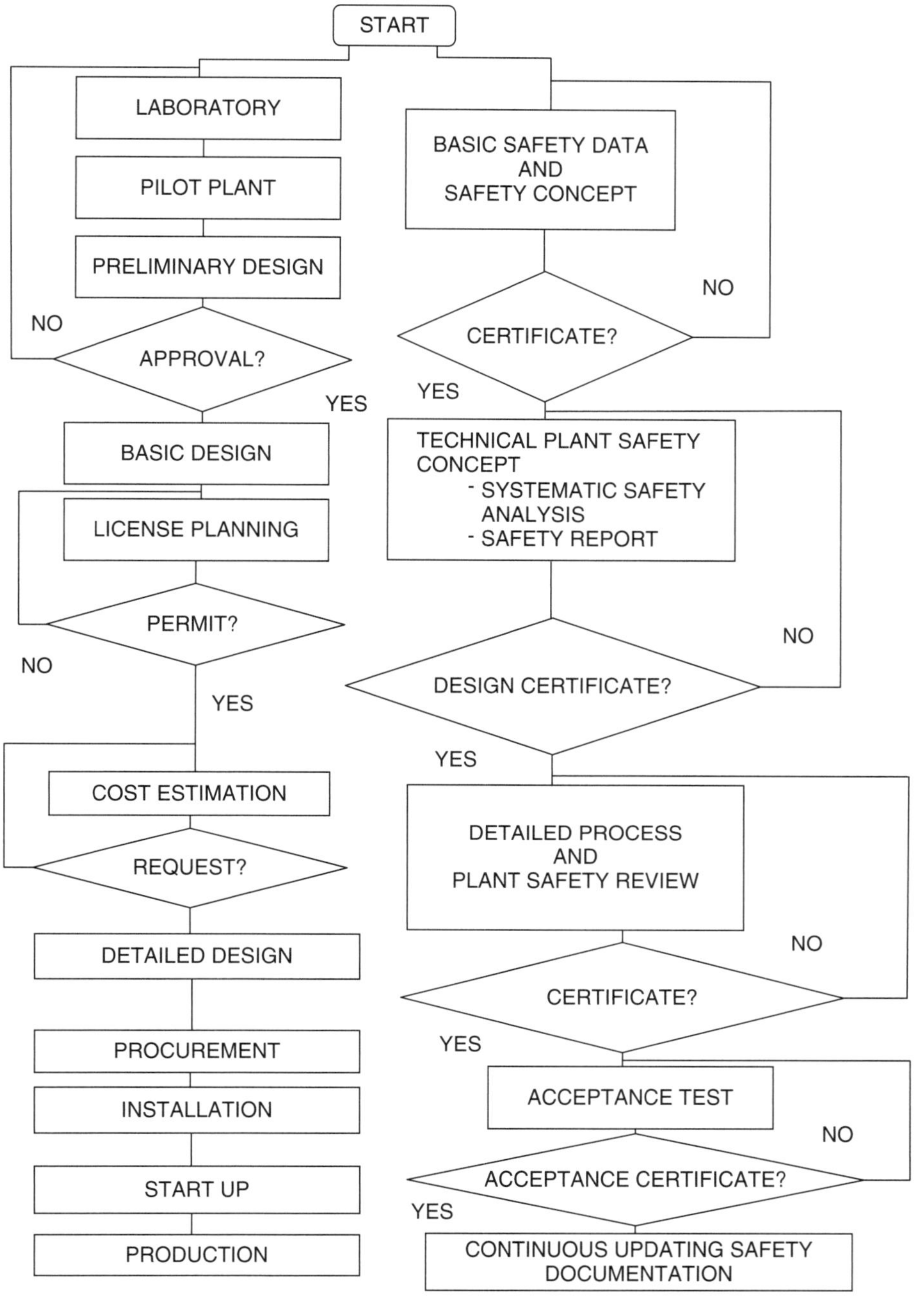

Figure 19-17 Plant design and safety development flowchart for a new plant.

In order that nothing be overlooked in the analysis, it is desirable to provide methodological aids to the specialists as they deal with various problems. The safety concept for a chemical plant must be complete, with an answer ready for any safety question. All necessary questions that can help reveal potential sources of danger must be posed ahead of time.

The problem of a comprehensive and complete analysis comes up twice in the course of process and plant development. First, safety figures for the various substances in a process (e.g., toxicity, flammability, explosibility) must be determined. Second, plant defects and system malfunctions that could activate hazard potentials must be identified. Appropriate safety engineering involves assessing the hazards as to both possible scope and probability of occurrence.

Methodological aids available for use in these tasks are all characterized by clear and easily understood structures and systematic procedures. The specific methods differ in:

- Objective of the analysis
- Functional principle employed
- Basic knowledge required
- Appropriate time of use
- Aids needed
- Results achievable
- Documentation
- Cost/benefit ratio

The principal tasks of safety analysis methods can be placed in four groups, two relating to the identification of problems and two to their assessment.

1. Completely identifying
2. Assessing hazards as to potential scope, determining the hazard potential
3. Identifying danger sources in the plant, identifying defects and malfunctions
4. Evaluating possible hazards with respect to probability of occurrence

Table 19-4 summarizes the most important methods. The literature refers to many more methods than those listed in the table. In most cases, the additional names do not refer to formal methods: "safety audit," "what-if method," and "preliminary hazard analysis"; or they are synonyms for known methods: "human error analysis," "action error analysis," and "human reliability analysis," all concerned with human errors and using checklists, keywords, tables, or event trees to study their effects; or they are a combination of two methods: "cause–consequence analysis," a combination of incident sequence analysis and fault tree analysis.

The methods of safety analysis noted in Table 19-4 increase in difficulty and cost from top to bottom. With the exception of consequence analysis, they are also arranged in roughly chronological order to use in process and plant development. In accordance with this ordering, the amount of input information required, the level of complexity, and the amount of special knowledge required also increase from top to bottom. Therefore, assessment methods must generally be carried out by a specialist, while the identification methods should be among the tools used by every chemist and chemical engineer in the process industries.

Identification methods used in safety analysis can be summarized as follows.

19.5.1.1 Checklists and Relationship Charts A checklist enumerates points that, according to experience, are associated with hazards in the handling of substances or in the performance of a process. It can be as detailed as desired and must be suitable for the type of analysis being carried out. For example, checklists can be developed and used for thorough analysis of all safety-relevant material properties. If the concern is whether explosions can be initiated by undesired reactions, the checklist comprises such questions as:

- Are the substances stable?
- What happens if foreign substances are admitted?
- What happens if the process temperature and concentration change?

The formal method is to work through such lists of questions and to determine whether:

TABLE 19-4 Methods for Safety Analysis

Task	Aim	Working Principle	Method
Identification of hazard	To complete the safety concept	Reasoning	Checklists Relationship charts
		Use of search aids and tables	Failure modes and effects analysis Action error analysis Hazard and operability studies
Evaluation of hazards according to probability of occurrence	To optimize safety systems with regard to reliability and availability	Representation of interconnections of failures in graphical form and evaluation of probability	Incident sequence synthesis Fault tree analysis
Evaluation of hazards according to consequences	To minimize the hazard potential and devise optimum protective measures	Mathematical analysis of physicochemical processes	Hazard consequence analysis

1. A hazard is possible (Yes or No)
2. Further study is needed
3. Safety measures are called for

The results of the analysis can be documented directly on the checklist or on a summary form reflecting its organization.

Checklists can be used in the design phase to ensure the completeness of the safety concept in later phases of a plant project, to ensure that possible events such as power, cooling, and stirred outages are included in the safety concept. The creation and use of checklists are also customary in plant operation. They have the obvious advantage that they can be adapted to any problem; their drawback is that things not included may not always be recognized and dealt with.

The same limitation applies to the second "reasoning" method, the use of compatibility or relationship charts. One use for this technique is to check the compatibility of all substances that can come into contact in a process (Fig. 19-18). If, for example, there are substances A to G, then half of a two-dimensional matrix gives all possible pairings of substances, as shown in Fig. 19-18. By systematical scanning over the interaction fields, it is possible to check whether mixing two substances creates a hazard and whether further experimental study is called for. The disadvantage of this method is obvious: Mixtures of three or more components do not appear. Caution is therefore indicated, and matrices are more often used simply to illustrate safety situations, causes and effects, or effects and corrective actions.

19.5.1.2 Hazard and Operability Studies (HAZOP) and Similar Methods The second group of problem-identifying methods includes:

- Failure modes and effects analysis

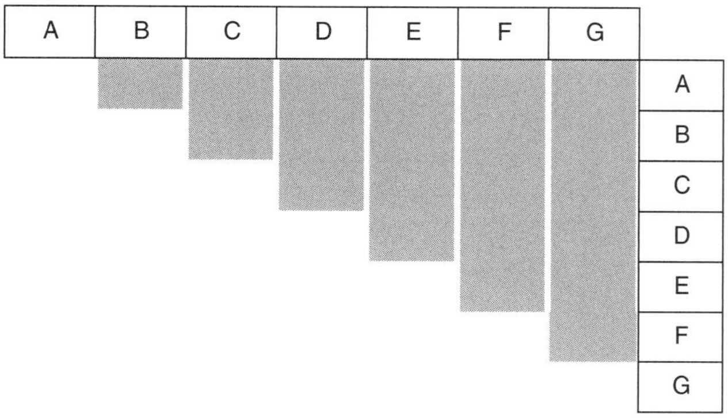

Figure 19-18 Reaction matrix showing all possible pairings of substances, and tests for unwanted reactions.

- Action error analysis
- HAZOP studies

These methods are so similar that they can be described together. All can be characterized as deviation analysis in that they look for possible hazards that can arise in the process, in the plant, or in plant operation if an error occurs or if the state or sequence of actions deviates from the state or sequence prescribed.

In this analysis of deviations, similar forms are employed for all three cases. As a rule, the form covers the following aspects:

- Deviation (error, failure)
- Cause
- Effect
- Corrective action

In much analysis, the following are also included:

- Error detection
- Frequency (often/seldom)
- Severity of effects

Also common to all three methods is the use of search aids. It is here and in the objects of study that the methods differ.

Failure modes and effects analysis is oriented to items of equipment and machinery, each having a certain function in a plant. It employs the working hypothesis that this function is not performed, that is, that each piece of equipment is examined for the effect its failure has and what corrective action may be required.

The search aids in action error analysis and HAZOP studies are keywords describing deviations from nominal conditions or a nominal sequence. Keywords used to characterize human errors include:

- Too early
- Too late
- Not
- Wrong object
- Wrong sequence

Typical keywords for the HAZOP method include:

- More
- Less
- No
- Different from

All that remains is to indicate the proper time for applying these methods. In the design of a new plant, they should come in at the detailed safety design phase, when most of the piping and instrumentation diagrams (P&IDa) are ready. These methods are also useful in the inspection of plants already built and onstream. In the author's view, the possibility of creating suitable keywords for all parameters, functions, and action sequences in a plant makes the HAZOP method so flexible that it can cover the full range of applications of the other techniques. So two identification methods are of practical importance in the development of a complete safety concept: checklist methods and the HAZOP method.

19.5.1.3 Assessment of Hazards by Consequences The concept of the hazards potential was introduced as a way of evaluating the magnitude of severity of a safety problem. If the scope and quality of safety practices are to be suited to the hazard potential so that a generally comparable and acceptable risk level is achieved, it is necessary to get at least a rough idea of the magnitude of the hazard potential. On the other hand, if effect-limiting safety devices are to be custom designed, such as an emergency pressure relief system or a scrubber to handle off-gases in case of an emergency, it is necessary to model the physical and chemical processes taking place during the accident and to investigate their effects. Protective systems can be tailormade, and hazard potentials can be minimized.

19.5.2 Evaluation of Hazards by Probability of Occurrence

Both methods from this class incident sequence analysis and fault tree analysis examine links between faults, represent them graphically, and can assign probabilities to them. Incident sequence analysis starts with a single fault and observes how it may develop. The working hypothesis is that every safety measure can succeed or fail with a certain probability.

Figure 19-19 shows a simple event tree for a temperature rise in an exothermic semibatch chemical reaction where three safety measures—heating shutdown, feed cutoff, and emergency cooling—have been instituted to keep the reaction from getting out of control. The figure illustrates the three-stage safety concept. The probability of the undesired event can be calculated if the probability of the initial event and the failure probabilities of the several safety measures are known.

More careful analysis of the sequence of events reveals that the model is far too simple for the process under discussion. Whether the safety actions succeed or fail depends heavily on the intensity and rate of temperature rise, as well as the dynamic response of each action, and these are quantities that cannot be described simply in probability terms. Therefore, the only utility of event tree analysis for such processes is that it can make relationships easy to understand in simple cases.

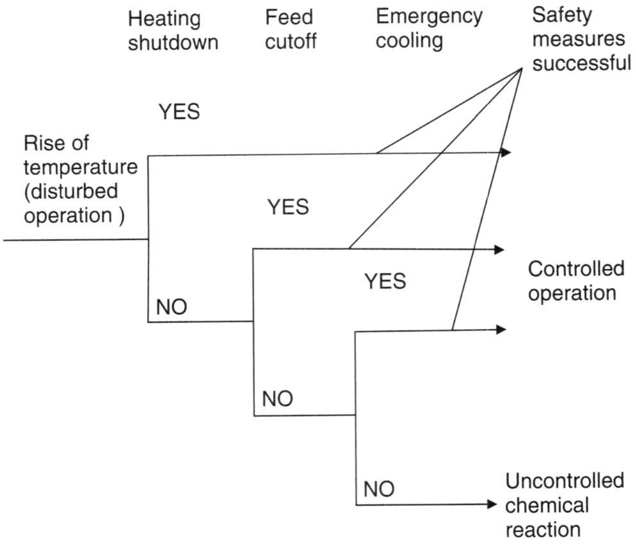

Figure 19-19 Event tree diagram showing success or failure of safety measures.

Similar limitations apply to fault tree analysis, at least when used for typical processes in chemical systems. The analysis goes in the direction opposite to that in incident sequence analysis deductively from an undesired event. Faults are connected from the top to the bottom in chains that can lead to the top event. As described in Section 19.3, there are two main types of link:

1. The blocking AND junction: More than one event must occur in order to open a path upward.
2. The passing OR junction: More than one event from a set of several events is sufficient to lead to the next-higher event.

Both fault tree analysis and event tree analysis pose stringent requirements on correct application and interpretation. Both call for experts.

The methods have been fully described in the literature. In this chapter, soybean and ammonia plants are the subjects of case histories. Their main area of application is in the assessment and optimization of complicated process systems in which the components display Yes/No failure behavior with assignable probabilities and where the failure rates are known.

Case History 19.1: Soybean Plant To study qualitative and quantitative safety aid building, the process separation of prolyl-tRNK sintetase from the soybean, *Phaseolus aureus*, was investigated. The process separation line for enzyme separation from the soybean consists of a container for the soybeans, a mill, an exchanger, pumps, extraction tanks, tanks for acid, vessels, and transporters, as shown in Fig. 19-20.

The study of fault detection and diagnostics in the soybean plant is concerned with designing a system that can assist a human operator in detecting and diagnosing equipment faults in order to prevent accidents. Many accidents have been caused by an operator's misjudgment or misoperation. There is a need to model a system that can also suggest appropriate action to be taken when a hazard occurs. Process safety analysis begins with system definition. Definition includes equipment, instruments and pipes, topology, input and output attributes, state variables, behavior rules, and initial scenarios. Process safety analysis includes hazard identification, risk analysis, frequencies, and probability analysis, consequence analysis, and hazard cost analysis.

Starting with the basic variables and their interrelations, the qualitative event model of the system can be formulated successively in the form of Boolean functions. The fault event of a system is generally in the first instance formulated in IF ... THEN form. This can be reformulated immediately using the operators AND, OR, and NOT in

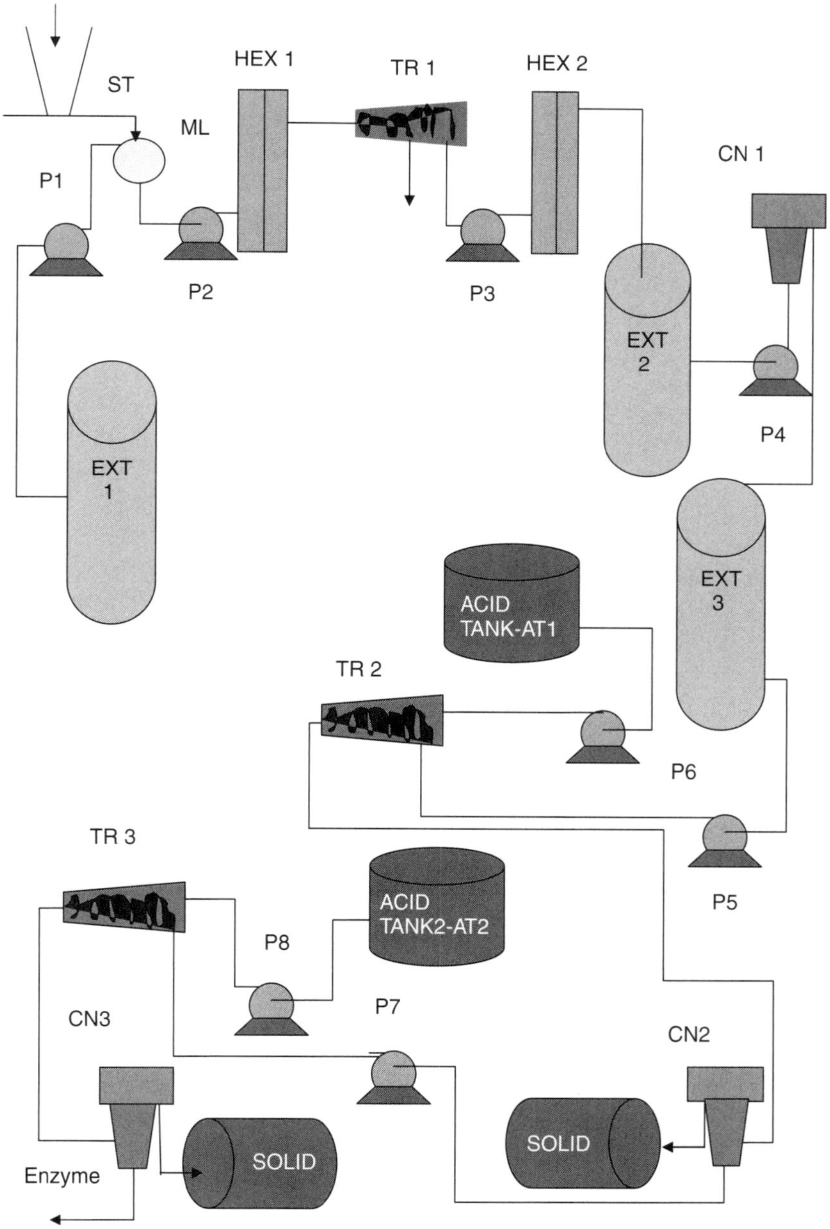

Figure 19-20 Enzyme extraction from a soybean plant. (From [53].)

Boolean form if one can assume that primary event have only two states: existence and nonexistence, as shown in Fig. 19-21.

The system can diagnose causes of faults associated with the state variables pressure, flow rate, and temperature. The qualitative variables are described in three discrete values (low, medium, and high). For diminishing the losses, a systematic cause–event analysis was made and the results of this summarized in the form of a fault tree. The attributes of the model are chosen to be pressure, supply, flow, and resistance. Supply is described in two discrete values (present, absent).

Equipment states are described in qualitative terms, such as closed, open, failed, blocked, and leaked. The following blocks are considered: blockage (B), leakage (L), and malfunction or mis operation (M) (Fig. 19-21). The study of fault detection and diagnostic is concerned with designing a system that can assist the human operator in detecting and diagnosing equipment faults to prevent accidents.

The original model generates various scenarios. In the aim of completion of the simulation runs, a resulting symptom scenario matrix is formed (Fig. 19-22). The interpretation and presentation involve monitoring system symptoms. In the symptom decomposition phase, the relational symptom/scenario matrix is decomposed by using a projection operation to produce elementary relations. This projection operation delineates which scenarios were found to have the same symptom values in their final state. Various scenarios of the process are considered.

Model for risk analysis and prevention of accidental situation for the soybean plant investigated is realized through development of logical frame and interface frame system equations. The knowledge base is composed of information streams and a database of symptoms and faults that occurred at a single unit.

The qualitative fault tree for soybean plant safety is shown in Fig. 19-21. The corresponding model is presented by system equations (19.19) (the operator ∪ is equivalent

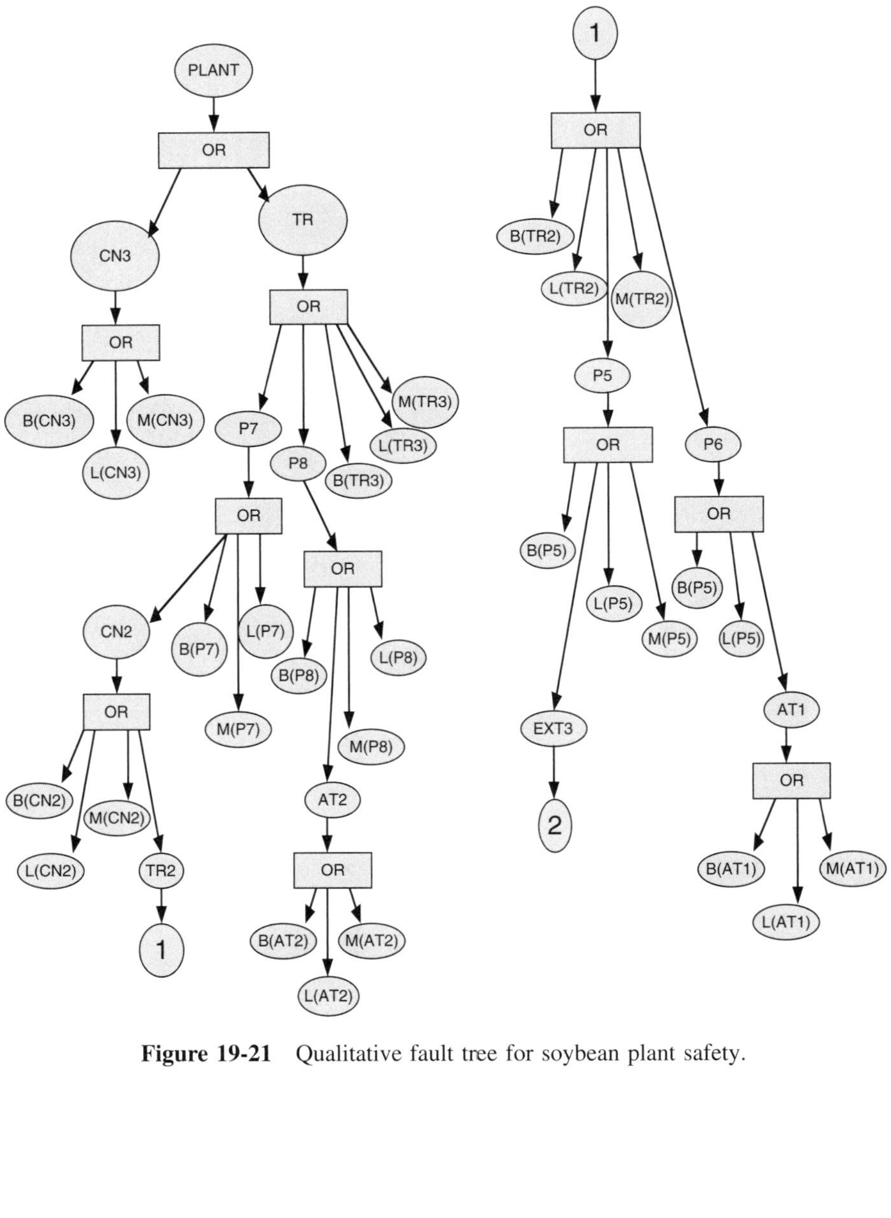

Figure 19-21 Qualitative fault tree for soybean plant safety.

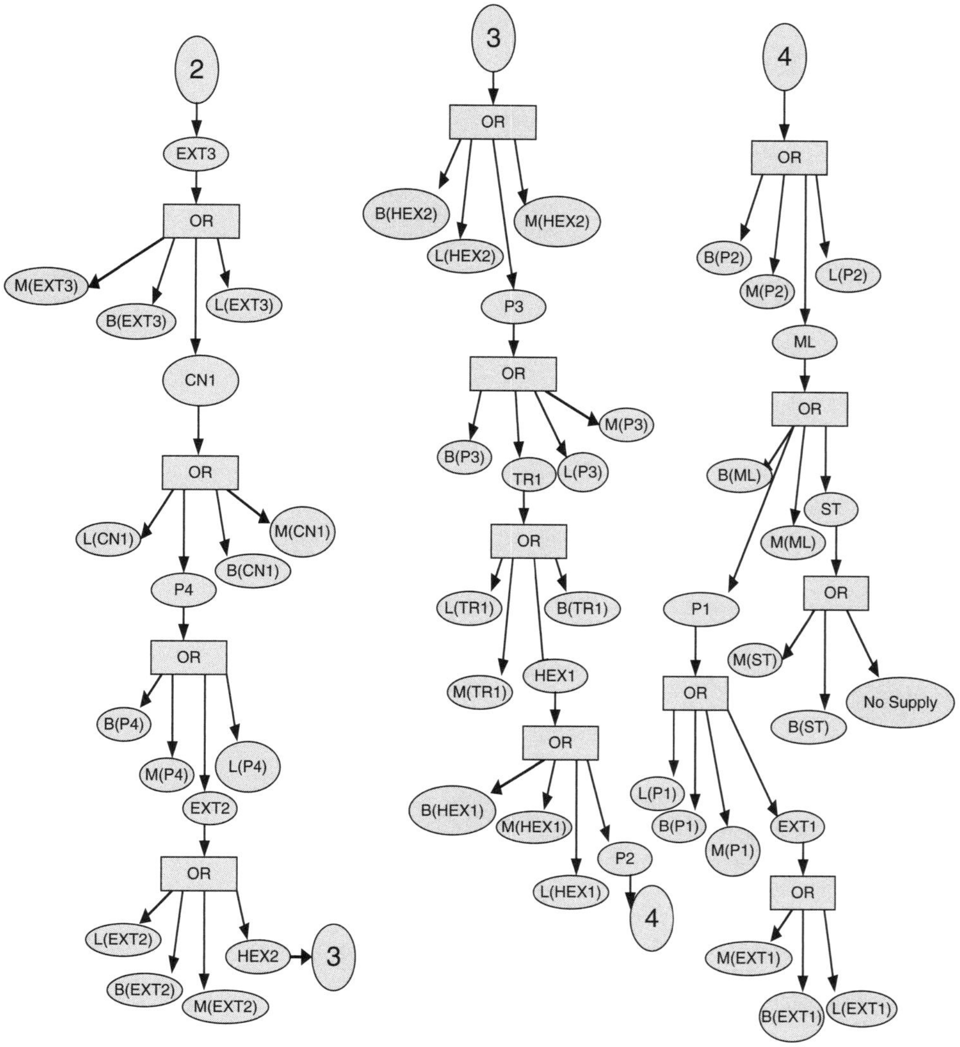

Figure 19-21 (*Continued*)

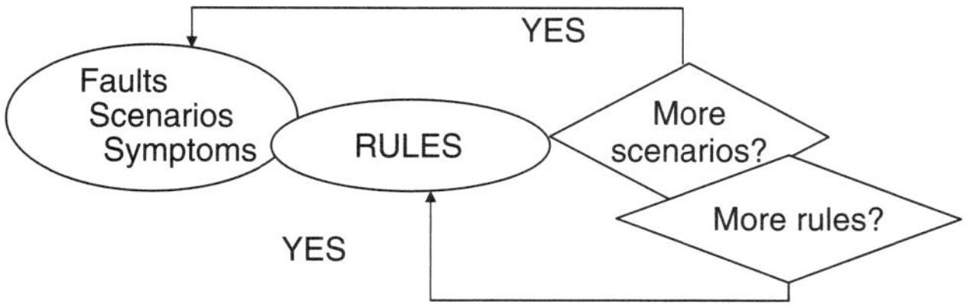

Figure 19-22 Fault diagnostic model executing.

to the operator OR in the fault tree).

$$\text{HEAD} = \text{ENZYME ABSENT}$$

$$\text{ENZYME} = \text{CN3} \cup \text{TR3}$$

$$\text{CN3} = \text{B(CN3)} \cup \text{L(CN3)} \cup \text{M(CN3)}$$

$$\text{TR3} = \text{B(TR3)} \cup \text{L(TR3)} \cup \text{M(TR3)} \cup \text{P7UP8}$$

$$\text{P8} = \text{B(P8)} \cup \text{L(P8)} \cup \text{M(P8)} \cup \text{AT2}$$

$$\text{P7} = \text{B(P7)} \cup \text{L(P7)} \cup \text{M(P7)} \cup \text{CN2}$$

$$\text{AT2} = \text{B(AT2)} \cup \text{L(AT2)}$$

$$\text{CN2} = \text{B(CN2)} \cup \text{L(CN2)} \cup \text{M(CN2)} \cup \text{TR2}$$

$$\text{TR2} = \text{B(TR2)} \cup \text{L(TR2)} \cup \text{M(TR2)} \cup \text{P6UP5}$$

$$\text{P6} = \text{B(P6)} \cup \text{L(P(6)} \cup \text{M(P6)} \cup \text{AT1}$$

$$\text{AT1} = \text{B(AT1)} \cup \text{L(AT1)}$$

$$\text{P5} = \text{B(P5)} \cup \text{L(P5)} \cup \text{M(P5)} \cup \text{EXT3}$$

$$\text{EXT3} = \text{B(EXT3)} \cup \text{L(EXT3)} \cup \text{M(EXT3)} \cup \text{CN1}$$

$$\text{CN1} = \text{B(CN1)} \cup \text{L(CN1)} \cup \text{M(CN1)} \cup \text{P4}$$

$$\text{P4} = \text{B(P4)} \cup \text{L(P4)} \cup \text{M(P4)} \cup \text{EXT2}$$

$$\mathrm{EXT2} = \mathrm{B(EXT2)} \cup \mathrm{L(EXT2)} \cup \mathrm{M(EXT2)} \cup \mathrm{HEX2}$$
$$\mathrm{HEX2} = \mathrm{B(HEX2)} \cup \mathrm{L(HEX2)} \cup \mathrm{M(HEX2)} \cup \mathrm{P3}$$
$$\mathrm{P3} = \mathrm{B(P3)} \cup \mathrm{L(P3)} \cup \mathrm{M(P3)} \cup \mathrm{TR1}$$
$$\mathrm{TR1} = \mathrm{B(TR1)} \cup \mathrm{L(TR1)} \cup \mathrm{M(TR1)} \cup \mathrm{HEX1}$$
$$\mathrm{HEX1} = \mathrm{B(HEX1)} \cup \mathrm{L(HEX1)} \cup \mathrm{M(HEX1)} \cup \mathrm{P2}$$
$$\mathrm{P2} = \mathrm{B(P2)} \cup \mathrm{L(P2)} \cup \mathrm{M(P2)} \cup \mathrm{ML}$$
$$\mathrm{ML} = \mathrm{B(ML)} \cup \mathrm{L(ML)} \cup \mathrm{M(ML)} \cup \mathrm{STUP1}$$
$$\mathrm{ST} = \mathrm{B(ST)} \cup \mathrm{ABSENT\_soybean}$$
$$\mathrm{P1} = \mathrm{B(P1)} \cup \mathrm{L(P1)} \cup \mathrm{M(P1)} \cup \mathrm{EXT1}$$
$$\mathrm{EXT1} = \mathrm{B(EXT1)} \cup \mathrm{(EXT1)} \cup \mathrm{M(EXT1)} \quad (19.19)$$

The Prolog programming language was used for the development of a computer simulation model. The procedure for model execution is shown in Fig. 19-22. The program begins with initial data following rules given for diagnostic aid. Table 19-5 shows the frequency of basic faults occurring.

The cost fault assessment was obtained by substituting for the Boolean variables using the appropriate event frequencies, linking faults with costs. Fault-occurring frequencies were estimated according to Eq. (19.20) and costs according to Eqs. (19.21) and (19.22):

$$\text{middle frequency} = \text{number of faults/year} \quad (19.20)$$

$$\text{cost} = (\text{middle frequency})(\text{cost unit}) \quad (19.21)$$

and remediation cost R_C[6U]:

$$R_C = a_0 P^{a_1} C^{a_2} \quad (19.22)$$

where P is the middle frequency of the fault, C is the cost of the fault occurring, and a_0, and a_1 are parameters [46] a_2 (see Appendix 19.2).

19.5.3 Reliability Analysis

Frequency and probability analysis involves frequency values of hazards, magnitude identification of each hazard, and development of a sound criterion for quantification of the logic reliability tree. All hazards, major and minor, need to be included. The relationship between hazard and risk must be defined [50]. Consequences modeling develops a troubleshooting system, formalized as a learning tool, and creates a recommendation to tolerant system building.

The fault event of a system is in the first instance generally formulated in an IF–THEN form. This can be reformulated immediately using the operators AND, OR, and NOT in Boolean form if one can assume that the primary events have only two states: existence and nonexistence.

TABLE 19-5 Frequencies of the Basic Fault Events as a Part of the Fault Events Database

Event Code	Middle Frequency[a]	Description
MF(1)	0.110	Mill
MF(2), MF(3)	0.095	Exchangers
MF(4), MF(5), MF(6)	0.033	Transporters
MF(7), MF(8), MF(9)	0.090	Centrifuges
MF(10), MF(11), MF(12)	0.095	Extractors
MF(13)–MF(20)	0.099	Pumps
MF(21), MF(22)	0.099	Acid tanks
MF(23), MF(24)	0.033	Solid tanks
MS(1)	0.150	Mill
MS(2), MS(3)	0.100	Exchangers
MS(4), MS(5), MS(6)	0.099	Transporters
MS(7), MS(8), MS(9)	0.090	Centrifuges
MS(10), MS(11), MS(12)	0.095	Extractors
MS(13)–MS(20)	0.099	Pumps
MS(21), MS(22)	0.055	Acid tanks
MS(23), MS(24)	0.033	Solid tanks
B(1)	0.100	Mill
B(2), B(3)	0.095	Exchangers
B(4), B(5), B(6)	0.050	Transporters
B(7), B(8), B(9)	0.100	Centrifuges
B(10), B(11), B(12)	0.089	Extractors
B(13)–B(20)	0.095	Pumps
B(21), B(22)	0.099	Acid tanks
B(23), B(24)	0.033	Solid tanks
L(1)	0.150	Mill
L(2), L(3)	0.100	Exchangers
L(4), L(5), L(6)	0.099	Transporters
L(7), L(8), L(9)	0.090	Centrifuges
L(10), L(11), L(12)	0.095	Extractors
L(13)–L(20)	0.099	Pumps
L(21), L(22)	0.099	Acid tanks
L(23), L(24)	0.900	Solid tanks

[a] Frequency is estimated: for example solid tank failed once per year (1/0.900 = 1 year).

Starting with the basic variables and their interrelations, the qualitative event model of the system can be formulated successfully in the form of Boolean functions. To make the qualitative model quantitative, the independent variables should be replaced by relative frequencies of the events p_i. The Boolean operators AND or OR should be replaced by the algebraic operators AND$(p_1, p_2, p_3, \ldots, p_n)$ and OR$(p_1, p_2, p_3, \ldots, p_n)$, producing the output frequency p_y from the input frequencies $p_1, p_2, p_3, \ldots, p_n$.

For the quantitative model the term *relative frequency* was used instead of *event probability*. Probabilistic variables must fulfill this.

$$\overline{p_i} = 1 - p_i \quad (19.23)$$

The AND $(p_1, p_2, p_3, \ldots, p_n)$ operator assigns for p_y the value n:

$$p_y = \prod_{i=1}^{n} p_i \tag{19.24}$$

Analogously, for the OR $(p_1, p_2, p_3, \ldots, p_n)$ operator:

$$p_y = \sum_{i=1}^{n} p_i \tag{19.25}$$

Equation (19.25) does not fulfill the requirement that the relative frequencies must lie in the range $0 \leq p_i \geq 1$. Therefore, Eq. (19.25) is transformed using Morgan's rule:

$$p_1 + p_2 + p_3 + \cdots + p_n = \bar{p}_1 \quad \bar{p}_2 \quad \bar{p}_3 \cdots \bar{p}_n \tag{19.26}$$

in the form

$$p_y = 1 - \prod_{i=1}^{n} (1 - p_i) \tag{19.27}$$

The event frequency can have a great variety of forms. Expressions (19.23) to (19.27) for the concatenation of the event frequencies in the fault tree can be viewed as satisfactory provided that the p_y's have crisp values.

The goal of process reliability analysis is to capture the benefits of commonsense reasoning about process malfunctions as displayed in human behavior [52,57]. The study of fault detection and diagnosis is concerned with design that can assist the human operator is detecting and diagnosing equipment faults to prevent accidents.

Case History 19.2: Ammonia Plant The plant consists of an ammonia converter, an ammonia storage tank, a centrifugal circulator, a nitrogen wash tower, four scrubbers, a liquid nitrogen/oxygen exchanger, a shift converter, two dryers, a gas-fired preheater, a gas generator, two chillers, and two coolers, as well as four reciprocating compressors, as shown in Fig. 19-23.

Methods for analyzing system safety and the synthesis for the construction of a fault-tolerant system are of elementary importance for equipment. Process safety analysis begins with plant, materials, and environmental definitions. It includes system components, topology, input and output attributes, state variables, behavior rules, and initial scenarios. The control features are qualitative variable description and logic rules for manipulating variable values between systematic states.

The system can diagnose for causes of faults associated with state variable pressures, flow rates, and temperatures. The qualitative variables are described in three discrete values: low, medium, and high. Equipment states are also described in qualitative terms, such as closed, open, failed, blocked, and leaked.

The following faults are considered: blockage, leakage, and malfunction or misoperation. The system topology or component interconnections are defined by the process connections of the working process model. The level of aggregation is defined by the modular component interconnections, which define propagation paths of attributes within the system. The study of fault detection and diagnostics is concerned with design that can assist the human operator in detecting and diagnosing equipment faults to prevent accidents. This system represents a qualitative events model expressed by logic algebra.

M, B, and *L* are independent logic variables representing the basic events malfunction, blockage, and leakage, respectively. Frequencies of the basic events are illustrated. The middle frequency of leakage for all components was taken to be approximately 0.099. The middle frequency of blockage for all components was assumed to be 0.089, and for malfunction or misoperation was 0.095.

Frequencies of induced events are derived based on frequencies of the basic events. The reliability model was derived by substituting for the logic variables using the appropriate event frequencies, and instead of the Boolean logic operators, using probability frequency operators, as shown in Eqs. (19.23) to (19.27).

The qualitative event model is given by expression (19.28). The fault event tree of the ammonia plant is shown in Fig. 19-24. The quantitative reliability model is given by expression (19.29). The system equations can be solved easily for all unknown frequencies using the values of the frequencies of the basic events given.

The reliability simulation and prevention of accidental situations is realized through development of a logical frame. Its knowledge base is composed of material streams and equipment units, and the database is composed of events and faults that occurred at a single unit and process variable state data.

The qualitative model event equations:

$$\begin{aligned}
\text{HEAD} &= \text{AST} \\
\text{AST} &= \text{B(AST)} \cup \text{L(AST)} \cup \text{M(C1)} \cup \text{M(C2)} \\
\text{M(C1)} &= \text{AC} \\
\text{AC} &= \text{B(AC)} \cup \text{L(AC)} \cup \text{M(CM4)} \\
\text{M(C2)} &= \text{M(P)} \cup \text{M(CM4)} \\
\text{M(CM4)} &= \text{CM3} \cup \text{N2T} \\
\text{N2T} &= \text{L(N2T)} \cup \text{B(N2T)} \cup \text{M(DR2)} \\
&\quad \cup \text{M(CC2)} \cup \text{SC4} \cup \text{SC3} \\
\text{SC4} &= \text{L(SC4)} \cup \text{B(SC4)} \cup \text{SC3}
\end{aligned}$$

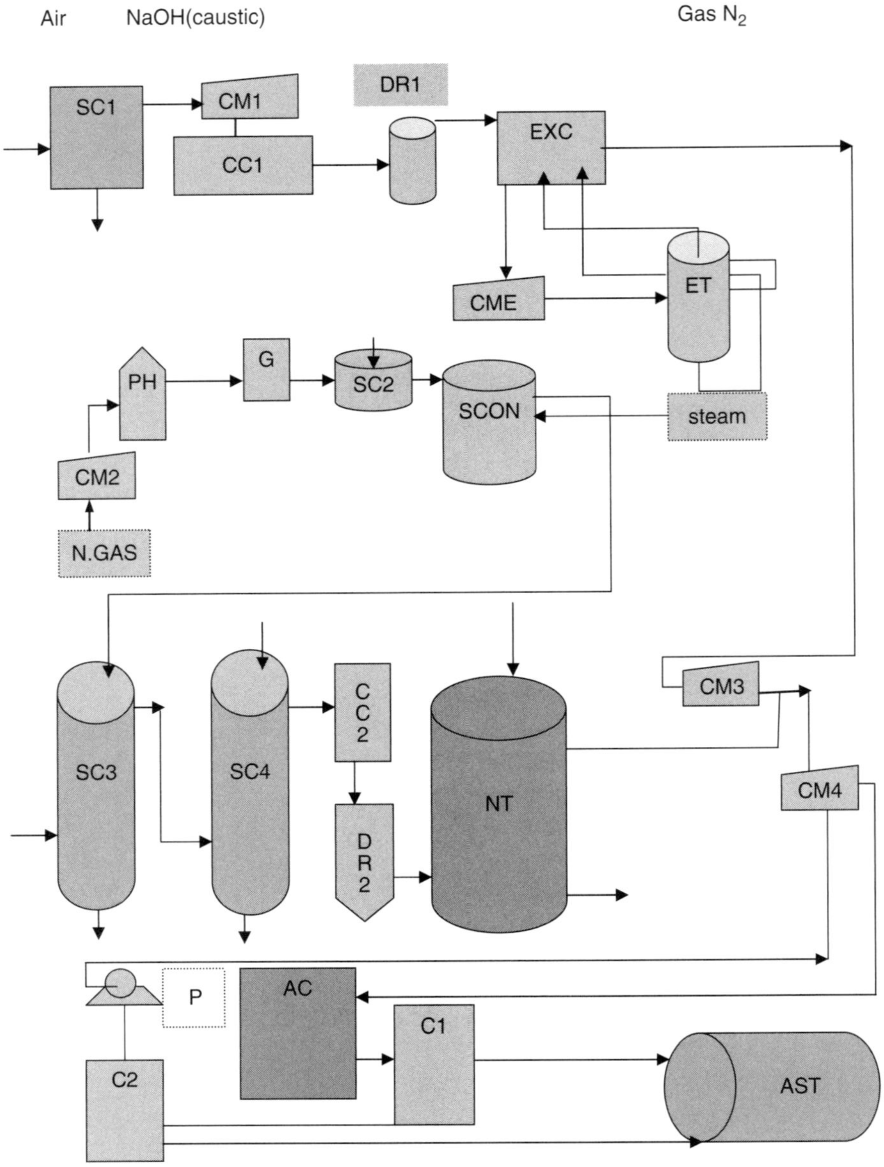

Figure 19-23 Ammonia production plant.

$$
\begin{aligned}
SC3 &= L(SC3) \cup B(SC3) \cup SCON \\
SCON &= L(SCON) \cup B(SCON) \cup S(STEAM) \cup SC2 \\
SC2 &= L(SC2) \cup B(SC2) \cup S(CAUSTIC) \cup M(G) \\
&\quad \cup M(PH) \cup CM2) \\
CM2 &= M(CM2) \cup S(NAT.GAS) \\
CM3 &= M(CM3) \cup EXC \\
EXC &= L(EXC) \cup B(EXC) \cup M(CM_E) \cup ET \cup DR1 \\
ET &= L(ET) \cup B(ET) \\
DR1 &= M(DR1) \cup M(CC1) \cup M(CM1) \cup SC1 \\
SC1 &= L(SC1) \cup B(SC1) \cup S(AIR) \cup S(CAUSTIC)
\end{aligned} \tag{19.28}
$$

The reliability model:

$$
\begin{aligned}
pHEAD &= 1 - pAST \\
pAST &= 1 - (1 - pB(AST))(1 - pL(AST)) \\
&\quad (1 - pM(C1))(1 - pM(C2)) \\
pM(C1) &= 1 - (1 - pAC) \\
pM(C2) &= 1 - (1 - pAC)(1 - pM(P))(1 - pM(CM4)) \\
pAC &= 1 - (1 - pB(AC))(1 - pL(AC)) \\
&\quad (1 - pM(CM4)) \\
pM(CM4) &= 1 - (1 - pCM3)(1 - pN2T) \\
pN2T &= 1 - (1 - pL(N2T))(1 - pB(N2T))
\end{aligned}
$$

$$(1 - pM(DR2))(1 - pM(CC2))(1 - pSC4)$$
$$(1 - pSC3)$$
$$pSC4 = 1 - (1 - pL(SC4))(1 - pB(SC4))$$
$$(1 - pSC3)$$
$$pSC3 = 1 - (1 - pL(SC3))(1 - pB(SC3))$$
$$(1 - pSCON)$$
$$pSCON = 1 - (1 - pL(SCON))(1 - pB(SCON))$$
$$(1 - pS(STEAM)(1 - pSC2)$$
$$pSC2 = 1 - (1 - pL(SC2))(1 - pB(SC2))$$
$$(1 - pS(CAUSTIC))(1 - pM(G))$$
$$(1 - pM(PH))(1 - pCM2)$$
$$pCM2 = 1 - (1 - pM(CM2))(1 - pS(NAT.GAS))$$
$$pCM3 = 1 - (1 - pM(CM3))(1 - pEXC)$$
$$pEXC = 1 - (1 - pL(EXC))(1 - pB(EXC))$$
$$(1 - pM(CM_E))(1 - pET)(1 - pDR1)$$
$$pET = 1 - (1 - pL(ET))(pB(ET))$$

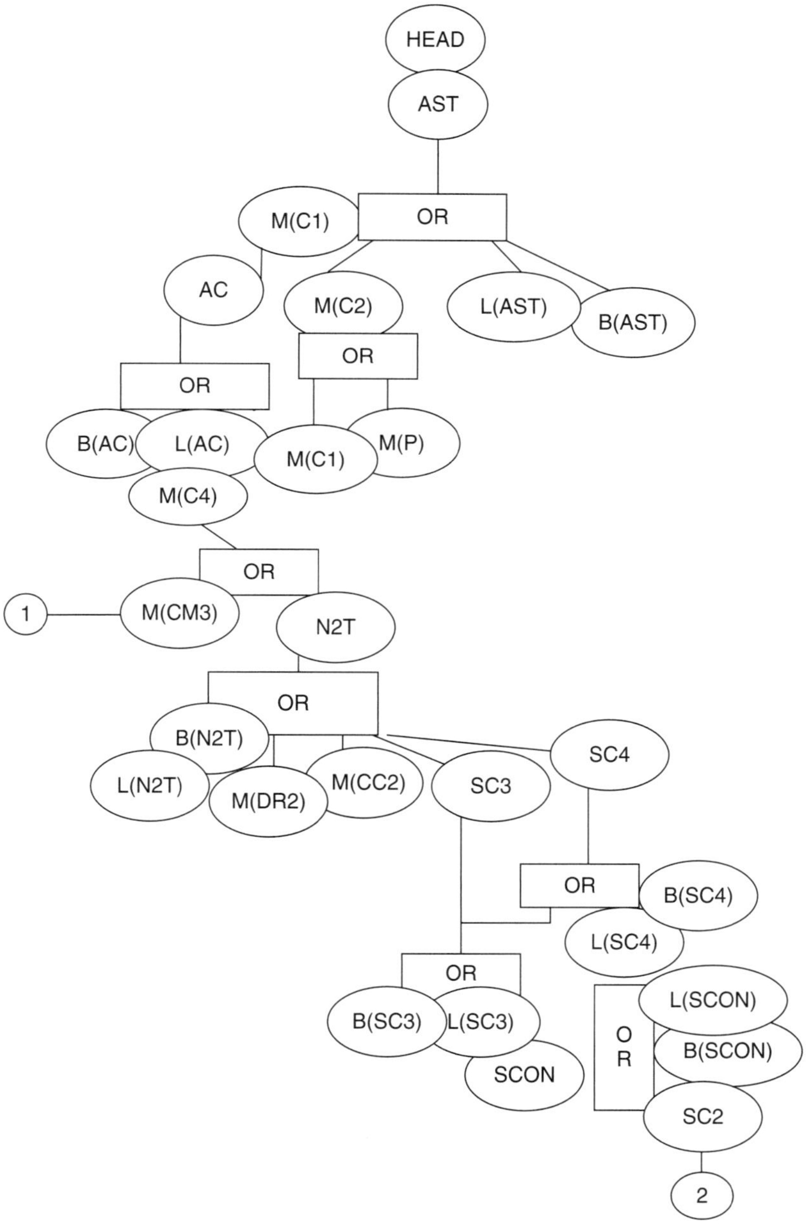

Figure 19-24 Fault tree diagnosis for the ammonia plant.

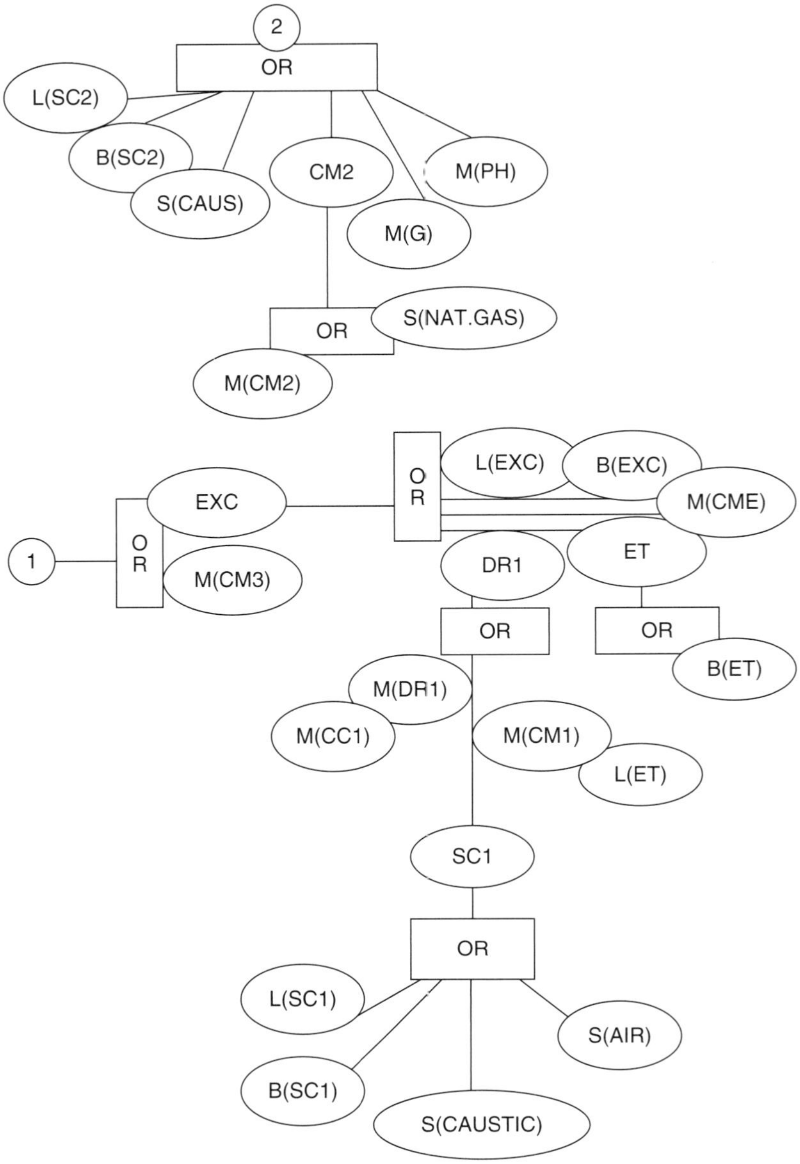

Figure 19-24 *(Continued)*

$$pDR1 = 1 - (1 - pM(DR1))(1 - pM(CC1))(1 - pM(CM1))(1 - pSC1)$$

$$pSC1 = 1 - (1 - pL(SC1))(1 - pB(SC1))(1 - pS(AIR)(1 - pS(CAUSTIC)) \qquad (19.29)$$

Safe Processing Access to the plant must be guaranteed by linking it to roads, streets, and fire service entries. Staircases (their number and placement dictated by the permissible lengths of escape and rescue routes) must give access to interior areas of the building. Escape routes serve simultaneously as lines of attack for firefighters and must therefore be built to remain safety usable for an extended period of time.

Flammable substances in the form of gas, mist, or dust mixed with a gaseous oxidizer can form an explosive mixture. The explosion of such systems is generally a fast, exothermic oxidation: a gas-phase chain reaction with chain branching. Special aspects are presented by substances that are not only oxidizable in the gas phase but can experience exothermic, explosive decomposition reactions without other reactants.

A gas-phase explosion is initiated by an ignition source delivering sufficient energy and having a suitable energy distribution. The reaction initiated propagates spontaneously through the mixture. Such combustion reactions are accompanied by the release of a large quantity of energy with increases in temperature and pressure and often also by the formation of dangerous reaction products. The hazards associated with an explosion are thus governed by three factors:

1. Occurrence of an explosive mixture
2. Presence of an effective ignition source
3. Effects of an explosion

These factors provide a logical structure for explosion prevention and protection:

1. Hazard identification
2. Risk assessment
3. Risk reduction and identification of countermeasures

In the chemical industry, a distinction is made between cases involving appreciable impacts and those where damage to plant and equipment is the most that can be expected. If appreciable impacts are not anticipated, economic considerations alone, plant availability, and product loss damage dictate that an analysis be done to determine how tolerable the probability of occurrence of an explosion. If appreciable impacts cannot be eliminated, the occurrence probability of events must be made low enough that such events are regarded as presented positively.

There are two options for lowering explosion risk measures: to prevent an explosion or measures to limit its effects. If explosion prevention is chosen, measures are instituted to reduce the explosion hazard or the ignition hazard; a combination of both approaches is possible. Limitation of the explosion hazard establishes a zone. If limitation of the ignition danger meets the requirements for this zone, prevention of an explosion is ensured.

19.5.4 Safety Based on Process Control

Safety objectives in chemical process plants can be achieved with systems based on process engineering and process control. Technical and organization measures, or a combination, can be employed. Alternative solutions that are equivalent from the safety standpoint can be found so that the most appropriate solution from the process engineering standpoint can be selected. Safety concepts for specific safety objectives set down in technical regulations must take first priority. In this section we deal with process control equipment for safety in a process plant.

The safety objective refers to preventing injuries to persons, major environmental damage, and major equipment damage to property. Safety precautions for electrical installations and equipment, occupational safety practices, or measures taken to safeguard machinery are excluded. Non-process control safety measures and equipment may include elimination of ignition sources, isolation to control explosions, pressure-proof design, safety valves, interception spaces, and retention systems.

Process control safety measures commonly reduce only a portion of the risk resulting from a unit. The risk on which the following discussion is based is just the portion to be covered by process control safety measures. The remainder can be dealt with by non-process control safety measures.

Safety tasks allotted to process control vary widely in character and significance. Similarly diverse are the safety requirements on process control equipment and practice. The use of process control equipment plant safety, the tasks assigned, and their performance are among the matters to be decided in a general safety review. The following are determined:

- The safety objectives (personnel, environment, property)
- The process control equipment to be used for plant safety
- The classification of process control equipment by task:
  - Operating systems
  - Monitoring systems
  - Safety systems
  - Damage-minimizing systems
- The required technical and organizational measures (cycle of functional testing)

The safety review is the basis for the task-driven planning, construction, and operation of process control safety systems at responsible expense and with a clearly defined functional scope. The selection of measures for maximum simplicity and direct impact generally leads to a safe and economic solution. Process control safety measures find use when other approaches are not applicable, are inadequate, or are uneconomic given a comparable level of risk reduction. The economic comparison takes in not only one-time investment costs, but also recurring maintenance expenses.

The requirements for process control safety and monitoring systems are derived from this definition. Inadvertent triggering of process control safety and monitoring systems during any phase of the process must not lead to an unacceptable fault condition of the plant.

Process control systems in chemical process plants are classified as operating systems, monitoring systems, and safety systems (Fig. 19-25) in relation to the ranges of a process variable:

- Specified operation, subdivided into:
  - Normal operating range
  - Admissible error range
- No specified operation and no admissible error range

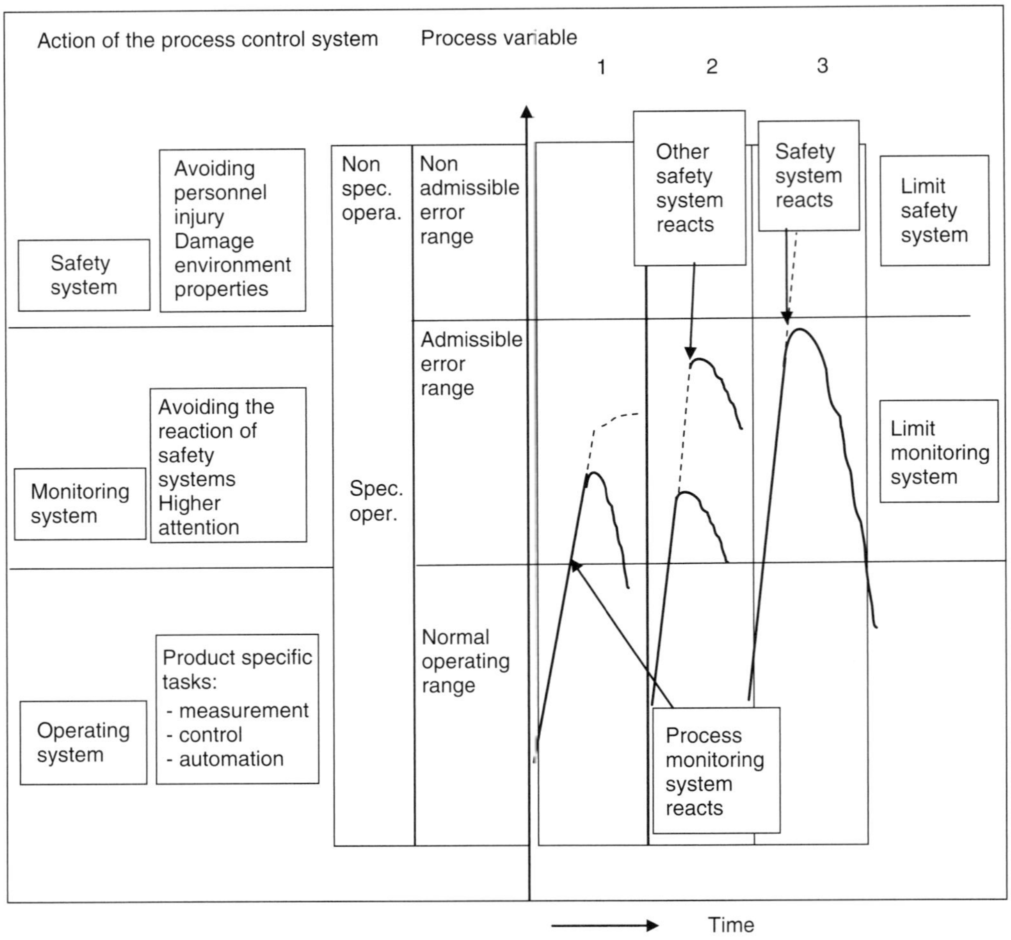

Figure 19-25 Functions of the process control system.

Region 1 in Fig. 19-25 corresponds to a situation in which factors inherent in the process keep the process variable from reaching the nonadmissible error range. The process variable is brought into the normal operating range by automatic or manual action. Region 2 represents the case where the process variable exceeds the limit for the nonpermissible range. Because another protective device is present (e.g., safety valve, rupture disk, rapid-opening or rapid-closing valve), a process control device connected ahead of it to signal or limit the increase in the process variable is classified as a monitoring system. In region 3, the process control system keeps the process variable from entering the nonpermissible range and is therefore classed as a safety system.

19.5.4.1 Process Control Operating System A process control operating system is used for the specified operation of a plant in its normal operating range. These devices implement the automation functions required for the production, measurement, and control of all variables relevant to operation, including related functions such as logging and report generation. High-level control algorithms, complex control sequences, automated recipe processing, and optimization strategies are employed increasingly. Many digital, binary, and analog signals have to be processed if all these tasks are to be performed. Because the functions of the process control operating systems are called on continuously or frequently in operation, these devices are subject to plausibility checking by plant personnel so that failures and malfunctions can be detected immediately.

19.5.4.2 Process Control Monitoring System When a plant is carrying out a specific operation, a process control monitoring system responds to conditions in which one or more process variables are outside the normal operating range but there is no safety reason to discontinue operation (i.e., these monitoring devices respond at the boundary between the normal operating and admissible error ranges of process variables).

Acceptable fault conditions of the plant are reported to evoke heightened awareness or direct action by operating personnel. The monitoring system may even initiate action itself to bring process variables back into the normal operating range. Also included are process control devices connected ahead of process control or non-process control safety systems, in order, if possible, to prevent them from responding.

19.5.4.3 Process Control Safety Systems In contrast to the functions of process control operating and monitoring systems, process control safety devices have the function of preventing no admissible error state of the plant. A process control safety system is necessary if its absence would allow the plant to reach states that could lead directly to personal injury, major environmental damage, or major equipment damage.

The task of a process control safety system is usually to monitor a process safety variable and to take one of the following actions if the variable goes outside the admissible error range:

1. Initiate a control process.
2. Notify the operating personnel (who are always present) so that the proper action can be initiated ahead of time.

The functions of process control safety systems always take priority over those of process control operating and monitoring systems, and must be executed at a level of minimum complexity. In contrast to the functions of process control operating equipment, those of process control safety systems are called upon very seldom, because the probability of occurrence of an undesired event is low and because process control engineering operating, monitoring, and safety systems are often in a staggered configuration. Because they are called upon so infrequently, it may be desirable to allow sharing of process control engineering safety components (e.g., actuators) by process control operating systems, to enhance availability and facilitate plausibility checking. Such common components must meet the requirements for process control safety equipment. Process control engineering safety aids can be useful for process plant supervision system development, as shown in Fig. 19-26.

19.5.5 Damage-Minimizing Systems

Damage-minimizing systems come into action when the plant is in no specific operational mode and when an undesired event occurs. They limit the impact on persons and the environment, limiting the extent of harm under these extremely rare circumstances. If process control equipment is used to detect an undesired event, what is monitored is not a process variable, such as pressure or temperature, but some other quantity, such as the concentration of gases released into the atmosphere.

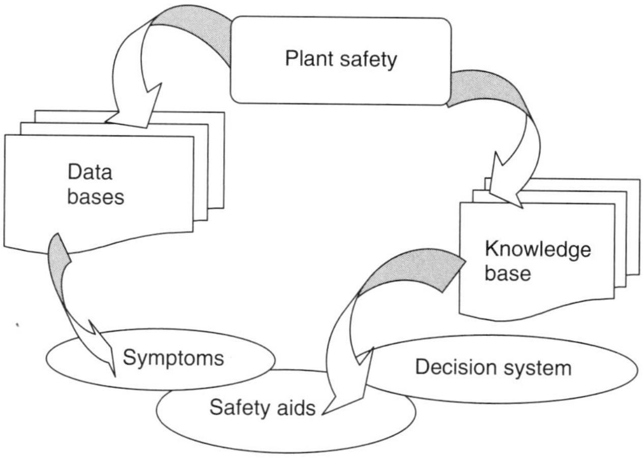

Figure 19-26 Supervision safety system.

Effectors put into action do not influence the process but have to do with the threatened region outside the vessels, apparatus, and piping, and initiation of a water curtain to prevent the spread of ammonia, for example.

There are no special requirements for the monitoring devices in a process control monitoring system; they are treated like operating equipment. If an interdisciplinary safety review establishes the need to employ a process control safety system, the following procedure is employed:

1. Qualitative estimation of risk
2. Setting requirements for process control safety system
3. Definition of technical and organizational measures

In particular, the following steps are identified and documented:

- Task statement, safety problem
- Function of process control safety system
- Technical design principles
- Nature and frequency of scheduled function testing
- Other organizational measures

Two points must be considered in the selection of process control safety devices: the minimum risk reduction aimed for, and the safety-related availability of process control safety systems. Process control safety systems must be designed and operated such that if one passive fault occurs, the protective function is still performed. Process control safety systems must be designed and operated such that if one passive fault occurs, the protective function is still performed. In the design of a process control safety system, the safety-relevant availability must therefore be chosen such that the risk is reduced to a residual level lower than the limiting risk, even if one passive fault occurs. The safety-relevant availability of process control safety systems depends on:

- The failure rate due to passive faults
- The mean time to detect and remedy passive faults
- The degree of redundancy of the process control safety system

The safety-relevant availability can thus be enhanced by the following practices or a combination of them:

- *Redundancy*: the presence of more operable technical means than are required to fulfill the basic functions intended.
- *Homogeneous redundancy*: redundant design of a device or parts thereof such that the redundant channels are structured identically and operate by identical physical processes.
- *Heterogeneous redundancy or diversity*: redundant design of a device or parts thereof such that the redundant channels operate by different physical processes or are structured differently.
- *Fail-safe quality*: the ability of a safety device to hold a process safety variable in a safe state, to modify the variable directly into another safe state, on the occurrence of certain faults in the safety device.
- *Self-monitoring*: a protective device in self-monitoring if, aside from certain faults, it is so constructed that all other faults are detected by the self-monitoring and thus the safe state is achieved.

If the requirements of process control safety systems are met after risk estimation, a distinction is made between high- and low-risk applications.

Organizational measures required for the operation of process control safety systems fall into four groups:

1. *Continuous monitoring*. In continuous monitoring, malfunctions of process control safety systems must be detected by regularly observing process variables and checking them for plausibility. Such observations are done by qualified personnel.

2. *Functional testing*. Functional tests are needed to reveal passive failure. Test instructions must be prepared, summarizing the nature and extent of periodic testing. These instructions must contain information on the nominal state and nominal performance of the safety device, along with a description of the properties and functions to be tested. This includes, in particular, statements of limiting values and measurement ranges, and other specified features to be inspected. The testing procedure must be described in test instructions and checklists that personnel can understand. The limiting values must be documented in writing by the operations manager.

The test cycle is set in the general safety review. Differences in availability may dictate that some parts of a protective device be tested more often than others. If no comparable experience exists, an appropriately short testing interval should be tested at first. If the tests reveal sufficient safety-related availability, the test interval can be lengthened as the time in service increases.

By analogy with pertinent guidelines, inspection of the entire process control safety system, from sensor to actuator, must take place at least once a year. The operations manager is responsible for seeing that functional tests are performed. It is desirable that testing be done under conditions corresponding to the demand case and with the least possible modification of the process control safety system. Whenever modifications are necessary to perform the check, special care should be taken to restore the process control safety system to its proper state. If manual overrides are installed in multichannel process control safety systems, only one channel at a time may be bridged.

Care must be taken that neither safety nor availability is substantially impaired during the test. The method used should start with the assumption that faults may be present in the safety device; thus, suitable measures should be taken to ensure that the plant remains in specified operation. In addition, tests should be performed after long shutdowns and repairs to the device.

3. *Maintenance*. When service conditions are severe, or in the case of certain measurement techniques or process analytical instruments, scheduled maintenance may be necessary. Work schedules must be prepared summarizing the nature and extent of periodic maintenance and the permits required from plant management. Maintenance operations are performed by qualified personnel in accordance with these work schedules, and the work must be documented. The formation of explosive mixtures inside apparatus and equipment can be prevented. If limitation of ignition danger meets the requirements for this zone, prevention of an explosion is ensured.

4. *Repair*. Repair of process control safety devices must be carried out without delay by qualified personnel whenever defects are found in devices and there is no alternative that will maintain the level of safety.

Testing, maintenance, and repair of process control safety and damage control devices must be documented. In particular, documentation of functional tests must include at least the following information:

- Identification of test objective
- Results of test, with detailed information on faults corrected

- Test data
- Signature of tester
- Signature of operations manager

The test report must be retained for at least five years so that the performance of specified tests can be proved.

To improve the reliability of protective devices, any faults discovered must be analyzed carefully. If longitudinal documentation of test results reveals weak points, there should be a call for improvement or testing at shorter intervals. In the technical safety concept, non-process control and process control safety measures employed to reduce the risk are evaluated qualitatively and the results used in grading protective measures. Technical measures are implemented as safety systems.

The task of a process control safety system is to reduce the risk or at least to limit the risk, possibly in cooperation with or without process control safety system or organizational measures. From the risk assessment and associated requirements, safety requirements of process control safety systems are first determined. In a second step, graded measures are devised to meet the requirements.

19.6 SAFETY PROCESS OPERATION

When hazardous substances are present, safety precautions must be taken so that concentration will stay below the limiting values in the workplace air, so that employees do not come into direct contact with hazardous substances, and so that explosions do not occur. Measures must be taken to prevent upsets in operation. If upsets come about anyway, the threat to employees must be limited.

Systematic safety analyses should be performed to determine the measures required. Hazard sources and the conditions for their activation are examined systematically and comprehensively so that the hazard potentials can be ascertained. The knowledge gained makes it possible to identify the necessary measures case by case.

Points to be considered in establishing safety practices include provisions of legislation, engineering codes, industry and factory standards, and safety information relating to specific processes. Measures, including the selection of processes with the lowest possible hazard potential, should be favored over administrative measures whenever possible. Preference should go to practices that safeguard employees or mitigate risks, regardless of employee behavior.

Despite many measures, ranging up to plant closure, contact of employees with hazardous substances cannot be prevented completely. Normal operation entails filling, transferring, and emptying media, taking samples, and cleaning and inspecting vessels. In abnormal operation, for example, a substance can get into work areas by leakage and when a pressure relief device operates.

1. *Sampling*. Samples are required for the following reasons:

- Determining identity and quality of feedstocks
- Monitoring and controlling chemical reactions and other processes in the plant
- Assessing quality of intermediate and end products

Employees conducting a sampling operation may be endangered by hazardous substances and the sampling conditions.

In the design of sampling stations, the following points must be considered:

- The sample should be withdrawn at a point in the plant where the pressure and temperature are as low as possible.
- The cross-sectional area of the sampling device must be kept as small as possible.
- Sampling must be designed such that large quantities of hazardous substances cannot escape from a plant or part of a plant because of a malfunctioning or damaged sampling device.
- The inevitable pre-sample liquid must be returned to the closed system, as far as possible.

2. *Sampling with vacuum*. A setup for taking samples without pre-sample flow is shown in Fig. 19-27. The procedure is described as follows.

a. Open ball valve V1.
b. Blow dip pipe P1 empty through fitting V2 (with nitrogen).
c. Close fitting V2.
d. Apply vacuum to fitting V3 to draw product into container C1.
e. Close ball valve V1.
f. Close fitting V3.
g. Apply nitrogen pressure through fitting V4 to transfer sample into suitable vessel.
h. Close fitting V4.

Nonreturn valves in the vacuum line prevent the working fluids in the vacuum generator from getting into the product vessel.

3. *Cleaning the vessel*. Cleaning methods should be designed so that worker health is not endangered by residues, contaminants, or cleaning agents. Before cleaning is begun, the status of the vessels to be cleaned must be determined. It must be known what substances, with what hazardous properties, are or have been present: solids deposits, liquids, vapors, or gases.

Hazards can arise from:

1. Reaction of the learning liquid with residues and contaminants

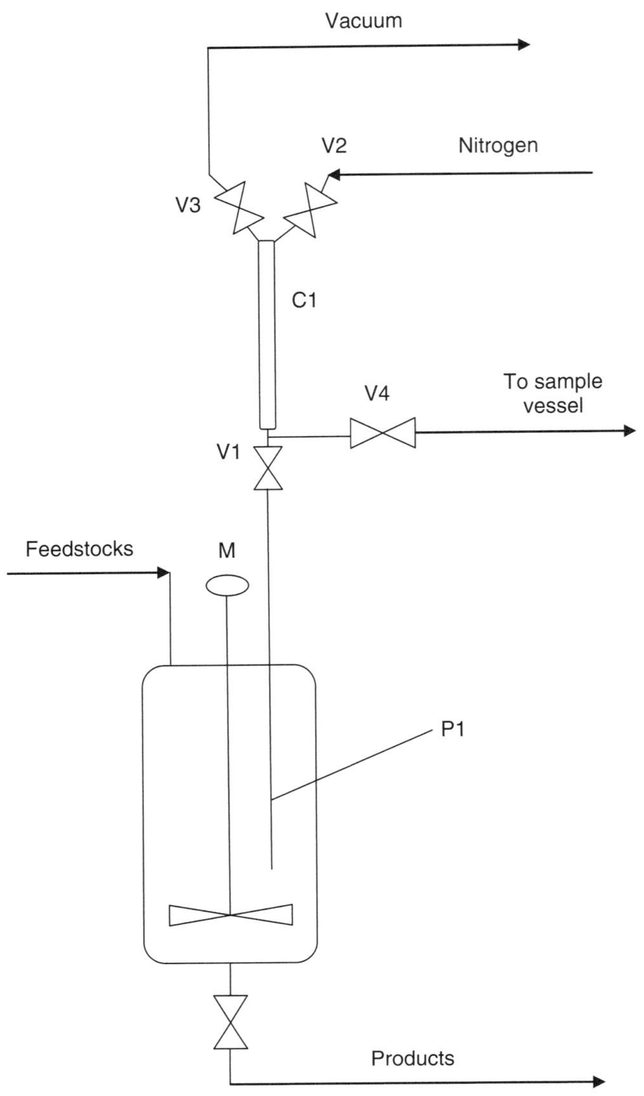

Figure 19-27 Sampling with vacuum.

2. Flammable cleaning agents
3. Static charging when liquids are sprayed under high pressure
4. Materials unstable with respect to the substances, pressures, temperatures, and mechanical stresses present

The proper precautions must be ascertained. Cleaning must be performed by trained personnel under expert supervision, in accord with written information. The cleaning method must be described in an instruction manual. Whenever possible, closed cleaning systems should be used.

Safety practices must prevent employees from being exposed to excessive concentrations of hazardous substances, even in abnormal or upset operation. Pressure vessels and pressure piping must be designed, constructed, and operated in such a way that personnel and third parties are not endangered. The action of pressure relief devices must not threaten employees or third parties. The quantity and nature of the substances present may make it necessary to conduct releases to the atmosphere or to provide supplementary features, scrubbers, flare systems, or blowdown tanks.

Leaks represent a further source of danger to personnel and the environment. Seal systems installed on shafts of pumps, drives, mixers, and stirrers must be selected in accord with the media present and the regulatory requirements. Along with proper design, the effectiveness of the seal systems must be tested regularly, and any leaks must be detected. When media are under pressure, splash deflectors must be used. Technical protective practices must also be coordinated with administrative and personnel practices, especially in the case of abnormal operation.

19.6.1 Batch and Continuous Processes

Safe production imposes stringent requirements on the owners and operators of chemical plants, who must observe and implement a body of legislation, rules, and regulations. What is more, the manufacturer of chemical products must make compromises among productivity, quality, safety, and environmental protection, but safety must take priority.

The principles of safety and environmental protection are established by management in accordance with the pertinent laws. Every chemical process involves special combinations of chemicals, equipment, and process conditions.

Safe operation makes it essential to study each process separately. The written set of instructions, the operating manual, occupies a central place in the safety effort. The operating manual is a comprehensive document that fully describes a particular chemical or physical process used in making a product. It must guarantee that the process can be carried on in a safe, environmentally benign, economical, quality-oriented way.

A fundamental requirement is that the manual must describe in full scope and depth the steps that must be performed in the process. It must be unambiguous in that the employee at the workplace knows at all times precisely what actions are to be taken, in what order. In extreme cases, every detail of a manipulation may have to be described. While the manual must be exact and complete, it must also be easily understood. It may be expedient to write separate standard operating procedures when complicated and extended actions must be repeated for each batch.

The operating manual is based on a wide range of documents and other information sources (Fig. 19-28). For reasons of bulk, it cannot include all this information down to the last detail. A vital task of the author is to set priorities and to pick out specific points about a process and set

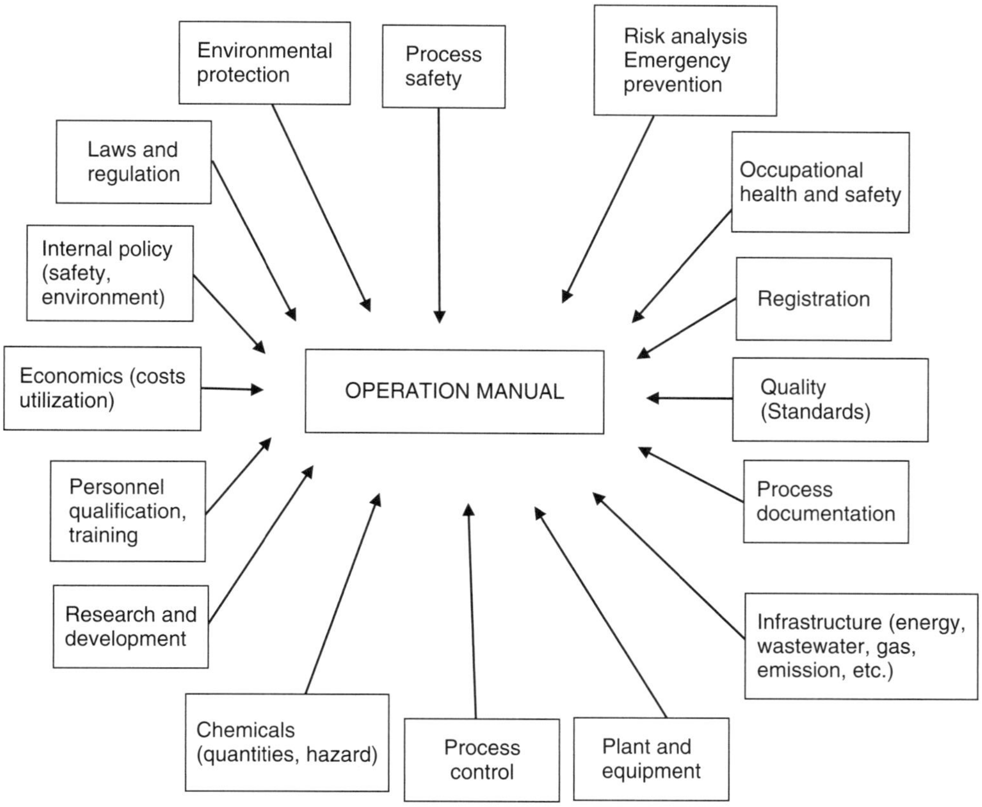

Figure 19-28 Documents and operating sources for operating manual.

them apart from what is obvious. Operators of multipurpose plants face a special challenge, for they must diagnose continuously varying conditions in the same apparatus and decide what process-specific information and actions are appropriate for each case. Electronic data processing systems are nearly universal today, and the use of standard forms makes it possible to prepare the operating manual in a consistent way and present it in an easily understood fashion. Modifications and additions must be made quickly if production is to be shifted to other plants.

The author of the manual must harmonize process data with plant-specific features while observing all guidelines on safety, environmental protection, occupational health, and so on. Many documents and other information sources can be used as aids in describing the process, but it is not enough merely to have these at hand. They must be interpreted and applied correctly. A plant manager often finds it difficult to read documents written by specialists in other disciplines. Close cooperation with these specialists is needed to ensure safe operation. The following information must be checked for completeness:

- Material and process safety data acquired and interpreted correctly
- Knowledge about the plant: cooling capacity, construction materials, plant status, and so on
- Problems of high-technology facilities: automation, measurement and control, process control engineering system
- Limitations of technical safeguards and safety equipment
- Utilization of infrastructure (energy supply, alarm systems, emergency systems, gas emission, wastewater, etc.)
- Correct and complete risk analysis process and plant, and transfer of results into practice

For some plants or chemical processes, additional points may have to be added to this list. A good operating manual is never the product of one person; it can only be created based on effective collaboration among specialists. As soon as the definitive operating manual is in being, the departments involved should apply their special knowledge to prepare opinions on the process. It is desirable to set up a fixed routing (toxicology/occupational health, safety, environment, registration, etc.) and have each activity signed off on the manual.

19.6.2 The Human Aspect of Safety

The chemical process employee works in an environment with a substantial hazard potential. Every day brings the need to work with harmful substances, dangerous reactions, and large quantities of chemicals. Each employee thus bears a high degree of responsibility for co-workers, the environment, and the plant. Even small mistakes can lead to grave accidents. In multipurpose plants, the situation is even more critical. Processes that differ in their demands on work methods and safety are performed in succession. Close concentration and an ability to adapt quickly to new conditions are required.

One of the most urgent tasks for management is to provide each employee with the conditions for safe work and the resources for performing the job. Occupational safety becomes a task for management, to ensure that workers are safety motivated and to aid them in carrying out their demanding tasks with discipline and self-confidence and without accidents.

The training of plant operating personnel is a continuing task for plant managers. Thorough training greatly enhances on-the-job safety by helping workers to deal better with the pressure of responsibility and to react calmly and correctly in stressful situations. Training must be related to practice in the plant. An ideal example is discussion of the operating manual before production is started. Special attention must be focused on the training of new employees who possess good technical knowledge but not the specific skills for their jobs. During their first months, these employees must be given technical orientation and support so that they can perform in accordance with safety rules.

A particularly important point in training is how to respond in abnormal situations. In the context of the operating manual, responses must be explained in the clearest possible way. Practical and theoretical emergency training with simulator programs help establish a routine for dealing with abnormal conditions. It is important that the worker have self-confidence that will lead to calm, considered actions in a real emergency. This training is documented and evaluated. Feedback, including worker self-assessment, helps reveal gaps so that the appropriate steps can be taken.

People are not inherently motivated to work safely, so external motivation must be provided. Repeated use, leading immediately to an accident, gives rise to bad habits. Administrative measures such as checkups and technical precautions can evoke safe working methods and make unsafe behavior difficult or impossible. Correct behavior should be associated with all possible advantages, such as recognition, easier job performance, and higher qualifications.

19.6.2.1 Internal Organization The processes that follow one another in a multipurpose facility call for intensified administrative policies. Without clear direction and clear structures, an operation where products are changed over, say, weekly, slides into chaos, and an accident becomes inevitable. In what follows, some points are listed to help avoid such problems.

The weekly program provides a quick survey of planning for the next week. All employees take part in a brief conference every week. This is an occasion for comments on the preceding week and the coming one, and is also very important for exchanging safety information. Absentees must be identified and informed later.

19.6.3 Safety in Production Practice

Central importance attaches to the production conference with all employees concerned. The operating manual is reviewed step by step and supplementary instructions are given by management. This production conference is an intensified form of the conference discussed with regard to introduction of a new operating manual. Its timing has much to do with its success. It should be held immediately before the start of production so that employees can concentrate on the new process without distraction and get themselves set for it mentally.

The preparations for a production changeover are extensive, but identical in their main features. It is worth standardizing these procedures, ideally with a checklist.

Checklist for production preparation

1. Information for leadership
   - Scope and timing of production
   - Study of process documentation
   - Leadership conference
2. Operating manual
   - Check operating manual (validity changes).
   - Make available report forms.
   - Post necessary documents (chemical datasheets) in plant.
3. Feedstock
   - Verify availability of reactants required.
   - Verify quality of reactants.
4. Training
   - Conduct process-specific training.
   - Discuss process with all employees concerned.
5. Plant/facilities
   - Check maintenance completed.
   - Carry out functional tests.
   - Install additional safety features.
   - Make piping hookups according to diagrams.
   - Perform leak tests.
   - Check cleanliness of equipment.
6. Trial run
   - Carry out and document trial run with solvent and prepare report.

7. Analysis
   - Prepare analytical sampling.
8. Administrative
   - Order necessary placards and labels.
   - Prepare folders and forms for process documentation.
9. Wastes
   - Check disposal system.
10. Storage
    - Prepare packages and containers.
    - Set aside storage capacity.
11. Safety
    - Verify completeness of safety documentation.
    - Consider safety and occupational health checks.

19.6.3.1 Continuous-Operation Production Plant A continuously operating production plant usually makes a single product. The steps involved can include both chemical reactions and physical separation operations. The quantity of material involved in both cases can be extraordinarily large, and special care is required, particularly when substances have dangerous properties (flammable, explosive, toxic, or environmentally harmful). Continuous-production plants are highly automated and are seldom run without process control systems. Apart from startup and shutdown modes, they operate according to fixed specified process parameters. Intervention by operating personnel is necessary only during indicated or infrequent deviations from normal operating conditions. Because the process does not change, the danger of neglecting to readjust safety-relevant process variables when there is a change of product is not crucial. A typical continuous process can operate in:

- Refinery processes
- Cracking processes
- Polymerizations
- Process for making organic and inorganic bulk products

Although the objectives, procedures, and plant sizes in the batch case are much different from those in the continuous case, quite a few of the basic statements can apply to continuous processes. For example, exact knowledge of the process steps is important for process safety, in particular the acceptable deviations from the process parameters specified. Safety parameters and their acceptable ranges, established before the process plant is designed, must be set forth in an operating manual and embodied in safe procedures for use by operating personnel. In the context of risk of safety analysis, conditions for the occurrence of conceivable upsets must be identified, the consequences must be analyzed in terms of occurrence probability and severity, and appropriate safety practices must be instituted and covered in detail by bulletins and training of operating personnel. The items studied include not only the failure of technical devices, compressors, pumps, mills, and centrifuges, along with malfunctions of instrumentation and control devices, but also errors made by operating personnel and the effects of partial and total power outages.

There must be a plant emergency plan, prescribing how workers should respond to a variety of abnormal conditions and accidents. A higher-order emergency and hazard control plan dealing with major accidents (i.e., those that can pose a serious danger to employees, the immediate neighborhood of the plant, or the environment) must establish internal report routing and hazard control measures. This plan must also designate those accidents in which external assistance (the public fire department or disaster response teams) must be requested, and what public institutions (i.e., police and regulatory authorities) are to be informed, immediately or later.

Before production is begun, operating personnel must receive training and orientation based on:

- Process documentation
  - Process flowsheet
  - Chemical reactions
  - Physical steps in the process
  - Parameter values to be maintained
  - Safety rules
  - Functioning of the process monitoring system
- Safety information
  - Working with dangerous substances
  - Use of personnel protective equipment
  - Periodic activities in the plant
- Operating and safety instructions:
  - Proper operation of apparatus and machinery
  - Repair procedures
  - Response to abnormal events

This information must be presented to all employees before they take up work in the plant, and at regular intervals thereafter. It is essential to test whether the knowledge presented has been understood and whether it can be put into practice promptly and correctly under stressful conditions. When a plant is being shut down on purpose, it is recommended that the shutdown process be combined with a simulated emergency.

Another vital activity in continuously operating plants is the keeping of a shift log. This helps inform later shifts about all important events in the plant. It is documentary in character and must contain the following items:

- Bulletins from management
- Process upsets
- Deviations from normal operation
- Defects found in the plant
- Repairs needed and those completed
- Availability of standby equipment and machinery
- Warnings from supplier and customer plants

Whole-plant inspection means a comprehensive system inspection and monitoring of a plant with regard to safety functions. This procedure takes into account material properties, process conditions, interactions between apparatus and other plant components, interactions between human begins and hardware, administrative policies, and the effects of external safety-relevant factors. In such a program inspections are concerned with whether the operation complies with legislative requirements (pollution control, occupational safety and health, water quality, spoil conservation), pertinent regulations, and recommendations.

These inspection and surveillance tasks involve all those concerned. The three pillars of whole-plant inspection are thus:

1. Supervision of the plant as a part of the operators' responsibilities
2. Inspection and surveillance by public agencies
3. Safety inspections by experts

Such procedures can also serve as a model for the inspection of plants for avoiding the dangerous consequences of serious accidents with hazardous substances.

19.6.4 Maintenance

The task of ensuring safety during maintenance work is to create the technical and administrative conditions for safe performance of the work. This requires a systematic procedure, as shown in Fig. 19-29. High reliability is obtained by designing and selecting components on the basis of their requirements. Key criteria are:

- Construction materials
- Functional principle
- Robustness

Secondary points are hardware safeguards against improper operating conditions, such as flow monitors on pumps, which help to prevent damage.

Because a component may fail anyway, the consequence of such a failure must be analyzed. Even if the probability

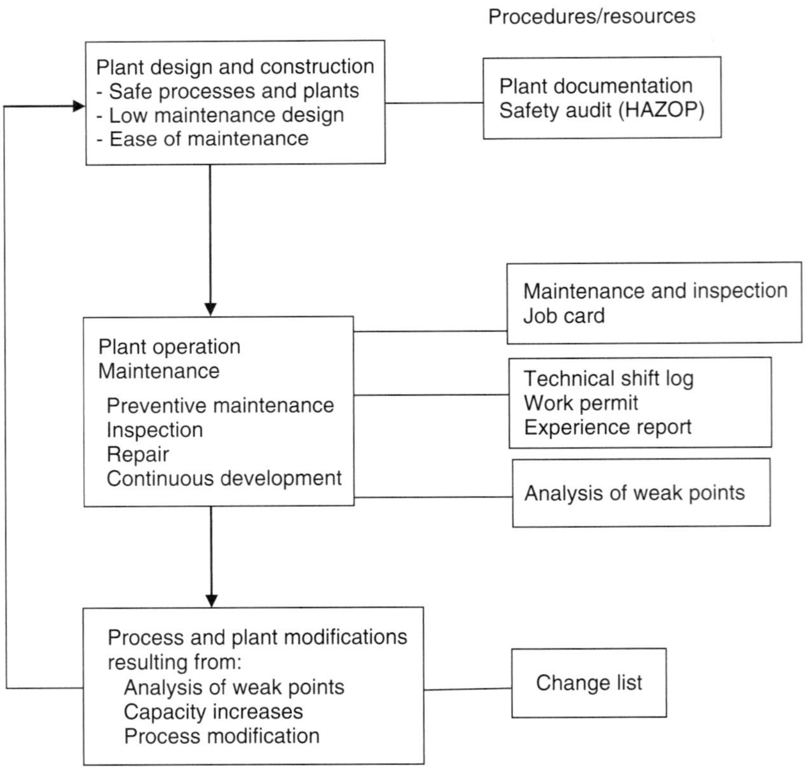

Figure 19-29 Plant maintenance procedure.

of occurrence is low, an event with serious consequences can influence the foregoing criteria and make it necessary to seek a different solution. Once these criteria are set, quality assurance systems must be instituted to monitor and guarantee compliance with the standards during subsequent fabrication and installation of apparatus and machinery.

Because maintenance and repair are inevitable, danger arising from them can be minimized most effectively by good installation and design practices. The criteria at this stage include clear layout, accessibility, easy and safe disassembly, and draining and pressure relief capabilities. In a chemical plant, special importance attaches to draining. Some maintenance jobs can be made safer by the use of assembly aids and a local spot vent system, for example.

The documents needed for reliable operation and safe maintenance should be created at the design stage. These documentations can usually be assembled from documents required for plant design and construction, manufacturers' datasheets, internal experience reports, and piping and instrumentation diagrams. Periodic inspections are supported by maintenance and inspection schedules, which must be present when the plant is commissioned. Test methods involve no opening of equipment or piping in the plant; ultrasonic inspection is preferred. Where safety-relevant process control equipment is not redundant, self-monitoring is recommended. Inspection, including functional tests and plausibility checks, help to discover concealed defects and to identify needed repairs.

19.6.4.1 Maintenance Activity The maintenance activity chart shown in Fig. 19-30 includes a procedure for scheduling and breakdown maintenance. The key step in minimizing the risk of breakdown is early detection of a deviation from nominal conditions, which can be apparent in:

- Obvious defects
- Concealed defects and slippage of process parameters

Regular rounds by shift personnel under a fixed plan can aid in problem detection and thus contribute to minimizing risks. The maintenance and inspection schedule for every technical component contains information on:

- Test criteria employed in maintenance and inspection
- Methods practiced
- Timing

The action ends with a finding documented in a report signed by the inspector. Descriptions should be oriented to concrete observations and should not contain speculation about the cause. The proper remedy cannot be established until the malfunction or damage has been jointly assessed by plant management, the shift supervisor, the shop supervisor, and in some cases a specialist. A meeting must be held to determine how and with what precautions the repair work is to be performed.

A further aid to proper reporting of technical problems by operating personnel to maintenance includes the following information:

- Trouble during the shift
- Exact location and identification of the component
- Nature of the trouble or malfunction observed
- Time of discovery

The log should also show what maintenance action was taken, and the person responsible must sign it. The technical shift log performs two other functions:

1. Identification of temporary and makeshift repairs
2. Evaluation in terms of recurrent problems (analysis of weak points)

Computer support is recommended so that:

- Temporary repairs are not left in place permanently.
- Statistical analysis can be done and appropriate countermeasures instituted.
- Service life can be extended and risk reduced by cutting the number of repair projects.

Safety procedures must be followed whether the maintenance job is:

- Anticipated, when preventive maintenance, inspection, and repairs are to be carried out during a scheduled shutdown, or
- Unplanned, caused by a failure (breakdown maintenance)

In either case, the procedures must be designed to cope with the risks occurring at every stage. If a problem arises in the failure or safety devices, it may affect plant safety directly. This aspect is not pursued here; the proper response is always set out in the context of safety analysis.

As a rule, the plant or sections of it must be put in a safe condition by a shutdown procedure when a problem occurs. This must not be confused with the emergency OFF or panic button for a section of the plant. While a planned or checklist plant shutdown is a frequent procedure, the failure of a technical component required for shutdown may lead to the isolation of specific apparatus or sections of the plant, or to overriding the safety interlocks. Examples include bypassing a leaking agitated vessel and realizing an

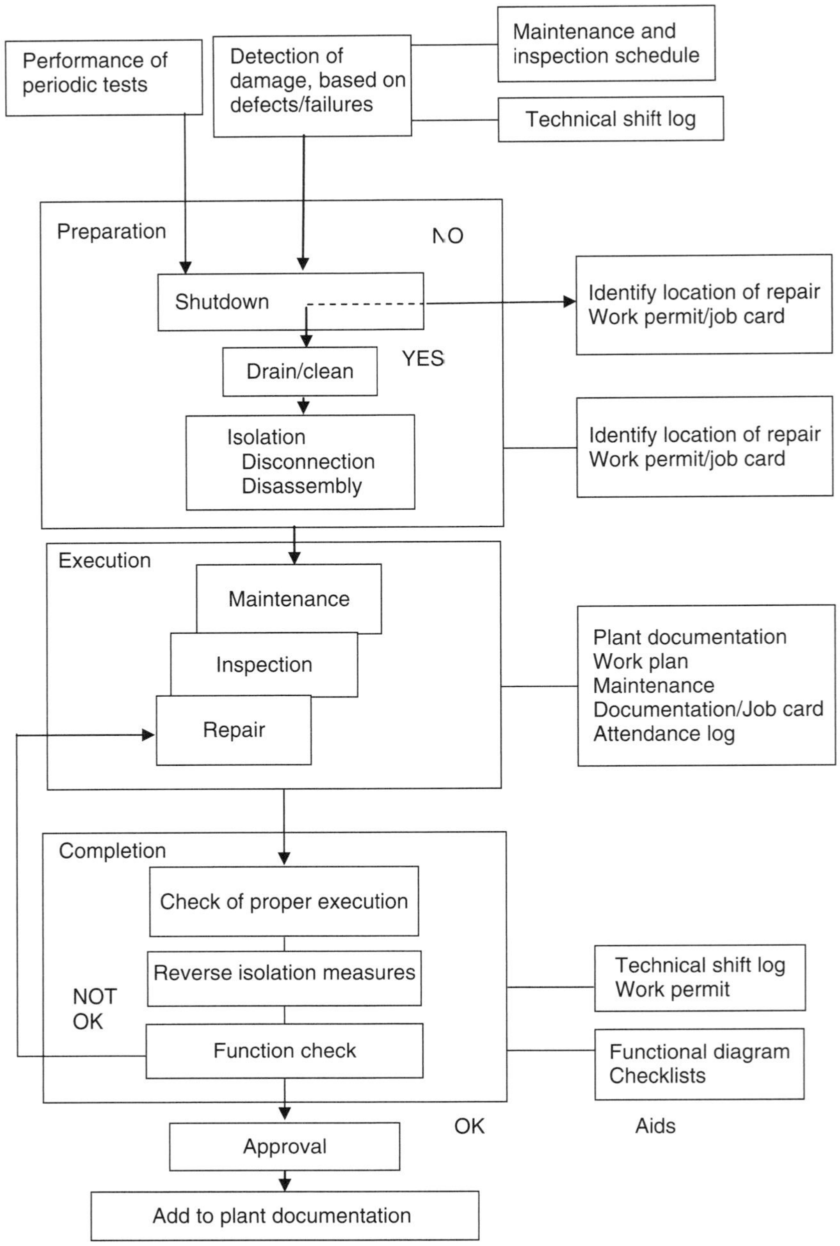

Figure 19-30 Chart of maintenance activity.

interlocked fitting that normally cannot be opened except at a certain point in the reaction.

Shutdown is not always necessary. In many instances, countermeasures can be taken that do not lead directly to a solution of the problem. There are three possibilities:

1. Postponing the proper repair (temporary plugging of leaks)
2. Using planned fallbacks (putting standby systems onstream)
3. Taking alternative measures (blowing a vessel down or pumping it dry)

Alternative actions are temporary modifications. Such makeshifts may be more desirable than immediate shutdown of the plant in a large continuous process where operation can be maintained until the next planned shutdown. Any modification must be checked first by a systematic safety analysis.

Scheduled and breakdown maintenance operations differ, especially in the documentation existing at the start of production, which describes the specific work to be done and takes the form of instructions.

- Scheduled maintenance is covered by preventive maintenance and inspection schedules describing the nature and scope of the work. Repair instructions (job cards) are generally used. Plans for shutdown maintenance are of this type, especially in continuous process plants.
- For breakdown maintenance, the problem and the corrective action taken are recorded in the technical shift log. Special repair instructions are not always available, and the documentation is then limited to the written plant documentation, including updates.

There are essentially two additional possibilities:

1. Maintenance work while the plant is onstream
2. Maintenance work during a shutdown

In what follows, only maintenance during shutdown is discussed. There is no fundamental difference between total and partial shutdown. The only routine actions that should be taken in an operating plant are those that raise no safety concerns.

Maintenance can be broken down into three phases:

1. Preparation
2. Execution
3. Post-completion measures

When danger can arise during the performance of this work, all workers involved must follow special safety practices, which are written up in a work permit. The purpose of this is two fold: to make certain that the proper safety procedures are followed, and to ensure that responsibility for the components of the work are clearly defined by signature.

Preparatory steps include single components or entire sections of the plant. This requires proper shutdown of the section covered, to a safe condition. Where redundant components exist for the malfunctioning components, plant management and/or the set operating instructions may allow these to be placed onstream. Other temporary actions, such as blowing down instead of pumping dry, changing the present process control switching or alarm values, or interfering with interlocks, are permitted only after a special safety check, unless they are provided for in special operating procedures.

After shutdown, the critical danger points are release of substances by opening up the system to drain it, and the possible introduction of undesired substances during cleaning. Ideally, a plant should have empting devices and holding tanks for the contents of the equipment so the plant does not have to be opened for drainage. This concept is desirable for substances with high hazard potentials, but the cost may be substantial. Regardless of whether fixed holding tanks are installed, emptying by gravity should be possible.

Normally, it is necessary to drain only certain portions of a plant, such as those that can be safety separated from other, product-containing sections by closing some fittings and inserting blanks. Residual quantities of product remain in the drained parts of the plant, however, especially in fittings, filters, and pumps. The individual draining features of these must be used to free them of product. Drainage is frequently aided by nitrogen pressure. Subsequent cleaning is not required with highly volatile substances. The vented air from such an operation must be considered. A vent system may be necessary together with downstream treatment of the gases.

When cleaning is required, the agents used most be compatible with the product and, physiologically, as innocuous as possible. If inert media cannot be used, the following become particularly important:

- Medium adequately diluted in the shortest possible time
- Agent reactive only with the small quantity of product found in the drained segment of the plant
- Cleaning liquid safety separated from sections of the plant containing larger amounts of product

Failure to observe these conditions has frequently led to problems, often because in a nonroutine cleaning operation, the flow direction, pressure, and valve configuration were different from those of normal conditions, and so were interpreted incorrectly. All connections to other sections of the plant must therefore be examined, including those not ordinarily fitted with product.

When maintenance work calls for disconnection or disassembly of a plant section, personal injury hazards can arise if the repair site is not rendered safe, if the electric power to drives is not disconnected, or if product-filled pipes are not closed safely with a blank flange after disassembly of a subsystem. The latter point is crucial when equipment such as pumps and fittings are disassembled, because the safety measures are carried out right at the disassembly site. In larger maintenance operations, entire piping systems are generally made product-free.

Engineering codes, accident prevention regulations, and special safety instructions aid in the proper execution of these operations. To avoid misunderstanding at the site, particularly in a complex installation, the defective item must be identified and its isolation must be indicated visually. The area can be marked further with a tag stating which department has jurisdiction and what type of work is being done.

The key requirement for sale repairs is up-to-date plant documentation. On the base technical specifications—complete and unambiguous descriptions of performance data, media, construction materials, and design data—it must be ensured that parts and materials used are such that no danger arises when the affected part of the plant is onstream.

Hazards in this sense are leaks due to improper materials or incorrect temperature/pressure design, unwanted reactions due to the catalytic action of construction materials or other media present, and parameters outside safe ranges due to the installation of unsuitable equipment (e.g., a pump with too large an impeller, generating excessive pressure). Similar hazards must be considered when replacing or repairing process control system components. A typical mistake involving an instrument is use of the wrong range.

Just as a repair or inspection site must be labeled to prevent confusion, replacement parts or newly delivered repaired components must be labeled so that they can be recognized as equipment meeting the specification. Often, there is no outward difference between two parts; only the materials inside the device differ.

These practices are intended to ensure that the on-site activity can be finished in the shortest possible time, without interruption or risk. Safety and economy are not in conflict.

Virtually every maintenance activity comprises many single activities performed by different departments. Because economics does not permit craft specialists to be employed especially for all such tasks, workers unfamiliar with the special hazards of the production plant have to be brought in. The requirement of up-to-data plant documentation must be supplemented by explicit instructions to the craftspeople at the work site, with clear task descriptions.

It is necessary to establish both the plant documentation needed for maintenance, and the scheduling of various maintenance activities in a plant area. Only by matching and skillfully coordinating maintenance activities to the rhythm of the process can they be prevented from threatening each other. An aid in this respect is a working plan that describes the sequence of tasks and their duration from their starting times: a bar chart.

In an area where welding and caulking work is done, for example, explosive mixtures are not permitted. Production not connected to the maintenance action may have to be interrupted. The same holds when flammable or explosive substances are present in the course of maintenance when a protective coating is applied to a surface.

Some types of work also require a second person to be present, even if the job requires only one worker. This is the case when a hazard such as the escape of toxic or explosive mixtures cannot be positively ruled out, or when work has to be done inside vessels.

Maintenance jobs can extend over more than a day, so interruptions cannot always be avoided. The area of the plant where repairs are being done should be marked off with signs throughout the job, especially to prevent an inadvertent return to service before the work is completed.

Approval to restart can be given only by persons whose functions include this task, such as the responsible plant engineer. Only a single person should be made responsible for maintenance. This person is the contact point throughout maintenance; otherwise, there is a danger that someone who does not have an overview of the work may make a wrong decision.

The danger of premature startup threatens, interruptions of the work. These must be recorded in an attendance log, and completion must be documented in the technical shift log. For work that requires a permit, the permit includes a space for approval after completion of the job.

A maintenance action cannot be considered finished until all the unit activities have been checked and accepted by the coordinator of the overall action. This includes both inspection for correct execution of a task, and functional testing. Proper execution means both use of the components provided (correct temperature and pressure ratings, material, etc.) and their functionally correct installation. Functional testing is then done to make certain that all technical components work together in the manner necessary for the process. The simplest example is leak testing of a repaired section of the plant; a very complicated one is the comprehensive testing of a process control system.

Where preparatory action has been taken to isolate sections of the plant, these must be reversed in the same way as temporary modifications. Approval can then be given for production. In some cases the plant is restarted a step at a time, extra functional tests with reduced throughput or an enlarged crew being done first. Such precautions are needed especially when extensive maintenance work has been carried out.

The elimination of problems by repair is a necessity, but it also offers plant operators a chance to analyze recurring problems and to use the results to improve the system. Analysis of maintenance actions for continuing development presupposes that the technology and the operating conditions are documented adequately. Entries in the technical shift log aid in evaluating the frequency of problems. If plant documentation is computerized, the cost of such a weak point analysis can be reduced substantially. The goal is to use technically feasible and economically acceptable means to optimize sections of the plant (i.e., to minimize the frequency and/or severity of trouble). Before such an analysis, the possibility of human error causing the problem must be eliminated.

This analysis of weak points must not be limited to the technology. Often, the technical defect can be eliminated more simply by changing the process conditions. The weak point found may not be the cause, but rather the effect, of defects in the process, process conditions, or product specifications. In such cases, the analysis of weak points becomes a process analysis.

The causes of problems may well be difficult to identify. A final statement of causes often has to be based on the vigilance of the operating personnel on the spot, as well as the years they have spent learning the details.

19.6.4.2 Wastewater Transport System Safety Wastewater from cities and from industrial and other manufacturers is transported in various sewerage systems. In an infrastructure system, process systems are used to receive, collect, and evacuate wastewater. Depending on how that is collected and evacuated, a wastewater system can be of the general, separation, or partial separation type.

In a general system all wastewater is transported in one channel. In a separation system, wastewater from industry, household, and the atmosphere travels in separate channels. In partial separation systems, industrial water is mixed with water from households or with industrial and atmospheric water. These systems can be found in various combinations in different parts of a city. Wastewater transport systems consist of a variety of elements, including collectors, pipelines, valves, and pumps (Fig. 19-31).

Maintenance of wastewater treatment and evacuation is very important. Safety of the wastewater transport system is ecologically significant, so a diagnostic system for risk analysis and supervision of wastewater systems has been developed. In the safety analysis and operation, simulation begins with data from the process components. For accident detection the model derived forecasts the future behavior of the system, and risk parameters are determined. The system consists of stream and process unit data as well as basic faults and symptoms. The results obtained constitute a wastewater transport safety protection system. In simulation, qualitative and quantitative analysis are often employed together. Usually, qualitative decisions are made efficiently using symbolic and graphic information, and quantitative analysis is performed more conveniently with numerical information. Diagnostic expert systems play an important role in risk analysis and accident prevention in production systems as in transport systems.

Operations create a historical database of object variables, symptoms, and scenarios. Likely scenarios are typically generated by instantiating parameter values in a parametric model according to a given situation. Modeling of the risk parameters involves uncertain processing.

The level of aggregation is defined by the modular component interconnections, which define propagation paths of attributes within a system. Research begins with the development of a conceptual framework that will facilitate the modular specification of models. The second phase develops a logic framework that will permit the use of attributes and simulation techniques linked to executable models. The fault events in a system are in the first instance generally formulated in IF–THEN form. These can be reformulated immediately using the operators AND, OR, and NOT in Boolean form if one can assume that the primary events have only two states: existence and nonexistence.

System identification involves identification of variables, elements and equipment as well as material supplies. System variables are defined in three discrete states: low, medium, and high. The equipment states are defined as blockage and leakage. The system states are "normal" and "does not work." The supply states are "exists" or "does not exist". System variables include pressure, flow, and level.

Figure 19-32 presents a safety information support system for wastewater transport system diagnosis. The system consists of 11 wastewater streams, four supply streams, and seven process units. A diagnostic system can be built as a support decision system. The database involves data streams and process units data as well as basic faults and symptoms connected by a semantic network. The diagnostic system can assist in maintaining wastewater system transport. Wastewater evacuation and treatment system safety is ecologically very important. The problem at hand is a problem of diagnosis, in which a major part of the solution consists of informing a supervisor and taking action. Such systems are used as plant maintenance aids.

19.6.5 Plant Safety Optimization

Safety objectives are isolated as shown in Fig. 19-33. Problem formulation is the most critical step in the safety problem solution, which includes optimization. Safety formulation requests identification of the main concept elements for a given specification. Needed are statement object functions, optimal criteria, and constraints.

Generally, two approaches, *static* and *dynamic*, can be applied in safety optimization. For process and plant safety, a dynamic system for an optimal policy needs to be developed:

$$\text{optimal policy} = f[e_1, e_2, e_3, e_4, q(e_1), q(e_2), q(e_3), q(e_4)] \qquad (19.30)$$

where e_1 is a set of processes, e_2 a set of hazards, e_3 a set of safety rules, e_4 a set of the optimal object functions, and $q(j)$ for $j = e_1, e_2, e_3, e_4$ is a changeable functionality. The basic objective function has economic criteria focusing on profit, material, energy, labor and risk costs, and yield. Product quality and plant capacity are technological criteria for optimal conditions.

While the objective of maintenance is to keep a plant in its nominal condition, modifications to the process and

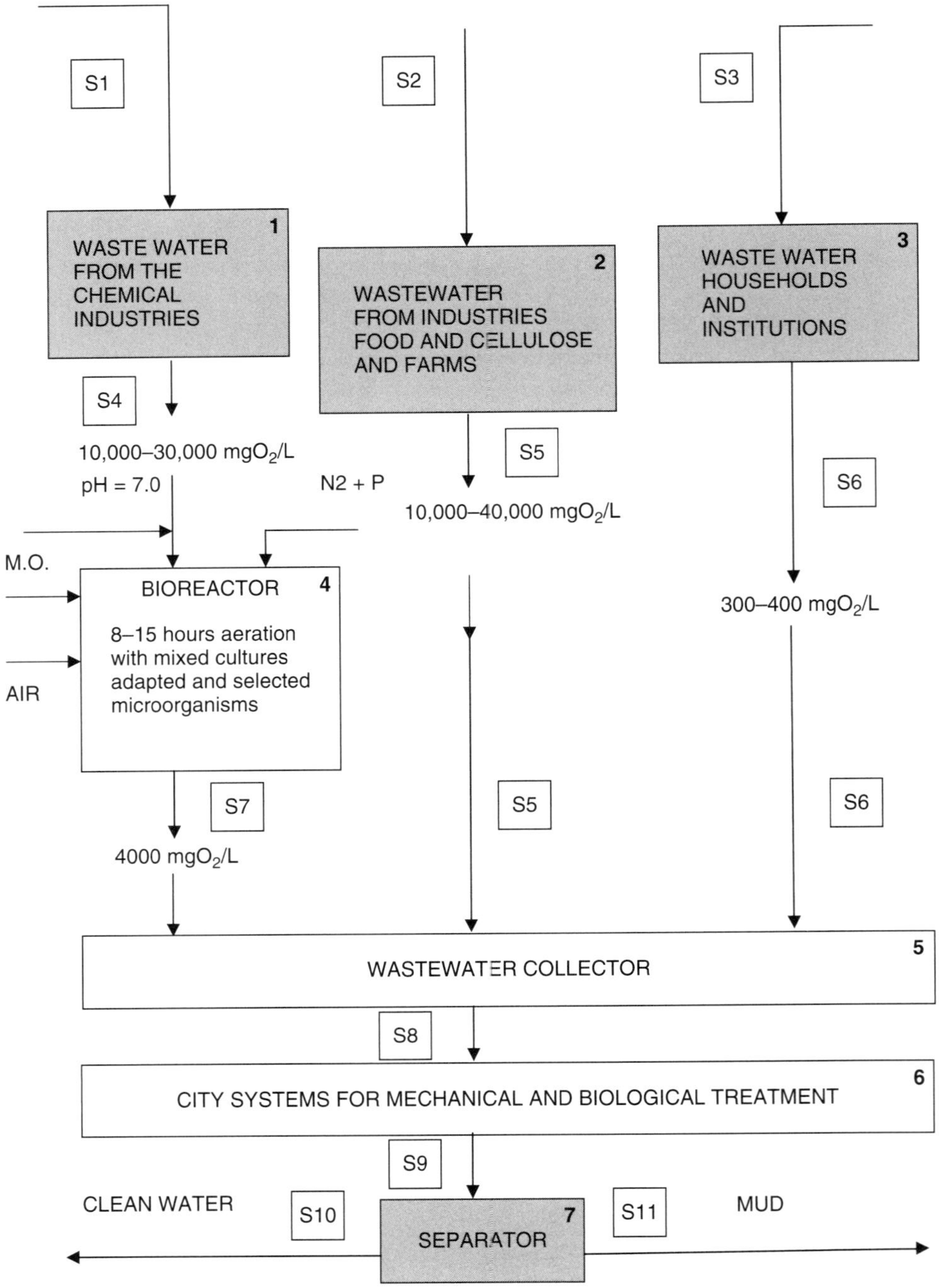

Figure 19-31 Wastewater system evacuation [56].

plant involve changing the documented nominal state of either or both. A variety of factors can lead to process and plant modifications:

- Increases in capacity
- Change in process to lower manufacturing costs and/or improve product quality or to modify the product
- Measures to eradicate weak points
- Upgrading to the state of the art (technology or safety)

Any change in the nominal state means that the corresponding parts of the process and plant documentation must be revised or rewritten. In this respect, modification differs in scope from a new design project but not in basic execution. The reason it is so important to treat a modification like a new design is that there is no way to know ahead of time how widely the modification will alter conditions. Even minor changes such as replacing one apparatus with another, or raising the temperature in a unit operation, can cause a violation of the design

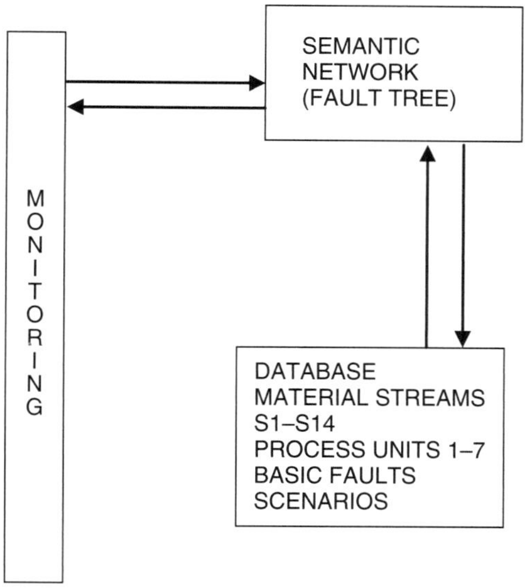

Figure 19-32 Structure of a wastewater diagnostic system.

or process control parameters. The systematic methods already described for evaluating new plants must be applied here, too, because they cover the entire field of possible hazards. First, all areas of concern, such as explosion hazards, are studied to determine whether the proposed change affects them at all. Because a change often relates to only a few danger spots, the number of areas of concern quickly becomes small, as does the number of aspects to be considered. Detailed analysis is needed for only a few aspects in most cases. Knowledge of the existing safety concept is essential for assessment of all new practices. Any change represents a "compatible" expansion of the entire concept.

19.6.6 Plant and Process Modification

A suitable procedure must be established if modifications are to be carried out correctly. As soon as a major change becomes necessary in the plant because a new unit operation is being introduced, the project is implemented in the same way as any other design project, right up to the construction of a new plant.

In the case of minor changes, the amount of investigative effort required is generally small, so a simplified procedure is useful in such cases. The principle remains the same, but the effort invested in safety analysis and documentation updating is matched to the scope of the planned change. A useful aid is the change sheet, which has two functions: to describe the objective and to describe implementation of the planned change. The change sheet also documents the work and its approval, and is used to identify the parts of the plant documentation affected by the change. A signature

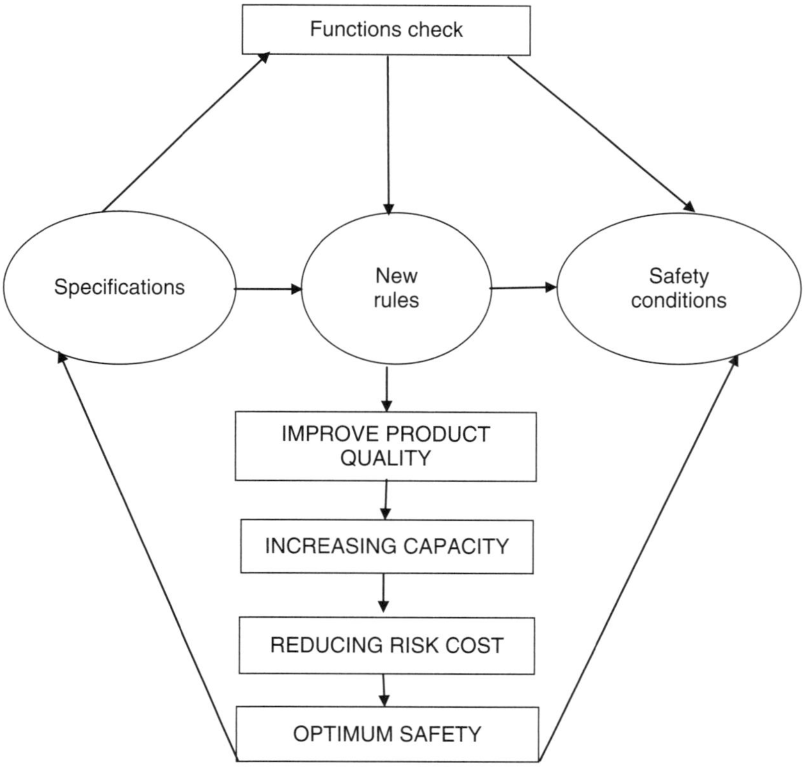

Figure 19-33 Optimization procedure.

is required for each approval and each inspection of each step in the process. The work permit complements this by listing actions to be taken during its execution.

The items contained in the change sheet are:

1. Identification of what is to be changed, including the nature of the change and other ordering criteria; suitable tools for this purpose are job cards, piping and instrumentation diagrams (P&IDs), and operating manuals.
2. Purpose of the change.
3. Description of the change.
4. Identification of essential danger spots, with the help of a checklist.
5. Discussion with specialist departments on the danger spots relevant to each.
6. Inquiry into necessary updating of safety analysis.
7. List of the plant and process documents affected, with the possibility of monitoring the progress of the change, testing steps after completion, an acceptance test, and a test run.

Important consultations with the specialist departments and other safety information must be recorded, and the safety analysis must be attached.

The systematic procedure with the associated aids contributes to plant safety from design to operation and process upgrades. By itself, however, it is not sufficient to guarantee plant safety. Although it cannot replace the knowledge of experienced workers, it offers a guide and support for daily work. The reasons for the increasing volume of technical documentation and administrative support are:

- Plants themselves have become far more complex, with process control and other advances.
- Maintenance work is being done by a larger number of specialized teams.
- There is increased public interest in readily understandable safety practices.

19.6.7 Hazard Impact Reduction

The responsible care of every plant operating company to take active measures to protect employees, third parties, and the environment against potential harm means that the operator must take responsibility for technical and administrative precautions to limit the impact of minor and major accidents in the facility. Legal and regulatory provisions independently require limitation of the consequences of accidents in plants.

In plants containing large amounts of products where the release of a substance would represent a high hazard potential, it may be useful to subdivide the plant with rapid-closing valves and fittings. These can be actuated after the detection of a release and after the scope of the release has been evaluated. They can be activated manually or automatically. The effect is to isolate the area of the leak from the rest of the facility. The quantity released at a leak can thus be limited to a valve estimated in advance. The use of a rapid-closing valve system presupposes detailed advance planning and design of the plant sections, knowledge of the amount of products contained, the way in which a leak is detected, and so on. The analysis may also include the emergency shutdown of the entire plant if dangerous conditions can be created by blocking individual sections.

The product in a plant or in an isolated portion of the plant may be drained or released into a retention system, catch tank, or blowdown tank permanently installed in the plant if a leak or other hazardous condition arises. The retention system may be fitted with a separator, a cyclone, if the properties of the substance and the system pressure demand separation of the phases of the substance. It is also useful to configure the system in such a way that the plant drains by gravity. A pressure relief system may also be needed to reduce the pressure in the gas phase.

Spray curtains are often employed in plants with critical gases or vapors, where a release at an unpredictable location cannot be ruled out and may lead to a hazard to employees, third parties, or the environment. The plant section in question, or the entire plant, is surrounded by a pipe system fitted with closely spaced spray nozzles. If a pollutant escapes, a high-pressure mixture of water and steam issues from the nozzles, surrounding the plant in a dense water mist that can condense or capture the released substances. Neutralizing agents can also be added to the water (e.g., ammonia for phosgene plants). Due to the high pressure and thermodynamic effects, the effectiveness of these systems can be maintained, even in windy weather. Mobile equipment such as articulated booms and water cannons operated by the plant's firefighting service can be used to create spray containment in smaller plants or as a supplementary measure in larger ones.

In what follows, plans established by a plant or works to control hazards inside the battery limits are referred to as plant and hazard control plans. An alarm and hazard control plan describes administrative and technical actions to minimize the impact of accidents, relates to various levels of organization, and is prepared with an eye to the immediate surroundings of each unit. The action prescribed includes:

- Management
- Personnel
- Technical resources

- Communications
- Procedures and jurisdictions

In facilities containing a large number of varied units, alarm and hazard control plans are prepared at the plant (smaller unit) and works (larger unit) levels.

19.6.7.1 Public Awareness The chemical industries worldwide *responsible care program* obliges companies to strive for steady improvements in health, safety, and environmental protection. It mandates that corporations seek a stronger dialogue with the public to enhance public knowledge of product and manufacturing safety. Many companies are bound by the following guiding principles:

1. To acknowledge the public interest in products and the activities of manufacturers, and to respect this interest.
2. To develop and manufacture only those products that can be produced, transported, and used safely.
3. In the planning of new products and production processes, to assign a high priority to health, safety, and environmental aspects.
4. In any chemical-related health and environmental activity, to provide full information to government agencies, employees, customers, and the public, and to recommended suitable protective practices.
5. To advise customers on the safe use and transportation of chemical products, as well as their safe disposal.
6. To operate production facilities so as to safeguard the environment as well as the health and safety of employees and the public.
7. To conduct research to increase knowledge about the possible health, safety, and environmental impacts of products, production processes, and waste products.
8. To cooperate with other companies in solving problems occasioned in the past by the handing and disposal of hazardous substances.
9. To cooperate with government and government agencies in devising responsible legislation, procedures, and standards that enhance the safety and protection of the public, workers, and the environment.
10. To promote the principles and practices of the responsible care program by sharing experiences with other companies manufacturing, using, transporting, and disposing of chemical products, and offering all types of support to them.

The six management practice codes deal with:

1. Information and emergency response
2. Process and plant safety
3. Environmental protection
4. Transportation safety
5. Occupational health and safety
6. Product responsibility

The operators of accident-relevant plants provide information on safety practices and proper accident response to persons who may be affected by an accident as well to the public at large. This information must be made available in suitable form, and without having to be solicited.

The objective is to create a bulletin, understandable to the general public, containing the following information:

- Name of plant operator and location of plant
- Name and title of the person giving information
- Brief description of the nature and purpose of the plant
- Identification of substances or preparations that can cause an accident, along with their essential hazardous attributes
- Nature of dangers in an accident, including possible impact on humans and the environment
- Nature of warning given to persons affected, and follow-up
- Correct behavior and actions of persons affected in the event of an accident
- Pertinent safety practices
- Internal and external hazard control plans
- Coordination of plans among plant operator, municipality, and agencies responsible for hazard control

In areas of high industrial concentration, a joint bulletin may be issued by the operators of plants where emergencies may occur.

19.7 SAFETY AND RELIABILITY ANALYSIS

A project process hazard analysis is one part of a comprehensive project safety program that will include pilot-plant safety reviews, preliminary hazard analysis, pre-process hazard analysis, implementation of management of change, confirmation of the "as built" status for piping and equipment, procurement and construction quality assurance, pre-startup safety reviews, and possibly a follow-up process hazard analysis before plant startup. The hazard identification method may be based on project timing, the study's purpose or intent, or even personnel availability.

19.7.1 Process Safety Information

Developing process safety information is regarded primarily as the engineering contractor's responsibility. Specific material safety datasheets may be the client's sole responsibility, due to the attendant liability from improper or inaccurate data. Other process safety information is, however, typically the design basis and product from the contractor. Most engineering firms have a standard data book which they offer on all projects. This should be reviewed as a standard contractual issue since the client must ultimately assume the responsibility for keeping process safety information current. The data book, following standards requirements, might include:

- Process technology
  - A block flow diagram or simplified process flow diagram
  - Process operation
    - Maximum intended inventory. In batch operations and for many vessels in continuous processes, this is the same as the maximum design capacity for a specific tank or vessel.
    - Safe upper and lower limits for items such as temperatures, pressures, flows, or compositions. Often, the limits may be the same as the design limitations on the equipment.
  - Evaluation of consequences of deviations from safe upper and lower limits, including those affecting employee safety and health.
- Equipment information pertaining to process equipment, especially materials of construction
  - Piping and instrumentation drawings
  - Electrical classification drawings
  - Relief system design and design basis
  - Ventilation system design
  - Design codes and standards used
  - Material and energy balances
  - Safety systems
- *Material and energy balances*: most commonly used to represent continuous process operations that are more readily modeled mathematically and where an energy balance is integral to process design. In certain batch operations where the client deems it inappropriate or where an energy balance has no impact on the process, the process safety management (PSM) requirements for a material and energy balance may be met sufficiently by preparing a material balance only. Documentation of such an agreement is critical to the contractor–client relationship.

19.7.2 Project Safety Information

Many operating companies rely on engineering/procurement/construction (EPC) contractors to provide needed services during expansions and other capital project ventures. These service providers are obligated to work under the same PSM guidelines as is the operating company. So what are the responsibilities for the operating company (client), contractor, and others under PSM?

EPC contractors and operating companies can develop and assign process safety responsibilities early in project planning. Under these conditions, all project parties involved are fully aware of their individual obligations. Fully disclosing individual party responsibility ensures dialogue between the client and the engineering procurement contractor and other involved service providers. Clearly defining and communicating project responsibilities may influence project execution and associated costs considerably. According to the checklist in Table 19-6, operating companies can include all PSM requirements in project planning.

Allocating responsibility may be considerably different when defining a relationship between other contractor types and the same client company. For clarification, major activities historically included in engineering, procurement, and construction are as follows:

1. *Engineering*: process design, conceptual/analytical design, production design, specifications, requisitions, and drawings
2. *Procurement*: inquiries, bid evaluations, purchase orders, expediting, inspection, and delivery to job site
3. *Construction*: temporary facilities, material receipt and erection of structures, equipment, piping, instruments, electrical, paint, and insulation

19.7.2.1 Project Planning Under Process Safety Management Project planning is complex but can be exacerbated if there are no clearly defining activities to be done under PSM requirements. Often it is assumed that PSM-related engineering activities are clear to an engineering procurement contractor because most activities are have been present historically in a sound design. Without a "road map," designers cannot be expected to provide a product that fits into the client's PSM program.

Procurement activities, along with construction activities, can be the linchpins between the engineering design intent and the final product as constructed. For procurement, process safety management–related activities may include:

- Assuring the suitability of equipment suppliers and subassembly equipment suppliers.

TABLE 19-6 Process Safety Management Checklist[a]

Description of PSM Compliance Issue for Capital Projects	PSM Requirement	Project Requirement	EPC Supply	Client Supply
	General Issues			
1. Written PSM over entire project plan		Y	J	J
2. Assign PSM coordinators for project		Y	Y	Y
3. Split of PSM responsibilities		Y	J	J
4. List of chemicals covered		Y	Y	
5. Description of boundaries of process covered		Y	Y	Y
6. Provide PSM training to appropriate project personnel		Y	J	J
7. Prequalification of vendors		Y	J	J
8. Prequalification of contractors	Y	Y	Y	J
9. Schedule periodic project PSM reviews		Y	J	J
	Employee Participation			
1. Written plan for implementation	Y	O	O	
2. Documentation of employee involvement in PSM program development and PHA	Y	N		Y
3. Other				
	Process Safety Information			
1. Material datasheets	Y	Y		Y
2. Process safety manuals to comply with standards	Y	Y	Y	
3. As-built P&IDs, PFDs, mass and energy balances	Y	Y	Y	
4. Equipment data files as-built interlock diagrams, electrical classification drawings, relief system design basis, other information				
5. Materials of construction	Y	Y	Y	
6. Documentation that equipment is in accordance with design specifications and that design complies with generally accepted good engineering practice	Y	Y	Y	
	Process Hazard Analysis			
1. PHA methodology acceptable to hazard standard	Y	Y	J	J
2. Performance and tracking of PHA to ensure that entire process covered has been studied and that recommendations have been approved, incorporating design	Y	Y	Y	A
3. Documentation that appropriate team attended the PHA	Y	Y	Y	
4. Documentation of PHAs retained for project/process history	Y	Y	Y	

(*continued*)

TABLE 19-6 (*Continued*)

Description of PSM Compliance Issue for Capital Projects	PSM Requirement	Project Requirement	EPC Supply	Client Supply
5. Approvals of team leader and team members		Y		Y
6. Documentation that engineering and controls have been considered as well as human factors, facility siting, and health effects	Y	Y	Y	
7. Final report to be suitable for company archieves		Y	Y	
	Operating			
1. Written procedures for process operating modes	Y	O	O	Y
2. Written procedures for all plant operating modes		Y	Y	
3. Safe work practices for control of hazards during nonroutine work activities	Y	Y	O	Y
	Training			
1. Training in operating procedures provided to all employees involved in operating the process	Y	Y	O	Y
2. Documentation of training for all employees involved in operating the process	Y	Y	Y	
	Operating			
1. Evaluation of contractor's safety programs	Y	Y	Y	
2. Company review of contractor's safety programs and performance	Y			Y
3. Generic contractor safety manual listing safe work practices to be followed while performing work on process covered		Y	J	J
4. Information related to surrounding processes if any safety hazards that might be encountered from them and emergency actions may be required	Y	Y		Y
5. Information related to the hazard of the process	Y	Y	J	Y
6. Control of entrance and exit of contractors for process covered	Y	Y	J	J
7. Emergency action plan training provided to each contractor	Y	Y	J	J
8. Documentation of training of contractor employees on hazards and emergency information	Y	Y	Y	
9. Documentation of actions when contractor employee safety infractions are encountered		Y	Y	

TABLE 19-6 ***(Continued)***

Description of PSM Compliance Issue for Capital Projects	PSM Requirement	Project Requirement	EPC Supply	Client Supply
10. Injury and illness logs for contractor employees	Y	Y	Y	
11. Process safety topics included in toolbox safety training		Y	J	J
12. Procedure to advise client of any unique hazards identified by the contractor		Y	Y	Y
	Pre-startup Safety Review			
1. PSSR procedures	Y	Y	O	Y
2. PSSR forms and checklists in place	Y	O	Y	
3. Documentation of PSSR and authorized to startup	Y	O	Y	
	Mechanical Integrity			
1. Written procedures to maintain the ongoing integrity of process equipment	Y	Y	O	Y
2. Training of maintenance personnal appropriate to meet the standard site safety emergency plans and safe work practices	Y	Y	Y	
3. Determination of appropriate frequencies of inspections and tests for process equipment covered	Y	Y	Y	
4. Documentation of initial inspections and tests and correction of equipment deficiencies prior to startup	Y	Y	Y	
5. Quality assurance procedures for new equipment and for maintenance materials and spare parts	Y	Y	Y	
6. Documentation and certification of training of contractor employees on appropiate craft skills	Y	Y	Y	
	Hot Work Permit			
1. Hot work permit procedures and permit forms must be in place		Y	Y	Y
	Management of Change			
1. Written procedures to MOC, including authorization, tracking, and documentation of final resolution	Y	Y	J	J
2. Documentation of training of employees on changes and their impact; updating of process safety information and operating procedures				

(continued)

TABLE 19-6 ***(Continued)***

Description of PSM Compliance Issue for Capital Projects	PSM Requirement	Project Requirement	EPC Supply	Client Supply
	Incident			
1. Procedures for performing incident investigations and other requirements of PSM standard	Y	Y	Y	Y
2. Assignment of incident investigation team member for any incident involving contractor employees	Y	Y	Y	
3. Documentation of initial inspections and tests and correction of identified equipment deficiencies prior to startup	Y	Y	Y	
	Emergency Planning and Response			
1. Emergency action plans in place	Y	Y	O	Y
2. Appropriate emergency actions and information regarding site safety issues transmitted to the contractor	Y	Y		Y
3. Documentation of training of contractor employees on safety issues and EAP	Y	Y	Y	
4. Additional OSHA training for contractor personnel as required, particularly under PSM standard	Y			Y
	Compliance Audits			
1. Compliance audits procedure in place	Y			Y
2. Certification that the facility has been evaluated for compliance with safety standard	Y			Y
	Trade Secrets			
1. Written policy for dealing with trade secrets	Y	Y	Y	
2. Completed confidentiality agreements		Y	J	J

[a]Y, yes; N, no; J, joint responsibility; O, optional; A, approval.

- Preparing a preapproved contractors' or subcontractors' list.
- Assuring that purchase requisitions address common safety-related issues before being released for fulfillment.
- Bringing appropriate process safety personnel to vendor review meetings before releasing final purchase orders.
- Performing shop inspections to conform that appropriate codes, standards, and accepted engineering practices are followed.
- Expediting suppliers to avoid out-of-sequence deliveries, thus avoiding out-of-sequence construction activities.
- Validating that vendors supply necessary standard operating and maintenance procedures for equipment.

- Assimilating and distributing vendor drawings, prints, and other pertinent information so that equipment data files may be comprehensive.

Process safety management–related construction activities include:

- Assisting in establishing and applying applicable codes, standards, and good engineering practices in construction efforts.
- Assuring that construction meets new equipment design (through field inspections, P&ID checks, punch lists, pre-startup safety review support, etc.).
- Documenting assurance of construction materials of new equipment, through confirmation of materials received, field checks versus P&IDs, positive material identification or other, and so on.
- Assisting in documenting that the new equipment adheres to design specifications and manufacturers' installation instructions by confirming that design specifications and manufacturers' instructions are implemented properly during a pre-startup safety review.
- Documenting any tests to assure that the equipment meets design intent and is properly installed (through hydro test sheets, electrical checks, instrument check-out lists, calibration sheets, etc.).
- Following all applicable client safe work practices during construction activities, especially when working near operating equipment of an existing covered process.
- Assisting in assuring that subcontractors are also following appropriate safe work practices by using safety analysis forms, job safety analysis forms, and so on.
- Assisting engineering and the client in performing a pre-startup safety review before initial startup of newly constructed equipment or facilities.
- Enhancing information transfer from construction to the client's maintenance department. This ensures that the documented equipment history begins with fabrication and includes inspection and construction information.

19.7.2.2 Process Safety Management Elements For major capital projects, contractors and clients must develop a process safety management plan. For example, the checklist in Table 19-6 can be used as a guide to start such a plan. All 14 elements are critical issues when planning a new or retrofit project. During the conceptual design stages, consider these key issues to maintain process safety management compliance during the project. The plan should clearly define the scope of activities under the plan and boundaries of the process covered.

19.7.3 Design and Control Safety

In process design and control safety, the system topology or component interconnections are defined by PFD and P&IDs as well as maps of the surrounding area (plant, people, computing infrastructure, alarm systems, etc.).

1. *Attributes*. Attributes include pressure, temperature, flow, and supply.

2. *Systematic state*. The criterion for state variables definition is that the system operate as close as possible to maximum product yields. The state variables are described in terms of low, medium, or high; present or absent; and so on. Equipment state, pipes, valves, and instrument control states are defined in terms of failed, blocked, leaked, open, closed, on, off, and so on. The environmental state can be defined as lighting, storm, flooding, and so on. The faults state can be identified in terms of blockage, leakage, misoperation, malfunction, and so on.

3. *Symptoms*. Symptoms are aids in diagnostic analysis. Possible hazards include explosion, fire, and flooding.

4. *Scenarios*. Scenarios are used to set the initial status of the system variables and attributes. In the symptom decomposition phase, the relational symptom/scenario matrix decomposes by using a projection operation to produce elementary relations.

5. *Diagnostic construction*. A systematic cause–event analysis give results which are summarized in the form of a fault tree. How the reliability of a construction element influences technological systems is described as a function of the system structure.

6. *Consequences analysis*. Formulating the cause–effect relationship of a system is performed taking into account all relevant antecedents and consequences (e.g., previous process hazard analysis, previous incident reports, off-site consequence analysis methodology such as look-up tables).

19.7.4 Operating Procedures

In general, operating procedures vary in form and content for individual operating companies. Unfortunately, many operating companies lack the resources to develop procedures for major engineering procurement contractor process projects. Most large engineering firms have prepared in-house policies for developing process safety information manuals, which may be used in writing operating manuals.

Despite differences in form and content, written operating procedures must address several issues as required by process safety management, including:

1. *Steps for each operation*: initial startup; normal operations; temporary operations; emergency shutdown, including conditions requiring assignment of shutdown responsibility; emergency operations; normal shutdown; startup following a turnaround or emergency shutdown.

2. *Operating limits*: consequences of deviation; steps to correct or avoid deviation.

3. *Safety and health considerations*: properties and hazards presented by process chemicals; precautions needed to prevent exposure, such as engineering controls, administrative controls, and personal protective equipment; control measures taken if physical contact or airborne exposure occurs; raw material quality control and control of hazardous chemical inventory levels; any special and unique hazards.

4. *Safety systems and their functions*: The operating procedures must be current, reflecting actual operating practices and up-to-date information on process chemicals, technology, and equipment. To fulfill this responsibility, the owner may have the contractor prepare and document the "as-built" status at project completion. Under these circumstances, all parties must understand the status of documents when the client takes custody.

Development and implementation of safe work practices are also required. They provide a means of controlling hazards during operations such as lockout/tagout, confined space entry, opening process equipment or piping, and entrance into a facility by maintenance, contractor, laboratory, or other support personnel. Safe work practices will apply equally to client and contractor employees, or others (e.g., subcontractors, vendors) who may provide services and have contact with process operations.

19.7.5 Training

Client companies usually provide training in-house. For large companies, the engineering/procurement/construction contractor will usually supply information to the trainers. For smaller client companies or for new personnel, this task may fall on the engineering/procurement/construction contractor.

In addition to operating procedures, training considerations may include safe work practices, emergency action plans, process overviews, and other topics in addition to basic job skills. Recent trends include annual job skill certification for trades being certified by local area contractor safety councils. Another trend is that client companies provide all safe work practices, emergency action plans, and process overview training directly to contractor personnel. By providing the training, the operating company can assure and document that the proper training has been performed.

When the engineering/procurement/construction contractor is providing or supporting training efforts, the presentation will typically represent the initial process safety management–required training. All employees who may be involved in operating the process must receive training. Emphasis should be placed on specific safety and health hazards emergency operations, including shutdown and safe work practices, applicable to the job tasks. Documentation of the training should also be kept.

19.7.5.1 Contractor Programs Process safety management contractor safety requirements are categorized as either the contractor's or the employer's (client's) responsibility. Interestingly, both parties must share in the effort. If a contractor does not meet the obligations, the client company does not have a safe workplace. Conversely, if the workplace hazards are not clearly identified and explained to the contractor's worker's, the client has an unsafe workplace. Cooperation must exist in developing emergency action plans and emergency response and responder training. If the contractor is involved in responding to an emergency, he or they should become part of the client's response team.

19.7.5.2 Pre-startup Safety Review This is an important issue for the contractor and client to resolve. Follow mechanical completion and precommissioning activities, and the responsibility split will depend on contractual obligations. For example, if the client company will do the pre-commissioning and startup, the engineering/procurement/construction contractor assumes unwarranted liabilities by merely lubricating rotating equipment. Conversely, if the engineering/procurement/construction contractor will start the plant, a client employee who modifies a control station or distributed control system function can nullify process guarantees. These issues must be addressed contractually if the relationships exacerbate communication and may cause failures in management systems and process safeguards.

Client companies and contractors can use checklists to ascertain and document equipment status prior to startup. If such a checklist is used in the custody transfer of completed facilities from the contractor to the client, it should be formalized in the project's contractual basis.

19.7.5.3 Mechanical Integrity The specialized nature of mechanical testing and inspection procedures requires up-to-date certification on state-of-the-art techniques, which most operating companies use infrequently. Some engineering/procurement/construction contractors have in-house expertise for baseline data development, but others may rely on equipment vendors and subcontractors. Consequently, responsibility delineation for the mechanical integrity baseline data gathering and development becomes more important. For example, who certifies field welds on pressure vessels or bench-test results for safety relief valves? For safety-critical electronic control systems, does the factory acceptance test suffice, or is a witness's test in the control room before plant startup needed? Close corporation between quality assurance personnel of the operating company and the engineering/procurement/construction contractor and subcontractors is crucial.

For older plants, where an engineering/procurement/construction expansion project must tie in with existing equipment, correction of equipment deficiencies is vital. The correction of deficiencies must meet requirements, and for existing equipment designed and constructed in accordance with codes, standards, or practices that are no longer in general use, the employer must determine and document that the equipment is designed, maintained, inspected, tested, and operated in a safe manner.

Issues that arise in this instance include understanding the original (out-of-data) codes and standards, knowing whether equipment revisions or upgrades meet the latest codes and standards, and verifying that the contractor is qualified to make any appropriate modifications. To avoid problems, clients should ask contractors to define and document needed qualification, as well as work closely with the contractor to ensure that administrative controls used in prior safe operations are fully understood and incorporated.

A final issue for the mechanical integrity element is developing written procedures to maintain equipment integrity. Procurement activities may include assuring that vendors supply necessary equipment and standard operating and maintenance procedures. Whenever the equipment is particularly unique in form or function or is new to the plant site, developing written procedures and proper personnel training take on added significance.

19.7.5.4 Incident Investigation State regulations require that contractors be included on incident investigation teams when incidents involve contractor employees. An engineering/procurement/construction contractor, when on site, should assign a site process safety management coordinator who is an incident investigation team member. This allows the client to train him or her as a team member and expedite investigations that involve contractor employees. One chemical company has a stated policy that any contractor who does not agree to be part of an incident investigation team when asked cannot enter the plant. This signifies the importance assigned to incident investigation by many companies.

19.7.5.5 Emergency Planning and Response Emergency action plans (including training of contractor employees) on alarms and notification procedures are the heart of the emergency planning and response element. Typically, client personnel are trained, and contractor personnel evacuate the premises during any major incident. At a large construction site, contractor personnel could be used as responders, but training and training certification often are cost prohibitive or ineffectual.

19.7.5.6 Compliance Audits The process safety management compliance audit, which is required at least every three years, is typically the client company's sole responsibility. However, contractor personnel are interviewed in a comprehensive site audit. Training contractor personnel on PSM issues, job skills, and safe work practices are issues, along with documentation of the training received.

19.7.6 Process Hazard Analysis Revalidation

Hazard identification methods can be used in different ways to model part of the incident scenario leading to a possible accident. So an effective model of the incident scenario is the key to a successful application. It follows the structure of a generic fault tree up to the release of materials, and of an event tree from this point to the impact of the release on people, the plant, and the environment. These techniques can be applied at any stage in the life cycle of a process. Such techniques are often classified as "pre-HAZOP" studies. Under process safety management, standard process hazard analysis must be updated and revalidated at least every five years.

Because safety standards only give guidance, hydrocarbon processing industry companies can select a revalidation method to use, and have several options:

- Redo the process hazard analysis as if it were the initial one.
- Retrofit, update, and revalidate the process hazard analysis by concentrating on deficiencies in the original analysis and incorporating process changes.
- Update and revalidate the original process hazard analysis for process changes only.

Using guidelines and the evaluation flow diagram in Fig. 19-34, companies can cost-effectively decide how much rework is needed to maintain process safety management compliance.

19.7.6.1 Regulatory Requirements The primary requirements to revalidate PHAs are described in safety standards. This topic explicitly addresses team composition, analysis frequency, and document retention. There are also implied requirements for analysis issues and revalidation report preparation. Specifically, the revalidation team must have the required engineering expertise and process operations knowledge. The revalidation must be performed within five years of the initial or previous process hazard analysis, and the revalidation results must be retained for the process's lifetime.

Safety standards require that PHAs be revalidated at least every five years. However, companies may decide that more frequent revalidations are more cost-effective, consistent with loss prevention goals or are necessary due to external factors such as the U.S. Environmental Protection Agency's (EPA's) requirements as to community relations concerns. For example, if a major process equipment/configuration revision is in progress, it may be more economical to revalidate the unaffected process portion while performing the modification process hazard analysis. In addition, some companies have established frequencies for revalidating process hazard analysis based on risk categorization (e.g., high, medium, low). For high-risk processes, revalidation can be more frequent than every five years. This approach, which uses variable revalidation frequencies based on perceived process risk, is consistent with American Petroleum Industry recommended practice. Although the maximum allowable interval between revalidations is five years, other factors may influence the frequency.

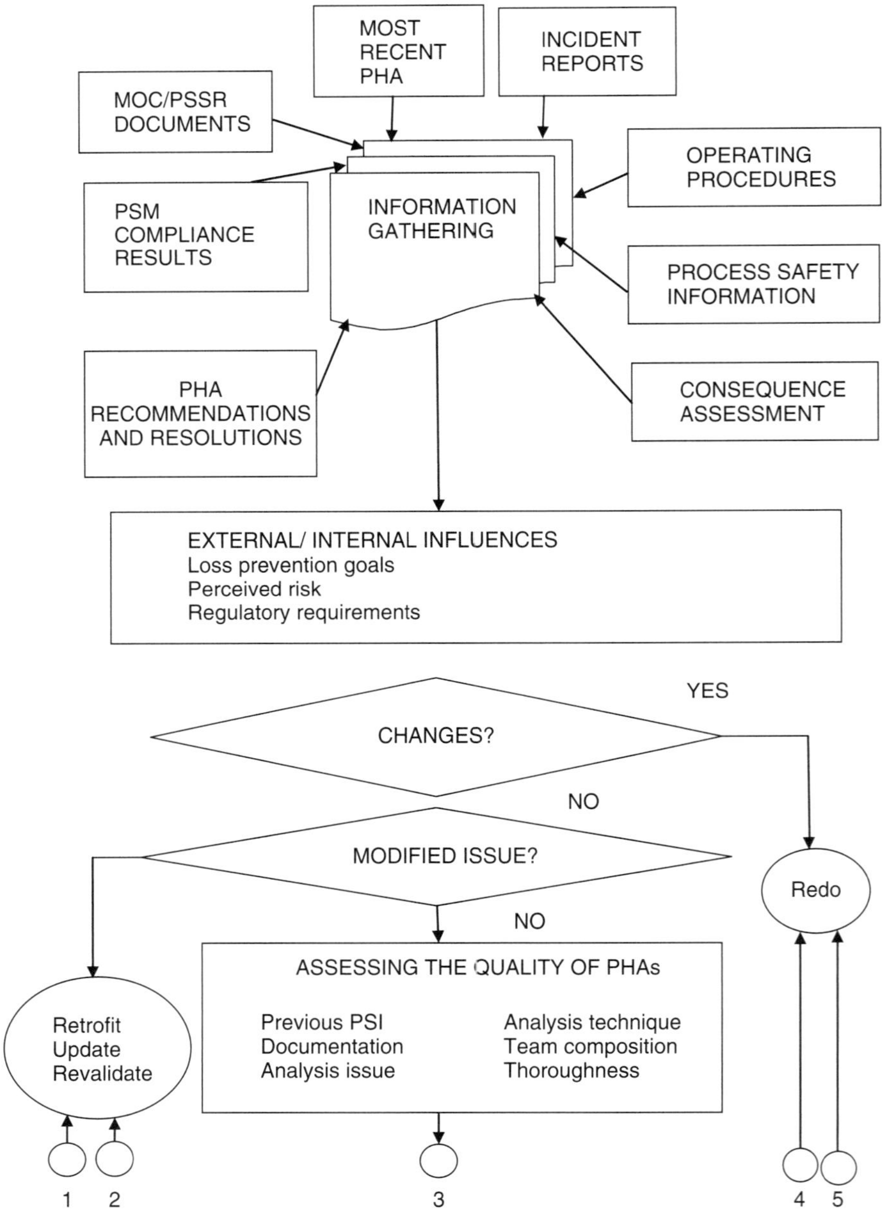

Figure 19-34 Process hazard analysis revalidation flowchart.

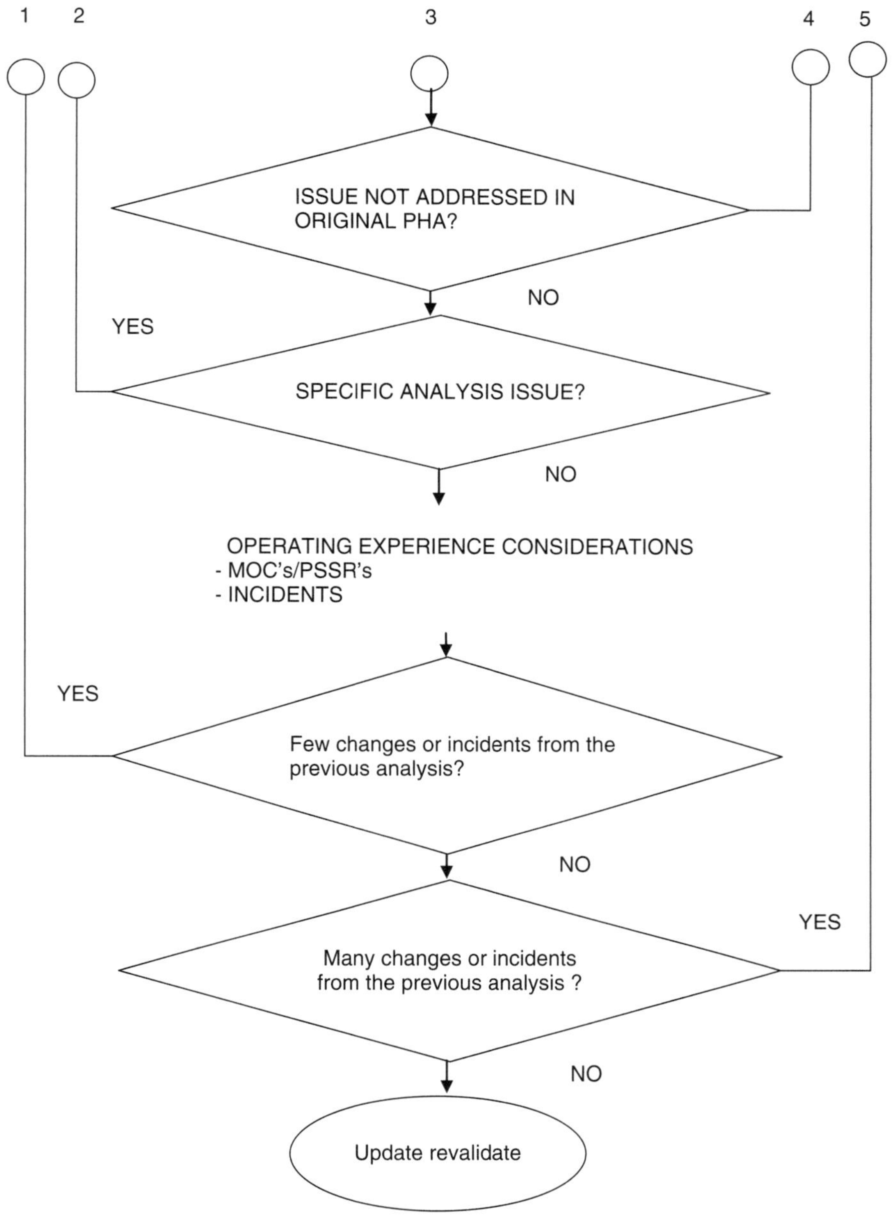

Figure 19-34 (*Continued*)

The revalidation approach is dependent on many factors, including the initial/previous PHA quality, goals and objectives for the revalidation, and process operating history. Therefore, a range of possible approaches may be appropriate to satisfy standard requirements. The corresponding process hazard revalidation terminology to be used is:

- *Redo*. Perform a process hazard analysis as if it were the initial process hazard analysis, using one or more of the techniques listed in the process safety management standard.
- *Retrofit, update, and revalidate*. Address specific analysis deficiencies (e.g., facility siting, human factors) and incorporate results into previous process hazard analysis, modify the initial PHA to address appropriately changes that have been made in the process, and verify that the initial PHA still accurately addresses process hazards and that adequate controls exist to manage them.

Figure 19-34 depicts a logical process to determine an appropriate PHA revalidation approach. Whereas actual activities done as part of a process hazard analysis revalidation can vary, the results desired are uniform and must ensure that the process hazard analysis is consistent with the current process, identifies the known process hazards, and confirms that existing controls can manage the hazards adequately. To achieve these results, three fundamental steps should be a part of every revalidation effort: collecting supporting documentation, determining issues to address during the revalidation, and conducting and documenting the revalidation analysis.

19.7.6.2 Supporting Documentation Before revalidating a PHA, certain support information should be collected, which may include:

1. Process safety information
2. Most recent process hazard analysis report
3. Operating procedures
4. Management of change (MOC) and pre-startup safety review (PSSR) documents
5. Previous incident reports
6. Resolution and status of recommendations from the previous process hazard analysis
7. Consequence assessments (facility siting studies)
8. Process safety mangement compliance audit results

This information can influence the selection of the revalidation approach. For example, process safety information can be used to identify new hazards or discover process modifications that inadvertently bypassed the management procedure. The most recent process hazard analysis report provides information about the quality of the previous process hazard analysis.

Operating procedures could be analyzed or referenced to identify nonroutine operating modes that should be addressed. Management of change and pre-startup safety review documents can provide valuable information about known modifications and controls that manage any associated hazards. Previous incident reports may identify new hazards or specific practices that warrant further evaluation. Process hazard analysis recommendation resolutions can help identify process and/or facility changes, confirm new control installation, and help ensure that hazardous accident scenarios identified previously have been adequately addressed. Consequence assessments performed for facility siting concerns or EPA requirements can help alert the team to how serious or minor the release effects could be. Process safety management compliance audit results can identify specific process hazard analysis deficiencies, indicate equipment and instrumentation reliability levels (the maturity of the mechanical integrity program), and substantiate confidence in related process safety management practices (emergency response, training).

Most of this information will be needed and used before PHA revalidation meetings to determine an appropriate, cost-effective revalidation approach and prepare properly for the meetings. Regardless of the approach selected, three fundamental factors should be considered when identifying process hazard analysis revalidation issues: internal and/or external influences, initial/previous process hazard analysis quality, and operating experience (Table 19-6).

19.7.6.3 Internal and/or External Influences Although a company may have performed and documented a previous process hazard analysis comprehensively, other factors may suggest redoing it. These factors may include new regulatory requirements or new interpretations of existing requirements, major process modifications occurring (or planned) since the previous analysis, and a full process hazard analysis not being done for an extended period. New regulatory requirements could be imposed by state or federal agencies (e.g., the EPA) or could be in response to settlement agreements with regulatory or local community/citizen groups. When major process modifications are planned, a company may choose to redo the process hazard analysis because it may be more cost-effective or necessary to adequately address potential upstream/downstream accident scenarios associated with modifications.

Furthermore, a company may decide to redo (as opposed to revalidate) a process hazard analysis due to concerns about:

- Possible hazardous interactions of multiple modifications
- Subtle changes that may not have been analyzed during the MOC process
- The practicality of revalidating future analysis due to the number of documents (MOC forms, PSSR documents) to be collected and retained.

In some cases, retrofitting a process hazard analysis can address specific issues, or a separate, less intensive study can be performed to address a specific concern (offsite effects for EPA).

19.7.6.4 Initial/Previous PHA Quality The process safety management standard does not specifically state that a process hazard analysis revalidation must address each initial process hazard analysis requirement. However, a reasonable interpretation is that safety standard would expect any recognized deficiencies in the initial process hazard analysis to be addressed, and certainly not later than the first revalidation cycle. Therefore, a prudent approach would verify that specific process hazard analysis issues listed in the process safety management standard have been addressed adequately. Although a process safety management compliance audit may not have identified specific deficiencies in a company's process hazard analysis program, an individual process hazard analysis could have specific deficiencies because it was not reviewed as part of the compliance audit and/or the process hazard analysis program changed over time and the current technique is more rigorous than the previous techniques.

Several quality issues should be considered when selecting a revalidation methodology. These issues are consistent with process hazard analysis requirements of process safety management regulation and can be used to do a quality self-evaluation of an initial/previous PHA. Assessing the quality of the initial/previous process hazard analysis, although not an actual part of the revalidation analysis, plays an integral part to determining the revalidation approach. It establishes the scope of work, which is important for establishing resource needs such as personnel and estimation of meeting time. Table 19-6 summarizes the issues that should be evaluated in a process hazard analysis quality assessment [13,17].

1. *Was the process safety information for the previous PHA adequate to assess the hazards?* Accurate and complete process safety information is essential to enable a thorough process hazard analysis. Incorrect or missing process safety information could result in the process hazard analysis team's inadvertently: (a) not identifying hazards (not being aware of a piping tie-in or chlorine use at the cooling tower); (b) acknowledging the adequacy of safeguards (administrative or engineering controls) that are not effective (e.g., that have been removed or disabled). Although experienced personnel on a process hazard analysis team may compensate for some process safety information shortcomings, a company should consider whether it is likely that known hazards were overlooked due to inaccurate or incomplete process safety information. Also, a company should consider how PHA recommendations were addressed. For example, if relief valve design basis information was developed in response to a process hazard analysis recommendation, was the design basis information assessed to ensure that identified hazards were addressed (e.g., are the design basis and relief valve set pressure adequate for scenarios identified in the process hazard analysis?). Redoing a process hazard analysis may the best approach to meeting a standard's revalidation requirement when a company believes that significant process hazards may have been overlooked or underestimated due to inaccurate or incomplete process safety information.

2. *Were the initial/previous PHA results documented adequately*? In addition, PHA revalidation teams can significantly benefit from comprehensive process hazard analysis documentation results. Consider the case where process hazard analysis results (i.e., deviations or questions evaluated and the team's judgment of the consequences) were documented only when the team recommended additional controls. This "documentation by exception" may not provide the revalidation team with sufficient evidence that many deviations (potential accident scenarios) were discussed. This makes completeness verification of the initial/previous process hazard analysis difficult at best and may not provide adequate information to properly evaluate future modifications.

Lack of documentation could compound itself, depending on the revalidation approach. Over time, people's memory of process hazard analysis discussions will fade or become distorted, making it nearly impossible to confirm exactly what potential accident scenarios were considered by process hazard analysis revalidation teams. The first process hazard analysis revalidation cycle provides an excellent opportunity to upgrade documentation shortcomings and provide a more comprehensive basis for future revalidations and management operation changes reviews [25]. Process hazard analysis is difficult at best and may not provide adequate information to properly evaluate future modifications.

Lack of documentation could compound itself, depending on the revalidation approach. Over time, people's memory of process hazard analysis discussions will fade or become distorted, making it nearly impossible to confirm exactly what potential accident scenarios were considered by process hazard analysis revalidation teams. The first process hazard analysis revalidation cycle provides an excellent opportunity to upgrade documentation shortcomings and provide a more comprehensive basis for future revalidations and management operation change reviews.

3. *Were all PHA requirements addressed adequately?* The process hazard analysis revalidation team should have confirmation that all process hazard analysis requirements of standard process safety management regulation were addressed adequately. This may be accomplished by evaluating the previous process hazard analysis documentation, reviewing any pertinent process hazard management compliance audit results, and/or interviewing previous process hazard analysis team members. Table 19-7 is a process hazard analysis checklist.

If one or more specific analysis requirements were not addressed adequately (e.g., process hazard, human factors, previous incidents, facility sitting), a company may be able to retrofit the process hazard analysis for the deficient analysis issues and update and revalidate the remainder. In some cases, deficient analysis issues may result in redoing the process hazard analysis because of significant quality concerns or to ensure overall cost effectiveness. For example, if significant process hazards were not addressed adequately, efficient retrofitting is probably impossible and redoing the process hazard analysis is more appropriate. In addition, deficiencies in multiple analysis issues may warrant redoing the process hazard analysis to ensure quality as well as cost-effectiveness.

TABLE 19-7 Evaluating the Process Hazard Analysis Checklist

Issue	Yes	No
1. Was the process safety information for the previous PHA adequate to hazards?	×	
2. Were the initial/previous PHA results documented adequately?	×	
3. Were all PHA requirements addressed adequately?	×	
4. Was the analysis technique appropriate?	×	
5. Was the team composition adequate?	×	
6. Were nonroutine operations addressed adequately?		×

4. *Was the analysis technique appropriate?* In the initial process hazard analysis, the analysis technique could have been inappropriate for the process's complexity or misapplied, resulting in overlooked hazards. In either case, the company should take reasonable measures to verify that all known hazards were addressed. It may be possible to retrofit the PHA to address specific known hazards or other specific analysis issues that are missing, depending on the deficiencies.

5. *Was the team composition adequate?* Analysis team composition is a critical part. The team needs personnel with knowledge and experience specific to the process being analyzed to:

- Identify hazards that are specific to the unit
- Provide insight on how the unit responds to upset conditions
- Relate actual operating, maintenance, and training practices
- Determine reasonable worst-case effects from upsets
- Judge the adequacy of existing safeguards

The process hazard analysis team may not identify unique hazards or assess appropriately potential causes, consequences, or safeguards of effectiveness in preventing or mitigating postulated accident scenarios. In addition, a person knowledgeable in the process hazard analysis technique used is required. Many companies require that certified process hazard analysis leaders facilitate the analyses.

6. *Were nonroutine operations addressed adequately?* A corresponding standard has issued citations for not addressing nonroutine operating modes in a process hazard analysis. For example, a citation was issued for not performing a hazard evaluation for a fired-furnace startup. Unfortunately, many initial process hazard analyses of continuous processes focus only on normal operations and may not address nonroutine, critical operating modes such as startup, shutdown, and emergency operations shutdown. Batch process process hazard analysis may neglect operations such as vessel cleanup or pressure testing that are considered normal operating procedures. To address this issue, many companies carry out a hazard evaluation of selected operating and/or maintenance procedures. A step-by-step hazard evaluation of nonroutine critical procedures can find such hazards as skipping steps or performing steps incorrectly. This analysis can identify reasons (root causes) for errors, potential consequences, and existing engineered and administrative safeguards against the errors or outcomes. From the analysis, recommendations for additional hardware or administrative safeguards to protect against critical errors can be developed. The analysis of nonroutine critical procedures can also identify procedure content and format improvements (highlighting critical steps) and enhance operator training by finding critical items that may require refresher training. Even if a step-by-step hazard evaluation of nonroutine critical operating procedures is not done, a company should still confirm that the associated hazards have been addressed.

19.7.6.5 Operating Experience Another important factor to consider is the process operating history. These factors include:

1. Operating years of process unit
2. Maturity of the MOC program
3. Number and/or significance of MOCs and PSSRs since the previous PHA
4. Number and/or significance of incidents or near misses that have occurred since the previous PHA, and how these changes may have affected the initial PHA.

1. *Was the initial PHA performed before the process unit was actually operating*? Standards require that a process hazard analysis be performed for any new process built. If an initial process hazard analysis was done before the unit became operational, the team could have: inadvertently overlooked hazards (it was based on design and not actual operation) or postulated hazards that did not materialize. In either case, the initial process hazard analysis results will need to be reviewed and revised to represent accurately process hazards and engineering and administrative controls that are in place. For this situation, redoing the process hazard analysis may be the must appropriate approach.

2. *When was an effective MOC program implemented?* Because MOC frequently requires a cultural change, many companies struggle setting up a formal and effective MOC program. Consequently, many process modifications may not have received a formal MOC review. This is particularly true for processes that had a process hazard analysis performed prior to the PSM standard (a MOC program was not required) or within the first year. To help ensure that process changes are not overlooked during revalidation, a company could:

- Compare P&IDs used for previous process hazard analysis to new P&IDs to identify modifications that bypassed the MOC process.
- Review and compare documentation from the previous process hazard analysis to current P&IDs.
- Redo the analysis.

In any event, a company should consider how they will try to identify and incorporate any modifications than escaped the MOC program.

Additional factors to be considered are facility, personnel, and procedural changes (particularly frequencies of performing tasks). All are related to MOC; however, many MOC programs do not specifically include these areas. For example, a process may have new buildings nearby that were not considered in the previous analysis. Also, the number of people in the process buildings could have changed. In both instances, a process hazard analysis team may decide that additional controls should be in place, because the consequence of control failure could be more severe than estimated originally. Companies should try to identify these subtle changes as part of the process hazard analysis revalidation efforts and ensure that process hazards are assessed and controlled appropriately.

3. *Have there been many or significant MOCs and/or PSSRs since the last PHA?* Process units may be modified over time for many reasons (technology improvements, optimization, etc.). MOC and PSSR are intended to maintain the integrity of original safety features designed into a process and ensure that new hazards are managed properly. However, potentially hazardous interactions may have been introduced by a process change. The likelihood that hazards exist due to unidentified interactions increases with the number of process modifications. Consequently, a company may consider redoing a process hazard analysis after a specified number of changes rather than waiting a specified time interval. This approach could results in redoing a process hazard analysis more frequently than regulatory revalidation requirements necessitate or redoing a process hazard analysis after a specified number of process hazard analysis revalidation have been done. The frequency of redoing a process hazard analysis could then be consistent with process modification frequency.

For example, a company has a large number of MOCs/PSSRs since the previous process hazard analysis and intends to update the process hazard analysis documentation (HAZOP tables) to reflect the current process. It may be more cost-effective to redo the process hazard analysis than to update and revalidate documentation to incorporate each MOC/PSSR. Reviewing all MOC and PSSR documents ensures that hazards were addressed adequately and may require more time than redoing the process hazard analysis. If a company intends to continue revalidation (not redoing the process hazard analysis) for an indefinite period, document retention and revalidation logistics (determining what was reviewed, that it is still valid, etc.) could become unmanageable after several revalidations. Therefore, updating process hazard analysis documentation at some interval to incorporate all applicable MOCs and PSSRs is more effective.

4. *Have there been many or individual major incidents since the previous PHA?* Numerous or major incidents can be an indicator of one or more weaknesses in a process safety management program. For example, although the process hazard analysis team may have identified the potential for an incident, the team may have judged the safeguards as adequate (alarms, training). On the other hand, the process hazard analysis team may not have identified the potential for an incident. Therefore, a company may decide that process areas that have experienced several or major incidents should receive special attention from the revalidation team. This may include:

- Step-by-step evaluations of relevant critical procedures
- Recommending that specific procedures be included in the refresher training program and an in-depth review of related training materials and practices be performed
- Redoing specific portions (sections of the process) of the previous process hazard analysis

In any case, the process hazard analysis team should always review incidents that have occurred since the previous process hazard analysis as well as the corrective actions recommended related to the incidents.

19.7.6.6 Conducting and Documenting the Revalidation Analysis After determining the revalidation scope, a company should identify and/or develop the necessary tools to conduct revalidation meetings and define desired revalidation results documentation.

1. *Redo*. By evaluating the issues listed in Fig. 19-34, a company may decide that redoing a previous process hazard analysis is the best choice. If so, a company should establish guidelines for conducting a process hazard analysis. Many

references are available on hazard evaluation techniques; therefore, documentation associated with redoing a process hazard analysis is not addressed here. However, companies should be cognizant of any documentation or analysis shortcomings in the original process hazard analysis and the end-user needs for revalidation results (reliability group, modelers of scenarios for EPA required hazards assessments).

2. *Retrofit, update, and revalidate*. Using Fig. 19-34, a company may determine that a retrofit, update, and revalidate approach is the most appropriate. A company could evaluate the previous process hazard analysis using a checklist to identify deficient or missing analysis issues (e.g., human factors, facility siting, range of effects on employees) that need attention during revalidation. The completed checklist, documentation, incident reports, current P&ID procedures, and PHA recommendation status could be reviewed in preparing for the revalidation meetings. This review preparation is intended to identify analysis issues and process unit areas that require retrofitting, as well as areas that can be updated and revalidated.

In areas suitable for revalidation, the team could confirm that results from the previous process hazard analysis and any subsequent MOC/PSSR reviews are still accurate and the consequences valid. The team could also review any required checklists and process sections to retrofit and/or update specific process sections and analysis issues. During these reviews, the team should consider whether additional safeguards or other actions are necessary to control and/or mitigate process hazards and develop recommendations if appropriate.

Updating and revalidation are intended for high-quality process hazard analysis of processes that have effectively managed hazards between hazard evaluations (MOC effectively implemented and documented).

Documentation from previous process hazard analysis, incident reports, MOC/PSSR documentation, and the process hazard analysis recommendation status can lead the team through a verification of results from previous process hazard analysis and MOC/PSSR reviews. The time required to perform this analysis will depend on the extent and number of process changes. A checklist can help the team identify other changes that may have occurred. Process hazard analysis must also address facility siting and human factors issues.

3. *Documentation*. The process safety management standard implies report document process hazard analysis revalidation meetings. The documentation will vary with the amount of updating or retrofitting required and needs of the end users. Also, the resources required (labor, scheduling, cost) for a process hazard analysis revalidation can be affected by these factors.

The most complete and common documentation form for process hazard analysis revalidations that are not redos is past documentation. With this style, the process hazard analysis revalidation report will:

- Identify the process unit examined
- List meeting participants
- Itemize documents examined in MOC' incidents since the previous process hazard analysis
- Describe the revalidation approach
- Document analysis results and findings

A report appendix could contain a detailed hazard evaluation table that has been updated to incorporate MOC analysis, applicable PSSR results, and any appropriate checklist results. This documentation style should result in hazard evaluation tables that are easier to revalidate in the future, demonstrate that all analysis issues were addressed, can be used to support other PSM activities (mechanical integrity, training), and accurately represent the process configuration, hazards, and applicable controls in one document. It can also help meet EPA requirements for process hazard analysis.

Process hazard analysis must be revalidated at least every five years for the life of the process. Due to the variety of starting points and company-specific goals, the actual revalidation methodology may vary between processes and/or from one revalidation to the next (for the same process). Following a logical, comprehensive process to identify appropriate tasks should help ensure that process hazards are adequately controlled.

19.7.7 Emergency Flaring Systems

To protect employees, equipment, and the public, pollution control systems such as flares should be properly designed and manufactured, without flaws or shortcuts, according to ISO-9000 standards. These systems must be thoroughly tested and applied correctly to ensure safety and reliability.

Due to mandated environmental regulations, collecting and controlling volatile organic compounds (VOCs) and other harmful emissions have increased safety hazards at operating facilities. For example, flares are essential safety systems. The routine disposal of refinery off-gases and emergency control of hydrocarbons, olefins, aromatics, and other VOCs from processing and storage areas is commonly accomplished by flaring. These systems, on call 24 hours a day, must provide reliable and dependable operation, regardless of wind or weather conditions. They must work right at all times or the consequences could be catastrophic.

In flaring hydrocarbons and other waste gases, most performance problems are due to poor design, manufacturing shortcuts, or improper operation, maintenance, and service. Serious safety problems to avoid are flame lift-off and flameouts of unburned volatiles or gases, flashback and burnback inside flare burner tips and stacks, flamelick down elevated stacks, high thermal radiation at grade level, or plugged or restricted waste gas flow.

19.7.7.1 ***Common Problems*** In the hydrocarbon-processing industry, common flaring problems involve the pilot and ignition systems, high operating costs for steam, and purge. For safety, there can be no shortcuts in a flare's design, manufacture, installation, operation, and maintenance. Meeting quality in these criteria ensures that the flaring system will completely destroy all volatile and carcinogenic vapors, minimize thermal radiation at grade level, and eliminate flame liftoff, flashback, burnback, flame lick, and "burning rain" (liquid carryover).

Many operating problems result from management "pinching pennies." When designing flare systems, areas not to be comprised include purchasing pilots without individual windshields, rain hoods, or protected well-type thermocouples to monitor the pilot flame; using pit flares with straight tips instead of T- or Y-shaped tips; and allowing improper welds and improperly sized flare tips. Other common problems are caused by operators deviating from instructions for pilot gas and air settings. In some cases operators are more comfortable when they can see highly visible and wasteful pilot flames.

19.7.7.2 ***Meeting ISO Standards*** Little documentation has been published about the proper design and development of combustion control systems to comply with ISO-9001 standards [69]. Even less information is available on manufacturing methods (e.g., full-penetration welds on inadequate gauge high-temperature alloys; the misuse of carbon steels, refractory linings, and other materials; and applying proven technologies). For maximum safety and reliability, the design of flare incinerators and other pollution control systems must incorporate proven technologies and practical developments. Innovative designs should be field-proven.

Case History 19.3 Serious safety problems at a refinery necessitated that maintenance personnel climb the derrick of a 76-m flare stack periodically to inspect and (frequently) repair a 1220-mm-diameter refractory-lined flare burner and molecular seal. Refractory pieces broke off the burner and accumulated in the seal base. Thus, the upper portion of the emergency flare burner was unprotected from 1200°C incineration temperatures. Searing heat further accelerated burnout of the flare tip.

Serious safety problems kept recurring as refractory pieces fell into the molecular seal, restricting the waste gas flow. Also, the broken refractory kept plugging a drain at the seal base. Moisture was trapped in the unprotected carbon steel base. Consequently, metal corrosion resulting from trapped moisture allowed air to enter the seal and fill the elevated stack below the flare burner, creating an explosive mixture.

Periodically, the entire refinery was shut down to patch the broken refractory, remove debris, and cut out and replace corroded sections of the molecular seal. Extreme caution was exercised during the cutting and welding operation 76 m above the refinery in an area where explosive fumes could be present.

To improve the flare's reliability and reduce processing downtime, the flaring system was revamped with new technology. A multibaffle kinetic seal was installed. The revamp affixed a conical windshield to prevent flames from licking down the outside of the tip and stack. To streamline installation, the windshield was custom-engineered to match existing piping. Thus, purge gas consumption has been reduced significantly without compromising safety considerations.

Case History 19.4 *Retrofitting Pilots* Pilots, like flare tips, require proven designs, proper manufacturing methods, and installation with "protective positioning" to ensure reliable flare ignition, regardless of wind or weather, including hurricane or typhoon force winds. A retrofit would be physically difficult for several reasons: space limitations, lack of dimensional drawings, and the possibility of burning up replacement pilots during their installation. To facilitate the actual installation, extra "eyes" were used. Boom-mounted, temperature-resistant TV cameras and a ground-mounted telescope helped technicians deal with constraints of installing new pilots. After operating nine months, the entire refinery was shut down for scheduled maintenance. During that outage, the flare tip was replaced with a redundant unit that had been in stock, complete with four flame-front pilots.

Within the next three months, three of the four flamefront pilots failed. Again, the situation was critical. To solve this serious safety/production problem, the system was retrofitted by an array of sparkler pilots to the operating flare.

Pilot and igniters are often installed improperly, despite specific instructions from manuals and the piping drawings. Another very common problem is unnecessary readjustment of a manufacturer's factory settings for pressure and flow. Instruction manuals and piping diagrams should be read and understood before operations or maintenance staff proceed with a startup. All equipment, including flare stacks, ladder platforms, piping and essential safety systems from detonation, flame arrestors, liquid seals and knockout drums to burner heads, and aircraft warning lights on elevated flares must be installed to ISO 9001 standards by the manufacturer or in compliance with instructions and drawings furnished by the manufacturer.

In the United States the EPA has endorsed combustion, incineration, and flaring as effective solutions to pollution problems. Many burners are installed in process heaters, furnaces, and boilers. As is true with all equipment types, there are performance differences between manufacturers. For example, "low nitrous oxides (NO_x) burners" were supplied to a Canadian refinery to replace 280 burners that were creating major maintenance and operational headaches. After installation, the new burners were not controlling the emissions of nitrous oxides and carbon monoxide [70].

None of the 280 burners was installed properly. These burners were not equipped with individual pressure gauges for steam and oil. Without these gauges, it is impossible to control oil automatization and thus prevent plugging and fouling of fuel lines. Thus, it is not possible to regulate excess air that affects carbon monoxide and nitrous oxide formation.

Uncontrolled fuel-oil buildups on heater floors and tubes and flames spilling over burner components did considerable damage throughout the refinery. These serious safety problems undetermined efforts to conserve fuel, reduce harmful emissions, and improve combustion operations. This refinery installation was also botched by not protecting the refractory blocks from rain and snow during storage and prior to mounting the burners. Because some burner guns were too short, they beat against the refractory, causing it to explode. Even small items such as the burner pilots were not uniform. Lengths, diameters, and basic pilot configurations were mismatched from burner to burner and even within individual burners. Everything had to be replaced systematically, including all heat-traced piping.

At the same refinery, while retrofitting the two most critical heaters with 36 low nitrous oxide burners, all refractory blocks had to be replaced along with several burner blocks. Each of the new burners was designed and manufactured to satisfy the most stringent regulations for nitrous oxide and carbon monoxide emissions and was equipped with individual steam and oil gauges. Piping on the steam side of each oil-fired unit included a steam bypass to blow out the heavy residue.

The same refinery has an elevated flare with a molecular seal and 765-mm-diameter refractory-lined flare tip. Commonly called "Old Smoky," this steam-assisted flare was plagued by refractory falling from the tip into the mole seal. The parent organization that manufactured the maintenance-prone flare and the poorly designed and manufactured and improperly installed burners is being sued by the EPA.

19.7.8 Computerized Hazard Identification

Process hazard analysis is expensive, but software can reduce a hazard identification study's overall cost by 10 to 20%. Consequently, an effective program can pay for itself in a few days. The search must be systematic and informative and the software must be proven, affordable, and usable. To begin, do some research and define your company's specific needs: what elements are essential in the hazard analysis software [24,25,83]. More important, ask the vendor specific and detailed questions about the product.

Manual process hazard analysis studies can be very time consuming and tedious. With the advent of Windows, however, unique, powerful interfaces can improve efficiency and communications, reduce the information gap, enhance previous documentation quality, and provide many new options. Computerization offers three major benefits:

1. Consistency of analysis
2. Access to stored data and information
3. Accurate and representative records

Use the purchaser's checklist along with the three popular myths about software (Section 19.7.8.5) when evaluating potential process hazard analysis programs.

19.7.8.1 Systems Check

- *What operating system should the PHA software use?* Find out what operating system your company uses and verify that the new software is compatible (e.g., Windows 3.1, Windows for Workgroups, Windows'97, Windows'00, or Windows NT-operating system). Windows systems are perhaps most common and offers a wide range of features.
- *Does my company operate on a local area network?* If your company system allows people to log into the same network, the process hazard analysis software should be able to operate on a network system as well. In this case, you probably would have to purchase a site license. The cost will be higher since the number of possible users increases considerably.
- *Is the PHA software compatible with management systems that help execute PHA recommendations?* Every company executes recommendations differently, but most use spreadsheets such as Excel or Lotus 1-2-3. Make sure that you can transfer recommendations for future data manipulation.
- *What programming language was used to write the software?* Never buy software at face value; it may look good on the surface, but how was it written? Software written in more powerful code, such as C++, reduces the execution time of commands and can offer greater program flexibility.
- *How much space should the software occupy?* To be effective, powerful programs do not have to take up vast amounts of space. An efficiently written program may take up 10% more space than another, still have more features, and run faster. Don't be deceived by a program's size; determine what it can do and how long it takes to do it. Many engineers like to bring laptops on site; therefore, find

a program that is written so that it takes up a relatively small amount of space. This ensures that the program will perform more effectively on the laptop, which tend to be less powerful than most desktop computers.

19.7.8.2 Technical Support

- *What technical support service does the vendor offer?* Buying process hazard analysis software should be like a good marriage: dedicated and long term. It is not unreasonable to pay a yearly support fee for technical support after a 90-day period. Make sure that you are supplied with comprehensive manuals, that help files are included, and that you can get online support. Another issue to consider is the availability of telephone technical support. Take a test before purchasing and call the technical support line. Test the speed, attitude, and general helpfulness provided by this service.
- *How often does the company update its software?* This is crucial. Software and hardware are similar to university textbooks—once published, they are probably out of date. If your prospective vendor publishes software only every five years, you must think carefully about purchasing. Software must be updated on a regular basis. The company must be kept technologically advanced to be competitive. Also, find out if the upgraded software remains compatible with previously created files.
- *Can the vendor provide support at your company's various sites?* If your organization is international, you should make sure that the vendor can support all pertinent locations. On-site training and risk-support services should also be affordable and available.

19.7.8.3 Licensing

- *Does the vendor offer multiple licensing options?* You should have the choice of a single local area network or corporate, plus academic, licenses where relevant. Prices will vary depending on the number of users. If you buy a corporate license, the software's unit cost will be greatly reduced.

19.7.8.4 Flexibility

- *What process hazards analysis methodologies does my company wish to use, and will the process hazard analysis software support them?* HAZOP, guide word, and knowledge-based approach "what-if" checklist, preliminary hazard analysis, and failure modes effects analysis are all acceptable process hazard analysis methodologies. The software should be flexible enough to cover the techniques that your company needs to comply with standards or other regulations. HAZOP is perhaps the must popular methodology. With the guide word HAZOP, make sure that the software can perform cause-by-cause hazard evaluation.
- *Is the PHA software compatible with other relevant software on the operating system?* If you are in the Windows operating system, for example, will your process hazard analysis software work with spreadsheets such as Excel or Lotus 1-2-3 or word processors such as Microsoft Word or Word Perfect? There are great benefits to having this compatibility. It allows you to create and store much additional information either before, during, or after process hazard analysis sessions. Also, exporting to the corporate database structure should be an option. This will be helpful in tracking all of the recommendations.

19.7.8.5 Features

- *Does the software have features that make my company's PHA studies more efficient?* All process hazard analysis software can only produce as good a result as your input (after all, garbage in, garbage out). Software can really only act as a tool to assist engineers. These tools must have features that make execution and documentation of the process hazard analysis more efficient and this more cost-effective.
- *What are some key features that will speed up PHAs?* Power features make a great deal of difference. Being able to input data without having to make too many keystrokes or mouse clicks will make the sessions flow smoothly and quickly. Make sure that you can easily:
    - Enter data and edit input.
    - Recall and copy data entered previously.
    - Access preestablished libraries of information and those you have created.
    - Update recorded data.
    - Export/import information to/from software.
    - Save and back up data entered.
    - Choose different printing options; one format rarely suits everyone.
    - Record/manage/track recommendations and resolutions.

- *What documentation features should I look for?* It is frustrating to conduct a meaningful process hazard analysis session only to find that the documentation fails to provide adequate records. Are you able to record the sessions fully? Can you print a separate list of recommendations according to different priorities? Other features that are important include session attendance records, key words used in the study, and P&IDs or drawings, reports.
- *What screen formats should be expected?* Manually based methodologies for performing PHAs have been based using the university blackboard ratio (i.e., much greater width than height). Horizontal scrolling computer monitors leave the user partly blind, since the full width is never available to the viewer. The vertical format (Fig. 19-35)

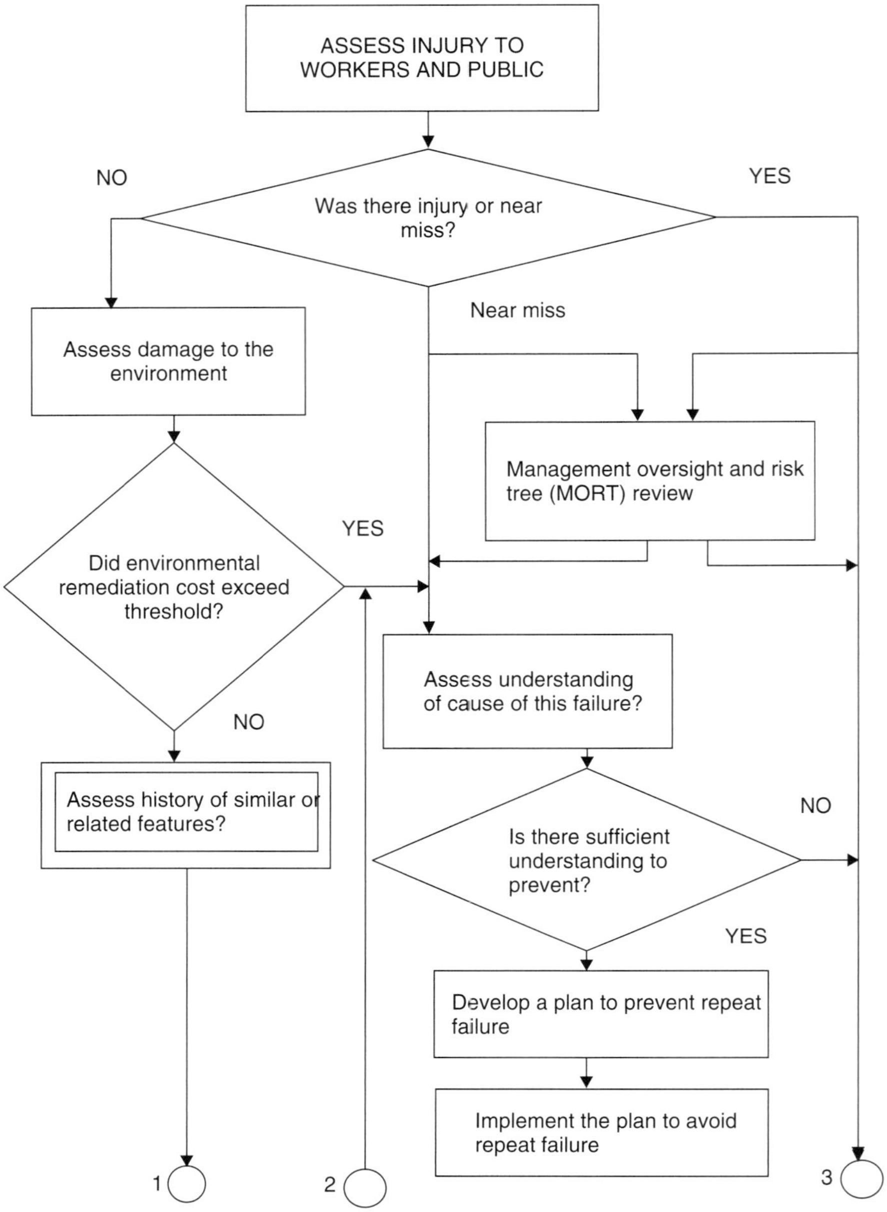

Figure 19-35 Assessment and decision tree provide cost-effective response to equipment failure.

allows the process hazard analysis team to view a larger data set and reduces information overload. Scrolling is a negative feature, especially when using display devices for team interaction.

Vertical organization of data eliminates scrolling, making the format ideal for group environments using projection systems or large monitors. Additionally, although the vertical format is used to input the data, the horizontal format can generate reports that are read easily.

• *Does the software have customizable options?* Customizable options such as a naming convention option permit the user to alter names of the key items, such as nodes, deviations, causes, consequences, and safeguards. A program can thus adopt a chameleon type of characteristic; it can adapt to many other applications beyond the standard process hazard analysis techniques. Such an option permits users to create and/or customize their own safety review techniques. Multiple alternative methods for entering data might include keyboard combinations, mouse, tool bar, and pull down/pop-up menus. These can allow for shortcuts and individual preferences, which may also change.

• *How does the software deal with filing and data storage?* Data integrity is paramount since lost files mean lost results and wasted human resources. If sufficient

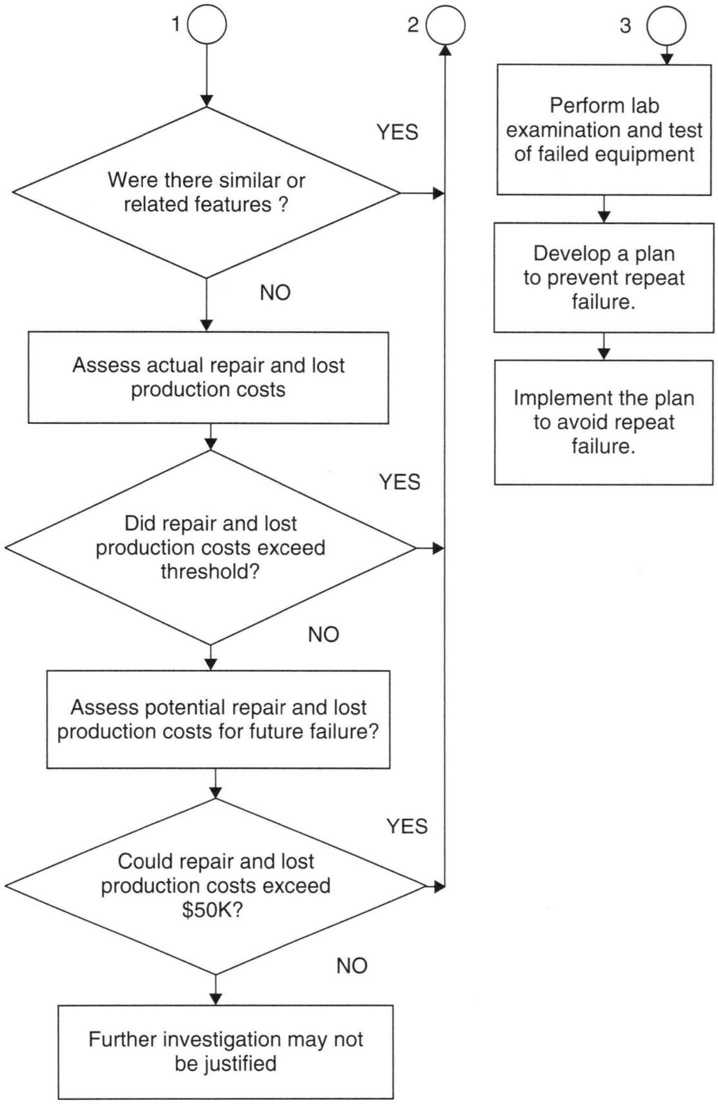

Figure 19-35 (*Continued*)

material is lost, it may not be possible to repeat the PHA process in its entirety. Make sure that you can save files on an automatic backup system so that the data loss is minimized. Additionally, being able to set the automatic backup timing sequence is a very valuable feature. Data loss from a study will not only make you unpopular but could also cost your company a considerable amount of money.

What risk matrix does my company use, and does the software support it? Since risk is defined as the consequence of an incident times the frequency, this evaluation provides some measure of risk (see Appendix 19.1). Computerization is well suited to using risk matrices. The software should support a range of alternatives. It should enable you to build your own risk matrix and assign values as well as to set these new risk matrices as a default, which will save you considerable time.

• *Can the software address additional needs, such as revamps and operating procedures?* Several companies have been caught offguard believing that they are not required to analyze revamps and operating procedures. Does the software address these issues and help you fully meet the needs of standards?

• *Does the PHA software have available knowledge-based libraries? Can you create your own*? Well-made PHA software uses knowledge-based libraries that contain vast amounts of data. Such libraries, held in special modules within the body of a program, supply information on node types, recommended deviations, and respective causes and safeguards (controls). Both predefined and user-constructed libraries are desirable. Such memory aids can greatly assist in preventing important topics from being overlooked. For less experienced teams, knowledge-based libraries provide a learning exercise. Since the plant may have very specific needs, you should be able to create and store your own libraries for future use. Also, the temporary library features (e.g., clipbooks) can be very useful. Remember that a knowledge-based library will grow and retain all data input previously.

- *What else should be considered?* Nothing is perfect; we need both quick fixes and time-saving features. Look for features such as duplication, insertion, search and replace, merging, moving, auto-renumbering, and deletion. If the user is not familiar with the process hazard analysis methodology HAZOP, online tutorials can be helpful. In the United States, process hazard analysis methodologies such as HAZOP are insufficient to cover all process hazard analysis demands according to their standards. Does the software provide assistance, such as additional checklists, in remediating this situation?

The following are three popular, although undesirable myths about process hazard analysis software:

1. *The most inexpensive PHA software will give you an adequate return*. Because all process hazard analysis software is not created equal, this is a complete fallacy. The cost of process hazard analysis software is minuscule compared to the cost of performing the process hazard analysis.
2. *One piece of PHA software isn't very different from another*. Let us suppose that you have two or three process hazard analysis software products written for Windows. Does this mean that one is as good as the other? No! In fact, a poorly written Windows product might be infinitely worse than a similar DOS product. Overall, avoid purchasing software that you cannot use.
3. *An organization that performs good PHAs must also produce good software*. Creation of good process hazard analysis software requires a combination of good programming skills and a working knowledge of process hazard analysis methodologies. There are great benefits to purchasing software from a company that focuses on software design. You are purchasing a technical tool and it should be written in a professional and expert manner.

There are still people conducting process hazard analysis using manual methods. The benefits to using process hazard analysis software exist only if the software meets your needs and provides more efficient preparation, execution, and recording of the PHA sessions. To use software, you must do some questioning and research, not accept it blindly. As with anything else, "try before you buy." Get a working demo and test it. If you have a problem, phone the vendor. The vendor may also offer training and assistance in performing process hazard analysis.

Computerized software tools have assumed major significance for the execution of process hazard analysis. They can offer better online presentations and performance to the user as well as providing better documentation and downstream tracking. The chances that something will be missed are thus greatly reduced. Process hazard analysis sessions are often arduous and painstaking.

19.7.9 Risk Assessment

When plant equipment such as rotating machinery, piping components, and pressure vessels can no longer perform their design function, they have failed. That failure may be so simple that the plant equipment can be restored to operation quickly and easily, or the failure may be so complete that the equipment must be replaced. In addition to the damage to capital facilities, plant equipment failure can injure workers or the public and damage the environment.

The total equipment failure cost can include the cost of the repair of existing equipment, equipment or component replacement, corrective actions for defective products, lost opportunity to produce products, reediting damage to the environment and personal injury liabilities, as well as the intangible costs of negative publicity or increased regulatory actions. When the total equipment failure cost is considered, significant economic incentive exists to examine each equipment failure incident with the goal of preventing recurrent failures. In some situations, regulations also mandate routine analysis of failures that are hazards to safety and the environment.

Risk assessment and plant reliability are considered as decision processes which can be used to identify those plant equipment failure incidents that warrant investigation of the cause [57,61]. Plant management can incorporate simple and systematic decision processes into the day-to-day activities of plant maintenance and engineering functions to improve the organization's response to equipment failure. Systematic review of each failure incident may be much more cost-effective than focusing attention on those failures that are especially costly. For example, safety management has recognized that systematic documentation and analysis of each reportable worker injury, however minor, is critical to avoiding major personnel injury. Similarly, systematic documentation and analysis of each equipment failure is useful in avoiding major equipment failure incidents.

A decision tree methodology can be applied to each equipment failure by plant maintenance with engineering's assistance. The decision tree is a series of six assessments with associated questions to determine the appropriate decision path (Fig. 19-35). The three from the decision tree are:

1. Further investigation may not be cost-effective.
2. Develop a plan to prevent repeat failure based on available information.

3. Perform a formal laboratory investigation of the failed equipment to determine the cause of failure and proper corrective action.

Routine decision tree execution can be incorporated into a maintenance work order fulfillment process that involves failed equipment repair or replacement. The strategic objective of the decision process is to understand the cause of equipment failure to the extent necessary to develop and implement a plan to avoid repetitive or similar failures. The decision tree can also help with prioritizing specific failures that should receive the greatest attention.

The relationship between hazard and risk must be defined. The intent of this decision process is to identify and focus on those plant equipment failures that place humans, the environment, and operating profits at the greatest risk. The incentive is to develop and implement plans to avoid the costs associated with equipment failure.

19.7.9.1 Decission Process Unit The decision tree includes six action blocks for assessing each equipment failure. Each assessment is linked to a specific question relating to the failure. Response to that question determines the path through the decision tree and eventually the exit and recommended action. With multiple paths through the decision tree, it is not always necessary to consider all six action blocks. In fact, the cause of some failures may be so obvious that it is possible to proceed directly to developing and executing a plan to prevent repeat failure. However, each action block can be considered for more significant failures to provide insight for prioritizing failure resources investigation [51].

The action blocks are:

- Assess injury to plant workers and the public, if any, that occurred as a result of the equipment failure.
- Assess damage to the environment, if any, that occurred as a result of the equipment failure and the remediation cost.
- Assess the history of other failures to determine if similar or related failures have occurred in this unit or plant.
- Assess the actual or estimated repair cost and lost production due to the equipment failure.
- Assess the potential worst-case repair and lost production should a similar failure occur in the future.
- Assess the current understanding of the failure cause to determine if sufficient knowledge exists to prevent a similar failure in the future.

19.7.9.2 Decision Tree Questions The six questions in the decision tree derive from the action block assessments. The answer to each is generally a simple yes or no. The decision process has been generalized for routine use in a wide variety of facilities, so there may be gray areas between different questions and decision paths on this generic decision tree. Users of the system should apply judgment to the assessment, the specific questions and the ultimate path through the decision tree when appropriate.

There are six questions corresponding to the action blocks:

- *Was there human injury or a near miss?* This is the only question with three exit paths: human injury, near miss, and neither injury nor near miss. Fortunately, most equipment failure does not result in injury to workers or the public. When injury occurs, there is a good economic reason for the equipment owner and operator to investigate the document carefully to understand the cause of that failure. One obvious goal of such an investigation is to prevent a recurrence of a similar event.

A second consideration when equipment failure results in human injury is to control, to the extent possible, the agenda for investigating and analyzing the findings. Even when the owner is relatively confident of the cause of the failure, there is a potential benefit in having the evidence documented by professional failure analysts, either within the operating company or by an independent laboratory. Rarely is it cost-effective for the owner and operator to voluntarily abdicate the responsibility for and control of failure investigation and documentation that caused human injury. In some cases it may be a legal requirement to conduct thorough investigations for plants covered by process safety management regulations.

An owner and operator should avoid the risks of allowing an injured party's legal representatives or other agency to take the initiative for and control of the investigation. There may be cases where regulatory agencies assume control of investigations for statutory reasons, but the owner should generally stay as involved in the investigation as possible.

The decision tree includes a separate path for a human injury near miss. Defining a near miss is subjective but includes those situations where slightly different circumstance could have led to human injury. A near-miss failure that actually involved human injury should be treated differently from the other classes due to the potential cost of a repeat failure that results in human injury. The decision tree user will determine when plant equipment failure should be considered to be on a near-miss path.

- *Did the environmental damage remediation cost exceed $10,000?* The environmental damage remedition and regulatory complications costs can be larger than the equipment repair and lost production costs. In fact, fairly mundane plant equipment failures such as a corrosion leak that spills process fluids into the environment for an extended period can cost far more for remediation than for repair.

The decision process recommends an arbitrary threshold cost for environmental remediation of \$10,000, which is four to five times the cost of a simple laboratory failure analysis. Decision tree users should adjust this \$10,000 threshold to suit their specific situation. However, for most equipment failures the environmental remediation costs and interaction cost will either be nil or well above this arbitrary threshold, especially when the remediation costs and interaction with regulatory agencies and negative publicity are included.

- *Have there been similar or related failures in the past?* Two repetitive failure classes need to be considered in the decision tree. The first is when the failure is almost identical to a previous failure or failures. It should be fairly obvious that once a pattern of repetitive failures in plant equipment exists, whether recognized or not, those failures will continue until the appropriate corrective actions are taken. Although the cost of a single small failure might not seem to justify the cost of a systematic investigation and corrective actions, it does. The cost of repetitive failures can be estimated by present-worth calculations. For example, the present worth of 10 failures once a year, each costing \$1500, is about \$−10,000. This present worth is more indicative of the magnitude of the problem than the \$1500 single failure cost.

The second repetitive failure class is when multiple plant equipment failures in the same facility appear to be unrelated but have some common thread. For example, there are situations in which apparently unrelated plant equipment failures are related to a local operating and maintenance culture that tolerates high risk, and thus to fundamentally questionable operating and maintenance practices. The operating and maintenance culture at a facility can gradually degrade to become much more tolerant of high-risk operating and maintenance practices.

Persons directly involved may not be aware of the significance of the shift in culture, especially when there is constant and long-term pressure to cut operating and maintenance costs. When a culture tolerant of high-risk operating and maintenance practices evolves, it can contribute to repetitive plant equipment failures in several ways. Just as a history of reportable minor injuries to workers is a red flag that more serious injuries may occur, a history of repetitive plant equipment failures can indicate that much more serious failures may occur, especially when the repetitive failures are due to high-risk practices. One incentive for investigating repetitive plant equipment failures by someone outside the local plant culture is that an outsider may be more likely to recognize a risk-tolerant operating and maintenance culture.

- *Did the actual repair cost and lost production exceed \$10,000?* The total actual plant equipment failure cost includes both the cost to repair or replace all the affected equipment and the lost production income. Both failure cost components must be included to manage response to equipment failure effectively. The recommended threshold of \$10,000 is an arbitrary value that is four to five times the cost of a typical laboratory failure investigation. Although \$10,000 is an arbitrary value, it is the right order of magnitude in that isolated failures that cost \$1000 may not justify formal laboratory investigations, whereas failures that cost \$100,000 clearly justify careful investigation.

- *Could the actual repair and lost production cost of a repeat failure exceed \$50,000?* Many plant equipment failures are analogous to the near miss case for human injury in that there was a near miss of substantial and much more expensive secondary equipment damage. An example is a pipe rupture that fills an area with flammable gas or fluid that does not ignite. In this example the actual repair and lost production costs could be nominal and not warrant a response beyond the repair and restart of the facility. However, had the fuel release ignited, the potential worst-case cost for repair and lost production could have been substantial.

The suggested threshold value of \$50,000 for potential worst-case cost is another arbitrary value that users can adjust to fit specific needs and management philosophy. The most important concept in this action block and question is that the potential for recurrent failure with much grater damage than has already occurred should be considered in responding to plant equipment failure.

Note that the cost for implementing a plan to avoid a single or repetitive failures is not considered in suggesting the \$10,000 and \$50,000 thresholds in this generic decision tree. The specific cause of an equipment failure must first be understood to estimate the cost of corrective actions needed to avoid recurrent failures. This decision tree focuses on the failure cost compared to the cost of determining how to avoid repeat failures without considering the cost to avoid recurrent failure.

- *Is the failure cause sufficiently understood to prevent a repeat failure?* Some equipment failures are caused by operating or maintenance practices that can be corrected once the cause is recognized. An example is when an operator allows rotating equipment to overspeed to failure during startup. Another example is when pipe or vessels rupture due to inadvertent overpressure. In these examples, investigating the failed components might add little to understanding the cause since the issues are probably human rather than mechanical.

On the other hand, when the cause of equipment failure is not understood, it is impossible to develop a meaningful plan to prevent a recurrent failure. When the cause is not understood and there is sufficient economic incentive to avoid repeat failures, a formal laboratory investigation should be conducted.

It can be costly and potentially dangerous when the cause of a particular failure is thought to be understood but really is not, since repetitive failures are likely to continue. When the cause is thought to be well understood, it may be useful to include some "what if" questions along with the decision tree. One important question is: What are the potential consequences if the presumed cause of this failure is misunderstood and goes uncorrected?

19.7.9.3 Managing Response to Equipment Failure

The maintenance work order and fulfillment system at each plant could include executing a decision tree. Figure 19-35 as a part of failed plant equipment repair. Executing the decision tree for each work order involving equipment failure could provide documentation of systematic process safety and mechanical integrity management. Including this decision tree or a similar decision process in the maintenance work order fulfillment system is a cost-effective way to achieve the strategic goal of avoiding injury, damage to the environment and capital equipment, and production loss.

Implementation could begin by preparing printed forms with the six action blocks and completed questions for each repair work order. This decision tree includes an unnumbered action block for executing a management oversight and risk tree.

As a facility gains experience with this simple decision tree, the decision process could be refined by substituting more quantified answers for the yes or no answers in the six questions. Adding more quantified answers relating to repair and lost production costs would better prioritize the effort that should be applied to avoiding future failures. This simple decision process should be installed and functioning before refinement is attempted.

While a plant's maintenance function should be responsible for executing the decision tree as a part of work order fulfillment, the plant's engineering function can and should provide significant assistance and input to the decision process. Also, maintenance and/or engineering personnel will need to interact with plant operations regarding the real and potential cost of lost production. A plant's safety, health, and environmental functions could also provide useful input into the decision process and champion its routine execution.

19.8 SUMMARY

This chapter moves from hazard identification to frequency and probability analysis and finally to consequence analysis. Considerable attention is given to hazard operation techniques, by the most important tool used in plant safety analysis. There is also an extensive discussion of failure and effects analysis, which has gained increasing recognition as a useful reliability analysis tool in the process industry.

Hazard identification methods can be used in different ways to model part of the incident scenario leading to a possible accident. So an effective model of the incident scenario is the key to a successful application. It follows the structure of a generic fault tree up to the release of materials, and of an event tree from this point to the impact of the release on people, the plant, and the environment.

Process safety analysis includes hazard identification, frequencies and probability analysis, consequence analysis, and hazard cost analysis. The initial phase is development of a conceptual framework that will facilitate the modular specification. The second phase is the development of a logic framework that will permit the use of attributes linked into executable models. Process safety design and control component interconnections are defined by process flow diagrams and piping and instrumentation diagrams as well as maps of the surrounding area (plant, people computing, infrastructure, alarm systems, etc.). Risk assessment and plant reliability are considered as decision processes that can be used to identify those plant equipment failure incidents that warrant detailed investigation of the cause.

The decision tree includes action blocks for assessing each equipment failure. Each assessment is linked to a specific question relating to the failure. Response to that question determines the path through the decision tree and eventually the exit and recommended action. The methodology should lead to a diagnostic system that can accommodate changes in the plant configurations. No limitation is placed on the number of equipment units or streams since it is a simple matter to create a new frame and working arrangement.

With multiple paths through the decision tree, it is not always necessary to consider all six action blocks. In fact, the cause of some failures may be so obvious that it is possible to proceed directly with developing and executing a plan to prevent repeat failure.

Project evaluating under process safety management is considered for engineering, procurement, and construction. Engineering, procurement, and construction contractors and operating companies can develop and assign process safety responsibilities early in project planning following the defined checklist. Clearly defining and communicating project responsibilities influences project execution and associated costs considerably. As a process safety management–related construction activities is considered a pre-startup safety review. A project process hazard analysis is one part of a comprehensive project safety program that will include plot-plant safety reviews, preliminary hazard analysis, pre-process hazards analysis, implementation of management of change, confirmation of the "as built" status for piping and equipment, procurement and construction quality assurance, pre-startup safety reviews, and possibly a follow-up process hazards analysis before plant startup. This techniques can be applied at any stage in the life cycle of a process. Such techniques are often classified as previous hazard operaion studies. Performing process hazard analysis is expensive, but using software can reduce the hazard identification study's overall cost by 10 to 20%. Consequently, an effective program can pay for itself in a few days. The major benefits of computerization are consistency of analysis, access to stored data and information, and accurate and representative records.

APPENDIX 19.1 METHOD FOR RISK MEASURE

This method uses the ratio of risk observation as a quality measure. That reduces the risk measure if states exist that have been faults observed, for example, eight times. The observed ratio Q_r is introduced as a contributor to the risk measure:

$$\text{risk measure} = HQ_r \tag{A19.1.1}$$

where Q_r is the ratio observed and H is uncertainty, entropy, related to one failure occurring. The ratio observed is defined as

$$Q_r = \frac{8n_8 + 7n_7 + 6n_6 + 5n_5 + 4n_4 + 3n_3 + 2n_2 + 1n_1}{n} \tag{A19.1.2}$$

where n is the total number of failure states; n_8 is the number of input failure states observed eight times, n_7 the number of input failure states observed seven times, and so on.

If every failure state has been observed at least eight times, Q_r is equal to 1. If any failure state has not been observed at all, Q_r is equal to zero. Thus, Q_r can be also defined as a quality measure. The overall risk measure is then defined as the product of H and Q_r, as expressed by Eq. (A19.1.1). H is related to one failure is defined as

$$H = -\sum_n p(\text{output/input}) \log_2 p(\text{output/input}) \tag{A19.1.3}$$

APPENDIX 19.2 METHOD FOR PARAMETER DETERMINATION

There are many software package and routines for parameter determination based on the least squares method. The function for remediation cost estimation can be defined as

$$R_C = a_0 P^{a_1} C^{a_2} \tag{A19.2.1}$$

If this expression is transformed into

$$\ln R_C = \ln a_0 + a_1 \ln P + a_2 \ln C \tag{A19.2.2}$$

the following equation (with linear parameters) is obtained:

$$y = \ln a_0 + a_1 x_1 + a_2 x_2 \tag{A19.2.3}$$

or

$$y = b_0 + b_1 x_1 + b_2 x_2 = \sum_0^3 b_k ({x_1}^1 {x_2}^1) \tag{A19.2.4}$$

The object function is the minimum sum of squares:

$$S = \sum_1^n w_i \left[Y_i - \sum_0^3 b_k ({x_1}^1 {x_2}^1) \right]^2 = \min \tag{A19.2.5}$$

This will be satisfied if

$$\frac{\partial S}{\partial b_k} = 0 \tag{A19.2.6}$$

and then the equations can be solved by

$$\overline{b} = (\overline{x})^{-1} \overline{Y} \tag{A19.2.7}$$

where w is the statistical weight, k the number of parameters, n the number of the incidents observed, y the value calculated, and Y the value observed.

REFERENCES

1. Ale, J. M., Risk analysis and risk policy in the Netherlands and the EEC, *J. Loss Prevent. Process Ind.*, vol. 4, 1991, pp. 58–64.
2. Austin, D. G., and Jeffreys, G. V., *The Manufacture of Methyl Ethyl Ketone from 2-Buthanol, IChemE/Godwin*, London, 1979.
3. Baker, W. E., et al., *A Short Course on Explosion Hazards Evaluation*, Soutwest Research Institute, San Antonio, TX, 1976.
4. Baker, W. E., et al., *Explosions in Air*, 2nd ed., Wilfried Baker Engineering, San Antonio, TX, 1983.
5. Balemans, A. W., Check lists: guidelines for safe design of process plants, in Bushmann, C. H. (Ed.), *Loss Prevention and Safety Promotion in the Process Industries*, Elsevier Science, Amsterdam, 1974.
6. Bartknecht, W., *Explosionsschuts, Grundlagen und Anwendung*, Springer-Verlag, Berlin, 1993.
7. Berthold, W., and Löffler, U., *Lexikon Sicherheitstechnischer Begriffe in der Chemie*, Verlag Chemie, Weinheim, Germany, 1981.
8. Bretherick, L., *Handbook of Reactive Chemical Hazards*, 4th ed., Butterworth, London, 1990.
9. Browning, E., *Toxicity and Metabolism of Industrial Solvents*, Elsevier Science, Amsterdam, 1965.

10. Casarett, L. J., Doull, J., Klaasen, C. D., and Amdur, M. O., *Casarett and Doull's Toxicology: The Basic Science of Poisons*, 4th ed., McGraw-Hill, New York, 1991.
11. Christen, H. R., *Thermodynamik a und Kinetik chemischer Reaktionen*, Diesterweg,u.Salle, Frankfurt am Main, Germany, 1974.
12. Crowl, D. A., and Louvor, J. F., *Chemical Process Safety: Fundamentals with Applications*, Prentice Hall, Englewood Cliffe, NJ, 1990.
13. Crumpler, D. K., and Whitte, D. K., How to effectively revalidate PHA, *Hydrocarbon Process*., 1996, pp. 75–60.
14. Cvetkovic, I., Perunicic, M., and Divac, Z., (2005) Properties and handling recommendation of tri-chloro-ethylene and benzene used in chemical laboratories, *Comput. Ecol. Eng*., vol. 1, no. 1, 2005, pp. 63–72.
15. Decant, W., and Neumann, H. G., *Tissue Specific Toxicity: Biomedical Mechanisms*, Academic Press, London, 1992.
16. Dow Chemical Co., The Dow safety guide, reprinted from *Chemical Engineering Progress*, 5th ed., AIChE, New York, 1980.
17. Early, W. F., Improve client-contractor working relationships under PSM, *Hydrocarbon Process*., vol. 75, October 1996, pp. 95–105.
18. Giesbrecht, H., et al., Analyse der potentiellen Explosionswirkung von kurzzeitung in die Atmosphäre freigesetzten Brenngasmengen, 1, *Chem. Ing.Tech*., vol. 52, no. 2, 1980, pp. 114–122.
19. Giesbrecht, H., et al., Analyse der potentiellen Explosionswirkung von kürzzeitung in die Atmosphäre freigesetzten Brenngasmengen, 2, *Chem. Ing. Tech*., vol. 53, no. 1, 1981, pp. 1–10.
20. Groothuizen, T. M., Hartgerink, J. W., Pasman, H. J., and Buschmann, C. H. (Eds.), Phenomenology, test methods and case histories of explosions in liquids and solids, *First International Loss Prevention Symposium*, Hague/Delft, The Netherlands, May 28–30, Elsevier Science, Amsterdam, 1974.
21. Hayes, W. A., *Principles and Methods of Toxicology*, Raven Press, New York, 1989.
22. Heinrich, H. J., *Zur Bemessung von Druckentlastungsoffnungen bei Gas und Staubexplosionen*, Wissenschafliche Berichte aus der Bundesanstalt fur Materialprüfung (BAM), Berlin, 1973, pp. 277–280.
23. Hessian, R. L., Jr., Greenberg, H. R., and Early, W. F., II, Utilization of PC-based hazards and operability study data, *AIChE-24 Annual Loss Prevention Symposium*, San Diego, CA, August 19–22, 1990.
24. Hyatt, N., The advantages of using software for process hazards analysis, *Computers IV*, Houston, TX, 1996.
25. Hyatt, M., and Spradle, J. S., Buy PHA software with confidence, *Hydrocarbon Process*., vol. 75, October, 1996, pp. 63–82.
26. Institution of Chemical Engineers, *Flowsheeting for Safety*, IChemE, London, 1976.
27. King, R., *Safety in the Process Industries*, Butterworth-Heinemann, London, 1990.
28. Kletz, T. A., *Hazard Analysis: A Quantitative Approach to Safety*, IChemE Symposium Series No. 34, IChemE, London, 1971, p. 75.
29. Kletz, T. A., Evaluate risk in plant design, *Hydrocarbon Process*, vol. 56, May 1977, p. 207.
30. Kletz, T. A., Inherently safer plants: an update, *Plant/Oper. Prog*., vol. 10, no. 2, 1991, pp. 81–84.
31. Lawley, H. G., Loss prevention, in *Operability Studies and Hazard Analysis*, AIChE, New York, 1974.
32. Lees, F. P., *A Review of Instrument Failure Data*, IChemE Symposium Series, No. 47, IChemE, London, 1976, p. 73.
33. Lowrance, W. W., *Of Acceptable Risk*, W. Kaufman, New York 1976.
34. Marshall, V. C., *Major Chemicals Hazards*, Ellis Horwoord Series in Chemical Engineering, Ellis Horwood, Chichester, UK, 1987.
35. Mitchell, J. R., and Horning M. G. and (Eds.), *Drug Metabolism and Drug Toxicity*, Raven Press, New York, 1984.
36. Nabert, K., and Shön, G., *Sicherheitstechnishe Kennzahlen brennbarer Gase und Dampfe*, 2nd rev. ed., Deutscher Eichverlag, Braunschweig, Germany, 1980.
37. Nagy, J., Explosion development in closed vessels, *Rep. Invest. U.S. Bur. Mines*, no. 7507, 1971, pp 1–50.
38. Napier, D. M., *Static Electrification in the Process Industries*, IChemE Symosium Series 34, IChemE, London, 1971, p. 170.
39. Napier, D. M., and Russell, D. A., Hazard assessment and critical parameters relating to static electrification in the process industries, *Proceedings of the First International Symosium on Loss Prevention*, Elsevier Science, Amsterdam, 1974.
40. Palmer, K. N., *Dust Explosion and Fires*, Chapman & Hall, London, 1973.
41. Pekalski, A. A., and Pasaman, H. J., Dinstinction between the upper explosion limit and the lower cool flame limit in determination of flammability limit at elevated conditions, *Process Safety Environ. Prot*., vol. 87, 2009, pp. 47–52.
42. Pilz, V., Grundlagen fur die Vorhersage der Auswirkungen von Störfällen, *VFDB*, vol. 2, no. 3, 1981, pp. 116–125.
43. Prugh, R. N., Applications of fault tree analysis, *Chem. Eng. Prog*., vol. 76, July 1980, p. 59.
44. Savkovic-Stevanovic, J., Framework for qualitative simulation in process engineering, *Comp. Chem. Eng*., vol. 16, 1992, pp. 127–134.
45. Savkovic-Stevanovic, J., A qualitative model for estimation of plant behaviour, *Comp. Chem. Eng*., vol. 18, 1994, pp. 713–720.
46. Savkovic-Stevanovic, J., *Process Modelling and Simulation*, Faculty of Technology and Metallurgy, Belgrade University, Belgrade, Serbia, 1995, Chap. 4.

47. Savkovic-Stevanovic, J., The chemical plant management qualitative support system, *Chem. Ind.*, vol. 58, no. 3, 2004, pp. 97–104.
48. Savkovic-Stevanovic, J., The process plant management risk reduction support system, *Proceedings of the 11th European Symposium on Loss Prevention*, Prague, Czechoslovakia, May 31–June 3, 2004, p. 223.
49. Savkovic-Stevanovic, J., Process hazards analysis and safety standards, *Comput. Ecol. Eng.*, vol. 1, no. 1, 2005, pp. 47–59.
50. Savkovic-Stevanovic, J., Risk assessment and reliability analysis, *Comput. Ecol. Eng.*, vol. 1, no. 1, 2005, pp. 69–75.
51. Savkovic-Stevanovic, J., Process safety management, *Comput. Ecol. Eng.*, vol. 1, no. 1, 2005, pp. 35–46.
52. Savkovic-Stevanovic, J., Cognitive reliability analysis of the process plant, in Puijagner, D., Espuna, A. (Eds.) *Computer Aided Process Engineering*, Vol. 22, Elsevier Science, Amsterdam, 2005, pp. 342–347.
53. Savkovic-Stevanovic, J., Process plant risk analysis and modelling, in Plesu, V., Agachi, P. S. (Eds.), *Computer Aided Process Engineering*, Vol. 24, Elsevier Science, Amsterdam, 2007, pp. 1229–1233,
54. Savkovic-Stevanovic, J., *Informatics*, Faculty of Technology and Metallurgy, Belgrade University, Belgrade, Serbia, 2007, Chap. 6.
55. Savkovic-Stevanovic, J., *Process Engineering Intelligent Systems*, 2nd ed., Srbisim, Belgrade, Serbia, 2008, Chap. 5.
56. Savkovic-Stevanovic, J., Waste water transport system safety, *Proceedings of the International Conference on Maritime and Naval Science and Engineering*, Brasov, Rumania, October 13–14, 2009, pp. 71–76.
57. Savkovic-Stevanovic, J., Reliability and safety analysis of the process plant, *Petrol. Coal*, vol. 52, no. 2, 2010, pp. 62–68.
58. Savkovic-Stevanovic, J., and Jovanovic-Jovic, V., Process safety analysis of the allyl chloride plant, *Proceedings of the International Symposium on Large Chemical Plants* (LPC10) Antwerp, Belgium, September 28–30, 1998.
59. Savkovic-Stevanovic, J., and Krstic, S., Process safety of the phenol plant, *II Regional Symposium on Chemistry and Environment*, IV-1, Krusevac, Serbia and Montenegro, June 18–23, 2003, pp. 245–246.
60. Savkovic-Stevanovic, J., and Krstic, S., A chemical plant risk reduction system modelling and application, *Proceedings of the 16th European Modelling and Simulation Conference* (CDROM), Riga, Latvia, May 23–27, 2005.
61. Savkovic-Stevanovic, J., and Krstic, S., The process plant hazard and risk reducing system, *ENENP-11*, Belgrade, Serbia and Montenegro, Sept. 25–28, 2005, p. 153.
62. Savkovic-Stevanovic, J., and Krstic, S., Risk reduction support system of the phenol recovery plant, *Petrol. Coal*, vol. 48, no. 6, 2006.
63. Savkovic-Stevanovic, J., Mosorinac, T., and Krstic, S., Process risk analysis operation modelling, *Comput. Ecol. Eng.*, vol. 2, no. 1, 2006, pp. 31–37.
64. Sax, N. I., *Dangerous Properties of Industrial Materials*, 4th ed., Reinhold, New York, 1975.
65. Schacke, H., Redundanz im Staubexplosionsschutz-Koncept komplementärer Schuzmassnahmen, *Staub. Reinhalt. Luft.*, vol. 53, 1993, pp. 453–459.
66. Seaton, W. H., *Chetan*, ASTM Data Series 51, ASTM, Philadelphia, 1974.
67. Sinnott, R. K., *An Introduction to Chemical Engineering Design*, Chemical Engineering Series, Vol. 6, Coulson, J. M., Richardson, J. F., (Eds.), Pergamon Press, New York, 1983.
68. Stojanovic, O., Stojanovic, N., and Kosanovic, D. J., *Harmful and Dangerous Substances*, Rad, Belgrade, Serbia, 1984.
69. Straits, J. F., Improve flare safety to meet ISO-9000 standards, *Hydrocarbon Process.*, vol. 75, June 1996, pp. 109–114.
70. Straits, J. F., Combat NOx with better burner design, *Environ. Eng., Suppl. Chem. Eng.*, November 1994.
71. *Ullmann's Encyclopedia of Industrial Chemistry*, Vol. B8, VCH Verlagsgesellschaft, Zurich, 1995.
72. United Nations, *Recommendations on the Transport of Dangerous Goods*, 8th rev. ed., UN, New York, 1993.
73. United Nations, *Recommendations on the Transport of Dangerous Goods: Test and Criteria*, 3rd ed., UN, New York, 1995.
74. Vitorovic, L. J., Skrlj, M., Mitic, N. V., and Levata, S., *Poisoning Chemicals in Yugoslavia*, Grmec, Belgrade, Serbia, 1996.
75. Webster, J. K., *Toxic and Hazardous Materials: A Source Book and Guide to Information Sources*, Bibliographical Indexes in Science and Technology No. 2, Greenwood Press, New York, 1987.
76. Wells, G. L., *Safety in Process Plant Design*, IChem/Godwin, London, 1980.
77. Wexler, P., *Information Resources in Toxicology*, Elsevier Science, New York, 1988.
78. Whiston, J., *Safety in Chemical Production*, Backwell Science, Oxford, UK, 1991.
79. Williams, R. T., *Biochem. Soc. Trans.*, vol. 2, 1974, pp. 359–367.
80. Lewis, D.J., *AIChE – Loss Prevention symposium*, Houston, April, The Mond fire, explosion and toxicity index: A development of Dow index, 1979.
81. Lewis, D.J., *Loss Prevention*, No. 13 (AIChE) 20. The Mond fire, explosion and toxicity index applied tp plant layout and spacing, 1979.
82. Freytagh, F.F., *Handbuch der Raumexplosionen*, Verlag, Chemie, Weinheim, 1965.
83. Gordon, R.L., Hessian Jr, R.T., Greenberg, H.R. and W.F. Early II, Utilization of PC based hazards and operability study data, *AIChE-24 Annual Loss Prevention Symposium*, San Diego, California, August 19–22, 1990.

SECTION III

PROCESS MEASUREMENT, CONTROL, AND MODELING

20

FLOWMETERS AND MEASUREMENT

GREGORY LIVELLI
ABB Instrumentation, Warminster, Pennsylvania
CHIKEZIE NWAOHA
Control Engineering Asia, Ten Alps Communications Asia, Aladinma, Nigeria

The accurate measurement of flow plays a vital role across all sectors of industry worldwide [10]. With the volatility of oil and gas prices, accurate measurement of product streams has a massive impact on revenues not only for operating companies but also for national governments that realize revenues from production [10]. Inaccurate flow measurements or failure to take measurements can cause serious or disastrous results. The principal question facing operators, regulators, and environmental bodies is: How can we be sure that the flow of streams is being measured to the accuracy and reliability required?

Flowing systems require energy, typically provided by pumps and compressors to produce pressure difference as the driving force, and flow sensors should introduce a small flow resistance, with minimal increase in process energy consumption. Most flowmeters require straight sections of piping designs, which can be partially compensated by straightening vanes placed in the piping.

Flowmeters are part of a control loop assembly. A control loop is a closed system consisting of selected instruments that work with the objective of monitoring and controlling fluid flow. These devices could be located on any set comprising a control valve assembly, pump, compressors, and so on, although they could be some distances apart. In good process operations, flow throughout the process must be regulated near their required values with small variability.

In addition, the job of the measurement personnel is to maximize the return on investment while utilizing the equipment in place. This in turn ensures that both the producer and the buyer are receiving fair value for the product being measured. There are many different methods of measuring fluid flow, which is useful but can be very confusing. The objective of this chapter is to unravel some of the mysteries of flowmetering technology selection, explain how different flowmeters work, and highlight the guidelines for the troubleshooting that needs to be performed to ensure accurate measurement results.

20.1 FLOW MEASUREMENT TECHNIQUES

A wide variety of different methods are available for metering flow rate and total flow. Each method has its own specific characteristics, which are directed toward individual installation requirements. Flow measurement can be classified as total flowmeters and flow rate meters.

20.1.1 Volumetric Totalizers

Volumetric totalizers with moving chambers which are driven by the fluid are known as *displacement meters* or *flow totalizers* and can be used for liquids or gases. Direct volumetric totalizers transport a fluid in measured chambers with defined geometrically bounded volumes, whereas indirect flow totalizers operate with moving metering chambers whose volume is known [4]. In these totalizers the enclosed chambers of a direct flow totalizer are missing

Process Plant Equipment: Operation, Control, and Reliability, First Edition. Edited by Michael D. Holloway, Chikezie Nwaoha, and Oliver A. Onyewuenyi.
© 2012 John Wiley & Sons, Inc. Published 2012 by John Wiley & Sons, Inc.

and the measurements are made either mechanically, utilizing vaned wheels through which defined volumes are transported, or electrically, using quantity proportional pulses. Some flowmeters utilize the fluid velocity or kinetic energy of the flow stream for flow rate determination, which is also an indirect method. The user encounters difficulty in selecting the technically best and most cost-effective meter for the particular process application. Volumetric totalizers can generally be classified as, but are not limited to: turbine meters, oval gear totalizers, lobbed impeller gas meters, vortex flowmeters, and swirl flowmeters.

20.1.2 Turbine Flowmeters

The history of gas turbine meters goes back more than a century. In 1901, Thomas Thorp applied for a British patent for an inferential gas meter based on a turbine principle [3]. In 1903 he also got a U.S. patent [3]. Turbine flowmeters are indirect volumetric totalizers in which the flow stream causes a vaned rotor to revolve. In a turbine flowmeter, a rotor is placed in the path of the flow. When a fluid flows through a turbine, it causes the turbine rotor to revolve with an angular velocity that is proportional to the flow rate. The speed of rotation of the rotor is proportional to the total flow and the frequency of the revolutions to the flow rate. The turbine rotors are light in weight, producing minimal friction in the bearings. Because of this, the flow ranges can be expanded because the system responds with greater sensitivity. Turbine flowmeters differ according to the design of the spinning rotor. Different types of turbine meters include axial, single-jet, multijet, paddlewheel, Pelton wheel, propeller, and Woltman [20]. Single- and multijet meters are widely used for municipal water measurement; paddlewheel, propeller, and Pelton wheel meters are used mainly to handle dirty liquids; and Woltman meters are water meters for large-volume applications [20].

Turbine flowmeters suit only low viscosities and are sensitive to flow profile and vibration, but recent technologies allow measurement of gases and liquids with elevated viscosities. A coil in the housing opposite the vanes of the rotor measures the signal by various methods. The three most common cases are as follows:

1. A magnet in one vane induces a voltage pulse in the coil during every revolution.
2. The coil encloses a magnet, and the vanes are made of permeable material. As the vanes pass the magnet, the field is distorted, inducing a voltage pulse.
3. A high-frequency ac voltage (10 kHz) is fed to the coil. The permeable vanes vary the amplitude of the supply voltage, resulting in a secondary frequency superimposed on the carrier frequency [4].

In all three cases, a frequency signal is generated that is proportional to the number of revolutions and therefore to the flow rate. The signal is fed to a preamplifier in the connected converter. In this manner, the totalizer, each of whose individual pulses represents a defined volume, becomes a flowmeter as a result of the time-based frequency that is generated. The fluid flow rate can then be used for utility and nonutility purposes and are generally used in larger line sizes, primarily those exceeding 4 in [4].

Turbine flowmeters are highly accurate and durable, but are restricted by the fact that they must be applied in noncorrosive and clean services and require a considerable amount of maintenance. Turbine meters make up about 8% of the installed base for flowmeters. Rangeability can reach 100:1 if a meter measures the rate of a single fluid under constant conditions. Expected accuracies of plus or minus 0.25% can be attained by certain turbine meters where proper stream conditions are maintained and the meter is properly installed [5]. To minimize installation effects, turbine flowmeters should have an integral flow conditioner. Figure 20-1 is an example of a turbine flowmeter.

Turbine flowmeters have both advantages and limitations.

Advantages

- No supply power is required for rotating vane and Woltman meters.
- Rotating vane and Woltman meters are certified for water.
- Turbine flowmeters are suitable for cryogenic fluids.
- Turbine flowmeters are usable at extreme temperatures and pressures.
- Turbine flowmeters are certified for gas.

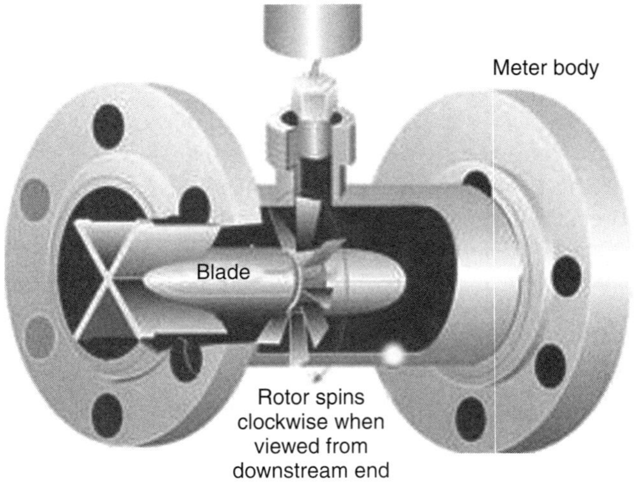

Figure 20-1 Example of a turbine flowmeter. Turbine meters are well accepted in the processing industries.

Limitations

- Limited material selection
- Reliable only for low viscosities
- Have moving parts; wear and erosion issues
- Sensitive to contamination
- Profile sensitivity in the case of axial flow totalizers
- Many require conditioning sections
- Affected by overloading and fast changes at high differential pressures; danger of overspeeding
- Vibration sensitive

20.1.3 Oval Gear Totalizers

Figure 20-2 shows a typical oval gear flowmeter. The metering element of this volume totalizer consists of two oval gears. The driving fluid produces the required torque, which varies as a function of the gear position, to rotate the gears. Therefore, the torques on the lower gear on the left side cancel each other, while the torque on the upper gear is one-sided and actually causes the rotation. Around the upper gear a bounded crescent-like volume exists which is pushed toward the outlet of the meter. Each rotation of the pair of oval gears transports a defined fluid volume. The number of rotations is therefore an exact measure of the quantity of fluid that has flowed through the meter. The precision teeth assure a good seal between the two gears. The play between the oval gears and the walls of the metering volume is so slight that the leakage flow is small.

The rotations of the pair of oval gears are transmitted without a stuffing box to an indicator either by a permanent magnet coupling or by a feedback-free magnetic field-controlled pulse transmitter. However, the gears and bearings are subject to mechanical wear. Through selection of materials for the housing, oval gears, and bearings as well as by design consideration of expansions due to high temperatures, oval gear meters are suitable for almost all operating conditions. For low viscosities, the flow range is appreciably less than that for higher viscosities [4].

Figure 20-2 Oval gear meter for industrial fluids, design with totalizer. (Courtesy of ABB, Inc.)

The advantages and limitations of oval gear meters follow.

Advantages

- High accuracy
- Suitable for fluids with high viscosity
- Operates in both flow directions (forward and reverse)
- No flow profile effects; therefore, no conditioning sections
- No supply power required
- Certification approvals

Limitations

- Volumetric totalizer
- Only for liquids
- High-pressure drop
- Moving parts; wear
- Accuracy decreases for lower viscosities due to leakage losses
- Sensitive to contamination; filter required
- Flow blockage at zero flow through soilage
- Sensitive to overloading
- Monitoring and maintenance required

20.1.4 Lobed Impeller Gas Meters

Two rotating impellers, designed with a figure eight cross section, rotate in opposite directions due to the forces exerted by the gas being metered. The shape of the impellers prevents contact while the gap between them remains constant. Figure 20-3 depicts the impeller movements in a lobed impeller gas meter during operation. A gear drive external to the measuring chamber synchronizes the impellers. During each rotation, four crescent-shaped volumes are moved through the measuring chamber. The number of rotations is proportional to the total flow. The rotation is coupled using an adjustable fine-tooth gear train to the totalizer.

An unmeasured flow, which is a function of the pressure drop, flows through the gaps. This negative error is compensated by an adjustment. The viscosity of gases increases at high pressures and reduces the losses in the gaps, which

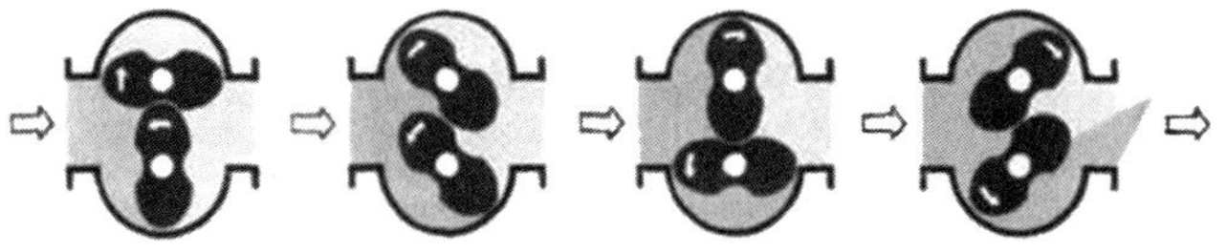

Figure 20-3 Operation of a lobed impeller meter. (Courtesy of ABB, Inc.)

compensates for the higher losses that would otherwise exist due to the higher-pressure drops. The pulsations in the gas discharge can cause the pipeline connected to the meter to vibrate. If resonance should occur, loud noises and sudden pressure drops can result. This condition should not be allowed to occur; if necessary, noise and pulsation dampers should be utilized. The pressure drop results from the mechanical and dynamic resistances in the meter. The dynamic portion increases appreciably with increasing flow. Lobed impeller meters are very susceptible to contamination. Since contamination affects the pressure drop, it must be monitored and the meter cleaned when required. Figure 20-4 is a typical impeller flowmeter from ABB.

The advantages and limitations of lobbed impeller flowmeters follow.

Advantages

- Exceptional accuracy for gas measurements
- No conditioning sections
- No supply power
- Certification approvals

Limitations

- Volumetric totalizer
- Only for gases
- Moving parts; wear
- Flow blockage at no flow due to contamination
- Sluggish toward quick changes
- Affected by quick changes at high pressure differentials; danger of overspeeding
- Monitoring

Figure 20-4 Lobed impeller meter. (Courtesy of ABB, Inc.)

20.1.5 Vortex Flowmeters

Vortex meters are currently entering a new growth phase after many years of relatively slow growth. They employ a principle called the *Von Kármán effect* [19]. According to this principle, flow will generate vortices when passing through a bluff body, a nonstreamlined object mounted at right angles to the fluid flow path [19]. As the fluid passes the obstruction, vortices are shed on alternating sides with a frequency that is linearly proportional to the velocity of the fluid. The spacing between the vortices remains constant regardless of the fluid velocity. Within a pipeline, the vortice effect attenuates a few pipe diameters downstream from the bluff body. Figures 20-5 and 20-6 show the principle of operation and a photo of a typical vortex flowmeter.

Technologies used to detect vortices include piezoelectric, ultrasonic, and capacitive. Flow rate is calculated by multiplying the cross-sectional area of the pipe by the flow velocity. During installation, it is necessary to install conditioning straight sections upstream (comparable to those required in a turbine flowmeter) of the meter, whose length is dependent on the type of distortion. Steam, gases, and liquids can be metered with a vortex flowmeter. Vortex meters may be installed vertically or horizontally. For high accuracy, vortex meters should be installed with proving connections.

The advantages and limitations of vortex flowmeters follow.

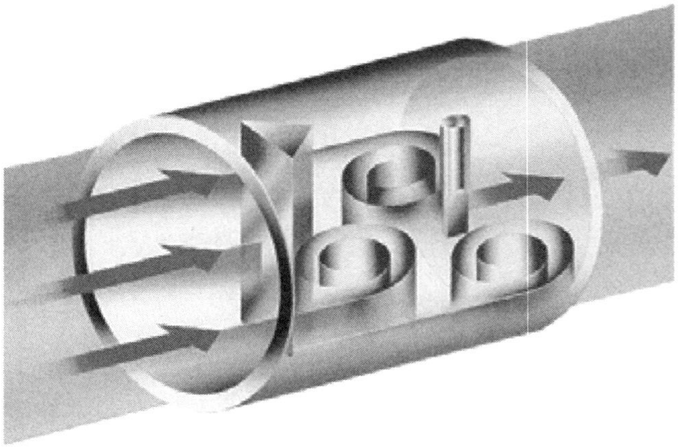

Figure 20-5 Principle of operation of a vortex flowmeter. (Courtesy of ABB, Inc.)

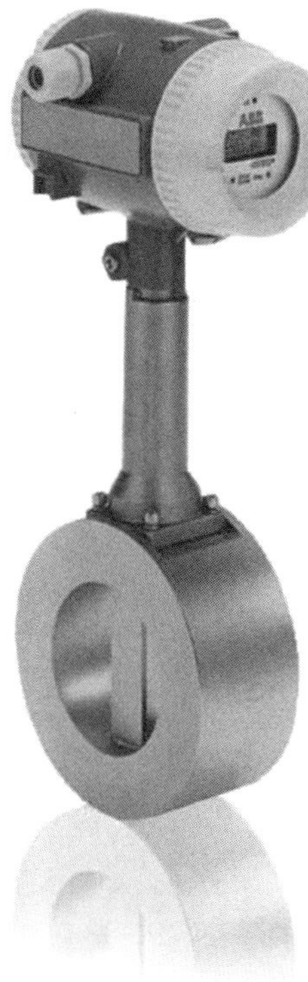

Figure 20-6 Compact design of a vortex flowmeter. (Courtesy of ABB, Inc.)

Figure 20-7 Swirl flowmeter. (Courtesy of ABB, Inc.)

Advantages

- No moving parts
- Rugged construction
- Suitable for liquids, gases, and steam
- Easily sterilized
- Unaffected by pressure, temperature, and density changes
- Linear relationship between flow rate and measurement signal

Limitations

- Conditioning sections required
- Minimum Reynolds number required

20.1.6 Swirl Flowmeters

A swirl flowmeter operates similar to that of a vortex flowmeter, however, a stationary guide body is located in the inlet of the meter (Fig. 20-7), whose shape is similar to that of a turbine rotor, which forces the fluid to rotate. The fluid flows through the meter pipe of the swirl flowmeter in a thread-like rotation. The flow stabilizes in the cylindrical cross section. Consideration of the cross sections in this region shows that the rotational velocity at the wall is relatively small and increases toward the pipe center until a stable vortex core is formed at the center. During the transition of the flow into the expanding section of the pipe, the vortex core is displaced because a backflow occurs in the expander section. The vortex core forms a spiral-like secondary rotation whose frequency is proportional to the flow rate over a wide range. This secondary rotation is measured with a detector. The detector utilizes the resulting pressure differences for its pulse measurements. The same sensors are used in both the swirl and vortex flowmeters. The rotation frequency is between 10 and 1500 Hz, the higher frequencies indicating higher flow rates.

The flow rate of liquids, gases, and steam can be measured with a swirl flowmeter. A definite advantage of the swirl flowmeter over other systems is the fact that conditioning sections up- and downstream of the meter are not required. A deswirler at the meter's exit eliminates any effects on downstream instruments.

The advantages and limitations of swirl flowmeters follow.

Advantages

- No moving parts
- No conditioning section

- Suitable for liquids, gases, and steam
- Exceptional repeatability
- Unaffected by pressure, temperature, and density changes

Limitations

- Pressure drop
- Minimum Reynolds number required

20.2 FLOW-RATE METERS

A wide variety of methods are available for flow-rate metering. Some flowmeters utilize the fluid velocity or the kinetic energy of the flow stream for flow-rate determination, which is also an indirect method. The various types of flow-rate meters are: differential pressure flowmeters, variable-area flowmeters, electromagnetic flowmeters, ultrasonic flowmeters, Coriolis mass flowmeters, and thermal mass flowmeters. These are described below.

20.2.1 Differential Pressure Flowmeters

Flowmeters based on differential pressure, also known as DP flowmeters, represent a popular choice in the processing industries, constituting nearly 30% of flowmeter installations [9]. They have good application flexibility since they can measure liquid, gas, and steam flows [9]. DP flowmeters take advantage of a constriction in the flow stream that creates a difference in pressure. This pressure differential is used to compute flow rate. The constriction in the line is created by various devices called *primary elements*. Types of primary elements are orifice plates, venturis, pitot tubes, and flow nozzles. Differential pressure meters can be used for liquid and gas metering at extremely high temperatures and pressures. The meters have been optimized by extensive research activities over decades, and the results published as standards. Differential pressure meters can be used without problems only under specific flow conditions. Nonuniform flow streams after disturbances prevent an axisymmetric velocity profile from forming in the throat and thereby alter the differential pressure values. That is why the differential pressure producer must be installed between two straight cylindrical pipe sections in which no disturbances or diameter changes may exist. Along these sections the required velocity profile for metering can form.

The arrangement and design of the installation is a function of the application. The minimum requirements for each metering location are pressure tap lines between the differential pressure producer and the differential pressure transmitter. Shutoff vales are installed in both pressure tap lines. For protection of the differential pressure transmitter, a combination of three to five valves are installed ahead of the transmitter converter to isolate and protect it by preventing the application of the pressure from one line only. If the differential pressure meter is to be used for gas metering, it should always be installed above the differential pressure producer to prevent any condensate from entering the pressure tap lines. Conversely, gas bubbles should not enter the pressure tap lines when metering liquids. Therefore, in those applications the converter should be installed below the differential pressure producer. For steam metering the pressure tap lines are filled with condensate from the condensate chambers. Figure 20-8 is an example of a DP flowmeter.

A particularly difficult flow condition is swirl, in which the fluid moves from side to side in the pipeline. The straight sections recommended are not by any means sufficient for conditioning such a flow profile. As a result, a flow straightener must be installed. A flow straightener can also be used to shorten the recommended straight lengths for the other types of disturbances.

When using a differential pressure flow measurement system, the density of the fluid varies with pressure and temperature changes. It is recommended that as a minimum when metering gases or steam, additional measurements be made of the process temperature and pressure and to calculate the state and condition of the fluid. With the advent of multivariable transmitters, process temperature and pressure can be measured directly at the flow metering point, thereby eliminating the needs for additional measurement devices and a separate flow

Figure 20-8 Differential pressure transmitter. (Courtesy of ABB, Inc.)

computer to calculate the mass flow output. This will assure a reliable measurement of the normal or mass flow rates even under varying conditions.

The advantages and limitations of differential pressure flowmeters follow.

Advantages

- Universally suitable for liquids, gases, and steam
- Useful in extreme situations (e.g., viscosity); suitable because of the variety of versions
- Calculations possible for unusual situations
- Suitable for extreme temperatures and pressures
- Range changes possible
- Low-pressure drop for nozzles and averaging pitot tubes

Limitations

- Square-root relationship between flow rate and differential pressure; therefore, a shorter range
- Affected by pressure and density changes
- Pressure drop for orifice plates
- Edge sharpness for orifice plates must be assured; therefore, no solids or contamination
- Very long conditioning sections
- Expensive installation requiring pressure lines, fittings, and transmitter
- Installation and maintenance experience advantageous
- Maintenance intensive

20.2.2 Variable-Area Flowmeters

Variable-area flowmeters, also known as rotameters, provide practical solutions for many flowmetering applications. They are simple and inexpensive. The flow rate of gases and liquids can be determined simply, yet relatively accurately, with variable-area flowmeters. Every variable-area flowmeter consists basically of two components: a tapered metering tube and a float that rides within the tube.The fluid flows upward through a vertical tube whose diameter increases in the upward direction. The upward-flowing fluid lifts a float located in the tube to a height so that the annulus has an area, which results in an equilibrium of the forces acting on the float.

An important requirement for metering is the exact centering of the float in the metering tube. Three methods have found acceptance.

1. Through slots on the float head, the flowing fluid forces the float to rotate and center itself. This principle cannot be utilized with all float shapes. Additionally, there is a strong dependence on the fluid viscosity.
2. The float is guided by three ribs or three flats (ball floats), which differ from the metering tube cone in that they are parallel to the tube axis. A variety of float shapes are possible. Even for cloudy, opaque fluids, the reading edge remains visible.
3. A guide rod (Fig. 20-9) in the middle of the meter tube is used to guide the float.

In addition, a wide variety of float shapes are available. The weight, shape, and materials are adapted to the individual installations. Examples are shown in Fig. 20-10.

When in service, the pressure drop occurs primarily at the float because the energy required to produce the metering effect is derived from the pressure drop of the flowing fluid and to a lesser degree in the meter fittings. The drop in pressure at the float is dependent on its largest outside diameter and its weight, and therefore is independent of its elevation in the meter tube. The pressure drop through the fittings, however, increases as the square of the flow rate. The resulting pressure drop is the reason for the requirement of a minimum upstream pressure. Its accuracy is relatively low ($\pm 2\%$) and depends on precise knowledge of the process and the fluid [4].

The advantages and limitations of variable-area flowmeters follow.

Advantages

- Inexpensive
- No supply power for local indication
- Suitable for liquids, gases, and steam

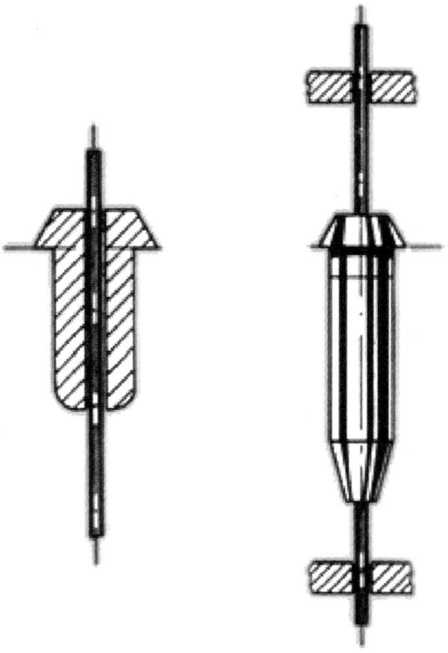

Figure 20-9 Float with a guided rod.

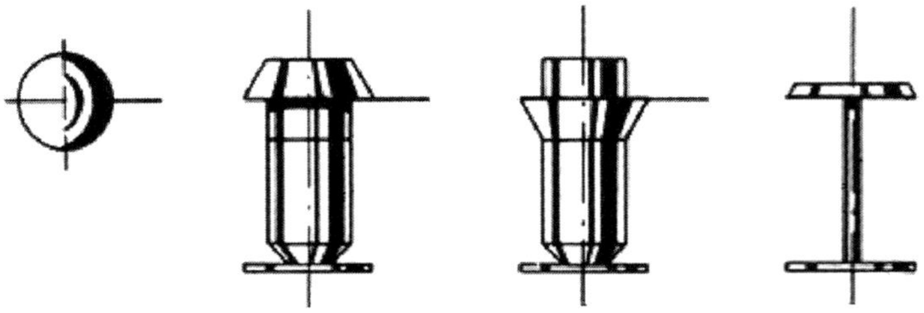

Figure 20-10 Float shapes: (a) ball float; (b) viscosity immune float; (c) viscosity nonimmune float; (d) float for low-pressure drop. (Courtesy of ABB, Inc.)

- No conditioning sections
- Simple meter construction; therefore, installation and maintenance friendly
- Indication with opaque fluids
- Metal meter pipe with converter
- Metal pipe meter can be sterilized and CIP tested

Limitations

- Vertical installation
- Constant pressure drop
- Affected by density and temperature changes
- Solids damage metering edge; otherwise, slight contamination allowed
- Affected by pulsation and vibration
- Expensive when exotic materials are required

20.2.3 Electromagnetic Flowmeters

Electromagnetic flowmeters, also known as magmeters, are a popular choice among instrument engineers, making up about 20% of flowmeter installations [9]. Faraday's law says that a conductor moving through a magnetic field produces an electric signal. In this case the fluid is the conductor, and electromagnetic coils surrounding the meter body generate the magnetic field. If an electrical conductor is moved in a magnetic field which is perpendicular to the direction of motion and to the conductor, an electrical voltage is induced in the conductor whose magnitude is proportional to the magnetic field strength and the velocity of the movement.

To utilize this operating principle, it is imperative that a magnetic field exist within the pipe and that the voltages induced can be measured without any form of interference [14]. Two coils generate the magnetic field that extends through the pipe only when it is not shunted by permeable pipe materials. For example, austenitic steel does not hinder the magnetic field; therefore, it is the most commonly used material for the meter pipe in an electromagnetic flowmeter. To prevent shorting out the signal voltage induced, the inner surface of the metering pipe must be electrically insulating. The signal voltage is measured at two electrodes which are in galvanic contact with the fluid. An additional requirement for the operation has already been mentioned: the fact that the fluid must be an electrical conductor. Therefore, a minimum conductivity of 20, 5, or 0.05 μS/cm is required. This is very dependent on the type of flowmeter [4].

Electromagnetic flowmeters are generally used to measure almost all electrically conducting liquids, pastes, slurries, and emulsions with excellent long-term stability and accuracy. Temperature, density, pressure, and viscosity have no major influence on measurement with this type of flowmeter [7]. Their various industry applications cover water and wastewater; the chemical and pharmaceutical industries; the food and beverage industries; the mining, aggregates, and cement industries; the pulp and paper industry; the steel industry; and the power, utility, and chilled water industries. Electromagnetic flowmeters have a rangeability up to 1000:1, depending on the maximum tolerable measurement error [9]. For flow velocities of 0.5 to 50 ft/s, accuracies are usually stated as a percentage of rate [9]. For lower velocities, accuracies are stated as a percentage of span. Typical velocities measured range from 3 to 15 ft/s for water and clean chemicals, 3 to 6 ft/s for abrasive fluids, and 6 to 12 ft/s for coatings and liquids with entrained air [9]. The major limitation of the magnetic flowmeter is that it cannot be used for hydrocarbon flow measurement, due to the low conductivity of hydrocarbons. Figure 20-11 shows a typical design of an electromagnetic flowmeter. Electromagnetic flow meters can be used in the food & beverage and pharmaceutical industries and is available with hygienic and flexible process connections. They are also applicable in waste water, sewage or sludge and chemical industries applications [6].

The advantages and limitations of electromagnetic flowmeters follow.

Advantages

- Unobstructed flow passage
- No moving parts
- No additional pressure drop

Figure 20-11 Standard design of an electromagnetic flowmeter. (Courtesy of ABB, Inc.)

- Essentially flow profile insensitive; therefore, no conditioning sections
- Unaffected by changes in temperature, density, viscosity, concentration, and electrical conductivity
- Favorable material selections for chemically aggressive and abrasive fluids
- Unaffected by contamination and deposits
- Especially suitable for hydraulic solids transport
- Can be sterilized and CIP tested
- Linear relationship between flow rate and measurement signal
- Meters in both flow directions (forward and reverse)
- Flow range setting can be optimized
- Minimum maintenance, but still maintenance friendly
- Certification approvals

LIMITATIONS

- For liquids only
- Lower conductivity limit of 0.05 μS/cm [4]
- Gas inclusions cause errors

20.2.4 Ultrasonic Flowmeters

Like the turbine flowmeter, the ultrasonic flowmeter (Fig. 20-12) is an inferential meter. An ultrasonic flowmeter

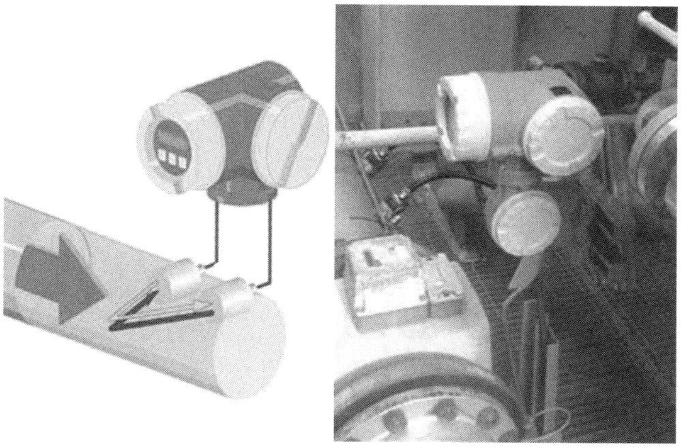

Figure 20-12 Ultrasonic flowmeter. (Courtesy of Endress & Hauser.)

sends a signal of known frequency across the flow stream and measures how the flow modifies it. This value is used to determine the flow rate. The liquid being measured must be relatively free of entrained solids or gas to reduce signal scattering. Ultrasonic meters are subdivided into two types:

1. *Doppler ultrasonic flowmeters* (Fig. 20-13) also send a signal across the pipe. However, with Doppler technology, the signal bounces off particles in the flow stream instead of off the other side of the pipe [21]. The flow particles are traveling at the same speed as the flow. As the signal from the transducer travels through the flow stream, its frequency shifts in proportion to the main velocity of the fluid. The reflected signal is detected by a receiver, which measures its frequency. The meter calculates flow by comparing the frequencies transmitted and those reflected.

2. *Transit-time meters* have both a sender and a receiver. A transducer sends an ultrasonic signal at an angle from one side of the pipe to the other side, and back [19]. The signal travels faster whe it travels with the flow than when it

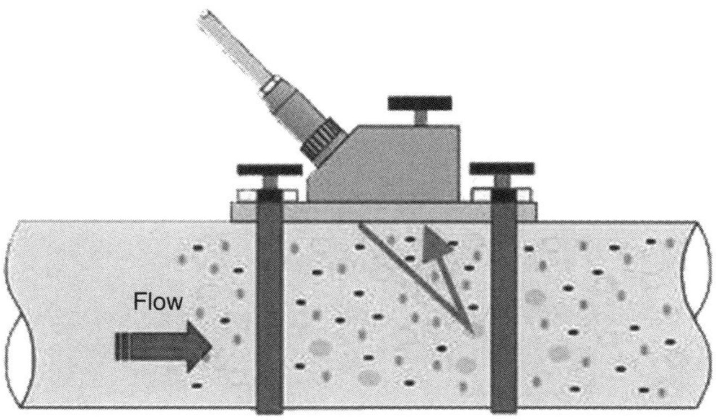

Figure 20-13 Doppler ultrasonic flowmeters work best on slurries and dirty liquids.

travels against the flow [19]. The flowmeter determines how long it takes for the signal to cross the pipe in one direction and how long it takes for the signal to cross the pipe in the reverse direction. The difference between these two times is proportional to the flow rate. In use, clamp-on transit-time technology represents a good fit for flow measurement applications where a challenging process liquid is the issue [11].

The advantages and limitations of ultrasonic flowmeters follow.

Advantages

- Unobstructed flow passage
- No moving parts [17]
- No additional pressure drop
- Favorable material selections for chemically aggressive fluids
- Linear relationship between flow rate and meter signal
- Minimum maintenance
- Meters in both flow directions (forward and reverse)
- For transit-time meters unaffected by temperature, density, and concentration
- Installation of transducers in existing pipelines possible, with on-site calibrations
- High rangeability in handling a variety of flow rates [17]

Limitations

- Sound beam must traverse a representative cross section; therefore, flow profile dependent, long conditioning sections required
- Errors due to deposits
- Transit-time meters require clean liquids
- Doppler meters only for slight contamination and few gas bubbles
- Doppler meters affected by sound velocity changes due to temperature, density, and concentration changes
- Unsuitable for heavily contaminated liquids
- Gas bubbles cause errors

20.2.5 Coriolis Mass Flowmeters

Coriolis flowmeters (Fig. 20-14) measure flow rates based on the mass of the fluid flowing through the pipe. Coriolis flowmeters are composed of one or more vibrating tubes. These meters have a sine-wave voltage applied to an electromagnetic drive, which in turn generates an oscillatory motion of the tube. This oscillatory motion causes the tubes to vibrate. When the tubes vibrate, it produces an angular rotation about its center. As the fluid accelerates away from the center, a resulting Coriolis force opposes the motion. This force generates the sine wave, which is measured and converted to the mass flow reading. The thickness of the tubing walls varies considerably from design to design. But even the sturdiest tubing will be thinner than the process piping. Long, bent tubes twist more easily than do short, straight tubes, so they will generate stronger signals under the same conditions. However, one problem is that corrosion and erosion cause thining of the tube walls inside a Coriolis flowmeter sensor. Curved-tube designs provide wider rangeability (up to 200:1), while straight-tube meters are limited to 30:1 to 50:1, with lower accuracy [4].

Figure 20-14 Coriolis mass flowmeter. (Courtesy of ABB, Inc.)

The operating principle of these flowmeters occurs in two different forms:

1. *Double-pipe measurement system*. The overwhelming majority of Coriolis instruments today are based on the double-pipe principle: with a flow splitter and two bent meter pipes (Fig. 20-15). This design has the advantage of temperature stability and, in particular, the decoupling of the meter pipe vibrations from external vibrations. The amplitudes of the vibrations which are required for determining the phase shift are measured between the two meter pipes and not relative to the housing. Possible vibrations of the housing therefore have no effect on the measurements [4]. Based on the appreciably more stable and defined signals, this system provides the most

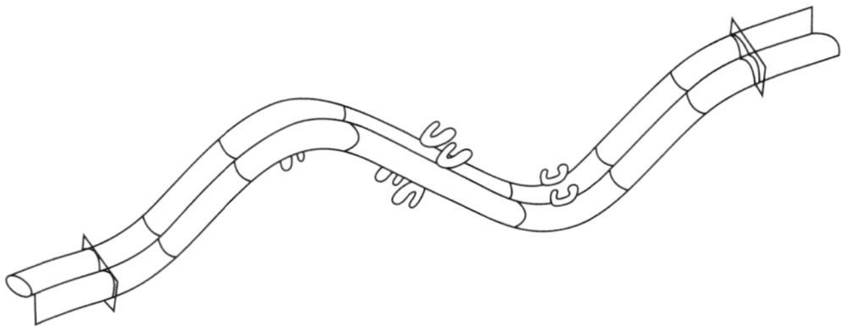

Figure 20-15 Double-pipe design.

accurate measurements coupled with insensitivity to outside influences. A well-designed double-pipe meter requires minimum energy to start and keep the system resonating and generates measurement signals for even the lowest flow rates.

2. *Single-pipe measurement system*. In addition to the double-pipe design there is a single-pipe design. To maintain the insensitivity to external vibrations, the meter pipes in this design are bent into loops (Fig. 20-16). The amplitudes of the vibrations, and thereby the phase shift, are measured between the pipe loops and not relative to the housing. This principle offers distinct advantages for the smaller meters because a flow splitter in not required. The straight single-pipe design has advantages in that it can be cleaned more easily, has a reduced drop in pressure, and is less harsh on the fluid itself. These advantages come with a lower accuracy and a higher sensitivity to external vibrations. Because of the straight meter pipe, the amplitude differences must be measured relative to the housing [4]. If the housing is also vibrating, the effects are difficult to compensate. In addition, the signals measured are appreciable smaller, which also contributes to the reduced accuracy, especially for the density measurement.

Figure 20-16 Bent single-pipe design.

It is somewhat difficult to start and keep a single pipe resonating. The elasticity of a pipe is directly related to its wall thickness. Therefore, vibrating straight pipes must be constructed thin and are available only for limited meter sizes. For abrasive or corrosive fluids the thin-walled sections of the meter pipe can add additional safety concerns.

Coriolis flowmeters are widely used in refineries, processing plants, and so on, for the transfer of petroleum products through trucks and the transfer of a charge stream to a reactor. In the food and beverage industries a deciding factor is the good cleanability of the instruments, even a double-pipe system. Furthermore, the highly accurate mass and density measurements of the materials are a great advantage. Compositions can be monitored online. The concentration of two-phase fluids can be determined from density measurements using special software. In the paper industry, Coriolis mass flowmeters were predestined for use in the coating and color kitchens. Problems always occurred due to the varying density of viscosity values. Due to its multivariability, flexibility, high accuracy, wear-free quality, and ruggedness, the Coriolis mass flowmeter continues to conquer new application fields. Although at first it may appear that the initial acquisition costs are higher, they often become negligible compared to the later savings due to more accurate and simpler fills. Compared to traditional instruments, the accuracy remains constant for a long period at minimum maintenance cost.

The advantages and limitations of Coriolis mass flowmeters follow.

Advantages

- True mass metering
- Multivariable, flexible, wear-free, and rugged
- Additional temperature and density measurements
- High accuracy (±0.1%), repeatability, and rangeability (100:1) [4]
- Unaffected by pressure, temperature, and density [1]
- No conditioning sections
- Meters in both flow directions (forward, reverse)
- Can be sterilized and CIP-tested
- Flow range settings can be optimized for flow rate and for density
- Self-draining

Limitations

- For liquids (and most recently, gas)
- Affected by gas inclusions (in liquid service)
- Vibration sensitive when installed improperly

- Material selections limited
- Large size limitations

Based on the advantages mentioned above, it is not difficult to understand why the Coriolis measurement principle is preferred to other measurement principles by more and more industries.

20.2.6 Thermal Mass Flowmeters

Thermal mass flowmeters are used almost entirely for gas applications. As the name implies, thermal mass flowmeters use heat to measure flow [18]. This method works best with gas flow measurement. It is difficult to get a strong signal using thermal flowmeters in liquids, due to considerations relating to heat absorption. This flow measurement requires additional measurements of pressure and temperature to calculate the mass flow rate. These corrective measures add cost and increase the complexity of the measurements; in addition, they decrease the measurement system accuracy. Thermal mass flowmeters provide mass flow rate directly without additional measurements or calculations. Using the normal density of a gas, the normal volume flow rate can be calculated. Two industrial methods are used for thermal gas mass flow-rate metering.

1. *Hot film anemometers*. These meters use the flow rate–dependent heat transfer of a heated body in a fluid. This flow rate–dependent cooling is not a function of the pressure and temperature but of the type and number of particles that affect the heated surface. The method determines the mass flow rate of the fluid directly. Figure 20-17 depicts the operating principle of a hot film anemometer. The sensor consists of two resistors, which are part of a bridge circuit. One resistor is at the temperature of the flowing gas, the other is heated electrically and simultaneously cooled by the flow. A control circuit applies heat to the resistor so that a constant temperature difference exists between the resistors. The power is then a measure of the mass flow rate of the gas.Typical applications for this type of thermal mass flowmeter (Fig. 20-18) are:

- Gas flow measurements in the chemical and process industries
- Compressed air balancing
- Gas burner control
- Digester gas and aeration measurements in sewage treatment plants
- Gas measurements in an air separation system
- Hydrogen measurements in processes
- Carbonization in breweries and soft-drink production

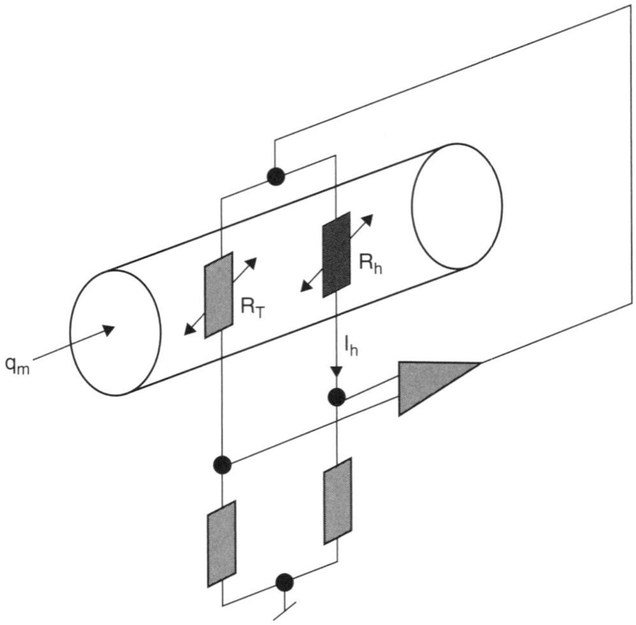

Figure 20-17 Operating principle of a hot film anemometer. [R_h, thin-film resistor, heated; R_T, thin-film resistor, fluid temperature; q_m, mass flow rate; I_h, heat flow; V, control circuit.

Figure 20-18 Thermal mass flowmeter. q_m, mass flow rate; P, constant electric heating power; L, thermal dissipation power. (Courtesy of ABB, Inc.)

2. *Calorimetric or capillary methods*. For very small pipe diameters or extremely low flow rates, which exist primarily in the gas analysis sector and in laboratories, the calorimetric method can be used. Figure 20-19 shows the principle of the calorimetric or capillary method of thermal mass flowmeters. The gas flows through a capillary that is heated with constant power. The mass flow rate can be calculated from the resulting temperature difference, the heat loss of the system, and an instrument constant. Typical applications of thermal mass flowmeters are in HVAC applications involving ventilation and temperature control, in flare gas measurements, in cogeneration plants, and in utility boilers.

The advantages and limitations of thermal mass flowmeters follow.

Advantages

- True gas mass metering
- No pressure and temperature corrections
- Neglible pressure drop [8]
- High accuracy
- Wide flow range
- No moving parts, shock resistant, dirt resistant [8]
- Rugged design
- Fast response time
- Easily sterilized

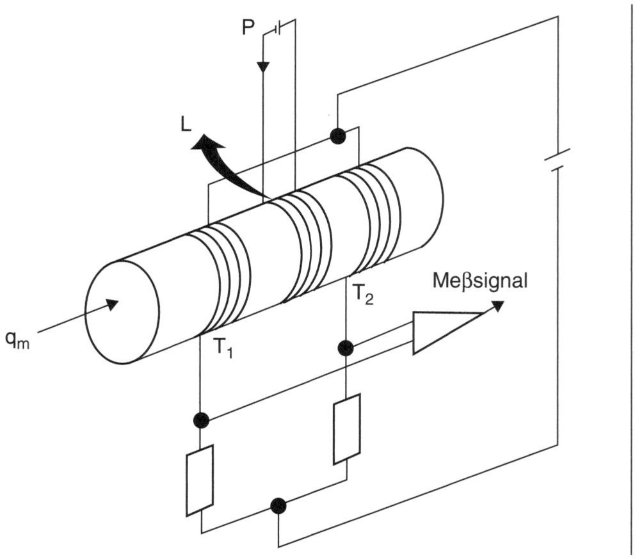

Figure 20-19 Operating principle of the capillary method.

Limitations

- Only for gases
- Conditioning sections

20.3 COMMON PROBLEMS OF FLOWMETERS

20.3.1 Liquid Carryover

When taking gas flow measurements, care must be taken to ensure that gas measurement is actually gas measurement and not gas and liquid measurement. During the transfer of gas through a pipeline, a significant amount of the gas being transferred must be a "dry pipeline-quality gas." This ensures that the gas is free from liquids. In gas flow measurements, free-flowing and entrained liquids distort the flow profile inside the meter tube, resulting in measurement errors. Liquids cause both high and low measurement readings and must be dealt with as soon as possible. The more free liquid that is eliminated in a gas stream, the more accurate the measurement. To prevent this, the gas–liquid separators must be working properly and in good condition. It is also important to make sure that the gas temperature is within the range required [15].

20.3.2 Dirt

Dirt in the fluid flow stream caused by corrosion or inadequate separation, for example, also poses a big threat to the measurement of flow. For example, erosion–corrosion produces pipe scrapes and debris that blocks the flowmeter filter efficiency, thereby restricting normal flow, and this alters the accuracies in the coefficient of discharge and the gas flowing profile measured [15]. To prevent errors in flow measurement, a filter of proper size (μm) should be installed upstream of the flowmeter.

20.3.3 Viscosity Effects

Viscosity is a fluid property. Through use of the Reynolds number (Re) it is possible to coordinate the viscosity effects and the sizing. For Re < 2300, laminar flow exists with a large viscosity effect. The transition region exists between approximately $2300 < \text{Re} > 3000$, above which turbulence exists [4]. In the turbulent region there are no limitations due to viscosity. Small Re values have varying effects, but this depends on the metering, method. For gas totalization or flow metering, the viscosity effect is hardly noticeable. Only for small variable-area flowmeters does the effect increase slightly in the lower flow ranges. The situation for liquids is quite different. The electromagnetic flowmeters and mass flowmeters are completely viscosity independent, although the pressure drop in the latter is affected as a function of the length of the flow path. Ultrasonic flowmeters have difficulties in the transition region from laminar to turbulent flow. In vortex, swirl, and turbine flowmeters, an increasing viscosity moves the minimum flow measurement to a higher value, thus reducing the flow range. Oval gear totalizers are ideal instruments for high viscosities. At higher viscosity they become more accurate because of the decrease in leakage losses. A special variable-area flowmeter float design decreases the effects of viscosity in this flowmeter.

20.3.4 Solids in a Fluid

Inherent solids in a fluid can mean various things. First, there is the usually undesirable contamination, then there are mixtures such as slurries, and finally, there is the hydraulic transport of solids. Contamination is undesirable because its amount and effects are difficult to predict. Gases can convey liquid or dust particles. Solid particles in a gas are dangerous because the gas flow velocities are high and therefore the kinetic energy of the particles can be appreciable, so that they may become more destructive. For differential pressure and variable-area flowmeters the dust particles damage the sensitive sharp metering edges. Unclean fluids affect lobed impeller meters by causing increased wear to their moving parts. Vortex and swirl flowmeters flush light contamination through the meter. If a nonconducting coating (e.g., oil, grease) in an electromagnetic flowmeter interrupts the electrical contact between the electrodes and the fluid, a meter with capacitively coupled electrodes must be installed. A special case is the ultrasonic flowmeter, because the Doppler principle requires foreign bodies as reflectors while the transit-time principle can tolerate only a minute number of particles and will not operate if gas bubbles are present. Deposits affect the sound path and cause errors. A prerequisite for proper functioning of a mass flowmeter is that the solid particles follow the vibrations, which again is a function of the fluid viscosity. Therefore, with an increasing mass and inertia of the particle, the danger of incorrect measurements increases.

20.3.5 Gas Content in a Liquid

The instruments used to meter liquids are volumetric meters which cannot differentiate between gas and liquid. Therefore, gaseous components in a liquid cause errors whose magnitude is equal to the proportion of gas in the mixture. A correct measurement should be expected from the Coriolis mass flowmeter. But here also errors occur because of the damping characteristics of the gases. Beyond these, there are side effects that must also be considered. For turbine flowmeters it might happen that larger gas bubbles could cause overspeeding. Cavitation is readily generated at higher flow velocities, particularly in vortex and swirl flowmeters. In the ultrasonic flowmeters using the transit-time principle, both the transit time and the damping are changed, so that even small gas bubbles can cause an effect.

20.3.6 Corrosion Risks with Aggressive Fluids

The effects of corrosion can be avoided only by proper material selection. Slight inattention in selecting the material for a gasket, for example, can result in an inoperative instrument. The complicated elements of volumetric totalizers are disadvantaged when it comes to material selections. Therefore, these meters are not preferred in corrosive applications. For differential pressure flowmeters it is important not only to consider that the orifice or nozzle must be corrosion resistant but also the pressure lines and fittings. In certain installations the injection of an isolating fluid or gas is used to prevent the entry of dangerous fluids. Variable-Area flowmeters made of special materials are expensive and are seldom installed; the same holds true for the mass flowmeters. Suitable solutions to the problems are offered only by electromagnetic and ultrasonic flowmeters. The smooth metering pipe can be lined with a corrosion-resistant liner such as PTFE. Transducers in ultrasonic flowmeters and electrodes in electromagnetic flowmeters are in direct contact with the fluid and therefore need to be protected against corrosion damage.

20.3.7 Vibration

Pipeline vibrations cause wear primarily on the moving parts and bearings in the volumetric totalizers. In the oscillating elements in vortex and mass flowmeters the vibration frequencies are superimposed on the measurement frequencies, causing errors. If resonance occurs, parts may fracture. The vortex flowmeter can be designed to be substantially insensitive to vibrations because of the separation of the sensor from the shedder and the oscillation compensation in the converter. The differential signals are essentially decoupled from external vibrations in the mass flowmeter because of its S-shaped double-pipe system. Additionally, the digital filter technology assures wide insensitivity to external vibrations. Should the vibrations be the same as the excitation frequency, the measurements could be affected. The relatively large mass of the variable-area float makes it sensitive to vibrations. Vibration dampeners and wall mounting should be utilized.

20.3.8 Pulsation

Pulsation effects are a function of the inertia of the metering system. Only when the metering element can follow the pulsations without lag will the measurement be unaffected. Meters that contain moving parts are therefore subjected to increased wear. The elimination of pulsation through the use of pulsation control devices is an important step to take. Pulsation has a tendency to introduce meter errors [12]. It is essential that a damping device be provided. Oval gear meters have so much inertia that they provide a degree of self-damping sufficient for averaging purposes. The metering error will increase somewhat. Turbine, vortex, and swirl flowmeters measure the flow essentially without inertia and in the connected converters include time-constant elements. Again the average is obtained with slightly larger errors. Gas turbine flowmeters are in danger when the rotational speed increases too quickly, resulting in more bearing wear should overspeeding occur. As a result of the nonlinear relationship between flow rate and differential pressure, the error in differential pressure flowmeters is also affected by pulsations. Therefore, a damping device is definitely recommended, preferably using storage volumes or expansion tanks in the fluid line. The Hodgson number can be used to calculate the volumes required. The damping elements in the differential pressure transmitter accommodate small pulsations. The variable-area float tends to dance when pulsations are present. Therefore, damping elements must also be used here. Metal pipe flowmeters can have damping elements installed as an option.

20.4 FLOWMETER INSTALLATION AND MAINTENANCE

The user of measurement information expects a problem-free installation of the meter and thereafter only wants to see exact measurement values; the meter itself is no longer of interest. Even though modern technology comes close to fulfilling these expectations, regular monitoring can prevent failures. Included in the selection of a meter is therefore a question regarding the capabilities of the maintenance personnel. The meter is installed in the pipeline in a relatively straightforward manner using flanged or threaded connections while considering the conditioning sections (e.g., swirl and thermal mass flowmeters). At this point, the requirement for stress-free installation must be stressed because very often the meters are installed in existing pipelines with existing stresses. Wafer designs (sandwich) require exact centering. There are ultrasonic flowmeters for which the transducers (transmitter/receiver) must be welded into the pipeline wall. This method produces error-free measurements only when an on-site calibration is conducted which takes into account the actual pipeline geometry. Table 20-1 provides an assessment of the flow regime effects for the applicability of various flow-measuring instruments. Table 20-2 provides guidelines on the proper installation and maintenance of flow-measuring instruments.

20.4.1 Flowmeter Installation

Improper flowmeter installation will in turn cause significant flow measurement problems. The first step leading to the optimum operating efficiency of a flowmeter is its proper installation. Therefore, strict measures should be taken into consideration before installing a flowmeter. The most important measure is to ascertain the function the flowmeter is to perform. The installation site is included

TABLE 20-1 Flow Regime Effects for Applicability of Various Flow-Measuring Instruments[a]

		Flow Regime			Conditioning Section		
	Instrument Type	Laminar	Transition	Turbulent	Upstream	Downstream	Shock-Type Operation
Direct	Oval gear	N	N	N	N	N	Wear
	Oscillating piston meter	N	N	N	N	N	Wear
	Lobed impeller meter	N	N	N	N	N	Resonance danger
Indirect volume totalizer	Rotary vane meter	N	N	N	N	N	N
	Woltman meter	N	N	N	$5 \times D$	$3 \times D$	N
	Turbine flowmeter	N	N	N	$15 \times D$	$5 \times D$	Overspeed wear
	Vortex flowmeter	U	U	N	$15\text{–}25 \times D$	$5 \times D$	N
	Swirl flowmeter	U	N	N	$3 \times D$	N	N
	Differential pressure flowmeter	N	N	N	ISO 5167	ISO 5167	Error
Flowmeter	Variable-area flowmeter	N	N	N	N	N	Error
	Electromagnetic flowmeter	N	N	N	$3 \times D$	$2 \times D$	N
	Ultrasonic flowmeter	N	U	N	$20 \times D$	$5 \times D$	N
	Coriolis mass flowmeter	N	N	N	N	N	Error
	Thermal mass flowmeter	N	N	N	$15 \times D$	$5 \times D$	N

[a]N, no effect; U, unsuitable; $1 \times D$, conditioning section $1 \times D$ in length.

TABLE 20-2 Installation Maintenance of Flow-Measuring Instruments

	Instrument Type	Installation Tasks	Pipeline Upstream from Meter	Maintenance During Operation	Self-Monitoring	Service
Direct	Oval gear	Connect flanges	Filter	None	No	Local disassembly possible
	Oscillating piston meter	Connect flanges or threads	Filter recommended, no conditioning section	None	No	
	Lobed impeller meter	Connect flanges	Filter	Monitor lubrication	No	
Indirect volume totalizer	Rotary vane meter	Connect flanges or threads	No conditioning section	None	No	Meter insert exchange possible
	Woltman meter	Connect flanges	No conditioning section	None	No	
	Turbine flowmeter	Connect flanges, electrical connections	No conditioning section	Monitor possible foreign lubrication	No	Exchange primary
	Vortex flowmeter	Connect flanges or install wafer design, electrical connections	No conditioning section	None	Continuous plausibility and error monitoring with error messages	Electronic control function and test values; sensor can be exchanged
	Swirl flowmeter	Connect flanges, electrical connections	No conditioning section	None		
	Differential pressure flowmeter	Left meter in flanges; connect pressure tap lines, fittings, transmitter, supply power	Long conditioning section	Regular inspection recommended	No	Direct measurement at transmitter
Flowmeter	Variable-area flowmeter	Connect flanges or threads	No conditioning section	None	Continuous plausibility and error monitoring with error messages	Glass flowmeter with Snallin design
	Electromagnetic flowmeter	Connect flanges, electrical connections	Short conditioning section	None		Electronic control function and test values, simulator
	Ultrasonic flowmeter	Connect flanges, or weld stubs, electrical connections	Long conditioning section	None	Indicate signal losses	
	Coriolis mass flowmeter	Connect flanges, wall mount, electrical connections	No conditioning section	None	Continuous plausibility and error monitoring with error messages	Electronic control function and test values
	Thermal mass flowmeter	Connect flanges or install wafer design and insert sensor	Conditioning section	None	Error messages	Sensor can be exchanged

in the selection of the meter because the ambient conditions can only be adjusted to the requirements of the particular instruments to a limited degree. If gas measurements are required, an electromagnetic flowmeter cannot be used, as the electrical conductivity required is not present.

To prevent improper flowmeter installation, there are four basic measures to consider: (1) temperature and pressure, (2) vibration and pulsation, (3) reverse flow metering, and (4) power supply.

20.4.1.1 ***Pressure and Temperature*** Housing strength due to its wall thickness and material selection; mechanical tolerances for temperature expansion; gasket type and material; and sensor limits and the effects on the transmitting element are some of the considerations that affect instrument selection at high temperatures and pressures. For variable-area flowmeters, differentiation must be made between glass and metal pipe meters. The pressure and temperature limitations for glass tube meters naturally lie far below those for metal pipe meters. Differential pressure flowmeter selection includes the fittings, pressure lines, and converter. These parts must be designed appropriately. The limiting values for the converter usually are controlling for the entire system. The upper temperature limit for the converters is usually around 120°F; higher temperatures must be reduced by isolators in the pressure lines [4].

20.4.1.2 ***Reverse Flowmetering*** An extreme form of pulsation can lead to reverse flow. Additionally many pipelines are designed specifically for bidirectional flow. Are there meters that can also measure reverse flow? The direct volumetric totalizers can naturally reverse their direction and totalize backward when the secondaries include appropriate provisions. The best solutions are provided by the electromagnetic and mass flowmeters, which provide all measurements for both flow directions and can switch automatically.

20.4.1.3 ***Power Supply*** Measurement and control technology operates with measurement signals that must be transmitted over long distances. Therefore, all totalizers and flow-rate meters provide appropriate output signals whose generation requires electrical power. The flowmeters additionally require a power supply for their operation. When only local indication is required, electrical lines are not required for the totalizers and the variable-area flowmeters. In very few instances these days, the standard 0.2 to 1 bar pneumatic signal is used. These meters must be designed with appropriate transmitters whose power is provided by a pressurized air supply of 1.4 bar. Pneumatic transmitters are available for differential pressure and variable-area flowmeters.

20.4.1.4 ***Vibration and Pulsation*** Considerations regarding vibration and pulsation effects have been discussed in Sections 20.3.7 and 20.3.8.

20.4.2 Flowmeter Maintenance and Operating Characteristics

Tables 20-3 and 20-4 provide the operating characteristics of flow-measuring instruments from the quantitative and qualitative perspectives. The key factors are (1) flow ranges, (2) error limits, (3) dynamic response, and (4) pressure drop.

20.4.2.1 ***Flow Ranges*** The flow ranges for oval gear meters are a function of the viscosity of the fluid. They are 10:1 for viscosities up to approximately 300 mPa·s and increase to 20:1 for viscosities of 1104 mPa·s [4]. The conditions are exactly the opposite for indirect totalizers and flowmeters. With increasing viscosity the start of the linear range increases and the flow ranges are reduced. Because of the square-root relationship that exists in differential pressure flowmeters between the flow rate and the differential pressure, the flow range is a function of the accuracy required. A flow range of 3:1 promises very high accuracy. Electromagnetic and ultrasonic flowmeters represent a special situation in that their range end values can be set and thereby optimized. Thermal mass flowmeters achieve flow ranges up to 150:1 [4].

20.4.2.2 ***Error Limits*** The error limits represent controversial specifications, because they are values published by the manufacturers for ideal boundary conditions. In practice, deviations from ideal conditions are common, so that additional stipulations must be made for instrument accuracies. It is important to realize that the accuracy may be based on the value measured (% of rate) or on the range end value (% of maximum). Through special calibrations, better accuracies can be achieved, which are generally valid only for a specific time period. Exactly how the error limits can be affected by contamination, wear, and physical changes is not always known. The values listed in the table are based on the values measured. Based on these values a classification system for variable-area flowmeters was established. VDI/VDE 3513, Sheet 2 represents a combination of error values based on both the measured and range end values. Of the error values for an accuracy class, 75% are a percentage of the rate values and the other 25% are a percentage of the maximum values.

20.4.2.3 ***Dynamic Response*** The term *time response* defines the output vs. time relationships that exist after a step change in the metered value has occurred (step response). The characteristic value is the time constant, the time required to reach the actual value. After a step change

TABLE 20-3 Operating Characteristics of Flow Instruments: Quantitative

	Instrument Type	Flow Range	Error Limits (% of rate)	Dynamic Behavior (time constants)	Pressure Drop $Q_{v,\max}$ (bar)
Direct	Oval gear	2:1 to 10:1	0.1 to 0.3	—	4
	Oscillating Piston meter	5:1 to 250:1	0.2 to 2	—	3
	Lobed impeller meter	20:1 (50:1)	1	—	0.003
Indirect volume totalizer	Rotary vane meter	100:1 to 350:1	2 to 3	—	0.25 to 0.75
	Woltman meter	100:1 to 1250:1, 0	2 to 3	—	0.005 to 0.5
	Turbine flowmeter	5:1 to 20:1	0.5 (liquid), 1 (gas)	—	0.5 to 1
	Vortex flowmeter	15:1 to 20:1	0.75 (liquid), 1 (gas)	0.2	0.7 (water), 0.007 (air)
	Swirl flowmeter	10:1 to 25:1	0.5	0.2	0.3 (water), 0.005 (air)
	Differential pressure flowmeter	5:1 (10:1)	0.75 to 0.3	—	0.005 to 1; function of DP and beta ratio
Flowmeter	Variable-area flowmeter	12:1	Class 1.6/2.5	—	0.005 to 0.2
	Electromagnetic flowmeter	100:1	0.25	0.2	Same as pipeline
	Ultrasonic flowmeter	10:1	0.5 to 1.0	1	Same as pipeline
	Coriolis mass flowmeter	100:1	0.15	0.2	1
	Thermal mass flowmeter	40:1 to 150:1	1	0.012	0.002

the indication reaches 63% of the actual value after time. After 5T, which corresponds to the time required to reach the actual value, the indication is essentially equal to the actual value. For totalizers, information regarding their time response is seldom given because both the moving masses and the viscosities of the fluids vary. For variable-area flowmeters the same comments apply, while for differential pressure flowmeters the pressure lines with their various fittings influence the time response.

20.4.2.4 Wear Long-term meter service life is understandably an important requirement of users. Therefore, the mechanical wear must be kept to a minimum. Wear is caused primarily by abrasion of the metering elements (by the fluid) and by the bearing friction of the moving parts. An electromagnetic flowmeter can be considered ideal because the smooth inside wall shows wear only when highly abrasive fluids are metered, such as lime slurries, sand–water mixtures, or coal–water mixtures in solids transport. If the walls are lined with polyurethane or soft rubber, even these fluids rarely cause difficulties. Even though vortex and swirl flowmeters have no moving parts, they do have parts that extend into the flow stream. A specific size, hardness, and edge sharpness of solid particles in the fluid cannot be exceeded. Small dust components in gases and plastic suspensions are permissible. The metering edge of the variable-area flowmeter floats is also machined precisely. It may not be damaged. Bearing wear in the totalizers and their rotating meter bodies can be added to the difficulties mentioned above. A small number of hard particles in the fluid can destroy these meters. Ideal are metered fluids, which have a tendency to lubricate.

20.4.2.5 Material Compatibility Material selection always requires information about corrosion problems. Some fluids are chemically harmless but can become corrosive when small amounts of other materials are present, perhaps only as contamination. Therefore, care must be exercised. To make reliable predictions about the compatibility between a given chemical and a material of construction, several fundamental questions must be answered [2]:

- What corrosive agents are in the process and in what concentration range?
- What is the process temperature range?
- What material is being used for the piping?
- What cleaning cycles exist, and what fluids are used in these cycles?

TABLE 20-4 Operating Characteristics of Flow Instruments: Qualitative

	Instrument Type	Moving Parts	Wear Parts	Material Selection for Wetted Parts	Cleaning/ Sterilization
Direct	Oval gear	Oval gears, gear train	Bearings, gear teeth	Oval gears and housing of gray cast iron, SST, bronze, bearings of hard carbon, SST	−
	Oscillating piston meter	Piston, gear train	Bearings, piston	Housing and metering chamber of gray cast iron, SS, bronze, Duroplast, piston of gray cast iron, hard rubber, carbon, PCTFE, tantalum, plastic	+
	Lobed impeller meter	Impeller, gear train	Bearings, impeller	Piston and housing of alum-alloy or gray cast iron, bearings of SS	−
Indirect volume totalizer	Rotary vane meter	Rotary vane, gear train	Bearings	Housing of brass, metering insert of plastic, axles of SS	−
	Woltman meter	Rotary vane, gear train	Bearings	Housing of gray or ductile cast, rotary vane and metering insert of plastic, brass, SS	−
	Turbine flowmeter	Rotor	Bearings	Rotor and housing of SS, bearings of sapphire, tungsten carbide	−
	Vortex flowmeter	None	Insignificant wear	Rotor and housing of SS, bearings of sapphire, tungsten carbide	+
	Swirl flowmeter	None	Insignificant wear	Housing of SS (1.4571[316Ti]); guide bodies of 1.4571[316Ti], Hastelloy C; sensor prot. of 1.4571[316Ti], Hastelloy C	+
	Differential pressure flowmeter	None	Metering edges	SS (1.4571[316Ti])	+
Flowmeter	Variable-area flowmeter	Float	Metering edges	SS (1.4571[316Ti]), Hastelloy C, PTFE, PVDF, glass	+
	Electromagnetic flowmeter	None	Insignificant wear	Liner of hard/soft rubber, PFA, PTFE, electrodes of 1.4571[316Ti], Hastelloy, tantalum, platinum, carbon	+ +
	Ultrasonic flowmeter	None	Insignificant wear	SS (1.4571[316Ti]), Hastelloy C	+ +
	Coriolis mass flowmeter	None	Insignificant wear	SS (1.4571[316Ti]), Hastelloy C, 1.4435	+
	Thermal mass flowmeter	None	Insignificant wear	SS, Hastelloy C, ceramic	+ +

- What is the velocity? (particularly important when handling sulfuric acid)
- What are the environmental cracking and degradation risks, such as risk of stress corrosion cracking, hydrogen embrittlement, and swelling of polymers and elastomers?

With answers to these questions it is possible to make a prediction regarding process compatibility. The variety of possible materials is especially restricted where complicated and difficult-to-manufacture parts are required. This is particularly true in volumetric totalizers and to a certain extent in swirl, mass and thermal mass flowmeters. At first glance differential pressure flowmeters would seem to present no problems. But not only must the orifice plate or the nozzle be manufactured of resistant metals, but the pressure lines, fittings, and pressure transmitter must also be made of suitable materials. Sometimes constant flushing of the pressure lines is advised. Now it becomes a question of cost. This is the case when variable-area flowmeters are made of resistant materials. They are installed only when supply power is unavailable at the metering location. The material selection problem is solved almost ideally for electromagnetic flowmeters because the PTFE liner material can be used for almost all fluids. Platinum is exceptionally good as an electrode material. It may be possible to solve the electrode material problem by using the capacitive electrode design. The ultrasonic flowmeter requires proper protection tube material for the sound transducers (transmitter and receiver).

20.4.2.6 Cleaning and Sterilization Why must a closed pipeline be cleaned? There are various answers: deposits due to sedimentation or adhesion reduce the cross sections, crystal formations block the flow, and residue can contaminate the product. Therefore, the ability to clean is a determining factor for meter selection. Where deposits are to be expected, meters with moving parts are seldom installed. In installations of swirl, vortex, and thermal mass flowmeters, difficulties can also be encountered. In the ultrasonic flowmeter, the sound path is altered so that erroneous measurements may occur. Nonconducting deposits affect electromagnetic flowmeters unless the same meter type with capacitive electrodes is installed. Electrically conductive deposits short out the signal and cause erroneous measurements. Deposits are removed by flushing, dissolving, or by mechanical cleaning using a brush or scraper. A prerequisite for using a scraper is a pipeline with a defined diameter without internal projections. This requirement can only be met by an electromagnetic flowmeter, or perhaps by an ultrasonic flowmeter. Among the residues remaining in the pipeline are bacteria in the food industry, which can spoil the product. Thorough cleaning and sterilization using steam, liquid cleaning agents, acids, and bases cannot be avoided. This is usually accomplished using CIP (clean in place) and SIP (sterilize in place) procedures in which all elements in the system remain in place. A CIP/SIP process is performed at high temperatures with cleaning agents such as dilute NaOH. However, CIP/SIP processes are often performed before a shutdown, such as overnight or before a weekend. If the piping is not designed to drain completely, cleaning liquids can puddle, giving them hours, or even days, to corrode the materials. CIP capability is determined by testing. Certificates have been granted for variable-area, electromagnetic, thermal, and Coriolis mass flowmeters.

20.5 CALIBRATION AND CERTIFICATION

20.5.1 Why Calibrate?

In a company, the need for rigorous traceability and compliance of test equipment to national standards to assure that everyone uses the "same measurement stick" is essential both for suppliers who manufacture the products and for users who install them. This is growing in importance due particularly to the increasing national and international division of work and the resulting requirements for interchangeability. In addition to the technical reasons, there are legal viewpoints. Relevant directives must also be observed together with contractual obligations with the purchaser of the products (assurance of the quality of the product) and the responsibility to bring to the marketplace only those products whose safety in regulated usage is not compromised by faults. As far as contractual responsibilities for the accuracy of measurement instruments are concerned, nonfulfillment of these requirements is a failure of a guaranteed property. Proof of the selection of adequate test equipment and its proper operation within the scope of product liability is of great importance, because systematic and completely documented test equipment monitoring, in combination with continuous proof of the adequacy of the test equipment for the task, is essential for possible exoneration [4].

20.5.2 Flow-Rate Calibration Methods

There are four common measurement calibration methods:

1. *Volumetric method*. The procedure is to measure the volume flow rate of a liquid using a calibrated volume tank and measuring the fill time.
2. *Gravimetric method*. This method is comparable to the volumetric method; however, the volume of the fluid is measured by weighing, taking into account the density and weight of displaced air. The calibrations can be conducted using either a standing start–stop operation (opening and closed a shutoff device) or a flying start–stop operation (operation of a diverter device). Furthermore, a distinction is made between a static and a dynamic method, the later not requiring a diverter device.
3. *Comparison method*. The liquid flows through the instrument being calibrated and a second flow-rate measurement instrument that was calibrated previously.
4. *Pipe test section method*. This method employs a volumetric measurement instrument consisting of a pipe section with a constant cross section and known volume. The flow rate is derived from the time required for a plug, driven either mechanically or by the fluid, to traverse the pipe section. In test stands with the highest accuracy requirements, a static weigh method is used almost exclusively in conjunction with a diverter using a flying start–stop operation [4].

20.5.3 Boundary Conditions and Measurement Fixtures [4]

During the design of a calibration test stand, the following series of requirements must be met:

1. The flow must be stationary.
2. The flow in the obstruction free inlet section must be axisymmetric and free of swirl and pulsation.
3. The reference flowmeter or the calibrated test section for measuring the flow rate or flow must satisfy the requirements of ISO 4185 and ISO 8316.
4. The flow range of the reference flowmeter of the calibration standard must be of the same range as the range of the instrument being tested. The error limits of the reference standard must not be larger than one-third of the error limits for the instrument being tested.
5. The measuring system and reference standard must be described in detail, including the traceability of the reference standard and the uncertainty of the measuring system. The calculation of the uncertainty of the flow-rate measurement must be in accord with ISO 5168, ISO 7066-1, and ISO 7066-2.190.
6. The flowmeter primary must be installed between a straight, unobstructed inlet section at least 10 × *D* long and an unobstructed outlet section at least 5 × *D* long. If swirl-free velocity profiles are required, a flow straightener must be installed.
7. During the calibration the flowmeter primary must be completely filled with fluid. In addition, the physical characteristics of the measurement fluid (e.g., density and viscosity) must also be considered. Analogy calculations can be made using the Reynolds number, which takes flow velocity conditions into account. Naturally, the other equipment on the test stand also has accuracies, which must be so good that they have only a minimal impact on the flow-rate measurement results.

For the test stand schematic shown in Fig. 20-20, the water flows through two comparison standards (2), which cross-check each other. The flowmeter being calibrated (3) is installed in a long, undisturbed pipe section. The control valve (4) and pump (1) determine the flow rate. The water can then either be returned directly to the supply reservoir (8) or by actuating the diverter (7) into the tank (5), where its mass can be measured. The scale arrangement (6) itself is checked in periodic intervals with weights by the certification agency.

20.6 LACT AND PROVER DESCRIPTIONS [16]

20.6.1 What Is a LACT Unit?

A LACT unit is an assembly of skid-mounted equipment and piping designed and configured to accurately measure

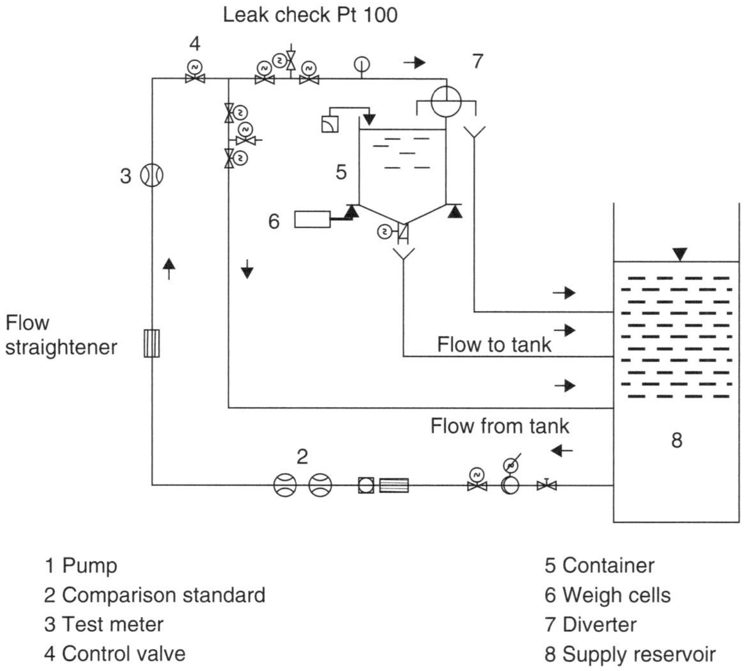

Figure 20-20 Test stand for gravimetric calibration and comparison calibration methods.

and prove the quantity and quality of a liquid. A LACT unit is a piece of equipment upon which both parties, buyer and seller, agree, which is used to transfer the custody of a liquid automatically from one responsible party or location to another. A LACT unit consists of piping and associated skid-mounted equipment that will measure the volume and quality of a fluid precisely.

20.6.2 What Is a Meter Prover Used For? [16]

A meter prover is used to verify the quantity of fluid measured by the LACT meter. A meter prover is made up of valves, detector switches, and a calibrated section that is used to check the quantity of fluid measured by the meter. The following are specific attributes of LACT and prover units:

1. Piping and pumps can be smaller because the liquid can be shipped continuously instead of in batches as required when tank measurement is used.
2. The inventory is smaller because only one small tank is required as a buffer for the irregularities of field production. This reduces cost, moves product to market faster, and reduces risk from loss due to a tank being damaged.
3. LACTs protect the environment because emissions to the atmosphere are reduced by reducing stored volume and reducing the time any volume needs to be stored.
4. One LACT unit is less expensive than one large measuring tank. Two tanks are required for tank measurement.
5. There is less maintenance on LACT units since they are smaller. Tank painting, repair, and cleaning are very expensive.
6. There is less impact on the environment since a LACT unit takes up less space and generally has fewer emissions than a tank.

20.6.3 Operation of a LACT Unit [16]

The operating guidelines for a LACT unit are as follows:

1. Before any liquid is directed to a LACT unit, all pumps should be off and all valves should be closed. Many components, especially the meters, could be damaged by the rush of air that is caused by the incoming fluid followed by the slug of oil. PD meters will overspeed on the rush of air, and when the slug of oil hits them, the speed will change dramatically and damage will occur.
2. Turn on the power to the LACT unit and prover if it has one. All equipment not needing power should be turned off at this point.
3. If the LACT unit is equipped with a pump, flood the pump slowly with fluid. Vent off at its high point any air trapped in the pump at the valve provided.
4. Open all vent valves on the LACT unit and prover if it is so equipped and slowly crack open the inlet valves and other valves that direct fluid to the prover and other runs. The discharge valves should remain closed.
5. Allow the LACT and prover to fill with fluid while checking for leaks and releasing the air to drain or atmosphere. Filling the equipment slowly protects the equipment from surges, and if there are any leaks, prevents a large amount of fluid from being lost.
6. Start the pump and slowly open the discharge valves and set the back pressure. As a rule of thumb, the backpressure minimum should be 20 psig above the vapor pressure of the fluid being measured. Check the meter to make sure that it is registering volume.
7. Observe the sample system to ensure that it is taking a sample at the proper intervals. Next to the meter that checks the quantity of fluid, the sampler is the next most important piece of equipment because it is the check on the quality of the fluid.
8. Increase the volume to the desired flow rate, set the pressure control and flow control valves, and open the inlet and outlet block valves to their fully open positions.
9. Check all remaining components for proper operation.
10. Calculate the volume and flow totals manually to ensure that the mechanical and flow computers are working properly.
11. The frequency of meter proving will be determined by the use of meters. For batch operations the minimum is once at the beginning of the batch run, in the middle, and toward the end. For continuous operation each meter should be proved at a minimum of once per day for the first 30 days to develop meter characteristics. After the first month the interval can be changed to weekly if the meter repeats continuously. After six months of weekly proves, the meter proves can be monthly. The conditions above are minimums. As a rule, meters should be proved

when either the flow rate, temperature, pressure, or density changes more than 10% in either direction for more than 10 minutes.

12. All calculations should be checked manually at regular intervals to ensure that no errors have been introduced into the computer. As the system becomes more complicated, the calculations should be checked more frequently.
13. All LACT equipment, BS&W monitors, temperature transmitters, pressure transmitters, and so on, should be checked monthly. Follow the manufacturer's instructions for calibration.
14. The prover should have a water draw (a check of the volume between the detector switches) before it is put into service on site. During transportation the prover piping could have gotten damaged or the detector switches could have been reset. After the first water draw on site, the prover should be checked for volume yearly. If any part of the prover is disassembled between the detector switches it should have a water draw preformed. If the prover switches are changed or the position of the switches are changed, the prover should have a water draw preformed.

20.6.4 LACT Unit Components [16]

20.6.4.1 Meters

20.6.4.1.1 Positive-Displacement Meters Positive-displacement (PD) meters for LACT units usually range in size from 2 to 16 in. and in flange ratings from 150 ANSI to 600 ANSI. Flow rates will be from 60 to 13,000 bbl/h. Larger sizes are available, but the cost of a meter over 10 in. tends to be very high. In cases of higher flow rates, more meter runs are used. PD meters are used for fluids that are too viscous for turbine meters.

The linearity of PD meters is normally ±0.25% and the repeatability is normally ±0.02%. The turndown range of a PD meter is usually 5:1. The temperature range for a PD meter is normally −25 to +125°F. The normal stackup (equipment used to record flow) for a PD meter in LACT service when no power is available is a direct stack with no compensation to a right-angle drive. The right-angle drive is used to attach the high-frequency pulser that is required for proving. Sometimes the high-frequency pulser is part of the stack, and then a right-angle drive is not required. On top of the high-frequency pulser a ticket printer is mounted to record the total flow over a period of time. With the ticket printers there are numerical displays that show the total flow in barrels and tenths of barrels. An accumulative barrel counter that is not resettable is also included as a safety measure. A low-frequency pulser, normally one pulse per barrel, is the last item in the stack and is used to detect meter or stack failure. When power is available, the stackup is the same as above, but normally the ticket printer is omitted and the high-frequency pulses are taken to a flow computer for flow calculations. Registers, both resettable and nonresettable, are recommended for backup.

20.6.4.1.2 Turbine Meters Turbine meter sizes range from 1 to 24 in. and ANSI 150 to ANSI 2500. The flow rates through a turbine meter range from 9 to 57,100 bbl/h. The temperature range of a turbine meter is −50 to +250°F for the smaller sizes and −50 to +700°F for the larger sizes. Turbine meters are used on less viscous fluids. The linearity of a turbine meter is ±0.025% and the repeatability is ±0.02%. The turndown is 10:1. For LACT unit application the turbine meter pulse output is sent to a flow computer, where the total flow and flow rate are computed. A totalizer powered by a backup battery is recommended on the turbine meter in case of power failure.

20.6.4.2 Strainers Fabricated in-line basket strainers are utilized in conjunction with flow meters, pumps, and compressors. Permanent in-line basket strainers are designed to allow flexibility in meeting material and construction requirements for a variety of applications. Basket strainers can be equipped with quick-opening closures or the closure can be a blind flange. Differential pressure gauges and/or switches are normally placed across the strainer basket to signal a high differential. An air eliminator is always placed on top of the strainer closure to allow trapped air to escape when filling and operating the strainer. This air eliminator will remove some air from the fluid as it passes through the strainer, but it is not a replacement for a deareator, which is designed to eliminate air and gas from the system. The basket for the strainer is constructed of carbon or stainless steel with $\frac{1}{8}$-in.-diameter holes on $\frac{3}{16}$-in. centers. Mesh wire cloth is available in mesh sizes of 20 to 325 for lining the strainer to remove smaller particles. Specification for strainers meet or exceed the following specifications:

Pipe ASTM A53 Grade B
Flanges ANSI B16.5
Studs A193, B-7
Nuts A194, 2H
Gaskets Flexataulic
Heads A516 Grade 70

Caps A234 WPB
Couplings A105

20.6.4.3 Prover Connection Tees All configurations of LACT units include two flanged prover connecting tees. These tees (one upstream and one downstream of the double block and bleed valve) are used to divert the flow through the prover. These tees can be furnished with connections for temperature and pressure transmitters. Valves are normally provided between the tee drop and the connection to the prover. These valves can be full port or reduced port, but care must be taken not to cause a high-pressure drop when the fluid is routed through the prover. When more than one LACT meter run is connected to a common prover, these must be block and bleed valves to ensure that no false fluid bypasses the prover. Optional hose connections with quick disconnects can be provided. Usually 6 in. is the largest size for hose connections. Caps are provided for these quick connects to prevent the entry of water or dirt when the LACT system is not connected to a prover. Due to the environmental concerns over spilled liquids, portable provers are becoming less and less popular.

20.6.4.4 Backpressure Valves and Flow Control Valves Backpressure valves are placed after the meter and meter prover connections to maintain a backpressure on the meter and prover system. The backpressure valves are normally set at 20 psi over the vapor pressure of the fluid that is being measured. Backpressure valves maintain backpressure on the meter and prover. These valves can also be flow control valves, which are used to maintain the desired flow rate. Proving must be done at the same flow rate as when the meter is in normal operation. When a prover is put online or taken off line or when a tank level that feeds the charge pump changes, a change in flow rate will take place if only a backpressure valve is used. Since a meter must be proved at the flow rate at which it operates, a constant flow rate must be maintained at the various conditions. Adjusting the flow control valve after the additional pressure drop of the prover is added is usually necessary.

20.6.4.5 Flanges Raised face weld neck forged steel flanges are normally used on LACT units and provers. These flanges allow for easy maintenance of the LACT equipment and provide for a tight, leakfree connection to other piping and components. The standard specification on flanges is ANSI B16.5. The maximum pressure ratings of the different flanges are listed in the ANSI specification.

20.6.4.6 Piping to Route the Fluid Horizontally and Vertically for Precise Measurement The pressure piping for a LACT unit is normally routed in a horizontal plane. However, for taking a sample and measuring the water cut of a fluid, the flow should be vertical. Vertical flow does not allow the fluid to stratify; that is, any mixture in the pipe (e.g., oil and water) will be moving on a vertical plane after the mixer and passing the sampler and BS&W (basic sediment and water) monitor in a completely mixed mode. When used, densitometers should be mounted in either a vertical or a 45° angle with the flow in a downward direction. Fluid flow in a downward direction through a densitometer keeps the vibrating tube clear of accumulated debris.

20.6.4.7 Static Mixers Uniform blending of the liquid in the line is accomplished at the full flow rate by the use of a static mixer. Flow rates through the mixer should not be lower than 4 ft/s or proper mixing will not occur. Some pressure drop is experienced across the static mixer, but a mixer is necessary for proper sampling and determining the water cut.

20.6.4.8 Deaerators In the LACT where the liquid is under pressure, free air or other gas may be present in the flow stream. An example is the unloading of petroleum tankers using shipboard pumps which discharge the liquid under pressure through piping to a shore installation. Due to operating procedures at pump startup and during shipping, large volumes of free air may be introduced into the piping. The presence of this free air in the discharge liquid makes accurate measurement impossible where positive-displacement or turbine-type flowmeters are used. In addition, the free air may cause damage to the metering apparatus. One method that has been used to eliminate free air from ship off-loading systems is to deliver the flow into relatively large storage tanks. To allow dissipation of free air, the liquid is permitted to stand under quiescent conditions at atmospheric pressure for periods of 2 to 4 h. The tank can then be gauged to determine the quantity of liquid transferred. This method involves the measurement of batch quantities as well as the associated inaccuracies of gauging large tanks. In addition, such large shore tanks are relatively expensive.

Another method that has been used is to employ relatively large retention or relaxation tanks into which the liquid is discharged from the shipboard pump, and from which the liquid is pumped through a measuring meter. This method requires the use of a second pump to deliver the liquid from the retention tank to the meter. Again, this involves the expense of a relatively large retention tank.

A superior method or apparatus for removing free gases from liquid flowing under pressure through piping is available. The deaerator allows immediate metering of the offloaded product while maintaining fluid line pressure. Furthermore, this deaerator eliminates the need for relatively large and costly storage and retention tanks. The gas deaertor can be a horizontal or vertical vessel, depending on the type of application and space limitations. The vessel is sized to have sufficient retention time for the gas and air bubbles to rise to the top of the vessel. To assist this gas and air removal, the interior of the deaerator will have a series of inclined steel scrubber plates to collect and segregate the bubbles, which then coalesce and pass to the top of the vessel.

20.6.4.9 Pumps Normally, the pumps used for LACT units are of the centrifugal or piston type. Pump sizing is critical for the proper flow of fluid through a LACT unit. Either ANSI or API pumps can be used on LACT units. A pumps on a LACT unit usually has pressure switches located on both the suction and discharge sides for protection. There is a misconception about the centrifugal pump on a LACT unit that it is a mixer for the fluid. Actually, because of the way the impeller throws out the fluid, the centrifugal pump acts as a separator. The water and oil are separated as in a centrifuge.

A piston or plunger pump can cause problems with fluid measurement by creating pulsations in the fluid flow. It is best to avoid using piston or plunger pumps on LACT units. If they must be used, they should be located as far up- or downstream as possible. Pulsation dampness should be used on all piston and plunger pump applications.

20.6.4.10 Sample Systems The sample system is a very important part of a LACT unit. The meter determines the quantity and the sample system is used to determine the quality of the fluid. A sample system is composed of a piece of equipment to take the sample out of the line and another piece of equipment to store the fluid. The location of the sampler is critical. API has made a recommendation that the sample be taken three pipe diameters downstream of the sample mixer. Although this location is logical and usually correct, the location should be proved to take a representative sample of the liquid. This is usually done by a third party that injects known quantities of liquid (e.g., oil and water) into the inlet of the LACT. The sample taken from the installed location should agree with the percentages injected. Not taking a true representative sample of the fluid will cause errors in quality measurement. Samplers (devices that take fluid out of the line) can be electric or pneumatic driven. Sample systems can be of the “grab” or beveled probe type. Sometimes the sampler is referred to as *isokinetic*, a description of which is as flows: Any technique for collecting a sample from an appropriately mixed flowing stream in which the sample collection chamber is so designed that the stream entering it has the same velocity as that of the stream passing around and outside the sample collection chamber.

The least expensive and adequate sample system is a tube cut at a 45° angle, with the open end facing the flow. The probe should be in the center one-third of the flow line after the mixer. The sample is directed to a spring-loaded adjustable-volume cylinder through a three-way valve. At regular intervals controlled by the flow or time, the solenoid opens to a vessel that holds the samples. A better and more expensive sampler is a device that grabs a sample out of the line as the fluid moves past it: called isokinetic sampling. The sampler can be actuated by time or can be proportional to flow. Flow proportional samplers are preferred for precise sampling. If it is time activated, the sample (normally 1.5 to 10 cm^3) is taken at regular adjustable timed intervals (e.g., every 30s). If the sampler is flow proportional, a sample would be taken every barrel or every 5 or 10 barrels. The rate would be adjustable, and on most samplers the amount of sample would also be adjustable. A standard sample system is composed of a sample probe that is installed in the process line. The probe is isolated from the sample pump by a valve that easily allows the pump to be removed from the line and serviced. The sample pump receives the fluid from the probe and pumps it into the sample cylinder. Line pressure is maintained on the sample cylinder by connecting the cylinder end on the opposite side of the piston to the process. The sample cylinder is connected with quick disconnects, which makes it easily removable.

The sample pump is activated by a regulated air supply that is controlled by an electric solenoid. The sample frequency is controlled by a counter mounted in a junction box. One pulse per barrel is sent to the counter from the flow computer. The counter is then set to the frequency required to fill a 500-cm^3 sample cylinder in the required amount of time. Another type of sample container is the sample pot. This container can be of any size and is mounted on the LACT skid. As the sample is collected, it is deposited into the sample container. At the end of day or week, depending on the amount of sample taken, a small volume is pumped into another container and transported to the lab to be analyzed. The remaining sample collected is then either pumped back into the line before the meter or discharged to a sump tank.

20.6.4.11 Basic Sediment and Water (BS&W) Monitors Historically, BS&W monitors have been used as indicators of excess water content in oil. They are able to detect to a reasonable degree of accuracy the amount of water in the oil. When a percentage of water is detected, a signal is sent to an alarm panel or to a solenoid on a three-way diversion valve that would divert the out-of-spec oil back to a separator tank for further removal of water. The oil is then routed back to the LACT, and if the water content is in spec, the oil is allowed to go down the pipeline. Net oil is always determined by the sampler and not by the water content meter.

BS&W monitors and the diversion valves are placed before the meter to keep "bad oil" from being sold down the pipeline. In some cases where the tanks that hold the diverted oil are in a different location than the LACT, the diversion skid is sold as a separate unit. Recently, new devices have been introduced on the market that have an accuracy of $\pm 0.05\%$ or better. As the measurement of water cut becomes more accurate, it will be possible to have a net oil meter within LACT specifications.

20.6.4.12 Densitometers Densitometers are devices that measure the density of a fluid, which is extremely important when determining crude oil quantities. The accuracies are 0.0002 to 0.0005 g/cm^3. A typical density range would be 0 to 3 g/cm^3. The density on a LACT unit is normally taken on a slipstream across a pressure drop. The densitometer fluid direction of flow is always downward, to keep the vibrating tube free of debris. The densitometer can be mounted vertically or at a 45° angle. The densitometer is checked with a device called a pycnometer, a sphere of precision volume, normally 1000 cm^3, which is connected in parallel with the densitometer, filled with fluid, and weighed. The weight of the fluid in the sphere compared to its volume is then compared to the reading on the densitometer. The densitometer should be checked occasionally against the measured fluid to ensure its accuracy.

20.6.4.13 Block and Bleed Valves Block and bleed valves are used when it is necessary to know for sure that no fluid is passing in the line. Normally, both the inlet and outlet of the valve are shut off. The cavity between the inlet and outlet can then be drained to verify that no fluid is passing from the inlet to the outlet.

20.6.4.14 Three-Way Diversion Valves A three-way diversion valve is used to send oil back to a tank if the water content is out of spec. Care should be taken to ensure a constant pressure on the system when the valve is in the diversion mode. Some three-way diversion valves are unbalanced and require a differential of no more than 40 psi across them. Some three-way diversion valves are also balanced but require piping configurations that result in high-pressure drops. By using two disk valves and a valve operator, the problems noted above can be avoided.

20.6.4.15 Check Valves Check valves are used at the discharge of the runs on a LACT unit to ensure that no fluid enters the metering system in a reverse direction.

20.6.4.16 Block Valves Block valves are used when it is not necessary to verify that the flow has been stopped. Block valves can be ball valves, plug valves, gate valves, or disk valves. Electric or pneumatic operators are sometimes used on valves either for remote operation or because the valves are too big to operate by hand. If an electric or pneumatic operator is used, position switches must be used to send a signal back to the control cabinet to indicate the position of the valve.

20.6.4.17 Relief Valves The description of a relief valve for a LACT unit is a thermal relief valve. Thermal relief valves are used when there is a possibility of pressure building between two shutoff points and damaging the equipment. These valves are sized to relieve the pressure due to thermal expansion only; they are not designed to relieve the total flow in the LACT if it should become overpressurized.

20.6.4.18 Vents Vents are used at high points of the LACT unit and on the strainers to release air that could become trapped in the high points. Although they will eliminate some air, these vents should not be used to eliminate large quantities of air. For large quantities of air, a deaerator must be used.

20.6.4.19 Skids Structural steel skids to mount the equipment are provided with lifting eyes to allow the equipment to be moved from place to place. Either steel or composition grating can be provided to cover the skid for walkways. Checkered steel plate can also be provided to cover the skid. Drip pans can be provided to route any leaks to a drain.

20.6.4.20 Heat Tracing Heat tracing is provided to keep the fluid warm in the piping when the fluid is not moving. Heat tracing is not used to heat fluid in a pipe that has liquid flowing through it. Heat tracing must be applied properly to ensure that the loads on each circuit and thermostat are not exceeded. Heat tracing is applied flat to the surface of the pipe and held on with high-temperature tape.

20.6.4.21 Insulation Heat tracing is always insulated to protect the wiring. Calcium silicate and mineral wool are the most commonly used forms of heat tracing. An outer coating of sheet aluminum is usually placed over the insulation. Insulation is used to reduce the transfer of heat either into or out of the piping and valves.

20.6.4.22 Protective Coating (Paint) All exposed steel surfaces are normally sandblasted to "white metal," followed by a painting system suitable for the service for which the equipment will be used. One-, two-, or three-coat systems are available, depending on whether the equipment is to be used on- or offshore.

20.6.4.23 NACE Compliance and Material Requirements [16]

1. The National Association of Corrosion Engineers (NACE) Standard Material Requirements are similar to the API standards in that they are recommended practices and not codes to follow, like ASME or NEC.
2. NACE Standard Material Requirement MRO175-90 was written to reduce the effect of sulfide stress cracking on metallic materials for oil field equipment.
3. NACE takes no responsibility for failed equipment due to sulfide stress cracking, even if their procedures are followed.
   a. When a request for a quote states that the package must be designed to fulfill the standards of NACE, it means that we should build the equipment following all the recommendations, including the scope and application recommendation of the standard. If we say that the equipment will be NACE, we are saying that regardless of the use of the equipment, it will meet NACE standards.
   b. When we purchase equipment that must meet NACE requirements, we can get that information in two forms:
      (1) We can ask the supplier to furnish us with a letter from the manufacturer stating that the item is built to NACE standards.
      (2) We can get from the supplier a list of the materials used, the certificates or MTRs on that material showing a Rockwell hardness of 22 HRC maximum, and a statement of heat treatment. With the information above we can certify that the material meets the NACE requirements.

Note: Tubing fittings (e.g., Swagelock or comparable fittings) cannot be NACE compliant because they have been cold-worked to harden the material to meet the material strength requirements.

20.6.4.24 Electrical Classifications The Classification of Areas for Electrical Installations at Drilling Rigs and Production Facilities on Land and on Marine Fixed and Mobile Platforms (API RP 500B) gives the National Electrical Code electrical classification for oil field equipment. The NEC has three areas of classification: (1) unclassified areas, where no danger of explosive gasses occur; (2) class 1 division 2 areas, where explosive gases could occur (e.g., during normal maintenance) but are not usually in the area; and (3) class 1 division 1 areas, where explosive gases are present. Each area of classification requires different electrical protection for the equipment.

20.6.5 Liquid Displacement Provers

The liquid displacement prover is a device that has reached great acceptance in the last 20 years as a method of obtaining a more accurate measurement with liquid meters used in hydrocarbon custody transfer measurements. It is defined as the "true measurement" in the definition of meter accuracy:

$$\text{meter accuracy} = \frac{\text{indicated measurement of meter}}{\text{pipe prover measurement}}$$

The concept of using an online, as-installed prover system offers the advantage of not interrupting the flow of the fluid during proving. This system eliminates static starting and stopping of the meter, which introduces error in any meter, and replaces it with a dynamic test, which is the manner in which the meter is normally used. With such a proving system, the amount of volume required to be displaced is dependent on:

1. The repeatability required
2. The resolution with which the meter register can be read
3. The resolution with which the displacer can be located at the extremities of the prover section

20.6.5.1 Description of Prover Systems The prover system is accomplished by several types of mechanical devices, which are dependent on a common basic principle of operation. The principle is the accurate and repetitive displacement of a precalibrated and known volume of liquid between two detector switches from a cylindrical container with a mechanical sealing displacing device driven through the container by fluid energy from the stream being measured. Simultaneously, the corresponding metered volume is indicated. These measurements provide a ratio of the known volume displaced, and the meter registration is determined and applied as a meter factor. Proper corrections must be made on the prover and meter for the effects of temperature and pressure as well as the effects on the liquid of these two variables. Since the prover is designed to be as small as possible consistent with the accuracy required, a proving counter is normally used in place of the standard register so that resolution can be increased for proving and the standard counter does not have to run at such a high rate. Similarly, the resolution of the detectors must be maintained high so that their operation does not add significantly to the test tolerance.

There are two basic designs of continuous-flow prover systems, the unidirectional and the bidirectional. The unidirectional prover allows the displacer to travel in only one direction through the prover system, whereas the bidirectional prover allows the displacer to move in either direction, since it incorporates a means of reversing the flow through the prover section. Two types of unidirectional provers have been used: (1) the manual in-line type, in which the displacer travels through a measured section of pipeline, where it is removed manually and returned to the launch site, and (2) the endless loop automatic displacer return device. Since the second device is more common, it is the only one described; however, the design considerations apply to both types of provers.

20.6.5.2 Endless Loop Provers The endless loop prover uses a spheroid in a prover section consisting of a closed loop of pipe in which a central control interchange unit incorporating a valve system acts as a launcher and receiver. The piping is arranged so that the downstream ends of a loop of pipe crosses over the upstream end of the section. The interchange is a vertical line connecting the upstream and downstream ends of the loop. The displacer detector switches are located at a distance from the interchange which allows launching and receiving to take place with the valve system in the proper position after stabilized flows have been established. The loop section of pipe is normally maintained in service with the metered fluid continually passing through the loop to keep the temperature and pressure stabilized and offer flushing action for variable products, which minimizes the stabilizing time necessary before proving starts and reduces the intermixing between different liquids.

20.7 TROUBLESHOOTING LACT AND PROVER SYSTEMS [16]

1. Make sure that all instruments on the LACT and prover are calibrated properly. All instruments and equipment used to do the calibration on the LACT and prover should have recently been calibrated to a standard and be in like-new condition.

2. Check to make sure that all the mechanical equipment operates properly, the four-way valve seats properly, the block and bleed valves do block and bleed, and so on.

3. Check the system for leaks, especially leaks to drain. Although a properly designed LACT and prover system requires all drains after the meter to be able to be inspected visually, some systems have hidden sources of leaks. Prover drain valves should be checked to ensure that they are not leaking.

4. The proper flowing conditions must be maintained through the entire prove. Make sure that there are no drastic changes in flow rate, temperature, and pressure. Many good proves have to be aborted because they ran out of fluid.

5. For a good prove the prove must be made at the same flow rate as the meter under operating conditions. All meters have curves that give them different pulse frequencies at different flow rates.

6. Always double check any component suspected of malfunction. For example, if while proving a meter the prover ball is suspected of not being inflated properly, check another meter, if possible, to see if the condition exists there also. Many times in a hunt to find the problem, other equipment is changed, so that even if the problem is found, other problems still exist. Although the following troubleshooting guide in Table 20-5 is very complete, it is not possible to have an answer to every possible problem. Good common sense is a necessity when troubleshooting mechanical, pneumatic, electric, and electronic problems.

Sometimes a combination of the above will make troubleshooting extremely difficult, but a LACT and prover are no more than a combination of mechanical, pneumatic, electrical, and electronic components. With common sense and a good logical procedure, you will always be able to identify the source of the problem and be able to remedy it.

20.8 TROUBLESHOOTING FLOWMETERS

To prolong the service life and high accuracy of flowmeters, the causes of its problems should be determined. The first step in troubleshooting a flowmeter obstacle is to make sure that it is selected and installed properly [13]. Table 6 provides guidance on troubleshooting four common problems for flowmeters.

TABLE 20-5 Troubleshooting Guide for the LACT and Prover Systems

Possible Problems/Causes	Remedies/Solutions
Repeatability cannot be achieved	
Entrapped air or gas	Open vent valves on the LACT and prover until all gas in the system is eliminated. Make several more proving runs while venting to ensure that all gas is out of the system.
Proving temperature not stable	Check the meter and prover differential. The temperature difference must not exceed 1°F unless the volumes are compensated for in both the prover and the LACT.
Unstable proving pressure	Check the meter and prover differential. The pressure difference must not exceed 1 psi unless the volumes are compensted for in both the prover and the LACT.
Unstable proving flow rate	Check the flow rate fluctuations. Limit the flow-rate variation to plus or minus 5% maximum.
Slippage due to improper sphere inflation	Check the sphere inflation by measuring around the seam of the sphere. Inflation should be 2 to 5% over the ID of the prover. Different sphere materials with different fluids require different inflations. Inflate the sphere to the smallest diameter that will not allow slippage and still not require a high-pressure drop across the ball to get it to move.
Defective displacer	Check for wear on the ball or piston. Long scratches will allow the fluid to bypass. Replace the sphere or the defective piston seals.
Pulse signals continue to accumulate after the second detector is passed	
Defective detector switch or pulse counter	Check the pulse counter first by simulating a switch closure during a prove run. If the counter is good, the detector is not operating properly. Replace or repair the sphere detector. *Caution:* An underinflated sphere can also cause this condition.
No pulse signals can be obtained	
Defective meter generator or defective wiring	Check the wiring for continuity. Simulate a detector switch pulse while connected to the pulse output to eliminate a defective sphere. Repair or replace the equipment as necessary. If the detector switches are removed or replaced, the prover will have to have a water draw done.
Calibration pulse signals become erratic at high flow rates	
Sphere hits the detector switch before the four-way switch is sealed off	Check the four-way differential switch to ensure that it engages before the first detector switch is reached. Speed up the action of the four-way valve, decrease the flow rate, or increase the prover pre-run.
Erratic proving results	
Random electrical noise affecting the pulse signal	Observe the pulse signal with an oscilloscope. Correct the source of the noise.
Defective pulse generator, amplifier, etc.	Check all devices one at a time by replacing or checking. Replace the defective parts.
Marginal pulse signals	Observe the signal with an oscilloscope. Replace the defective parts.
Unexpected or out-of-pattern meter factor	
Leaking block and bleed valves	Check all valves in the meter string, especially valves to other meter runs in the system. Manually operate the valves and repair or replace as necessary.
Strainer has become plugged	Check for high-pressure drop across the strainer. Clean out the strainer and check the flow rate.
On turbine meters, a blade may have broken off or has become shorted out	Check the meter by removing it from the line. Action 6C = repair or replace the defective meter.
Rotor blade damaged in a PD meter or meter housing damaged	Inspect the body and blades of the PD meter. Repair or replace the defective meter.
Unexpected out-of-pattern decrease in meter factor	
Air or gas slugs in the line	Check the line by opening the air vents. Eliminate air or gas in the line.
Leakage of meter run outlet block and bleed valve	Check the meter run outlet block and bleed valve. Repair or replace the block and bleed valve.
Leakage in line, flanges, vents, etc.	Visually check all lines, flanges, vents, etc. for leaks. Stop all leaks.

(continued)

TABLE 20-5 ***(Continued)***

Possible Problems/Causes	Remedies/Solutions
Leakage of the four-way valve	Observe the differential across the four-way valve when it seats. Make sure that the differential goes to zero when the valve is off its seat. Repair or replace the four-way seals.
Flashing in the meter run	Make sure that the operating pressure is above the vapor pressure by at least 20 psig. Increase the pressure on the system.
Linearity exceeds the manufacturer's tolerance	
Viscosity change	Check the fluid in a lab to determine the viscosity. Determine the reason for the viscosity change, and correct.
Bearing wear	Inspect the meter. Repair or replace as necessary.
Meter corrosion	Inspect the meter. Repair or replace as necessary.
Long-term one-way drift-of-meter factor	
Bearing wear	Inspect the meter. Repair or replace as necessary.
Pressure and/or temperature instruments out of calibration	Check pressure and/or temperature instruments for calibration. Recalibrate or replace as required.
Change in base prover volume	Open the prover, clean, and inspect. Clean the prover, recheck, or water-draw the prover.

TABLE 20-6 Troubleshooting Guide for Flowmeters [13]

Possible Problems/Causes	Remedies/Solutions
Flowmeter does not respond	
Is the pump off?	Check for voltage at the control box. Verify that the control box has not tripped. Verify that the supply breaker is on. Start the pump.
Is the flow sensor not connected properly to the flowmeter?	Verify that the flow sensor is connected to the flowmeter. Connect the flow sensor to the flowmeter.
Is the flowmeter faulty?	Verify that the pump is running by using a clamp-on ammeter on one lead of the pump power cable. Install a spare flowmeter.
Is the flow sensor faulty?	Verify that the pump is running by using a clamp-on ammeter on one load of the pump power cable. Install a spare flowmeter to check for a response. If no indication, replace the flow sensor.
Flowmeter indicates higher flow than normal	
Is new a pump and/or filters installed?	Check when the pump and/or filters were lined up in service.
Is a filter not installed?	Check that the filter is installed properly. Install the filters (the flow rate decrease must not be greater than 15% with new filters).
Is the flowmeter off-scale high?	Check if the flow sensor is on greater than the flow indicated, but not hard pegged. If not, the flowmeter should be recalibrated. If the flow sensor is hard pegged and recalibration does not correct the problem, install a new filter.
Is there pulsation in the flow line?	If so, install a process pulsation dampener.
Flowmeter filters load-up too quickly	
Is the filter selection for the application incorrect?	Check the filter status and install the proper size (μm), based on the application.
Flowmeter indicates lower flow than normal	
Is the pump phased incorrectly, with the pump running?	Reverse two of the three motor leads at the control box. Record the flow rate. Check for an approximate flow of one-third (if the pump is phased incorrectly, the system will run at approximately one-third flow). If so, maintain normal flow.
Is the flowmeter faulty?	Verify pump is running by using a clamp-on ammeter on one lead of the pump power cable. Install a spare flowmeter.
Is the flow sensor faulty?	Verify that the pump is running by using a clamp-on ammeter on one load of the pump power cable. Install a spare flowmeter to check for a response. If no indication, replace the flow sensor.
Are there liquid droplets in the gas line?	If so, install a demister and/or heat gas upstream of the sensor.

REFERENCES

1. Berge, J., Time for calibration, *Control Engineering Asia*, vol. 4, no. 8, October 2009, pp. 24–27.
2. Berry, G., Strategies for proper material selection: *lessons from 30 years of application experience*, *Flow Control*, vol. xiii, no. 3, March 2007.
3. Bluvshtein, L., Uncertainties of measurement instruments, *Pipeline and Gas Journal*, vol. 234, no. 5, May 2007, pp. 28–32.
4. Frenzel, F., et al., 2006 *Industrial Flow Measurement Practice*, ABB Automation Products, Ladenburg, Germany.
5. Gas Processors Suppliers Association, 1998 *Engineering Data Book, SI Version*, 11th ed., Vol. 1, Sec. 3, GPSA, Tulsa, OK,
6. http://www.automation.siemens.com/w1/automation-technology-electromagnetic-18627.htm#.
7. http://www.rshydro.co.uk/Electromagnetic-c-316.html.
8. Kurz, J., Theorizing on thermal flow: evolution of mass thermal convention sensor tecnology, *Flow Control*, vol. xv, no. 2, February 9, 2009, pp. 26–30.
9. Livelli, G., Matching the flowmeter to the application: selection strategies for systems designers, *Flow Control*, vol. xiii, no. 8, August 2007.
10. McGillvray, A., The flow data factor, *World Pipelines*, vol. 9, no. 8, August 2009, pp. 131–133.
11. Milford, S., Outside the pipe: understanding the benefits of clamp-on transit-time flow measurement, *Flow Control*, vol. xv, no. 5, May 2009, pp. 16–20.
12. Mokhatab, S., and Lamberson, G., Fundamentals of gas pipeline metering stations, *Pipeline and Gas Journal*, vol. 236, no. 1, January 2009, pp. 34–38.
13. Nwaoha, C., Flow meters: minimising flow measurement setbacks, *Everything About Water*, no. 6, June 2009, pp. 46–48.
14. Nwaoha, C., Flow meters: minimizing flow measurement setbacks, *Petro Min*, March–April 2009, pp. 42–46.
15. Nwaoha, C., Flow meters: proper selection minimises set-backs, *Pipeline and Gas Journal*, vol. 236, no. 7, July 2009, pp. 43–44.
16. Rudroff, D. J., *L.A.C.T. Units Metering Systems and Proving Systems*, Welker Flow Measurement Systems, Sugar Land, TX, December 28, 2005, pp. 1–56.
17. Yoder, J., A growing need for oil flow measurement, *Pipeline and Gas Journal*, vol. 233, no. 7, July 2006, pp. 36–37.
18. Yoder, J., A market prime for boom: thermal flow meters positioned to gain from environmental push, *Flow Control*, vol. xv, no. 3, March 2009, pp. 16–23.
19. Yoder, J., High pices drive flow measurement in the energy market, *Pipeline and Gas Journal*, vol. 235, no. 7, July 2008, pp. 26–28.
20. Yoder, J., Something to be said for tradition: legacy flow measurement systems live on, *Flow Control*, vol. xiv, no. 11, November 2008.
21. Yoder, J., Ultrasonic flow meters in the energy measurement spotlight, *Pipeline and Gas Journal*, vol. 236, no. 7, July 2009, pp. 41–42.

21

PROCESS CONTROL

JOHN A. SHAW
Process Control Solutions, Cary, North Carolina

The measurement of a process variable, the comparison of that variable with its respective set point, and the manipulation of the process in a way that will hold the variable at its set point when the set point changes or when a disturbance changes the process is known as *process control*. Process control is used to maintain a variable in a process plant at a set point or to cause it to respond to a setpoint change. The most common method used in process control is the PID (proportional, integral, derivative) control algorithm. This algorithm and how it is used are discussed in this chapter. The chapter focuses on assisting the new practitioner, perhaps a young engineer starting a career or a person experienced in other fields who has taken on process control as part of his or her responsibility. The objectives are to:

- Provide an understanding of the basic common control method: PID
- Assist the reader in making use of the PID control algorithm.

Suppose that the temperature of the heated water leaving a heat exchanger is to be held at its set point by manipulating the flow of steam to the exchanger using the steam flow valve. In this example the temperature is known as the *measured* or *controlled variable* and the steam flow (or the position of the steam valve) is the *manipulated variable* (see Fig. 21-1).

Processes can contain many variables that need to be held at a set point and many variables that can be manipulated. Usually, each controlled variable is affected by more than one manipulated variable, and each manipulated variable will affect more than one controlled variable. However, in most process control systems manipulated variables and control variables are paired so that one manipulated variable is used to control one controlled variable. Each pair of controlled variable and manipulated variable, together with the control algorithm, is referred to as a *control loop*. The decision of which variables to control is based on both knowledge and operation of the process.

21.1 CONTROL SYSTEM COMPONENTS

A simple control loop is composed of the following four elements:

1. *Field devices*. A sensor (or transmitter) measures some variable in the process, such as temperature, liquid level, pressure, or flow rate, and converts that measurement to a signal (typically, 4 to 20 mA) for transmission to the controller or control system.
2. *Control algorithm*. A mathematical algorithm inside the control system is executed at some time period (typically, every second or faster) to calculate the output signal to be transmitted to the final control element.
3. *Final control element*. A valve, airflow damper, motor speed controller, or other device receives a signal

Process Plant Equipment: Operation, Control, and Reliability, First Edition. Edited by Michael D. Holloway, Chikezie Nwaoha, and Oliver A. Onyewuenyi.
© 2012 John Wiley & Sons, Inc. Published 2012 by John Wiley & Sons, Inc.

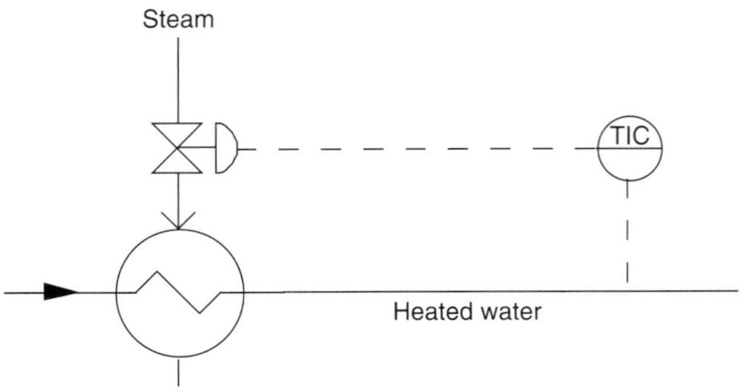

Figure 21-1 Typical control loop with controlled variable (temperature) and manipulated variable (steam valve).

from the controller and manipulates the process, typically by changing the flow rate of some material.

4. *Process*. The process responds to the change in the manipulated variable, with a resulting change in the measured variable. The dynamics of the process response are a major factor in choosing the parameters used in the control algorithm and are covered in detail in this chapter.

21.2 CONTROL SYSTEM REQUIREMENTS

The control algorithm does not "know" the correct output to send to a valve. Instead, the algorithm merely continues to move the output in the direction that should move the process toward the set point until the process reaches the set point. The algorithm must have feedback (process measurement) to perform. The control algorithm must be tuned for any particular process loop. To be able to tune a PID loop, each of the terms of the PID equation must be understood.

21.3 SENSOR RESPONSE

Most often the sensor will respond to the process rapidly, with only a short dead time and small lag. This typically will not affect the loop response. However, there are some situations where it may have an effect. Two examples are:

- Large thermowells, heavy steel devices used to protect thermocouples from the process, may add a small lag to the process, affecting tuning.
- The dead time inherent in digital sensors or in digital transmission can affect fast-acting control loops such as flow loops.

21.3.1 Process Response

Loops are tuned to match the response of the process. We first discuss the responses of the process to the control system. The dynamic and steady-state response of the process are related to changes in the controller output. These responses are used to determine the gain, reset, and derivative of the loop. The process response is the effect on the controlled variable caused by a change in the controller output.

21.3.1.1 Steady-State Response The steady-state process response to controller output changes is the condition of the process after sufficient time has passed so that the process has settled to new values. It is characterized primarily by process action, gain, and linearity. Action describes the direction the process variable changes following a particular change in the controller output. A direct-acting process increases when the final control element increases (typically, when the valve opens); a reverse-acting process decreases when the final control element increases.

For example, if we manipulate the inlet valve on a tank to the control level, an increase in the valve position will cause the level to rise. This is a direct-acting process. On the other hand, if we manipulate the discharge valve to control the level, opening the valve will cause the level to fall. This is a reverse-acting process.

Next to action is process gain, which is the most important process characteristic. The *process gain* (not to be confused with *controller gain*) is the sensitivity of the controlled variable to changes in a controller output. Gain is expressed as the ratio of change in the process to the change in the controller output that caused the process change. The process gain is affected by the valve itself, by the process, and by the measurement transmitter. Therefore, the size of the valve and the span of the transmitter will affect the process gain.

The gain of the process often changes based on the value of the controller output. That is, with the output at one value, a small change in the output will result in a larger change in the process measurement than the same output change at some other output value. The process shown in Fig. 21-2 is nonlinear. With controller output very low, a 1% increase in the output causes the measured variable to increase by 2% (Fig. 21-3). When the output is very high, the same 1% output increase causes the process to increase by only 0.5%. The process gain decreases when the output increases.

From the standpoint of controller tuning, the process linearity includes the linearity of the process, the final control element, and the measurement. It also includes any control functions between the PID algorithm and the output

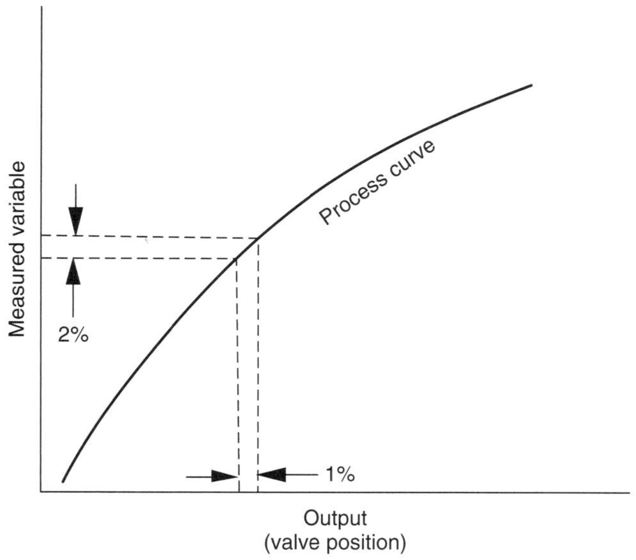

Figure 21-2 Nonlinear process.

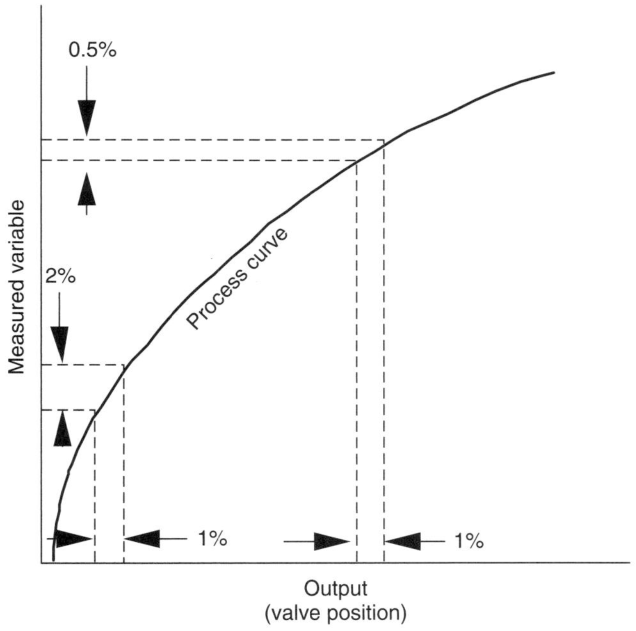

Figure 21-3 Nonlinear process: gain changes with measured variable.

to the valve. Valves may fall into three classes: linear, equal percentage, and quick opening (see Fig. 21-4).

Linear valves have the same gain regardless of the valve position. That is, at any point a given increase in the valve position will cause the same increase in the flow as at any other point. Equal percentage valves have a low gain when the valve is nearly closed and a higher gain when the valve

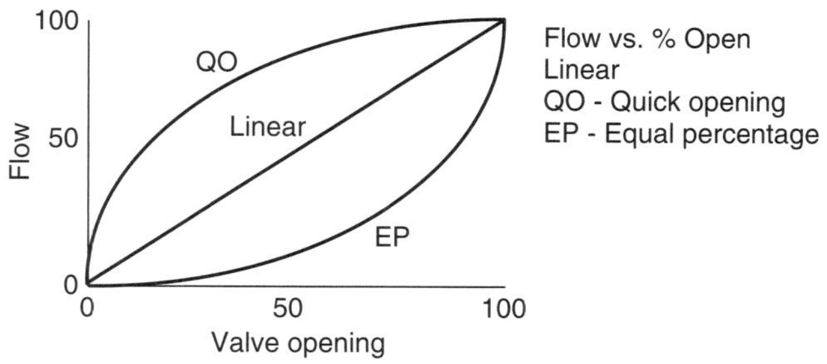

Figure 21-4 Flow vs. valve opening for quick opening, linear, and equal percentage valves.

is nearly open. Quick-opening valves have a high gain when the valve is nearly closed and a lower gain when the valve is nearly open.

21.3.1.2 Dead Time *Dead time* is the delay in the loop due to the time it takes material to flow from one point to another. For example, in the temperature control loop shown in Fig. 21-5, it takes some amount of time for the liquid to travel from the heat exchanger to the point where the temperature is measured. If the temperature at the exchanger outlet has been constant and then changes, there will be some period of time before any change can be observed by the temperature measurement element. Dead time is also called distance velocity lag and transportation lag. Dead time is often considered to be the most difficult dynamic element to control. This will become apparent later in the discussion of controller tuning.

21.3.1.3 Lags The most common dynamic element is the simple lag. If a step change is made in the controller output, the process variable will change as shown in Fig. 21-6. An example of a process dominated by one loop is the level in a vessel (Fig. 21-7). The flow of the liquid out of the vessel is proportional to the level. If the inlet

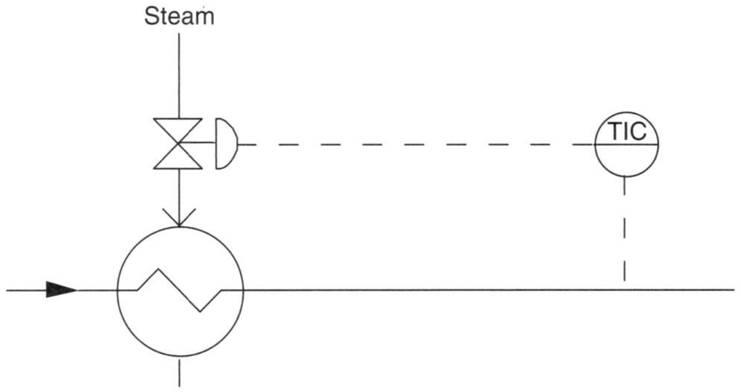

Figure 21-5 Control loop with dead time caused by distance between process and measurement.

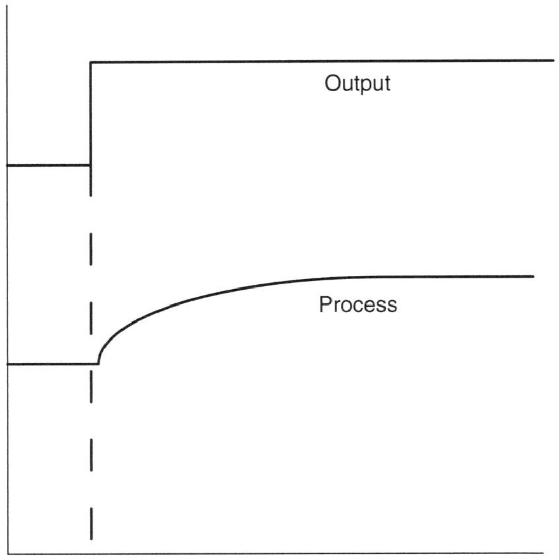

Figure 21-6 Process with single lag.

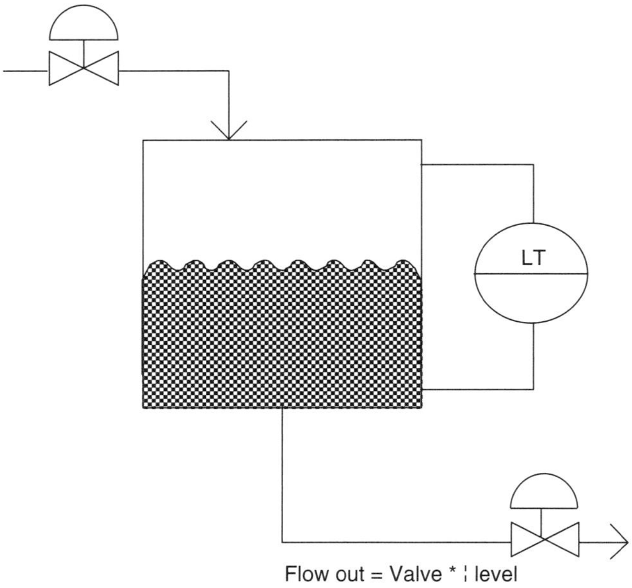

Figure 21-7 Level loop: typical single lag process.

valve is opened, increasing the flow into the vessel, the level will rise. As the level rises, the flow output will rise, slowing the rate of increase in the level. Eventually, the level will be at the point where the flow out will be equal to the flow in.

Figure 21-8 illustrates the response of a simple dead time and a process with dead time and a lag (more typical). The output from the process controller and the response of the process, over time, are shown. Most processes have more than one lag, although some of the lags may be insignificant. Lags are not additive. A response of a multiple lag is illustrated in Fig. 21-9. The process measured variable begins to change very slowly, and the rate of change increases up to a point, known as the *point of inflection*, where the rate of change decreases as the measurement approaches its asymptote. The first part of the curve, where the rate of change is increasing, is governed primarily by the second largest lag. The second part of the curve, beyond the point of inflection, is governed primarily by the largest lag.

Often, processes are described as first order, second order, and so on, based on the number of first-order linear lags included in the process dynamics. It can be argued that all processes are of a higher order, with a minimum of three lags and a dead time. These lags, which are present in all processes, include the lag inherent in the sensing device, the primary lag of the process, and the time that the valve (or other final control element) takes to move. However, in many processes the smaller lags are so much smaller than the largest lag that their contributions to the process dynamics are negligible.

Dead time is also present in all processes. With pneumatic control, there is some dead time due to the transmission of the pressure signal from the process to the controller, and from the controller to the valve. This is eliminated by electronic controls. With digital controls, there is an effective dead time equal to one-half the loop scan rate. In most cases, the loop will be scanned fast enough so that this dead time is insignificant. In some cases, such as liquid flow loops, the dead time is significant and affects the amount of gain that can be used.

Rather than consider a process to be first order, second order, and so on, it may be better to consider all loops to be higher order to a degree. We can characterize processes by the degree to which one first-order lag dominates the other lags in the process (not considering any true dead time). *Dominant-lag processes* are those that consist of a dead time plus a single significant lag, with all other lags small compared to the major lag. *Multiple-lag* or *non-dominant-lag processes* are those in which the longest lag is not significantly longer than the next-longest lag (see Fig. 21-10). One measure of the dominance of a single lag is the value of the process measurement at which the point of inflection (POI) occurs. In the most extreme case (only a single lag), the POI occurs at the initial process value. With about three equal, noninteracting lags, the POI occurs at about 33% of the difference between the initial and final process values.

Process dynamics usually consist of several lags and dead time. The dynamics differ from one loop to another. The dynamics can be expressed by a detailed list of all of the lags and the dead time of the loop, or they can be approximated using a simpler model. One such model consists of a dead time and a first-order lag. Graphically, the process response of such a model is shown in Fig. 21-11. The dynamics can be approximated by two numbers:

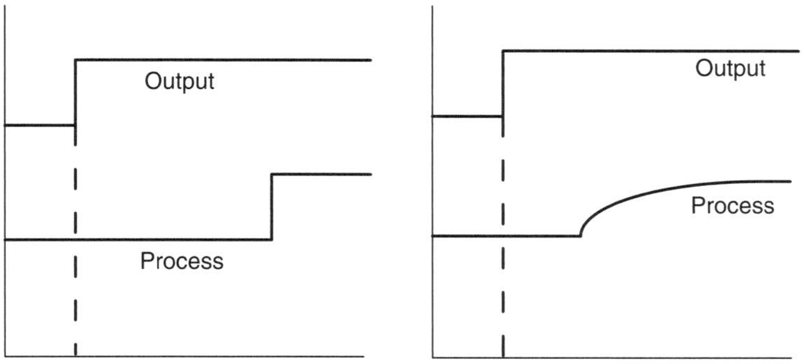

Figure 21-8 Process with dead time only: process with dead time and single lag.

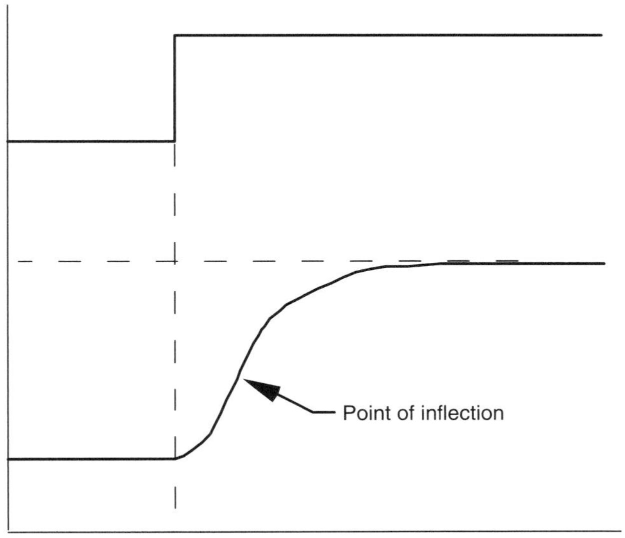

Figure 21-9 Point of inflection for process with lags.

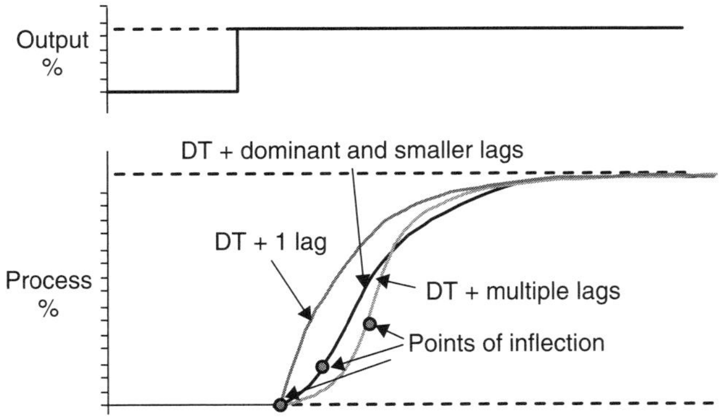

Figure 21-10 As the number of lags increase, the value of the process at the point of inflection increases. DT, dead time.

τ, the process time constant, is approximately equal to the largest lag in the process. T_d is the pseudo dead time and approximates the sum of the dead time plus all lags other than the largest lag.

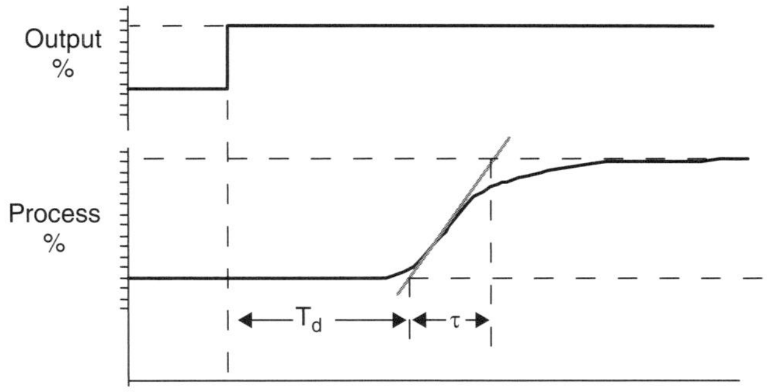

Figure 21-11 Calculation of dead time (T_d) and lag (τ).

Several tuning methods, such as the Ziegler–Nichols open-loop method [1], are based on an approximation of the process as a combination of a single first-order lag and a dead time, known as the first order plus dead time (FOPDT) model. These methods identify the process by making a step change in the controller output. The process trend is recorded, and graphical or mathematical methods are used to determine the process gain, dead time, and first-order lag.

Process gain is the ratio of the change in the process to the change in the controller output signal. It depends on the range of the process measurement and includes the effects of the final control element.

Pseudo dead time (T_d) is the time between the controller output change and the point at which the tangent line crosses the original process value. The pseudo dead time is influenced by the dead time and all of the lags smaller than the longest lag in the process.

Process time constant (τ) is the rate of change of the process measurement at the point at which the rate of change is the highest. The time constant is strongly influenced by the longest lag in a multiple lag process.

The ratio of the pseudo dead time to the process time constant is often referred to as an *uncontrollability factor* (*Fc*), which is an indication of the quality of control that can be expected. The gain (for a P, PI, and PID controller) at which oscillation will become unstable is inversely proportional to this factor.

21.3.2 Controller/Actuator Response

The most important configuration parameter of the control algorithm is the *action*, as it determines the relationship between the direction of a change in the input and the resulting change in the output. If a controller is *direct acting*, an increase in its input will result in an increase in its output. With *reverse action*, an increase in its input will result in a decrease in its output. The controller action is always the opposite of the process action.

21.4 CONTROL ALGORITHMS

21.4.1 On/Off Switch

The most common algorithm is the simplest: an on/off switch. For example, most appliances use a thermostat to turn the heat on when the temperature falls below the set point and then turn it off when the temperature reaches the set point. This results in a cycling of the temperature above and below the set point but is sufficient for most common home appliances and some industrial equipment.

21.4.2 PID Algorithm

To obtain better control, there are a number of mathematical algorithms that compute a change in the output based on the controlled variable. Of these, by far the most common is known as the *PID* (proportional, integral, and derivative) *algorithm*, on which we focus in this chapter. The PID control algorithm is made of three basic responses: proportional (or gain), integral (or reset), and derivative. In the next several sections we discuss the individual responses that make up the PID controller.

The term *error* means the difference between the process and the set point. If the controller is direct acting, the set point is subtracted from the measurement; if reverse acting, the measurement is subtracted from the set point. Error is always a percentage.

1. *Proportional response*. The most basic response is proportional, or gain, response. In its pure form, the output of the controller is the error times the gain added to a constant known as "manual reset":

$$\text{output} = EG + k \tag{21.1}$$

where output is the signal to the process, E the error (the difference between the measurement and the set point), G the gain, and k the manual reset, the value of the output when the measurement equals the set point.

2. *Integral response*. If we look only at the reset (or integral) contribution from a more mathematical point of view, the reset contribution is

$$\text{output} = g\left(K_r \int e\,dt\right) \tag{21.2}$$

where g is the gain and K_r is the reset setting (repeats per minute). At any time the rate of change of the output is the gain × the reset rate × the error. If the error is zero, the output does not change; if the error is positive, the output increases.

3. *Derivative response*. Derivative is the third and final element of PID control. Derivative responds to the rate of change of the process (or error). Derivative is normally applied to the process only. The derivative contribution can be expressed mathematically as

$$\text{output} = g\left(K_d \frac{de}{dt}\right) \tag{21.3}$$

where g is the gain, K_d the derivative setting (min), and e the error. The open-loop response of the controller with proportional and derivative is shown graphically in Fig. 21-12.

4. *Complete response*. If we combine the three terms (proportional gain, integral, and derivative), we obtain the complete PID equation (Fig. 21-13):

$$\text{output} = G\left(e + R\int e\,dt + D\frac{de}{dt}\right) \tag{21.4}$$

where G is the gain, R the reset (repeats per minute), D the derivative (min), and e the error. This is a general form of the PID algorithm and is close, but not identical, to the forms actually implemented in industrial controllers.

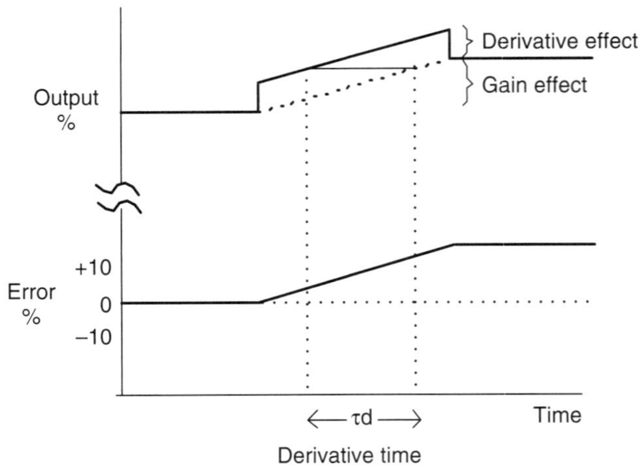

Figure 21-12 Calculation of the amount of derivative.

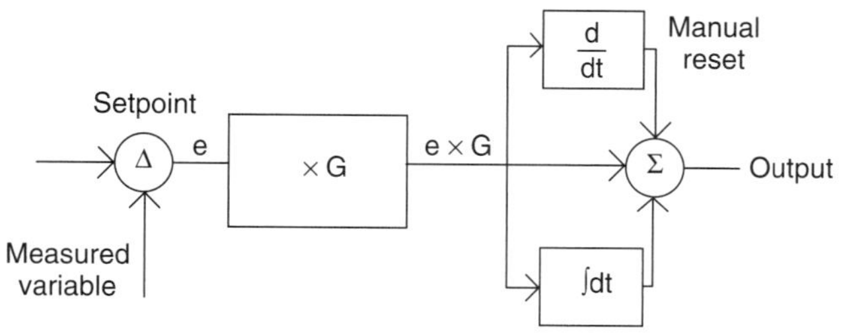

Figure 21-13 Combined PID block diagram (see the text for definitions of terms).

Most commercial controllers allow the user to specify proportional-only controllers, proportional-reset (PI) controllers, and PID controllers that have all three modes. The majority of loops employ PI controllers. Most control systems also allow all other combinations of the responses: integral, integral–derivative, derivative, and proportional–derivative. When proportional response is not present, the integral and derivative is calculated as if the gain were one.

21.4.3 Control Modes

Most control systems allow the operator to place individual loops into either manual or automatic mode. In manual mode (Fig. 21-14) the operator adjusts the output to bring the measured variable to the value desired. In automatic mode the control loop manipulates the output to hold the process measurements at their set points.

In most plants the process is started up with all loops in manual mode. During the process startup loops, are individually transferred to automatic. Sometimes during the operation of the process, certain individual loops may be placed in manual mode for periods of time.

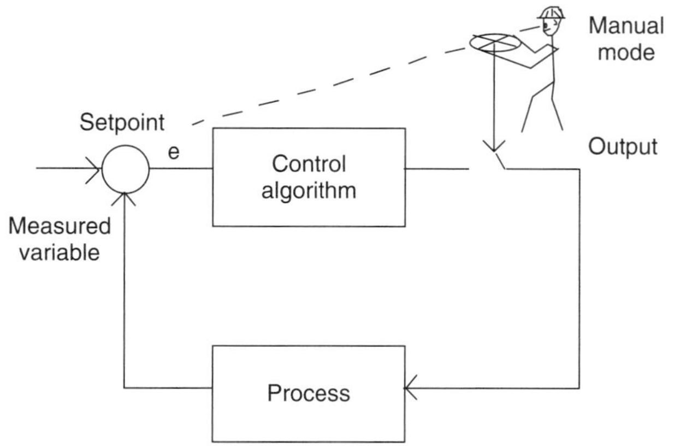

Figure 21-14 Manual mode.

21.5 LOOP TUNING

Once a loop is configured and started up, for it to work correctly someone has to put the correct gain, reset, and derivative values into the PID control algorithm. This is known as *tuning the loop*. One of the most important, and most ignored, facets of loop tuning is the determination of whether the loop is tuned properly.

The loop performance must fall between two extremes. First, the loop must respond to a change in set point and to disturbances. That is, an error must eventually result in the manipulation of the output so that the error is eliminated. If the gain, reset, and derivative of the loop are turned to zero, there will be no response.

The other extreme is instability. An unstable loop will oscillate without bound. A set point change will cause the loop to start oscillating, and the oscillations will continue. At worst, the oscillations will grow (or diverge). Proper tuning of a loop will allow the loop to respond to set point changes and disturbances without causing instability. There are several informal methods or rules of thumb for determining the quality of the tuning of a control loop.

21.5.1 Quarter-Wave Decay

Traditionally, quarter-wave decay (Fig. 21-15) has been considered to be the optimum decay ratio. This criterion is used by the Ziegler–Nichols tuning method [1], among others. There is no single combination of tuning parameters that will provide quarter-wave decay. If the gain is increased and the reset rate decreased by the correct amount, the decay ratio will remain the same.

Quarter-wave decay is not necessarily the best tuning for either disturbance rejection or set point response. However, it is a good compromise between instability and lack of response. For some loops the objective of the tuning is to minimize the overshoot following a set point change, as illustrated in Fig. 21-16. For other loops the primary concern is the reduction of the effect of disturbances, as illustrated in Fig. 21-17. The choice of methods depends on the loop's place in the process and its relationship with other loops.

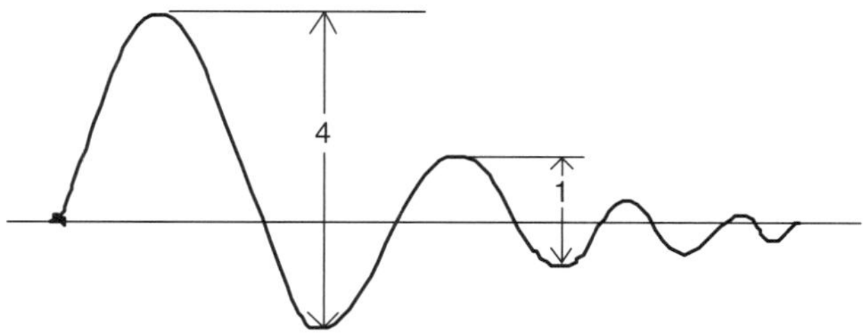

Figure 21-15 Quarter-wave decay.

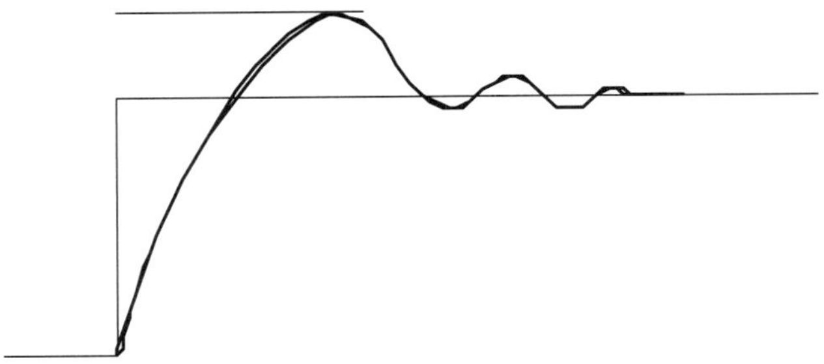

Figure 21-16 Minimizing the overshoot.

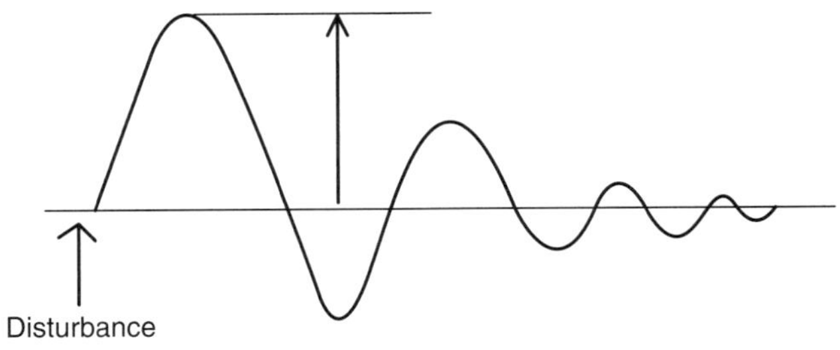

Figure 21-17 Minimizing effect of disturbance.

There are several criteria for evaluating tuning that are based on integrating the error following a disturbance or set point change. These methods are not used to test control loops in actual plant operation because the usual process noise and random disturbances will affect the outcome. They are used in control theory education and research using simulated processes. The indices provide a good method of comparing various methods of controller tuning and different control algorithm. The two most common methods are [1]:

1. The integral of the absolute value of error,

$$\mathrm{IAE} = \int |e| dt$$

2. The integral of the error squared,

$$\mathrm{ISE} = \int e^2 dt$$

21.5.2 Ziegler–Nichols Tuning Methods

In 1942, J. G. Ziegler and N. B. Nichols, both of the Taylor Instrument Companies (Rochester, NY), published a paper that described two methods of controller tuning that allowed the user to test a process to determine its dynamics. Both methods assume that the process can be represented by the model (described above) comprising the process gain, a pseudo dead time, and a lag. The methods provide a test to determine process gain and dynamics and equations to calculate the correct tuning.

The Ziegler–Nichols methods provide quarter-wave decay tuning for most types of process loops. This tuning does not necessarily provide the best ISE or IAE tuning but does provide stable tuning that is a reasonable compromise among the various objectives. If a process does actually consist of a true dead time plus a single first-order lag, the Ziegler–Nichols methods will provide quarter-wave decay. If the process has no true dead time but has more than two lags (resulting in a pseudo dead time), the Ziegler–Nichols methods will usually provide stable tuning, but the tuning will require online modification to achieve quarter-wave decay. Because of their simplicity and because they provide adequate tuning for most loops, the Ziegler–Nichols methods are still widely used.

The Ziegler–Nichols methods, as well as several other methods for controller tuning, rely on a model of the process that comprises one first-order lag plus dead time. The FOPDT model parameters can be determined from the actual process using a simple process reaction test. The output from the controller is increased (or decreased) in a step change, and the reaction of the process is recorded. The process gain is the ratio of the change in the process (as a percentage) to the change that had been made in the controller output.

Several methods have been proposed to calculate the pseudo dead time and time constant from the reaction curve. Ziegler and Nichols proposed a graphical method using a tangent line drawn through the steepest part of the curve (the point of inflection). The line continues below the original process value. The time between the output change and the point at which the tangent line crosses the original process line is called the *lag* (the term *pseudo dead time* is used in this chapter). The slope of the line is then calculated (see Fig. 21-18). The original Ziegler–Nichols formulas used the slope or the rate of change rather than the time.

Another graphical method, which is the mathematical equivalent of the original Ziegler–Nichols method, is

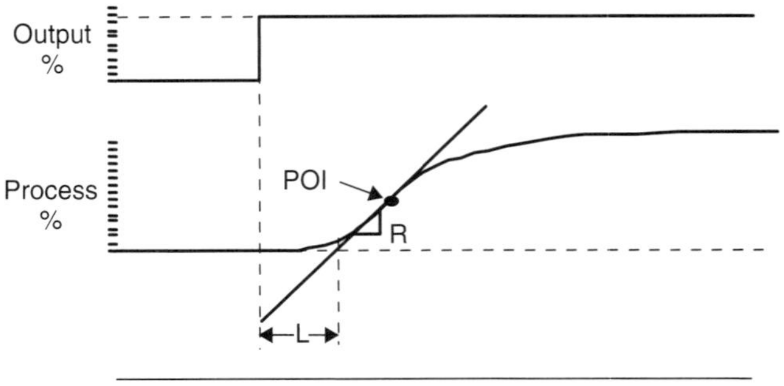

Figure 21-18 Finding slope of tangent line.

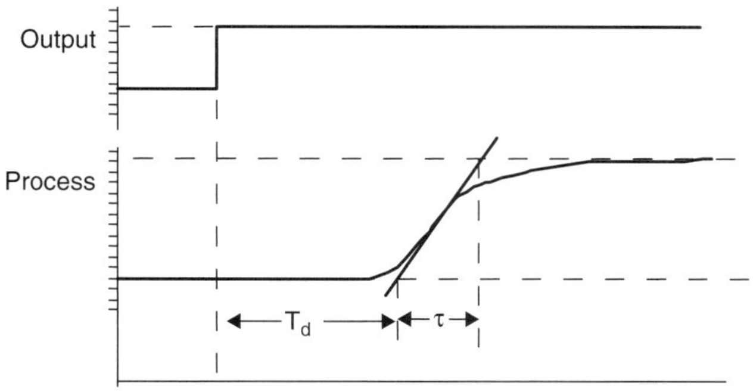

Figure 21-19 Depiction of T_d and τ using the tangent line.

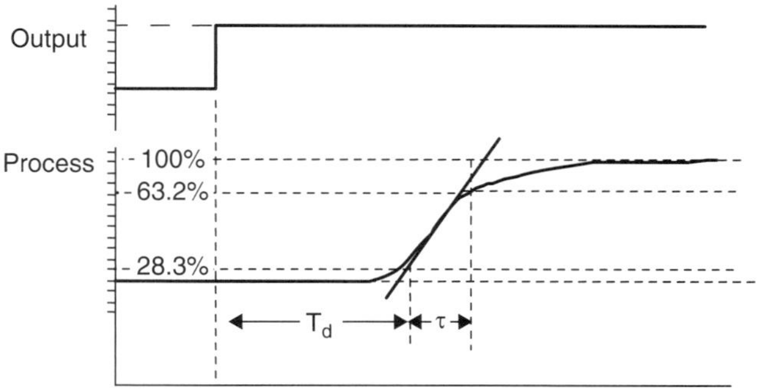

Figure 21-21 Two-point method to find T_d and τ.

known as the *tangent method* (see Fig. 21-19). In this method the same tangent line is drawn, but the process time constant is the time between the interception of the tangent and the original process value line and the eventual process value line. The formulas that are commonly provided for the Ziegler–Nichols open-loop method use the process time constant. These two methods will provide identical results when applied perfectly: that is, no error in the drawing of the line and no noise in the process signal.

Two additional methods are the mathematical equivalent of the previous two only if the process dynamics really did comprise only a single-order lag and a dead time. When the process differs from this model, the following two methods will provide different results that may actually provide better tuning. The first of these is sometimes known at the *tangent-and-point method* (Fig. 21-20). In this method the same tangent line is drawn as before and used to calculate the pseudo dead time as before. However, a point equal to 63.2% of the value between the original and the ultimate process measurement is made on the tangent line. The time between the end of the pseudo dead time and the time at which the tangent line goes through the 63.2% point is the process time constant. This method will give the same results as the first two when the process is truly a dead time plus first-order lag. As the values of the smaller lags

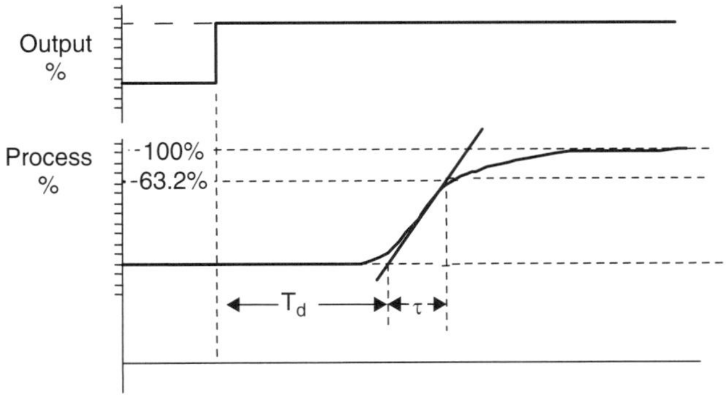

Figure 21-20 Tangent and point method.

increase, the tangent and point method gives a smaller time constant than that given by the two graphical methods.

Another variation is the *two-point method* (Fig. 21-21) proposed by Smith (2009). This method does not require the drawing of a tangent line but measures the times at which the process changes by 28.3% (t_1) and 63.2% (t_2) of the total process change. Then two formulas are used to calculate the pseudo dead time and the process time constant [1].

$$\text{Process time constant:} \quad \tau = 1.5(t_1 - t_2) \qquad (21.5)$$

$$\text{Pseudo dead time:} \quad T_d = t_1 - \tau$$

One advantage of the two-point method is that it does not require drawing the tangent line. This improves the accuracy, particularly for processes with noise. This method will provide the same pseudo dead time and time constant when the process dynamics are dominated by a true dead time and single lag. Like the tangent and point method, the two-point method provides different values for the pseudo dead time and the process time constant when the process dynamics comprise multiple lags. When used with the multiple-lag processes, the one- and two-point methods also provide better tuning. Tests within each set all result in the same pseudo dead time and time constant using the tangent line method. However, using the two-point method, the pseudo dead time and time constant are different. For multiple-lag processes the two-point method results in a longer dead time and shorter time constant than those of the tangent methods. This provides more conservative tuning using any of the FOPDT tuning methods.

21.5.2.1 Ziegler–Nichols Open-Loop Tuning Method The first method is the *open-loop method*, also known as the *reaction curve method*. This method calculates the actual values of the assumed process model (the gain, pseudo dead time, and lag). For this method to work the process must be "lined out," that is, not be changing. With the controller in manual, the output is changed by a small amount. The process is then monitored.

The following values are calculated using one of the methods described above: K_p, process gain; τ process time constant; and T_d, pseudo dead time. The gain, reset, and derivative are calculated using Table 21-1.

21.5.2.2 Ziegler–Nichols Closed-Loop Tuning Method The *closed-loop* (or *ultimate gain*) *method* determines the gain that will cause the loop to oscillate at a constant amplitude. Most loops will oscillate if the gain is increased sufficiently. The following steps are used:

1. Place the controller in automatic mode with low gain and no reset or derivative.
2. Gradually increase the gain, making small changes in the set point, until oscillations begin.
3. Adjust the gain to make the oscillations continue with a constant amplitude.
4. Note the gain (ultimate gain, G_u) and period (ultimate period, P_u). G_u is the gain of the controller, resulting in constant-amplitude oscillation. P_u, shown in Fig. 21-22, is the period of the oscillation.

The gain, reset, and derivative are calculated using Table 21-2.

TABLE 21-1 Gain, Reset, and Derivative for Open-Loop Method

	Gain	Reset	Derivative
P	$\frac{t}{T_d Kp}$		
PI	$0.9\frac{t}{T_d Kp}$	$\frac{0.3}{T_d}$	
PID	$1.2\frac{t}{T_d Kp}$	$\frac{0.5}{T_d}$	$0.5T_d$

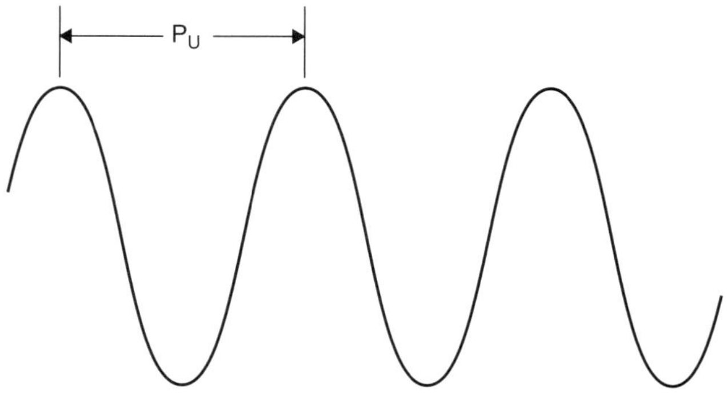

Figure 21-22 Closed-loop method: constant oscillation with P_u noted.

TABLE 21-2 Gain, Reset, and Derivative for Closed-Loop Method

	Gain	Reset	Derivative
P	$0.5G_u$		
PI	$0.45G_u$	$\frac{1.2}{P_u}$	
PID	$0.6G_u$	$\frac{2}{P_u}$	$\frac{P_u}{8}$

TABLE 21-3 Gain, Reset, and Derivative for the Cohen–Coon Method

	Gain $\times$ *Kp*	Reset Rate/τ	Derivative/τ
P	$\frac{t}{T_d} + 0.333$		
PI	$0.9\frac{t}{T_d} + 0.082$	$\frac{3.33(t/T_d)[1 + (t/T_d)/11]}{1 + 2.2(t/T_d)}$	
PID	$1.35\frac{t}{T_d} + 0.27$	$\frac{2.5(t/T_d)[1 + (t/T_d)/5]}{1 + 0.6(t/T_d)}$	$\frac{0.37(t/T_d)}{1 + 0.2(t/T_d)}$

21.5.3 Other Methods

The *Cohen–Coon method* is similar to the Ziegler–Nichols reaction rate method in that it makes use of the FOPDT model to develop the tuning parameters. The parameters (shown in Table 21-3) are more complex, involving more arithmetic operations. As can be seen from the tables, the Cohen–Coon method will result in a slightly higher gain than that obtgined using the Ziegler–Nichols method. For most loops it will provide tuning closer to quarter-wave decay and with a lower ISE index than are provided by, the Ziegler–Nichols open-loop method.

Integral absolute error and integral squared error are two methods of judging the tuning of a control loop (see below). A method of selecting tuning coefficients to minimize the IAE or ISE criteria for disturbances was developed by Lopez et al. [1]. This method uses the FOPDT model parameters and a set of equations to calculate the tuning parameters (Table 21-4). Tests show that the parameters provide results close to the minimum IAE or

TABLE 21-4 Gain, Reset, and Derivative for the Lopez et al. Method

	Gain $\times$ *Kp*	Reset Rate	Derivative
P	$1.411T_d/\tau^{-0.917}$		
PI	$1.305T_d/\tau^{-0.959}$	$(\tau/.492)(T_d/\tau)^{0.739}$	
PID	$1.495/T_d\tau^{-0.945}$	$(\tau/1.101)(T_d/\tau)^{0.771}$	$0.56\tau(T_d/\tau)^{1.006}$

ISE, particularly when the actual process dynamics are similar to those of the FOPDT model (the process contains a true dead time and one lag). When the process has multiple lags, the equations do not provide the best possible tuning, but they still provide better tuning (lower IAE and ISE indices) than do the other methods.

21.5.4 Controllability of Processes

The gain at which a loop will oscillate depends on the dynamics of the loop. In general, a loop that has no dynamic elements other than one first-order lag will not oscillate at any gain. A loop with dead time or with multiple lags will oscillate at some gain.

If we refer to the model used for the Ziegler–Nichols open-loop test, the gain at which a loop will exhibit undamped, sustained oscillations (the ultimate gain in the Ziegler–Nichols closed-loop test) will depend on the ratio of the process time constant (τ) to the pseudo dead time (T_d) (Fig. 21-23). The importance of this fact goes beyond finding the best tuning parameters. There is an advantage to a loop that can have a higher gain. If a loop can have a higher gain, it will have greater rejection of disturbances and will respond more rapidly to set point changes. Therefore, it is advantageous to be able to increase the gain if doing so will not cause the loop to become unstable.

Remember that the time constant is proportional to the largest lag in the system. The pseudo dead time is based on the dead time and all other lags. The allowable gain (and the gain required for quarter-wave decay) can be increased by increasing the ratio τ/T_d. An increase in either dead time or in any lag other than the longest lag will decrease the ratio and therefore decrease the allowable gain. The loop scan period has the effect of process dead time. Increasing the scan period will decrease the ratio and the allowable gain. Also, adding any lag smaller than the longest lag (e.g., adding a large well to a thermocouple or a filter to a noisy loop) will decrease the allowable gain

Flow loops are too fast to use the standard methods of analysis and tuning. The speed of the loop compared to

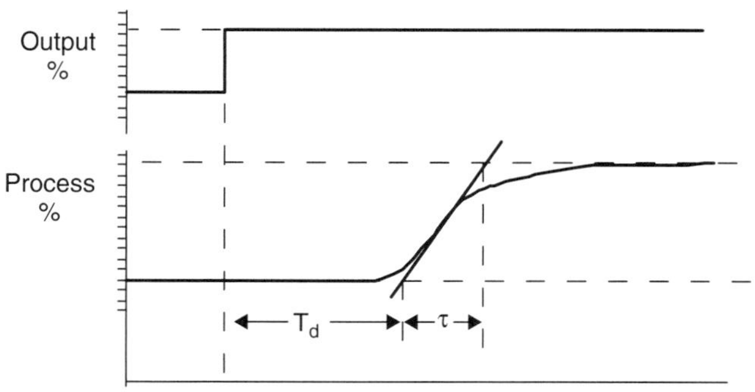

Figure 21-23 Ratio of τ to T_d is an indication of controllability.

the update rate of the display prevents the collection of data for either the open- or closed-loop methods. However, there are some rules of thumb regarding the tuning of flow loops. Typically, the tuning of a flow loop using digital control is gain = 0.5 to 0.7, reset = 15 to 20 repeats/min, and no derivative.

21.5.4.1 Digital Versus Analog Control The guidelines described above are based on digital controller such as that used in distributed control systems. Some flow loops using analog controllers are tuned with higher gain. However, the same loop may go unstable if a digital controller is used. Although normally a digital control system will provide response very similar to that of an analog control system, the performance can be quite different with fast loops such as flow. With an analog controller, the flow loop has a time constant, τ of a few seconds and pseudo dead time much smaller than 1 s. However, with a digital controller, the scan rate of the controller can be considered dead time. Although this dead time is small, it is large enough when compared to τ to cause instability when the gain is higher than 1.

21.6 MULTILOOP CONTROL

21.6.1 Cascade Control

Cascade is the most common form of multiple variable control. In *cascade control* the controller manipulates the set point of another controller rather than a final control element.

21.6.1.1 Cascade Control Basics In many processes there some process variables that we want to hold at a specific value, and other intermediate variables that change as necessary to hold the first type of variable at its desired set point. For example, if the temperature of liquid leaving a heat exchanger is the controlled variable, the one we want to hold at a set point, we do so by manipulating the steam valve. The flow of steam to the exchanger affects the temperature directly, but we do not care how much is flowing as long as the temperature is controlled (see Fig. 21-24). The amount of steam required will depend on the flow rate of process fluid and the difference between the inlet temperature and the set point of the outlet temperature (see Fig. 21-25).

We can control the temperature using a single PID controller with the temperature as its input and its output connected to the steam valve. This arrangement will control the temperature. However, there are some problems in the control. The steam header pressure may change, causing a sudden reduction in the steam flow to the exchanger. The temperature controller will bring the temperature back to

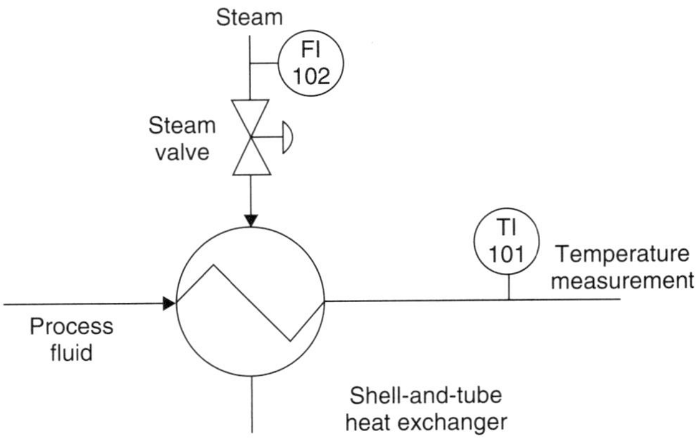

Figure 21-24 Simple shell-and-tube heat exchanger.

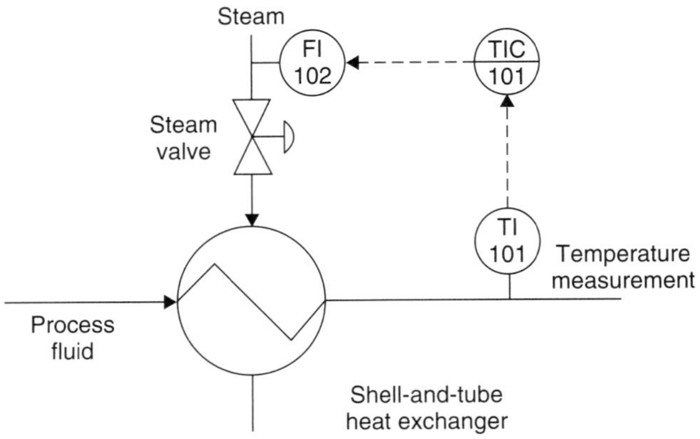

Figure 21-25 Simple control loop for heat exchanger.

its set point, but because of the slow tuning required of the temperature controller, the correction will take longer than desired.

The temperature loop, which probably contains multiple lags and dead time, is a more difficult loop to tune. Nonlinearities in the valve will further complicate the tuning.

The use of cascade control will correct both of these problems (see Fig. 21-26). If the header pressure changes, causing a change in the flow, that change will be detected by the flow measurement and corrected immediately.

21.6.1.2 Cascade Structure and Terminology The "outer" loop of a cascade loop pair is called the *primary loop*. This loop controls the variable that must be held at a specific set point. The "inner" loop of the pair is called the *secondary loop*. This loop controls the variable that will influence the primary variable. The primary and secondary loops are sometimes referred to as the *master* and *slave loops*.

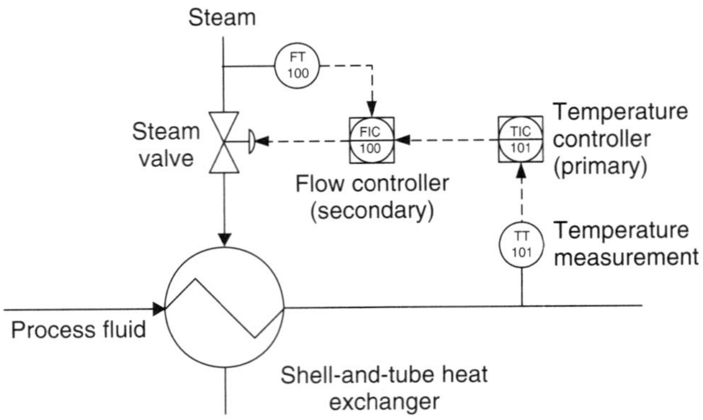

Figure 21-26 Cascade control loop for heat exchanger.

21.6.1.3 Guidelines for Using Cascades Cascade should be used when the following conditions are met:

- A variable exists that has a great influence over the primary variable.
- This variable has a dynamic time at least four times faster than the primary loops dynamics.
- The variable can be controlled easily.

Otherwise, cascade control probably will not work well. Other factors also influence the decision to use cascade:

- If the secondary variable is measured for other reasons, such as data recording or operator information, there is little if any additional cost to cascade. If the secondary variable would not be measured except to allow cascade control, the costs of installing the measurement will need to be justified.
- If there are times when the operator will need to control the secondary variable directly (such as during startup), the secondary loop will be needed, so as long as the guidelines above are met, cascade control may as well be implemented.

At one time cascade control had a higher cost and caused more complexity for the operator. However, with modern DCS and other digital control systems, the only cost of cascade control is the cost of measuring the secondary variable, and thus the operational complexity can be reduced.

21.6.1.4 Cascade Implementation Issues A cascade loop does not always operate in full cascade operation, just as a simple PID loop is not always in the automatic mode. To implement a cascade control scheme, the secondary controller must have the ability to operate in the remote set point mode. This option was required when purchasing analog controllers but is generally a user-selectable option on digital controllers. In DCS controls the user may have to select a PID function block or select the remote set point option on a more general PID function block.

Some manufacturers provide a single mode "switch" in the PID block with the options of manual, auto, and cascade (or remote set point). Other manufacturers provide two switches, manual/auto and local set point/remote set point. Either way, the operator will have the ability to place the control loop into (1) *manual*, and manipulate the output to the valve directly; (2) *auto*, with the ability to manipulate the set point of the secondary loop; and (3) *cascade* (or remote set point) to allow the primary controller to operate in automatic made and control the process.

This is illustrated in Fig. 21-27, which shows three possible modes of a cascade control scheme. In A, manual mode, the operator is adjusting the valve directly while monitoring the flow. In B, the operator has placed the flow loop (secondary) into automatic and is adjusting its set point. In C, the operator has placed the flow loop into cascade or remote set point mode and may adjust the set point of the temperature controller, if desired. The secondary controller can be in manual, with the secondary set point tracking the secondary measurement; the secondary controller may be in automatic, with the primary output tracking the secondary set point; or the secondary controller may be in cascade mode, with the primary controller in automatic.

An important consideration for the implementation is that the transfer between one mode and another must be bumpless. That is, the act of changing the mode must have no immediate effect on the output to the valve. The most common technique to deal with this need is tracking. In Fig. 21-27A, while the operator adjusts the output to the valve, the set point of the flow loop tracks the actual flow.

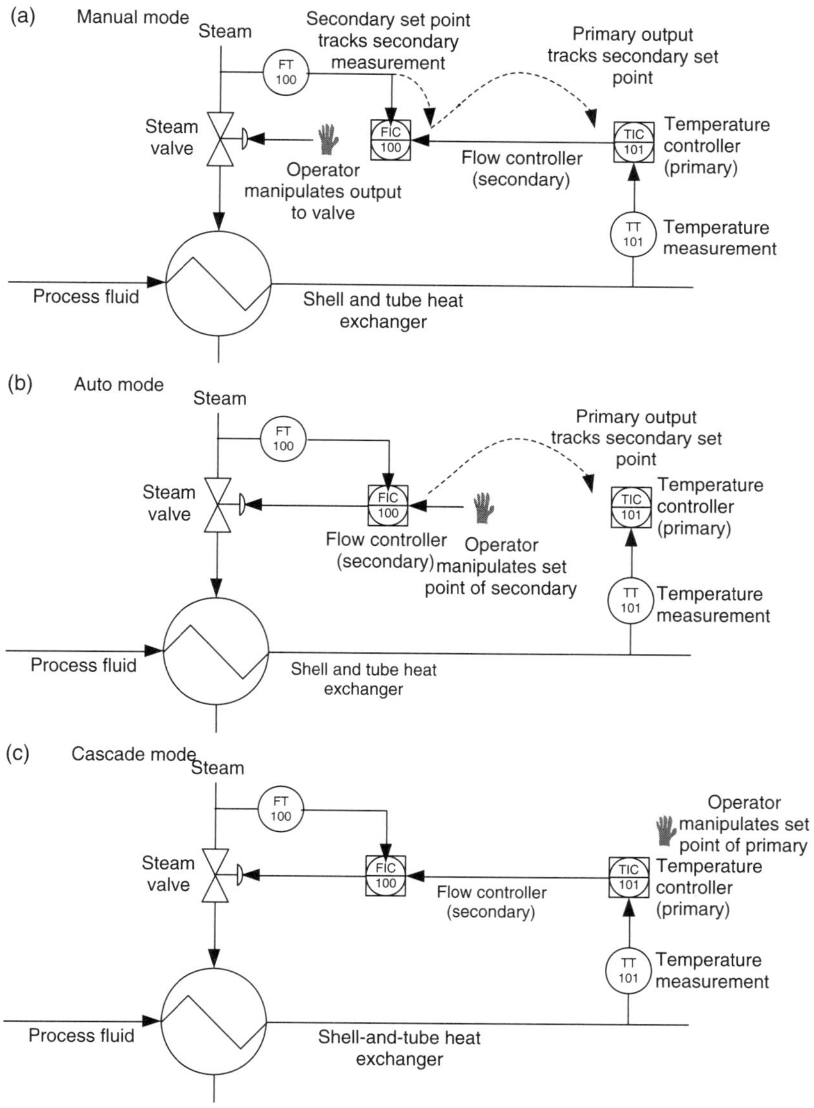

Figure 21-27 Manual, auto, and cascade modes.

At the same time, the output of the temperature controller tracks the set point of the flow controller (with the flow value converted to percent). Therefore, when the operator switches the flow controller to automatic, the set point is the same as the actual flow, and there is no sudden change in the valve.

In Fig. 21-27B, while the operator adjusts the set point of the flow controller, the output of the temperature controller continues to track the flow set point. So when the operator switches the flow control to cascade (see Fig. 21-27C), the output of the temperature controller is already at the correct value for that time, and there is no immediate change in the flow set point or the output to the valve. After the switch to cascade, if the operator adjusts the temperature set point or there are load changes, the temperature controller will manipulate the flow controller to increase or decrease the steam flow as needed.

21.6.1.5 Tuning Cascade Loops The secondary loop is tuned first, using any of the tuning methods. When the loops are tuned, careful attention should be paid to the response to set point changes. The primary loop is left in manual while the secondary loop is tuned. After the secondary loop is tuned, the primary loop is tuned with the secondary loop in automatic.

21.6.2 Ratio Control

Ratio is another common form of multiple-loop control. The object of *ratio control* is to maintain two or more flows at a constant ratio even when the flows are changing. Most often, the flows being controlled are blended, that is, mixed together. However, there are some situations where flows not mixed together are controlled by ratio control. For example, we may wish to maintain a flow of steam to the reboiler of a distillation column in ratio with the feed to the column, even though the two never mix. For most examples and discussion here we use a flow mixing example.

21.6.2.1 Ratio Control Basics Ratio control usually involves the control of one flow (known as the *controlled flow*) in ratio to another flow (known as the *wild flow*). The wild flow may be controlled; if so, it is controlled by an unrelated loop (Fig. 21-28). In this example the wild flow is measured by FT-101. The controlled flow is measured by FT-102 and controlled by controller FIC-102, which is a remote set point (similar to the secondary loop of a cascade control pair). The signal from FT-101 is multiplied (in block FF-102) by the desired ratio, which can be adjusted by the operator.

In actual implementation on most control systems, the remote set point to the controlled flow controller (FIC-102) is expecting a percentage signal (100% will result in a set point at the maximum range of the controller). Therefore,

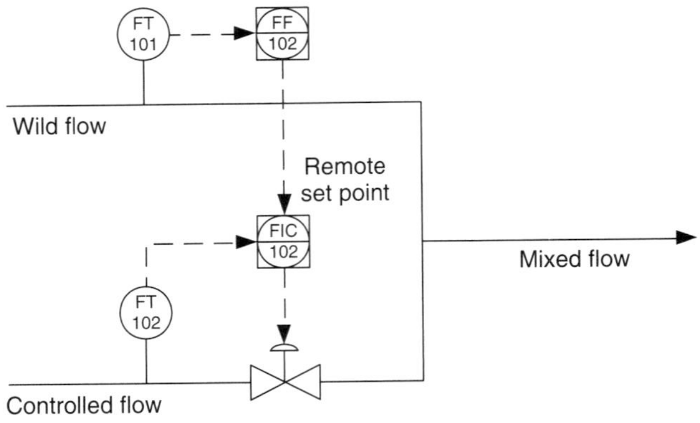

Figure 21-28 Ratio loop.

the ratio unit (FF-102) will multiply the actual flow by the ratio and then convert the signal to a percentage based on the range of FIC-120. In some cases the controller may be able to accept a remote set point in engineering units, requiring no normalization. In most cases the operator will be able to adjust the ratio in FF-102.

21.6.2.2 Mode Change Just as in the case of cascade loops, the operator will have a choice of operating the controlled flow controller by (1) manual, adjusting the valve position directly; (2) placing the flow controller in automatic and adjusting the set point; or (3) placing the controlled flow controller into remote set point (cascade) mode and allowing the ratio control to take place. This mode change will probably take place during startup of the process. To prevent a "bump" or sudden change in the set point of the flow controller when the operator switches from local set point to remote set point (ratio control), the ratio is often recomputed, either continuously while the controller is in local set point or when the switch to ratio control is made, so that the ratio at the time of the switch is the actual ratio. After the switch to ratio control, the operator can adjust the ratio.

21.6.2.3 Ratio Manipulated by Another Control Loop In blending applications the purpose of ratio control is to achieve some desired mix of the two flows, usually to result in a particular physical property, such as density. To ensure that the correct ratio is used, the physical property may be measured using an analyzer. The analyzer signal can then be used as the input of a PID controller that adjusts the ratio. In this case, the ratio is a secondary loop and the analyzer controller is the primary loop in a cascade pair of loops. The output of the analyzer loop is converted from 0 to 100% to the number to be multiplied by the wild flow signal to provide the controlled flow set point (see Fig. 21-29).

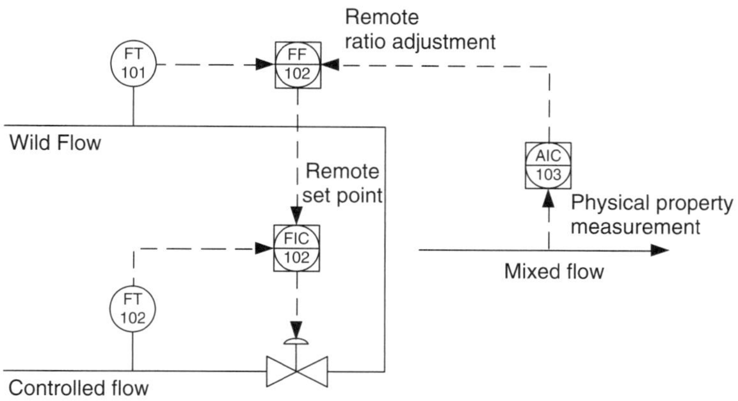

Figure 21-29 Ratio set by external loop.

21.6.3 Feedforward Control

In *feedback* control we measure the controlled variable, compare it to a set point, and then take corrective action after the controlled variable has departed from its set point, usually as a result of a disturbance (change in a load variable). If, instead, we can measure the loads and change the manipulated variable to anticipate the effect of the load change, we might keep the controlled variable from departing from the set point (or, at least, keep the error short). This strategy is called *feedforward control* (Fig. 21-30).

In the example above, the feedforward calculation block calculates the amount of steam needed to heat the process fluid to the set point based on the flow and inlet temperature. This calculation is known as the *feedforward calculation*. Because there are usually disturbances that cannot be measured and the feedforward calculation is not perfect, the calculation will not result in the exact amount of steam flow. Therefore, the output of the feedback controller is added to the feedforward calculation to trip the steam flow set point continuously to keep the temperature at its set point.

21.7 FINAL CONTROL ELEMENTS

The term *valve* has been used for *final control elements*. Although most common, valves are controlled from a signal from the controllers. Some other types of final control elements are dampers, time-proportional heating elements, and time-proportional solenoid valves.

21.7.1 Time-Proportional Heating Elements and Solenoid Valves

A heating element may be turned on or off, or a solenoid valve supplying cooling water may be opened or closed for a period of time based on the output of the controller. For example, a 50% output of the controller may cause the element to be on or the valve to be open for 5 s of a 10-s cycle period, while a 0% output would cause the element to

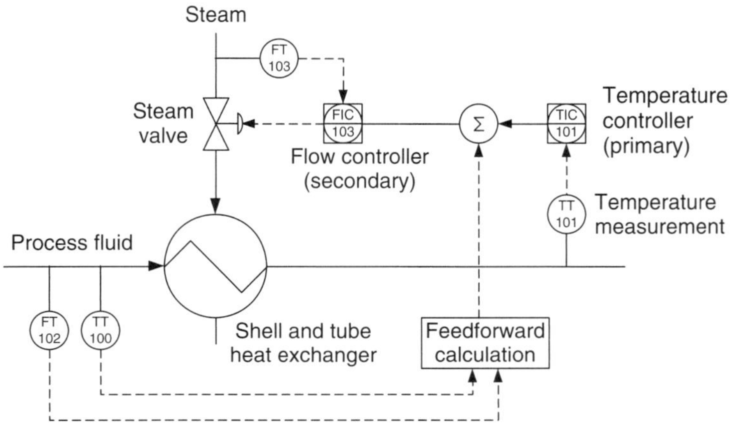

Figure 21-30 Feedforward control.

remain off and the valve to remain closed the entire cycle. A separate device or software program would typically be used to convert the output signal, in percent, to the cycle time.

21.8 PROCESS CONTROLLERS

The information provided above can apply to any controller, whether pneumatic (based on air pressure), electronic analog, single-loop digital, or multiloop digital.

21.8.1 Distributed Control Systems

A distributed control system (DCS) is a control system built with a number of microprocessor modules. Some of the microprocessors will provide the controls for a number of control loops, from a dozen to several hundred. Other microprocessors provide displays for operators and for engineers and technicians. Typically, the vendor of the DCS will provide both control software and the displays.

21.8.2 Programmable Logic Controllers

Programmable logic controllers (PLCs) were first used to replace relays for logic (on/off) control. PLCs and DCSs are now very similar, with many of the same features. One difference that remains is that with PLCs, the PLC vendor typically does not provide the displays. Other companies provide display software that will run on personal computers.

REFERENCE

1. Lopez, A. M., Miller, J. A., Smith, C. L., and Murrill, P. W., Controller tuning relationships based on integral performance criteria, *Instrumentation Technology* November 1967, pp. 57–62.
2. Smith CL (2009), Practical Process Control: Tuning and Troubleshooting, Wiley, 2009, p. 40.

22

PROCESS MODELING AND SIMULATION

MATHEW CHIDIEBERE ANEKE
Northumbria University, Newcastle upon Tyne, England

Before the advent of and advances in computer technology, most engineering designs were usually carried out manually, which entails solving numerous equations (algebraic, differential, partial differential, numerical, etc.) to evaluate the parameters of a given process unit and operation, as well as to investigate or examine the behavior of a unit when there were variations in the process design variables. This approach is time consuming and is prone to human error. Furthermore, to unravel some of the mysteries of the design that cannot be achieved through manual calculations, the design engineer may resort to construction of pilot plants to investigate the behavior of a process or unit operation. This is time consuming and very expensive.

However, with the advances made in computer technology, the work of the design engineer has been made easier. Although the design equations are still complex, solving them is now easier through the use of computer codes. This has given rise to a new style of engineering design and has reduced the level of investment in the construction of pilot plants. It has also brought more flexibility into engineering design. Therefore, it can be affirmed that advances in computer technology have brought a lot of benefits to engineering design, including minimal design cost, shorter product development cycle, reduced time, and advancements in competitiveness [11]. Also with the help of the knowledge and information embedded in engineering design model tools, less experienced engineers can comfortably solve complex design problems [1].

Engineering design cuts across many fields of engineering. Each field has modeling tools peculiar to it, and this depends on the nature of the design. In this chapter we will focus primarily on modeling in chemical/process engineering design. Our objectives are to:

- Establish the relationships among process modeling, simulation, and optimization
- Introduce some commercial process simulation tools
- Present process modeling and simulation case studies

22.1 PROCESS MODELING

Process modeling is one of the principal steps in process design. Modeling involves the development of a set of equations that help us to gain insight into the behavior of a given process or unit operation. A good model for this is one that is capable of showcasing behavior similar to that of the real process that it represents. The complexity of any given model is synonymous with the amount of information or detail that the process designer wants to capture from the model. For example, at the preliminary process design and evaluation stages, simplified models are developed that could be solved by manual calculations or with the use of spreadsheets [2]. This might be useful for conducting simple mass and energy balances on preliminary process flowsheets. However, for more detailed design, computer programs or process simulations tools are necessary to produce more rigorous mass and energy balances. There are several classifications of mathematical models [2]; (1) steady-state models vs. dynamic models, (2) lump-sum model vs. distributed models, and (3) shortcut model vs. rigorous model.

Process Plant Equipment: Operation, Control, and Reliability, First Edition. Edited by Michael D. Holloway, Chikezie Nwaoha, and Oliver A. Onyewuenyi.
© 2012 John Wiley & Sons, Inc. Published 2012 by John Wiley & Sons, Inc.

22.1.1 Steady State Versus Dynamic Models

Steady-state models are models in which the process parameters are assumed to be constant with respect to time. In other words, they are models in which the rate of change of any process parameter with respect to time is assumed to be zero. Steady-state modeling is commonly used to investigate process behavior during normal operation. It is represented by a set of algebraic equations.

Dynamic modeling is concerned with the transient behavior of a process or unit operation. It is usually represented by a set of algebraic and differential equations combined. Its application is primarily in the study of process startup, shutdown, and for controller design.

22.1.2 Lump-sum Versus Distributed Models

When a model is homogeneous (i.e., exists in the same state throughout the system), the parameters are lumped. However, if the model is heterogeneous (varying states within the system), the parameters are distributed [3]. In real systems, the operating conditions (e.g., temperature, pressure, composition) inside the process equipment vary with time; however, since systems operate at conditions of almost perfect mixing, they can be assumed to be homogeneous and in such cases are always modeled as lump-sum equipment. A *lump-sum model* can be either a steady-state or a dynamic model [2]. A steady-state lump-sum model is represented by a set of algebraic equations, while a dynamic lump-sum model is represented by a set of differential and algebraic equations.

Distributed models are used to model systems in which imperfect mixing occurs. Imperfect mixing results in different operating conditions within a piece of equipment [2]. Distributed models are represented using partial differential equations.

22.1.3 Shortcut Versus Rigorous Models

A third classification is based on the level of complexity of a model. Model complexity usually involves a trade-off between simplicity and accuracy [2]. *Shortcut models* are generally used during the preliminary design stage and usually involve manual calculations. Although manual calculations are less accurate, they help to provide a quick assessment of a design problem. On the other hand, a *rigorous model* is used during detailed modeling and provides much more accurate results. Special modeling tools or computer codes are used.

22.2 PROCESS SIMULATION

Simulation simply means solving a set of process modeling equations that define a process or unit operation to obtain a solution for unknown process variables. A process may contain many unit operations that are connected together with process streams. This assembly of the process units connected together by process streams makes up the process flowsheet. Hence, a simulation flowsheet consists of (1) a large number of nonlinear equations that describe the connectivity of the units of the flowsheet through process streams, (2) specific equations for each unit which describe the internal mass and energy balance, and (3) an equilibrium relationship as well as physical property equations that define densities, enthalpies, and other transport and thermodynamic properties of process streams [4].

Most unit operations, such as heat exchangers, pumps, and valves, are conventional process units, and thus their equations are basically the same in different processes. Hence, since the model equations for these conventional units are the same, the behavior of such units in any given process model depends on the nature of the process stream that they handle. Thus, for such conventional units, one or more models are usually developed and used to solve the energy and material balances.

When the unit operations of a process are connected together using the process streams and the degrees of freedom of the process have been satisfied, the assembled process can be solved simultaneously by an equation solver. Some of the equation solvers used in process simulation tools are the Dormand–Prince, the Bogacki–Shampine, the Adams, and the Newton–Raphson.

22.3 PROCESS OPTIMIZATION

Process simulation provides a solution for the performance of a process under certain operating conditions. Variations in process operating conditions give rise to variations in process performance. In most engineering applications, the aim is always to obtain the best process performance at the least possible cost. This is the problem that process optimization seeks to solve. It involves the systematic improvement of a process [2]. Process optimization finds the best solution to a given process within certain defined operating constraints. To optimize a process, an objective function has to be specified which serves as the performance indicator of the process. The objective function may be the operating cost, desired product yield, or any other parameter that the process engineer wishes to investigate. The best operating point is obtained by varying the manipulated variables of the process (e.g., pressure, temperature, flow rate). The constraints are usually imposed on a system to avoid operating the system in unsafe regions or to achieve a specific product requirement, model validity, or process operation. Process optimization is therefore an iterative process simulation to achieve the set objective function by varying the manipulated variables while satisfying the constraints imposed on the process or unit operation.

Process modeling, simulation and optimization are related. To solve any given process design, operation, or control problem, the process had to be modeled first; then the model is solved through simulation either by the use of process simulation tools or by hand calculation. The model solved is then validated. The validated model can then be used to optimize the objective function or design controllers for the system.

22.4 COMMERCIAL TOOLS FOR PROCESS MODELING, SIMULATION, AND OPTIMIZATION

Due to the increasing complexity of engineering design and the advancement in computer technology, the majority of the process modeling, simulation, and optimization problems are handled with special process modeling, simulation, and optimization tools known as *process simulators*. Many commercial process simulators are currently available in the market, and they can be broadly classified into two groups: (1) modular mode process simulators, and (2) equation-oriented mode process simulators.

22.4.1 Modular Mode Process Simulators

In modular mode process simulators, the modeling equations are grouped according to the individual units in the process, and specialized solution strategies are applied to each unit [5]. The sequence of the calculation then goes from one unit in the process flowsheet to the next through the process streams that connect the units. The direction of flow of information usually follows the direction of material flow [6]. Modular process simulators are easily initialized. Also, solution procedures are unit specific and locally robust [5].

Modular mode process simulators are the most common process simulator tools employed in the industrial environment. The individual unit operations are usually represented as a black box using process equipment symbols. Inside each symbol is embedded the model equations of the unit. Each symbol also has input(s) and output port(s). Some examples of process modular simulators are shown in Table 22-1.

22.4.2 Equation-Oriented Process Simulators

For equation-based simulators, the process model equations are considered as a single large set of equations to be solved with a large-scale nonlinear algorithm [5]. It saw considerable industrial development and applications in the 1980s and 1990s, especially in the area of online modeling and optimization [2]. They are more flexible but more difficult to use than modular simulators. The user is allowed access to the model equations and he or she can modify the model equations of the units. Some of the equation-oriented process simulators currently in use are listed in Table 22-2. The reader is referred to the individual developer Web sites for further information on any process simulators that may be of interest.

In carrying out any process modeling work for any given process there are some basic procedures that must be followed to achieve the modeling objectives. These procedures are described below.

1. *Understand the process to be modeled and the modeling objective*. Before carrying out any modeling, there is always a set of objectives that you would like your model to achieve. A good understanding of the process to be modeled is essential in developing a very good process model of the process or unit operation. This can be done through a relevant literature review and study of similar operations for the given process. This will be helpful in understanding the relevant factors that will be considered as part of your model.

2. *Decide whether the model will be dynamic, steady state, or both*. A good understanding of the process and the modeling objective will help you to decide what aspect of the model to consider for the given process. If you are interested in carrying out transient analysis of any given process, dynamic modeling should be used, while for steady-state operation analysis, steady-state modeling has to be used.

TABLE 22-1 Examples of Modular Process Simulator Tools

Modular Simulator	Developer	Web Site
Aspen Hysys	Aspen Technologies, Inc.	http://www.aspentech.com
Aspen Plus	Aspen Technologies, Inc.	http://www.aspentech.com
PRO/II	Simulation Science, Inc.	http://www.simsci.com
CHEMCAD	Chemstations, Inc.	http://www.chemstations.com
ProMax	Bryan Research and Engineering	http://www.bre.com
IPSEpro	SimTech Simulation Technologies	http://www.simtechnology.com
OLGA	Spt Group	http://www.sptgroup.com

TABLE 22-2 Examples of Equation-Oriented Process Simulator Tools

Equation-Oriented Simulator	Developer	Web Site
Aspen Custom Modeller	Aspen Technologies, Inc.	http://www.aspentech.com
gProms	Process Systems Enterprise Ltd	http://www.psenterprise.com
IPSEpro	SimTech Simulation Technologies	http://www.simtechnology.com
MATLAB and Simulink	MathWorks	http://www.mathworks.com

3. *Decide on the complexity of the model*. This has to do with the amount of detail you want your model to capture. Complex models capture more detailed information about a given process but are very difficult to develop and are time consuming. The required level of complexity can be achieved by making practical and relevant assumptions.

4. *Choose the modeling principles and process simulator*. At this stage, you have to decide whether your model will be carried out using a modular-mode or equation-oriented approach. If a modular-mode approach, you can choose from the numerous modular-mode process simulators; however, with an equation-based approach, you have to develop a set of model equations to describe your process. (Note: For a steady-state process, the model equations comprise a set of algebraic equation, while for a dynamic process, the model equations comprise a set of algebraic and differential equations.) Most modeling equations are based primarily on the law of conservation of mass, energy, and momentum.

5. *Be sure to satisfy the degrees of freedom*. For you to be able to solve the set of model equations developed, the degrees of freedom for the system must be satisfied. This means that the number of equations must be equal to the number of unspecified variables. Hence, some of the variables have to be specified prior to solving the set of equations, whereas others have to be determined by solving the equations. The specified variables are regarded as the *parameters of the model or process*. Apart from satisfying the degrees of freedom, you should also make sure that the parameters you specify do not cause any of the model equations to be redundant.

6. *Simulate or solve the set of model equations*. When the degrees of freedom have been satisfied and none of the equations is redundant, the set of model equations can be solved or simulated using the process simulator.

7. *Verify and validate the simulation results*. This involves making sure that the result of your model has practical meaning. A model will have to show relevance to a real-world situation. For example, it makes no real-world sense to obtain a negative flow rate of a cold fluid to a heat exchanger system. The steps above are shown diagrammatically in Fig. 22-1.

22.5 PROCESS MODELING CASE STUDIES

Case Study 22.1: Steady-State Modeling of a Splitter Unit Using a Modular Mode Process Simulator

(a) *Process description*. Figure 22-2 shows an arrangement for a simple splitter system. The feed stream F enters the splitter and is divided into two streams, W and D. The feed contains three components, benzene, toluene, and diphenyl, with mole fractions x_{FB}, x_{FT}, and x_{FD}, respectively. One of the outlet streams of the splitter is represented as D. Its mole composition with respect to benzene, toluene, and diphenyl is given as x_{DB}, x_{DT}, and x_{DD}, respectively. Similarly, the remaining outlet stream flow is given as W, with the mole composition of benzene, toluene, and diphenyl given as x_{WB}, x_{WT}, and x_{wD}, respectively.

(b) *Principles of the model*. Since a splitter is used to divide fluid streams, the model will be based on the law of conservation of mass and energy.

(c) *Modeling assumptions*. The model is based on the following assumptions:
   1. The splitter works at steady state (given)
   2. The process is an isobaric process (i.e., there is no pressure drop in the splitter: $P_F = P_D = P_w = P$).
   3. The process is an isothermal process (i.e., there is no temperature change in the system: $T_F = T_D = T_W = T$. This is because the splitter is used to divide the fluid, and thus the fluid temperatures are supposed to be constant.
   4. The process is an adiabatic process (i.e., there is no heat transfer between the system and the environment).

(d) *Modeling equations*. At steady-state operation, the independent material and energy balance equations are

$$Fx_{FB} = Dx_{DB} + Wx_{WB} \quad (22.1)$$

$$Fx_{FT} = Dx_{nT} + Wx_{WT} \quad (22.2)$$

$$Fx_{FD} = Dx_{DD} + Wx_{WD} \quad (22.3)$$

$$Fh_F = Dh_D + Wh_W \quad (22.4)$$

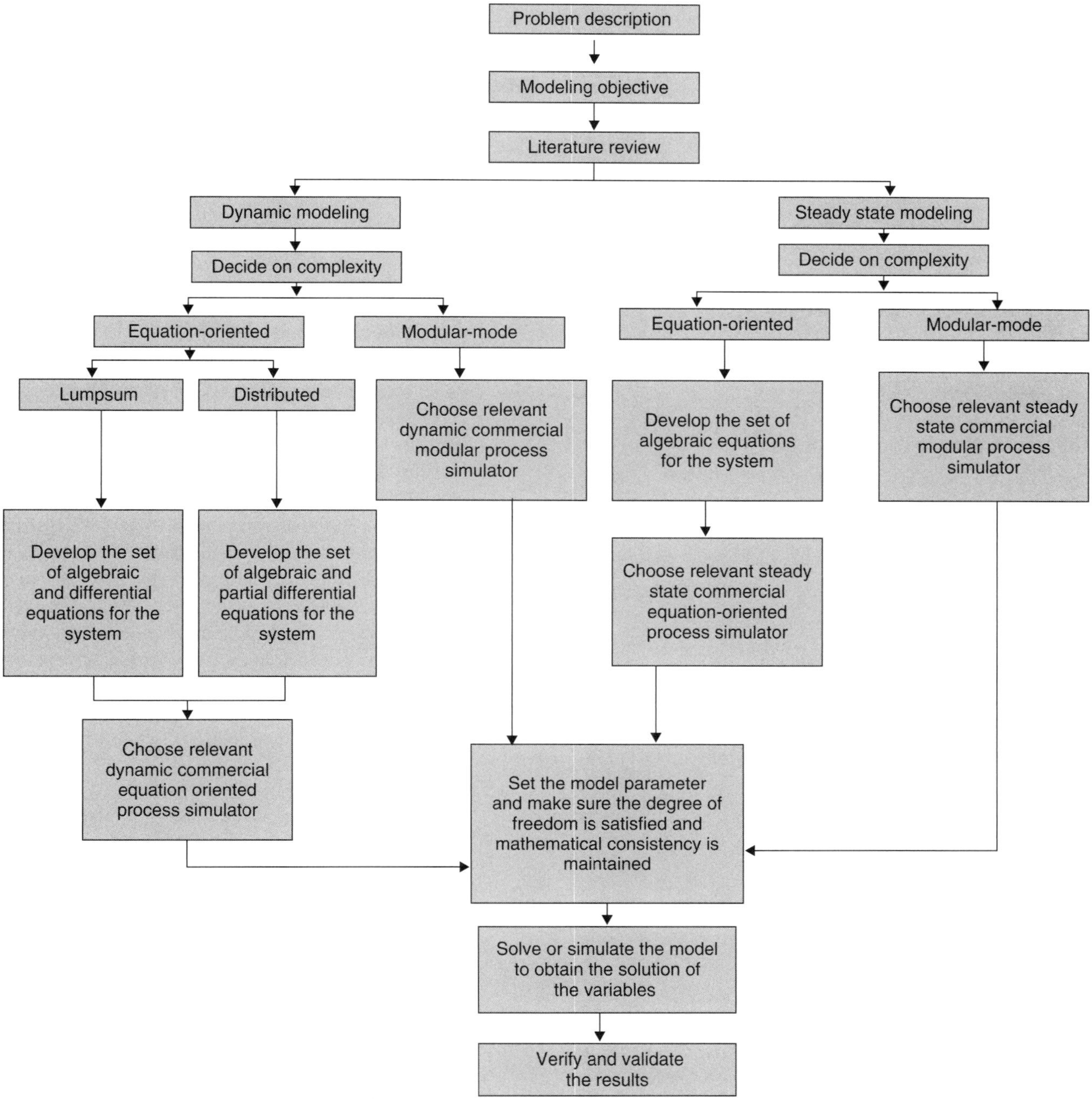

Figure 22-1 Basic procedures in developing process models.

The enthalpy h of each stream can be expressed as follows:

$$h_i = h(T, P, x_{iB}, x_{iT}, x_{iD}) \qquad (i = F, D, W) \tag{22.5}$$

where T and P are the temperature and pressure of the system, which is assumed to be constant, and

$$x_{iB} + x_{iT} + x_{iD} = 1 \qquad (i = F, D, W) \tag{22.6}$$

(e) *Analysis of the Degrees of Freedom*. From the model equations formulated above, the number of independent equations $N_E = 10$. The number of variables $N_v = 17(F, D, W, x_{iB}, x_{iT}, x_{iD},$ and $h_i(i = F, D, W)$, P, and T). The number of degrees of freedom $N_D = N_v - N_E = 7$. Therefore, to solve the steady-state model of the splitter, seven variables must be specified. This has been implemented in one of the commercially available modular-mode process simulators, Aspen Hysys version 7.0, and

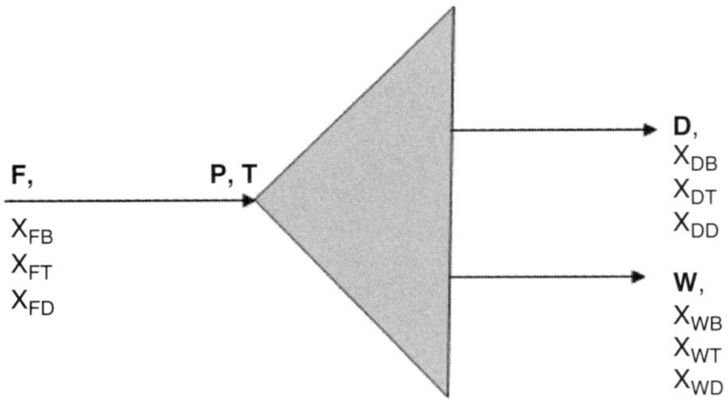

Figure 22-2 Simple splitter arrangement.

TABLE 22-3 Variables Specified

Variable	Value Assigned
F	100 kgmol/h
T	25°C
P	101.3 kPa
X_{FB}	0.5
X_{FT}	0.3
X_{FD}	0.2
D	40 kgmol/h

the results are shown in Table 22-3. The seven variables specified include the composition of the feed (x_{FB}, x_{FT}, x_{FD}), the temperature, pressure, and flow rate of the feed stream (T, P, F), and the flow rate of one of the product streams, D.

The variables specified are shown in lighter grey (columns F and D) in Figs. 22-3 and 22-4. The results of the simulation obtained after satisfying the degrees of freedom are shown in Fig. 22-5.

Case Study 22.2: Dynamic Modeling of Condenser, Reflux Drum, and Splitter Used at the Top of a Distillation Column Using a Modular-Mode Process Simulator

(a) *Process description*. The arrangement of the column overhead is as shown in Fig. 22-6. It contains the condenser, reflux drum, and the splitter used in the top of the distillation column. The vapor coming out of the top tray of the column passes through the condenser, which is usually a shell-and-tube heat exchanger. The heat contained in the vapor is extracted by the cold condenser liquid (usually, water). The extraction of the heat from the vapor makes the vapor condense into liquid, which subsequently flows into the reflux drum. As the liquid accumulates in the reflux drum, it leaves from the drum to the splitter, where the splitter splits the liquid flow into two parts. One part is returned as reflux to the top tray of the column, and the other part is removed as distillate (top product). From the literature, there are different methods of modeling a condenser, reflux drum, and splitter arrangement for the column overhead, and the method used for any purpose depends on the aim of the model and the availability of resources to solve the model equations.

To model the column overhead arrangement adequately, as shown in Fig. 22-6, the following procedure was adopted.

(b) *Condenser modeling*. Condensers are modeled as either total or partial condensers. In total condensers, all the vapor is condensed into liquid and there is no vapor product, while in partial condensers, the cooling duty is such that only a fraction of the vapor is condensed, while the rest leaves as vapor. According to Luyben [7], most commercially available distillation columns employ total condensers, so the modeling of the condenser in this chapter will be based on total condensation.

Different models have been developed for column overhead by different authors. For example, Ramirez [8] modeled the column overheads as a single unit. He claimed that the accumulation of liquid in the condenser is negligible. Luyben [7] modeled both the condenser and the reflux drum as one unit. Pantelides [3] modeled the condenser and the reflux drum separately. He also admitted that the accumulation of liquid in the condenser is negligible compared to that in the reflux drum, and hence no accumulation term was considered in the condenser model. Again he assumed that the liquid leaving the condenser is at its bubble point, the composition of vapor entering the condenser is the same as that of the liquid leaving the condenser, and that the pressure in the condenser and that of the reflux drum are approximately equal.

However, the assumption that the composition of the inlet vapor to the condenser is the same as that of the liquid from the condenser means that no subcooling takes place in the condenser, which is ideal. In real processes it is almost inevitable to avoid subcooling [3], and as a result of this assumption, the product purity level obtained from this model will definitely be higher than that obtainable in real processes.

(c) *Definition of process variables for condenser modeling*. The subscript "con" represents the condenser variables, the subscript "1" represents variables of the vapor stream from tray 1 (based on the top-down tray numbering method), the subscript "n" represents

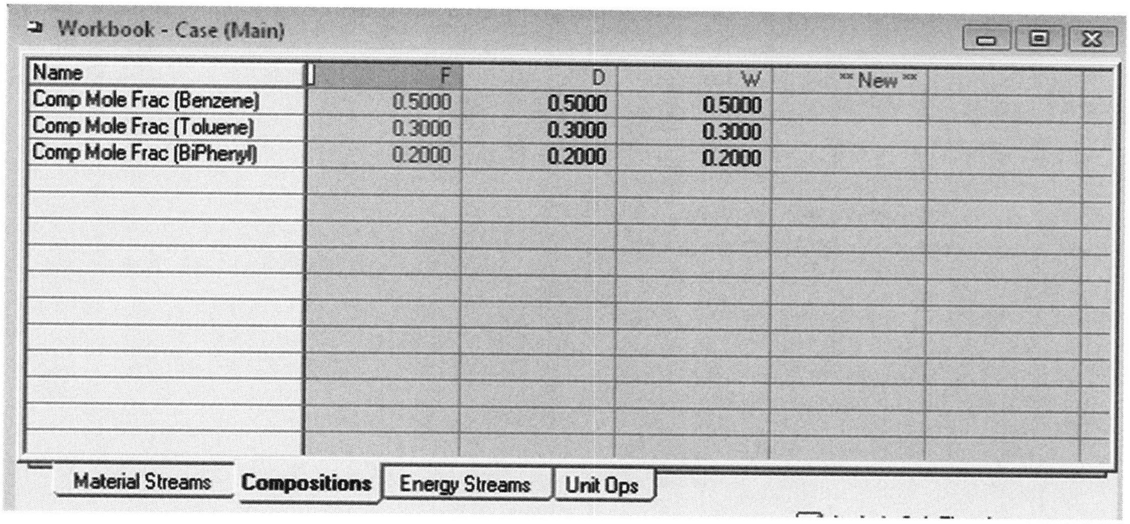

Workbook - Case (Main)

Name	F	D	W	** New **
Comp Mole Frac (Benzene)	0.5000	0.5000	0.5000	
Comp Mole Frac (Toluene)	0.3000	0.3000	0.3000	
Comp Mole Frac (BiPhenyl)	0.2000	0.2000	0.2000	

Material Streams | Compositions | Energy Streams | Unit Ops

Figure 22-3 Workbook showing the specified feed composition for the splitter model in Aspen Hysys.

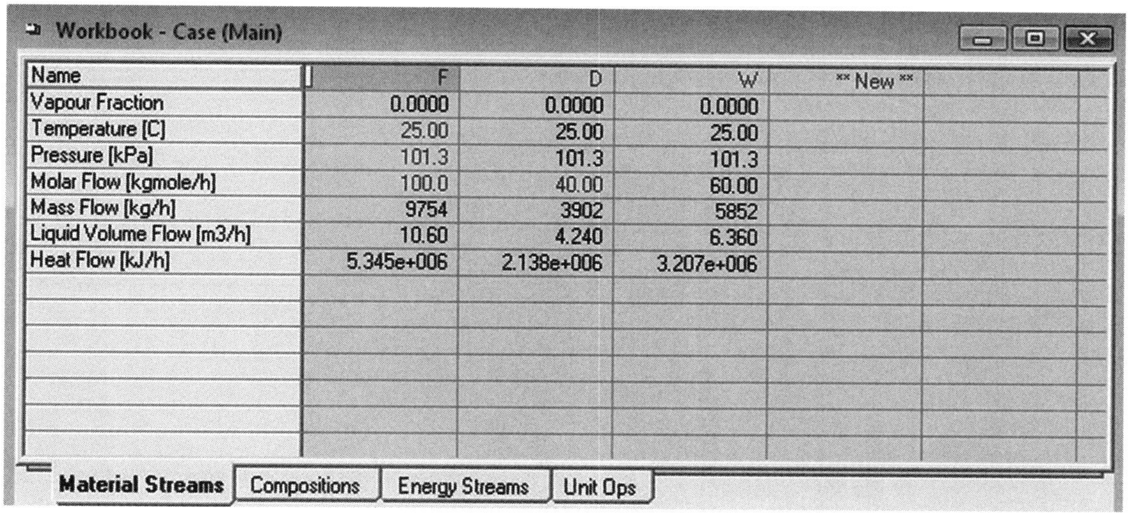

Workbook - Case (Main)

Name	F	D	W	** New **
Vapour Fraction	0.0000	0.0000	0.0000	
Temperature [C]	25.00	25.00	25.00	
Pressure [kPa]	101.3	101.3	101.3	
Molar Flow [kgmole/h]	100.0	40.00	60.00	
Mass Flow [kg/h]	9754	3902	5852	
Liquid Volume Flow [m3/h]	10.60	4.240	6.360	
Heat Flow [kJ/h]	5.345e+006	2.138e+006	3.207e+006	

Material Streams | Compositions | Energy Streams | Unit Ops

Figure 22-4 Workbook showing the specified feed stream temperature, T, pressure, P, flow rate, F, and the product stream flow rate, D, for the splitter model in Aspen Hysys.

the number of components, and the subscript D represents the reflux drum and splitter variables.

The modeling approach adopted in this work is that of Pantelides [3], and it is based on the following assumptions.

(d) *Principles of the model*. The model is based on the law of conservation of mass and energy.

(e) *Condenser modeling assumptions*. The condenser model is based on the following assumptions:

- The condenser is modeled as a total condenser.
- No accumulation occurs in the condenser.
- Liquid leaving the condenser to the reflux drum is at its bubble point.
- Composition of the vapor entering the condenser is the same as that of the liquid leaving the condenser.
- The pressure in the condenser and that of the reflux drum are approximately equal.
- The composition of the distillate and the reflux are the same.

Based on the assumptions listed above, the condenser model proposed by Pantelides [3] is as given below. The material balance around the condenser gives

$$V_1 y_{11} = L_{con} x_{i,con} \qquad (i = 1, \ldots, n) \qquad (22.7)$$

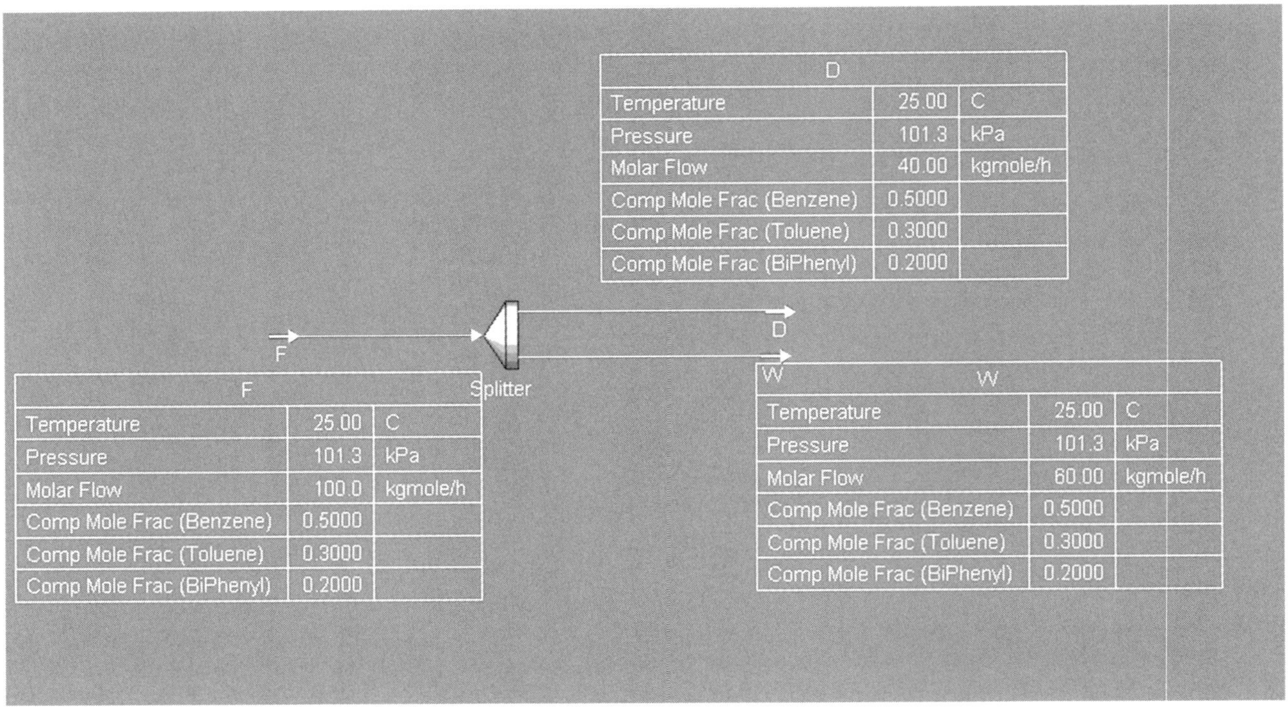

Figure 22-5 Aspen Hysys simulation result for the splitter system.

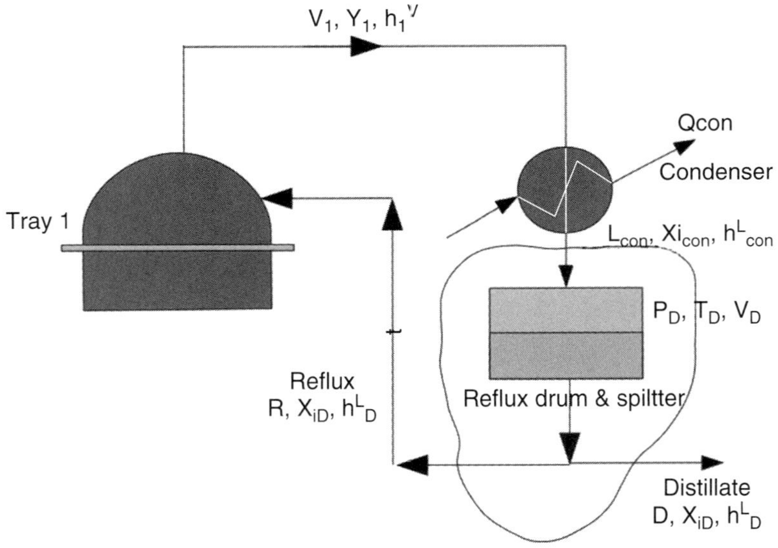

Figure 22-6 Column overhead.

Energy balance around the condenser gives

$$V_1 h_1^v = L_{con} h_{con}^L + Q_{con} \tag{22.8}$$

Enthalpy is a function of temperature, pressure, and composition:

$$h_1^v = h^v(T_1, P_1, y_1) \tag{22.9}$$

$$h_{con}^L = h(T_{con}, P_{con}, x_{con}) \tag{22.10}$$

Since the liquid leaving the condenser is assumed to be at its bubble point, the liquid and vapor are in equilibrium, and hence for any component,

$$\overline{y_{i,con}} = K_i xi, \text{con} \qquad (i = 1, \ldots, n) \tag{22.11}$$

K_i are complex functions of temperature, pressure, and phase composition:

$$K_i = k(T_{\text{con}}, P_{\text{con}}, x_{\text{con}}, y_{\text{con}}) \qquad (i = 1, \ldots, n) \tag{22.12}$$

$$\sum_{i=1}^{n} y_{i,\text{con}} = 1 \tag{22.13}$$

$$\sum_{i=1}^{n} x_{i,\text{con}} = 1 \tag{22.14}$$

(f) *Reflux drum and splitter modeling*. In the reflux drum, there is an accumulation of liquid in the drum.

(g) *Modeling principles*. The model is based on the law of conservation of mass and energy.

(h) *Modeling assumptions*

- The temperature of the liquid in the reflux drum is uniform.
- The composition of the reflux stream and of the distillate leaving the splitter is the same.
- The fluid leaving the splitter is at the same pressure as that of the reflux drum.

Based on the assumptions listed above, the following model equations are developed. The material balance around the reflux drum and splitter envelope in Fig. 22-7 gives

$$\frac{dM_{i,D}}{dt} = L_{\text{con}} x_{i,\text{con}} - (R + D) x_{iD} \qquad (i = 1, \ldots, n) \tag{22.15}$$

The energy balance around the reflux drum and splitter envelope gives

$$\frac{dU_D}{dt} = L_{\text{con}} h_{\text{con}}^L - (R + D) h_D^L \tag{22.16}$$

$$M_{i,D} = M_D^L x_{i,D} + M_D^V y_{i,D} \qquad (i = 1, \ldots, n) \tag{22.17}$$

$$U_D = M_D^L h_D^L + M_D^V h_D^V - P_D V_D \tag{22.18}$$

$$V_D = \frac{M_D^L}{\rho_D^L} + \frac{M_D^V}{\rho_D^V} \tag{22.19}$$

$$y_{i,D} = K_{i,D} x_{i,D} \qquad (i = 1, \ldots, n) \tag{22.20}$$

$$\sum_{i=1}^{n} y_{i,D} = 1 \tag{22.21}$$

$$\sum_{i=1}^{n} x_{i,D} = 1 \tag{22.22}$$

$$h_D^L = h^L(T_D, P_D, x_D) \tag{22.23}$$

$$h_D^V = h^v(T_D, P_D, y_D) \tag{22.24}$$

$$\rho_D^L = \rho^L(T_D, P_D, x_D) \tag{22.25}$$

$$\rho_D^V = \rho^V(T_D, P_D, y_D) \tag{22.26}$$

$$K_{i,D} = K(T_D, P_D, x_D, y_D) \qquad (i = 1, \ldots, n) \tag{22.27}$$

$$P_D = P_{\text{con}} \tag{22.28}$$

$$T_D = T_{\text{con}} \tag{22.29}$$

(i) *Analysis of the degrees of freedom*. From the equations listed above, we have the following variables: $V_1, L_{\text{con}}, , h_1^V, h_{\text{con}}^L, Q_{\text{con}}, T_1, P_1, T_{\text{con}}, P_{\text{con}}, R, D, U_D, h_D^L, M_D^L, M_D^V, h_V^D, T_D, P_D, V_D, \rho_D^L, \rho_D^V, y_{i1}, , y_{i\text{con}}, x_{i\text{con}}, K_i, M_{iD}, x_{iD}, y_{iD}, K_{iD} (i = 1, \ldots, n)$. The number of variables $N_V = 8n + 21$. The number of equations $N_E = 7_n + 16$. The degrees of freedom $N_D = N_V - N_E = n + 5$. Therefore, for a unique solution of the system to be obtained, $n + 5$ variables have to be specified. This has been implemented in one of the commercially available modular-mode process simulators, Aspen Hysys version 7.0, as shown in Table 22-4.

From the table, the number of components $n = 12$. The degrees of freedom $N_D = 12 + 5 = 17$. Hence, 17 variables have to be specified to obtain a unique solution of the process. The 17 variables specified are as shown in Table 22-5 and are

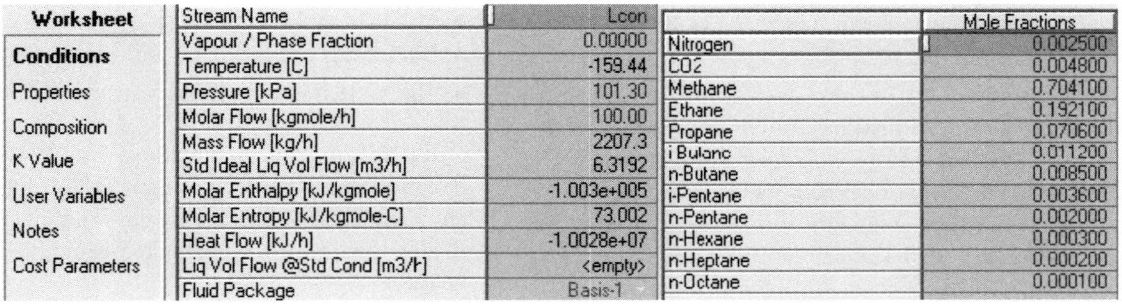

Worksheet	Stream Name	Lcon		Mole Fractions
Conditions	Vapour / Phase Fraction	0.00000	Nitrogen	0.002500
Properties	Temperature [C]	-159.44	CO2	0.004800
Composition	Pressure [kPa]	101.30	Methane	0.704100
K Value	Molar Flow [kgmole/h]	100.00	Ethane	0.192100
User Variables	Mass Flow [kg/h]	2207.3	Propane	0.070600
Notes	Std Ideal Liq Vol Flow [m3/h]	6.3192	i-Butane	0.011200
Cost Parameters	Molar Enthalpy [kJ/kgmole]	-1.003e+005	n-Butane	0.008500
	Molar Entropy [kJ/kgmole-C]	73.002	i-Pentane	0.003600
	Heat Flow [kJ/h]	-1.0028e+07	n-Pentane	0.002000
	Liq Vol Flow @Std Cond [m3/h]	<empty>	n-Hexane	0.000300
	Fluid Package	Basis-1	n-Heptane	0.000200
			n-Octane	0.000100

Figure 22-7 Worksheet showing the specified composition, vapor-phase fraction, pressure, and molar flow for stream L_{con} and some of the solved variables.

TABLE 22-4 Composition of Stream L_{con}

Component	Mole Fraction
Nitrogen	0.0025
CO_2	0.0048
Methane	0.7041
Ethane	0.1921
Propane	0.0706
i-Butane	0.0112
n-Butane	0.0085
i-Pentane	0.0036
n-Pentane	0.0020
n-Hexane	0.0003
n-Heptane	0.0002
n-Octane	0.0001

TABLE 22-5 Variables Specified

Component	Mole Fraction
Stream L_{con}	
Nitrogen	0.0025
CO_2	0.0048
Methane	0.7041
Ethane	0.1921
Propane	0.0706
i-Butane	0.0112
n-Butane	0.0085
i-Pentane	0.0036
n-Pentane	0.0020
n-Hexane	0.0003
n-Heptane	0.0002
n-octane	0.0001
Vapor-phase fraction	0
Pressure	101.3 kPa
Molar flow rate	100 kgmol/h
Stream V_1	
Vapor-phase fraction	1
Stream *D*	
D	60 kgmol/h

shown in light grey in Figures 22-7 and 22-8. The results of the simulation obtained after satisfying the degrees of freedom is shown in Fig. 22-9. The implementation of the model in Aspen Hysys version 7.0 is shown in Fig. 22-10.

(j) *Concluding remarks*. From the work carried out so far, it can be concluded that the degrees of freedom have to be satisfied before a unique solution of any given model can be obtained. When using the modular-mode process simulators, it is not necessary to write down the model equations (as we did in this case study) of the system since these equations are already embedded as a black box in the process simulator software. It is only carried out here to establish the fact that for any model to have a unique solution, the degrees of freedom must be satisfied. Hence, when modeling with a modular-mode process simulator, it is required only that you specify the parameters of the model to satisfy the degrees of freedom of the system as well as avoiding introducing redundancy in the system.

Case Study 22.3: Dynamic Modeling of a Cooling Tank System Using an Equation-Oriented Process Simulator

(a) *Process description*. In this cooling tank system (Fig. 22-11), a hot liquid flows into the tank at flow rate F and temperature T_0. The liquid in the tank is cooled by cooling water which flows into the tank via a coil with a flow rate of F_0 and an inlet temperature of $T_{c,\mathrm{in}}$. As the cooling water passes through the coil located inside the tank, heat exchange takes place between the hot liquid in the tank and the cooling water inside the coil. This results in the transfer of heat from the hot fluid contained in the tank to the cooling water in the coil. Hence, the temperature of the water in the tank decreases while that of the cooling water increases to $T_{c,\mathrm{out}}$. The liquid temperature in the tank is T.

(b) *Cooling tank modeling*. From the process description it can be observed that variations in any of the process variables, such as feed temperature and flow rate, cooling water inlet temperature and flow rate will affect the temperature of the fluid in the tank. The tank dynamics can be obtained through dynamic modeling.

(c) *Modeling principles*. The tank is modeled based on the law of conservation of mass and energy.

(d) *Modeling assumptions*. The model of the tank is based on the following assumptions:

- The tank is well insulated, so that there is negligible heat transfer from the tank to the surroundings.
- There is no accumulation of heat in the tank wall and the wall of the cooling water coil.
- The tank is well mixed.
- Both fluids are the same and incompressible.
- The system is initially at a steady state.

Based on the assumptions listed above, the dynamic modeling of the tank is developed using an equation-oriented approach. The mathematical model of the tank cooling tank system is as explained below.

The rate at which the tank temperature varies depends on the rate at which heat is transferred from

Worksheet	Stream Name	D
Conditions	Vapour / Phase Fraction	0.00000
	Temperature [C]	-159.44
Properties	Pressure [kPa]	101.30
Composition	Molar Flow [kgmole/h]	60.000
	Mass Flow [kg/h]	1324.4
K Value	Std Ideal Liq Vol Flow [m3/h]	3.7915
User Variables	Molar Enthalpy [kJ/kgmole]	-1.003e+005
	Molar Entropy [kJ/kgmole-C]	73.002
Notes	Heat Flow [kJ/h]	-6.0166e+06
Cost Parameters	Liq Vol Flow @Std Cond [m3/h]	<empty>
	Fluid Package	Basis-1

Figure 22-8 Worksheet showing the specified molar flow rate for stream D and some of the solved variables.

Workbook - Case (Main)

Name	V1	Lcon	V2	Lreflux	D
Vapour Fraction	1.0000	0.0000	1.0000	0.0000	0.0000
Temperature [C]	-38.22	-159.4	-159.4	-159.4	-159.4
Pressure [kPa]	101.3	101.3	101.3	101.3	101.3
Molar Flow [kgmole/h]	100.0	100.0	2.765e-004	100.0	60.00
Mass Flow [kg/h]	2207	2207	4.794e-003	2207	1324
Std Ideal Liq Vol Flow [m3/h]	6.319	6.319	1.426e-005	6.319	3.792
Heat Flow [kJ/h]	-8.445e+006	-1.003e+007	-20.21	-1.003e+007	-6.017e+006
Molar Enthalpy [kJ/kgmole]	-8.445e+004	-1.003e+005	-7.309e+004	-1.003e+005	-1.003e+005
Name	R	CONDENSER I	** New **		
Vapour Fraction	0.0000	<empty>			
Temperature [C]	-159.4	<empty>			
Pressure [kPa]	101.3	<empty>			
Molar Flow [kgmole/h]	40.00	<empty>			
Mass Flow [kg/h]	882.9	<empty>			
Std Ideal Liq Vol Flow [m3/h]	2.528	<empty>			
Heat Flow [kJ/h]	-4.011e+006	1.582e+006			
Molar Enthalpy [kJ/kgmole]	-1.003e+005	<empty>			

Streams | Unit Ops

Figure 22-9 Workbook showing some of the specified variables and the simulation result.

the hot fluid to the tank and the rate at which heat leaves the tank to the cooling water. Also, the rate at which the cooling water temperature varies depends on the rate at which it absorbs heat from the hot water contained in the tank. This can be represented mathematically as shown below.

The rate of change of the tank temperature is given by the equation

$$\frac{dT}{dt} = \frac{F}{V}(T_0 - T) - \frac{Q}{V\rho C_p} \tag{22.30}$$

Similarly, the rate of change of the coil fluid temperature is given by

$$\frac{dT_{c,\text{out}}}{dt} = \frac{F_c}{V_c}(T_{c,\text{in}} - T_{c,\text{out}}) + \frac{Q}{V_c\rho C_p} \tag{22.31}$$

The heat transfer from the hot water in the tank to the cold fluid in the coil is given by

$$Q = UA(T - T_{c,\text{out}}) \tag{22.32}$$

The equations listed above represent the temperature dynamics of the entire system.

(e) *Analysis of the degrees of freedom*. The model equation for the system contains the following variables: T, F, V, T_0, Q, ρ, C_p, $T_{c,\text{out}}$, T_{cin}, F_c, V_c, and UA. The number of variables $N_V = 12$. The number of equations $N_E = 3$. The degrees of freedom $N_D = N_V - N_E = 9$. Hence, nine variables must be specified in order to obtain a unique solution to the cooling tank system. Of the nine variables to be specified, two will be the initial conditions for

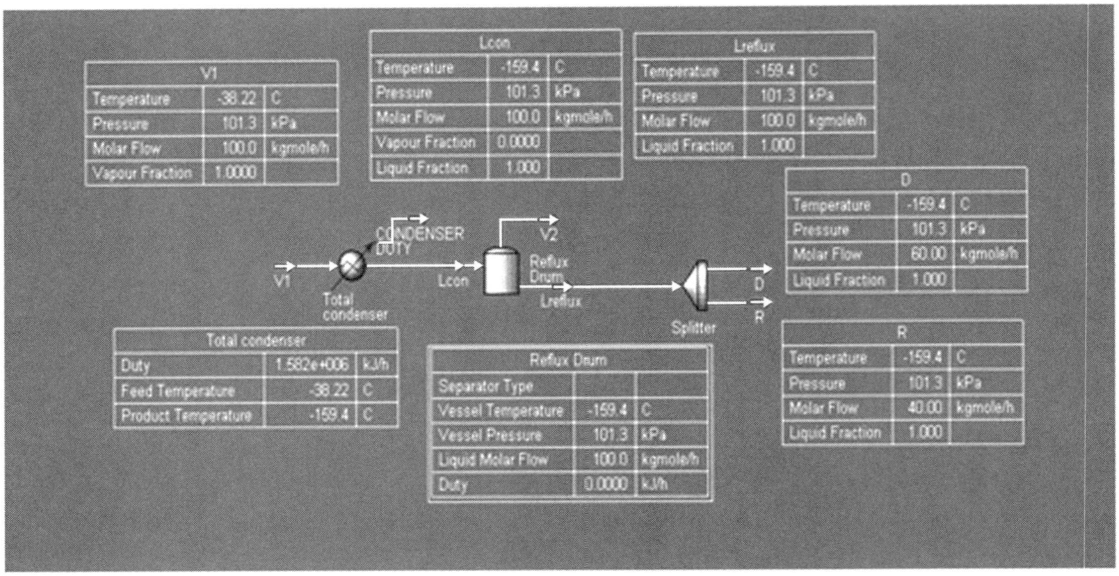

Figure 22-10 Aspen Hysys model of the condenser, reflux drum, and splitter arrangement for the column overhead.

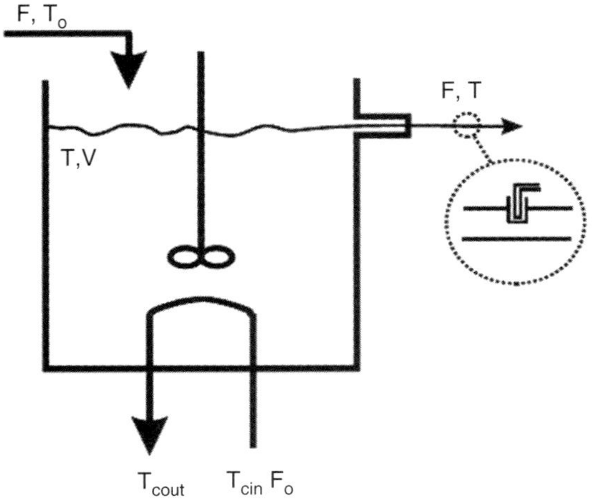

Figure 22-11 Cooling tank system.

the two differential equations. Assume the following specified variables:

Tank:

Volume, $V = 2.1\text{m}^3$

Initial temperature (steady state) : $T_s = 85.4^\circ\text{C}$

Tank inlet flow:

$$F = 0.085\frac{\text{m}^3}{\text{min}}$$

Temperature : $T_0 = 150^\circ\text{C}$

$$\text{Density} : \rho = \frac{10^6\text{g}}{\text{m}^3}$$

Specific heat capacity, C_p : $1\text{cal/g} \cdot \text{C}$

Cooling coil:

Inlet temperature, $T_{c,\text{in}} = 25^\circ\text{C}$

Initial flow rate (steady-state)$F_{cs} = 0.5\text{m}^3/\text{min}$

Coil volume : $V_c = 0.2\text{m}^3$

At steady state, the derivative terms in model equations (22.30) and (22.31) become equal to zero.

$$\frac{dT}{dt} = \frac{dT_{c,\text{out}}}{dt} = 0 \tag{22.33}$$

Therefore, Eq. (22.30) reduces to

$$\frac{F}{V}(T_0 - T) = \frac{Q}{V\rho C_p} \tag{22.34}$$

Solving for Q in Eq. (22.34) gives

$$\begin{aligned} Q &= F\rho C_p(T_0 - T) \\ &= 0.085\frac{\text{m}^3}{\text{min}} \cdot \frac{10^6\text{g}}{\text{m}^3} \cdot 1\frac{\text{cal}}{\text{g} \cdot \text{C}}(150 - 85.4)^\circ\text{C} \\ &= 5{,}491{,}000\ \text{cal/min} \end{aligned} \tag{22.35}$$

Similarly, Eq. (22.31) reduces to

$$\frac{F_c}{V_c}(T_{c,\text{in}} - T_{c,\text{out}}) = -\frac{Q}{V_c \rho_c C_p} \tag{22.36}$$

Substituting 5,491,000 cal/min for Q in Eq. (22.36) and solving for $T_{c,\text{out}}$ yields

$$\begin{aligned} T_{c,\text{out}} &= T_{c,\text{in}} + \frac{Q}{F_c \rho_c C_P} \\ &= 25^\circ\text{C} + \frac{5{,}491{,}000 \text{ cal/min}}{0.5\text{m}^3/\text{min} \cdot 10^6 \text{ g/m}^3} \cdot 1\text{cal/(g}^\circ\text{C)} \\ &= 35.982^\circ\text{C} \end{aligned} \tag{22.37}$$

Substituting 5,491,000 for cal/min and Q for 35.982°C for $T_{c,\text{out}}$, in Eq. (22.32) and solving the heat-transfer coefficient UA gives us

$$\begin{aligned} UA &= \frac{Q}{T - T_{c,\text{out}}} = \frac{5{,}491{,}000\text{cal/min}}{(85.4 - 35.982)^\circ\text{C}} \\ &= 111{,}113.3595 \text{ cal/min} \cdot \text{C} \end{aligned}$$

The model equations have been solved to obtain some of the process parameters. The model can now be developed in an equation-oriented process simulator in order to observe the dynamics in the tank temperature. The equation-oriented process simulator adopted for this case study is Simulink. The steps involved in the development of the Simulink model of the cooling tank system using the model equations listed above are explained below.

The Simulink model for a cooling tank system is built using the model equations (22.30) to (22.2) developed earlier for the cooling tank system.

The step-by-step procedure is as follows.

Step 1: The tank temperature model is built using tank temperature model equation (22.30):

$$\frac{F}{V}(T_0 - T) - \frac{Q}{V\rho C_p} = \frac{dT}{dt}$$

From the equation, the Simulink blocks are linked as described below. The signal from the input parameters, F and T_0 are linked to a gain of magnitude $1/V$ and a summing point, respectively. At the summing point, the difference $T_0 - T$ between the input temperature signal T_0 and the output temperature signal T is obtained. This difference and output signals from the gain are both sent to the product block, where the product signal, $(F/V)(T_0 - T)$ is obtained. The product signal is sent to the second summing point block, where the difference between the product signal $(F/V)(T_0 - T)$ and the rate of change of temperature signal $Q/V\rho C_p$ obtained by passing heat-transfer rate signal Q [from Eq. (22.32)] through a gain of $1/V\rho C_p$. The signal from the second summing point block is then passed to an integrator to integrate the differential temperature signal to obtain the output temperature signal T.

Table 22-6 shows at a glance the blocks from the Simulink library used in building the Simulink

TABLE 22-6 Building Blocks for the Tank Temperature Model

Block Description	Reason It Is Needed	Magnitude	Quantity Needed	Symbol
Constant	For inputs to the system (F, T_0)	$F = 0.085$ $T_0 = 150$	2	1
Gain	For evaluation of the inverse functions $(1/V)$, $1/(V\rho C_p$	$\frac{1}{V} = \frac{1}{2.1}$ $\frac{1}{V\rho C_p} = \frac{1}{2.1 \cdot 10^6 \cdot 1}$	2	1▷
Summing point	For the evaluation of addition and subtraction of signals $(T_0 - T)$, $\frac{F}{V}(T_0 - T) - \frac{Q}{V\rho C_p}$	$T_0 - T$ $\frac{F}{V}(T_0 - T) - \frac{Q}{V\rho C_p}$	2	+−
Product	For the multiplication of parameters $\frac{F}{V}(T_0 - T)$	$\frac{F}{V}(T_0 - T)$	1	×
Integrator	For integration of the derivative dT/dt	Initial condition, 85.4°C	1	$\frac{1}{s}$

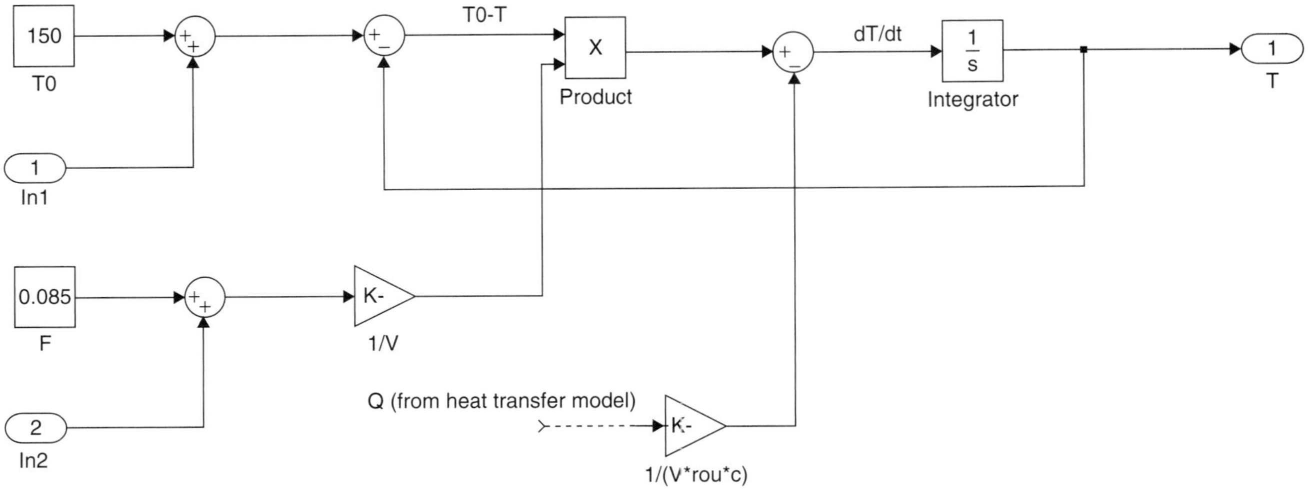

Figure 22-12 Simulink model for the tank temperature.

model for the tank temperature differential equation [Eq. (22.30)] under consideration. The Simulink model for the tank temperature differential equation is shown in Fig. 22-12.

Step 2: The coil temperature model is built using the coil temperature model equation [Eq. (22.3)]:

$$\frac{F_c}{V_c}(T_{c,\text{in}} - T_{c,\text{out}}) + \frac{Q}{V_c \rho_c C_p} = \frac{dT_{c,\text{out}}}{dt}$$

From the equation, the Simulink model is developed as explained below. The signal from the input parameters F_c are $T_{c,\text{in}}$ are linked to a gain of magnitude $1/V_c$ and a summing point, respectively. At the summing point, the difference $(T_{c,\text{in}} - T_{c,\text{out}})$ between the coil input temperature signal $T_{c,\text{in}}$ and the output temperature signal $T_{c,\text{out}}$ is obtained. This difference signal and the output signal from the gain are both sent to the product block, where the product signal, $(F_c/V_c)(T_{c,\text{in}} - T_{c,\text{out}})$ is obtained. The product signal is sent to the second summing point block, where the sum of the product signal $(F_c/V_c)(T_{c,\text{in}} - T_{c,\text{out}})$ and the rate of change of coil temperature signal $\frac{Q}{V_c \rho_c C_p}$; obtained by passing heat transfer rate signal Q [from Eq (22.32)] through a gain of $1/V_c\rho, C_p$. The signal from the second summing point block is then passed to an integrator to integrate the differential temperature signal to obtain the output temperature signal, $T_{c,\text{out}}$.

Table 22-7 shows at a glance the blocks from the Simulink library used in building the Simulink model for the coil temperature differential equation. The Simulink model for the coil temperature differential equation is shown in Fig. 22-13.

Step 3: This step shows the modeling of the heat transfer from the tank hot liquid to the coil cold liquid across the coil surface. It serves as a link between the tank temperature equation model and the coil temperature equation model. The Simulink model is obtained based on the heat-transfer model equation [Eq. (22.32)]:

$$Q = UA(T - T_{c,\text{out}})$$

From the model equation, the Simulink model is built. The difference in temperature between the two output temperature signals from step 1, T, and step 2, $T_{c,\text{out}}$, is obtained by passing the two signals through a summing point block. The output signal from the summing point block is then passed through a gain of magnitude UA to obtain the heat-transfer rate signal. The signal from the heat-transfer rate is then sent to the Simulink model for the tank and coil temperature to complete the model. Table 22-8 shows the building blocks for the model, and Fig. 22-14 shows the Simulink model for the heat-transfer rate.

The overall Simulink model for the entire cooling tank system is obtained by combining the individual Simulink models obtained in steps 1, 2, and 3. The Simulink model for the cooling tank system is shown in Fig. 22-15.

TABLE 22-7 Building Blocks for the Coil Temperature Model

Block Description	Reason It Is Needed	Magnitude	Quantity Needed	Symbol
Constant	For inputs to the system (F_c, $T_{c,\text{in}}$)	$F = 0.5$ $T_{c,\text{in}} = 25$	2	[1]
Gain	For the evaluation of the inverse functions $1/V_c$, $1/V_c\rho_c C_p$	$\frac{1}{V_c} = \frac{1}{0.2}$	2	1▷
		$\frac{1}{V_c\rho_c C_p} = \frac{1}{0.2 \cdot 10}$		(+ −)
Summing point	For the evaluation of addition and subtraction of signals $(T_{c,\text{in}} - T_{c,\text{out}})$, $\frac{F_c}{V_c}(T_{c,\text{in}} - T_{c,\text{out}})$	$T_{c,\text{in}} - T_{c,\text{out}}$ $\frac{F_c}{V_c}(T_{c,\text{in}} - T_{c,\text{out}}) + \frac{Q}{V_c\rho_c C_p}$	2	(+ +)
Product	For the multiplication of parameters $\frac{F_c}{V_c}(T_{c,\text{in}} - T_{c,\text{out}})$	$\frac{F_c}{V_c}(T_{c,\text{in}} - T_{c,\text{out}})$	1	[×]
Integrator	For integration of the derivative $\frac{dT_{c,\text{out}}}{dt}$	Initial condition, (obtained by simulation)	1	[1/s]

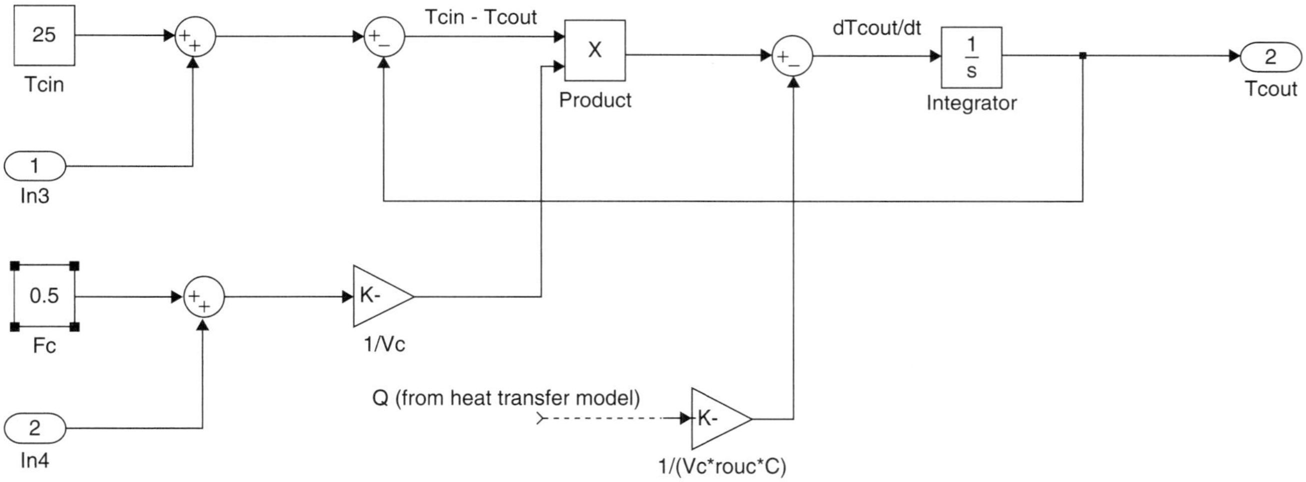

Figure 22-13 Simulink model for the coil temperature.

TABLE 22-8 Building Blocks for the Heat-Transfer Model

Block Description	Reason It Is Needed	Magnitude	Quantity Needed	Symbol
Gain	For the evaluation of the inverse functions, $1/UA$	$\frac{1}{UA} = \frac{1}{1.11113 \cdot 10^5}$	1	1▷
Summing point	For the evaluation of addition and subtraction of signals, $T - T_{c,\text{out}}$	$T - T_{c,\text{out}}$		(+ −)

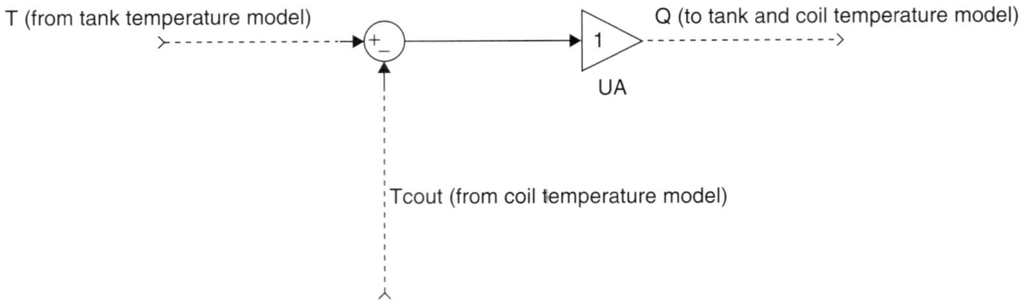

Figure 22-14 Simulink model for the heat transfer.

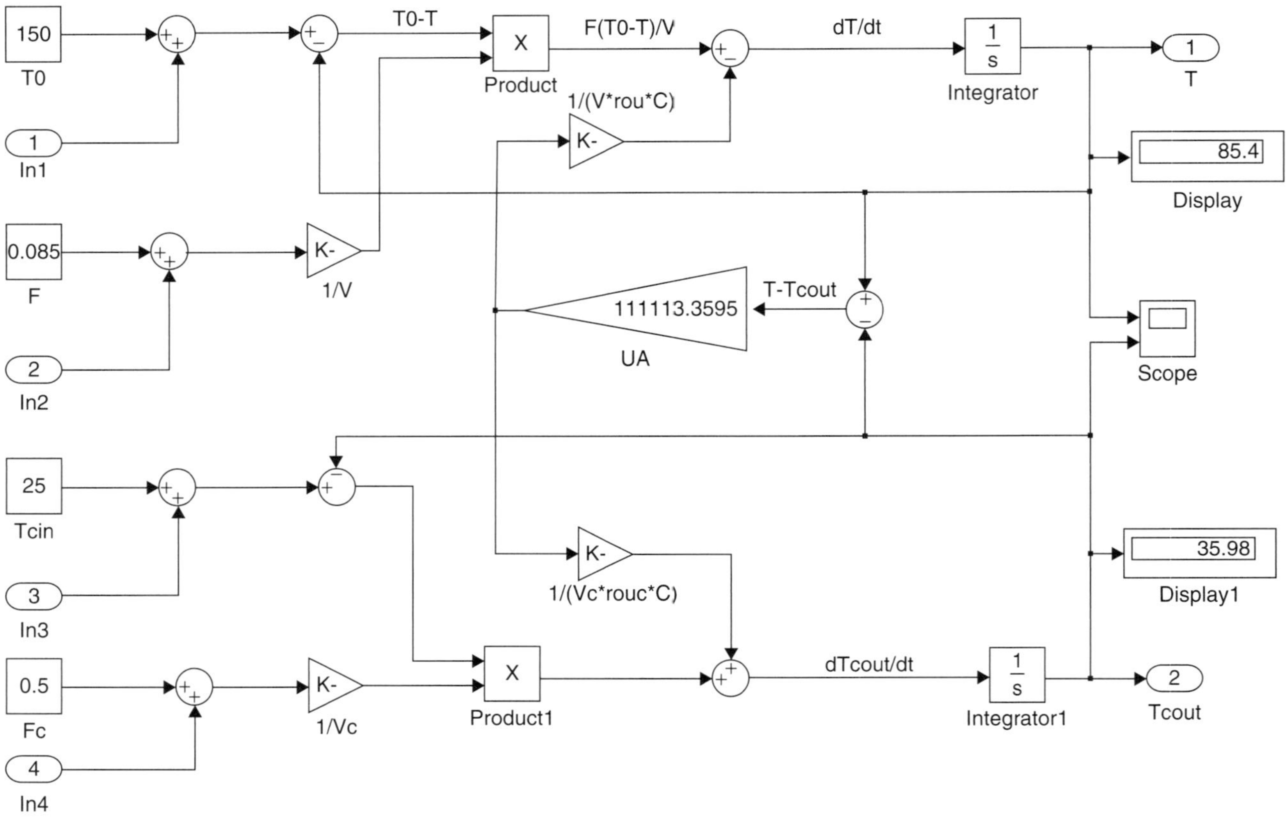

Figure 22-15 Simulink model for the cooling tank system.

22.6 CONCLUDING REMARKS

Equation-oriented modeling involves a great deal of development of mathematical equations. Unlike the modular-mode approach, the equation-oriented approach requires the design engineer to have a prior knowledge of the model equations that define the process. The set of model equations had to be solved using an equation-oriented process simulator. This can be carried out using a signal flow diagram (as shown in the case study) or by the use of computer codes written in a particular programming language.

REFERENCES

1. Horvath, L. and I.J. Rudas, *Modeling and Problem Solving Techniques for Engineers*. 2004: Elservier Academic Press, USA.
2. Wang, M., *Process Simulation and Design*. 2008, Cranfield University, United Kingdom.
3. Pantelides, C.C., *The Mathematical Modelling of the Dynamic Behaviour of Process Systems*. 2003, Imperial College, London.
4. Biegler, L.T., I.E. Grossman, and A.W. Westerberg, *Systematic methods for Chemical Process Design*. 1997: Prentice Hall, New Jersey.

5. Alkaya, N., S. Vasantharajan, and L.T. Biegler, *Successive Quadratic Programming*: Application in the Process Industry, in Encyclopedia of Optimization, C.A. Floudas and P.M. Pardalos, Editors. 2001, Kluwer Academic Publishers.
6. Smith, R., *Chemical Process Design and Integration*. 2005: John Wiley.
7. Luyben, W.L., Process Modelling, *Simulation and Control for Engineers*. 2nd ed. 1990: McGraw Hill, New York.
8. Ramirez, W.F., *Computational Methods for Process Simulation*. 2nd ed. 1997: Butterworth-Heinemann, Oxford, USA.
9. ISA Committee SP51, *Control Valve Terminology*, Technical Report ANSI/ISA-75.3-1996, Instrument Society of America, Research Triangle Park, NC, 1996.
10. Jejeli, M., and Huang, B. (Eds.), *Detection and Diagnosis of Stiction in Control Loops*, Springer-Verlag, New York, 2010.
11. Nwaoha, C., Extending control valve life by proper selection and maintenance, *Pipeline and Gas Journal*, vol. 235, no. 11, November 2008, pp. 78–80.

APPENDIX I

METHODS FOR MEASURING PROCESS TEMPERATURE

CHIKEZIE NWAOHA
Control Engineering Asia, Ten Alps Communications Asia, Aladinma, Nigeria

Temperature measurement of a substance, regardless of its state, is a very critical aspect of process measurement in the oil and gas, water and wastewater, pharmaceutical, food and beverage, and power industries, and its accuracy is very pertinent to proper operation of a plant. Determining the temperature of a substance contained in equipment or passing via the equipment, temperature measurement devices usually form part of a control loop assembly: a closed system of selected instruments that have the objective of monitoring fluid flow, controlling fluid flow, and so on. In this case, their role is to monitor and control the temperature of the fluid within the design limits.

In most cases, a computer is included in the temperature measurement loop to handle functions such as data gathering and transmission, bulk data storage, display, alarms, and control. This has given rise to the manufacture of temperature measurement instruments that are computerized and intelligent.

I.1 TYPES OF INSTRUMENTS

A variety of industrial temperature measurement instruments are used to measure the temperature profile of a fluid, gas, or solid (see Table I-1). The most important of these are detailed and described below.

I.1.1 Liquid-Filled Glass Thermometers

A reliable and low-cost device, the liquid-filled glass thermometer (Fig. I-1) is composed of a glass bulb with a capillary. When it expands, the liquid in the bulb forms a thin thread in the capillary, with the length of the thread forming a measure of the existing temperature. The space above the liquid contains only the saturated vapor of the fluid. The liquids used regularly are pentane, −200 to 20°C; alcohol, −110 to 50°C; toluene, −70 to 100°C; mercury (vacuum), −35 to 280°C; and mercury (gas-filled), −35 to 750°C [1]. Ocassionally, the liquid is pressurized to increase the measuring limit. The thickness of the material used in manufacturing a liquid-filled glass thermometer usually affects the time of response. To achieve high accuracy, requirements such as high thermal conductivity, high coefficient of thermal expansion for the fluids, and linear expansion must be considered.

I.1.2 Bimetallic Thermometers

A bimetallic thermometer consists of a strip of two dissimilar metals with different coefficients of heat expansion joined together and attached to an indicator. As a result of the metals' different rates of expansion, the bimetal is caused to warp, indicating a temperature change. One end of the metal strip is attached to the housing and the other end to an indicator. The metal strips can be riveted together but they are usually rolled or soldered jointly. To attain a high degree of deflection, a bimetallic strip is wound to form a spiral. Bimetalic thermometers can come in different designs: cone-shaped spiral, cylindrical spiral, flat spiral, and a combination of flat and cylindrical spiral. For high accuracy, bimetallic thermometers must be calibrated at or close to the conventional

Process Plant Equipment: Operation, Control, and Reliability, First Edition. Edited by Michael D. Holloway, Chikezie Nwaoha, and Oliver A. Onyewuenyi.
© 2012 John Wiley & Sons, Inc. Published 2012 by John Wiley & Sons, Inc.

TABLE I-1 Comparison of Various Types of Sensors [1]

Sensor	Advantages	Limitations
Thermocouple	Self-powered Simple Rugged Inexpensive Wide variety Wide range	Nonlinear Low voltage Reference required Least stable Least sensitive
RTD	Most stable and accurate Area sensing More linear than thermocouple Most repeatable Contamination resistant	Expensive Current source required Slow response time Low sensitivity to small temperature change Self-heating
Infrared	No contact required Very fast response time Good stability over time High repeatability No oxidation/corrosion to affect sensor	High initial cost More complex support electronics Spot size restricts application Emissivity variations affect readings Accuracy affected by dust, smoke, and background radiation
Bimetallic	Simple, robust, and inexpensive Has good accuracy Can measure temperature in the range −40 to 550°C Can withstand 50% overage temperature measurement	Not recommended for measurement of temperature above 550°C The metals undergo permanent warp distortion Use limited to local mounting
Liquid-filled glass	More economical, versatile, widely used Rugged in construction, low maintenance Can be used for remote indication Stable in operation System provides enough power to drive the control mechanism	Compensation necessary for ambient (surrounding) temperature changes and long capillary tube For accuracy the pressure bulb should be large In case of error the entire system has to be replaced

operating temperature being monitored. Although these devices are portable and do not require a power supply, they are usually not as accurate as thermocouples or resistance temperature devices. Bimetallic thermometers tend to be used where relative changes need to be monitored, and applications include use in oil refineries and in hot-work wire heaters, and they also work effectively in tempering tanks.

I.1.3 Thermocouples

Thermocouples represent a type of temperature measurement technique that allows direct electronic monitoring of temperature. A thermocouple consists of two different types of wires (dissimilar metals) joined together at one end. The wire metals can be an alloy or a very pure metal, and the measurement technique works on the principle of a thermo voltage being created when two different metals are brought into contact with one another.

A variety of temperature limits and materials of construction are available to measure temperature for different applications: copper/constantan; chromel/alumel; platinum/rhodium–platinum; and iron/constantan. To achieve high accuracy, thermocouples used for process measurements must be protected by a thermowell. For thermal contact with the thermowell, a physical contact or thermally conductive lubricant is placed between the thermocouple and the thermowell. For installations at low temperatures, extreme precautions must be taken to reduce sources of moisture. With their ability to measure

Figure I-1 Liquid-filled glass (capillary) thermometers offer low maintenance and enable remote indication.

a wide range of temperatures, thermocouples are used in furnace, salt bath, heat treating, molten metal, and ceramic applications.

I.1.4 Infrared Thermometers

Infrared (IR) thermometers (Fig. I-2) are noncontact temperature measurement devices that have many industrial applications. They measure temperature by detecting the amount of radiation emitted by a surface. To function properly, an IR instrument must take into account the emissivity of the surface being measured. The most basic design consists of a lens to focus the infrared energy onto a detector, which converts the energy to an electrical signal that can be displayed in units of temperature after being compensated for ambient temperature variation. With the ability to measure temperature measurement from a distance without contact with the object, an IR thermometer is useful for measuring temperature under circumstances where

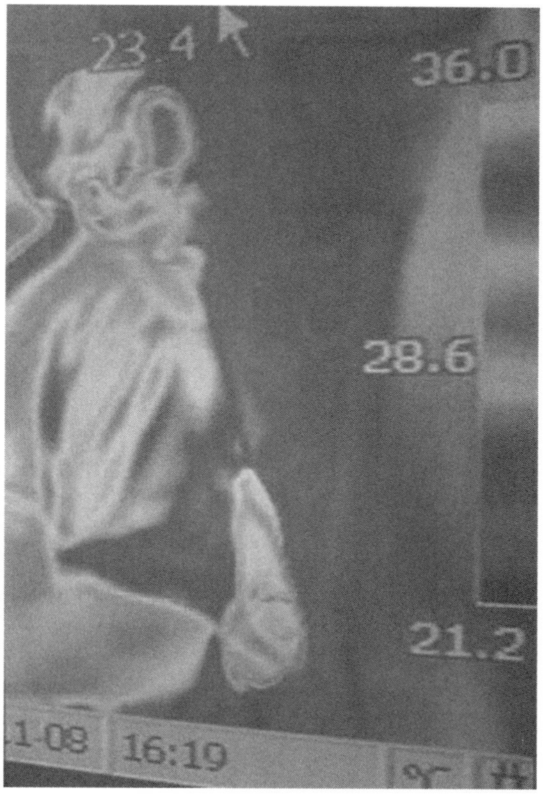

Figure I-2 Infrared thermometers deduce temperature by detecting the amount of radiation emitted by the surface.

thermocouples or other probe-type sensors cannot be used or do not produce accurate data. The noncontact advantage is particularly relevant for food industry applications.

I.1.5 Resistance Thermometers

Resistance thermometers, or RTDs (resistance temperature devices), work on the principle of the resistance of a material changing with temperature. Platinum, nickel, and copper are regularly used for RTDs because the change in resistance with these metals is large. RTDs are used in applications where faster response and greater accuracies are required than may be obtained with thermocouples. Their relatively high electrical output also makes them suitable for direct connection to indicators, controllers, recorders, and so on, and RTDs may also be more economical than thermocouples in some installations since the extension wires can be made of copper. In addition, a reference temperature source is not required for calibration [1].

I.2 SELECTION STRATEGY

To maintain the efficiency and prolong the service life of temperature meters, proper selection strategy must be followed. Indeed, improper selection strategy is regarded as the most likely cause of measurement failure, which can also result in significant plant downtime. In a worst-case scenario, an improperly specified temperature meter can fail to measure temperature during an over- and under-temperature event, compromising plant safety and damaging valuable equipment. To improve temperature meter operating efficiency and service life, there are certain basic selection criteria end users should consider, as we discuss next.

I.2.1 Accuracy

The closeness of a reading of an measurement device to the actual value of the quantity being measured, the accuracy, goes a long way toward prolonging the efficiency of a process. Usually expressed as $\pm$ a percent of the full-scale output or reading, accuracy specifications can be misleading because they are values usually published by the original equipment manufacturer for ideal boundary conditions. In practice, however, deviations from the ideal conditions are common, so that additional stipulations must be made for instrument accuracy. It is also important to establish if the accuracy is based on the measured value (% of rate) or on the range end value (% of maximum) [1].

I.2.2 Process Media

In selecting a particular temperature meter for installation, the phase media compatibility with the device must be taken into consideration; otherwise, corrosion and wear are likely to occur (Fig. I-3). Corrosion of parts will eventually cause meter failure, possibly compromise the safety of plant personnel and equipment, and risk the environment. Similarly, some fluids are chemically harmless but can become corrosive when small amounts of other materials are present (perhaps only contamination); therefore, care must be exercised. The variety of possible materials is especially restricted where complicated and difficult-to-manufacture parts are required [1]. A meter removed from operation with one process medium and placed into service with a second medium that is incompatible with the first can cause unwanted chemical reactions that may result in contamination of process media or even a catastrophic explosion. Corrosion of the temperature element assembly in the first installation may also be sufficient to cause early failure in the second installation. Therefore, it is not advisable that temperature meters be moved from one application to another. Should it be necessary, however, it is vital to establish the chemical compatibility of the two media. Refer to NEMA-12 specifications for more detailed information on protection against corrosion.

Figure I-3 Improper temperature device selection can lead to significant plant downtime, equipment damage, and risks to plant safety.

I.2.3 Temperature Range

Range represents a vital selection criterion because it stipulates the maximum and minimum allowable operating temperatures that a gauge can detect. Therefore, any temperature meter that does not meet the required temperature range requirements for an application should not be considered (Fig. I-4). For example, thermocouples typically have a range of -200 to 1600°C [1].

I.2.4 Plant History

The plant history of the temperature meter in the same or a similar scenario is a good measure to use when

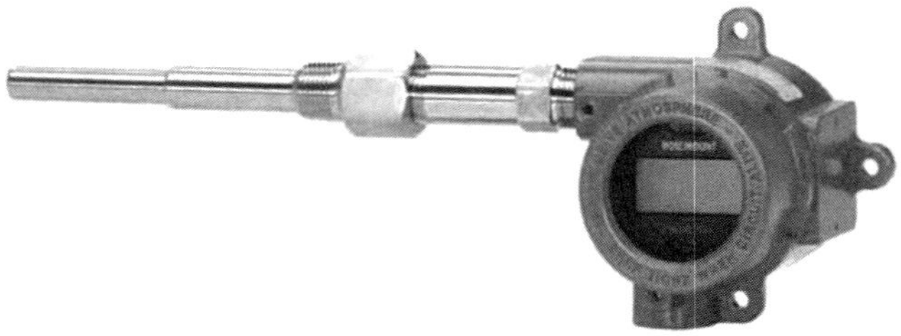

Figure I-4 Because of potential incompatibility between process media, it is not advisable that temperature meters be moved from one application to another. (Courtesy of Emerson.

specifying a particular instrument. The end user should evaluate such plant history, including lifetime cost and ease of maintenance and interchange. Note that at this stage, the expertise of the temperature meter manufacturer does not really affect the final decision as to the meter design to use in any specific application.

REFERENCE

1. Nwaoha, C., Temperature techniques, *Control Engineering Asia*, March 2011, pp. 26–29.

APPENDIX II

AIRFLOW TROUBLESHOOTING

Chikezie Nwaoha

Control Engineering Asia, Ten Alps Communications Asia, Aladinma, Nigeria

Many industries utilize air for production purposes, as in mixing, pressurizing, atomizing, and agitating applications. Applications include oil and gas, food, pharmaceutical, and others. By managing compressed-air systems and controlling airflow in such scenarios, users can improve production efficiency by up to 50%. Compressed air flows as a result of pressure differential (Fig. II-1). This implies that pressure drop is the major cause of insufficient airflow. To maintain adequate airflow, much attention should be focused on the pressure losses that are caused by obstacles in the compressed-air systems [1].

II.1 COMMON OBSTACLES TO PROPER AIRFLOW

Obstacles in a compressed airflow system alter the pressure of the flowing air. As the push to optimize production continues, it becomes imperative for production personnel to identify and combat such problems.

1. *Air quality*. Air cleanliness affects the airflow required. Atmospheric air contains a large amount of airborne contaminants: dust, dirt, water vapor, and in an oil-related industry, oil vapor in the form of unburned hydrocarbons. In many applications, dirt and dust particles can pass through the air compressor and gradually form deposits on the interior surface of the compressor. As these deposits accumulate, friction increases and the compressor loses it ability to generate the head required for airflow [2].

While many operators are concerned about the risk of dirt and dust particles, oil vapor and water vapor in the airstream also pose a problem. During compression, oil and water vapor escape with the compressed air, and after compression the air is cooled in the interstage cooler, resulting in condensed vapors. If this condensate is not removed, it causes corrosion and blockage to the compressed-air systems, hence reducing airflow and production efficiency. To prevent this, filters must be located properly in the system, and an interstage cooler with automatic drain traps must be fitted to the air compressor.

2. *Air compressor type and operation*. The air compressor can also be an obstacle to airflow. Air compressors make use of lubricating oil for sealing and lubrication and use cooling water (mostly applied) for cooling (Fig. II-2). During operation, cooling water and lubricating oil may seep into the compressed air as a liquid or aerosol. The resulting leakage causes contamination problems similar to those of condensed oil and water vapors [1]. The type of compresssor also affects proper airflow. A turbine-driven air compressor, for example, utilizes more lubricating oil and cooling water for operation than does an electrically driven compressor, so a turbine-driven compressor is more susceptible to lubricating oil and/or cooling water seepage.

3. *Improper configuration of distribution systems*. The main objective of proper sizing and configuration of distribution systems is to transport the maximum expected volumetric airflow from the compressor to the point of use with minimum pressure drop. Poor distribution system configuration can lead to insufficient airflow, and

Process Plant Equipment: Operation, Control, and Reliability, First Edition. Edited by Michael D. Holloway, Chikezie Nwaoha, and Oliver A. Onyewuenyi.
© 2012 John Wiley & Sons, Inc. Published 2012 by John Wiley & Sons, Inc.

Figure II-1 Newly installed air compressor. (Courtesy of Port Harcourt Refining Company, Nigeria.)

Figure II-2 The author adjusting cooling water in an air compressor. (Courtesy of Port Harcourt Refining Company, Nigeria.)

TABLE II-1 Troubleshooting Insufficient Airflow [1]

Airflow Problems	Remedies
Low air pressure at point of use	
Is the distribution piping leaking?	If so, refer piping for inspection and maintenance.
Is the filter element clogged?	Clean the filter element.
Is the pressure regulator faulty or not installed properly?	Refer pressure regulator for maintenance, or check manufacturer's manual for installation procedures. If not, install pressure regulators at various points in the system.
Is the air receiver leaking?	Request inspection. Check status and refer for repairs.
Liquid in the air lines	
Is the drain trap too small?	If so, install a properly sized drain trap.
Is the air compressor cooling system leaking?	Refer cooling jackets for maintenance, or install a new one.
Is the drain trap clogged?	Clean, repair, or replace the drain trap.
Is the compressed air dryer undersized or faulty?	Check the status of the air dryer. Refer the dryer for maintenance.
Is the compressor sealant leaking?	Refer for sealant replacement.
Low air pressure at compressor discharge	
Is the suction pressure adequate?	Maintain suction pressure.
Is it leak free?	Check the compressor for leakage. Refer for maintenance.
Is the discharge valve worn out?	Install a new valve.
Is the air capacity system adjusted improperly?	Refer to the manufacturer's recommendation for adjustment of air capacity systems.
Dirt or scale in the air lines	
Is the filter application correct?	Check the filter status, and install a proper-sized filter based on the application.
Is the distribution piping aging or corroded?	Install a new distribution piping. Check for moisture content in the airstream, and install a filter along the line.
Low air level in the receiver	
Is the level indicator in good working condition?	Confirm status. Refer for repairs or install a new indicator.
Are the air compressor loading valves functional?	Refer faulty valves for maintenance.
Is the air receiver leak-free?	If so, request inspection and refer for maintenance.

thus affects the discharge pressure, robbing the user of expensive compressed air power. This is not limited to the interconnecting piping from the discharge of the air compressor to the header. It also applies to the air storage system and the distribution line conveying air to production areas. To prevent this, international quality standards and guidelines must be adhered to strictly when sizing distribution systems [2].

II.2 TROUBLESHOOTING

Quickly diagnosing and correcting application issues will help ensure that small problems don't become big ones. Thus, it is important always to keep in mind the old truism: "A problem identified is half solved." Table II-1 provides guidance on troubleshooting five common problems.

REFERENCES

1. Nwaoha, C., Airflow troubleshooting strategy: achieving production efficiency through proper air systems maintenance, *Flow Control*, vol. xv, no. 7, July 2009, pp. 28–29.
2. Nwaoha, C., Process optimization: controlling air flow, *Steam and Boiler Review*, vol. 3, no. 11, November 2009, pp. 22–24.

APPENDIX III

MIG SHIELDING GAS CONTROL AND OPTIMIZATION

JERRY UTTRACHI
WA Technology, Florence, South Carolina

Question: What is the best way to deliver shielding gas to a MIG system?

Answer: For systems with up to 50 ft (for typical pipeline pressure it can even be longer) from the gas supply to the wire feeder, the best system (Fig. III-1) consists of a rotameter flowmeter (one with a flow indicator ball) and the WA Technology gas saver system (GSS). This system will work for any gas supply. The benefits include being able to see the actual gas flow while the GSS reduces associated gas waste by about 80% and controls the gas surge velocity, improving start weld quality.

Option: For a pipeline shielding gas supply, an orifice can be used to set the flow.

Question: Why is a MIG gas hose a wasteful $\frac{1}{4}$-in. ID? (*Hint:* It's not for the nonexistent pressure drop!)

Answer: Why is most MIG gas delivery hose $\frac{1}{4}$-in. ID? This large size causes excess gas to be stored in the hose when welding stops and is wasted every time welding starts. Fabricators find it causes from 30% to over 60% wasted gas! Is this large hose size needed to handle pressure drop? No! The typical 35 ft^3/h shielding gas flow rate creates very little pressure drop. A 100-ft $\frac{1}{4}$-in. ID hose, operating with pressures needed to flow 35 ft^3/h, has a pressure drop of less than 1 psi! The answer is MIG welding, developed in the 1950s by two of the dominant U.S. industrial gas producers. The major equipment product line manufactured at the time that MIG was introduced was oxyfuel welding and cutting apparatus. The flow rate of oxygen required when cutting is quite high. It can use 250-ft/h and higher flows. At a 50-psi regulator pressure setting, 250 ft^3/h will produce an 11-psi pressure drop in a 100-ft $\frac{1}{4}$-in. ID hose versus the flow of less than 1 psi at 35 ft^3/h that is used for MIG welding. (Note that the pressure drop at 35 ft^3/h will be less than 1 psi at either 50 or 5 psi hose pressure.) The highest-volume gas hose used for the largest-sales-volume hoses sold (25 ft) was $\frac{1}{4}$-in. ID, which easily handled the pressure drop. Cutting and fitting hose of this size was automated. In addition, the Compressed Gas Association (CGA) committee developed gas hose fittings designed for various size hoses (Fig. III-2). The inlet end of these fittings handles hoses up to $\frac{3}{8}$-in. ID. Using CGA designs, $\frac{1}{4}$-in. inert gas fittings are relatively easy to make. Production can start with heavy wall tubing, making drilling the gas passage hole quick and economical. A minimum amount of material is required to be removed with this design approach. Functional hose clamps are also readily available for $\frac{1}{4}$-in. hose fitted to a hose barb.

Therefore, 1/4-in. hose, hose fittings, and hose clamps were readily available and lowest in cost. Production economics are the reason that MIG gas delivery hose was $\frac{1}{4}$-in. ID!

The special fittings required for use with our very heavy-walled small-ID *GSS* hose are much more difficult and costly to manufacture. The hose does not expand over the fitting as do thinner-walled hoses used for TIG, and special hose clamps must be used. See Fig. III-3 to view the GSS fitting on the gas-supply side of the hose versus a $\frac{1}{4}$-in. fitting. As the saying goes, "Follow the money!"

Question: Are there other things that we can do other than purchase GSSs to reduce gas waste?

Process Plant Equipment: Operation, Control, and Reliability, First Edition. Edited by Michael D. Holloway, Chikezie Nwaoha, and Oliver A. Onyewuenyi.
© 2012 John Wiley & Sons, Inc. Published 2012 by John Wiley & Sons, Inc.

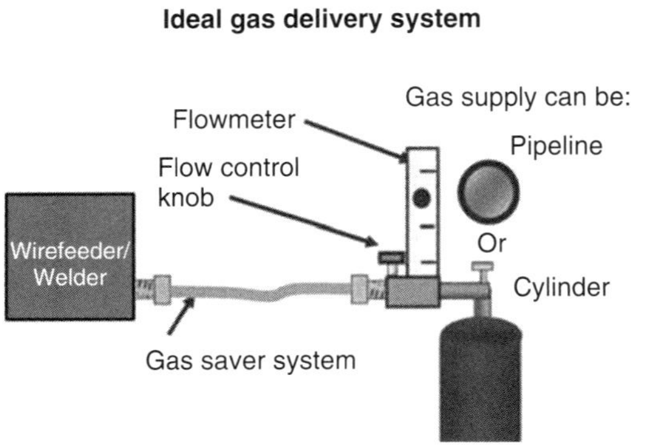

Figure III-1 Schematic of a typical MIG/MAG shielding gas delivery system.

Figure III-2 Patented Flow Rate Limiter shown mounted on flowmeter, which prevents increase in flow after it is set.

Answer: Yes.

1. We have found that many welders set flow rates too high. We have a patented *flow rate limiter* that can be installed on most flowmeters. You set the maximum desired flow rates and install the lock, which prevents the flow rate from being increased. See the details of this recently patented product at www.netWelding.com/Flow_Rate_Limiter.htm.
2. Another thing to consider is leaks. These can be very wasteful and also allow moisture-laden air to enter the shielding gas lines.

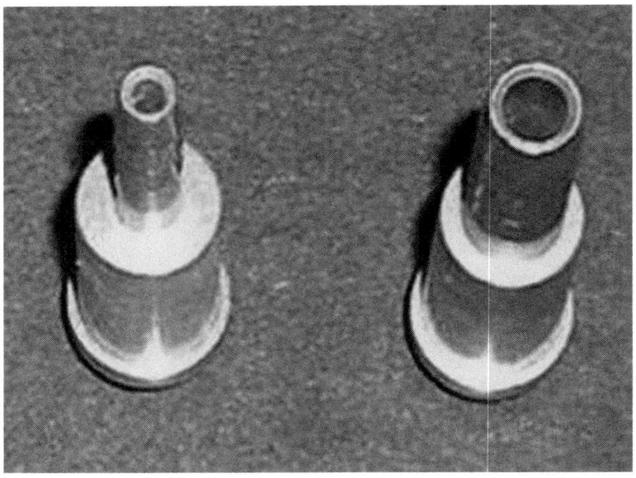

Figure III-3 Patented custom extruded small ID gas saving delivery hose requires a special small OD hose barb, shown on left.

	A	B	C	D	E	F
1	Insert Data; Leave 0 if None		Answer	Insert Special Pipe Diamter - if Needed		
2						
3	Pipe Size ID inches	ID Volume ft³ Per Foot of Pipe	Input Number of Feet of Pipe in System	Total Physical ft³ in Piping System	Total ft³ of Gas at STP At 50 psi	Total ft³ of Gas at STP At 40 psi
4	1.0	0.09	150	13	58	49
5	1.5	0.20	0	0	0	0
6	2.0	0.35	400	140	615	520
7	2.5	0.53	0	0	0	0
8	3.0	0.79	0	0	0	0
9	4.0	1.40	20	28	123	104
10	0.0	0.00	0	0	0	0
11	Gas Deliver Hose Size inches	ID Volume ft³ Per Foot of Hose	Total Piping =	181	795	672
12	0.375	0.0123	0	0	0	0
13	0.250	0.0055	300	2	7	6
14	0.125	0.0014	0	0	0	0
15	Total Pipe and Hose Volume; ft³ =			182	802	678
16						
17			ft³ Gas Reduced with 10 psi Pressure Drop =			124
18			Input Time to Reduce Pressure in Minutes =			5.5
19			Cubic Feet per Hour (CFH) Leak Rate =			1353

Figure III-4 An Excel spread sheet allows simple calculation of leak rate in a shielding gas pipeline supply by using a timed leak-down test.

In addition to leaks, there needs to be ongoing vigilance to check the gas delivery systems in a systematic way. We offer several training programs that can help. One is called the "Lean Welding Manufacturing Learning Program Optimizing Shielding Gas Use and Eliminating Waste." This 75-page program can be used as a self-study leaning program. It now includes an Excel spreadsheet (Fig. III-4), which provides a suggested method of quantifying and monitoring leak rates in a pipeline gas supply.

A shorter 27-page program that simply defines shielding gas flow rates (minimums and maximums) and can be used to educate welders and welding foreman is also available. See www.NetWelding.com/prod03.htm.

APPENDIX IV

RUPTURE DISK SELECTION

CHIKEZIE NWAOHA
Control Engineering Asia, Ten Alps Communications Asia, Aladinma, Nigeria

It has been a practice in the process industry to fit rupture disks in pressure equipment to provide means of relieving excess pressure. The disk ensures the safety of life and plant equipment in case of overpressure (Fig. IV-1). A rupture disk is a non-reclosing pressure relief device designed to provide overpressure relief in chemical, petrochemical, and sanitary applications. It can also be defined as a thin metal membrane designed to burst at a certain pressure and temperature to prevent overpressurization of the attached vessel. A rupture disk can be scored, machined, or injection molded. Rupture disks are also known as *burst disks*, due to their characteristic way of bursting when relieving pressure buildup [1].

Rupture disks are similar to pressure relief valves due to their function of reducing overpressurization but differ in the sense that while a pressure relief valve will reseat after discharging, a rupture disk remains open after discharging. A rupture disk consists of a thin metal disk held at both ends by specially machined flanges. The disk can be made of aluminum, silver, copper, nickel, or other metal. It is because of the disk material and thickness that it will burst at a predetermined pressure. Since the rupture disk metal is thin, the stresses are high, and noncorrosion allowance is allowed. However, the disk material must be more resistant to corrosion than the metal from which the pressure vessel is constructed. As a preventive alternative, linings, coatings, or other materials can be employed to protect the disk against corrosion.

A rupture disk with accessories consists of selected instruments that work as a unit with the single objective of controlling pressure buildup. A disk's accessories consist of burst checks, burst indicators, clamps, and ferrules [1]. A rupture disk can be actuated thermally or mechanically. Regardless of the disk design, it should be made with a *safety factor*, which means that a rupture disk designed to burst at 80 psi must burst at 80 psi (it must burst within the agreed tolerance) [3].

IV.1 RUPTURE DISK TYPES

Various types of rupture disks are used in the oil and gas, sanitary, and other industries: forward- and reverse-acting rupture disks. The type employed depends on its suitability for the process.

IV.1.1 Forward-Acting Rupture Disk

The forward-acting rupture disk is dome shaped and is oriented into a system with the process medium (phase application) pressure against the concave side of the disc (Fig. IV-2). When there is an increase in the process pressure beyond the designed or allowable operating pressure, the tensile strength of the material is reached and rupture occurs. This type of rupture disk is employed in systems that typically have operating ratios at about 80% or less. They are usually designed to act in tension, somewhat like a balloon. There are different types of forward-acting rupture disks: forward-acting composite disk, forward-acting solid metal disk, forward-acting scored metal disk, and graphite disk. They have the advantage of being cheaper than reverse-acting rupture disks.

Process Plant Equipment: Operation, Control, and Reliability, First Edition. Edited by Michael D. Holloway, Chikezie Nwaoha, and Oliver A. Onyewuenyi.
© 2012 John Wiley & Sons, Inc. Published 2012 by John Wiley & Sons, Inc.

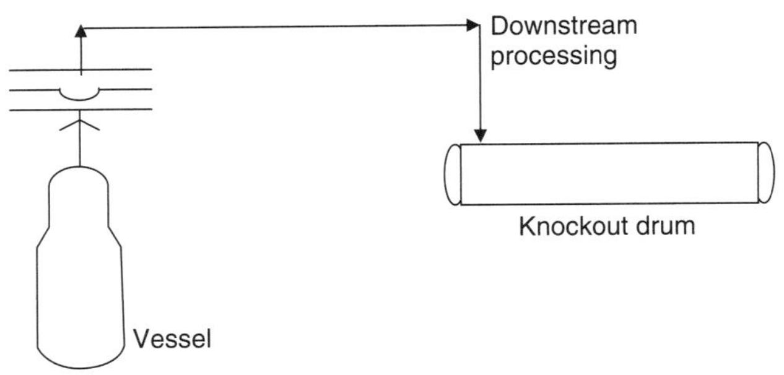

Figure IV-1 Vessel being protected by a rupture disk.

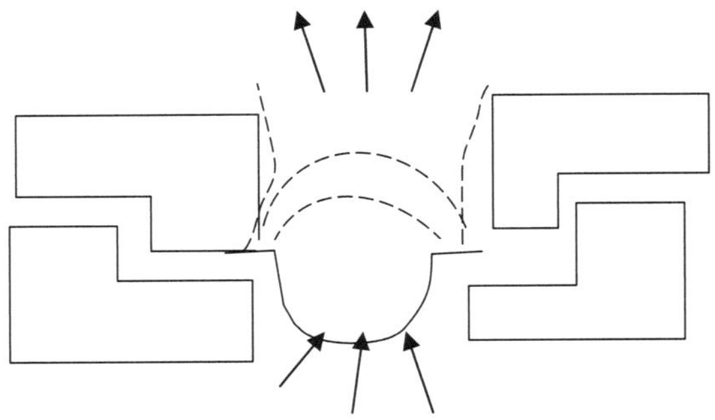

Figure IV-2 Forward-acting rupture disk.

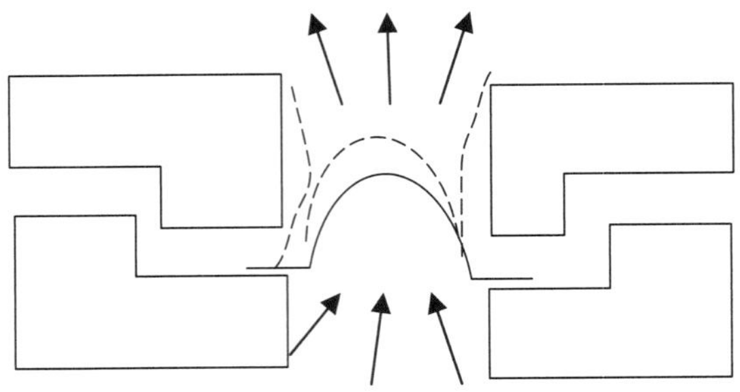

Figure IV-3 Reverse-acting rupture disk.

IV.1.2 Reverse-Acting Rupture Disk

The reverse-acting rupture disk is also dome shaped, but is installed with process pressure against the convex side of the disk, thereby placing the rupture disk in compression (Fig. IV-3). It is designed such that as the designed burst pressure of the rupture disk is attained, the compression loading on the disk causes it to reverse back into a forward-acting disk and then burst. This causes the disk to open by a predetermined knife blade or scoring pattern penetration. Reverse-acting rupture disks are employed in systems where operating ratios are as high as 95% or less [4]. They can be used in combination with relief valves, which means that they can be nonfragmenting. Reverse-acting rupture disk can also be used in vacuum or larger backpressure applications without special supports. They have the advantage of controlling burst pressure at close tolerances.

IV.2 OPERATING A RUPTURE DISK

While using a rupture disk, backpressure buildup in the space between the disk and the pressure relief valve should be controlled. The contrary will lead to the disk not bursting at its designed pressure. This backpressure buildup could be caused by leakage, as a result of physical damage or corrosion. Proper inspection for corrosion and physical damage of the rupture disk will give satisfactory service and enhance process safety [2].

Rupture disks are used on vessels, piping, and pressure relief valves, where the pressure buildup is formed as a result of unavoidable mechanical malfunctions and runaway chemical reactions. They are also used to protect heat exchangers and large equipment such as compressors and pumps. Rupture disks can be termed "nonreturn" when the relief is discharged to the atmosphere, and can be termed "return" when they discharge to some downstream collection and treatment systems, such as the knockout drums (Fig. IV-1).

IV.3 RUPTURE DISK SELECTION CRITERIA

Process industry professionals claim that over 60% of rupture disks installed do not perform satisfactorily, and improper selection strategy accounts for over 80% of these problems. Proper selection of a rupture disk is more than performing sizing calculations to make sure that it is adequately sized for the emergency event. Failure to select the correct rupture disk for an application can result in significant plant downtime due to "nuisance failures." In a worst case, an improperly specified disk can fail to open during an overpressure event, resulting in a catastrophic failure.

There are five basic selection criteria to consider to improve rupture disks operating efficiency and service life:

1. *Phase application*. In selecting a particular rupture disk for installation, its phase application must be considered. This implies that gas-only disks should be used under gas-only conditions, because a disk may not open at all under liquid conditions. For example, many reverse-buckling disks are applied under gas conditions; its opening is actuated by the energy stored in the compressed gas; and since liquids are incompressible, stored energy due to compression is not available [1]. Therefore, all liquid-only disks are eliminated in this stage if the application is gas.

2. *Rupture disk operating ratio*. The rupture disk operating ratio indicates the pressure at which the disk can be operated with a prolonged service life. The operating ratio is gotten by dividing the maximum operating pressure by the rupture disk burst pressure. Rupture disk selection with a higher-than-required operating ratio will be a waste of revenue and failure, but selecting a rupture disk with a lower operating ratio than required will cause reduced production and frequent change-outs. Rupture disks have a recommended maximum operating ratio of about 50 to 95%, depending on the construction method and materials. They should not be selected for applications where the disk will be subjected to conditions above the specified maximum operating ratio. Therefore, any rupture disk that does not meet the required operating ratio requirements must not be considered.

3. *Does the rupture disk withstand full vacuum*? Rupture disks are required to be vacuum resistant. In some cases, a rupture disk needs an additional vacuum support, whereas in others they withstand the full vacuum as its standard condition. There are forward- and reverse-acting rupture disk that don't need optional vacuum supports and at the same time don't withstand full vacuum. Such disks are not considered when the need for vacuum resistance exists [2].

4. *Does the rupture disk require being nonfragmenting?* Rupture disks used upstream of a pressure relief valve are normally specified to be the nonfragmenting type. This is because when a disk fragment lodges in the relief valve, it obstructs the valve from closing properly [4]. In some applications it is advisable not to contaminate the process with pieces of rupture disks. This is the case in sanitary applications. If fragmentation is not required for the application, all fragmenting rupture disks should be eliminated.

5. *The soft criterion*. Once the final selection list is established, this criterion is used to evaluate the final suitability of a rupture disk design for a given application. The most important point is that time-intensive evaluation is carried out on a smaller set of rupture disks suited for the application. At this stage, the expertise of the rupture disk manufacturer does not really affect the final decision on the type of disk design to use in the application [3]. Such a soft criterion is previous plant history with the rupture disk design in the same or similar applications (e.g., interchangeability, ease of maintenance, cycle life).

IV.4 TROUBLESHOOTING

Understanding the causes that make a rupture disk fail will help ensure that they don't fail. When the cause of a rupture disk failure is unknown, it is imperative that the disk manufacturer be consulted for further analysis. To prolong the service life and high accuracy of rupture disks, the causes of its problems should be determined. The first step in troubleshooting a rupture disk problem is to make sure that it is selected and installed properly. Table IV-1 provides guidance on troubleshooting four common problems.

TABLE IV-1 Common Rupture Disk Problems and Remedies [1]

Possible Problems	Remedies
Disk is not discharging	
Is scale lodged in the disk?	Clean the disk.
Is the disk applied in the wrong phase media?	If so, check for adequate rupture disk for the particular phase medium: gas or liquid.
Is the disk scored or machined to the vessel properly?	Check coupling.
Is the pressure gauge functioning?	Replace a faulty gauge.
Is the rupture disk small?	Install a properly sized disk.
Disc is experiencing excess pressure	
Is valve opening and/or closing too quickly?	Install slow-closing valves upstream of the rupture disk.
Is the pressure meter installed far from the disk?	Check for proper disk and pressure meter distance.
Is there a hammer effect along the line leading to the disk?	If so, install air-filled dampers to the line.
Disk leaks	
Is the rupture disk installed incorrectly?	Check the manufacturer's manual.
Is the rupture disk assembly worn out?	Replace the rupture disk assembly.
Is there residual porosity in the disk?	Refer the disk for impermeability test.
Disk is not installed correctly	
Is the disk installed with the wrong holder?	Specify the proper disc holder.
Is a damaged disk installed?	If so, replace the disk, and refer for maintenance.
Are the installation instructions complied with?	Maintain design conditions.

REFERENCES

1. Nwaoha, C., When the pressure is on: improving process safety through proper rupture disc selection, *Flow Control*, vol. xv, no. 5, May 2009, pp. 26–28.
2. Nwaoha, C., Enhancing plant safety via rupture disc selection, *Hydrocarbon Asia*, vol. 19, no. 3, July–September 2009, pp. 42–45.
3. Nwaoha, C., Controlling process pressure through proper selection of rupture, *Petroleum Africa*, October 2009, pp. 52–54.
4. Nwaoha, C., Achieving process safety via proper selection of rupture disc, *Oil Review Africa*, no. 1, January–February 2010, pp. 76–78.

APPENDIX V

PRESSURE GAUGE SELECTION

CHIKEZIE NWAOHA
Control Engineering Asia, Ten Alps Communications Asia, Aladinma, Nigeria

To maintain the efficiency of process gauges, the proper selection strategy must be followed. An improper selection strategy is regarded as the most likely cause of pressure gauge failures. As such, end users should be very careful to make the necessary steps to ensure effective pressure gauge selection (Fig. V-1), as failure to select the correct pressure gauge for an application can result in significant plant downtime. Further, in a worst-case scenario, an improperly specified pressure gauge can fail to measure pressure during an overpressure event, compromising plant safety and damaging valuable equipment [1].

V.1 SELECTION CRITERIA

To improve pressure gauge operating efficiency and service life, there are seven basic selection criteria that users should consider: phase media/wetted parts, accuracy, case style/material, connection size and location, pressure ranges, dial size, and plant history.

1. *Phase media/wetted parts*. In selecting a particular pressure gauge for installation, the phase media compatibility with the pressure gauge wetted parts (e.g., bourdon tube, socket) must be taken into consideration. If not compatible with the wetted parts of the gauge, corrosion most will probably occur. Corrosion of gauge wetted parts will eventually cause gauge failure, possibly compromise the safety of plant personnel and equipment, and pose a risk to the environment. Similarly, some fluids are chemically harmless but can become corrosive when small amounts of other materials are present (perhaps only contamination) [1]. Therefore, care must be exercised. The variety of materials possible is especially restricted where complicated parts that are difficult to manufacture are required. On the other hand, a gauge removed from operation with one process medium can be placed into service with a second medium that is incompatible with the first. This can cause unwanted chemical reactions that may result in contamination of process media or a catastrophic explosion. This incompatibility can also cause corrosion of the pressure gauge wetted parts. Therefore, it is not recommended that pressure gauges be moved from one application to another. Should it be necessary, however, the following must be considered:

- *Chemical compatibility*. The consequences of an incompatibility can range from contamination to explosive failures.
- *Corrosion*. Corrosion of the pressure element assembly in the first installation may be sufficient to cause early failure in the second installation.
- *Partial fatigue*. The first installation may involve pressure pulsation that has expended most of the gauge's life, resulting in early fatigue in the second installation.

To minimize such problems, when a gauge wetted part is not compatible with the process medium, a diaphragm should be considered.

2. *Accuracy*. Pressure gauge accuracy represents controversial specifications, values usually published by the manufacturers for ideal boundary conditions. In practice, deviations from ideal conditions are common, so that additional stipulations must be made for instrument accuracy. It is imperative to realize that accuracy may by based on

Process Plant Equipment: Operation, Control, and Reliability, First Edition. Edited by Michael D. Holloway, Chikezie Nwaoha, and Oliver A. Onyewuenyi.
© 2012 John Wiley & Sons, Inc. Published 2012 by John Wiley & Sons, Inc.

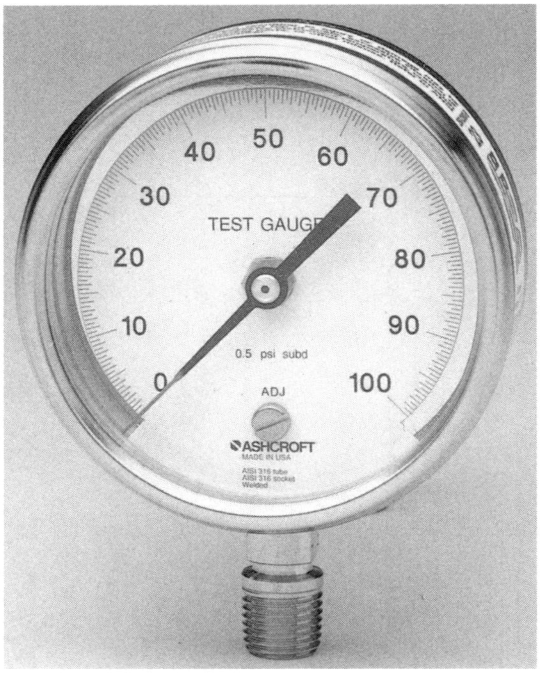

Figure V-1 A pressure gauge: the Duragauge. (Courtesy of Ashcroft, Inc.)

the measured value (% of rate) or on the range end value (% of maximum). For a mechanical pressure gauge, accuracy is identified as a percentage of the full-scale range. Although requirements differ from one industry to another, the following are general guidelines:

- *Critical processes*: 0.5% full-scale accuracy
- *Test gauges and standards*: 0.25% through 0.10% full-scale accuracies
- *General industrial processes*: 1.0% accuracy; less critical commercial uses: 2.0% accuracy

For more detailed information on accuracy, refer to ASME B40.100 or the DIN specifications.

3. *Case style/material*. In specifying the case style/material for a particular pressure gauge, environmental considerations must not be overlooked. Such considerations include ambient temperature, airborne particulates, condensation, humidity, water, and chemicals, all of which can affect the overall performance of the gauge. Ambient temperature, particularly, may affect the accuracy and integrity of a pressure gauge. Gauges are available either temperature compensated or non-temperature compensated. Ambient conditions may require that the gauge be isolated from extreme temperatures. When required, the gauge should be isolated from temperature extremes with a flexible line assembly. When ambient conditions are corrosive, contain a large number of particulates, or if the gauge will be exposed to a wet or humid environment such as humidity, rain, or wash-downs, it is recommended that a weatherproof, hermetically sealed, or liquid-filled gauge be specified.

4. *Connection size and location*. Pressure gauges are usually available with a variety of connections including NPT, DIN, JIS, BSP, and SAE. Process pressure gauges with $4\frac{1}{2}$-in. dial sizes or larger are most often supplied with a $\frac{1}{4}$-in [1]. NPT connection to provide the best support to the gauge. There are some factors to be considered when selecting a pressure gauge connection: gauge size, space limitations, leak integrity, process pressures, and past experience. After the selection of the appropriate pressure gauge connection size, the connection location is then considered. In selecting the required pressure gauge connection location, there are various mounting options and requirements to be considered:

- Well/surface mount lower connect
- Stem mount lower connect
- Panel mount back connect
- Front flange flush mount back connect for panel mounting
- U-clamp flush mount back connect for panel mounting

5. *Pressure ranges*. The pressure gauge pressure ranges represent controversial specifications because they stipulates the maximum and minimum pressure the gauge can detect. ASME B40.100 recommends that normal operating pressure be confined to 25 to 75% of the scale. If pulsation effects are present in the process, the maximum operating gauge pressure should not exceed 50% of the full-scale range. Therefore, any pressure gauge that does not meet the required pressure range requirements for an application should not be considered [1].

6. *Dial size*. Pressure gauge dial sizes range from $\frac{1}{2}$ to 16 in. in diameter. Generally, readability requirements, required gauge accuracy, and space limitations are used to determine the dial size. Measurement accuracies of 0.25% or 0.5% generally have a dial size of $4\frac{1}{2}$ in. or larger since more graduations are required.

7. *Plant history*. The plant history of a pressure gauge under the same or similar conditions is a good measure to use when specifying a particular gauge. The user should evaluate such plant history as lifetime cost, ease of maintenance, and interchangeability. Note that at this stage, the expertise of the pressure gauge manufacturer does not really affect the final decision as to the gauge design to use in the application [1].

V.2 CONCLUSIONS

To ensure prolonged service life and the operating success of any pressure measurement technique, proper gauge selection must be followed. Key factors to consider include dial size, accuracy, and plant history. If the cause of a pressure gauge failure is unknown, the user should consult the manufacturer of the gauge for further analysis.

Acknowledgment

I wish to thank John Carissimi and Lou Altieri of Ashcroft Inc. for their permission to make use of an excerpt from material entitled "Guide: Seven Steps to Select a Pressure Gauge."

REFERENCE

1. Nwaoha, C., Pressure Gauge Selection Strategy: 7 Key Considerations to Ensure Application Success, *Flow Control*, vol. xvi, no. 9, September 2010, pp. 14–17.

APPENDIX VI

CORROSION AND ITS MITIGATION IN THE OIL AND GAS INDUSTRIES

KRUPAVARAM NALLI
Tebodin & Co., Muscat, Oman

In exploration and production operations in the oil and gas industries, the field operators would normally like to have an uninterrupted supply of oil and gas to the export or processing points. The lines and component fittings of the lines undergo material degradation with the varying conditions of a well due to changes in fluid composition, souring of wells over a period, and changes in pressures and temperatures under operating conditions. The material degradation results in the loss of mechanical properties such as strength, ductility, and impact strength, leading to loss of materials, reduction in thickness, and at times ultimate failure. Hence, it is imperative for field operators, pipeline engineers, and designers to have *corrosion awareness* concerning the oil and gas industries in their day-to-day activities to combat and mitigate corrosion and to ensure smooth and uninterrupted flow of oil and gas to users.

A wealth of information is available on corrosion and its mitigation in the oil and gas industry through case histories, technical papers, published literature, and corrosion institutes. The present appendix is a consolidation of information pooled from various sources for the benefit of process, operations, maintenance, and integrity engineers involved in oil and gas activities.

VI.1 WELLHEAD FLUIDS

The reservoir fluids in oil and gas fields around the world vary greatly in composition. In some cases the fluid is in a gaseous state; in others, it is in a liquid state. Frequently, gas and liquid coexist in a given reservoir. The reservoir fluids basically comprise one or a combination of the following types of fluids with dissolved and suspended solids, based on the reservoir formation and location:

- Liquid hydrocarbons and multiphase systems
- Gas and gas condensates
- Formation waters
- Seawaters
- Brackish waters

Aging reservoirs will be souring (increasing in acid gas–hydrogen sulfide level) and with the increase in water cut, the corrosivity of the fluids increases as the well is being drained of fluids.

Figure VI-1 is a general process scheme indicating the various functions in a typical oil and gas exploration and production facilities.

VI.2 CORROSION AND CORROSION MORPHOLOGY

Corrosion is defined as a physiochemical reaction between a metal and the environment resulting in material degradation and thus leading to impairment of the intended function of the metal, environment, or the integrity of the system. This can be general corrosion or regular loss of metal on the exposed surface, or can be localized corrosion where only a limited portion of the surface is in contact.

Process Plant Equipment: Operation, Control, and Reliability, First Edition. Edited by Michael D. Holloway, Chikezie Nwaoha, and Oliver A. Onyewuenyi.
© 2012 John Wiley & Sons, Inc. Published 2012 by John Wiley & Sons, Inc.

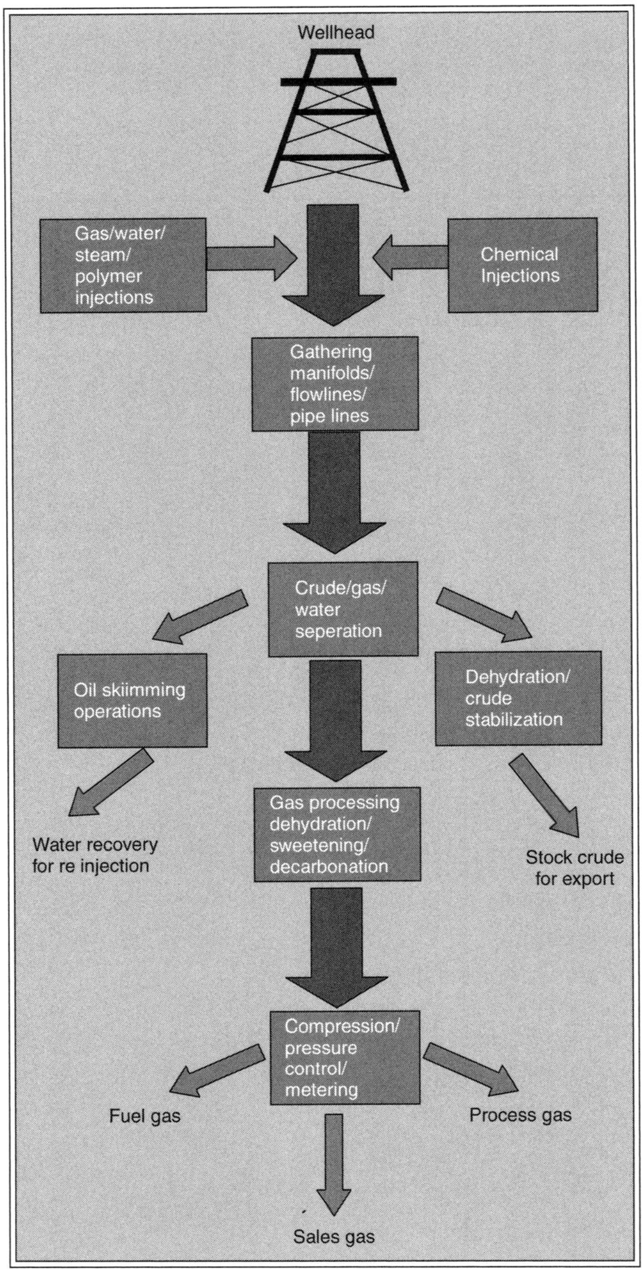

Figure VI-1 Typical processing scheme for an oil and gas facility.

VI.3 CORROSION IN THE OIL AND GAS INDUSTRY

Corrosion in oil and gas industry is due primarily to the reservoir and well fluids tapped during exploration and production operations. Some of the contents present in the well fluid are represented in Table VI-1. Few or all of these constituents may be present in varying compositions. Each component in the fluid will have an influence on the corrosivity of the fluid and will determine the performance of the materials in contact. Some of the corrosive mechanisms generally observed in the oil and gas industry and the impact of corrosion agents in the fluid are discussed briefly in the following sections.

TABLE VI-1 Typical Well Fluid Components: Multiphase Liquid Hydrocarbons

Component	Symbol	Unit
Carbon dioxide	CO_2	Mol/mol%
Hydrogen sulfide	H_2S	Mol/mol%
Oxygen (dissolved)	O_2	Mg/L or ppb
Chlorides	Cl^-	Mg/L (ppm)
Bicarbonates	HCO_3^-	Mg/L (ppm)
Iron	Fe^{2+}	Mg/L (ppm)
Magnesium	Mg^{2+}	Mg/L (ppm)
Potassium	K^+	Mg/L (ppm)
Sodium	Na^+	Mg/L (ppm)
Calcium	Ca^{2+}	Mg/L (ppm)
Sulfates	SO_4^-	Mg/L (ppm)
Sulfur	S	Mg/L (ppm)
Mercury	Hg	Mg/L (ppm)
Lead	Pb	Mg/L (ppm)
Zinc	Zn	Mg/L (ppm)
Chromium	Cr	Mg/L (ppm)
Barium	Ba	Mg/L (ppm)
Total hardness	As $CaCO_3$	Mg/L (ppm)
Total suspended solids	TSS	Mg/L (ppm)
Total dissolved solids	TDS	Mg/L (ppm)
pH value	pH	Number

VI.3.1 CO_2 Corrosion (Sweet Corrosion)

CO_2 is a principal corroding agent in oil and gas production systems. CO_2 will mix with the water, forming carbonic acid and making the fluid acidic (reducing the pH value). CO_2 corrosion is influenced by temperature and an increase in pH. At elevated temperatures, iron carbide (siderite) scale will form on the material as a protective scale, reducing the corrosion rate. The metal begins to corrode under these conditions. Forms of CO_2 corrosion—ringworm corrosion, mesa corrosion, and pitting corrosion—are shown in Fig. VI-2. CO_2 corrosion is enhanced in the presence of oxygen and organic acids, which dissolve the protective iron carbide scale and prevent further scale formation. The presence of bicarbonates in the phase improves the alkalinity of the fluid and thus reduces the corrosivity of the environment. Acetates in well fluids are observed to affect the corrosion rate, which is observed to be low when the acetate concentration in the fluid is low.

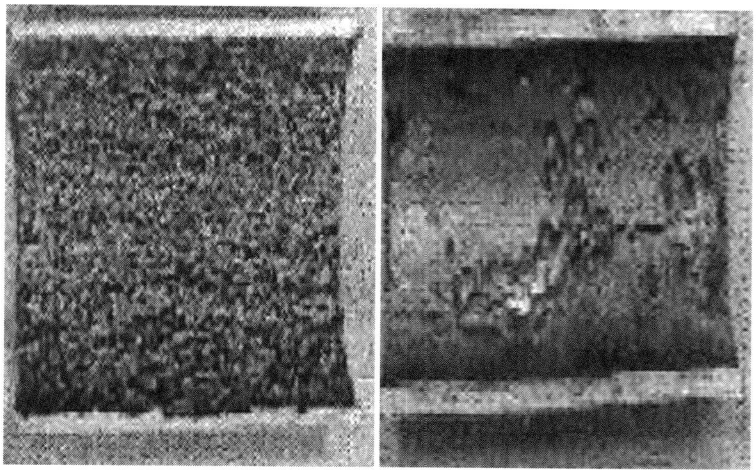

Uniform corrosion Mesa corrosion

Figure VI-2 Corrosion due to CO_2 (sweet corrosion).

VI.3.2 Hydrogen Sulfide Corrosion (Hydrogen Attack and Sour Corrosion)

Although the presence of hydrogen sulfide corrosion is less significant, the primary concern with hydrogen sulfide is hydrogen attack on the metal, causing the metal embrittlement. Fluids with high levels of H_2S are termed *sour*, and NACE defines H_2S levels above 0.05 psi partial pressure as sour conditions.

The hydrogen attack mechanism is complex and is caused by absorption of atomic hydrogen in steel, depending on certain pressure, temperature, and pH values of the fluid. The forms of sour corrosion include uniform, pitting, and stepwise cracking, as shown in Fig. VI-3.

VI.3.3 Corrosion Due to Oxygen

Oxygen is a strong oxidant and reacts with the metal very quickly. The dissolved oxygen in the formation or water produced is a primary causes of corrosion in production equipment. Although oxygen is not present in well fluids, oxygen ingress takes place in well fluids through leaking pump seals, casing and process vents, and open hatches. The forms of corrosion associated with oxygen are mainly uniform and pitting corrosion, as shown in Fig VI-4.

VI.3.4 Chlorides: Stress Corrosion Cracking

Chlorides in well fluids attack the material through the depassivation effect induced by chloride ions and is quite aggressive on 300 series austenitic stainless steels. High-nickel alloys are practically immune to this attack, and resist chloride corrosion. The chloride stress corrosion is influenced greatly by the temperature, chloride concentration, and residual stresses in the metal. The presence of oxygen and low pH accelerates the attack on the metal. Chloride corrosion is normally indicated by pitting on the metal surface, as shown in Fig. VI-4.

VI.3.5 Elemental Sulfur Corrosion

Elemental sulfur is present in some reservoir fluids and is a very strong oxidant. It mixes with the water in

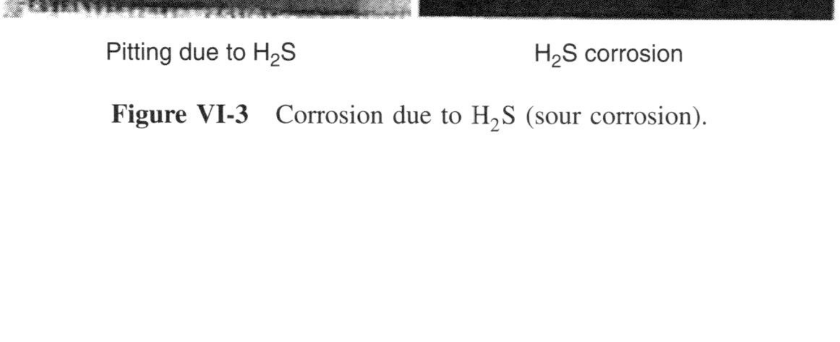

Pitting due to H_2S H_2S corrosion

Figure VI-3 Corrosion due to H_2S (sour corrosion).

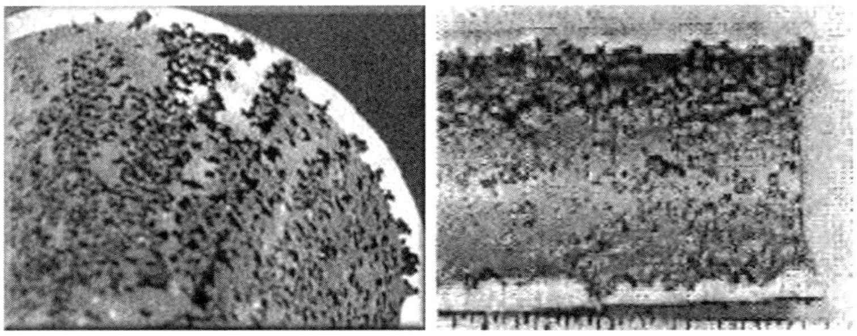

Figure VI-4 Corrosion due to oxygen and chlorides.

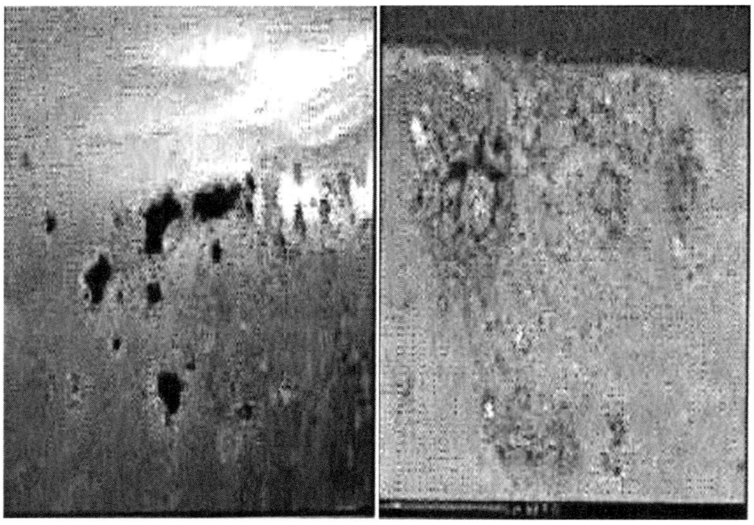

Figure VI-5 Galvanic and MIC corrosion.

the fluid and forms sulfuric acid and reacts with carbon and low-alloy steels to form sulfides. Corrosion due to elemental sulfur increases with temperature. Low-alloy materials such as CRA are quite susceptible to elemental sulfur attach; however, highly alloyed nickel CRAs such as Inconel/Incoloy are resistant to sulfur attack. Sulfur attack is by and large localized and is similar to pitting corrosion.

VI.3.6 Galvanic Corrosion

Galvanic corrosion occurs when two metallic materials with different nobility (electrochemical potential) are in contact and are exposed to an electrolytic environment. In such a situation the metal with less or more negative potential becomes anode and begins to corrode. The most prominent type of galvanic corrosion occurs in a coupling between CRAs (stainless steel or nickel alloys) with carbon or low-alloy steels in deaerated surroundings. Hydrogen ebrittlement is also possible on the more noble metal if it is susceptible. The presence of H_2S and low temperatures encourages galvanic corrosion (Fig. VI-5).

VI.3.7 Erosion Corrosion

Erosion corrosion is a form of corrosion with mechanical removal of metal due to high flow rates of the medium and is common to all metallic materials. The rate of corrosion increases with sand or solid suspended particles in the fluid and is dependent on fluid flow rate and on the density and morphology of solids present in the fluid.

VI.3.8 Crevice Corrosion

Crevice corrosion is normally localized corrosion that takes place in the narrow clearances or crevices in the metal, with fluid becoming stagnant in the gap. This is normally in the form of pitting formed on the surface of the metal and gradually extending into the metal itself.

VI.3.9 Microbial-Induced Corrosion (Sulfate-Reducing Bacteria)

Microbial-induced corrosion is due to the presence of a sulfate-reducing bacterium that grows in anaerobic conditions. These bacteria are normally present in reservoir fluids, formation waters, and soils. In a hospitable environment, the microbes tend to form colonies that enhance corrosion. The formation of colonies is promoted by neutral water, especially when stagnant. The form of microbial-induced corrosion is pitting, and the products of corrosion involve iron sulfates, slime, plugging, and bacterial growth (Fig. VI-5).

VI.4 METHODS TO COMBAT CORROSION IN THE OIL AND GAS INDUSTRY

Fighting corrosion to improve equipment and plant life continues to be a nightmare for many plant operating staff personnel. Although many methods have been suggested to arrest corrosion, they can be classed broadly into three main categories:

1. Changing the material of construction for the specific application.
2. Reducing the intensity of corrosive attack by modifications in corrosive media.
3. Creating a barrier layer between the material and the medium to avoid direct contact.

VI.4.1 Change in Materials of Construction

When it is observed that the existing material of construction is prone to corrosive attack, it is normally decided to change the materials of construction and select alternative material to suit the specific need. Generally, the materials used in hydrocarbon applications can be grouped broadly as metals or nonmetals. Each variety of these materials has its specific applications and limitations. At times a change of material may add to the cost. However, it is worthwhile to think in terms of life-cycle costing, which may show a longer equipment life and lower maintenance cost despite high initial cost. A detailed study of process and operating conditions has to be carried out before selection of a new material.

The new generation of stainless steels—duplex stainless steels, and super duplex stainless steels—by and large resist almost all types of corrosion. Steel mills all over the world continue to develop new materials with different metallurgies to resist almost any type of corrosion in the oil and gas industry. Although exotic materials such as titanium and zirconium will operate in almost all corrosive and high-temperature environments, the initial cost is prohibitive unless equipment downtime is critical to a process or a facility. Nonferrous materials such as copper, nickel, and copper–nickel alloys have also found good use in seawater environments, where normal materials such as carbon and stainless steels generally perform badly and fail. Selecting a suitable grade of stainless steel such as a low-carbon variety and stabilized grades will avert intergranular corrosion or weld decays if welding operations are involved.

Table VI-2 shows some of the commonly used materials in the hydrocarbon and oil and gas industries. The table is for guidance only. Detailed study of flow conditions, corrosion mechanisms involved, and the expected life of material is important before selecting a specific metal for an application. It is always more important to understand that no single material is a cure for all the corrosion evils. A material that is good for stress corrosion cracking may fail due to fatigue. Another material that can resist high-temperature corrosion may fail due to pitting.

VI.4.2 Change in Corrosive Media or the Environment

At times it is necessary to reduce the intensity of corrosive attack by adding *inhibitors* to reduce the aggressiveness of the media. These chemicals are injected into a medium point to point in the process. The chemicals, their concentration, and the frequency of injection depend on the process medium and, normally, on the recommendations of the inhibitor manufacturer, since these chemicals, although generic in nature, are generally proprietary items. The inhibitors used are normally chromates, phosphates, and silicates, added following the recommendations of the manufacturer.

The removal of oxygen from a fluid medium improves the chances of corrosion resistance by materials in contact with the fluid. Controlling and stabilizing the pH value of the medium is another method of combating corrosion. Many varieties of corrosion inhibitors are available in the market, but a judicious approach and the manufacturer's recommendations must be followed while injecting these chemicals in well fluid streams.

VI.4.3 Intermediate Barrier to Avoid Direct Contact with Media

A protective layer or barrier on material to prevent direct contact with process media will enhance material and equipment life. The barrier layer can be a paint, coating, or lining, or a metallic lining or metallic sheets. There are also nonmetallic linings, such as fiberglass, glass flake, epoxy, and rubber, which are normally employed on such equipment as separators, knockont drums, and storage tanks. Nickel, zinc, and cadmium coatings are also preferred at times on certain components, such as flanges and bolting. It must be clearly understood that such arrangements are not permanent cures and will only extend the life of the bare materials underneath the barrier to some extent. Following prolonged exposure to the atmosphere, heat, and sunlight, the paint may flake off. The tape or lining on a pipeline

TABLE VI-2 General Material Selection for the Hydrocarbon Industries[a]

Sl No.	General Material Specification	Material Designations: Plates	Pipes	Forgings	Applicational Use in Hydrocarbon Industry	Use in Hydrocarbon Industry	Oil and Gas Applications
1	**Carbon–manganese–silicon steels**						
	Carbon steels	A283 A285 A515 A516 A537	A53 A106 API5L A333	A105 A694	General-purpose, medium-corrosion, medium-temperature (to 200°C) applications; also low-temperature (up to −45°C) applications with 516, 537, and 333, grades	Pressure vessels, heat exchangers, tanks, spheres, and piping	Bulk fluids, crude pipelines, flow lines, water and steam injection lines; production and test separators, knockont drums, storage tanks
2	**Carbon–chrome–molybdenum steels**						
	Low- and medium-alloy steels	A204 A387	A335	A182	Medium corrosive and high temperature to 600°C media applications; an economic compromise between CS and SS	Furnace tubing, heat exchangers, re-boilers, pressure vessels, high-temperature piping	Wellhead items, chokes, manifolds, and well components with sour and high-temperature applications
3	**Straight chromium steels**						
	Chromium >12% and <18%	A240	A312 A213	A182	Highly corrosive and very high temperature above 800°C applications	Furnace/heater tubing, high-temperature vessels, columns, high-temperature heat exchangers	Christmas trees, wellheads, down-hole rods,valves, and casing pipes (OCT components)
4	**Chromium–nickel steels**						
	Stainless steels Chromium >18%, Nickel >8%	A240	A312 A213	A182	Highly corrosive, high-temp. media up to 800°C and strong oxidizing media; also, cryogenic applications	Pressure vessels, columns, heat exchangers, alloy claddings, piping and cryogenic applications	Valve trims, instruments, and internals of separators and tanks; low chloride levels
5	**Nickel steels**						
	2.5% nickel, 3.5% nickel, 9% nickel	A203 A353	A333	A420	Mildly corrosive and very low temperature (up to −100°C) media	Cryogenic storage vessels, heat exchangers, and piping especially for LNG applications	Rarely used in oil and gas sectors, LNG storage tanks, piping, and pumps
6	**Duplex stainless steels**						
	22% chromium—duplex, 25% chromium—superduplex	A240	A790	A182	Saline and highly chloride concentrated media and moderate temperatures (up to 60°C)	Pressure vessels, exchangers, piping with saline and chloride environments	Piping, vessel and tank internals where very high level of chlorides are present

TABLE VI-2 (*Continued*)

Sl No.	General Material Specification	Material Designations: Plates	Pipes	Forgings	Applicational Use in Hydrocarbon Industry	Use in Hydrocarbon Industry	Oil and Gas Applications
7	**Nickel–chrome (Inconels)**						
	Nickel–chromium–iron alloys	B443	B444	B564	High-corrosive, high-temperature, high-chloride, and high-sour media	Piping, tubing, instruments normally for high-temperature and high-sour environments	Wellhead and flow lines; manifolds with high-sour and high-temperature applications
8	**Nickel–Iron (Incoloys)**						
	Nickel–iron–chromium alloys	B424	B423	B564	High-corrosive, high-temperature, high-chloride, and high-sour media	Piping, tubing, instruments normally for high-temperature and high-sour applications	Wellhead and flow lines; manifolds with high-sour and high-temperature applications

[a]This table is to be used for guidance only. Every application must be considered separately and independently.

may become physically damaged, crack, and delaminate, exposing the bare material beneath to corrosive media. However, this method of combating corrosion is cheap and less expensive than opting for a costly material of construction.

VI.5 CORROSION MONITORING AND MANAGEMENT

The best way to check corrosion is by visual inspection and periodical checkups for material degradation. However, it may be possible to check the material condition externally, and it is impractical, if not impossible, to check the internal surfaces now and then. One method is to carry out on-stream inspection through periodic wall thickness measurements on fixed and vulnerable locations on equipment, piping, and pipelines to assess material conditions and corrosion rates. However, this method has its own limitations, since the checkpoint under investigation may show a level of corrosion lower than that at some other point, which may be heavily corroded and have gone undetected.

Corrosion is also monitored by placing electronic probes in the pipelines and by measuring the change in the electrical resistance in the probe coil. However, this method is more indicative of the process fluid than of a wetted materials condition. Cross-country pipelines are normally checked with intelligent pigging operations such as those using magnetic flux or ultrasonic pigs. These pigs will detect the internal conditions of the pipeline and corrosion conditions on the pipe wall thickness, and will indicate the thickness available on the pipe wall.

Most equipment, including separators, drums, and heaters, is checked for corrosion during annual shutdown and turnaround operations. Based on the physical assessment of the material conditions, corrective action is initiated to change the material or replace the equipment, or at times to do temporary repair work before replacement. In practice it is observed that physical inspection is the best method of monitoring corrosion and assessing material conditions.

VI.6 CONCLUSIONS

Undoubtedly, understanding the corrosion mechanism is very important before considering various material options for applications. However, it should be clearly understood that no particular material is the cure for all the evils of corrosion. Each case has to be considered in its totality before a decision is made as to the proper materials to use. Consultations with process, operations, materials, and corrosion engineers can save millions in the fight against the corrosion menace.

APPENDIX VII

MIXERS

Jayesh Ramesh Tekchandaney
Unique Mixers and Furnaces, Thane, India

VII.1 ADVANCES IN MIXING TECHNOLOGY

Mixing technology has made rapid advances in recent times. In addition to the laboratory and pilot scale trials, design engineers have access to tools such as Computational fluid mixing (CFM), Digital particle image velocimetry (DPIV), Laser doppler anemometry (LDA), Laser induced fluorescence (LIF), for better understanding of mixing systems. CFD provides detailed analysis of fluid flow early in the development process. The effects of vessel geometry and internals such as baffles can be factored into the system design. Mixing problems, such as low-velocity areas, circulation zones, staging and anomalies in the fluid flow and other potential problems can be identified using mixing simulation programs such as *VisiMix* and quickly resolved, instead of adopting the conventional and time consuming trial and error process. Using DPIV technology, it is possible to measure the fluid velocity field in the mixing vessel at all points almost instantaneously thereby making it possible to study large-scale, time-dependent phenomena which are responsible for the mixing process. LDA is probably the best method of non-intrusively determining mean velocity and turbulence data to high level of accuracy. This information may be used with CFD to accurately determine the flow in the mixing vessel. LIF is a measurement technique, which provides both qualitative and quantitative assessment of mixing.

The modern approach to the analysis of physico-chemical phenomena and processes, as well as the prediction of their course and results, is based on mathematical modeling. A number of sophisticated mathematical models and several advanced simulation software programs

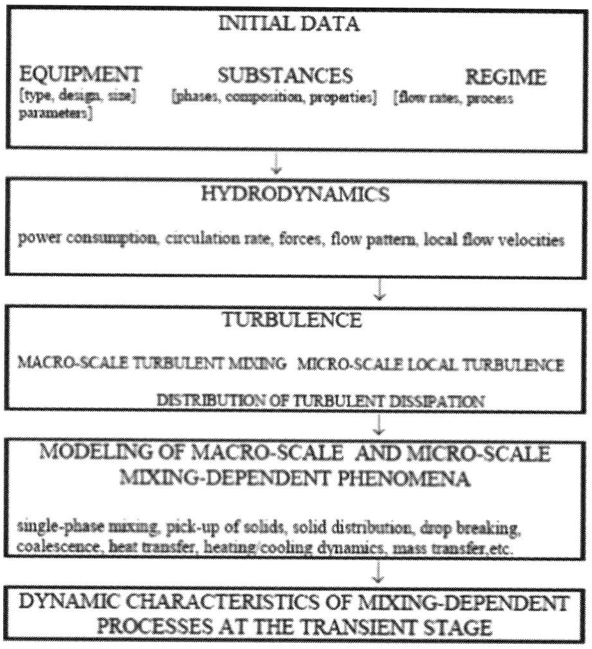

Figure VII-1 Flowchart of mathematical model and calculation algorithm used by *VisiMix*, Source [1].

have been developed. *VisiMix* is a computer program for calculation, analysis, scale-up and development of mixing processes and equipment based on mathematical simulation. The software allows chemical and process engineers to visualize mixing processes and calculate the most important process parameters for single and two-phase systems as power consumption, circulation rates, local concentrations of solutes and suspended particles, drop size,

Process Plant Equipment: Operation, Control, and Reliability, First Edition. Edited by Michael D. Holloway, Chikezie Nwaoha, and Oliver A. Onyewuenyi.
© 2012 John Wiley & Sons, Inc. Published 2012 by John Wiley & Sons, Inc.

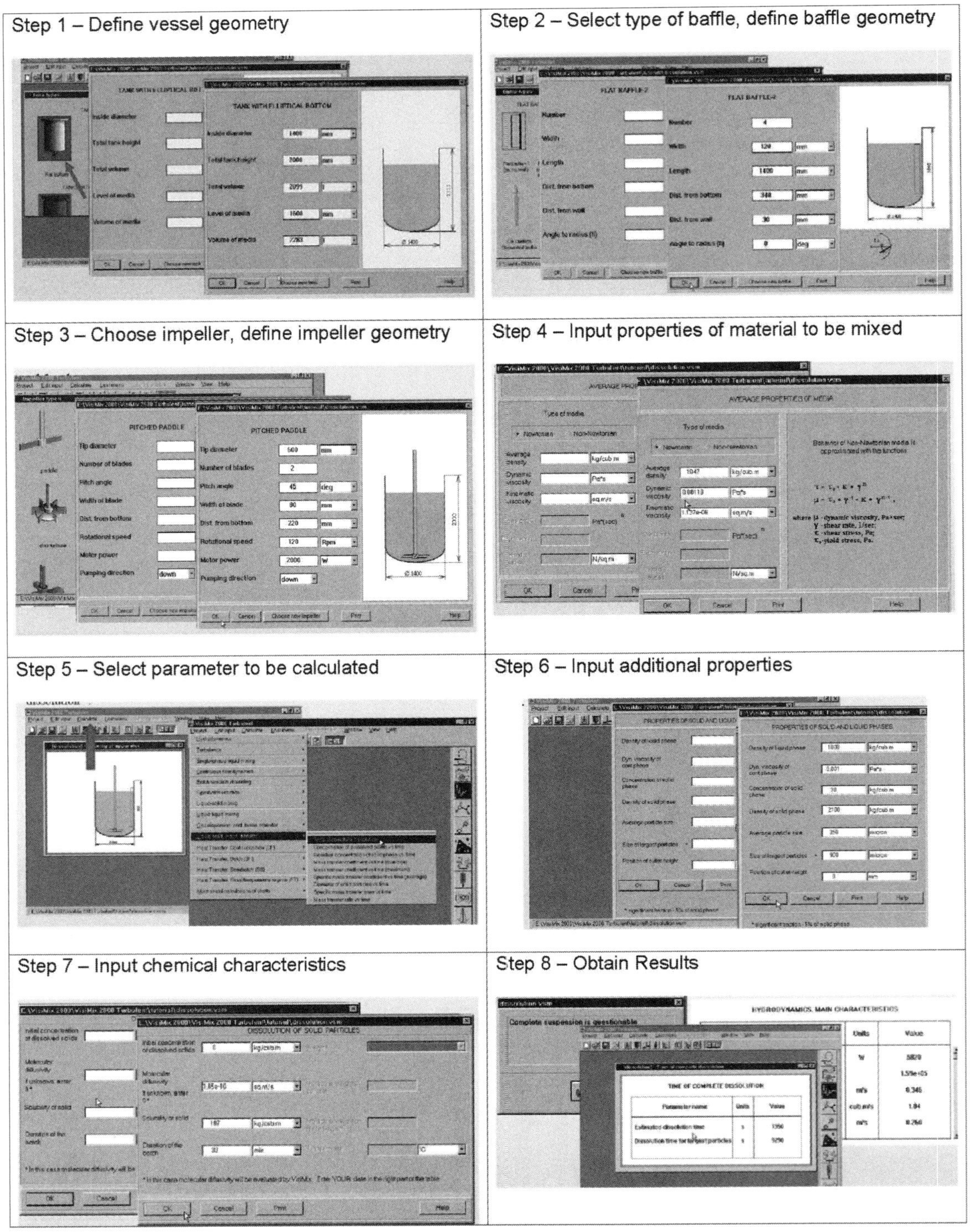

Figure VII-2 Mixing simulation using *VisiMix*.

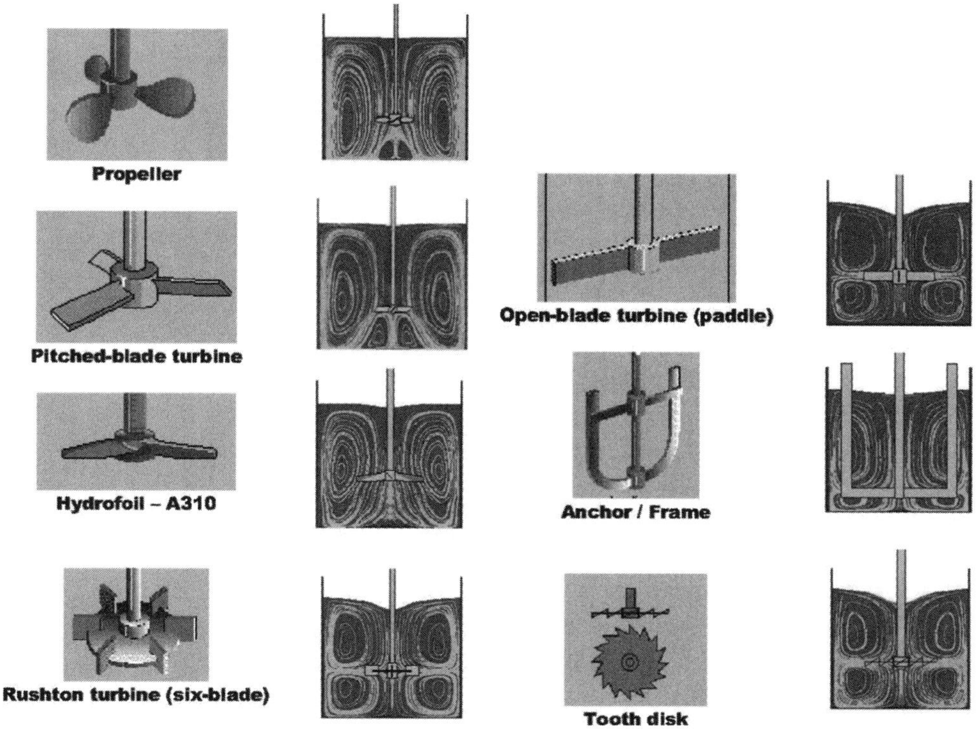

Figure VII-3 Impeller flow patterns, visualizations generated by *VisiMix*.

concentrations of reactants in chemical reactors, and so on. The mathematical models and calculation methods used in the *VisiMix* software have been subject to experimental verification, and most models have been tested on industrial scale and used in engineering practice for many years. With the application of mathematical models through computer simulation programs, it is possible to evaluate and select the most suitable mixing condition corresponding to the specific process requirement. These mathematical models and methods of calculations form a single system which allows for performing a chain of consecutive steps of mathematical simulation shown schematically in Fig. VII-1.

VisiMix input consists of data on mixing equipment and properties of the media. The program presents the data on the actual dynamic characteristics of mixing equipment, and a wide set of data on local flow velocities, intensities of turbulence, micro-mixing, etc., that are required for mathematical modeling of different unit operations. Figure VII-2 explains the operating steps of the software, along with some results. Figure VII-3 depicts the flow patterns generated by VisiMix for different type of impellers in an agitated vessel.

Advances in mixing technology have enabled mixer manufacturers to develop and improvise agitator designs customized for end user applications. Moreover, the improvements and the technological advances in the mechanical, electrical, and electronic components, instrumentation and process control, have accelerated the development of mixing equipment that maximize mixing performance at lower energy consumption.

REFERENCE

1. Braginsky L, Kokotov Y. *Review of the Main Mathematical Models Used in VisiMix*, VisiMix Ltd, Har Hotzvim, Jerusalem, 2000. (http://visimix.com/PROF/Reviewof theMainMathematicalModels.pdf)

GLOSSARY OF PROCESSING TERMS

Garlock Sealing Technologies (A Garlock Family of Companies)

Actuator: A device attached to a valve that moves the stem, usually by means of a motor or pneumatics. A valve fitted with an actuator is usually referred to as a *control valve*.

Adhesion: The clinging, binding, or sticking of two material surfaces to one another, such as rubber to rubber, rubber to metal, or rubber to cloth. It also refers to the bond strength between two surfaces.

Ambient temperature: The temperature of the surrounding environment.

Anchor: The terminal point or fixed point from which directional movement occurs.

Angle: An L-shaped steel member used either as a duct flange or as the fastening connection of an expansion joint. Angles are also used for bolting the joint to the mating flange surfaces of the ductwork or adjacent equipment.

Angular movement: Movement that occurs when one flange of the expansion joint is moved to an out-of-parallel position vis-à-vis the other flange. Angular movement is measured in degrees.

ANSI: Acronym for "American National Standards Institute."

Antiextrusion ring: A ring of packing used at one or both ends of a packing set to prevent extrusion of the packing into clearances. Also called a *bull ring*. *See also* End ring *and* Junk ring.

API: Acronym for "American Petroleum Institute."

ASME: Acronym for "American Society of Mechanical Engineers."

Assembled spring inside diameter: The inner diameter of the garter spring, with the ends securely joined.

ASTM: Acronym for "American Society for Testing and Materials."

Baffle/flow liner: A metal shield that is designed to protect an expansion joint from the abrasive particles in a gas stream and to reduce flutter caused by air turbulence in a gas stream. Baffles and flow liners may be welded or bolted into position.

Bellows: That portion of an expansion joint which accommodates movement of the joint convoluted or flat.

Belt-type expansion joint: An expansion joint in which the flexible bellows portion of the joint is made flat and bolted or clamped to metal adapter flanges or frames.

Blocking agent: A coating applied to braided packing to fill the area between fibers, blocking the passage of a medium through the body of the braid, improving its sealing characteristics. This may be a grease, an oil, or a PTFE dispersion.

Burst: A rupture caused by internal pressure.

Bushing: A metallic or carbon ring used to take up excess space in a stuffing box.

Butt cut: A straight cut at the seam (or joint) of a packing ring.

Case OD Sealer: A coating applied to the outside diameter of a seal's case to prevent leakage between the case and the housing bore.

Cavitation: An undesirable phenomenon that sometimes occurs in pumps, in which small vapor bubbles are created in the area of the impeller. As the vapor bubbles move along the vanes of the impeller to an area of higher pressure, they rapidly collapse. This collapse or "implosion" is so rapid that it may be heard as a rumbling noise or felt as vibration. The forces generated

Process Plant Equipment: Operation, Control, and Reliability, First Edition. Edited by Michael D. Holloway, Chikezie Nwaoha, and Oliver A. Onyewuenyi.
© 2012 John Wiley & Sons, Inc. Published 2012 by John Wiley & Sons, Inc.

as a result of cavitation may damage the impeller or even the packing set.

Centering ring (or guide ring): A solid piece of metal used on the outside diameter of a gasket winding to center the gasket properly on the sealing surface by locating the ring against the studs in the connection. The ring also acts as a compression limiter to prevent overcompression of the winding.

Centrifugal pump: A type of pump that relies on the rotation of an impeller to generate pressure and cause flow.

Checking: The short, shallow cracks on the surface of a rubber product resulting from damaging action or environmental conditions.

Chevron® Packing: Also known as V or *vee packing* and *stack packing*, an automatic hydraulic and pneumatic seal designed for sealing rods, rams, pistons, and plungers. Consisting of a male and female adapter and at least three chevron "vee" rings, which utilize a distinct hinge area to allow automatic reaction to pressure. Chevron-type packing seals are lip-type seals that are gland sensitive.

Cold flow: Permanent and continual deformation of a material that occurs as a result of prolonged compression or extension at or near room temperature.

Compressed height: The height of a packing ring or packing set after it has been compressed in a stuffing box.

Compressed sheet: A gasketing material containing primarily fibers, rubber, and fillers, manufactured on a special calender (known as a *sheeter*) in such a manner that the compound is caused to build up under high load as an oriented sheet on one roll of the sheeter.

Compressibility: The quality or state of being compressible. In the case of gasketing, it is the percentage loss of thickness when subject to a given load.

Compression limiter: A ring used in conjunction with a seal to control or limit compression of the seal.

Compression packing: A deformable material used to prevent or control the passage of a pressurized fluid between surfaces that move in relation to each other.

Compression set: The deformation that remains in the gasketing after it has been subjected to, and released from, a specific compressive stress for a definite period of time at a prescribed temperature. (Compression set measurements are for the purpose of evaluating the creep and stress relaxation properties of elastomeric constructions.)

Compressive stress: The Force per unit area compressing (squeezing) the seal.

Cross section: (1) The view of a part as if it were cut to show its internal structure. (2) The distance between the shaft or stem surface and the bore of a stuffing box. *See also* Packing space. (3) The shape of a packing ring at a cut. A packing ring may have a square, rectangular, or round cross section.

Cryogenic: Of or relating to a very low temperature.

Cup and cone: The shape of certain types of die-formed ring sets. In a cup and cone set, the rings have conical surfaces that nest into one another.

Cup packing: A specific type of hydraulic or pneumatic sealing element that seals primarily on its outside diameter.

Cure: The act of vulcanization. *See also* Vulcanization.

Cut-ring set: A braided material cut into individual rings for a specific stem or stuffing box size and packaged as a set.

Cycle and adjust procedure: A procedure used to consolidate a packing set after it has been installed in a stuffing box. This procedure helps to reduce the amount of gland load relaxation that occurs after a valve has been put into service.

Deformation: A stress-induced change of form or shape.

Density: The ratio of the mass of a body to its volume, expressed as lb/ft^3 or g/cm^3. It is common for die-formed graphite foil rings to be produced to a specific density.

Design pressure or vacuum: The pressure or vacuum condition that exists during system startup and/or shutdown operations. During this cyclic phase in the system, both pressure and vacuum conditions may occur.

Design temperature: The maximum or most severe temperature expected during normal operation caused by equipment failure. *See also* Excursion temperature.

Discharge pressure: The fluid pressure measured on the discharge (or outlet) side of the pump, where the fluid exits the volute.

Ductility: The ability to be drawn. Example: The stretching of metal wire to a smaller cross section.

Eccentricity: The distance that the central axis of a shaft is offset from the center of the stuffing box through which it passes.

Elastic limit: The limiting extent to which a body may be deformed and yet return to its initial shape after removal of the deforming force.

Elastomer: An elastic rubber-like material which, in the vulcanized state and at room temperature, can be stretched repeatedly to at least twice its original length. Upon release of the stress, this material will return immediately to approximately its original length.

Elongation: The increase in length expressed numerically as a fraction or percentage of the initial length.

Emissions: Gaseous or liquid leakage given off by a piece of equipment. This usually refers to volatile organic hydrocarbons monitored by government agencies. It is typically expressed in parts per million volumetric (ppmv, or simply ppm).

End ring: A ring used at the top or bottom of a packing set, usually functioning as a wiper ring and/or an antiextrusion ring. *See also* Antiextrusion ring, Junk ring, *and* Wiper ring.

EPA: Acronym for "Environmental Protection Agency," the U.S. government agency responsible for protecting the public's health and the environment by developing and enforcing regulations and studying environmental issues.

Excurision temperature: The temperature the system could reach during an equipment failure such as an air heater failure. Excursion temperature should be defined by maximum temperatures and time duration of excursion.

Expansion: The increase in any linear dimension or volume.

External pressure: The pressure (lb/in^2) exerted on the outside diameter of a seal.

Extrusion: The distortion, under pressure, of a portion of a seal into the clearances between mating metal parts.

Gasket: A deformable material which when clamped between essentially stationary faces prevents the passage of matter through an opening or joint.

Gasket spacer: A gasket material cut to fit in a stuffing box between braided packing rings. Gasket spacers may be used to provide protection against abrasive particles, to increase the pressure resistance of some packing sets, or to reduce the flow of fluid through the body of the braid.

Gasketing: Material in bulk form from which gaskets may be cut.

Gauge: The thickness of a material; also the device used to measure material thickness.

Gland: *See* Packing gland.

Gland follower: A part that protrudes into a stuffing box to compress a packing set or packing ring.

Gland load: The amount of load applied to a packing set. This may be expressed in terms of force (lb, N) or in terms of pressure (psi, kPa).

Gland pressure: The amount of pressure applied to a compression packing set by the gland follower, usually expressed in psi or kPa.

Gland stud: A threaded rod or eye bolt extending from a valve or pump body that is used to compress a packing set.

Hand wheel: A wheel located at the top of a manually operated valve that is used to actuate the valve stem.

Hat (flange) packing: An old form of sealing rod which seals on the inside diameter only.

Head: The pressure at any point in a fluid can be thought of as being caused by a vertical column of the liquid that, due to its weight, exerts pressure at the point in question. The height of this column is called the *static head* (or sometimes simply *head*) and is expressed in terms of feet or meters of liquid.

Heat exchanger: A piece of equipment designed to transfer thermal energy from one medium to another. The unit is typically a long cylindrical body with multiple tubes passing within the body. A medium at one temperature flows within the body but external to the tubes, while a medium of another temperature flows within the tubes, thus allowing thermal energy to be exchanged without cross-contamination of the products.

Helical motion: The circular, screw-like, up and/or down motion of a rod or valve stem.

Homogeneous: Of uniform composition throughout.

Hydrodynamic seal: A dynamic sealing device that utilizes the viscous shear and inertia forces of the fluid, imparted by a helically grooved or ribbed seal lip, to generate a pressure differential that opposes fluid flow.

ID: Acronym for "inside diameter." Used when denoting the inside dimension of a packing set, usually equal to the stem or shaft diameter.

Impeller: The part of a centrifugal pump that when rotated will generate pressure and cause flow.

Inclusion: Foreign matter included in seal material.

Inner ring: A solid ring used on the inside diameter of a spiral wound gasket in higher-pressure applications. This ring also acts as a compression limiter and provides protection from the media stream.

Inside face: That surface of the inner case which faces, and is usually in contact with, the fluid being sealed.

Internal pressure: The pressure (lb/in^2) exerted on the inside diameter of a seal.

ISO: Acronym for "International Standards Organization."

Jacketed gasket: Type of gasket with a metallic cover either partially or completely surrounding a filler material to improve temperature, pressure, chemical, and crush resistance.

Joint: An interstice (crevice) between rigid members of a fluid container.

Leakage: An escape of gases or liquids from a piece of equipment.

Leakage rate: The quantity of fluid passing through (or around) a seal in a given period of time.

Lip diameter: The inner diameter of a seal lip, measured with the spring installed.

Lip force: The radial force exerted by an extension spring and/or lip of a seal on the mating shaft. Lip force is expressed as force per unit of shaft circumference.

Lip height: The axial distance from the outside seal face to the toe face.

Lubricant starvation: Lack of proper lubrication at the seal interface, which may cause premature wear and early failure.

Lubrication: The use of a liquid or fluid to reduce the friction, heat, or wear between two surfaces where movement is taking place.

***M* factor:** Gasket property. The factor that provides additional preload capability in the flange fasteners to maintain sealing pressure on a gasket after the internal pressure is applied to the joint. Used in conjunction with the Y factor to perform flange design.

Medium: The fluid that is being sealed.

Metal C-ring: A C-section metal seal with low-load characteristics. Plating and coating are added for improved performance.

Metal O-ring: A hollow tubular seal with a circular cross section. Various shapes and sizes can be fabricated and different plating and coating can be added for better performance.

Metal seal: A seal composed entirely of metal components which when compressed between two stationary faces prevents the passage of a particular media through a specified space or opening.

Misalignment: The out-of-line condition that exists between the adjacent face of equipment or system flanges.

MOV: Acronym for "motor-operated valve," a motor-actuated control valve.

Movements: The dimensional changes that an expansion joint is required to absorb, such as those resulting from thermal expansion or contraction.

MRO: Acronym for "maintenance and repair organization."

MSS: Acronym for "Manufacturers Standardization Society" of the valves and fittings industry.

MTI: Acronym for "Material Technology Institute" of the chemical process industry.

Nonmetallic expansion joint: An expansion joint that utilizes flexible nonmetallic material to accommodate joint movements.

O-ring: An elastomeric seal of homogeneous composition molded in one piece to the configuration of a torus, with circular cross section; more simply, a round ring with a round cross section.

OD: Acronym for "outside diameter," used to denote the outside dimension of a packing set (usually equal to the stuffing box bore).

Oil seal: A seal designed primarily for the retention of oil.

Oil swell: The change in volume of a gasketing product resulting from contact with oil.

Omnidirectional strength: Strength in all directions.

Operating pressure/vacuum: The temperature at which a system generally will operate during normal conditions.

Outer case: The outer thin-wall rigid structure of a lip-seal assembly, which contains the inner case, the primary-seal ring, the spring parts, and the secondary seal.

Packing gland: The space into which a compression packing is inserted. Also known as a *stuffing box*.

Packing groove: A groove machined into a flange or joint to accommodate a packing ring.

Packing hook: A tool similar to a corkscrew for removing packing from a stuffing box.

Packing space: The distance between the shaft or stem surface and the bore of a stuffing box. Packing space (x) can be calculated by the following equation: $x = (\text{OD}-\text{ID})/2$.

Permanent set: The degree to which an elastic material fails to return to its initial form after deformation.

Permeability: The quality or condition of allowing passage of fluid, gases, or air through a material such as rubber.

pH: The measure of the strength of an acid or base. On the pH scale, a neutral solution (neither acidic or basic) has a pH of 7. Solutions with a pH below 7 are acidic; the smaller the pH value, the more acidic the solution. Solutions with a pH above 7 are basic.

Piston: A cylindrically shaped part that fits within a cylinder and transmits or receives motion by means of a connecting rod (a cylinder rod).

Pitting: Surface cavities that occur on a metal as a result of galvanic corrosion or mechanical erosion.

Plunger: A cylindrically shaped part that has a uniform diameter and is used to transmit thrust (as in a hydraulic cylinder) or develop pressure and cause flow (as in a reciprocating pump).

Pneumatic seal: An extruded or molded elastomeric seal designed to expand and retract to provide a secure, reliable seal that can hold, position, or handle objects in a wide range of applications.

Precompression: Compression of the expansion joint during installation so that in the cold position the joint has a given amount of compression in addition to the rated amount of compression. The purpose of precompression is to allow for unexpected or additional axial extension.

Pressure: Force per unit area, usually expressed in pounds per square inch (psi) or kilograms per square centimeter (kg/cm^2).

Pressure cycling: The variation of pressure in a system.

Pressure, absolute (PSIA): The pressure above zero absolute: the sum of atmospheric and gage pressure.

Pressure, atmospheric: The pressure exerted by the atmosphere at any specific location. Sea level pressure is approximately $14.7\,lb/in^2$ absolute.

Pressure, gauge (PSIG): The pressure differential above or below atmospheric pressure, expressed as lb/in^2 or psig.

Primary lip: The normally flexible elastomeric component of a lip seal assembly, which rides against the rotating surface and affects the seal.

Protective cover: Outer cover material used to protect boot material during shipment and installation.

psi: Abbreviation for "pounds per square inch," a unit of pressure.

PTFE: Acronym for "poly(tetrafluoroethylene)," a polymer having excellent chemical resistance. PTFE dispersion is used as a coating for many styles of packing. Some packing styles are constructed of PTFE fibers. Teflon® is DuPont's trade name for their PTFE materials.

Pump shaft: The metal rod connecting the impeller of a pump to the motor.

Purge fluid: A clean liquid (usually, water) which is injected through a flush port to flush solid particles from the stuffing box area to minimize abrasive wear.

Purge port: A hole in the side of a stuffing box through which a flushing fluid is injected.

Quarter-turn valve: A valve that will fully open or close with a 90° rotation of the stem.

Radial: In the direction perpendicular to a shaft axis.

Radial expansion: The ability of a packing material to spread out in the radial direction of a stuffing box when it is compressed.

Radial lip seal: A type of seal that features a flexible sealing member referred to as a lip. The lip is usually of an elastomeric material. It exerts radial sealing pressure on a mating shaft in order to retain fluids and/or exclude foreign matter.

Radial load: The total force (load) acting on a seal lip which tends to maintain contact of the lip on the shaft. It is the sum of the forces developed from seal interference and the garter spring.

Reciprocating: The motion of a shaft back and forth in the direction of its axis.

Reciprocating motion: An oscillating, back-and-forth motion as it normally pertains to a piston rod or valve stem.

Reciprocating pump: A type of pump that relies on the reciprocating motion of a plunger, or series of plungers, to generate pressure and cause flow.

Recovery: The degree to which a product returns to its normal dimensions or shape after being distorted.

Recovery (gasketing): The increase in thickness of a gasket after a load is removed.

Remote handling: Handling without direct individual contact, usually done with robotic arms.

Resilience: The ratio of energy given up by the elastomer upon releasing it from a definite deformation to the energy required to produce the deformation. Unless the deforming is momentary, the result obtained will be complicated by the partially plastic nature of the elastomer.

Retaining rings: Segmented metal rings installed directly against the back of the expansion joint flange and bolted through to the metal flange of the pipe.

Rising stem valve: A valve in which the movement of the stem is simply reciprocating, with no rotation.

Rising/rotating stem valve: A valve in which the movement of the stem is both reciprocating and rotating at the same time, usually following a helical path.

Rod: The metal shaft that extends from the piston to outside the cylinder.

Roll: Sheet rubber and gasketing material of a uniform width rolled up on itself from which gaskets and other products of various shapes may be cut.

Rotary: The motion of a body turning on an axis.

Rotary service: A dynamic type of sealing application in which one member remains stationary while the other moves past it in a rotating mode. This type of service generally involves a continuous motion between members and higher surface speeds than those encountered in reciprocating applications. The generally higher degree of frictional heating, combined with the fact it is concentrated in one area, sometimes leads to heat-related seal problems.

Rotation (flange): The warping, bowing, or bending of a flange that can occur from too much bolt load.

Rough trim: A trimmed surface with irregularities on the outside and inside lip surfaces in the immediate vicinity of the contact line.

Roughness: Irregularities in shaft surface texture that result from the production process. See SAE J448a, (June, 1963) for more information.

rpm: Abbreviation for "revolutions per minute," a measurement of the rotary speed of a rotating shaft.

Runout: A measurement of how far a shaft moves in the radial direction.

Scoring: Gouges on the surface of a shaft, stem, or bore due to mechanical wear.

Scratch: A shallow discontinuity in seal material whereby no material is removed.

Seal: Any device designed to prevent or control the movement of fluid from one chamber to another to exclude contaminants.

Seal assembly: A group of parts, including sealing surfaces, provisions for initial loading, and a secondary sealing mechanism, which accommodates the radial movement necessary for installation and operation.

Seal cage: *See* Lantern ring.

Seal case: A rigid member to which the seal lip is attached.

Seal gasket: A gasket that is placed between two adjacent metal parts to make a gastight connection.

Seal outer diameter: The external diameter of a lip-seal assembly, which normally corresponds to the outer diameter of the outer seal case.

Seal solution: Any gasket or seal, which when properly installed prevents the escape of matter.

Sealability: The measure of fluid leakage both through and across both faces of a gasket. Measured either by using ASTM F-37 or DIN 3535 equipment and procedures.

Seating load: The load required to compress a seal properly. This will vary depending on the size and shape of the seal. (This does not include operating hydro end forces.)

Self-energizing O-ring: An O-ring with a hole drilled on the pressure side, allowing pressure to equalize inside the O-ring; prevents collapse of the ring.

Setback (stand of height): The distance the expansion joint is set back from a gas stream to allow for lateral movements and to prevent the joint from protruding into the stream or rubbing on the baffle when operating under negative pressure. Setback also reduces the heat input and prevents abrasion from solids or particles in the stream.

Shaft: The metal rod connecting the impeller of a pump to the motor.

Shaft eccentricity: The radial distance that the geometric center of a shaft is displaced from the axis of shaft rotation.

Shaft lead: Spiral grooves on a shaft surface caused by relative axial movement of grinding wheel to shaft.

Shaft runout: *See* Dynamic runout.

Shaft surface finish: *See* Shaft surface texture.

Shaft surface texture: A term used to describe the quality, appearance, or characteristics of the shaft surface resulting from such operations as grinding, polishing, and burnishing. See SAE J448a (June 1963) for additional information.

Sheeter: A special calender used to make compressed sheet.

Shipping straps or bars: Braces that are located between the two expansion joint flanges to prevent overcompression or distortion during shipment and joint assembly.

Single-end coating: The process of applying a coating to the individual yarns (or "ends") of a packing before they are braided. This process results in a very thorough uniform coating throughout the braid.

Size, actual: Actual dimensions of a part, including tolerance limits.

Size, nominal: The approximate size of a part in fractional dimensions.

Skim coat: A layer of rubber laid on a fabric but not forced into the weave. Normally laid on frictioned fabric.

Slab: A thick sheet.

Sleeve: A metal cylinder that is placed over a pump shaft in the sealing area. In pumping applications, certain media and packing materials can cause abrasive wear on the rotating surface. A sleeve is a relatively inexpensive, replaceable component that protects the pump shaft from wear.

Slip stick: A friction-related phenomenon in which the sealing element tends to adhere and rotate with the shaft surface momentarily until the elastic characteristics of the sealing element overcome the adhesive force, causing the seal lip to lose sufficient connection with the rotating shaft long to allow leakage. This cycle repeats itself continuously and is normally associated with nonlubricated and boundary-lubricated conditions.

Slurry: A fluid mixed with solid particles. In packing applications handling slurries, abrasion is a major concern. Steps must be taken to minimize abrasive wear of the packing materials.

Specific contact pressure: Seating load divided by seal contact area.

Specific gravity: The ratio of the weight of a given substance to the weight of an equal volume of water at a specified temperature.

Spiral trim: A trimmed surface that has a spiral pattern.

Splicing: A procedure for making an endless boot or bellows from open-ended material. Splicing may be accomplished by one or more of the following: cementing, bonding, heat sealing, stitching, vulcanizing, or mechanical fasteners.

Spool packing: Packing material that is braided and sold on a spool, as opposed to cut-ring sets or die-formed ring sets.

Spring groove: A depression formed in the head section of a seal. It is generally semicircular in form and serves to accommodate and locate the garter spring.

Spring outside coil diameter: The outer diameter of an individual helical coil of a garter spring.

Spring rate: The force in pounds required to deflect an expansion joint 1 in. in compression and elongation or in a lateral direction.

Spring retaining lip: The portion of the primary lip that restricts the axial movement of the extension spring from a predetermined position.

Spring witness marks: A series of indentations or depressions remaining in the spring groove of a radial lip seal after the spring has been removed or dislocated.

Square braid: A type of braiding construction that yields a soft, flexible packing material having a square cross section. Also referred to as *regular braid*.

Squeeze seal: Also known as *installation activated*, a squeeze seal relies on the squeeze or compression it achieves at installation to create a seal. The squeeze or compression is due to its high degree of interference. This type of seal generally seals well at low pressure; however, in dynamic applications, it is characterized by a high wear rate and friction.

Stack height: (1) The combined height of all the rings of a packing set. (2) The combined height of all the components in a stack of Belleville washers used to live-load a packing set.

Static seal: A seal whose sealing surface sees no motion—the opposite of dynamic seal.

Stem: The meral rod that connects the internal components of a valve to a handwheel, handle, or actuator.

Stock: In the papermaking industry, the wet pulp mixture at any point in the papermaking process.

Strain: The unit change, due to force (stress), in the size or shape of a body compared to its original size or shape—it is a nondimensional quantity but is frequently expressed in inches per inch, centimeters per centimeter, and so on.

Stress: The force applied to a unit area.(lb/sq in.).

Stress relaxation: The loss of an initial bolt load or bolt stress, often accelerated by thermal cycling.

Stress–strain: The relationship of load and deformation in a body under stress. In rubber this is most commonly the relationship of tension (stress) and elongation (strain).

Strong oxidizer: In packing applications, strong oxidizers cause the degradation of organic packing materials such as carbon, graphite, and cellulosic fibers. PTFE packing materials are generally used in these applications, due to their oxidation resistance.

Stuffing box: The space into which compression packing is inserted. Also known as a *packing gland*.

Suction pressure: The fluid pressure measured on the suction (or inlet) side of the pump where the fluid enters the volute.

Surface finish: A measure of the roughness of a surface, usually expressed in microinches or micrometers.

Surface speed: The linear speed of a point on the surface of a rotating shaft, usually expressed in ft/min or m/s.

Surge pressure: Operating pressure plus the increment above operating pressure that the expansion joint will be subjected to for a very short time period. Surge pressure is typically due to pump starts, valve closing, and the like.

Swelling: The increase in volume or linear dimension of a specimen immersed in liquid or exposed to a vapor.

Symmetrical seal: A seal that has the same shape on either side of its centerline.

Tacky (rubber surface): Sticky material that will adhere to itself, usually to ply up layers.

Tensile strength: The amount of tensile stress that causes a specimen to rupture; measured in psi.

Thermal barrier: A layer of insulating material designed to reduce the surface temperature at the gas sealing layer to a level compatible with its heat resistance capability.

Thermal conductivity: The ability of a material to transmit heat. A measure of the rate at which thermal energy is transferred through a substance. High thermal conductivity is an advantage in pump packing applications, where it is important to transfer frictional heat away from the shaft–packing interface so that the packing does not burn.

Thermal cycling: The repeated heating and cooling of a tank, vessel, or piping system. The duration of the cycle can vary tremendously.

Thermal expansion: The increase in volume or length of a material that occurs as a result of a temperature increase.

Thermal movements: Movements created within a system caused by a thermal change. Can be axial, lateral, or torsional.

Tolerance: The upper and lower limits between which a dimension must be held.

Torque: Informal definition: A measure of "twisting force." In packing applications, one might be concerned with the torque applied to the gland stud nut or the torque required to overcome friction between the packing and the valve stem.

Torsional rotation: The twisting of one end of an expansion joint with respect to the other end about its longitudinal axis, Such movement being measured in degrees, as is angular rotation.

Trim: The removal of superflous parts from a molded product, usually removal of parting-line flash or feed sprues.

Trim cut: Damage to the elastomeric portion of a seal during trimming.

Tube: The inner ply of the expansion joint that is in direct contact with the system medium.

U-seals: A packing in which the element has a U-shaped cross section.

Ultimate elongation: The maximum elongation prior to rupture.

Unbonded flash: Flash that does not adhere properly to the mating material to which it is intended to be bonded.

Uncompressed height: The height of a packing set or packing ring before being compressed in the stuffing box.

Unirotational seal: A seal designed for applications having a single direction of shaft rotation.

Unitized seal: A seal assembly in which all components necessary for accomplishing the complete sealing function are retained in a single package.

Vacuum: Negative pressure; when a gasket or seal is used to prevent the passage of external media or air into a system.

Valve body: The part of a two-piece valve that houses the internal workings of the valve. The flanges that attach the valve to a pipeline are also part of the body.

Valve bonnet: The part of a two-piece valve that attaches to the valve body. It houses the stuffing box and provides support to guide the valve stem.

Valve stem: *See* Stem.

Vibration: The ability of a flexible connector to absorb mechanical oscillations in a system.

Viscosity: A manifestation of internal friction opposed to mobility. The property of fluids and plastic solids by which they resist an instantaneous change of shape (e.g., resistance to flow).

Volume swell: An increase in physical size caused by the swelling action of a liquid.

Volute: The internal area (housing) of a centrifugal pump where the fluid comes in contact with the impeller.

Wear sleeve: A replaceable metal ring generally used in assemblies to eliminate expensive shaft replacement caused by grooving that may occur at the seal–shaft interface.

Weathering: The surface deterioration of a rubber article during outdoor exposure, such as cracking, crazing, or chalking.

Weepage: A minute amount of liquid leakage by a seal.

Weld in baffle: A baffle that is designed to be welded to the duct wall. This design can be of either single- or double-acting type.

Welding blanket: A fire-resistant blanket that is placed over an expansion joint to protect it from weld splatter during a field welding operation.

Whip: Deflection of a shaft (usually, on a mixer or pump) due to a rotating mechanical load. A long shaft that is not supported by bearings is more susceptible to whip than is a short shaft or one that is supported by bearings.

Wicking: Leakage through a gasket, not around it.

Wiper ring: A ring of braided packing used in conjunction with graphite foil rings to wipe a reciprocating valve stem clean of graphite particles and keep the graphite contained in the stuffing box.

Wire-reinforced: A product containing wire(s) to give added strength, increased dimensional stability, or crush resistance.

Work pressure: The maximum operating pressure encountered during normal service.

***Y* Factor:** Gasket property. The minimum design seating stress in either psi or megapascal over the contacted area of a particular gasket. The stress required to provide a sealed joint with a 2-psig (virtually 0 psig) internal pressure in the joint. Used in conjunction with the M factor to perform flange design.

Yield strength: The amount of stress that causes a material to lose its elasticity permanently.

INDEX

Process Plant Equipment: Operation, Control, and Reliability, First Edition. Edited by Michael D. Holloway, Chikezie Nwaoha, and Oliver A. Onyewuenyi.
© 2012 John Wiley & Sons, Inc. Published 2012 by John Wiley & Sons, Inc.